EUROPE'S ENVIRONMENT

The Dobříš Assessment

NOTE TO THE READER

This book is a copy of the first full version of the Dobříš Assessment, printed and published in August 1995 and made available to the public. In the final stages of preparation of this report, a limited number of black and white and colour preprints were prepared, to promote the early dissemination of the information.

The report was written and first published in English. Commonly used English equivalents for place names and English spellings are therefore used throughout the report.

Translations into other European languages are planned. First to be available will be the conclusions and highlights – Chapters 39 and 40.

The overview of the report, first published in English, is being prepared in all languages of the European Environment Agency and Russian.

A summary of the report will also be published.

The statistical compendium to this report, produced by Eurostat, is also available under Catalogue Number CA-82-94-488-EN-C (+DE, FR).

This report was co-financed by the European Union's PHARE programme, which provides grant finance to support its partner countries in central and eastern Europe to the stage where they are ready to assume the obligations of membership.

The European Union's PHARE Programme

EUROPE'S ENVIRONMENT
The Dobříš Assessment

Edited by David Stanners and Philippe Bourdeau

The report on the state of the pan-European environment requested by the environment ministers for the whole of Europe at the ministerial conference held at Dobříš Castle, Czechoslovakia, June 1991.

**Prepared by the European Environment Agency Task Force
(European Commission: DG XI and Phare)**
in cooperation with:

United Nations Economic Commission for Europe
United Nations Environment Programme
Organisation for Economic Cooperation and Development
The Council of Europe
IUCN – The World Conservation Union
World Health Organization
together with Eurostat
and individual European countries.

**European Environment Agency
Copenhagen**

This report was designed, copy-edited and produced for the European Commission and the European Environment Agency by

Earthscan Publications
120 Pentonville Road
London N1 9JN
Tel: 44 171 278 0433
Fax: 44 171 278 1142

Publishing Director
Jonathan Sinclair Wilson
Project Manager
Jo O'Driscoll
Project Assistant
Andrew Young

Design and typesetting
Paul Sands,
S&W Design

Cartography and illustrations
Gary Haley,
PCS Mapping & DTP

Editorial work
Nina Behrman
Caroline Richmond
Audrey Twine
Wolfgang Klar
Georgina Klar
Don Shewan

Index
Frank Pert

Picture research
Brooks Krikler Research

European Environment Agency
Kongens Nytorv 6
DK-1050
Copenhagen K
Denmark
Tel: +45 33 36 7100
Fax: +45 33 36 7199

Cataloguing data can be found at the end of this publication

Luxembourg: Office for Official Publications of the European Communities, 1995

ISBN 92-826-5409-5

© EEA, Copenhagen, 1995

Printed in Belgium

In Memoriam
Josef Vavroušek
(1944–1995)

It was Josef Vavroušek who, as Czechoslovak Minister of the environment, suggested holding the first pan-European Conference of Environment Ministers in 1991, effectively paving the way for the 'Environment for Europe' process and the preparation of this document: *Europe's Environment: The Dobříš Assessment.*

These actions arose out of the new horizons which emerged from the collapse of the Iron Curtain in the autumn of 1989. In Josef Vavroušek's own words: '....there are not many fields where effective cooperation in continental or even global scale is so necessary, like in the field of environmental protection and restoration. Not a single group of countries can solve the severe legacy of environmental deterioration and find ways towards sustainable future alone. Environmental cooperation is thus not only the necessary condition for substantial improvements of our ecological situation but also the most natural first step of European integration.'

From the initial idea, Josef Vavroušek went on to organise the first European environment ministers conference which was held at Dobříš Castle near Prague from 21 to 23 June 1991. The Czechoslovak president Václav Havel opened the historic meeting together with the Swiss president Flavio Cotti. The meeting was attended by 36 environment Ministers or deputies from almost all European countries and representatives of many international institutions and environmentally oriented European non-governmental organisations. Following suggestions from Josef Vavroušek and others, the conference requested the preparation of the current report and the development of an environmental programme for Europe. Human values and environmental ethics were of particular importance to Josef Vavroušek. The conference underlined the importance of these ideals in the search for ways of sustainable living. Josef Vavroušek's initiative and vision thus set the basic framework of long-term European environmental cooperation which, after Dobříš, was followed by the Lucerne Conference held in April 1993, and Sofia in October 1995.

Although Josef Vavroušek did not live to see the final publication of this report and the development of the environmental programme for Europe, the European Environment Agency had the privilege of his presence at a seminar in Copenhagen on 10 March 1995, one week before he died, where the results of the report were discussed and ideas for follow-up were considered. At this seminar, Josef Vavroušek emphasised the importance of broadly disseminating this report and its results. In addition to the usual concerns about improving data quality and standards, and the need to develop and maintain a broad and developing picture of the state of Europe's environment, Josef returned again to the social aspects prevalent at the Dobříš conference and reminded the seminar of the need to address human values for a sustainable future for Europe.

That Josef Vavroušek died in the 'nature' he loved and worked to protect is an ironic demonstration of how subject to and dependant on the environment we are. For his contribution to European cooperation and understanding in the field of environment and sustainable living, Josef Vavroušek will justifiably be remembered as the 'father' of the 'Environment for Europe' process and 'grandfather' of *The Dobříš Assessment,* where his work and influence will not be forgotten.

Josef Vavroušek (1944–1995) died in an avalanche in the Roháče West Tatry Mountains in Slovakia on 18 March 1995 while hiking with his daughter Petra, who also died in the accident.
Dr Vavroušek was the first minister of the Environment of Czechoslovakia (1990–1992), one of the leaders of the Czechoslovak 'Velvet Revolution' in 1989, the founder and President of the Society for Sustainable Living (1992–1995), a member of the Czech and Slovak Association of the Club of Rome, as well as a member of the student expedition 'Lambaréne' which brought medicaments to Albert Schweitzer's hospital in Lambaréne in 1968.

Contents

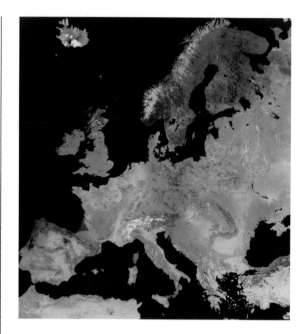

PART III PRESSURES

PART IV HUMAN ACTIVITIES

PART V PROBLEMS

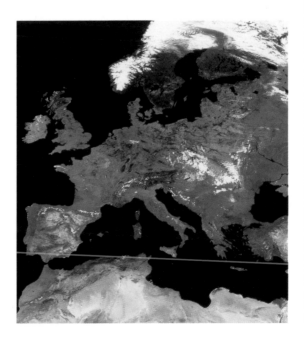

PART VI CONCLUSIONS

Foreword

What you have in your hands is the result of more than three years' work following the request of the first Pan-European Conference of Environment Ministers which took place at Dobříš Castle, near Prague, from 21 to 23 June 1991. This conference called for the preparation of a State of the Environment Report for Europe and invited the European Commission to take responsibility for the work. This was accepted by the then Commissioner responsible for Environment, Carlo Ripa di Meana, and mostly accomplished during the mandate of Commissioner Yannis Paleokrassas who strongly supported the work.

To make progress in solving common environmental problems, a common information base had to be compiled, based as far as possible on comparable and reliable data - as we say now the best available data.

Many goals were identified to be supported by this report:

- to develop a comprehensive Environment for Europe Programme addressing in particular transboundary envronmental problems;
- to provide a sound basis for effective measures, strategies and policies to address environmental problems nationally and regionally; and
- to inform the public and raise awareness about our common responsibility for the environment.

Above all it was clear that the report had to become a base-line and reference, for periodic updating and review, to contribute to the "Environment for Europe" pan-European co-operation process. This process is institutionalised by the Conference of Environment Ministers that meets periodically - first at Dobříš Castle in 1991, then in Lucerne in April 1993, and in October 1995 in Sofia. The preparation of this report has been a key element to the preparation of the Sofia conference. Its findings were already made available to governments and international organisations concerned, interested parties, officials and NGOs, and the report was broadly distributed in proof form, in the months preceding final publication.

Within the European Commission the task of producing the report was given at the time to DG XI (Directorate General for Environment, Nuclear Safety and Civil Protection) then headed by Laurens Jan Brinkhorst who assigned the work to the Task Force preparing for the European Environment Agency. The Agency officially came into being on 30th October 1993 when the decision was made to locate it in Copenhagen. When the Task Force completed its work in August 1994 the Agency took over from the Commission to finalise and publish the report.

It is a satisfaction for the European Commission and the European Environment Agency to present now this final report for broad public distribution. The report identifies, through the data it makes available, for the first time the key environmental issues which face our continent as a whole. The report certainly has its limitations. It has not been possible in all cases to obtain appropriate data or data of sufficient quality. However, it has provided not only for the compilation of what is considered to be the best available data, and therefore for the best possible base to support the mentioned goals, but also for the building up during the exercise of a network of institutions, official as well as non-governmental bodies, working together. Exposing data and aggregating information, identifying deficiencies and lacuna, is not only an unavoidable task but the guarantee of more sound decision making, now and in the future. The report also forms the basis for periodic reviews and updates of the state of the environment and the supporting information networks. From now on we should only expect to make progress against this base-line.

The report confirms the poor quality of the European environment, particularly, but not exclusively, in parts of Central and Eastern Europe. The report identifies serious threats and stresses on the environment, natural resources and human health which have to be tackled more efficiently. Above all, by performing, for the first time, a global structured review of the situation and trends, and by casting light on the shadows, the report allows the dimensions of the problems to be realised. This is a strong basis for identifying priorities and alternatives for action without losing the global perspective and respecting an integrated approach, something which is paramount in the interrelated environment.

When we evaluate progress in terms of environmental quality, the dimensions of the problems are shown to be much larger than assumed up to now when focusing on specific issues and their progress. While reductions are being achieved in some cases, such as with emission controls, these actions are often insufficient on their own to achieve recovery and improvement of natural resources and environmental quality. This is due to the level of degradation and the sensitivity and the often limited carrying capacity of the environment. It is a question of scale of action. If we take seriously our responsibility of not increasing the debt of future generations, there is no room for complacency. This concerns not only "European space" but also the abusive European contribution to global degradation or problems.

As for environmental threats, in Europe these still mostly follow the ongoing trend of their coupling to economic development - the economies in transition representing a special case because of the lower performance of parts of their production sectors. Increase in domestic, agricultural and industrial wastes of all kinds, and of energy and transport intensity, follow, in general - with some notable exceptions - the level of economic activity, while the reduction of final environmental pressures is still mostly achieved by end-of-pipe measures.

These are the main challenges, but they are also the opportunities for Europe, where appropriate policies have to be established to progress in the process towards sustainable development. Europe created the world economy, and its agricultural and industrial revolutions of the 18th and 19th century are still transforming the earth; a Europe driven by sustainable environmental values could be leading the so-called third industrial revolution to enter the 21st century approaching a more sustainable pattern.

These opportunities are also reflected in the report. Europe is still a very diverse continent, an environmental asset in its own right. Nature is still a valuable resource of large dimensions in many Southern, Central and Eastern European regions where in parts it faces many threats. A correspondingly large effort, and commitment of financial resources, should be devoted to protecting and enhancing these common natural infrastructures; a

most rewarding enterprise if conducted in a convincing and professional manner. The pollution problems in many parts of Central and Eastern Europe, largely related to poor industrial performance and to old and decaying installations, should be considered a common challenge for Europe to renew its production system. This appeals to the use of more efficient, clean and, in general, best available technologies. Such an enterprise is difficult to apply at the scale required in the more established and highly performing western parts of Europe, but in Central and Eastern Europe there is the opportunity to develop new and leading experiences and win-to-win strategies on a broad front for the benefit of the whole of Europe.

The information contained in this report, though still incomplete and able to be improved, already provides the basis for efficient action. The longer we wait the harder and more costly it will be.

We wish to express our appreciation and thanks to the United Nations Economic Commission for Europe, the United Nations Environment Programme, OECD, the Council of Europe, WHO and IUCN as well as government authorities of Central and Eastern Europe, EFTA and EU for their cooperation and for producing a successful result. We further thank the large number of individuals who are named in the list of acknowledgements for their valuable work.

All of us will have to continue to work together for the benefit of the European environment in the years to come.

Copenhagen, August 1995

Ritt Bjerregaard
Member of the European Commission responsible for Environment, Nuclear Safety and Civil Protection

Domingo Jiménez-Beltrán
Executive Director European Environment Agency

Preface

This report, requested in 1991 at the Dobříš conference of European environment ministers, presents an assessment of the state of Europe's environment and the pressures and human activities impacting on it, and analyses prominent environmental problems of concern to Europe as a whole. The report was prepared by the European Commission's task force for the European Environment Agency (EEA-TF, DG XI) in cooperation with the United Nations Economic Commission for Europe (UNECE), calling on the assistance of other relevant international organisations and individual European countries.

To facilitate this task a Project Group was formed consisting of representatives from the European Commission, UNECE, UNEP, OECD, the Council of Europe, IUCN, WHO, and countries preparing the Environment for Europe conferences and/or associated with the EU presidency. In addition, each country nominated a National Focal Point to act as a national interlocutor throughout the project. Beginning in December 1991 a series of Project Group meetings were held to plan and organise the work. In order to keep all countries informed of progress, a series of orientation meetings were held during the period of preparation of the report, attended by National Focal Point representatives. In the last phase, drafts were circulated to these participants as well as relevant experts in EC Directorate Generals, non-governmental organisations and scientific institutes, for corrections, comments and review.

To respond to the needs arising out of the Dobříš request for this report (see Foreword and page 2) a strong factual basis on the state of the environment was required for decision making and awareness raising covering the whole scope of Europe's environment. Preparing this in a single volume has resulted in a large report, and therefore great importance has been given to the presentation and structure of the information to improve accessibility (see Chapter 1 and Summary). To lighten the report from the details of the data on which it is based and yet give access to them, a statistical compendium has been prepared to accompany this report in which the background data and statistics to the main assessment are presented. Much reference is made to this companion volume in the current report, referred to as the *Statistical Compendium*. This compendium should help serve the main objectives of the report and lay down a bench-mark for follow-up activities.

From the start, it was in fact recognised that a single report would not be able to serve all the given objectives. Although policy makers were identified as the primary target audience, the results of this European-level assessment, if presented in different forms, are expected to serve a variety of uses and a variety of audiences. In particular, additional publications are required to inform the public adequately, and to this end other products are envisaged, including a popular version, CD-ROM and specially designed educational material.

The starting point for data collection was information held by the European Commission (especially DG XI and Eurostat), and by well established international environmental and socio-economic sources such as UNECE, UNEP and OECD. This was supplemented with existing data from national or subnational levels as well as from non-governmental bodies using a list of data requirements distributed to all countries. In many cases appropriate data were found to be unavailable. For essential items, supplementary information was directly requested of National Focal Points, using, in a few cases, specially designed questionnaires (eg for inland waters and urban areas). Environmental data appropriate to a full appraisal of European environmental problems are in fact widely scattered. To gather the most up-to-date information in all fields at European level thus takes time, and in this report the most recent data presented vary widely depending on the field and location, ranging from the late 1980s to 1993.

Non-governmental organisations have been involved in many aspects of the project preparing this report, sometimes acting as alternative suppliers of information, being involved in the review process and as sources of an independent constructive critique to the whole process. As pointed out at the Dobříš conference, much benefit can be expected from cooperation with the independent sector. Considerable resources of time, money and personnel, as well as much organisation, are required to support this. Since these resources were at a premium during the preparation of this report, the full benefits expected from such cooperation were not able to be fully realised during this exercise. Future similar activities should, however, aim to build and improve on the current experience.

An important aspect of such an international effort is the coordination with other complementary activities for efficient data collection, and to ensure up-to-date and verifiable results. Noteworthy is the close cooperation pursued with the World Health Organization (WHO). In more or less the same period as the current report was being prepared, the WHO Regional Office for Europe was implementing a project called *Concern for Europe's Tomorrow* (CET) to collect and analyse information on the effects of the environment on human health in the European region. The first findings were presented to the second European Conference on Environment and Health held in Finland in June 1994. Clearly there are many areas of common interest between CET and the Dobříš assessment and, in order to ensure complementarity of information in the complex field of human health and the environment, a mutually beneficial exchange of information has taken place.

Without the extraordinary cooperation among all the individuals and organisations involved (see Acknowledgements), this report would never have been put together, let alone completed. For this commitment the editors gratefully acknowledge and thank all co-workers involved in the preparation of this report. Although every effort has been made by the EEA-TF to consult and obtain the agreement of all contributors to the report, the responsibility for the content and the way in which these contributions have been used rests with the Editors.

The findings of the report make clear the enormity of

the tasks still to be tackled. Data availability and quality throughout Europe are patchy; there are important gaps which make it difficult to draw meaningful conclusions. Moreover, data (and even terminology and methodology) are by no means always comparable, let alone standardised. No attempt has been made to disguise these deficiencies; indeed, their identification (see Appendix 2) provides the first elements for future action plans. Improving these inadequacies and providing future continuity in this work is now the responsibility of the newly established European Environment Agency, inheritors of the information and database built up throughout the course of this project. Nevertheless, reporting on the environment and more generally monitoring progress towards sustainability is a cooperative enterprise and it is to this that the editors dedicate the report.

Acknowledgements

This report was prepared and written with the collaboration of a large number of individuals. To acknowledge their work the following two lists have been prepared. First, there is a chapter-by-chapter listing of coordinators, principal authors and other important contributors to the chapters in question.

The second is a comprehensive list of all contributors involved in the preparation of the report, grouped according to teams and affiliations including the Project Group and the National Focal Points, and those who contributed more on a generic national and thematic basis.

All chapters were reviewed by the Project Group and National Focal Points, but for many chapters special reviews were made. The names of the people who carried out these additional reviews are included in the first list by chapter.

The whole project was managed and coordinated throughout by David Stanners of the European Environment Agency Task Force (EEA-TF, European Commission, DG XI) on secondment from and supported by the European Commission's Joint Research Centre, DG XII, and under the guidance of Philippe Bourdeau (Head of EEA-TF, 1992–94). Roy Gibson (Head of EEA-TF, 1990–91) initially launched the project, and continued through 1993 to provide guidance and assistance as it developed. As the project was being completed it was inherited in August 1994 by the European Environment Agency, under the directorship of Domingo Jiménez Beltrán who facilitated the report's completion.

The Danish National Environmental Research Institute, the Norwegian water resources and energy administration, the Finnish Meteorological Institute and the UK Department of the Environment each supported a national expert to work with the EEA-TF to help prepare this report.

The European Commission's 1991 PHARE Regional programme provided financial support for the development of the report for work covering Central and Eastern European PHARE countries.

Apart from the people cited in the acknowledgements there are further individuals at national, regional and city levels who have contributed to this report through the collection, processing and supply of data, in particular for Chapter 5, 9 and 10. These people are also gratefully acknowledged.

With the large number of people involved in a project such as this it is difficult to be exhaustive. Consequently, the Editors apologise for the involuntary omission of any individual who contributed to this assessment.

CONTRIBUTORS BY CHAPTER

Chapter 1: Reporting on Europe s environment
Concept, coordination and editing:
David Stanners (EEA-TF)
Principal author:
David Stanners (EEA-TF)
Other contributors:
Marina Alberti (EEA-TF)
Sylvain Joffre (EEA-TF)

Chapter 2: Environmental Changes and human development
Concept, coordination and editing:
Marina Alberti (EEA-TF)
Sylvain Joffre (EEA-TF)
David Stanners (EEA-TF)
Principal authors:
Marina Alberti (EEA-TF)
Sylvain Joffre (EEA-TF)

Chapter 3: Europe: The continent
Concept, coordination and editing
Sylvain Joffre (EEA-TF)
David Stanners (EEA-TF)
Principal authors:
Sylvain Joffre (EEA-TF)
David Stanners (EEA-TF)
Other contributors:
Jonathan Parker (EEA-TF)

Chapter 4: Air
Concept, coordination and editing:
Sylvain Joffre (EEA-TF)
Arne Tollan (EEA-TF)
Barbara Lübkert (EEA-TF/WHO- ECEH)
Principal authors:
F A M de Leeuw, Hans Eerens, Rob Sluyter, Roel van Aalst and Joe Alcamo (RIVM), Arne Semb, Steiner Larssen, Frode Stordal and Ulf Pederson (NILU), Sylvain Joffre (EEA-TF)
Other contributors:
Arne Tollan (EEA-TF)
Additional reviewers:
Peringe Grennfelt (IVL,S), Erne Meszaros (Inst Atm Phys, H), M L Williams (DoE, UK), M Nowicki (Min Env Protection, PL)

Chapter 5: Inland waters
Concept, coordination and editing:
Niels Thyssen (EEA-TF)
Principal authors:
The water resource
Niels Thyssen (EEA-TF), Nigel Arnell (Institute of Hydrology, UK);

Groundwater
Ron Franken, Luc Kohsiek, Karel Kovar, Ton van der Linden, Kees Meinardi, Rob van de Velde and Jaap Willems (RIVM, NL), Bon van Kappel, Nico Pellenbarg, Eric Sprokkereef and Frans van de Ven (RIZA, NL)

Rivers, reservoirs and Lakes
Peter Kristensen, Hans Ole Hansen and Kristian Ditlev Frische (NERI, DK)

Other parts
Niels Thyssen (EEA-TF)

Other contributors:
Victor Oancea, Guy Oberlin (CEMAGREF, F), Aage Rebsdorf (NERI, DK)

Additional reviewers:
Pertti Heimonen (National Board of Waters and the Environment, Helsinki), Michel Maybeck (Université Pierre et Marie Curie, Paris), John Seager (National Rivers Authority, Bristol), Hanna Soszka (Institute of Environmental Protection, Warsaw), Genrich S Vartanyan (All-Union Research Institute for Hydrology and Engineering Geology, Moscow), Jonathan Moore (DG XI), Hartmut Barth (DG XII)

Chapter 6: The seas

Concept, coordination and editing:
Niels Thyssen and Joanne Davies (EEA-TF)

Principal authors:
Steve Nixon (WRc, UK), except for the North Sea (J-P Ducrotoy, North Sea Task Force) and the Baltic Sea, (National Board of Waters and the Environment, Helsinki)

Additional reviewers:
Hartmut Barth (DG XII)
H Charnock (Southampton University),
Georges Verriopoulos (DG XI)

Chapter 7: Soil

Concept, coordination and editing:
Anne Teller (EEA-TF)

Principal authors:
Anne Teller (EEA-TF), Luc Kohsiek, Rob van de Velde, Adi Cornelese, Jaap Willems, Dico Fraters and Frank Swartjes (RIVM, NL), Ben van der Pouw, Dethmer Boels, Wim de Vries (The Winand Staring Centre), Godert van Lynden (ISRIC, NL)

Additional reviewers:
Winfried Blum (A), Martin Desaules (CH), Alexander Gennadiev (Russian Federation), A Gomez (F), Mike Hornung (UK), Rudy Dudal (B), Peter Reiniger, Denis Peter (DG XII)

Chapter 8: Landscapes

Concept, coordination and editing:
David Stanners and Dirk Wascher (EEA-TF)

Principal authors:
Johan Meeus (NL), David Stanners, Dirk Wascher (EEA-TF)

Other contributors:
Nicol Bischoff (EEA-TF/Agricultural University, Wageningen), Adrian Phillips (UK), Hans-Werner Koeppel (Bundesamt für Naturschutz, Bonn), Margareta Ihse (Stockholm University), Graham Drucker (WCMC)

Additional Reviewers:
Margareta Ihse (Stockholm University), J Primdahl (Den Kgl Veterinæ-og Landbohøjskole, Copenhagen), F Diaz Pineda (Universidad Complutense de Madrid), Adrian Phillips (UK), Angel Velchev (University of Sofia, Bulgaria), Zdenek Lipsky (Agricultural University of Prague), Rob Jongman (Agricultural University, Wageningen)

Chapter 9: Nature and wildlife

Concept, coordination and editing:
Dirk Wascher (EEA-TF)

Principal authors:
Forests and mountains
Gerhard Heiss (CNPPA, IUCN),
Dirk Wascher (EEA-TF)

Scrub and grasslands
Paul Goriup
(The Nature Conservation Bureau, Berkshire)

Rivers, lakes and wetlands
Edward Maltby and David Hogan (University of Exeter)

Coast and marine
Denny Elder and Sara Humphrey (IUCN, Gland),
Dirk Wascher (EEA-TF)

Desert and tundra
Paul Goriup (NCBB), Dirk Wascher (EEA-TF)

Mammals
Francois de Beaufort (Musée Nationale d' Histoire Naturelle, Paris), Dirk Wascher (EEA-TF)

Birds
Melanie Heath (BirdLife International),
Dirk Wascher (EEA- TF)

Amphibians and reptiles
Anton Stumpel (Institute for Forestry and Natural Resources, Wageningen)

Fish
Peter Maitland (Fish Conservation Centre, Stirling)

Invertebrates
Martin Speight (National Parks and Wildlife Service, Dublin)

Plants
Martin Schnittler (Bundesamt für Naturschutz, Bonn), Dirk Wascher (EEA-TF)

Nature conservation
Adrian Phillips (UK), Dirk Wascher (EEA-TF)

Other components
Dirk Wascher (EEA-TF)

Network organiser for ecosystems:
Zbigniew Karpowiez (IUCN)

Other contributors:
Udo Bohn (Bundesamt für Naturschutz), Keith Corbett, Graham Drucker and Richard Luxmoore (WCMC), Hanns Langer (IEB), Patric Haffner (Musée Nationale d' Histoire Naturelle, Paris), Jean Pierre Ribaut (Council of Europe), Ian Hepburn (The Nature Conservation Bureau), Ward Hagemeijer (SOVON), Gaynor Whyles (WWF-UK), Roland Tarr (UK), Roland Wirth (IUCN), Didier Vangeluwe (IRSNB), Desmond Kime (B), Sheila MacDonald (University of Essex), Pertti Uotila and Tapani Lahti (Botanical Museum Helsinki), Georges Verriopoulos (DG XI), Marc Roekaerts, Niels Thyssen (EEA-TF)

Additional reviewers:
Claus Helweg Ovesen (Miljövårdsenheten, Malm), Gérard Collin (Cévennes National Park, Florac) Rob Jongman (Agricultural University, Wageningen) Michael Succow (Ernst-Moritz-Arndt-Universität, Greifswald), Angheluta Vadineanu (University of Bucharest)

Chapter 10: The urban environment

Concept, coordination and editing:
Marina Alberti (EEA-TF)

Principal author:
Marina Alberti (EEA-TF)

Other contributors:
Jonathan Parker, Nicolas Sifakis, Niels Thyssen (EEA-TF), Tjeerd Deelstra, Ad Wouters, R uit de Bosch (IIUE), Ute Enderlein (UNECE), Alex von Meier (M+P, Aalsmeer), Jaroslav Beneš , Miroslva Hátle (Czech Republic), Herbert Müller (DG XI), Luigi Zanda (Consorzio Venezia-Nuova)

Additional sources of data:
Ariel Alexandre (OECD), Patricia Brown and David Huthinson (LRC), Xavier Bonnefoy and J-F Collin (WHO), Rob Sluyter (RIVM), The International Council for Local Authorities (ICLEI)

Additional reviewers:
Arian Alexandre (OECD), Ian Clark (DG XI), Joseph Convittz (OECD), Tjeerd Deelstra (IIUE), Erich Den Hamer (DG XI), Vyacheslav L Glazichev (European Academy of the Urban Environment, Moscow), Ekhart Hahn (Gesellschaft für Ökologischen Städtebau und Stadtforschung, Berlin), Nick Hanley (DG XI), Ulf Christiansen (Danish Building Research Institute), John Zetter (UK DoE), John Whitelegg (Lancaster University)

Chapter 11: Human health
Concept, coordination and editing:
Michael Krzyzanowski (WHO- ECEH), David Stanners (EEA-TF)
Principal author:
Michael Krzyzanowski (WHO- ECEH)
Additional reviewer:
Kathleen Cameron (DG XI)

Chapter 12: Population, production and consumption
Concept, coordination and editing:
Marina Alberti (EEA-TF)
Principal author:
Marina Alberti (EEA-TF)
Additional reviewers:
Brian Wynne (Lancaster University), Nick Robins (IIEE), Sylvain Joffre (EEA-TF)

Chapter 13: Exploitation of natural resources
Concept, coordination and editing:
Marina Alberti (EEA-TF)
Principal authors:
Marina Alberti (EEA-TF), Deborah Chapman (UNEP)
Additional reviewers:
Brian Wynne (Lancaster University, Sylvain Joffre (EEA-TF)

Chapter 14: Emissions
Concept, coordination and editing:
Marina Alberti, Gordon McInnes, Niels Thyssen (EEA-TF)
Principal authors:
Marina Alberti (EEA-TF)
Air
Gordon McInnes (EEA-TF)
Water
Niels Thyssen (EEA-TF)
Other contributors:
Peter Kristensen (NERI, DK)

Chapter 15: Waste
Concept, coordination and editing:
Marina Alberti (EEA-TF)
Principal author:
Marina Alberti (EEA-TF)
Other contributors:
Nuclear waste
Serge Orlowski (DG XI)
Transfrontier movements
Andreas Bernstorff (Greenpeace)
Additional reviewers:
Harvey Yakowitz (OECD), Steve Wassersug (USEPA), Brian Wynne (Lancaster University, Eusebio Murillo-Matilla, Seth Davies, Andrew Piavaux, Etienne Le Roy, Nadia Vink (DG XI), Maria Elizabeth Wolf (Eurostat)

Chapter 16: Noise radiation
Concept, coordination and editing:
Jonathan Parker and David Stanners (EEA-TF)
Principal authors:
Alex von Meier (M+P, Aalsmeer), Peter Maxson (Institute for European Environmental Policy, Brussels)
Other contributors:
Herbert Müller (DG XI), Jaroslav Beneš (Czech Ecological Institute, Prague)
Additional reviewers:
Noise
Jon Berry (DG VII), Herbert Müller and Gunnar Jordfald (DG XI), Per Kruppa (DG XII)
Ionising radiation
George Fraser (DG XI), Karl Schaller (DG XII), Marc De Cort (JRC)

Chapter 17: Chemicals and genetically modified organisms
Concept, coordination and editing:
David Stanners and Niels Thyssen (EEA-TF)
Principal authors:
Niels Thyssen (EEA-TF), Anne- Marie Prieels (Tech-Know, B)
GMOs
Joanna Tachmintzis (DG XI)
Other contributors:
Walter Karcher (JRC), Maurice Lex (DG XII), Ferdinand Kaser (EEA-TF)
Additional reviewer:
Goffredo Del Bino (DG XI)

Chapter 18: Natural and technological hazards
Concept, coordination and editing:
David Stanners (EEA-TF)
Principal authors:
Paul Wenman, Jenny Baer (ERM)
Other contributors:
Serge Orlowski (DG XI), Marc De Cort (JRC)
Additional reviewers:
Alessandro Barisich, George Fraser, Karl Schaller (DG XI)

Chapter 19: Energy
Concept, coordination and editing:
Chris Groom (EEA-TF)
Principal authors:
Ray Tomkins, Diana Hill (ERM)
Additional reviewers:
Manfred Decker and Kevin Leydon (DG XVII), Jorgen Henningsen and Henry Mangen (DG XI), Rosemary Montgomery (Eurostat), Gaynor Whyles (WWF-UK), Jonathan Parker (EEA-TF)

Chapter 20: Industry
Concept, coordination and editing:
Chris Groom (EEA-TF)
Principal author:
Chris Groom (EEA-TF)
Other contributors:
Kevin Flowers (ERM), County of Ringkøbing (Denmark), CEFIC
Additional reviewers:
Kevin Flowers (ERM), Paul Schreyer (OECD), Sylvain Joffre (EEA-TF), Dominique Bé (DG III), Jonathan Parker (EEA-TF)

Chapter 21: Transport

Concept, coordination and editing:
  Jonathan Parker (EEA-TF)
Principal authors:
  Henning Arp (DG XI), Jonathan Parker (EEA-TF)
Other contributors:
  Zissis Samaras (Aristotle University, Thessaloniki),
  Per Kågeson (Transport and Environment, Stockholm),
  Steve Peake (University of Cambridge/Royal Institute for
  International Affairs, London)
Additional reviewers:
  Chris Groom, Sylvain Joffre (EEA-TF), Wim Blonk,
  Wolfgang Elsner, Ronny Rohart (DG VII), Ian Clark (DG XI),
  Malcom Fergusson (Institute for European Environmental
  Policy, London), Miriam Linster (OECD), Jack Short (ECMT),
  Gaynor Whyles (WWF-UK)

Chapter 22: Agriculture

Concept, coordination and editing:
  Chris Groom, Jonathan Parker, Anne Teller (EEA-TF)
Principal authors:
  Chris Groom (EEA-TF), Anne Teller (EEA-TF)
Other contributors:
  Niels Thyssen (EEA-TF), Frank Raymond (ERM),
  Kevin Flowers (ERM)
Additional reviewers:
  Bertrand Delpeuch (DG XI), Dominique Guenat
  (ACADE, Fontanezier, Switzerland), Denis Peter (DG XII),
  J J Pacheco and Pierre Ruel (DG VI), Heino von Meyer
  (OECD), Gaynor Whyles (WWF-UK)

Chapter 23: Forestry

Concept, coordination and editing:
  Anne Teller (EEA-TF)
Principal authors:
  Kevin Flowers (ERM), Arnold Grayson (UK),
  Anne Teller (EEA-TF)
Additional reviewers:
  Jeremy Wall (DG III), Hans Andresen (Eurostat)

Chapter 24: Fishing and aquaculture

Concept, coordination and editing:
  Jonathan Parker (EEA-TF)
Principal authors:
  Kevin Flowers and Joe Jarrah (ERM),
  Jonathan Parker (EEA-TF)
Other contributors:
  David Cross (Eurostat), Mike Holden (ERM),
  Dominique Levieil, Alessandro Piccioli (DG XIV)
Additional Reviewers:
  David Armstrong, Armando Astudillo-Gonzalez,
  Dominique Levieil (DG XIV), Gaynor Whyles (WWF-UK),
  Roger Bailey, Melodie Karlson and Janet Pawlak (ICES),
  John Caddy, C H Newton and R Welcomme (FAO),
  Henrik Gislasson (Danish Fisheries and Marine Research Institute,
  Copenhagen), W Wijnstekers and Georges Verriopoulos (DG XI),
  Kirstan Sundseth (Ecosystems Ltd, Brussels), Gaynor Whyles
  (WWF- UK), Chris Groom (EEA- TF)

Chapter 25: Tourism and recreation

Concept, coordination and editing:
  Jonathan Parker (EEA-TF)
Principal author:
  Anna MacGillivray (ERM)
Other contributors:
  Gaynor Whyles (WWF-UK), Dirk Wascher (EEA-TF)
Additional reviewers:
  Lucio D' Amore and Robert Kremer (DG XI),
  Richard Dickenson (DG XXIII), Robert Bentley and
  Peter Shackleford (WTO), Gaynor Whyles (WWF-UK),
  Chris Groom (EEA-TF)

Chapter 26: Households

Concept, coordination and editing:
  Chris Groom and Jonathan Parker (EEA-TF)
Principal author:
  Raymond Colley (ERM), Chris Groom (EEA-TF)
Other contributors:
  Carolyn Stokoe (ERM)

Chapter 27: Climate change

Concept, coordination and editing:
  Sylvain Joffre (EEA-TF)
Principal author:
  Sylvain Joffre (EEA-TF)
Other contributors:
  Arne Tollan (EEA-TF)
Additional reviewers:
  Roel van Aalst and H F van der Woerd (RIVM)

Chapter 28: Stratospheric ozone depletion

Concept, coordination and editing:
  Sylvain Joffre (EEA-TF)
Principal author:
  Sylvain Joffre (EEA-TF)
Other contributors:
  Arne Tollan (EEA-TF)
Additional reviewers:
  Roel van Aalst and H F van der Woerd (RIVM)

Chapter 29: Loss of biodiversity

Concept, coordination and editing:
  Jeffrey McNeeley (IUCN), Dirk Wascher (EEA-TF)
Principal author:
  Jeffrey McNeeley (IUCN)

Chapter 30: Major accidents

Concept, coordination and editing:
  David Stanners (EEA-TF)
Principal authors:
  Paul Wenman and Jenny Baer (ERM)
Other contributors:
  Serge Orlowski (DG XI)
Additional reviewers:
  Alessandro Barisich (DG XI), Karl Schaller (DG XI)

Chapter 31: Acidification

Concept, coordination and editing:
  Sylvain Joffre (EEA-TF)
Principal author:
  Sylvain Joffre (EEA-TF)
Other contributors:
  Anne Teller and Niels Thyssen (EEA-TF),
  Arne Tollan (EEA-TF), Roel van Aalst (RIVM),
  Arne Semb (NILU)

Chapter 32: Tropospheric ozone and other photochemical oxidants

Concept, coordination and editing:
  Sylvain Joffre (EEA-TF)
Principal author:
  Sylvain Joffre (EEA-TF)
Other contributors:
  Roel van Aalst (RIVM), Arne Tollan (EEA-TF)
Additional reviewers:
  Roel van Aalst, Frank de Leeuw and J P Beck (RIVM)

Chapter 33: The management of freshwater

Concept, coordination and editing:
Arne Tollan and Niels Thyssen (EEA-TF)

Principal authors:
Arne Tollan and Niels Thyssen (EEA-TF)

Additional reviewer:
Peter Gammeltoft (DG XI)

Chapter 34: Forest degradation

Concept, coordination and editing:
Anne Teller (EEA-TF)

Principal author:
Anne Teller (EEA-TF)

Other contributors:
Thomas Haussmann and Michéle Lemasson (DG VI),
Pietro Ducci (DG XI)

Chapter 35: Coastal zone threats and management

Concept, coordination and editing:
Niels Thyssen and Jonathan Parker (EEA-TF)

Principal authors:
Steve Nixon (WRc, UK), Niels Thyssen and
Jonathan Parker

Other contributors:
Susan Gubbay (Marine Conservation Society, UK)

Additional reviewer:
Robert Kremer (DG XI)

Chapter 36: Waste production and management

Concept, coordination and editing:
Marina Alberti (EEA-TF)

Principal author:
Marina Alberti (EEA-TF)

Other contributors:
L Barniske (UBA)

Additional reviewers:
Harvey Yakowitz (OECD), Steve Wassersug (USEPA),
Brian Wynne (Lancaster University), Eusebio Murillo-Matilla,
Seth Davies, Andrew Piavaux, Etienne Le Roy and
Nadia Vink (DG XI)

Chapter 37: Urban stress

Concept, coordination and editing:
Marina Alberti (EEA-TF)

Principal author:
Marina Alberti (EEA-TF)

Other contributors:
Sylvain Joffre (EEA-TF)

Additional reviewers:
Ian Clark and Etrich Den Hamer (DG XI)

Chapter 38: Chemical risk

Concept, coordination and editing:
David Stanners and Niels Thyssen (EEA-TF)

Principal authors:
Niels Thyssen (EEA-TF),
Anne- Marie Prieels (Tech-Know, B)

Other contributors:
Walter Karcher (JRC), Ferdinand Kaser (EEA-TF)

Additional reviewer:
Goffredo Del Bino (DG XI)

Data processing and mapping

Jan Bliki, Wim Devos, Jef Maes (team leader),
Vital Schreurs (GISCO), Kris Segers,
Gert Vanhaverbeke

ALL CONTRIBUTORS, MAIN GROUPS AND TEAMS

Project group

Jean-Pierre Ribaut, L Hyver Yésou (Council of Europe),
Jaroslav Beněs (former Czechoslovakia), Torben Moth Iversen
(Denmark), Philippe Bourdeau (EEA-TF), Gertrud Hilf (Eurostat),
Liz Hopkins, Zbigniew Karpowicz, Tina Rajamets (IUCN),
Klaas van Egmond, Roel M van Aalst, Adriaan Minderhoud
(The Netherlands), Christian Avérous, Chris Chung, Paul Schreyer
(OECD), Maria Leonor Gomes (Portugal), Alexandra M Goudyma,
Dmitri Kolganov (Russian Federation), Thomas Litscher
(Switzerland), Jock Martin (UK), Andreas Kahnert (UNECE),
Anastassios Diamantidis, Boris Ivanov, Sipi Jaakkola (UNEP),
Ian Waddington, Janos Zákonyi (WHO)

National focal points

Lirim Selfo, Rikkard Shllaku (Albania), Günter Liebel (Austria),
Andrei Dopkyunas, Oleg Ivanov, Alexander Rachevsky (Belarus),
John Huylebroeck, K De Brabander (Belgium), V Beschkov,
Ivan Filipov (Bulgaria), Viktor Simončič , Margita Mastrovi (Croatia),
I Shekeris (Cyprus), Jiři Dlouhý (Czech Republic), Torben Moth
Iversen (Denmark), Tõnu Miller, Leo Saare (Estonia), Guy S derman
(Finland), Jacques Varet (France), Karl Tietmann (Germany),
Miltos Vassilopoulos (Greece), Miklós Bulla, Tibór László, Nandor
Zoltai (Hungary), Karitas Gunnarsdóttir, Thórir Ibsen (Iceland),
Brendan Linehan, Luke Griffin (Ireland), Patrizia De Angelis (Italy),
Martins Belickis, Aivars Berner, Andris Krikis, Valts Vilnitis (Latvia),
Helmut Kindle, Felix NŇscher (Liechtenstein), Inesis Kiškis,
Arunas Kundrotas (Lithuania), Charles Zimmer (Luxembourg),
Lawrence Micallef (Malta), Mumitru Drumea, Gheorghe Sprinceanu,
Margarita Petrushevskaya (Moldova), Klaas van Egmond,
Hubert M van Schouwenburg (The Netherlands), Kristian
Kristjansson, Berit Kvæven, Vitek Skopal (Norway), Pawel Blaszczyk,
Janusz Radziejowski (Poland), Maria Leonor Gomes (Portugal),
Adriana Gheorghe, Dumitru Mihu, Lavinia Toma (Romania),
Alexandre Groudyma, Dmitri Kolganov (Russian Federation),
Juraj Cúth (Slovak Republic), Emil Ferjančič, Jorg Hodalic,
Jozef Skultety (Slovenia), Antonio Garciá Alvarez (Spain),
Lars-Erik Liljelund (Sweden), Charles Emmenegger, Patrick Ruch
(Switzerland), Esra Fatma Karadag, Tunç Ügdûl (Turkey),
Volodymyr Demkin, Andriy Demydenko, Viacheslav Lipinsky,
Yuriy Ruban (Ukraine), Jock Martin (UK), Vesna Vjadiković
(former Yugoslavia)

Other national contributors

A Dorofeev, Alexei Mozhukhov (Belarus), Andrija Randić (Croatia),
Miroslev Hátle, Jaroslav Mejar, J Pleskaulca, Radim Srám,
Josef Vavroušek (Czech Republic), Bjarne Moeslund, Claus
Lindegaard, Jørgen Lund Madsen, Peter Gravesen (Denmark),
Andrews Kratovits, Jaan Saar, Matti Viisimaa (Estonia),
Tarja Söderman, Erik Wahlstöm (Finland), Márti Bagi, Gábor Lányi,
Gábor Nechay, Elmér Szabó, Gyula Vigh (Hungary), Kozlova Tatjana
(Latvia), Viktorija Maceikaite (Lithuania), Neville Ebejer (Malta),
Arcadie Capcelea, V S Snegovoy (Moldova), Gabriel Kielland,
L O Reiersen (Norway), Zdzislaw Cichocki, Alina Ruszczycka-
Jakubiak, Teresa Pawlowska, Jadwiga Sienkiewicz, Barbara Walkowiak
(Poland); E S Dmitriev, M G Karpinsky, Pyotr M Rubtsov, Valeriy
Rumjanstsev, Vadim Voronin, Dmitri Zimin (Russian Federation);
Pavel Petrovic, J Richtorcik, Dana Syihlová (Slovak Republic),
Vilibald Premzl, Ducan Vuković (Slovenia), Manuel Soler (Spain),
Kjell Andersson, Anders Berntell, Gunnar Blychert,
Carl-Elis Boström, Kjell Grip, Ingrid Hasselsten, Bert Karlsson,
Gunnar Zettersten (Sweden), H Völkle (Switzerland), Sema Acar,
Sveda Zeytinpğlu (Turkey); Alexandre Belov (Ukraine)

UNECE statistical team

Ute Enderlein, Andreas Kahnert (team leader), Kari Nevalainen,
Dimitra Ralli

Eurostat statistical team and GISCO

John Allen (team leader), Els Janssens, Dietmar Maaß, Rosemary Montgomery, Josephine Oberhausen, Maria Pau, Vital Schreurs, Leo Vasquez, Hilda Williamensen, Mattias Wohlén

European Environment Agency Task Force

Marina Alberti, Patrizia Baldi, Nicol Bischoff, Jan Bliki, Philippe Bourdeau (Head of EEA-TF, 1992-94), Michel-Henri Cornaert, Joanne Davies, Wim Devos, Eric Evrard (DG I), Anne-Marie Fynbo, Roy Gibson (Head of EEA-TF, 1990-91), Nathalie Grisard, Chris Groom (UK, Department of the Environment), Brian Grzelkowski, Sylvain Joffre (Finnish Meteorological Institute), Ferdinand Kaser, Bruno Kestermont, Barbara Lübkert-Alcamo, Jef Maes, Gordon McInnes (NETCEN, AEA Tech., UK), Elise Milan, Jonathan Parker, Denis Peter, Marc Roekaerts, Régine Roy, Kris Segers, Nicolas Sifakis, David Stanners (Project Manager), Catherine Stevens, Anne Teller, Niels Thyssen (NERI, DK), Arne Tollan (Norwegian water resources and energy administration), Carine Van Cauter, Jane Van Den Plas, Geert Vanhaverbeke, Dirk Wascher

Other European Commission services

Claudine Elisée, Andrea Fennesz, Richard Moxon, Hans Stausboell (DG I); Dominique Bé, Jeremy Wall (DG III); Thomas Haussmann, Michèle Lemasson, Joao Jose Pacheco, Pierre Ruel (DG VI); Jon Berry, Wim Blonk, Wolfgang Elsner, Ronny Rohart (DG VII); Marc Arnett, Henning Arp, Alessandro Barisich, Philippe Bourel de la Roncière, Ilona Bräunlich, Kathleen Cameron, Ian Clark, Lucio D'Amore, Seth Davies, Goffredo Del Bino, Bertrand Delpeuch, Erich Den Hamer, Tanino Dicorrado, Pietro Ducci, George Fraser, Peter Gammeltoft, Nick Hanley, Jorgen Henningsen, Gunnar Jordfald, Robert Kremer, Henry Mangen, Jonathan Moore, Eusebio Murillo-Matilla, Herbert Müller, Serge Orlowski, Prudencio Perera Manzanedo, Andre Piavaux, Tue Rohrsted, Etienne Le Roy, Karl Schaller, Joanna Tachmintzis, Georges Verriopoulos, Willem Wijnstekers, Nadia Vink (DG XI); Hartmut Barth, Augustin Janssens, Per Kruppa, Maurice Lex, Canice Nolan, Denis Peter, Peter Reiniger (DG XII); Marc De Cort, Walter Karcher, Julian Wilson (JRC - DG XII); Manfred Decker, Kevin Leydon (DG XVII); David Armstrong, Armando Astudillo-Gonzalez, Dominique Levieil, Alessandro Piccioli (DG XIV); Richard Dickinson (DG XXIII); John Allen, Hans Andresen, David Cross, Theo van Cruchten, Gertrud Hilf, Els Janssens, Rosemary Montgomery, Josephine Oberhausen, Maria Pau, Vital Schreurs, Leo Vasquez, Hilda Williamensen, Mathias Wóhlen, Maria Elizabeth Wolf (Eurostat)

Other thematic contributors

Winfried Blum (A), Bertrand Delpeuch, Dominique Guenat (ACADE, Fontanezier, Switzerland), Genrich S Vartanyan (All- Union Research Institute for Hydrology and Engineering Geology, Moscow), Rudy Dudal (B), Alex von Meier (M + P, Aalsmeer), Melanie Heath (BirdLife International), Tapani Lahti, Pertti Uotila (Botanical Museum Helsinki), Udo Bohn, Hans-Wener Koeppel, Martin Schnittler (Bundesamt für Naturschutz, Bonn), Angheluta Vadineanu (University of Bucharest), Angel Velchev (University of Sofia, Bulgaria), Steve Peake (University of Cambridge/Royal Institute for International Affairs, London), Victor Oancea, Guy Oberlin (CEMAGREF, F), Gérard Collin (Cévennes National Park, Florac), Martin Desaules (CH), Luigi Zanda Consorzio Venezia-Nuova), Karin Dubsky (Coastwatch Europe), Miroslva Hátle (Czech Republic), Ulf Christiansen (Danish Building Research Institute), Henrik Gislasson (Danish Fisheries and Marine Research Institute, Copenhagen), J Primdahl (Den Kgl Veterinæ-og Landbohøjskole, Copenhagen), M L Williams, John Zetter (DoE, UK), Jack Short (ECMT), Kirstan Sundseth (Ecosystems Ltd, Brussels), Jenny Baer, Raymond Colley, Alister Fulton, Diana Hill, Mike Holden, Joe Jarrah, Anna MacGillivrary, Kevin Plowers, Frank Raymond, Carolyn Stokoe, Ray Tomkins, Paul Wenman (ERM), Sheila MacDonald (University of Essex), European Academy of the Urban Environment, Moscow), David Hogan, Edward Maltby, (University of Exeter, UK), A Gomez (F), John Caddy, C H Newton, R L Welcomme (FAO), Peter Maitland

(Fish Conservation Centre Scotland), R Karottki (Folkcenter for Renewable Energy, Denmark), Ekhart Hahn (Gesellschaft für Ökologischen Städtebau und Stadtforschung, Berlin), Slava Korzhenko (Global Eco Reform), Andreas Bernstorff (Greenpeace), Michael Succow (Ernst-Moritz-Arndt-Universität, Greifswald), Roger Bailey, Harry Dooley, Melodie Karlson, Janet Pawlak (ICES), Hanns Langer (IEB), Nick Robins (IIEE), Tjeerd Deelstra, Ad Wouters, R uit de Bosch (IIUE), Erne Meszaros (Institute Atm Phys, H), Hanna Soszka (Institute Environmental Protection, Warsaw), Malcom Fergusson (Institute for European Environmental Policy, London), Peter Maxson (Institute for European Environmental Policy, Brussels), Anton Stumpel (Institute for Forestry and Natural Resources, Wageningen), Nigel Arnell (Institute Hydrology, UK), Harald Loeng (Institute Marine Res, N), Torgeir Bakke (Institute Wat Res, N), The International Council for Local Environmental Initiatives (ICLEI), Didier Vangeluwe (IRSNB), Godert van Lynden (ISRIC, NL), Danny Elder, Sarah Humphrey, Jeffrey McNeely, Roland Wirth (IUCN), Peringe Grennfelt (IVL,S), John Whitelegg, Brian Wynne (Lancaster University, UK), Patricia Brown and David Hutchinson (LRC), F Diaz Pineda (Universidad Complutense de Madrid), Susan Gubbay (Marine Conservation Society, UK), Claus Helweg Ovesen (Miljövårdsenheten, Malmø), M Nowicki (Ministry Environmental Protection, PL), François de Beaufort, Patric Haffner, Hervé Maurin (Musée Nationale d'Histoire Naturelle, Paris), Pertie Heinonen (National Board of Waters and the Environment, Helsinki), John Seager (National Rivers Authority, Bristol), Martin Speight (National Parks and Wildlife Service, Dublin), Paul Goriup, Ian Hepburn (Nature Conservation Bureau, UK), Nikolai Friberg, Kristian Frische, Hans Ole Hansen, Peter Kristensen, Aage Rebsdorf (NERI, DK), Harald Dovland, Steiner Larssen, Jozef Pacyna, Frode Pedersen, Arne Semb, Bjarne Sivertsen, Frode Stordal (NILU), J-P Ducrotoy (North Sea Task Force), Ariel Alexandre, Joseph Convittz, Heino von Meyer, Harvey Yakowitz (OECD), Bo Libert, Miriam Linster (OECE), George Neuhof (Oslo & Paris Commission), Michel Meybeck (Université Pierre et Marie Curie, Paris), Zdenek Lipsky (Agricultural University of Prague), Roel van Aalst, Joe Alcamo, Jan Bakkes, J P Beck, James Burn, Adi Cornelese, Robert Downing, Frank de Leeuw, Hans Eerens, Ron Franken, Dico Fraters, Jean Paul Hettelingh, Luc Kohsiek, Karel Kovar, Kees Meinardi, Harrie Slaper, Rob Sluyter, Frank Swartjes,Ton van der Linden, Rob van de Velde, H F van der Woerd, Jaap Willems, (RIVM, NL), G Arnold, Bob van Kappel, Erick Sprokkereef, Frans van de Ven (RIZA, NL), Nikita Glazovsky, Vladimir Kolossov (Russian Academy – Science), Alexander Gennadiev (Russian Federation), H Charnock (Southampton University), Ward Hagemejier (SOVON, F), Margareta Ihse (Stockholm University), Per Kågeson (Transport and Environment, Stockholm), Finn Bro-Rasmussen (Technical University, DK), Bogdan Paranici (TER), Zissis Samaras (Aristotle University, Thessaloniki), L Barniske (UBA), Arnold Grayson, Mike Hornung (UK), Yinka Adebayo, A Aluse, F Belmont, Deborah Chapman, H Gopalan, S Patwary, Peter Peterson Walter Rast, G Schneider, Peter Schröder, Philomène Verlaan (UNEP), Steve Wassersug (US-EPA), Rob Jongman (Agricultural University, Wageningen), Keith Corbett, Gerhard Heiss, Graham Drucker, Richard Luxmoore (WCMC), Xavier Bonnefoy, J-F Collin, Michael Krzyzanowski, Barbara Lübkert-Alcarno, Robert Mérineau, Bent Fenger (WHO/EURO), Wim de Vries, G J Reinds, Dethmer Boels, J J H van den Akker, Ben van der Pouw, (Winand Staring Centre NL), Tony Long, Gaynor Whyles (WWF), Peter Newman, Steve Nixon (WRc,UK), Robert Bentley, Peter Shackleford (WTO), Desmond Kime, Johan Meeus, Adrian Phillips, Anne-Marie Prieels, Roland Tarr (independent consultants)

Other contributors

Sally Cavanaught (Climate Action Net), Annie Roncerel (Climate Network Eur), Raymond Van Ermen (EEB) Malcolm Fergusson (Earth Resources Res), Robert Atkinson (E-W Environment), Nick Robins (IIED), Jean-Michel Decroly, Christian Vandermotten (Free Univ Brussels), Herbert Hamelt (Seft), Mikuláš Huba (SZOPK), Anne Blommaert (translations) Christina Pannocchia (WISE), Robin Clarke, Nigel Jones (Words and Publications)

Summary

PART I THE CONTEXT

1 Reporting on Europe's environment

Explains how the report has been developed and organised and the functions it fulfils. The coverage, constraints, information selection and assessment methods used are presented and discussed. The analytical structure of the report is explained and the interrelationships between environmental assessment, policy making and implementation are specified.

2 Environmental changes and human development

Provides the context for the assessment of Europe's environment where human action is now altering the global environment on an unprecedented scale and where Europe often plays a disproportionately large role in contributing to environmental change. The interactions between environment and development are examined in relation to changes in human induced interferences with biogeochemical cycles. Demographic and economic growth are key to understanding human-induced pressures on the environment, where sustainability and carrying capacity are important concepts against which to view these problems.

3 Europe: the continent

Europe is the second smallest continent, comprising little more than 7 per cent of the Earth's land area. The continent is surrounded by nine major seas, including the world's two largest land-locked seas, the Caspian and the Black Sea. The area includes 46 states, of which 19 are in the EU and EFTA area, 21 in Central, Eastern and Southeastern Europe, 4 are small continental states and 2 are small island territories. The continent spans three climatic zones: the circumpolar, the temperate and the subtropical. The chapter discusses the climatic, geological and biogeographic factors which give the continent its soil and vegetation patterns. It concludes with an account of landuse in Europe.

PART II THE ASSESSMENT

The environment is assessed for each basic environmental medium – air, water and soil – separately, as well as in more integrated functional units – landscapes, nature and wildlife and urban areas. A summary of human health in Europe completes the picture.

4 Air

Presents an overview of the state and trends of air quality in Europe, examining sources, impacts and responses for air pollutants on local, regional and global scales. Although air quality is improving with respect to some pollutants (SO_2), it is deteriorating with respect to others. The impacts of air pollution on human health and the environment are seen as major European problems which require regulations and conventions that, for example, set limits on emissions.

5 Inland waters

Reviews the state of groundwater, rivers and lakes, evaluates water quantity and quality trends, and relates both the state and the trends to natural processes and human activities. Where possible, the results highlight the condition of inland water in each European country and provide comparisons of water issues in different areas of Europe. The data were obtained from many sources, including national water resource surveys, state of the environment reports and the scientific literature, from the results of a specially prepared questionnaire and as a result of EU legislation.

6 The seas

Evaluates a series of problems common to each of Europe's nine major seas – Mediterranean, Black Sea, Caspian Sea, White Sea, Barents Sea, Norwegian Sea, Baltic Sea, North Sea, North Atlantic Ocean. The problems are: lack of effective catchment management; coastal zone pollution; eutrophication; conflict of uses in the coastal zones; introduction of non-indigenous species; lack of control of offshore activities; over-exploitation of resources; and sea level rise as a result of global warming.

7 Soil

Highlights the major role of soil in the functioning of ecosystems and the importance of soil protection for the maintenance of a healthy environment. The functions performed by soil and how they are affected by human activities are reviewed and assessed. The most severe soil degradation processes are discussed. For each threat, the main causes, magnitude, impacts and remedies are presented. Since there is little quantitative information on soil degradation, most of the assessments are qualitative. Some important quantitative assessments were estimated from updates of available models, or derived from case studies.

8 Landscapes

Provides an overview of the values and functions that characterise cultural landscapes. Thirty major European landscape types are differentiated and their geographical locations presented on a map. Typical landscape stresses are identified and illustrated with case studies. A review of legal and strategic measures for landscape conservation is given.

9 Nature and wildlife

Analyses the states of ecosystems, fauna and flora, and nature conservation measures. The main habitat types are described and the ecological functions of and environmental threats to eight natural (or quasi natural) major ecosystem types are examined. The geographic distribution, management qualities and prevailing stresses are analysed and illustrated. Data for this assessment derive from a review performed by an expert network. The assessment of Europe's fauna and flora deals with seven groups of species. Special attention is given to threatened species according to Red Data lists. For both ecosystems and species, a number of case studies portray typical examples and give detailed information to illustrate the overall findings. Existing as well as potential legal and strategic measures in nature conservation are reviewed on national and international scales.

10 The urban environment

Examines the quality of urban environments in Europe and their regional and global impacts. Experimental urban environmental indicators are used to identify major problems in selected European cities. The assessment focuses on urban environmental quality, flows and patterns. The chapter stresses the need for an integrated approach to urban areas and examines planning and management strategies for improving the urban environment.

11 Human health

Summarises the main issues related to the health status of Europeans and the possible links between health and the environment. The review is based on the results of a contemporaneous assessment of environment and health performed by WHO-EURO – Concern for Europe's Tomorrow. Indicators of health are examined, including self-assessment, life expectancy and infant mortality. The major causes of death in Europe – circulatory diseases, cancers, respiratory diseases, communicable diseases, and injury and poisoning – are reviewed. A review of the potential for environment-related health problems in Europe is presented.

PART III PRESSURES

This section examines the agents and means by which Europe's environment is altered by focusing on the pressures induced by human activities.

12 Population, production and consumption

Examines the complex and still poorly understood relationship between people, resources and development in an attempt to clarify the issues involved. The chapter stresses the importance of accurate assessments of the environmental implications of long-term economic and development programmes.

13 Exploitation of natural resources

Explores the differences between renewable (water, forests and crops) and non-renewable (fossil fuels and metal ores) resources and explains how international trade and the growing interdependence of nations have made the sustainable management of these resources a global issue. The chapter focuses on major developments in the use of resources in Europe and reviews the statistics used to monitor resource use.

14 Emissions

Presents an overview of emissions to air and water in Europe: their physico-chemical characteristics, magnitude, pathways and sinks. The review of atmospheric emissions of major pollutants in European countries is based on data reported to UNECE and, where available, on CORINAIR 1990. The analysis of emissions to water is based on the limited quantitative information available, and focuses on emissions from agriculture and wastewater. The contribution of industry as a source of emissions to the aquatic environment is illustrated. Existing emission inventories in European countries are examined, highlighting the need for an integrated means of collecting data on emissions and waste to all media, and the need to harmonise emission inventory methodologies at the European level.

15 Waste

Analyses present trends in waste production in Europe, and assesses the potential threats to the environment resulting from current waste management practices. Current patterns of transfrontier movements of hazardous waste across European countries are examined. The options to reduce waste and recycle materials through integrated processes are presented for a number of waste streams. The assessment is based on information from a joint OECD/Eurostat survey and state of the environment reports. The limited availability, quality and comparability of existing waste statistics highlight the importance of harmonising waste classification systems at the European level.

16 Noise and radiation

Reviews the main impacts of the important 'physical fields' in Europe – environmental noise, ionising and non-ionising radiation – and relates them to main sources. Noise data from international reviews (OECD and WHO) are presented and analysed to assess the overall situation in Europe and in individual countries. Non-ionising radiation – electromagnetic fields and ultraviolet radiation is reviewed, and the main sources and effects of natural and artificial ionising radiation are described.

17 Chemicals and genetically modified organisms

The manufacture, marketing and use of chemicals result in the release of many compounds to the environment, often with undesirable effects on human health, welfare and ecosystems. The sources of these compounds are described, as are the impacts of selected chemicals of concern. The use of genetically modified organisms in EU countries is examined and their potential undesirable effects and the procedures in place to control their safe use discussed.

18 Natural and technological hazards

Examines the characteristics and importance of accidents and natural hazards as causes of environmental impacts; the different types of damage that can be produced are identified. Examples are given of industrial, transport, marine and nuclear accidents. Natural hazards such as storms and floods, heatwaves, fires and droughts which also impact the environment and their potential exacerbation by human activities are described.

PART IV HUMAN ACTIVITIES

Human activities are the sources of pressures on the environment. This section analyses eight key sectors, providing overviews of their environmental impacts, outlooks and driving forces.

19 Energy

Analyses energy-related activities at three stages – production of primary energy, conversion to derived energy (electricity and heat) and end-use – on local and pan-European scales. Environmental impacts from fossil fuels, nuclear power and renewable energy are briefly reviewed. Factors determining future energy use and projected changes are presented.

20 Industry

Presents an overview of the environmental impact of pan-European industry, and highlights differences between various parts of Europe. The importance of industry as a whole in terms of emissions, waste production and the use of natural resources is examined and an evaluation is made of the environmental performance of certain industrial sectors. Ways in which business practice has changed in response to environmental challenges are also considered. There is little industrial data that can be directly related to environmental impact, hence production and energy use data from international sources and national and state of the environment reports for specific industries are used.

21 Transport

Reviews the impacts of transport on the environment and provides an overview of the pan-European transport situation with regional differences. Trends in transport activities and their implications for the environment are examined, as are some underlying forces. The future outlook for transport in Europe is also assessed. The main data used are drawn from international sources, especially the European Conference of Ministers for Transport (ECMT), Eurostat and UNECE. Data from the scientific literature are used for illustrating some specific environmental impacts.

22 Agriculture

Analyses the trends in agricultural structure and practice which have evolved to meet demand, and indicates the potential impacts of agriculture on the environment. Although much data are available at national level, especially in the EU and EFTA countries, most are related to production, employment structure, fertiliser and pesticide use, livestock demographics and farm sizes; it is difficult to assess the size and contribution to impacts on the environment of agricultural production and changes in agricultural systems.

23 Forestry

Examines the situation of Europe's forests and how they are being used. The chapter summarises how associated activities and practices can result in impacts on the environment, and identifies the main driving forces influencing these changes. The data used include traditional forest inventories as well as qualitative information on specific environmental effects and non-wood production forestry.

24 Fishing and aquaculture

Reviews the nature and importance of the impacts of fishing on the environment, and provides an overview of the pan-European situation with regard to fishing and aquaculture, with regional differences. The effectiveness of existing fishing policies are evaluated. The main data are drawn from international organisations (FAO and Eurostat) and examples are taken from national state of the environment reports and the scientific literature.

25 Tourism and recreation

Presents an overview of the pan-European situation with regard to tourism and recreation and highlights local differences. The impacts of tourism and recreation are assessed in six key settings: protected areas; rural zones; mountains; coastal areas; cities and heritage sites; and theme and leisure parks. General trends are presented using data from the World Tourism Organisation, but since tourism statistics do not adequately reflect the pressures of tourism and recreation on the environment, a case-by-case approach using national or local data is adopted.

26 Households

Reviews the environmental impacts of households in terms of resource use and emissions, assesses underlying driving forces and evaluates possible control measures. Data relating households directly to the environment are not available. Socio-economic data are available from international organisations such as Eurostat. Case studies using national data are used to help create a more complete picture.

PART V PROBLEMS

Environmental problems require an integrated approach for their appraisal, cutting across the media, pressures and human activities. This section highlights 12 problems of particular European concern, focusing on their causes and the goals and strategies being adopted for their resolution.

27 Climate change

Deals with the potential impacts in Europe of the enhanced greenhouse effect caused by a rise in CO_2 levels in the atmosphere, which are already 50 per cent more than in pre-industrial times. The chapter discusses the causes of the problem, the consequences (in terms of changed climatic patterns, sea level rise, effects on hydrology, threats to ecosystems and land degradation), and the international strategies being used to try to limit effects.

28 Stratospheric ozone depletion

Analyses the problem of stratospheric ozone depletion caused by the release to the atmosphere of the class of chemicals known as chloro- and bromofluorocarbons, used as refrigerants, industrial cleaners, foaming agents and fire extinguishers. Consequences include possible changes in atmospheric circulation and increased UV-B radiation on the Earth's surface which may lead to increased levels of skin cancer, eye cataracts and effects on ecosystems and materials. The measures necessary to minimise ozone depletion are discussed.

29 The loss of biodiversity

Reviews the extent of biological diversity in Europe, and the reasons for its decline on a continent where human influences are particularly pervasive. The chapter outlines a series of goals that should lead to the conservation of biodiversity and the sustainable use of biological resources, and strategies for achieving these goals, including implementation of the Convention on Biodiversity.

30 Major accidents

Reviews the environmental problems caused by accidents and the attention that has been given to trying to set acceptable risk levels, for both human health and the environment. Risk management is analysed, focusing on the magnitude of the consequences of an accident and the probability that it will occur. The need for industry to assess its own risks and to use integrated safety management systems and audit tools is discussed. Emergency response or contingency plans are discussed in both national and transboundary situations. The chapter concludes with a special section on the causes of nuclear accidents, and strategies for avoiding them.

31 Acidification

Discusses how the combustion of fossil fuels emits sulphur and nitrogen oxides into the atmosphere where these gases are converted into acids which, after deposition, lead to a series of undesired changes in terrestrial and aquatic ecosystems. The chapter focuses on the adverse chemical and biological effects found in lakes, soils and forests as a result of deposition of acidifying substances in amounts exceeding critical loads. Possibilities for reducing emissions through international agreements are discussed.

32 Tropospheric ozone and other photochemical oxidants

Reviews the complex reactions in the lower atmosphere that produce oxidants such as ozone from the main precursors – nitrogen oxides, volatile organic compounds, methane and carbon monoxide. Levels of these oxidants which can have adverse effects on human health are increasing. They can also affect materials such as paint and plastics, crops and possibly forests. Tropospheric ozone concentrations in the northern hemisphere are expected to keep rising at 1 per cent a year. No limiting goals have yet been set and the actions already undertaken are not thought to be sufficient in Europe.

33 The management of freshwater resources

The regional distribution of problems concerning European water resource – such as the imbalance of water availability and demand, the destruction of aquatic habitats, and water pollution – is highlighted and discussed in relation to the pressures arising from human activities in the catchment areas. A series of sustainable goals for water resource management has been proposed, along with the means of reaching them. Particular attention is devoted to the necessity of international cooperation for management of transboundary rivers.

34 Forest degradation

This chapter focuses on the two most important causes of forest degradation across Europe: air pollution, which seriously threatens the sustainability of forest resources in Central, Eastern and, to a lesser extent, Northern Europe; and fire, a major concern in Southern Europe. The analysis of the damage is derived from large-scale remote sensing observations of European-wide surveys, however they do not readily permit cause-effect relationships to be identified. Detailed monitoring could improve understanding. For fires, causes are often related to socio-economic factors which render the control of the causes complex since they often stem from conflicts and tensions in the overall system of land management.

35 Coastal zone threats and management

Highlights the importance of coastal zones as a buffer between the land and the sea, and examines how human activities which create physical modifications of the coastline and emissions of contaminants have led to the deterioration of habitats and water quality. In order to alleviate the serious environmental problems found in many coastal areas, a strategy for integrated coastal zone management has been proposed. This strategy takes into account the importance of coasts for human well-being and, at the same time, provides for habitats which plants and animals require.

36 Waste production and management

Analyses the increasingly severe problem of waste disposal and processing caused by steady increases in the quantity of wastes and their toxic component. Despite increased emphasis on waste prevention and recycling, most European waste is still disposed of by landfill and incineration. Waste control options are discussed; in spite of progress achieved, most waste still escapes control or avoids strict regulations by transfrontier movement. Strategies to minimise waste generation and ensure safe management are seen as crucial to achieve sustainable patterns of production and consumption.

37 Urban stress

Urban areas in Europe show increasing signs of environmental stress, notably in the form of poor air quality, excessive noise and traffic congestion. Cities also absorb increasing amounts of resources and produce increasing amounts of emissions and waste. This chapter analyses the causes of urban stress and their link to the rapid changes in urban lifestyles and patterns of urban development which have occurred in the last few decades. A series of goals and means to achieve sustainable urban patterns in Europe are discussed including: improved urban planning; integrated transport management; efficient use of water, energy and materials; the setting of new standards and improvement of information.

38 Chemical risk

Few environmental problems in Europe cannot be traced to some form of excessive chemical loading, and this chapter reviews the resulting problems and the ways of reducing the danger. The goal is to reduce levels of chemicals in the environment to a low-risk, target level where only negligible effects occur to the population and the environment. The far-reaching EU programme, designed to reduce risks from chemicals in the environment, is discussed.

PART VI CONCLUSIONS

39 General findings

It is found that there are no simple answers to the question 'how healthy is Europe's Environment?', and that a full analysis of responses and the state of actions to protect the environment was beyond the scope of the report. Twelve prominent European environmental problems were identified for which there is clear evidence for justifiable concern and for action to be taken. Although their choice depends on the criteria used it is found that the list is robust and one to which others could be added. The importance of a multi-media cross-sectoral integrated assessment has been demonstrated. A major obstacle however is the lack of comparable, compatible and verifiable data at European level. In spite of the comprehensive approach taken, some areas, such as rural environments and the service sector are seen to have been neglected. To disseminate the results broadly, a line of products such as a CD-ROM version of the report and the underlying databases, popular versions and short summaries in various languages will be produced utilising the processes developed during the preparation of this report.

40 Highlights and responses

An analysis of the highlights in terms of findings, responses and policy options for each of the 23 priority thematic or sectoral issues are presented. The review illustrates the wide range of problems facing Europe's environmental managers and policy makers, and the types of responses and policy options which are currently contemplated. No attempt is made at setting priorities or estimating costs or benefits of actions, alternative possibilities, or inaction. These and the definition of detailed methodologies were beyond the scope of this report.

Appendix 1: An inventory of 56 environmental problems

A non-prioritised listing of the 56 major environmental problems identified in the course of preparation and analysis of Europe's environment for the report, from which the twelve consolidated prominent European environmental problems were identified.

Appendix 2: Information strengths and weaknesses

A listing, for each of 20 sectoral issues and themes, of the information strengths and weaknesses found during the preparation of the report and analysis of available information. In general there is adequate coverage for certain areas (national emission inventories, inland water quality, soil mapping, census data, etc) for large parts of Europe. Harmonised data however are lacking for an extremely wide range of issues, and some significant data gaps exist.

Part I
The Context

Introduction

This report is organised in six parts over a series of forty chapters. The main assessment of the state of the environment is presented in Part II (Chapters 4 to 11). Part III (Chapters 12 to 18) describes the pressures. Part IV (Chapters 19 to 26) provides an appraisal of the major human activities or socio-economic sectors which are the source of these pressures, and Part V (Chapters 27 to 38) presents twelve prominent European environmental problems. The main highlights and conclusions are summarised in Part VI (Chapters 39 and 40).

This first part of the report sets the context for the assessment. There are three aspects to this, each developed in succeeding chapters.

Reporting on the environment has become an important activity as environmental policies and programmes have developed. The transition to sustainable development requires among other things an adequate supply of reliable environmental information. State of environment reports can help in the assessment of alternative policy options, improve political accountability and meet the public right to know. The conclusions of the Dobříš Castle ministerial conference (June, 1991) saw these needs when they called for the preparation of this report. The conference concluded that the report should '... facilitate the development of an environmental programme for Europe' (Conclusion 32), be '... a basis for the effective implementation of environmental policies and strategies' (Conclusion 5) and act '... as a useful tool to inform the public and raise awareness about environmental problems' (Conclusion 5). The declaration to the Lucerne conference (April, 1993) endorsed the work going into preparing the report and stated that the report will '... serve as the basis for the further development of the Environmental Programme for Europe (EPE)'.

These conclusions formed the objectives and guidelines for producing this report. How these were interpreted and applied is the subject of Chapter 1. Here, an explanation is given as to how the report was developed, organised and presented to serve the aspirations expressed in the ministerial declarations and to supply decision makers and the public with a comprehensive knowledge base on the state of the environment in Europe.

Chapters 2 and 3 provide relevant details of the settings for this. Chapter 2 gives a summary of the current understanding of environmental change, how it relates to human development and responses at international level. Chapter 3 describes the specific characteristics and definitions of the geographical setting – the European continent.

Photo overleaf shows satellite mosaic image of the whole of Europe. The colours approximate to natural tones.
Source: Worldsat Productions/NRSC/ Science Photo Library

Monitoring pollution
Source: Martin Bond,
Science Photo Library

1 Reporting on Europe's environment

INTRODUCTION

Europe's environment is changing under the influence of
human activities. This is no recent phenomenon – for many
hundreds, and even thousands, of years humankind has
deliberately changed the surrounding environment to serve
both immediate and long-term needs. The extent to which
this has occurred in Europe now means that our entire
surroundings, even including most of what is perceived as
natural, has been moulded in some way by human activities.
Any 'natural balance' that might have existed at one time
between the earth and *Homo sapiens* has long been broken,
with signs of the current imbalance being evident in the state
of the air, water and soil. There have been many unwanted
side-effects as a result of these changes; some were
predictable, others came as a surprise. This report aims to
describe and explain these changes and effects, to provide a
comprehensive picture of the state of Europe's environment
in the early 1990s, to assist sound decision making and to help
raise public awareness about environmental problems. This
chapter outlines the approach taken in the report to do this
and explains how the large amount of diverse information
that needs to be presented has been selected and ordered.

SCOPE AND COVERAGE OF THIS REPORT

Since the environment is involved with and touched by most
of what we do, the scope of a report on the state of the

environment is potentially very wide. Almost all human
activities depend on, or affect, the environment in one way or
another and thus it is difficult to decide where the boundaries
to environmental reporting lie. It is the very nature of the
environment, its complexity, its multiple interconnections
and in-built delay mechanisms, which render environmental
problems difficult to understand and thus easy to ignore.

This inherent breadth in the concept 'environment'
makes environmental problems important to everybody and
explains the increasing attention being given them. This has
also led to uncertainties and disagreements about the fields of
greatest importance and conflicts over the boundaries of
interest. Views vary between individuals and organisations,
depending upon their particular concerns and interests, and
the time-scale of their outlook. Furthermore, these views
have changed with time as our understanding has increased
and capacities to alter the environment have expanded.

The call and expectations for this report, described in the
Foreword, and the evolving European and international
situation, provide the general context for this current
assessment. Further specific criteria (described below) were
used to give the report focus and meaning. Two important
requirements emerge from the general context:

1   the need for an integral view of Europe, for the first time
    combining information from Central and Eastern Europe
    with that of Western Europe; and
2   an appreciation of Europe's place and role in a global context.

Appraising the pan-European environment together as a unit
for the first time, it was deemed more appropriate to adopt a

comprehensive approach than one which was particularly selective. The comprehensive approach provides a fuller and more independent basis for building the expected Environment Programme for Europe, while also facilitating the laying down of a baseline against which future reporting at this level can be compared. Counting against this all-inclusive approach is the fact that few collections of environmental data have been made on a pan-European basis and that a comprehensive appraisal at this level would therefore be faced with many difficulties. Thus it was recognised that part of the task of preparing the report was to identify and highlight the problems with the data, pointing out the inadequacy of the coordination of the activities surrounding it. In this way the report was conceived as a comprehensive snapshot of the state of Europe's environment, and, as far as possible, a description of the state of environmental information.

Environmental reporting activities have been carried out in European countries since the early 1970s. Some European countries produce state-of-the-environment reports on an annual basis, others less regularly. The European Commission has already published four state-of-the-environment reports (1977, 1979, 1986 and 1992). Other international organisations such as OECD and UNEP prepare state-of-the-environment reports which also feature Europe to some extent. Equally, reports are available from non-governmental organisations such as the World Resources Institute and the Worldwatch Institute which feature the European area. The UNCED conference held in Rio de Janeiro in 1992 requested that each participating country submit a National Report. These reports provide a broad review of environmental policy and initiatives in different countries. However, often there is also a useful description of the state and trends in the environment in that particular country.

CONSTRAINTS AND LIMITATIONS

The nature of environmental reporting can best be appreciated, and the expectations arising from it made more realistic, if it is seen as part of a wider process ranging from research, measurement, analysis and interpretation, through information dissemination to awareness raising and action taking. Each report fits into a wide field of activities and, at best, can be only as good as these activities are in providing a good understanding of the problems involved and appropriate quality data to back it up.

Preparing a report on the state of the environment is hampered by difficulties with the information itself, its objectivity, reliability and comparability, as well as its availability. The very nature of the environment is the underlying cause of many of these problems, which might make it difficult, for example, to set in place adequate and appropriate monitoring. The inadequacy of what is known combined with an often incomplete appreciation of its importance may inappropriately bias what is reported on and lead to erroneous conclusions being drawn.

Research has a crucial role to play ensuring that environmental information properly reflects scientific findings. Particularly significant here are the understanding and results provided by research concerned with the mechanisms of environmental processes, their interaction with human-induced pressures and natural variabilities. Particular difficulties are often encountered connecting the results of research to monitoring – improvements in knowledge and understanding arising from research usually take time to pass to routine applications, and new findings also take time to be disseminated and accepted. To be useful for environmental reporting, companion data are often required on a scale previously unavailable. In the face of such difficulties, there is an understandable tendency to fall back on the presentation of 'official' data and statistics.

The transformation of data into information is at the heart of environmental reporting. Data are observations obtained by various measuring methods, and become information when they are put into context. Information implies interpretation, which is ultimately the basis for action.

A substantial amount of environmental data and information are available. In recent years, environmental data have increased in availability as environmental monitoring and research has become more extensive. Despite this, there is still a paucity of sound environmental information. The fact is, much relevant data are still missing and those which do exist are not always known or accessible. Insufficient use is made of existing data, but those available are often inadequate – spatially and temporally patchy, incomplete and inconsistent. There are many factors contributing to this situation. Environmental monitoring is generally driven by sectorial or thematic considerations. Monitoring is usually confined within strict geographical boundaries, information needs at European or global levels are usually not included, and long term considerations are often lacking. This results in a profusion of often inflexible data collection initiatives in all environmental sectors from which a single comprehensive reliable picture of the environment cannot be formed.

The underlying cause of many of these problems is the absence of an appropriate institutional framework to address environmental concerns directly and fully, leading to a fragmentation of data collection and assessment activities. Across Europe, this is exacerbated by uneven technologies and resources for implementing monitoring activities in the different countries.

Assessment and reporting are carried out in the company of other limitations. The treatment of uncertainty is of key importance in a responsible presentation of 'facts' about the state of the environment. Uncertainty is found at all stages. Early on, when a problem is first being recognised, there is the uncertainty or ignorance as to whether or not a problem exists at all. Further along, there is uncertainty arising from the imperfect understanding of the mechanisms driving the problem. Later, uncertainty and ignorance can also be associated with how much is known about the geographical extent or degree of a problem, being the result, for example, of the scientific or technical limitations of monitoring, or simply the lack of data being collected. Outside these considerations, uncertainty also arises from the unpredictability of many natural processes, which no amount of extra research or monitoring can remove.

Reporting on the environment must proceed in the face of these many difficulties and limitations. It necessarily begins with reference to current scientific findings and with the information which is already available – ideally only that with a clearly defined pedigree and level of data quality. This was the starting point for the preparation of the current report. Information was gathered from government, research and independent sectors, at national and international levels. In special cases, when significant data were missing, an attempt was made to generate the information during the project. The origin of the data was clearly marked and, when uncertainty was a primary issue, reference was made to the current scientific findings. An appraisal of data quality formed part of the assessment and the findings included. Finally, areas were identified where additional or improved data are on a critical path for better understanding of an environmental problem.

Improved reporting on Europe's environment can be properly realised over time only with a fully coordinated data collection system at European level. This is an ongoing process and requires good cooperation between all partners. This is still a long way off, but the results of this exercise, and especially the work of the newly established European Environment Agency (EEA), are expected to contribute to this significantly.

INFORMATION SELECTION AND ASSESSMENT

What is involved in measuring the 'state' of the environment, and how is 'quality' defined? These points are considered below, elucidating what to cover, how to select the relevant material and how to invest the information with meaning.

What to report

When environmental problems first come to light, this is usually the result of information which has been gathered on environmental conditions and often by chance observations. This is closely followed by an analysis of the immediate causes of the recorded environmental changes, both natural (eg, pests, floods or drought) and artificial (eg, pollutant releases), tracing back to the original sources of the environmental pressures – human activities. At this point it begins to be possible to build cause and effect stories by which the fuller consequences of the environmental problems and options for tackling them become clearer.

Four types of information are thus needed in a state-of-the-environment report to give it relevance and purpose:

1 information on the state and trends of different environmental compartments and systems;
2 information about the agents stressing the environment in the context of natural changes;
3 information about the sources of these stresses;
4 a description of the environmental problems themselves.

These elements dictate the main structure of the current report (see below).

The information needed to describe the first three elements (the state, pressures and human activities) is mainly of an objective nature: for example, the actual condition of a particular river is a fact – although the figures may be disputed. A description of environmental problems goes beyond this. The information provided by objective observation of state and pressure, when combined with knowledge of physics, chemistry, biology and of the Earth's biogeochemical and climatic systems, can lead to propositions about causes and effects which may or may not prove to be correct. Long periods of research and debate usually separate the time of the first proposition and its general acceptance as fact, or its rejection. During this time opinions will vary as to whether a particular interpretation is correct or not. Further, even when such a proposition is entirely proven within the scientific community, opinions will differ as to its practical importance or significance. This is the realm of individual and group perception, ethical and social codes, politics and vested interests – influences which affect everybody. Including descriptions of environmental problems therefore means that the fields of theorising, uncertainty and perception cannot be avoided. The most objective presentation includes a description of the most accepted current international thinking and disputes on each problem, explanations of causes and consequences, and outlines of the different options and response strategies being adopted.

Criteria for selection

Within these broad outlines of what should be reported on, there are many variables and details which could be covered. A selection had to be made, identifying what is important and significant for this report. An assessment based on the outline above requires an enormous amount of quantitative data which is not available in a sufficiently comprehensive and reliable form compatible across Europe to produce a full assessment. Nevertheless, large quantities of environmental

data are available which should certainly be employed to their maximum degree. The very existence of such data means that caution needs to be exercised when selecting what to include in the report: bias will be introduced if mere availability of information dictates importance.

There is in fact a close liaison between what information is available and the degree of recognition of an environmental problem. Since environmental monitoring has necessarily to be efficient and selective, data collection is heavily influenced by recognised problems. This in turn affects what is available to report on. If it were certain that the environment was being systematically assessed and the problems of importance were being acted upon with all the necessary supplementary data being collected, then concentrating reporting on those data sets which are available, consistent and complete for the territory of interest would not introduce bias into the results. However, as discussed above, this is far from the case. Furthermore, throughout the different countries and regions of Europe, different sets of environmental problems are recognised as important and, except for a few major issues, few consistent approaches are being adopted to tackle them.

The developments in environmental indicators potentially provide a way out of this problem. Once established they would provide a base set of reliable and comparable information on the state of the environment (see Box 1A). Environmental indicators are designed to provide a succinct and concise summary of the state of the environment and can act as a guide to environmental reporting.

In the absence of an established set of environmental indicators which could be used to report on the state of Europe's environment, a series of criteria was used to help assess which information, out of all that exists, should be included. Nine criteria were defined, derived from the original mandate and objectives of the report:

1 threat to sustainability;
2 impact of global problems in the European area;
3 prominence of a European aspect;
4 transboundary aspect;
5 long-term character;
6 risk for human health;
7 social or cultural impacts;
8 ecological damage;
9 economic loss.

Principally, these criteria were used for identifying prominent European environmental problems chosen for more detailed treatment in Part V of the report (Chapters 27 to 38). However, these criteria were also used as guidelines when selecting which material to focus on in each chapter.

How to give the information significance

When it is stated that the environment is degraded and requires restoration, this involves a quality judgement and implies that something is known about what the environment should be like. In heavily degraded areas, such considerations may seem too obvious or academic. Nevertheless, it is often overlooked that in order to know whether the environment is degraded, to determine the level and importance of that degradation or the amount of restoration required, a clear, concise and understandable measure of environmental quality or health is needed.

The natural environment is a varied background against which to assess and compare the quality of the environment influenced by humans. An appropriate level of understanding of the natural variability of the environment is one way, however, to assess the level of degradation and, from this, the amount of restoration required. In all but a few cases, this approach is neither feasible nor satisfactory due both to lack of information, and to the fact that so often the environment

Box IA Environmental indicators

Attempts to monitor the state of environmental resources have increased the volume of environmental data and information available. However, much relevant data are still missing and those which are usable are patchy and incomplete. Many data are collected in order to meet specific regulatory purposes, rather than the broader aims of environmental policy making.

To enhance the ability to monitor environmental change, sets of environmental indicators have been put forward by international and national organisations. The minimum level of organisation of data that is required (coordination and compatibility) is a list of selected descriptive indicators needed to produce an environmental quality profile. These are particularly useful where it is necessary to present social and economic indicators along with environmental data in order to provide a full picture of the state of the environment in a particular country or region.

In order to provide policy-relevant information, however, it is necessary to select environmental indicators which can help explain changes in the environment and their relationships with human activities. Indeed, this reasoning dictated the choice of data to be included in this report. The lack of sufficient understanding and agreement at all levels as to what constitutes a core set of indicators can partly explain the inadequacy of the results (data availability and quality being two other important confounders).

Three main groups of environmental indicators can be distinguished, and these have their counterparts in the data presented in this report:

I   state-of-the-environment indicators;
2   stressor indicators;
3   pressure indicators.

Environmental indicators are useful for several reasons, including:

● their ability to catch the 'essence' of a state of the environment in a few statistics;
● their capability for allowing comparisons between countries and between regions and the global environment;
● their capability for deriving trends over time (eg, between successive state-of-the-environment reports);
● their potential for measuring the success of environmental programmes and policies.

Further refinements can be made to provide more useful information to policy makers. *Response* indicators can be developed to describe what society is doing to deal with environmental problems (such as environmental abatement expenditures) or for public opinion. *Performance* indicators (such as those currently being developed by the OECD) can be devised which relate to clearly defined norms (legislative or scientifically based). In the longer term, indicators of *sustainability* are needed, as called for in Agenda 21; essentially this means addressing the question of what level of socio-economic development is compatible with a sustainable use of natural resources, which in turn requires some notion of appropriate norms of sustainability.

is profoundly and irreversibly altered from its natural state.

Alternative yardsticks are therefore required to invest measures of the state of the environment with significance. Such 'assessment criteria' might include: limit values, environmental quality standards (EQSs), air quality guidelines (AQGs), maximum permissible concentrations (MPCs), or the use of the critical loads and the critical levels concept. For those problems which are not strictly 'polluting' (eg, loss of biodiversity, landscape degradation, urban problems) more subjective criteria are needed (eg, conservation targets or zoning regulations) but are more difficult to establish and often lacking. Such yardsticks arise out of the application of different environmental principles (such as those described in Box 2B) and depend on the level of understanding of the environmental problem of concern. Thus, some approaches can be strongly policy orientated and target based, arising out of technological and economic considerations, or they might arise directly out of, and adhere strictly to, a particular understanding of the mechanisms of an environmental impact and its environmental significance.

In the current report, the approach adopted varies with the problem of concern. Generally, significance is given to the data presented by referring to the best scientific authorities and experts in the problems presented, and adopting their assessment criteria. Thus, for example, maximum permissible concentrations may be referred to or, in other cases, natural values or conditions may be used. Where this has not been possible due, for example, to problems with data, the available information has sometimes been presented simply as reported, in the form made available, with its significance and interpretation left open for future analysis. The comparison with target or limit values is included where appropriate and contributes towards assessing the success of implementing policies and strategies.

PRESENTATION STRATEGY

For the significance of the material being presented to be properly realised, and for the causes and consequences of environmental problems to be discussed (from which priorities for action can be decided), a simple presentation strategy is required which nonetheless accommodates the complexities. Previous state-of-the-environment reports have tackled this in various ways. A common approach, first proposed and adopted by the OECD, is one that covers: information on the pressures acting on the environment; descriptions of the actual condition or state of the environment (including trends); and descriptions of the responses taken in order to check and control environmental degradation. A different configuration of these information blocks has been adopted here. Special emphasis is given to integrating diverse findings and confronting environmental problems of particular concern to Europe.

A description of the physical, chemical and biological attributes of the environment is fundamentally what a report on the state of the environment is all about. There is no unique way, however, to treat these, since individually or in different combinations they provide alternative views of the environment, their importance and relevance being relative to the use being made of the information. In order to cover the diversity of perspectives for the present assessment, different views of the environment were chosen. Thus, descriptions of environmental attributes have been presented by individual media (the air, water and soil), and in more spatial or functional units (landscapes, ecosystems and wildlife, and urban areas). This is completed by a brief overview of human health in Europe, based on a recent WHO assessment.

A description of the pressures is usually arranged around the sectors of human activities. In this report, the pressures are analysed as the 'agents' which are potentially stressing the environment (emissions to air and water, waste, noise and radiation, chemicals and natural and technological hazards).

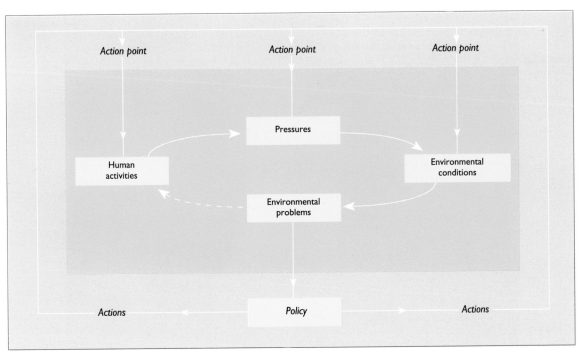

Figure 1.1
The report model

The 'sources' of these agents, the human activities themselves, are dealt with separately by main economic sectors and activities (energy, industry, transport, agriculture, forestry, fishing, tourism and households). This separation helps to demonstrate that it is not necessarily the satisfaction of human 'needs' *per se* which is the cause of environmental damage, but rather the manner and extent by which they are pursued and fulfilled.

Environmental problems link this three-way presentation of media, pressures and human activities (Figure 1.1). Many problems are discussed in the individual chapters of these three sections. However, special attention is given to a series of 12 prominent environmental problems identified to be of particular concern for Europe. These presentations integrate information from the previous chapters in full cause and effect stories, emphasising in particular the goals and strategies being applied to tackle them, or options for action. These four groupings of information are not mutually exclusive, and sometimes create redundancies. Overlaps contribute different visions of the environment and the redundancies are necessary for clarity.

Underlying this presentation is the importance given to demonstrating the interlinkages between the information being presented, and the consequent importance of having comparable data between and among all relevant fields and sectors. This cannot be fully realised in one exercise, particularly given the current inadequate situation with respect to environmental information in Europe. Nevertheless, the approach sought here should be more appropriate for understanding the source of environmental problems, particularly at a heterogeneous European level, as well as for more easily identifying alternative courses of action needed for their resolution. Furthermore, it should also help to clarify data requirements, in particular identifying common needs, leading to more effective data collection and consequently improved environmental assessments.

The analytical structure of the report illustrated

schematically in Figure 1.1 shows a simplified chain from the sources of environmental pressures (human activities), through the pressures themselves to the effects (environmental conditions and problems). The origins and the multiple contributory pathways which lead to environmental problems cannot be adequately illustrated by the model in the figure since the way in which 'causes' lead to 'effects' are many and complex. Responses can be made or, more specifically, actions can be taken, at different points in the chain, treating the sources or the agents of pressure, or treating the environment directly. The figure specifies the inter-relationships between the environmental assessment *per se* (the central part of the figure, and the task of the current report), and the area where policy is made and actions can be taken (the role of society at large and politicians in particular). Although descriptions of each step in this model do have their place in this report, determining or prescribing responses is not the purpose or intention. Instead, the report attempts to present the most objective analysis possible of the current environmental situation in Europe, including the role of current initiatives being used to tackle them and options for future actions. This is expected to form a reliable and rigorous basis for decision making.

In order to make an effective presentation, the storyline illustrated in Figure 1.2 has been followed. This recalls the all-too-familiar retro-active order by which environmental problems normally come to light and steps are taken to tackle them. The linkages between the different parts of this story are of key importance for a proper application of the results of this report. By enumerating the linkages between the different parts presented, sectors deserving priority action can be determined by the degree of influence they have across all fields and problems. Similarly, this approach can help define the multiple contributions to the environmental problems of concern and thus help design better strategies to tackle them across sectors.

The broad way that information has been structured in

What is the state of Europe's environment?	What is causing it to be like this?	What are the causes/ sources of these pressures?	What prominent European environmental problems can be identified, and what is being done to tackle them?
The assessment	*Pressures*	*Human activities*	*Problems*

Figure 1.2
The storyline adopted for the report

this report, separating human activities, pressures and environmental conditions, is expected to be of benefit to policy makers by showing how actions can be directed at different points in the chain (Figure 1.1). Thus, human activities can be targeted directly, changing or eliminating them; or the pressures themselves can be addressed, reducing emissions or altering the manner in which a particular intervention or intrusion in the environment is carried out. Alternatively, the environmental conditions can be tackled directly, by cleaning up and restoring degraded areas, but without addressing the underlying causes this is not a final solution. Finally, the descriptions of environmental problems act to inform the reader about which combination of circumstances and actions is relevant to address both causes and effects and where the interlinkages lie.

Opencast mining
Source: David Preutz

2 Environmental changes and human development

ENVIRONMENTAL CHANGE

Environmental change occurs as a result of both natural and human processes. Environmental systems and human activities contribute to environmental changes through the transformation and transportation of large quantities of energy and materials. Natural systems transform the sun's energy into living matter and cause changes by cycling materials through geological, biological, oceanic and atmospheric processes (the biogeochemical cycles described below). Human activities, on the other hand, transform materials and energy into products and services to meet human needs and aspirations.

Compared with natural processes, human transformation of materials and energy has for the most of human history been relatively small. Nowadays, human activities are altering these flows at unprecedented scales; human-induced consumption and transformation of net primary productivity is estimated to be about 40 per cent of that carried out by the Earth's terrestrial ecosystems (Vitousek et al, 1986). Humans fix almost as much nitrogen and sulphur in the environment as does nature (Graedel and Crutzen, 1989). We are also altering the carbon cycle by releasing large quantities to the atmosphere from the burning of fossil fuels. Human emissions of trace metals such as lead exceed natural flows by a factor of 17. The human contribution of other metals such as cadmium, zinc, mercury, nickel, arsenic and vanadium is twice or more than that of natural sources (Nriagu and Pacyna, 1988).

The scale of planetary changes induced by human activities is also evident in the modification of the physical landscape. Since the eighteenth century, the planet has lost 6 million km^2 of forests – an area larger than Europe (Clark, 1989). In addition, the degradation of land to the point that its biotic function is damaged has increased. According to a recent study from the United Nations Environment Programme (UNEP) the extent of vegetated soil degradation has reached 1964.4 million hectares (17 per cent of the Earth's land area) in the last 45 years, due to overgrazing, deforestation, overexploitation, and improper agricultural and industrial practices (UNEP, 1993). In Europe, the portion of degraded vegetated land reached about 23 per cent of the total over the same period (Oldeman, et al, 1991).

Human – biosphere interactions

Humans belong to the biosphere – the thin layer surrounding the Earth's surface which embraces the whole living world. Nevertheless, *Homo sapiens* has not excelled in conserving the planet for either current or future generations. This is particularly significant given the minuscule length of time that humans have lived on the planet, compared with geological time (see Box 2A).

Four ecological phases of human existence can be described (Boyden, 1992): hunter-gatherer, early farming, early urban, and modern high-energy. During the hunter-gatherer phase, humanity utilised no more than one ten thousandth of one per cent (0.0001 per cent) of the then available photosynthesised solar energy (solar energy powers the biosphere and is available to humans through photosynthesis). One of the turning points of human development occurred around 12 000 years ago (Boyden, 1992) when humankind started manipulating biological systems to its own (perceived) advantage, beginning the agricultural transition. Thereafter, populations grew and proliferated. Cultural development added a technological dimension to the metabolism of human populations, increasing food production on the one hand, and lengthening lifespans on the other. Civilisations also developed in parallel with the advance of military innovation, each reinforcing the other in the domination of the local environment, people or the globe.

Since humans began systematically to utilise their closed environment, the natural equilibrium of continental ecosystems has been affected. This had already started when prehistoric tribes regularly abandoned their camps because of the surrounding accumulation of wastes. The degradation of the environment was noticed by Plato who wrote 2400 years ago:

what now remains [of forested lands in Attica] *is like the skeleton of a sick man, all the fat and soft earth having been wasted away, and only the bare framework of the land being left.* (Critias III, section B)

Although the acuteness of problems started to emerge only in the eighteenth century with the intensification of agriculture, demographic development and the start of industrialisation, it is clear that there has always been an interaction between humankind and the biosphere. These interactions led initially to changes which were local; they have now resulted in profound changes in both the biophysical environment and the human condition, becoming global in character.

Today, humanity is seen as the main force influencing ecosystems. We are not distant or external harvesters of Nature's surplus, but intimately part of the natural world. We are subject to most of the same constraints as other organisms, but also have the power to alter our environment. Until relatively recently we have not hesitated to exercise this ability; the consequences for Europe's environment are described in this report.

Biogeochemical cycles

The interactions between humans and the environment can best be understood by considering the Earth's natural cycles. In general terms, the environment can be seen as a system of reservoirs or stores, with fluxes linking them. The global biogeochemical cycles are essential to global ecology, and the critical role of living systems in the Earth's geochemical cycles is becoming increasingly understood. Living organisms may provide regulatory control of the natural metabolism responsible for maintaining the Earth's atmosphere, waters and soils.

Each element follows a loop or cycle as it is incorporated into the living cells of organisms and then released as the organisms die and decompose. These cycles are an expression of life; they are the metabolic system of the planet. Their various patterns are the consequence of myriad biological, chemical and physical processes that operate on all time-scales. In the absence

---

Box 2A Humans and geological time

The time for which humans have lived on Earth is relatively insignificant on a planetary time-scale. This can be illustrated by shrinking the Earth's historical 4600 million years into the span of a single year, with the present day being midnight on 31 December.

On this scale the oldest rocks were formed on 25 February, the first multi-cellular organisms appeared on 29 October and the first plants started growing at the beginning of November. By 30 November the current composition of the atmosphere had nearly been reached (with 90 per cent of its present oxygen level). The rest of evolution is an ever-accelerating spiral: the first mammals appearing in mid-December; the first primates appearing on December 29 and the first *Homo habilis* emerging at 6 pm on December 31. The Lascaux caves were painted one and a half minutes before midnight while the industrial revolution is only two seconds old.

These two seconds represent the period of time over which humans have carried out most of the exploitation of natural resources relative to the age of the Earth (that is, during the past few centuries). This perspective allows a better understanding of the rapidity of this exploitation with respect to the slow accumulation of natural resources over hundreds of millions of years of geological time.

---

of significant disturbances, these processes define a natural cycle for each element with approximate balances in their sources and sinks. This results in a quasi-steady state for the cycle, at least on time-scales less than a millennium. Our knowledge of the way these biogeochemical cycles relate to each other still compares poorly with our general understanding of the individual cycles themselves.

The study of these cycles has at least two fundamental goals. One is the characterisation of the natural elemental cycles and the linkages among the four elements especially significant to Earth biota – carbon, nitrogen, phosphorus and sulphur. The other is the identification and quantification of changes in these cycles due to anthropogenic activity.

Disturbances in major cycles due to human activities

Since the industrial revolution, human activity has increased to such an extent that it must now be regarded as a significant perturbation of the critical biogeochemical cycles of the planet. Indeed, it can be argued that some activities, such as mining and agriculture, have initiated new cycles. The magnitude of human activity is global and the effects are vast.

Certain indicators of the state of particular cycles, such as levels of atmospheric carbon dioxide (CO_2), carbon monoxide (CO) and methane (CH_4) for the carbon cycle, have moved well outside their recent historical distributions. Similarly, changes in the nitrogen and sulphur cycles are reflected by the onset of what is now called 'acid rain'. It is difficult to identify any major river or estuary that has not been affected in some way by the addition of phosphate from agricultural, urban or industrial sources.

The direct use of natural 'reservoirs' can affect their sustainability if the rate of extraction is greater than their regenerative capacity. For instance, fossil fuels and raw materials, such as ores, can be replaced only in geological time-scales; biological reservoirs can, on the other hand, be regenerated provided certain limits are not exceeded. It is important to note that although fluxes of matter or energy may be small in comparison to the amount in reservoirs, many processes and equilibria are non-linear or rate-limited so that even small disturbances can alter the prevailing equilibrium.

The huge amount of gases and particles released by anthropogenic emissions affects the general equilibrium of the various reservoirs. Sulphur anthropogenic emissions, for instance, represent approximately 25 per cent of the global emissions, and they are concentrated in only 2 per cent of the global area. Disturbing one main cycle will also affect other cycles since they are interconnected, as, for example, are the closely linked iron and sulphur cycles in the atmosphere and oceans (Zhuang et al, 1992).

Carbon

A sketch of the carbon cycle is given in Figure 2.1. The two primary sources of the observed increase in CO_2 are combustion of fossil fuels and landuse change; cement production is a further important source. The latest data available from the IPCC (Intergovernmental Panel on Climate Change) (IPCC, 1992a) give estimates of global fossil fuel emissions in 1989 and 1990 of 6.0 ($\pm$0.5) Gt C (10^9 tonnes of carbon). The annual average net flux to the atmosphere from landuse change during the 1980s was about 1.6 ($\pm$1.0) Gt C. The IPCC estimates the global ocean sink to be 2.0 $\pm$0.8 Gt C per year. The terrestrial biospheric processes which are suggested as contributing to sinks are due to forest regeneration and fertilisation, but these have not been adequately quantified. This implies that the imbalance between sources and sinks (of the order of 1 to 2 Gt C per year), that is the so-called 'missing sink', has not yet been adequately resolved (IPCC, 1992b), although recent research indicates that boreal and temperate forests in the northern hemisphere may be responsible.

Nitrogen

Humans are modifying the nitrogen cycle in four main ways:

1 agriculture and the production of artificial fertilisers place more than half the global nitrogen fixation process within human influence;

2 industrial agriculture has increased the rate of decomposition of organic matter in the soil and, thus, the return of nitrogen to the atmosphere;

3 anthropogenic fossil fuel combustion and biomass burning contributes to 50 to 80 per cent of the total flux of nitrogen oxides to the atmosphere (IPCC, 1992a), which are having an important impact on atmospheric chemistry, biological productivity and the acidity of precipitation in the northern mid-latitudes;

4 the concentration of human populations in coastal zones and the routine discharge of associated wastes and sewage are having important effects on coastal aquatic ecosystems.

Phosphorus

Phosphorus also is essential to growth in terrestrial ecosystems and yet, unlike carbon, it is frequently in short supply. Its relative insolubility limits its availability to organisms in soils, rivers and oceans, while sedimentary deposits provide its major reservoir. Because it is not volatile, phosphorus plays little role in atmospheric chemistry. Without human activities, these characteristics would normally limit this element's involvement in global biogeochemical cycles. However, human interventions have altered the availability of phosphorus both directly and indirectly. The application of phosphorus fertiliser is a direct perturbation, but even more subtle alterations of the phosphorus cycle may influence the dynamics of other cycles. Fire, either natural or as a management technique, may increase the available stocks of phosphorus, because oxidation of plant litter transforms organically bound phosphorus into more available forms. Increased levels of available phosphorus can, in turn, raise the rate of restoration of nitrogen to soils.

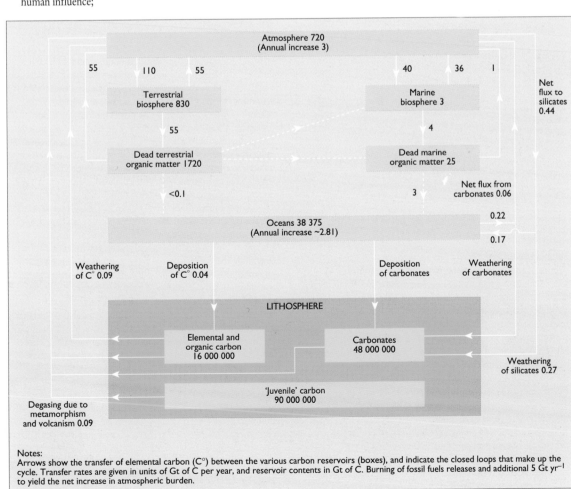

Figure 2.1
The global carbon cycle
Source: From Holland, 1978, and updated from data by Clark, 1982 and Bolin and Cook, 1983

Notes:
Arrows show the transfer of elemental carbon (C°) between the various carbon reservoirs (boxes), and indicate the closed loops that make up the cycle. Transfer rates are given in units of Gt of C per year, and reservoir contents in Gt of C. Burning of fossil fuels releases and additional 5 Gt yr⁻¹ to yield the net increase in atmospheric burden.

Figure 2.2
The global sulphur cycle
Source: NASA

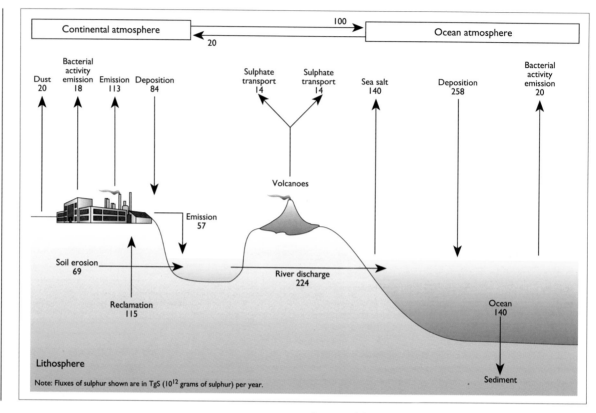

Figure 2.3
World population growth
Source: UN, 1993

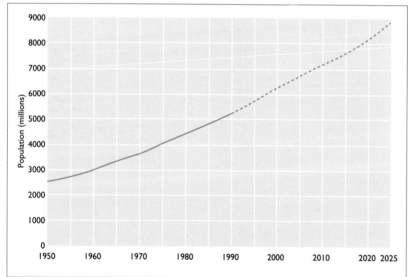

Sulphur

Sulphur plays a vital role in maintaining biological systems because it is an essential nutrient for all plants, animals and simple organisms such as bacteria, fungi and algae. In addition, since sulphur is so widely available in sea water (as sulphate), it strongly influences the carbon cycle and the flux of energy in marine ecosystems. The sulphur cycle is shown in Figure 2.2. Current understanding of the global sulphur cycle is limited. Most of the reduced sulphur gases are believed to be biogenic in origin, but little data exist regarding the nature or strengths of the sources or their spatial and temporal distributions. Furthermore, the mechanisms and rates of the processes which oxidise these gases to sulphur dioxide (SO_2) remain poorly understood. In contrast to human contributions to the global cycles of carbon, nitrogen and phosphorus, where human activity represents relatively minor components of their respective natural fluxes, anthropogenic sulphur emissions may be comparable to releases from natural systems. Recently, this has become a key issue for assessing future climate trends, because of the role that ocean-derived reduced sulphur (mainly in the form of dimethylsulphide, DMS, which oxidises, first to SO_2 and then to particulate sulphate) might play in dampening temperature rises by promoting cloud formation.

Demographic and economic trends

The relative magnitude of natural and human processes has changed enormously throughout different stages of human development. The nature and size of environmental changes can be understood by examining the contemporary demographic and economic trends worldwide. The last 100 years has seen the number of people inhabiting the planet multiply by a factor of more than three, the world economy expand by more than 20 times, the consumption of fossil fuels grow by a factor of 30, and industrial production expand by a factor of 50. Most of these increases have occurred in the last 40 years (MacNeill, 1989), in which same time period the world population has more than doubled (Figure 2.3).

A set of macro-indicators is given in Table 2.1 to illustrate the changes at the global level in the population, the economic growth and the human-induced pressures on the planet's resources. The increase in material and energy flows that contribute to human activities is the result of the exponential growth of population and economic activities. Production and consumption have increased globally at a rate far beyond the growth of human population (Figure 2.4), explaining the widespread rise in living standards. Global emissions of carbon dioxide and CFCs (chlorofluorocarbons) have grown rapidly in the same period, although reduction in the growth rate of the latter has recently been achieved through international action (Figure 2.5).

The rate and distribution of growth has not been uniform in time and place. Population, production and consumption have particularly increased in the last two decades. Population growth has shifted to developing countries, whereas growth in production and consumption has occurred in the most developed part of the planet. The unbalanced distribution of economic and population growth is clearly expressed by the fact that industrialised countries, which represent only one quarter of the world's population, consume about 80 per cent of the world's resources (UNEP, 1992).

These growth trends are expected to continue into the future for the next 20 to 50 years at least. Today's global annual economic output of approximately $ 20 trillion is projected to grow three and a half times over the next 40 years reaching about $ 69 trillion by the year 2030 (World Bank, 1993). If this is achieved according to current development models, environmental impacts will be enormous.

Medium- and long-term projections of the state of the world and national economies are generated by bodies such as the United Nations (UN LINK project), the World Bank, the International Monetary Fund (IMF) and the European Commission, and research institutes such as the London Business School. The latest scenarios available (from the UN LINK project and the World Bank) for the EU are for a modest recovery from 1994 to 2002, with annual growth ranging from 2.5 to 2.7 per cent (UNESC, 1993); this compares with a current annual rate of growth in the EU of just over 2 per cent (CEC, 1993a, p 44). In 1992, the European Commission indicated that annual growth rates for EFTA between 1995 and 2000 will be 2.6 per cent (CEC, 1992, p 25). The period of transition brought a drop in GDP in 1992 for most Central and Eastern European countries. However, projections for Central and Eastern Europe (excluding the former USSR) indicate annual growth rates between 3 to 4 per cent for the period 1994–97, and 3.3 to 4.0 per cent for the period 1998–2002. For the republics of the former USSR, projections indicate annual growth rates between 1.3 and 2 per cent for the period 1994–97, and 2.1 to 5.2 per cent for the period 1998–2002 (UNESC, 1993).

SUSTAINABILITY

The world economy is dependent on the natural environment in a number of significant ways. On one side the environment is a source of energy and materials which are transformed into goods and services to meet human needs. On the other, it is a sink for the wastes and emissions generated by producers and consumers. In addition, the environment provides a number of basic conditions for human life and the economy, such as, for example, a stable climate. Environmental economists refer to this threefold contribution (as source, sink and service provider) of the environment to economic development as 'natural capital'.

The capability of natural systems to provide energy and materials, and to absorb the interference of pollution and waste, is the critical threshold within which the world economy can expand. But how much human-induced change can the global environment sustain? Although knowledge of environmental systems is still too limited to answer this question with any certainty, a host of warning signals provides us with increasing evidence that the impact of human activities might have already gone beyond the capability of maintaining the integrity and productivity of natural resources.

One perspective from which this can be assessed is that of 'carrying capacity'. In ecology, this is the maximum impact that a given ecosystem can sustain, or:

the maximum population that can be supported indefinitely in a given habitat without permanently impairing the productivity of the ecosystem(s) upon which that population is dependent.
(Rees, 1988, p 285)

Thus, the critical threshold for the planet is the carrying capacity of the Earth's ecosystem. The concept of carrying capacity provides an objective basis for defining the sustainability of economic activities. For the human population, the carrying capacity is the maximum rate of resource consumption and waste and emissions production that can be sustained in the long term globally and regionally without impairing ecological integrity and productivity. In order to be sustainable, the level and rate of natural resources depletion and pollution emissions should be no greater or faster than the level and rate of regeneration or absorption of environmental systems.

To assess the sustainability of current patterns of

Indicators (units)	1950	1970	1990
Population (billion)	2.5	3.7	5.3
Urban population (per cent)	29.3	36.6	43.1
GDP ($ billion)	–	2040[a]	22 300
Life expectancy (years)[b]	46.4	57.9	64.7
Energy consumption (million toe)	–	3752[c]	5571
Aluminium consumption (thousand tonnes/year)	–	11 350[d]	17 909
Water consumption (km^3/year)	1360	2590	3240
Fertiliser consumption (thousand tonnes/year)	–	69 308	137 547
Passenger cars (million)[e]	–	193.5	444.9
Passenger kilometres travelled (billion veh-km)[e]	–	2584.3	4892.8
Freight kilometres travelled (billion veh-km)[e]	–	656.1	1702.4
Road network (thousand km)[e]	–	11 701	12 454
CO_2 emissions (million tonnes/year)	6000	15 000	27 870
NO_x emissions (million tonnes/year)[f]	6.8	18.1	89–125
SO_2 emissions (million tonnes/year)	60	114	142–166
CFC emissions (million tonnes CFC-11 equivalents)	–	–	1.4
Municipal waste (million tonnes/year)[e]	–	302[g]	420

See the *Statistical Compendium* for full details on the indicators in this table, and Table 4.7 of this report.

a data refer to 1965
b data on life expectancy represents annual mean values for the periods 1950–55, 1970–75 and 1990–95, respectively
c data refer to 1971
d data refer to 1975
e OECD countries only
f data for 1950 and 1970 refer to million tonnes of nitrogen per year
g data refer to mid-1970s

development, both the level of demand for natural resources and the transformation processes required by human activities should be considered. These are influenced by the size and characteristics of human activities but also by the processes and technologies employed.

The interdependence between the demand and supply of natural resources across world regions is also crucial to assess the sustainability of human activities. In the last two centuries, radical changes in production and transportation techniques have produced an interdependent global economy. Nowadays, almost no region sustains itself by drawing on the resources available within its regional boundaries. Instead, most world regions depend on the supply of resources from other areas.

Table 2.1
Selected world indicators, 1950–90
Source: Based on data compiled by the Monitoring and Assessment Research Centre, see UNEP, 1993

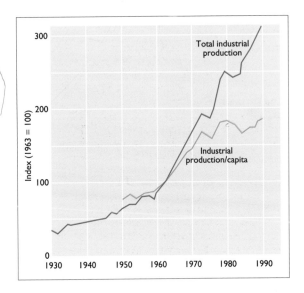

Figure 2.4
Growth of world industrial production (base year, 1963)
Source: Meadows et al, 1992

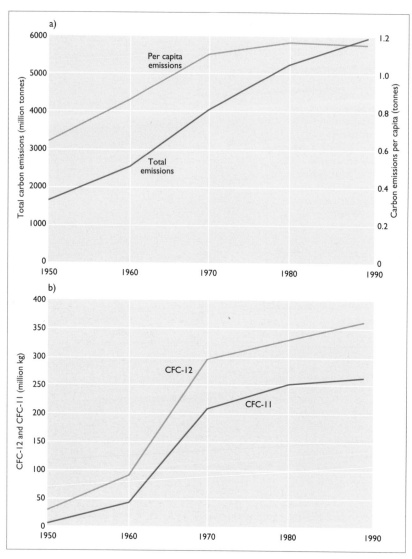

Figure 2.5
Global emissions of
a) CO₂ (measured as
carbon) and
b) CFC-11 and
CFC-12
Source: WRI, 1992

To develop and implement this approach, conceptual work is required to define the factors involved and how they interact. To meet this challenge, international and national initiatives have been undertaken for revising the system of national accounting. The UN statistical division and the World Bank have prepared a handbook, *Integrated Environmental and Economic Accounting*, to accompany the revised SNA (UN/WB, 1993).

The development of key environmental indicators is a parallel complementary initiative which can contribute to a better measure of sustainability by providing base sets of reliable and comparable information on the state of the environment (see Box 1A).

EUROPEAN AND GLOBAL DIMENSIONS

Europe's environment is tightly interlinked with the global environment, natural processes and economic activities. Environmental changes in different parts of the world are also linked together through the interdependence of the demand and supply of natural resources. During the last few decades, increased evidence has emerged showing the interdependence between the environmental conditions of different countries, and between these conditions and the demand and supply of natural resources worldwide. Economic development in Europe not only depends on the trade of natural resources from other regions, but also results in the production of pollution and wastes, which often have global consequences.

Irrespective of political and geographical boundaries, once released into the environment, pollutants are transported by natural processes. Air pollution is one of the most demonstrable areas where the significance of the European and global dimensions can be clearly illustrated. The radionuclides transported across Europe as a result of the Chernobyl accident in 1986 is a powerful reminder of this.

Accidental releases of pollutants on this scale are rare, but emissions are routinely and continuously occurring from myriad sources across Europe as part of the normal functioning of the economic infrastructure. Thus, as air masses pass over Europe they also carry a host of airborne pollutants, dispersing and depositing them at the surface in rain, snow and by simple transfer.

Meteorological variabilities and fluctuations facilitate the dispersion of atmospheric pollutants. Eventually these give them the chance to reach any location in Europe and beyond, exposing humans and ecosystems to the transported contaminants, either directly, or after they have been deposited. Taking sulphate (SO₄) in air as an example, Map 2.1 shows the results of a model simulation of the global mean SO₄ concentration distribution, and the fraction that is due to European anthropogenic emissions. The figure shows a region extending from the North Pole to the Sahel region in Africa, and from central Asia to the mid-North Atlantic, where more than 50 per cent of the sulphate in air derives from Europe.

The disproportionate role that Europe often plays in environmental problems is found in many areas, not only in those related to air pollution. With the European population representing approximately 12.8 per cent of the world's total (1990), atmospheric emissions from Europe contribute comparatively greater proportions of global values: approximately 30 per cent for CO₂, and 36 per cent for CFCs. Figures for industrial waste production show that Europe is responsible for 38 per cent of the total; indeed OECD countries, which account for about 16 per cent of the world's population, produce about 93 per cent of the world's total industrial waste (OECD, 1991).

Europe's interactions with the globe should not, however, be seen solely in terms of sharing out the pressures in the right proportions. The consequences of Europe's impact on the global environment go far beyond this. Thus, for example, some European activities require the extraction of natural resources from non-European countries; and certain compounds, now

Trade enables regions to exceed their local carrying capacity by importing materials and energy, and exporting emissions and waste. Trade is not based on real ecological 'surplus' and generates an ecological deficit in the exporting regions. Worldwide trends in trade of raw materials and waste show exceedance of the carrying capacity in importer countries and an ecological deficit in those countries exporting their natural capital (Rees, 1992). Due to increased integration among production systems, trade among the world's various regions is expected to increase, especially in view of the conclusion of the Uruguay Round of the General Agreement on Tariffs and Trade (GATT) in December 1993.

Measuring sustainability

The UN Conference on Environment and Development (UNCED, 1992) pointed out the inadequacy of indicators of economic development, such as Gross Domestic Product (GDP), Gross National Product (GNP), and of the System of National Accounts (SNA) more generally, to reflect sustainability. Current approaches to national accounting fail to take into account pollution and the depletion of natural resources. Indicators used to assess development do not reflect the conditions and trends in environmental resources. Agenda 21 thus concludes:

Indicators of sustainable development need to be developed to provide solid bases for decision-making at all levels and to contribute to a self-regulating sustainability of integrated environment and development systems.

(UNCED, 1992, Chapter 40)

forbidden or restricted in Europe (eg, DDT and other pesticides) are exported to other countries, not only potentially causing environmental problems there, but also possibly returning in agricultural or processed products through importations to Europe.

RESPONSES

How environmental problems emerge and become recognised depends on society, its organisation, its values and objectives, as well as on the level of environmental awareness. The past two decades have seen an increasing awareness of environmental problems, and in this period some important progress towards their resolution has been made.

The last 20 years

The United Nations conference on the Human Environment held in Stockholm in 1972 was a turning point for better understanding of the extent of existing and expected future impacts of humankind on the earth. Also in 1972, *The Limits to Growth*, the first report of the Club of Rome (Meadows et al, 1972), called attention to the constraints of natural resources, and, although many of its projects and assumptions came under detailed criticism, it contributed to building the concept of sustainability. The Action Plan for the Human Environment adopted by the Stockholm Conference, the establishment of the United Nations Environment Programme (UNEP), the creation of environmental ministries in many countries of the world, and the enthusiasm of non-governmental organisations gave further impetus to the environmental movement and gave it effective expression in the international community for environmental awareness.

This growing awareness did not develop evenly across Europe. From the beginning of the 1970s, a strong divergence appeared between Eastern and Western Europe (Kara, 1992). Most Western European countries began to pursue environmental strategies to tackle environmental problems, achieving some improvements in the process. However, environmental concerns did not achieve the same attention in Eastern European countries where different economic and political conditions gave the environment little priority until the end of the 1980s. This growing tension contributed to the changes in Eastern Europe at the end of the 1980s and eventually advanced the desire, forcefully endorsed at the Dobříš ministerial conference in 1991, to pursue a pan-European approach to environmental restoration and improvement.

From problems to institutions

Recent years have seen the development of another phase of the environmental movement, characterised by the concern evinced at the national and international level around some important, complex, and widespread problems. Major issues such as climate change, stratospheric ozone depletion, acid rain, the disposal of hazardous wastes, desertification and the destruction of tropical rain forests, have seen an increasing application of science in the identification and better understanding of the problems and in the search for strategies to help tackle them. At the same time environmentalism and environmentalists have increasingly entered the political arena.

The United Nations Conference on Environment and Development (UNCED), held in Rio de Janeiro in June 1992, was convened in this context to underline that a new responsibility for environmental protection has to be shared by all countries if sustainable development at the global level is to be ensured. The principal achievement of UNCED was the action plan for the 1990s and into the 21st century, commonly referred to as 'Agenda 21'. This agenda details strategies and an integrated programme of measures to halt and reverse environmental degradation and to promote environmentally sound and

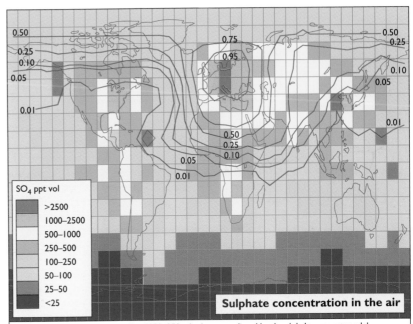

SO₄ ppt vol

▓	>2500
▓	1000–2500
□	500–1000
▨	250–500
▨	100–250
▨	50–100
▓	25–50
▓	<25

Sulphate concentration in the air

Note: This concentration map (for 1000–950 mbar) was predicted by the global transport model MOGUNTIA-S. The contours indicate the fraction of this concentration field that is attributable to anthropogenic sulphur emissions from Europe.

Map 2.1 Annual average sulphate concentration in air close to the surface, and Europe's contribution to it
Source: Environment Institute, CEC Joint Research Centre, Ispra, after Langer and Rodhe (1991)

sustainable development in all countries (UNCED, 1992). A direct result of UNCED was the establishment of the high level Commission on Sustainable Development (CSD) (Agenda 21: Chapter 38). Particularly relevant to environmental information were Chapters 8 (Integrating Environment and Development in Decision-Making), 35 (Science for Sustainable Development) and 40 (Information for Decision-Making).

A series of initiatives before this conference paved the way for this Agenda. The World Conservation Strategy of 1980 (IUCN/UNEP/WWF, 1980) promoted the idea of sustainable development. In 1987 the World Commission on Environment and Development reported on the integration of environment and economic activities, firmly establishing this approach and calling for relevant policy and institutional reforms (WCED, 1987). A new World Conservation Strategy known as 'Caring for the Earth', was published in 1991 elaborating the principles and actions for a sustainable society (IUCN/UNEP/WWF, 1991).

At European level these considerations have been the subject of a series of initiatives. The Bergen Ministerial Conference (May 1990) on sustainable development in the ECE region emphasised the need to improve reporting on the state of the environment and to encourage debate on the environmental implications of national policies. Prompted by the social and political changes beginning at that time in Central and Eastern Europe, a special joint meeting of environment ministers from this region and the EU was held in Dublin in June 1990. A first 'pan-European' conference of environmental ministers took place in Dobříš Castle in the former Czechoslovakia in June 1991 at which the present report on the state of the European environment was requested.

Attention on the environment, particularly in Central and Eastern Europe, has often been triggered by concerns over health. This complex and important relationship between environment and health was discussed for the first time at ministerial level for the whole of Europe in Frankfurt in December 1989. At this meeting, organised by the World Health Organisation (WHO), the European Charter on Environment and Health (WHO, 1989) was adopted, incorporating the basic philosophy of the World Commission on Environment and Development. In preparation for the second environment and health conference (Helsinki, June 1994) the report 'Concern for Europe's Tomorrow' was compiled to assess the health-environment issues of concern across Europe (WHO, in press).

Box 2B
Concepts for environmental protection

Sustainable development: development which, '...meets the needs of the present without compromising the ability of future generations to meet their own needs.' (WCED, 1987, p 43)

Precautionary principle: this broadly demands that if an activity or substance carries a significant risk of environmental damage it should either not proceed or be used, or should be adopted at only the minimum essential level and with maximum practicable safeguards (UNEP, 1992, pp 706–7). Article 130r.2 of the new *Treaty on European Union* provides that EU environmental policy should be based upon the precautionary principle.

Polluter pays principle (PPP): that polluters should bear the full costs of pollution-reduction measures decided upon by public authorities to ensure that the environment is in an acceptable state. More recently the PPP has been extended to accidental pollution (OECD, 1991, p 257).

Shared responsibility: this principle involves not so much a choice of action at one administrative level to the exclusion of others, but rather a mixing of actors and instruments at different administrative levels, enterprises or indeed the general public or consumers (CEC, 1993c, p 78).

Environmental impact assessment (EIA): the necessary preliminary practice of evaluating the risks posed by a certain project before granting permission for a development (UNEP 1992, p 707, and Council Directive 85/337/EEC). A UNECE convention sets certain provisions for the EIA process in the case of transboundary impacts (Espoo, Finland, February 1991).

Best available technology (BAT): signifies the latest or state-of-the-art techniques and technologies in the development of activities, processes and their methods of operation which minimise emissions to the environment (CEC, 1993b, p 35).

Environmental quality standard (EQS): the set of requirements which must be fulfilled at a given time by a given environment or particular part thereof, as set out in legislation (CEC, 1993b, p 35).

Integrated pollution prevention and control (IPC): to provide for measures and procedures to prevent (wherever practicable) or to minimise emissions from industrial installations so as to achieve a high level of protection for the environment as a whole. The IPC concept arose when it became clear that approaches to controlling emissions in one medium alone may encourage shifting the burden of pollution across other environmental media. This concept requires that emission limit values are set with the aim of not breaching EQSs: only when EQSs or relevant international guidelines do not exist can emission limit values be based on BAT (CEC, 1993b, pp 34, 32 and 38–9).

The way forward

Solutions to environmental problems require a new approach and view of societal life, requiring the cooperation and participation of all sectors of society in the identification of problems, the search for solutions and their implementation. Improved environmental education and awareness raising are particularly vital for achieving greater cooperation. Europe, together with other industrialised countries, has a particular responsibility to establish policies which take better account of sustainability. The EC Fifth Environmental Action programme, *Towards Sustainability*, is one contribution to a more participatory form of environmental policy making (CEC, 1993c). The programme promotes the integration of environmental concerns into the different sectors of the economy and encourages closer cooperation between them.

Worldwide, many different strategies are being adopted to tackle environmental problems. These include: regulation (through legal and economic instruments), control and management, international cooperation and agreements, and monitoring and assessment. There are a number of philosophies, principles and concepts which are employed by some of the adopted strategies, many of which are implicitly or explicitly included in international conventions, environmental legislation or in the formulation of environmental standards. Some of the most important of these are briefly described in Box 2B.

The supply of objective, reliable and comparable information on the European environment to inform the public and policy makers about environmental problems and to improve the way they are tackled, is now the task of the European Environment Agency (EEA) and its network. The Regulation establishing this Agency, published in May 1990, came into force at the end of October 1993 with the decision to place its headquarters in Copenhagen. The EEA's commitment to meet the 'public right to know' will be important in providing timely environmental information and for encouraging openness and public access. The activities of the Agency are also expected to improve the coordination of European data systems with global ones, strenthening the European partnership towards the solution of global environmental problems.

REFERENCES

Bolin, B and Cook, R B (Eds) (1983) *The major biogeochemical cycles and their interactions.* SCOPE 21, John Wiley, Chichester.

Boyden, S (1992) *Biohistory: The Interplay between Human History and the Biosphere, Past and Present.* Parthenon.

CEC (1992) *Energy in Europe: a View to the Future.* Special Issue September 1992, DG XVII Energy, Commission of the European Communities, Luxembourg.

CEC (1993a) Growth, competitiveness, employment: the challenges and ways forward into the 21st century, white paper. *Bulletin of the European Communities,* supplement 6/93, Commission of the European Communities, Luxembourg.

CEC (1993b) Proposal for a Council Directive on integrated pollution prevention and control. Commission of the European Communities, OJ No C 311/06, Luxembourg.

CEC (1993c) *Towards sustainability: a European Community programme of policy and action in relation to the environment and sustainable development.* Commission of the European Communities, Luxembourg.

Clark W C (Ed) (1982) *Carbon dioxide review: 1982.* Oxford University Press, Oxford.

Clark, W C (1989) Managing planet Earth. *Scientific American* 261(3), September, 46–54.

Graedel, T E and Crutzen, P J (1989) The changing atmosphere. *Scientific American* 261(3), 28–36.

Holland, H D (1978) *The chemistry of the atmosphere and oceans.* John Wiley, Chichester.

IPCC (1992a) *Climate change 1992: the Supplementary Report to the IPCC Scientific Assessment.* World Meteorological Organisation (WMO) and UNEP, Cambridge University Press, Cambridge.

IPCC (1992b) *1992 IPCC Supplement: Scientific Assessment of Climate Change.* Submission from Working Group I, Cambridge University Press, Cambridge.

IUCN/UNEP/WWF (1980) *The World Conservation Strategy: Living resource conservation for sustainable development.* Gland, Switzerland.

IUCN/UNEP/WWF (1991) *Caring for the Earth: A Strategy for Sustainable Living.* Earthscan Publications, London.

Kara, J (1992) *Geopolitics and the Environment: The Case of Central Europe.* Environmental Politics, 1(2), pp186–95.

Langer, J and Rodhe, H (1991) A global three-dimensional model of the tropospheric sulfur cycle. *Journal of Atmospheric Chemistry* 13, 225–63.

MacNeill, J (1989) Strategies for sustainable economic development. *Scientific American* 261(3), September, 155–65.

Meadows, D H, Meadows, D L and Randers, J (1972) *The Limits to Growth.* Universe Books, New York.

Meadows, D H, Meadows, D L and Randers, J (1992) *Beyond the Limits: Global Collapse or a Sustainable Future?* Earthscan Publications, London.

Nriagu, J O and Pacyna, J M (1988) Quantitative assessment of worldwide contamination of air, water and soil by trace metals. *Nature* (London) 333(6169), 134–9.

OECD (1991) *The State of the Environment.* Organisation for Economic Cooperation and Development, Paris.

Oldeman, L R, Hakelling, R T A and Sombroek, W G, *World Map of the Status of Human-Induced Soil Degradation: an Explanatory Note (revised second edition).* International Soil Reference and Information Centre, Wageningen, and United Nations Environment Programme, Narobi.

Rees, W E (1988) A role for environmental assessment in achieving sustainable development. *Environmental Impact Assessment Review* 8(4), 273–91.

Rees, W E (1992) Ecological footprints and appropriated carrying capacity: what urban economics leaves out. *Environment and Urbanization* 4(2), 121–30.

UN (1993) *World Population Prospects: the 1993 Revision,* United Nations, New York.

UNCED (1992) *Agenda 21.* United Nations Conference on Environment and Development, Conches, Switzerland.

UNESC (1993) Medium- and long-term perspectives for the world economy: synopsis of available projections. Paper prepared by UN Economic and Social Council for UNECE Senior Economic Advisers to ECE Governments, 7–11 June 1993, (EC.AD/R.69 7 April 1993), UNECE, Geneva.

UNEP (1990) *Global Assessment of Soil Degradation: World Map on Status of Human-Induced soil degradation: Sheet 2: Europe, Africa, and Western Asia.* United Nations Environment Programme, Nairobi.

UNEP (1992) *The World Environment 1972–1992.* Tolba, M K, El-Kholy, O A, El-Hinnawi, E , Holdgate, M W, McMichael, D F and Munn, R E. (Eds). United Nations Environment Programme, Chapman and Hall, London.

UNEP (1993) *Environmental Data Report, 1993–94.* United Nations Environment Programme. Blackwell, Oxford.

UN/WB (1993) Handbook of National Accounting. Integrated Environmental and Economic Accounting. United Nations Department of Economic and Social Development Statistical Division, and the World Bank, New York.

Vavroušek J (1992) *A proposal: a better European Environmental Systems Concordare,* 1, 1–4. The International Environmental Negotiation Network, MIT-Harvard Public Disputes Program, Harvard Law School, Cambridge, MA, USA.

Vitousek, P M, Ehrlich, P R, Ehrlich, A H and Matson, P A (1986) Human appropriation of the products of photosynthesis. *BioScience* 36, 368–73.

World Bank (1993) *The World Bank Annual Report 1993.* The World Bank, Washington DC.

WCED (1987) *Our Common Future.* World Commission on Environment and Development. Oxford University Press, Oxford and New York.

WHO (1989) *European Charter on Environment and Health.* First European Conference on Environment and Health, Frankfurt 7–8 December 1989, WHO Regional Office for Europe, Copenhagen.

WHO (in press) *Concern For Europe's Tomorrow.* World Health Organisation, Copenhagen.

WRI (1992) *World Resources 1992–93: a guide to the global environment.* Oxford University Press, Oxford.

Zhuang, G, Yi, Z, Duce, R A and Brown, P R (1992) Link between iron and sulphur cycles suggested by detection of Fe(II) in remote marine aerosols. *Nature* 355, 537–9.

Sunset, Renney Rocks near Plymouth
Source: Spectrum Colour Library

3 Europe: the continent

EXTENT AND BOUNDARIES

Europe is a continent apparently distinct and precise on its western boundaries, but less well-defined in the east (see Map 3.1). Europe stretches from tundra in the north to Mediterranean and desert climes in the south. It abuts Asia in the east, shares the Atlantic with the Americas, and the Mediterranean with Africa and the Middle East. The air it shares with the globe.

The exact boundaries of what constitutes Europe is a controversial matter. It is neither well defined, nor the subject of common agreement. Particularly difficult to decide whether or not they are part of Europe are outlying islands and countries such as Greenland, the arctic islands, Turkey and the trans-Caucasian republics. In the main, this report uses the common geographical definition which excludes the trans-Caucasian republics, Greenland and Anatolia. The border with Asia is taken to follow the Ural mountains, the river Ural to the Caspian Sea, the Manych valley to the Sea of Azov and the Black Sea, and the Bosporus. Being an environmental report, however, it focuses on environmentally significant geographical units which may straddle the often artificial borders created by this definition, for example when considering river catchments or sea areas.

GEOGRAPHY AND GEOLOGY

Europe forms little more than 7 per cent of the land surface of the Earth. It is the second smallest continent, being nearly 30 per cent larger than Oceania, but less than a quarter of the size of

Asia. Its length from Nordkapp (71°12'N) in Norway, to Cape Matapan, in the south of Greece (36°23'N) is about 3850 km, and its breadth from Cabo da Roca in Portugal to the Urals is about 5050 km.

Four major morphological zones can be distinguished:

1 the old Fennoscandian Shield and Caledonian range, consisting of the Scandinavian highlands, and the north and west of Ireland and the UK;
2 the North European Plain from France to Russia;
3 the central and southern European Highlands, comprising the Sierra Nevada, Pyrenees, Alps, Apennines, Carpathians and the Balkans, and;
4 the littoral zone of the Mediterranean.

On the fringe of the bulky Eurasian landmass, Europe appears like an appendage. As a continent it is very fragmented; its mountain ranges and hydrographic network divide up the landscape, and its peninsulas, islands and archipelagos give it complexity and interest. Indeed, 'an aggregation of peninsulas' (from Iberia to Scandinavia, Italy and Greece) describes the continent well. These diverse characteristics combine to give Europe a rich variety of landscapes in a comparatively small area.

Much of Europe's history, economy, landscapes and traditions have been tightly connected to the close presence of the sea. In the western part of the continent there are few areas more than 200 km from the coast. The continent is surrounded by nine identifiable sea basins, including the world's two largest land-locked seas, the Caspian and the Black Sea.

Europe's morphology has been shaped from north to

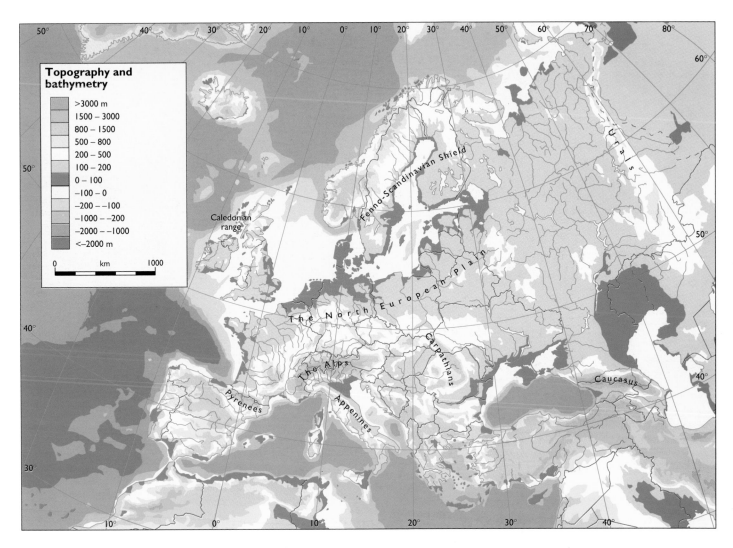

south by successive periods of mountain building (see Map 3.2). The oldest rocks found in Europe – the Precambrian shields (approximately 3500 to 540 million years old) and rocks of the Palaeozoic period (approximately 540 to 245 million years old) – cover most of the northern part of the continent including Scandinavia, the Russian Federation and Ukraine, large parts of Britain and Ireland, Brittany and parts of Spain. The more recent rocks of the Mesozoic period (245 to 65 millions years ago) and the Cenozoic period (from 65 to 2 million years ago) form a continuous belt across north and central Europe from the North Sea into Russia.

Until about 60 to 70 million years ago, Greenland and Scandinavia were parts of the same continent. Northern Europe, of old geological formation, has stabilised, while parts of the south are still unstable as shown by the active volcanoes and the frequency of significant earthquakes. Most types of important mineral resources are found in Europe, distributed unevenly across the continent according to geology. Thus, important deposits of coal are found in Belgium, the Czech Republic, France, Germany, Poland, the Russian Federation and the UK, while these same countries, plus Austria, Spain and Sweden, have deposits of iron ore. Copper, lead, zinc, gold, platinum, silver, mercury, sulphur, and graphite are all found in Europe. There are large oilfields in the Caucasus, Ukraine and Romania and both oil and natural gas offshore beneath the North Sea and Caspian Sea.

Large parts of the continent have been shaped during the Quaternary era (the last 2 million years) and in particular by the last glaciation which ended approximately 10 000 years ago. A great deal of Europe's soils were formed in this era in connection with the many inland ice periods. A characteristic feature of Europe is the Alpine ridge, a permanent divide in the continent for climate, history and trade. The plains are home to most economic and social activities; the so-called 'cradle' of European welfare and power.

This complex and fragmented geography and geology has to a large extent driven Europe's history and its political and territorial complexity. Geography and geology have contributed significantly to the political subdivisions found in Europe today, as well as to the conflicts and changes which have occurred and continue to influence the configuration of European countries and regions. Nowhere in the world is there such a multitude of independent states in such a small area. This provides both richness and also scope for enormous disparities arising from the variety and differences in culture, religion, language, lifestyle and habits. The potential for transboundary problems with environmental consequences is thus of particular significance in Europe.

Europe covers an area of 10.2 million km². At the end of 1993, geographical Europe embraced the whole of 43 independent states, and included parts of three others (the Russian Federation, Turkey and Kazakhstan) (Map 3.3). These states vary enormously in size, from 4.25 million km² for the European part of the Russian Federation, to 0.44 and 1.95 km² for the two smallest independent states in Europe, the Vatican City and the Principality of Monaco respectively.

Two major political groupings of European countries, the European Union (EU) and EFTA, comprise 19 of these independent states, making up 35.3 per cent of Europe's total surface area. In Central and Eastern Europe (including the European parts of the Russian Federation, Kazakhstan and Turkey) a further 21 states can be found (64.5 per cent of the surface of Europe). The remaining area of Europe is made up of 4 small continental states (Andorra, Monaco, San Marino and Vatican City) and 2 island territories (Cyprus and Malta).

Map 3.1
A topography of Europe
Source: CEC/Eurostat

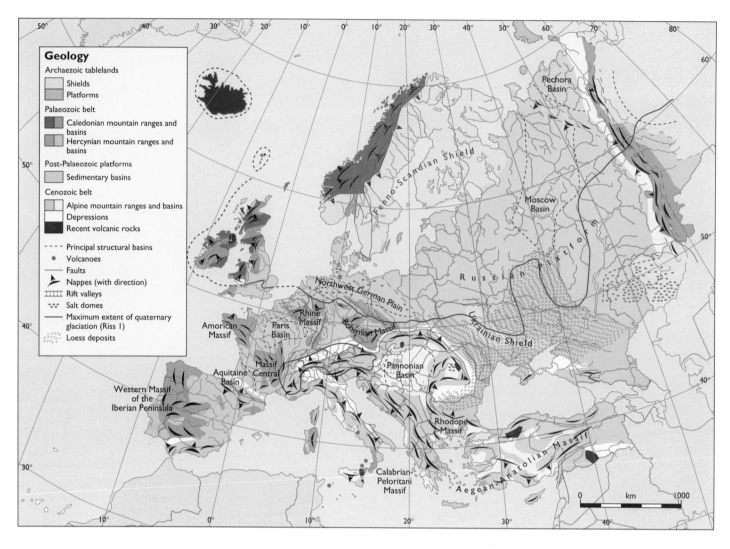

Map 3.2
Geological structure
of Europe
Source: De Agostini, 1992

Since this report was requested (June 1991), 3 states have ceased to exist in Europe (Czechoslovakia, the USSR and Yugoslavia) and 14 new countries have come into being; 7 from the European part of the former USSR (involving approximately 52 per cent of the land surface of Europe), 2 from former Czechoslovakia, and 5 from Yugoslavia.

The country groupings mentioned above are frequently referred to in the report to present information. In some cases the data necessitate alternative arrangements and then the country coverage is explained. Furthermore, some chapters use particular physical or geographical regions in which to present the data, these are either dictated by available statistics or more relevant to the topic under discussion and largely related to important landuse characteristics. These are defined explicitly at the appropriate place in the report.

CLIMATE

The complex and fragmented structure of Europe contributes to the variable meteorological and climatological conditions: from Mediterranean in the south, through temperate oceanic in the west and temperate continental in the east, to boreal in the north. These main types contain many subdivisions linked to altitude (eg, Alpine climate), position and aspect in plains or valleys, and the distance from the sea. Temporal variations of climate have also influenced historical and economic development. The following succinct description is based on the monograph of Martyn (1992).

Europe lies across three climatic zones:

1 the circumpolar zone – the northern tip of Scandinavia, the Svalbard archipelago and northern Iceland;

2 the subtropical zone – the area south of the Pyrenees, Alps, Dinaric Alps and Balkans;

3 the temperate zone – warm in the central south and west, and cool further north especially in eastern Scandinavia and Finland.

The climatic features of Europe also change in an easterly direction: the greater the distance from the Atlantic, the more the climate becomes continental (ie, drier, with greater seasonal and diurnal temperature regimes).

The influence of the Atlantic is strongly felt owing to the warm North Atlantic Current or Drift originating from the Gulf Stream which reaches the shores of western Europe beyond 45°N (mainly fetching up on the coasts of Ireland, Britain and Norway, as well as all the way to the Russian island of Novaya Zemlya). Thanks to this current, the whole area experiences a positive temperature anomaly in winter which gradually decreases south and eastwards across the continent.

Europe's latitudinal position on the globe, between 36 and 71°N, determines the duration of incoming light and heat from the sun. Annual sunshine totals vary from 1000 hours in the cloudiest areas of eastern Iceland, the Faeroes and western Scotland, 1400 to 1600 hours in northwest central Europe to 2800 to 3000 hours in the southern Mediterranean, and even over 3400 hours in Portugal and southeast Spain.

The atmospheric circulation over Europe is established in the presence of permanent cyclonic centres stationed over the North Atlantic (the Icelandic Low), and of the Azores High. In winter, Europe may come under the influence of the seasonal Asian high, and lows moving in from the Atlantic into the Mediterranean basin or forming further north. Summer is characterised by the presence of the Arctic High in the Svalbard region and the South Asian Low. This pressure system gives rise

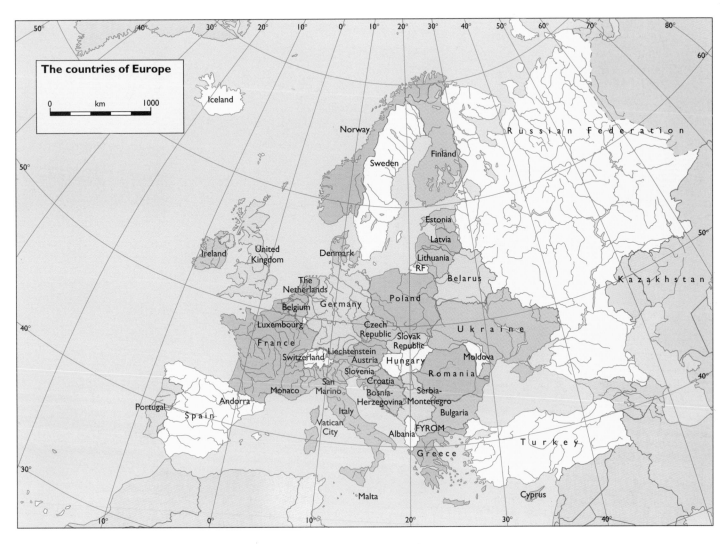

to prevailing southwesterlies over Northern, Western and Central Europe, and northwesterlies and westerlies in Southern Europe (see Map 3.4). The characteristic feature of the climates of Northern, Western and Central Europe is the changeability of the weather. Summer in Southern Europe is much more reliable under the influence of tropical air masses.

These insolation and circulation conditions are well reflected by the air temperature distribution. In January, when thermal conditions are determined by the atmospheric circulation, the isotherms run latitudinally. The 0°C isotherm can be regarded as the boundary between the warm south and west of Europe and its cool centre and north. In July, insolation affects the continent far more than the atmospheric circulation does, although all along the Atlantic fringe the reverse is still the case. The shape of Europe and the dominant westerlies cause the continentality of the climate to increase eastwards, as shown by the annual temperature amplitude. It increases from 8 to 10°C in Iceland, along the Irish and British coasts and Portugal, and from 14 to 16°C along the Atlantic coast of Scandinavia to 20°C in the Iberian Meseta, the Lombardy plain, the lee of the Scandinavian mountains, the Baltic States, central Poland, the Middle Danube plain, the Dinaric Alps and southern Greece. A more continental climate is found in the Middle Danube and Wallachian plains (25 to 26°C), Lapland and eastern Finland (26 to 28°C), as well as central Russia.

Moist air masses from the Atlantic cause the largest amounts of precipitation to fall on western coasts and mountain slopes. In such places the annual precipitation (see Map 3.5) may exceed 1000 to 2000 mm and, in particularly exposed spots, may even reach 3000 to 4500 mm (western parts of Britain and Ireland, Norway, the Iberian and Balkan peninsulas, southern and northern slopes of the Alps and the western Pyrenees).

Precipitation is scantiest (300 to 500 mm per annum) in the southeast of the Mediterranean peninsulas, chiefly southeast and central Spain, southeast Italy and eastern Greece. Under 500 mm per annum falls in Sweden, in the Baltic States, central Poland, the Middle Danube plain, the Dinaric Alps and southern Greece.

Although small, Europe has a wide mosaic of seasons of maximum precipitation, reflecting the influence of orography. Southern Europe has a dry season (spring and summer) which lengthens eastwards and southwards. The rest of Europe has precipitation all the year round, western coasts receiving more in autumn and winter. As a sign of increasing continentality, the proportion of more abundant summer rain grows eastwards from southeast Sweden and west-central Europe. In the Alps and the interior lowlands, however, rainfall is more plentiful in spring. Both the particular climate and the orography contribute to the accumulation of waters in the Alps. Thus Switzerland is the water tower of Europe. The main watercourses leaving the country flow either towards the North Sea (Rhine), or towards the Mediterranean Sea (Rhone). A smaller area of territory belongs to the catchment areas of the Po (Adriatic) and the Danube (Black Sea).

The interplay between natural conditions and cycles and anthropogenic influences can sometimes lead to undesirable conditions being reinforced. In winter, for example, maximum emissions of pollutants often combine with periods of maximum precipitation leading to washout and greater than expected environmental impacts. Under drier, colder, conditions (due to latitude or altitude), many pollutants have longer residence times, leading to a build-up in the atmosphere during winter; at the start of spring – a critical phase in biospheric cycles – the pollutants can still be abundant.

Map 3.3
The countries of Europe
Source: CEC/Eurostat

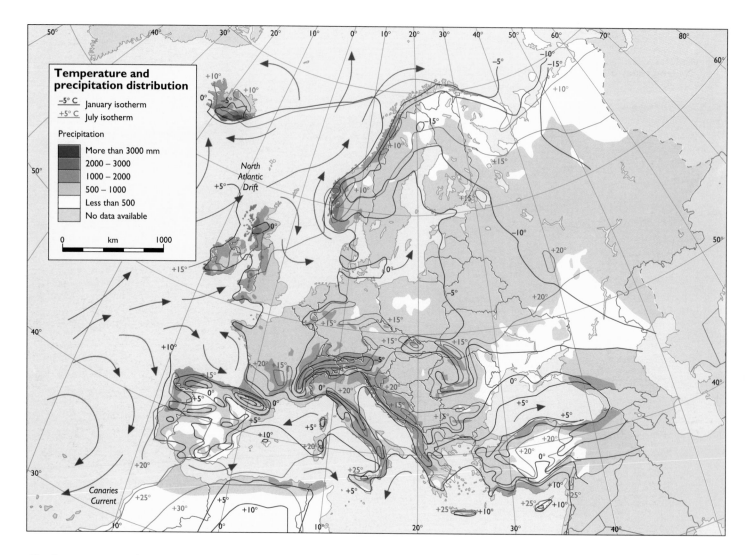

Map 3.5
Annual temperature
and precipitation
distribution in
Europe
Source: De Agostini, 1992

BIOGEOGRAPHY

In addition to the differences in geology and climate across Europe presented above, there are also large variations in vegetation. The combination of geological, morphological and climatological features determines (and is partly determined by) the distribution of soil and vegetation patterns (see Chapter 7, for a generalised soil map of Europe). Climatic conditions modulate the close relationships between soil and vegetation. Distinct geographical areas having their own specific character and originality can be identified by the composition of natural plant and animal life. These can be grouped into different biogeographic provinces as illustrated for Europe in Map 3.6.

Within the northernmost borenomoral zone (subpolar belt) can be found permafrost and tundra (a permanently frozen soil layer reaching down to 400 m below the surface) maintained by long cold winters lasting from September to June. Running alongside and to the south of the borenomoral zone is the taiga, a wide cool and moist temperate climate belt dominated by coniferous forests.

The Boreonemoral province resembles a slightly mixed coniferous forest. The southern limit in Sweden is the northern limit of the beech (*Fagus sylvatica*) forests. The Pontian province covers the true steppes of Eastern Europe (grasslands) and the steppe-woodland belt of Ukraine.

Central and Southern European vegetation is not distributed in clear tracts across the continent; instead, due to the complex morphological and climatic patterns, a mosaic of various types is created. The Pannonian province is recent in origin and secondary in nature; it has a steppe-like appearance and covers the plains of the Middle Danube in the Carpathian basin. This latter belongs to the woodland-belt with riparian oak-forests mixed with Central European forests. The Continental

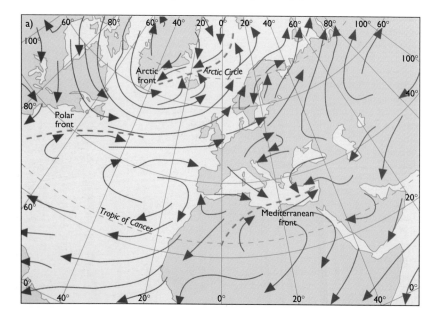

Map 3.4 *Prevailing wind directions and atmospheric fronts over Europe and the North Atlantic, in a) January and b) July (opposite)*
Source: Martyn, 1992

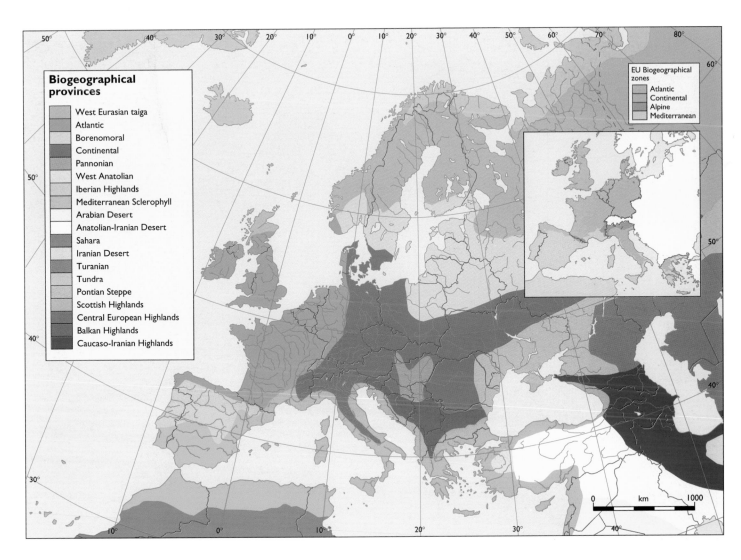

Biogeographical provinces

- West Eurasian taiga
- Atlantic
- Borenomoral
- Continental
- Pannonian
- West Anatolian
- Iberian Highlands
- Mediterranean Sclerophyll
- Arabian Desert
- Anatolian-Iranian Desert
- Sahara
- Iranian Desert
- Turanian
- Tundra
- Pontian Steppe
- Scottish Highlands
- Central European Highlands
- Balkan Highlands
- Caucaso-Iranian Highlands

EU Biogeographical zones
- Atlantic
- Continental
- Alpine
- Mediterranean

province is the heartland of the West Palaearctic broadleaved deciduous forest. The Central European Highlands include the Pyrenees in the west, the Alps proper, the Caucasus in the east, as well as northern outliers. In this province there is much Mediterranean influence and endemism.

The Atlantic province, covering western Denmark, southwest Norway, Ireland, the UK and western France, is under the influence of a mild oceanic climate. Evergreen dwarf shrubs grow where forests cannot establish themselves for climatic or cultural reasons.

The area of typically winter-rain, broadleaved evergreen, sclerophyllous vegetation is the Mediterranean province bordering, in most parts, the Mediterranean Sea. Endemics abound in Mediterranean flora and fauna and are spread throughout the region. In the warm temperate climatic belt, Mediterranean types of vegetation (forests, scrubs and brushwoods) with resistant and hard leaves are prominent.

The semi-arid regions of Central, Southern and Eastern Europe, with few or no trees, are essentially dominated by grass-like (graminaceous) species with fertile chernozem soil (see Chapter 7).

LAND-COVER AND LANDUSE

In the absence of human interference, the prevailing conditions defined by the various biogeographical provinces described above would determine the land-cover in Europe. This would result in a vegetation cover very different to that which is found today; a view of what this might look like is illustrated in a map of 'potential natural vegetation' (see Map 9.1).

Few areas of Europe remain in their 'natural' conditions.

Those that do are mainly found at the extremities of the continent (mountains, lakes, islands) and in northern and eastern regions. Actual land-cover depends upon the level of human habitation and the degree and type of human activity and exploitation of the land. In most parts of Europe this has created entirely artificial landscapes principally as a result of the practices of agriculture, forestry or the construction of settlements (particularly, large urban areas). Although artificial,

Map 3.6
Biogeographical provinces of Europe
Source: Udvardy, 1975, with EU-approved biogeographical regions inset

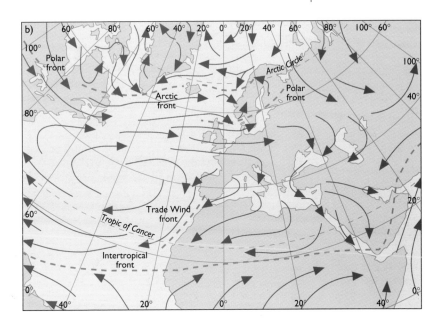

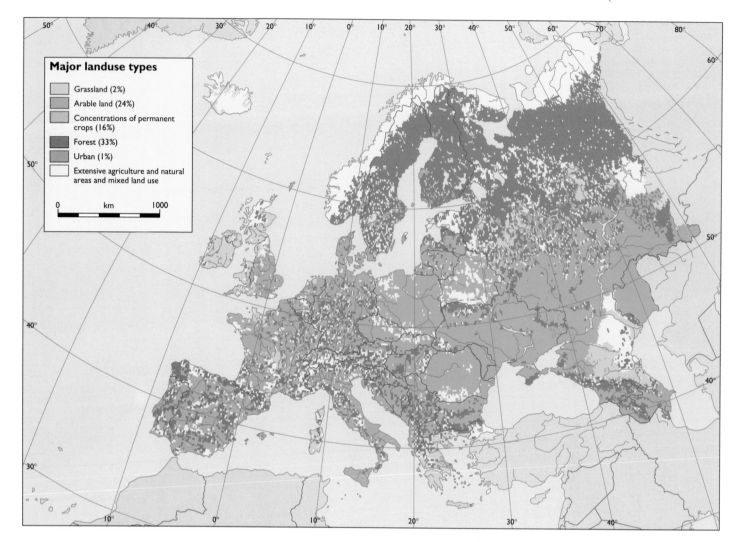

Major landuse types

Grassland (2%)

Arable land (24%)

Concentrations of permanent crops (16%)

Forest (33%)

Urban (1%)

Extensive agriculture and natural areas and mixed land use

0   km   1000

Map 3.7
Landuse in Europe
Source: Van de Velde et al, 1994

Table 3.1
*Population density
and available land
per inhabitant for
some world regions*
Source: Based on FAO,
1990, and CEC/Eurostat
Note: 1 excluding
permanent crops
2 including forest,
mountains, tundra, etc

such areas are not necessarily poor in biodiversity. Land is subject to many competing uses or functions and, being a finite resource, conflicts often arise between them. These pressures on land as a resource are addressed in Chapter 13 wherein the exploitation of Europe's natural resources is examined.

The way that land is used has a primary influence on the type of pressures which are allowed to act on the environment. An assessment of the actual landuse in Europe is illustrated in Map 3.7. This assessment (Van de Velde et al, 1994) calculates that about 42 per cent of total land area in Europe serves some agricultural purpose, 33 per cent is covered by forest and 24 per cent by mountains, tundra, etc.

These percentages, however, vary greatly between countries:

● The proportion of land devoted to agriculture in Europe varies between less than 10 per cent in Finland, Sweden and Norway, to 70 per cent or more in Hungary, Ireland, Ukraine, and the UK. Farming is certainly the landuse which has most significantly shaped, and in many ways enhanced, the landscape to which most Europeans are accustomed, and if agricultural landuse changes significantly (for instance under 'set aside' arrangements in the EU) this could alter the landscape considerably (see Chapter 22).

● Forest cover varies from 6 per cent in Ireland up to 66 per cent in Finland. Unlike agriculture areas, forest areas are expanding in Europe. Recent trends indicate that there has been a 10 per cent increase of forest cover in Europe, over the last 30 years (see Chapter 23).

● The land surface devoted to urban areas in Europe is also increasing. To give an example, it is calculated that each decade 2 per cent of agricultural land is lost to urban areas in Europe (see Chapter 10).

Some comparisons of land and landuse statistics per inhabitant between Europe and other world regions are found in Table 3.1. These show Europe to be 2 to 3 times more densely populated than the USA or Africa, but about half as much as Asia (excluding the republics of the former USSR). Recent figures show that, per inhabitant, only Africa and Asia (as defined above) have less arable land for food production than Europe. With regard to forests and woodlands, only the areas of Asia outside the former USSR have less than Europe.

REFERENCES

De Agostini (1992) Europe, *Le Grand Atlas*. Istituto Geografico De Agostini, Novara.

FAO (1990) *FAO Production Yearbook 1989*. FAO Statistical Series No 94. Food and Agriculture Organisation, Rome.

Martyn, D (1992) *Climates of the World*. Elsevier Publications, Developments in Atmospheric Sciences 18, Amsterdam.

Udvardy, M D F (1975) *A Classification of the Biogeographical Provinces of the World*. IUCN Occasional Paper No 18. IUCN, Gland.

Van de Velde, R J, Faber, W, van Katwijk, V, Scholten, H J, Thewessen, Th, Verspuy, M and Zevenbergen, M (1994) *The Preparation of a European Land Use Database*. RIVM report no 712401001, Bilthoven.

	Europe	USA	Africa	Asia (excl former USSR)
Population per km^2 total land	67	27	21	114
Population per km^2 arable land	267	132	374	725
Arable land[1] per inhabitant (ha)	0.38	0.76	0.27	0.14
Forest & woodland per inhabitant (ha)	0.83[2]	1.1	1.1	0.17

Part II
The Assessment

Introduction

The assessment of environmental conditions in Europe is presented in the following eight chapters, each adopting a different focus and perspective on the environment. The first four chapters concentrate on environmental media (air, water – inland and marine – and soil) and the next three on more integrated or functional environmental units (landscapes, nature and urban areas). Chapter 11 completes the picture with an assessment of human health.

Each chapter is presented in a form appropriate to the nature of the environment being considered, the type of issues of importance and the state of information about the subject.

Cross references to other chapters are used to reduce repetition but some overlap between chapters is unavoidable, and sometimes desirable, for a clear presentation. A proper understanding of the linkages between environmental 'compartments' and with human activities and the pressures they exert on the environment is key to developing appropriate and effective strategies to tackle environmental problems.

Photo overleaf shows coloured satellite view of Europe. This is a mosaic of several cloud-free images of parts of Europe taken by NOAA weather satellites. The colours approximate to natural tones and indicate different surface cover. The mosaic shows Europe in summer so very little snow is seen.
Source:
GEOSPACE/Science
Photo Library

Air pollution
Source: Gamma

4 Air

INTRODUCTION

Good air quality is a prerequisite for the health and well-being of humans and ecosystems. Polluted air will affect human health, ecosystems and materials in a variety of ways. The atmosphere can act as a means for transporting local pollution emissions to other locations, even long distances away and to other media (land and water). In this chapter, the mechanisms behind the build-up of atmospheric pollution at different scales are described, to explain the causal chain between anthropogenic pollutant emissions and the status of air quality in Europe. Present and future trends of air quality are discussed in relation to implemented and planned policy measures and technological developments. The consequences and impacts of changed air quality are described and assessed using common and consistent procedures at three spatial scales extending across national borders. There are, however, important issues which cut across these scales, where, for example, control policies optimal at one scale may not be optimal at another, and air pollution emission reductions targeted in one area may lead to pollution of other parts of the environment.

It was for long believed that air pollutants, once released, were eventually diluted to negligibly low concentrations in the atmosphere. However, measurements taken during the last 20 to 30 years have shown this belief to be erroneous and incomplete. There are three main reasons for this:

1 Not all of the troposphere is available to dilute released pollutants. Most pollutant emissions occur at or close to the Earth's surface in the lowest layer of the atmosphere, the so-called 'mixing layer' (Box 4A). Depending on the meteorological conditions, and especially when the mixing layer corresponds to a temperature inversion, pollutants can accumulate in locally restricted zones leading to high concentrations and 'smog'. Constructing stacks higher than temperature inversions was a short-sighted solution that brought some local relief, but shifted pollution problems to different areas and even to larger scales.

2 Emitted pollutants undergo changes in the atmosphere: the many different anthropogenic and natural compounds (see Box 4B) disperse, mix, are transported and undergo chemical and physical reactions. Sooner or later, nearby or remote to the original release, ingredients of this pollution 'cocktail' are returned to the Earth's surface in one of a number of forms, where they can have adverse effects on ecosystems, humans and buildings.

3 Compounds remain in the atmosphere for differing lengths of time. This duration, the residence time, is determined by the processes of deposition and chemical conversion (see also Box 4B). If atmospheric residence times are of the order of 30 days, vertical mixing may extend to the whole troposphere, and hemispheric transport will be important. Only if residence times are considerably longer, between 6 and 12 months, will exchange between the northern and southern hemispheres take place; after 12 months, exchange between the troposphere and the stratosphere becomes important. It is in this way that European emissions of so-called 'greenhouse gases' and of CFCs contribute to the global problems including the 'enhanced greenhouse effect' and stratospheric ozone depletion.

The fate of airborne pollutants is determined mainly by the release height and the prevailing weather. Thus, the scale of distribution patterns and effects will range from local (up to a few tens of kilometres), through regional (up to a few hundred kilometres), to continental (up to a few thousand kilometres) and global. Concentrations of air pollutants will vary greatly with time (daily, weekly, and seasonally) and in space.

Box 4A The structure of the atmosphere

The Earth's atmosphere consists of several layers (see Figure 4.1) extending from the Earth's surface to outer space. However, the part of the atmosphere that affects weather and climate, and in which most air pollutants are emitted and dispersed, is just a thin layer, from the surface up to a height of 10 to 15 km, called the troposphere. (If the Earth were the size of an apple this layer would be less than the thickness of its skin.) Above the troposphere lies the stratosphere, between 15 and 50 km above the Earth's surface. The stratosphere is rather isolated from the weather in the troposphere because vertical transfers between these two layers are very slow. The stratosphere holds the ozone layer (at a height of 20 to 40 km), which protects the Earth from the sun's ultraviolet radiation, an essential feature of the Earth's atmosphere for the continuing sustenance of life.

The so-called 'mixing layer' of the atmosphere is generally turbulent, varying in depth with meteorological conditions, ranging from a few hundred metres (the height of the Eiffel tower) to about 2 km. Pollutants released into this layer are thus effectively dispersed into a larger volume of air within a few hours, including some transfer to higher levels of the troposphere. If turbulence in the mixing layer is impeded, pollutants accumulate in an even smaller volume, giving rise to high concentrations ('smog'). This occurs when wind speeds are low and when the Earth's surface temperature is lower than the air above (a temperature inversion). Contrary to the troposphere, where on average the temperature decreases with height, the inversion layer (typically 100 to 300 m in depth) is characterised by a temporary increase of temperature with height, with weak wind and slow mixing, and practically no exchange with the overlying background troposphere.

Figure 4.1
Average vertical structure of the atmosphere

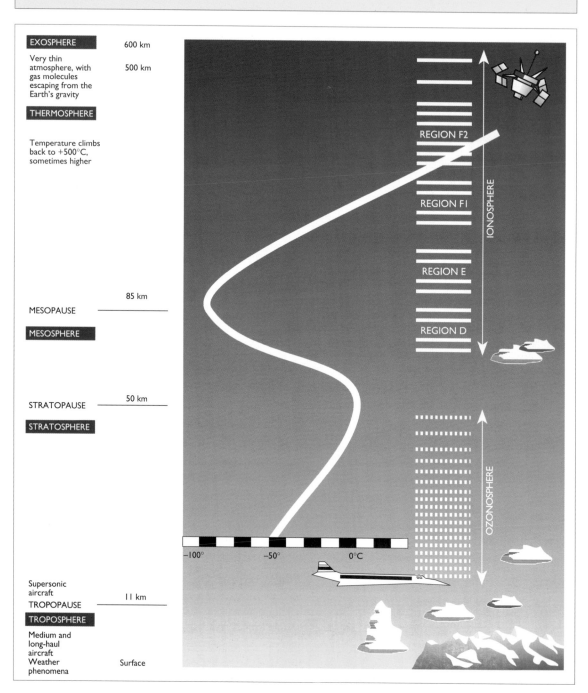

Box 4B Atmospheric constituents and their residence times

The chemical composition of the atmosphere has been evolving since the Earth was formed; it has now reached an equilibrium which sustains life. The 'natural' atmosphere is made up almost entirely of nitrogen, oxygen, argon and water vapour (Table 4.1). Air pollutants are very minor constituents of the atmosphere, but nevertheless can disturb the equilibrium, and thus affect human health, ecosystems and climatic conditions on Earth.

Atmospheric constituents are removed from the atmosphere through deposition and chemical reactions. Deposition can occur in both wet – with rain or snow – and dry forms. Wet deposition is an efficient but episodic cleansing mechanism of the atmosphere for both gases and particles. Particles can be trapped in raindrops or snowflakes during the process of condensation or by the impact of falling drops. Particles are also deposited by impaction, sedimentation (for large particles), or filtration. Many gases are taken up by vegetation through stomata (pores in leaves), or absorbed on wet surfaces. Dry deposition is a continuous process depending on the properties of the surface and of the depositing species.

The atmospheric residence times of the gases in the atmosphere vary widely (Table 4.1): from less than one hour (eg, reactive volatile organic compounds such as cyclopentadiene), to several decades (for chlorofluorocarbons or CFCs). Winds can easily transport pollutants over continental-scale distances even if their residence times are as short as 1 or 2 days, and the pollutants will generally be confined to the lowest 1 to 2 km of the troposphere. Thus, air pollution levels in Europe are dominated by the contribution of European emissions.

Chemical transformations convert primary – directly emitted – pollutants (such as sulphur dioxide or nitrogen oxides) into secondary pollutants (such as acids and particulate sulphates and nitrates). These can be formed through chemical oxidation (eg, by the very reactive hydroxyl molecule OH or by ozone and other oxidants). Photo-oxidants (eg, ozone and peroxyacetylnitrate or PAN) are formed by photochemical degradation of organic compounds in the presence of nitrogen oxides. Chemical transformation processes do not remove material from the atmosphere, but may convert it to substances with different deposition properties, and effects. For example, only a very small proportion of the gaseous nitrogen oxides (NO_x) is removed by wet deposition. However, once NO_x has been converted to nitric acid or the nitrate aerosol, wet deposition becomes one of the major removal processes.

Gases	Mean percentage	Mean residence time*	Present change (per cent per year)
Nitrogen (N_2)	78	10^6 years	
Oxygen (O_2)	21	10^3 years	
Argon (Ar)	0.9	–	
Water vapour (H_2O)	variable: 0–3	8–10 days	
Carbon dioxide (CO_2)	0.035	50–200 years	+0.4
Methane (CH_4)	0.00017	7–10 years	+1 (decreasing to 0.6)
Hydrogen (H_2)	0.00006	–	+0.6
Nitrous oxide (N_2O)	0.000033	130 years	+0.3
Carbon monoxide (CO)	$4–20 \times 10^{-6}$	0.4 year	+1–2
Ozone (O_3)			
– tropospheric	$10^{-6} – 10^{-5}$	weeks–months	+1.5
– stratospheric	$10^{-5} – 5 \times 10^{-5}$	months	–0.5
Ammonia (NH_3)	$10^{-8} – 10^{-6}$	3 days	
Sulphur dioxide (SO_2)	$10^{-7} – 5 \times 10^{-5}$	3 days	
Nitrogen oxides (NO_x)	$10^{-8} – 5 \times 10^{-5}$	3 days	
CFCs (Freons)	10^{-7}	50–150 years	+5–10
PAN	$10^{-7} – 5 \times 10^{-6}$		
Volatile organics (VOC)	$10^{-5} – 10^{-4}$	–	

*Note: For pollutants with short residence times, the indicated range goes from clean areas to polluted cities.

Table 4.1
Average percentage of main atmospheric constituents under clean air conditions, with mean residence times and current relative changes
Source: Compiled by EEA-TF from multiple sources

example of this phenomenon.)

Local and national emissions of air pollutants may thus have implications at regional (transboundary) and global scales. The assessment of air pollution cannot therefore be limited to the local or the national level. Three levels of assessment are considered appropriate:

1 the local level (time-scale less than one day) with the related exposure of population and materials;
2 the regional level (between a day and a week) with the related atmospheric deposition input into terrestrial, marine and surface water systems in relation to critical loads;
3 the European contribution to global air pollution (time-scale longer than a month).

URBAN AND LOCAL AIR POLLUTION

Europe today is a highly urbanised continent with more than 70 per cent of Europeans living in urban areas (see Chapter 10). Many urban activities (eg, traffic, combustion processes, industrial production) are accompanied by emissions into air yielding elevated concentrations of pollutants. This is especially significant when a large number of activities are concentrated together, as in an urbanised area.

Air pollution on the urban scale is the source of a range of problems: health risks mostly associated with inhalation of gases and particles, accelerated deterioration of building materials, damage to historical monuments and buildings, and damage to vegetation within and near the cities. In order to tackle these problems it is necessary to understand: what characterises urban air quality; what are the sources of urban air pollutants; and which exposures are associated with the high concentrations of air pollutants occurring in cities.

The actual occurrence and frequency of increased air pollution concentrations depends primarily on the magnitude and the distribution of emission sources, on local topography (eg, flat terrain or basin or valley) and local meteorology (eg, average wind speed, frequency of calm weather conditions, occurrence of inversion layers). The significance of any air pollution depends ultimately on the type of pollutants, the resulting exposure and the health and other effects associated with this exposure. In this section, the World Health Organization air quality guidelines (WHO-AQGs) have been taken as reference values to assess where

High concentrations of primary pollutants, including enhanced deposition, can occur within and around emission areas. Practically all large particles are deposited locally. Local weather is an important factor determining short-term pollution levels; in Southern Europe, systems of local air circulation (such as land–sea breezes) are particularly influential. Continental-size weather patterns (cyclones and anticyclones), usually lasting a few days, can suddenly increase pollution loads on the regional scale, resulting in 'pollution episodes'. The transfer of pollutants from the mixing layer to the upper troposphere and stratosphere increases residence times which may have far-reaching global impacts on the properties of the atmosphere. (CFCs emitted at the surface but which destroy the ozone layer at an altitude of 20 to 30 km are an important

ambient concentrations may possibly cause effects on human health and where further study may be necessary (WHO, 1987). The AQGs are given in Table 4.2, together with the effects they aim to prevent. As a precaution, some national limit values are even lower than WHO-AQG values.

A considerable safety margin is often built into AQGs with respect to the lowest observed effect level, to help protect the sensitive part of the population. However, AQGs have only been specified for a limited number of the thousands of possible constituents of urban air and for a limited number of averaging times. The number of constituents covered by AQGs, and the number of averaging times specified for each, increase as evidence of possible adverse effects accumulates.

Table 4.2
WHO air quality guidelines for relevant pollutants and expected effect concentrations
Source: Based on WHO, 1987

Pollution type	Indicator	AQG ($\mu g/m^3$)	Averaging time	Effect level	Effects
Short term effects					
Summer smog	O_3	150–200	I hour	200 $\mu g/m^3$; classification: mild	Lung function decrements, respiratory symptoms, inflammation
Winter smog	SO_2 + PM[1]	125 + 125	I day	400 $\mu g/m^3$; classification: moderate	Decreased lung function; increased medicine use for susceptible children
Urban/traffic	NO_2	150	I day	–	–
Long term effects					
Traffic/industry	Lead	0.5–1.0	I year	–	Effects on blood formation, kidney damage; neurologic, cognitive effects
Combustion	SO_2 + PM[1]	50 + 50	I year		Respiratory symptoms, chronic respiratory illness

Note:
I Particulate matter measured as black smoke

Source category	SO_2	NO_2	CO	PMI	Organics	Pb	Heavy metals[2]
Power generation (fossil fuel)	•	Δ	Δ				Δ/•
Space heating							
– coal	•	Δ	•	•	•/Δ		Δ/•
– oil	•	Δ					
– wood				•	•/Δ		
Road transport							
– petrol		•	#		•	#	
– diesel	Δ	•		•	•		
Solvents					Δ		
Industry	Δ	Δ	Δ	Δ	Δ	Δ	•/#

Notes:
Δ Between 5 and 25% of total emissions in commercial non-industrial cities
• Between 25 and 50% of total emissions in commercial non-industrial cities
More than 50% of total emissions in commercial non-industrial cities
I Particulate matter
2 With the exception of lead

Table 4.3
Main emission sources for the different pollutants
Source: RIVM

Table 4.3 gives the main categories of sources that contribute to the total emissions of the components mentioned in Table 4.2 (see also Chapter 14). In non-industrial cities the largest contributions come from local traffic and domestic heating when oil, coal or wood is used.

Most large cities in Europe operate monitoring networks in order to assess the air quality in their city. However, the design and the technical procedure (components measured, methods used, number and location of the stations) of the monitoring activities vary widely within Europe. The most advanced networks have incorporated air dispersion models and emission inventories in order to determine the geographical distribution of air pollution in the city and to determine the contribution from the different sources. In such cities the authorities are able to use this information to decide on the measures needed to reduce air pollution to an acceptable level, and to assess the cost of these measures (cost-efficient abatement measures).

Consistent information about air quality in large European cities is not yet collected systematically except by the European Commission under Council Decision 82/459/EEC (see also CEC, 1990) and WHO/UNEP (UNECE, 1992). In general, air pollution problems may occur in all the 2000 cities in Europe with more than 50 000 inhabitants (WHO, in press). For this report, only a limited number of cities were evaluated. Questionnaires were sent out to all cities or conurbations in Europe with more than 500 000 inhabitants, as well as to the largest city in each country (105 cities in all). (See Map 10.1 for city locations.) Approximately 22 per cent of the European population, or 150 million people, live in these cities. A summary of the information from these questionnaires covering emissions, air concentrations and exposure situations in these European cities is presented in this chapter, and the details elsewhere (RIVM/NILU, 1994).

Air quality in large European cities

An overview of the most relevant air quality indices in the selected European cities is given in Table 4.4. The table gives five groups of indices derived according to the methodology of RIVM/NILU (1994). The indices are:

- environmental pressure (combination of city population and population density);
- emissions (for winter and summer smog);
- climatological impact: average dispersion (wind speed) and smog forming potential (frequency of adverse dispersion conditions);
- exceedances (maximum pollution concentrations relative to AQG);
- exposure (percentage of population exposed to concentrations above AQG).

These indices are presented to help guide the interpretation and comparison of the data in order to obtain some understanding of the relative severity of three major air pollution situations that may occur in the selected European cities:

1 winter-type smog by sulphur dioxide (SO_2) and particulate matter (PM) measured by the black smoke or gravimetric methods;
2 summer-type smog by ozone (O_3) as resulting from emissions of volatile organic compounds (VOCs) and nitrogen oxides (NO_x);
3 high annual average concentration levels (including benzene, benzo(a)pyrene (BaP), and lead, in addition to SO_2 and PM).

The indices for the exceedance and exposure situations are based mainly on results from air pollution measurements in each city. The indices for environmental pressure, emissions and dispersion may be used to explain the differences in air quality from city to city. The quality of the data can differ considerably between cities and many of the data are uncertain (as marked). However, this is the best comparison that can be made with existing data and those collected specially for this exercise. Table 4.4 is based mainly on measurement data from 1990, although for some cities data were given for other years, between 1984 and 1991. (See *Statistical Compedium*).

City (country code)**	Environmental pressure1	Emission2 of smog-forming pollutants		Climatological impact1			Exceedances3		Exposure4
					Potential conditions for smog			City background	City background
		Summer smog	Winter smog	Average dispersion	Summer	Winter	O3	SO2+PM	SO2+PM
Tirana (AL)	4	1*	2*	4				2	1
Vienna (AT)	4	4	2*	4	3	3		2	2
Minsk (BY)	4	2	2*	4	3	4		2*	3
Sofia (BG)	3	3	5	5	3	3		3*	3
Brussels (BE)	3	4*	3*	3	3	2	2	1	1
Antwerp	3	3*	3*	3	3	2	2	1	3*
Zagreb (HR)	3	4	2	5	4	4		2	3*
Prague (CZ)	3	2	3*	3	3	3		3	4
Copenhagen (DK)	3	4	3	1	1	3			
Tallinn (EE)	3	1*	3*	2	2	4			
Helsinki (FI)	3	2	2	2	2	4	1	2	3*
Paris (FR)	5	5*	3	3	3	2	2	2	2
Lille	3	3*	2*	2	3	2	1	2	4
Lyons	4	3*	3*	4	4	3	1	3*	3
Marseilles	4	2*	3*	3	4	1	1	1	1
Toulouse	2	2*	2*	4	4	2	2	1	1
Berlin (DE)	4	5	5	3	3	3	2	4*	4
Bremen	4	2*	2*	2	3	3	4	4*	3*
Cologne	4	3*	2*	4	3	2	2	1*	4
Dortmund	3	2*	1*	3	3	2	2	1	4
Dresden	2	2	4*	5	3	3		2	3
Duisburg	4	2*	1*	4	3	2	2	2	4
Düsseldorf	3	2*	1*	4	3	2	2	1	3
Essen	3	2*	1*	4	3	2	2	1	4
Frankfurt am Main	3	2*	2*	4	3	3	2	1*	3
Hamburg	3	4*	2*	3	2	3	2	2*	4
Hanover	2	2*	1*	3	3	3	3	3*	2*
Leipzig	3	3*	5*	3	3	3		5	4
Munich	3	4*	2*	4	3	3	1	0.5	1
Nuremberg	3	2*	2*	4	3	3		1	1*
Stuttgart	2	2*	2*	5	3	3	2	1	3
Athens (GR)	4	4	2*	4	4	1		2	3
Thessaloniki	3	2*	3*	4	4	3			
Budapest (HU)	5	4*	5*	4	4	3	2	2*	4
Dublin (IE)	3	2	2	3	1	2		3	4
Reykjavik (IS)	3	1*	1	2	1	3		0.5	1
Rome (IT)	5	4*	5*	3	4	2		3	3*
Genoa	4	2*	4*	3	4	1			
Milan	4	3*	3*	4	4	5			
Naples	4	3*	4*	4	5	3			
Palermo	4	2*	3*	3	4	1			
Turin	5	4*	4*	4				2	4
Riga (LV)	3	2	1	2	2	4		4*	4*
Vilnius (LT)	3	2*	2*	4	2	3			
Luxembourg (LU)	3	1*	1	4	3	3		0.5	1*
Kishinev (MD)	3	2*	3*	5	3	4			
Amsterdam (NL)	3	3*	2*	2	2	2	2	0.5	2
Rotterdam	3	3*	2*	2	2	2	2	1*	2
The Hague	3	2*	1*	1	2	2	2	0.5	1
Oslo (NO)	2	2	1	5	2	5	1	2	2
Warsaw (PL)	3	4	3*	3	3	3	1		1*
Gdańsk	3	2*	4	2	1	3			
Katowice		2	5*	2	3			3	4
Cracow	4	2	4*	5	3	4		2	3*
Łódź	3	2*	4*	3	3	3		1	3*
Poznań	3	2*	4*	3	3	3		2	3*
Wrocław	4	2*		3	3	3			
Lisbon (PT)	5	3*	4*	3	4	1		1	1*
Oporto	5	2*	3*	3	3	1		2	4
Setubal	3	2*	2*	2	4	1		1	1*
Bucharest (RO)	4	3	3	4	4	5		2*	4
Moscow (RU)	5	4	2	4	3	4		2*	2*
Izhevsk (Ustinov)	3	2	2	3	3	4		2*	3*
Kazan	3	2	2	3	3	4		2*	2*
Krasnodar	2	2	2	3	4	4		3*	4*
Nizhny Novgorod	3	2	3	4	3	4		2*	3*
Perm	3	4	3	4	3	4		2*	3*
St Petersburg	5	2	3	3	3	4		1*	2*
Rostov-on-Don	4	2	2	3	4	3		2*	3*
Samara	3	3	3	4	4	3		4*	4*
Sratov	3	3	2	3	3	5		2*	3*
Togliatti	3	2	3	3	3	4		3*	4*
Tula	3	2*	2	4	3	4		1*	2*
Ufa	3	2	3	4	3	4		2*	3*
Volgograd	3	3	3	4	4	4		2*	3*
Voronezh	4	1	3	4	3	4		2*	3*
Yaroslavl	2	4	3	4	3	5		4*	2*
Bratislava (SK)	3	1	3	3	4	3		2	4
Ljubljana (SI)	3	1*	3	5	4	4		2	4
Madrid (ES)	5			3	5	4			
Barcelona	4			4	4	1		1*	3*
Seville	3			4	5	1		0.5	1*
Valencia	4			3	4	2		2	1
Zaragoza	4			3	4	3		1	4
Stockholm (SE)	3	3	2	2	2	4		1	1*
Gothenburg	3	2	1	2	2	3		0.5	1*
Zurich (CH)	4	1	2	5	3	3		0.5	1*
Istanbul (TR)	4			4	3	3		5*	4
Kiev (UA)	3	2	2	2	3	3		2	3
Dnepropetrovsk	4	2	4	2	3	4		2	3
Donetsk	4	1	3	2	3	4		2	3
Kharkov	4	2	2	3	3	4		2	3
Krivoj Rog	4	2	4	3	3	3		4	4
Lvov	3	1	1	3	3	4		3	3
Mariupol	3	2	4	4	3	4		2	3
Odessa	4	2	2	4	3	3		4	4
Zaporozhye	4	2	3	3	3	3		2	3
London (UK)		5*	2	3	2	2	2		
Belfast		2*	5		3	4		3	4
Birmingham	4	5*	5*	3	2	3			
Glasgow	3	2*	4*	3	1	3			
Leeds	4	2*	2*	3	1	3		1	4
Liverpool	3	2*	3*	1	1	2		3	4
Manchester	5	2*	2*	3	2	2		2	4
Sheffield	4	2*	3*	3	2	3	2	1	3*

Notes: * = uncertain data.

1 Indices have been ranked in five classes ranging from lowest (1) to highest (5) for 'environmental pressure' or 'climatological impact'. The number '1' corresponds to least pressure or most favourable climatological conditions, and '5' corresponds to most pressure or most unfavourable climatological conditions. For 'environmental pressure', class intervals have been chosen on the basis of the standard deviations from the population, and population density averages for the 105 selected cities. For 'climatological impact', class intervals have been chosen on the basis of the standard deviations from the average dispersion (measured as wind speed) and smog-forming potential conditions (measured as the frequency of adverse dispersion conditions for summer and winter smogs separately) for the 105 selected cities. In each case the class intervals used are defined as follows, (where a = the average, and s = the standard deviation): 1 = < (a–1.5s); 2 = (a–1.5s) to (a–0.5s); 3 = (a–0.5s) to (a+0.5s); 4 = (a+0.5s) to (a+1.5s); 5 = > (a+1.5s). The class values given for each city and category in these columns are calculated as the mean of the classes obtained from the two indicators used in each case separately (ie, each of the two pairs of indicators are assumed to be equally significant as measures of environmental pressure and climatological impact respectively).

2 The emission indices have been ranked in five classes from 'least unfavourable conditions' (1), to 'most unfavourable conditions' (5), relative to the average emission conditions in the 105 selected cities for compounds with summer (VOCs and NOx) and winter (SO2 and PM) smog-forming potential. Class intervals have been chosen on the basis of the standard deviations from the averages in the ranges described in note 1 above.

3   0.5: < 0.5 AQG   1:   0.5–1 AQG   2:   1–2 AQG
    3:   2–3 AQG   4:   3–4 AQG   5:   4–5 AQG

4   1: 0–5% population   2: 5–33% population   3: 33–66% population
    4: >66% population   ** See Table 5.9 for country codes

Table 4.4
Indices describing outdoor air pollution conditions in large European cities, 1990
Source: Specially compiled data from multiple sources by RIVM (See *Statistical Compendium*)

Winter smog

Results in Table 4.4 indicate that winter smog concentrations exceed WHO-AQGs in 61 of the 87 cities with data on SO_2 and/or PM. Extrapolation to all 105 cities gives an estimate of exceedance of the AQG for winter smog in 70 per cent of the surveyed cities. Such exceedances occur in general over a large part of the city during periods (days) with poor dispersion conditions. In about 24 of the 87 cities (28 per cent), concentrations around or higher than twice the AQG have been measured during recent years. Among these cities are: Ljubljana, Łódź, Sofia, Istanbul, Milan, Turin, Stuttgart, Belfast, Dublin and Prague. Urban winter smog can be found in all parts of Europe.

According to the reported measurements, the eight cities of Istanbul, Leipzig, Berlin, Donetsk, Krivoj Rog, Odessa, Riga and Yaroslavl have the highest measured concentrations of winter smog, because of high emission densities and the topographical situation of the cities (valley or basin). Coal is widely used for domestic heating in many of these cities, and they experience the same type of pollution as in London's notorious smog episodes of the 1950s, but fortunately the present concentration levels in the affected cities are lower than in those London smogs.

High concentration of suspended particles is a more widespread urban air pollution problem than high SO_2 concentrations. Most of the surveyed cities have significantly higher PM than SO_2 concentrations. Exceptions are the 'coal cities' already mentioned, and also Antwerp, Lille, Lyons, Bremen, Leeds, Liverpool, Sheffield and Oporto, where SO_2 is higher than PM.

Summer smog

The occurrence of photochemical oxidants, as indicated by the ozone concentration, is a regional phenomenon in Europe (see below). In most cases this regionally elevated ozone concentration will be partly suppressed in urban areas due to the emission of nitrogen oxide (NO), which instantaneously reacts with ozone to produce nitrogen dioxide (NO_2). Further reactions, involving VOCs and nitrogen oxides, will result in production and build-up of ozone and oxidants in the city 'plume', but these are slow processes and the additionally produced ozone will usually be found downwind of the area where the emissions occurred. An exception is when cities are located in confined valleys, or when the polluted air is trapped in land–sea breeze circulation systems, where the residence times are long enough for significant photo-oxidation to take place. For example, particularly high ozone concentrations, up to 400 $\mu g/m^3$, are measured in Athens and Barcelona.

Long-term exposure

Long-term exposure to SO_2 and PM exceeds WHO-AQGs in 24 of the 61 cities with winter smog problems, particularly in the city centres of Prague, Turin, Donetsk, Krivoj Rog, Krasnodar, Lvov, Samara, Togliatti and Istanbul.

For long-term exposure to benzene and BaP, a systematic overview of the concentrations is not possible as these are measured in only a few cities, except for BaP in the former USSR. These compounds are indicators of organic compounds associated with volatile organic hydrocarbons and with soot and polycyclic aromatic hydrocarbons (PAHs) from small-scale combustion processes, respectively. From available data and model calculations, benzene will probably exceed the lifetime cancer risk level of 10^{-4} in most cities, and BaP probably even up to a lifetime risk level of 10^{-3} (RIVM, 1992), especially in those cities which still use coal or oil for domestic heating or which are unfavourably located (see Table 4.4).

Countermeasures

Successful measures for reducing air pollution in cities in Europe over the past 20 to 30 years have included the regulation of fuel for domestic heating (low sulphur content in oil and coal, introduction of natural gas instead of coal and oil), development of district heating, reduced lead content in petrol and restriction on the use of cars in city centres (eg, Cologne, Dortmund, Zurich). Furthermore, dust removal from industrial sources and large boilers has resulted in major improvements. These measures have been particularly successful in reducing SO_2, particulate and lead emissions, and hence daily and annual average urban atmospheric concentrations, by significant amounts in many cities. However, there has been little or no evidence of similar downward trends in NO_x concentrations. During smog conditions, emergency actions are taken by some cities (eg, Rome, Milan, Athens, Prague) involving, for example, restrictions on car traffic and fuel substitution in power plants/industries. Cost-effective energy conservation measures and substitution of fossil fuels with renewable energy resources have a large potential for reduction of emissions.

'Hot spot' air pollution

The term 'hot spot' pollution may be used to describe the high short-term pollution concentrations to which the population may be exposed when located close to pollution sources. 'Hot spot' pollution includes the case of urban streets with busy traffic, and the pollution impact from industrial stacks in cities.

Air pollution from road transport

Urban street pollution is monitored in many cities, and the measurements show that short-term maximum concentrations of carbon monoxide (CO), NO_2 and particles may exceed AQGs by a factor of 2 to 4, depending upon the actual traffic and dispersion conditions of the street. It has been estimated (Mage and Zali, 1992) that there are between 9.5 and 18 million people in Europe who spend a considerable part of their working day in and near roadside settings (eg, commuters, those working in roadside buildings), although only some of them experience exceedances of AQGs. However, this population exposure is important, since, undoubtedly, for many regular commuters in city centres this exposure constitutes the main contribution to their total exposure to air pollution. Such exposure occurs in all cities, but high concentrations occur more frequently in cities with unfavourable winter smog dispersion conditions (Table 4.4).

In all European cities, smog occurrences and long-term average concentrations of harmful compounds such as lead, benzene, PM and BaP receive a significant contribution from emissions from road transport. Road transport contributes on average more than half to the nitrogen oxides emissions in Europe, and about 35 per cent of the VOC emissions. Within most cities, the relative contribution from road transport to the overall city emissions is considerably higher. Moreover, most of the summer smog formation in Europe, both in individual cities and on the regional scale in Europe, originates from the potential ozone-forming capacity of VOCs together with relatively high concentrations of nitrogen oxides, both being emitted from road transport. Diesel-engine vehicles also produce very fine particulate material, less than 10 μm aerodynamic diameter (PM_{10}), although their per cent contribution has not yet been quantified.

While NO_x emissions from stationary sources have decreased in many urban areas, emissions from motor vehicles have increased because the growth in the number of vehicles and the distance travelled per vehicle has been much larger than the reduction in emission factors. The fitting of three-way catalyst exhaust-cleaning devices is a major improvement, but is limited to new vehicles. Since the lifetime of vehicles is typically more than 10 years and in some countries more than 20 years, reductions in emissions will be slow even if vehicle numbers and distances travelled were to stop increasing. Further improvements are possible, particularly with respect to reductions of VOC emissions, for example through new EC regulations for reducing emissions of petrol vapours from service stations. More rapid improvements in air quality in the short term can only be achieved by limitations on road transport in urban areas.

A particular and significant air pollution problem in Nordic countries results from the use of studded tyres during the winter. These cause major wear of the road surface, producing grit which is suspended in the air through the action of car turbulence. This results in high levels of suspended particulate material close to roads and generally in city-centre air during dry winter conditions.

Urban pollution from industry

In some cities, air pollution from road transport and domestic heating is supplemented by that from local industry. However, depending on the height of the stack and the prevailing wind directions, the impacted area is usually a few kilometres from the emission sources. In many European cities, monitoring stations are operated in areas where principal industries are expected to have a large impact. Only in some cities (eg, Bratislava, Duisburg, Rotterdam, Athens, Antwerp, Turin, Ljubljana and Helsinki) do these 'hot spot' stations indicate a possible significant impact from industrial emissions, ie, air concentration of SO_2 and/or PM are considerably higher than in the city centre. Other large European cities with important sources of industrial air pollution and exposure include: Sofia, Bucharest, Zagreb, Perm, Togliatti, Budapest, Cracow, Lille, Lyons, Poznań, Lisbon, Donetsk, Odessa and Valencia.

The data collected specifically for this report do not allow an evaluation of the actual magnitude of 'hot spot' industrial pollution exposure in large European cities. (Local industrial pollution in areas outside the largest cities is treated below.)

Population exposure to air pollution

Health effects which arise from exposure to air pollution can be classified as: irritation and annoyance, loss of organ functions (eg, reduced lung capacity), morbidity and mortality (see Chapter 11). Some of these effects can be acute and reversible, while others develop gradually into irreversible chronic conditions. The respiratory system and the eyes are the main organs affected by air pollution, while systemic effects may also be evoked. Table 4.2 also shows expected effects upon exposure to given levels of the pollutants (WHO, 1991; 1992). For the carcinogens, risk estimates are given for lifetime exposure to the indicated concentrations.

The actual exposure of the urban population to air pollutants is difficult to estimate. Besides estimating the spatial distribution and time variation of the pollutant concentration, the location and physical activity level (in relation to inhalation volume rate) of the population should be known. Since detailed data about the activity and actual location of the population are not available, the description of exposure has been limited here to the description of ambient air concentrations in relation to population density.

The role of indoor air pollution to the overall population exposure must not, however, be overlooked. This is a complex problem which involves the entrapment of outdoor air pollutants inside, as well as the indoor release of compounds, such as combustion products, formaldehyde, solvents and tobacco smoke (see Chapter 11 and WHO, in press).

As a rough indication of air pollution exposure, pollutant concentrations have been compared to WHO-AQGs as given in Table 4.2. Qualitative results of exposure estimates are presented in Table 4.4. Data were made available for cities which contained about 116 million of the 150 million people covered by the questionnaires. Of this population, it is estimated that about 56 million people (48 per cent) live in surroundings where at least once a year (actually in 1990) the concentration level is above a short term AQG level for SO_2 and/or PM (winter smog conditions). Extrapolation to all 105 cities gives an estimate of 71 million people exposed to winter smog conditions. There is not sufficient city data to calculate exposure to high ozone concentrations. Regional background estimates (see below) suggest that more than 80 per cent of the people of these cities are exposed to high ozone concentrations (above the AQG) at least once a year.

Exposure of materials, buildings and cultural heritage to air pollution

Air pollution in urban and industrial areas increases the deterioration of many buildings and construction materials. For structural metals such as steel, zinc, copper and aluminium, quantitative dose–response relations are available describing the corrosion rate as a function of, inter alia, sulphur dioxide concentrations, chloride deposition rates, and climatic factors such as time of wetness. Pollution effects on other materials have been documented, but general quantitative assessment may still be difficult to express. Extensive documentation exists of the effect of acid pollutants, particularly SO_2, on the deterioration of marble, and other calcareous stone used in buildings and monuments (Coote et al, 1991; ECOTEC, 1986; Kučera et al, 1992).

There is a strong correlation of the weight decrease of calcareous sandstone with ambient sulphur dioxide concentrations (Figure 4.2, p 35). This deterioration depends on exposure, particularly wetting of surfaces and also on the mineralogical quality of the stone material. Different types of marble and limestone have quite different deterioration characteristics.

In sheltered positions, a surface layer known as 'black crust' forms on calcareous materials. Black crust does not act as a protection patina and it will often conceal advanced stages of decay until larger parts begin to peal away leaving a surface with a bad visual appearance. On several materials, nitrogen oxides and ozone increase the effects of sulphur dioxide attack; moreover, nitrogen compounds are thought to encourage the growth of mosses, lichens and algae on buildings.

Various attempts have been made to estimate the costs of material degradation and maintenance due to air pollution. Extrapolation of the data from one study suggests that costs for damage by sulphur dioxide to buildings and construction materials might be in the order of ECU 10 billion per year for Europe as a whole (Kucera et al, 1992).

Many historic monuments and buildings are affected by air pollutants, especially those made from marble, calcareous sandstone, or other materials susceptible to damage. Many of these objects are situated in heavily or moderately polluted areas and are thus subject to serious deterioration. Examples are the Acropolis in Athens, Cologne Cathedral, and whole cities, such as Cracow and Venice, on the UNESCO cultural heritage list.

The deterioration of historic buildings and monuments may be seen as something to be avoided at all costs. However, materials cannot be expected to last for ever; repairs, and therefore renovation and protective measures, would be necessary even if air pollution levels were reduced. Cost figures are not available. Much of the restoration work which is now being carried out is due to damage which has occurred in the past 50 to 100 years and the accumulated need for repair and restoration is not known.

Exposure to toxic air pollution from industry

Some of the most severe exposure to harmful air pollutants in Europe occurs near individual industrial plants or industrial areas, usually located outside large centres of population. Metal smelters, oil- and coal-fired power plants, chemical industry (eg, production of fertiliser, pesticides) and oil refineries/petrochemical industries are industrial sectors responsible for some of the major local air pollution problems in Europe (RIVM/NILU, 1994). The concentration level of harmful air constituents is obviously dependent upon the type of industry, the level of process technology and emission control, as well as emission conditions, mainly the height of stacks. The technological level of process and emission control is less developed in Eastern Europe, and in parts of Central and Southern Europe as well, than in Western and Northern Europe. This is probably due to a multiplicity of factors including differences in economic development and perhaps also the rigour of national policies on air pollution management and abatement.

Sulphur dioxide, particles and heavy metals are overall the most significant pollutants emitted from industrial sources. Indeed, SO_2 and PM (measured by the black smoke or other methods) are the compounds most often monitored near industrial sites. Measurements of heavy metals, fluoride,

ammonia and other compounds have also been reported, but this information has not been assessed in this study.

High air pollution exposure from industrial stacks typically occurs under unstable atmospheric dispersion conditions (ie, high insolation and strong vertical mixing) with moderate winds, as opposed to high general urban air pollution which occurs under calm stable weather. An exception is the situation near industries with low-level diffuse emissions (eg, iron and steel works, nickel or aluminium smelters), where air pollution is also highest under calm weather conditions. For both low-level emissions and stack emissions, the exposure situation is the most severe in valleys, when residential areas are located downwind from the plant in the prevailing wind direction(s).

The principal air pollution impact near SO_2-emitting industrial sources is typically short term (one to a few hours, although industrial plants may also give extremely high long-term impact), as opposed to general urban air pollution, where the 24-hour average concentrations are often more significant, relative to AQGs.

Comprehensive data on air pollution concentrations and exposure near industrial plants and areas in Europe do not exist. Data have been collected unsystematically, and often not at all. However, selected data on pollution in industrial areas in EU countries have been collected as part of the Council Decision 82/459/EEC. For the present report, information on local industrial air pollution was collected

Table 4.5
Examples of available data on local industrial air pollution problems in Europe
Source: NILU

Country	Maximum concentration measured in µg/m3	Area, type of industry	Country	Maximum concentration measured in µg/m3	Area, type of industry
Eastern and Central Europe			**Southern Europe**		
Bulgaria	2200 (SO_2, 1 hr)	Sofia, non-ferrous metals	Greece	377 (SO_2, 1 hr-98%ile)	Thessaloniki
	485 (SO_2, year)	Asenovgrad	Italy	168 (SO_2, 1 hr-98%ile)	Venice industrial zone
	530 (PM, year)	Dimitrovgrad, cement		220 (PM, 1 hr-98%ile)	Venice industrial zone
	870 (SO_2, max month)	Pirdup Zlatitza, non-ferrous metals	Portugal	686 (SO_2, 1 hr-98%ile)	Barreiro-Seixal
	Very high H_2SO_4	Pirdup Zlatitza		466 (PM, 1 hr-98%ile)	Barreiro-Seixal
	248 (NH_3, max month)	Devnia, chemical industry, power plants	Spain	500 (SO_2, 1 hr-98%ile)	Cartagena, iron/steel, metallurgy
	290 (H_2SO_4, max month)	Devnia, chemical industry, power plants		499 (PM, 1 hr-98%ile)	Cartagena, power plants, chemical industry
Croatia	90 (SO_2, year)	Rijeka, petrochemicals refinery	**Western Europe**		
	381 (SO_2, 24 hr)	Zagreb, power plants (oil)	Austria	780 (SO_2, 0.5 hr)	Lenzing, viscose, pulp/paper
	608 (PM, 24 hr)	Zagreb, cement industry	Belgium	2 000 (SO_2, 30 min)	Antwerp, oil refinery, power plants, chemical industry
	817 (NH_3, 24 hr)	Zagreb, fertiliser		1.6 (Pb, year)	Hoboken, lead smelter
Czech Republic	1500 (SO_2, 24 hr)	Sokolov, brown coal industry/power	France	955 (SO_2, 1 hr-98%ile)	Salsigne
Hungary	94 (SO_2, year)	Dorog, chemical industry/mining		150 (PM, 1 hr-98%ile)	Le Havre, H_2SO_4, petrochemicals, oil refinery, metallurgy
	146 (SO_2, winter)	Tata, aluminium smelter	Germany	571 (SO_2, annual av.)	Merseburg
Latvia	100 (PM, year)	Daugavpils, chemical industry		790 (SO_2, 1 hr-98%ile)	Klingental
Poland	584 (SO_2, year)	Toruń		2.4 (Pb, year)	Brauback, lead smelter
	477 (PM, year)	Katowice, steel, coking, power plants	The Netherlands	6.3 (HF, 24 hr)	Delfzijl, aluminium industry
	0.17 (BaP, year)	Katowice, steel, coking, power plants	UK	122 (SO_2, 24 hr-98%ile)	Newry
Romania	285 (PM, year)	Bucharest, non-ferrous metals, power plants, chemical industry		418 (black smoke, 24 hr-98%ile)	Sunderland
	128 (SO_2, year)	Zlatna, aluminium smelter		3.6 (Pb, year)	Walsall, lead smelter
	3020 (SO_2, max month)	Zlatna, aluminium smelter	**Northern Europe**		
	8130 (SO_2, max 24 hr)	Baia Mare, non-ferrous metals	Finland	630 (SO_2, 1 hr-98%ile for max month)	Harjavalta, copper smelter
	70 (Pb, max 24 hr)	Baia Mare, non-ferrous metals		0.040 (Cr, 24 hr)	Tornio, FeCr smelter
	280 (NH_3, max 24 hr)	Bacau, fertiliser, chemicals, petrochemicals, power plants	Norway	500 (SO_2, 1 hr)	Ålvik, ferroalloy
	901 (HCl, max 24 hr)	Bacau, fertiliser, chemicals, petrochemicals, power plants		4 (PAH, 24 hr)	Årdal, aluminium industry
	1350 (SO_2, max month)	Copsa Mica, Sibin, non-ferrous metals			
Slovenia	1400 (SO_2, 0.5 hr)	Ljubljana, thermal power plants (coal), various small industries (plus domestic coal use)			

Notes: Measurement data from recent years.
Information in brackets indicates the pollutant and the period over which the concentration was measured.

through the return, by more than 20 countries, of a simple questionnaire (RIVM/NILU, 1994).

Data provided by this questionnaire are not complete regarding the number of industrial areas, and data from individual areas are often not very detailed. Nevertheless, an overview is presented in Table 4.5, giving, for each country from which reliable information is available, maximum reported concentrations, location of polluting site, and the types of industries involved.

By far, the most severe exposures to high pollution concentrations occur in Eastern European countries, where, in several industrial areas, largely uncontrolled industrial emissions from old-type processes result in severe exposure of nearby population and vegetation to harmful air pollution. In most of the EU and EFTA countries, efforts to clean up industrial emissions have substantially reduced such problems, but here also industrial areas with air pollution exposure exceeding WHO-AQGs still exist. The following countries indicated no significant local industrial air pollution problems: Denmark, Iceland, Ireland, Sweden and Switzerland. No specific information was available from seven countries.

Table 4.5 shows that high exposure to SO_2, PM and other compounds associated with specific industries occurs in all parts of Europe. Sulphur dioxide concentrations above 1000 $\mu g/m^3$, one-hour average, are not uncommon. Extreme SO_2 exposure occurs, for example, in Zlatna and Baia Mare (Romania), Asenovgrad (Bulgaria), Sokolov and Teplice regions (Czech Republic) and in Toruń (Poland), with annual averages around 500 $\mu g/m^3$ (Toruń, Asenovgrad) and 24-hour averages up to 8000 $\mu g/m^3$ (Baia Mare). Non-ferrous metal (cadmium and aluminium) industries and coal-fired power plants are those industries most often responsible for very high local industrial pollution in Europe. An estimate of the size of the population affected by local industrial emissions in Europe cannot be made at this time.

Conclusions

Considerable improvements in local air quality in recent decades have been achieved in many cities through substitution of coal and heavy fuel oils by cleaner fuels such as gas oil and natural gas, and by electricity and heat supplied from large electric power plants and district heating plants. Emission controls have been particularly successful in the reduction of SO_2, dust and fly ash emissions to the atmosphere, and in the reduction of the emissions of various gases from the process industries.

On the other hand, it is estimated that in 70 to 80 per cent of European cities with more than 500 000 inhabitants, air pollution levels of one or more pollutants exceed the WHO-AQGs at least once in a typical year. Similar conditions may occur in many smaller cities. These exceedances are taken in this report as indicators of where possible human health risks and health effects due to air pollution in these cities may occur and where further study may be necessary. This is part of a wider problem concerning the urban environment (see Chapter 10). The main pollution sources are combustion and industry in many Central and Eastern European countries, and road transport, particularly in Western cities. The deterioration of historic monuments and buildings due to air pollution may be seen as a destruction of priceless value. Urban air pollution is therefore regarded as a prominent environmental problem of concern to Europe (see also Chapter 37).

In many cities, consistent information on pollution sources, emissions, air concentrations, exposure and health effects, and their interconnections is not available. To rectify this it will be necessary to collect, harmonise and evaluate data on urban air pollution, human exposure and health effects on a European scale. Data for smaller cities and data on exposure to indoor air pollution, occupational exposure and exposure to

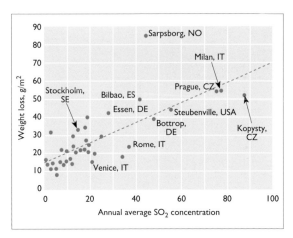

Figure 4.2
Observed annual weight loss of exposed tablets of dolomitic sandstone (White Mansfield) as a function of annual mean SO_2 concentrations at the exposure sites
Source: Coote et al, 1991

pollutants in food, soil and water should also be considered (see Chapter 11 and WHO, in press).

REGIONAL AND TRANSBOUNDARY AIR POLLUTION

Air pollutants can often be transported over considerable distances, affecting air quality and deposition of pollutants in adjacent or distant countries. Such effects can extend over hundreds or several thousands of kilometres. The pollutants involved have residence times in the air of between half a day to one week (see Box 4B). The effects from such regional or continental scale transportation of pollutants have received considerable attention, both scientifically and politically, and have led to the UNECE convention on Long-Range Transboundary Air Pollution (LRTAP), an important framework for environmental assessment and policy in Europe.

Atmospheric processes important on the regional scale are, in general, confined to the lowest few kilometres of the troposphere. However, processes at higher levels of the troposphere may affect concentrations at ground level (see global air pollution, below). Specific weather conditions or increased emissions and often both simultaneously can result in enhanced levels of air pollution. Such pollution episodes may last for several days or even weeks. During winter smog episodes concentrations of a variety of pollutants, including sulphur dioxide and aerosol particles, are high, while during summer episodes concentrations of ozone and other photochemical oxidants may be increased over large areas in Europe. The combination of meteorological patterns and spatial distribution of air pollutant sources (Map 4.1a to 4.1d) explains to a great extent the deposition and concentration fields in Europe. An overview of these emissions in European countries is presented in Chapter 14.

The presentation below of regional scale air pollution in Europe is based on a discussion of the pollutants introduced in Box 4C (p 40) grouped under eight different headings. Air pollutants important at this scale constitute a major input/source of pollution to other media (freshwaters, seas, soils; see Chapters 6 and 7) and are summarised at the end of this section.

To help characterise the impacts of acid deposition on surface waters and terrestrial ecosystems (notably forests), and of air pollutants on natural and agricultural ecosystems, 'critical loads' and 'critical levels' have been formulated. Critical loads are defined as those deposition loads below which no adverse effects are to be expected (Downing et al, 1993). Critical levels are defined as those concentrations above which adverse effects on sensitive receptors are expected.

The concept of critical load was developed during the 1980s. Values for critical loads of sulphur and nitrogen for sensitive receptors have been developed and values for critical levels of the atmospheric pollutants SO_2, NO_x, O_3, NH_x are under discussion. Although it was realized that the definition

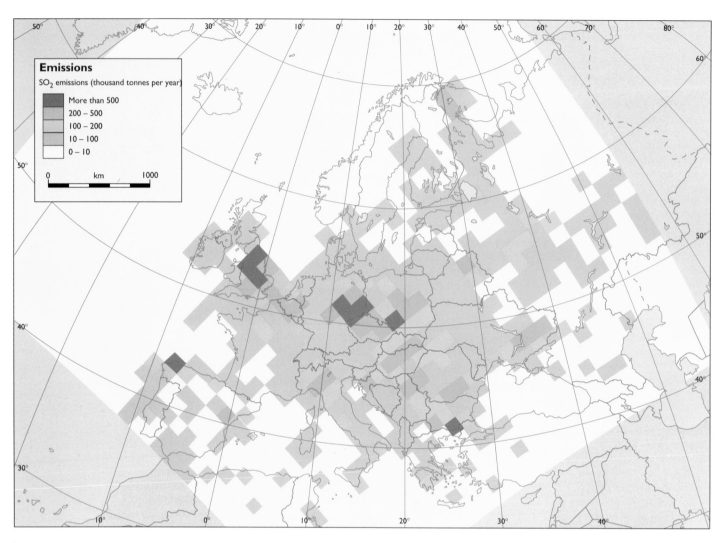

Emissions

SO$_2$ emissions (thousand tonnes per year)

- More than 500
- 200 – 500
- 100 – 200
- 10 – 100
- 0 – 10

0   km   1000

Map 4.1a)
Spatial distribution
of emissions of SO$_2$,
1990
Source: EMEP

of critical loads and levels is an iterative process and that further research was still needed, it was believed that mapping areas where the values presented are currently exceeded would be useful in a first estimate of areas at risk in different regions in Europe. This approach is adopted below. It should be stressed that the exceedance areas thus mapped do not, at present, necessarily correspond to those with recognised adverse effects.

Acid deposition

When the air pollutants SO$_x$ (SO$_2$ and sulphate), NO$_y$ (NO$_x$, nitric acid and nitrate) and NH$_x$ (NH$_3$ and NH$_4$, ammonium) are scavenged from the atmosphere and deposited at the surface, a series of problems result, often collectively referred to as acidification. The recognition of acid deposition as a threat to ecosystems has resulted in major research projects and international negotiations and agreements to reduce emissions. Present understanding emphasises that it is not the acidity of the precipitation itself that matters (as suggested by the familiar term 'acid rain'). Rather, what is important is the acidification of soils and waters, due to the total deposition of sulphur and nitrogen compounds by precipitation (rain, snow and mist) and dry deposition, in relation to the soil's or the ecosystem's capacity to accommodate and utilise these compounds, as well as the direct exposure of vegetation to atmospheric pollution. In the past decade, it has been shown that, next to sulphur and nitrogen oxides and their oxidation products, ammonia (NH$_3$) and its atmospheric conversion products may act as acidifying agents due to conversion to nitric acid in soils and

waters. The combined deposition of SO$_x$, NO$_y$ and NH$_x$ is therefore hereafter referred to as 'potential acid deposition', representing the maximum atmospheric acid load that could lead to acidification.

The actual potential acid deposition pattern in Europe derived from model calculations is shown in Map 4.2 (Sandnes and Styve, 1992; Sandnes, 1993). The highest deposition values are found in the industrialised, densely populated zone that runs from Poland and the Czech Republic, over Germany and the Benelux countries to the UK. Maximum values of more than 10 000 eq/ha/year are found here. This translates into exceedances of critical loads in these areas by several hundred per cent. Map 4.2 is based on model calculations using the EMEP model; the spatial resolution in the model is 150 km by 150 km. The model results are corroborated by deposition measurement. Considerable variation can occur within this grid size, and local deposition may be much higher.

Acid deposition in Europe originates mainly from air pollutants coming from European emissions. Emissions of acid components result largely from combustion of fossil fuels (SO$_2$ and NO$_x$) and agricultural activities (NH$_3$) (see Box 4C). Today, more than 70 per cent of total atmospheric SO$_2$ emissions stem from coal combustion in thermoelectric power plants while motor vehicles account for about 50 per cent of total atmospheric NO$_x$ emissions in Europe. Because the spatial distribution of the emissions and the chemical properties of the compounds involved are different, the relative contribution of the three components to the total acid deposition varies over Europe. In Central and Eastern Europe sulphur dominates, whereas in Western and Southern Europe NO$_x$ may be relatively more important. In countries

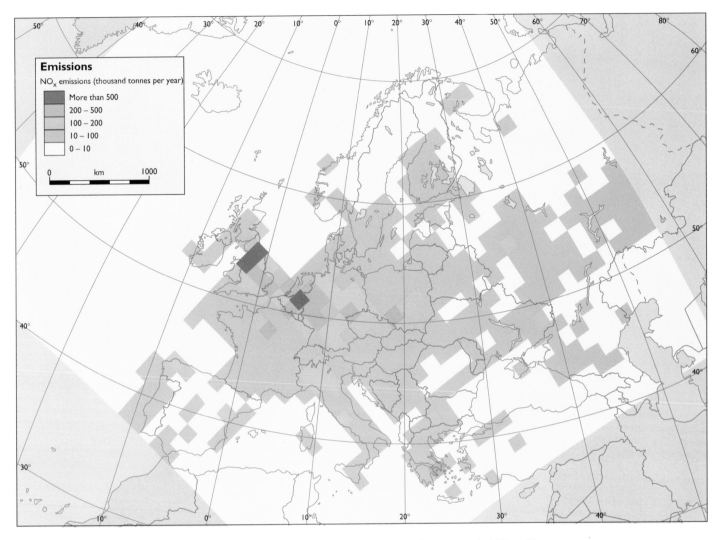

Emissions

NO$_x$ emissions (thousand tonnes per year)

More than 500

200 – 500

100 – 200

10 – 100

0 – 10

where there is intensive cattle breeding (such as Denmark, The Netherlands, and parts of the UK) ammonia makes a significant contribution. Model calculations (Tarrason, 1992) indicate that anthropogenic and biogenic emissions from North America contribute less than 5 per cent to total SO$_x$ deposition on Europe's west coast. The influence of Asian sources is estimated as less than 1 per cent. European emissions from power plants contribute most to total acid deposition (30 to 55 per cent) (RIVM, 1992). Next most important is the contribution of NO$_x$ emissions from transport; the range is from 7 per cent in Poland to about 30 per cent in Finland and Norway (see Chapter 31). In the UNECE LRTAP convention, 'budget matrices' have been constructed from which the contribution of emissions in each of 36 European countries and areas of Europe to deposition in these countries and areas can be derived.

Long-term trends in total acid deposition are not yet available because of the lack of 'historical' data on the dry deposition contribution. However, measurements of concentrations in air and precipitation indicate a growing contribution of NO$_x$ in environmental acidification. Long-term measurements of the nitrate and sulphate contents of the atmospheric aerosol in southern England show an increasing nitrate to sulphate ratio through a 30-year period (Atkins and Law, 1984). Precipitation networks continue to record an upward trend in the nitrate content of European precipitation. Although the analysis of ice cores from the south Greenland ice sheet is showing a decreasing trend in sulphate, the nitrate content is growing (Mayewski et al, 1986). The nitrate concentrations of over 300 lakes in Norway have shown a doubling over the period 1974 to 1986, despite there being little change in sulphate and pH levels

(Henriksen et al, 1988). The impacts of acid deposition are discussed in Chapters 5 (Inland waters), 7 (Soil), 31 (Acidification) and 34 (Forest degradation).

Photo-oxidants and photochemical smog episodes

Under conditions of high insolation and low winds, concentrations of ozone and photochemical oxidants can build up to high levels which are damaging both to human health and to vegetation. Usually such situations take several hours to develop. 'Photochemical smog' is formed particularly in land basins, and in coastal areas where air pollutants may be contained in a land–sea breeze circulation system, as in Barcelona or Athens. However, under high-pressure conditions, ozone concentrations often reach serious levels over much larger areas in Europe, and may reach Scandinavia and other more remote areas following transport of ozone and its precursors from the more central source areas. Because of the strong link to particular meteorological situations, the occurrence of 'ozone episodes' in Europe is very variable from year to year, in both space and time.

NO$_x$ and VOC emissions from motor vehicles are the main cause of photochemical oxidant and ozone formation on a regional scale. Volatile organic compounds (VOCs) present in the air from anthropogenic and natural emissions are oxidised in the atmosphere in the presence of sunlight. The process generally starts through reaction with the very reactive hydroxyl (OH) radical, by reaction with ozone, or by direct action of sunlight. The oxidation processes are complex, and NO$_x$ play an important role as necessary catalysts. Oxidation takes place on time-scales ranging from a

Map 4.1b)
Spatial distribution of emissions of NO$_x$, 1990
Source: EMEP

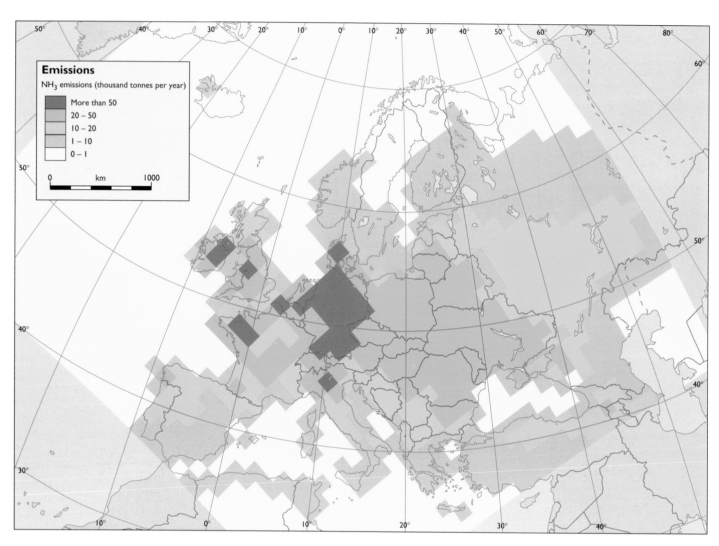

Emissions

NH₃ emissions (thousand tonnes per year)

More than 50
20 – 50
10 – 20
1 – 10
0 – 1

0   km   1000

Map 4.1c)
Spatial distribution of emissions of NH₃, 1990
Source: EMEP

Table 4.6
Guideline values for exposure to ozone as set by WHO, EC and under discussion in UNECE for the protection of human health and vegetation (ppb)
Sources: WHO, 1987; EC Directive 92/72/EEC; UNECE, 1988

Exposure duration	Protection of human health		Protection of vegetation		
	WHO	EC	WHO	EC	UNECE
1 hour	76–100	88	100	100	75
8 hours	50–60	55*			30
24 hours			33	33	
Growing season (100 days)			30		
Growing season (183 days), 9.00–16.00 average					25

Notes:

* The 8-hour average is calculated as a moving average over the day, ie, it is calculated at every hour based on the 8-hourly values between minus one hour and minus nine hours before the reference time.

For ozone (O₃), 1 ppb ~ 2 μg/m³

few hours to several months. A host of secondary products is formed during this oxidation; they are usually called photo-oxidants. Ozone is by far the most important of these in terms of the adverse effects it has on human health and ecosystems. The role of less reactive, relatively long-lived VOCs such as methane in tropospheric chemistry, and their contribution to the global background levels of ozone, is discussed below with other 'global' pollutants. Here, only the more reactive organic compounds which lead to photo-oxidant formation will be considered.

Several international bodies such as WHO, UNECE and EC have discussed, proposed or set guideline values for ozone for the protection of human health and/or vegetation (Table 4.6). These guideline values indicate levels combined with exposure times during which no adverse effects are currently expected. However, progress in scientific understanding may lead to revisions of these guideline values.

The 75 ppb hourly concentration level, chosen here as a reference, was exceeded in summer 1989 at all stations in the EUROTRAC-TOR and ECE-EMEP networks, except for a few stations in Scandinavia. In Switzerland and Austria this level was exceeded on more than 20 per cent of days during April to September (Beck and Grennfelt, 1993). These networks of background measurements have a fair coverage in Northern and Western Europe but the information is incomplete and largely lacking in Southern, Central and Eastern Europe. Therefore, model estimates are presented here in order to give a European picture of the problem.

In order to quantify exposure to high levels, excess ozone is defined as the sum of the concentrations minus a given limit value (here 75 ppb), summed over all hours in a period. The modelled values of excess ozone show a large variation over Europe; for summer 1989 the highest values were calculated in Western Europe (Map 4.3) (Simpson and Styve, 1992). Excess ozone depends strongly on number and severity of summer smog episodes in a particular year; at individual stations this quantity can vary by a factor of 100 or more. In heavily populated areas the ozone concentrations may be lower due to chemical scavenging by local nitrogen oxide emissions.

Calculations with a variety of atmospheric transport models indicate that emissions from Europe are the main

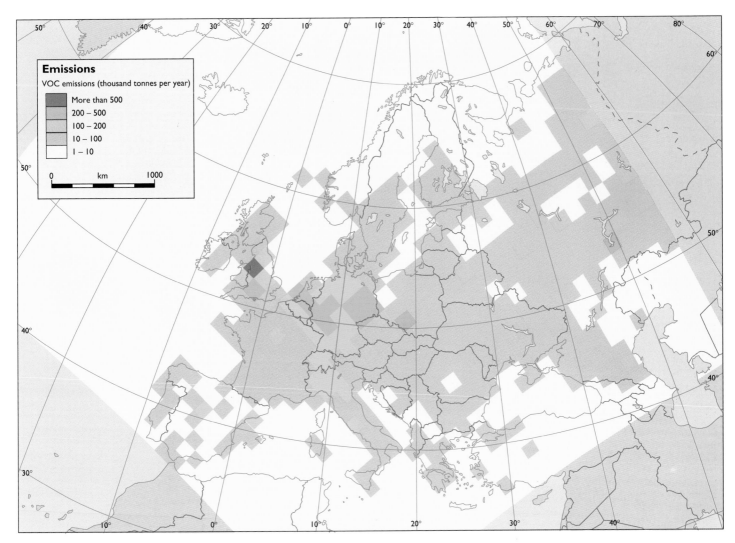

Emissions

VOC emissions (thousand tonnes per year)

More than 500
200 – 500
100 – 200
10 – 100
1 – 10

0 — km — 1000

cause of the occurrence of high ozone peak values (Simpson and Styve, 1992; de Leeuw and Van Rheineck Leyssius, 1991). Model calculations show that implementation of the 1991 VOC protocol under the UNECE LRTAP convention would result in most areas in a 40 to 60 per cent reduction in ozone levels in excess of 75 ppb. The effect on the annual average ozone concentration is much less: 4 to 8 per cent in northwest Europe, but 1 to 4 per cent on average (Simpson and Styve, 1992).

Winter smog episodes

High levels of SO_2 and suspended particulate matter (SPM) lead to winter air pollution episodes when low winds and a strong temperature inversion impede the vertical mixing and dilution of pollutants in the lowest atmospheric layers. Low temperatures increase the demand for energy, resulting in increased emissions and the further accumulation of pollutants. During these episodes, emissions related to domestic heating can be up to 70 per cent higher than the winter season average (de Leeuw and Van Rheineck Leyssius, 1990).

As well as SO_2 and PM, the winter smog mixture contains compounds such as carbon monoxide (CO), nitric acid, nitrogen dioxide (NO_2), and inhalable particles of variable content such as soot, sulphate, nitrate, ammonium, metals and organic compounds, such as polycyclic aromatic hydrocarbons (PAH). The concentrations of strong oxidants such as ozone are low during winter episodes.

Winter smog episodes occur most frequently and are most severe in Central Europe. For the densely populated parts of the Czech Republic, eastern Germany and southern

Poland, model calculations, supported by measurements, show yearly averaged concentrations up to 10 times higher than yearly averaged values in Western Europe. Based on these data it is expected that, under winter smog conditions, daily average SO_2 concentrations in rural areas will be well above $400 \, \mu g/m^3$. Even higher episodic values can be measured in cities.

As an illustration, a recent winter smog episode from 2 to 13 February 1993 in Northern Bohemia is considered (J Beneš Czech Ecological Institute, personal communication) (Figure 4.4). During this episode, daily average SO_2 and PM concentrations reached maximum values of 825 and $480 \, \mu g/m^3$ respectively. These values may be compared to the WHO-AQG for the evaluation of ambient concentrations with respect to their probable effects on human health of $125 \, \mu g/m^3$ for SO_2 and $120 \, \mu g/m^3$ for PM as 24-hour averages, and EC guideline values of 100 to $150 \, \mu g/m^3$ for SO_2 and 100 to $150 \, \mu g/m^3$ for PM. Maximum 30-minute average concentrations were $1850 \, \mu g/m^3$ for SO_2, $2600 \, \mu g/m^3$ for PM and $760 \, \mu g/m^3$ for NO_x. Similar or even more severe episodes are still expected to occur 1 to 2 times a year in the future. The most severe smog episode ever reported was in London in December 1952 when SO_2 and PM daily average concentrations reached values of about $5000 \, \mu g/m^3$ each (see eg, Brimblecombe, 1987). In the two-week period during and immediately after this smog episode, a total of approximately 4000 excess deaths were observed compared to a similar period in previous years (Ministry of Health, 1954).

Winter smog episodes often extend across several European countries due to long range transport across the continent. This is evident from measurements and model

Map 4.1d)
Spatial distribution of emissions of VOC, 1990
Source: EMEP

Box 4C Pollutants leading to regional air pollution

Primary pollutants with residence times corresponding to the regional scale include the acidifying species sulphur dioxide (SO_2), nitrogen oxides (NO_x), and ammonia (NH_3), and aerosol-bound pollutants (such as dust, heavy metals, and persistent organic pollutants). Secondary pollutants that are formed through chemical reactions in the atmosphere on these time-scales have to be considered as well. Examples are the oxidation products of SO_2, NO_x and NH_3, sulphate, nitrate and ammonium respectively and photochemical oxidants such as ozone, aldehydes and peroxides (eg, PAN) resulting from atmospheric reactions of emissions of volatile organic compounds (VOCs) and NO_x. These air pollutants lead to the following typical regional air pollution problems: atmospheric deposition of sulphur and nitrogen compounds leading to acidification of soils and surface waters; photochemical oxidant formation; and occurrence of summer and winter type pollution episodes. Air pollution problems (including visibility degradation) are also associated with aerosol particles, metals and persistent organic pollutants (including their deposition in marine waters).

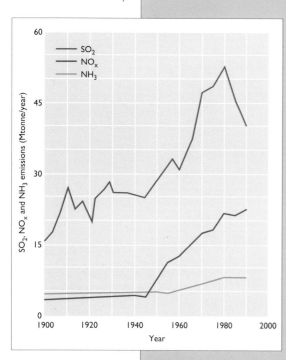

Figure 4.3
Temporal evolution of European emissions of acidifying compounds since the beginning of the century
Source: RIVM

Sulphur dioxide, nitrogen oxides and ammonia

Anthropogenic emissions of SO_2 are due largely to combustion of sulphur-containing fossil fuels (oil, coal and lignite). Nitrogen oxides are also emitted mainly from combustion sources: especially transport, power generation and domestic heating. Agriculture, particularly the storage and spreading of animal manure, is the main source of ammonia emissions to the atmosphere. In Europe, anthropogenic emissions outweigh emissions from natural sources: for SO_2 and NO_x natural emissions are less than about 4 per cent of total emissions (Tarrason, 1991; Builtjes and Hulshoff, 1991), and for NH_3 the natural contribution is estimated with a large uncertainty to be approximately 9 per cent (Asman, 1992).

Volatile organic compounds

Road transport and industry (in particular, combustion of fossil fuels, solvent use and evaporation losses) are the most important sources of VOC. Natural sources are vegetation (especially forests), soil and oceans. VOC emissions consist of a mixture of many organic species, each having its own reactivity and time-scale in ozone formation. VOC emissions, both total amount and speciation over the various chemical reactivity classes, are less well known than SO_x or NO_x emissions. The potential of the individual VOC components in ozone formation varies widely. Although ozone formation depends on the actual weather situation and the physico-chemical conditions (eg, concentrations of VOC and NO_x), in general terms the ozone-creation potential decreases in the order: aromatics (excluding benzene), olefins, aliphatics, methane.

Aerosol particles

Aerosol particles are microscopic solid or liquid particles suspended in air. This includes wind-blown sand and dust and a host of air pollutants. The chemical and physical properties of atmospheric aerosols vary widely. The size of the aerosol varies from smaller than 0.1 up to a few hundred microns (μm) in diameter. Aerosol particles smaller than about 10 μm are considered to be mainly responsible for health effects. Typical examples of aerosol-bound pollutants that have adverse effects are heavy metals, persistent organic pollutants with a high molecular mass (such as PAHs and dioxins), and soot or black smoke. Atmospheric aerosols consist partly of primary (directly emitted) aerosols, and partly of aerosols formed in the atmosphere by chemical conversion of primary pollutants such as SO_2, NO_x, NH_3 and VOCs. These secondary aerosols generally have a fine particle size of less than 2.5 μm aerodynamic diameter (standard borderline between large and small/fine particles). No official data for European emissions of aerosols exist, although several estimates are available (RIVM, 1992; 1993). The most important source categories for aerosol emission are the combustion of (fossil) fuels by industry, road transport and households. Reduced visibility is one of the most obvious effects of air pollution by secondary aerosols.

Atmospheric visibility is limited by the attenuation of light. In the case of fog, this attenuation is effected by atmospheric water droplets. In clean, dry air, the molecules of air limit visibility to about 300 km, but the presence of small particles generally reduces visibility to much shorter distances, particularly in areas affected by air pollution. Studies, in both North America and Europe, indicate a strong causal relationship between the concentration of small particles and visibility (WMO, 1983). In urban air, soot and other primary particles are important contributors to visibility impairment, whereas in rural areas the visibility reduction is caused mainly by sulphates and nitrates (Leaderer et al, 1979; Barnes and Lee, 1978) (see also Regional and transboundary air pollution).

Both sulphates and nitrates take up water and transform to liquid droplets at relative humidities above 70 to 80 per cent. This water uptake increases the particles' effect on visibility. However, for all relative humidities the optical attenuation is nearly proportional to the air's content of sulphates and nitrates. The relative contribution of elemental carbon, or soot, to visibility reduction in rural areas can be estimated at 10 to 20 per cent on the basis of estimated emissions and limited measurements (Heintzenberg and Meszaros, 1985).

Metals

Metal emissions result predominantly from combustion (coal, oil and waste incineration), the metal industry and transport (lead and copper) (Pacyna and Munch, 1988). However, emission data for metals are rather limited and uncertain. Metal emissions pose a direct health hazard to people through inhalation. When deposited they can be taken up by vegetation and may enter the food chain, resulting in exposure to humans and animals.

Trends

The emissions of all these pollutants have strongly increased in Europe since the beginning of this century, although the increase has been much more dramatic for NO_x and SO_2 than for ammonia (see Figure 4.3). During this century, the 'hot spots' of SO_2 emissions have moved from West to East and South: from the Black Country in England to the Ruhr area and thence to the 'Black Triangle' – the border area between Germany, Poland and the Czech Republic (Mylona, 1993). Due to the growing industrialisation of Southern Europe, emissions in this area are expected to become relatively more important. The largest NO_x emission densities have until now been mostly located in Western Europe (UK, Germany, Benelux); these are expected to shift eastwards in the immediate future. For ammonia, the largest emission densities are in The Netherlands and Denmark as a result of intensive cattle and pig farming.

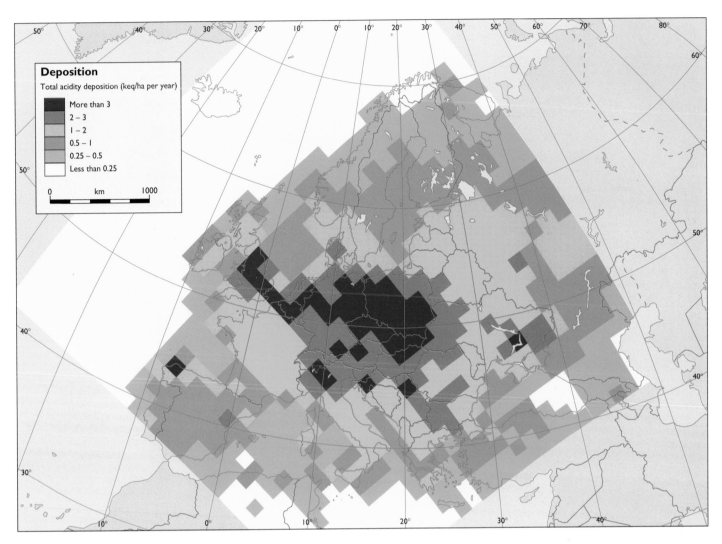

Deposition

Total acidity deposition (keq/ha per year)

- More than 3
- 2 – 3
- 1 – 2
- 0.5 – 1
- 0.25 – 0.5
- Less than 0.25

0   km   1000

calculations. Map 4.4 shows the transboundary spread of an episode in January 1987, with maximum 24-hour SO₂ and PM concentrations of 900 and 700 μg/m³ in northwest Europe. Instead of the normal West-to-East transport in Europe, East-to-West transport was important during some of these episodes (see also Chapter 2).

For continental Western Europe (western part of Germany, Benelux countries, France) it has been estimated that during the 1985 and 1987 episodes at least 50 per cent – but probably up to 75 per cent – of the SO₂ was of Eastern European origin. For sulphate, and probably also PM, the Eastern European contribution is probably of the order of 80 to 90 per cent. In contrast, the contribution of Eastern Europe to the annual average sulphur deposition in northwest Europe is of the order of only 10 per cent (Lübkert, 1989; de Leeuw and Van Rheineck Leyssius, 1990).

Aerosol particles

Atmospheric aerosols can be primary – directly emitted – or secondary – formed in the atmosphere by chemical conversion of primary pollutants (see Box 4C). Aerosol particle measurements available at present are insufficient to prepare a European map of aerosol concentration. Only a few of the national monitoring stations report their data to international data collecting programmes. As for other pollutants, particulate measurements are concentrated in urban areas and, being influenced by local sources, are often not representative for larger areas. Moreover, the variety of measuring techniques applied makes it difficult to interpret the data.

A first estimate of average large-scale aerosol

Figure 4.4 *Daily average concentration of SO₂ during a winter smog episode in Northern Bohemia, Czech Republic, February 1993 (μg/m³)*
Source: ČHMÚ

Map 4.2
Total potential acid deposition in 1990 (EMEP/MSC-W acid model)
Source: Sandnes and Styve, 1992; Sandnes, 1993

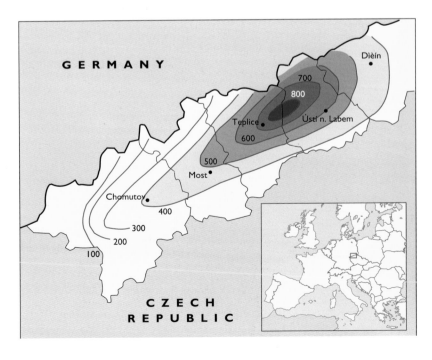

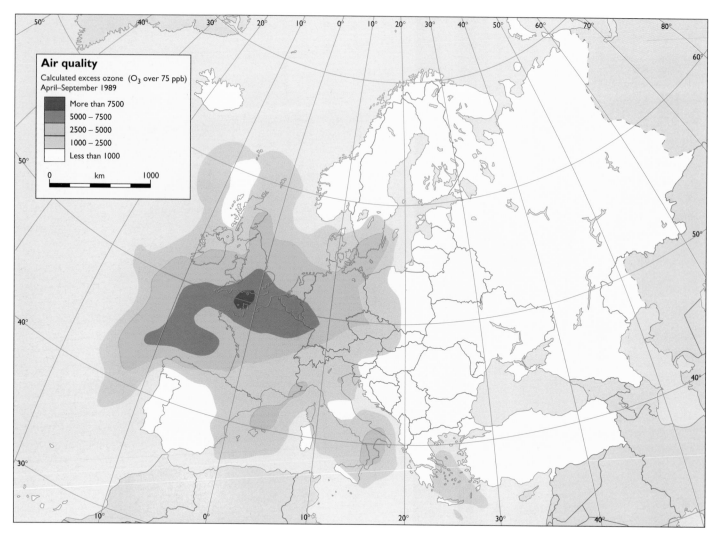

Air quality

Calculated excess ozone (O_3 over 75 ppb)
April–September 1989

- More than 7500
- 5000 – 7500
- 2500 – 5000
- 1000 – 2500
- Less than 1000

0   km   1000

Map 4.3

Excess ozone (above threshold value of 75 ppb) during summer 1989 (EMEP/MSC-W model)

Source: Sandnes and Styve, 1992

concentrations for Europe has been made by means of the TREND atmospheric dispersion model (Map 4.5) (RIVM, 1992). It is estimated that, in the whole of Europe, secondary aerosols form a substantial contribution to total aerosol concentrations in rural areas. In urban/industrial areas the contribution of primary aerosol is high. In the Upper Silesia area, yearly averaged PM concentrations of 180 to 230 $\mu g/m^3$ are measured, with episodic background values of the order of 500 $\mu g/m^3$ or higher. For the Czech Republic, PM yearly average values up to 220 $\mu g/m^3$ PM are reported. The WHO-AQG for total suspended particulates both for maximum daily values (120 $\mu g/m^3$) and for yearly average values (100 $\mu g/m^3$) are probably exceeded in these areas.

Sulphate and nitrate aerosol levels are expected to decrease by 20 to 30 per cent between 1990 and 2000. For aerosol particles, including both emitted dust and soot, present emissions are not well known and future developments are uncertain.

Concurrent with technological changes and shifts in fuel use, primary dust emissions from power generation and industry in Central and Eastern Europe are expected to decrease in the next 20 years by at least 30 per cent. Strong emission reductions for dust and aerosol particles can be achieved by implementing various filtering techniques (such as electrostatic precipitation) on power plants and baghouse filters on industrial installations, fuel change from coal to gas in power production and residential heating, and energy conservation. While industrial particulate emissions are expected to decrease, the trend in future transport emissions is uncertain, particularly with respect to soot, since traffic volume is expected to rise strongly in Central and Eastern Europe.

Metals and persistent organic pollutants

Heavy metals – mostly aerosol-bound – result from combustion processes, industry and transport. The long-range atmospheric transport of heavy metals is well documented, and the influence of anthropogenic sources has been observed as far as the polar regions. Measurements in southern Scandinavia indicate that the concentrations of many trace elements may be an order of magnitude higher in air masses that have passed the European continent than in air masses coming from the North Atlantic region (Lannefors et al, 1983).

Since the 1970s the deposition and concentration of heavy metals have fallen. The decrease is most pronounced for lead, due to a reduction of the lead content in petrol, as illustrated for rural (Cottered), suburban (North Tyneside), urban (Cardiff) and motorway edge (Manchester) measurement stations in the UK in Figure 4.5 (McInnes, 1991).

The deposition of metals in Europe is generally much less than the European emissions, indicating that Europe contributes to deposition outside its borders, to the oceans and to the metal content of the atmosphere. For instance, in a recent study (Petersen, 1992), the deposition of mercury in Europe is estimated at 120 to 300 tonnes/year, while European emissions are 730 tonnes/year. Europe is probably contributing to the background concentration level of 2 to 4 ng/m^3. The accumulation of mercury in soils from atmospheric deposition is considered to be the dominant source of high contents of mercury in freshwater fish from lakes in Scandinavia (Chapter 5).

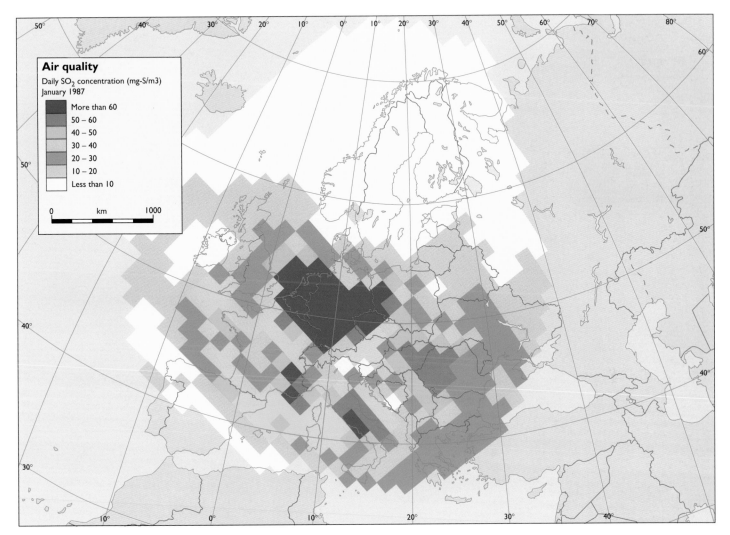

Adverse effects of lead and cadmium are known from a number of specific studies carried out in both West and East Europe (WHO, 1987). Only a few air quality standards have been defined for some metals (for example lead, cadmium, zinc) by national governments and by international bodies. Generally, these are exceeded only at the local scale, close to sources.

For heavy metals, few target or limit values have been defined for atmospheric deposition to prevent accumulation in soil, groundwater and food. Research is ongoing to assess exceedances of target values for soil and sea and related critical deposition load values in Europe for various heavy metals and organic pollutants. The modelled spatial distribution of the deposition of cadmium (Map 4.6) is generally in agreement with the available measurements (RIVM, 1992). The model resolution is not sufficient to resolve the contributions of sources or the precipitation patterns on the local scale. Measurements of heavy metals in mosses in Scandinavia indicate various areas where deposition is increased by local emissions (see Maps 7.5 and 7.6 for lead and cadmium respectively) (Rühling et al, 1992). Deposition maps are available for other metals as well. Uncertainties in emissions, particle size distribution and measurements contribute to the remaining discrepancies between model estimates of deposition and depositions inferred from measurements.

Organic pollutants which have a long environmental residence time are generally called persistent organic

pollutants (POPs). These include polycyclic aromatic hydrocarbons (PAHs) and a host of chlorinated organic compounds. Among these, the chlorinated dibenzodioxins and dibenzofurans (PCDD/PCDFs) and the polychlorinated biphenyls (PCBs) are of special concern because of their toxicity and their accumulation in food-chains. These pollutants are present in the atmosphere owing to their use and release in a wide variety of activities. Major sources of PCDD/PCDF are waste incinerators, metal recycling, wood preservatives, transport and the metal industry.

Map 4.4
Daily SO$_2$
concentration field,
19 January 1987
Source: EMEP

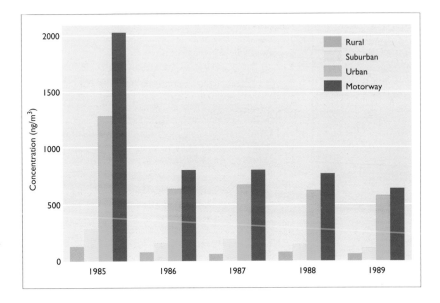

Figure 4.5 *Annual average concentration of lead in air at four locations in the UK (ng/m^3)*
Source: McInnes, 1991

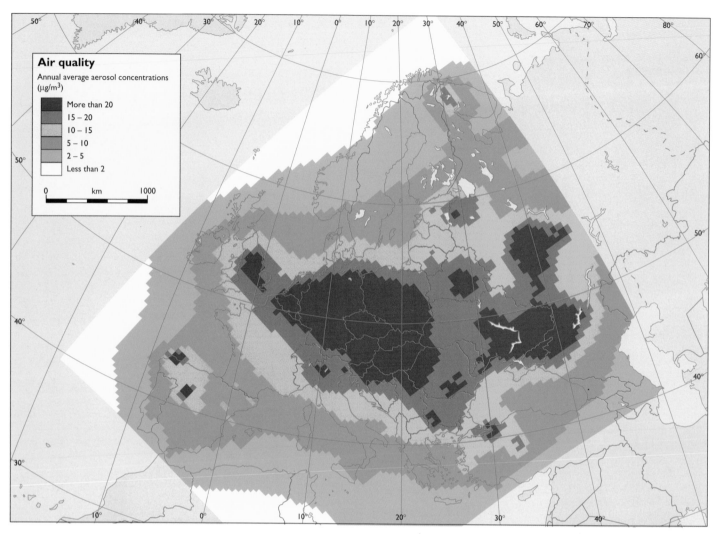

Map 4.5
Yearly averaged
aerosol
concentrations, based
on model
calculations
(RIVM/TREND-
model) (µg/m³)
Source: RIVM, 1992

Polychlorinated biphenyls have been extensively used in electric equipment such as transformers and capacitors, and as fire-resistant hydraulic fluid.

Concentration levels of dioxins in ambient air are of the order of 1 to 100 femtogrammes (fg) i-TEQ/m³ (fg=10^{-15} g; i-TEQ = international 2,3,7,8 TCDD toxic equivalents). Model studies indicate deposition loads in northwestern Europe of the order of 1 to 20 ng i-TEQ/m² per year. There is no doubt that there is a significant transboundary transport of these and related compounds (Van Jaarsveld and Schutter, 1993).

Important sources for PAH are combustion, including transport. For benzo(a)pyrene (BaP), emissions in Europe are dominated by domestic fuel combustion, especially of coal and wood. Highest concentrations are found in Central and Eastern European cities in wintertime. High concentration levels of BaP at more remote locations due to long-range transport have been documented (Lunde and Bjorseth, 1977). Map 4.7 shows the deposition of BaP over Europe from model calculations (Van Jaarsveld, 1994).

Concentrations of PCB in air in Europe are of the order of 10 to 10 000 picogrammes (pg)/m³ (pg=10^{-12} g). Regulations on the production and use of PCBs were introduced already in the 1970s (OECD, 1973) because of their recognised harmful effects. However, due to releases both before and after these regulations were passed, PCBs are still present in the atmosphere, and their concentrations are being reduced only slowly (Jones et al, 1992).

Visibility degradation by air pollution

Visibility degradation can be caused by primary pollutants – soot and other particles – emitted directly into the atmosphere, and by secondary pollutants – mainly sulphates and nitrates – formed in the atmosphere by reactions of primary pollutants (see Box 4C). Using measured sulphate aerosol concentrations from the EMEP network, 'average' visibility under dry conditions may be quantified, and is usually in the range 50 to 100 km. Relative humidities of more than 80 per cent will reduce expected visibility to less than 50 km, depending on the concentration of sulphates and other aerosol particles.

Even in dry conditions, visibility will be reduced greatly if, for example, aerosol concentrations are as high as those shown for 21 July 1990 in Map 4.8. The high concentrations of sulphate aerosols in this case were caused by a combination of rapid oxidation of SO_2 and NO_x, due to photochemical oxidant formation, and low wind speeds. The reported visibilities shown in Map 4.8b indicate reductions down to 5 km for wide areas across Europe including the UK, the Benelux countries and the Po Valley on this day.

Radioactivity

Airborne artificial radioactivity originates from the operation of nuclear facilities and the past testing of nuclear weapons in the atmosphere. Until the Chernobyl accident (see Box 18E) the fall-out resulting from the airborne testing of nuclear weapons was the largest source of artificial radioactivity in air across Europe, forming a background concentration in air in the order of 10^{-6} Bq/m³ for caesium-137 (Cs-137, the most

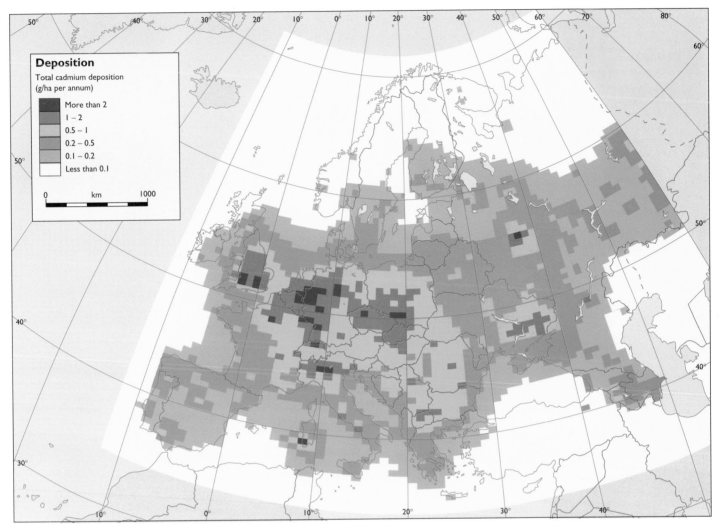

Map 4.6
Cadmium deposition
based on model
calculations
(RIVM/TREND-
model)
(g/hectare/year)
Source: Van Jaarsveld,
1994

significant radionuclide of interest in the long term). This level is many orders of magnitude below atmospheric radiation levels of natural origin predominated by radon (1 to 2 Bq/m^3 in outdoor air). Other natural radionuclides present in aerosol particles include beryllium -7 (^{7}Be) and potassium-40 (^{40}K) (see Chapter 16).

The occurrence of airborne radioactivity levels above natural levels and of significance from the point of view of health is always the result of an event, an accident or an unauthorised release. Radioactive decay and the same natural scavenging processes which act on other airborne pollutants usually mean that elevated air concentrations last only over relatively short periods of time (days to weeks). However, this can be sufficient for significant levels of radiation exposure to be experienced and, as a result of the scavenging processes, can lead to deposition and elevated radiation levels in soil and water (see Chapters 5 and 7).

Authorised routine emissions from nuclear installations are based on the exposure limits recommended by the ICRP (International Commission on Radiological Protection) and, more significantly, within these limits on the ALARA (as low as reasonably achievable) principle. In addition to the on-site control of the quantities of radioactivity actually emitted, routine monitoring of radioactivity is carried out in the vicinity of nuclear facilities. This monitoring acts as a check that these limits are adhered to, giving a direct measure of the environmental consequences of the authorised routine emissions, as well as a warning of any unexpected increase in ambient radioactivity, from whatever source.

Away from nuclear facilities, regional scale levels of radioactivity are monitored by networks of radioactivity measuring stations. Data for the period 1984 to 1991 for 12 of the 18 stations across Europe, where highly accurate measurements are continuously performed (Map 4.9), clearly shows the influence of the 1986 Chernobyl accident superimposed on the normal background levels. In the weeks following the accident, levels in air diminished very rapidly through atmospheric dilution, washout of the radioactive plumes and dry deposition. However, the data clearly indicate that it took several years for concentrations to fall back to the pre-Chernobyl level of 10^{-6} Bq/m^3, mainly due to resuspension of deposited activity and to the early dispersion of the caesium-contaminated particles in the upper atmosphere, which acted thereafter as a reservoir from which subsequent leakage occurred.

Atmospheric inputs to other media

Input to coastal seas

Deposition of pollutants from the air plays an important role in polluting coastal seas in Europe. It is estimated that of the total anthropogenic emissions of heavy metals in Europe about 1 to 15 per cent is deposited on the North Sea and 4 to 20 per cent is deposited over the northwestern Mediterranean (see Chapter 5). For parts of the Mediterranean Sea (the Ligurian Sea) atmospheric inputs of heavy metals may exceed the input of rivers, especially for lead (Migon et al, 1991).

About 90 per cent of all the lead deposited annually in the North Sea (some 1120 tonnes in 1991) originates in Belgium, France, western Germany, The Netherlands and the UK. Relative to the total pollution load to the North Sea the atmosphere contributes about 35 per cent for lead, 14 per cent for cadmium and 5 per cent for zinc (Warmenhoven et al, 1989).

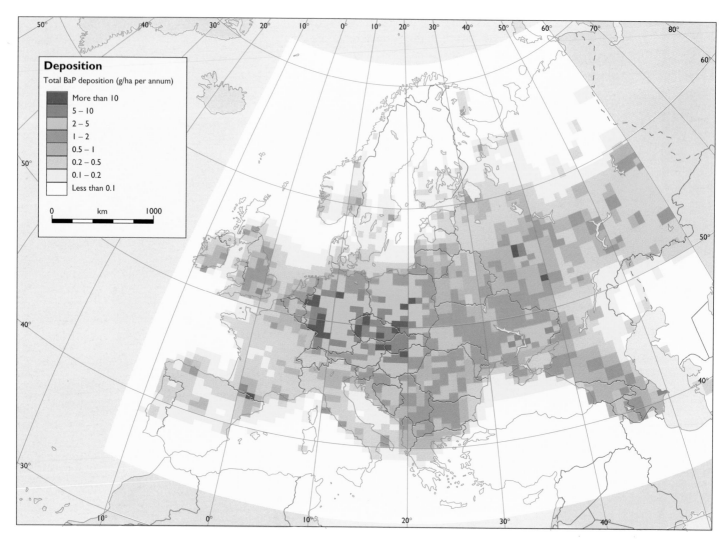

Deposition

Total BaP deposition (g/ha per annum)

- More than 10
- 5 – 10
- 2 – 5
- 1 – 2
- 0.5 – 1
- 0.2 – 0.5
- 0.1 – 0.2
- Less than 0.1

Map 4.7
Benzo(a)pyrene deposition based on model calculations (RIVM/TREND model) (g/hectare/year)
Source: Van Jaarsveld, 1994

Model calculations for the Baltic Sea reveal that, for the second half of the 1980s, the total deposition of lead appears to be approximately 1300 tonnes/year, with 70 per cent of this coming from countries directly surrounding the Baltic Sea; the remainder is due to long-range transport from other regions in Europe (HELCOM, 1991).

The total nitrogen deposition (sum of reduced nitrogen NH_x, and oxidised nitrogen NO_x) originates mainly from surrounding countries as well. It contributes about 60 to 70 per cent of the total atmospheric nitrogen deposition of 380 to 510 kilotonnes/year to the North Sea (Van Jaarsveld, 1994). For the Baltic Sea, adjacent countries contribute about 65 per cent to the total annual atmospheric nitrogen flux of about 300 kilotonnes of nitrogen (HELCOM, 1991). This accounts for one third of the total load of nitrogen to the Baltic Sea, which has increased by a factor of four since the beginning of the century. This has resulted in increased biological production followed by sedimentation and biological destruction, giving rise to reductions in the levels of dissolved oxygen in the sea (see Chapter 6).

Deposition in coastal areas is substantially greater than in more remote areas. The influence of distance from major pollution sources is illustrated by measurements from the Baltic Sea, where annual nitrogen deposition decreases from about 1100 kg/km² in the south to around 650 kg/km² in the northern parts (HELCOM, 1991).

Inputs to forest soils, surface waters and ecosystems

Critical loads for acidity and for sulphur and nitrogen are shown in Map 4.10 (Downing et al, 1993). Comparing these with actual atmospheric deposition over Europe (Sandnes and Styve, 1992; Sandnes, 1993) by calculating the difference gives the exceedance map (Map 4.11). The results show that the greatest excess (more than 2000 acid equivalents per hectare per year) occur in Central and northwestern Europe. Present loads of acidic deposition are higher than critical loads in roughly 60 per cent of Europe. Almost all European countries have areas where current levels of acidic deposition exceed critical loads, and many countries have areas where critical loads are significantly exceeded (by 1000 to 2000 acid equivalents per hectare per year). Central parts of Europe receive 20 or more times acidity than the ecosystems' critical loads, thus affecting the long-term sustainability of these ecosystems at the present levels of deposition (see Chapter 7).

Maps of critical levels are still under development. Analysis of direct effects of atmospheric concentrations of particular pollutants is complicated by the difficulty of pollution measurements in the field and of realistic manipulation of air pollution in laboratory experiments. This situation is further complicated by the fact that, although critical levels are set for individual pollutants (eg, SO_2, O_3, etc), in reality vegetation is exposed to a complex mixture of potentially damaging substances in parallel with climatic and biological stresses; all these interact with one another in poorly understood ways (but which can be additive, synergistic or antagonistic).

More attention is now being paid to the critical level for ozone with the introduction of a VOC emission protocol under the UNECE LRTAP convention. The original critical level for chronic exposure to ozone for vegetation (see Table 4.6) is now considered unsatisfactory, since it did not take account of exposure outside the 0900 to 1600 'daytime' period, nor does it give greater weight to episodic concentrations. At a UNECE workshop on critical levels for ozone held in Bern in 1993, new advances in scientific understanding, stemming especially from the European Open-Top Chamber Programme, led to two new considerations for critical ozone values. It was recommended that critical ozone levels be defined separately for forest trees and crops, and that those for crops be based on significant crop damage (at least 10 per cent) and those for trees on significant decreased biomass growth. The workshop recommended for crops a critical level of 5300 ppb-hours excess ozone above a reference level of 40 ppb (at daylight, during 3 months), and for forests 10 000 ppb-hours above 40 ppb (24 hours during 6 months).

Whether the critical level for ozone is expressed as an average or as a cumulative dose makes little difference to the situation with regard to exceedances. For most of Europe, ozone concentrations, particularly in recent years, are well above the values set by either method. Even taking into account the uncertainties mentioned above, this suggests that present levels of ozone in the lower atmosphere are damaging forests, crops and natural vegetation, and that efforts are required to reduce these effects by reducing emissions of precursors.

Conclusions

Concentrations of acidifying compounds in the atmosphere lead, after deposition, to acidification problems in soils and freshwaters as well as eutrophication in fresh and marine waters. Acidification is a major problem in Europe. The deposited acidifying compounds originate from emissions of sulphur and nitrogen oxides and of ammonia. The highest depositions are found in the highly populated and industrialised zone extending between Poland and the UK. In 60 per cent of Europe's area, critical loads for acidification are being exceeded and are expected to continue to be exceeded, resulting in prolonged risk for ecosystems. In many European countries these critical loads are exceeded by several hundreds of per cent. Studies indicate that considerable emission reductions of all acidifying compounds are needed to avoid exceedance of critical loads in Europe (reductions of about 90 per cent for SO_2 and NO_x and more than 50 per cent for ammonia).

In summary, in the next ten years, acid deposition in Europe is expected to decrease following the expected decrease in European emissions of SO_2 and NO_x (while ammonia emissions will probably not change significantly). But deposition will still exceed critical loads in some areas.

Concentrations of ground-level ozone have exceeded the WHO air quality guideline level of 75 ppb one-hour average for human exposure at nearly all existing European stations. In view of their transboundary character and harmful effects to natural and agricultural ecosystems, as well as to human health, activities leading to acidification and tropospheric photochemical oxidant formation are addressed as prominent environmental problems in Chapters 31 (Acidification) and 32 (Tropospheric photochemical oxidants).

Smog episodes continue to occur during the winter when weather conditions prevent the dispersion of pollutants and emission rates are increased. These episodes occur most frequently and are most severe in Central Europe, where maximum daily concentrations of SO_2 and PM can exceed the WHO-AQGs by factors of 4 or more. Both local sources and long-range transport can contribute to these episodes.

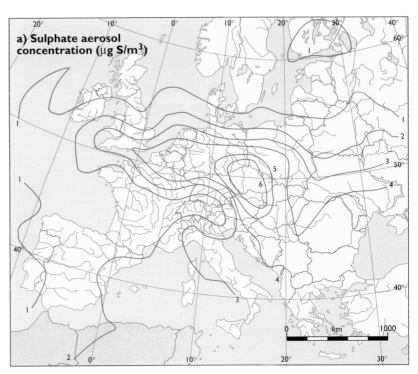

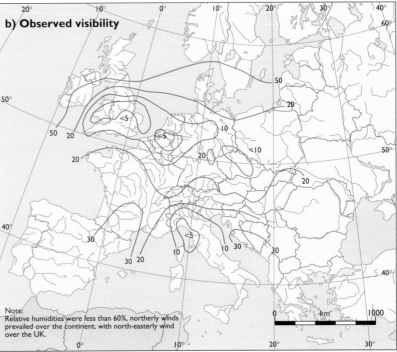

Information on atmospheric aerosol concentrations, particularly secondary aerosols and rural concentrations, is limited. Emission reductions have been achieved for some dust and aerosol particles, particularly from large stationary sources, but the trend in transport emission is not yet certain.

Heavy metal depositions and concentrations have decreased since the 1970s, particularly airborne lead concentrations following reductions in the lead content of petrol. Heavy metals such as lead, cadmium, mercury and zinc have been a problem mainly in the immediate vicinity of industrial sources and road transport but are also a subject for concern in terms of long-range transport into the Arctic region and beyond Europe.

Persistent organic pollutants, including dioxins and PAHs, are released from a number of diverse sources and, as for metals, are of concern on the local and transboundary scale.

Visibility reduction is caused by both primary and

Map 4.8
a) Concentration of sulphate aerosol in Europe on 21 July 1990 (μg S/m³);
b) Observed visibility (km) in Europe on 21 July 1990, 12 noon GMT
Source: EMEP

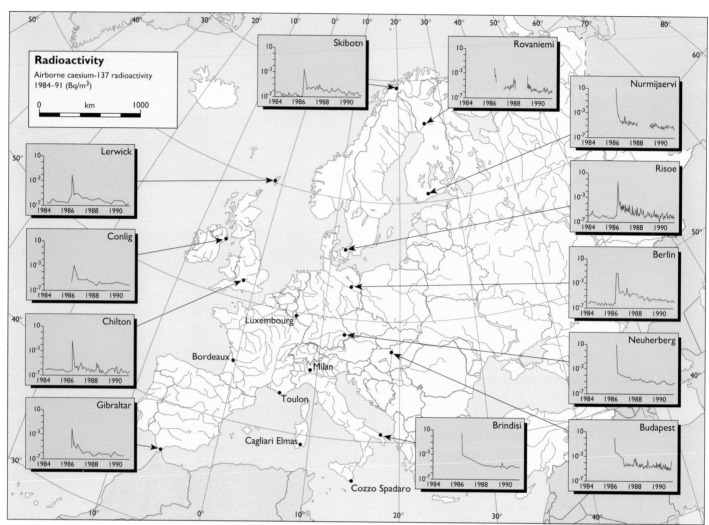

Map 4.9
Airborne caesium-137 radioactivity at 12 monitoring sites across Europe, 1984–91
Source: CEC, 1989 and CEC (in preparation)

secondary particulate pollutants. Secondary aerosols produced during photochemical oxidant formation at low wind speeds can be particularly effective in reducing visibilities.

The occurrence of airborne radioactivity levels significantly above the natural background is always the result of an event, an accident or an unauthorised release. Data from a series of monitoring stations across Europe clearly show the accidental release from Chernobyl. The initially high airborne Cs-137 levels from the accident diminished rapidly through washout and deposition in the weeks following, but took several years to fall back to pre-Chernobyl levels because of resuspension and the initial dispersion into the upper layers of the atmosphere.

Atmospheric emissions can contribute significant proportions of the heavy metals reaching the North and Baltic seas, with most coming from countries bordering the seas but some also arriving after long-range transport.

Improved monitoring and emission inventories are required for assessing these mechanisms consistently:

- For many compounds, the inventories are uncertain (eg, for VOCs, ammonia), not up to date (heavy metals) or missing (aerosols, persistent organic compounds).
- Monitoring data coverage is still insufficient for ozone and VOCs, and especially for compounds such as toxic aerosol components, metals and persistent organic pollutants.
- The EMEP network for assessing acid deposition is still too coarse in southwestern and Eastern Europe.
- Dry deposition is still estimated indirectly through unharmonised methods.
- A framework for assessment of suspended particulate matter is lacking.

GLOBAL AIR POLLUTION

Scientific understanding and public concern have recently been rising about potential global impacts such as increased exposure to ultraviolet solar radiation, sea level rise, more frequent droughts and storms with all the subsequent economic and societal damage. It is becoming increasingly clear that human activities can now disturb or alter processes at the planetary level. In effect, long-lived gaseous releases from anthropogenic activities can accumulate in the atmosphere, thus destabilising the prevailing equilibrium beyond Europe, and affecting global-scale processes. The main effects known so far are the enhanced greenhouse effect, the depletion of stratospheric ozone and the changing composition of the atmosphere.

Long-lived air pollutants

Air pollutants (eg, CO_2, CH_4, N_2O, CO, O_3 and CFCs) with residence times longer than one week have a significance on a global scale. In terms of population or surface area, Europe contributes disproportionately to the global budget of most of these pollutants (Table 4.7).

The main sources of carbon dioxide (CO_2) in Europe are combustion of fossil fuel in power generation, road transport and landuse changes. CFC emissions originate from industrial and household applications of refrigeration, packaging and cleaning. In the 1992 Copenhagen Amendments to the Montreal Protocol it was agreed that CFC emissions should be phased out by 1996. Anthropogenic emissions of nitrous oxide (N_2O) are

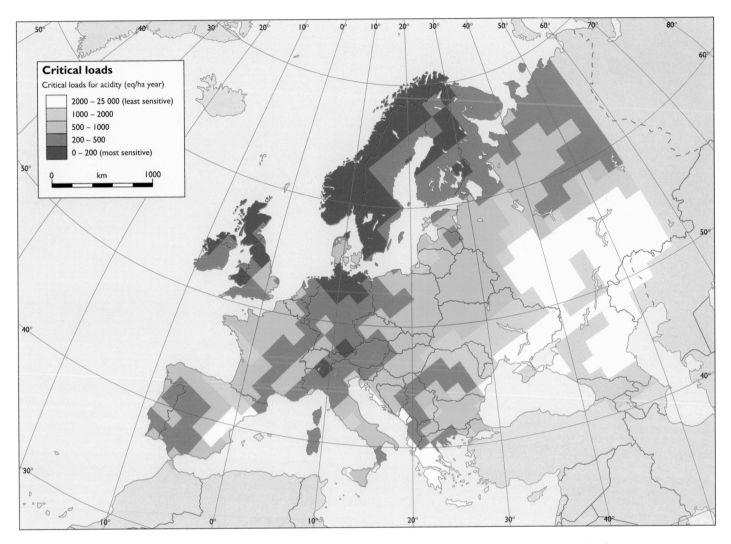

Critical loads

Critical loads for acidity (eq/ha year)

- 2000 – 25 000 (least sensitive)
- 1000 – 2000
- 500 – 1000
- 200 – 500
- 0 – 200 (most sensitive)

0 km 1000

Map 4.10
Critical loads
(5-percentile) for
acidity
Source: RIVM/CCE

probably connected to agriculture, landuse change and combustion; the sizes of these emissions are quite uncertain. Major anthropogenic sources for methane (CH_4) in Europe are ruminants (cows, goats, sheep, etc), agriculture (wetland rice cultivation), landfills and energy systems. Nitrogen oxides and carbon monoxide (CO) are emitted predominantly by road transport, power production and industry. Volatile organic compounds (VOCs) are emitted mainly by road transport and industry, with important contributions from industrial and domestic uses of solvents and evaporative losses from the distribution and transport of motor spirit (IPCC, 1990b; 1992; RIVM, 1992).

The changing troposphere

The chemical composition of the global troposphere, the lowest 10 to 15 km of the atmosphere, is changing. Concentrations of both the greenhouse gases that are chemically inert in the troposphere, such as CO_2, N_2O and the CFCs, and those of more reactive gases such as CH_4, CO and O_3 are increasing (see Table 4.1). The largest increase (4 per cent per year) has been noted for CFCs, and there have been even greater increases for some individual CFCs (WMO-UNEP, 1991). Recent data verify that the rate of increase of methane has fallen from about 1 per cent in the late 1970s to about 0.6 per cent per year at present (Steele et al, 1992). Carbon monoxide concentrations are increasing by 1 per cent per year in the northern hemisphere. Ozone concentration is increasing in the northern hemispheric troposphere at a rate of 0.5 to 1 per cent per year (Figure 4.6). The increased concentration of all

these trace gases contributes to the global warming potential (see below).

Ozone, as already mentioned, is not directly emitted, but is formed in the troposphere by photochemical reactions involving methane, carbon monoxide, VOCs and NO_x. A higher average ozone concentration in the troposphere has led to higher background levels of ozone at the Earth's surface. In Europe, the average ozone concentration at ground level has risen from 10 ppb at the end of the last century to about 25 ppb today (Volz and Kley, 1988). Regional episodes of ozone in Europe are superimposed on these higher background levels, with the result that higher episodic levels of ozone occur, aggravating ozone's adverse impacts on human health and vegetation (see above).

Ozone concentrations in the northern hemispheric troposphere are expected to keep rising at a rate of about 0.5 to 1.0 per cent per year, as a consequence of the expected increase during the next decade of global emissions of carbon monoxide, methane, and nitrogen oxides by 0.9, 0.9, and 0.8 per cent per year, respectively. This will lead to increased exposure of humans and ecosystems to ozone.

Another large-scale change of importance is the decline of the self-cleansing ability of the troposphere because of a lowering of the level of the hydroxyl radical concentration. Hydroxyl levels have probably decreased by 5 to 20 per cent since 1900 as a consequence of the build-up of methane and carbon monoxide stemming from anthropogenic sources (Thompson, 1992). Although hydroxyl is present in very small concentrations in the atmosphere, it reacts with virtually all trace gases, including those that would otherwise be inert. As a result, lower concentrations of hydroxyl

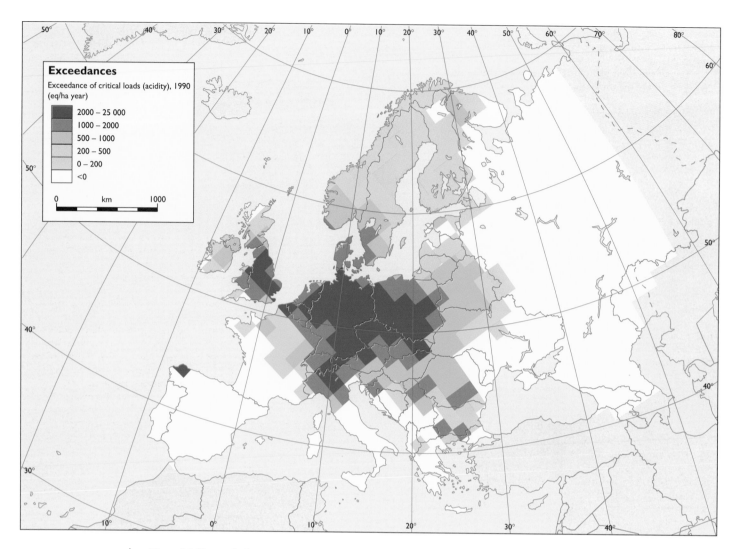

Figure 4.6 *Tropospheric ozone concentration at various mountain sites in Western Europe, 1880–1990 (ppb)*
Source: Marenco et al, 1994

Map 4.11 Exceedances of critical loads (acidity), 1990
Source: RIVM/CCE

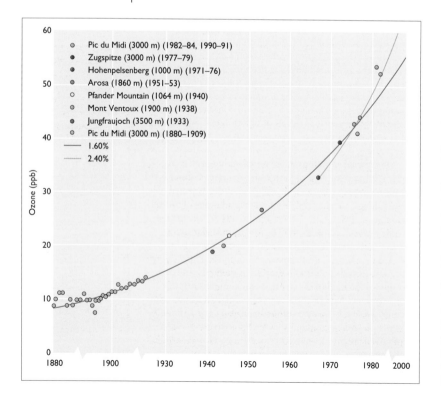

increase the atmospheric residence times and hence the concentrations of trace gases. The very complex chemistry of the troposphere provides some compensation mechanisms (eg, increased concentration of hydrogen peroxide concurrent with lower hydroxyl levels could speed up oxidation of SO_2 in clouds). Generally, however, this lowered reactivity of the troposphere may have important impacts on trace gas concentrations – including, for example, methane, which contributes to the global warming potential, and substitutes for CFCs such as HCFCs, which could reach the stratosphere and deplete the ozone layer.

For trace gases with very long residence times, such as CO_2, N_2O and CFCs, the European contributions to tropospheric change are approximately those given in Table 4.7. This does not apply to methane, for which the observed concentration trend may be towards higher levels than can be explained from the increased global emissions, as a consequence of the reduced hydroxyl concentration (IPCC, 1992). Simplified model calculations indicate that if global emissions of methane and carbon monoxide were reduced by 10 per cent – about equal to Europe's share – methane levels would stabilise rather than increase (Rotmans et al, 1992). For trace gases such as ozone, CO, NO_x and VOCs the residence time is less than the time needed for global mixing, and Europe's contribution to tropospheric change cannot be derived simply from Table 4.7. Model calculations indicate that European contributions to tropospheric ozone concentrations are of the order of 10 to 15 per cent of present levels, with considerable uncertainty (Roemer and Van den Hout, 1989).

Aircraft emissions play an important role in changes in the upper parts of the troposphere, the level at which aircraft cruise. Model calculations suggest that NO_x emissions from aircraft are responsible for 5 to 12 per cent of the observed ozone concentrations in the northern hemisphere's upper troposphere (Beck et al, 1992).

Recently, other greenhouse gases have also been discussed, especially fluoridised compounds such as carbon tetrafluoride (CF_4) and sulphur hexafluoride (SF_6). Globally the contribution from these gases to the global warming potential is modest. Nationally, however, such emissions contribute significantly to total greenhouse gas emissions.

Depletion of stratospheric ozone

The observed depletion of the stratospheric ozone layer during the last decades is a problem of global scope (WMO-UNEP, 1991). The major concern is that certain wavelengths of ultraviolet solar radiation penetrate the atmosphere down to the Earth's surface, damaging human health and ecosystems. Another important consequence could be a change in global circulation and global climate, as absorption of solar radiation by ozone leads to heat formation, which is a key factor in stratospheric climate. It is now evident that the principal cause of this problem is the substantial increase of industrially produced unreactive CFCs and halons which have found their way to the upper atmosphere and are expected to cause ozone depletion up to at least the year 2100.

The annual average depletion of beneficial ozone in the stratosphere, which provides protection against excess ultraviolet-B (UV-B) radiation, is not compensated by the average increase in destructive tropospheric ozone, since the stratospheric ozone layer contains about 90 per cent of all the ozone in the atmosphere (Figure 4.7).

Figure. 4.7 Average ozone concentrations in the stratosphere and troposphere, 1969–88
Source: Staehelin and Schmid, 1991

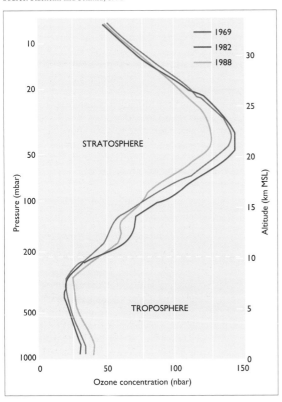

Species	European emissions	Global emissions		European contribution to global anthropogenic emissions (%)[5]
		Anthropogenic	Natural	
CO_2	8070[3]	27 870[1,2]	–[4]	30
CH_4	55[3]	335[2]	155[2,7]	16
CFCs[8]	0.5[3]	1.4[3]	–	36
N_2O	0.5[3]	7[6]	15[6]	7
NO_x (as NO_2)	22[10]	89–125[2]	23–131[2]	21*
CO	125[3]	1100[3,11]	70–280[2,7]	11*
VOC (NMHC)**	25[12]	100[1]	1100[1]	25*
SO_2	39[10]	142–166[2]	14–20[9]	25*

Notes:
This table does not represent the large differences in per capita or unit area emission rates, particularly found between European countries of the OECD and those with former planned economies. Furthermore, recent developments are modifying these differences.
** VOC = volatile organic compounds; NMHC = non-methane hydrocarbons
1 IPCC (1990a).
2 IPCC (1992); anthropogenic sources include fossil fuel combustion, cement production and deforestation.
3 SEI (1992). These are country estimates for Europe + (0.5 x the former USSR). The global estimates of SEI (1992) agree with the IPCC (1992) assessment for fossil combustion related emissions and for CFCs. The global estimates for CH_4 from landfills and from a number of land-use related sources of CH_4 and N_2O of SEI (1992) show a disagreement with IPCC (1992). Apparently, the uncertainty in the European estimates is greater than in global estimates.
4 Natural sinks and sources of CO_2 far outweigh anthropogenic emissions. Note that deforestation is included in the anthropogenic emission estimate.
5 Where indicated with a * the European contribution is given, but global estimates are highly uncertain.
6 Khalil and Rasmussen (1992).
7 CH_4 and CO sinks have not been included.
8 Indicated as CFC-11 equivalents.
9 This estimate includes only volcanic emissions, and not demethylsulphide (DMS) from oceans and DMS and H_2S from soils and vegetation.
10 Sandnes and Styve (1992).
11 Crutzen and Zimmermann (1991), also presented in IPCC (1992).
12 Simpson and Styve (1992). Estimate includes mainly non-methane anthropogenic volatile organic compounds. It may, however, include minor amounts of CH_4 and natural emissions for parts of Europe.

The annual average ozone column abundance (the sum of stratospheric and tropospheric ozone) has now been decreasing on a global scale for more than a decade, apparently at an increasing rate. The trend is larger nearer the poles, with no detectable change at the tropics. In the northern hemisphere the largest depletion has been at mid-latitudes at an average rate of 0.4 to 0.5 per cent per year over the period 1979–90 (Stolarski et al, 1991). WMO has reported that seasonal average values in winter 1991–92 over the entire 50°–60° N latitudinal belt, including Northern Europe, were 11 per cent lower than multi-year averages, an unprecedented minimum in 35 years of observations. Preliminary evaluations (Bojkov et al 1993) show even larger depletions (13 per cent) in winter 1992–93. These recent large ozone depletions coincide with high stratospheric aerosol loadings arising from the eruption of the Pinatubo volcano in 1991.

The decrease is dramatically manifest in spring at polar latitudes. The Antarctic ozone hole that started to appear after 1975 tends to become deeper, last longer and extend to a larger area each year. In 1992, WMO reported up to 65 per cent ozone destruction in the Antarctic spring. For the first time, inhabited parts of South America experienced a 50 per cent reduction in total ozone amount, causing approximately doubled UV-B radiation levels (WMO press releases, 1992). In the Arctic region, a recent European research campaign has demonstrated strongly perturbed atmospheric chemistry characterised by high levels of active chlorine compounds (Pyle, 1992).

Table 4.7
European anthropogenic emissions and global anthropogenic and natural emissions for selected atmospheric compounds, 1990/91 (megatonnes/year)
Sources: See table

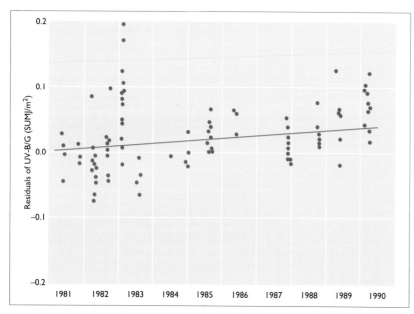

Figure 4.8
Long-term average ratio of UV-B and global radiation measured at Jungfraujoch observatory, Switzerland, 1981–90
Source: Blumthaler and Ambach, 1990

Figure 4.9
Land air temperatures of the northern hemisphere expressed as anomalies relative to the average, 1861–1990
Source: IPCC, 1992

still contain chlorine or bromine atoms and, moreover, have an elevated global warming potential and thus should be replaced themselves rather rapidly.

As a consequence of ozone layer depletion, an increase of harmful UV-B radiation is expected to occur in Europe. Since adequate and well calibrated UV-B trend measurements are in short supply, it has not yet been possible to detect a positive trend in UV-B radiation in Europe, possibly also because of other compensating atmospheric factors. Figure 4.8 shows an example where a positive trend is indeed evident. An increase in UV-B radiation levels, if it does occur, is likely to produce increases in skin cancers, effects on the human immune response system and effects on aquatic and terrestrial ecosystems, although these effects cannot yet be quantified. The effects on elements of aquatic ecosystems, such as phytoplankton, could have severe consequences for biomass production and ocean food-chains (UNEP, 1991).

Given the global consequences and the significant European contribution to this problem, stratospheric ozone depletion is considered as a prominent environmental problem (see Chapter 28).

Global climate change

Levels of greenhouse gases regulate the temperature of the Earth and its atmosphere. The most important of these gases are water vapour and carbon dioxide; of lesser importance but still significant are CFCs, methane, nitrous oxide, and ozone. According to climate change theory, an increase in the concentration of these gases should cause an increase in global average surface temperatures (IPCC, 1990b;1992). Indeed, palaeoclimatological data provide some support for this theory (Schneider, 1989). However, it should be noted that emissions of yet another group of gases, SO_2 and dimethylsulphide (DMS), lead to the build-up of sulphate aerosols in the atmosphere, which may affect cloud structure and earth albedo (reflectivity). In this way, these aerosols may partly offset the effect of greenhouse gases by reflecting solar radiation back into space (Charlson et al, 1992).

The growing interest in the greenhouse effect has led to more intense study of historical and current climatic data. Some major findings have emerged (see, eg: ECGB, 1992):

- the trend of global mean surface temperatures indicates an increase of 0.45°C over the past hundred years (see Figure 4.9);
- the six warmest years during this period were 1989, 1987, 1983, 1988, 1991, and 1990 (in order of global annual average temperature);
- between 1949 and 1989, tropical ocean surface temperatures increased by 0.5°C;
- in the last 20 years mean wind velocities have increased at all latitudes;
- since 1983 there has been approximately an 8 per cent decrease in annual snow cover on the continents of the northern hemisphere.

Stratospheric levels of chlorine have increased from a pre-industrial level of less than 1 ppb to more than 3 ppb, and are not expected to decrease before the end of the century, even if emissions could be halted now (RIVM, 1992). In the 12-year period 1976 to 1987, the EU countries produced 3.9 Mtonnes of CFCs, 38 per cent of world production. Europe consumes about 0.4 Mtonnes per year of CFCs and halons, some 34 per cent of the world's total. Consumption in Eastern Europe is only 1 per cent of the world's total (RIVM, 1992). Assuming CFC emissions to be the dominant cause of ozone depletion, Europe's present contribution is probably close to one third of the global total.

The Copenhagen Amendment to the Montreal Protocol will put a stop to global emissions of CFCs, carbon tetrachloride and methyl chloroform before 1996. The EC has already adopted a faster mandatory reduction scheme aiming at phase-out before 1995.

There are many alternatives for replacement of the CFCs in such uses as aerosol propellants, foam blowing agents and cleaning of electronic equipment. Replacement of refrigerating, freezing, and air conditioning equipment will have to be made in stages. However, substitute products do

Although these observations are broadly consistent with theoretical predictions of climate models, it is still rather uncertain when and how much global warming to expect. This is because large gaps remain in our knowledge of the global climate system, as in the strength of feedbacks to climate from the atmosphere, ocean, and biosphere (the role of clouds, for example), or in the influence of particles on the Earth's energy balance and global warming potential. The assessment (derived from global climate models) of the Intergovernmental Panel on Climate Change (IPCC) is that a doubling of pre-industrial CO_2 would lead to an increase of global average temperature of 1.5 to 4.5°C (IPCC, 1992).

To evaluate the consequences of climate change, it is common to take as a reference case the doubling of CO_2 (or an equivalent amount of total greenhouse gases) in the

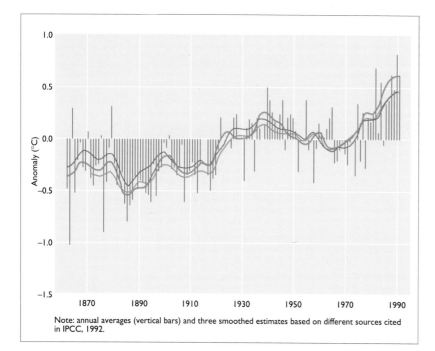

Note: annual averages (vertical bars) and three smoothed estimates based on different sources cited in IPCC, 1992.

atmosphere relative to its pre-industrial level (before the years 1750–1800). As to the likely timing of a doubling of CO_2 in the atmosphere, the IPCC scenario (IS92a) suggests this will occur by about 2050 (IPCC, 1992).

Model predictions of climate change are that the temperature increases will be greater at high latitudes than low latitudes. Hence the extent of temperature increases in Nordic countries is expected to be larger on average than those in the Mediterranean area. The IPCC reports predictions for Southern Europe from pre-industrial times to the year 2030, under the assumption that average global warming would be 1.8°C by that year. Temperature was estimated to be 2°C warmer in winter and 2 to 3°C warmer in summer, precipitation 0 to 10 per cent greater in winter and 5 to 15 per cent lower in summer, with a –5 to +5 per cent change in soil moisture in winter and a –15 to –25 per cent change in summer (IPCC, 1990a).

The main biophysical consequences of climate change in Europe will be (see Chapter 27): a rise in the global mean sea level creating threats to coastal areas, and risks to ecosystems in connection with rapid changes of temperature and hydrological conditions. These in turn will alter agricultural potential, hydrological processes leading to increased land degradation and threats to the food supply. The overall European contribution to the processes of global climate change is significant, although unevenly distributed across the continent. Annual CO_2 emissions are estimated to be about 2200 million tonnes of carbon (Table 4.7) from fossil fuel combustion and cement manufacturing, which is equivalent to approximately 30 per cent of the world's total. Carbon dioxide emissions from landuse changes are relatively minor compared with other world regions. European nations also emit 55 million tonnes a year of methane, another important greenhouse gas (Table 4.7); this is about 16 per cent of the world's total anthropogenic emissions of methane. Carbon dioxide taken up by Europe's forests may partly offset emissions from the region (see Box 23F).

An effective CO_2-equivalent doubling of greenhouse gas concentrations is expected around 2050 and will result in rapid changes in surface temperature (about 0.3°C/decade), precipitation and sea level (3 to 10 cm/decade), leading to high risks and to substantial costs. For most trace gases contributing to climate change, the global emissions are expected to rise in the next decade: carbon dioxide at 2.10 per cent/year, methane at 0.9 per cent/year, and nitrous oxide at 0.6 per cent/year. Tropospheric ozone concentrations are also expected to keep on increasing. In contrast, CFC emissions are expected to decrease to near zero around 2000.

Apart from the obvious way in which Europe affects climate change by emitting and absorbing greenhouse gases, there are less obvious linkages between the European continent and the regional and global climate. These linkages are sometimes called climate feedbacks, with one example being the feedback between land-cover and climate. In the future, a combination of economic development and climate changes may alter the proportion of Europe's land that is covered by forests, crops, settlements, and so on. Moreover, the amount of snow and ice coverage will also be modified by changes in the region's climate patterns. Together, the changes in land-cover and ice will alter the amount of solar radiation reflected by Europe, as well as the amount of heat and moisture that it gives off or absorbs. This in turn will affect the Earth's heat balance, which will lead to further changes in the region's climatic patterns. Hence, a change in land-cover and ice brought about by global warming or other factors will feed back to the climate system, possibly accelerating (or dampening) the rate of climate change. (For further detail see Chapter 27.)

Conclusions

Long-term changes to the composition of the atmosphere may potentially lead to changes in the world's climate. Europe plays important direct and indirect roles in these changes. Europe's contribution to most anthropogenic emissions of greenhouse gases and ozone depleting compounds is proportionately far larger than its geographical extent and population size relative to the rest of the world. At the same time, Europe may suffer consequences of global changes from non-European sources of emissions.

If fully realised, impacts could be far-reaching with: increased occurrence of skin cancer due to exposure to enhanced UV-B radiation; flooding and salinisation of groundwater in coastal areas due to sea level rise; alterations to ecosystems, land degradation, and changed runoff patterns leading to modification of agricultural potential. Regional predictions of these global effects are still very uncertain.

From the perspectives of both cause and effect, climate change and stratospheric ozone depletion have been ranked as prominent environmental problems facing Europe as a threat to sustainability, and a risk to human health and ecosystems.

SUMMARY AND CONCLUSIONS

Urban air pollution

Urban and local air pollution is a potentially important factor affecting human health, especially in Central and Eastern Europe. It was estimated that in 70 to 80 per cent of the 105 European cities with more than 500 000 inhabitants studied for this report, levels of one or more pollutants exceed the WHO air quality guidelines (WHO-AQG) at least once in a typical year. Long-term exposure to SO_2 and particulate matter exceeds WHO-AQG in 24 of the 61 cities with winter smog problems. Especially in Eastern Europe, estimated long-term exposure to benzo(a)pyrene and particulates is above agreed risk levels, in some cases by a factor of 100 or more.

Short-term peak levels of ozone during summer smog episodes are estimated to exceed AQG values in 60 large cities, and sulphur dioxide and particulates or black smoke levels in 79 cities, in each case concerning over 100 million inhabitants. The main pollution sources are combustion and industry in many Central and Eastern European countries, and road transport, particularly in western cities (see Chapter 10).

Some local improvements in urban air quality have occurred since the 1970s as a consequence of technical and structural improvements. These improvements have been particularly dramatic for SO_2, particulate material and lead in many areas. Pollution peaks in urban/industrial areas can, however, occur unnoticed by existing monitoring networks due to: the sparsity of the network, unsampled pollutants, inappropriate location and obsolete or ill-maintained sampling apparatus. Monitoring systems need improvement. In many cities, consistent information on pollution sources, emissions, air concentrations, exposure and health effects, and their interconnections is not available. Assessing urban air quality throughout Europe for this report has required a new collection of data. To obtain a consistent and regular assessment of air quality in Europe there is a need to monitor, collect and evaluate data on urban air pollution, human exposure and health effects in a harmonised manner at the European scale.

A desire to reduce the health effects of air pollution and the associated costs can be considered as one of the major reasons for air pollution emission reductions. Furthermore, annual maintenance costs due to air pollution are of the order of ECU 10 billion. For these reasons urban air pollution is recognised as an important environmental problem in Europe. The processes leading to urban air pollution are described in Chapter 10. Its links with the wider issue of urban stress are addressed in Chapter 37.

Regional air pollution

Pollutants originating from sulphur and nitrogen oxides and ammonia emissions can be transported long distances downwind and deposited onto natural surfaces, causing major problems including acidification and contamination of soils (Chapter 7) and surface waters (Chapter 5). This has dramatic consequences for the diversity and conditions of ecosystems including forests and crops. Fish stock die-back is a serious problem in regions susceptible to freshwater acidification. The highest deposition levels are found in the highly populated and industrialised zone extending beween Poland and the UK. In 60 per cent of Europe's area, critical loads for acidification are being exceeded, and in many European countries these critical loads are exceeded by several hundreds of per cent, for example in Norway and Sweden. Acidification has already been identified as a priority problem for which a European abatement strategy has been developed in the UNECE Convention on Long-Range Transboundary Air Pollution. Because of the effects of acidification on forest and aquatic ecosystems it is also addressed as a prominent environmental problem in Chapter 31 of this report. It has been estimated that very significant emission reductions of all acidifying compounds are needed to avoid exceedance of critical loads in Europe (about 90 per cent for SO_2 and NO_x and more than 50 per cent for ammonia).

Photo-oxidant formation from increased NO_x and VOC emissions has led to elevated ozone concentration values exceeding the one-hour 75 ppb maximum level proposed by the WHO at nearly all existing European stations. Photochemical ozone, both in the free troposphere and at ground level, has been recognised as a priority problem by the EU and UNECE, in view of its transboundary character and its harmful effects on human health and on natural and agricultural ecosystems. Photochemical ozone is also considered as a prominent environmental problem (see Chapter 28).

Severe extensive winter smog episodes creating health hazards occur frequently in Central and Eastern Europe, and less frequently also in other parts of Europe. In the densely populated parts of the Czech Republic, eastern Germany and southern Poland, levels of sulphur dioxide and particulate matter can exceed short-term WHO-AQGs by a factor of four or more. Combustion processes, including residential heating, and power production and industry are the main sources for this problem. Concentrations of lead and other heavy metals in air have decreased since the 1970s (lead decreases were mostly in Western Europe following the introduction of unleaded petrol).

The total input of pollutants, especially of heavy metals and nutrients, to the Baltic Sea, the North Sea or the Mediterranean Sea is estimated to have a significant atmospheric contribution (5 to 70 per cent). Most of this atmospheric input is coming from adjacent countries. These estimates of atmospheric inputs into regional seas are rather preliminary, but still have been valuable for supporting protection measures for the North Sea and the Baltic Sea.

For many compounds, emission inventories are uncertain (eg, for VOCs and ammonia), not up to date (heavy metals) or even missing (aerosols and persistent organic compounds). Future assessments of atmospheric pollution and reduction measures will need improved and quality assured European emission inventories (see Chapter 14). For pan-European transboundary air pollution problems the UNECE LRTAP convention and the EMEP network have formed the most important source of data and assessments. Once fully operational, the CORINAIR system, available to all countries in the UNECE area, will provide more detailed data relating to the type and structure of sources and their emissions.

Air quality monitoring coverage is still insufficient for ozone and VOCs but especially for toxic substances such as aerosols, metals and persistent organic pollutants for which no European network exists. No formal European network for supporting smog alert systems is available yet. The EMEP network for acidifying compounds and photo-oxidants is still too sparse to cover regional air pollution and deposition problems in certain regions, especially southwestern and Eastern Europe. Impacts of air pollutants on natural and agricultural ecosystems are starting to be monitored, for example, through integrated ecological monitoring. Critical levels for a number of air pollutants are under discussion and conclusions are still evolving.

Global air pollution

The build-up of long-lived pollutant compounds in the atmosphere alter the composition, chemistry and dynamics of the atmosphere with possible climatic changes and depletion of the protecting shield against solar ultraviolet radiation provided by the stratospheric ozone layer. Europe's contribution to most anthropogenic emissions of greenhouse gases and ozone-depleting compounds is disproportionately large in relation to its geographical extent and population size.

Tropospheric ozone concentration in the northern hemisphere is increasing continuously at a rate of 0.5 to 1.0 per cent per year. European emissions contribute some 10 to 15 per cent to the present levels. High ground level concentrations contribute to effects on humans and ecosystems. There is concern about the concurrent decline of the self-cleansing ability of the atmosphere that may lead to longer residence times of many trace gases, enhanced greenhouse effects and more extensive stratospheric ozone depletion.

While 'bad' tropospheric ozone is increasing, 'good' stratospheric ozone has been depleted in the northern hemisphere at an average rate of 0.4 to 0.5 per cent per year over the period 1979 to 1990. Winter 1991/92 average values over the entire 50–60°N latitudinal belt were 12 per cent lower than multi-year averages. The contribution of European emissions to this problem is around 35 to 40 per cent. Exposure to enhanced UV-B radiation in Europe would lead, according to model calculations, to increased occurrence of skin cancer and eye cataract, and depletion of sea plankton, thus affecting the food-chain.

Observations of climate changes are broadly consistent with theoretical predictions of climate models, but it is still rather uncertain how much global warming can be expected in Europe. If realised, impacts could be far-reaching:

- sea level rise, leading to flooding of lowland areas, increased salinity of estuaries and groundwater resources;
- changed agricultural potential in various areas in Europe with a shift of vegetation zones;
- risks to ecosystems and loss of biodiversity;
- soil erosion and desertification; and
- changed hydrological patterns.

Europe's contribution to climate change includes emission of greenhouse gases, but also absorption by its forests. Changes in landuse may also affect climate change.

Tropospheric change, stratospheric ozone depletion and climate change are global problems with a relatively large contribution from Europe, and possibly large impacts in Europe. These characteristics combined with the risk to human health and the damage they inflict on ecosystems is the basis for treating them as prominent environmental problems (see Chapters 27, 28 and 32).

Overall

Some local improvements in urban air quality and regional deposition levels have been achieved in Europe as a consequence of international conventions and technical improvements, as well as economic and societal incentives. However, urban air concentrations exceeding AQG, and hence potentially detrimental to human health, or causing damage to material and buildings, are still observed in many places. Moreover, deposition of acidifying compounds and photo-oxidants exceeds critical loads for protection of natural ecosystems over large parts of Europe. Technological possibilities exist for greatly reducing many urgent air pollution problems. However, following past attention on reducing the impact from stationary combustion emission sources, emissions from mobile sources – in particular from road transport – are now increasing in absolute terms or as a proportion of total emissions.

There is a need to establish harmonised monitoring networks and assessment methodologies (analysis, models, inventories, description of uncertainty) across countries to assist in optimising the control strategies which are required.

Finally, scenario studies are needed to assess the impact of economic development and implementation of new technologies on pollutant emissions (RIVM 1992, 1993). Such studies have not been incorporated here since they go beyond the scope of this report.

REFERENCES

Asman, W A H (1992) *Ammonia emission in Europe: updated emissions and emission variations.* Report 228471008, RIVM, Bilthoven.

Atkins, D H F and Law, D V (1984) *Retrospective trend analysis of atmospheric sulphate and nitrate at a site in the United Kingdom between 1954 and 1982.* AERE report R-11144, HMSO, London.

Barnes, R A, and Lee, D O (1978) Visibility in London and the long distance transport of atmospheric sulphur. *Atmospheric Environment* **12**, 791–49.

Beck, J P, Reeves, C E, de Leeuw, F A A M and Penkett, S A (1992) The effect of aircraft emissions on tropospheric ozone in the northern hemisphere. *Atmospheric Environment* **26A**, 17–19.

Beck, J P and Grennfelt, P (1993) Distribution of ozone over Europe. *Proceedings of EUROTRAC Symposium '92* (Borrell, P M et al, Eds) pp 43–58. SPB Academic Publishing bv, The Hague.

Blumthaler, M and Ambach, W (1990) Indication of increasing solar ultraviolet-B radiation flux in alpine regions. *Science* **248**, 206–8.

Bojkov, R D, Zierefus, C S, Balis, D M, Ziomas, I C, and Bais, A F (1993) Record low total ozone during northern winters of 1992 and 1993. *Geophysical Research Letters* **20**, 1351–4.

Brimblecombe, P (1987) *The big smoke: a history of air pollution in London since medieval times.* Cambridge University Press, Cambridge.

Builtjes, P J H and Hulshoff, J (1991) *The LOTOS project: long period modelling of photochemical oxidants in Europe using the Lotos model.* Report P91/021, TNO, Delft, The Netherlands.

CEC (in preparation) *Environmental radioactivity in the EU, 1987-90.* De Cort, M, Janssens, A and de Vries, G (Eds), EUR 15699 EN. Commission of the European Communities DG XI/JRC Ispra.

CEC (1989) *Environmental radioactivity in the European Union, 1984-1986,* EUR 12254 EN. Fraser, G and Stanners, D (Eds), CEC Radiation Protection Report 46. Commission of the European Communities, Luxembourg.

CEC (1990) *Exchange of Information Concerning Atmospheric Pollution in the European Community,* EUR 12477 EN. Commission of the European Communities, Luxembourg.

Charlson, R J, Schwartz, S E, Hales, J M, Cess, R D, Coakley, J A Jr, Hansen, J E and Hofman, D J (1992) Climate forcing by anthropogenic aerosols. *Science* **255**, 423–30.

Coote, A T, Yates, T J S, Chakrabarti, S, Bigland, D J, Ridal, J P and Butlin, R N (1991) *UN/ECE international cooperative programme on effects on materials, including historic and cultural monuments. Evaluation of decay to stone tablets: Part 1. After exposure for 1 year and 2 years.* Building Research Centre, Garston, Watford, UK.

Crutzen, P J and Zimmermann P H (1991) The changing photochemistry of the troposphere. *Tellus* **43**, 136–51.

Downing, R P, Hettelingh, J-P and de Smet, P A M (1993) *Calculation and mapping critical loads in Europe. Status Report 1993.* Coordination Centre for Effects, RIVM, Bilthoven.

ECGB (1992) Enquete Commission ('Protecting the Earth's Atmosphere') of the German Bundestag (Ed), *Climate change: a threat to global development.* Economica Verlag, Bonn.

ECOTEC (1986) *Identification and assessment of materials damage to buildings and historic monuments by air pollution. Final report.* ECOTEC Research and Consulting Ltd, Birmingham.

Heintzenberg, J, and Meszaros, E (1985) Elemental carbon, sulfur and metals in aerosol samples at a Hungarian regional air pollution station. *Időjárás* (Journal of the Hungarian Meteorological Service) **89**, 313–19.

HELCOM (1991) Airborne pollution load to the Baltic Sea 1986–1990. *Baltic Sea Environmental Proceedings No 39.* Baltic Marine Environment Protection Commission – Helsinki Commission.

Henriksen, A, Lien, L, Traaen, T S, Sevaldrud, I H, and Brakke, D F (1988) Lake acidification in Norway – present and predicted chemical status. *Ambio* **17**, 259–66.

IPCC (1990a) *Climate Change. The IPCC Scientific Assessment.*

IPCC (1990b) *The IPCC response strategies.* WMO/UNEP.

IPCC (1992) *Climate Change (1992). The supplementary report to the IPCC scientific assessment,* pp 25–46. Cambridge University Press, Cambridge.

Jones, K C, Sanders, G, Wild, S R, Burnett, V and Johnston, A E (1992) Evidence for a decline of PCBs and PAHs in rural vegetation and air in the United Kingdom. *Nature* **356**, 137–40.

Khalil, M A K and Rasmussen, R A (1992) The global sources of nitrous oxide. *Journal of Geophysical Research* **97**, 14651–60.

Kucera, V, Henriksen, J F, Knotkova, D and Sjöström, C (1992) *Model for calculations of corrosion cost caused by air pollution and its application in 3 cities.* Paper 084, 10th European Corrosion congress, 5–8 July 1993, Barcelona.

Lannefors H, Hansson H-C and Granat, L (1983) Background aerosol composition in southern Sweden: 14 micro and macro constituents measured in 7 particle size intervals at one site during one year. *Atmospheric Environment* **17**, 87–108.

Leaderer, B P, Holford, T R and Stolwijk, J A (1979) Relationship between sulfate aerosol and visibility. *Journal of the Air Control Association* **29**, 154–7.

de Leeuw, F A A M and Van Rheineck Leyssius, H J (1990) Modeling study of SO_x and NO_x transport during the January 1985 smog episode. *Water, Air and Soil Pollution* **51**, 357–71.

de Leeuw, F A A M and Van Rheineck Leyssius, H J (1991) Calculation of long-term averaged and episodic oxidant concentrations for the Netherlands. *Atmospheric Environment* **25A**, 1809–18.

Lübkert, B (1989) Characteristics of the mid-January 1985 $SO2$ smog episode in Central Europe. Report from an international workshop. *Atmospheric Environment* **23**, 611–23.

Lunde, G and Bjorseth, A (1977) Polycyclic aromatic hydrocarbons in long-range transported aerosols. *Nature* **268**, 518–9.

Mage and Zali (1992) *Motor vehicle air pollution: public health impact and control measures.* WHO Division of Environmental Health, Geneva.

Marenco, A, Gouget, H, Nédélec, P, Pagès, J-P and Karcher, F (1994) Evidence of a long-term increase in tropospheric ozone from Pic du Midi data series. Consequences: Positive radiative forcing. *Journal of Geographysical Research.* **99**(D8), 16617-32.

Mayewski, P A, Lyons, W B, Spencer, M J, Twickler, M, Dansgaard W, Koci, B, Davidson, C I and Honrath, R E (1986) Sulphate and nitrate concentrations from a South Greenland ice core. *Science* **232**, 975–7.

McInnes, G (1991) *Airborne lead concentrations in the United Kingdom 1985–1989.* Report LR814 (AP) Warren Spring Laboratory, Stevenage, UK.

Migon, C, Morelli, J, Nicolas, E and Copin-Montegut, G (1991) Evaluation of total atmospheric deposition of Pb, Cd, Cu and Zn to the Ligurian Sea. *Sci Total Environ* **105**, 135–48.

Ministry of Health (1954) *Mortality and morbidity during the London fog of December 1952*. HMSO, London.

Mylona, S (1993) *Trends of sulphur dioxide emissions, air concentrations and depositions of sulphur in Europe since 1880*. EMEP/MSC-W Report 2/93. Norwegian Meteorological Institute, Blindern, Norway.

OECD (1973) *Council Decision on Protection of the Environment by Control of PCBs*. OECD, Paris.

Pacyna, J M and Munch, J (1988) *Atmospheric emissions of arsenic, cadmium, mercury, and zinc in Europe in 1982*. NILU report 17/88. Norwegian Institute for Air research, Lillestrøm.

Petersen, G (1992) *Numerische Modellierung des Transport und der chemischen Umwandlung von Quecksilber über Europa*. GKSS Forschungszentrum Geesthacht, externer Bericht GKSS 92/E/51, Geesthacht, Germany.

Pyle, J (1992) *Presentation at the quadrennial ozone symposium*. 4–23 June 1992. Charlottesville, USA.

RIVM (1992) *The environment in Europe: a global perspective*. Report 481505001, RIVM, Bilthoven.

RIVM (1993) *Scenarios for economy and environment in Central and Eastern Europe*. Document prepared for the World Bank, RIVM, Bilthoven.

RIVM/NILU (1994) *Urban and local air quality: scientific background document for Europe's Environment 1993*. RIVM, Bilthoven (in preparation).

Roemer, M G M, Hout, K D Van den (1989) *De invloed van Europese en mondiale emissiereducties op mondiale ozonconcentraties*. Report R89/073. TNO, Delft, The Netherlands.

Rotmans, J, der Elzen, M G J, Kiol, M S, Swart, R J and Van der Woerd, H J (1992) Stabilizing atmospheric concentrations: towards international methane control. *Ambio* **21**, 404–13.

Rühling, A, Brumelis, G, Goltsova, N, Kvietkus, K, Kubin, E, Liiv, S, Magnússon, S, Mäkinen, A, Pilegaard, K, Rasmussen, L, Sander, E and Steinnes, E (1992) Atmospheric heavy metal deposition in Northern Europe 1990. *Nord 1992* **12**. Nordic Council of Ministers, Copenhagen.

Sandnes, H (1993) *Calculated budgets for airborne acidifying components in Europe, 1985, 1987, 1988, 1989, 1990, 1991 and 1992*. EMEP/MSC-W Report 1/93, Norwegian Meteorological Institute, Blindern, Norway.

Sandnes, H and Styve, H (1992) *Calculated budgets for airborne acidifying components in Europe 1985, 1987, 1988, 1989, 1990 and 1991*. EMEP/MSC-W Report 1/92. Norwegian Meteorological Institute, Blindern, Norway.

Schneider, S (1989) The changing climate. *Scientific American* **261(3)**, 38–47.

SEI (1992) *National greenhouse gas accounts: current anthropogenic sources and sinks*. Stockholm Environment Institute, Stockholm.

Simpson, D and Styve, H (1992) *The effect of the VOC protocol on ozone concentrations in Europe*. EMEP/MSC-W Note 4/92. Norwegian Meteorological Institute, Blindern, Norway.

Staehelin, J, and Schmid, W, (1991) Trend analysis of tropospheric ozone concentrations utilizing the 20 year data set of ozone balloon soundings over Payerne (Switzerland). *Atmospheric Environment* **25A**, 1739–1749.

Steele, L P, Dlugokencky, E J, Lang, P M, Tans, P P, Markin, R C and Masarie, K A (1992) Slowing down of the global accumulation of atmospheric methane during the 1980s. *Nature* **358**, 313–16.

Stolarski, R S, Bloomfield, P, McPeters, R D and Herman, J R (1991) Total ozone trends deduced from Nimbus 7 Toms data. *Geophys Res Let* **18**, 1015.

Tarrason, L (1991) *Biogenic sulphur emissions from the North Atlantic ocean*. EMEP/MSC-W Note 3/91. Norwegian Meteorological Institute, Blindern, Norway.

Tarrason, L (1992) *Contributions to sulphur background deposition over Europe: results for 1988*. EMEP/MSC-W Note 5/92, Norwegian Meteorological Institute, Blindern, Norway.

Thompson, A M (1992) The oxidizing capacity of the Earth's atmosphere: probable past and future changes. *Science* **256**, 1157–65.

UNECE (1988) *Critical Levels Workshop*, Bad Harzburg, Germany, March 1988. Unweltbundesamt, Berlin.

UNECE (1992) *The environment in Europe and North-America: annotated statistics 1992*. United Nations Statistical Commission and Economic Commission for Europe, Conference of European Statisticians, Statistical Standards and Studies No 42, New York.

UNEP (1991) *Environmental effects of ozone depletion: 1991 update*. Panel Report Pursuant to Article 6 of the Montreal Protocol on Substances that Deplete the Ozone Layer. Under the auspices of UNEP, November 1991, Nairobi.

Van den Hout, K D, Van Aalst, R M, Besemer, A C, Builtjes, P J H and de Leeuw, F A A M (1985) *Koolwaterstoffen in relatie tot de luchtkwaliteit*. Report CMP 85/03, TNO, Delft, The Netherlands.

Van Jaarsveld, J A and Schutter, M A A (1993) Modelling the long-range transport and deposition of dioxins: first results for NW Europe. *Chemosphere* **27**, 131–9.

Van Jaarsveld, J A (1993) Atmospheric inputs to the North Sea and coastal waters. *J IWEM* **7**, 24–31.

Van Jaarsveld, J A (1994) *Atmospheric deposition of cadmium, copper, lead, benzo(a)pyrene and lindane over Europe and its surrounding marine areas*. Report 222401002, RIVM, Bilthoven.

Volz, A and Kley, D (1988) Evaluation of the Montsouris series of ozone measurements made at the nineteenth century. *Nature* **332**, 240–42.

Warmenhoven, J P, Duyzer, J A, De Leu, Th and Veldt, C (1989) *The contribution of the input from the atmosphere to the contamination of the North Sea and the Dutch Wadden Sea*. Report R89/349A, TNO, Delft, The Netherlands.

WHO (1987) *Air quality guidelines for Europe*. WHO Regional Publications, European Series No 23. World Health Organization, Regional Office for Europe, Copenhagen.

WHO (1991) *Impact on human health of air pollution in Europe*. EUR/HFA TARGET 21 report EUR/ICP/CEH 097 1499n. World Health Organisation, Regional Office for Europe, Copenhagen.

WHO (1992) *Acute effects on health of smog episodes*. WHO Regional Publications, European Series No 43. World Health Organization, Regional Office for Europe, Copenhagen.

WHO (in press) *Concern for Europe's tomorrow (CET)*. World Health Organization, Regional Office for Europe, Copenhagen.

WMO (1983) Effects of sulphur compounds and other air pollutants on visibility: report prepared by the World Meteorological Organization (WMO) in cooperation with the ECE Secretariat. *Air Pollution Studies* **1**, 125–54. United Nations, New York.

WMO-UNEP (1991) *Scientific assessment of ozone depletion, 1991*. World Meteorological Organisation Report no 25, Geneva.

Unpolluted river,
River Ticino,
Switzerland
Source: Michael
St Maur Sheil

5 Inland waters

INTRODUCTION

Human health and development are threatened in many places because of insufficient or poor quality water. Flooding is a serious problem in some countries, and destruction of aquatic habitats by channelisation, rough maintenance schemes and damming of rivers can lead to an overall impoverishment of native plant and animal species.

By tradition most evaluations of the quality of the aquatic environment have been based on measurements of a set of concentrations, speciations, and physical partitions of inorganic or organic substances in the water. Other elements such as the amount of water and the physical conditions of the waterbody have only recently been considered to be of equal importance as the water quality *per se* for determining the ecological quality of the aquatic environment.

Demand for water of good quality has increased with the advent of industrialisation and rapid population growth. This trend has continued over time and has become more widespread geographically. In addition to domestic and industrial use of water, other requirements have become increasingly important. These include improved personal hygiene, agricultural irrigation and livestock supply, hydropower generation, cooling water for power plants and industry, as well as recreational purposes such as boating, swimming and fishing. Each of these intentional water uses affects, more or less, the quality of the water. Together with increased intensity of water use, discharge of untreated domestic and industrial wastes, excessive application of fertilisers and pesticides in agriculture, and accidental spills of harmful substances (including radioactive substances) have led to increasing pollution of many European waterbodies: groundwaters, rivers, lakes, coastal areas and seas. Figure 5.1 illustrates in chronological order the sequence of water pollution problems found in European freshwaters since

1850. Most pollution problems have evolved unrecognised over time until they have become apparent and measurable. Recognition of a problem, therefore, took considerable time and control measures took, in most cases, even longer.

Not all water quality problems are due solely to human impacts. Locally, natural geochemical conditions may cause high content of reduced iron (eg, in the Russian Federation and Denmark), fluoride (in Moldova and Bavaria, Germany), arsenic and strontium (in some mountainous countries), and salts in the groundwater, reducing its use as a source of drinking water. Natural events like volcano eruptions and subsequent mud flows, floods and droughts can lead to serious local and regional deterioration of the aquatic environment. The impact of some of these events, however, can be made worse by human activities; for example, by landuse changes, deforestation and river channelisation.

Water pollution used to be primarily a local problem, with identifiable sources of pollution by liquid waste. Up to a few decades ago most of the wastes discharged to waters came from animal and human excreta and other organic components from industry. In areas with low population density without sewerage systems, such problems are to a

Figure 5.1
Consecutive
occurrence and
perception of
freshwater pollution
problems in
industrialised
countries
Source: Meybeck and
Helmer, 1989

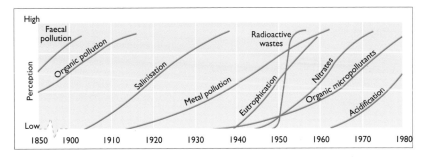

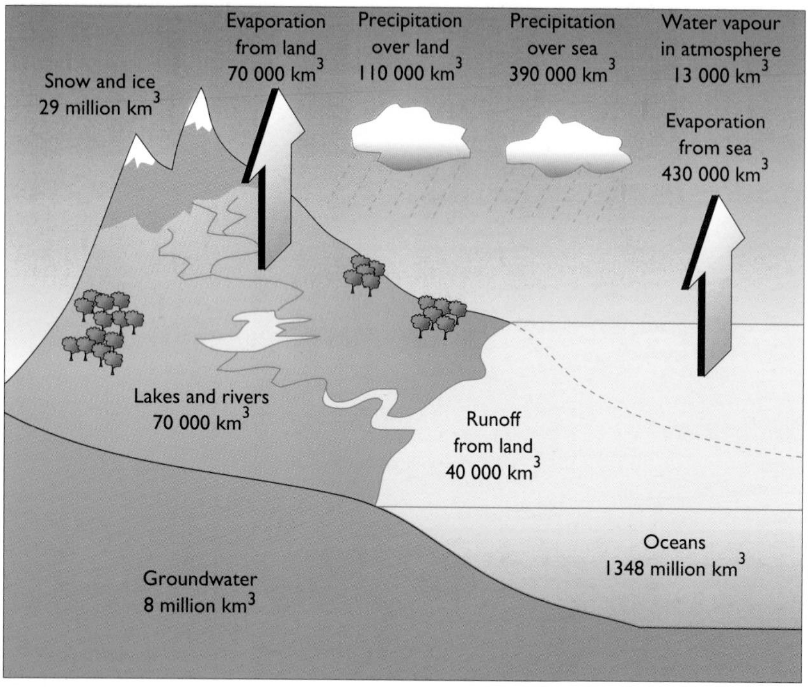

Snow and ice
29 million km³

Evaporation
from land
70 000 km³

Precipitation
over land
110 000 km³

Precipitation
over sea
390 000 km³

Water vapour
in atmosphere
13 000 km³

Evaporation
from sea
430 000 km³

Lakes and rivers
70 000 km³

Runoff
from land
40 000 km³

Groundwater
8 million km³

Oceans
1348 million km³

Figure 5.2
The hydrological cycle with major reservoirs
Source: UNEP, 1991

Note: The fluxes of evaporation, precipitation and annual runoff are in km³/year.

great extent alleviated by the natural self-purification capacity of the receiving water. However, with the increasing urbanisation of the 19th and early 20th centuries, and subsequent expansion of sewerage systems without any or adequate treatment, liquid waste loads have become so large that the self-purification capacity of receiving waters downstream of large human settlements can no longer prevent adverse effects on water resources. The results of discharges of such materials include dying fish, offensive smells and the risk of infection. In addition, the widespread channelisation of rivers that took place over this period contributed significantly to the reduction of the natural self-purification capacity of rivers.

Over the years, the pollution load of most receiving waters has further increased. In addition to impacts from point sources, pollution from non-point (diffuse) sources, for example leaching and runoff from agricultural areas and long-range transported air pollutants, have become increasingly important. Consequently, the associated problems are no longer just local or regional, but have become continental in scope.

For Europe, no general overview of water quality exists. Therefore, it is the main objective of this chapter and the following one to compile a picture, as comprehensive as possible, of the present state of European groundwaters, rivers, lakes, coastal areas and seas. Although the information available in many cases is anecdotal, an evaluation of water quality trends has been undertaken, and the present state of the waterbodies has been related to natural processes, and to human activities in the catchments overlying the aquifers or draining to the waterbodies in question.

THE WATER RESOURCE

A thorough understanding of water in the hydrosphere is necessary if we are to appreciate its role in the Earth system and provide a solid basis for rational water management to meet human demands, while at the same time preserving the integrity of the environment. The International Hydrological Programme and the Hydrology and Water Resources Programme of UNESCO and WMO, respectively, have provided important contributions to knowledge about the hydrology and water resources of the Earth and its

continents. Other significant contributions to the understanding of global water resources have been provided by Lvovich (1973), Baumgartner and Reichel (1975) and Shiklomanov (1991).

The hydrological cycle

The Earth's salt water and freshwater have been formed in the course of the evolution of the planet as a by-product of numerous chemical processes transforming rock matter at large depths. It penetrated to the surface of the Earth as water vapour in particular as a result of volcanic activity, after having leaked from the interior of the Earth.

In total, it has been estimated by UNEP (1991) that the Earth's total water amount equals approximately 1360 million km³ of which less than 3 per cent is present as liquid freshwater or locked up temporarily in ice caps. As shown in Figure 5.2, 8 million km³ are present as groundwater, 0.2 million km³ as fresh surface water, and 29 million km³ form the polar ice caps and minor glaciers elsewhere. For Europe the approximate figures are, according to original estimates and UNESCO (1978): 1 million km³ of groundwater (very approximate), 2580 km³ surface water (131 km³ in rivers, 2027 km³ in lakes and 422 km³ in reservoirs) and 4090 km³ of water locked up in glaciers.

Although the total amount of water on Earth is fixed on a short time-scale, the physical state of the water is continuously changing between the three phases (solid in ice, liquid, and as atmospheric water vapour), circulating through the different environmental compartments (ocean, atmosphere, glaciers, rivers, lakes, soil moisture and groundwater) and renewing the resources. The typical average renewal rates of water resources show considerable differences, ranging from thousands of years for the ocean and polar ice down to fortnightly or even weekly renewal of water in rivers and the atmosphere.

Average annual global evaporation from the ocean is six times higher than the evaporation from land (0.43 versus 0.07 million km³), whereas the ratio between precipitation over ocean and land is 3.5 (0.39 against 0.11 million km³), showing that approximately 40 000 km³ of water each year is transported from the ocean via the atmosphere to renew the freshwater resources. This amount of water constitutes the natural water resource that appears as global river runoff and which is potentially available for consumption each year.

Water resources

As shown in Figure 5.3, the annual average renewable water resources per capita show very large variability between different geographical regions in Europe. The populations of Nordic countries have in general between six and eight times as much water available for consumption per capita than the population of the other three geographical regions: Eastern, Southern and Western Europe.

Regional averages, however, provide only a very general measure of the adequacy of water availability, and obscure any variation between and within countries. It is interesting to examine the per capita distribution of available water resources in Europe, which is highly uneven, as shown in Map 5.1.

Relative water availability can be classified in the seven categories shown in Table 5.1.

Low, very low or extremely low water availabilities are found in 36 per cent of the European countries, in particular in the southern countries, with Malta having the lowest amount of water available of all European countries – 100 m³/capita per year. Densely populated Western countries with moderate precipitation (eg, Belgium, Germany, Denmark and the UK) also fall in these

Box 5A Definitions and sources

The **renewable freshwater resource** available for use in a country is the amount moving in rivers and aquifers, originating either from local precipitation over the country itself, or by water received from neighbouring countries in transboundary rivers and aquifers (Brouwer and Falkenmark, 1989). River runoff represents the dynamic component of the total water resource, as compared with the considerably less mobile water volumes in aquifers, lakes and glaciers. The groundwater contribution to river runoff is, according to Shiklomanov (1991), about 25 per cent of total runoff. **Water abstraction (or water withdrawal)** means water physically moved from its natural site of occurrence. The amount of abstracted water which is not returned to the site of abstraction after use is referred to as **water consumption. Water use** means all use of water, both in and outside the river, lake or aquifer.

National estimates of renewable water resources, including internal and external contributions, and water abstractions per capita are available for most European countries. Very often, however, the estimates vary between different information sources. For the purpose of this report, estimates of renewable water resources and water abstractions have been obtained for each country by updating the water resource database at Eurostat, with recent water resource statistics taken primarily from national reports prepared by many countries for presentation at the 1992 UNCED conference in Rio de Janeiro.

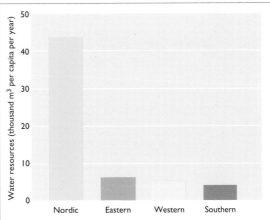

Note: Country groupings refer to WHO/UNEP (1991).
Nordic: Finland, Iceland, Norway, Sweden (Iceland excluded from this figure).
Eastern: Belarus, Bulgaria, Czech Republic, Estonia, Hungary, Latvia, Lithuania, Moldova, Poland, Romania, Russian Federation, Slovak Republic, Ukraine.
Southern: Albania, Bosnia-Herzegovina, Croatia, Cyprus, Greece, Italy, Malta, Portugal, Serbia-Montenegro, Slovenia, Spain and FYROM.
Western: Austria, Belgium, Denmark, France, Germany, Ireland, Liechtenstein, Luxembourg, The Netherlands, Switzerland, UK.

Figure 5.3
Average total renewable water resources in Nordic, Eastern, Western and Southern European countries
Source: Multiple sources compiled by EEA-TF and Eurostat (see *Statistical Compendium*)

Map 5.1
Per capita available water resources (total renewable water resource) and water abstractions for each country in Europe
Source: Multiple sources compiled by EEA-TF and Eurostat (see *Statistical Compendium*)

categories. Water availability is also low in some Eastern countries (Moldova and Ukraine), mainly due to low precipitation.

Thirty-two per cent of European countries fall in the category of medium availability.

For Europe as a whole, above medium, high or very high water availabilities are found in another 32 per cent of the

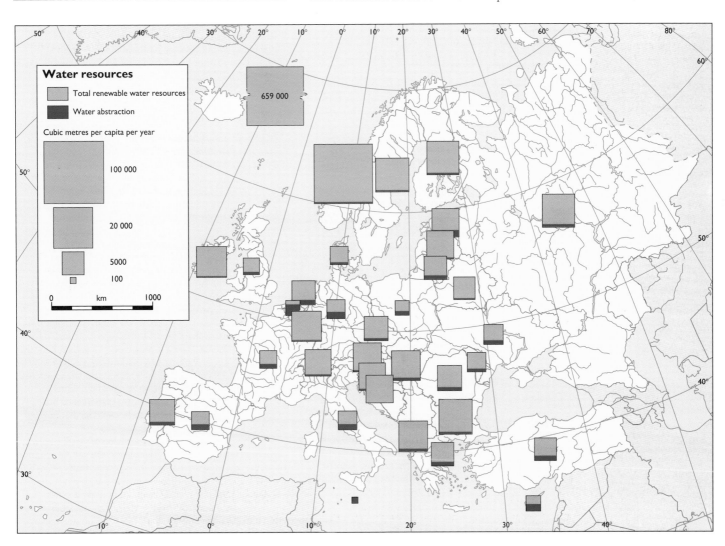

Table 5.1
Classification of relative per capita water availability
Source: Shiklomanov, 1991

Category	Water availability (m³ per capita per year)
Extremely low	<1000
Very low	1000–2000
Low	2000–5000
Medium	5000–10 000
Above medium	10 000–20 000
High	20 000–50 000
Very high	>50 000

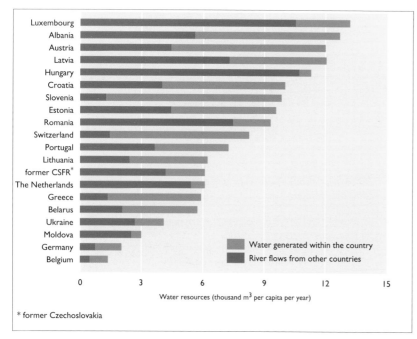

* former Czechoslovakia

Figure 5.4
Internal and external contributions to the total renewable water resource in selected European countries
Source: Multiple sources compiled by EEA-TF and Eurostat (see *Statistical Compendium*; OECD, 1992; Arnell et al, 1993)

countries. Plenty of water is found either in sparsely populated countries where precipitation is very high, like the Nordic countries (Iceland, for example, has more than 600 000 m³ of water available per year for each of its citizens), or in countries with large transboundary rivers running through them. For example, the per capita water availability in Spain is only one third of that for Portugal, which receives 48 per cent of its water in transboundary rivers from Spain. Many other countries are dependent on external contributions of water for meeting their demands. The importance of transboundary rivers as water suppliers is summarised in Figure 5.4 for those countries which receive a significant proportion (more than 10 per cent) of their water from other countries.

In this context, 20 European countries receive a high proportion of their water from transboundary rivers. Five countries – Hungary, Moldova, Romania, Luxembourg and The Netherlands – are heavily dependent on external inputs, receiving more than 75 per cent as external contributions, while three countries – Latvia, Ukraine and former Czechoslovakia (CSFR) – receive between 50 and 75 per cent of their renewable water resources from abroad. Bulgaria and Romania share the water resources of their boundary river, the Danube.

For countries with major rivers running through them, estimates of total renewable water resources tend to overestimate sustainable water resources. In humid, temperate climates, a good estimate of a country's sustainable freshwater resources is the long-term average river runoff generated within the country itself. Therefore, consideration of the spatial and temporal variability of river flows is of utmost interest for the assessment and management of European water resources.

River flow characteristics in Europe

River runoff is one of the main sources of freshwater from which the various water demands are satisfied.

There is considerable spatial and temporal variation in river flow across Europe. The seasonal flow regime of a Mediterranean catchment, with its winter high flows, dry periods in summer and frequent flash floods, is very different from that of an English lowland chalk stream, where flows vary relatively little over the year. They both have a very different seasonal regime from a catchment in Poland, which has a maximum flow following the spring snow-melt. Two factors, climatic and physical properties of the catchment, lead to such variations in seasonal flow regimes. In addition, European rivers also experience considerable variation from year to year. Finally river flow regimes are also affected by human activities, such as river impoundment and landuse changes.

European flow regimes are described here, with special emphasis on:

● average annual river runoff;
● monthly river flow regimes; and
● time variations in seasonal and annual runoff totals.

The information for this section of the report has been extracted primarily from four databases:

1 the FRIEND database for Northern and Western Europe;
2 the AMHY database for Alpine and Mediterranean Europe;
3 the CORINE river flow database of the EC; and
4 the GRDC database operated by the WMO.

Supplementary information has been obtained from hydrological yearbooks and monographs (Korzun et al, 1977; Stancik et al, 1988), or has been prepared by hydrological agencies across Europe specifically for this report.

Average annual runoff

The average annual runoff in Europe (Map 5.2) follows very closely the pattern of average annual rainfall – and topography. Annual runoff is greater than 4500 mm in western Norway, and decreases to less than 25 mm in parts of Spain, central Hungary and eastern Romania, and in large regions of Ukraine and the southern part of the Russian Federation. The greater variation of runoff in Western Europe, compared with Eastern Europe, reflects the greater variability in topography, and hence rainfall.

Across most of lowland Europe, between 25 and 45 per cent of rainfall runs off into waterbodies. In high rainfall areas, such as the Alps, western Norway and western Scotland, over 70 per cent of the rainfall may become runoff. In drier regions, particularly southern Spain, runoff may amount to less than 10 per cent of the annual rainfall.

The annual average runoff for Europe is estimated from Map 5.2 at 3100 km³ over a territory of 10.2 million km², that is, 304 mm/year, or 9.6 l/second per km². This is the equivalent of 4560 m³/capita per year for a population of 680 million. Compared to the present average total water abstraction of 700 m³/capita per year (1920 l/capita per day) in Europe, the immediate conclusion would be that Europe faces no water shortage problems.

Even when average runoff, evenly distributed, would suffice to cover the needs of water users, the natural variation from one season to another, or from year to year, can set constraints. The part of runoff which can serve as a continuous supply source, even under drought conditions, is obviously much less than average runoff, and depends primarily on the local hydrological regime. For most of

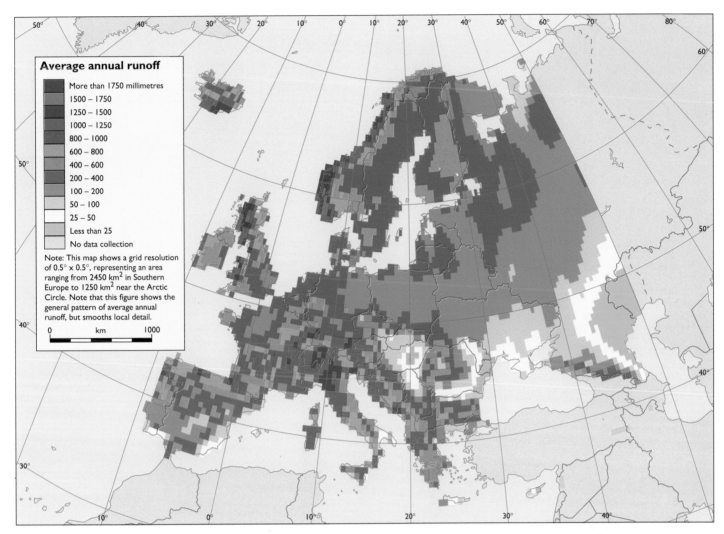

Average annual runoff

More than 1750 millimetres
1500 – 1750
1250 – 1500
1000 – 1250
800 – 1000
600 – 800
400 – 600
200 – 400
100 – 200
50 – 100
25 – 50
Less than 25
No data collection

Note: This map shows a grid resolution of 0.5° x 0.5°, representing an area ranging from 2450 km² in Southern Europe to 1250 km² near the Arctic Circle. Note that this figure shows the general pattern of average annual runoff, but smooths local detail.

0   km   1000

Western Europe, it has been estimated that this runoff, taken as the mean annual minimum 10-day flow, is between 1 and 3 l/second per km².

Seasonal river flow regimes

Variations in the distribution of river flows through the year define different flow regimes. All are based, to a large extent, on the timing of maximum and minimum flows, the number of flow seasons, and the origin of the river flows (Gottschalk et al, 1979; Krasovskaia and Gottschalk, 1992). Climate is usually the dominant influence on river flow regimes, but there are cases where the effect of physical characteristics of the catchment can dominate. Such 'atypical' regimes, with very little variability in flow from month to month, include those of catchments with very large lakes, or where river runoff is controlled by seepage from an aquifer.

River flow regimes vary considerably across Europe, but insufficient information is available to produce maps showing their geographical distribution.

The river flow regimes of large catchments (Map 5.3) can be different from those of small catchments. In particular they are much less variable because they integrate runoff over a large area, and can include sub-catchments with very different characteristics.

The differences in river flow regimes are apparent between Western Europe (where flows are at a minimum in summer and late autumn), the mountain-fed catchments (where flows are at a maximum in summer), and Eastern and Northern Europe (where most runoff occurs during the spring snow-melt period).

Inter-annual variability

The droughts of the early 1990s, which have been affecting several parts of Europe, illustrate clearly the variability in river runoff from year to year. From an analysis of runoff data from 14 gauging stations with long records (between 40 and 150 years) across Europe, three important conclusions can be drawn:

1 Consistent patterns in the variability of river flow regimes from year to year exist across Europe. Droughts, for example, tend to affect large parts of Europe at the same time. The regional consistencies and differences between wide areas reflect the correlation with large-scale weather patterns.
2 It appears that extreme years tend to cluster. Several low flow years occurred in succession in the early 1970s in Western Europe, followed by a period in the 1980s where several years were above average.
3 There is no evidence of any consistent trend in river flow characteristics. Some records show upward or downward trends for a part of the record, but in each case these periods are followed by periods with no trend, or an opposing trend.

Human influences on river flow regimes

Many European river flow regimes are heavily affected by human activities, ranging from direct abstraction of water through regulation of flow regimes by reservoirs, to changes in catchment landuse. It is, however, difficult to generalise about the impacts of human activities on flow regimes. However, it is important to recognise that these impacts are

Map 5.2
Average annual runoff in Europe
Source: Arnell et al, 1993

Note: This map shows a grid resolution of 0.5° x 0.5°, representing an area ranging from 2450 km² in Southern Europe to 1250 km² near the Arctic Circle. Note that this figure shows the general pattern of average annual runoff, but smooths local detail.

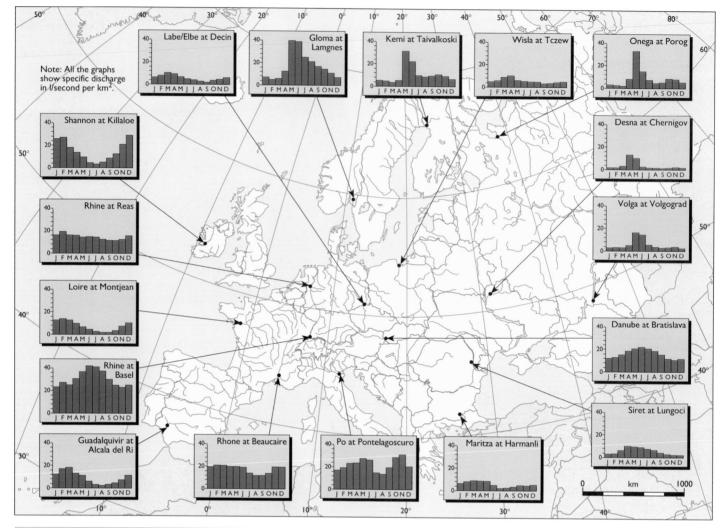

Map 5.3
Mean monthly flow for several major river catchments in Europe
Source: Arnell et al, 1993

Note: All the graphs show specific discharge in l/second per km².

not limited to areas with high population densities (examples are reservoirs and river channelisation).

Main sources of water for abstraction

Surface water is the main source for water abstraction by all utilisation sectors in Europe. On average, 70 per cent of total abstraction is drawn from this source, but with large variation between countries (Figure 5.5). Groundwater is by far the next most important source, and other sources include desalinated sea water.

Countries like Spain, Belgium, The Netherlands, Finland and Moldova, with insufficient groundwater supplies, abstract more than 90 per cent from surface water sources. However, in Cyprus, Switzerland, Slovenia, Iceland and Denmark, which are countries with extensive groundwater reservoirs, more than 75 per cent is drawn from groundwater.

The public water supply systems serve primarily domestic users and some industrial demands, and the source of water for this sensitive use is predominantly groundwater. As a source of public drinking water groundwater is of enormous importance. For Europe as a whole, about 65 per cent of the public supply is provided from groundwater (Zektser et al, 1992) which

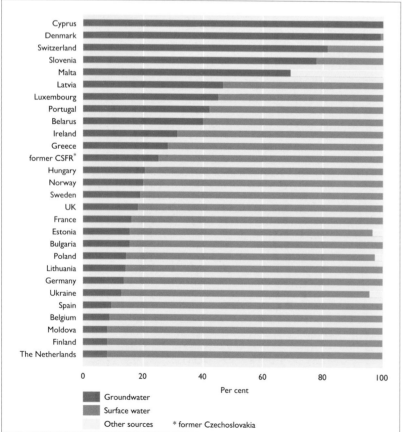

Figure 5.5
Total water abstraction (by all utilisation sectors) by source in European countries
Source: Multiple sources compiled by EEA-TF and Eurostat (see *Statistical Compendium*)

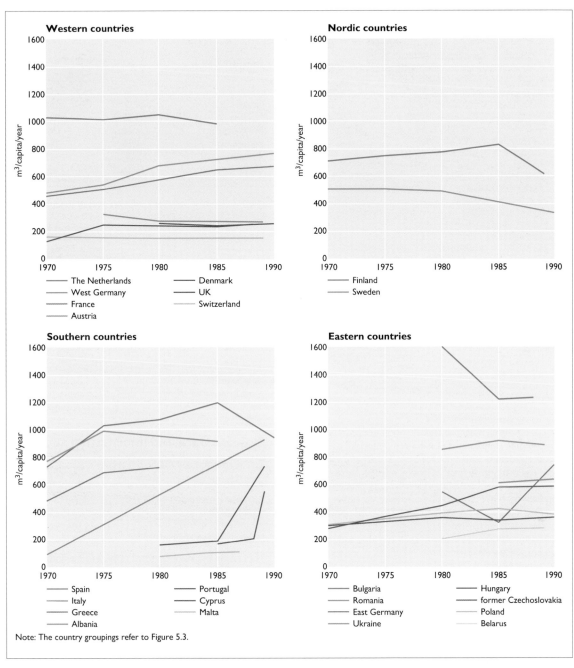

Figure 5.6
Trends in total water abstraction in various countries in different parts of Europe
Source: Multiple sources compiled by EEA-TF and Eurostat (see *Statistical Compendium*)

normally is of a better quality than surface water. On average, the total water abstraction in Europe amounts to approximately 480 km³/year (or 700 m³/capita per year), ranging from below 200 m³/capita per year in Luxembourg, Malta and Switzerland to above 1000 m³/capita per year in Bulgaria, Estonia, Lithuania and Spain.

Over the last two decades total water abstraction has in general increased in Europe (Figure 5.6). This trend, however, masks great variability between countries. Abstraction increases have been particularly marked in Southern European countries, but also in the majority of countries in Eastern and Western Europe more water was abstracted in the late 1980s than before. Stabilisation or even an abstraction decrease has occurred in some of these countries, including Austria, Bulgaria, The Netherlands and Switzerland. This is also the case in the Nordic countries: Sweden and Finland.

For Europe as a whole, 53 per cent of the abstracted water (surface and groundwater) is used for industrial purposes, 26 per cent in agriculture and only 19 per cent for domestic purposes (WRI, 1990).

There is large variability in sectoral water abstraction between countries (Figure 5.7). Interpretations of such statistics, however, should be carried out with caution because it is not always clear how they have been derived and what they comprise. Water use in the industrial sector is particularly difficult to evaluate, since it is not always clear whether cooling water and water used for power generation is included in the industrial share. Water used for these purposes often accounts for about 70 to 80 per cent of industrial water use, and most of this is used in power generation (see Chapters 19 and 20).

Agriculture's share is highest (30 to 70 per cent) in a number of Southern and Eastern European countries with low net precipitation and in Denmark and The Netherlands where agriculture is very intensive (see Chapter 22). Water use in agriculture is mainly for irrigation and livestock.

Water abstraction for public use shows less variability between European countries than other sectoral abstractions, ranging between 15 and 25 per cent of total abstractions in the majority of countries to more than 40 per cent in Luxembourg and the UK.

The intensity of water use in a country can, according to the OECD (1992), be represented as the percentage abstraction of total available water resources, which includes internally generated water and inputs from neighbouring

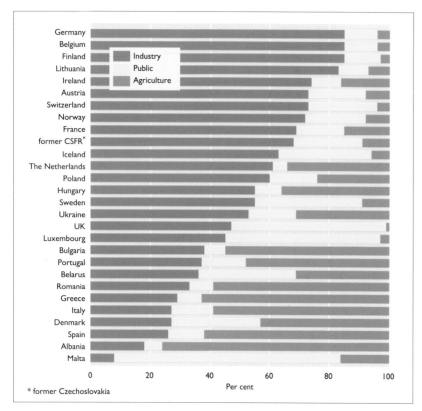

Country	Water use indicator (per cent)	
	(1)[a]	(2)[b]
Belgium	72	91
Bulgaria	6	46 – >100[c]
Lithuania	19	31
Hungary	5	96
Moldova	13	89
The Netherlands	16	136
Portugal	10	26
Romania	9	49
Ukraine	16	48

Table 5.2 *National water use intensity indicators*
Source: (a) EEA-TF and Eurostat (see *Statistical Compendium*); (b) internally generated water resources from Arnell et al (1993); (c) Bulgarian Ministry of Environment (pers comm)

Note: Indicators are calculated as the country's water abstraction in percentage of (1) total available resources and (2) internally generated water resources.

most difficult water resources to determine. Natural groundwater resources are stored in aquifers, which are permeable rock formations or unconsolidated deposits, chiefly gravels, sands and silts.

The main characteristics of groundwater systems are:

- invisible and relatively inaccessible location;
- very low flow rates, resulting in long residence times and slow reaction to changes on the surface; and
- huge extent of aquifers.

Figure 5.7
Total water abstraction (surface and groundwater) by economic sectors in European countries
Sources: If not otherwise specified, data are from WRI (1990) as presented by Veiga da Cunha, 1993. Data from Germany, UK, Belarus, Ukraine and Lithuania are extracted from Umweltbundesamt, 1992; UK DoE, 1992, Eurostat, pers comm and national reports prepared for the UNCED Conference by Ukraine, 1992 and Lithuania, 1992, respectively.

countries via transboundary rivers (see Figure 5.4).

The calculated water use intensity indicators cover a very wide spectrum, from less than 0.1 per cent in Iceland to over 70 per cent in Belgium, with a general average of around 15 per cent. If this indicator were calculated by relating abstractions exclusively to internal resources (Table 5.2), its values would be appreciably higher in those countries where total resources are, to a considerable extent, accounted for by external resources.

This is particularly true for Belgium, Bulgaria, Hungary, Moldova and The Netherlands, where water abstraction approximately equals total internal resources, or even exceeds it, as in the case of The Netherlands. These countries, as well as Romania, Lithuania, Portugal and Ukraine, are therefore particularly sensitive to upstream impacts affecting the quantity and quality of the water they receive through transboundary rivers.

GROUNDWATER

Characteristics and distribution

Groundwater is an important element in the Earth's hydrological cycle. It remains one of the least-studied and

---

Box 5B How to read the maps

The maps in this section on groundwater (prepared by RIVM and RIZA) should not be taken as giving an assessment of the situation at the local or sub-regional scale. The degree of detail of geographically distributed input data varied greatly between countries, and was often available only for big regions within a country or for whole countries. Therefore, it is important to bear in mind that, where maps are presented in this chapter, they should be evaluated primarily on a European scale. The main purpose of the maps is to outline a particular feature (for example nitrate leaching) and to show the relative differences between countries. For bigger countries the maps can also be used to

---

Box 5C Data and information sources

The information and data in this section are based on:

- questionnaires sent to national focal points;
- various national and regional state-of-the-environment reports; and
- a review of scientific literature concerning the quantitative and qualitative aspects of groundwater in Europe.

The first questionnaire consisted of three forms. Form 1 requested qualitative information about toxic contamination (such as pesticides), nitrate, acidification and overexploitation for the country as a whole. Form 2 requested quantitative regional information about a) landuse and b) groundwater resources (eg, rates of groundwater abstraction and lowering of the groundwater table). The selection of regions (territorial units) to report on, within a country, were decided by the responding country itself. The landuse information asked for included: application of nitrogen and pesticides, and number of domestic animals and manure production. Form 3 requested site specific measured data for nitrate and pesticide concentration in the groundwater and groundwater level. An extended version of this questionnaire was forwarded to some countries to obtain more detailed information.

Not all countries completed and returned the questionnaires and the degree of detail of the information received varied greatly. Some countries provided only aggregated qualitative information, while other countries provided very detailed information, documented by measurements and maps. Consequently, in many cases additional information from national and regional surveys and other literature had to be used.

Especially important was information on the actual (observed) concentrations of nitrate and pesticides in groundwater, and the quantitative aspects describing use of the resource (abstractions and overexploitation). Table 5.3 summarises the sources of information used for each country in the preparation of the section on groundwater.

In spite of its non-visibility, groundwater has very important functions, including economic, ecological and those relating to public health – which are not always fully recognised. For example, groundwater is an important source for drinking water. Human activity, however, can have a great effect on quantity and quality of the available groundwater resources. Due to the aforementioned characteristics, groundwater systems are normally very stable, in both quantity and quality. However, the effects of pollution and overexploitation will accumulate over time. In general, the periods of recovery will be centuries and decades, respectively.

The availability of groundwater (as a natural resource) is limited by three factors:

1 the total amount of recharge (renewal of groundwater), resulting from precipitation, evapotranspiration, infiltration and seepage from rivers and lakes;
2 the quality of the recharged water; and
3 the properties of the soil and aquifer (permeability, porosity, etc).

As with surface water, there is an uneven distribution of natural groundwater resources within Europe. The geographical distribution of different types of aquifers is shown in Map 5.4. A note on how to interpret the maps appears in Box 5B.

Recharge and loads

The renewal of groundwater resources occurs through natural and artificial recharge. The natural recharge is the amount of water available for percolation into the aquifer as a result of excess precipitation and runoff. The precipitation excess is the amount of precipitation less evapotranspiration.

Artificial recharge is the result of excess irrigation or of human-induced recharge (forced feeding) of aquifer systems. Although irrigation is common in large parts of Europe, systematic data on water sources and quantities used are not available. As a result, it was not possible to quantify the effects of irrigation on the groundwater systems. The essential objective of forced feeding recharge is to transform surface water with a periodically unreliable quality into a safe source for water supply. It can be done by applying natural gravity forces, for example, by infiltration canals, or by damming rivers, or via injection wells. The importance of forced feeding could not be assessed due to the lack of systematic data, but projects are in general limited to a local scale. Forced feeding of aquifers is currently being used in a number of European countries (eg, river bank infiltration in Germany and The Netherlands, and surface water infiltration in the Uppsala area of Sweden, and in dunes in The Netherlands). In Germany, for example, these techniques together account for 15.7 per cent of public water supply (Umweltbundesamt, 1992).

All groundwater contains natural chemical constituents in solution. During its passage underground, the water dissolves and deposits various substances, while other solutes are being transformed or degraded. Most of these changes tend to be slow but long-lasting. Pollution of the percolating water due to human activities sometimes has a severe impact on the constituent load reaching the groundwater.

In many places, the natural quality of groundwater is degraded because of human activities. Basically, two different sources of groundwater pollution can be recognised: non-point (diffuse) sources and point sources. The diffuse sources are mainly from agriculture (eg, irrigation excess, animal wastes, fertilisers, pesticide residues), urban areas (storm drainage) and contaminant deposition from the atmosphere. The point sources comprise municipal and industrial activities, such as surface disposal of liquid and solid waste, improper storage of materials used in manufacture, sewer, tank and pipeline leakages, and activities related to mining (especially mine tailings) and oil exploitation.

Groundwater quantity

During recent centuries, human-made changes in the hydrological cycle have had enormous impacts on groundwater levels and flows. These have changed for many reasons, including pumping of groundwater for domestic, industrial and agricultural water supply, provision of cooling water, mine pit drainage, intensified and deeper drainage of agricultural lands, changing natural land into agricultural and urban areas, and regulation of surface waters.

These changes have important economic and ecological effects. Economic effects can include crop and industrial production as well as the production of drinking water. Groundwater-related terrestrial ecosystems are notably affected by changing groundwater levels, while at the same time wetlands are becoming more scarce. Moreover, groundwater is the last remaining water source for rivers and small surface waters during dry periods. If the base flow in the rivers is not maintained, severe damage can be caused to aquatic ecosystems, and the economic functions of the rivers – such as navigation and water supply – will also suffer.

Country	Nitrate Form	Nitrate Other	Pesticides Form	Pesticides Other	Quantity Form	Quantity Other
Albania						+
Austria	+		+	+	+	
Belarus				+		
Belgium				+		
Bulgaria	+		+		+	+
Croatia	+		+		+	+
Czech Republic						+
Denmark	+	+	+	+	+	+
Estonia	+				+	+
Finland	+				+	
France		+		+		+
Germany*		+		+		+
Greece						+
Hungary	+			+	+	+
Iceland	+				+	+
Ireland	+				+	+
Italy				+		
Latvia	+				+	
Lithuania	+			+	+	
Luxembourg						
Moldova	+		+		+	
The Netherlands		+		+		
Norway	+				+	+
Poland	+				+	+
Portugal					+	+
Romania	+				+	+
Russian Federation		+				+
Slovak Republic	+		+		+	
Slovenia	+		+	+	+	+
Spain	+	+				+
Sweden	+			+	+	+
Switzerland	+		+	+	+	+
Ukraine	+		+		+	+
UK		+		+	+	

Notes: 'Form' indicates the questionnaires. 'Other' implies reports, surveys, etc.
* refers to West Germany only.

Table 5.3
Sources of information for actual concentrations of nitrate and pesticides in groundwater, and for the quantitative aspects such as abstractions and overexploitation

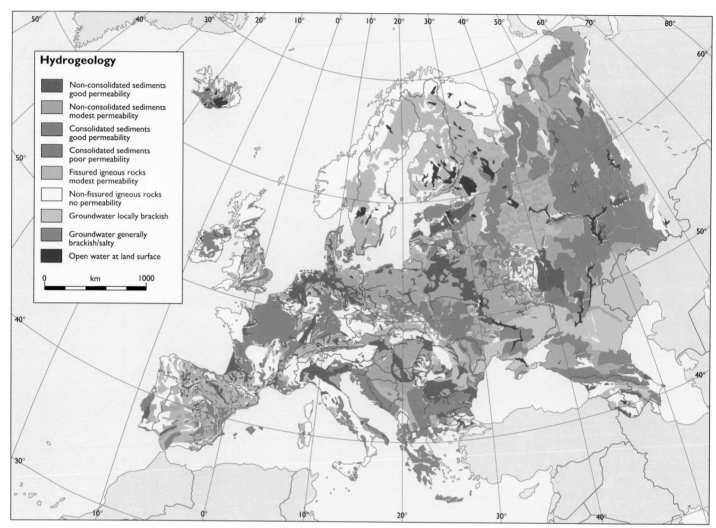

Hydrogeology

- Non-consolidated sediments good permeability
- Non-consolidated sediments modest permeability
- Consolidated sediments good permeability
- Consolidated sediments poor permeability
- Fissured igneous rocks modest permeability
- Non-fissured igneous rocks no permeability
- Groundwater locally brackish
- Groundwater generally brackish/salty
- Open water at land surface

0   km   1000

Map 5.4
General typology of aquifers in Europe
Sources: RIVM, using data from IAH, 1974–87; VSEGINGEO, 1985; UN, 1990 and 1991

Abstractions

In many areas groundwater abstractions exceed the recharge, and the aquifer becomes overexploited (Map 5.5), leading to a systematic and in many cases ongoing lowering of the groundwater levels. This results in a wide range of problems: damage to wetlands, drying up of springs and upper river reaches, reduction of river flows, salt water intrusion (the flow of salt water into an aquifer) and settling phenomena and damage to buildings.

Reported cases of overexploitation include abstraction for drinking and industrial water supply, irrigation water and drainage of mines. Drainage of waterlogged agricultural soils (see Chapters 7 and 9) and some river engineering works also lead to lowering of the groundwater level. Many cases of lowering of the groundwater table up to several dozens of metres have been reported, and in a number of cases the lowerings are in the order of hundreds of metres, affecting the groundwater level in an area up to 500 km away from the overexploited site. Major lowerings of the groundwater table caused by mining activities have, for example, been reported for the area around Kharkov (Ukraine), Lille (France) and in the Ruhr basin (Germany).

However, due to their requirement for water, most overexploited sites are found in or near large urban and industrial centres. Thus, about 60 per cent of the European cities with more than 100 000 inhabitants (or a total of approximately 140 million people) are located in or near areas with groundwater overexploitation. The consequences of this can be far-reaching, as shown by the recent severe problems in Spain and Greece, where cities have been given restrictions on the use of water. Rivers and marshes also have dried out, leading to loss of wetlands. An estimated 6 per cent of the area

with aquifers suitable for abstractions is presently overexploited. This was estimated by overlaying the aquifer typology map (Map 5.4) with the overexploitation map (Map 5.5).

Salt water intrusion in aquifers may result from groundwater overexploitation along the coast. As urban, tourist and industrial centres are commonly located in this zone, intrusion of salt water is a problem of many coastal regions, especially along the Mediterranean, Baltic and Black Sea coasts.

A systematic European inventory of human-induced lowerings of groundwater levels and other changes in the groundwater systems is not available. Therefore, the information presented in Map 5.5 is based largely on problem cases. The risk of underestimating the extent of the problem is thus considerable.

Potential wetland damage

Wetlands, or wet ecosystems, are considered very important nature reserves, with a high ecological value (see Chapter 9). Many wetlands are located in areas prone to flooding such as river floodplains and the systems are highly dependent on the shallow depth of the water table, making them very sensitive to minor changes in the groundwater level. A tragic example is the Tables de Daimiel nature reserve in Spain (Custodio, 1991). The two main rivers feeding the area are fully dependent on groundwater but the groundwater resources in the area have been severely overexploited for irrigation, and at the beginning of 1993 the nature reserve had almost completely dried out.

An estimate of the major European wetlands being threatened by lowering of the groundwater level is shown in Map 5.6, obtained by combining a map of these areas with the

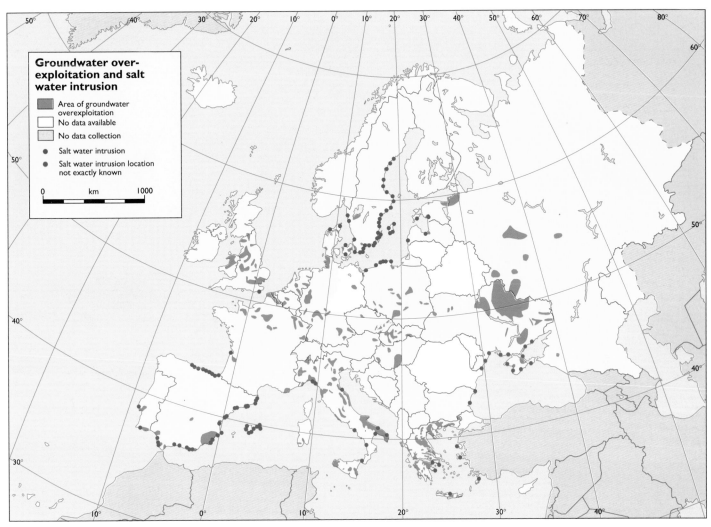

Map 5.5
Overexploitation of groundwater resources and salt water intrusion
Source: Compiled by RIZA from multiple sources listed in Box 5C or referred to in the main text

inventory of overexploited areas.

Map 5.6 is based on data for major bogs, mires and waterbird sanctuaries in Europe (Goodwillie, 1980; Grimmett and Jones, 1989), and forms the basis of the inventory of the most important wetlands (total area 55 000 km²). As can be seen, about 25 per cent of the wetlands are potentially endangered.

Raising of groundwater levels

The water table has been rising due to reduced groundwater abstractions by industry in some cities and industrial areas (eg, Paris, London, Birmingham, Liverpool). Rises in groundwater level can cause problems like flooding of tunnels and basements, chemical attack on structural materials and increased humidity in houses (UK DoE, 1992). Also, increased volumes in sewers may decrease their effective capacity. Similar effects may result from large hydro-technical constructions changing the groundwater level in the drainage basin (eg, in the Russian Federation and Ukraine) and from artificial recharge and irrigation. Another example can be found in the Russian Federation around the Caspian Sea, where the rise in groundwater level was caused by a sea level rise of a few metres. In Ukraine raising groundwater level, in an area of approximately 25 000 km² in the northern and central parts of the country, has caused problems in 2000 towns and settlements. The groundwater rise, which was caused by excess irrigation, produces unfavourable geological processes like displacements, landslides and soil salinisation.

Groundwater pollution

The main pollution threats to groundwater are:

- a wide range of inorganic and organic contaminants from point sources in urban, industrial, mining, military and landfill areas;
- leaching of nitrates;
- leaching of pesticides; and
- acidification.

Point sources

Soil pollution on urban, industrial, mining, military and landfill areas can affect the quality of groundwater, either directly or after a delay in time. In Table 5.4 potential contaminants are presented with respect to point sources. Severe problems of groundwater pollution have already occurred (Brömssen, 1986).

Actual data on polluted groundwater sites in Europe are very scarce. The area of groundwater potentially polluted by industry, mining, military activities and landfills (Table 5.5) can be estimated using various assumptions and by extrapolating available data (Franken, forthcoming). It is thus estimated that potential groundwater pollution by point sources occurs in less than 1 per cent of the European territory. Although this seems to be a rather insignificant portion, it should be borne in mind that most groundwater abstraction takes place in the vicinity of the areas potentially contaminated, posing a threat to the public water supply.

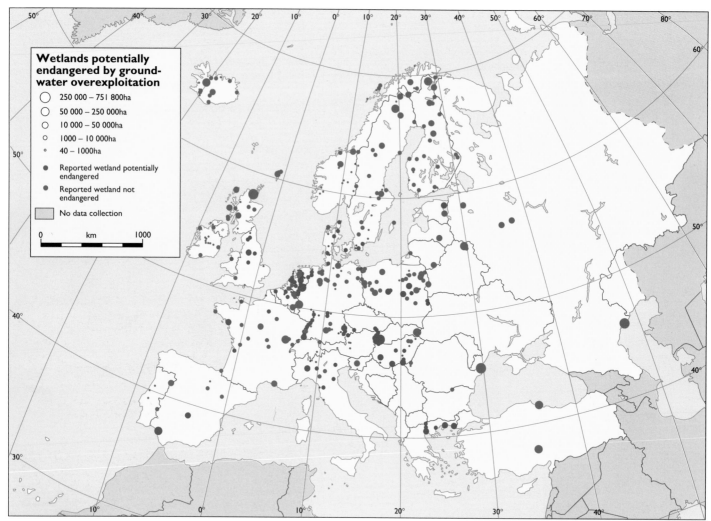

Wetlands potentially endangered by ground-water overexploitation

○ 250 000 – 751 800ha
○ 50 000 – 250 000ha
○ 10 000 – 50 000ha
○ 1000 – 10 000ha
○ 40 – 1000ha
● Reported wetland potentially endangered
● Reported wetland not endangered
▨ No data collection

0   km   1000

Map 5.6
Wetlands potentially endangered by overexploitation of groundwater
Source: Compiled by RIZA from multiple sources listed in Box 5C or referred to in the main text

Table 5.4
Point sources and the expected contamination of groundwater
Source: Prepared by RIVM

Nitrate leaching

The use of manure and fertilisers can lead to leaching of nitrate, ammonium, sulphate, potassium and, to a lesser extent, phosphorus into the groundwater, and hence into surface water. Because nitrate in water above certain concentrations can be a danger to human health and adversely affects the stability of aquatic ecosystems, this report focuses only on this contaminant. Whether or not nitrate will leach to the groundwater depends primarily on the time and rate of application, crop uptake, the soil type and climatological conditions. In Europe, there is a lack of nitrate monitoring data for groundwater. Therefore, a model was used to generate a European overview of nitrate leaching.

Taking into account the present rate and spatial distribution of nitrogen application (see Chapter 22), together with the soil type and climate, a model (based on Drecht, 1991) was used to predict the nitrate leaching from the root zone of agricultural land in Europe. Agricultural land is taken as the total of arable land, grassland and land used for permanent crops. The results of the calculations are presented in terms of the values 10, 25 and 50 mg NO_3/l. The latter two values are specified in the Council Directive (80/778/EEC) relating to the quality of water intended for human consumption as guide level (25 mg NO_3/l) and maximum admissible concentration (50 mg NO_3/l), respectively.

Map 5.7 shows the areas with computed nitrate concentrations exceeding the guide level (GL) and the maximum admissible concentration (MAC) at one metre below the soil surface where concentrations would be at their maximum. Model computations indicate that over 85 per cent of the agricultural area in Europe has nitrate levels above the GL, and also that the MAC is exceeded below approximately 20 per cent of the agricultural area. The red spots on the Kola Peninsula are due to lacking information on regionally distributed nitrogen input to agricultural land in that area.

Several processes influence the fate of nitrate as it percolates from the root zone to the groundwater. Denitrification is particularly important in this respect, leading to a decrease of nitrate concentrations between the root zone and groundwater level (see Chapter 14 on emissions to water) as well as in the deeper groundwater, provided that (bio)chemical conditions are favourable.

The physical processes affecting the fate of nitrate involve mixing of 'old' resident groundwater with percolating 'young' water from upper soil layers. As a result of mixing of two types of water with different nitrate concentrations (lower in groundwater than in the percolating water), the nitrate concentrations in abstraction wells and seepage areas supporting wetlands, rivers and lakes are generally well below

Source	Inorganic contaminants	Organic contaminants
Urban areas	heavy metals (Pb, Zn) and salts	oil products, chlorinated hydrocarbons
Industrial areas	heavy metals	PAHs, chlorinated hydrocarbons, hydrocarbons, oil products and synthetic organic compounds
Mining areas	heavy metals, salts (SO_4^{2-}, Cl^-) and arsenic	PAHs, hydrocarbons, oil products
Military sites	heavy metals	hydrocarbons (oil products)
Landfills	salts (Cl^-, NH_4^+) and heavy metals	biodegradable and non-biodegradable organics

the leachate concentrations and/or their increase is retarded in time.

In order to obtain an overview in terms of a 'hot spots' map of the areas in Europe where groundwater is expected to be particularly vulnerable to nitrate contamination arising from agricultural activities, the nitrate leaching map (Map 5.7) was superimposed on an 'aquifer-risk map' prepared for, but not shown in, this report. The rationale behind the calculation of the risk map is explained in Box 5D. This analysis indicates that 'hot spots' for nitrate pollution (Map 5.8) can be expected primarily in areas where intensive cultivation of sandy soils takes place on top of unconsolidated, unconfined aquifers. For instance, this is the case for extended areas of several north-western European countries, the Czech Republic, the Slovak Republic, Hungary, Ukraine and Belarus.

Monitoring data on nitrate in groundwater supplied by countries were very heterogeneous. About 60 per cent of the countries replying to the questionnaires indicated that they had serious nitrate problems and the majority of these countries gave some regional details showing the geographical distribution of the problem areas and actual concentration measurements. Some countries (eg, Poland) reported quality data for the upper groundwater, while for others (eg, Spain) concentration values for deep groundwater have been used. In other cases, no specifications at all were provided about the monitoring depth, and quality data for both upper and deeper groundwater may therefore have been mixed. For large parts of Europe there is a complete lack of monitoring data on

Source	Landuse (km²)	Potentially contaminated area (km²)
Built-up area (incl housing, industry, traffic area)	100 000–200 000[1]	24 000–48 000[2]
Mining area (incl mining industries & spoil dumps)	50 000–100 000[3]	15 000–30 000[4]
Military sites	12 000–24 000[5]	7200–14 400[6]
Landfills	2000–4500[7]	1200–2700[8]
Total		47 400–95 100

The area of potentially contaminated sites has been estimated assuming the following:

1. 1 per cent of the European territory (which is 10.2 million km²) is urban (Velde et al, in press). The total built-up area (including industry, roads, airports, etc) is taken to be twice that.
2. 20 per cent of the built-up area is industrial and 20 per cent of this is seriously polluted. Further, dispersion of contaminants increases this area by a factor of six.
3. Estimated from extrapolation of a few national landuse statistics available (including metal, salt and coal mines and mining waste disposal sites).
4. 5 per cent of the mining area has polluted groundwater and dispersion increases this area by a factor of 6.
5. Extrapolated from national statistics of The Netherlands and Switzerland.
6. The groundwater is polluted below 10 per cent of the military sites. Dispersion assumption as above.
7. Dutch figures indicate that there are 100 km² of landfills in The Netherlands (population 15 million). For the European population of 680 million this equals 4500 km². This estimate is taken as the maximum due to generation of less municipal waste in Eastern European countries as compared to Western countries.
8. The groundwater is polluted below 10 per cent of landfills. Dispersion characteristics as above.

Map 5.7 Calculated concentration of nitrate in the leachate from agricultural soils (at 1 metre depth)
Source: Compiled by RIVM from multiple sources listed in Box 5C or referred to in the main text

Table 5.5 Estimated area of point sources and contaminated land in Europe
Source: See table

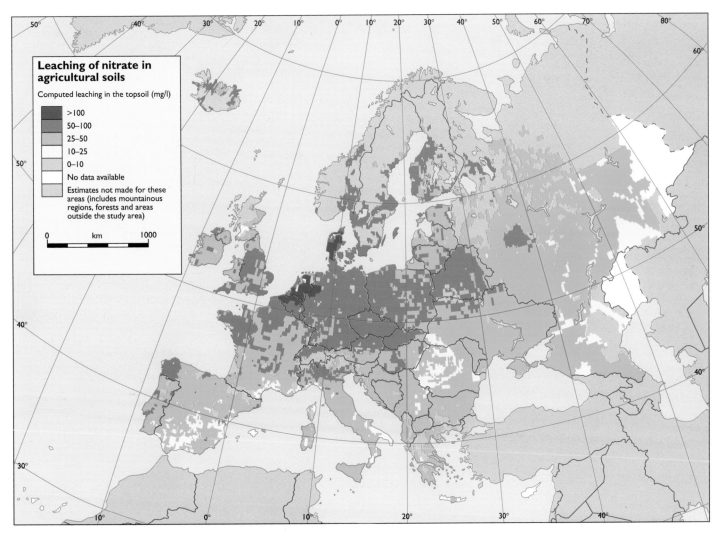

groundwater pollution by nitrate. A standardised groundwater monitoring network across Europe (based on unified and agreed principles for network design, site selection, sampling frequency, choice of determinants, etc) would have enabled a much clearer picture of the problems to be presented.

A comparison of the nitrate 'hot spots' map based on model calculations (Map 5.8) with the data on nitrate pollution leads to the following conclusions:

- the 'hot spots' map suggests that 24 countries face nitrate problems (18 serious, 6 minor);
- from the available data on the nitrate problem it follows that in 26 countries the groundwater pollution by nitrate can be considered as a problem (18 serious, 8 minor).

Therefore, it is concluded for nitrate that the 'hot spots' map (Map 5.8) is in reasonable agreement with the available information on groundwater quality.

Further comparison of the nitrate 'hot spots' map with the nitrate problem data suggests that:

- for Western and Central European countries the data show a reasonable fit with the calculated 'hot spots' map (eg, France, The Netherlands, Denmark, the western part of Germany and Hungary);
- regional differences occur which are related mainly to the lack of detailed information on the regionally distributed nitrogen input to agricultural land – this is the case in Bulgaria, Romania, Moldova and the southern part of the Russian Federation;
- in some countries, the groundwater recharge is greater than was accounted for in the model due to irrigation practices; this in particular holds true for areas with

intensive horticulture, such as Spain and probably the southern part of the Russian Federation, leading to higher nitrate leaching than predicted from model computations.

In general it may be concluded that, despite the generalisations and assumptions attached to the preparation of the nitrate 'hot spots' map, the application of nitrogen (manure and fertilisers) in agriculture does lead to nitrate contamination (occasionally severe contamination) of European groundwater.

Pesticides

Approximately 600 different pesticides (as broadly defined to include herbicides as well as fungicides and insecticides) are applied in European agriculture, silviculture and horticulture (see Chapter 17). On their passage through the subsurface environment, these 600 active ingredients are transformed into an (often) unknown number of degradable products (residues). The effects of active ingredients and their residues on non-target terrestrial organisms (side effects), their fate in the soil and their undesired effects in groundwater and surface waters are far from known with certainty for every substance.

The extent to which pesticides in groundwater have reached concentrations limiting the use of groundwater for drinking water abstraction is essentially unknown. Only a few pesticide measurements in groundwater are available for a restricted number of constituents. The information about pesticide use per country (amounts of pesticide) available for this assessment was also limited. The modelled applications of pesticides to agricultural soils are shown in Chapter 22. Taking into account the available data on the present rate and spatial distribution of pesticide application, the organic matter

Map 5.8
The nitrate 'hot spots' for groundwater
Source: Data compilation and model calculations by RIVM

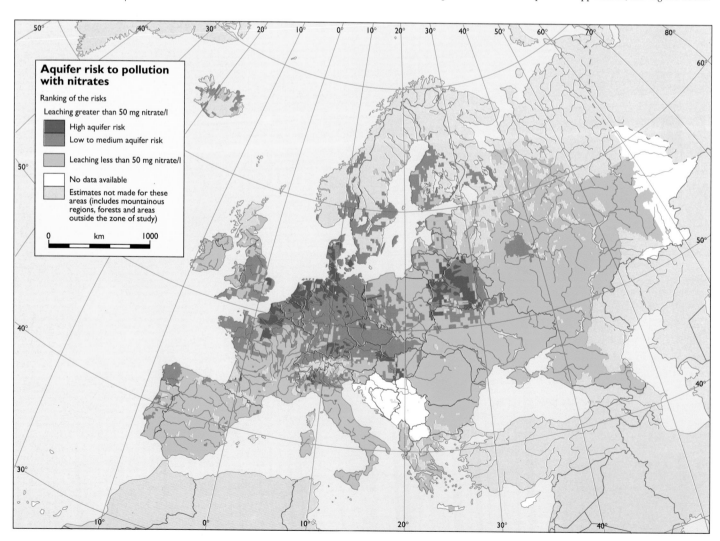

Aquifer risk to pollution with nitrates

Ranking of the risks

Leaching greater than 50 mg nitrate/l

High aquifer risk

Low to medium aquifer risk

Leaching less than 50 mg nitrate/l

No data available

Estimates not made for these areas (includes mountainous regions, forests and areas outside the zone of study)

0   km   1000

Box 5D Estimating aquifer risk to pollution

The 'aquifer-risk map' combines various processes (risk factors) affecting the fate of nitrate as it percolates down from the topsoil to the aquifer. For nitrate (and also for pesticides) these factors are:

B1 the groundwater recharge in terms of annual averages;
B2 the thickness of the unsaturated zone, ie, the layer
 between the upper soil and the groundwater table;
B3 the composition (texture) of layers in the unsaturated zone;
B4 the aquifer type (Map 5.4).

Groundwater recharge (B1) has been calculated as precipitation excess less water flow components (surface runoff and interflow). The water flow components have been estimated by taking into account the slope of the land surface, the composition (texture) of the upper soil (infiltration capacity), the air temperature (frost periods) and the aquifer type. This information has been derived from existing databases. The pedology of the upper soil layers and the thickness of the unsaturated zone were taken from the FAO World Soil Map.
 Determination of the aquifer risk for each grid cell (10 x10 minutes) was carried out by ranking the individual risk factors and adding them using the following equation:

$$\text{Aquifer risk} = WF_{B1} \times R_{B1} + WF_{B2} \times R_{B2} + WF_{B3} \times R_{B3} + WF_{B4} \times R_{B4}$$

where WF is the weighting factor, and R is the value of the risk factor. The weighting and risk factors (Table 5.6) have been selected so as to calculate aquifer risks ranging from no risk to maximum risk.

Risk factor (R)	Weighting factor (WF)	Value of risk factor (R)
B1: Groundwater recharge	3	range: 1–10 10 = groundwater recharge >500 mm/year 1 = groundwater recharge < 50 mm/year
B2: Thickness of unsaturated zone	1	range: 8–10 10 = groundwater < 3 m below surface 8 = groundwater > 5 m below surface
B3: Composition (texture) of layers in unsaturated zone	2	range: 0–9 9 = coarse soils 0 = histosols (peat and muck soils)
B4: Aquifer type	4	range: 1–10 10 = sedimentary aquifer with good porosity 1 = hardrock with poor porosity

Table 5.6 Weighting and risk factors used to generate the 'aquifer-risk map' Source: RIVM

Map 5.9 Calculated concentration of pesticides in the leachate from agricultural soils (at 1 metre depth) Source: Compiled by RIVM from multiple sources listed in Box 5C or referred to in the main text

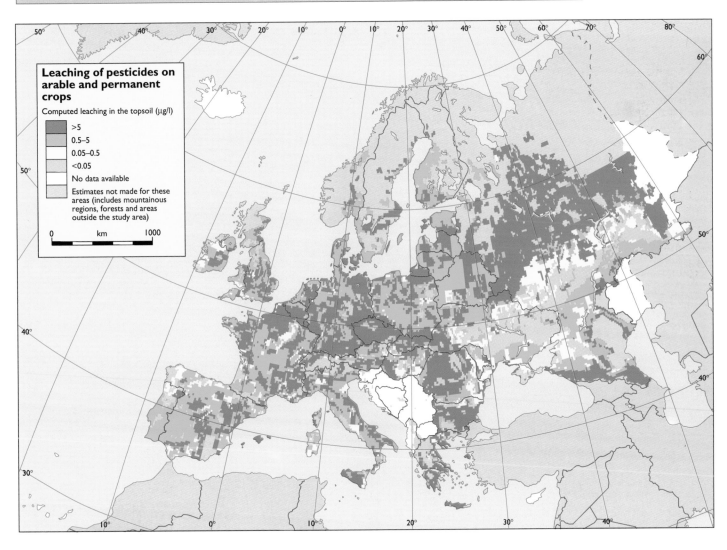

Leaching of pesticides on arable and permanent crops

Computed leaching in the topsoil (μg/l)

- >5
- 0.5–5
- 0.05–0.5
- <0.05
- No data available
- Estimates not made for these areas (includes mountainous regions, forests and areas outside the study area)

0 km 1000

content of the soil, and climate, a model (Boesten and van der Linden, 1991) was used to predict the pesticide leaching from the root zone of agricultural soils. Based on this model, the computed area with pesticide levels exceeding the EU drinking water standard (maximum admissible concentration of 0.5 µg/l for the total amount of pesticides and their residues as specified in Council Directive (80/778/EEC)) at one metre below the soil surface is shown in Map 5.9.

Based on the model, the EU standard is exceeded in the leachate under 75 per cent and 60 per cent of the total arable and permanent crop land in EU/EFTA and Central and Eastern European countries, respectively. The use of pesticides on grasslands was not taken into account because (a) application on grassland does not in general occur and (b) if it does occur the application rate is low.

The pesticide 'hot spots' map for groundwater (Map 5.10) was produced following the procedure as described above for nitrate, that is, using an overlay of the pesticides leaching map (Map 5.9) and the 'aquifer-risk map' (not shown). The expected 'hot spots' are in areas where intensive cultivation of coarse textured soils, low in organic matter, occurs on top of unconsolidated, unconfined aquifers. This is the case in (parts of) Denmark, northern France, The Netherlands, Lithuania and Belarus. The most extensive risk of high pesticide contamination is estimated to occur in Belarus.

There is an almost complete lack of systematic monitoring data for pesticides in Europe. Where measurements are available, they are mostly incidental (not part of a systematic monitoring network) and the density of monitoring points is very low. Often no specifications are given about the exact location of the measurement site and

the monitoring depth. In general, only a few compounds (mostly fewer than ten) have been included in monitoring programmes.

Comparison of the calculated pesticide 'hot spots' map (Map 5.10) and the areas with 'observed' pesticide problems leads to the following considerations. For the Western European countries, the calculated 'hot spots' (eg, in France, The Netherlands, Denmark, the UK, Germany and Italy) show a reasonable fit with observations. Moldova and Ukraine have reported serious pesticide problems, but neither of these countries has indicated the specific areas where the problems occur. The 'hot spots' map shows only a small part of Ukraine having a high aquifer risk for pesticide pollution. One obvious explanation for the apparent mismatch is the lack of regionally distributed data on observed pesticide problems within Moldova and Ukraine. Another explanation is that in these countries very mobile pesticides have been used, which may lead to groundwater pollution not accounted for by the model used to generate the pesticide 'hot spots' map, which is based on the average mobility of pesticides. Finally, Map 5.10 shows that a large 'hot spot' area was calculated for Belarus. Unfortunately, no observations were available for Belarus for verification of the calculated 'hot spot' area. Taking into account the above considerations, it is assumed that the pesticides 'hot spots' map (Map 5.10) is in reasonable agreement with the available information on groundwater quality with respect to pesticides.

In summary, it may be concluded that, despite the generalisations and assumptions attached to the preparation of the pesticide 'hot spots' map, the application of pesticides leads to pesticide contamination of groundwater in many European countries.

Map 5.10
The pesticides 'hot spots' for groundwater
Source: Data compilation and model calculations by RIVM

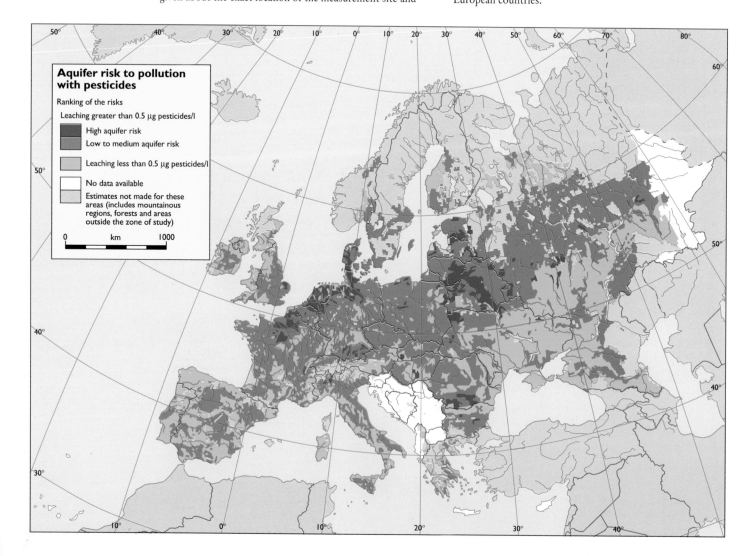

Aquifer risk to pollution with pesticides

Ranking of the risks

Leaching greater than 0.5 µg pesticides/l

■ High aquifer risk

■ Low to medium aquifer risk

■ Leaching less than 0.5 µg pesticides/l

□ No data available

■ Estimates not made for these areas (includes mountainous regions, forests and areas outside the zone of study)

0   km   1000

As previously indicated, very few measurements of pesticides are available for groundwater. The main reason is the high cost of chemical analyses, combined with the high number of chemical compounds. Out of approximately 600 pesticides, only about 30 have been monitored. These 30 pesticides are neither the most mobile, nor representative of all pesticides. The availability of suitable analytical methods is also an important limiting factor (see also Chapters 7 and 17).

Acidification

Acidification of the soil is a well documented and serious problem in large parts of Europe (see Chapters 7 and 31). In addition to natural acidification, sandy and poorly buffered (low alkalinity) soils are subject to enhanced acidification caused by atmospheric deposition of sulphur and nitrogen compounds, fertiliser application and land drainage. Because water must filter through the soil before it reaches the aquifer it is likely that groundwater below those areas may also have become acidified. Unlike surface waters (treated in detail later in this chapter) there are very little data available to show the present extent of groundwater acidification and its trends in Europe and only a few examples of documented groundwater acidification can be presented.

Acidification of forest soils is, in particular, a well described and generally recognised problem in Northern and Central Europe (Last and Watling, 1991). It is caused primarily by atmospheric deposition of acidifying compounds emitted from human activities (see also Chapters 4, 7 and 31). As a result, increased concentrations of aluminium, sulphate, hydrogen ions, and sometimes nitrate are reported in the upper groundwater layer below sandy soils in Norway, Sweden, Denmark, Finland, the Czech Republic, the Slovak Republic, Germany and The Netherlands.

In Sweden, for example, a national groundwater survey showed severe influence of acid inputs on wells in forested areas (Bernes and Grundsten, 1992). Thus more than 50 per cent of surveyed wells in 14 municipalities (in total 52 municipalities were surveyed) showed very strong or strong acidification effects in the mid-1980s.

Acidification of deeper groundwater under agricultural land has also been reported (Targbill, 1986; RIVM/RIZA, 1991; Brömssen, 1986; Overgaard, 1986), but in these cases acidification cannot be explained solely by atmospheric deposition. Other processes (such as oxidation of ammonia to nitrate by nitrification) can also contribute to acidification (RIVM, 1992a; Rebsdorf et al, 1991), as can denitrification if nitrate is reduced by ferrous sulphide (see Chapter 14).

In Denmark, for example, an analysis of a 30-year time series (period 1952–82) of pH, alkalinity and nitrate in groundwater from unconfined aquifers below predominantly agricultural land in the western, sandy part of the country showed that pH and alkalinity had declined steadily over time, whereas nitrate had increased. The average yearly decrease was 0.023 units for pH and 33 μmol/l for alkalinity. For nitrate the increase was 14 μmol/l (Rebsdorf et al, 1991). If alkalinity continues to decrease at the present rate the groundwater in that part of Denmark would completely lose its buffering capacity within 20 to 25 years. This would make the groundwater unsuitable for human consumption and almost all other purposes, unless expensive water treatment processes are introduced.

The most serious consequences of acidification of groundwater are the increased mobilisation of trace elements, especially aluminium, and the increased solubility of some metals in water distribution systems, both resulting from a lowering of the pH (Chapter 7).

RIVERS, RESERVOIRS AND LAKES

Introduction

This section focuses on European inland surface waters, that is, rivers, lakes and reservoirs, and assesses their environmental state and the environmental problems caused by human activity. It is one of the first attempts at an all-European assessment aimed at providing a general overview as a basis for implementing measures to improve the environmental state of rivers and lakes, and to identify areas with environmental problems.

Considerable environmental information on rivers and lakes is currently collected and reported by various regional or national authorities. However, as the European continent covers about 10 million km² and there are several million kilometres of flowing waters and more than a million lakes, this information is very heterogeneous, and therefore difficult to collate on a pan-European basis. Nevertheless, a first attempt at doing so has been made in this section. The primary focus is on frequently measured water quality parameters (eg, organic matter in rivers, nutrients in rivers and lakes, and acidification of rivers and lakes) since the wide geographical coverage makes these variables well suited to illustrate the general environmental state of European inland surface waters. Some environmental problems related to more rarely monitored variables are also described.

Characteristics of European rivers, reservoirs and lakes

A river is a system comprising both the main course and all the tributaries that feed into it; the area that the river system drains is known as the catchment. The main characteristic of rivers is their continuous one-way flow in response to gravity. In addition, because of changes in physical conditions such as slope and bedrock geology, rivers are dynamic and may change nature several times during their course (eg, from a fast-flowing mountain stream to a wide, deep, slowly flowing lowland river).

When assessing river characteristics and water quality it is important to bear in mind that a river comprises not only the main course, but also a vast number of tributaries. Thus although the main course of Europe's largest river, the Volga, is 3500 km long, it receives water from ten tributaries each longer than 500 km, and more than 151 000 tributaries each longer than 10 km (Fortunatov, 1979).

Rivers are greatly influenced by the characteristics of the catchment area (Figure 5.9). The climatic conditions influence the water flow, as does bedrock geology and soil type. The latter also affects the mineral content of the river water. Human activity affects river systems in numerous ways, for example, through afforestation or deforestation, urbanisation, agricultural development, land drainage, pollutant discharge, and flow regulation (dams, channelisation, etc). The lakes, reservoirs and wetlands in a river system attenuate the natural fluctuation in discharge and serve as settling tanks for material transported by the rivers. For example, whereas the water of the Rhine is very muddy and turbid when entering the Bodensee, it is clear and transparent when leaving. Water flow and water quality are therefore the net result of the various characteristics of the catchment.

Lakes are bodies of standing water that is usually fresh, but which may also be brackish. Although lakes may be characterised by physical features of the lake basin, such as lake area and water depth, the characteristics of the catchment are important when describing the lake environment. Nutrient loading of a lake is determined not only by the bedrock geology and soil type in the catchment, but also by

Box 5E Data and information sources

Information in this section is based on:

- questionnaires forwarded to national focal points;
- various national and regional state of the environment reports; and
- a review of the scientific literature concerning the environmental state of European rivers, lakes and reservoirs.

Four different questionnaires were used. Two were designed to provide general information about the rivers and lakes in each country, while the other two were designed to provide more specific information about 10 to 20 rivers and lakes in each country. In addition, time-series data for rivers were obtained from the EU according to Council Decision (77/795/EEC), the UNEP/GEMS database (in Burlington, Canada), the INFODANUBE database (Regional Environmental Centre for Central and Eastern Europe, in Budapest, Hungary), and from a supplementary questionnaire sent to countries outside the EU. The questionnaires were used as a tool to ensure that comparable information was obtained on the various environmental problems in the different countries. Many persons in many countries have made a great effort to answer the questionnaires. A total of 28 countries, especially Eastern European, Nordic, and EFTA countries, were kind enough to answer the questionnaires (Table 5.7). However, the information supplied differed greatly; some countries provided only sparse information about a few waterbodies, while others supplied information from national surveys that included more than 500 or 1000 rivers or lakes. When a response to the questionnaires was lacking or incomplete, national or regional environmental reports were reviewed and analysed to ensure geographical coverage as wide as possible. In these cases the country information shown on maps in this chapter may not always accurately represent the actual situation.

Major additional sources of information were as follows: Austria: Wasserwirtschaftkataster, 1992; the Belgian Flanders: Vlaamse Milieumaatschappij, 1992; Estonia: Milius, 1991; 116 large rivers in the EU countries: Council Decision (77/795/EEC); Ireland: Flanagan and Larkin, 1992; Italy: Istituto di Ricerca sulle Acque 1990, Gaggino et al, 1987; Northern Ireland: Gibson, 1986, 1988; Norway: NIVA, 1990a,b;

Country	Rivers		Lakes	
	Form I	Form II	Form I	Form II
Albania	X	5	–	(3)
Austria	(X)	4	(X)	–
Bulgaria	X	10	(X)	5
Croatia	X	26	–	4
Czech Republic	X	15	–	–
Denmark	X	70	X	37
Estonia	–	12	X	3
Finland	X	21	X	19
France	X	–	–	–
Georgia	X	14	(X)	6
Hungary	X	4	–	3
Iceland	X	2	X	1
Latvia	X	15	X	–
Lithuania	X	15	–	–
Luxembourg	X	–	X	–
Moldova	X	7	X	9
The Netherlands	X	3	X	7
Norway	X	19	(X)	12
Poland	X	19	X	16
Romania	X	26	–	16
Russian Federation	X	16	–	4
Slovenia	X	21	–	–
Spain	X	3	(X)	7
Sweden	X	21	X	9
Switzerland	X	4	X	9
Turkey	–	–	X	–
UK	X	–	X	–
Ukraine	–	37	–	6

Note: Form I is a general questionnaire and Form II gives information about specific rivers and lakes. The number of questionnaires with information about specific rivers and lakes is listed. X forms completed; (X) only sparse information; – no form information.

Table 5.7
Countries that have answered the river and lake questionnaires

the human activity.

Reservoirs are human-made lakes created to serve one or more purposes. As their water residence time is generally relatively short, and as the water level fluctuates much more widely and frequently than in natural lakes, they can be regarded as hybrids between rivers and lakes.

European rivers

On average, European rivers discharge a total of 3100 km³ of freshwater to the sea each year, about 8 per cent of total world discharge . Because Europe has a temperate humid climate and a high percentage of limestone in the surface rock, the weathering rate is the highest of all the continents; as a result, 12.6 per cent of all dissolved solids discharged to the oceans are derived from Europe (Kempe et al, 1991). That Europe is relatively densely populated and has a high proportion of agricultural areas also affects the concentration of dissolved substances in river water; thus the median nitrate level is 1.8 mg N/l in European rivers as compared with only 0.25 mg N/l in non-European rivers (Meybeck et al, 1989).

Major European river catchments
In proportion to its land area, Europe has the longest coastline of all continents. As it is a relatively young and

structured continent, geologically, river catchments are numerous but relatively small and rivers are short (Map 5.11). About 70 European rivers have a catchment area exceeding 10 000 km², and only rivers arising deep inside the continent are relatively large. The three largest rivers in Europe, the Volga, the Danube and the Dnepr, drain one quarter of the continent, but are only small by world standards, their catchments ranking 14th, 29th and 48th, respectively (Showers, 1989).

The 31 largest European rivers, all of which have catchments exceeding 50 000 km², drain approximately two thirds of the continent. More than half of these rivers have their catchment area in the European part of the former USSR. The major rivers flowing north into the Barents Sea and the White Sea are the Severnaya (Northern) Dvina and the Pechora. The Volga and the Ural which flow south and the Kura which flows east drain into the Caspian Sea while the Dnepr and the Don drain south into the Black Sea. The largest river to discharge into the Black Sea is the Danube, which has its catchments in 16 countries of Central Europe and the Balkans. The main rivers to discharge into the Baltic Sea are the Neva, the Wisla, the Oder and the Neman. Ten rivers with catchments larger than 50 000 km² drain into the Atlantic and the North Sea, with the Rhine, the Elbe, the Loire and the Douro being the largest. The European rivers

Portugal (river Douro and tributaries): DGQA, 1991; Spain: Dirección General de Obras Hidráulicas del MOPU, 1989; the UK: Institute of Hydrology and British Geological Survey, 1990 and UK DoE, 1992; West Germany: Umweltbundesamt, 1992. The very small European countries are only briefly considered.

In all, information has been collected from 800 river stations in more than 550 European rivers, as well as from more than 1500 lakes. Information was obtained about the largest rivers in nearly all countries (Table 5.9), this being supplemented by data from numerous smaller rivers for which land usage and human activity in the catchment area are known. Descriptive statistics for frequently measured physical and chemical variables in European rivers are given in Table 5.8. Information provided about the general environmental state of European lakes was less extensive, only 18 countries having completed the general lake questionnaire; this information was therefore supplemented with data from national lake surveys, as well as information from the scientific literature.

The European countries dealt with in this section have been divided into four regions: Nordic, Eastern, Western, and Southern (Figure 5.3).

The results on chemical concentrations in this section are generally based on annual mean values (rivers) or summer mean values (lakes) from the period 1988–91. Several results are presented as summary statistics for a group of rivers or lakes by means of frequency distributions (pie-charts) and box-cox diagrams. The frequency distribution for organic matter and nutrients in rivers and lakes (maps with pie-charts) are based on information supplied by each country on the percentage of rivers or lakes having organic matter or nutrient levels within specified classes, for example, the percentage of rivers with annual mean phosphorus levels below 25 μg P/l, between 25–50 μg P/l, 50–125 μg P/l, and so on.

The box-cox diagrams that illustrate, for instance, the relationship between total phosphorus and population density in river catchments (eg, Figure 5.8) have been compiled by grouping the rivers in population density classes, for example, 30 rivers with fewer than 15 inhabitants per km^2 and 19 rivers with more than 100 inhabitants per km^2. The median value (the middle bar), the upper and lower quartiles (the upper and lower box), and the 10 and 90 per cent percentiles (the bar at the end of the lower and upper lines) have been calculated for each class. In rivers with a catchment population density exceeding 100 inhabitants per km^2, 50 per cent of the rivers have annual mean total phosphorus levels above and below 300 μg P/l, and 75 and 25 per cent of the rivers have phosphorus levels above and below approximately 100 μg P/l respectively (lower quartile).

Table 5.8
Descriptive statistics of annual mean physical and chemical variables at European river stations

	Number of river stations	Percentage of river stations with concentrations not exceeding		
		25%	50%	75%
pH	717	7.5	7.8	8.0
Total alkalinity (meq/l)	274	1.0	2.5	4.0
Chloride (mg Cl/l)	442	17.3	26.5	68.3
Organic matter and oxygen level (mg O$_2$/l)				
Biochemical oxygen demand	645	1.9	2.8	4.7
Chemical oxygen demand	470	7.8	15.0	25.0
Dissolved oxygen	620	8.4	9.7	10.7
Nitrogen (mg N/l)				
Ammonium	580	0.1	0.2	0.4
Nitrate	654	0.7	1.8	3.9
Total nitrogen	329	0.8	2.1	4.5
Phosphorus (μg P/l)				
Dissolved orthophosphate	412	45.0	124.0	286.0
Total phosphorus	546	59.0	170.0	366.0
Heavy metal (μg/l)				
Copper	192	1.0	4.0	8.0
Zinc	176	5.0	10.0	

Note: The following example for dissolved orthophosphate shows how to read the table. For this compound annual mean concentrations have been gathered for 412 European river stations. Of these, 25 per cent (103 stations) had concentrations below 45 mg P/l; 50 per cent (206 stations) were below 124 mg P/l and 75 per cent (309 stations) were below 286 mg P/l, respectively.

that drain into the Mediterranean are relatively small, the Rhone, the Ebro and the Po being the largest. Nevertheless, since the damming of the Nile, the Rhone has become the Mediterranean's most important freshwater source (Kempe et al, 1991).

Major rivers in European countries

Countries whose coastline is long in relation to their area, for example Iceland, the UK, Ireland, Norway, Sweden, Denmark, Italy and Greece, are usually characterised by having a large number of relatively small river catchments and short rivers, the three to four largest of which drain only 15 to 35 per cent of their area (Table 5.9). The population tends to congregate in towns along the coastline and wastewater is consequently discharged directly into coastal areas rather than into the river systems.

Many European countries are drained by only a few river catchments; thus the Wisla and Oder drain more than 95 per cent of Poland, and the Danube drains most of Hungary, Romania and Slovenia (Table 5.9).

European lakes and reservoirs

Many natural European lakes appeared 10 to 15 thousand years ago, being formed or reshaped by the last glaciation period, the Weichsel. The ice sheet covered all of Northern Europe, but in Central and Southern Europe it was restricted to the mountain ranges. As a rule the regions that have many natural lakes are those that were affected by the Weichsel ice. Norway, Sweden, Finland and the Karelo-Kola part of the Russian Federation have numerous lakes that account for approximately 5 to 10 per cent of national surface area. Large numbers of lakes were also created in the other countries around the Baltic Sea, as well as in Iceland, Ireland and the northern and western parts of the UK. In Central Europe most natural lakes lie in mountain regions, those at high altitudes being relatively small and those in the valleys being the largest, for example Lac Léman, Bodensee, Lago di Garda, Lago di Como and Lago Maggiore in the Alps and Lake Prespa and Lake Ohrid in the Dinarian Alps. Exceptions are the two large lakes, Lake Balaton and Neusiedler See, that lie on the Hungarian Plain.

In contrast to glaciation, processes such as tectonic and volcanic activity have played only a minor role in the formation of European lakes. Numerous lakes have been created by natural damming of rivers and coastal areas, however.

Countries that were little affected by the glaciation period, such as Portugal, Spain, France, Belgium, southern England, central Germany, the Czech Republic, the Slovak

Table 5.9
The major European rivers, their catchment areas, and the percentage of the country area drained
Source: Compiled by NERI from multiple sources listed in Box 5E

Notes: River numbers 1–31 refer to Map 5.11. Symbols refer to tributaries to: #, Danube; †, Rhine; ◊, Neva; ‡, Kura; ¶, Neman.

Same rivers with different names: Strimon/Struma; Daugava/Zapadnaya Dvina.

Country (country code)	River		Catchment Area (10³ km²)	Per cent of country	
Albania (AL)	64	Drin	~7.0	~23	
	87	Seman	5.6	20	
	82	Vijosë	~4.5	~16	
Austria (AT)	2	Danube	80.6	96	
	#	Inn	15.9	19	#
	11	Rhine	2.4	3	
Belarus (BY)	3	Dnepr	~130.0	~63	
	15	Neman	45.5	22	
	18	Zapadnaya Dvina	~30.0	~14	
Belgium (BE)	39	Meuse	13.5	44	
	52	Schelde	~10.0	~33	
Bosnia-Herzegovina (BA)	#	Sava	~37.5	~74	#
	70	Neretva	~10.0	~20	
Bulgaria (BG)	2	Danube	48.2	43	
	47	Maritza	21.1	19	
	55	Struma	10.8	10	
Croatia (HR)	2	Danube	~37.0	~65	
	#	Sava	~24.5	~44	#
	#	Drava	~8.0	~14	#
Czech Republic (CZ)	12	Elbe	51.4	64	
	#	Morava	~25.0	~30	#
	13	Oder	4.7	6	
Denmark (DK)	99	Gudenå	2.6	6	
	101	Skjern Å	2.3	5	
	103	Storå	1.1	3	
Estonia (EE)	35	Narva	~20.0	~45	
	81	Pärnu	6.9	15	
	102	Jägala	~1.5	~3	
Finland (FI)	◊	Vuoksa	61.1	18	◊
	31	Kemijoki	51.1	15	
	38	Kymijoki	37.2	12	
	42	Kokemäenjoki	27.0	9	
former Yugoslav Republic of Macedonia (MK)	48	Vardar	~20.5	~81	
	64	Drin	~3.5	~13	
France (FR)	14	Loire	117.5	21	
	17	Rhone	85.6	16	
	19	Garonne	85.0	16	
	22	Seine	79.0	14	
Georgia (GE)	10	Kura	~42.5	~61	
	66	Rioni	13.4	19	
	‡	Alazani	7.5	11	‡
Germany (DE)	11	Rhine	102.1	29	
	12	Elbe	97.0	27	
	2	Danube	59.6	17	
	33	Weser	45.8	13	
Greece (GR)	73	Aliákmon	~9.5	~7	
	80	Piniós	7.1	5	
	55	Strimon	6.0	5	
Hungary (HU)	2	Danube	93.0	100	
	#	Tisza	44.6	48	#
	#	Drava	6.2	7	#
Iceland (IS)	77	Jökulsá-á-Fjöllum	7.8	8	
	78	Thjórsá	7.5	7	
	84	Ölfusa	6.1	6	
Ireland (IE)	65	Shannon	~14.0	~20	
	88	Barrow	~5.5	~8	
	96	Suir	3.6	5	
Italy (IT)	26	Po	69.0	23	
	56	Tevere	17.2	6	
	68	Adige	12.2	4	
Latvia (LV)	18	Daugava	23.6	37	
	75	Lielupe	8.8	14	
	74	Gauja	7.9	12	
Lithuania (LT)	15	Neman	46.6	71	
	¶	Vilnya	13.8	71	¶
	69	Venta	5.2	8	
Luxembourg (LU)	†	Süre	~2.0	~77	†
	†	Mosel	~0.5	~19	†
Moldova (MD)	25	Dnestr	~18.0	~53	
	#	Prut	~12.0	~36	#
	94	Kogel'nik	3.9	11	
The Netherlands (NL)	11	Rhine	~25.0	~60	
	39	Meuse	6.0	14	
Norway (NO)	36	Glomma	41.4	13	
	56	Drammens-elva	17.1	5	
	59	Tana	10.9	3	
Poland (PL)	9	Wisla	191.8	61	
	13	Oder	114.2	37	
	100	Rega	2.6	1	
Portugal (PT)	16	Douro	~20.0	~22	
	21	Tajo	~18.0	~20	
	24	Guadiana	~14.0	~15	
Romania (RO)	2	Danube	232.2	98	
	#	Muresul	27.8	12	#
	#	Oltul	24.0	10	#
Russian Federation (RU)	1	Volga	1360.0	35	
	4	Don	~380.0	~10	
	5	Severnaya Dvina	357.0	9	
	6	Pechora	322.0	8	
	7	Neva	~220.0	~6	
	8	Ural	~110.0	~3	
	3	Dnepr	~105.0	~3	
Serbia-Montenegro (SB)	2	Danube	~95.0	~93	
	#	Sava	~20.0	~19	#
	64	Drin	~4.5	~4	
Slovak Republic (SK)	2	Danube	~46.5	~97	
	#	Váh	~17.5	~37	#
	#	Tisza	~16.0	~33	#
Slovenia (SI)	#	Sava	10.8	53	#
	#	Drava	~5.0	~24	#
Spain (ES)	20	Ebro	84.2	17	
	16	Douro	~78.0	~15	
	21	Tajo	~62.0	~12	
	24	Guadiana	~58.0	~11	
Sweden (SE)	32	Göta älv	~41.0	~9	
	37	Torne älv	34.1	8	
	40	Ångermanälven	30.6	7	
	41	Dalälven	29.0	6	
Switzerland (CH)	11	Rhine	28.0	68	
	†	Aare	17.8	43	†
	17	Rhone	10.4	25	
Ukraine (UA)	3	Dnepr	293.0	48	
	27	Yuzhnyy Bug	63.7	11	
	25	Dnestr	52.7	9	
United Kingdom (UK)	62	Thames	15.0	6	
	71	Severn	11.6	5	
	72	Trent	10.5	4	

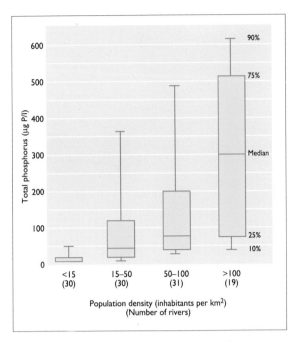

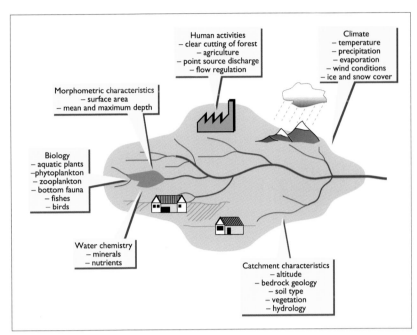

Figure 5.8 An example box-cox diagram showing the relationship between annual mean total phosphorus and population density in river catchments

Republic and the Central European part of the Russian Federation, have few natural lakes. In these areas human-made lakes such as reservoirs and ponds are often more frequent than natural lakes. Many river valleys have been dammed to create reservoirs, and a large number have been built in mountain ranges for use by the hydroelectric industry. In several countries, for example The Netherlands, Germany, France and the former Czechoslovakia, numerous artificial small lakes have been created by other human activities such as peat and sand quarrying, and for use as fish ponds.

Natural lakes in Europe
There are more than 500 000 natural lakes larger than 0.01 km² (1 ha) in Europe; of these about 80 to 90 per cent are small, with a surface area between 0.01 and 0.1 km², and only about 16 000 have a surface area exceeding 1 km² (Table 5.10). Three quarters of the lakes are located in Norway, Sweden, Finland and the Karelo-Kola part of the Russian Federation.

The approximate number and size distribution of natural lakes is shown for each country in Table 5.11; however, the number of small lakes is somewhat uncertain, and the figures given are generally minimum estimates.

Human-made lakes
Reservoirs are the most important human-made lakes in Europe, there being more than 10 000 major reservoirs covering a total surface area of more than 100 000 km². The numbers of relatively large reservoirs are greatest in the Russian Federation (ca 1250), Spain (ca 1000), Norway (ca 810) and the UK (ca 570). Other countries with a large number of reservoirs are Hungary (ca 300), Italy (ca 270), France (ca 240) and Sweden (ca 225). Many European countries have numerous smaller human-made lakes, for example Latvia, Bulgaria and Estonia, which have about 800, 500 and 60, respectively.

Large lakes and reservoirs in Europe
There are 24 natural lakes in Europe that have a surface area larger than 400 km², the largest being Lake Ladoga, which covers an area of 17 670 km² (Map 5.12). The latter is located in the northwestern part of the Russian Federation, together with Lake Onega, the second largest lake in Europe. Both are

considerably larger than other European lakes and reservoirs, but nevertheless rank only 18th and 22nd in world order (Herdendorf, 1982). The third largest European freshwater body is the 6450 km² Kuybyshevskoye reservoir on the Volga. Another 19 natural lakes larger than 400 km² are found in Sweden, Finland, Estonia and the northwestern part of the Russian Federation, and three in Central Europe – Lake Balaton, Lac Léman and Bodensee, the surface areas of which are 596, 584, and 540 km², respectively (Map 5.12).

The six largest reservoirs are located in the Volga river system in the Russian Federation, the two largest being the 6450 km² Kuybyshevskoye and the 4450 km² Rybinskoye reservoirs. Of the 13 European reservoirs with an area exceeding 1000 km², only the Dutch reservoir Ijsselmeer lies outside the Russian Federation and Ukraine.

Deep and shallow lakes and reservoirs
Lake water depth is an important parameter with which to characterise the lake environment. It is determined largely by the surrounding topography, lakes in mountainous regions generally being deeper than those in lowland areas. In two lowland countries, Finland and Poland, most lakes have a mean depth of 3 to 10 m; lakes with a mean depth greater than 10 m are rare. In Austria and Switzerland, in contrast, large shallow lakes are virtually absent, and most lakes have a mean depth greater than 25 m. As with natural lakes, the deepest reservoirs are located in mountainous regions of countries such as Norway, Spain, France, Scotland and Greece. Examples are the 190 m deep Spanish reservoir Almendra, the 132 m deep Greek reservoir Kremasta, and the 125 m deep Norwegian reservoir Blåsjø (these are maximum depths).

Regulation of European rivers

River regulation is a general term describing the physical changes that people impose on watercourses. Various human activities that physically influence rivers are listed in Box 5F.

Many of the rivers in Europe have now been regulated; in some countries there are very few unregulated rivers (Brookes, 1987; Petts, 1988; Garcia De Jalon, 1987). River regulation has been undertaken to the greatest extent in Western and Southern Europe. Thus in countries such as Belgium, England, Wales and Denmark, the percentage of river reaches that are still in a natural state is low, typically 0 to 20 per cent. By contrast, in countries such as Poland,

Figure 5.9
Conceptual diagram describing river catchment relationships
Source: Prepared by NERI

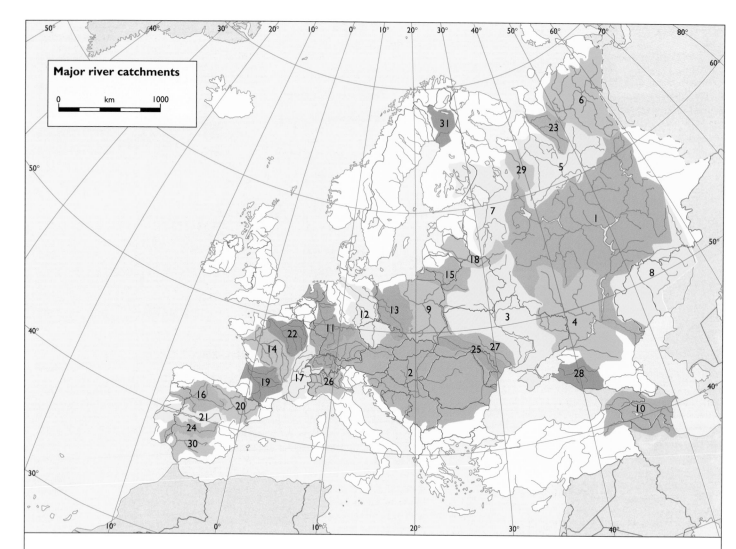

River	Country (The country code refers to Table 5.9)	Catchment area (10^3 km^2)	Mean discharge (km^3/yr)	Length (km)
1 Volga	RU	1360	230	3530
2 Danube	DE, AT, SK, HU, HR, SB, RO, BG, UA, CH*, PL*, IT*, CZ*, SI*, BA*, AL*, MD*	817	205	2850
3 Dnepr	RU, BY, UA	558	53	2270
4 Don	RU, UA*	422	38	1870
5 Severnaya Dvina	RU	358	148	740
6 Pechora	RU	322	129	1810
7 Neva	RU, FI*, BY*	281	79	75
8 Ural	RU, Kazakhstan	270	-	2540
9 Wisla	PL, SK*, UA*, BY*	194	31	1050
10 Kura	GE, TK, AZ, AR*, Iran*	188	18	1360
11 Rhine	CH, AT, DE, FR, NL IT*, LU*, BE*	185	69	2200
12 Elbe	CZ, DE, AT*, PL*	148	24	1140
13 Oder	CZ, PL, DE	119	16	850

River	Country (The country code refers to Table 5.9)	Catchment area (10^3 km^2)	Mean discharge (km^3/yr)	Length (km)
14 Loire	FR	118	32	1010
15 Neman	BY, LT, RU, PL*	98	22	960
16 Douro	ES, PT	98	20	790
17 Rhône	CH, FR	96	54	810
18 Zapadnaya Dvina	RU, BY, LV, LT*	88	21	1020
19 Garonne	FR	85	21	575
20 Ebro	ES	84	17	910
21 Tajo	ES, PT	80	6	1010
22 Seine	FR	79	16	780
23 Mezen	RU	78	26	970
24 Guadiana	ES, PT	72	-	800
25 Dnestr	UA, MD	72	10	1350
26 Po	IT, CH*	69	46	670
27 Yuzhnyy Bug	UA	65	3	860
28 Kuban	RU	58	13	870
29 Onega	RU	57	18	420
30 Guadalquivir	ES	57	2	675
31 Kemijoki	FI	51	17	510

***Map 5.11** Larger European river catchments exceeding 50 000 km²*

Note: * The country includes part of the catchment area but the main river does not run through it.
Source: Compiled by NERI from multiple sources listed in Box 5E or referred to in the main text

Estonia and Norway, many rivers still have 70 to 100 per cent of their reaches in a natural state.

River regulation often causes major changes in river processes, primarily the flow regime and the transport of dissolved and particulate matter. The effects are seen not just locally, but may be extensive, with downstream reaches nearly always being affected, and upstream reaches and the surrounding areas often being affected as well. Some types of river regulation, for example land drainage, may affect much if not all of a catchment area, but that which has the most widespread and marked effect is the construction of reservoirs. Nevertheless, many river systems have also been changed by channelisation, especially those in lowland areas.

Reservoirs

Because reservoirs usually have a relatively short water residence time – often less than a year and sometimes just a few days – they can be regarded as a hybrid between a river and a lake that is internally divided in three zones: a river-like zone at the inflow end, a lake-like zone in the outflow end, and a transitional zone in between. Another prominent feature of many reservoirs is that the water level fluctuates, so that the littoral zone is biologically poorly developed where the fluctuations exceed that of the natural lakes in the region.

Reservoirs have been constructed in Europe for thousands of years; the earliest were relatively small, and used mainly for domestic water supply and crop irrigation. During the last two centuries there has been a marked increase in both reservoir size and number, with large storage capacity reservoirs constructed in many countries, especially the former USSR. Thus there are currently about 3900 large reservoirs with dams higher than 15 m in Europe (not including the territory of the former USSR), half of which have been built since 1961 (Boon, 1992). To this must be added the many large reservoirs in the European part of the former USSR, and the thousands of smaller reservoirs and ponds spread throughout Europe.

The development of reservoir construction in Europe can be illustrated using the UK and Spain as examples (Figure 5.10). In the UK the number of reservoirs grew rapidly during the second half of the 19th century from about 50 to about 200; from then until the mid-1970s, the rate was about six new reservoirs per year (Boon, 1992). In Spain the number of reservoirs grew at the rate of about two per year between 1900 and World War 2, but at a rate of about 20 per year in the post-war period from 1950 to 1980 (Garcia De Jalon, 1987; Riera et al, 1992). Reservoir construction in Europe has now fallen off and growth in total reservoir area seems to be stagnant. This is mainly because of the lack of suitable sites (Williams and Musco, 1992).

Reservoir usage

Reservoirs are usually built to serve several purposes, the primary uses typically being the generation of hydroelectric power, irrigation, flood control, and domestic and industrial water supply. Other uses are commercial fishery and various recreational activities.

Environmental problems related to reservoirs

Reservoir construction leads to a number of environmental problems, both during the building phase and following completion. As the water level in the reservoir rises upon the closing of the dam, major changes often take place in the area to be inundated; farmland can be lost, settlements flooded, and the groundwater table elevated. Once the reservoir has been established, the environmental problems can be divided in two groups: those that render the reservoir unsuitable for its purpose, for example algae and toxic substances in reservoirs used for drinking water, and those that induce ecological deterioration of the river system, especially downstream of the reservoir (Box 5G).

Surface area (km²)	Total in Europe	Norway, Sweden, Finland, and Karelo-Kola (part of the Russian Federation)
>400	24	21
>100	150	125
>10	2000	1500
>1	16 000	12 500
>0.1	100 000	75 000
>0.01	500 000–700 000	>450 000

Table 5.10
Estimated number of natural lakes in different surface area classes
Source: Compiled by NERI from multiple sources listed in Box 5E

Country	Surface area (km²) 0.01–0.1	0.1–1	1–10	10–100	>100
Albania[1]	–	–	–	>3	3
Austria[1]	some 100s		19	7	2
Bulgaria[1]	53	175	288	14	0
Croatia[1]	–	1	3	0	0
Denmark[1]	354	256	74	6	0
England and Wales[1]	1665		50	2	0
Estonia[1]	750	209	41	1	3
Finland[1]	40 309	13 114	2283	279	47
France[2]	–	128	23		1
Georgia[1]	799	58	21	14	0
Germany[3]	–	–	~100	~20	2
Greece[3]	–	–	–	>16	1
Hungary[1,4]	–	–	–	2	2
Iceland[1]	~7000	1650	176	17	0
Ireland[3]	–	–	~100	14	3
Italy[5]	–	>168	>82	13	5
Latvia[1]	2164	740	122	20	0
Moldova[1]	>3300	48	30	6	0
The Netherlands[1]	–	–	–	47	3
Norway[1]	208 000		2000	450	7
Poland[1]	6050	2627	545	32	2
Russian Federation[1]	471 000		4626	412	51
Spain[1]	–	–	–	800	
Sweden[1]	59 500	19 374	3990	358	22
Switzerland[6]	1300		10	15	5
Ukraine[1]	950			>4	2
former Yugoslavia[6]	>200		>10	15	4

Notes:
– Missing information.
There are no or only a few lakes in Belgium, the Czech and Slovak republics, Luxembourg and Portugal. There is no information from Belarus, Romania and Lithuania.
Information obtained from: 1 Questionnaires; 2 Ministère de la Culture et de l'Environnement, 1978; 3 Estimated by NERI; 4 Biró, 1984; 5 Gaggino et al, 1987; 6 Dill, 1990.

Table 5.11
Number of lakes in different European countries
Source: See table

Note: Country estimates may include some reservoirs

The water quality of a reservoir, as reflected by the content of pathogens, toxic chemicals and poisonous algae, is of primary concern when the reservoir is used for drinking water, commercial fishery, industrial processes, and recreational activities such as bathing and water sports. At present it is not possible to give a general overview of the extent to which European reservoirs are contaminated with toxic substances. However, contamination of reservoirs with oil, organic solvents, heavy metals and radionuclides has been reported to be a problem in the Russian Federation and in Ukraine (Mnatsakanian, 1992; Gavrilov et al, 1989). Many of the environmental problems to which reservoirs are subject, such as eutrophication and heavy metal, organic and thermal pollution, also occur in natural lakes and rivers.

Since reservoirs interrupt the natural continuity of a river, the ecological consequences can be manifold. Access to

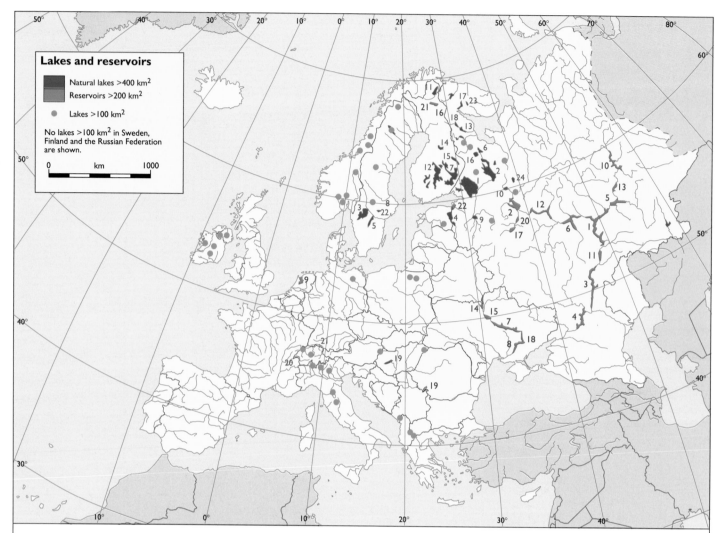

Lakes and reservoirs

■ Natural lakes >400 km²
■ Reservoirs >200 km²
● Lakes >100 km²

No lakes >100 km² in Sweden, Finland and the Russian Federation are shown.

Natural lake	Country (code refers to Table 5.9)	Area (km²)	Mean depth (m)	Max depth (m)
1 Ladoga (Ladozhskoye)	RU	17 670	51	258
2 Onega (Onezhskoye)	RU	9670	30	120
3 Vänern	SE	5670	27	106
4 Peipus	RU, EE	3570	23	47
5 Vättern	SE	1912	39	128
6 Vygozero	RU	1285	7	19–24
7 Saimaa	FI	1147	12	82
8 Mälaren	SE	1140	13	61
9 Il'men'	RU	1124	2.6	10
10 Beloye	RU	1120	4.2	20
11 Inari	FI	1102	14	96
12 Päijänne	FI	1054	17	98
13 Topozero	RU	1025	15	56
14 Oulujärvi	FI	893	7.6	35
15 Pielinen	FI	867	9.9	60
16 Segozero	RU	781–910	23	97
17 Imandra	RU	845	16	67
18 Pyaozero	RU	660–754	15	49
19 Balaton	HU	596	3	11
20 Lac Léman	CH, FR	584	153	310
21 Bodensee (Constance)	DE, CH, AT	540	90	252
22 Hjälmaren	SE	478	6.1	22
23 Umbozero	RU	422	30	115
24 Vozhe (Charonda)	RU	420	1.4	5

Reservoir	Country (code refers to Table 5.9)	Area (km²)	Mean depth (m)	Max depth (m)
R1 Kuybyshevskoye	RU	6450	12.6	40
R2 Rybinskoye	RU	4450	5.6	30
R3 Volgogradskoye	RU	3320	10.1	41
R4 Tsimlyanskoye	RU	2702	8.8	–
R5 Nizhnekamskoye	RU	2650	4.9	–
R6 Cheboksarskoye	RU	2270	6.1	–
R7 Kremenchugskoye	UA	2250	6.0	–
R8 Kakhovskoye	UA	2150	8.5	–
R9 Ijsselmeer	NL	2000	–	–
R10 Kamskoye	RU	1915	6.4	29
R11 Saratovskoye	RU	1830	7.3	32
R12 Gor'kovskoye	RU	1591	5.5	21
R13 Votkinskoye	RU	1120	8.4	28
R14 Kiyevskoye	UA	922	4.0	–
R15 Kanevskoye	UA	582	4.3	–
R16 Lokka	FI	417	–	–
R17 Ivankovskoye	RU	327	3.4	–
R18 Dnepr	UA	320	–	60
R19 Djerdap	SB, RO	253	–	92
R20 Uglich	RU	249	5.0	–
R21 Porttipahta	FI	214	–	–
R22 Narva	EE, RU	200	1.9	9

Map 5.12 Large European lakes and reservoirs Source: Compiled by NERI from multiple sources listed in Box 5E or referred to in the main text

spawning sites for migratory fish is prevented, the problem being especially acute for fish such as salmon, trout and sturgeon. As reservoirs trap the suspended matter flowing into them, they reduce the suspended matter load to downstream reaches; in contrast, because of their high biological productivity, the organic load to the downstream reaches may increase. Since reservoirs regulate the water flow, sudden flow fluctuations may occur downstream of reservoirs used for hydroelectric power generation. In other cases reservoirs may have a stabilising effect on the downstream flow regime, thereby benefiting downstream floral and faunal communities. Still other reservoirs warm up the river water, thereby elevating the downstream water temperature. Finally, the water flowing from reservoirs with bottom water outlets comes from the deeper layers and is consequently colder and, due to poor mixing and decomposition of organic matter, occasionally low in oxygen content. The changes in flow regime and water temperature detrimentally affect the downstream aquatic community. To quote Casado et al (1989), these are: 'a reduction of macrophytes, a reduction in faunistic richness both of fish and invertebrates, and a reduction of fish biomass, density and growth'.

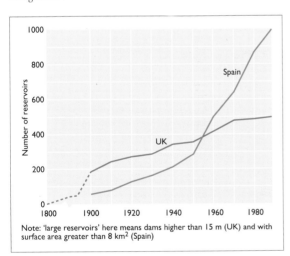

Figure 5.10 The increase in total number of large reservoirs in the UK and in Spain
Sources: Boon, 1992, and Riera et al, 1992

During the last 10 to 15 years there has been growing public opinion against the construction of dams and reservoirs. For example, in Norway, the Lappish population protested against the damming of the river Alta, and, in Spain, a 101 m high dam built on the river Esla in 1970 to form the Riano reservoir remains unused because the inhabitants of the valley to be expropriated have refused to move (Garcia De Jalon, 1987). A more recent example is the construction of the Gabcikovo hydroelectric power plant on the Danube; although both Hungary and the former Czechoslovakia were jointly involved in the original project, environmental groups in Hungary forced the Hungarian Government to withdraw. The Hungarians are now calling attention to an environmental catastrophe in the area.

River channelisation

The objective when modifying the course of a river is to improve certain features, for example flood control, drainage of the surrounding land, navigation and erosion prevention. River channelisation comprises a number of physical measures, each of which is related to hydrological parameters; hence straightening changes the slope, dredging changes the depth and width, and dredging and weed cutting change the roughness. Other more radical methods of river channelisation are culverting, lining, and piping.

In many countries where there is intensive agricultural production, many of the rivers have been regulated. In Denmark, for example, 85 to 98 per cent of the total river network has been straightened (Brookes, 1987; Iversen et al, 1993).

Physical and biological effects

Channelisation has great impact on a river because it disrupts the existing physical equilibrium of the watercourse; to compensate for the alteration in one or more of the hydraulic parameters, and to establish a new, stable equilibrium, other parameters will change. Because straightening of a river increases its slope, the energy in the moving water has to be dispersed over a smaller surface; as a result the water is able to move larger particles and sediment discharge increases through bank erosion. If the river is not repeatedly manipulated or stabilised by culverting, lining etc, this will eventually lead to widening of the river channel and to a subsequent reduction in water velocity. River channelisation generally changes a heterogeneous system into a homogeneous one such that flow becomes uniform, pools are lost and the substrate becomes uniform throughout the channel (Figure 5.11).

Channelisation can also have great impact on riparian vegetation; trees are often logged to allow channel maintenance (eg, machine dredging) and scrubs are cut to ensure sufficient drainage. This increases solar radiation at the stream surface, thereby increasing the water temperature, reducing the concentration of dissolved oxygen, and increasing the in-stream primary production. In nutrient-rich watercourses this results in enhanced growth of benthic algae and macroalgae.

Another effect of the channelisation of rivers and drainage of wetlands may be increased nutrient and organic matter loading of rivers and the marine environment. The reason is that while the annual nitrogen removal capacity of wetlands and natural rivers can be as much as several hundred kg/ha, that of channelised rivers and drained wetlands may be negligible. Naturally meandering riparian zones alongside rivers may therefore play an important role in balancing intensive agricultural and ecological interests.

The velocity of river water is one of the major factors regulating the structure of riverine plant and animal communities (Brookes, 1988; Westlake, 1973). The uniform and often unstable sediment found in channelised watercourses is suitable for few, if any, plant species. Furthermore, as the uniform water flow precludes areas with little or no flow, resting sites for fish and invertebrates are virtually absent. The general effect of channelisation is therefore a reduction in habitat number and diversity and a consequent reduction in species number and diversity. The latter may be further reduced by the above-mentioned decrease in oxygen concentration. Hence, the biomass of organisms such as fish and invertebrates is usually lower in channelised watercourses, as illustrated in Figure 5.12.

It is not only animals and plants living within the watercourse that are affected by channelisation, however. Thus animal species which depend on the bank for foraging and/or breeding decline in number, with the consequence that species diversity on the river banks also decreases (Figure 5.13). In addition, a number of plant species that are confined to the more or less water saturated soil adjacent to the river are also affected.

Organic pollution of rivers

Organic matter derived from diverse human activities is a major source of pollutant discharge to rivers. The decomposition and breakdown of this organic matter is mediated by microorganisms and takes place mainly at the surface of the sediment and vegetation in smaller rivers, and in the water column in larger rivers. As the process requires the consumption of oxygen, severe organic pollution may

lead to rapid deoxygenation of the river water and hence to the disappearance of fish and aquatic invertebrates. The habitat then becomes uniform with only a few robust species able to tolerate the low oxygen concentration. Decomposition of organic waste also results in the release of ammonium, which, although not in itself toxic, may, depending on the pH and temperature of the water, be converted to ammonia, which is poisonous to fish.

The most important sources of the organic waste feeding into rivers are domestic and industrial sewage. Immediately downstream of a sewage effluent, organic matter decomposition reduces the oxygen content of the water and results in the release of ammonium (Figure 5.14). Further downstream the concentration of organic matter decreases as a result of dilution and continuing decomposition. As the distance from the effluent increases, bacteria oxidise the ammonium to nitrate, and oxygen enters the water via the water surface, thereby increasing its oxygen content. Eventually the levels of organic matter, oxygen and ammonium reach those present immediately upstream of the sewage effluent; this process of recovery is called self-purification. An example is the Danube, which is already polluted by organic matter when it enters Hungary. As it winds its way through the country the river receives large amounts of organic matter from tributaries and cities, especially from Budapest. However, by the time it leaves the country and enters Croatia, an amount of organic matter equal to that discharged in Hungary has been decomposed (Varga et al, 1990; Benedek and Major, 1992). Nevertheless, this does not imply that rivers can take up an unlimited amount of organic matter without suffering as a result; the pollution may be so severe, widespread and long lasting that self-purification is insufficient. Thus the Danube is still polluted when it leaves Hungary, and the Rhine was polluted with such excessive amounts of organic matter between World War 2 and the early 1970s that there was very serious oxygen depletion in its middle and lower course and the river virtually died (Friedrich and Müller, 1984).

Organic matter content in European rivers

Because decomposition of organic matter requires oxygen, the amount of organic matter in a river can be measured in terms of the biochemical oxygen demand (BOD) or the chemical oxygen demand (COD), the units of which are mg O_2/l. River reaches little affected by human activities generally have a BOD below 2 mg O_2/l whereas a BOD

Figure 5.11
The difference in physical variation between a) a natural and b) a channelised watercourse
Source: Prepared by NERI

Figure 5.12 *Difference in the amount of fish and invertebrates in natural and channelised watercourses*
Source: Adapted from Brookes, 1988

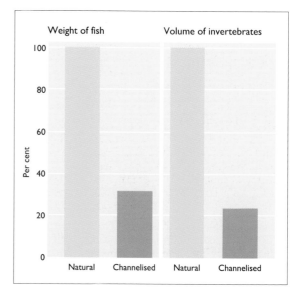

exceeding 5 mg O_2/l generally indicates pollution. Measurement of BOD is the most widespread method in Europe, but many countries also measure the COD, and some use only the COD. Although both BOD and COD indicate the potential oxygen demand of the organic matter in the water, there is not necessarily a correlation between the two measurements. In large rivers suffering from severe eutrophication, elevated BOD values can occur due to decomposition of phytoplankton, and in these cases high BOD values are not necessarily indicative of organic pollution.

BOD, COD and oxygen content data from a large number of river stations in 33 European countries are collated in Table 5.12. Median BOD, COD and oxygen are 2.8, 14.5, and 9.7 mg O_2/l, respectively. As can be seen, annual mean BOD was below 5 mg O_2/l and the oxygen content above 8 mg O_2/l at more than 75 per cent of the river stations. Extremely high BOD is generally seen only in smaller rivers polluted with raw sewage or animal slurry, a BOD exceeding 500 mg O_2/l then being possible.

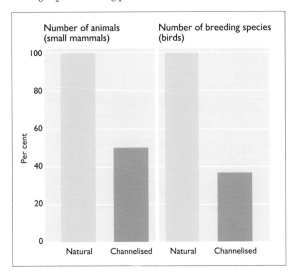

Figure 5.13 *Difference in the numbers of small mammals and birds alongside natural and channelised watercourses*
Source: Adapted from Brookes, 1988

The information obtained from the 30 European countries concerning the percentage of rivers with BOD or COD levels in specified classes is summarised in Map 5.13, and the BOD and COD concentrations of a large number of European rivers is presented in Map 5.14.

In Iceland, Norway, Sweden and Finland organic matter content is measured only as COD. In these countries discharge into rivers of organic waste derived from human activity is negligible and COD levels therefore are generally low (Map 5.13). In terms of BOD, rivers in Ireland, Georgia, Estonia, Latvia, Austria, Switzerland, The Netherlands, the UK, Denmark and Croatia are least affected, with less than 25 per cent of the rivers having a BOD exceeding 3.5 mg O_2/l. In Hungary, Lithuania, Portugal, France, Ukraine, Germany, Slovenia and Italy the rivers are moderately affected, with less than 25 per cent having a BOD exceeding 5 mg O_2/l. More affected rivers are found in Albania, Poland, the Czech Republic, Moldova, the Russian Federation and Spain, where more than 25 per cent of the rivers have a BOD exceeding 5 mg O_2/l. BOD is highest in Bulgaria, Belgian Flanders and Romania, exceeding 5 mg O_2/l in 60 per cent, 69 per cent and 80 per cent of the rivers, respectively.

Human activities and organic matter

The organic matter naturally occurring in rivers originates from soil erosion and from dead plants and animals and is

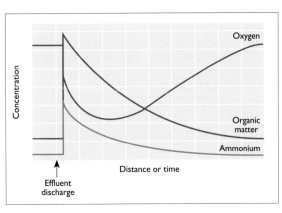

Figure 5.14
Impact of an organic matter effluent on river concentrations of organic matter, oxygen, and ammonium
Source: Modified from Meybeck et al, 1989

normally relatively insoluble and only slowly decomposed. In contrast, organic matter derived from human activities is generally soluble, finely divided and rapidly decomposed, the result being a marked and abrupt increase in oxygen consumption in the river.

As population density in catchments increases, the level of organic matter in the rivers generally increases, and the oxygen content decreases (Figure 5.15). Thus whereas BOD concentration is lower than 2 mg O_2/l in catchments with fewer than 15 inhabitants/km², it generally exceeds 5 mg O_2/l in catchments with more than 100 inhabitants/km². The great variation found in extensively populated areas is attributable mainly to variation in the extent of wastewater treatment – well-functioning treatment plants can decompose up to 90 per cent of the organic matter in the wastewater.

Assessment of river quality

In most European countries the environmental state of the rivers has been monitored for many years. However, as pollution has both physico-chemical and biological effects on the receiving water, the quality of the water can be assessed in many different ways, and there are numerous methods in use throughout Europe. The definition of river quality used in this chapter is given in Box 5H.

The quality of European rivers is summarised in Table 5.13. About a quarter of the river reaches are classified as having poor or bad water quality. Most of the countries classify 50 per cent or more of their river reaches as having good or fair quality. Iceland, Scotland, Northern Ireland and Ireland have the highest proportion of rivers classified as having good water quality, while the Russian Federation, Finland, England and Wales, West Germany, Austria, Croatia, Lithuania and Latvia classify more than 75 per cent of their rivers as having good or fair water quality. More than 25 per cent of the rivers have poor or bad water quality in Bulgaria, Romania, the Czech Republic, Poland, Denmark, The Netherlands, Luxembourg, Slovenia and Italy. Furthermore, the percentage of river reaches classified as having bad water

	Number of river stations	Percentage of river stations with organic matter and oxygen concentrations (not exceeding mg O_2/l)					
		Mean	10%	25%	50%	75%	90%
All rivers							
BOD	645	4.5	1.4	1.9	2.8	4.7	7.9
COD	470	18.5	4.5	7.8	15.0	25.0	36.6
Oxygen	620	9.4	6.4	8.4	9.7	10.7	11.6
Near pristine rivers							
BOD	11			1.2	1.6	2.7	
COD	23			5.1	13.3	29.9	
Oxygen	8			10.2	10.6	11.1	

Note: Guidance as how to read the table is given in Box 5E (Table 5.8).

Table 5.12
Descriptive statistics for annual mean organic matter and oxygen at European river stations
Source: Compiled by NERI and EEA-TF from multiple sources listed in Box 5E or referred to in the main text

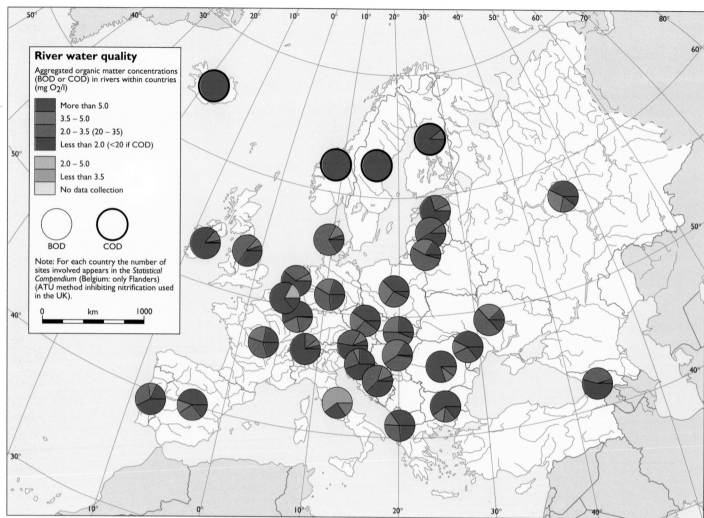

Map 5.13
Frequency distribution of annual mean concentrations of organic matter in European rivers by country 1988–91
Source: Compiled by NERI and EEA-TF from multiple sources listed in Box 5E or referred to in the main text

quality is highest in the Czech Republic and Poland (28 per cent) and in Belgian Flanders (37 per cent).

The greater the amount of organic matter present in river water, the lower the oxygen concentration, and the higher the ammonium concentration. These two relationships are illustrated in Figure 5.16 using data from 482 and 365 river stations, respectively. A low oxygen concentration influences the river fauna. To maintain a salmon or trout population in a river, a minimum oxygen concentration of 6 mg O_2/l is necessary (Council Directive 78/659/EEC) (Figure 5.16a). To reduce the risk of fish kill, the ammonium concentration should not exceed 1 mg N/l (Figure 5.16b). Although ammonium is not in itself toxic to fish, it becomes toxic when converted to ammonia. Thus if the ammonia concentration exceeds 0.025 mg/l, trout growth is prevented, and if it exceeds 0.25 mg/l, the trout die. Rivers with large sewage discharges exceed these limits.

Trends in organic matter discharge to rivers

After World War 2 riverine discharge of organic waste increased in many European countries with resultant severe oxygen depletion. During the last 15 to 20 years, however, biological treatment of domestic and industrial wastewaters has intensified, and organic matter loading of rivers has consequently decreased in many parts of Europe, the result being that many rivers are now fairly well oxygenated.

An example is the river Rhine. Rebuilding of industry after World War 2 led to high loads of poorly or untreated sewage being discharged into the river. This caused oxygen contents of the water to fall considerably (Figure 5.17), and long stretches became biologically devastated. In the 1960s the deterioration in river quality became so apparent that

countermeasures such as increasing the number of sewage treatment plants were implemented. In the eight-year period 1973–81, the percentage of sewage water that was treated before being discharged into the Rhine rose from about 30 per cent to 80 to 90 per cent (Dijkzeul, 1982 (in Wolff, 1987)). The consequent marked reduction in organic pollution led to a clear improvement in water oxygen content and the return of several of the animal species that had disappeared. Thus, although the general pollution of the Rhine is still serious, there are signs that conditions are improving.

A comparison of BOD levels at 223 river stations throughout Europe reveals signs of improving conditions. From the period 1977–82 to the period 1988–90 the organic matter concentration decreased at almost 75 per cent of the river stations (Figure 5.18), the reduction being greater than 25 per cent at 41 per cent of the river stations. An increase of more than 25 per cent was seen at 8 per cent of the river stations, however.

The improvement was greatest in Western Europe, where about half of the river stations show a decrease in organic matter concentration of more than 25 per cent, primarily as a result of intensified wastewater treatment. The improvement was less pronounced in Eastern Europe, where about 30 per cent of the river stations show a decrease of more than 25 per cent and about 15 per cent show an increase of more than 25 per cent (Figure 5.18).

Organic pollution is still a serious problem in many European rivers, and will continue to be so for as long as large amounts of sewage water are discharged into the rivers without being treated. Countries presently experiencing industrial depression should be especially attentive to this problem and develop their sewage plant

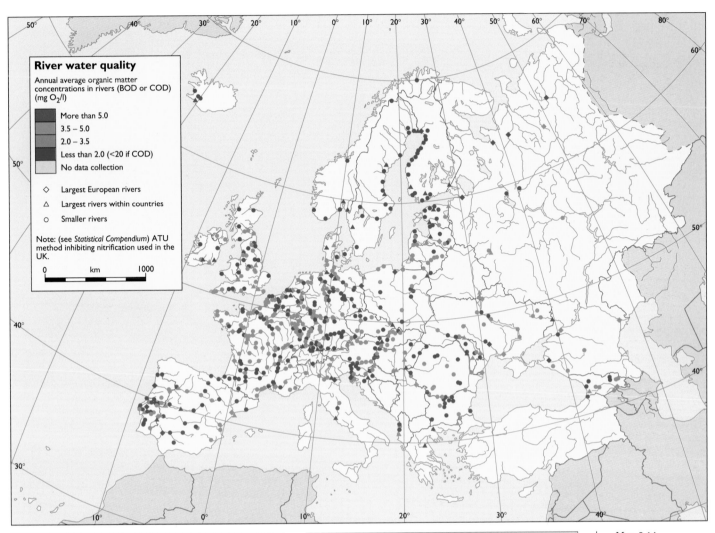

River water quality

Annual average organic matter concentrations in rivers (BOD or COD) (mg O₂/l)

- More than 5.0
- 3.5 – 5.0
- 2.0 – 3.5
- Less than 2.0 (<20 if COD)
- No data collection

◇ Largest European rivers
△ Largest rivers within countries
○ Smaller rivers

Note: (see *Statistical Compendium*) ATU method inhibiting nitrification used in the UK.

0 ———— km ———— 1000

Map 5.14
Annual mean concentrations of organic matter at specific stations in European rivers, 1988–91
Source: Compiled by NERI and EEA-TF from multiple sources listed in Box 5E or referred to in the main text

Box 5H River quality classification

Good quality: river reaches with nutrient-poor water, low levels of organic matter; saturated with dissolved oxygen; rich invertebrate fauna; suitable spawning ground for salmonoid fish.

Fair quality: river reaches with moderate organic pollution and nutrient content; good oxygen conditions; rich flora and fauna; large fish population.

Poor quality: river reaches with heavy organic pollution; oxygen concentration usually low; sediment locally anaerobic; occasional blooming of organisms insensitive to oxygen depletion; small or absent fish population; periodic fish kill.

Bad quality: river reaches with excessive organic pollution; prolonged periods of very low oxygen concentration or total deoxygenation; anaerobic sediment, severe toxic input; devoid of fish.

capacity, as their industrial production can be expected to increase in the future.

Nutrients in European rivers and lakes

Precipitation that falls on the surface of the land dissolves and absorbs minerals as it drains through the soil. In pristine areas, the chemical composition of river and lake water is determined mainly by that of the soil and underlying bedrock, nutrient levels in such waters generally being low.

Human settlement and associated clearance of forest,

	Water quality (%)			
	Good	Fair	Poor	Bad
Austria 1991[1]	14	82	3	1
Belgium 1989–90[2]	17	31	15	37
Bulgaria 1991[1]	25	33	31	11
Croatia[1]	15	60	15	10
Czech Republic 1990[3]	12	33	27	28
Denmark 1989–91[4]	4	49	35	12
England and Wales 1990[8]	65	23	10	2
Finland 1989–90[1]	45	52	3	0
Germany 1985[5]	44	40	14	2
Iceland[1]	99	1	0	0
Ireland 1987–90[6]	77	12	10	1
Italy[7]	27	31	34	8
Latvia[1]	10	70	15	5
Lithuania[1]	2	97	1	0
Luxembourg[1]	53	19	17	11
Netherlands 1990[1]	5	50	40	5
Northern Ireland 1990[9]	72	24	4	0
Poland 1990[1]	10	33	29	28
Romania[1]	31	40	24	5
Russian Federation[1]	6	87	5	2
Scotland 1990[9,10]	97	24	4	0
Slovenia 1990[1]	12	60	27	1

Note: For definitions see Box 5H.

Table 5.13
Quality of European river reaches
Sources: 1 Questionnaires; 2 Vlaamse Milieumaatschappij, 1991 (Flemish part only); 3 Kinkor, 1992; 4 Danish Ministry of the Environment, 1992; 5 Toner, 1987 (West Germany only); 6 Clabby et al, 1992; 7 De Pauw et al, 1992; 8 NRA, 1991; 9 UK DoE, 1992; 10 The Scottish Office, Environment Department, 1992

Figure 5.15
Relationship between
population density
and the annual mean
BOD, COD and
oxygen levels in more
than 100 European
river catchments
*Figure 5.15
Relationship between
population density
and the annual mean
BOD, COD and
oxygen levels in more
than 100 European
river catchments*
Source: Compiled by
NERI and EEA-TF from
multiple sources listed in
Box 5E or referred to in
the main text

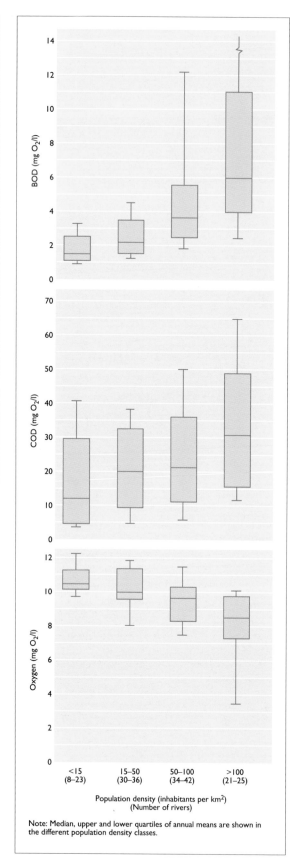

Note: Median, upper and lower quartiles of annual means are shown in
the different population density classes.

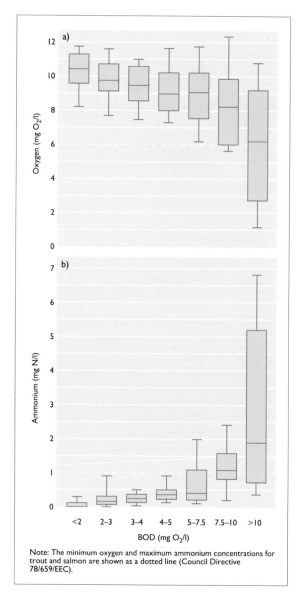

Note: The minimum oxygen and maximum ammonium concentrations for
trout and salmon are shown as a dotted line (Council Directive
78/659/EEC).

Figure 5.16 *Relationship between BOD levels and
a) annual mean oxygen concentration; and
b) annual mean ammonium concentration*
Source: Compiled by NERI and EEA-TF from multiple sources listed in Box
5E or referred to in the main text

agricultural development and urbanisation greatly accelerate
the runoff of materials and nutrients into rivers and lakes.
This stimulates the growth of phytoplankton and other
aquatic plants and, in turn, that of organisms higher up the
aquatic food chain. The process is usually known as 'cultural
eutrophication'. Enhanced biological production and other
associated effects of eutrophication are generally more
apparent in lakes, reservoirs, coastal areas, and large, slowly
flowing rivers, than in small rivers. Although phosphorus
tends to be the nutrient that most limits plant growth in lakes
and reservoirs, increased nitrogen levels can also lead to
higher biological production (OECD, 1982), especially in
marine areas, where it is usually the nutrient that most limits
plant growth.

The ecological consequences of cultural eutrophication
can be significant. High nutrient input to standing waters
may provoke a shift in the biological structure; clearwater
shallow lakes with submerged plants become dominated by
phytoplankton, and the water therefore becomes so turbid
that the submerged plants disappear. In addition, excessive
growth of blue-green algae may sometimes occur, leading
to the formation of surface scum and the production of
toxins potentially poisonous to fish, cattle, dogs and
humans. The fish community also changes and becomes
dominated by species more tolerant to the turbid
environment. When the phytoplankton sink to the lake
bottom their decomposition may reduce the oxygen
concentration in the water to levels too low to support fish
and benthic invertebrates, the result being fish kill and the
disappearance of benthic invertebrates. Low oxygen levels
may also enhance the release of phosphorus from the lake

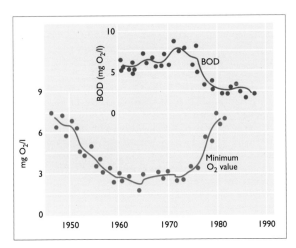

Figure 5.17 Dissolved oxygen and BOD in the Rhine at the German/Dutch border
Sources: Modified from: 1 RIWA, 1983 (in Friedrich and Müller, 1984); RIWA, 1983; 2 Umweltbundesamt, 1992

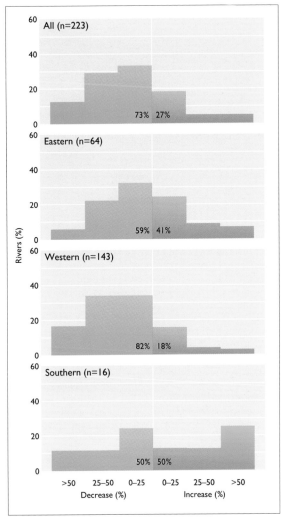

Figure 5.18
Percentage change in BOD concentration from the period 1977–82 to the period 1988–90 at 223 European river stations
Source: Compiled by NERI and EEA-TF from multiple sources listed in Box 5E or referred to in the main text

sediment, thereby further enhancing phytoplankton production.

Excessive phytoplankton growth is easy to detect and it significantly affects both the use and the aesthetic quality of the lake. In cases where lake water is used for domestic water supply, eutrophication leads to taste and odour problems, and necessitates treatment and filtration of the water before use; this is an expensive and time-consuming process, especially in the case of very eutrophic water.

Nutrient concentrations in European rivers

Phosphorus in rivers

Phosphorus is measured both as total phosphorus and soluble reactive phosphate (dissolved orthophosphate). In 321 European rivers dissolved orthophosphate was found to average 59 per cent of total phosphorus. In some countries total phosphorus is not measured and data are available only for dissolved orthophosphate. In such cases the data have not been converted to total phosphorus, but are presented as dissolved phosphate and the fact noted in the figures and tables.

Information concerning annual mean phosphorus concentrations at 546 river stations in 34 countries is collated in Table 5.14. Median annual mean total phosphorus and dissolved phosphate were found to be 173 and 126 μg P/l, respectively, annual mean phosphorus levels being below 50 μg P/l at only 25 per cent of the stations. In catchments with little or no human activity phosphorus levels in rivers are generally lower than 25 μg P/l. Phosphorus levels exceeding 50 μg P/l indicate an anthropogenic influence, for example sewage effluent and agricultural runoff. In rivers heavily polluted by sewage effluent, phosphorus levels may exceed 500 to 1000 μg P/l.

The frequency distribution of phosphorus levels in rivers has been compiled for 34 European countries, each country having supplied information about the percentage of rivers with annual mean phosphorus levels in specified classes: below 25 μg P/l, 25 to 50 μg P/l, 50 to 125 μg P/l, etc (Map 5.15). This provides an overview of the percentage of rivers with low and high phosphorus levels. The phosphorus level at a large number of specific European river stations is presented in Map 5.16 as well.

The lowest phosphorus levels are found in the rivers in Norway and Iceland. Although many rivers in Sweden, Finland, Ireland, Austria, France, the Russian Federation, Slovenia, Albania, Georgia and Bulgaria also have a low annual mean phosphorus concentration (Map 5.15), the concentration exceeds 50 μg P/l in 10 to 40 per cent of the rivers in Sweden, Finland and Ireland, and exceeds 125 μg P/l in 10 to 50 per cent of the rivers in France, the Russian Federation, Slovenia and Bulgaria.

In many other countries only 10 to 20 per cent of the rivers have a phosphorus concentration below 50 μg P/l. These relatively unpolluted rivers are generally situated in catchments in mountainous and forested regions where the population density is low. In Estonia, Latvia, Lithuania, Switzerland, Austria, Croatia, Italy and Portugal, more than 40 per cent of the rivers have a phosphorus level below 125 μg P/l; however, many of the rivers in these countries have a high phosphorus level.

The highest phosphorus levels are found in a band stretching from southern England, across the central part of Europe through Romania and Moldova to Ukraine; in these countries more than 80 per cent of the rivers usually have a phosphorus concentration exceeding 125 μg P/l. In Poland, Belgium (Flanders), Luxembourg and the UK, about 50 per cent or more of the rivers even have a phosphorus level exceeding 500 μg P/l (Maps 5.15 and 5.16).

With the exception of large rivers in the Russian Federation and the Nordic countries, the phosphorus levels generally exceed 100 μg P/l in the downstream reaches in all of the largest rivers in Europe (Map 5.16).

Bearing in mind the regional differences described above, Europe can be divided into four main regions, the river phosphorus levels of which are shown in Figure 5.19. The phosphorus levels are lowest in the Nordic countries, medium in both the Southern and Eastern European countries (although slightly lower in Southern European rivers), and highest in the Western European countries.

Nitrogen in rivers

Dissolved inorganic nitrogen, particularly nitrate and ammonium, constitutes most of the total nitrogen in river water; thus inorganic nitrogen was found to constitute 88 per

Table 5.14
Descriptive statistics of annual mean phosphorus concentrations in European rivers
Source: Compiled by NERI and EEA-TF from multiple sources listed in Box 5E or referred to in the main text

	Number of river stations	Percentage of river stations with concentrations not exceeding (µg P/l)					
		Mean	10%	25%	50%	75%	90%
All rivers							
Total phosphorus	546	280	17	59	170	366	590
Dissolved phosphate	412	329	15	45	124	286	770
Near pristine rivers							
Total phosphorus	39	26		8	12	30	

Note: Guidance as how to read the table is given in Box 5E (Table 5.8).

Map 5.15
Frequency distribution of annual mean phosphorus concentrations in European rivers by country, 1988–91
Source: Compiled by NERI and EEA-TF from multiple sources listed in Box 5E or referred to in the main text

cent (nitrate 78 per cent and ammonium 4 per cent) of the total nitrogen in 240 to 420 European rivers. In some countries data are available only for nitrate and occasionally ammonium. The results presented in this chapter are therefore either total nitrogen, inorganic nitrogen, or only nitrate nitrogen, as indicated in the figures and tables.

Annual mean nitrogen concentration at 654 river stations in 34 countries is summarised in Table 5.15. Whereas ammonium levels are generally below 0.5 mg N/l, nitrate and total nitrogen concentrations exceed 1 mg N/l in most of the rivers. The average level of ammonium and nitrate in pristine rivers is reported to be 0.015 mg N/l, and 0.1 mg N/l, respectively (Meybeck, 1982). Although rivers in a strictly pristine state are rarely found in Europe because of high atmospheric nitrogen deposition, the levels of nitrogen in relatively unpolluted streams ranged from 0.1 to 0.5 mg N/l

(Table 5.15). Nitrogen levels exceeding 1 mg N/l indicate an anthropogenic influence, for example agricultural runoff and sewage effluent. The ammonium level normally rises and the oxygen level falls when rivers receive sewage effluent or effluent from animal husbandry farms; in heavily polluted rivers the ammonium level may rise to as much as 1 to 5 mg N/l, which, when converted to ammonia, may be toxic to fish and other river fauna.

The nitrogen concentration is generally lowest in rivers in Iceland, Norway, Sweden, Finland and Albania, being below 0.75 mg N/l in 60 to 100 per cent of the rivers (Map 5.17). In southern Sweden and Finland, and in Latvia, Lithuania, the Russian Federation, Ireland, Austria, Switzerland, Slovenia, France, Portugal and Spain, nitrogen concentrations are higher, ranging from 1 to 3 mg N/l in the majority of the rivers (Maps 5.17 and 5.18).

The highest nitrogen levels are found in Eastern and Western European rivers, particularly in the Czech Republic, Denmark, Estonia, Germany, Luxembourg, Moldova, The Netherlands and Romania, the concentration exceeding 2.5 mg N/l in more than two thirds of the rivers (Map 5.17).

Whereas the median nitrate concentration is only 0.18 mg N/l in Nordic rivers, it is 3.5 mg N/l in the Western European countries, where rivers with nitrogen levels below 1 mg N/l are rare (Figure 5.20). The nitrate levels in the rivers of Southern and Eastern European countries are basically the same, although nitrate levels in the Southern European countries are rather homogeneous whereas there are large regional differences in Eastern Europe; thus the concentration is high in Central European countries like Romania, the Czech Republic and Poland, but low in the new Baltic states and large parts of the Russian Federation.

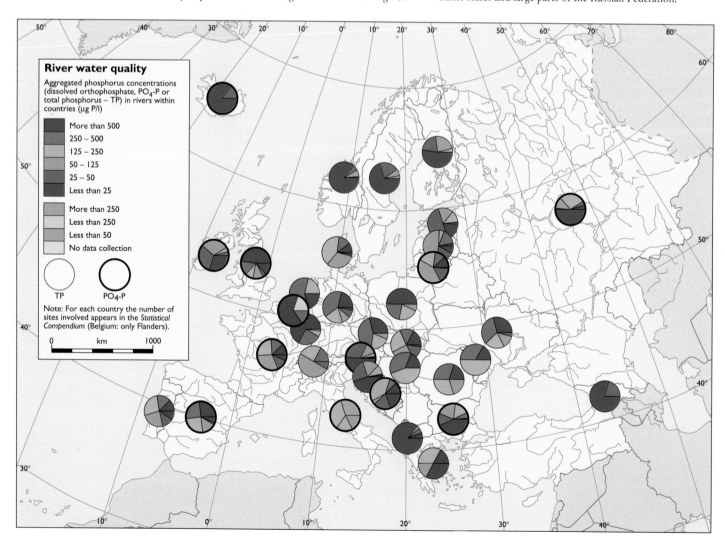

In Chapter 14 the relationship between human activities and nutrient concentrations in rivers is described.

Nutrient concentrations in European lakes

The nutrient concentration of the inflowing water generally determines the lake water nutrient concentration (Vollenweider, 1976; OECD, 1982). However, nutrients may be partly lost from the lake water, either to the sediment, especially in the case of phosphorus, or to the atmosphere through denitrification, in the case of nitrogen. The proportion of nutrients lost is directly related to water residence time. Nutrient levels in lakes are therefore generally lower than in their tributaries. If external phosphorus loading of a lake is high for a period, a large amount of phosphorus will accumulate in the lake sediment; when external loading is subsequently reduced, some of the accumulated phosphorus may be released to the water and thereby delay lake recovery (Sas, 1989; Cullen and Forsberg, 1988; Marsden, 1989).

In relatively unpolluted areas such as remote mountain regions and national parks, lake nutrient levels are usually low, the total phosphorus concentration being below 10 to 20 μg P/l and the nitrogen level below 0.5 mg N/l (Table 5.16). Higher concentrations generally imply an anthropogenic influence on the lake catchment.

The frequency distribution of lake phosphorus and nitrogen concentrations gives an indication of lake water quality in each country. However, although the presence of a high percentage of lakes with low nutrient levels suggests that the water quality is generally good, the frequency distribution may be dominated by clearwater lakes in sparsely populated areas and as a consequence may not be fully representative of the relatively more important lakes in the most inhabited areas.

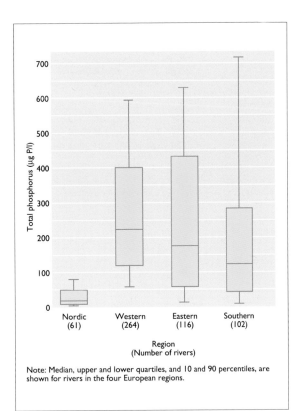

Note: Median, upper and lower quartiles, and 10 and 90 percentiles, are shown for rivers in the four European regions.

Figure 5.19
Annual mean phosphorus levels in rivers in four European regions
Source: Compiled by NERI and EEA-TF from multiple sources listed in Box 5E or referred to in the main text

Map 5.16
Annual mean phosphorus concentrations at specific stations in European rivers, 1988–91
Source: Compiled by NERI and EEA-TF from multiple sources listed in Box 5E or referred to in the main text

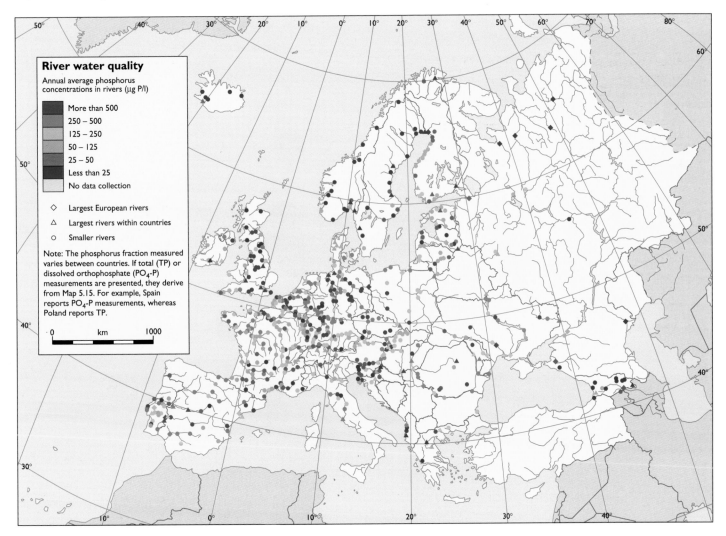

River water quality

Annual average phosphorus concentrations in rivers (μg P/l)

- More than 500
- 250 – 500
- 125 – 250
- 50 – 125
- 25 – 50
- Less than 25
- No data collection

◇ Largest European rivers
△ Largest rivers within countries
○ Smaller rivers

Note: The phosphorus fraction measured varies between countries. If total (TP) or dissolved orthophosphate (PO$_4$-P) measurements are presented, they derive from Map 5.15. For example, Spain reports PO$_4$-P measurements, whereas Poland reports TP.

0   km   1000

Table 5.15
Descriptive statistics of annual mean nitrogen concentrations in European rivers
Source: Compiled by NERI and EEA-TF from multiple sources listed in Box 5E or referred to in the main text

	Number of river stations	Percentage of river stations with concentrations not exceeding (mg N/l)					
		Mean	10%	25%	50%	75%	90%
All rivers							
Total nitrogen	329	3.34	0.30	0.80	2.12	4.50	7.32
Nitrate	654	2.63	0.24	0.70	1.80	3.90	5.73
Ammonium	580	0.66	0.03	0.07	0.18	0.45	1.40
Near pristine rivers							
Total nitrogen	43	0.40		0.19	0.33	0.39	
Nitrate	39	0.30		0.05	0.10	0.22	

Note: Guidance as how to read the table is given in Box 5E (Table 5.8).

Map 5.17
Frequency distribution of annual mean nitrogen concentrations in European rivers by country, 1988–91
Source: Compiled by NERI and EEA-TF from multiple sources listed in Box 5E or referred to in the main text

Phosphorus in lakes

The European lakes with the lowest phosphorus concentration are found in Norway, the concentration of total phosphorus being less than 10 µg P/l in more than 70 per cent of lakes (Map 5.19) and only exceeding 25 µg P/l in approximately 10 per cent of lakes. In nearby Sweden and Finland, as well as in Latvia and Austria, lake phosphorus levels are also relatively low, being below 25 µg P/l in 75 to 90 per cent of lakes. In Switzerland, Italy and Germany, the lake phosphorus levels are higher, approximately 50 per cent of the lakes having a concentration below 25 µg P/l, and the other 50 per cent having a concentration between 25 and 125 µg P/l. German lakes with low phosphorus concentrations tend to be located in the Alpenvorland, while those with high concentrations are located in the northern part of the country. In Italy there are

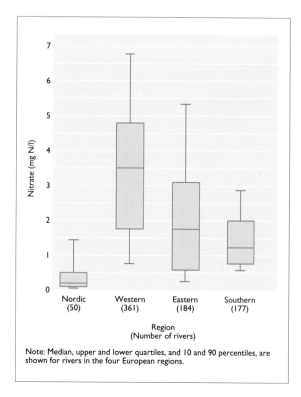

Note: Median, upper and lower quartiles, and 10 and 90 percentiles, are shown for rivers in the four European regions.

Figure 5.20 Annual mean nitrate concentrations in rivers in four European regions
Source: Compiled by NERI and EEA-TF from multiple sources listed in Box 5E or referred to in the main text

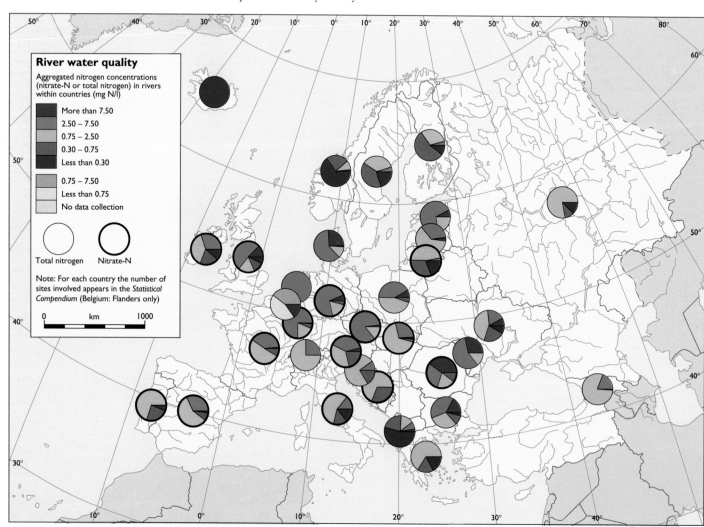

also many lakes with low phosphorus levels in the Alpine regions. In Estonia and Northern Ireland, phosphorus levels in the majority of the lakes are below 50 and 125 μg P/l, respectively.

In Spain, England and Wales, Romania, Denmark, Poland, The Netherlands and Moldova, lake phosphorus levels are generally high, exceeding 50 and 125 μg P/l in more than 80 per cent and 45 per cent of lakes, respectively.

Nitrogen in lakes

The frequency distribution of total nitrogen in lakes shows the same tendency as for phosphorus. The percentage of lakes with low nitrogen levels is highest in Norway, where 50 per cent and 90 per cent of the lakes have nitrogen levels below 0.3 and 0.75 mg N/l, respectively. In Sweden, Finland, Latvia and Estonia, nitrogen levels are lower than 0.75 mg N/l in the majority of the lakes. The picture is the opposite in Denmark, Poland and Moldova, where 85 to 90 per cent of the lakes have nitrogen levels higher than 0.75 mg N/l.

Phosphorus concentrations in large lakes

Extreme nutrient levels are rarely seen in the large European lakes, total phosphorus and total nitrogen generally being below 125 μg P/l and 1.0 mg N/l, respectively.

In the two largest European lakes, Lake Ladoga and Lake Onega in the Russian Federation, total phosphorus concentrations are currently 26 and 10 μg P/l, respectively (Map 5.20), levels three to five times higher than in the 1950s (Gutelmacher and Petrova, 1982; Petrova 1987). Although information about nutrient levels in many of the other large European lakes and reservoirs that lie in the Russian Federation is limited, the phosphorus concentrations in Lake

Peipus and Lake Il'men' are 40 and 70 μg P/l, respectively. In 13 large reservoirs on the Volga and the Dnepr, the total phosphorus concentration ranges from 49 to 201 μg P/l, the mean value being 87 μg P/l (Datsenko, 1990).

In Sweden, Finland and Norway the large lakes usually have a phosphorus concentration below 15 μg P/l. Nevertheless, a few lakes located in more densely populated areas have higher phosphorus levels, examples being the Swedish lakes Mälaren and Hjälmaren, the concentrations of which are around 30 and 50 μg P/l, respectively. The largest lake in Ireland and the UK, Lough Neagh in Northern Ireland, has a phosphorus level of approximately 100 μg P/l, while other large lakes in Ireland and the large lochs in Scotland generally have low phosphorus concentrations. The majority of large lakes in The Netherlands, northern Germany, Denmark and Poland have phosphorus levels around or exceeding 100 μg P/l.

Large lakes in the Alps have phosphorus concentrations ranging from 10 to 60 μg P/l, with concentrations in the largest approximately as follows: 60 μg P/l in Lac Léman, 50 μg P/l in Bodensee, 10 μg P/l in Lago di Garda, 30 μg P/l in Lake Neuchâtel, and 18 μg P/l in Lago Maggiore. The two large lakes at the Hungarian Plain, Lake Balaton and Neusiedler See, have phosphorus levels of 40 to 100 μg P/l, while the lakes in the mountains between Albania, Greece and the former Yugoslavia generally have low phosphorus levels.

Impact of nutrients on lake water quality

Phosphorus is usually the nutrient that most limits plant growth in lakes and reservoirs. Many studies have shown that the amount of phytoplankton (generally measured as the level

Map 5.18
Annual mean nitrate concentrations at specific stations in European rivers, 1988–91
Source: Compiled by NERI and EEA-TF from multiple sources listed in Box 5E or referred to in the main text

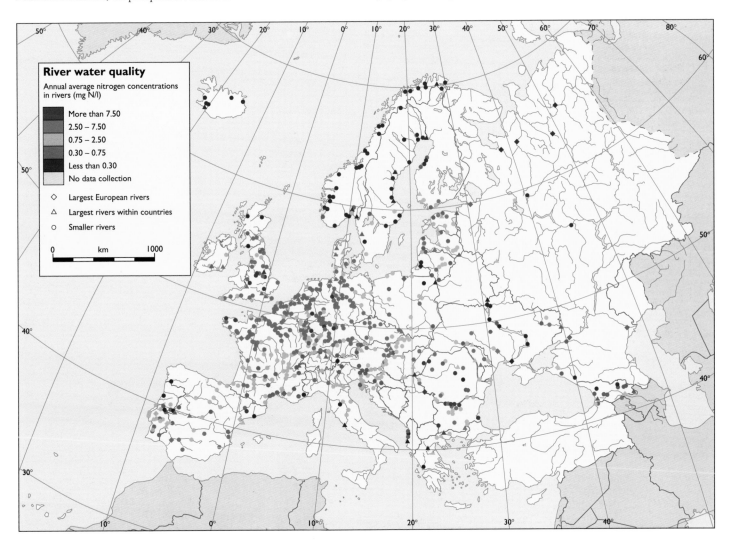

River water quality
Annual average nitrogen concentrations in rivers (mg N/l)

- More than 7.50
- 2.50 – 7.50
- 0.75 – 2.50
- 0.30 – 0.75
- Less than 0.30
- No data collection

◇ Largest European rivers
△ Largest rivers within countries
○ Smaller rivers

0   km   1000

Site	Country†	Number of lakes	Total phosphorus µg P/l	Nitrate µg N/l	Ammonium µg N/l
Sierra Nevada[1]	ES	10	3.9*	6	–
Pyrenees[2]	ES	102	~15	~139	~25
Tatra Mountains[3]	SK	10	5.7	400	14
Northern Apennines[4]	IT	43	14	112	28
Southern Alps[5]	IT	50	19	327	42
Italian Alps[6]	IT	320	<10 (85%)	14	18
Italian Alps, Aosta Valley[7]	IT	100	–	140	–
Sweden, reference-lakes[8]	SE	154	<15 (80%)	<750# (87%)	–
Northern Sweden, forest lakes[9]	SE	59	13.2	350	–
Finland, forest lakes[10]	FI	135	10	–	–
Black Forest Lakes[11]	DE	6	–	429	–

Notes: Values shown are mean values or percentages with concentrations below a specified limit.
* dissolved orthophosphate   # total nitrogen   † for country codes see Table 5.9.

Map 5.19
Frequency distribution of summer mean total phosphorus concentrations in European lakes by country, 1988–91
Source: Compiled by NERI and EEA-TF from multiple sources listed in Box 5E or referred to in the main text

Table 5.16
Phosphorus and nitrogen levels in lakes in relatively unpolluted areas such as remote mountain regions and national parks
Sources: 1 Morales-Baquero et al, 1992; 2 Catalan et al, 1992; 3 Fott et al, 1992; 4 Viaroli et al, 1992; 5 Mosello, 1981; 6 Mosello, 1984; 7 Mosello et al, 1991; 8 SCB Sweden, 1990; 9 Borg, 1987; 10 Heitto, 1990; 11 Schwoerbel, 1989

of the photosynthesis pigment, chlorophyll) increases in lakes with increasing phosphorus levels (eg, OECD, 1982; Canfield and Bachman, 1981; Berge, 1987; Jeppesen et al, 1991). The increase in phytoplankton concentration reduces water transparency and at high nutrient levels the lake water becomes very turbid.

In lakes with summer mean total phosphorus levels lower than 10 µg P/l, chlorophyll concentrations are low and the water is clear, that is, water transparency is high (Figure 5.21). With increasing phosphorus concentrations chlorophyll levels rise and water transparency declines. Thus in lakes with phosphorus levels between 50 and 125 µg P/l, the chlorophyll level ranges from 10 to 50 µg/l, and water transparency is reduced to 1 to 2 m. At phosphorus concentrations exceeding 125 µg P/l the chlorophyll concentration may reach very high levels, and water transparency becomes reduced to less than 1 m. However, in lakes with extremely high phosphorus concentrations – above 500 µg P/l – the increase in chlorophyll levels out, the biological structure becomes very unstable, and fish kill may occur (Søndergaard et al, 1990).

Blue-green algae and phosphorus
In lakes with high phosphorus levels excessive growth of blue-green algae may lead to surface scum and the production of toxins potentially poisonous to fish, cattle, dogs and humans. Data from 300 shallow Danish lakes indicate that blue-green algae are of only minor importance in the summer at phosphorus levels below 50 µg P/l, but become increasingly important at higher phosphorus levels (Figure 5.22), and become the dominant phytoplankton group at phosphorus concentrations from 200 to 1000 µg P/l. In deeper lakes, however, blue-green algae tend to dominate at even lower phosphorus levels, and the concentration generally has to be below 20 µg P/l for blue-green algae not to become dominant in the summer (Sas, 1989).

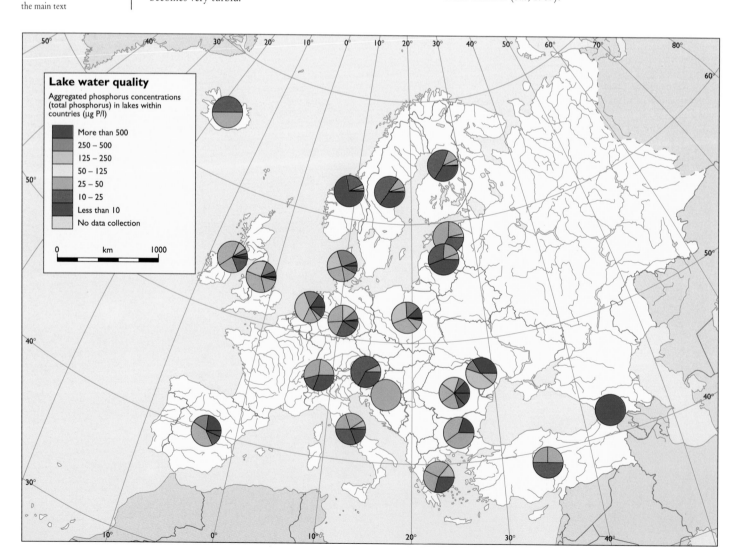

Lake water quality
Aggregated phosphorus concentrations (total phosphorus) in lakes within countries (µg P/l)

More than 500
250 – 500
125 – 250
50 – 125
25 – 50
10 – 25
Less than 10
No data collection

0   km   1000

Trends in nutrient concentrations

Time series data for the period 1977–90 concerning nutrient concentrations in more than 250 European rivers have been collated for this report. Most of the time series cover the whole period but, especially in the case of Southern European rivers, some of the time series cover only the last five to eight years.

Phosphorus trends

The increase in phosphorus levels that occurred in many lakes during the 1960s and 1970s caused severe eutrophication problems and led several countries to take measures to reduce phosphorus discharge to rivers and lakes. This was achieved primarily by reducing discharge from point sources, especially wastewater, by improving wastewater treatment. In some countries part of the reduction was achieved by reducing or banning the phosphorus content of detergents.

The reduction in phosphorus loading is reflected by a marked decrease in the phosphorus levels of many European rivers (Figure 5.23); the phosphorus concentration in the majority of rivers (64 per cent) having decreased between the periods 1977–82 and 1988–90. In one third of the rivers, the reduction was greater than 25 per cent.

While the phosphorus concentration decreased in the majority of rivers in the Nordic countries and in Western and Southern Europe, it rose in many Eastern European rivers, largely because of very limited construction in the Eastern European countries of sewage plants incorporating phosphorus removal. The reduction in phosphorus levels was most marked in Western Europe, where a reduction of more than 25 per cent was observed in half of the rivers. Nevertheless, the highest phosphorus levels are still to be found in the rivers of Western Europe (Figure 5.19).

A reduction in phosphorus loading and concentration has also been observed in several European lakes during the last 10 to 20 years (Table 5.17). During the 1960s and early 1970s an increase in phosphorus concentration caused environmental deterioration in many of the large lakes in the Alps (Figure 5.24). Action plans were consequently implemented in the 1970s with the objective of reducing phosphorus loading. As a result there was a marked reduction in phosphorus levels during the 1980s. In the two largest alpine lakes, Lac Léman and Bodensee, the phosphorus level has been reduced from about 90 μg P/l to 60 and 50 μg P/l, respectively (Stabel, 1991; Lang, 1991). However, these concentrations are still high, and little improvement has been observed in the water quality (Tilzer et al, 1991; Lang, 1991).

In Denmark there has been a general reduction in the phosphorus level of many lakes (Figure 5.25a), but without any significant improvement in water transparency (Figure 5.25b); this discrepancy may be explained by the fact that the phosphorus level still exceeds 125 μg P/l in 70 per cent of the lakes, and so water transparency is therefore generally less than 1 m (Figure 5.22).

Nitrogen trends

Nitrate levels have tended to increase in many European rivers over the last 20 to 40 years (Figure 5.26), and the same applies to European lakes (Barbieri and Mosello, 1992; Tilzer et al, 1991; Mosello, 1989; Henriksen et al, 1988). The increase is attributable mainly to a corresponding increase in the use of nitrogen fertilisers in most European countries (see Chapter 22). However, the increase in lake nitrate levels in areas with relatively low human activity, for example Norway,

Map 5.20
Total phosphorus concentrations in large European lakes and reservoirs, 1988–91
Source: Compiled by NERI and EEA-TF from multiple sources listed in Box 5E or referred to in the main text

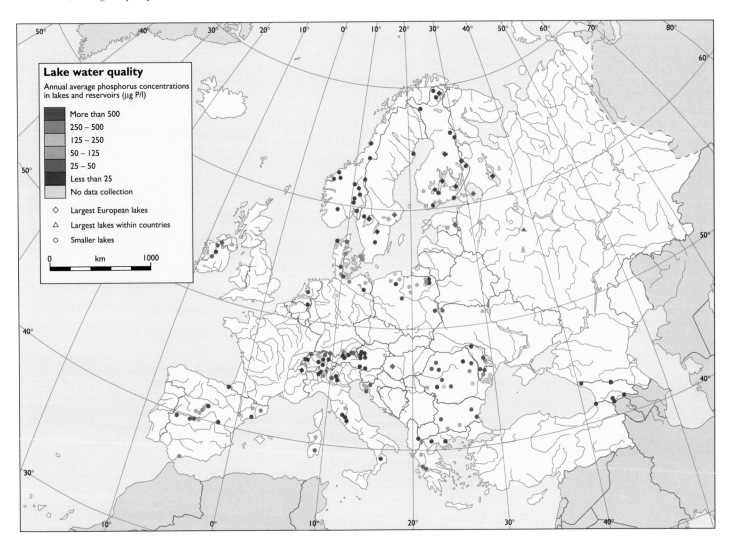

Figure 5.21
Relationship between lake total phosphorus concentration and chlorophyll and water transparency
Source: Compiled by NERI and EEA-TF from multiple sources listed in Box 5E or referred to in the main text

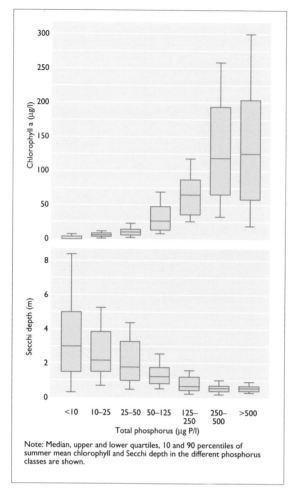

Note: Median, upper and lower quartiles, 10 and 90 percentiles of summer mean chlorophyll and Secchi depth in the different phosphorus classes are shown.

Southern European countries, respectively. The increase in nitrate concentration was most marked in Eastern and Southern European rivers, because of the marked increase in the use of nitrogen fertilisers during the same period. In many Western European countries, in contrast, the increase in the use of nitrogen fertilisers peaked in the late 1970s and remained at that high level during the 1980s (see Chapter 22).

The ammonium level in 70 per cent of 230 European rivers decreased markedly between 1977–82 and 1988–90 (Figure 5.28). The decline was most significant in rivers in which the ammonium level was between 0.5 and 1 mg N/l in the period 1977–82, the ammonium level falling to below 0.5 mg N/l in most of these rivers. The decreasing trend in ammonium levels is in concert with the decline in organic matter content of the water (Figure 5.17), and can be explained by improved wastewater treatment and the consequent reduction in organic matter discharge. The reduction in ammonium levels was most significant in the Western

Figure 5.23 Percentage change in annual mean river total phosphorus concentration from the period 1977–82 to 1988–90
Source: Compiled by NERI and EEA-TF from multiple sources listed in Box 5E or referred to in the main text

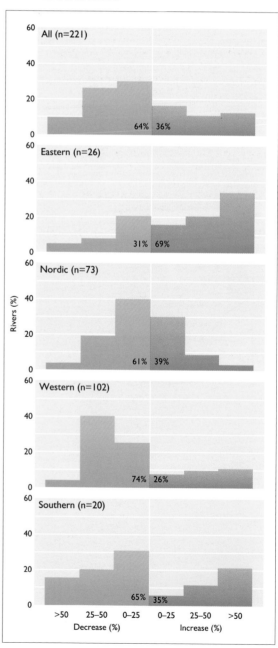

is attributable mainly to increased atmospheric deposition of nitrogen caused by the combustion of fossil fuels and agricultural activity (Henriksen et al, 1988). A decrease in the ammonium level in European rivers is expected as a consequence of general improvement in wastewater treatment and a growing awareness of the problem of agricultural pollution with manure and silage.

Nitrate levels increased between 1977–82 and 1988–90 in more than two thirds of European rivers, the median concentration increase being 0.14 mg N/l, or 13 per cent (Figure 5.27). The percentage of rivers in which the concentration increased was nearly identical in the four regions of Europe, but the median concentration increase varied, being 0.013 mg N/l in Nordic rivers and 0.18, 0.49, and 0.35 mg N/l, in the rivers of Western, Eastern and

Figure 5.22
Relationship between total phosphorus concentration (May–September) and blue-green algae dominance in 300 shallow Danish lakes
Source: Redrawn from Jeppesen et al, 1991

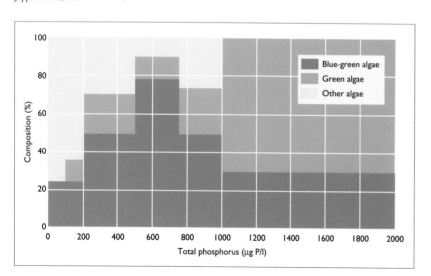

Country[*]	Lake(s)	Period	Reduction in in-lake phosphorus level (%)
Europe[1,2]	9 deep lakes	Mid-1970s to late 80s	51
Europe[1,2]	9 shallow lakes	Mid-1970s to late 80s	63
DK[3]	46 lakes	1972–79 to 1985–89	33
SE[4]	Vättern	1962 to 86	57
CH[5]	Zürichsee	1979 to 88	31
CH[5]	Walensee	1979 to 88	92
AT, CH, DE[5,6]	Bodensee	1979 to 89	45
CH, FR[7]	Lac Léman	1976 to 89	36
CH, IT[8]	Lago di Lugano	1984 to 90	32
IT[9]	Lago Maggiore	1977 to 88	46
AT[10]	Mondsee	1979 to 89	70
NO[11]	Mjøsa	Early 1970s to 1991	39

1 Sas, 1989; 2 Seip et al, 1990; 3 Kronvang et al, 1993; 4 Persson et al, 1989; 5 Stabel, 1991; 6 Tilzer et al, 1991; 7 Lang, 1991; 8 Barbieri and Mosello, 1992; 9 Mosello, 1989; 10 Dokulil and Jagsch, 1992; 11 NIVA, 1992
* See Table 5.9 for country codes.

Table 5.17 Examples of European lakes in which phosphorus concentration has decreased during the last 10 to 20 years

European and Nordic countries, while the concentration tended to increase in many Eastern European rivers.

Acidification of surface waters

Surface water acidification has been of public concern since the early 1970s, when awareness of the problem was aroused

Figure 5.24 Trends in the phosphorus concentration of seven European lakes (a and b different scales)
Sources: Walensee, Zürichsee and Bodensee after Stabel, 1991; Mondsee after Dokulil and Jagsch, 1992; Vänern after Olsson, 1991; Lac Léman after Lang, 1991; Lake Lugano after Barbieri and Mosello, 1992

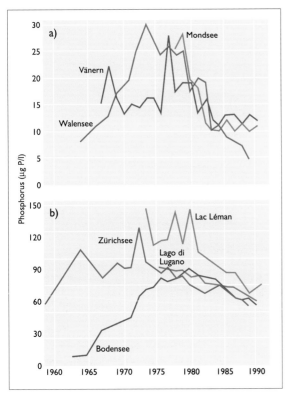

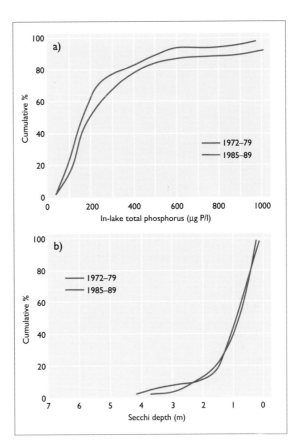

Figure 5.25
a) The cumulative percentage of in-lake total phosphorus concentration in 46 Danish lakes during the periods 1972–79 and 1985–89;
b) the cumulative percentage of summer water transparency (Secchi depth) in 46 Danish lakes during the periods 1972–79 and 1985–89
Source: Redrawn from Kronvang et al, 1993

by episodes of severe fish kill in rivers and lakes in the southernmost part of Norway, and along the west coast of Sweden (Overrein et al, 1980; Monitor 12, 1991). These episodes of fish kill were attributed to 'acid rain', that is, precipitation acidified by the combustion products of fossil fuels. Norwegian and Swedish observations have indicated that the emission of acidifying gaseous by-products such as sulphur dioxide and nitrogen oxides was probably the main cause of acidic precipitation in areas situated up to hundreds of kilometres from the source.

Studies were initiated to evaluate the extent, the cause/effect relationships, and the chemical and biological impact of surface water acidification in Norway and Sweden (Overrein et al, 1980; National Swedish Environment Protection Board, 1983). It was found that, of a total of 5000 lakes in a 28 000 km[2] region of southern Norway, 35 per cent had lost their fish populations. In southern and central Sweden, damage to fish communities was observed in 2500 lakes; however, on the basis of lake pH, and from the knowledge that many fish species are unable to tolerate pH levels below 5.5, it was estimated that the fish populations of about 18 000 lakes were affected.

Surface water acidification and damage to freshwater life forms have since been documented in numerous studies (eg Muniz, 1991). Over the last two decades, surface water acidification as a consequence of the atmospheric emission of sulphur dioxide and nitrogen oxides has been recognised as a serious environmental problem in most other European countries, as well as in North America (Merilehto et al, 1988; Howells, 1990; Skjelkvåle and Wright, 1990). In the mid-1980s a further acidifying air pollutant was identified. Laboratory studies in The Netherlands documented that the deposition of ammonia volatilised from agricultural areas with large concentrations of livestock could be a major cause of acidification in softwater lakes (Schuurkes et al, 1986; van Dam, 1987; van Breemen and van Dijk, 1988).

The acidifying effect of atmospheric deposition has been documented from studies in areas such as nature reserves and remote mountain lakes (Mosello et al, 1992b), where the only possible influence of human activity is through atmospheric

Figure 5.26
Trends in the nitrate concentration of European rivers
Sources: Eight English rivers from José, 1989; River Rhine from van der Weijden and Middelburg, 1989; six Czech rivers from Czechoslovakia, 1992; six Northern Irish rivers from Smith et al, 1982; five Hungarian rivers from Hock and Somlyódy, 1989; three Danish rivers from Kristensen et al, 1990

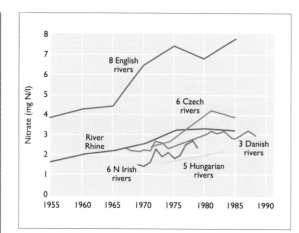

deposition. When considering areas with industrial and agricultural activity, several factors that can positively or negatively affect surface water acidity have to be taken into account: for example mining, afforestation, clear cutting, draining, crop removal, fertilisation and liming. Internal processes in lakes (eg, reduction and oxidation processes) may also influence their acid/base status. These various factors are described in standard textbooks, for example Cresser and Edwards (1987); Howells (1990); Monitor 12 (1991).

Acidification of surface waters in Europe

Surface water acidification can be expected in areas where acidic deposition is high and the catchment soil or bedrock is poor in lime and other easily weatherable minerals that buffer against acid precipitation. Small high altitude lakes and streams are generally affected more severely than larger lowland surface waters, the latter usually being acidified only in well leached, sandy areas covered by heath or forest. Dune lakes and pools may also be acidified. Even naturally acidic bog lakes (eg, in the Campine region of Belgium) may be further acidified by acid precipitation, thereby causing a change in their ecology.

Geological characteristics render large areas of Europe sensitive to surface water acidification if exposed to acidic atmospheric deposition (Map 5.21). The areas of Europe where surface water acidification has actually been observed are shown in Map 5.22 and the current state of surface water acidification in Europe is briefly summarised below. A more comprehensive summary can be found in Merilehto et al (1988), Howells (1990), and Skjelkvåle and Wright (1990).

Figure 5.27 (left)
Percentage change in annual mean river nitrate concentration between the periods 1977–82 and 1988–90
Source: Compiled by NERI and EEA-TF from multiple sources listed in Box 5E or referred to in the main text

Figure 5.28 (right)
Percentage change in annual mean river ammonium concentration between the periods 1977–82 and 1988–90
Source: Compiled by NERI and EEA-TF from multiple sources listed in Box 5E or referred to in the main text

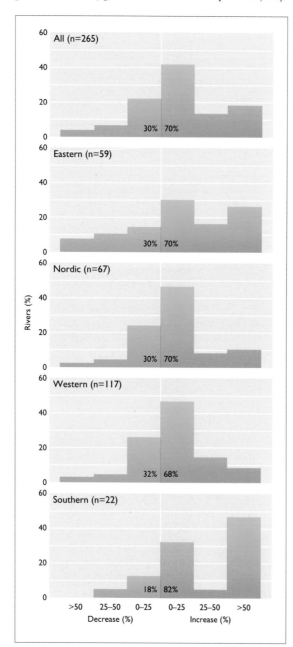

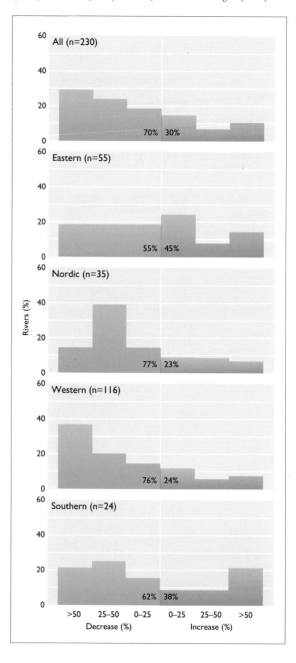

In southern Finland, Sweden and Norway, lake and river acidification increased slowly from 1915 to 1950 and rapidly from 1950 to 1980. Since then conditions have remained relatively unchanged, possibly as a result of a reduction in acidic deposition (Brodin and Kuylenstierna, 1992). In southern Norway lake sulphate concentration tended to decrease between 1974 and 1986 in line with the decrease in sulphur emission in Western Europe. Surprisingly, however, this was not accompanied by a decrease in lake acidity (Norwegian State Pollution Control Authority, 1987). The reason could be the two-fold increase in nitrate concentration that occurred during the same period as a result of an increase in the emission of nitrogen oxides. According to a Swedish survey undertaken in the winter of 1990, 45 per cent of the 85 000 lakes in Sweden had a winter pH below 6.0, and 7 per cent had a pH below 5.0 (Monitor 12, 1991). Most of the acidic lakes were small, however: 23 per cent of lakes smaller than 0.1 km², but only 2 per cent of lakes larger than 1 km², had a pH below 5.5 (Monitor 12, 1991). In Finland about 10 per cent of 1000 lakes were found to be seriously affected (Merilehto et al, 1988). Lake acidification has also been found in the Russian Federation in Karelia and in the Kola Peninsula. As surface waters in Karelia have a high humus content, mean pH is 6 to 7 (Merilehto et al, 1988).

Most surface waters in Western and Central Europe are not as seriously affected by acidification as those in Finland, Sweden and Norway, despite the fact that acidic deposition is greater; this is because the soil of Western and Central Europe is generally well buffered, and therefore able to neutralise the acid deposited. In the UK acidified surface waters are found in various regions, but primarily in areas dominated by acidic soils, especially Scotland, northern England and Wales (Howells, 1990). In parts of Scotland (Battarbee, 1989) and Wales (Jenkins et al, 1990) afforestation of areas heavily loaded by sulphur oxides has exacerbated river acidification, while in more pristine areas, such as the northernmost part of Scotland, acidification has not occurred despite an increase in seasonal input through canopy interception. In Ireland there is little evidence of surface water acidification attributable to acidic precipitation. However, rivers draining evergreen afforested catchments of soil poor in buffering capacity were found to be more acidic when the catchment was covered with evergreen forest than when not, the increased acidity being attributable to the scavenging of acidifying air pollutants by the canopies of the evergreen trees, and therefore affected by acid rain (Bowman, 1991). In The Netherlands 50 to 60 per cent of about 5000 lakes are acidic (Leuven et al, 1986). Furthermore, of 187 randomly selected softwaters, mainly small isolated moorland and dune pools and lakes, 35 per cent were extremely acid with pH below 4.0. Although the main cause of this acidification is believed to be deposition of acidifying air pollutants, oxidation of sulphides is a contributing factor under certain geological conditions.

Surface water acidification has also been recorded in Belgium, Denmark, Germany, Poland, the Czech Republic, the Slovak Republic, Austria, France, Switzerland and Italy (Merilehto et al, 1988). Acidification of high altitude lakes has been reported at a workshop in 1991 on remote mountain lakes (Mosello et al, 1992b). Studies of such lakes are important because many of them are acid-sensitive and, since they are situated far from human settlements, atmospheric deposition is the main source of pollution (Wathne, 1992). Acidification has been recorded in mountain lakes in Italy

Map 5.21
Areas of Europe geologically sensitive to surface water acidification
Sources: Redrawn from Skjelkvåle and Wright, 1990; Merilehto et al, 1988

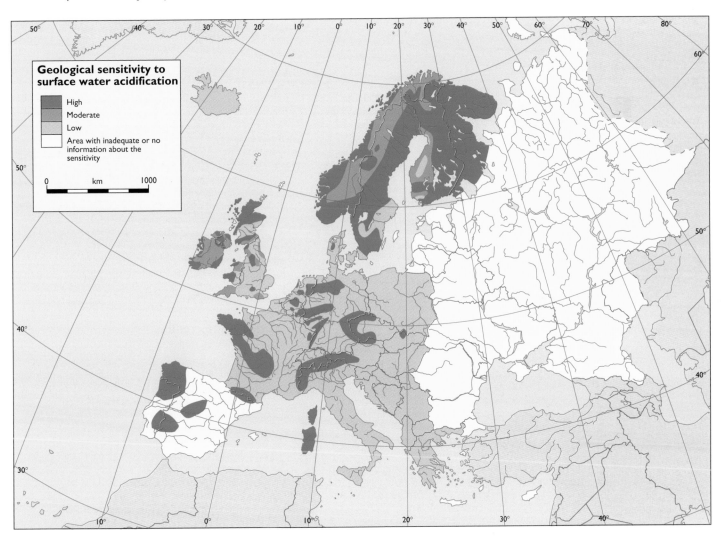

Geological sensitivity to surface water acidification

High
Moderate
Low
Area with inadequate or no information about the sensitivity

0   km   1000

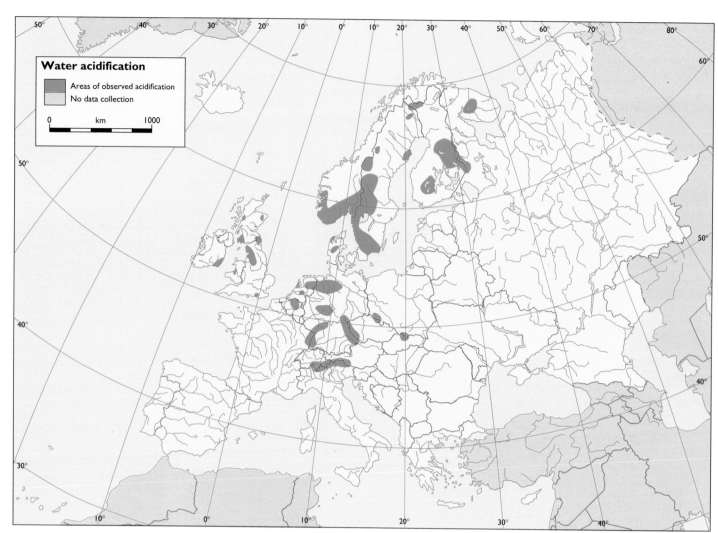

Map 5.22
Areas of Europe
where surface water
acidification has been
observed
Sources: Redrawn from
Skjelkvåle and Wright,
1990; Merilehto et al, 1988

(Mosello et al, 1992a; Schmidt and Psenner, 1992), the Slovak
Republic (Fott et al, 1992) and the Czech Republic (Vesely
and Majer, 1992). In contrast, no acidified lakes have been
recorded in the Pyrenees (Catalan and Camarero, 1992) and
in the northern Apennines (Viaroli et al, 1992). Acidification
of lakes and rivers in forested areas has been recorded in the
Vosges Mountains, the Schwarzwald, the Bavarian Forest,
Thüringerwald and Erzgebirge. Acidification of small,
isolated seepage lakes has been recorded in lowland areas of
The Netherlands, Belgium, Denmark and northern Germany,
mainly in regions dominated by sandy soils of low buffering
capacity (Merilehto et al, 1988).

Impact of acidification

Acidification affects aquatic ecosystems at all levels and has a
profound impact on both plant and animal communities.
Aquatic organisms are influenced both directly, because of
the resulting toxic conditions, and indirectly, because of the
loss of suitable, acid-sensitive prey. One of the first recorded
effects of acidification caused by human activity was a
decrease in fish diversity in the Nordic countries (Overrein et
al, 1980; Drabløs and Tollan, 1980).

Aquatic organisms have extremely different tolerance
levels towards acidification, as illustrated for various
species/groups of benthic invertebrates and fish in Figure
5.29. The toxicity of acid freshwaters towards fish is only
partly attributable to the low pH. In most cases, the crucial
toxic factor is the content of inorganic aluminium ions that
have been leached from minerals in the catchment by acid
deposition and subsequently transported to the lakes and
streams (Drabløs and Tollan, 1980; Muniz, 1983).

That acidification can severely affect aquatic

ecosystems is especially well documented in Sweden and
Norway. For example, during 1940–75, 1750 of 5000 lakes in
southern Norway became completely devoid of fish as a
result of acidification, with another 900 lakes being seriously
affected (Overrein et al, 1980). Recent Norwegian results
(T. Hesthagen, NINA, personal communication) indicate
serious and ongoing damage. Of 13 600 surveyed fish stocks
in lakes in southern Norway, 2600 had become extinct and
3000 had reduced their number of individuals. In rivers in
southern Norway, Atlantic salmon are badly affected. In
25 rivers, the salmon is now virtually extinct due to
acidification (Hesthagen and Hansen, 1991). The total
annual number of adult salmon lost was estimated to be
between 92 000 and 305 000 individuals weighing some 345
to 1150 tonnes. Fish kill on such a scale is detrimental to
both commercial fishery and recreation and quality of life.

Acidification trends

There is good evidence that many lakes have become acidified
during the past 100 years; this has been documented both
chemically and by palaeolimnological studies of pH-indicator
species such as diatoms in lake sediment cores (Charles et al,
1990; Howells, 1990). Such studies indicate, for example, that
whereas the pH of lake Gårdsjön in Sweden fell slowly from
pH 7 to 6 over the 12 500 years from the end of the last ice
age until 1960 by natural acidification processes, it has
subsequently fallen rapidly from pH 6 to 4.5 (Figure 5.30).
The pH values determined in this way agree reasonably well
with lake water chemistry, the measured pH of lake Gårdsjön
having been 6.3 in 1948, and 4.7 in 1977 (Wright, 1977). A
trend towards acidification has also been recorded in
freshwaters in Finland, Sweden, Norway, the UK, The

Netherlands, Belgium, Denmark and Germany (Charles et al, 1990; Monitor 12, 1991).

Critical loads

As a result of international cooperation on reduction of emissions of sulphur and nitrogen oxides (see Chapter 31) the concept of 'critical load' emerged in order to establish reduction strategies for emissions of these substances, which, in contrast to the preceding strategy, should be based on scientific knowledge of the capacity of different ecosystems and sites to withstand various types of acid deposition.

Since the early 1980s both 'target load' (ie, a political goal for loading reduction) and 'critical load' have been extensively explored. Since 1988 the most widely used definition of the critical load for surface waters has been: 'A quantitative estimate of the loading of one or more pollutants below which significant harmful effects on specified sensitive elements of the environment are not likely to occur according to present knowledge' (Kämäri et al, 1992).

Operational criteria for assessing the effects of acid deposition on surface waters (rivers and lakes) have, for example, been developed in Scandinavia. These criteria are based upon selected organisms and populations (fish and invertebrates) which are sensitive to chemical changes in the water as a result of changes in atmospheric inputs of acidifying substances (Figure 5.31).

The emission reduction needed to reach critical loads for nitrogen and sulphur is estimated to be as high as 70 to 80 per cent for southern Sweden and southern Norway, and even higher for other European countries (Brodin and Kuylenstierna, 1992).

The critical load concept appears to be a rational method for quantifying how much acidic deposition and hence emission must be reduced in order to protect sensitive sites or ecosystems. Furthermore, the concept has now found acceptance in political negotiations concerning reduction strategies. European critical load maps have been prepared by the Coordination Centre for Effects (Hettelingh et al, 1991), and Kämäri et al (1992) have described in more detail how the critical load concept should be employed.

Figure 5.29
Acid tolerance of various benthic invertebrates and fish
Source: Redrawn from Monitor 12, 1991

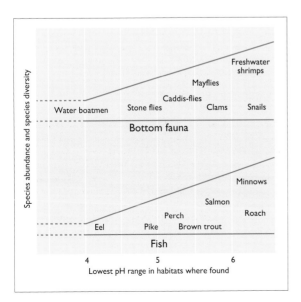

Heavy metals, organic micropollutants, radioactivity and other hazards

Human activity leads to chemical compounds being discharged into the aquatic environment in various ways, for example in domestic sewage water and industrial effluent, through mining and through atmospheric deposition. While some of these compounds are known to be toxic, the environmental impact of many others remains to be elucidated. Nevertheless, these contaminants together represent a threat both to the aquatic ecosystem *per se*, and to human health – a threat that is intensified by the vast number of new synthetic organic compounds being produced and released into the environment.

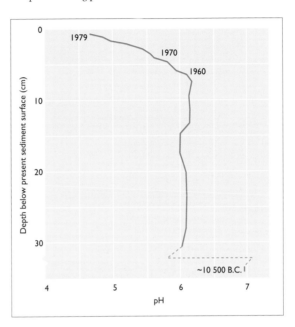

Figure 5.30
pH in lake Gårdsjön, Sweden, calculated from the sediment subfossil diatom flora
Source: Redrawn from Renberg and Wallin, 1985

Heavy metals

Production and use of heavy metals has increased markedly in Europe during the 19th and 20th centuries. As a result heavy metal contamination of inland surface waters has also increased, the main sources being mining and industrial activities. Other important sources are sewage discharge, runoff from the land and atmospheric deposition. In the 1970s surface water concentrations of heavy metal levels reached alarming proportions and national and international regulations (eg, Council Directive 76/464/EEC and daughter directives on pollution caused by certain dangerous substances discharged into the aquatic environment of the Community) were therefore implemented to control heavy metal release at the source. Although this has led to reduced levels of harmful metals in many Western European rivers during the last decade (RIVM, 1992b; UK DoE, 1992; Umweltbundesamt, 1992), the level is still high in some European rivers.

General assessment of the state of heavy metal pollution of European rivers and lakes is difficult, primarily because measurement of metals is rarely included in monitoring programmes, but also because concentration levels are usually so low that problems arise with sample preparation and methodological precision. Until recently metals have generally been analysed either as total water concentration or on the dissolved fraction. Recent evidence, however, suggests that anthropogenic inputs generally can be better evaluated from particulate associated metals (Meybeck et al, 1989). Comparison and assessment of the state of heavy metals in European rivers is therefore even more difficult than for most of the other water quality variables. The following evaluation of heavy metals in European rivers must be looked at with these reservations

Figure 5.31
Critical loads for fish
Source: After Henriksen
et al, 1990

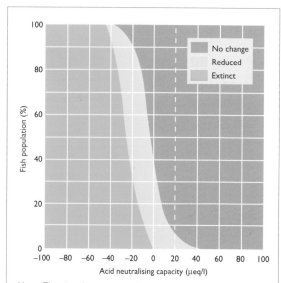

Notes: There is a close connection between the concentration of ANC (acid neutralising capacity) in a lake and damage to fish populations. The ANC is the difference between the sum of base cations (calcium, magnesium, sodium, potassium) and strong acid anions (sulphate, nitrate, chloride). At an ANC above 20 μeq/l there is little probability of damage to fish. Below −40 μeq/l, fish cannot survive.

Table 5.18
Descriptive statistics of annual mean concentrations of total heavy metals in European rivers
Source: Compiled by NERI and EEA-TF from multiple sources listed in Box 5E or referred to in the main text

section has led to increased mobilisation of metals from the soil (see Chapter 7), and hence to elevated levels in the aquatic environment; thus increased levels of aluminium in surface waters and elevated levels of mercury in fish have been found in areas of Europe known to be acidified (Henriksen et al, 1989; Andersson et al, 1987).

The problems associated with heavy metal contamination of inland surface waters are illustrated below using cadmium, copper and zinc as examples.

Cadmium

Cadmium ranks among the most hazardous metal pollutants, cadmium-induced Itai-Itai disease in the Japanese Jinzu River basin being a frightening reminder of its toxicity (Nogawa, 1981). The cadmium concentration in natural waters unaffected by humans is generally less than 1 μg Cd/l (Meybeck et al, 1989). Although cadmium levels in European rivers rarely exceed 1 μg Cd/l, there are a few rivers with high cadmium concentrations, especially the Iskar in Bulgaria, the rivers Guadiana and Tajo in Portugal, and the Danube in Romania. The cadmium that is discharged into inland surface waters accumulates in the sediment; as a result, the sediment of several European rivers and lakes is contaminated (Förstner, 1980).

A study of the cadmium cycle in the Rhine basin showed that the most important sources of cadmium pollution during the period 1983–87 were industrial point sources and runoff from the land (RIVM/GLOBE, 1992). Emission reduction was implemented in the 1980s, especially at point sources, and as a result cadmium levels are presently declining; however, the levels are still high, and additional measures may be needed to reduce cadmium concentrations further.

Copper

Because of its widespread use, copper is both an actual and a potential pollutant of the aquatic environment. In catchments with no human activity the river copper concentration is generally lower than 2 to 5 μg Cu/l (Nriagu, 1979; Hodson et al, 1979), a level well below the drinking water standard of 100 to 3000 μg Cu/l. Such a high concentration is rarely found, and then primarily in connection with mining. It is well known that copper is highly toxic to aquatic biota – copper compounds are often used as insecticides, fungicides, algicides and molluscides. Its toxicity is generally higher in water with a low mineral content than in water with high alkalinity and hardness. Although the Council Directive (78/659/EEC) on the quality of freshwaters needing protection or improvement in order to support fish life recommends that the copper level be below 40 μg Cu/l in water with a hardness of 100 mg CaCO₃/l, fish and invertebrates have been reported to be affected by copper levels as low as 10 to 20 μg Cu/l (Hodson et al, 1979). Since about 20 per cent of the river stations analysed in this report have copper levels exceeding 10 μg Cu/l, the aquatic biota may be affected. However, the copper levels rarely exceed 40 μg Cu/l.

Zinc

Like copper, zinc is also widely used. It is produced in 12 European countries, with what is now the Russian Federation being the single most important producer (Cammarota, 1980). There are also, for instance, at least 140 zinc mining sites in Belgium, England, Wales, France, Ireland and northern Spain (Whitton, 1980). In addition, mining of many other metals also results in zinc being released into the environment. It can thus be assumed that several European rivers are being polluted by zinc.

In natural water unaffected by human activity the zinc concentration is generally below 5 μg Zn/l. Although human tolerance to zinc is generally high, zinc may be rather

in mind. It has been possible in this report to collate data on total heavy metal concentrations in more than 200 European rivers (Table 5.18). In general, the concentrations are well below standards for drinking water; only cadmium and mercury exceed drinking water standards in some rivers. It should be remembered, however, that the aquatic biota are usually affected by much lower concentrations: thus while the drinking water standard for copper is 100 to 3000 μg Cu/l, salmonoid fish are affected at 10 to 50 μg Cu/l (Hodson et al, 1979). Statistical descriptive variables – the median and 10 and 90 percentiles – of heavy metals in European rivers were found to be nearly identical to those for a river dataset covering the whole world (Meybeck et al, 1989). Heavy metal levels were generally lowest in the rivers of the sparsely populated Nordic countries, while those in the rivers of Western, Eastern, and Southern European countries were roughly the same.

High concentrations of heavy metals were recorded at some of the river stations, in particular in the upstream reaches of the Wisla, Oder and Elbe (where the majority of the Polish and Czech heavy industry is located), the Danube and many of its tributaries, and some of the rivers of Ukraine. These heavy metal 'hot spots', which are generally located near mining areas and industries using large quantities of metals, are only examples: a more extensive survey would undoubtedly reveal many more locations with high levels of many metals.

The increase in acid deposition described in the previous

	Number of river stations	Drinking water standard (μg/l)	Percentage of river stations with concentrations not exceeding (μg/l)			
			25%	50%	75%	90%
Copper	192	100-3000**	1.6	4.8	8.0	16.0
Zinc	176	100-5000**	5.0	10.4	36.0	91.0
Cadmium	145	5*	0.0	0.0	0.4	1.8
Mercury	163	1*	0.0	0.0	0.2	1.1
Lead	72	50*	0.8	2.7	6.7	11.0
Chromium	56	50*	1.3	4.3	11.5	17.0
Nickel	48	50*	0.8	4.3	11.5	17.0

Note: * Maximum Admissible Concentration, and
** Guide level as specified in Council Directive (80/778/EEC).

poisonous to the aquatic biota. The Council Directive (78/659/EEC) concerning water standards for fish recommends that zinc levels be below 300 μg Zn/l in water with a hardness of 100 mg CaCO$_3$/l but below 30 μg Zn/l in water with a hardness of 10 mg CaCO$_3$/l. The annual mean zinc concentration exceeded 35 μg Zn/l at 25 per cent of the river stations analysed in this report.

Organic micropollutants

Some organic micropollutants, for example the pesticide DDT and polychlorinated biphenyls (PCBs), are well known, as are their toxic effects. Others are suspected of having effects of various types, including carcinogenic effects, while the environmental effects of others remain to be elucidated (see Chapter 17).

Traditional monitoring for organic micropollutants in the aquatic environment has focused on organochlorine compounds (eg, DDT, PCBs), and polycyclic aromatic hydrocarbons (PAHs). However, organic micropollutants have rarely been measured on a large-scale basis in national monitoring programmes. In recent years there has been a growing awareness of the problem of aquatic pollution by other dangerous compounds, such as non-organochlorine pesticides. A number of surveys aimed at defining the magnitude of the problem have therefore been undertaken. Nevertheless, it is extremely difficult to quantify the risk presented by organic micropollutants, firstly because the biological effects of most of them are poorly known, and secondly, most of them occur at levels too low to be analytically determined. In addition, the behaviour of many of the compounds in freshwater – their adsorption, degradation, temporal and spatial variability, as well as their bioaccumulation and combined biological effects – are virtually unknown.

Although some of the very toxic organic micropollutants have been either restricted or banned in several European countries during the last 20 to 30 years, they remain in use in other parts of Europe. In the late 1960s and early 1970s several European countries banned the use of DDT; as a result, there has been a subsequent marked reduction in DDT levels in their surface waters (Olsson and Reutergårdh, 1986; RIVM, 1992b). Similarly, since the use of PCBs was restricted in several European countries about 15 years ago, PCB levels in their surface waters have also declined (Olsson and Reutergårdh, 1986; RIVM 1992b). High levels of PCBs are still found in some rivers, however (Chevreuil et al, 1987).

Micropollutants in inland surface waters originate primarily from industrial and urban activity and from agriculture. Only a few European studies of the pollution sources, their flow pattern and their environmental impact exist. Galassi et al (1992) studied the toxic effect of organic micropollutants on the zooplankton *Daphnia magna* in the large river Po, in Italy, and found that toxicity was highest in May, this being related to the use of pesticides in the river catchment area. The European aquatic environment is currently exposed to growing numbers and quantities of pesticides as a consequence of the marked increase in the use of pesticides during the last three decades (see Chapter 22). The level of pesticide is particularly high in rivers and lakes located in intensively cultivated areas with crops requiring large quantities of water soluble pesticides together with high surface runoff during the pesticide spraying season. Assessment of the environmental risk of pesticides must be based on a combination of studies of their impact on the aquatic biota and model calculations of potential pesticide runoff.

In the EU a long-term strategy has been adopted (Council Regulation 93/793/EEC) to control the environmental risk from chemicals (see Chapter 38). This, among others, involves testing the ecotoxicity of a list of priority chemicals, including some pesticides.

Accidents and leakage from waste disposal sites

Of the many threats posed by organic micropollutants, industrial accidents and leakage from industrial or agricultural storehouses also need to be considered. For example, the well publicised fire in 1986 at a chemical warehouse in Basle, Switzerland, resulted in a significant discharge of organic micropollutants into the Rhine; the spill caused fish kill in more than 250 km of the river, and macroinvertebrates were affected more than 600 km downstream (Stumm, 1992).

The leakage of substances from the numerous European waste sites, for example landfills and military sites, is another potential source of organic micropollution of inland surface waters.

Radioactivity

Although most radiation stems from natural sources (background radiation), various human activities have increased the potential for contaminating the aquatic environment with radionuclides (see Chapter 16). Natural radionuclides in river and lake waters include: tritium (H-3) (0.02 to 0.1 Bq/l), potassium-40 (K-40) (0.04 to 2 Bq/l), radium, radon and their short-lived decay products (<0.4 to 2 Bq/l).

The effect of human activities can be to enhance these levels (such as through the production of phosphate fertilisers), or to introduce into inland waters artificial radionuclides (such as caesium-137 (Cs-137) and strontium-90 (Sr-90)) by the testing of nuclear weapons, the production of nuclear power (including the operation of the full nuclear fuel cycle) as well as through scientific and medical uses of radionuclides.

Nuclear power production and radioactive waste

At present there are more than 200 nuclear power plants in operation in Europe (see Chapter 19). They are usually built close to a major water source, either inland or coastal, the water being needed for cooling purposes. Besides a reduction in flow and thermal pollution of the water, authorised discharges of low-active effluents from the nuclear power plant will lead to low-level contamination by radionuclides.

The radioactivity levels (total alpha and residual beta – ie total beta less K-40 activity) at 35 locations in 11 large EU rivers which have been monitored since 1984 lie mainly between 0.04 and 0.4 Bq/l (CEC, 1989; in press). Sparse information about radioactivity in surface waters was obtained from the countries questioned for the present report. However, the information available indicates levels ranging from 0.01 to 0.8 Bq/l, the highest levels being found in rivers in Ukraine and the Czech Republic, and the lowest levels generally in northern Finland. In some rivers, eg, the Rhone, increased levels of radioactivity have been recorded in the lower reaches (Figure 5.32). This is attributable to discharge from nuclear installations. However, the radiation levels observed are well below those considered of any significance from the point of view of human health (CEC, 1989; in press). While it may be theoretically possible that aquatic biota could be affected in some areas, in practice there is no evidence of such effects.

Nuclear waste disposal sites are another potential source of contamination of inland surface waters. There have been several reports from Eastern European countries of careless handling of radioactive waste products leading to contamination of lakes and rivers.

In the southern Urals, an area of severe contamination is found between Chelyabinsk and Sverdlovsk resulting from the operation of the Mayak nuclear facility and the Kyshtym accident (see Chapters 16 and 18). Whilst operating this plant between 1949 and 1956, high levels of liquid waste were released into the Techa river (about 10^{17} Bq in total) causing severe contamination and radiation exposure (Burkart and Kellerer in WHO (in press)). The river is still largely fenced off, although the remaining radioactive contamination is substantially reduced.

Figure 5.32
Annual mean gross beta activity in the river Rhone
Source: Redrawn from CEC, in press

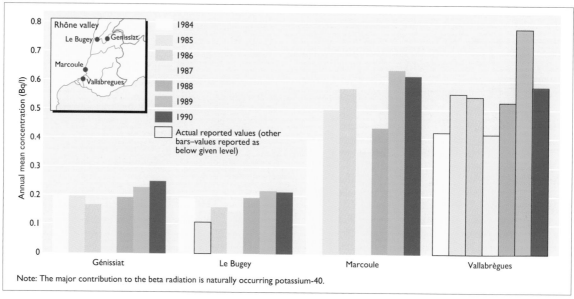

Note: The major contribution to the beta radiation is naturally occurring potassium-40.

Accidents

The Chernobyl accident in 1986 was the most severe nuclear accident to affect almost the whole of Europe (see Chapter 18). It resulted in a large amount of radioactive material being released into the atmosphere. Fall-out spread to many countries, with Ukraine, Belarus, the Russian Federation, Finland, Sweden, Norway and Alpine regions being affected most seriously. The year after the accident the highest effective radiation levels recorded in nearby countries were up to 30 per cent greater than background levels (RIVM/GLOBE, 1992).

In the areas affected, part of the radioactive material was washed out in the rivers and lakes. In the rivers situated near the nuclear plant, the Pripyat and the Dnepr, very high radiation levels (approximately 3500 Bq/l) were observed in the first days following the accident. Within a month the levels decreased by a factor of 100, but high radiation levels can still be observed (Voitsekhovitch et al, 1993). In Finland, the southern part was most heavily affected by the nuclear fall-out and high radiation levels were observed in the two large rivers draining the area, Kokemäenjoki and Kymijoki (Figure 5.33a). In the following years the level decreased, but it is still much higher than before the accident. In areas not so severely affected directly by fall-out, for example northern Finland and around St Petersburg, rivers nevertheless showed appreciable effects of runoff within their drainage areas (Figure 5.33b).

The effect of this fall-out on the biota of Swedish rivers and lakes was studied shortly after the accident (Petersen et al, 1986), but no immediate damage to aquatic organisms was found. Nevertheless, the possibility of a long-term biological effect cannot be ruled out because the effects of the increase in background radiation and the bioaccumulation of radioactivity through the aquatic food chain remain unclear.

Other hazards

Pathogens

Inland surface water polluted by faecal discharge from humans and animals may transport a variety of pathogens (bacteria and viruses). Pathogens in rivers and lakes are generally observed in the most densely populated areas where domestic and animal excreta are not adequately treated. In Europe increasing awareness of this pollution problem during the last century resulted in more efficient wastewater collection and treatment. Additionally waterbodies have been divided into those used for waste removal and those used for drinking water. Today, pathogens in European rivers and lakes generally do not pose a problem for public water supply. However, pathogens may restrict the use of a waterbody for bathing. Council Directive (76/160/EEC) sets standards for the microbiological quality of bathing waters. For the 1992 bathing season the microbiological quality of 5266 freshwater bodies in the EU was assessed, of which 63 per cent complied with standards for total and faecal coliforms (CEC, 1993).

Salinisation

In many dry parts of the world rivers have become saline because of high concentrations of dissolved salts, which may impair the use of water for human drinking and livestock water, just as it affects water use for irrigation and even industry. Salinisation may impoverish the ecological state of the rivers as well.

The downstream areas of the river Volga and Don catchments in the southern part of the Russian Federation are very dry and affected by salinisation, which has also become a problem in other dry parts of Southern Europe. In these places salinisation is aggravated by the increasing use of water for irrigation, which allows rapid surface water evaporation (see Chapter 7).

Mining activity, particularly salt, potash and iron mines as well as ore fields, releases brines, which contain high concentrations of dissolved salts, into inland surface waters. In Poland, the daily load to the upstream reaches of river Wisla amounts to 7000 tonnes of salt. Several other European rivers, such as the Rhine and the Elbe, have also been seriously affected by mining for decades. Chloride impacted rivers are also found in the southern Harz potash mining area in East Germany (LAWA, 1990) where, for example, concentrations reach values above 30 000 mg Cl/l in tributaries to the river Saale.

As an example, Figure 5.34 shows the development of the chloride concentration in the river Rhine. The pollution first became a problem in the 19th century. It then increased gradually until the Second World War, after which it rose steeply (Wolff, 1987). The high salt content of the Rhine water has negatively affected the crops grown by the market gardeners of the Dutch Westland district, and they have fought many juridical battles with the potash mining industry to seek compensation for their economic losses (Wolff, 1987).

CONCLUSIONS

Recent geological evolution in Europe is responsible for creating the complex pattern of aquifers and the multitude of short rivers and small lakes that form the European freshwater system. Thus, only three rivers (the Volga, the Danube and the Dnepr) and two lakes (the Russian lakes

Ladoga and Onega) rank among the 30 largest rivers and lakes of the world. There are more than 500 000 natural lakes with a surface area larger than one hectare, and more than 10 000 artificial lakes.

The water resource

Overall, there is no water shortage problem in Europe. However, the amount of water available for sustained consumption is very unevenly distributed across the continent. Total renewable water resources, including water generated internally in a country from precipitation, as well as external contributions from transboundary rivers, show extreme variations that range from less than 100 m^3/capita per year in Malta to more than 630 000 m^3/capita per year in Iceland.

Internal water resources ranging from under 25 mm to over 4500 mm, taken as average annual river runoff, are determined by rainfall and topography. The highest runoff is in Northwestern Europe and in mountain ranges, decreasing towards the east and south. The annual average runoff for Europe is estimated to be approximately 3100 km^3 or 4560 m^3/capita per year for a population of 680 million.

Many countries are heavily dependent on external contributions of water through transboundary rivers to meet their demands. Ten countries receive more than 50 per cent of their total water resources from neighbouring countries. Countries located downstream on large rivers (Moldova, Romania, Hungary, Luxembourg and The Netherlands) receive more than 75 per cent of their water in river flows from other countries. This means that there is a source of possible dispute over transboundary water pollution, or disposal of water resources.

Limited analysis of long flow records from throughout Europe indicates that:

- there are consistent patterns in the variability of river flow from year to year across Europe (eg, droughts, for example, tend to affect large parts of Europe at the same time), and
- there is no evidence of a consistent upward or downward trend in river flow characteristics.

On the average, the total annual European abstraction of water amounts to approximately 480 km^3 (or 700 m^3/capita per year). This is about 15 per cent of the total renewable resources, ranging from 0.1 per cent in Iceland to over 70 per cent in Belgium.

For a number of countries, water abstraction is approximately equal to, or even in excess of, the amount of water generated within the country. This leaves those countries dependent on upstream impacts which can affect the quantity and quality of the water they receive from transboundary rivers.

Groundwater

Overexploitation of aquifers takes place in about 60 per cent of the industrial and urban centres in Europe. About 140 million inhabitants of major cities are supplied with water from overexploited sources, sometimes leading to restricted supply. About 6 per cent of the area of aquifers suited for abstractions suffers from overexploitation, and this portion is increasing. Consequently, severe drawdown of the groundwater level in most of these aquifers has occurred and is probably continuing.

In many European coastal areas, salt-water intrusion due to overexploitation seriously affects soils and the drinking water, or will affect them in the near future. This may threaten the development of urban and tourist areas. This problem is most prominent along the coasts of the Mediterranean, the Baltic and the Black Sea.

The amount of abstractions for agricultural irrigation,

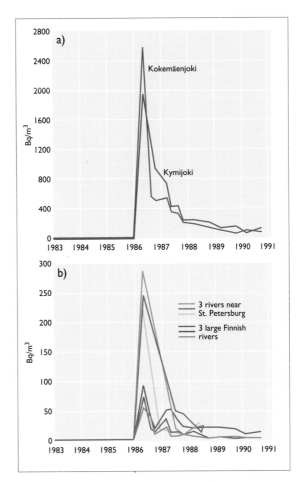

Figure 5.33
Observations of Caesium-137 in a) two large rivers in southern Finland; and b) three rivers running near St Petersburg and three rivers from northern Finland
Sources: Finnish rivers from Stuk (1984–1992) and rivers near St Petersburg from Gavrilov et al, 1989

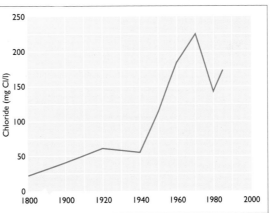

Figure 5.34
Chloride concentration in the Rhine near Lobith
Source: Redrawn after van der Weijden and Middelburg, 1989

the irrigation excess and related effects on the groundwater budget are uncertain due to an almost complete lack of data.

Economic consequences of lowering groundwater tables, such as reduced crop production, deterioration of well-water quality or pumping wells falling dry, are already considered to be important problems. However, ecological consequences of groundwater level drawdown (impoverishment of groundwater-related vegetation and deterioration of aquatic ecosystems due to diminished groundwater supply to rivers) are still not emphasised enough. The potential loss of wetlands is considerable; about 25 per cent of major wetlands are suffering from overexploitation of the groundwater.

Groundwater quality is threatened by high concentrations of nitrate due to excessive use of manure and artificial fertilisers on agricultural soils. Model computations of nitrate leaching from agricultural soils indicate that 87 and 22 per cent of the agricultural area in Europe has nitrate levels above the EU target value (25 mg NO_3/l) and the drinking water standard (50 mg NO_3/l), respectively. Typical 'hot spots' are found where a high load of nitrogen is applied to sensitive soils overlying sensitive aquifers, as in, for example, parts of Germany, The Netherlands, the UK and Denmark.

Using model calculations and the sparse monitoring data available it has been demonstrated that groundwater quality is threatened by pesticides in almost all European countries despite the fact that there is great spatial variability in both pesticide loads (from 2 to more than 10 kg/hectare for active ingredients) and sensitivity to leaching.

The EU standard of 0.5 μg/l for the total amount of pesticides in drinking water is (according to model calculations) exceeded on 75 and 60 per cent of the total arable and permanent crop land in EU/EFTA and Eastern Europe, respectively. The expected hot spots are in areas with intensive agriculture on top of unconsolidated, unconfined aquifers, covered by soil with low organic matter content – for example, in parts of Denmark, northern France, The Netherlands, Lithuania and Belarus.

In the longer term, if no future action is taken to reduce the leaching of nitrate and pesticides to groundwater, it will be necessary to carry out additional purification measures to produce good quality drinking water, and the functioning of groundwater-fed ecosystems might become threatened.

At present, 10 000 to 20 000 km² is thought to be seriously polluted by point sources. Since much industrial activity takes place in urbanised areas, near which large amounts of groundwater are abstracted, the probability of polluted groundwater being abstracted is relatively high. It has been estimated that, if no remedial action is taken, 100 000 to 200 000 km² (1 to 2 per cent of the total European area) will be potentially polluted by industry, mining, military activities and landfills within a period of 50 years.

Rivers and lakes in Europe

Monitoring of nutrient and organic concentrations in rivers and lakes is included in most national programmes for surveying the state of the environment. Although the selection of determinants and methods of measurement differ somewhat between countries, it has been possible (by carefully evaluating the available data) to assess the present state of pollution by these substances on a European scale. However, at this stage, it has not been possible to include a general European assessment of water pollution by heavy metals, organic micropollutants and radionuclides, mainly due to lack of comparable data.

Excessive discharges of organic matter from human activities to waterbodies can lead to undesirable changes in the composition of aquatic biota. In addition, the water can become unsuitable for human consumption. Generally, the lowest levels of organic matter are found in rivers in the Nordic countries, and the highest in Central and Southern Europe. There is an evident relationship between the concentration of organic matter in a river and the population density of its catchment.

The concentration of organic matter declined markedly in many European rivers from the end of the 1970s to the end of the 1980s, with the concentration level decreasing at 75 per cent of monitored river sites. The improvement is greatest in Western European rivers, primarily because of intensified wastewater treatment.

Excessive inputs of nutrients (nitrogen and phosphorus) to waterbodies can result in a series of adverse effects known as cultural eutrophication. This can cause significant ecological changes, and negatively affect the quality of water for human consumption and other uses. Phosphorus is usually the primary nutrient responsible for freshwater eutrophication, whereas nitrogen is the main nutrient causing eutrophication of coastal areas and seas.

In general, nutrient levels are lowest in the rivers of the sparsely populated Nordic countries, and highest in those in a band stretching from the southern part of the UK to the Balkan area and Ukraine. There is a close relationship between phosphorus concentrations and the catchment population density and between nitrogen levels and the percentage of the catchment that is agricultural land. Most of the phosphorus loading of inland surface waters is attributable to discharge from point sources, especially municipal sewage water and industrial effluent, while the nitrogen loading is derived primarily from agricultural activity, especially the use of nitrogen fertilisers and manure.

Many European lakes also have high nutrient concentrations, especially those lakes in densely populated areas. Because of the high nutrient levels, many of the lakes have high concentrations of phytoplankton and therefore very turbid water.

The phosphorus concentration in most European rivers and many lakes has decreased during the last 10 to 15 years as a consequence of improved wastewater treatment and substitution for phosphorus in detergents. The ammonium level in many European rivers has also declined, mainly because of improved wastewater treatment and increased awareness of the importance of correct handling and storage of manure and silage. The improvement has been most marked in Western European rivers; phosphorus levels have increased in many Eastern European rivers. In contrast to the phosphorus and ammonium levels, the nitrate level in most European rivers has increased during the last 10 to 15 years, mainly as a result of the increasing use of nitrogen fertilisers.

The nutrient levels in many areas of Europe are still too high, and unless drastic efforts are made to reduce inputs of nutrients, eutrophication is likely to continue to be an important European environmental issue. In many cases, it is essential to remove phosphate in wastewater treatment plants and to reduce the phosphorus content of detergents, as well as to reduce the nitrogen (in particular) and phosphorus loadings from agricultural areas.

Their geological characteristics make rivers and lakes in large parts of Europe sensitive to acidification, especially the Nordic countries, parts of Central Europe and the northwestern part of the Russian Federation. Atmospheric deposition of acidifying substances in these areas has led to an impoverishment of flora and fauna in many rivers and lakes, resulting in decreased biodiversity and fish kills (see Chapter 31).

In general, the concentration of heavy metals in European rivers is well below drinking water standards. However, aquatic biota are usually affected at much lower concentrations. Relatively high concentrations of heavy metals are observed in rivers receiving mine effluents and discharges from metal processing industries.

The use of toxic substances (eg, DDT, PCBs) has been either restricted or banned in several European countries during the last 25 years. As a result, there has been a marked reduction in their levels in surface waters. However, other organic micropollutants suspected of having detrimental environmental effects, and many with unknown effects, are still being discharged into freshwater bodies. Nuclear accidents, due to low safety standards at nuclear power plants and careless handling of nuclear waste, are another potential source of water contamination.

A river with water of good quality does not necessarily support a diverse flora and fauna. For rivers to be of high ecological quality it is also necessary that they do not fall dry and that their natural physical conditions (such as meandering course, pool-riffle sequences) are maintained (where not already destroyed) or re-established. River regulation works, in particular reservoir construction and channelisation, have in almost all European countries severely degraded the physical conditions of rivers, leading to a reduction of their self-purification capacity and to hindrance of migration of important fish species like trout, salmon and sturgeon to their spawning sites in upstream river reaches. The percentage of river reaches still in a natural state is in many countries very low, ranging from below 20 per cent in densely populated countries with intensive agriculture like Belgium and Denmark, to above 70 per cent in Norway, Estonia and Poland, with lower population densities or less intensive agriculture.

To improve the physical conditions, and hence the ecological quality of watercourses, river restoration projects, covering a broad spectrum of river sizes and restoration works, are currently being initiated in a number of countries (including Sweden, Denmark, the UK, Germany and France).

Endpiece

At present, European freshwater resources, although plentiful and of good quality in some regions, are still under threat from a multitude of human impacts. This can reduce the amount and quality of the water available for human consumption and other intentional uses. Because of these problems, better integrated water management is urgently needed to halt and reverse deterioration of water, taking into account the acuteness of the problems, their transboundary nature, their linkage with human activities in the catchments, and their cross-media importance. With this background, the condition of European freshwater resources and their management has been identified as a prominent environmental problem, to be treated in detail in Chapter 33.

REFERENCES

Andersson, T, Nilsson, Å, Håkanson, L and Byrdsten, L (1987) *Kvicksilver i Svenska Sjøer.* Rapport 3291, Statens Naturvårdsverk, Solna, Sweden. [In Swedish with an English summary].

Arnell, N, Oancea, V and Oberlin, G (1993) *European river flow regimes.* Report to the European Environment Agency Task Force. Institute of Hydrology, Wallingford, and CEMAGREF, Lyons.

Barbieri, A and Mosello, R (1992) Chemistry and trophic evolution of Lake Lugano in relation to nutrient budget. *Aquatic Sciences* 54, 219–37.

Battarbee, R W (1989) Geographical research on acid rain I. The acidification of Scottish lochs. *The Geographical Journal* 155, 353–77.

Baumgartner, A and Reichel, E (1975) *The World Water Balance.* R. Oldenburg Verlag, Munich.

Benedek, P and Major, V (1992) In: *Point source pollution in the Danube Basin. Country technical reports. Bulgaria, the CSFR, Hungary, and Romania. Volume III.* Water and Sanitation for Health Project, USAID, Washington DC.

Berge, D (1987) *Fosforbelastning og respons i grunne og Middelgrunne Insjøer.* NIVA rapport O–85110.

Bernes, C and Grundsten, C (1992) *The environment national atlas of Sweden.* The National Environmental Protection Agency, Stockholm.

Biró, P (1984) Lake Balaton: a shallow Pannonian water in the Carpathian Basin. In: Taub, F B (Ed) *Ecosystems of the world – lakes and reservoirs,* pp 231–45. Elsevier, Amsterdam, Oxford, Tokyo.

Boesten, J J T I and van der Linden, A M A (1991) Modelling the influence of sorption and transformation on pesticide leaching and persistence. *Journal of Environmental Quality* 20, 425–35.

Boon, P J (1992) Essential elements in the case for river conservation. In: Boon, P J, Calow, P and Petts, G E (Eds) *River conservation and management,* pp 11–33, Wiley, New York.

Borg, H (1987) Trace metals and water chemistry of forest lakes in northern Sweden. *Water Research* 21, 65–72.

Bowman, J (1991) *Acid sensitive surface waters in Ireland.* Environmental Research Unit, Dublin.

Breemen, N van and van Dijk, H F G (1988) Ecosystem effects of atmospheric deposition of nitrogen in The Netherlands. *Environmental Pollution* 54, 249–74.

Brodin, Y W and Kuylenstierna, J C J (1992) Acidification and critical load in Nordic countries, a background. *Ambio* 21, 332–8.

Brömssen, U von (1986) *Acidification of drinking water-groundwater,* pp 251–61. Elsevier, Amsterdam.

Brookes, A (1987) The distribution and management of channelized streams in Denmark. *Regulated Rivers* 1, 3–16.

Brookes, A (1988) *Channelized rivers: perspectives for environmental management.* Wiley, New York.

Brouwer, F and Falkenmark, M (1989) Climate-induced water availability changes in Europe. *Environmental Monitoring and Assessment* 13, 75–98.

Burkart, W, and Kellerer, A M (Eds) (in preparation) A first assessment of radiation doses and health effects in workers and communities exposed since 1948 in the southern Urals. *Science of the total environment.*

Cammarota, V A (1980) Production and uses of zinc. In: Nriagu, J O (Ed) *Zinc in the environment. Part I: Ecological cycling,* pp 1–38. Wiley, New York.

Canfield, D E and Bachman, R W (1981) Prediction of total phosphorus concentration, chlorophyll a, and Secchi depth in natural and artificial lakes. *Canadian Journal of Fisheries and Aquatic Sciences* 38, 414–23.

Casado, C, de Jalon, D G, del Olmo, C M, Barcelo, E and Menes, F (1989) The effect of an irrigation and hydroelectric reservoir on the downstream communities. *Regulated Rivers* 4, 275–84.

Catalan, J and Camarero, L (1992) Limnological studies in the lakes of the Spanish Pyrenees (1984–90). *Documenta Ist Ital Idrobiol* 32, 59–67.

Catalan, J, Ballesteros, E, Camarero, L, Felip, M and Garcia, E (1992) Limnology in the Pyrenean Lakes. *Limnetica* 8, 27–38.

CEC (1989) *Environmental radioactivity in the European Community 1984–86.* Fraser, G and Stanners, D (Eds). Radiation Protection report 46. EUR 12254EN, Commission of the European Communities, Luxembourg.

CEC (1992) *Research and technological development for the supply and use of freshwater resources.* SAST Project No. 6. Strategic Dossier prepared by H Williams and D Musco. Commission of the European Communities, Brussels.

CEC (1993) *Quality of bathing water 1992.* EUR 15031EN, Commission of the European Communities, Luxembourg.

CEC (in press) *Environmental radioactivity in the European Community, 1987-90.* De Cort, M, Janssens, A, de Vries, G (Eds) EUR 15699EN. Commission of the European Communities, Luxembourg.

Charles, D F, Battarbee, R W, Renberg, I, van Dam, H and Smol, J P (1990) Palaeoecological analysis of lake acidification trends in North America and Europe using diatoms and chrysophytes. In: Stephen, A, Lindberg, S E and Salomon, W (Eds) Soils, aquatic processes and lake acidification. *Advances in Environmental Sciences, Vol 4, Acidic Precipitation.* Springer Verlag, New York.

Chevreuil, M, Chesterikoff, A and Letolle, R (1987) PCB pollution behaviour in the river Seine. *Water Research* 21, 427–34.

Clabby, K J, Lucey, J, McGarrigle, M L, Bowman, J J, Flanagan, P J and Toner, P F (1992) *Water quality in Ireland 1987–1990. Part I, General assessment.* Environmental Research Unit, Dublin.

Cresser, M and Edwards, A (1987) *Acidification of freshwaters.* Cambridge University Press, Cambridge.

Cullen, P and Forsberg, C (1988) Experiences with reducing point sources of phosphorus to lakes. *Hydrobiologia* 170, 321–36.

Custodio, E (1991) *Characterization of aquifer overexploitation: comments on hydrogeological and hydrochemical aspects: The situation in Spain.* IAH XXIII International Congress, Puerto de la Cruz, Tenerife, Spain.

Czechoslovakia (1992) *National report of the Czech and Slovak Republic for UNCED.* Czechoslovak Academy of Sciences and the Federal Committee for the Environment, Prague.

Dam, H van (1987) *Acidification of moorland pools: a process in time.* Doctoral thesis, Wageningen.

Danish Ministry of the Environment (1992) *Environmental indicators 1992.* Ministry of the Environment, Copenhagen.

Datsenko, Y S (1990) Some problems of studies the reservoirs eutrophications. Unpublished paper, Department of Geography, University of Moscow.

De Pauw, N, Ghetti, P F, Manzini, P and Spaggiari, R (1992) Biological assessment methods for running water. In: Newman, P J, Piavaux, M A and Sweeting, R A (Eds) *River water quality. Ecological assessment and control.* pp 217–48. Commission of the European Communities, Brussels.

DGQA (1991) *Compêndo de dados da qualidade da aqua 1990.* Direcção-general da Qualidade da Ambiente, Portugal.

Dijkzeul, A (1982) *De waterkwaliteit van de Rijn in Nederland in de periode 1970–1981.* Rijksinst, Zuivering Afvalwater, Nota 82–061.

Dill, W A (1990) *Inland fisheries of Europe.* EIFAC technical paper No 52. Food and Agricultural Organization of the United Nations, Rome.

Dirección General de Obras Hidráulicas del MOPU (1989) Análisis de calidad de aguas 1987–88. Dirección General de Obras Hidráulicas del MOPU.

Dokulil, M T and Jagsch, A (1992) The effects of reduced phosphorus and nitrogen loadings on phytoplankton in Mondsee, Austria. *Hydrobiologia* 243/244, 389–94.

Drabløs, D and Tollan, A (Eds) (1980) Ecological impact of acid precipitation. *Proceedings of the International Conference on the Ecological Impact of Acid Precipitation,* SNSF project report, Norway.

Drecht, G van (1991) *Model NLOAD.* RIVM, Bilthoven.

Flanagan, P J and Larkin, P M (1992) *Water quality in Ireland. Part II, River quality data.* Environmental Research Unit, Dublin.

Förstner, U (1980) Cadmium in polluted sediments. In: Nriagu, J O (Ed) *Cadmium in the environment. Part I, Ecological cycling,* 306–63. Wiley, New York.

Fortunatov, M A (1979) Physical geography of the Volga basin. In: Mordukhai-Boltovskoi, E D (Ed) *The River Volga and its life,* Junk Publishers, The Hague, 1–29 Monographiae Biologicae 33.

Fott, J, Stuchlík, E, Stuchlíková, Z, Straskrabova, V, Kopácek, J and Simek, K (1992) Acidification of lakes in Tatra Mountains (Czechoslovakia) and its ecological consequences. In: Mosello, R, Wathne, B M, and Guissani, G (Eds) *Limnology on groups of remote lakes: ongoing and planned activities,* pp 69–81. *Documenta Ist Ital Idrobio* 32.

Franken, R O G (forthcoming) *Point-sources of soil and groundwater pollution.* Report 712401002, National Institute for Public Health and Environmental Protection (RIVM), Bilthoven.

Friedrich, G and Müller, D (1984) Rhine. In: Whitton, B A (Ed) *Ecology of European rivers,* pp 265–315. Blackwell, Oxford.

Gaggino, G F, Cappelletti, E, Marchetti, R and Calcagnini, T (1987) The water quality of Italian lakes. European Water Pollution Control Association. International Congress, *Lake Pollution and Recovery,* Rome 15–18 April, 1985.

Galassi, S, Guzzella, L, Mingazzini, M, Vigano, L, Capri, S and Sora, S (1992) Toxicological and chemical characterization of organic micropollutants in river Po waters (Italy). *Water Research* 26, 19–27.

Garcia De Jalon, D (1987) River regulation in Spain. *Regulated Rivers* 1, 343–8.

Gavrilov, V H, Gritchenko, Z G, Ivanova, L M, Orlova, T E and Tishkova, N A (1989) Strontium-90, caesium-134, and caesium-137 in water reservoirs of the Soviet Unions' Baltic region (1986–1988). In: Three years observation of levels of some radionuclides in the Baltic Sea after the Chernobyl accident. Seminar on the radionuclides in the Baltic Sea. 29 May 1989. Rostock-Warnemünde, German Democratic Republic. *Baltic Sea Environment Proceedings* 31, 62–80.

Gibson, C E (1986) A contribution to the regional limnology of Northern Ireland: the lakes of County Down. *Record of Agricultural Research* 34, 75–83.

Gibson, C E (1988) A contribution to the regional limnology of Northern Ireland (II): the lakes of County Fermanagh. *Record of Agricultural Research* 36, 121–32.

Goodwillie, R (1980) *European peatlands.* European Committee for the Conservation of Nature and Natural Resources. Council of Europe. Nature and Environmental Series No 19.

Gottschalk, L, Jensen, J L, Lundquist, D, Solantie, R and Tollan, A (1979) Hydrological regions in the Nordic Countries. *Nordic Hydrology* 10, 273–86.

Grimmett, R F A and Jones, T A (1989) *Important bird areas in Europe.* International Council for Bird Preservation, Technical Publication No 9.

Gutelmacher, B L and Petrova, N A (1982) Production of individual species of algae and its role in the productivity of phytoplankton in Ladoga Lake. *International Revue der Gesamten Hydrobiologie* 67, 613–24.

Heitto, L (1990) A macrophyte survey in Finnish forest lakes sensitive to acidification. *Verhandlungeu des Internationalen Vereins Limnologie* 24, 667–70.

Henriksen, A, Lien, L, Traaen, T S, Sevaldrud, I S, and Brakke, D F (1988) Lake acidification in Norway – present and predicted chemical status. *Ambio* 17, 259–66.

Henriksen, A, Lien, L, Rosseland, B O, Traaen, T S and Sevaldrud, I S (1989) Lake acidification in Norway, present and predicted fish status. *Ambio* 18, 314–21.

Henriksen, A, Kämäri, J, Posch, M, Lövblad, G, Forsius, M and Wilander, A (1990) *Critical loads to surface waters in Fennoscandia.* Environmental report 1990: 17. Nordic Council of Ministers, Copenhagen.

Herdendorf, C E (1982) Large lakes of the world. *Journal of Great Lakes Research* 8, 379–412.

Hesthagen, T and Hansen, L P (1991) Estimates of the annual loss of Atlantic salmon, *Salmo salar* L, in Norway due to acidification. *Aquaculture and Fisheries Management* 22, 85–91.

Hettelingh, J-P., Downing, R J and de Smet, P A M (1991) *Mapping Critical Loads for Europe.* CCE Technical Report 1, RIVM Report 259101001, Bilthoven.

Hock, B and Somlyódy, L (1989) Water quality in Hungarian rivers. *River Basin Management V,* pp 19–27. International Association on Water Pollution Research and Control.

Hodson, P V, Borgmann, U and Shear, H (1979) Toxicity of copper to aquatic biota. In: Nriagu, J O (Ed) *Copper in the environment Part II: Health effects,* pp 308–72. Wiley, New York.

Howells, G (1990) *Acid rain and acid waters.* Ellis Horwood series in environmental science. Ellis-Horwood Limited, New York.

IAH (1974–87) *International hydrogeological map of Europe* (1:1 500 000) (Eds W Struckmeier and H Karrenberg). BGR (Hanover) and UNESCO (Paris).

Institute of Hydrology and British Geological Survey (1990) *Hydrological data UK, 1989 yearbook.* Institute of Hydrology, Oxfordshire.

Istituto di Ricerca sulle Acque (1990) *Un sistema informativo per la gestione della qualità delle acque, un'applicazione ai corsi d'acqua italiani.* Istituto di Ricerca sulle Acque.

Iversen, T M, Kronvang, B, Madsen, B L, Markmann, P and Nielsen, M B (1993) Re-establishment of Danish streams, restoration and maintenance measures. *Aquatic Conservation* 3, 1–20.

Jenkins, A, Cosby, B J, Ferrier, R C, Walker, T A B and Miller, J D (1990) Modelling stream acidification in afforested catchments: an assessment of the relative effects of acid deposition and afforestation. *Journal of Hydrology* 120, 163–81.

Jeppesen, E, Kristensen, P, Jensen, J P, Søndergaard, M, Mortensen, E and Lauridsen, T (1991) Recovery resilience following a reduction in external phosphorus loading of shallow, eutrophic Danish lakes: duration, regulating factors and methods for overcoming resilience. *Mem Ist Ital Idrobiol* 48, 127–48.

José, P (1989) Long-term nitrate trends in the river Trent and four major tributaries. *Regulated Rivers* 4, 43–57.

Kämäri, J, Amann, M, Brodin, Y-W, Chadwick, J, Henriksen, A, Hettelingh, J-P, Kuylenstierna, J, Posch, M and Sverdrup, H (1992) The use of critical load for the assessment of future alternatives to acidification. *Ambio* 21, 377–86.

Kempe, S, Pettine, M and Cauwet, G (1991) Biogeochemistry of European rivers. In: Degens, E T, Kempe, S and Richey, J E (Eds) *Biochemistry of major world rivers,* pp 169–211. SCOPE 42, Wiley, New York.

Kinkor, J (1992) River management in the Czech Republic. In: Newman, P J, Piavaux, M A and Sweeting, R A (Eds) *River water quality: ecological assessment and control,* pp 607–25. Commission of the European Communities, Brussels.

Korzun, V I et al (1977) *Atlas of world water balance.* UNESCO, Paris.

Krasovskaia, I and Gottschalk, L (1992) Stability of river flow regimes. *Nordic Hydrology* 23,137–54.

Kristensen, P, Kronvang, B, Jeppesen, E, Græsbøll, P, Erlandsen, M, Rebsdorf, A, Bruhn, A and Søndergaard, M (1990) *Vandmiljøplanens overvågningsprogram – ferske vandområder – vandløb, kilder og søer.* Report 5. National Environmental Research Institute, Denmark.

Kronvang, B, Ærtebjerg, G, Grant, R, Kristensen, P, Hovmand, M and Kirkegaard, J (1993) Nationwide monitoring of nutrients and their ecological effects, State of the Danish aquatic environment. *Ambio* 22, 176–87.

Lang, C (1991) Decreasing phosphorus concentration and unchanged oligochaete communities in Lake Geneva, how to monitor recovery. *Archiv für Hydrobiologie* 122, 305–12.

Last, F T and Watling, R (Eds) (1991) *Acidic deposition: its nature and impacts.* Proceedings of the international symposium held in Glasgow, Scotland, 16–21 September 1991. Royal Society of Edinburgh.

LAWA (1990) *Die Gewässergütekarte der Bundesrepublik Deutschland 1990.* Länderarbeitsgemeinschaft Wasser (LAWA)

Leuven, R S E W, Kersten, H L M, Schuurkes, J A A R, Roelofs, J G M and Arts, G H P (1986) Evidence for recent acidification of lentic soft waters in the Netherlands. *Water, Air and Soil Pollution* 30, 387–92.

Lithuania (1992) *National Report to UNCED*. Environmental Protection Department of the Republic of Lithuania, Vilnius.

Lvovich, M I (1973) The global water balance. *Transactions of the American Geophysical Union* 54, 28–42.

Marsden, M W (1989) Lake restoration by reducing external phosphorus loading, the influence of sediment phosphorus release. *Freshwater Biology* 21, 139–62.

Merilehto, K, Kenttämies, K and Kämäri, J (1988) *Surface Water Acidification in the ECE Region*. Miljörapport 1988: 14. Nordic Council of Ministers, Copenhagen.

Meybeck, M (1982) Carbon, nitrogen, and phosphorus transport by world rivers. *American Journal of Science* 282, 402–50.

Meybeck, M and Helmer, R (1989) The quality of rivers: from pristine stage to global pollution. *Palaeogeography, Palaeoclimatology, Palaeoecology* 75, 283–309.

Meybeck, M, Chapman, D V and Helmer, R (Eds) (1989) *Global freshwater quality – a first assessment*. GEMS: Global Environmental Monitoring System. WHO/UNEP – Blackwell, Cambridge.

Milius, A (1991) Chlorophyll a in Estonian lakes. *Proceedings of the Estonian Academy of Sciences* 40, 199–206.

Ministère de la Culture et de l'Environnement (1978) *Environnement et cadre de vie – dossier statistique – Tome 2*. La Documentation Française, Paris.

Mnatsakanian, R A (1992) *Environmental legacy of the former Soviet Republics*. Centre for Human Ecology, University of Edinburgh, Edinburgh.

Monitor 12 (1991) *Acidification and liming of Swedish waters*. National Swedish Environment Protection Board, Stockholm.

Morales-Baquero, R, Carrillo, P, Cruz-Pizarro, L and Sanches-Castillo, P (1992) Southernmost high mountain lakes in Europe (Sierra Nevada) as reference site for pollution and climatic change monitoring. *Limnetica* 8, 39–47.

Mosello, R (1981) Chemical characteristics of fifty Italian alpine lakes (Pennine-Lepontine Alps), with emphasis on the acidification problem. *Mem Ist Ital Idrobiol* 39, 99–118.

Mosello, R (1984) Hydrochemistry of high altitude lakes. *Schweizerische Zeitschrift für Hydrologie* 46, 86–99.

Mosello, R (1989) The trophic evolution of Lake Maggiore as indicated by its water chemistry and nutrient loads. *Mem Ist Ital Idrobiol* 46, 69–87.

Mosello, R, Marchetto, A, Tartari, G, Bovio, M and Castello, P (1991) Chemistry of alpine lake in Aosta Valley (N. Italy) in relation to watershed characteristics and acid deposition. *Ambio* 20, 7–12.

Mosello, R, Barbieri, A, Marchetto, A, Psenner, R and Tait, D (1992a) Research on quantification of the susceptibility of alpine lakes to acidification. *Documenta Ist Ital Idrobiol* 32, 23–30.

Mosello, R, Wathne, B M and Giussani, G (Eds) (1992b) Workshop on: *Limnology on groups of remote mountain lakes, ongoing and planned activities*. Pallanza, March 15th 1991. *Documenta Ist Ital Idrobiol* 32.

Muniz, J P (1983) The effects of acidification on Norwegian freshwater ecosystems. In: *Ecological effects of acid deposition*, pp 299–322. Nat Swed Environ Proj Bd Report PM 1636.

Muniz, J P (1991) Freshwater acidification: its effect on species and communities of freshwater microbes, plants and animals. *Proceedings of the Royal Society Edinburgh*, 97B, 227–54.

National Swedish Environment Protection Board (1983) *Ecological effects of acid deposition*. Report and background papers 1982 Stockholm Conference on the Acidification of the Environment. Expert Meeting I. Report SNV PM 1636.

NIVA (1990a) *Landsomfattende undersøkelse av trofitilstanden i 355 innsjøer i Norge*. Norsk Institutt for Vannforskning, Oslo.

NIVA (1990b) *Oppfølging av 49 av de 355 undersøkte innsjøer i 1989*. Norsk Institutt for Vannforskning, Oslo.

NIVA (1992) *Tiltaksorientert overvåkning av Mjøsa med tilløpselver. Årsrapport for 1991*. Rapport 490/92. Norsk Institutt for Vannforskning, Oslo.

Nogawa, K (1981) Itai-Itai disease and follow-up studies. In: Nriagu, J O (Ed) *Cadmium in the environment, Part II, Health effects*, pp 1–37. Wiley, New York.

Norwegian State Pollution Control Authority (1987) *1000 lake survey 1986 Norway*. Report 283/87. Norwegian Institute for Water Research, Oslo.

NRA (1991) The quality of rivers, canals and estuaries in England and Wales. Report of the 1990 survey. *Water Quality Series No 4*, National Rivers Authority, Bristol.

Nriagu, J O (1979) The global copper cycle. In: Nriagu, J O (Ed) *Copper in the environment, Part I, Ecological cycling*, pp 1–17. Wiley, New York.

OECD (1982) *Eutrophication of waters. Monitoring, assessment and control*. OECD, Paris.

OECD (1992) *Environmental indicators. (In-depth analysis of individual indicators: intensity of use of water resources)*. OECD Secretariat, Group of the State of the Environment, Paris.

Olsson, H (1991) Materialbalancer for Vänerns fosfor och kvävetransporter 1970–1989. *Vatten* 47, 263–72.

Olsson, M and Reutergårdh, L (1986) DDT and PCB pollution trends in the Swedish aquatic environment. *Ambio* 15, 103–9.

Overgaard, K (1986) *Nitrate and pH in drinking water*. Miljøprojekt 75, National Agency for Environmental Protection, Copenhagen.

Overrein, L N, Seip, H M and Tollan, A (1980) *Acid precipitation – effects on forests and fish*. Final Report of SNSF Project, 1972–1980, FR/80. Norwegian Council for Scientific and Industrial Research, Oslo.

Persson, G, Olsson, H, Wiederholm, T and Willén, E (1989) Lake Vättern, Sweden: a 20-year perspective. *Ambio* 18, 208–15.

Petersen, R C, Landner, L and Blanck, H (1986) Assessment of the impact of the Chernobyl reactor accident on the biota of Swedish streams and lakes. *Ambio* 15, 327–31.

Petrova, N A (1987) The phytoplankton of Ladoga and Onega lakes and its recent successional changes. *Archiv für Hydrobiolologie* 25, 11–18.

Petts, G E (1988) Regulated rivers in the United Kingdom. *Regulated Rivers* 2, 201–20.

Rebsdorf, A, Thyssen, N and Erlandsen, M (1991) Regional and temporal variation in pH, alkalinity and carbon dioxide in Danish streams, related to soil type and land use. *Freshwater Biology* 25, 419–35.

Renberg, I J and Wallin, J E (1985) The history of acidification of Lake Gårdsjön as deduced from diatoms and Sphagnum leaves in the sediment. *Ecological Bulletin* 37, 47–52.

Riera, J L, Jaume, D, de Manuel, J, Morgui, J A and Armengol, J (1992) Patterns of variation in the limnology of Spanish reservoirs: a regional study. *Limnetica* 8, 111–23.

RIVM (1992a) *Milieudiagnose 1991. I. Integrale rapportage lucht, bodem en grondwaterkwaliteit*. Report 724801004, 33–34, RIVM, Bilthoven.

RIVM (1992b) *National environmental outlook 2: 1990–2010*. National Institute of Public Health and Environmental Protection, Bilthoven.

RIVM/GLOBE (1992) *The environment in Europe: a global perspective*. RIVM, National Institute of Public Health and Environmental Protection, Bilthoven.

RIVM/RIZA (1991) *Sustainable use of groundwater*. Report 600025001, RIVM Bilthoven.

RIWA (1983) Jahresbericht 1982. *Teil A Der Rhein*. The Hague.

Sas, H (Ed) (1989) *Lake restoration by reduction of nutrient loading: expectation, experiences, extrapolation*. Academie Verlag Richardz Gmbh, Sankt Augustin.

SCB Sweden (1990) *The natural environment in figures*. Third edition. Statistics Sweden, Stockholm.

Schmidt, R, and Psenner, R (1992) Climate changes and anthropogenic impacts as causes for pH fluctuations in remote high alpine lakes. *Documenta Ist Ital Idrobiol* 32, 31–57.

Schuurkes, J A A R, Heck, I C C, Hesen, P L G M, Leuven, R S E W and Roelofs, J G M (1986) Effects of sulphuric acid and acidifying ammonium deposition on water quality and vegetation of simulated soft water ecosystems. *Water, Air and Soil Pollution* 31, 267–72.

Schwoerbel, J (1989) Waters in the Black Forest. In: Lambert, W and Rothhaupt, K-O (Eds) *Limnology in the Federal Republic of Germany*, pp 46–52. Report SIL 1989, Kiel.

Seip, K L, Sas, H and Vermij, S (1990) The short term response to eutrophication abatement. *Aquatic Sciences* 52, 199–220.

Shiklomanov, I A (1991) The world's water resources. In: *Proc International Symposium to Commemorate the 25 Years of IHD/IHP*, pp 93–126. UNESCO, Paris.

Showers, V (1989) *World facts and figures* (3rd edn). Wiley, New York.

Skjelkvåle, B L and Wright, R F (1990) *Overview of areas sensitive to acidification: Europe*. Acid Rain Research, Report 20/1990. Norwegian Institute for Water Research, Oslo.

Smith, R V, Steven, R J, Foy, R H and Gibson, C E (1982) Upward trend in nitrate concentrations in rivers discharging into Lough Neagh for the period 1969–1979. *Wat Res* 16, 183–8.

Søndergaard, M, Jeppesen, E, Kristensen, P and Sortkjær, O (1990) Interaction between sediment and water in a shallow hypertrophic lake: a study on phytoplankton collapses in Lake Søbygård, Denmark. *Hydrobiologia* 191, 149–64.

Stabel, H–H (1991) Irregular biomass response in recovering prealpine lakes. *Verhandlungen des Internationalen Vereins für Limnologie* 24, 810–15.

Stancik, A et al (1988) *Hydrology of the River Danube*. Publishing House Priroda, Bratislava.

Stuk (1984–1992) *Radioactivity of surface waters and freshwater fish in Finland 1983–1990*. Five annual reports from the Finnish Centre for Radiation and Nuclear Safety, Helsinki.

Stumm, W (1992) Water, an endangered ecosystem, assessment of chemical pollution. *Journal of Environmental Engineering* 118, 466–76.

Targbil, G (1986) *The Norwegian monitoring programme for long range transport of air pollutants, results 1980–1984*. The Norwegian State Pollution Control Authority.

The Scottish Office, Environment Department (1992) *Water quality survey of Scotland*, Edinburgh.

Tilzer, M M, Gaedke, U, Schweizer, A, Beese, B and Wieser, T (1991) Interannual variability of phytoplankton productivity and related parameters in Lake Constance: no response to decreased phosphorus loading. *Journal of Plankton Research* 13, 755–77.

Toner, P F (1987) Freshwater pollution in Ireland: does it present a problem? In: Proceedings of third annual conference: *Lake, river and coastal pollution – can it be contained?*, pp 15–26. Sherkin Island Marine Station, Ireland.

UK DoE (1992) *The UK environment*. UK Department of the Environment HMSO, London.

Ukraine (1992) *Ukraine National Report to UNCED*. Ministry of Environmental Protection of Ukraine, Kiev.

Umweltbundesamt (1992) *Daten zur Umwelt*. Erich Schmidt Verlag, Berlin.

UNEP (1991) *Freshwater pollution*. UNEP/GEMS Environment Library, No 6, Nairobi.

UNESCO (1978) *World Water Balance and Water Resources of the Earth*. UNESCO Series Studies and Reports, No 25, UNESCO, Paris.

United Nations (1990) *Groundwater in Eastern and Northern Europe*. Natural Resources/Water Series No 24, UN, New York.

United Nations (1991) *Groundwater in Western and Central Europe*. Natural Resources/Water Series No 27, UN, New York.

van der Weijden, C H and Middelburg, J J (1989) Hydrogeochemistry of the River Rhine: long term and seasonal variability, elemental budgets, base levels and pollution. *Water Research* 23, 1247–66.

Varga, P, Abraham, M and Simor, J (1990) Water quality of the Danube in Hungary and its major determining factors. *Water Science and Technology* 22, 113–18.

Veiga da Cunha, L (1993) Water resources in Europe In: Sesimbra (Ed) *The European common garden. Based on reports prepared for the "East-West Interparliamentary Meeting"*. Strasbourg, 17–20 May 1992, pp 159–204, Sesimbra, Brussels.

Velde, R J van de, Verspuy, M, Faber, W, Katwijk, V van, Scholten, H J, Thewessen, T and Zevenbergen, M (in press) *The preparation of a European land–use database*. RIVM, Bilthoven.

Vesely, J and Majer, V (1992) The major importance of nitrate increase for the acidification of two lakes in Bohemia. *Documenta Ist Ital Idrobiol* 32, 83–92.

Viaroli, P, Ferrari, I, Mangia, A, Rossi, V and Menozzi, P (1992) Sensitivity to acidification of Northern Apennines lakes (Italy) in relation to watershed characteristics and wet deposition. In: Mosello, R, Wathne, B M and Guissani, G (Eds) Limnology on groups of remote lakes: ongoing and planned activities. *Documenta Ist Ital Idrobiol* 32, 93–105.

Vlaamse Milieumaatschappij (1991) *Jaarverslag biologisch meetnet 1989–1990. De biologische waterkwaliteit in het vlaamse gewest. Rapport over de jaren 1989 and 1990*. Vlaamse Milieumaatschappij V M M, Erembodegem.

Vlaamse Milieumaatschappij (1992) *Jaarverslag fysico-chemisch meetnet oppervlaktewater 1990. De fysico-chemische kwaliteit van de oppervlakteren in het vlaamse gewest in 1990*. Vlaamse Milieumaatschappij V M M, Erembodegem.

Voitsekhovitch, O V, Kanivets, V V, Laptev, G V and Biley, I Y (1993) *Hydrological processes and their influence on radionuclide behavior and transport by surface water pathways as applied to water protection after the Chernobyl accident*. A paper by the Ukrainian Hydrometeorological Research Institute, Kiev.

Vollenweider, R A (1976) Advances in defining critical loading levels for phosphorus in lake eutrophication. *Mem Ist Ital Idrobiol* 33, 53–83.

VSEGINGEO (1985) Groundwater runoff map of Central and Eastern Europe (1:1 500 000). USSR Ministry for Geology, All-Union Research Institute of Hydrology and Engineering Geology (VSEGINGEO).

Wasserwirtschaftkataster (1992) *Gewässergüte in Österreich*. Jahresbericht 1990. Bundesministerium für Land-und Forstwirtschaft. Vienna.

Wathne, B M (1992) Acidification of mountain lakes, palaeolimnology and ecology. The ALPE project. In: Mosello, R, Wathne, B M and Giussani, G (Eds) Limnology on groups of remote mountain lakes. *Documenta Ist Ital Idrobiol* 32, 7–22.

Westlake, D F (1973) Aquatic macrophytes in rivers: a review. *Polskie Archivum Hydrobiologii* 20, 31–40.

Whitton, B A (1980) Zinc and plants in rivers and streams. In: Nriagu, J O (Ed) *Zinc in the environment, Part II, Health effects*. Wiley, New York.

WHO/UNEP (1991) *Water quality. Progress in the implementation of the Mar del Plata action plan and a strategy for the 1990s*, London.

Williams, D and Musco, H (1992) *Research and technological development for the supply and use of freshwater resources – SAST Project 6. Strategic Dossier*. Report published by the Commission of the European Communities, DG XII, Brussels, Luxembourg.

Wolff, W J (1987) Rhine water in The Netherlands: effects on lakes, rivers and the North Sea. Proc 3rd Ann Conf: *Lake, river and coastal pollution: can it be contained?*, pp 49–56. Sherkin Island Marine Station, Ireland.

WRI (1990) *World resources 1990–91*. Oxford University Press, World Resources Institute, Oxford and New York.

Wright, R F (1977) *Historical changes in the pH of 128 lakes in Southern Norway and 130 lakes in Southern Sweden over the period 1923–1976*. SNSF project – TN 34/77. Norwegian Institute for Water Research, Oslo.

Zektser, I S, Lorne, G E, and Cullen, S J (1992) Groundwater pollution: an international perspective. *European Water Pollution Control*, Vol 2 (6).

Europe is surrounded by nine sea basins
Source: Stock Imagery

6 The seas

INTRODUCTION

Europe's seas, as part of the world's oceans, play a key role in maintaining the natural balance of the Earth's biosphere. Covering over two thirds of the Earth's surface, the oceans contain about 97 per cent of the world's water. Interaction between the atmosphere and the oceans and seas has a large influence on climate and weather patterns. Furthermore, plant life in the oceans is an important part of the 'lungs' of the planet, and photosynthesis in the seas and oceans is responsible for the removal of a large amount of carbon dioxide from the atmosphere.

The sea regions covered in this report are: the Mediterranean Sea, Black Sea and Sea of Azov, Caspian Sea, White Sea, Barents Sea, Norwegian Sea, Baltic Sea, North Sea and North Atlantic Ocean. Map 6.1 shows Europe's seas with the main subsidiary seas and bays, and their respective catchment and drainage areas.

Of the countries included in this report all except Austria, Belarus, the Czech Republic, Hungary, Luxembourg, Moldova, the Slovak Republic and Switzerland share coastlines of at least one of these European sea regions. Countries such as Germany, France, the UK, Denmark and Norway have coastlines with at least two major seas while the Russian Federation shares the shores of five European seas.

How human activities affect a given sea area depends very much on the sea's ability to dilute, disperse and assimilate pollutants or other influences. The characteristics which determine this include the circulation of the water, the strength and movement of the currents, the 'flushing' or retention time of waterbodies and the geological diversity of the sea floor. A comparison of some of these characteristics and physical features is shown in Figures 6.1a to 6.1e.

The open oceans, such as the North Atlantic Ocean, due to their vast areas, great depth and efficient water circulation, are still relatively unaffected by human activities compared with coastal areas and enclosed or semi-enclosed seas. The smaller, shallower seas can be classified into those which are nearly completely enclosed by landmasses and those which have a large water exchange with the oceans. The state of the seas which have only limited exchange of water with adjacent oceans is strongly influenced by their water circulation, which reflects the local dominant wind patterns (eg, the Black Sea).

The general nature of the enclosed and semi-enclosed seas is essentially dependent on whether or not the freshwater lost through evaporation is more or less than the amount of freshwater input from precipitation and direct runoff from land. In the case where evaporation is higher than the freshwater input, the surface waters become denser and sink, resulting in considerable vertical mixing of the waters. The oxygen acquired from the atmosphere thus becomes available in deeper waters to foster life. An example of this type of sea is the Mediterranean. In the reverse situation where the freshwater inflow exceeds evaporation the lighter freshwater remains on the surface. When such a sea is also isolated from the ocean, for example by a sill, oxygen becomes depleted and even absent in the deeper water. Often oxygen cannot be replenished and the variety of life-forms becomes much reduced. This is the case for the Black Sea and to some extent the Caspian Sea and the Baltic Sea.

In all seas, estuaries exist where freshwater runoff and sea water mix. They are smaller and shallower than seas, and are more influenced by the neighbouring landmass. Areas of exceptionally rich biological production, they are especially sensitive to the activities of local human populations such as excavation, construction and the discharge of domestic, industrial and agricultural pollutants.

Another important physical feature is the retention or turnover time within the seas. This is more relevant to the enclosed and semi-enclosed seas (eg, the Mediterranean and the Baltic), rather than the more open 'oceanic' seas, for example, the Barents and Norwegian seas and the North Atlantic Ocean. Retention time has a direct influence on how contaminants are retained or accumulated in the marine ecosystem. Values range from 0.1 to 3.9 years in the North Sea to 140 years in the Black Sea (Figure 6.1c). The Caspian Sea is completely land-locked and would largely retain non-degradable contaminants.

The environmental quality and status of a sea is also influenced by the amount or load of contaminants entering the sea and their degradability, persistence and toxicity to aquatic organisms. Total loads entering a sea are not only a function of the population size and industrialisation within its catchment but also on the level of treatment or control, of contaminants in discharges. The type of human activity – for example, agricultural or industrial – within the catchment is also important in determining the type as well as the quantity of load. The Black Sea and the Sea of Azov have not only the largest catchment of Europe's seas but also the largest population within the catchment (Figures 6.1d and 6.1e): contaminant loads would therefore be potentially higher than in other seas. Europe's northern seas, the White, Barents and Norwegian seas, have by far the smallest populations living in their catchments, relatively small catchment areas, and, in the case of the latter two, relatively large surface areas: these features contribute to the relatively uncontaminated nature of these seas.

An attempt to compare loads of contaminants entering each sea has been made in this report: this should be considered as only approximate as there are likely to be variations in data collection methodology, and for some

seas data were not available for all of the different sources or do not exist at all. It should also be noted that the data do not necessarily relate to the same year for all seas; in all cases the most recent available data were used. Figure 6.2 compares the riverine loads of selected contaminants into Europe's seas – this source has the most complete dataset. Where riverine data were not available, land-based (riverine and direct discharges) load data have been used. It should also be noted that for some seas, and for the various contaminants, riverine and land-based sources may not necessarily be the most important. For most seas, it has not been possible to differentiate between the contaminant load associated with particulate material and that in the dissolved phase, or to take account of non-conservative estuarine processes that may add to or decrease loads entering the sea. Total gross loads have therefore generally been used. An example of how allowance for contaminant form and estuarine processes may affect total or gross loads entering seas is given for the Mediterranean in terms of mercury, lead and zinc (Figures 6.2b, 6.2d and 6.2e, respectively).

Within Europe there are examples of regional conventions which deal with pollution at a regional or individual sea basis. These will be treated in the sections on individual seas. In addition, there are two global conventions that address marine pollution (Nauke and Holland, 1992). The two conventions are the London Dumping Convention (now the London Convention) and the MARPOL Convention: the former deals with direct disposal of waste into the sea and the latter with ship-borne operations. Both are administered by the International Maritime Organisation. Compared with the other sources of

Map 6.1
Europe's seas with subsidiary seas and bays and catchments
Source: Eurostat-GISCO

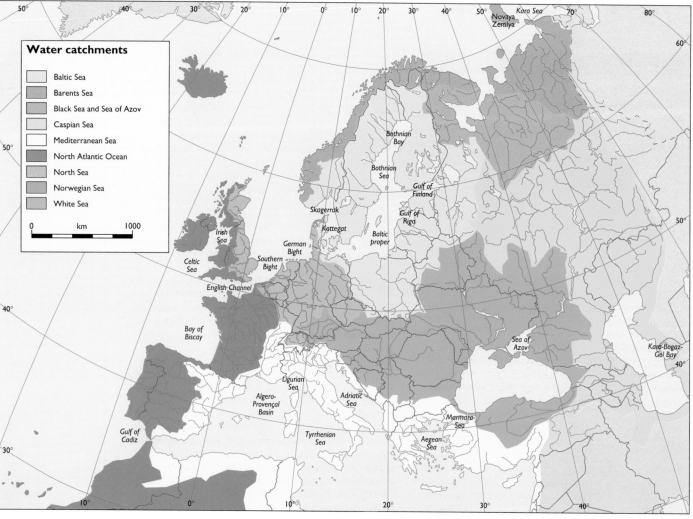

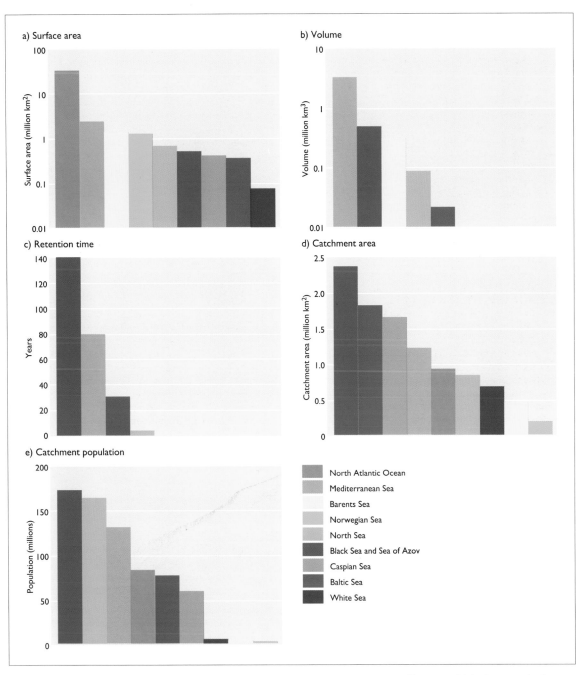

Figure 6.1
Comparison of Europe's seas by
a) surface area;
b) volume;
c) retention times;
d) catchment areas; and
e) catchment populations
Source: Data compiled by WRc from the information sources listed in the text

pollution (such as land-based and the atmosphere) these two are generally relatively small. The London Dumping Convention came into force in 1975; enforcement is undertaken through national legislation by the contracting parties. Sixty-seven countries have signed and ratified the convention. MARPOL is now known as the International Convention for the Prevention of Pollution from Ships 1973, as modified by the protocol of 1978 (MARPOL 73/78). The regulations contained in Annexes I (oil) and II (bulk liquid chemicals) are mandatory and must be applied by all parties, while those contained in Annexes III (packaged goods), IV (sewage) and V (garbage) are optional. By June 1992, 72 states, representing 90 per cent of total ship tonnage, had accepted Annexes I and II of the Convention.

COMMON PROBLEMS

When reading the text on the individual seas it will become apparent that there are problems which are recurrent or common to many of the seas in Europe. These problems, briefly outlined here, not only arise from the direct or

indirect consequence of human activities but may also be compounded by the natural variability and changes within the marine environment itself. Most of the problems are associated with, or manifest themselves more noticeably within, the coastal zone of each sea (which is closest to humankind's direct influence). The degradation of the coastal zone has, therefore, been identified as a prominent environmental problem of concern to Europe to be treated in more detail in Chapter 35.

Lack of effective catchment management, control and regulation

Many of Europe's seas have large multinational catchments which may include states with no seaboard. Contaminants within the seas may originate hundreds or even thousands of kilometres away from the sea. Hence, effective catchment management and pollution control measures require concerted international efforts and cooperation. Not only water quality, but also manipulation of river flows can affect the receiving seas. There is also exchange of water, and hence

Figure 6.2
Riverine loads of
selected chemical
elements in Europe's
seas:
a) cadmium;
b) mercury;
c) copper;
d) lead;
e) zinc;
f) total nitrogen; and
g) total phosphorus
Source: Data compiled by
WRc from the
information sources listed
in the text
Note: The pink-shaded
area for the Mediterranean
Sea indicates re-assessed
loads taking into account
where indicated the form
of the chemical elements
(dissolved or particulate)
and estuarine process
which may remove a
portion of the total load
before discharge into
coastal waters

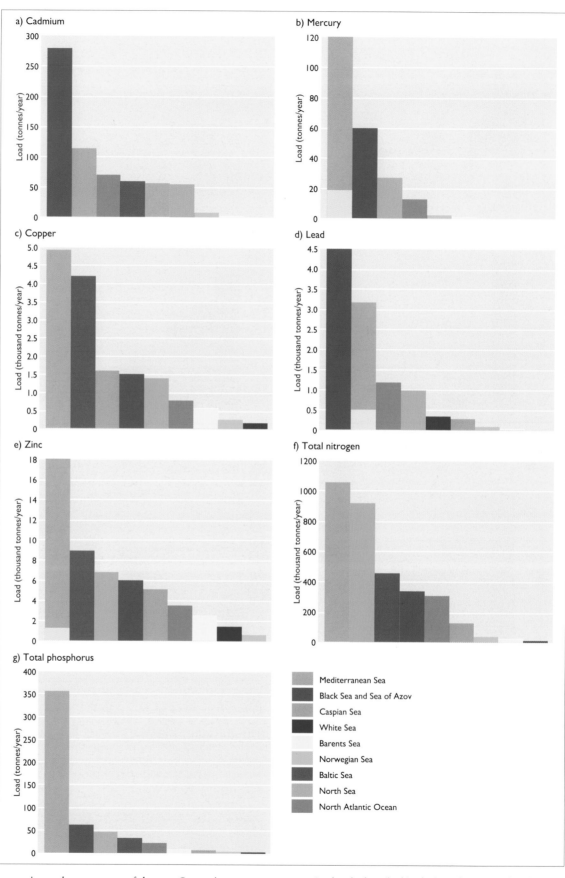

contaminants, between many of the seas. Contaminants can also be transported by air and in some seas atmospheric deposition can be very significant in relation to total contaminant loads (for example, inorganic nitrogen and some heavy metals, such as cadmium, in the Mediterranean and the North Sea).

As already described in the introduction to this chapter, there are regional conventions covering a number of Europe's seas. Acceptance of these conventions often requires a periodic assessment of loads of contaminants entering the relevant areas from different sources, for example, direct, riverine and atmospheric inputs. These data give not only a quantitative indication of the magnitude of the problem and where control and improvement measures should be targeted, but also of how loads change with time as controls and catchment management are introduced. Standard methodology for the collection of both load and quality data is defined so that results should be directly comparable. For example, sample locations, analytical methods and limits of detection for contaminants are all stipulated. Where standard methodology is not applied, large differences in the estimates of loads and quality measurements may occur.

Coastal zone pollution

The effect of any contaminant on marine water quality (and hence on sediment and biological quality) depends on its concentration, physico-chemical form (eg, dissolved or associated with particulate material), fate, behaviour, persistence and toxicity in the environment. Chemical contaminants are often placed on lists for priority control and reduction (with the exception of nutrients) on the basis of their persistence, toxicity and bioaccumulation. Nutrients are selected because of concerns over eutrophication. The most important contaminants in the coastal zone are synthetic organic compounds (eg, PCBs, and pesticides such as DDT), microbial organisms, oil, nutrients, litter, and, generally to a lesser extent, heavy metals (eg, cadmium, mercury and lead) and radionuclides.

- As well as being an important source of nutrients, sewage contributes to the microbial contamination of coastal waters – other sources include agricultural runoff and sea-birds. People bathing in contaminated waters or eating contaminated seafood may develop gastro-intestinal upsets or, in grossly polluted water, more serious illnesses such as typhoid and hepatitis (see Chapter 11). In the EU the quality of bathing water is regularly assessed for up to 19 microbiological and physico-chemical parameters according to Council Directive (76/160/EEC). For the 1992 bathing season, 10 977 coastal bathing waters were assessed for compliance with microbiological standards for total and faecal coliforms. Overall, 89 per cent of the waters complied (CEC, 1993). Further, to increase awareness about the coastal environment, the Foundation for Environmental Education in Europe in 1986 launched the Blue Flag Campaign. In order to be awarded the blue flag a beach must fulfil certain criteria regarding its environmental quality (eg, bathing water quality), management and safety provisions. In 1993 1200 beaches in 13 countries could fly the blue flag.
- Oil is a very noticeable contaminant which can adversely affect the quality of the environment. A major source to the marine environment is shipping (eg, from cleaning operations and bilge water) and it is heavy shipping routes and ports which are often most contaminated. Accidental spills cause localised and sometimes major damage to the marine environment, with sea-birds often being the most conspicuous victims. Riverine inputs and other land-based sources often make the most significant contribution to total oil loads (GESAMP, 1990).
- Metals enter the marine environment from a variety of sources, but in general rivers and deposition from the atmosphere predominate. Metals occur naturally in sea water and sediments at levels reflecting local geology and geochemistry, and are generally present in such low concentrations in the sea that they do not constitute a major threat to marine organisms except at some contaminated sites.
- Chlorinated hydrocarbon pesticides, PCBs and other synthetic organic compounds reach the sea mainly from the same sources as metals and are known to accumulate in the sediments, as do some heavy metals. From here they can be recycled through the food-chain long after their initial introduction. Some marine mammals and sea-birds are believed to have been harmed by a build-up of such organic compounds.
- Nuclear weapons testing, the dumping of radioactive waste, and discharges from nuclear power stations and reprocessing plants add to the inventory of naturally occurring radionuclides. The latter two sources are the most important to the coastal zone.
- Sediments and suspended particles in shallow coastal waters act as both sinks and sources for many environmental contaminants, and, therefore, play an important role in determining the impact of chemical contaminants in the marine environment (see also Chapter 35).
- Litter, in particular that made of synthetic materials, such as ropes, nets, plastic bags and packaging rings and straps, is increasingly being introduced to the marine environment, and is particularly noticeable in the coastal zone. Such litter often floats in the sea, entangling and killing fish, birds and marine mammals. The accumulation of some litter on beaches is potentially dangerous (eg, medical litter) while all of it aggravates beach users. The high prevalence of litter in the coastal zones of Europe's seas has been highlighted by the annual surveys undertaken by Coastwatch Europe (Coastwatch, 1994), which has played a part in raising awareness of the complex problems that affect coastal areas (see Box 6A for more details). Indeed without this volunteer effort little information would be available and it is difficult to know how such data could otherwise be collected for the cost.

Eutrophication

The enrichment of natural waters by nutrients (eutrophication), primarily nitrogen in marine waters but also phosphorus in low salinity waters, has been associated with increased primary productivity and nuisance algal growth in the coastal zones and semi-enclosed and enclosed areas of seas. Increased loads of nutrients to coastal waters of a number of Europe's seas have caused increasing eutrophication (Chapter 35). The consequences of eutrophication can be an increased frequency of algal blooms (sometimes toxic), increased water turbidity, slime production, oxygen depletion in deep waters and mass fish and benthic fauna kills. The major sources of nutrients to coastal waters are from sewage disposal, runoff from agricultural land and atmospheric deposition.

In freshwaters, phosphorus is considered to be limiting to algal growth; in marine waters this role is more likely to be played by nitrogen. Algal production in transitional waters between fresh and fully saline waters (eg, in relatively enclosed coastal waters) would be limited by either nitrogen or phosphorus depending upon whether the transitional area receives important freshwater supplies (eg, the Emilia-Romagna area of the northwest Adriatic Sea (Vollenweider et al, 1992)). In estuaries, factors such as water turbidity, light availability and retention time would be at least of equal importance as nutrient concentrations and availability.

Conflict of uses in the coastal zone

In many European seas the coastal zone is an important area for human habitation, industry and recreation (see Chapter 35). This inevitably leads to a conflict of use, not only of water (eg, for bathing, surfing, scuba diving and shell fisheries) but also of landuse (eg, harbours and marinas), which can have important secondary effects on the quality of marine environment. The lack of effective coastal zone management can lead to the loss of important components of the ecosystem and habitats, for example dunes and wetlands.

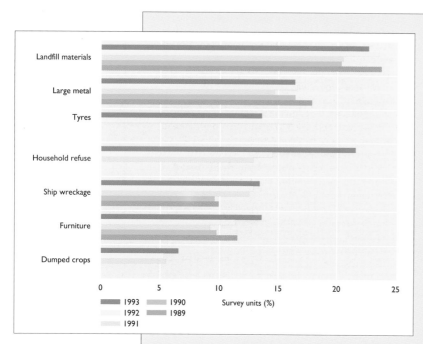

Figure 6.3 Coastwatch Europe: larger litter items in the splash zone (collective data), 1989–93
Source: Coastwatch, 1993, 1994

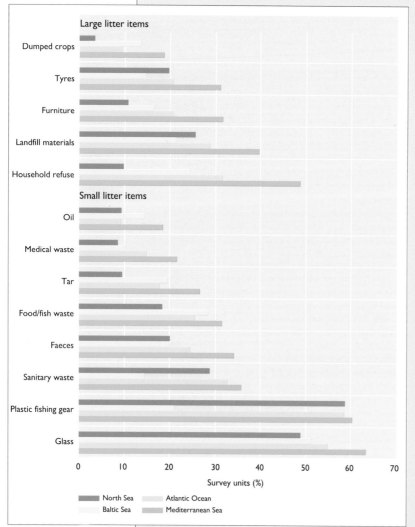

Figure 6.4 Coastwatch Europe: large and general litter items in the splash zone of four seas (collective data), 1993
Source: Coastwatch, 1993, 1994

Box 6A Coastwatch Europe surveys

The coastal surveys coordinated by Coastwatch Europe began in 1987 and ran in six countries as an international survey in 1989. They aim to collect comparable information on the coast, its characteristics and threats, effluent discharges, quality indicators, litter pollution, changes to the 'coastal rim' and destruction of habitats. The surveys are carried out by volunteers guided by regional and local coordinators and instructed by local authority representatives, teachers, wildlife officers and others knowledgable in the field. For the fifth survey completed in September/October 1993 (Coastwatch, 1994), over 100 000 volunteers took part in 21 European countries.

Each year since 1989 the sample size has been over 10 000 survey units (each unit is equal to 500 m of coastline): in 1993 there were approximately 15 000 survey units. Two countries provided more than 50 per cent of the returns: UK (34 per cent) and Spain (26 per cent). The percentage of coastline covered in each country is very variable and has been changing since 1989. In 1993 more than 7.5 per cent of the national coastline of seven countries was surveyed: Ireland (9 per cent), UK (12 per cent), Germany (15 per cent), Spain (16 per cent), The Netherlands (20 per cent), Lithuania (30 per cent) and Poland (55 per cent). Of these countries, all except The Netherlands believe the results to be representative of the accessible part of the coastline.

The Coastwatch information on coastal litter pollution provides the most comprehensive data on this subject available for Europe. A summary follows:

- In 1993, approximately one third of units surveyed were thought to be clean or almost clean from litter; this ranged from 50 per cent in the North Sea to 20 per cent in the Mediterranean. The results from 1989 to 1993 suggest that the general litter situation is worsening in the Baltic States and Poland, while improving marginally in two of the countries surveyed – Germany and The Netherlands. In most countries, packaging made of non- or slow-degradable materials, such as plastic and polystyrene, is increasingly found on coasts.
- Litter is divided into large and small categories. The most common large litter item in the 'splash zone' in 1993 was landfill materials and household refuse (Figure 6.3). Of the four large litter categories which increased in 1993, household wastes increased most significantly. A big improvement in large litter items was noticed in particular in Poland.
- Among the small or general litter items in the last three survey periods, about 80 per cent of shores were reported to have some form of plastic litter: the most prevalent types were hard plastic containers (mainly plastic bottles), foamed polystyrene/polyurethane, plastic fishing gear and other plastic materials. Glass and cans were found on 53 and 66 per cent of survey units respectively in 1993. The occurrence of plastic bottles or hard plastic containers increased over the last two surveys, while cans and other plastic, paper/card and wood items have stabilised or reduced in 1993. Counts of beverage containers yielded an average of 23 bottles per survey unit. Medical and other sanitary wastes have increased consistently over the past three years, being found in 13 and 29 per cent of survey units respectively in 1993.
- The results of the collective data are strongly influenced by the returns from Spain and the UK, which have the largest sample sizes. Comparing the results by sea area compensates for this and reveals regional differences (Figure 6.4). Most significantly, Mediterranean coasts consistently show the highest returns for all categories of litter, large and small, compared with the North Sea, Baltic and Atlantic.

Changes such as the construction of ports and tourist facilities result in the loss of habitats. The damming of rivers, resulting in a reduction of freshwater flow, alters the hydrological regime and can have serious consequences. This reduction in freshwater flow generally means a reduced sediment load which may induce coastal erosion.

Offshore activities

Offshore activities such as oil and gas exploration and exploitation, disposal of waste (sewage sludge, dredged spoil and radioactive waste) and shipping are responsible for the further release of contaminants, especially oil, into Europe's seas. Offshore oil and gas activities give rise to oil pollution of the seabed in the vicinity of the rig when drilling with oil-based muds, and of the water column due to the discharge of oil-contaminated water used in the production process. Oil-contaminated cuttings also cause smothering of biota. In addition, oil contamination may be related to tainting of fish. Solid radioactive waste has been disposed of offshore in a number of seas, though it has been estimated that the risk from contamination from past dumping is very small. Control of these activities requires both international agreements (eg, MARPOL and the London Dumping Convention) and national control measures.

Introduction of non-indigenous species

There are examples of the introduction of non-indigenous species of flora and fauna, usually accidentally (for example by shipping or by escapes from aquaculture and aquaria), to Europe's seas. In some cases the introduced species have been able to out-compete indigenous species and, if there are no predators or control, may dominate an ecological niche and in extreme cases change the structure, function and balance of the ecosystem (see Chapter 9). Examples of this in the marine environment are the introduction of the tropical alga *Caulerpa taxifolia* in the Mediterranean, and the predatory comb jelly *Mnemiopsis leidyi* (a ctenophore) in the Black Sea and Sea of Azov.

Overexploitation of resources

Europe's seas are important as a source of food (fisheries, shellfisheries and various algae) and also for other resources such as oil, gas, sand and gravel. Overfishing of commercial fish stocks causes changes in the dynamic balance between species. New technologies such as purse seining has added further pressure by increasing catches as well as killing non-target species (see Chapter 24). Shellfisheries may also detrimentally affect the marine environment. Dredging activities disrupt the sea floor habitats, and waste associated with aquaculture farms can cause local deoxygenation and eutrophication problems. Oil and gas exploration and extraction may not only introduce pollutants but may also cause physical disturbance, for example from seismic testing, which might have some detrimental impact.

Sea-level rises and climate changes

Climatic change and potential global warming are discussed in Chapter 27. One of the predicted consequences of global warming is an increase in global mean sea level. Eustatic (real) changes in the ocean level are affected by many factors, including differences in atmospheric pressure, winds, ocean currents and density of sea water; all cause spatial and temporal variations in sea level in relation to land. Over the last 100 years global mean sea level has been rising, the estimated increase ranging from 0.5 mm per year to 3 mm per year, with most estimates lying in the range 1 to 2 mm per year. This historical rise has been due to the thermal expansion of the oceans, and increased melting of glaciers and the margins of the Greenland ice sheet.

The Intergovernmental Panel on Climate Change (IPCC) has tested a number of climate scenarios, including the worst-case 'business-as-usual emissions' scenario (IPCC, 1990). This scenario predicted that the average global mean sea-level rise from 1990 to 2100 would be about 6 cm per decade (with an uncertainty range of 3 to 10 cm per decade). The predicted rise is about 18 cm by 2030, and 44 cm by 2070. Although over the next 100 years the effect of the Antarctic and Greenland ice sheets is expected to be small, they make a major contribution to the uncertainty in predictions. Any rise in sea level is not expected to be uniform over the globe. Thermal expansion, changes in ocean circulation and surface air pressure will vary from region to region as the world warms, but in an as yet unknown way (IPCC, 1990).

The potential effects of sea-level rise include: permanent inundation of low-lying land; increased frequency of temporary flooding from high tides or storm surges; changes in rates of beach, dune or cliff erosion; and salinisation of groundwater and surface water supplies, wetland ecosystems or agricultural soils. The financial consequences of sea-level rise are potentially great. For example, it has been estimated that the cost to adapt sea defences against high water and to adapt water management systems over the next century to account for a 1 m sea-level rise along the coast of The Netherlands would be about $10 billion (de Ronde, 1993).

Climate change will also potentially affect ocean circulation and mixing patterns, which could have secondary effects on nutrient availability and phytoplankton productivity. At high latitudes warming would result in diminished temporal and spatial extent of sea ice, indeed some models predict an ice-free Arctic. A significant reduction in the extent and persistence of sea-ice would have profound consequences for marine ecosystems, notably of the Nordic seas (IPCC, 1990). A temperature rise or fall of only one degree can have a significant impact on fish migration and reproduction patterns. It is thought that natural changes of these kinds have affected the success of fishing in the Nordic seas, particularly for herring, since the Middle Ages (Bernes, 1993).

Relative importance of common problems

An indication of the occurrence and relative importance of the common problems within the individual seas (not between seas) is given in Figure 6.5. For example, eutrophication is seen to be the, or a, major concern in the Baltic, Black and North seas, and a smaller relative or more localised problem in the Mediterranean and Caspian seas. In the more open seas, such as the Norwegian Sea and the North Atlantic Ocean with low relative nutrient inputs, eutrophication is not a problem. Coastal zone pollution is seen to be a problem in all of Europe's seas (although very localised in the Norwegian Sea, which has a sparsely populated coastline and a relatively small catchment population). This is discussed in further detail in Chapter 35 concerning coastal zones. Sea-level changes (as a result of climate change) would potentially affect all of Europe's seas.

THE EUROPEAN SEAS: OVERVIEW

In the rest of this chapter each individual sea is described under the following broad headings:

● location and general geographical and sociological features, and any relevant international conventions and agreements (general situation);

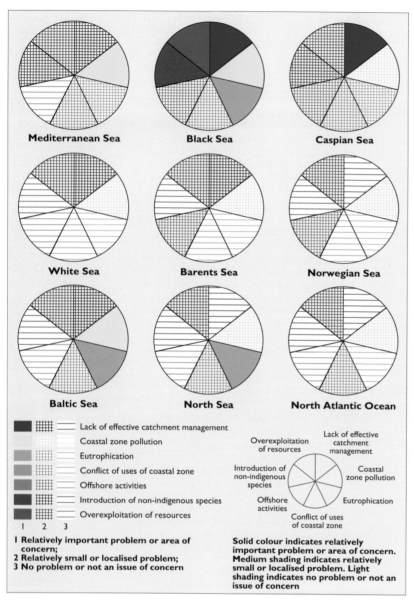

Mediterranean Sea

Black Sea

Caspian Sea

White Sea

Barents Sea

Norwegian Sea

Baltic Sea

North Sea

North Atlantic Ocean

Lack of effective catchment management
Coastal zone pollution
Eutrophication
Conflict of uses of coastal zone
Offshore activities
Introduction of non-indigenous species
Overexploitation of resources

1 2 3

1 Relatively important problem or area of concern;
2 Relatively small or localised problem;
3 No problem or not an issue of concern

Overexploitation of resources

Introduction of non-indigenous species

Offshore activities

Conflict of uses of coastal zone

Lack of effective catchment management

Coastal zone pollution

Eutrophication

Solid colour indicates relatively important problem or area of concern. Medium shading indicates relatively small or localised problem. Light shading indicates no problem or not an issue of concern

Figure 6.5
Relative importance of the 'common problems' in each sea
Source: WRc
Note: This assessment does not attempt to compare the relative scale or magnitude of the problems <u>between</u> the seas

- physical and hydrological nature (physical features);
- biological characteristics (biological features);
- quantity and source of contaminants entering the seas (inputs);
- level and distribution of various marine contaminants (contaminant levels);
- effects of these contaminants and other human activities on the ecosystem (biological effects); and
- the identification of the most important problems of the sea region (conclusions).

THE MEDITERRANEAN SEA

General situation

The Mediterranean Sea is the largest of the semi-enclosed European seas (see Map 6.1). It is surrounded by 18 countries and has shores on three continents (Europe, Africa and Asia) with a combined population of 129 million people in the catchment draining into sea, and sharing a coastline of 46 000 km. It is one of the leading tourist areas in the world, hosting 100 million visitors every year. This influx of people increases the waste discharges from domestic and industrial sources.

The Mediterranean Sea has an average depth of 1.5 km, though more than 20 per cent of the total area is covered by

water less than 200 m deep (UNEP, 1989). The sea consists of two major basins, the eastern and the western. There are also smaller regional seas within the Mediterranean: the Ligurian, Tyrrhenian, Adriatic and Aegean seas. It is linked to the Atlantic by the Strait of Gibraltar, with the Black Sea and Sea of Azov by the Dardanelles, the Sea of Marmara and the Bosporus, and with the Red Sea by the Suez Canal. The Mediterranean Sea is characterised by low precipitation, high evaporation, high salinity, low tidal action and relatively low nutrient concentrations outside the inner coastal zone and parts of some regional seas.

The coasts of the northwestern Mediterranean are the most affected by pollution because of the concentration of urban populations, industrial activities and discharges of major rivers including the Ebro and the Rhone. The Adriatic receives the discharge of the River Po. The North African coast, in contrast, is for most part arid with little urbanisation or industrialisation. Pressures on the marine environment therefore vary widely depending on the local or regional situation.

Countries bordering the Mediterranean Sea met in Barcelona in 1975, under the auspices of UNEP, to draw up a programme of action to protect the Mediterranean – this became the Barcelona Convention for the Protection of the Mediterranean Sea against Pollution. The UNEP Regional Seas programme had been initiated in 1974 when the Mediterranean had been selected as a 'concentration' area where coastal states would be assisted in the implementation of an action plan.

The Mediterranean Action Plan (MAP) consists of three main components: legal; environmental assessment; and environmental management. The legal component is contained within the Barcelona Convention. The long-term Mediterranean Pollution Monitoring and Research Programme (MEDPOL) was launched in 1975 under MAP, Phase 1 lasting until 1980. Phase II, endorsed in 1981, foresees monitoring at four levels: sources of pollution; nearshore areas; offshore areas; and transport of pollutants from the atmosphere. MEDPOL monitoring started in 1983 through the implementation of National Monitoring Programmes and at present 16 countries have ongoing programmes and are submitting data. The Blue Plan, which forms a part of the environmental management component of MAP, was launched in 1979 to assist the Mediterranean countries in making appropriate practical decisions for the protection of their marine and coastal environment.

Physical features

Surface water, entering the Mediterranean from the Atlantic, migrates generally towards the east (Map 6.2). Evaporation processes transform this surface water into denser, deep water which flows east to west back into the Atlantic. In fact, water loss by evaporation exceeds the water input from runoff and precipitation, resulting in the Mediterranean's characteristic high salinity (average 38.5 per thousand, ranging from 37 in the west to 39 in the east). Strong vertical currents in winter ensure mixing of the water column and oxygenation of the deep waters. It takes on average about 80 years for the water in the Mediterranean to be completely exchanged.

Biological features

The Mediterranean Sea has a high species diversity but its biological productivity, while being extremely varied, is among the lowest in the world due to extremely low nutrient concentrations (UNEP, 1989). The total number of species of animals and plants has been estimated to be around 10 000 (Boudouresque, 1993). Its fauna includes many endemic species and is notably richer than the fauna of the Atlantic coasts. The eastern and western Mediterranean, separated by the relatively shallow straits between Sicily and Tunisia, show differences in resident fauna and flora, indicating a degree of isolation between

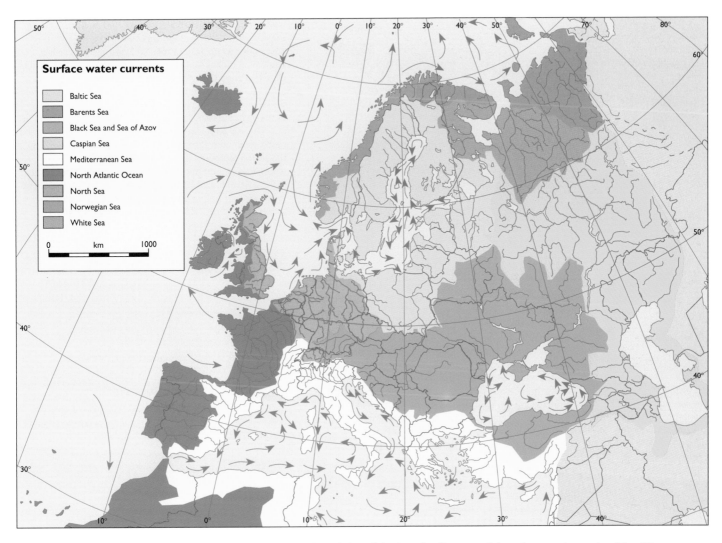

Map 6.2
Europe's seas with
main surface water
currents
Source: Eurostat-GISCO
and WRc

the two regions (Clark, 1986). The biodiversity of the western Mediterranean is also greater than that of the eastern (Boudouresque, 1993).

The most threatened species in the Mediterranean is the monk seal, *Monachus monachus*, which is one of the ten most threatened species of mammals in the world (IUCN, 1988) (see Chapter 9).

The loggerhead turtle, *Caretta caretta*, and the green turtle, *Chelonia mydas*, nest regularly and in significant numbers in the Mediterranean (COE, 1990a) – both are recognised by the IUCN as globally threatened species, the former being ranked as 'vulnerable' and the latter as 'endangered' (IUCN, 1988). There are on average 2000 female loggerhead turtles nesting annually in the Mediterranean, the majority in Greece and Turkey. Green turtles, as far as is known, nest only in the extreme southeast of Turkey and in Cyprus (COE, 1990a).

There are around nine species of whales and dolphins regularly found in the Mediterranean (Tethys Research Institute, 1991). The biggest populations are found in the particularly rich pelagic zone of the western Ligurian Sea.

The extensive seagrass beds of *Posidonia oceanica* are an important part of the Mediterranean marine ecosystem, often occupying a considerable part of the littoral zone. A characteristic feature is the formation of *Posidonia* seagrass beds parallel to the shore in sheltered and shallow bays, isolating a coastal lagoon (Augier, 1982). These beds have suffered greatly from physical modifications of the coast (see Chapter 9). *Posidonia* plays an important role in the ecosystem: through the production of organic material at the base of the food-chain; as a primary oxygen producer; as a feeding and nursery area for numerous species of fish (many commercially important);

through the stabilisation of sediments; and through attenuation of wave and swell (protection of beaches) (Boudouresque, 1993).

A tropical, non-indigenous alga, *Caulerpa taxifolia*, has been observed in the Mediterranean since 1984. First seen in the Monaco area (perhaps as a result of an accidental release from an aquarium in Monaco (Meinesz and Hesse, 1991)), this alga had, by 1990, been found up to 150 km from Monaco at Toulon. At some locations it inhabits a wide range of substrates, including rock, mud and sand, and a wide range of depths, 3 to 35 m, and has achieved 100 per cent coverage in some places. Wherever it becomes established it considerably modifies the vegetal communities in the infralittoral zone. It also contains a toxin which may inhibit some other organisms such as grazers, epiphytes and competitors. It appears to be consumed by only a few fish species.

There are important fisheries in the Mediterranean, with fish such as mullet (*Mugilidae*) and hake (*Merluccius* spp) being in most demand (see Chapter 24). Other fish such as anchovy (*Engraulis encrasicolus*), sardines (*Sardina pilchardus*) and mackerel (*Trachurus* spp) in the northwest are also intensively fished. Oil pollution in some parts of the sea has led to tainting of a variety of fish and bivalves.

Inputs

Contaminants enter the Mediterranean from rivers, direct discharges (land-based and offshore), atmospheric deposition, and through water exchange primarily with the Atlantic Ocean and Black Sea. A major assessment of land-based inputs into the Mediterranean, reported by UNEP et al

Figure 6.6
Mediterranean Sea:
proportions of oil
load from different
sources
Source:
UNEP/IMO/IDC, 1987

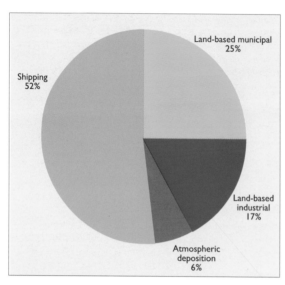

tankers during ballasting and loading operations and during the washing of the bilge and tanks. Land-based sources account for 42 per cent of total oil loads, and the atmosphere accounts for 6 per cent (Figure 6.6). Natural seeps have also existed over geological time-scales, particularly in the northeastern parts of the sea (UNEP, 1989). The coastal area off Libya receives the highest oil input. Oil spills are also a problem, but are infrequent and generally localised.

Nuclear power stations are operating in three countries on the Mediterranean: Spain, France and Slovenia. However, it is the nuclear reprocessing plant at Marcoule in France that represents the most important source of artificial radioactivity in the sea (see Figure 5.32). It is also apparent that atmospheric inputs are a significant proportion of the total input load of some radionuclides such as Cs-137 (Martin et al, 1989).

Contaminant levels

In the Mediterranean Action Plan status report of 1989, it was reported that the concentrations of trace metals and chlorinated hydrocarbons in sea water and sediment should be considered (at that time) with caution, and typical concentrations could not be identified – this was because of inadequate analytical quality control and different analytical methods. However, concentrations in biota were considered to be more reliable because extensive laboratory intercalibration had been carried out (UNEP, 1989). Over the last five years the CEC has funded research through a number of projects implemented in the western Mediterranean and some regional seas (eg, EROS 2000) as reported by Scoullos (1993).

Heavy metals

Only a few data have been found on the open-sea metal concentrations in the Mediterranean (Fowler, 1990). For mercury these concentrations are similar to concentrations in the adjacent North Atlantic, whereas for cadmium the concentrations (1 to 7 ng/l in the open sea), while still being well below accepted values for pollution, are generally higher than in the Atlantic. Recent analyses of cadmium show some relatively high values for certain coastal areas of Spain and Italy (5 and 10 ng/l, respectively). Lead values are also slightly higher than for Atlantic water – between 20 and 40 ng/l in the northwest Mediterranean, compared with oceanic levels of 5 to 15 ng/l. Offshore concentrations of zinc are reported to range between 150 and 240 ng/l (Morley and Burton, 1991); higher levels of 410 ng/l are found in the north Adriatic (Scoullos, 1993). A gradient of increasing concentrations of cadmium, lead, copper and zinc from the south to the north of the Adriatic Sea is also reported.

Elevated concentrations of mercury, cadmium, zinc and lead in sediments are found at 'hot-spots', which are generally in the coastal zones receiving industrial effluents, solid waste and domestic sewage. For example, concentrations of up to 37 mg/kg dry weight of mercury have been reported, compared with a typical background of 0.05 to 0.1 mg/kg dry weight (Fowler, 1990). Zinc concentrations are reported to be as high as 6480 $\mu g/g$, 5930 $\mu g/g$ and 2550 $\mu g/g$ (dry weight) at 'hot spots' along the coasts of Spain, at Venice and at Marseilles, respectively (Scoullos, 1993).

The MEDPOL programme has used two indicator species for monitoring contaminants, the mussel (*Mytilus galloprovincialis*) and the red mullet (*Mullus barbatus*). Mercury has been given special attention because recent data show that Mediterranean fish (eg, bluefin tuna, *Thunnus thynnus*, sardines, *Sardina pilchardus*, anchovy, *Engraulis encrasicolus*, and scads, *Trachurus* spp) and other marine animals generally have higher levels than those of the North Atlantic (Figure 6.7) (FAO, 1986). However, results are very variable and are available mainly for the northern parts of the sea.

(1984), relied upon data collected by national monitoring networks. More recent work, such as that undertaken within the CEC EROS 2000 project, has indicated that these earlier estimates are too high (eg, CEC, 1992; Martin et al, 1989 and Dorten et al, 1991). In particular, when non-conservative processes, which reduce gross loads passing through estuaries, are taken into account, and when the partitioning of contaminants between the dissolved and particulate phase are considered, the net riverine loads of metals, such as mercury, copper, lead and zinc, entering the main body of the sea are now calculated (Dorten et al, 1991; Martin et al, 1989) to be much lower than estimated in the earlier compilation. In Figure 6.2(a–e) both the UNEP and the more recent estimates of loads for some heavy metals are illustrated.

The Mediterranean basin is unusual in that it is rich in mercury deposits, for example cinnabar and metallic mercury (Zafiropoulos, 1986) – 65 per cent of the world's mercury mineral resources are located in the region (Scoullos, 1993). Natural inputs of mercury (via rivers) can be locally very significant, for example into the Tyrrhenian Sea (Baldi, 1986), compared to anthropogenic sources. Dissolved mercury riverine inputs to the Mediterranean are now estimated to be between 1.7 and 20 tonnes per year (Dorten et al, 1991), compared with the original estimate of 120 tonnes per year (UNEP et al, 1984) from rivers.

Recent work in the northwestern basin has assessed the relative contributions of atmospheric and river inputs of about 40 elements including heavy metals, radionuclides and nutrients. For the majority of the elements analysed the proportion of atmospheric deposition relative to total deposition (from rivers and atmosphere) did not exceed 20 per cent (CEC, 1992). However, for heavy metals atmospheric input generally appears to predominate: this derives from both the heavily industrialised northern boundary and dust loads originating in the Sahara region (Dorten et al, 1991). For cadmium, lead and copper (in the dissolved phase) atmospheric inputs are much greater than riverine (50, 200 and 5 times respectively) (Martin et al, 1989). For inorganic nitrogen, atmospheric and riverine inputs are roughly equivalent, and in the case of inorganic phosphorus, riverine inputs are most important. Major anthropogenic contaminant loads are discharged from the rivers Nile, Rhone, Ebro and Po.

There is also an exchange of dissolved trace metals between the Mediterranean and the Atlantic Ocean through the Strait of Gibraltar, and also with the Black Sea via the Bosporus. Recent mass balance calculations (Martin et al, 1993) indicate that in the case of dissolved copper, nickel and cadmium there may be a net export to the Atlantic.

Inputs of oil to the sea are estimated at 635 000 tonnes per year (UNEP, 1989). Of this, around half is spilt from

Sea gooseberry in the Mediterranean
Source: Frieder Sauer, Bruce Coleman Ltd

Synthetic organic compounds

A three-fold decrease in PCBs has been detected in coastal waters of the northwestern Mediterranean between the mid-1970s and the period 1978 to 1982. This trend has also been confirmed by measurements made along the French coast in 1984.

The limited number of measurements of organochlorines in sediments in the Mediterranean show a few 'hot spots'. For example, elevated sediment concentrations of PCBs have been found near the Athens sewage outfall; in the Bay of Naples; near the Marseilles outfall; and offshore from Nice (UNEP, 1989).

Measurements on biota are scattered and variable, but again the industrialised regions and major estuaries stand out. Mussels containing elevated levels of PCBs have been observed at Toulon and Marseilles near the Rhone estuary (UNEP, 1989). Levels of PCBs and other organochlorines in red mullet and mussels show a general decrease from the more northern parts of the sea to the south and east. There are as yet not enough reliable data for trend analysis.

Oil

Oil pollution in the Mediterranean Sea is associated mainly with shipping routes, ports and oil and gas exploration activities. Data on concentration levels of hydrocarbons show increases over recent years, especially with regard to concentrations in water and on beaches. In general, concentrations of dissolved/dispersed petroleum hydrocarbons in open waters are between 0 and 5 $\mu g/l$, and values above 10 $\mu g/l$ have been observed near the shore, particularly near industrialised areas and river mouths (UNEP, 1989).

The amounts of petroleum hydrocarbons in marine organisms and sediments in the area are poorly known. The available data cover mainly the coastal zone, and thus the contamination of the open waters is less well known. Results also show an increased level of petroleum hydrocarbons in

sediments compared with the concentrations in water, which indicates that they may be accumulating in the sediments. Aliphatic and aromatic petroleum hydrocarbon concentrations in sediments range from 1 to 62 $\mu g/g$, and 2 to 66 $\mu g/g$, respectively, along the Spanish coast outside harbours, oil terminals and river mouths (Scoullos, 1993). Polycyclic aromatic hydrocarbons (PAHs) in the northwest Mediterranean range from 0.4 to 0.7 $\mu g/g$ in the deep sea basin, from 0.3 to 0.5 $\mu g/g$ on the continental shelf, and from 0.4 to 5 $\mu g/g$ off Barcelona (Tolosa et al, 1993).

Almost no observations exist about the effect of petroleum hydrocarbons on Mediterranean marine organisms (UNEP 1989).

Microbiological contamination

The results from the 1992 bathing season (CEC, 1993) indicate that 97 per cent of the designated bathing waters around Greece complied with the mandatory standard of 2000 faecal coliforms per 100 ml, as specified in the EC Directive on the quality of bathing water, and 95 per cent were within the guideline value of 100 per 100 ml. The bathing water quality around the Italian coastline was also reported to be of relatively high quality, with 92 per cent complying with the mandatory standard and 85 per cent with the guideline value. The main areas of non-compliance included the Naples and Caserta districts, around Genoa in northern Italy and along the north coast of Sicily around Palermo. Along the Spanish Mediterranean coastline, approximately 95 per cent of bathing waters complied with the mandatory standard: non-compliant waters included those around the Granada, Malaga and Valencia areas. In France, 95 per cent of bathing waters complied with the Directive.

Nutrients

The open Mediterranean Sea is nutrient-depleted. Typical 'background' concentrations of nitrate nitrogen are 7 μg N-NO_3/l, in moderately eutrophic areas 21 μg N-NO_3/l, and in heavily eutrophic, 70 to greater than 110 μg N-NO_3/l (GESAMP, 1990). Corresponding values for phosphate are 0.93, 4.7 and 9.3 μg P-PO_4/l, respectively. It is generally Mediterranean shores adjacent to urban agglomerations and tourist resorts (such as the Adriatic coast) which show the highest nutrient concentrations.

Figure 6.7
Mediterranean Sea: mean and range of concentrations of mercury
Source: UNEP, 1989

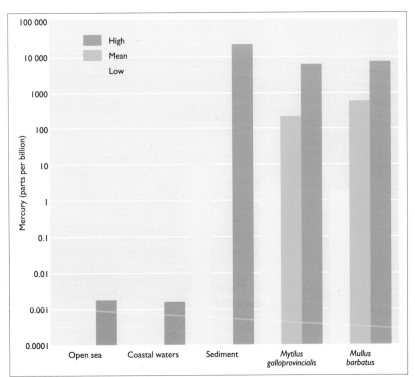

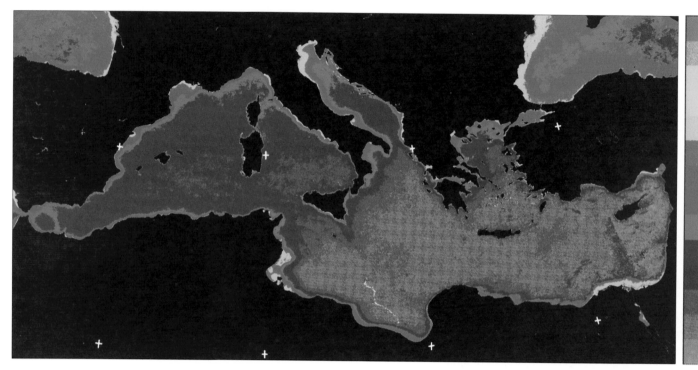

Map 6.3
Satellite image of the Mediterranean: the colour bars show concentrations of chlorophyll-like pigments, indicating eutrophication
Source: JRC/CEC

Note: The image is a four-year composite for the period 1979-82.

Monitoring in the Mediterranean
Source: Frank Spooner Pictures

Radionuclides

Immediately following the Chernobyl accident, the surface water and sediment concentrations of caesium (Cs)-137 increased by one to two orders of magnitude and by a factor of 2 to 4, respectively (UNEP, 1989). The distribution of the fall-out was very heterogeneous due to variations in local conditions and, in particular, was concentrated around river discharges, reflecting the fall-out that occurred within the drainage catchments. Areas of higher concentrations included the Rhone estuary and the Italian coast. The high concentrations decreased again relatively rapidly and were close to pre-Chernobyl values by the end of 1989. It has been estimated that the Chernobyl accident resulted in a 25 to 40 per cent increase of the total amount of Cs-137 in the Mediterranean Sea (UNEP, 1989). The effects of artificial radionuclides on living organisms in the Mediterranean are considered to be negligible.

Biological effects

The Mediterranean Sea is generally considered to be oligotrophic (poor in nutrients) and does not have widespread problems arising from nutrient enrichment. However, coastal areas which receive anthropogenically enhanced nutrient loads from rivers and the direct discharge of untreated domestic and industrial wastewater are most susceptible to eutrophication. Examples can be observed in many coastal lagoons, estuaries and semi-enclosed bays, particularly in northern areas (eg, bays of the Ebro delta, the Albufera of Valencia, the coastal lagoons of southeastern France, the lagoon of Tunis, the Kastela Bay in Croatia and the Izmir Bay in Turkey (Estrada, 1993)). This is clearly illustrated in the satellite image in Map 6.3.

Two areas show extensive cultural eutrophication: the Gulf of Lions and the Northern Adriatic Sea. In the Northern Adriatic extensive dinoflagellate and diatom blooms occur in the spring and autumn: on occasions massive quantities of mucilage are produced (Estrada, 1993). However, the species responsible for this mucilage or gel production or the trigger mechanism is not yet fully known (Barth and Fegan, 1990). The key features of the North Adriatic which make it vulnerable to gel production are believed to be: shallowness (under 35 m); low turbulence during the summer and the high riverine nutrient input (Estrada, 1993). Studies in the Emilia-Romagna coastal waters of the northwest Adriatic (to the south of the River Po delta) have indicated that phosphorus rather than nitrogen is the prevailing limiting nutrient (Vollenweider et al, 1992). However, in some circumstances in this area, such as in the summer and when nitrogen supply is low, nitrogen may also become limiting. The most severe eutrophic conditions are restricted to semi-enclosed bays and to the areas within the estuarine plume of the Po, which dominates the freshwater inputs. Numerous point sources are also important nutrient inputs. In response to the increased nutrient load, primary productivity increases with a maximum in the Emilia-Romagna area. During summer this area has suffered from persistent heavy algal blooms whose eventual decomposition causes anoxic conditions and mass kills of fish and benthic fauna. In 1990 the accumulation of gelatinous material produced by *Phaeocystis* spp on the beaches of Benicasim (Spain) had an adverse effect on the tourist trade.

'Red' tides also occur in the Mediterranean, often

associated with the dinoflagellate *Noctiluca scintillans* (eg, frequently along the Catalan coast since the 1970s) (Estrada, 1993). Toxic algal episodes are also reported, for example, paralytic shellfish poisoning (from *Alexandrum minutum* and *Gymnodinium catenatum*) and diarrhetic shellfish poisoning (from *Dinophysis* spp). There is increasing evidence to suggest that mass development of toxin-producing unicellular algae is related to eutrophication. However, a cause–effect relationship between increased productivity and toxin production has not been proven (Barth and Fegan, 1990).

There has been a marked reduction in the *Posidonia oceanica* (seagrass) beds over the last decades. The reduction has been particularly noticeable around large industrial ports (eg, Barcelona, Marseilles, Toulon, Nice, Genoa, Naples, Athens and Algiers). The depths at which *Posidonia* is able to grow has also decreased. This has been associated with an increase in water turbidity reducing the amount of light exposure on the sea bed. Because of the role *Posidonia* plays in the Mediterranean ecosystem (see above, on biological features), its reduction is considered to have severe economic consequences (Boudouresque, 1993).

Conclusions

- The Mediterranean Sea has the highest species diversity of all of the European seas. It is the 'home' of a number of species endangered by human activities including: pollution of the environment, uncontrolled fishing, and construction causing the loss of the natural habitats.
- The areas of the Mediterranean most severely affected by pollution are the coastal areas, in particular the northern part of the coast of Spain, those of France and Italy and the Adriatic. Other coastal areas in the vicinity of major cities (eg, Athens) are also considered to be polluted.
- Pollution is caused mostly by river flow draining agricultural land and the discharge of inadequately treated domestic and industrial wastewater.
- Because the Mediterranean Sea is one of the most popular tourist destinations in the world (see Chapter 25) these problems are further aggravated during the tourist season. Oil pollution is also a relatively important problem for the Mediterranean.

THE BLACK SEA AND THE SEA OF AZOV

General situation

The Black Sea, the world's largest land-locked and anoxic sea, is located in a semi-arid climatic zone (see Map 6.1). It is bordered by Bulgaria and Romania to the west, Ukraine to the north, the Russian Federation and Georgia to the east and Turkey to the south. Although only these six countries surround it, the total catchment area draining into the Black Sea is over five times the size of the actual sea and includes parts of 21 countries. The northwestern basin receives the heaviest impact from human activities because, of the 171 million people living in the Black Sea catchment, 81 million live in the River Danube basin alone. The human activities in the catchments have had major effects, not only on the rivers, but also on the receiving water body, the Black Sea. The Black Sea coasts are important tourist areas, with up to 40 million visitors during the summer.

The Black Sea has until recently been unprotected by any common policy or legal regime. In 1992 a legal Convention for the Protection of the Black Sea, based on the Barcelona Convention, was signed by Bulgaria, Georgia, Romania, the Russian Federation, Turkey and Ukraine. The Global Environment Facility (GEF – co-managed by the World Bank and UNEP) is assisting the countries in drawing up an action

plan for implementation of the Convention. A recent common policy declaration (the Odessa Declaration, April 1993) by all six environment ministers of the Black Sea Convention calls for improved assessment and monitoring of contaminants, and the 'development of comprehensive and co-ordinated plans for the restoration, conservation and management of living natural resources'. A project for environmental management of the Black Sea was recently approved by GEF. This project should provide the urgently needed short-term support for environmental assessments, institutional capacity building, pre-investment feasibility studies and an emergency investment portfolio of actions to complete unfinished key pollution control projects such as sewage treatment works.

There is also a Convention on Fishing in the Black Sea implemented in 1959 by Bulgaria, Romania and the former USSR. However, it was not signed by Turkey, an important Black Sea fishing nation. Also of direct relevance to the Black Sea is the Bucharest Declaration which was signed by eight countries in 1985 and under which Danube water quality data are collected and the information exchanged. The control of pollution in the Danube catchment is very important as it is a major source of contaminants to the Black Sea. Building on this Declaration, the riparian states decided in February 1991 to elaborate a convention on the protection and management of the river Danube and an ecological agreement for the entire basin (signed in Sofia in June 1994).

The main sources of information for the following sections have been Balkas et al (1990), Dechev (1990), Mee (1991 and 1992) and Mnatsakanian (1992). Additional sources are identified in the text where appropriate.

Physical features

The Black Sea has a surface area of 461 000 km^2 and an average depth of 1240 m. About 25 per cent of its area is occupied by its northwestern continental shelf which is less than 200 m deep. This northwestern area is subject to the discharge of the largest rivers (the Danube, Dnepr, Dnestr and Yuzhnyy (Southern) Bug – see Map 5.11) entering the Black Sea. The sea's only link to other seas is with the Mediterranean through the Bosporus, the Sea of Marmara and the Dardanelles. In the north the Kerch Strait connects the Black Sea to the shallow Sea of Azov. This latter sea has an area of 39 000 km^2, an average depth of 8 m and a maximum depth of 12 m. The major rivers flowing into the Sea of Azov are the Don and the Kuban.

The link of the Black Sea to the Mediterranean Sea via the Dardanelles is shallow (50 m) and narrow, and the inflow of salty water from the Mediterranean is outweighed by the outflow of surface water from the Black Sea. This water deficit is compensated by freshwater discharge. These two inflow sources have resulted in a stratification of the water column with fresher water (salinity 17.5 to 19 per thousand) at the surface and denser water (salinity average about 22 per thousand) at depth. This permanent halocline (change in salinity) located between 100 and 200 m is a distinguishing characteristic of the Black Sea. The difference in density between these two water masses, coupled with the absence of well-defined vertical currents, prevents the mixing of these two water layers and the subsequent penetration of oxygen from the surface to the bottom. Over the years organic matter has been sinking and decomposing in the deep waters of the Black Sea. For this reason the Black Sea is permanently anoxic below a depth of 150 to 200 m, and 90 per cent of its total water volume is anoxic.

Under these anoxic conditions, further degradation of organic matter takes place using oxygen bound in nitrates and especially in sulphates. The latter chemical reduction results in the formation of hydrogen sulphide. Hydrogen sulphide has been produced for millennia and has contaminated 90 per cent of the total waterbody of the Black Sea, making it fit only for anaerobic bacteria. The depth of the aerobic and anaerobic layers in the water column reflects the steady-state balance of the flows of organic matter and oxygen that has been established in the past centuries. The depth of the oxygenated layer varies spatially. It is greatest near coasts and less over the deeper basin. Whether the oxygenated layer is still decreasing is currently a question of scientific debate.

The surface water temperature varies on both a seasonal and a regional basis, with a range from 0°C in the northwestern coastal and shelf area during winter up to 25°C in the west during summer. The deep water remains constant at 9°C at 1000 m depth all year round.

The main feature of the water currents is an anti-clockwise boundary current which runs approximately parallel to the shore, two cyclonic gyres (spiral flows) that almost divide the basin in two and a series of smaller and anti-clockwise eddies originating from the larger features (see Map 6.2). The freshwater input and its associated contaminants from the large rivers on the northwestern shelf are transported by gyres throughout the Black Sea.

During the last 30 years the damming of the Danube and other major rivers, particularly in the former USSR, has resulted in an estimated decrease of freshwater input to the Black Sea of up to one fifth. This reduction in freshwater and associated sediment load has resulted in coastal erosion, particularly of the Romanian coasts, and an increase in salinity. Some Romanian beaches have been eroded at a rate of 12 m per year. Partly because of this erosion, extensive coastal engineering work has been carried out along the Romanian coast to protect the shoreline and to enlarge Constanţa harbour. As a consequence a large amount of material has been dumped in the coastal waters, reducing the water quality. The most severely affected biota have been filter feeding organisms and rocky zone macroflora.

Heavy pollution and increased agricultural activity coupled with limited mixing results in oxygen depletion in the coastal waters, especially during summer, further exacerbating the naturally anoxic nature of the Black Sea.

Biological features

The most biologically productive areas of the Black Sea include the northwestern and northeastern parts of the basin, including the Azov Sea. This is because these areas receive enhanced nutrient loads from the inflowing rivers as well as being subject to efficient vertical mixing.

There are 180 species of fish in the Black Sea, more than

half of which are also present in the Mediterranean Sea. There are three species of dolphin. The brackish nature of the Black Sea restricts the number of species of organisms present. Because the water below about 150 m contains no oxygen it is largely devoid of life, except for anaerobic bacteria.

The Ukrainian part of the Black Sea has four conservation/nature reserves which comprise extensive marine areas (COE, 1990b). The Danube Delta Protected Area has a rich and varied marine fauna, with 92 fish species including salmon, sturgeon and *Clupeidae* (including many endemic species). The area also has great ornithological value with 225 species of aquatic birds. Chernomosk National Reserve includes large marine areas (11 000 ha) of the Black Sea and Sea of Azov: its waters are productive with extensive mats of *Zostera* and *Phyllophora*. Its rich fauna includes the very rare Black Sea herring. Karadag Park in the Crimea (809 ha) also has extensive beds of *Zostera*, rocky shores and many nesting sea-birds. Mys Martjan Park, Crimea, holds many species that have disappeared from other areas.

Inputs

The nutrient load to the Black Sea has increased markedly in recent decades, probably as a consequence of the widespread use of phosphate detergents and intensification of agriculture. This has been reflected in an increase in the concentrations of nitrogen and phosphorus compounds. Between 1970 and 1991 a two- to three-fold increase in the nitrate maximum, just above the halocline, has been reported. During the same period a seven-fold increase in phosphate concentration was observed along the Romanian shelf (Mee, 1992).

A recent study (Ludikhuize, 1992) commissioned by UNEP (for the Global Environment Facility) gave a preliminary assessment of inputs to the Black Sea. It suggests that 65 per cent of the nitrogen input to the Black Sea is via rivers, 40 per cent alone from the Danube River (that is, 340 kilotonnes total inorganic nitrogen). This study also estimated the inputs of nitrogen from different sources (Figure 6.8a) of which agriculture and domestic wastewater contributed the largest share (31 and 26 per cent, respectively). Other important inputs were from industry and atmospheric deposition. However, data about the different sources were very limited in this study and values are rough estimates. There are few data on phosphorus inputs but the same study estimates that the Danube River is again the most imporant source, contributing some 60 kilotonnes of total phosphorus input to the Black Sea (Figure 6.8b).

Even though nutrient loads may be reduced in the

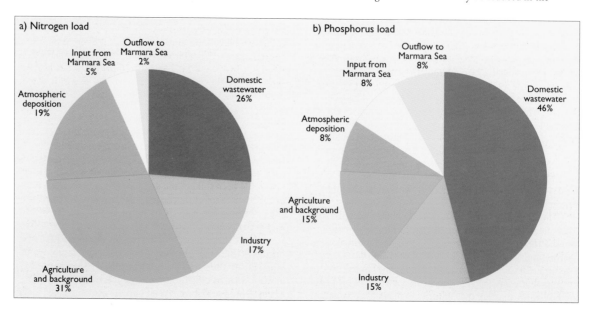

Figure 6.8
Black Sea:
proportion of
a) nitrogen and
b) phosphorus loads
from different sources
Source: Ludikhuize, 1992

a) Nitrogen load

Input from Marmara Sea 5%
Outflow to Marmara Sea 2%
Domestic wastewater 26%
Atmospheric deposition 19%
Industry 17%
Agriculture and background 31%

b) Phosphorus load

Input from Marmara Sea 8%
Outflow to Marmara Sea 8%
Domestic wastewater 46%
Atmospheric deposition 8%
Agriculture and background 15%
Industry 15%

future, release of sediment-bound nutrients (particularly from the continental shelf areas) would fuel eutrophication for many years or decades with gradually diminishing intensity.

There is limited information available on the loads of heavy metals being discharged into the Black Sea: it is, however, reported that the Danube alone (based on the Bucharest Declaration's 1989 figures) is responsible for discharging annually up to 280 tonnes of cadmium, 60 tonnes of mercury, 900 tonnes of copper, 4500 tonnes of lead, 6000 tonnes of zinc and 1000 tonnes of chromium (Mee, 1992). More recent water quality data on the rivers Danube, Dnepr, Don, Kuban and Belaya were obtained during the collection of data for inclusion in Chapter 5 on inland waters (EEA-TF questionnaire). These indicate generally similar total loads (to the Danube) from these rivers – 87 tonnes of cadmium, 1500 tonnes of copper, 825 tonnes of lead, 2600 tonnes of zinc, 207 kilotonnes of nitrate and 47 kilotonnes of total phosphorus. There appear to be no recent data from other Black Sea rivers though loads for three Turkish rivers have been quoted (Balkas et al, 1990). These data indicate that in 1984 between 300 and 1200 tonnes of cadmium, 30 to 300 tonnes of copper and up to 3000 tonnes of lead were discharged. Heavy metal loads must be considered to be minimum values, as additional sources such as direct discharges and atmospheric deposition have not been quantified.

The annual oil load of the Danube River in the period 1988 to 1989 has been estimated at 50 000 tonnes. Few data exist about sea-based sources, which are expected to account for the oil pollution along shipping lanes and in ports. Oil drilling along the Romanian coastline may be another potential source of oil pollution. However, several active oil rigs have been subject to protective measures and monitoring has not yet detected adverse effects (Balkas et al, 1990). Inadequate port reception facilities and deballasting activities add further inputs of oil.

There are also 16 official waste disposal sites within the western Black Sea on the continental shelf: dredging spoils are disposed of at these sites and these would add an unquantified load of contaminants to the offshore zone of the sea.

Contaminant levels

Heavy metals

There are few published data on heavy metals in the Black Sea, and those available reflect high and very variable concentrations. This is probably due to analytical difficulties caused by their low concentration in sea water. Concentrations of metals in water are generally higher in the nearshore zones than further offshore and in the open sea. In particular, high levels of cadmium are found along the coastline of Romania (up to 1.6 μg/l compared with open sea levels of 0.05 μg/l), and high levels of mercury and copper along the Bulgarian coast (up to 2.6 μg/l mercury and 83 μg/l copper, compared with open sea levels of 0.1 μg/l and 0.5 μg/l, respectively (Dechev, 1990)).

Synthetic organic compounds

There are generally no validated data on pesticides in the Black Sea, though levels of up to 200 to 300 ng/l total organochlorine pesticides have been reported in the River Don, compared with general levels of 10 to 30 ng/l in water of the Sea of Azov, and 5 ng/l in the open Black Sea. In addition, mean and maximum water column concentrations of αHCH and γHCH (lindane) are reported to be 5 ng/l and 23 ng/l, and 1 ng/l and 5 ng/l, respectively.

Total DDT water concentrations in the Kerch Strait range from 8 to 20 ng/l, approximately two orders of magnitude higher than in the Mediterranean Sea (Mee, 1992). Measurements made in 1981 and 1982 indicated DDT water levels of 32 to 486 ng/l in the region of the Danube delta and 20 to 550 ng/l of lindane. Total DDT levels in two fish species were reported to be in the order of 1 μg/g (dry weight) for *Gobius* spp and 4.6 μg/g (dry weight) for *Sprattus* spp.

Surfactants (synthetic detergents) are reported to be widespread in the surface waters of the Black Sea and Sea of Azov, ranging from 0.6 mg/l around Odessa to 0.1 mg/l in the open waters of the Black Sea, and 1.2 mg/l in the coastal zone to 0.1 mg/l in open waters of the Sea of Azov (Mnatsakanian, 1992). There are also 'hot spots' of phenol in water, such as up to 14 ng/l in the Odessa area and 18 ng/l in other northern coastal areas. Levels of phenol in water up to 3 ng/l have been found along the western shores around the Danube, along eastern shores and in the Sea of Azov.

Oil

Areas of the Black Sea are severely polluted with oil, particularly those areas subject to river discharge and ports. Sevastopol Bay, which serves as the major port for the Black Sea navy, is the most polluted, with an average annual concentration of 5 mg/l, over 100 times higher than the maximum permissible concentration (MPC) allowed by the Russian Federation water standards. Even the average open sea oil concentration of 0.1 mg/l exceeds the MPC by a factor of two, and is two orders of magnitude higher than oil concentrations in the open North Sea (1 to 3 μg/l). Oil pollution along shipping lanes is especially heavy (typically around 0.3 mg/l) and is suggested to be caused by deballasting and bilge discharges. Some ports (Odessa and Novorossiysk) have facilities for this but others do not (eg, Varna, Burgas and Sevastopol).

Nutrients

Associated with the increasing loads of nutrients into the Black Sea over the last 25 years, there has been a long-term change in nitrogen and phosphorus concentrations. For example, between 1960 and 1975 the nitrate and phosphate concentrations along the Romanian coast have increased between 5-fold and 10- to 20-fold respectively (Mee, 1991). Ammonia and phosphate concentrations increase, and nitrate decreases, with depth in the Black Sea; this is caused by the decomposition of organic matter under anoxic conditions. Mean surface water concentrations of ammonia, nitrate and phosphate were reported to be 2.8 mg N/l, 0.8 mg N/l and 0.4 mg P/l, respectively (Dechev, 1990). At 500 m nitrate is no longer present but ammonia concentrations are some 21 times greater (58.5 mg N/l) than at the surface, and phosphate some 14 times greater (5.6 mg/l).

Microbiological contamination

There are few data on the bacterial level of municipal wastewater and the level of bacteriological pollution of coastal waters, but a large number of beaches have been known to suffer from bacteriological contamination. This led to the closing of many Ukrainian beaches during the summer of 1989. The beaches of the easternmost part of the Sea of Azov and Taganrog Bay were also closed in 1989 because of alleged bacterial contamination.

Radionuclides

Deposition of Chernobyl radionuclides occurred in the Black Sea, but there are limited data on the resultant concentrations within sea water, sediments and biota. Available information indicates that sea water concentrations in the Crimea area ranged between 190 and 650 Bq/m^3 Cs-137, and 79 and 320 Bq/m^3 Cs-134 during 1986: these had decreased to between 90 and 120 Bq/m^3 Cs-137 and to between 20 and 55 Bq/m^3 Cs-134 during 1987 at the same place (Stepanets et al, 1992). Other sources of information report that Cs-137 concentrations in the Black Sea ranged from 41 to 165 Bq/m^3, and Cs-134 from 17 to 78 Bq/m^3, following the Chernobyl accident (Balkas et al, 1990).

Biological effects

Eutrophication

The evidence is overwhelming that a major part of the Black Sea is critically eutrophic (Mee, 1992). Eutrophication is especially apparent in the northwestern shelf area because of the heavy anthropogenic nutrient load carried by the rivers. Other areas suffering from eutrophication include Crimea, Kavka and near Batumi. The consequences of eutrophication in the Black Sea and Sea of Azov include:

- increased algal blooms;
- gradual basin-wide shallowing of the euphotic zone (eg, Secchi disc depths have decreased from 50 to 60 m in the early 1960s, to present values of 35 m, or as little as 10 m in parts of the coastal zone);
- a massive loss of shallow water macrophytes (large algae and rooted plants);
- widespread reduction in dissolved oxygen concentrations;
- occasional formation of anoxic benthic layers of hydrogen sulphide distinct from the permanent anoxic layer in the main basin;
- changes in the food-chain; and
- severe reductions in fish stocks.

Eutrophication has caused the base of the food-chain to change, resulting in an increase in the development of monospecific blooms of plankton. Higher levels within the food-chain have subsequently also been changed with massive basin-wide biomasses of jellyfish (*Aurelia aurita*) and a predatory comb jelly (*Mnemiopsis leidyi*). The decomposition of the massive quantities of these two species has resulted in widespread hypoxia and has caused a large reduction in the number of macrobenthic marine species. These changes to the marine ecosystem of the Black Sea have contributed to the demise of the fisheries and reduced its tourist potential.

Opportunistic species

A number of opportunistic species, brought into the Black Sea in the ballast water of ships, have found ecological niches in which they thrive. Some of these species have had important ecological consequences for the Black Sea and Sea of Azov:

- *Rapana thomasiana* (a predatory sea snail from Japan) has been blamed for a decrease in commercially harvested oyster populations and a decrease in biodiversity.
- *Mya arenaria* (a soft-shelled clam) successfully established itself on the Romanian shelf and reached densities of over 1000 individuals/m². This species has actually been beneficial to the environment in this area by improving the ecosystem's capacity for self-purification.
- *Mnemiopsis leidyi* (a predatory comb jelly) has attained a gigantic biomass in the Sea of Azov. It feeds on zooplankton and fish larvae and is itself at the end of the food-chain. The decomposition of the large numbers of individuals is exacerbating the problem of hypoxia, especially in shallow waters of the Sea of Azov and the northwestern shelf area, although it is also affecting the entire basin area.

Fisheries

The combination of reduced river discharge, increased water pollution, recruitment failure and overexploitation has adversely affected fish stocks in the Black Sea and Sea of Azov. Commercial fishing in the Dnepr and Dnestr estuaries has been much reduced and some valuable species such as pike (*Esox lucius*), perch (*Perca fluviatilis*), roach (*Rutilus rutilus*), bream (*Abramis brama*) and vimba (*Vimba vimba*) have disappeared altogether. In fact, of the 26 commercial fish species abundant in 1970, only five are left in commercial

quantities today: the anchovy (*Engraulis encrasicolus*), sprat (*Sprattus sprattus* and *Clupeonella cultriventris*) and horse mackerel (*Trachurus mediterraneus* and *Trachurus trachurus*). Depletion of the stocks of the dominant predators such as bonito (*Sarda sarda*), bluefish (*Pomatomus saltatrix*) and dolphins in the 1970s coupled with the increase in primary productivity and zooplankton biomass has resulted in this increase in biomass of small pelagic fish.

Conclusions

- The Black Sea and the Sea of Azov are subject to several very major ecological problems. They have changed over the last 30 years from highly diverse ecosystems supporting a highly productive fishery into a hyper-eutrophic state which leads to environmental conditions unsuitable for most higher organisms.
- Eutrophication caused by the input of nutrients from rivers and untreated domestic, industrial and agricultural waste is resulting in algal blooms and the flourishing of jellyfish and a comb jelly which represent essentially the end of the food-chain.
- The decomposition of the large number of individuals of these species is further exacerbating the Black Sea's natural anoxic condition and causing the widespread demise of benthic organisms.
- Oil pollution is relatively high in the Black Sea and microbiological contamination of beaches from sewage waste is posing a human health threat.
- Dam and irrigation projects on the major rivers draining into the northwestern part and the Sea of Azov are reducing the freshwater and sediment inflow to these areas, resulting in erosion of coastal areas, particularly along Romanian coastlines, and changes to the salinity of the areas affected.
- Overfishing coupled with the problems caused by eutrophication has resulted in a fundamental change in the structure of the food-chain and the major commercial species are now small pelagic fish such as anchovies, mackerel and sprats, which are themselves overexploited.

THE CASPIAN SEA

General situation

The Caspian Sea is situated in an arid and semi-arid zone and is surrounded by the Russian Federation, Azerbaijan, Kazakhstan, Turkmenistan and Iran (see Map 6.1). It is the largest relatively low-salinity, land-locked body of water in the world. The main ports along the coasts of the Caspian Sea are Baku (Azerbaijan), Astrakhan and Makhachkala (Russian Federation), Krasnovodsk (Turkmenistan), Shevchenko (Kazakhstan), and Bandar Anzali in Iran. The Caspian Sea is an important source of oil and gas and extensive drilling operations are located at the Baku Archipelago and Neftyanyye Kamni Fields in the Baku district.

Physical features

The Caspian Sea has a surface area of 436 000 km² and is structurally divided into three parts: northern, middle and southern. The major rivers flowing into the Caspian are the Volga, the Ural, the Kura, the Terek and the Samur. Of these, the Volga, to the north of the sea, is alone responsible for three-quarters of the riverine freshwater input. Salinity varies little with depth but widely from north to south. In the north it ranges from 2 per thousand close to the mouth of the Volga, to 5 to 10 per thousand elsewhere; the whole of this part of the sea is like a gigantic estuary inhabited by fresh and brackish-water organisms. The north is also very shallow, with a depth of 10 to 12 m. In contrast, the middle and southern Caspian are deep waterbodies with maximum depths of 788 m and 1025 m, respectively, divided by the pronounced Aspheron Ridge, with a marine type of water circulation and a salinity of 12 to 13 per thousand.

The Caspian Sea has been influenced by long-term sea level variations. In the early 1930s the level of the sea was 26 m lower than the world ocean. By 1977 this level had dropped to −29 m, from where it started to rise to its present level of −27.5 m. The reasons for these sea-level changes are presently unknown but one could be the Volga drainage rate. Dams constructed on the Volga and Kura rivers for hydroelectric plants and irrigation schemes have reduced the flow of freshwater into the Caspian. This has been only partly mitigated by the construction of canals to divert water from the Siberian rivers, which discharge into the Arctic, into the Volga catchment. This sea-level rise has caused a problem for beaches and cities in the North Caucasus region. The construction of large reservoirs on the Syr-Dar'ya and Amudar'ya rivers to support rice and cotton growing has also seriously depleted the water in the Aral Sea. Between 1965 and the late 1980s the Aral Sea had lost two thirds of its volume and its water level fell by 17 m (Tursunov, 1989). Because it is situated in Asia, a discussion of the serious environmental problems concerning the Aral Sea is beyond the scope of this report.

In summer, the surface temperature of the Caspian Sea is 24 to 27°C and in winter it falls to 9°C in the south and to 0°C in the north, where sea-ice forms. Vertical water circulation is limited, resulting in serious oxygen depletion at depth. This is particularly significant below 200 to 300 m and results in the production of hydrogen sulphide in deep sediments (Clark, 1986).

Rivers add 350 km³ of freshwater per year into the Caspian Sea and a further 100 km³ comes from direct precipitation onto the sea; all of this volume (450 km³) is lost by evaporation. The eastern side of the sea receives little precipitation and negligible river input; there is also a high rate of evaporation. This results in some bays and semi-enclosed areas of the sea having very high salinities, for example up to 200 per thousand in the Kara-Bogaz-Gol Bay, which is almost completely separated from the Caspian proper. The bay is shallow and joined to the main sea through a very narrow channel which was closed off by a dam at the end of the 1970s. This area was used for the evaporation of water for the collection of several sea-water minerals.

Biological features

Lack of oxygen at depth has resulted in an impoverished benthic fauna in the central and southern parts of the Caspian Sea; in some parts it is completely missing (Clark, 1986). The northern Caspian benthic fauna is dominated by molluscs and there was a previously conspicuous absence of polychaetes (bristleworms). Between 1939 and 1941 some 65 000 *Hediste diversicolor*, a polychaete, were introduced to restructure the benthic ecosystem (Clark, 1986). This has become the dominant food for sturgeon, of which there are two main species: the Russian sturgeon, *Acipenser gueldenstaedti*, and the stellate sturgeon, *Acipenser stellatus*. However, the sturgeon population has been much reduced and is still diminishing rapidly. There is a reported trend of decreasing species diversity of phytoplankton, zooplankton and macrobenthos; for example, the latter showed a decrease in species numbers from 90 in 1980, to 63 in 1990 (Izrael and Tsyban, 1992). This trend towards a decrease in benthic biomass and species diversity is more strongly marked in the central and southern regions than in the northern Caspian Sea. The species diversity has been reduced to almost the minimal level in some stretches of water.

Inputs

About 12 km³ of wastewater per year is discharged into the Caspian Sea from the Volga basin and 8 km³ comes from the coastal towns: 2.5 km³ out of that is discharged without any treatment, a further 7.5 km³ is partially treated. The Volga is the largest European river in terms of catchment area (1 360 000 km²), mean discharge (230 km³ per year) and length (3530 km), and, with 15 major cities and major industry in its catchment, represents the principal source of pollution in the northern Caspian. There is also major oil industry along the Azerbaijan coast with numerous offshore oil wells with associated refineries and petrochemical plants. In addition, there is a large urban population around Baku.

An estimate of the minimum gross load of some contaminants entering the Caspian has been made from water quality data from the River Volga obtained from the EEA-TF's inland waters questionnaire (see Chapter 5). No data have been found on the loads from other sources such as other rivers and direct inputs. From the available data it has been calculated that 98 kilotonnes (kt) of nitrate, 126 kt of total nitrogen and 5 kt of phosphates entered the Caspian from the Volga in 1991. In addition, there are some limited data on metal concentrations in Volga river water (Karpinsky, 1992). These have been used, with the flows provided by the Hydrochemical Institute, Rostov-on-Don, to estimate the annual load of cadmium (114 tonnes (t)), copper (1600 t), lead (297 t) and zinc (5100 t).

Contaminant levels

Heavy metals

The concentrations of copper, zinc, lead and cadmium in Volga water are reported to be 7 μg/l, 22.5 μg/l, 1.3 μg/l and 0.5 μg/l, respectively, and all are reported to have increased significantly over the last 15 years, exceeding maximum permissible pollutant levels designated by the former USSR (Karpinsky, 1992). There are apparently no other data reported or available.

Synthetic organic compounds

Organochlorine pesticide concentrations in water are reported to be as high as 27 ng/l around the Volga delta compared with 1.8 to 1.9 ng/l in more open sea locations (Mnatsakanian, 1992). The average annual concentration of phenol in 1988 in the northern Caspian and in the Dagestan and Far Eastern coastal areas were 4 to 6 μg/l. There were, however, high phenol contamination levels, 5 to 16 μg/l, along the Azerbaijan coast.

Oil

Oil pollution has a very long history in the Caspian Sea. Natural benthic oil springs have long caused some contamination, but it was not until the middle of the twentieth century that the threat of oil pollution increased from offshore oil production. Oil pollution is also associated with petroleum extraction operations, river runoff and along shipping routes. The level of oil in coastal and open sea waters of the Caspian is relatively high, with coastal waters having in general reported concentrations of between 100 and 300 μg/l (Mnatsakanian, 1992). However, some coastal waters are even more polluted. The average annual concentration of oil compounds in 1988 in the northern Caspian region and in the Dagestan and Far Eastern coastal areas were between 100 and 150 μg/l. Higher levels of 200 to 600 μg/l were also found along the Azerbaijan coast, caused not only by the runoff from large cities but also from offshore oil rigs.

Turkmenistan also has many oil wells in the

Krasnovodsk-Cheleken region, where frequent accidents at wells and pipelines have occurred. In the region of oil fields the oil concentrations along Turkmenistan coasts is between 100 and 200 μg/l. The highest observed concentrations in the Turkmenistan coastal region are in the Kara-Bogas-Gol Bay, with concentrations of up to 750 μg/l.

In 1989 four accidents were recorded which resulted in more than 300 tonnes of oil being discharged into the sea. One accident alone resulted in oil being spread over more than 200 km^2. These spillages have had extremely adverse effects for the marine environment because the Caspian Sea has no official organisation which is responsible for cleaning up spilled oil. In 1989 the Azerbaijan government decided to stop all new deep sea drilling until new equipment was available to prevent these oil spillages.

Nutrients

There were no available data on the levels of nutrients in the Caspian Sea.

Microbiological contamination

In the North Caucasus region there is a problem due to a lack of sufficient wastewater treatment facilities in the cities (Mnatsakanian, 1992). For example, in Makhachkala all of the sewage goes into the sea without any treatment. In 1990 no seasonal fluctuations were observed in the bacterial populations in the west, middle and south Caspian: total numbers of bacteria varied from 0.5 to 3.5 million per ml (Izrael and Tsyban, 1992). In the Bay of Bakinsky the numbers of bacteria were a little lower than in the other regions.

Radionuclides

It has been recently reported that surface water concentrations of Sr (strontium)-90 in the Caspian Sea in 1990 and 1991 were 15 Bq/m^3, and from 12 to 87 Bq/m^3, respectively (Yablokov et al, 1993).

Biological effects

It is reported that there is a trend of decreasing ecological quality of the Caspian Sea, with the middle and southern parts being the most impacted (Izrael and Tsyban, 1992). In 1988–89 hundreds of thousands of sturgeon were found dead with gross changes in their bodies (destruction of muscles): overall the stocks of sturgeon are believed to be falling (Mnatsakanian, 1992). A phenomenon first recorded in 1984 and observed on a mass scale in 1987–88 (in up to 90 per cent of the stock) is the lamination of the muscle tissue and egg membrane in the sturgeon, which was given the name 'myopathy' (Karpinsky, 1992). Associated with this is metabolic disturbance, atrophy and adiposity of muscle fibres, blood changes, and elevated content of heavy metals and organochlorine pesticides in the blood and tissues – a possible cause of this disease is cumulative toxicosis.

Significant declines have been observed in the breeding and nursery grounds for sturgeon, bream, pike, crucian carp (*Carassius carassius*) and other commercial fish species. The decrease in fish catches has been linked with the variations in sea level, increased salinities and pollution levels, and is possibly compounded by overfishing. The total fish catch from the Caspian was 0.3 million tonnes per year in the mid-1930s; this had dropped to 0.1 million tonnes by 1970.

Conclusions

- The condition of the Caspian Sea varies from north to south. While the northern part is not considered to be in a satisfactory ecological state, it is the central and southern Caspian Sea which are considered to be particularly at risk from, for example, the oxygen deficiency at depth (caused by limited vertical water circulation) and subsequent changes in the food-chain.
- Apart from this, the major problems faced by the Caspian Sea are:
  – oil exploitation and pollution;
  – damming of the bay of Kara-Bogas-Gol;
  – lack of sewage treatment;
  – general contamination levels.
- The Caspian has undergone considerable long-term variations in sea levels, which may be due partially to changes in the discharge volume of the Volga (Karpinsky, 1992). Climatic changes that induced changes in freshwater flows into the sea may add to this variation in the future and possibly change the nature of the ecosystem.

THE WHITE SEA

General situation

The White Sea is a semi-enclosed inland sea of the Arctic Ocean situated on the northern coast of the Russian Federation between the Kanin and Kola peninsulas. In the north it joins the Barents Sea. The coastline topography varies, with predominantly high, rocky outcrops along the northwest coasts and flat low-lying land along the southeast coasts. The sea has a number of gulfs or bays: the Gulf of Mezen, the Gulf of Onega, the Gulf of Kandalaksha and the Gulf of Dvina. In winter the sea is completely covered with ice, thus limiting its use for shipping. The main coastal cities are Arkhangelsk, Onega, Kem and Kandalaksha. A large nuclear submarine shipyard is located in the city of Severodvinsk, near Arkhangelsk.

At a Ministerial Conference in Rovaniemi, Finland, in June 1991, the ministers from the eight arctic countries, including the Russian Federation, agreed to develop an Arctic Monitoring and Assessment Programme (AMAP), which will be implemented by an Arctic Monitoring and Assessment Task Force (AMAP, 1993). The initial priorities are work on persistent organic contaminants, selected heavy metals and elements, and radionuclides. The White Sea is included within the programme.

Physical features

The White Sea surface covers 90 000 km^2 and is comparatively shallow, with a maximum depth of 350 m. There are three major rivers discharging into the White Sea (see Map 5.11): the Severnaya (northern) Dvina; the Mezen and the Onega. These three rivers are all within the 30 largest European rivers in terms of catchment area. The largest, the Severnaya Dvina, has a catchment of 358 000 km^2, a mean annual discharge of 148 km^3 and a length of 740 km. These catchments are relatively sparsely populated.

Biological features

The formerly rich White Sea fisheries, numerous seal herds and natural seagrass mats have become depleted due to overexploitation and pollution (Mnatsakanian, 1992). Seagrass is very intensively used for the production of agar and iodine. It has been estimated that the amount of seagrass in the White Sea in 1989 was only half of that present in 1975

(Mnatsakanian 1992). The population of Greenland seals is also thought to have halved in the four years between 1985 and 1989.

Inputs

Very few data have been found on the input loads of contaminants entering the White Sea. In 1989 it was estimated that 2600 tonnes of oil, 107 tonnes of phenol and 382 tonnes of surfactants were discharged into the White Sea (Mnatsakanian, 1992).

Loads of some other contaminants have been estimated from 1991 data on river water quality and flow obtained by the EEA-TF inland waters questionnaire (Chapter 5). The data relate to the three major rivers discharging into the White Sea: the Severnaya Dvina, Mezen and Onega. From these data it was estimated that, in 1991, 187 tonnes (t) of copper, 314 t of lead, 1656 t of zinc, 276 t of nickel, 8.8 kilotonnes (kt) of nitrate nitrogen, 17.2 kt of total nitrogen and 1.5 kt of orthophosphate were discharged from these rivers. These values must be considered as minima, since no data are available on direct discharges into the sea, other riverine inputs or atmospheric deposition.

Radioactive waste has also been discharged into the White Sea. According to a recent report, between 1959 and 1984, a total of 3.7 TBq of liquid waste was discharged (1 TBq = 10^{12} Bq or 27 Curies). In 1984 the Murmansk Marine Shipping Line (the organisation responsible for radioactive waste disposal) ceased dumping liquid waste (Yablokov et al, 1993).

A considerable problem in the White Sea is the effect of the large number of logs transported down river. Not all of the logs are successfully collected from the rivers and as a result log jams are sometimes formed in the river mouths. All shores along the White Sea often resemble timber stock yards.

Floating logs can create problems for navigation and fishing and pollution of the coast of the receiving sea
Source: Stock Imagery

Contaminant levels

There are limited data published or available on the levels of contaminants in water, sediment and biota of the White Sea:

- Open sea concentrations of oil (average 0.02 mg/l) in 1989 were lower than those found in nearshore bays and gulfs: for example up to 0.15 mg/l in the Gulf of Kandalaksha, and 0.35 mg/l in the Gulf of Dvina.
- Relatively high levels of phenols have also been found in the enclosed bays and gulfs, such as Dvina (7 µg/l) and Onega (8 µg/l), compared with open sea levels of less than 3 µg/l.
- Levels of the pesticides αHCH and γHCH were reported to be 2.1 and 1.1 ng/l, respectively, for open sea areas and 4.5 and 3.7 ng/l, respectively, in the Gulf of Dvina (Izrael and Tsyban, 1992).
- Average sea surface water concentrations of Sr-90 in the White Sea were reported to be 9.3 Bq/m³ in 1990 and 1991 (Yablokov et al, 1993). This value can be compared with average surface water concentrations of Sr-90 of 16 to 17 Bq/m³, and 6 Bq/m³ in the Baltic and Barents seas, respectively, in 1991.
- From microbiological figures, the open part of the sea has been classified as clean, the Gulf of Kandalaksha, the Gulf of Onega and the Gulf of Dvina are considered to be 'fairly polluted' (Mnatsakanian, 1992).

Biological effects

In the spring of 1990 there was a massive death of starfish, crabs, mussels and marine mammals, which the Soviet Inter-governmental Commission attributed to sulphuric compounds (such as mustard gas) or the leakage of missile fuel from a submarine (Izrael and Tsyban, 1992).

Conclusions

- The water in the open White Sea was classed as clean from the 1990 USSR monitoring results (Izrael and Tsyban, 1992).
- The most heavily polluted waters are in the Onega and Dvina gulfs (Bruggeman et al, 1991).
- The primary pollution source is river runoff, and the majority of pollutants originate from the enterprises of the lumber and woodworking industries, the former RSFSR State Agroindustrial Committee, the former USSR Ministry of the Medical and Microbiological Industries, and others.
- The major problems of the White Sea are:
  – contamination levels of synthetic organic compounds, organics and oil are locally high;
  – the depletion of natural seagrass beds;
  – a decrease in the seal population;
  – floating logs.

THE BARENTS SEA

General situation

The Barents Sea is a marginal sea of the Atlantic Ocean and is situated at the outermost bounds of the polar ocean between the coasts of Europe and Svalbard, Franz Josef Land and Novaya Zemlya (see Map 6.1). It reaches the coasts of the Russian Federation and Norway. The hydrological characteristics of the Barents Sea are strongly affected by its location between the Atlantic Ocean and the arctic basin. In the northwest region which is under the influence of the warm Atlantic waters the sea never freezes, whereas in the northern area it is partially covered by polar ice for part of the year. The coastline of the Barents Sea is strongly indented with gulfs and fjords. The most important river flowing into the Barents Sea is the Pechora.

The Barents Sea is relatively unpolluted as it is situated a long way from dense human populations and industrialised areas. The coastal areas of the Barents Sea are not heavily populated. The largest town is Murmansk (the Russian Federation) which has a large naval base and heavy industry. Apart from that there are just a few small Norwegian towns and fishing villages. The most important industry is fishing.

The Barents Sea is included within the area covered by the Paris Convention, and, as Norway is a signatory to the Convention, input and quality data are submitted annually to the Paris Commission for the Norwegian sector of the Barents Sea. However, as the Russian Federation is not a signatory to the Convention, comparable data for the Russian Federation sector of the sea are not available. More details on the Paris Convention are given below in the section on the North Sea. The Barents Sea and the Kara Sea (the sea to the east of Novaya Zemlya) are also included in the Arctic Monitoring and Assessment Programme, with both Norway and the Russian Federation being signatories – for more details on AMAP see the section on the White Sea.

This section on the Barents Sea has been compiled largely from the following references (specific points are referenced as appropriate): Gjosaeter (1992), Loeng (1989 and 1991), Northern Europe's Seas (1989), and Saetre et al (1992).

Physical features

The Barents Sea covers an area of 1 425 000 km^2 and has an average and maximum depth of 230 m and 600 m, respectively. More than 20 per cent of its area is shallower than 100 m, but troughs deeper than 400 m enter the area from the west and northeast. The deepest waters are found in the western part, between Norway and Bear Island; the shallowest water is on Svalbard Bank and in the southern part where depths are less than 50 m. The bottom topography strongly influences the current conditions.

There are large long-term variations in the properties of the Atlantic inflow into the Barents Sea which may influence the properties of locally formed water masses. In the Barents Sea there is warm Atlantic water in the south and cold Arctic water in the north. The transition zone between the Atlantic and Arctic water masses is called the polar front, and is formed by the mixing of these two water masses. During winter strong vertical mixing gives rise to a water column with very low stability. Nutrient concentrations are therefore homogeneously distributed with depth during winter, and so are available for phytoplankton utilisation in spring. During spring and summer vertical stability develops differently in ice-covered and ice-free areas. In ice-covered areas, the vertical stability in the surface layers increases as soon as the ice starts to melt and a surface layer of low salinity water is formed. In ice-free areas, that is in areas with Atlantic water, vertical stability is formed through heating of the surface layers by the sun. The vertical mixing during winter and the formation of layered surface water during spring and summer are important for primary production. Maximum ice coverage occurs between March and May, and minimum in August to September.

There are three main current systems linked to the three different water masses (see Map 6.2). The Norwegian coastal current flows along the whole Norwegian coast, bringing relatively low salinity and high temperature water from the Baltic, the North Sea and the Norwegian fjords. The Atlantic current has a high salinity (over 35 per thousand) and temperature (between 3.5 and 6.5°C) at its inflow area between Norway and Bear Island. Atlantic water flows north along the western coast of Svalbard and also into the Barents Sea. The influx of Arctic water occurs through two main routes: between Svalbard and Franz Josef Land, and, more importantly, through the opening between Franz Josef Land and Novaya Zemlya.

Biological features

The Barents Sea is an arctic sea with high biological production and relatively low species diversity. During the six summer months the Barents Sea experiences almost optimal conditions for life, with intense summer sunlight and adequate nutrient supply. One of the most important production areas is the edge of the polar ice cap. Phytoplankton grow very rapidly and in large quantities and these are eaten by the zooplankton which are the staple food of herring (*Clupea harengus*) and capelin (*Mallotus villosus*). This high density of phytoplankton and zooplankton and plankton-eating fish forms the basis for the rich fauna of the area. The sea contains some of the world's largest sea-bird populations, and is also an important area for fisheries and sea mammals. As the biological production is very limited in space and time the ecosystem is potentially vulnerable to the influence of human activity.

The Barents Sea contains some of the world's largest fish stocks of capelin, cod (*Gadus morhua*) and the Norwegian spring-spawning herring (*Clupea harengus*). There are strong interactions between these stocks, and variations in the year-class sizes have a marked influence on other components of the ecosystem. The capelin stock reached a maximum biomass in 1975 of 7.3 million tonnes. Since then it has generally declined, with the lowest estimate of 0.02 million tonnes in 1987. The Barents Sea is a nursery area for several fish species including the herring, cod, haddock (*Melanogrammus aeglefinus*) and saithe (*Pollachius virens*). The Norwegian spring-spawning herring stock was the largest fish stock in European waters until its collapse in the 1960s. The total fish biomass landed over the last 40 years was 2 to 3.5 million tonnes, which is the same as from the North Sea. However, the proportion of fish of the cod family in total catches has decreased from about 50 per cent in the 1950s to around 30 per cent in the 1980s. Pelagic species like capelin and herring have increased to about 60 per cent of total catch.

The Barents Sea contains several species of migratory whales: the minke whale (*Balaenoptera acutorostrata*) is the most important species during the summer and autumn. There is also the humpback (*Megaptera novaeangliae*) and fin whale (*Balaenoptera physalus*). In addition there are populations of grey seal (*Halichoerus grypus*), common seal (*Phoca vitulina*) and harp seal (*Pagophilus groenlandicus*). Walrus (*Odobenus rosmarus*) and polar bears (*Ursus maritimus*) are also common.

Practically all of the Barents Sea's shores are ice-bound from late autumn onwards, and during the summer thaw the shores are scoured by drift ice. As a result few animals and plants inhabit the shore and nearshore environment, but brown and red algae flourish from a depth of 5 to 10 m.

Inputs

Direct and riverine inputs of contaminants to the Barents Sea from Norway are reported annually to the Paris Commission. Information on discharges from the Russian Federation into the Barents Sea have been obtained from two sources: the Knipovich Polar Research Institute of Marine Fisheries and Oceanography (PINRO), Murmansk, and from information obtained from the EEA-TF inland waters questionnaire (Chapter 5). The estimates of loads are gross values and do not take into account estuarine processes which might reduce the net loads discharged to coastal waters. Compared with some other European seas, such as the North Sea, the load of contaminants to the Barents Sea is relatively low. A comparison of the data from Norway and the Russian Federation indicates that loads from Russian rivers can greatly exceed riverine and direct discharges from Norway; for example, the zinc load is 26 times greater, copper 18 times, and total nitrogen 5 times. In addition, some 3840 tonnes of oil and 3.4 tonnes of synthetic organic compounds (eg, DDT, HCH and PCBs) are discharged annually from rivers of the Russian Federation.

In addition to local sources, there are chemical contaminants transported by ocean currents and by long-range atmospheric transport. The atmospheric south–north poleward transport of contaminants from more urbanised and industrialised areas plays an important role in the Barents Sea.

A limited part of the Norwegian coast was opened to oil exploration in 1980: between 1980 and 1992, 54 exploratory wells were drilled in the Norwegian sector. In addition, there are approximately 17 wells in the Russian Federation sector. The main impacts from drilling are: seismic activity; pollution; and area conflicts. Seismic testing can lead to the death of larval fish, and can frighten adult fish away (for example, adult cod can be affected 20 nautical miles away). Though not reported in other sections, the effects of seismic testing could also be relevant to other European seas where oil exploration is undertaken.

Radioactive inputs to the Barents Sea arise from a number of sources, including:

- inputs from surrounding seas, in particular the Gulf Stream system (Saetre et al, 1992)
- atmospheric nuclear weapon tests in the 1950s and 1960s (this source is common to all of Europe's seas);
- fall-out from the Chernobyl accident;
- discharge of low-level liquid waste;
- possible runoff from Siberian rivers;
- possible leakage from underground nuclear weapon tests on Novaya Zemlya; and
- potential leakage of dumped solid nuclear waste in the vicinity of Novaya Zemlya.

The first two sources are predominant, as demonstrated by a recent assessment of the anthropogenic radionuclide inventory of the Barents and Kara seas over the period from 1961 to 1990 (Yablokov et al, 1993). This indicated that, in the Barents Sea, 63 per cent of anthropogenic radioactivity originated from surrounding seas, 31 per cent from atmospheric fall-out, 4 per cent from the dumping of solid and liquid radioactive waste and 2 per cent from river runoff.

Inputs from surrounding seas arise principally from the nuclear reprocessing plant at Sellafield in the Irish Sea (UK), with smaller contributions from the reprocessing plant at Cap de la Hague (France) and the nuclear power stations discharging to Northern European waters.

There is an exchange of water between the Barents and Kara seas through the Kara Gate or north of Novaya Zemlya. The present levels of radioactivity in the Kara Sea are, in addition to the sources mentioned for the Barents Sea, also attributed to runoff from Russian rivers, in particular the Ob and Yenisey, as well as to a potential release from dumped radioactive solid waste, especially in bays at the eastern part of Novaya Zemlya. The potential future importance of this latter source is indicated by an inventory of anthropogenic radioactivity in the Kara Sea (Yablokov et al, 1993). This indicates that the major contribution (95 per cent) is from the sinking of solid radioactive waste contained within nuclear reactor chambers. The remaining inputs have arisen from atmospheric fall-out (3 per cent), river runoff (1.3 per cent), and the dumping of liquid and other solid radioactive waste (0.7 per cent).

The discharge of liquid radioactive waste into the Barents and Kara seas from the former USSR began in 1960, and, by 1991, 450 TBq had been discharged into the former and 315 TBq (Sr-90 equivalent) into the latter. Five main disposal areas were used in the Barents Sea, the largest quantity being discharged at three sites to the west of Novaya

Oil exploration
Source: L Lee, Robert
Harding Picture Library

Zemlya. Smaller amounts were disposed of at the two sites closest to the mainland coast of the Russian Federation. There was one main disposal site in the Kara Sea off the southern part of the east coast. The dumping of liquid waste by the Murmansk Marine Shipping Line was stopped in 1984. The naval fleet continued the dumping of low-level liquid waste until 1991.

Solid radioactive waste dumped in the Kara Sea has arisen from the Russian Federation nuclear powered naval and ice-breaking fleets, and from ship-repairing and shipbuilding factories. Solid, low - and medium-level radioactive waste was contained in metallic containers; large sized waste was sunk separately or inside vessels such as barges and tankers. According to available data the total activity of the low and medium solid radioactive waste sunk between 1957 and 1991 in the Kara Sea was 574 TBq (Sr-90 equivalent) and in the Barents Sea 1.5 TBq (Sr-90 equivalent). This waste was disposed of at eight sites along the eastern coast of Novaya Zemlya in the Kara Sea, with the largest volume in the deep trench (380 m deep) of Novaya Zemlya and the largest radioactive quantity in the Gulf of Sedov (13 to 33 m deep).

In addition to the solid radioactive waste, seven reactors with spent nuclear fuel and ten reactors without fuel have been dumped in the Ambrosimov, Tsivolki and Stepovogo bays, along the eastern coast of Novaya Zemlya, as well as in the Kara Sea trench. It is estimated that there could be up to 85 000 TBq of radioactivity within this waste. These reactors are considered by the Russian Federation to represent the greatest radiological risk to the Kara Sea (Yablokov et al, 1993). The potential future impact on the Barents Sea would depend on a number of factors, including water currents and the exchange of water between the Kara and Barents seas. As a result the relevant areas are being closely monitored by the Norwegian and Russian Federation authorities.

Contaminant levels

Heavy metals

There are few data on trace metals in sea water and sediment: concentrations are generally at natural 'background' levels, and generally lower than those found in the North and Baltic seas.

Levels of metals in fish are generally found to be at background levels (Saetre et al, 1992). However, measurements of mercury in the liver and kidney of harbour porpoises showed a decreasing mercury concentration gradient from south to north along the Norwegian coast: concentrations ranged from 0.26 to 9.9 μg/g in liver samples and from 0.15 to 3.5 μg/g in kidney. Polar bear livers were found to have concentrations between 0.4 and 6.0 μg/g mercury, from less than 0.1 to 1.2 μg/g cadmium and from less than 0.5 to 1.6 μg/g lead. Levels of copper and zinc were at normal physiological levels.

Synthetic organic compounds

There are limited data on trace organic compound concentrations in sea water. In 1985, HCH (α and γ) concentrations ranged from 4.8 to 6.2 ng/l in the North Sea, from 2.3 to 3.8 ng/l along the coast of Norway and from 1.2 to 1.8 ng/l in the Barents Sea. Other chlorinated organics are usually at lower concentrations than HCH.

The analysis of sediments is part of the AMAP programme (see the section on the White Sea). Preliminary results indicate that aromatic hydrocarbons are present in sediments from all the sampling stations, indicating that they are influenced by anthropogenic inputs (Saetre et al, 1992). Mean concentrations of NPD (sum of naphthalene, phenanthrene, dibenzothiophenes) and PAHs (polycyclic aromatic hydrocarbons) were found to be 0.14 and 0.26 μg/g dry weight, respectively: these are 4 to 5 times lower than in the Skagerrak and 2 to 3 times lower than sediment in parts of the Norwegian Sea. Mean PCBs concentration was found to be 0.5 ng/g dry weight, which is 8 to 10 times lower than found in the Skagerrak.

Persistent organic compounds have been measured in a range of different biota, and results indicate that fish, sea-birds, marine and terrestrial mammals accumulate persistent organic compounds from different sources (Saetre et al, 1992). For example, the levels of PCBs in ringed seal (*Phoca hispida*) blubber from Svalbard ranged from 0.39 to 9 μg/g wet weight. With a few exceptions, birds from the Barents Sea region contain low levels of chlorinated hydrocarbons compared with more southerly populations; the highest concentrations of PCBs were found in the predatory glaucous gull (*Larus hyperboreus*).

Oil

In recent years there has been an increase in oil and gas exploration on the continental shelf of the Barents Sea by both Norway and the Russian Federation. In 1988 the Kola Gulf was the most polluted of the gulfs in the Barents Sea. Murmansk is located in Kola Bay and the bay is especially polluted near the mouth of the Severomorsk River and the settlement of Roslyakovo. In this bay concentrations of oil and phenol were up to 0.05 mg/l and 0.007 mg/l, respectively (Bruggeman et al, 1991). Petroleum product concentrations in sea water of up to 0.5 mg/l have been recorded in areas of petroleum and gas exploration off the sea shelf.

Radionuclides

The most widely measured artificial radionuclide in the coastal waters of North Europe is Cs-137. Before Chernobyl its source was largely the reprocessing activities at Sellafield in the UK together with a smaller contribution from fall-out

due to weapons testing (Camplin and Aarkrog, 1989). The most pronounced radioactive contamination in the Barents Sea was detected during atmospheric nuclear weapon tests (Saetre et al, 1992). The highest concentrations in fish were observed in the early 1960s, followed by a rapid decrease from about 80 Bq/kg wet weight in 1963, to below 10 Bq/kg wet weight by 1968. More recent levels are reported to be in the range of 0.5 to 1 Bq/kg wet weight, compared with levels of 30 Bq/kg in the Irish Sea and Baltic Sea (NRPA, 1993): both the latter values are well below the Norwegian action level for human consumption of 600 Bq/kg. Based on a joint Norwegian–Russian expedition in 1992 to the Barents and Kara seas, the concentrations of Cs-137 in surface waters were within 3 to 8 Bq/m^3, with the highest concentrations of 20 Bq/m^3 in bottom waters (Foyn and Semenov, 1992). The results indicate a significant decrease in the concentration of Cs-137 during the last ten years. A significant contribution of Sr-90 from the Ob and Yenisey rivers was also identified.

Biological effects

In general there are no major problems of nutrients and eutrophication in the Barents Sea, although locally increased nutrient levels may be found (eg, in some fjords).

The Barents Sea is a marginal area for several fish species. Changes in climatic conditions influence the distribution of fish, recruitment and growth. Global changes in climate may, therefore, have drastic biological consequences regardless of whether a warming or cooling takes place (Saetre et al, 1992).

The greatest human impact on the Barents Sea is from fishing, for example:

- direct mortality of target fish and other biota;
- food availability from discarding unwanted fish and benthos, from discarding wastes and by killing or damaging animals from fishing gear deployment;
- disturbance of the seabed by fishing gear; and
- litter.

Conclusions

- The Barents Sea can be considered to be relatively unpolluted since, with few exceptions, the levels of organic contaminants and trace metals in the sea are lower than those closer to domestic and industrial sources. However, persistent chlorinated organic compounds accumulating in arctic marine biota are of concern.
- The sinking of solid radioactive waste (eg, nuclear reactors) and the uncertainty associated with this as a potential future source of contamination is an issue of concern for people living in the area. This has led to the establishment of an extensive monitoring programme by the governments of Norway and the Russian Federation.
- Current radioactive contamination in the Barents Sea is, however, orders of magnitude lower than defined 'safe' limits, and generally originates from other parts of Europe (such as the Irish Sea and fall-out from the Chernobyl accident) rather than from the direct dumping and sinking of radioactive waste into the sea.
- The Barents Sea is highly productive and is able to maintain large stocks of pelagic fish which have inherent tendency to fluctuate between good and bad periods of recruitment and abundance. Fisheries for pelagic stocks may play, and have played, a major role in the ecosystem by intensifying these natural fluctuations.
- Global changes in climate may have drastic biological consequences regardless of whether a warming or cooling take place. This latter threat is being monitored as part of AMAP, the Arctic Monitoring and Assessment Programme.

THE NORWEGIAN SEA

General situation

The Norwegian Sea is bordered to the east by the coast of Norway between 61°N and North Cape (of Norway), 25° 45'E, to the northeast along a line from North Cape over Bear Island to the south point of West Svalbard (see Map 6.1). Its western limit extends from this point along a line to Jan Mayen and further to Gerpir, the easternmost point of Iceland. The southern limit extends from there to the Faeroe Islands and on to the position 61°N, 00° 53'W. The coastal areas of the sea are scantily populated and the major occupations are fishing, aquaculture, mining and some heavy industry. The port of Narvik is especially subject to shipping of ore vessels. Freight vessels and fishing boats are also important in the area, as is transit traffic from the Russian harbours in the north. The Norwegian Sea continental shelf is a site of oil and gas exploration activities.

Norway is a signatory to the Paris Convention, which covers the Norwegian Sea. The part of the Norwegian Sea north of the Arctic Circle is also included in the Arctic Monitoring and Assessment Programme (AMAP), to which Norway is a signatory (for more details on AMAP see the section on the White Sea).

This section on the Norwegian Sea has been compiled largely from the following references (specific points are referenced as appropriate): Aksnes et al (1993), Blindheim (1989) and Northern Europe's seas (1989).

Physical features

The Norwegian basin has two main depressions separated by a ridge between the Voring Plateau and Jan Mayen, through which only a narrow passage is deeper than 3000 m. The larger basin to the south contains wide areas with depths between 3500 and 4000 m. The Norwegian Sea is separated from the North Atlantic Ocean by the Scotland–Iceland ridge: the deepest passage is the channel to the southwest of the Faeroes with a depth of 850 m; the channel between the Faeroes and Iceland is shallower than 500 m. To the north the connection to the Greenland Sea is deeper than 2500 m off the Barents Sea shelf, which forms the northeastern margin of the basin.

There are four major water masses within the Norwegian Sea. The warm and saline Atlantic water (temperature higher than 8°C and salinity greater than 35 per thousand) flowing in from the North Atlantic is confined to the upper layers over most of the area. Arctic water from the east Icelandic current (temperature less than 3°C and salinity between 34.7 and 34.9 per thousand) occupies upper layers in the southwestern part of the sea: at intermediate depths its distribution is much wider. Bottom water (salinity close to 34.92 per thousand and temperature below 0°C) originating from the Greenland Sea and the Arctic Ocean fills the deeper layers and represents the largest water volume. Within the area these three different water masses are continuously being mixed and modified. Coastal water along the coast of Norway is the fourth major water mass in the Norwegian Sea. It is made up of a mixture of freshwater from land and more oceanic water, and is generally of low salinity. Locally salinity can be very variable, reaching a minimum in spring when runoff is at a maximum due to snow-melt. As a result the coastal waters are strongly stratified, and their temperature may be subject to large seasonal variations, particularly in surface layers. Most of the Norwegian Sea is ice-free during winter, with sea-ice occurring only in the border areas between Bear Island and West Svalbard, and along the borderline between West Svalbard and the island of Jan Mayen.

Biological features

In the Norwegian Sea winter convection currents extend to about 300 m depth, varying somewhat in different areas. Thus waters with high concentrations of nutrients are brought into the surface layer. In early spring the waters of this layer are rich in nutrients and have a large potential for primary production. As a result the Norwegian Sea is amongst the most productive seas in the world. Phytoplankton production starts along the Norwegian coast, where the wide salinity range in the coastal waters maintains some stratification throughout the year. In the open sea the mixed layer in early spring is too deep to support primary production and consequently the spring bloom occurs one to two months later than in coastal waters. The spring bloom consists mainly of diatoms but there are considerable yearly variations in the qualitative and quantitative aspects of the bloom in all parts of the Norwegian Sea. Summer production is relatively poor but there is more diversity with regard to species.

Herring (*Clupea harengus*) has traditionally been the most commercially important pelagic fish in the Norwegian Sea. Herring spawn along the Norwegian coast, mainly in March. The hatched larvae are carried northward by the coastal current and most are transported into the Barents Sea, which is the main nursery area. The herring return to the Norwegian Sea when sexually mature. By the end of the 1960s the herring stock was almost depleted. Since then the blue whiting (*Micromesistius poutassou*), which feeds in the Norwegian Sea and spawns west of the British Isles, has formed the largest pelagic plankton-consuming fish population in the area. Three species of redfish (*Sebastes marinus, Sebastes mentella* and *Sebastes viviparus*) are also found in the Norwegian Sea, the first two having commercial value. In addition, cod (*Gadus morhua*), saithe (*Pollachius virens*) and haddock (*Melanogrammus aeglefinus*) are found along the continental shelf.

Seals and whales are important components of high latitude ecosystems. Bowhead (*Balaena mysticetus*), blue (*Balaenoptera musculus*), fin (*Balaenoptera physalus*), sei (*Balaenoptera borealis*), humpback (*Megaptera novaeangliae*), sperm (*Physeter catodon*), and minke (*Balaenoptera acutorostrata*) whales all use the Norwegian Sea as an important feeding area. The bowhead, blue, sei and minke whales have a more or less clear seasonal migration pattern with a northward movement in the spring and a return to more southern areas in autumn. It has been estimated that up to 10 000 individual whales live in the area during the summer season, with fin and sperm whales being the most common. The world population of minke whale has been estimated to represent about 86 000 individuals.

Hundreds of fish farms are located along the Norwegian coasts between Sogn og Fjordane and Tromsø. These farms have produced increasing quantities of organic matter, resulting in local nutrient and eutrophication problems (see below).

Inputs

Norway submits data to the Paris Commission on direct and riverine inputs into the Norwegian Sea. It is considered that discharges to the Norwegian Sea from riverine and direct sources are amongst the lowest when compared with those to other seas in the Convention's area (OSPARCOM, 1992a). For example, in 1990, cadmium loads to the Norwegian Sea were estimated to be 8.9 tonnes, compared with 55 tonnes to the North Sea and 70 tonnes to the North Atlantic. In the case of total nitrogen the loads were 31 kilotonnes (kt), 920 kt and 314 kt, respectively. No data were available on the contribution of other sources of contaminants, for example atmospheric deposition. There are also inputs of contaminants via ocean currents flowing into the Norwegian Sea.

The largest scale salt-water fish farming industry in the Nordic region is located in the fjords of the west coast of Norway, including the Norwegian Sea coastline. Such operations have marked effects on the nutrient status of the waters where exchange with the open sea is limited (eg, in some fjords). It has been estimated that at present aquaculture releases some 13.5 kt of nitrogen and 2.7 kt of phosphorus into Norwegian coastal waters annually, equivalent to the nutrient load in untreated sewage from a population of 2 million people (Bernes, 1993).

Contaminant levels

The concentrations of trace metals in sea water in the Norwegian Sea are reported to range from 12 to 23 ng/l cadmium, 88 to 120 ng/l copper, 187 to 195 ng/l nickel and 25 to 35 ng/l lead. There are very limited data available on trace organic compound concentrations in sea water. For example, mean concentrations of αHCH and γHCH in the Norwegian Sea were reported to be 0.2 to 1.0 ng/l, respectively. Levels of synthetic surface active substances and hydrocarbons in sea water were reported to be very low, and concentrations of chlorinated hydrocarbons did not exceed 1 ng/l (Izrael and Tsyban, 1992).

Data were submitted from two coastal sites along the Norwegian Sea for inclusion in the Oslo and Paris Commissions' 1990 supplementary baseline study of contaminants in fish and shellfish. Neither showed 'elevated concentrations' of metals in tissues or 'higher concentrations' of organochlorines in one or more species (OSPARCOM, 1992a). The concentration of PCBs in zooplankton in the southern part of the sea have been found to be up to 55.6 μg/kg, and HCH levels were between 0.3 and 1 μg/kg in molluscs and 2.2 μg/kg in zooplankton – these concentrations were considered to be very low (Izrael and Tsyban, 1992).

Biological effects

It has been postulated that the ecosystem of the Norwegian Sea has been affected by climatic variability and, in particular, variability in temperature and salinity (Blindheim, 1989). For example, during the years 1965–68 waters of below 0°C in the East Iceland current extended further eastward into the Norwegian Sea than previously observed. This change is believed to have brought about a complete alteration in the feeding migrations of herring. In addition, a succession of several poor year classes of blue whiting during the late 1970s is also a possible effect. Even though the herring stock was subject to heavy fishing pressure in the 1960s, it is likely that the climatic variability was also an important factor for the depletion of stock around 1970. The biomass of the herring is at present estimated to be 2 million tonnes compared with 1950s values of 10 million tonnes (Aksnes et al, 1993).

Conclusions

- The Norwegian Sea is generally considered to be of high quality in terms of water and biological quality.
- Local problems in the coastal zone have been caused by the nutrient and industrial discharges in fjords and aquaculture.
- Oil discharges come from shipping, particularly that heading for Narvik and further north.
- Oil and gas exploratory operations are also a potential threat to the sea.

THE BALTIC SEA

General situation

The Baltic Sea (see Map 6.1) is a shallow, relatively young and sheltered inland sea, created after the last ice age some 10 000 years ago. It is the largest brackish-water area in the world. The present salinity conditions have remained largely unaltered for the last 3000 years. The sea is surrounded by Germany, Poland, the Russian Federation, Lithuania, Latvia, Estonia, Finland, Sweden and Denmark. The drainage basin of the Baltic Sea is over 1 800 000 km², which is more than four times larger than the entire sea surface area. It is densely populated (77 million people) and heavily industrialised, and there are large areas of intensive agriculture.

The area of the Baltic Sea is 412 560 km², and the volume is 21 000 km³; the mean depth is only 55 m although the maximum depth is 459 m (the Landsort Deep). Roughly 17 per cent of the area is shallower than 10 m. The Baltic Sea consists of a series of basins, most of them separated by shallow areas or sills, except for the Gulf of Finland which is a direct continuation of the Baltic proper. The Belt Sea, including the Danish Straits and the Sound, forms together with the Kattegat a transition zone and the only link between the Baltic Sea and the North Sea.

Due to its location in high latitudes, relative poor mixing and low salinity, ice-cover is a characteristic feature of the Baltic Sea. The Bothnian Bay (the north part of the Baltic) can be ice-covered for up to 6 months a year. Therefore, the productive season in northern parts of the Baltic Sea is only 4 to 5 months, compared with that of 8 to 9 months in southern sounds and the Kattegat.

There is high river discharge into the Baltic Sea, giving it a positive freshwater balance. The rivers Neva, Wisla and Oder are the largest rivers discharging to the Baltic Sea (see Map 5.11). Together they constitute about 23 per cent of the total freshwater inflow. The river discharges vary seasonally, the maximum runoff generally occurring in the spring during the snow-melting period.

The fact that the Baltic Sea is a land-locked basin with a relatively small water volume and a long residence time of water (30 years) makes it very susceptible to any kind of pollution load. The Baltic Sea ecosystem is subject to severe natural fluctuations as well as anthropogenic disturbances. During the recent decades there has been a clear tendency towards eutrophication.

Concern about the environmental problems of the Baltic Sea led in 1974 to the signing of the Baltic Marine Environment Protection Convention (Helsinki Convention) by the states surrounding the sea, and also resulted in the formation of the Helsinki Commission (HELCOM). A new Convention, revising the 1974 version, was signed in April 1992 in Helsinki and adds the states within the catchment area of the Baltic Sea: the Czech Republic, Norway (though not a member of HELCOM), the Slovak Republic and Ukraine. The Baltic Republics (Estonia, Latvia and Lithuania) also acceded. Today, monitoring is coordinated in all the Baltic Sea countries and periodic evaluations of the state of the sea are carried out. There have been two pollution load compilations, in 1987 and 1990. In the Ministerial Declaration of 1988 it was agreed that loads of nutrients, persistent organic substances and heavy metals should be reduced by 50 per cent. In 1991 the first periodic assessment of all coastal waters was undertaken by an expert group of HELCOM. The 14 states in the Baltic Sea catchment area, together with international financial institutions, have (in 1992) started the implementation of the Baltic Sea Joint Comprehensive Action Programme in order to improve the state of the sea during the next two decades.

The main contribution to this report on the Baltic Sea

was made by the National Board of Waters and the Environment, Helsinki, and, unless otherwise stated, the main sources of information for this section were HELCOM (1987, 1990, 1991, 1992), Saxén et al (1989) and Voipio (1981).

Physical features

The three important factors regulating the characteristics of the Baltic Sea ecosystem are salinity, temperature and oxygen concentration. The variation in salinity is large in the Kattegat, between 15 and 30 per thousand, but much smaller in the Baltic Sea itself. In the open areas of the Bothnian Sea the salinity is almost stable. In the Baltic proper the surface salinity varies between 6 and 8 per thousand, in the Gulf of Finland, between 2 and 7 per thousand, in the Bothnian Sea, between 4 and 6 per thousand and, in the Bothnian Bay, between 1 and 3 per thousand.

The surface temperature of the Baltic Sea shows a strong annual cycle. A spring and summer thermocline is developed over most of the Baltic Sea at depths between 15 and 20 m. The summer thermocline may be very sharp, and the stability at the thermocline layer can thus effectively suppress vertical mixing. During autumn and early winter the surface layers become uniform in temperature because of thermohaline convection and wind-induced mixing.

A clear natural trend of decreasing oxygen content of deep and bottom waters in the Baltic Sea basins has been demonstrated. The oxygen content of these waters is influenced by weak mixing across the halocline, by stagnant periods and by oxidation of organic matter. The lack of oxygen in the bottom is accompanied by a deterioration of the benthic community and a disappearance of higher forms of life. In the Baltic Sea, a continuous oxygen deficiency exists occasionally below the permanent halocline, except in the Bothnian Bay. The area of bottom water and sediments with reduced conditions for life varies naturally and was largest during 1968 and 1969, the current area being about 100 000 km^2.

The freshwater supply to the Baltic Sea generates a brackish surface layer of outflowing water, and incoming subsurface flow from the North Sea forms layers of more saline deep and bottom waters. A dominating feature of the Baltic oceanography is, therefore, the marked permanent salinity stratification at approximately 60 m. The salt water inflow from the North Sea is made up of two principal components: a frequently occurring deep-water stream, and an irregular, intensive inflow connected with prevailing southwesterly winds. Usually the intensive inflows cause a successive renewal of the bottom water in the deep basins. A large inflow occurred in 1975 and 1976, followed by a period of stagnation which lasted until January 1993, when the last inflow of high saline water into the Baltic Sea occurred.

Biological features

The Baltic Sea has a flora and fauna very poor in species due to its brackish-water. Most of the species are immigrants from neighbouring regions. Three different types of immigrants can be distinguished: marine organisms from estuaries and shallow coastal areas in the North Sea and the Atlantic, freshwater organisms from lakes, and glacial relicts. The genuine brackish-water species are also characteristic to the Baltic Sea. The number of marine species decreases steeply as one goes through the Danish Straits into the Baltic proper, and continues to decrease further up to the Gulf of Finland, and finally to the Bothnian Bay.

Of the 1500 macroinvertebrate species on the adjacent Norwegian coast, only some 70 species are found in central Baltic Sea, and of the several hundred marine macroalgal species on the Norwegian coast, only 24 are found on the Finnish coast. Several organisms originating from freshwater or marine areas are reduced in size or altered in their morphology in the Baltic Sea.

Among the most threatened species in the Baltic Sea area are the three resident species of seals, the grey seal (*Halichoerus grypus*), the ringed seal (*Pusa hispida*), and the harbour seal (*Phoca vitulina*). The populations of these species have greatly reduced during the last decades, although some recovery of grey seal populations has been found recently. The number of white-tailed eagles (*Haliaeetus albicilla*) has also decreased. Today, only about 200 nesting pairs live in the Baltic Sea. As top predators, the seals and the sea eagles have been affected by organic micropollutants such as PCBs and other organochlorine compounds. The populations of seals have also been affected by hunting and modern fishing gear.

Inputs

Contaminants enter the Baltic Sea from many sources: domestic sewage from about 77 million people; industrial and agricultural effluents; traffic; atmospheric deposition; and inflow from neighbouring seas.

The most significant industrial inputs come from: the paper and pulp processing industry; mining; steel and metal manufacture; and fertiliser production. Paper and pulp production in the Russian Federation and Sweden takes place mainly along the coastline, whereas it is more widespread throughout Finland and Poland. This area is responsible for 25 per cent of the world's total production of pulp, paper and board. The pulp mills were formerly major sources of mercury discharges, and the associated bleaching plants are the dominant source of organochlorine compounds arising from Sweden and Finland. Nitrogenous and phosphate fertilisers are produced in large quantities, leading to major discharges of nutrients and cadmium. In addition, Sweden is a major producer of iron and zinc ore, Poland of copper, lead and zinc, and Finland of iron, zinc and some silver.

Atmospheric deposition of some heavy metals to the Baltic Sea accounts for over 50 per cent of the total load. The upper sediment layers of the Baltic Sea have been found to contain concentrations of cadmium, mercury, lead, zinc, arsenic, copper and cobalt, several times higher than the pre-industrial layers. It is evident from data submitted to HELCOM that the dominating portion of the river-borne metal emissions originate in the Russian Federation, the Baltic States and Poland. The atmospheric deposition of many synthetic organic compounds is a more important source than riverine discharges, runoff or wastewaters. It is also believed that the acidification of freshwaters has increased the quantity of some metals entering the Baltic Sea.

The use of DDT has been banned in most industrialised countries since the 1970s. Now the main source of input of DDT into the Baltic Sea is through the atmosphere from areas where it is still in use. The manufacture and use of PCBs have decreased markedly since the 1970s because of regulatory action.

Petroleum hydrocarbons enter the Baltic Sea environment in many different ways. The total amount of oil products released into the Baltic Sea has been estimated to range from 21 kt to 660 kt per year. There are no natural seeps of oil. Transportation of oil and intensive shipping activities may be a larger source of petroleum hydrocarbons (10 per cent of total) than in other seas, but the highest input figures are from land (62 per cent) and atmospheric deposition (14 per cent) (Figure 6.9). As in other seas, oil spills in the Baltic are a danger to ecosystems, the effects depending, among other things, on the magnitude and location of the spill.

The major nutrient inputs are from runoff from

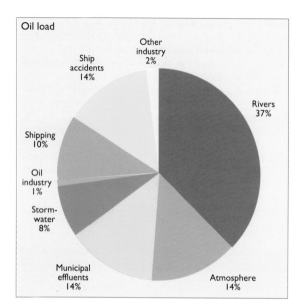

agricultural land, from domestic and industrial wastewaters and from atmospheric deposition. This results in nutrient concentrations being higher in the coastal waters than in the open sea. About 40 per cent of the nitrogen load comes from atmospheric inputs, and about 90 per cent of the phosphorus input is from land areas. The total input of nutrients to the Baltic Sea from land-based sources, including the Danish Straits and the Kattegat, is estimated to be 46 kt of phosphorus per year, and 630 kt of nitrogen per year. The total loads of nitrogen and phosphorus into the Baltic have increased two-fold and four-fold, respectively, since 1950. It has also been estimated that 75 per cent of the total nitrogen load and 90 per cent of the total phosphorus load from land-based sources is anthropogenic in origin. Of the total nitrogen load arising from land-based sources (in 1990), it was calculated that approximately 43 per cent arose from the former USSR, 17 per cent from Poland, 15 per cent from Sweden and 11 per cent from Denmark. The corresponding figures for total phosphorus loads were 43 per cent from the former USSR, 27 per cent from Poland, 11 per cent from Denmark and 9 per cent from Sweden.

Agriculture is important in all the Baltic countries: generally high amounts of artificial fertilisers are used, and surplus amounts of by-products, such as manure, are produced. These are diffuse sources of pollution which are now recognised as a significant component of the total load entering the Baltic Sea.

Contaminant levels

It has been estimated that even if all the riparian countries of the Baltic were immediately to reduce their discharges of contaminants by 50 per cent, it would still take 10 to 15 years before a measurable reduction in the concentration of some contaminants occurred in the marine environment.

According to the Second Periodic Assessment of the State of the Baltic Sea, the area most affected by pollution is the eastern part of the Gulf of Finland, which is heavily polluted by the city of St Petersburg and the Neva River. For example, 50 to 70 per cent of the wastewater load of the city is estimated to be discharged into the Neva River or the Neva Bay without any purification whatsoever. Other areas heavily affected by pollution are the Belt Sea and the Kattegat, where heavy blooms of diatoms and dinoflagellates, with subsequent oxygen deficiency in the deep water, have been reported.

In the northern and western riparian countries of the Baltic the application of modern technology has significantly reduced the emissions from industries over the last 10 to 15 years. On the other hand, in Central and Eastern European

countries, many industrial plants are technologically outdated, and need to be rebuilt to meet the present environmental demands (see Chapter 12).

Heavy metals

Investigations show that mercury concentrations increase from north to south in the Baltic, which supports the assumption that sources on the European continent are important contributors to the atmospheric mercury emissions in the Nordic countries. Total mercury content in water samples ranges from 2 to 630 ng/l in various sub-regions of the Baltic Sea, being the highest in the Bay of Gdańsk (277–630 ng/l). The concentrations in Baltic fish (cod and plaice) normally range from 20 to 500 μg/kg wet weight.

In general, the regional variations in the cadmium content of water are fairly small. The mean concentration is approximately 30 ng/l, with some higher values reported from nearshore regions. The concentrations in fish (cod and herring) and in shellfish (mussels) have been reported to range from 2 to 200 μg/kg wet weight and 1300 to 10 800 μg/kg wet weight, respectively.

The concentrations of lead in fish and blue mussels in the Baltic Sea have shown a decreasing trend since the early 1980s. The concentration of lead in fish (cod and herring) from the Baltic Sea has been found to range from 50 to 1300 μg/kg wet weight. These findings are supported by decreasing dissolved lead concentrations in Baltic Sea water.

Typically the concentrations of copper and zinc in muscle tissue of fish are below 1 mg/kg and 10 mg/kg wet weight, respectively, whereas the corresponding liver concentrations are about 5 to 10 times higher. The levels of copper and zinc are not significantly different in the Baltic Sea from those in the North Sea.

Synthetic organic compounds

The decreased use of PCBs is reflected by a clear downward trend in their concentrations in biota from the 1970s to the 1980s, and the concentrations now seem to have stabilised at approximately one third of the maximum level of the 1970s. The highest values of PCBs have been recorded in the northern part of the Baltic proper. Between 1980 and 1988 the mean levels of PCBs in liver of cod collected from the Baltic proper declined from approximately 8.5 mg/kg (lipid weight) to less than 2.5 mg/kg.

Pesticides

Investigations in 1983 revealed quantifiable concentrations of DDT in the water of the western Baltic proper and the Belt Sea. Since then the concentration in water has fallen below detection limits. There was also a clear downward trend for DDT concentration in biota of the Baltic Sea from the 1970s to the 1980s, but during the last decade the levels in biota have remained almost unchanged.

Toxaphene was introduced as a substitute after the use of DDT was prohibited. Today, the concentration of toxaphene in the Baltic Sea environment is of the same magnitude as that of DDT. Lindane (γHCH) has also been used as a substitute for DDT and today it can be found in all parts of the Baltic Sea environment. In 1983 the concentration of lindane in surface water from different parts of the Baltic Sea varied between 1.9 and 7.2 ng/l, compared with between 1.5 and 3.2 ng/l in 1988. This indicates a decline in surface water concentrations of lindane in the Baltic Sea.

Microbiological contamination

The various control systems used for determination of the microbiological conditions, standards and categories applied differ between the countries in the Baltic Sea area.

The relative number of beaches where bathing is

temporarily or frequently prohibited is markedly higher in the southern and eastern parts of the Baltic Sea than in the northern parts. For example, in Estonia the internationally known beaches in the Pärnu Bay have been closed because of microbiological pollution. In summer, values exceeding 100 000 *E. coli* per 100 ml have been recorded. Bathing was also banned from 25 bathing areas along the Baltic coast of Denmark in 1992 (CEC, 1993). However, overall in 1992 Denmark achieved compliance with the EU mandatory standard of 2000 faecal coliforms per 100 ml in 93 per cent of its designated bathing waters. Along the Baltic coast of Germany, 74 per cent of designated bathing waters were found to comply with the mandatory standard in 1992: this represented a marked improvement over 1991, when 63 per cent complied. In 1992, 15 bathing waters along the German Baltic coast were also closed for swimming.

Oil

The petroleum content in the surface water is at about the same level in all parts of the Baltic Sea but, compared to the North Sea and the North Atlantic, the levels are two and three times higher, respectively.

Nutrients

The Baltic Sea was for a long time considered to be an oligotrophic sea (nutrient poor), but lately that opinion has changed. In the Baltic proper, the Gulf of Finland and the Kattegat, nitrogen plays a key role as a limiting nutrient, whereas phosphorus is the limiting nutrient in the Bothnian Bay. In many areas of the Baltic Sea, the strong increase of phosphorus and nitrogen concentrations observed in the 1970s has stopped, with the exception of the Kattegat, the Gulf of Riga and the eastern Gulf of Finland, where the nitrogen concentrations are still increasing. Phosphorus and nitrogen concentrations have been at such high levels that the increasing biological production and sedimentation, followed by the microbial decomposition of the biogenic organic material, cause further deterioration of the oxygen conditions in basins of the Baltic Sea.

Radionuclides

The areal radionuclide fall-out from the Chernobyl accident was unevenly distributed. In 1988 the surface water concentrations of Cs-137 were about 150 to 170 Bq/m^3 in the Baltic proper, 120 to 230 Bq/m^3 in the Gulf of Finland, 220 to 320 Bq/m^3 in the Bothnian Sea, and about 120 Bq/m^3 in the Bothnian Bay. In the gulfs of Finland and Bothnia the total amounts of Cs-137 were about five to six times higher in 1988 than before the Chernobyl accident, but only three times higher in the Baltic proper. The post-Chernobyl values of Cs-137 in pike were in 1988 about 10 to 40 times higher than the pre-Chernobyl values.

Biological effects

Eutrophication

Regardless of whether anthropogenic activities or natural variations play the dominating part in the nutrient increase, eutrophication has been identified as one of the most serious problems in the Baltic Sea area. Phytoplankton primary production seems to have doubled within the last 25 years in the area from the Kattegat to the Baltic proper. Today, unusually intensive algal blooms appear to occur more frequently in the Kattegat, in the Gulf of Finland and in the Belt Sea. Signs of eutrophication are also evident in the pelagic ecosystem of the Baltic Sea, although no drastic changes in species or population numbers have occurred there.

The increase in primary production and the subsequent increase in algal sedimentation and decomposition in the benthic system have decreased the oxygen content of the deep water. The decrease is no longer restricted to the deep basins but increasingly affects the shallower parts, and has resulted in drastic changes in the composition of benthic communities below the halocline. The phytoplankton composition has not changed drastically, but there is a tendency towards more frequent, and sometimes toxic, algal blooms.

The mass development of microscopic algae over large areas of the Baltic Sea drastically reduces the water transparency and sometimes also creates surface scums and odours. The present transparency shows pronounced decrease (by 2 to 3 m, generally) in the open coastal waters compared with that during the first half of the century. The depth at which *Fucus vesiculosus*, the bladder wrack, occurs has also decreased since the 1940s, apparently due to the increased turbidity of the waters. In addition, an increased dominance by filamentous benthic algae has been reported from the Swedish coast, leading to local adverse effects on the hatching of herring eggs.

Within the southern Kattegat, mortality of benthic macrofauna, mainly bivalves, has been observed in most seasons and years. This has been attributed to oxygen deficiency from the degradation of algal biomass. In 1988 the area of the Kattegat affected was 5600 km^2 over which fish and the Norway lobster, *Nephrops norvegicus*, were either absent or found only in low numbers.

Fisheries

The Baltic fish stocks are a major living resource, with herring (*Clupea harengus membras*), sprat (*Sprattus sprattus*) and cod (*Gadus morhua*) representing about 90 per cent of the total catch. The catch of fish has increased steadily during the last 50 years, due mainly to increased fishing efforts and improved techniques. The annual yield has grown to about one million tonnes.

The environmental conditions in the Baltic influence the fish stocks considerably. For example, the stocks of cod are hampered by the increasing frequency of oxygen depletion and water containing hydrogen sulphide. The hatching of cod is affected by the gradually decreasing salinity in the spawning areas in the Bornholm Sea, the Gotland Deep, and the Gdańsk Deep. Pelagic fish, like herring and sprat, benefit from moderate eutrophication.

Conclusions

- The environmental situation in many parts of the Baltic Sea has deteriorated rapidly since the 1950s. The areas affected by pollution have grown, and the type of pollutants has changed, as wastes contain more toxic substances. Together with stagnation of the deeper water, the pollution of the Baltic Sea has now become a threat to its living resources.
- Human activity is also responsible for the increase of certain natural substances in the Baltic Sea, such as nutrients. Since the beginning of this century the levels of nitrogen and phosphorus in the Baltic Sea have increased by about four and eight times, respectively.
- The eutrophication affects the whole Baltic Sea, particularly the coastal areas. Low oxygen concentrations in late summers and autumns in the 1980s were often observed in the southern Kattegat, the Belt Sea, the Sound and the Arkona Basin (a sub-region of the southern Baltic proper), resulting in fish mortalities and damage to bottom fauna.
- As a result of increasing eutrophication, the intensity of toxic blooms has increased. Together with the microbiological contamination, eutrophication decreases the value of coastal areas for recreational purposes. A decline of benthic vegetation and a changed composition of species, for example from perennial brown and red algae

to filamentous green algae, is reported from all coastal zones in the Baltic Sea area.

- The pelagic fish stocks partly benefit from the moderate eutrophication. Currently, all other herring and sprat stocks are either well exploited or still underfished except the herring stock between the Kattegat and the Arkona Basin.
- As a result of the ban on certain harmful substances, some positive changes have been observed. The levels of DDT and PCBs have decreased since the 1970s. However, the concentrations of organochlorine residues in fish from the Baltic proper are still three to ten times higher than in the fish from the northern Atlantic. The Baltic Sea seal populations are slowly recovering from the reproductive failures, possibly caused by the accumulation of toxic substances. The relatively high concentrations of toxic organic compounds probably represent an even greater threat than eutrophication to the ecosystem in the long term.

THE NORTH SEA

General situation

The North Sea is situated on the continental shelf of northwest Europe. It is open to the Atlantic Ocean in the north, to the English Channel in the south and to the Baltic Sea in the east (see Map 6.1). The climate of the North Sea is dominated largely by the Atlantic Ocean, and is therefore characterised by a large variety of wind directions and speeds, a high rate of cloudiness, and relatively high precipitation compared with other marine areas (425 mm on average per year for the North Sea proper). Its coastlines are shared between the UK, Norway, Sweden, Denmark, Germany, The Netherlands, Belgium and France. The catchment area of the rivers that flow into the North Sea (including the English Channel, Skagerrak and Kattegat) is about 850 000 km² (165 million people). The catchment areas of the rivers Elbe, Weser, Rhine, Thames, Humber and Seine are densely populated, highly industrialised and farmed intensively. As a consequence these river systems are among the largest sources of contaminants and nutrients flowing from land into the North Sea.

The Oslo Convention (Convention for the Prevention of Marine Pollution by Dumping from Ships and Aircraft) and the Paris Convention (Convention for the Prevention of Marine Pollution from Land-Based Sources) provide the regulatory framework for the protection of the maritime area of the northeast Atlantic (including the North Sea) against pollution. The Oslo Convention came into force in 1974 and has been ratified by 13 European states; the Paris Convention came into force in 1978 and has also been ratified by 13 states. Finland has ratified only the Oslo Convention and Luxembourg has signed only the Paris Convention. The commissions established by the Conventions carry out programmes to assess the state of the marine environment and formulate policy to eliminate or reduce existing pollution and prevent further contamination of coastal waters and open seas. The commissions adopt appropriate measures to implement these policies, and assess their effectiveness on the basis of reports on their implementation and on the results of monitoring, thus adapting their policies and introducing new measures as appropriate.

In 1984, the perceived slow progress in reducing pollution of the North Sea resulted in a series of ministerial conferences on the North Sea attended by the environment ministers of all riparian countries. There have been three conferences to date; the last was in The Hague in 1990. At the end of each conference the ministers from the participating countries agreed objectives by way of declaration. These declarations are not legally binding and it is up to each government to decide how to achieve the stated objectives.

Following the London Ministerial Conference Declaration, the North Sea Task Force (NSTF) was set up during 1988 with the following membership: Belgium, Denmark, France, Germany, The Netherlands, Norway, Sweden, the UK, the Commission of the European Communities and the International Council for the Exploration of the Sea (ICES). The NSTF set up an extensive monitoring master plan, providing the various countries bordering the North Sea with a responsibility for specific areas. The primary objective of the NSTF was to produce a new quality status report by the end of 1993, and to achieve this objective a monitoring master plan was drawn up in 1989 for implementation during 1990/91 (Portman, 1991). Each North Sea riparian state was to carry out a field monitoring programme at sea and in coastal waters for an agreed list of substances using agreed techniques.

Ministers from 15 European countries agreed in September 1992 a new Convention for the Protection of the North East Atlantic. Together with its action plan (which is subject to annual review) and the accompanying Ministerial Declaration, the new Convention will provide a comprehensive framework for the protection of the Convention's area. The Convention covers pollution from vessels, offshore installations, dumping, monitoring and assessment. The Convention's area is the North Sea and the North Atlantic Ocean.

Those North Sea states which are members of the European Union have also to comply with relevant directives aimed at controlling and eventually eliminating the discharge of the most toxic substances to the aquatic environment.

Unless otherwise stated, the source of information and contribution on the North Sea is from the secretariat of the North Sea Task Force (Ducrotoy, personal communication).

Physical features

Together with the English Channel, the Kattegat and the Skagerrak, the North Sea covers 750 000 km² for a volume of 94 000 km³. The sea has an average depth of 85 m, with depths increasing towards the Atlantic Ocean to about 200 m at the edge of the continental shelf. The Norwegian Trench reaches a maximum water depth of 700 m. The English Channel is shallow and deepens gradually from about 30 m in the Strait of Dover to about 70 m in the west. In the shallow North Sea, the entire waterbody derives from North Atlantic water and freshwater runoff in different admixtures. Most of the North Sea water flows through the Skagerrak before leaving through the Norwegian Coastal Current. The flow from the English Channel is from west to east, feeding a salty core of Atlantic water through the Strait of Dover.

Tidal currents are the most energetic feature in the North Sea, stirring the entire water column in most of the southern North Sea, and the English Channel. Tidal energy from the Atlantic Ocean also forces a persistent current with an anti-clockwise circulation (see Map 6.2). Tidal heights are greatly amplified in the bays of the French coast of the Channel, where they can reach 12 m.

Most areas of the North Sea are vertically well mixed in winter. In late spring, a thermocline is established over large areas; this separates the lower from the upper layer and a self-stabilising stratification develops (50 m in the northern North Sea, 20 m in the western English Channel). The deeper parts (Norwegian Trench and Kattegat) are permanently stratified. A coastal strip along the southern part of the North Sea, stretching from northern France to the German Bight, remains vertically mixed during the whole year.

Residence times vary greatly between areas and layers in the North Sea. In the Skagerrak, the residence time is much longer for the deeper water (one to three years) than for the near surface water (one month), and times of up to 3.9 years are observed along the British coast (Reid et al, 1988).

Biological features

The distribution patterns of macroalgae and shore animals show that the North Sea is a transition area between the warm region in the southwest and the cold boreal region in the east and the north. A number of saltmarsh plants reach their northern limit of distribution in the Firth of Forth (Scotland) and their eastern limit in The Netherlands. In some coastal areas several types of plant populations have been reported to have declined, such as the seagrass *Zostera marina* in the 1930s and during the 1960s. Other plants, on the contrary, have dramatically increased in biomass in recent years. Green algae have developed on many tidal flats, for example *Enteromorpha* and *Ulva* species in the Wadden Sea and on the French Channel coast and in some British and Danish estuaries and lagoons.

In 1988, the phytoplankton bloom of the toxic flagellate *Chrysochromulina polylepis* resulted in the massive kill of epibenthic fauna along the Kattegat and Skagerrak coasts of Sweden and Norway. *Phaeocystis* has also caused nuisance foams on beaches of Germany, The Netherlands, Belgium and France. An increase of macrofauna in the Wadden Sea has been observed; for example, on tidal flats, the biomass doubled between 1990 and 1991.

The fish communities today reflect the impact of more than a century of intensive utilisation by the fisheries industry. In general, smaller, short-lived plankton-feeding species dominate; for example, sand eel (*Ammodytes tobianus*), dab (*Limanda limanda*), Norway pout (*Trisopterus esmarki*), herring (*Clupea harengus*) and mackerel (*Scomber scombrus*).

Many of the sea-birds in the North Sea are present in numbers that represent substantial proportions of their world population. Counts of bird populations breeding on British coasts indicate an increase in numbers during the last decades, and, in some cases, expansion of range; for example fulmar (*Fulmarus glacialis*), gannet (*Sula bassana*), lesser black-backed gull (*Larus fuscus*), guillemot (*Uria aalge*), great skua (*Stercorarius skua*), and Arctic skua (*Stercorarius parasiticus*). Some species, such as roseate tern (*Sterna dougallii*), herring gull (*Larus argentatus*) and black-headed gull (*Larus ridibundus*), have declined.

There is a long-term trend of increasing seal populations, but in 1988, phocine distemper virus caused a major mortality of mainly common seal (*Phoca vitulina*) and, to a lesser extent, the grey seal (*Halichoerus grypus*). For example, there was a 50 per cent reduction in the common seal population of the Wash on the east coast of England. Other populations along the UK's North Sea coast were probably less severely affected (Northridge, 1990). These two species of seal are the only ones to breed along the coasts of the North Sea. Since the disease outbreak, the number of seals has remained generally unchanged or has been increasing. At present, current estimates of the North Sea populations give a number of about 40 000 common seals and 47 000 grey seals.

The harbour porpoise (*Phocoena phocoena*) is the most common coastal water cetacean in the North Sea. Another species, the bottle-nose dolphin (*Tursiops truncatus*), is found in small resident populations at certain sites such as in the Moray Firth, Scotland. Other cetaceans show varying degrees of migratory behaviour, for the most part dependent on the movement of their food supplies. White-beaked dolphins (*Lagenorynchus albirostris*), for example, which breed off the Dutch coast in June and July, migrate to British waters to feed on herring and mackerel. Other species, particularly those in the northern and southwestern sectors of the North Sea, such as the minke whale (*Balaenoptera acutorostrata*) and the common dolphin (*Delphinus delphis*), visit these waters seasonally from the North Atlantic.

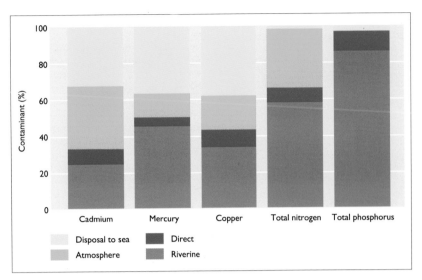

Inputs

Rivers, the atmosphere and dumping of dredged material are the major input sources of contaminants into the North Sea (Figure 6.10). In 1990 riverine inputs were generally greater than inputs from direct land-based discharges. Where comparable data were available, land-based inputs in general exceeded atmospheric inputs and inputs via dumping, though for several contaminants (eg, cadmium and lead) the atmospheric inputs were of the same order of magnitude as the other land-based inputs. Riverine discharges of heavy metals such as cadmium and mercury into the North Sea are generally lower than those into other European seas; in the case of total nitrogen, loads are close to the highest. Figure 6.11 shows a country-by-country breakdown of riverine and direct inputs in 1990.

It is not yet possible to estimate how much of the riverine and direct loads are retained within estuaries or how much reaches the open sea. It should be noted that riverine loads (metals and nutrients) have both natural and anthropogenic contributions.

An important step towards the reduction and termination of dumping of industrial waste was taken in 1987 when France ceased its dumping of phosphogypsum. A further reduction occurred in 1989/90 when Belgium and Germany phased out the dumping of waste from the production of titanium dioxide.

Figure 6.10
North Sea: proportion of selected contaminants from different sources
Source: OSPARCOM, 1992a

Figure 6.11
North Sea: proportion of riverine loads of selected contaminants from different states, 1991
Source: OSPARCOM (1992a) for 1990 figures and unpublished data for 1991 obtained by WRc
Note: Loads from Belgium, Germany and The Netherlands include loads from countries upstream

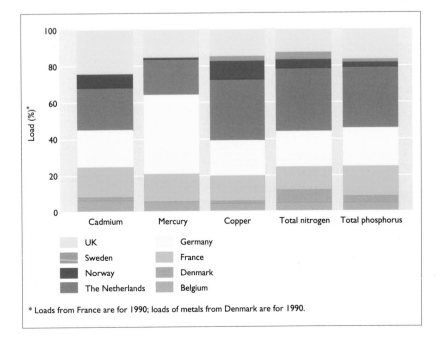

* Loads from France are for 1990; loads of metals from Denmark are for 1990.

By the end of 1992, the UK had ended the dumping of industrial waste at sea. Disposal of sewage sludge at sea will be terminated by the end of 1998. The metal load of dredged material from estuaries, harbours and navigation channels is considerable. However, this load is made up of generally unquantified contributions from both natural and anthropogenic sources. The anthropogenic metal load from this source is, therefore, overestimated.

Besides the global fall-out from nuclear weapon tests, primarily from the 1960s anthropogenic inputs of radioactivity in the North Sea originate mainly from authorised releases from the reprocessing plants of Sellafield (UK, to the Irish Sea) and Cap de la Hague (France, to the English Channel). Soluble radioactivity from Sellafield is initially transported southwards and then northwards before leaving the Irish Sea via the North Channel between Ireland and Scotland. It then travels northwards along the west coast of Scotland, around the northern coast of Scotland and then southwards into the North Sea. In addition there are 14 other nuclear sites that discharge relatively small amounts of radioactivity into the tidal waters of the North Sea and English Channel. The amount of deposition of radioactivity in North European waters from the reactor accident at Chernobyl in 1986 was highest in the Baltic Sea, though the North Sea, the northwest coast of Scotland and the Irish Sea also received a significant input (Camplin and Aarkrog, 1989). As is the case for routine discharges from the nuclear industry, the dominant radionuclide was found to be Cs-137, accompanied by Cs-134 and other radionuclides.

An estimated maximum of 260 000 tonnes of oil enters the North Sea from all sources annually (GESAMP, 1993). The major source of oil inputs from the activities of the offshore oil and gas industry is the discharge of cuttings arising from the use of oil-based and water-based drilling fluids. Particular concentrations of drilling sites occur in the major production areas of the northern North Sea, in particular in the East Shetland basin (primarily oil-fields) and the south/central North Sea (primarily gas-fields). The annual input of oil from cuttings was in the range of 15 000 to 25 000 tonnes per year over the period 1984 to 1990. There are some indications of a general increase of sediment hydrocarbon levels in the East Shetland basin as a result of the concentration of production platforms in that area, whereas no such general elevations have been observed in the southern gas-field areas. Operational discharges of oil from shipping are estimated at 1000 to 2000 tonnes per year, about 0.5 to 2.0 per cent of the total input to the North Sea.

The cultivation of fish in coastal waters provides a direct input of organic matter, nutrients and other chemicals (eg, therapeutants) to the North Sea. Shellfish cultivation does not normally involve the addition of material such as feed.

The main areas of fish farming activities around the North Sea are Shetland and Norway. The inputs of nutrients are small in relation to the natural flux of nutrients through the area by water movements.

Ninety to ninety-five percent of the nutrients entering the North Sea from land (excluding atmospheric deposition) originates from a few main rivers, more than half of them from the river Rhine. Nutrient inputs were maximal in the late 1970s and early 1980s for the Southern Bight, where they reached four times the input levels for total nitrogen and seven times those for total phosphorus, compared with 1930–40 levels. The importance of the second pathway of eutrophication by airborne nitrogen loads is now appreciated and can be in the same order of magnitude as the riverine sources in summer. In the central parts of the North Sea aerial transport is more important than riverine transport. Measurements suggest that atmospheric inputs of ammonia and nitrogen oxides are higher than previously thought.

Contaminant levels

Heavy metals

From sea water monitoring carried out by the North Sea Task Force (NSTF) in 1990 and 1991, the highest average values for lead, mercury and cadmium were in the central North Sea. A periodic or seasonal variation in the concentration of mercury and cadmium in sea water over the Dogger Bank has been described. Stormy conditions have been observed to lead to remobilisation of manganese, iron, cadmium and zinc following disturbance of the sediments. A plume of elevated and variable concentrations to the northeast of the Humber estuary was also described. The high metal concentrations found over the Dogger Bank and off the coast of England have been tentatively ascribed to airborne transport from land. Concentrations of lead in sea water were found to be lower in the eastern part than in the rest of the English Channel in springtime only, possibly due to removal of lead by phytoplankton during the spring blooms but coupled with reduced river inputs.

In 1990 the central parts of the North Sea had sediments with background levels of mercury. This implies that most of the mercury is trapped in estuaries and near coastal waters. Some elevated levels occur in the inner Skagerrak, the Norwegian Trench, the German Bight, to the north of the Dogger Bank and in estuaries surrounding the North Sea. The most notable areas of higher concentrations of lead in sediment are the Hardangerfjord (Norway), the Scheldt area, the Baie de Seine and the northeast coast of England. The northeast coast of England is heavily impacted from old lead mines. The concentrations of chromium are higher near the English coast, in the Skagerrak and in the Norwegian Trench.

The results of the Oslo and Paris Commission's (OSPARCOM) 1990 supplementary baseline study of contaminants in fish and shellfish identified the following areas with 'elevated concentrations' of metals in tissues (OSPARCOM, 1992a):

- cadmium – off the Humber estuary, the Dogger Bank, the Sørfjord/Hardangerfjord;
- copper – the western and eastern English Channel, the Ems estuary, the Outer Silver Pit (relatively deep water to the south of the Dogger Bank), off the Tees estuary;
- lead – east coast of England, the Sørfjord/Hardangerfjord area;
- mercury – off Calais, the Ems estuary, off the Thames estuary, Stenungsund area, Hvaler/Singlefjord, and Sørfjord/Hardangerfjord;
- zinc – Roscoff, western English Channel, Outer Silver Pit, southeast of Dogger Bank, west Dogger Bank, off the Tees estuary, the Tyne estuary, the outer Oslofjord, the Sørfjord/Hardangerfjord area.

Dumping of dredged contaminated mud adds harmful substances (such as heavy metals) to the marine environment.
Source: Michael St Maur Sheil

Synthetic organic compounds

The highest inferred concentrations of tributyl tin (TBT) in 1991 and 1992 (through the study of abnormalities in the sexual organs of female dogwhelks, *Nucella lapillus*) were found along the coast from Denmark to Belgium (from 40 to greater than 128 ng/l). These can be compared with the relatively low levels found along the coasts of Sweden and Norway (1 to greater than 4 ng/l). Since controls were instituted, TBT concentrations have decreased markedly at estuarine locations frequented by yachts, but in harbours and dry-dock areas concentrations are variable and still often high.

The occurrence of numerous other compounds even in very remote regions shows that some trace organics are degraded so slowly, if at all, that long-range transport is possible. High concentrations (mainly of PCBs) have been observed in predators. Relatively high DDT concentrations, particularly in biota, still occur. Clear gradients occur in several estuaries (eg, Thames, Western Scheldt, Weser, Elbe, Göta). Work is required to determine the source of these compounds. Determination of the origin and the toxicological significance of relatively high levels of toxaphene observed in river fish and mammals' blubber is also required.

The highest levels of lindane (γHCH) in fish were from the southern North Sea and ranged from 2 μg/kg wet weight in cod muscle to 10 μg/kg wet weight in herring. Corresponding levels in cod liver were higher: between 20 and 50 μg/kg wet weight.

The results of the OSPARCOM 1990 supplementary baseline study of contaminants in fish and shellfish identified the following areas with 'higher concentrations' of organochlorines in one or more species/tissues (OSPARCOM, 1992a):

- Dieldrin – off the Thames estuary (liver tissue only);
- γHCH – East Frisian coast;
- PCBs – French coast of the English Channel, western English Channel off Brixham, Thames estuary, Southern Bight of the North Sea, Western Scheldt estuary, Ems estuary, inner Oslofjord.

On different occasions, slicks of alkylphenols (nonylphenols and dodenylphenol), probably originating from ships' illegal discharges, have caused the death of sea-birds on the north coast of the Frisian Isles.

Microbiological contamination

The North Sea riparian countries who are members of the EU have to submit data annually on the quality of designated bathing waters in terms of microbial contaminants and physico-chemical parameters.

- Of the designated bathing waters along the UK's North Sea coastline, 85 per cent complied with the mandatory faecal coliform standard in 1992, with non-compliant areas including the Thanet coast, the northeast coast of England and some waters along the English Channel, and the east coast of central Scotland (CEC, 1993).
- Along the North Sea coast of Germany, 80 per cent of bathing waters complied with bacterial standards in 1992; non-compliance was found around the Ems, Weser and Elbe estuaries.
- Bathing was banned from two beaches along the North Sea coast of Denmark in 1992; overall, though, around 99 per cent of bathing waters did comply with the faecal coliform mandatory standard.
- In France approximately 74 per cent of bathing waters along its North Sea and Channel coasts complied with bacterial standards in 1992; non-compliance was recorded along the Normandy and Picardy coast and the Cherbourg peninsula.

Outflow from chemical plant, West Hartlepool, UK
Source: Michael St Maur Sheil

- In 1992, 90 per cent of Belgium's bathing waters complied with mandatory bacterial standards and 21 per cent complied with guideline values – non-compliance generally occurred near harbour mouths, for example at Ostende.
- The Netherlands reported that 86 per cent of its coastal bathing waters complied with mandatory standards in 1992, and 73 per cent met guideline values. Non-compliant areas included Harlingen on the Wadden Sea and Groede in the south of the country.

Oil

The highest concentrations of oil hydrocarbons in sediments are found around installations where oil-based drilling muds have been used. The concentrations in the sediments rapidly decline from the platforms out to a distance of about 500 m. High concentrations of polycyclic aromatic hydrocarbons (PAHs) in sediments were found in the Skagerrak and the Norwegian Trench. But PAHs appear to be widespread in sediments, including offshore areas such as the Dogger Bank and the Oyster Ground (off the Dutch coast).

Nutrients

Winter background concentrations of total inorganic nitrogen (TIN) ($NO_3 + NO_2 + NH_4$) and soluble reactive phosphorus (SRP) in the central North Sea are around 0.14 mg N/l and 0.019 to 0.022 mg PO_4-P/l, respectively, and in the Kattegat 0.056 mg N/l and 0.013 mg PO_4-P/l, respectively (OSPARCOM, 1992b). Between 1985 and 1990, relatively high winter TIN and SRP concentrations were found along many coasts of the North Sea, including those of The Netherlands, Belgium, Germany, Denmark, Norway (in the Skagerrak and Oslofjord) and the UK (southeast coast). In these areas winter nutrient concentrations could be up to 11 times and 5 times greater than background levels of TIN and SRP, respectively.

Oil spills can cause major damage to the environment, with sea-birds often the most conspicuous victims
Source: The Environmental Picture Library

Radionuclides

The concentration of most of the artificial radionuclides shows a declining trend in sea water and biota in the North Sea. A number of different radionuclides have been detected in biota during recent years. The radionuclide of greatest (though still small) radiological significance is Cs-137. In cod liver, Co (cobalt)-60, Ag (silver)-110m and Ag-108m have been detected in extremely low concentrations. Elevated concentrations of Po (polonium)-210 have recently arisen in some localities from the phosphate ore processing industry. The concentration of the Chernobyl derived nuclides Cs-134 and Cs-137 were almost negligible in biota in 1990.

Biological effects

Eutrophication

Enhanced nutrient concentrations are associated with the large continental European rivers, and their areal extent depends on the influence of river water. The areas of concern are the continental coastal waters from France to Denmark, where the river water influence is confined to a narrow coastal water zone with water exchange offshore limited by hydrographical fronts. The major extension of the coastal water and hence of eutrophication is found in the German Bight and the Kattegat. Also, estuaries, fjords and bays with reduced water exchange may be showing signs of eutrophication. However, transport of nutrients bound to organic particles can reach further out into the sea, accumulating in sedimentation areas, such as the Norwegian Trench.

Potential effects of eutrophication include the enhancement of growth of phytoplankton with its consequences for the food-chain, including the zooplankton and the benthos. The dominance of certain species of *Phaeocystis*, with increasing cell numbers and duration of blooms, has become an issue of concern in some parts of the North Sea. For example, massive blooms of

Phaeocystis gelatinous colonies are observed each year during spring in the area extending from the Strait of Dover to the German Bight. Dinoflagellates, like *Dinophysis* and *Alexandrium* species, which can cause shellfish poisoning, have developed large blooms since 1980. In 1988 a bloom of *Chrysochromulina polylepis* gave rise to serious fish kills in the Kattegat and Skagerrak. The 1988 bloom was a result of a specific combination of biological, physical and chemical conditions which could happen again in the future, although the probability is low.

Deficits in oxygen have been reported as a consequence of eutrophication in areas with a temporal density stratification of the water column which reduces turbulent water (and oxygen exchange): the Oyster Ground (temporarily), the German Bight, the Danish Jutland current, some Norwegian fjords (others are naturally anoxic) and Danish bays, and the southeastern Kattegat. Oxygen deficiencies have also been found in large areas in the bottom waters of the southeastern North Sea and in deep waters in the Kattegat.

Fisheries

The most obvious direct effect of fishing is the removal of fish from the ecosystem (for details see Chapter 24). At present, between 30 and 40 per cent of the total biomass of commercially exploited fish is caught in the North Sea each year; the stocks are therefore heavily exploited. The effect of fishing on such a complex ecosystem as the North Sea is, however, difficult to define with respect to the inherent natural variability.

Introduction of non-indigenous species

In the course of the past century, a number of alien species have found their way into the North Sea. The accidental spreading of exotic species due to aquaculture or by ballast water from ships causes a potential environmental hazard. Some species, such as the slipper limpet, *Crepidula fornicata*, and the brown macroalga, *Sargassum muticum*, can be a threat (eg, by out-competing) to native species and can cause problems (eg, to boat users, through excessive growth).

Conclusions

- The occurrence of exceptional algal blooms in the eastern North Sea, leading to fish kills and other detrimental effects (eg, shellfish poisoning), has generated considerable concern, and has been instrumental in instigating extensive research into eutrophication and its relationship with nutrient loads, hydrographic factors and weather conditions.
- Elevated concentrations of contaminants in the coastal zone of the North Sea are also of concern, particularly in relation to the accumulation of heavy metals and synthetic organic compounds within the higher trophic levels of the food-chain.
- Within the coastal zone there is also concern about the effects of pathogens arising from sewage and other sources on the health of bathers.
- General increases in sea levels arising from climate changes would potentially have a serious impact along parts of the North Sea coastline, for example in The Netherlands and along the east coast of the UK.

THE NORTH ATLANTIC OCEAN

General situation

For the purpose of this report the North Atlantic Ocean is defined as the area of the ocean lying between Europe's western seaboard and 42°W, and between 36°N and the Arctic Circle (see Map 6.1). It therefore includes the southeast coastline of Greenland, the coast of Iceland, the Faeroe Islands and the western coasts of the UK, Ireland, France, Spain and Portugal. It also includes the relatively shallow coastal waters of the semi-enclosed Irish Sea, the Celtic Sea, the Bay of Biscay and the Gulf of Cadiz. The Irish Sea in particular is subject to extensive monitoring by the UK and Irish authorities, and an overview of its current status has been recently compiled by the Irish Sea Study Group (ISSG, 1990).

The continental shelf ends at varying distances off the mainland, being relatively close to the northern shores of Spain and Portugal (less than 10 km in places) and more distant from the southwest peninsula of England (300 km). Within the continental shelf, water depths are generally less than 200 m; within the main body of the ocean, depths of 5700 m are found.

Major rivers discharging into the North Atlantic (see Map 5.11) include: the Guadiana and Guadalquivir in southwest Spain (Gulf of Cadiz); the Douro and Tajo on the west coast of Portugal; the Garonne and Loire on the west coast of France (Bay of Biscay); the Severn on the southwest coast of Britain (Celtic Sea); the Mersey (Irish Sea) and Clyde on the west coast of Britain; and the Shannon on the west coast of Ireland. The rivers on the mountainous north coast of Spain have relatively small catchments, though many are heavily industrialised. Major cities along the coastal zone include: Seville, Lisbon, Oporto, Bilbao, Bordeaux, Nantes, Dublin, Liverpool and Glasgow. Many river catchments are heavily industrialised, for example the River Mersey and the River Oria, whereas others have largely rural or agricultural catchments, for example the Shannon and the Loire. Human influence on the North Atlantic is most readily detectable within the relatively shallow coastal zone.

The Oslo Convention and the Paris Convention provide the regulatory framework for the protection of the maritime area of the northeast Atlantic against pollution (for details see the section on the North Sea). As for the North Sea, those EU countries with Atlantic coastlines have to comply with the requirements of the relevant EU Directives (76/464/EEC with six 'daughter directives') that concern the discharge of dangerous substances to surface waters and the subsequent quality of those waters.

Physical features

The North Atlantic exchanges water with: the Arctic Ocean; the Norwegian Sea; the North Sea through the English Channel and to the north of the UK; and the Mediterranean through the Strait of Gibraltar. Surface water movements in the north of the ocean are dominated by the relatively warm and saline water from the main body of the Atlantic, flowing in the North Atlantic Current. This current flows on both sides of the Faeroes, directly and indirectly towards the Norwegian Sea with its lower water temperature. Here a sharp front is formed with the colder and less saline water from the East Iceland Current. Warm water moves along this front into the Norwegian Sea while the colder water from the north passes into the deep Atlantic troughs and flows southwards. Around Iceland the Arctic Current branches off from the warmer Gulf Stream. Iceland's northern and northwestern coasts are occasionally blocked by drift-ice from the north. The warmer Atlantic water moves west, beneath and outside the polar water. To the east, the Arctic water gives way to the East Iceland Current which flows in the direction of the Faeroes.

Open ocean salinity remains relatively constant at between 35 to 36 per thousand, any variation reflecting regional precipitation and evaporation patterns and the effects of ocean currents. Salinity of the water around Iceland, the UK and the Bay of Biscay is generally between 34 and 35 per thousand.

Biological features

Extensive fisheries exist in the northern part of the ocean. The waters around Iceland are particularly productive, with the total annual catch of fish being approximately 1.5 million tonnes. The most important species caught are cod (*Gadus morhua*), redfish (*Sebastes* spp), haddock (*Melanogrammus aeglefinus*) and saithe (also known as coalfish) (*Pollachius virens*). There is growing concern over the deteriorating state of fish stocks in Icelandic waters and around the Faeroes. The reasons for the deterioration are not fully understood but could include fishing and periodic climatic variability. Aquaculture (eg, of salmon, *Salmo salar*, and trout species), is becoming more important in Iceland. Fisheries are also generally important within the other parts of the ocean, for example off the coasts of Spain, Portugal, France and the UK.

Inputs

The atmospheric inputs of many metals (excluding mercury) to the North Sea, Baltic Sea and Mediterranean Sea are three to ten times higher than those to the open North Atlantic. In open ocean basins, however, the atmospheric flux of man-made contaminants is generally more important than the net input from rivers; this is partially because riverine materials are generally captured within the coastal zone. An estimate of the direct and riverine inputs into the northeast Atlantic Ocean during 1990 can be obtained from data submitted to the Paris Commission by the riparian Atlantic states. It was considered, however, that the loads are underestimates (OSPARCOM, 1992a) as, most significantly, there were no data from the Atlantic coasts of France and Spain, and, less significantly, none from Iceland and Greenland. (There is very little marine pollution arising from Iceland and Greenland.) Direct and riverine loads of metals to the North Atlantic are generally of the same order of magnitude as those to the North Sea, although the impact of the loads away from the coastal zone is generally very small compared with that of other seas, because of the North Atlantic's very large volume and energetic mixing and dispersion characteristics.

The disposal of solid, low-level radioactive waste in the North Atlantic ceased in 1982, by which time some 54 PBq (1.5 million Curies) of radioactive waste in 140 000 tonnes of packaged material had been disposed of at ten sites around 46°N, 17°W (Camplin and Aarkrog, 1989). The present and future risk to individuals from past oceanic dumping is considered to be extremely small. Nuclear waste reprocessing plants are sited at Sellafield (UK, on the Irish Sea) and Cap de la Hague (France, on the English Channel). In 1990, it was reported that 71 TBq (1900 Curies) of total beta radioactivity, 2 TBq of total alpha and 23 TBq of Cs-137 were discharged from the Sellafield site (Kershaw et al, 1992). Discharges from Sellafield have decreased significantly in recent years because of a progressive introduction of more advanced treatment measures (ISSG, 1990). Overall, this trend is expected to continue into the future. In addition, there are a total of ten nuclear power stations bordering the northeast Atlantic: one in France and nine in the UK. They generally discharge smaller amounts (than Sellafield) of radioactivity into the Atlantic.

Contaminant levels

The areas considered by the Oslo and Paris Commissions in their 1990 quality report (OSPARCOM, 1992a) to be generally more contaminated were: the Gironde and Loire estuaries (France); the Spanish coast near Santander, La Coruna and the Navia estuary; and the Portuguese rias de Arosa, Pontevedra and Vigo.

Heavy metals

Lead concentrations in the open North Atlantic are between 5 to 50 ng/l in surface samples – that is, eight to ten times greater than in deeper water layers. Lead concentrations in sea water may have decreased over the last few years because of the increasing use of unleaded petrol. Cadmium and mercury concentrations in the open ocean are reported to range from 1.1 to 11.2 ng/l, and 0.6 to 1.2 ng/l, respectively (UNEP, 1990).

The results from the 1990 supplementary baseline study of contaminants in fish and shellfish (OSPARCOM, 1992a) identified the following areas with 'elevated' concentrations of metals:

- cadmium – Gironde estuary;
- copper – Gironde and Loire estuaries;
- lead – coast of northern Spain, the west, southwest and south coasts of Ireland;
- mercury – the Portuguese south coast, Spanish convention waters;
- zinc – Gironde, western English Channel.

Synthetic organic compounds

'Higher' concentrations of organochlorines (PCBs) in one or more species were found by OSPARCOM along the west coast of France and the north coast of Cornwall (UK) . Neither of the two criteria, 'elevated' and 'higher' concentrations, however, implied any hazard to human health or the environment.

Chlorinated hydrocarbon concentrations in water are in the order of a few ng/l and are fairly uniformly distributed at all depths (UNEP, 1990), but with highest concentrations occurring in surface microlayers naturally enriched with lipids.

Microbiological contamination

The only areas of the North Atlantic that are contaminated with microbial contaminants are along the coastal zone.

- Bathing waters in the Vizcaya, Guipuzcoa and Asturias areas of the north coast of Spain, for example, failed to comply with EU bacterial mandatory standards in 1992 (CEC, 1993); however, overall there was 89 per cent compliance.
- In Portugal, 83 per cent of designated bathing waters complied with the mandatory standard, and 75 per cent with the guideline value.
- Along the French Atlantic coast, around 88 per cent of designated waters complied; waters that did not comply included some around La Rochelle, at the mouth of the Gironde and along the southern Brittany coast. Non-compliance was often associated with poor quality discharges from sewage outfalls, from storm overflows and from rivers.
- There was an overall 66 per cent compliance in the bathing waters along the UK Atlantic coast, with the main non-compliant areas occurring around Liverpool Bay in the Irish Sea and along the northern coast of Devon and Cornwall.

Radionuclides

Within the North Atlantic Ocean the highest concentrations of Cs-137 (one of the most widespread and abundant artificial radionuclides arising from the nuclear industry) in sea water are found in the Irish Sea. Between 1980 and 1985 the average sea water concentration of Cs-137 in the Irish Sea was 730 Bq/m^3 (arising from the Sellafield reprocessing plant), with relatively high levels off the west coast of Scotland (190 Bq/m^3) and the North Sea (10 to 49 Bq/m^3). Average concentrations in the open Atlantic, and for comparison, the Norwegian Sea, Barents Sea and Baltic proper, were 3.3 Bq/m^3, 23 Bq/m^3, 27 Bq/m^3 and 18 Bq/m^3, respectively (Camplin and Aarkrog, 1989). More recent surveys indicate

lower levels of Cs-137 in the surface waters of the Barents and Kara seas: 7 and 6 Bq/m^3, respectively (NRPA, 1993). As a comparison, the usual content of naturally occurring radionuclides in sea water is about 12 000 Bq/m^3 – mainly due to K (potassium)-40 but also from the uranium and thorium radionuclide decay series. The mean concentration of radionuclides arising from Sellafield in the Irish Sea is about 2000 Bq/m^3 (Camplin and Aarkrog, 1989).

It has been estimated that the average annual radiation dose arising from natural sources in the UK is 1.87 mSv a year (Hughes and Roberts, 1984). As a comparison, the radiation doses to the most exposed groups of people are estimated to be: from six times less to twice 'natural' levels for the Sellafield discharges; 20 times less for Cap de la Hague; from 20 000 times less to six times less from the other nuclear sites; and 90 000 times less from the previous solid waste disposal in the northeast Atlantic Ocean (Camplin and Aarkrog, 1989) (see also Chapter 16).

Biological effects

Most of the North Atlantic ecosystem is relatively unpolluted, and any adverse biological effects are limited to the coastal zones, often in semi-enclosed areas and bays, and estuaries bordering the main ocean. For example, there are concerns about the possible effects of eutrophication along the Irish Sea coasts, and the destruction of coastal habitats along the eastern coasts of the Atlantic.

Conclusions

- The open Atlantic Ocean is relatively unaffected by human activities; levels of contaminants, though detectable, are biologically insignificant. Oil slicks and litter are often found along shipping routes but are of little consequence to the biological communities of the open ocean.
- The situation is different along the coastal zone and in the Irish Sea, where contamination levels and biological effects (such as eutrophication) are similar to those found in the more enclosed seas of Europe.
- Overfishing may also be partially responsible for the depletion of stocks in some local areas, for example around Iceland and in the Irish Sea.

CONCLUSIONS AND THE WAY FORWARD

This chapter has reviewed the current quality and status of Europe's seas. It has introduced and highlighted some problems common to Europe's seas and these have been expanded upon in the detailed description of each sea.

The review has established, as far as possible, the current status of the individual seas within Europe. It is clear that some of the seas have environmental problems associated with human activities. However, for many seas, the scale of the problems has not been fully quantified or understood. In order to improve this situation, detailed quality and status assessments are required. These are already periodically undertaken in many seas, for example the North and Baltic seas. Such assessments establish the scope of any environmental impacts and effects, and provide a quantitative baseline against which future quality can be compared and progress monitored. In other seas very few baseline data exist, or if they do exist they have not been collated in a coordinated way. Thus, for some of Europe's seas, establishing a baseline of quality in terms of input loads, contamination levels and biological status and effects would be a useful starting point for improvement. Indeed the Action

Plan of the new Convention of the north east Atlantic requires a quality assessment of the Convention area by the year 2000, and for reports on constituent parts of the area before 2000.

As described in the introduction to this chapter, a comparison of riverine (or for some seas, land-based, if separate riverine data are not available) loads of contaminants has been made from available information. It is re-emphasised here that this should be considered as only approximate as there are likely to be differences in methodologies by which the data were derived, and, for some seas, data were unavailable for some of the different sources, or did not exist at all. Also for some seas, riverine or land-based sources may not necessarily be the most important in terms of total loads. The comparison undertaken is of gross loads, as it does not take into account non-conservative estuarine processes which would remove a proportion of the riverine loads before reaching the main body of the sea. Studies to quantify this loss are under - way, for example in the Mediterranean and North seas. In addition, no attempt has been made to separate the dissolved component from the particulate component of the load. This distinction has been defined for some seas, particularly for the Mediterranean Sea. The effect of allowing for non-conservative estuarine processes and the form of the contaminant on the assessment of loads reaching any sea is illustrated by the pink-shaded areas for the Mediterranean Sea in Figures 6.2b (mercury), 6.2d (lead) and 6.2e (zinc). It can be clearly seen that net riverine loads are potentially considerably less than gross loads.

Bearing in mind all the qualifying statements made above, it would appear from the available information (Figure 6.2a–g) that:

- the Mediterranean Sea receives the highest riverine or land-based loads of mercury, copper, zinc, total nitrogen and total phosphorus (these gross loads are now considered by some experts to be overestimates, for example, Dorten et al (1991) estimate that dissolved mercury loads from rivers are in fact six times lower than original estimates);
- the Black Sea receives the highest riverine loads of cadmium and lead;
- the Mediterranean, Baltic, Black, and North seas are consistently in the 'top' four of the seas receiving the highest loads of land-based or riverine contaminants;
- the northern seas (White, Barents and Norwegian seas)

consistently receive relatively small loads of contaminants – loads to these seas are often 10 or even 100 times less than those receiving the highest loads.

The keystone to the collection of baseline data is the application of appropriate standard methods and quality control procedures. To help improve the situation, a joint initiative taken by marine institutes of EU and EFTA countries has, in 1993, led to the launching of QUASIMEME, a programme for quality assurance of information for marine environmental monitoring in Europe (Wells, 1993). This is essential for the provision of quantitative and comparable data on which judgements and decisions can be based with some confidence. It is evident that such baseline data have not been collected, or are not available, for many of Europe's seas, thus making quantitative comparisons of, for example, contamination levels between seas very difficult and somewhat qualitative.

To achieve quality improvements, there is a clear need for international collaboration and agreement, not only among the riparian states around a particular sea, but also involving the states within the catchment, and even those upwind in the case of atmospheric transport of contaminants. As has been described in the text, such international cooperation has been formalised in a number of conventions which cover a number of Europe's seas, for example, the Barcelona and Helsinki conventions.

The quality assessments presented here, based on the available information, indicate that the seas most at risk, or most affected by human activities, are the Black Sea and the Sea of Azov, and the Caspian Sea; and those least affected are the North Atlantic Ocean, the Norwegian Sea and the Barents Sea.

The key environmental issues and problems facing Europe's seas identified in this review are:

- coastal zone pollution;
- eutrophication;
- overexploitation of resources (biological and physical); and
- the longer-term, but potentially very serious, effects of climate change and sea-level rise.

Problems of the coastal zone are identified here as a prominent environmental problem of concern for Europe, and discussed in Chapter 35. Marine eutrophication is linked to similar problems in freshwaters, and discussed in Chapter 33. Climate change is discussed in Chapter 27.

REFERENCES

Aksnes, D, Blindheim, J, Christensen, I, Giske, J, Hopkins, C C E, Mork, M, Noji, T T, Rottingen, I, Sakshaug, E, Skjoldal, H R and Sundby, S (1993) *Science plan of Mare Cognitum – the ecology of the Nordic Sea, 1994–2000. A regional program of the global ecosystem dynamics.* Preliminary draft report produced by the Institute of Marine Research, Bergen.

AMAP (1993) *The Arctic monitoring and assessment programme (AMAP).* Draft June 1993. Arctic Monitoring and Assessment Programme, Oslo.

Augier, H (1982) *Inventory and classification of marine benthic biocoenoses of the Mediterranean.* Council of Europe Report, Nature and Environment Series No 25.

Baldi, F (1986) The biogeochemical cycle of mercury in the Tyrrhenian Sea. In: *Meeting on the biogeochemical cycle of mercury in the Mediterranean,* pp 29–43. FAO Fisheries Report No 325, Supplement.

Balkas, T, Dechev, G, Mihnea, R, Serbanescu, O and Unluata, U (1990) *State of the marine environment in the Black Sea region.* UNEP Regional Seas Reports and Studies No 124.

Baltic Marine Environmental Protection Commission, Helsinki Commission (1987) Seminar on oil pollution questions. Baltic Sea Environment Proceedings No 22, Helsinki.

Barth, H and Fegan, L (Eds) (1990) *Eutrophication-related phenomena in the Adriatic Sea and in other Mediterranean coastal zones.* Commission of the European Communities Water Pollution Research Report 16.

Bernes, C (1993) *The Nordic Environment – present State, trends and threats.* Nord 1993, 12. Nordic Council of Ministers, Copenhagen.

Blindheim, J (1989) Ecological features of the Norwegian Sea. In: Rey, L and Alexander, V (Eds) *Proceedings of the Sixth Conference of the Comité Arctique International,* 13–15 May 1985, pp 366–401. E J Brill, Leiden, New York, Copenhagen, Cologne.

Boudouresque, C F (1993) Etat actuel de la biodiversité marine en Méditerranée. In: Briand, F (Ed) *Pollution of the Mediterranean Sea,* pp 75–90. STOA project. Technical working document for the meeting of 10–11 September 1993, Corfu.

Bruggeman, W A, Jost, B, Huau, M C, Karaszi, K, Myers, S D, Nawalany, M, Pravdic, V, Storfelder, P B M, Teichgraber, B, van der Gaag, M A and van der Kooij, L A (1991) The state of the aquatic environment in the USSR, 1988. *European Water Pollution Control* 1, 37–43.

Camplin, W C and Aarkrog, A (1989) *Radioactivity in north European waters: report of Working Group II of CEC Project MARINA.* Fisheries Laboratory, Lowestoft, UK. Fisheries Research Data Report No 20.

CEC (1992) *The European river ocean system – EROS 2000.* Commission of the European Communities DG XII Environment, Brussels.

CEC (1993) *Quality of bathing water 1992.* Commission of the European Communities EUR 15031 EN, Luxembourg.

Clark, R B (1986) *Marine pollution.* Oxford University Press, Oxford.

Coastwatch (1993) *Coastwatch Europe: International results of the 1992 survey*. Coastwatch Europe Network, Trinity College, Dublin.

Coastwatch (1994) *Coastwatch Europe: International results summary of the autumn 1993 survey*. Coastwatch Europe Network, International Coordination, Trinity College, Dublin.

COE (1990a) *Marine turtles in the Mediterranean: distribution, population status, conservation*. Council of Europe Nature and Environment Series, No 48.

COE (1990b) *Marine reserves and conservation of Mediterranean coastal habitats*. Council of Europe Nature and Conservation Series, No 50.

Dechev, G (1990) *The ecological state of the Black Sea*. Bulgarian Academy of Sciences, Scientific Information Center, Vol XVII, N8, 5–61.

Dorten, W S, Elbaz-Poulichet, F, Mart, L R and Martin, J-M (1991) Reassessment of the river input of trace metals into the Mediterranean Sea. *Ambio* **20**, 2–6.

Estrada, M (1993) Harmful algal blooms in the Mediterranean. In: Briand, F (Ed) *Pollution of the Mediterranean Sea*, pp 91–102. STOA project. Technical working document for the meeting of 10–11 September 1993, Corfu.

Food and Agriculture Organization of the United Nations (FAO) (1986) *Meeting on the biogeochemical cycle of mercury in the Mediterranean*. FAO Fisheries Report No 325 Supplement.

Fowler, S W (1990) Concentration of selected contaminants in water, sediments and living organisms. In: *Technical annexes to the report on the state of the marine environment*, pp 143–208. UNEP Regional Seas Reports and Studies No 114/1.

Foyn, L and Semenov, A (1992) *Cruise report. Norwegian/Russian expedition to the Kara Sea. August–September 1992*. Institute of Marine Research, Bergen.

GESAMP (IMO/FAO/UNESCO/WMO/IAEA/UN/UNEP Joint Group of Experts on the Scientific Aspects of Marine Pollution) (1990) *Review of potentially harmful substances: Nutrients*. Reports and Studies No 34. London.

GESAMP (IMO/FAO/UNESCO/WMO/IAEA/UN/UNEP Joint Group of Experts on the Scientific Aspects of Marine Pollution) (1993) *Impact of oil and related chemicals and wastes on the marine environment*. Reports and Studies No 50. London.

Gjosaeter, H (1992) Pelagic fish and the ecological impact of modern fishing industry. In *Proceedings of symposium on: 'Man and the Barents Sea ecosystem'*. Groningen, The Netherlands 19–20 November 1992.

HELCOM (1987) Baltic Marine Environment Protection Commission, Helsinki Commission 1987. *Seminar on oil pollution questions*. Baltic Sea Environment Proceedings No 22.

HELCOM (1990) Baltic Marine Environment Protection Commission, Helsinki Commission 1990. *Second periodic assessment of the state of the marine environment of the Baltic Sea, 1984–1988; general conclusions*. Baltic Sea Environment Proceedings No 35 A and B.

HELCOM (1991) Baltic Marine Environment Protection Commission, Helsinki Commission 1991. *Interim report on the state of the coastal waters of the Baltic Sea*. Baltic Sea Environment Proceedings No 40.

HELCOM (1992) Helcom ad hoc high level Task Force 1992. *The Baltic Sea Joint Comprehensive Environmental Action Programme (Preliminary version)*. Conference Document No 5/3. Helsinki, April 1992.

Hughes, J S and Roberts, G C (1984) *The radiation exposure of the UK population – 1984 review*. National Radiological Protection Board, UK, Report No NRPB-R173.

Intergovernmental Panel on Climate Change (IPCC) (1990) *Climate change: the IPCC scientific assessment*. World Meteorological Organisation/United Nations Environment Programme, Cambridge University Press, Cambridge.

Irish Sea Study Group (ISSG) (1990) *The Irish Sea: an environmental review. Part 2: waste inputs and pollution*. Liverpool University Press, Liverpool.

IUCN (1988) *1988 IUCN red list of threatened animals*. IUCN, Gland and Cambridge.

Izrael, Y A and Tsyban, A V (Eds) (1992) *Review of the ecological state of the seas of the USSR and several regions of the world ocean for 1990*. Committee of Hydro-meteorology, Academy of Science of the USSR, Institute of Global Climate and Ecology, St Petersburg (in Russian).

Karpinsky, M G (1992) Aspects of the Caspian Sea benthic ecosystem. *Marine Pollution Bulletin* **24**, 384–9.

Kershaw, P J, Pentreath, R J, Woodhead, D S and Hunt, G J (1992) *A review of radioactivity in the Irish Sea: a report prepared for the Marine Pollution Monitoring Management Group*. Directorate of Fisheries Research, Lowestoft, Aquatic Environment Monitoring Report No 32.

Loeng, H (1989) Ecological features of the Barents Sea. In: Rey, L and

Alexander, V (Eds) *Proceedings of the Sixth Conference of the Comité Arctique International, 13–15 May 1985*, pp 327–65. E J Brill, Leiden, New York, Copenhagen, Cologne.

Loeng, H (1991) Features of the physical oceanographic conditions of the Barents Sea. In: Proceedings of the Pro Mare Symposium on Polar Marine Ecology, Trondheim 12–16 May 1990. *Polar Research* **10**, 5–18.

Ludikhuize, D (1992) *Environmental management and protection of the Black Sea*. Background document prepared for Technical Experts Meeting, 20–21 May 1992, Constanţa; Romania. UNDP/UNEP/World Bank.

Martin, J-M, Elbaz-Poulichet, F, Guieu, C, Loye-Pilot, M-D and Han, G (1989) River versus atmospheric input of material to the Mediterranean Sea: an overview. *Marine Chemistry* **28**, 159–82.

Martin, J-M, Huang, W W and Yoon, Y Y (1993) Total concentration and chemical speciation of dissolved copper, nickel and cadmium in the Western Mediterranean. In: Martin, J-M and Barth, H (Eds) *EROS 2000 - Fourth Workshop on the North West Mediterranean Sea*, pp 119–28. Commission of the European Communities Water Pollution Research Report 30.

Mee, L D (1991) *The Black Sea: a call for concerted international action*. Report of a UNEP/IOC/IAEA mission.

Mee, L D (1992) The Black Sea in crisis: a need for concerted international action. *Ambio* **21**, 278–86.

Meinesz, A and Hesse, B (1991) Introduction of the tropical alga *Caulerpa taxifolia* and its invasion of the Northwestern Mediterranean. *Oceanologica Acta* **14**, 415–26.

Mnatsakanian, R (1992) *Environmental legacy of the former Soviet republics*. Centre for Human Ecology, University of Edinburgh, Edinburgh.

Morley, N H and Burton, J D (1991) Dissolved trace metals in the northwestern Mediterranean Sea: spatial and temporal variations. *Water Pollution Research Report* **28** 279–92.

Nauke, M and Holland, G L (1992) The role and development of global marine conventions. *Marine Pollution Bulletin* **25** (1–4), 74–9.

Northern Europe's Seas (1989) *A Report to the Nordic Council's International Conference on the Pollution of the Seas, 6–18 October 1989 Stockholm* (translated from Swedish).

Northridge, S (1990) Seals. In: *1990 North Sea Report*, pp 85–8. The Marine Forum for Environmental Issues, London.

Norwegian Radiation Protection Agency (NRPA) (1993) Press release from NRPA on radioactivity in marine fish, dated 18 August 1993.

OSPARCOM (1992a). *Monitoring and assessment*. Oslo and Paris Commissions, London.

OSPARCOM (1992b). *Nutrients in the Convention area*. Oslo and Paris Commissions, London.

Portman, P (1991) The implementation of the Monitoring Master Plan. *North Sea Task Force News*, November 1991, London.

Reid, P C, Taylor, A H and Stephens, J A (1988). The hydrography and hydrographic balances of the North Sea. In: Salomons, W, Bayne, B L, Duursma, E K and Förstner, U (Eds) *Pollution of the North Sea: an assessment*, pp 3–19. Springer-Verlag, London.

de Ronde, J G (1993) What will happen to the Netherlands if sea level rise accelerates. In: Warrick, R A, Barrow, E M, and Wigley, T M L (Eds) *Climate and sea level change: observations, projection and implications*, pp 322–34. Cambridge University Press, Cambridge.

Saetre, R, Foyn, L, Klungsoyr, J and Loeng, H (1992) *The Barents Sea ecosystem – impact of man*. Report submitted for publication in 'Arctic' by the Institute of Marine Research, Bergen.

Saxén, R, Ikäheimonen, T K and Ilus, E (1989) *Monitoring of radionuclides in the Baltic Sea in 1988*. STUK-A90. Finnish Centre for Radiation and Nuclear Safety, Helsinki.

Scoullos, M (1993) Mediterranean pollution: chemical and biological aspects. In: Briand, F (Ed) *Pollution of the Mediterranean Sea*. STOA project. Technical working document for the meeting of 10–11 September 1993, Corfu.

Stepanets, O V, Karpov, V S, Komarevsky, V M, Borisov, A P, Farrahov, I T, Solov'eva, G Y, Pilipets, L A, Batrakov, G F and Chudinovskyh, T A (1992) Peculiarities of the distribution of man-made radionuclides in several European Seas. *Analyst* **117**, 813–16.

Tethys Research Institute (1991) *Annual Report 1991*. Tethys Research Institute, Milan.

Tolosa, I, Bayona, J M and Albaiges, J (1993) Spatial and temporal distribution of aliphatic and aromatic hydrocarbons in the NW Mediterranean. In: Martin, J-M and Barth, H (Eds) *EROS 2000 – Fourth Workshop on the North West Mediterranean Sea*, pp 79–84. Commission of the European Communities Water Pollution Research Report 30.

Tursunov, A A (1989) The Aral Sea and the ecological situation in central Asia and Kazakhstan. *Hydrotechnical Construction* **3**, 319–25.

UNEP, EEC, UNIDO, FAO, UNESCO, WHO, IAEA (1984) *Pollutants from land-based sources in the Mediterranean*. UNEP Regional Seas Reports and Studies No 32, United Nations Environment Programme, Geneva.

UNEP, IMO, IOC (1987) Assessment of the present state of pollution by petroleum hydrocarbons in the Mediterranean Sea. UNEP/WG 160/11, Athens.

UNEP (1989) *State of the Mediterranean marine environment*. MAP Technical Reports Series No 28, Athens.

UNEP (1990) *Technical annexes to the report on the state of the marine environment*. UNEP Regional Seas Reports and Studies No 114/1.

Voipio, A (Ed) (1981) *The Baltic Sea*. Elsevier Oceanography Series, Amsterdam.

Vollenweider, R A, Rinaldi, A and Montanari, G (1992) Eutrophication, structure and dynamics of a marine coastal system: results of ten-years monitoring along the Emilia-Romagna coast (Northwest Adriatic Sea). *Science of the Total Environment*, Supplement 1992 on 'Marine Coastal Eutrophication'.

Wells, D E (1993) Quasimeme: an introduction. Quasimeme Bulletin No 1, pp 2–5, Marine Laboratory, Aberdeen.

Yablokov, A V, Karasev, V K, Rumyantsev, V M, Kokeyev, M Y, Petrov, O I, Lystsov, V N, Yemelyanenkov, A F and Rubtsov, P M (1993) *Facts and problems related to radioactive waste disposal in seas adjacent to the territory of the Russian Federation*. Office of the President of the Russian Federation, Moscow. (Translated from Russian.) Small World Publishers Inc, Albuquerque.

Zafiropoulos, D (1986) The biogeochemical cycle of mercury: an overview. In: *Meeting on the biogeochemical cycle of mercury in the Mediterranean*, pp 168–87. FAO Fisheries Report No 325, Supplement, Food and Agriculture Organisation, Rome.

Ploughed field,
Adinkerke, Belgium
Source: J Herbauts

7 Soil

Figure 7.1
The soil ecosystem

INTRODUCTION

Soil is often seen as an inert medium, merely a support for human activities. However, soil is more than that: it is a dynamic, living system which comprises a matrix of organic and mineral constituents enclosing a network of voids and pores which contain liquids and gases. The structural arrangements of these components determine the main soil types in Europe; 320 different major soil types can be identified in the EU (CEC, 1985). In addition, soils contain populations of biota ranging from bacteria and fungi to worms and rodents (Figure 7.1); the chemical, physical and biological properties of soils vary both vertically and horizontally at a variety of scales. Soil is formed from the combined effects of climate, vegetation, soil organisms and time on rocks and parent materials (Jenny, 1941). So, any alteration in one of these components may result in changes in the soils. Soil genesis is a long process – the formation of a layer of 30 cm of soil takes from 1000 to 10 000 years (Häberli et al, 1991). It is formed so slowly that soil can be considered as a non-renewable resource.

Soil is a complex system where crucial biogeochemical processes occur. In the top 30 cm of one hectare of soil, there are on average 25 tonnes of soil organisms, that is: 10 tonnes of bacteria and actinomycetes,
10 tonnes of fungi, 4 tonnes of earthworms and 1 tonne of other soil organisms such as springtails, mites, isopods, spiders, coleoptera, snails, mice, etc. (Blum, 1988). Earthworms, alone, can represent from 50 to 75 per cent of the total weight of animals in arable soils; in one hectare, 18 to 40 tonnes (that is a layer 1 to 5 mm deep) of soil is ingested each year by the earthworms and passed onto the surface. The number of worms varies with the soil and its management. The soil fauna and flora recycle organic matter to form humus and mix it with

Box 7A Data and information sources

The information and data in this chapter are based on the following: work performed by the International Soil Reference and Information Centre (ISRIC), the Dutch National Institute of Public Health and Environmental Protection (RIVM) and the Winand Staring Centre; the draft version of the The European soil resource: current status of soil degradation in Europe: causes, impacts and need for action, published in 1994 by the Council of Europe (Van Lynden, 1994); various national and regional state of the environment reports; and a review of scientific literature.

Soils of Europe
For the purpose of this report, the FAO-UNESCO Soil Map of the World (FAO, 1981) has been generalised. According to the FAO classification (FAO, 1974) there are 24 major soil groups in Europe, which can be further subdivided in 75 soil units. In this chapter, the 75 soil units have been aggregated into 11 groupings, presented under 4 main headings (Map 7.1), on the basis of their similarity of behaviour with regard to degradation processes. Each of the groupings is characterised by specific soil or land parameters that have a major influence on the use and management of the soil and its vulnerability to degradation processes. These parameters are organic matter content, texture, acidity, soil depth, stoniness, slope and drainage. Although an attempt has been made to ensure a maximum of homogeneity within each of the groupings (for example, all organic soils are deep and non-stony; almost all 'black earths' are loamy and non-acid) it must be realised that the small scale of the map, and hence the limitation in representing separate soil units, entails a certain variability within each of the groupings and allows for only a crude generalisation to be made. For further detail, reference is made to the CEC Soil Map of Europe at scale 1:1 000 000 (CEC, 1985).

Problems and threats
The information and data on erosion and physical degradation are based on an updated version of the European part of the Global Assessment of Soil Degradation map (GLASOD) carried out by ISRIC and UNEP (Oldeman et al, 1991). For this update, questionnaires were sent to scientific teams in each European country for comments and additions on the GLASOD map. Not all countries completed and returned the questionnaires and the degree of detail of the information received varies greatly. It must be kept in mind that the scale

of the maps (1:10 000 000) does not allow detailed information to be shown; therefore, these maps must be interpreted with care (see Box 5B).

Data on pollution from heavy metals, pesticides, nitrates and phosphates were compiled by RIVM. Maps of nitrogen supply from manure and fertiliser loads and of pesticide loads and accumulation were prepared by using RIVM (1992) information and new data collected in the framework of the present project. The actual maps are presented in Chapter 5 for groundwater (Maps 5.7–5.10) and Chapter 22 for fertilisers and pesticides applications.

Data on acidification and compaction are based on the work done by the Winand Staring Centre. For acidification, critical loads and their exceedance on European forest soils have been calculated using a one-layer steady-state model (START). The same model has been used to calculate aluminium (Al) concentration and aluminium/base cation (Al/BC) ratio in a steady-state situation using present deposition rates. Estimates of the present Al concentration and Al/BC ratio have been derived from the dynamic model (SMART). For soil compaction, it was not possible to make an inventory for the whole of Europe due to the lack of data; instead, for regions where soil compaction problems are well known (southern Netherlands, eastern Germany, Poland, mid-western Russian Federation, mid-southern Sweden, northern France, southern Spain, southern Portugal and northern Italy), case studies were used to help quantify the effects.

Availability
An evaluation of existing databases on soils in Europe in the context of environmental threats suggests that data on soil fauna and flora, organic matter and heavy metals are inadequate. Experts consider that a general purpose map at a scale of 1:250 000, supported by national databases is missing for Europe. It was also recommended that the Soil Map of Europe (1:1 000 000) be revised to incorporate new knowledge.

Quality
In most European countries, the available data on soils are qualitative and their interpretation requires expert judgement; survey methods and classification systems vary between countries and sometimes within a country (eg, Germany, Spain). Information on soil and terrain attributes that influence environmental processes is often missing.

the mineral material; they also create and maintain the airways within the soil that are essential to plant roots; some species found in soil control others that are pests to crops.

The carbon stored in soils is nearly three times that in the above-ground biomass and approximately double that in the atmosphere (Eswaran et al, 1993). Replenishing carbon is a slow process not easily achieved; soil protection should therefore, apart from its other benefits, assist in maintaining this reservoir of soil carbon (see also Chapter 27).

In contrast with the earlier concern for the atmosphere and hydrosphere, the need to protect the soil has been appreciated only more recently. Soil is static and thus acts as an enormous receptacle for any type of pollutant which can be mobilised under different triggers (such as acidification) and finally released to the environment. Since the residence time of these substances is far longer in the soil than in air or water, effects are often hidden for a long time. Soil-related problems are also site-specific, which makes any attempt at generalisation very difficult. The unique capability of the soil for recycling biodegradable waste, and the delay until effects are usually detected, has led to complacency, and concern for the well-being of soils has been slight. But soil is necessary

for the growth of crops of food, fibre and timber, and it is an essential component of all terrestrial ecosystems. Consequently, the role of the soil is of vital importance to humankind and the maintenance of a healthy natural environment. The 1972 European Soil Charter (Council of Europe, 1972) recognised for the first time that any biological, physical or chemical degradation of soil should therefore be of primary concern and that appropriate measures to protect soils should be implemented without delay. Unlike air and water, soil can be owned as personal property, which renders soil conservation or protection policies difficult to enforce and requires acceptance by landowners and managers. The data and information sources used in this chapter are described in Box 7A.

Soil functions

Soil protection policies currently being discussed in Europe emphasise consideration of the functions which soils perform, rather than the uses of soils. This represents a fundamental change in approach since it seeks to identify any possible conflicts between uses and forms a sound basis for the assessment of the impact of a given use on particular soils. Soils have six basic functions: basis for biomass production; filtering; buffering and transformation medium; habitat and gene reservoir; foundation; raw material and a historical medium (Blum, 1990). The first three are primarily ecological while the last three highlight the technical/industrial, socio-economic and cultural aspects.

Biomass production

Most production of food, fodder or renewable raw materials is soil-related, with soil providing nutrients, air, water, and a medium in which plant roots can penetrate. Serious degradation of the soil by human activities may cause a loss of production of food and wood. Erosion and compaction are the most important threats to a sustainable agriculture and in (1989–90) affected 15 to 20 per cent of the European land area (Oldeman et al, 1991). Production losses due to these degradation processes may amount to 30 per cent or more yield reduction (as detailed below in this chapter). In Northern Europe, soil acidification may be the main factor impacting on timber production.

Filtering, buffering and transformation

It is only in the last few decades that the full significance of these soil functions has properly been recognised. They enable soils to deal with harmful substances, preventing them from reaching the groundwater or the food-chain. Substances can be mechanically filtered, adsorbed or precipitated, and, for organic substances in particular, even decomposed and transformed. This is crucial since about 65 per cent of the inhabitants of Europe depend on groundwater for their drinking water (see Chapter 5). Soil organic matter is an important factor in determining the buffer capacity of soil.

The soil buffers chemical substances as well as temperature; external inputs of chemicals, such as acidifying compounds, are buffered by the basic cations (of sodium, calcium, potassium and magnesium) present in the soil and derived from the weathering of clay minerals. The soil thus acts as a sink in which pollutants accumulate until the buffer capacity is depleted. In optimal conditions, more than 99 per cent of pesticides are transformed into non-toxic compounds within the plough layer of arable soils (Canton et al, 1991; Boesten and van der Linden, 1991). Nevertheless, although small, the remaining part which is not decomposed is sufficient to threaten drinking water in areas with high inputs of pesticides (see Chapters 5 and 22). The presence of other substances, such as copper, may, by reducing soil biological activity, indirectly hinder pesticide degradation; this is an important problem in areas of vine cultivation. Moreover, the buffer function changes with time and, when the buffer capacity is depleted, the soil can turn into a source of chemicals, and pollutants can begin to leach into the groundwater. The metaphor 'chemical time bomb' is being used for a chain of events resulting in the delayed and sudden occurrence of harmful effects due to the mobilisation of chemicals stored in soils and sediments in response to slow alterations of the environment (Stigliani et al, 1991). Soil microorganisms are responsible for the decomposition of organic matter and the transformation of other substances, such as sulphates and nitrates. Environmental changes may reduce substantially the capacity of the soil to withhold pollutants; consequently, soil is a central factor in determining critical loads for natural ecosystems.

Biological habitat and gene reserve

Soil provides habitats for numerous organisms and microorganisms (see Figure 7.1); it is also a gene reserve. Relatively little is known about the biology of soils, and many soils under agriculture are currently being subjected to new and alien forms of management. Decline of the quality of soils will generally contribute to the decline in biodiversity (Blum and Prieur in Council of Europe, 1990). Loss of biological activity and of animal species may be caused by removal or burning of vegetation, by excessive application of fertilisers and/or biocides or by acidification from atmospheric pollution. Frequently, biological soil degradation is associated with physical and chemical soil degradation. For example, compaction leading to reduced aeration of the soil may result in a decrease in activity of earthworms, which in turn will cause further degradation of soil structure. Acidification can lead to a loss of cast-forming worms and mycorrhizae. Unfortunately, the degree of biological soil degradation is difficult to estimate. In view of its strong links with other types of degradation it may be assumed that biological degradation is occurring whenever other types of soil degradation are significant, especially erosion, compaction and chemical pollution.

Foundation

Soil is also a physical medium for the development of infrastructure – houses, industrial premises, roads, recreational and leisure facilities – and for waste disposal. Today, built-up areas cover about 2 per cent of the total land area in Europe (see Table 5.5), ranging from 0.5 per cent in Iceland (1990) up to 12 per cent in Hungary, 13 per cent in Italy and 14 per cent in The Netherlands (*Statistical Compendium*), urban areas alone covering 1 per cent of the total land surface of Europe. In many European countries the built-up area has increased by 25 to 75 per cent during the period 1950–80. Soil sealing through urbanisation dominates in the more densely populated regions and major industrial areas of Western Europe; in West Germany, for instance, about 120 hectares of soil are lost every day (*Statistical Compendium*). This results in an increase of impermeable surface and a decrease of water infiltration into the soil. As soil sealing is bound to increase in the next century in Europe, a number of countries have taken the view that the best agricultural soils should not be used for building or infrastructure developments and there are several examples of new roads being re-routed to avoid fertile soils.

Source of raw materials

The ground is a source of raw materials such as clays, gravels, sands and minerals, as well as fuels such as peat. The land used for opencast mining is estimated at 0.05 to 0.1 per cent of the whole of Europe. However, mining activities can have an important impact locally. Lignite mines of the West German border region (North Rhine Westphalia) are the largest opencast mining operations in Western Europe. By 1990, about 250 km² of land had been excavated for lignite extraction (see Chapter 20). In the Czech Republic, the opencast mining of brown coal is concentrated in a few areas and covers almost 10 000 hectares.

Historical medium

Soil is a historical medium, concealing archaeological artefacts and palaeontological materials which are a unique source of historical information (see Chapter 8).

Interaction of functions

Functions may interact in space or time, creating conflicts for different uses of the land. Interactions in space may be due to agricultural practices which have an impact not only on agricultural soils but also on adjacent non-agricultural soils and groundwater. Drift of pesticides may affect sensitive species, and the transport of fertilisers by wind or water from agricultural fields to bordering nature reserves can lead to eutrophication of nutrient-poor ecosystems. The supply of nutrients to these ecosystems usually leads to an increase in biomass and a significant decrease in the number of species or the loss of rare species. On the other hand, the deterioration of natural ecosystems may lead to a decrease of their water-holding capacity and an increase of soil erosion. This can threaten agricultural soils by deposition of eroded soil, or through increased runoff. If runoff increases, infiltration decreases and the replenishment of groundwater is not ensured, thus threatening the ultimate replenishment of the drinking water reservoir. Soil functions may also change in time; for instance, agricultural land may be abandoned or deliberately changed to semi-natural land as a result of set-aside policy.

Functions at risk

For centuries, all soil functions were maintained without much difficulty. Problems arose at the beginning of this century, when increasing development started to conflict with the ecological functions of the soil. Expansion of settlements and infrastructure, especially for industries and transport, waste dumping, raw materials mining and intensive agriculture, exert pressure on soil. Sometimes these modifications are positive and enhance certain soil functions (the construction of terraces, for example, has led to reductions in the natural rates of erosion in Southern Europe), but more often a deterioration of soil characteristics takes place as a result of human activities, leading to a degradation of one or more soil functions.

Impairment of any soil function diminishes soil quality and value, and the capacity to provide the basic

requirements to support ecosystems. Since degradation of soil functions results from a pattern of competition and overutilisation, there is a need to strike a balance among all the interests concerned and to harmonise soil use at regional level, to enable the essential soil functions outlined above to be performed concurrently. Thus, soil suitability/vulnerability assessment is the logical forerunner to landuse policy making.

Soils of Europe

A glance at any landscape in Europe shows that even a small area may have a wide variety of soils, depending on the local climate, geology, relief, altitude, exposure, slope, vegetation, hydrology and human influence. As a result, landuse, soil management and degradation hazards differ considerably within the European context. Thousands of years of human activity have marked the soils throughout Europe. Yet these soils can be distinguished by differences in their most important features: their organic and inorganic components, their biological activity and their structure. In Map 7.1, which shows the four most important soil classes in Europe, soil types have been grouped in classes of broadly similar vulnerability to different degradation processes (Fraters, 1994).

The main soil groupings are briefly described below. The soil names used here do not belong to any formal nomenclature, and therefore, to allow for easier comparison, corresponding names from the FAO classification are given in italics (FAO, 1974; 1981).

Map 7.1
Soils of Europe
Source: Fraters, 1994

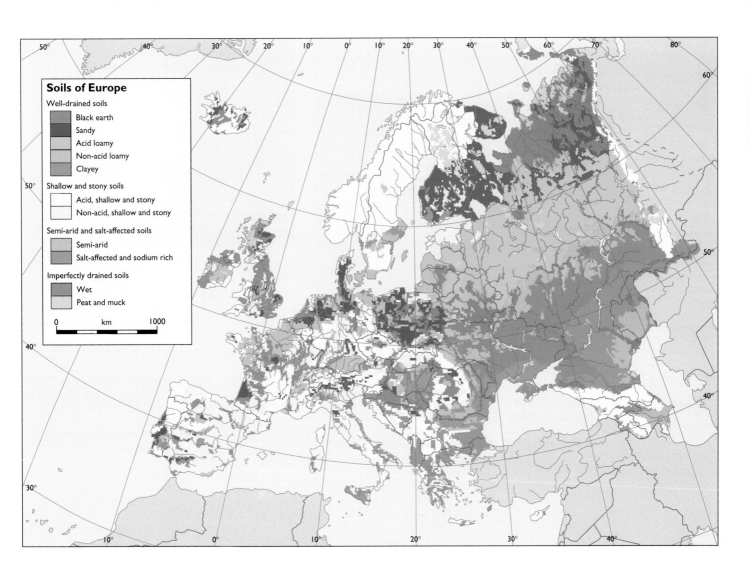

Soils of Europe

Well-drained soils
- Black earth
- Sandy
- Acid loamy
- Non-acid loamy
- Clayey

Shallow and stony soils
- Acid, shallow and stony
- Non-acid, shallow and stony

Semi-arid and salt-affected soils
- Semi-arid
- Salt-affected and sodium rich

Imperfectly drained soils
- Wet
- Peat and muck

0   km   1000

Well-drained soils

Well-drained soils cover 44 per cent of the European land area. They occur generally in an undulating relief. Their formation was conditioned by climate and natural vegetation, hence they show a 'zonal' distribution from north to south in relation to rainfall and temperature variations. The well-drained soils differ from each other by their texture, organic matter content and acidity (pH) from which five soil categories can be identified.

'Black earths'

From an agricultural viewpoint, the 'black earths', or *chernozems*, are of major importance. They are characterised by a thick humus-rich topsoil (50 to 100 cm deep), overlaying a lighter horizon containing calcium carbonate. They occur in the continental climate zone of the Central European plain, the steppe, in a belt extending from Poland and Hungary to the Urals, and cover about 9 per cent of Europe's land area. They are very fertile and suitable for arable farming. However, erratic rainfall often limits yields. In the Russian Federation and Ukraine, black earths are notable for extensive cereal production.

Sandy soils

The sandy soils are mainly *podzols*. They develop under the influence of a slowly decomposing acid humus which moves through the soil with percolating rainwater and precipitates in the subsurface in combination with iron and aluminium oxides. This accumulation layer limits rooting depth. In Europe, sandy soils cover over 9 per cent of the land, mainly in Northern Europe where they occur under forest. In Western Europe they are currently used intensively for agriculture with the application of fertilisers to overcome the inherent low fertility. The sandy soils also comprise *arenosols*, which are confined to limited areas, mainly in Poland, Spain and the UK.

Black earth
Source: R Dudal

Sandy soil
Source: J Herbauts

Acid loamy soils

Acid loamy soils are assepresented mainly by the *podzoluvisols*. They occur most extensively under the cold continental conditions of the central taiga of the north and the centre of the Russian Federation, and cover about 14 per cent of the land area. In the northern latitudes forestry is the dominant landuse. In the more central areas, grassland and crops occupy a larger portion of the land. Low fertility and a short growing season are limitations to high production.

Non-acid loamy soils

Non-acid loamy soils are widespread in Western and Central Europe, and cover about 8 per cent of the land. They comprise the *orthic luvisols* on loessic parent materials, which are among the most productive agricultural areas. In more northern latitudes, in Ireland and Sweden, they also include *cambisols*, which are used for livestock production, and the *haplic* and *luvic phaeozems*.

Acid loamy soil
Source: R Dudal

Non-acid loamy soil
Source: R Dudal

Clayey soil
Source: R Dudal

Clayey soils

In the Mediterranean countries, red and reddish-brown soils of heavy texture have developed over limestone, the *chromic cambisols* and the *chromic luvisols*. Despite their high clay content, these soils are well drained because of their texture and the presence of a permeable subsoil. They are associated in level areas and in depressions with heavy dark clay soils, the *vertisols*, which, although fertile, present management problems because of their unfavourable physical properties. The *luvisols* are sensitive to erosion. They are used for vineyards, olives and citrus. The *vertisols* are preferably used for the production of cereals. Clayey soils cover about 4 per cent of Europe's land area.

Shallow and stony soils

These soils – or *lithosols*, *cambisols* and *rendzinas* – cover 30 per cent of the total land area of Europe, and usually occur on steep slopes. They are predominant in the major mountain and hill ranges of Europe, such as the Pyrenees, the French Massif Central, the Alps, the Apennines, the Carpathians, the Scandinavian mountain ranges and the Caucasus. These soils are not suitable for intensive arable farming and use of heavy machinery. They are used mainly for non-intensive grazing and wood production. Among them, the acid, shallow and stony soils which occur on non-calcareous parent material (such as in Scandinavia and Germany) are more vulnerable to acidification than the non-acid shallow and stony soils on calcareous bedrock, for example in France, Greece, Italy and Spain.

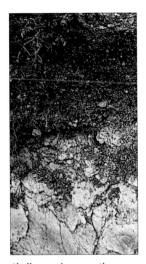

Shallow and stony soil
Source: J Herbauts

Semi-arid and salt-affected soils

In specific areas, such as the eastern part of Southern Europe, high evaporation results in soil salinisation, which leads to the formation of *solonchaks*, with excess of salts, or *solonetz*, with excess of sodium. The semi-arid and salt-affected soils cover 9 per cent of Europe, mostly in the southern part of the Russian Federation, Ukraine and Romania. The lack of rainfall combined with the excess of salt or sodium permits only non-intensive agriculture. The semi-arid soils (*xerosols* and *kastanozems*) do have potential for more intensive agriculture, with irrigation.

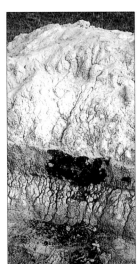

Semi-arid and salt-affected soils
Source: R Dudal

Imperfectly drained soils

These soils represent 17 per cent of the land surface of Europe. The imperfect drainage can be due to surface water stagnation (for example, with *gleyic luvisols* and *planosols*) or to fluctuating groundwater (for example, with *gleysols* and *fluvisols*). Wet soils occur in level relief and in areas with high groundwater tables. They are very extensive in the northern part of Europe (the Russian Federation, Scotland and Ireland), where they occur in association with organic soils. Imperfectly drained soils due to surface waterlogging are extensive in England and in Germany. Organic soils – peat and muck – are confined mainly to the boreal parts of the Russian Federation, Finland, Ireland and the uplands of

Imperfectly drained soil
Source: J Herbants

Britain. Organic soils – or *histosols* – are developed under the influence of high water tables and low temperatures. Remnants of former extensive peat areas occur elsewhere, for instance in The Netherlands and northern Germany. The dominant feature of these soils is their prolonged waterlogging, which strongly influences their use and management. Depending on the climate under which they occur, they are used either for extensive wood production and/or grazing (such as in the boreal parts of the Russian Federation and Finland), or, after reclamation (drainage), for arable cropping, dairy farming or horticulture (in Western Europe).

Soil vulnerability

As explained in the introduction to this chapter, the soils in Europe do not all respond similarly to soil degradation processes. The vulnerability of soils to certain contaminants or degradation processes depends upon those characteristics which enable soils to resist alteration and which maintain the biological functions expected of a soil (Bridges, 1991). Soil vulnerability is then the capacity for the soil system to be harmed in one or more of its ecological functions.

To assess the degree of soil vulnerability, the following soil-environmental parameters have been identified as critical for future soil monitoring programmes: pH, clay content, clay mineral type, texture, and organic matter content. These criteria are readily available and are combined with environmental features such as the site relief, total depth of soil, depth of groundwater table, water regime and length of growing season. Vulnerability must be specified with respect to agents, causes and effects (Desaules, 1991). The classification adopted in Map 7.1 and described in the preceding section can facilitate a general understanding of the areas of Europe which are most vulnerable to a certain type of soil degradation (Fraters, 1994).

A qualitative appraisal of the vulnerability of the main soil types to the main soil threats – erosion, soil acidification, soil pollution, compaction, organic matter losses, salinisation and waterlogging – has been developed and is presented in Table 7.1 (Fraters, 1994). A rough estimation of the area in Europe affected by the main soil threats is given in Table 7.2. The vulnerability is given with respect to soil characteristics assuming that, unless explicitly stated, there are no differences in land characteristics. Land comprises the total physical environment, including climate, relief, soils, hydrology and vegetation. All these characteristics determine whether or not degradation occurs.

Table 7.1 shows that sandy soils are much more vulnerable to soil pollution than 'black earths'. Pesticides are

Table 7.1
Vulnerability of European soils to degradation
Source: Fraters, 1994

Degradation type / Soil type	Erosion	Compaction	Acidification	Pollution
Black earth	Moderate			
Sands	Moderate	High	High	High
Acid loams	High	High	High	Moderate
Non-acid loams	High	High		Moderate
Clays	Moderate	Moderate		Moderate
Acid shallow	High		High	High
Non-acid shallow	High			Slight
Semi-arid	High			
Salt-affected	Moderate			
Wet		Moderate		High
Peat and muck	Depends on landuse and/or reclamation type			

Slight · Moderate · High

adsorbed by the organic matters of the 'black earths' and degradation is usually fast in the fertile topsoil. Heavy metals too are adsorbed strongly by the soil, which prevents their uptake by plants or soil fauna. In contrast to 'black earths', sandy soils have mostly acid, humus-poor topsoil with limited capacity for adsorption of pesticides and heavy metals. These contaminants are therefore more available for uptake by soil fauna and flora and for leaching from the topsoil than 'black earths' (see Chapter 5). Consequently in sandy soils heavy metals and pesticides threaten the groundwater as well as the soil flora and fauna. Loamy and clayey soils have an intermediate position between the sandy soils and the 'black earths'. The increase in adsorption capacity from sandy to clayey soils, due to the increase in clay content, often accompanied by increasing organic matter content, means increasing buffer capacity and decreasing vulnerability. However, soils with a relatively high concentration of heavy metals may become a potential threat if soil characteristics are changed, for instance if soils are acidified. In that case, the buffer capacity of the soil decreases and heavy metals may be released.

PROBLEMS AND THREATS

The soil functions described above are at risk because of the misuse of soil by humans. Soil degradation occurs when human-induced phenomena lower the current and/or future capacity of the soil to support human life (Van Lynden, 1994). Soils are generally resilient within certain limits, but outside those limits the soil will not recover naturally even if the load or stress is removed. In this section, emphasis is placed on the most severe soil degradation processes in Europe in terms of their irreversibility: soil erosion, acidification, pollution (by heavy metals, pesticides and other organic contaminants, excess of nitrates and phosphates, artificial radionuclides) and soil compaction. The particular problems of sites contaminated by waste disposal and of derelict industrial land are also developed. Other important but less extensive threats to soil functions are organic matter loss, salinisation and waterlogging. For each of these, the causes, magnitude, impact on the soil functions and remedies are discussed.

Badlands in Abruzzo, Italy
Source: A Giordano

Threat	Area affected (million ha)	Percentage of total European land area
Water erosion	115	12
Wind erosion	42	4
Acidification	85	9
Pesticides	180	19
Nitrates and phosphates	170	18
Soil compaction	33	4
Organic matter losses	3.2	0.3
Salinisation	3.8	0.4
Waterlogging	0.8	0.1

Note: Different threats can affect the same land area so that numbers cannot be added up.

Table 7.2
Estimated areas affected by major soil threats in Europe
Source: Oldeman et al, 1991

Soil degradation processes are not mutually exclusive, operating in strictly separated compartments; interactions will often occur with one type of degradation leading to another. The combined effect of pollution and erosion can illustrate this where pollutants from upstream areas can be transported with eroded sediment, accumulating downstream where the sediment is deposited.

Soil erosion

Causes

Soil erosion consists of the removal of soil material by water or, to a lesser extent, by wind. It is a natural phenomenon but it has been accelerated by human activities. It is one of the most widespread types of soil degradation worldwide. Unlike other forms of degradation, erosion removes soil from one area and takes it to another (this may imply important clean-up costs in deposition areas). It may be caused by any human activities that expose the soil to the impact of raindrops or wind, or that increase the amount and speed of surface runoff water. Some farming practices, such as ploughing up-and-down slopes, removal of vegetative soil cover and/or hedgerows, increased field size (open fields), abandonment of terraces, late sowing of winter cereals, overstocking, poor crop management, untimely farming operations and compaction by heavy machinery, may have these effects. The same holds true for improper forestry practices. In the Mediterranean area, deforestation and overgrazing in the past, on well-drained loamy and clayey soils (see Map 7.1), have led to severe soil erosion and degraded vegetation forms such as *maquis* and *garrigue*. The occurrence of frequent fires and summer droughts on these systems has led to irreversible desertification (UNEP, 1987 and Box 7B). Tourism, sport and recreational activities (walking, skiing, mountain bikes, off-road vehicles, etc) can also be important causes of erosion in areas which are already fragile and need protection, such as mountain areas, forests and parks (see Chapter 25).

Magnitude

Soil erosion occurs mainly where land is used over-intensively, and the resulting losses are irreversible over time-scales of tens or hundreds of years. Soil erosion is increasing in Europe (Blum, 1990), but figures are not available on the amount of erosion which takes place each year in different areas. However, models are being developed to predict areas at risk (Box 7C and Map 7.2). Maps 7.3 and 7.4 show the occurrence of water and wind erosion, respectively. An estimated 115 million ha, or 12

Box 7B Desertification in Europe

Desertification is now considered to be a major environmental problem affecting the most fragile areas of the Earth. Although it has been questioned whether this phenomenon is occurring in Europe, it is now accepted that desertification is affecting the Mediterranean region. It is caused by a combination of human exploitation (population pressure and landuse) that oversteps the natural ecological potential of the land, and the inherent fragility of the resource system.

In the Mediterranean, desertification has arisen for a variety of reasons:

- renewed pressure on land resources through migration;
- changes in agriculture in terms of both what is produced and the mode of production;
- increased demands on water through more intensive agriculture;
- impact of land degradation on flooding, groundwater recharge, salt water intrusion and soil salinisation.

The most obvious symptoms of desertification relate to a reduction in the biological and economic productivity or value of a piece of land (UNEP, 1992) and include:

- reduction of crop yields on irrigated or rain-fed farmland;
- reduction of biomass production and water availability due to a decrease in river flow or groundwater resources;
- encroachment of sand-bodies that may overwhelm productive land, settlements or infrastructure;
- deterioration of life-support systems, leading to societal disruption.

In the soil, desertifications lead to the lowering of organic matter content, deterioration of soil structure, changes in salt and water balances, higher surface runoff and higher erosion risks.

According to the *World Atlas of Desertification* (UNEP, 1992), the lands in Europe with the largest dry zones susceptible to desertification are found mainly in Spain, Sicily and the former USSR. Desertified areas are mainly uncultivated and abandoned lands (15 million hectares), dry lands (13 million hectares) and irrigated lands (1.6 million hectares).

The problem of desertification is manageable in principle, mainly through the use of proper irrigation systems, water management practices and reforestation. If these actions are technically feasible and practically implementable, they need special political and economic support, with due consideration given to social aspects, to be effective and sustainable.

Despite the severity of the problem, data are poor on recent trends in the extent of the desertifying areas in the northern Mediterranean region, as well as on the factors that are causing desertification and the processes involved. The EU has established programmes of investigation designed to identify, understand and mitigate the effects of desertification in Southern Europe. More research is particularly needed with regard to climate change and in the context of the World Convention to Combat Desertification (presently being negotiated in the framework of the United Nations). Changes in climatic conditions are likely to be exacerbated as a result of global warming due to the enhanced greenhouse effect. The risk of a northward shifting of the nearest desert belt, which might then cover Southern Europe, cannot be excluded and should receive the closest attention.

per cent of Europe's total land area, is affected by water erosion. Forty-two million ha are affected by wind erosion, of which 2 per cent is severely affected. Soil erosion occurs almost anywhere in Europe on sloping land, but particularly in the Mediterranean region (Box 7D), in parts of Central and Eastern Europe and in Iceland, due to a combination of climatic factors, soil properties and grazing practices over many centuries. Iceland is subject to wind erosion, as is the southeastern European part of the Russian Federation in spite of the very different soils and climate (Jansson, 1988).

Impacts

There is a variety of impacts of soil erosion, and the economic consequences are considerable. Erosion by water usually affects crop production through a decrease in plant rooting depth, removal of plant nutrients and organic matter, and, sometimes, uprooting of plants and/or trees together with dissection of the terrain by rills and gullies. Erosion may cause serious loss of fertile topsoil at rates of several millimetres per year (Morgan, 1986). In the Russian Federation, it is estimated that the humus content of agricultural soils decreases each year by about 1 per cent, while on irrigated land, far higher rates of humus losses occur; long-term land productivity is endangered since the annual erosion rates exceed the rate of humus formation (Karavayeva et al, 1991). The water-holding capacity of the soil can also be lowered through erosion. Agricultural production losses may add up to 30 per cent of normal yields (Follet and Stewart, 1985). Costs in terms of the quantity of fertilisers and manure required to replace the nutrients and organic matter lost with the removed topsoil are difficult to assess and deserve more attention. Decreases in crop yields due to erosion are not always clearly visible, but the costs of amelioration required to maintain yield levels could give a good indication of the damage done (Morgan, 1980). Arable fields may also be damaged by erosion upstream through deposition of excessive stream sediment loads originating in eroding areas.

In the mountainous Alpine regions, soils are predominantly shallow and usually covered with forest or non-intensive grassland. In the 1970s, large areas were developed into ski-runs. Intensive use of these slopes in the 1970s and 1980s (the Alps receive some 100 million tourists every year – see Chapter 25) has led to physical deterioration of the soils, principally by compaction. In addition, the loss of protective soil cover has resulted in erosion, and sometimes, with heavy rainfall, in landslides, causing deaths and devastation of buildings and houses. For example, in Val Pola (Italy) in 1987, a huge rock landslide occurred following a period of heavy rainfall killing 28 people (Agostoni et al, 1991).

Soil erosion may also threaten an ecosystem through the removal of the fertile topsoil, especially in areas where the vegetative cover of the soil has been damaged or removed. Soil material eroded from agricultural land can physically disturb natural ecosystems. The indirect (off-site) negative effects of erosion on the ecosystem include deposition of excessive amounts of sediments, possibly including pollutants (pesticides, fertilisers, heavy metals, etc) transported from upstream areas. In Southern Europe erosion is the major pathway for pollutant transport, while in Northern Europe leaching prevails on sandy and loamy soils. Deposition of sediments in waterbodies can have severe implications for aquatic life and human health.

Practically all soils on sloping land are vulnerable to erosion, but in particular sandy or silty soils are at risk (acid and non-acid loams on Map 7.1). Other factors like organic matter, infiltration rate, structure and surface roughness also play an important role, as well as external factors such as topography, climate, vegetation and management practices.

Old terraces in Peloponnisos, Greece
Source: A Nychas

Remedies

Since technological solutions to curb soil erosion abound, it is surprising that this type of soil degradation is still so widespread. One reason for this could be that other aspects of soil conservation – socio-economic and ecological factors – have been ignored for too long. An integrated approach is required to remedy this in which non-technological factors (population pressure, social structures, economy, ecological factors, etc) determine the most appropriate technical solution. A wide variety of possible technical solutions can be chosen, such as strip and alley cropping, rotation farming, contour ploughing, agro-forestry practices, adjusted stocking levels, mulching, use of cover crops, or construction of mechanical barriers such as terraces, banks and ditches. Though severe soil erosion is nearly always entirely irreversible, in less severe cases damage can still be prevented.

Box 7C Potential and actual soil erosion risk in southern EU countries

The CORINE soil erosion risk maps (Maps 7.2a and b) are the result of an analysis by a geographical information system where soil, slope, climate, vegetation layers and irrigation have been superimposed, enabling the evaluation of the soil erosion risk category. In order to estimate soil erosion risk information on a range of soil properties is required (soil texture, soil depth, stoniness, drainage, organic matter, pH, carbonates and salinity). The main source of information used was the soil map of the European Communities (CEC, 1985). *Potential* soil erosion risk is defined as the inherent risk of erosion, irrespective of current landuse or vegetation cover. The map of *potential* erosion risk (Map 7.2a) therefore represents the worst possible situation that might be reached. The area of land in this region with a high erosion risk totals 229 000 km[2] (about 10 per cent of the rural land surface). The largest area is found in Spain, mainly in the southern and western parts. In Portugal, areas of high erosion risk cover almost one third of the country. About 20 per cent of the land surface in Greece, 10 per cent in Italy and 1 per cent in France is subject to high erosion risk. The difference between the areas of *potential* and *actual* erosion risk (compare Maps 7.2a and b) reflects the protective influence provided by present land-cover, and the dangers inherent in changes in landuse practices.

Source: CORINE, 1992

Map 7.2
a) Potential and
b) actual erosion risk
in southern EU
countries
Source: CORINE, 1992

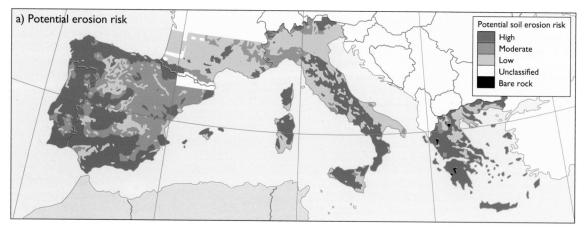

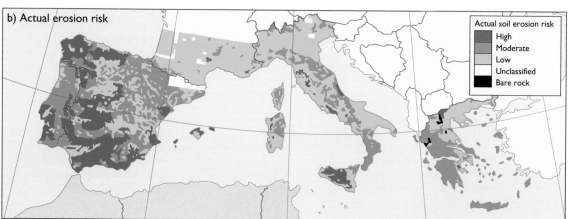

Box 7D Erosion in Spain

In Spain, it has been estimated (ICONA, 1991) that about 44 per cent of the territory is affected by some kind of erosion and 18 per cent (about 9 million hectares) is losing more than 50 tonnes of soil per hectare per year (t/ha/year), a level considered as the critical load for erosion. More than 90 per cent of the eroded land is located in the Mediterranean zone of Spain, where fires aggravate the situation. The major landuses concerned are essentially degraded woodlands (which lose about 95 t/ha/year of soil), non-irrigated grassland (which lose about 36 t/ha/year), sclerophyllous vegetation or the so-called *maquis* or *garrigue* (which lose about 21 t/ha/year), and to a lesser extent cultures of vine, almond or olive trees. According to ICONA, the average soil loss in Spain is about 27 t/ha/year. Considering that soil formation in this area is about 2 to 12 t/ha/year, the desertification rate through soil degradation is alarming. The cost of the direct impact of erosion on the environment (considering the reduced lifetime of reservoirs, the loss of agricultural production and the damage due to flooding) is estimated at 280 million ECU per year. The cost of rehabilitation to restore plant cover, improve water retention and protect soil is estimated at about 3 000 million ECU over a period of 15 to 20 years.

Gully erosion in Andalucia, Spain Source: J M Moreira

Map 7.3
Water erosion of soils in Europe
Source: Van Lynden, 1994

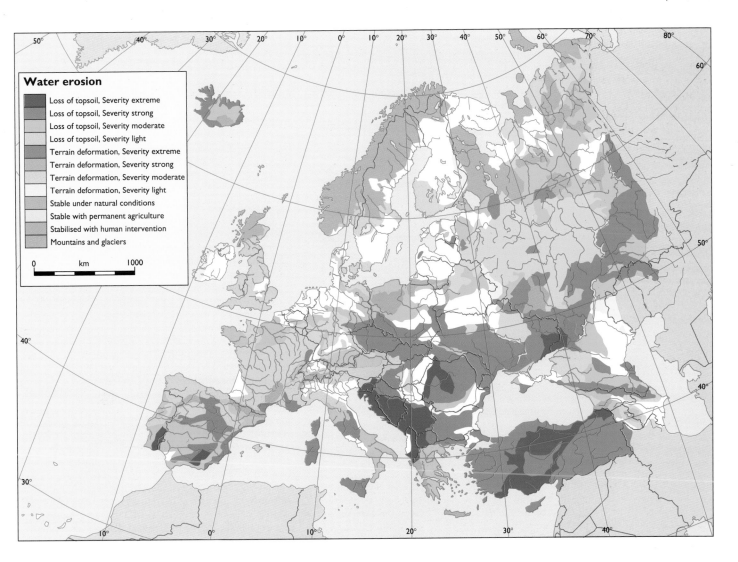

Water erosion

- Loss of topsoil, Severity extreme
- Loss of topsoil, Severity strong
- Loss of topsoil, Severity moderate
- Loss of topsoil, Severity light
- Terrain deformation, Severity extreme
- Terrain deformation, Severity strong
- Terrain deformation, Severity moderate
- Terrain deformation, Severity light
- Stable under natural conditions
- Stable with permanent agriculture
- Stabilised with human intervention
- Mountains and glaciers

0   km   1000

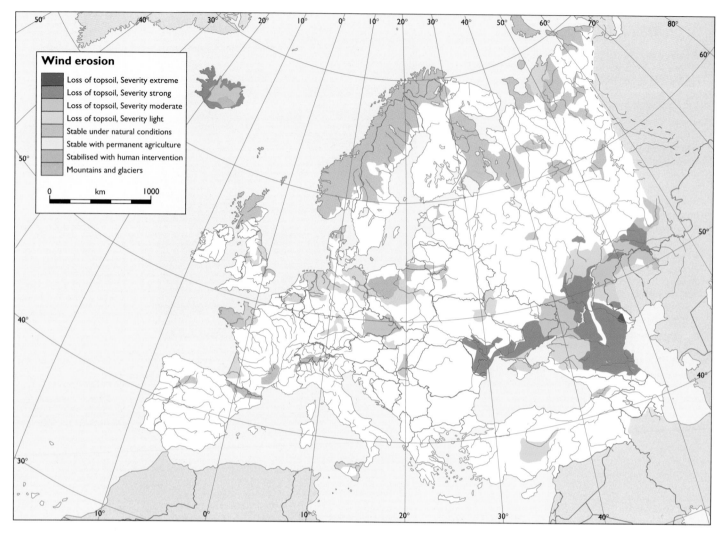

Map 7.4
Wind erosion of soils in Europe
Source: Van Lynden, 1994

Soil acidification

Causes

Soil acidification is a natural process, but it has been enhanced recently by human action through the emission of sulphur and nitrogen compounds from the combustion of fossil fuels and from industrial processes, subject to long-range transboundary dispersion (see Chapter 31). Acidification may also be caused by fertilisers and soil drainage. In Central and Western Europe, acidic deposition is by far the most important cause of soil acidification (de Vries and Breeuwsma, 1986). The Scandinavian countries are among the most severely affected areas in Europe, while the main sources of acidic emissions are the Northwestern and Central European countries. This well known discrepancy arises from the preferred track of low pressure systems over Europe giving rise to prevailing southwesterly winds over the north of the continent. Soil acidification is an important trigger for releasing cations such as iron, aluminium, calcium, magnesium and heavy metals (which are present in the soil in significant amounts but usually have little mobility). This has the effect of depleting the soil buffering capacity. Soils vary greatly in their ability to neutralise acidity; sandy soils have a low buffering capacity and thus only a small change in pH may be sufficient to change the soil from a sink to a source of pollutants. Six mechanisms contribute to soil acidification:

1 natural processes, such as the dissociation of carbonic and organic acids together with leaching of bases by rainfall;
2 landuse through the removal of base cations from the soil by harvesting;
3 careless use of nitrogen fertilisers;
4 afforestation with conifers;
5 drainage of wetland soils;
6 atmospheric deposition of sulphur dioxide (SO_2), nitrogen oxides (NO_x) and ammonia (NH_3) mainly from energy, industry, transport and animal husbandry.

Magnitude

Soil acidification is not a visually obvious feature, but it is an important, irreversible threat to the vitality of ecosystems. In agriculture, the effects of natural soil acidification, acidic deposition and acidification through landuse are hidden by liming; in the Russian Federation, however, it is estimated that 5 million hectares of arable land are strongly acidified despite liming (National Report of Russia, 1992). In most non-agricultural areas (forests, extensive grasslands, semi-natural areas), fertilisers are rarely used; in forests, tree canopies filter and concentrate larger quantities of pollutants from the atmosphere than do other vegetation types.

Important indicators of the acidification status of a soil are the aluminium (Al) concentration and the molar ratio of Al to base cations (Al/BC ratio) in soil solution, since the Al ion is toxic to roots. For both parameters, critical values, related to effects on fine roots, have been derived. Model predictions of the Al concentration and Al/BC ratio in current conditions, and in the steady-state situation that will be reached if present deposition rates of acidifying compounds continue, show that the area currently at risk may increase by 50 per cent in the future (see eg Table 7.3). Critical values of Al or molar Al/BC ratios are estimated (by numerical modelling) to be exceeded in 30 per cent of the forested area of Europe, or about 75 million ha. If acidic deposition does not change in future, up to 45 per cent, or about 110 million ha, of the forested area in Europe may come under risk (de Vries et al, 1992, 1994). The scientific threshold concentration for adverse effects on the environment is called the critical load (see Chapters 4, 5 and 31) and depends on the receptor (air, water or soil).

Impacts

The most important impact of acidification on the environment is the leaching of acidifying compounds from the soil to the surface water and groundwater (see Chapter 5). Soil acidification is the prerequisite to water acidification, direct atmospheric deposition only contributing to the process. Water draining from acidified soils contains increased levels of aluminium (Van Breemen and Verstraten, 1991) which can have serious impacts on the quality of surface waters (especially aquatic life) and on groundwater (and possibly therefore on the drinking water supply). Indeed, a major effect of acidification is the mobilisation of aluminium from clay minerals which the soil might have accumulated.

Acidification may also cause a depletion of the soil's buffering capacity through the breakdown of clay minerals. The most vulnerable soils are the ones with a low buffering capacity, that is acid and sandy soils (Map 7.1). Acidification considerably reduces the fertility of the soil, mainly by affecting its biology, by breaking up organic matter, and by causing loss of plant nutrients. There is also a risk of heavy metals being released where pH values are low (see eg Box 7E). On non-agricultural land where no chemical countermeasures are applied, acidification may affect plant vitality by causing loss and discolouration of leaves or needles (see Chapter 34); this can lead ultimately to forest die-back (Roberts et al, 1989). Furthermore, it may cause a decrease in the species diversity of the vegetation, followed by changes in plant and soil organisms, eventually favouring the more acid-tolerant and acidophilous species. However, certain acid soil habitats may be worth protecting for their biodiversity.

Remedies

Apart from emission reductions through coordinated abatement strategies (soil acidification can be slowed down if acidic inputs are reduced), soil acidification and its effects on forest vitality can be counteracted by fertilisation. However, buffering capacity of the soil cannot be restored. This makes soil acidification one of the most severe environmental threats in Europe, the full effects of which cannot be reversed. Liming will increase soil pH, but will also have an impact on soil biota and groundflora which is not necessarily desirable. In any case, liming should be done in such a way as to induce only a gradual increase in pH. This can be achieved by using slowly dissolving limestone, for example dolomite. A rapid pH increase may cause increased mineralisation of the humus layer, thus causing increased leaching of nitrates and base cations to groundwater.

Soil pollution by heavy metals

Causes

Soil contamination by heavy metals, such as cadmium (Cd), lead (Pb), chromium (Cr), copper (Cu), zinc (Zn), mercury (Hg) and arsenic (As), is a problem of concern. Although heavy metals are present naturally in soils, contamination comes from local sources (see Table 7.4), mostly industry (mainly non-ferrous industries, but also power plants, iron and steel and chemical industries), agriculture (irrigation with polluted water, use of mineral fertilisers, especially phosphates, contaminated manure, sewage sludge and pesticides containing heavy metals), waste incineration, combustion of fossil fuels and road traffic (see Chapter 14). Long-range transport of atmospheric pollutants adds to the metal load and is the main source of heavy metals in natural areas.

Heavy metals accumulate in the soil where they are fixed on mineral particles. From there they can be mobilised by 'triggers' such as acidification, and released to soil solution from which they can be taken up by soil organisms and plant roots, or leached into groundwater, thus polluting the food-chain or affecting drinking water quality.

Magnitude

It is difficult to give an evaluation of heavy metal pollution on a European scale since appropriate data are not available. Survey results which are available include those from Northeastern Europe where mosses here have been used to make a survey of heavy metal deposition (Box 7F, Maps 7.5 and 7.6). From the results of a study reviewing the environmental fate and effects of cadmium in Europe over the last 10 years (Jensen and Rasmussen, 1992), the average cadmium concentrations in European soils are estimated to range from 0.06 to 0.5 mg/kg with a tendency towards lower concentrations in Scandinavian soils. Unfortunately, this study refers essentially to Northern European countries due to the lack of information in Southern Europe. In contaminated soils (essentially mining areas), concentrations are 10 to 100 times greater than the average found in surface soils. In Europe, fertilisers produced from rock phosphate give the most important but highly variable contribution to cadmium in soils (from 0.3 to 38 g/ha/year); farmyard manure can give a similar and higher input than commercial fertilisers; atmospheric deposition adds to the cadmium load which over the past 50 years has brought about 2 to 7 g/ha/year to soils. For Europe as a whole, increasing cadmium concentrations have been found in European soils, especially during the last 20 to 30 years. However, only in a few cases has a significant long-term increase of cadmium concentration in harvested crops also been found.

Parameter	Critical value	Forested area exceeding critical values (%)	
		Present situation[1]	Steady-state situation[2]
Al concentration	0.2 mol$_c$ m^3	28	43
Al/BC ratio	1.0 mol/mol	14	30
Al and Al/BC	as above	30	45

Table 7.3
Predicted forested areas exceeding critical values of aluminium and/or aluminium/base cations in soil solution at present and in a (future) steady-state situation
Sources: *(based on current deposition levels)*
1 de Vries et al, 1994;
2 de Vries et al, 1992

	Cadmium	Copper	Lead	Zinc
Atmospheric deposition (local and regional)	+	+	+++	+
Pesticides	–	+/++[1]	–	+/++[1]
Manure	+	++	+	++
Mineral fertilisers and liming	+++	+/++[1]	+	+
Sewage sludge	+	+	+	+

Notes: – no contribution, + slight contribution, ++ moderate contribution, +++ high contribution.
1 Large differences between countries occur: copper compounds, for instance, are sometimes registered as fertiliser and sometimes as pesticide; differences also occur between and within countries in distinctions between arable land and grassland.

Table 7.4
Main anthropogenic sources of diffuse pollution by heavy metals to agricultural land
Source: RIVM, 1992

Box 7E The problem of methylmercury in Sweden

The environmental history of mercury in Sweden vividly illustrates the problem of hidden pollution. In the 1960s, the toxic effects on birds and other animals from the use of fungicides were obvious and the use of mercury compounds as fungicides was stopped. In the 1970s, emissions from industrial sources (for instance, chlor-alkali plants) were decreased to almost zero. Thus, the mercury problem was thought to be solved in Sweden. In the 1980s, the problem re-emerged as a delayed response. The present mercury content in pike exceeds the level of 1 mg/kg in fish in at least 10 000 lakes in Sweden. Part of this can be attributed to the fact that, even though the source has been drastically reduced, mercury which has accumulated over time, especially in forest soils (mainly bound to organic matter), is able to be mobilised through chemical and biological processes.

Recent results indicate that a substantial amount of the mercury flux is in the methylmercury form which is highly toxic and bio-available. Currently, however, the present leaching of methylmercury forms only a small part of the actual amount stored in Swedish soils (the average methylmercury transport to the lakes per year is less than 0.2 per cent of the total methylmercury stored in soils – see Figure 7.2). In addition, the soil stores are still increasing due to atmospheric deposition inputs and, unless the driving forces for the methylation and leaching processes are stopped (by reduced deposition of acidifying compounds - see Chapters 4 and 31), the soil will act as a supplier of mercury for a long time to come (Hultberg et al, 1994).

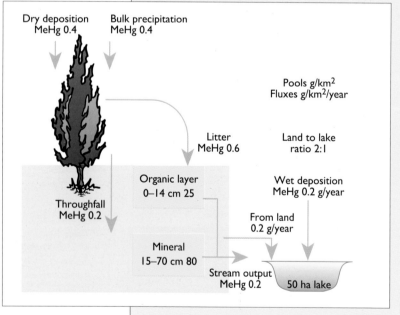

Impacts

Pollution of agricultural soils by heavy metals may lead to reduced yields (growth inhibition of leafy vegetables is reported at cadmium concentrations of 0.9 to 1.5 mg/kg in sandy soils and about 3 mg/kg in clay soils). Pollution can also lead to elevated levels of these elements in agricultural products, and thus to their introduction into the food-chain (Alloway, 1990). Heavy metals deposited on grassland soils remain predominantly in the top few centimetres and are directly ingested with soil by grazing animals. Soil ingestion is a major pathway of heavy metals to livestock grazing

contaminated land (Thornton and Abrahams, 1983). Heavy metals are toxic and it cannot be discounted that low levels may have long-term effects which may not yet have become apparent. They may have negative impacts through the inhibition of soil microorganism activity; for example, surface accumulation of heavy metals in acid forest soils can inhibit litter decomposition by influencing fungal and faunal communities. Cadmium concentration in soil exceeding 4 mg/kg dry weight can inhibit soil microbial processes, growth and reproduction of living organisms. Heavy metals may accumulate in soil, especially in soil with a large binding capacity for heavy metals (for instance in non-acid soils and 'black earths'). The most important factors governing plant uptake of heavy metals from soil are the soil pH, the clay content (texture), the organic matter content, and the metal concentration. This is illustrated in Figure 7.3. In soils with a low binding capacity (the acid and sandy soils of Map 7.1), there will be no accumulation of heavy metals even under high input load, which implies a relatively high concentration of heavy metals in the groundwater, important leaching, and high plant uptake.

Forest soils seem to accumulate more heavy metals than grasslands because forest crowns constitute receptors with a larger deposition surface, and forest litter is not exported (De Temmerman et al, 1987). Metal concentrations in the soil solution can remain low for decades or even centuries, as long as the binding capacity of the soil is not saturated. This reservoir of metals in the soil may turn into a source if the binding capacity of the soil is lowered. This may occur through a change in pH (pH<4) due, for example, to atmospheric deposition, changes in landuse, or climatic changes. Figure 7.4 illustrates how the behaviour of heavy metal cations adsorbed onto soil undergoes an abrupt change over a narrow range of pH.

Remedies

Reduction in the emissions of heavy metals is the most direct way to decrease the atmospheric deposition of these elements and their build-up in soil. Despite the great increase in traffic, substantial reductions in lead emissions have already been achieved through incentives to use unleaded petrol. Nevertheless, emissions of heavy metals from industrial plants (especially from smelters of Eastern Europe) are still very important. Many of the reduction measures already being considered and implemented for NO_x and SO_2 emissions will also reduce heavy metal emissions. Indeed, a decrease in acidification brought about by a reduction in the emission of these compounds can lead to a restoration of the binding capacity of soils.

On agricultural land, the heavy metal load can be decreased by:

- using low-metal-content resources for the production of fertilisers (especially low-metal-containing phosphate rock);
- using fewer animal feed supplements and feed with low metal contents (such as those containing copper used to improve the assimilation of foodstuffs);
- replacement of inorganic pesticides (such as Bordeaux mixture used in vineyards, Pb-arsenate, etc) by organic products;
- reducing fertiliser use and the number of animals per hectare; and
- reducing the use of other fertilisers such as manure and sewage sludge (the disposal of sewage sludge has become a soil-related problem in Europe, and particularly in The Netherlands).

Cadmium balance studies in European agricultural soils with moderate to high cadmium contamination confirm that a zero accumulation could be achieved by a combination of measures aimed at reducing the present cadmium inputs to the following rates: 50 to 70 per cent for fertilisers, 65 to 75 per cent for sewage sludge and 50 to 65 per cent for atmospheric deposition.

Box 7F Heavy metal monitoring in Northern and Eastern Europe

Atmopheric heavy metal deposition in Northern and Eastern Europe, including Denmark, Estonia, Finland, Iceland, Latvia, Lithuania, Norway, Sweden and adjacent areas of the Russian Federation, was monitored in 1990 by using mosses. This technique is based on the fact that some lichens and bryophytes are capable of trapping airborne heavy metal pollution so that the concentrations of heavy metals in the moss are closely correlated to atmospheric deposition.

Results for lead and Cadium are illustrated in Maps 7.5 and 7.6 (Nordic Council of Ministers, 1992). For all metals (cadmium, chromium, copper, iron, lead, nickel, vanadium and zinc) the regional deposition shows a decreasing gradient from relatively high values in the southern parts of Scandinavia to low values towards the north. The pattern of deposition in the Baltic region is due to long-range transport and local sources.

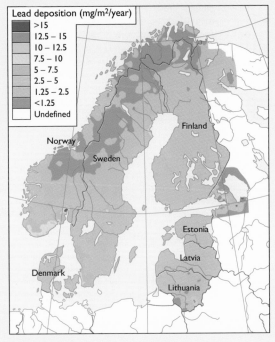

Lead deposition (mg/m²/year)
- >15
- 12.5 – 15
- 10 – 12.5
- 7.5 – 10
- 5 – 7.5
- 2.5 – 5
- 1.25 – 2.5
- <1.25
- Undefined

Map 7.5
Lead deposition, inferred from analysed moss samples
Source: Nordic Council of Ministers, 1992

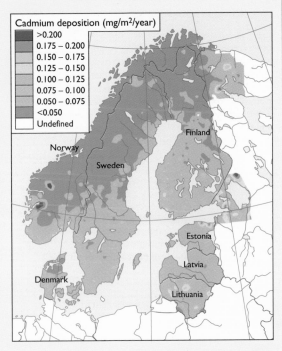

Cadmium deposition (mg/m²/year)
- >0.200
- 0.175 – 0.200
- 0.150 – 0.175
- 0.125 – 0.150
- 0.100 – 0.125
- 0.075 – 0.100
- 0.050 – 0.075
- <0.050
- Undefined

Map 7.6
Cadmium deposition, inferred from analysed moss samples
Source: Nordic Council of Ministers, 1992

Soil pollution by pesticides and other organic contaminants

The development and widespread use of synthetic and natural organic compounds in industry and agriculture has led to concern over the possibility of environmental pollution by these chemicals (see Chapters 5, 17, 22 and 38). Not only are pesticides known to pollute soil directly by affecting soil organisms, but also soil acts as a vector for the pollution of water. Organic pollutants enter soils through atmospheric deposition, direct spreading onto land (sewage sludge), or contamination by wastewaters and waste disposal. Besides pesticides, organic contaminants include many other components, such as oils, tars, chlorinated hydrocarbons, PCBs and dioxins. There is such a wide variety of organic substances that their detection and monitoring in the soil is practically impossible. Long-term studies at Rothamsted Experimental Station in the UK (ITE, 1989) have shown that, since the 1860s, there have been large increases in the soil burden of polynuclear aromatic hydrocarbons (PAHs), which are derived from combustion of fossil fuels. Their effects on soil microbial processes are unknown. PAHs accumulate on or within herbage, where they become part of the dietary intake of herbivores. In the Russian Federation, concentrations of PCBs in soils in industrial areas exceed by 20 to 30 times the national reference value (National Report of Russia, 1992).

Causes

Pesticides (mainly fungicides, herbicides and insecticides) are used in agriculture to protect crops and to ensure good quality of the harvest. The use of pesticides is also very cost-efficient compared with mechanical or other alternatives for pest control. Many pesticides, especially the older ones, have

Vineyards in Bouches du Rhone, France
Source: R Dannau

Figure 7.3
Governing factors for plant uptake of mobile trace elements from soil, as cadmium
Source: Tjell et al, 1983

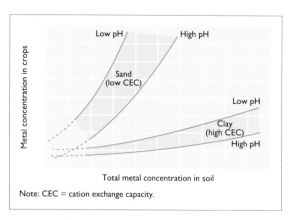

Note: CEC = cation exchange capacity.

a rather broad activity spectrum; that is, not only are the target organisms affected, but also non-target organisms, and side-effects to soil organisms or human beings are likely to occur. Persistence of pesticides in the plough layer and their leaching to groundwater are dependent, for the most part, on their intrinsic properties of transformation (degradation) and mobility in the soil. The most important problems occur for pesticides which are very persistent or very mobile. These problems are not related to the mode of action of a pesticide.

Magnitude

Intensive use of pesticides in agriculture is recent – after the Second World War in Western and Northern Europe, the amount used increased enormously up to about 1980 and is still increasing in Southern Europe (see Chapter 22 for the data on consumption and load of pesticides in Europe).

Impacts

The use of pesticides may unintentionally lead to:

- destruction of part of the soil microflora and micro- and mesofauna, which in turn may cause both physical and chemical deterioration;
- effects on the availability of organic matter, especially by herbicides;
- severe yield reductions in crops which follow in rotation due to the presence of residual herbicides – approximately 10 per cent of all herbicides have a persistency that may affect crops following those to which they were applied, and over 5 per cent of the applied amount stays in the plough layer for more than one year, which may influence the germination and growth of the succeeding crops (Cornelese et al, 1994);
- leaching of pesticides to groundwater, potentially threatening drinking water resources (see Chapter 5).

Up to now, only the threat of pesticides to groundwater as a resource for drinking water has received much attention. In the EU, Council Directive 80/778/EEC sets a maximum admissible concentration of 0.5 μg/l for the total amount of pesticides and metabolites in drinking water (see Chapters 5 and 22). The fact that there are about 1000 different compounds (active ingredients) on the market (Promochem, 1993), showing different behaviours in the soil, makes it extremely difficult to identify and evaluate the threats on soil posed by pesticides (see Chapter 17). Effects on non-target organisms in soil are expected to occur very frequently; they may influence not only the soil biotic functions, but also the abiotic functions. Until very recently, only the impact on major soil functions such as mineralisation and nitrification had been studied. Most pesticides only hamper these functions for a relatively short period after application (up to a few months) after which the soil may completely recover from this loss of activity. Side-effects in the soil are very much dependent on the specific pesticide used.

Remedies

The use of pesticides has been, and still is, widespread. Consequently, the reduction of side-effects can be obtained only through improved application and legislation (UNECE, 1992a). This might include:

- integrated pest management, using biological control;
- selection of pest-resistant species for crops;
- purification of active ingredients (separation of active and non-active isomers) which may reduce the application rates for several pesticides by 50 to 75 per cent;
- banning of persistent and very mobile pesticides;
- banning of broad-spectrum pesticides (having effects on many organisms, both target and non-target);
- improvement of application machinery which may lead to better targeting and reductions in application rates;
- development of biotechnology.

Soil pollution by nitrates and phosphorus

Causes

Nitrogen and phosphorus are elements essential to all forms of life and are therefore relevant to soil systems and to food-producing crops. They are important plant nutrients, but overapplication may lead to nitrogen or phosphate saturation in the soil, causing losses of nitrates into shallow groundwater, and saturation of the soil with phosphate, which may also move into the groundwater. Nitrate and phosphate pollution originates mainly from diffuse sources such as high manure and fertiliser application levels on agricultural soils and by atmospheric deposition on forests and natural areas. Unlike nitrate, where the major source comes from agriculture, phosphorus losses from sewage water (and animal manure) are the major factors contributing to increased content in water (see Chapter 14).

Magnitude

Nitrogen input levels to agricultural soils differ significantly within Europe. This is illustrated in Chapter 22 (Map 22.3), where the combined contribution of nitrogen from fertiliser and manure is depicted. Fertilisation inputs and related effects on water are usually less important in Southern Europe, with the exception of areas with intensive fruit and vegetable production; in these areas, the interaction of fertilisation and irrigation causes severe soil problems.

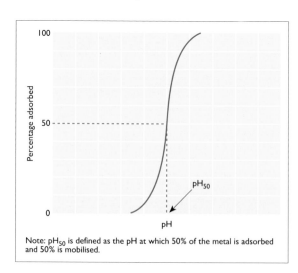

Note: pH_{50} is defined as the pH at which 50% of the metal is adsorbed and 50% is mobilised.

Figure 7.4 *Typical pH adsorption curve for hydrous metal oxides (pH_{50} ranges from 3 to 8)*
Source: IIASA, 1991

Natural areas and forests generally receive nitrogen input only from the atmosphere (deposition, and fixation by vegetation). In parts of Europe, ammonia generated by decomposition of manure-slurry in the soil, or that vented from sheds holding large numbers of livestock, is transported long distances, causing damage to forests and certain types of moorland vegetation over quite wide areas. Some landuse practices, such as drainage of peat soils, can enhance nitrogen leaching.

Unlike nitrogen, phosphorus is strongly fixed in soils – on average, only 3 per cent of total phosphorus in soil is extractable and available to plants (Heck and Hanotiaux, 1976) – often leading to deficiency problems in agricultural systems. The main soil components responsible for phosphate bonding are organic matter, clay minerals, and hydrous oxides of aluminium and iron. It is only in areas where phosphate is applied in large amounts under wet conditions on soils with low adsorption capacity (eg, sandy soils with low organic matter content under intensive agriculture in Northern Europe) that the load may reach levels where phosphorus is released to the groundwater; on slopes, runoff can contribute to large quantities of phosphorus to surface waters. This is particularly the case for intensive crops (such as maize) on sandy soils with a shallow groundwater table, because the phosphate fixing capacity is much lower under anaerobic conditions. These soils are common in Denmark, The Netherlands, the UK, France and Germany. In The Netherlands, it is estimated that 80 per cent of soils are saturated with phosphorus. The problem with phosphorus is that it can take 15 to 30 years before the soil is saturated and effects are seen, as in the Po valley in northern Italy. Where there are high phosphate concentrations in freshwater, eutrophication may occur. A level of 0.02 mg phosphorus per litre is sufficient to allow eutrophication and many freshwater bodies in Europe have higher concentrations than this (see Chapter 5).

Impacts

Together with an intensification of landuse in general, the main effects of nitrogen and phosphorus application in excess of plant uptake are the loss of nutrient-poor habitats within agricultural areas throughout Europe, the leaching and runoff to surface water of nitrogen and phosphorus, and the leaching to groundwater mainly of nitrogen in the form of nitrate. The most vulnerable soils are the sandy and loamy soils (see Map 7.1).

The amount of leaching depends on the soil, the local climate and crop management (see Box 7G). Well-drained soils with their plentiful supply of air favour the formation of nitrate so that the risk of leaching is increased. In waterlogged conditions (which are common for peat and muck soils and many wet soils, see Map 7.1), nitrate is denitrified by bacteria into nitrogen or nitrous oxide (N_2O) which dissipate to the atmosphere (N_2O is a greenhouse gas which contributes to climate change problems as described in Chapter 27); however, the amount of nitrate available for leaching is proportionally reduced. In intensive horticultural systems, interaction between high fertiliser inputs and irrigation enhances nitrate leaching. Most nitrate leaching occurs in the autumn when fields are either fallow, recently sown, or awaiting sowing. The impacts of (over)application of nutrients on surface water caused by runoff and leaching are treated in detail in Chapter 5.

In regions with high livestock densities, accumulation of phosphorus in the upper soil layers occurs. This is the case in the central and southern sandy regions of The Netherlands, and in the Po valley in Italy, where some of the highest livestock densities in Europe are found (Breeuwsma and Silva, 1992). In soils saturated with phosphorus, especially those with shallow groundwater, high phosphorus concentrations occur in the upper groundwater layer and surface water causing eutrophication.

Remedies

The problem of nitrate pollution is recognised internationally, and is usually associated with intensive agricultural practices. In the EU, Council Directive 91/676/EEC on nitrates imposes on Member States the establishment of mandatory codes of good agricultural practice and the designation of vulnerable areas. Codes of good agricultural practice try to narrow the imbalance between fertiliser input and plant uptake, and the following measures are favoured:

- selection of less nutrient-demanding crops;
- application of fertiliser at the proper time (in the growing season), ensuring that the efficiency of the application can be substantially increased;
- improvement of methods of manure application by incorporating slurry directly into the soil – this prevents the volatilisation of ammonia;
- better adjustment of fertiliser application and crop demands by using modern field methods to test the amount of available nitrogen in the soil – information systems have been developed in Bavaria (Germany) and Denmark which have permitted fertiliser applications to be decreased by one third without yield losses;

Box 7H Inventory of contaminated sites in Denmark

In Denmark, a register of contaminated sites is issued once a year by the Danish Environment Protection Agency since the adoption in 1990 of the law regulating waste dump sites. The data collected by each regional council comprise: the location of each site, its size, present and former use and main contaminants present; also included is an estimation of the risk the site represents for the groundwater, the surface water and the people who live nearby. On the basis of this information, priorities are selected for remediation plans which are submitted to the Danish Environment Protection Agency for financial support. From this register, it appears that the majority of contaminated sites are former waste dump sites, and approximately 15 per cent are used for residential purposes (Figure 7.5). Leachates contain mainly organic substances, heavy metals, solvents, chlorinated organic solvents, oil, petrol and other substances coming from tars. Over the period 1982 to 1992, 2547 sites have been recorded (Figure 7.6), and about 1250 projects and 460 protective measures have been carried out. Based on the conclusions of the registration activities, it is estimated that there should be about 11 000 contaminated sites in Denmark. By the end of 1993, ECU 95 million had been spent for the registration of contaminated sites, the implementation of protection measures against contaminants dispersion, and clean-up activities. The total costs of cleaning up would be about ECU 3 billion; this figure includes voluntary actions estimated at ECU 1 billion .

Source: Miljøstyrelsen, 1993

Figure 7.5
Inventory of contaminated sites in Denmark by pollution type and polluted area
Source: Miljøstyrelsen, 1993

Figure 7.6
Trend in the number of registered waste dump sites in Denmark (1982–92)

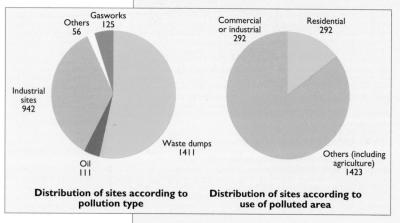

Distribution of sites according to pollution type

Distribution of sites according to use of polluted area

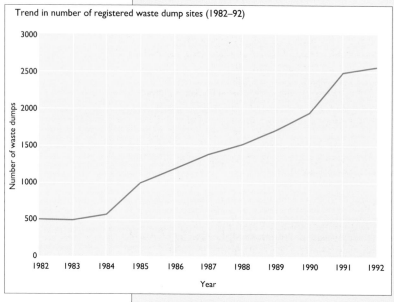

Trend in number of registered waste dump sites (1982–92)

- minimising leaching losses of arable land by sowing winter catch-crops in autumn;
- shortening the length of the grazing season;
- less intensive use of grassland, for example by lowering cattle density.

In general, nitrogen leaching cannot be completely prevented, especially in areas with precipitation excess and where high production levels are achieved on permeable soils.

Phosphate is more a local problem in Europe. It has been partly limited since the use of phosphates in detergents was controlled. However, unlike nitrate, phosphate binds to soil particles and if a soil becomes phosphate-saturated then phosphate will remain in solution along with nitrate. In The Netherlands, the capacity of the soils to adsorb phosphate is already exhausted at several locations. For these soils saturated with phosphate, rehabilitation would then involve the replacement of topsoil, which is not practicable.

Soil pollution by artificial radionuclides

Contamination of soil with artificial radionuclides has given rise to much public concern since the Chernobyl accident. The radionuclide currently causing most interest is caesium-137 (Cs-137) although others may also be detectable in some areas (eg, strontium-90 and plutonium isotopes). Contamination from atomic bomb tests, nuclear accidents and/or explosions all have a high risk of causing widespread contamination through the transport of artificial radionuclides by wind and water. Radioactive contamination of soils can also arise from leakage of waste dumps or during production, use and transport of radionuclides. Although these may not necessarily lead to effects over extensive areas, impacts may be locally highly hazardous (see Chapter 18).

Radionuclides behave chemically in soil as their stable isotope counterparts. The half-lives of the radionuclides (or the time to decay to half their original activity) and the retention properties of soils vary considerably; for example, clayey soils tend to retain radionuclides (as well as other contaminants) better than sandy soils. Since most sources of contamination come from the surface, most artificial radionuclides are concentrated in the upper few centimetres of mineral soils. In organic soils, they tend to concentrate at greater depths.

Radionuclides in the upper layers of soil may expose plants and animals to radiation, and may in extreme cases present a threat to humans from direct ionising radiation (see Chapter 16). From the soil, radionuclides can be available to enter the food-chain, leading to intake by humans and animals through ingestion. Exceptionally, wind-blown contaminated soil material may lead to exposure through inhalation. Some plants concentrate radionuclides and can act as indicators of soil contamination. At very high levels of exposure, flora can also sustain deleterious effects from radionuclide contamination, as noticed for example after the Chernobyl accident (see Box 18E in Chapter 18). Fauna living on or in the soil can equally be affected; the direct ingestion of soils by ruminants can be a significant pathway for artificial radionuclides to enter the food-chain and ultimately to reach humans.

A full European survey of radionuclides in soils does not exist. The impacts and radioactive contamination of the land surface resulting from fall-out after the Chernobyl nuclear accident in April 1986 are presented in Box 18E. Radioactivity levels of more than 40 GBq/km^2 (1 GBq = 10^{9}Bq) were still reported in a zone of 55 000 km^2 around the Chernobyl power station five years after the accident (National Report of Russia, 1992). In a zone of 10 000 km^2 (Mnatsakanian, 1992) radioactivity levels of more than 550 GBq/km^2 were found.

Country	Number of registered contaminated sites	Number of contaminated sites in critical condition	Remediation expenses already spent (ECU)	Estimated remediation costs in million ECU for crucial sites (15-year programme)
Belgium	8300	2000	incomplete information	1000
Denmark	3600	3600	incomplete information	200
France	incomplete data	incomplete data	incomplete information	4000
Germany	32 500	10 000	228 million	7000
Greece	incomplete data	incomplete data	incomplete information	200
Ireland	incomplete data	incomplete data	incomplete information	180
Italy	5600	2600	89 million	3000
Luxembourg	incomplete data	incomplete data	incomplete information	50
The Netherlands	5000	4000	1300 million	1000
Portugal	incomplete data	incomplete data	incomplete information	no information
Spain	4300	incomplete data	incomplete information	1000
United Kingdom	incomplete data	incomplete data	267 million/year	9000
TOTAL EU (estimation)	more than 55 000	more than 22 000		26 630

Table 7.5
Soil clean-up in the EU
Source: Carrera and Robertiello, 1993

In the framework of the CEC/CIS collaborative programme (see Box 18E), a project was started at the end of 1993 with the aim to establish a first European atlas of radioactive contamination and external exposure resulting from the Chernobyl accident (Joint Study Project 6).

Contaminated land

Past and present economic activities have often resulted in contamination of the soil underlying the place where these activities took place. Contaminants affecting soil may be in the form of solids, liquids or gases. The most common toxic soil pollutants include metallic elements and their compounds, organic chemicals, oils and tars, pesticides, explosive and toxic gases, radioactive materials, biologically active materials, combustible materials, asbestos and other hazardous minerals. These substances commonly arise from the disposal of industrial and domestic waste products in designated landfills or uncontrolled dumps. The existence of unauthorised dumps and the improper disposal of an extremely wide variety of wastes and sludges is one of the environmental problems which deserves the highest priority in Europe (see Chapter 15); in Central and Eastern Europe, it may be a limiting factor for foreign investment. Since the second half of the 1980s, attempts have been made to register contaminated land in the EU. According to a recent study (CEC, 1992), out of 200 000 hectares of derelict industrial land, about 25 per cent was from the coal and steel industry, of which more than 90 per cent was concentrated in five countries only (Belgium, France, Germany, Spain and the UK). In East Germany, 70 000 hectares have been abandoned following lignite mining. The process of site assessment is difficult, but some countries, such as Denmark and The Netherlands, have started a systematic recording, monitoring and clean-up programme of contaminated sites (see Box 7H).

A tentative list of contaminated sites in the EU has been compiled (Table 7.5). More than anything else, this inventory reveals the inconsistency of definitions (in many countries, contaminated sites have been restricted to landfills) rather than the actual number of contaminated sites. Better information can be obtained only if contaminated sites are properly registered and if reference levels for all potential contaminants are established.

The number of polluted sites is already huge and remediation costs are very high, roughly estimated at more than ECU 100 billion (Carrera and Robertiello, 1993). It is expected that the situation in Central and Eastern Europe will be at least as bad, especially in industrial areas.

Considering the rehabilitation costs, a compromise needs to be found between economically sustainable remediation policies and potential health effects.

Causes

In addition to the diffuse contributions to degradation described in the preceding sections, point sources also have important effects, mainly at a local level. Soil pollution can be created either directly by spillage, leakage, handling, or disposal and leaching of waste. Contamination from such sources often becomes detectable after relatively long periods when pollutants have already begun to be released, contaminating food supplies and drinking water. Uncontrolled dumping can threaten human health and the environment. Leaching processes in landfills can contaminate soil and groundwater. Almost all municipal solid wastes contain hazardous and toxic substances which can be leached to water.

Magnitude

The pollution of soil at both former and present industrial sites represents potentially serious threats to human health and soil biology. These problems have appeared, for example, with the expansion of cities, where industrial sites have been transformed into residential areas. Since there is no registration of such areas, the extent of soil pollution is difficult to assess. When such data do exist, they are often unreliable because many of them are collected unsystematically and are often based on qualitative information. Many sites of former polluting industries and landfill dumps are forgotten until contamination re-emerges to pollute water supplies or redevelopment surveys reveal the hazard. Preliminary inventories in The Netherlands indicate that at 20 per cent of present and former industrial sites, serious soil pollution will occur (Meeder and Soczó, 1992). For Germany it is estimated that 10 to 20 per cent of the 135 000 suspected sites (industrial areas, landfills and military sites) will potentially be confirmed as contaminated (NATO/CCMS, 1992).

Impacts

Because of the wide variety of soil pollutants and concentrations, impacts are imprecisely known. The effects of waste disposal, past and present, are often not confined to the actual site but may influence a large surrounding area (tens of square kilometres) including agricultural land, dwellings and/or nature reserves. As indicated in Table 5.4, a wide range of pollutants is involved and many are unknown.

Remedies

Various technologies are available for treatment. These include soil excavation, washing and disposal. However, this type of treatment is very expensive (more than ECU 500 per tonne of soil) and sometimes clean-up of soils is not possible although the actual threat to human health and the environment is high. In case of heavy metal pollution, the metal ions can be immobilised in the soil by increasing the chemical binding capacity. Cost-effective 'soft technologies' exist (ECU 3 to 350 per tonne of soil) which can be used *in situ* and where soil is treated with immobilising additives such as lime, 'Thomas basic slag' or hydrous iron and manganese oxides (Mench et al, 1993).

Strategies to tackle problems concerned with point sources of pollution require both prevention and remediation of soil and groundwater contamination. The prevention can be focused on:

● reducing industrial emissions and reducing the amount of generated waste; and

● soil and groundwater protection.

Box 7 I Geochemical characteristics of river and floodplain sediments in Europe

Work carried out by FOREGS (Forum of European Geological Surveys) shows that sediments from streams and floodplains of numerous rivers in various parts of Europe have been severely contaminated by heavy metals in waste from mining and other human activities over several centuries. For example, parts of the Innerste River floodplain in Germany, which is now used for agriculture, contain up to 2 per cent lead and several grams per tonne (ppm) of cadmium and mercury in the upper 50 cm as a result of mining in the Harz mountains. Such floodplains constitute chemical threats that may be triggered by future catastrophic floods or through the action of acid rain. Downstream, harmful effects on ecological systems may cross political borders and eventually reach the sea. This type of pollution needs to be surveyed in order to quantify the extent of the problem, evaluate health effects and develop plans for mitigation and abatement strategies. FOREGS has developed methods comprising chemical analyses of regionally distributed samples of pre- and post-industrial deposits of floodplain sediments, which facilitate an appraisal of the pollution against a varying natural geochemical background. Examples of the type of data that could be obtained by such geochemical mapping are given in Figure 7.7.

Soil and groundwater protection includes technical and legal measures, and management to avoid (new) contamination or to reduce further dispersion of contaminants by insulation (sealing). To verify the quality of soil and groundwater and to check the efficacy of insulation measures, monitoring networks need to be systematically set up (see Box 7 I). For the rehabilitation of priority sites, a remediation strategy is required.

Great difficulty surrounds determining threshold values for pollutants above which soil may be regarded as seriously contaminated and for which a clean-up procedure is necessary. Indeed, as mentioned in the introduction to the present chapter, soils are composed of mineral and organic constituents which vary in concentration and content geographically, as well as in depth down the soil profile. In some cases, natural concentrations in clean soils and waters can exceed the threshold trigger level for pollutants. For instance, the concentration range in unpolluted sites varies from 0.01 to 1 mg/kg for cadmium, 2 to 200 for lead and 2 to 100 for copper (Bridges, 1991), while in the EC Directive on sewage sludge (86/278/EEC), limit values for concentrations of heavy metals in soil range from 1 to 3 mg/kg for cadmium, 50 to 300 mg/kg for lead and 50 to 140 mg/kg for copper (see Box 7J). In comparison, concentrations of heavy metals found in polluted sites in Poland reach 290 mg/kg for cadmium, 4650 for lead and 1200 for copper. It is therefore difficult to define a minimum concentration above which a soil may be considered contaminated by any particular pollutant.

Legislation concerning contaminated land

Only a small number of countries such as The Netherlands and Denmark have elaborated comprehensive legislation with respect to the registering and cleaning up of polluted soils. In most countries, reference has to be made to general rules of law or with other provisions relating to the removal of abandoned waste, or pollution of groundwater.

Whether or not clean-up operations are undertaken, polluted soils raise a number of legal and administrative issues.

Where are the polluted sites?
To answer this it is necessary to establish a systematic examination of polluted areas. The transfer of property, or the closure of an installation, offer good opportunities for an assessment of soil quality. A legal obligation which imposed on the operators of industrial installations a duty to report spills and other causes of new soil pollution would also be useful. A mandatory soil investigation could also be required before any construction permit is delivered, at least in former industrial areas (Bocken, 1993).

How clean is clean? How dirty is dirty?
Many countries have guidelines or standards related to target or threshold values of certain potentially hazardous substances. However, the values differ from country to country. At EU level, several directives exist with regard to maximum admissible values – for nitrates in drinking water, heavy metals in sewage sludge and air pollutants and other chemical substances, for example. However, none of these addresses soil directly.

The use of critical loads of contaminants for soils is the subject of research in many European countries. The idea is to set up acceptable concentrations of contaminants which are permissible in soils, particularly those soils used to produce foodstuffs.

● In The Netherlands, in 1987, the government passed a Soil Protection Act, following the experience gained under the provisions of the Soil Clean Up (Interim) Act of 1983. To accompany this legislation, three threshold values were determined for concentrations of seven

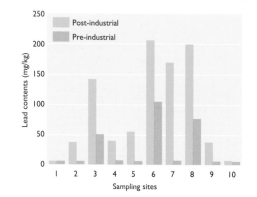

Figure 7.7
Contents of lead in samples of pre- and post-industrial floodplain sediments in Limburg, The Netherlands
Sources: De Vos et al (in preparation) Hindel et al (in press)

Note: At each site the pollution can be evaluated against the local but regionally varying natural background as the difference between the height of the two columns. The pollution is caused mainly by waste from old mining. The ten sample sites are within an area of 5000 km².

Box 7J EC Directive on sewage sludge

In the EC, Council Directive (86/278/EEC) regulates the use of sewage sludge in agriculture and sets limit values for concentrations of heavy metals in sludge and soils. The maximum values permitted by the EU are in nearly all cases higher or at least equal to the maximum values permitted in the different European countries. Recent research findings on the effects of heavy metals on soil microbiology (Chaudri et al, 1993) show that significant reductions in *Rhizobium* numbers occur at metal concentrations well below the current EU upper limit for all metals. Recommendations for an update of the Directive have also been made by an independent scientific review on the use of sewage sludge in agriculture (UK Ministry of Agriculture, Fisheries and Food, 1993).

classes of pollutants (metals, inorganic pollutants, aromatic compounds, polycyclic hydrocarbons, chlorinated organic compounds, pesticides and others): an uncontaminated 'reference' value (A value); a figure which indicates that further investigation is required (B); and a third figure, 'intervention' value (C), above which a clean-up is necessary. Recently, a new set of risk-based threshold values, taking into account potential risk for humans and ecosystems, has been defined for integration into Dutch legislation in 1994. The objective of the Dutch legislation is to maintain, or if clean-up is necessary to retrieve, the multifunctionality of the soil. Thus, whenever the multifunctionality of the soil is endangered, soil clean-up will be required.

- In the UK, attempts have been made to determine threshold values for pollutants below which soil may be regarded as uncontaminated, or above which it is seriously contaminated so land reclamation is necessary. Between these two figures lies a range of concentrations where interpretation is required before the land may be safely re-developed or used for particular purposes. The UK system differs from that of the Dutch by defining contamination in relation to the end use of the site.

- In Germany, water protection is the driving force; risk levels are classified in a number of categories which varies from one Land to another (a 'U' value below which a soil may be considered uncontaminated and an 'S' value above which treatment is necessary).

Whatever approach is used (multifunctionality or end use), it is first important to know where seriously polluted soils are found so that control over their end use can be ensured. The potentially severe effects that such control might have on the real estate market cannot be ignored.

Soil clean-up itself presents a number of serious problems. Clean-up costs are often so high that the only practical approach is to share the cost between the owner of the polluted property and society. The waste resulting from the clean-up has itself to be treated and disposed of in some way (see Chapter 15). Since soil pollution can affect areas or people far from where the pollution originated (migration of pollutants to the groundwater), clean-up activities need to cover the whole area affected by pollution even if parts belong to different owners (eg, heavy metal pollution from smelters near residential areas).

Finally, a distinction must be made between past and new pollution. The level of financial resources required means that the decision to clean up past pollution needs to be based on the actual danger to human health and the environment and be subject to a list of established priorities. At the same time, new pollution must be avoided and clean-up required as soon as the soil quality objectives are exceeded.

Soil compaction

Causes

Soil compaction is caused by the repetitive and cumulative effect of (heavy) machinery on the same piece of land, or to a lesser extent by the trampling of cattle when overstocked, in wet conditions or on wet soils. But it is not confined solely to agricultural soils. Building sites and intensively used recreational areas, for instance, are also susceptible. Compaction may occur at shallow soil depths as well as deeper in the subsoil. Shallow soil compaction occurs during seed bed preparation, spreading of fertilisers and pesticides. Subsoil compaction occurs when tractor wheels pass through the open furrow during ploughing, creating a persistent 'plough pan'. Severe subsoil compaction is also caused by heavy machinery used during harvest and the spreading of slurry with high capacity tankers having heavy axle loads. Soil compaction is potentially a major threat to agricultural productivity.

Magnitude

Soil compaction mostly affects highly productive soils and heavily mechanised agricultural land or wet grasslands with high cattle density (see Map 7.7). According to the *World map of the status of human-induced soil degradation* (Oldeman et al, 1991) about 33 million ha or 4 per cent of European land is at risk for soil compaction. In Map 7.7, it is mainly shallow compaction which is shown. However, due to the increasing weights and axle loads of farm vehicles, subsoil compaction is becoming an urgent soil conservation problem, especially in Central and Eastern Europe (Håkansson and Petelkau, 1994). Deep soils with less than 25 per cent clay are most sensitive to subsoil compaction (Hébert, 1982). Such soils cover large areas in the Russian Federation, Poland, Germany, The Netherlands, Belgium and northwestern France. These correspond on Map 7.1 to the acid and non-acid loamy soils, and to a lesser degree the sandy soils. Because subsoil compaction is not easily detected without specific measurements, the area affected by this type of soil degradation tends to be underrated.

Impacts

Compaction changes the quantity and quality of soil biochemical and microbiological activities. The major physical impact of soil compaction is to reduce soil porosity, which means less air and water are available to plant roots. At the same time the roots have more difficulty in penetrating the soil and have reduced access to nutrients in the soil. Biological activity is substantially reduced. Another effect of compaction is the increase in surface runoff, since less rainwater is able to percolate. This increases the risk of water erosion and loss of topsoil and nutrients.

Recent calculations (Boels and Van der Akker, in press) show that surface soil compaction may cause yield reductions of up to 13 per cent (average 5 per cent), whereas subsoil compaction may reduce yields by 5 to 35 per cent (average 12 per cent). In the Russian Federation, decreases in yield up to 50 per cent have been recorded on compacted soils (National Report of Russia, 1992).

Remedies

Surface soil compaction is easily countered by reworking the soil and eventually disappears after a couple of years if biological processes remain undisturbed. Subsoil compaction however is persistent and cannot be restored easily since present techniques do not provide long-term solutions. Since compaction is mostly caused by heavy machinery, it can be prevented by increasing the number of axles and wheels under agricultural machinery, increasing

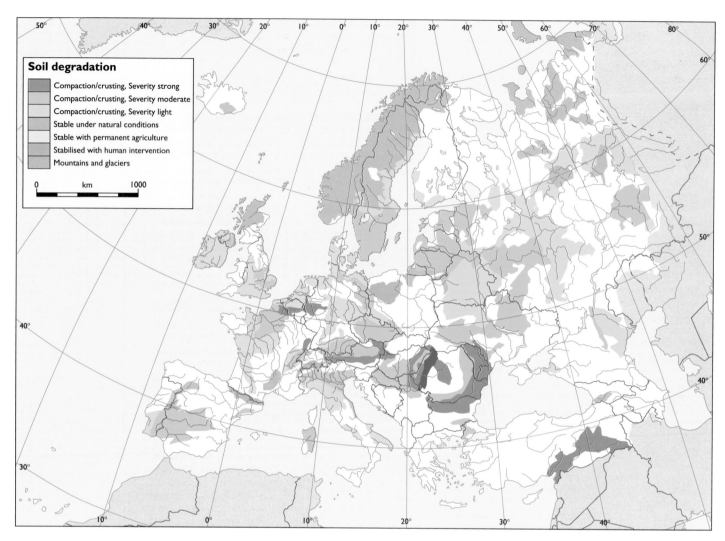

Map 7.7
Physical degradation of soils in Europe
Source: Van Lynden, 1994

tyre width, and reducing tyre pressures (Håkansson and Petelkau, 1994). The use of smaller and lighter vehicles does not necessarily help since it may require more frequent passages, which may counter the preventive effect. In general the use of large machines is preferable as they require fewer passages on the land, but the ground pressure under wheels or tracks should be below a certain threshold; less than 100 kPa has been proposed as critical value (Soane, 1982). It is the timing of operations more than the design of the equipment which is the crucial factor; for example, using machinery during wet conditions should be avoided.

Other threats

Organic matter losses

As mentioned in the first section of this chapter, soil organic matter is important in maintaining soil structure, in retaining water, and as a nutrient reserve and chemical buffer. Certain landuses adversely affect the amount of organic matter in soils. Intensive arable cultivation, especially in Western Europe, has led in certain cases to a decline in organic matter and a loss of biological activity and diversity (in Belgium, France, the UK, etc); in Beauce (France), organic matter has decreased by half in 16 years (Secrétariat d'Etat auprès du Premier Ministre, Chargé de l'Environnement, 1988). The ploughing up of grassland, the abandonment of rotations in the agricultural system and the burning of crop residues all reduce the amount of vegetation matter returning to the soil. Unfortunately, it is not possible to determine an ideal or target level of organic matter required in a healthy soil. Nevertheless, there is a threshold

value under which the soil is less able to hold moisture, store nutrients and adsorb pollutants; when its biological activity is reduced, the soil is less stable and more prone to erosion, leaching and runoff. Loss of organic matter is particularly apparent in regions where peat soils have been drained and farmed intensively. It is estimated that 3.2 million hectares suffer from losses of nutrients or organic matter in Europe.

Various measures could be taken to increase the organic content of soils. These include the introduction of grass into agricultural rotations; the ban on straw and stubble burning (Denmark, the UK), which is likely to result in crop residues being ploughed back into soils; and the increase of land left fallow or under grass as a consequence of the EU set-aside policy, which could lead to improvement in organic matter content. Monitoring of the current extent of organic matter loss is also needed.

Salinisation

Salinisation arises in a number of ways: by irrigation with improper drainage, intrusion of salt water from the sea or from saline fossil sources, and through evapotranspiration of saline soil moisture. It has direct negative effects on soil biology and crop productivity, and indirect effects leading to loss of soil stability through changes in soil structure (alkalinisation). In Europe, the surface area affected by salinisation is estimated to be 3.8 million ha (Oldeman et al, 1991; Szabolcs, 1991).

Salinisation is most strongly tied to site-specific soil properties and climatic conditions and therefore its distribution is restricted to Southeastern Europe where semi-arid conditions prevail, that is on semi-arid and salt-affected soils (Map 7.1). The irrigation of the Hungarian plain, for instance, has caused salinisation and alkalinisation of more than 20 per cent of the region. In the Russian Federation, about 7 per cent of agricultural soils are saline, part of them naturally so – for example the *solonchaks*. In Romania, where huge irrigation schemes were introduced 25 years ago (about 3.2 million ha), it is estimated that 200 000 ha have been salinised through irrigation, which represents about 6 per cent of total irrigated land. Salinisation is reversible but reclamation of saline/alkali soils is expensive, as it requires complex amelioration techniques.

Waterlogging

Waterlogging due to accidental or deliberate flooding, increased runoff from higher areas resulting from lower infiltration rates, or raising of the water table (as a result for example of irrigation) may also negatively affect crop productivity. Waterlogging, by driving the air from the soil, reduces the redox potential to anoxic conditions. This has an impact on many biochemical processes in the soil and may even trigger the release of hazardous substances previously safely stored in the soil. Waterlogging also increases the risk of compaction and, in drier areas, the risk of salinisation through capillary rise of saline groundwater. The most vulnerable soils are the wet soils and parts of the semi-arid and salt-affected soils (Map 7.1).

Waterlogging occurs mainly in the north of the Russian Federation and at a few scattered sites in the rest of Europe (foremost along the Black Sea coast and the lower Danube Valley). The area involved is estimated at 0.8 million ha (Oldeman et al, 1991).

RESPONSES

Ecological responses

The reversibility of the effects of various threats on soils is an important concept in soil protection. Irreversible changes should be avoided if they impair any of the soil functions. In some cases, soil properties may return naturally to their normal range of values once the threat is removed. Then the main question is the time-scale required. In many cases, the soil will not return naturally to its original condition, but would do so under appropriate management. In other cases, the soil will not return to its original condition but could, under management, be converted to some other desirable state. Therefore, to achieve a sustainable landuse policy, those uses leading to effects which can be reversed only by technological means should be avoided or minimised.

It is inevitable that some uses of soils will result in damage to the soil, or in the reduction of the ability of the soil to perform various functions; the most important soil threats have been extensively discussed above. A number of forms of damage can, to a greater or lesser extent, be reversed by a variety of management measures, either by the use of an ecological approach to management, or by application of technology. For example, surface soil compaction can be reversed naturally (the bulk density will decrease once the loading is reduced). Soil compaction can also be reversed by suitable cultivation techniques, but severe subsoil compaction will require deep ripping with large machines and is reversible only technologically. If a soil has been acidified by atmospheric deposition, reversibility would express the ability of the soil to recover (to increase its base saturation) as the inputs were decreased or as basic substances were added. In some cases, rehabilitation may not be practicable.

Critical load: a useful policy tool

As developed above, it is possible to rank soils in terms of their vulnerability to a particular stress, such as their sensitivity to acidification, wind or water erosion or nitrate leaching, or their sensitivity to compaction through excessive animal load and agricultural or civil engineering traffic.

The critical-load concept follows the same type of approach. It has been developed in connection with assessments of the impact of acidic deposition (see Chapters 4 and 31) but the approach could be applied more broadly. The concept aims at defining the maximum load of a given stress which will not produce adverse changes in soils. The load, which could be pollutant inputs, physical loading or fertiliser inputs, has much in common with the maximum permissible loads calculated for heavy metal inputs in sewage sludge. The definition of critical load will vary with the particular threat: for instance in relation to compaction it might be a quantitative estimate of the maximum physical loading of a soil which will not produce increases in bulk density, or decreases in permeability resulting in adverse effects on the functioning of the soil system. The critical-load approach provides a means of linking controls on the load of given stresses to soil protection by cause–effects models, but the necessary models are available for only a few stresses at the moment. Nilsson and Grennfelt (1988) have determined critical loads of sulphur and nitrogen for forest soils. Studies by RIVM in The Netherlands have produced models defining the maximum inputs of nitrogen fertiliser, and the timing of these inputs, to limit leaching of nitrate to groundwaters (see above and Chapters 5 and 22). In effect, these models are defining a critical load of nitrogen fertiliser with respect to impacts on water quality. Considerably more research is needed, however, before quantitative models can be developed to allow the determination of rates of recovery (reversibility) for any particular threat. Reference levels for all potential contaminants should be available in future.

Technical responses

The technical solutions to specific pollutant problems have been described above, together with the magnitude, impacts and causes of the different problems, and preference has been given to prevention methods. However, in cases of severe soil contamination, clean-up methods may be necessary. There is a variety of techniques with very varying divergent costs, and decisions should be taken case by case.

It has been estimated that, for the EU in 1988, the total expenses for the implementation of a 15-year remediation programme for those sites requiring immediate intervention would have been ECU 27 billion, an average of ECU 5 per inhabitant and per year (Carrera and Robertiello, 1993).

In any overall risk assessment of any given use, or of the growth of a given crop under a given management regime, it is desirable to balance soil suitability with sensitivity before making any decision. Thus, the combined suitability/vulnerability approach would take into account the risks of erosion, or nitrate leaching, for example, as a consequence of growing a particular crop at a given location and under a specific management scenario. The aim would be to balance suitability against any consequent risks – an optimisation procedure.

Policy and legislation

The term 'soil protection' is used here to indicate a broad protection of soils which is not linked to any particular end use (ITE, 1989). Indeed, the difficulty when dealing with soil protection legislation is that there is a wide variety of soil types, functions and sensitivities to various degradation types. Because of the broad definition of soil used in this chapter (see the Introduction), soil protection implies, therefore, the protection of a complex dynamic, heterogenous system. This renders the setting of realistic targets very difficult. The purpose of this section is not to make a comparative analysis of the rules applied in soil protection in different legal systems but rather to identify the major questions which could be usefully addressed in this context and to indicate a number of policy implications of some of the possible solutions.

Perspectives for future policies and approaches

- An analysis of the processes by which soils are polluted or severely degraded shows that international as well as national action has to be taken to protect them, as some problems arise across national borders and can be solved only through the joint effort of all countries, especially where the harmful substances are taken in from the atmosphere. Various activities are under way in Europe to this end (involving the OECD, Council of Europe, UNEP, etc).
- National and international efforts to protect the soil need to conform to certain guidelines or principles, which have been presented throughout this chapter, and could be summarised as follows:

  1 the first principle laid down in the European Conservation Strategy (Council of Europe, 1990) could form the basis of any soil protection strategy: the emphasis in soil protection *should move from remedial and reactive strategies to preventive and pro-active strategies. Past damage needs to be repaired but more and more we should seek to prevent such damage*, and not impair soil potential for future generations;
  2 the emphasis should be on protecting soil *functions* rather than uses;
  3 landuses should be related as much as possible to the most suitable soils.

- A wider application of the concept of multifunctionality, which already plays a large part in the soil protection policies of The Netherlands, Germany and Switzerland, would improve landuse policy and represent a marked change from the idea of protecting a soil for a given use.
- Consideration of the functions exploited by a given use of soils identifies any possible conflicts between uses, and forms a sound basis for assessing the impact of a given use on a particular soil. Soil types differ not only in the range of functions which they can perform, but also in their responses to different types of stress. For this reason it is important to assess soil vulnerability to different stresses, and to follow appropriate landuse practices accordingly. In this way the impact of human activities on soil can be minimised.
- To ensure soil protection, an integrated approach is required beyond pure 'environmental' policies. Sectoral approaches focusing on specific types or aspects of soil degradation are useful, but should be integrated in broader issues at policy and decision-making levels. Each soil degradation problem should be treated within an

overall policy on land suitability which would need to cover the different landuses such as agriculture, forestry, industry, town and countryside planning and tourism.
- Information on the pollution state of soils is fundamental to future policy development. The supply of such information must go hand in hand with the development of standards for the sampling and analysis of soils. The collection and analysis of data needs to be strengthened and should include the building of databases related to environmental risks (integrating information on pollutants, their content in soils, soil characteristics, landuses, concentrations in groundwater or the lower atmosphere, plants and food-chains).
- The evidence of soil degradation shows that there is an urgent need for action even if the degradation processes are not yet fully understood or scientifically explained (the 'precautionary principle'). Policies and strategies for soil protection should be flexible enough to allow a variability in soil responses to external disturbances to occur (soil being part of a dynamic environmental system). More cognisance should be taken of interactions in order to ensure that remedies are not worse than the disease.

Current situation

Up to now, soil conservation policies and legislation have aimed at ensuring the maintenance of the production potential of soils, mainly for agricultural and forestry uses, to protect watercourses and reservoirs and to prevent, or reverse, soil contamination occurring mainly through human activities (industrial activities and waste disposal).

In a number of countries (Austria, Italy, The Netherlands, Switzerland) laws exist that focus specifically on soils or on causes of soil degradation (such as overgrazing). Statutory protection of soils against erosion is probably the oldest type of legislation (eg, the Act for Resolution on Sand Erosion and Reclamation of 1895 in Iceland). Legal measures to control soil pollution are very recent in most countries, whereas legislation relating to other types of soil degradation is far less common, and is difficult to implement because soil, unlike air and water, is a personal property; its protection can be achieved only through the negotiation of voluntary management agreements between the relevant authorities and the land owners or occupiers.

In a report for the Council of Europe (1990), six types of legal treatments of soil protection are not mutually exclusive, which, have been identified:

1 drafting of specific legislation to cover all soil functions, which is most innovative but also least common (The Netherlands, Liechtenstein);
2 specific legislation concerning only certain soil functions (Cyprus, Iceland, Malta and Spain with respect to erosion control, Austria and Italy);
3 integration of soil protection considerations into various sectoral environmental legislation (Germany, Denmark, Greece, Norway, France, UK, Belgium, Luxembourg);
4 specific soil regulations within other legislative fields (agriculture, physical planning) – these regulations concern soil fertility (in Switzerland), soil as a natural resource (in Portugal), or are part of physical planning acts and/or other legislative instruments (Denmark, Greece, Norway, Sweden, Turkey, UK);
5 preference for negotiated agreements or for incentives (UK, France, Switzerland, Spain – codes of practice, incentives, subsidies);
6 economic measures which may consist of incentives or subsidies, of levies (as in the case of over-production of organic manure), or of fines in case of negligence, for

instance if soil erosion is due to a lack of proper conservation practices (Austria, The Netherlands). Despite these tools, soil degradation continues on a wide scale in Europe. Soils in many areas are failing to protect water resources from contamination; acidity, aluminium, nitrates, phosphates, pesticides, pathogens and contaminated sediments originating from the soil are polluting water resources; some soils are eroding at rates that exceed their formation; organic matter levels tend to decrease in arable soils; and peat soils are degraded through drainage. Comprehensive information is not available on the extent of such effects in Europe. Better data about the extent of all types of soil degradation is the prerequisite to any soil protection policy in Europe. Policies need to be coordinated in Europe and when legislation is required the protection of soil resources should be explicitly addressed.

CONCLUSIONS

Soil degradation is a crucial environmental problem today, a problem that is increasing, and one that tends to get too little attention (Barrow, 1991). Soil is a dynamic, living system which is formed so slowly that it can be considered a non-renewable resource. Compared with air and water, soil has the particularity to act as a sink in which pollutants are filtered or transformed until the buffer capacity is depleted. The soil can then turn into a source of chemicals, and pollutants can start leaking into the groundwater. The fact that pollution effects are hidden for a long time has led to complacency, and until more recently concern for the well-being of soils has been slight. Finally, unlike air and water, soil is a personal property which renders any soil protection measure difficult to implement.

In Europe, the three most severe soil degradation processes in terms of their irreversibility are erosion, acidification and pollution by heavy metals, pesticides and other organic contaminants, nitrates and phosphates, and artificial radionuclides. Other important threats are soil compaction, losses of organic matter through improper management practices, salinisation and waterlogging.

To establish the magnitude of soil degradation requires a qualitative and quantitative assessment but there is still little accurate information for all of Europe, measurement is difficult and existing data are inconsistent.

Since soil-related problems are site-based, any attempt at generalisation is also very difficult. There is a need for development of models for environmental impact prediction and risk assessment. For each main threat identified, an assessment has been made of the causal factors, the magnitude and severity of the damage, and remedies. Reversibility is the main criterion by which the different problems affecting soils in Europe have been selected and ranked.

The analysis of the processes by which soils are polluted and degraded shows that international as well as national action needs to be implemented to protect soils, as some problems arise across national borders and can be solved only through the joint efforts of all countries – particularly where the various substances are taken in from the atmosphere. To implement any soil protection programme, it is essential:

1 to identify the problems, to determine the main causes and effects and to gain better understanding of the links between the two; an appraisal of the current situation is the first step to be achieved (for instance, there is a need to register degraded sites in Europe);
2 to check the damages which have occurred and record trends, through the establishment of targeted and integrated monitoring networks involving vulnerable soils, and to assess methodologies (analysis, models, inventories, description of uncertainty) which should be harmonised between countries, data collection not being an aim in itself;
3 to take prevention measures to protect the soil from irreversible damages and to allow the soil to serve only for uses which cannot irreversibly alter or damage other uses – this precautionary principle leads to the concept of soil suitability and vulnerability;
4 to use technical solutions, when they exist, for greatly reducing or remedying certain soil pollution problems bearing in mind that they entail enormous expenses – ultimately, the question is whether these expenses are to be financed mainly or in part by governments or by the owners of the polluted land; moreover, some crucial contamination problems still remain without solutions.

Soil protection requires an integrated approach within the larger context of sustainable development, taking into account socio-economic factors (UNCED, 1992).

REFERENCES

Addiscott, T (1988) Farmers, fertilisers and the nitrate flood. *New Scientist* **120** (1633), 50–54.

Agostoni, S, De Andrea, S, Lauzi, S and Padovan, N (1991) Sistemi di monitoraggio nell'area della Val Pola (SO): Studi Trentini di Scienze Naturali. *Acta Geologica* **68**, 249–302.

Alloway, B J (1990) Soil processes and the behaviour of metals. In: Alloway, B J (Ed) *Heavy Metals in Soils*, pp 7–28. Blackie, London.

Barrow, C J (1991) *Land Degradation*, Cambridge University Press, Cambridge.

Blum, W E H (1988) Problems of soil conservation. *Nature and Environment Series* **39**. Council of Europe, Strasbourg.

Blum, W E H (1990) The challenge of soil protection in Europe. *Environmental Conservation* **17**, 72–4.

Bocken, H (1993) Key-issues in the legislation with respect to soil rehabilitation. In: Eijsackers, H J P and Hamers, T (Eds) *Integrated soil and sediment research: a basis for proper protection*, pp 725–32. Kluwer Academic Publishers, Dordrecht, The Netherlands.

Boels, D and Van den Akker, J J H (in press) A simulation procedure to estimate the effect of subsoil compaction on crop yield in Europe. *DLO The Winand Staring Centre Report* **81**. Wageningen.

Boesten, J J T I and van der Linden, A M A (1991) Modelling the influence of sorption and transformation on pesticide leaching and persistence. *Journal of Environmental Quality* **20**, 425–35.

Breeuwsma, A and Silva, S (1992) Phosphorus fertilisation and environmental effects in the Netherlands and the Po region (Italy). *Winand Staring Centre (SC-DLO) report* **57**. Wageningen.

Bridges, E M (1991) Dealing with contaminated soils. *Soil Use and Management* (7) 3, 151–8.

Canton, J H, Linders, J B H J, Luttik, R, Mensink, B J W G, Panman, E, van de Plassche, E J, Sparenburg, P M and Tuinstra, J (1991) *Catch-up operation on old pesticides: an integration.* RIVM report no. 678801002. Bilthoven, The Netherlands.

Carrera, P and Robertiello, A (1993) Soil clean up in Europe – feasibility and costs. In: Eijsackers, H J P and Hamers, T (Eds) *Integrated soil and sediment research: a basis for proper protection*, pp 733–53. Kluwer Academic, Publishers, Dordrecht, The Netherlands..

CEC (1985) *Soil map of the European Communities, scale 1:1 000 000.* Commission of the European Communities (DG VI), Luxembourg.

CEC (1991). *Pesticides in ground and drinking water.* Fielding, M, Barcelo, D, Helweg, A, Galassi, S, Torstensson, L, Van Zoonen, P, Wolter, R and Angeletti, G (Eds). Water Pollution Research Report no. 27, Commission of the European Communities (DG XII), Brussels.

CEC (1992) *Etude des problèmes posés par les sites abandonnés et pollués des industries charbonnières et sidérurgiques, volet A: Ampleur du problème.* Association des Régions Européennes de Technologie industrielle (RETI). Commission of the European Communities (DG XI-B) Brussels.

CEC (1993) *Feasibility study on the creation of a soil map of Europe at a scale of 1:250 000.* Institute for Land and Water Management, K U Leuven, Belgium, and The Winand Staring Centre for Integrated Land, Soil and Water Research, Wageningen, The Netherlands. Commission of the European Communities (DG XI-Task EEA Force), Brussels.

Chaudri, A M, McGrath, S P, Giller, K E, Rietz, E and Sauerbeck, D R (1993) Enumeration of indigenous *Rhizobium leguminosarum* Biovar *Trifolii* in soils previously treated with metal-contaminated sewage sludge. *Soil Biology and Biochemistry* **25** (3), 301–9.

CORINE (1992) *Soil erosion risk and important land resources in the southern regions of the European Community.* EUR 13233. Commission of the European Communities, Luxembourg.

Cornelese, A A, Malkoç, N and Van der Linden, A M A (1994) *Pan-European state of the environment: pesticides in soil and groundwater.* RIVM report 71240300. RIVM, Bilthoven, The Netherlands.

Council of Europe (1972) *European Soil Charter.* B(72)63. Council of Europe, Strasbourg.

Council of Europe (1990) *European conservation strategy: 6th European Conference on the Environment, Brussels, 11–12 October 1990,* Council of Europe, Strasbourg.

Desaules, A (1991) The Soil Vulnerability Mapping Project Group for Europe (SOVEUR): Methodological considerations with reference to conditions in Switzerland. In: Batjes, N H and Bridges, E M (Eds) *Mapping of soil and terrain vulnerability to specified chemical compounds in Europe at a scale of 1:5M,* pp 23–8. Proceedings of an International Workshop held at Wageningen, The Netherlands (20-23 March 1991) ISRIC, Wageningen.

De Temmerman, L O, Istas, J R, Hoenig, M, Dupire, S, Ledent, G, Van Elsen, Y, Baeten, H and De Meyer, A (1987) Définition des teneurs 'normales' des éléments en trace de certains sols belges en tant que critère de base pour la détection et l'interprétation de la pollution des sols en général. *Newsletter from the FAO European Cooperative network on trace elements.* Fifth issue, pp 57–86. State University Ghent.

De Vos, W , Ebbing, J , Hindel, R, Schalich, J , Swennen, R and Van Keer, I (in preparation) Geochemical mapping based on overbank sediments in the heavily industrialized border area of Belgium, Germany and The Netherlands. *Journal of Geochemical Exploration.*

de Vries, W and Breeuwsma, A (1986) Relative importance of natural and anthropogenic proton sources in soils in the Netherlands. *Water, Air and Soil Pollution* **28**, 173–84.

de Vries, W, Posch, M, Reinds, G J and Kämäri, J (1992) *Critical loads and their exceedance on forest in Europe.* The Winand Staring Centre for Integrated Land, Soil and Water Research. Report 58, Wageningen.

de Vries, W, Reinds, G J, Posch, M and Kämäri, J (1994) Simulation of soil response to acid deposition scenarios in Europe. *Water, Air and Soil Pollution* **78** (in press).

Eswaran, H, Van den Berg, E and Reich, P (1993) Organic carbon in soils of the world. *Soil Sci Soc Am J* **57**, 192–4.

FAO (1974) *Legend, FAO/Unesco soil map of the world, scale 1:5 000 000* Volume I. UNESCO, Paris.

FAO (1981) *Europe, FAO/Unesco soil map of the world, scale 1:5 000 000.* Volume V, UNESCO, Paris.

Follet, R F and Stewart, B A (Eds) (1985) *Soil erosion and crop productivity.* ASA, CSSA, SSSA, Madison (USA).

Fraters, D (1994) Generalized soil map of Europe: Aggregation of the FAO-Unesco soil units based on the characteristics determining the vulnerability to soil degradation processes. Report 71240300, RIVM, Bilthoven, The Netherlands.

Häberli, R, Lüscher, C, Praplan Chastonay, B and Wyss, C (1991) *L'affaire sol. Pour une politique raisonnée de l'utilisation du sol.* Rapport final du programme national de recherche 'Utilisation du sol en Suisse' (PNR 22), Georg Editeur, Geneva.

Håkansson, I and Petelkau, H (1994) Benefits of reduced vehicle weight. In: Soane, B D and van Ouwerkerk, C (Eds) *Soil compaction in crop production.* Elsevier, Amsterdam.

Hébert, J (1982) About the problems of structure in relation to soil degradation. In: Boels, D, Davies, D B and Johnston, A E (Eds) *Soil degradation,* pp 67–72. A A Balkema, Rotterdam.

Heck, J P and Hanotiaux, G (1976) Le phosphore minéral du sol et son utilisation spécifique par quelques plantes de grande culture. Actes 4e Coll sur le *Contrôle de l'alimentation des plantes cultivées,* pp 510–18. Ghent.

Hindel, R, Schalich, J, De Vos, W, Ebbing, J, Swennen, R and Van Keer, I (in press) Vertical distribution of elements in overbank sediment profiles from Belgium and The Netherlands. *Journal of Geochemical Exploration.*

Hultberg, H, Iverfeldt, Å and Lee, Y-H (1994) Methyl mercury input/output and accumulation in forested catchments and critical loads for lakes in south-western Sweden. In: Watras, C and Huckabee, J (Eds) *Mercury as a global pollutant – toward integration and synthesis.* Lewis Publishers (in press).

ICONA (1991) *Plan national de lutte contre l'érosion.* Ministère de l'Agriculture, de la Pêche et de l'Alimentation. Institut National pour la Conservation de la Nature, Madrid.

IIASA (1991) *Chemical time bombs: definition, concepts, and examples,* Stigliani, W M (Ed) Executive Report 16. IIASA, Laxenbourg, Austria.

ITE (1989) *An assessment of the principles of soil protection in the UK. Volume I; Concepts and principles,* Institute of Terrestrial Ecology, Soil Survey & Land Research Centre, HMSO, London.

Jansson, M B (1988) A global survey of sediment yield. *Geografiska Annaler* **70** A (1–2), 86.

Jenny, H (1941) *Factors of soil formation.* McGraw Hill, New York.

Jensen, A and Bro-Zasmussen, F (1992) Environmental cadium in Europe. *Reviews of Environmental Contamination and Toxicology.* **125**, 101 – 81.

Karavayeva, N A, Nefedova, T G and Targulian, V O (1991) Historical land use changes and soil degradation on the Russian Plain. In: Brouwer, F M, Thomas, A J and Chadwick, M J (Eds) *Land use changes in Europe: processes of change, environmental transformations and future patterns,* pp 351–77. Kluwer Academic Publishers, Dordrecht, The Netherlands.

Meeder, T A and Soczó E R (1992) Aanpak van bodemsanering in Europa en Noord-Amerika *Bodem* **3**, 103–6.

Mench, M J, Didier, V, Gomez, A and Löffler, M (1993) Remediation of metal-contaminated soils: an assessment of mobile and bioavailable metals in soil after treatment with possible immobilizing additives. In: Eijsackers, H J P and Hamers, T (Eds) *Integrated soil and sediment research: a basis for proper protection,* pp 270–71. Kluwer Academic Publishers, Dordrecht, The Netherlands.

Miljøstyrelsen (1993) *Depotredegørelse 1993.* Miljøstyrelsen, Copenhagen.

Mnatsakanian, R A (1992) *Environmental legacy of the former Soviet Republics,* p 57. Centre for human ecology, University of Edinburgh, Edinburgh.

Morgan, R P C (Ed) (1980) *Soil conservation: problems and prospects.* Wiley, New York.

Morgan, R P C (Ed) (1986) *Soil erosion and conservation.* Longman, Harlow (UK).

National Report of Russia (1992) Status of the Environment in 1991: 'Eurasia – monitoring', no **5**.

NATO/CCMS (1992) First International Conference. (Phase II). Summary Report. *The 1992 NATO/CCMS pilot study meeting on evaluation of demonstrated and emerging technologies for the treatment and clean up of contaminated land and groundwater,* October 18–22, 1992. Hungarian Academy of Sciences Facility, Budapest.

Nilsson, J and Grennfelt, P (Eds) (1988) Critical loads for sulphur and nitrogen. Report from a workshop held at Skokloster, Sweden 19–24 March 1988. UN/ECE and Nordic Council of Ministers. *Nordic Council of Ministers, Miljørapport 1988,* **15**.

Nordic Council of Ministers (1992) *Atmospheric heavy metal deposition in Northern Europe 1990.* Nord 1992, **12**. Copenhagen, Denmark.

OECD (1993) Group on Economic and Environment Policy Integration – Working group on soil and land management. *Public policies for the protection of soil resources.* ENV/EPOC/GEEI/SR (93).

Oldeman, L R, Hakkeling, R T A and Sombroek, W G (1991) *World map of the status of human-induced soil degradation, an explanatory note* (second revised edition). ISRIC, Wageningen; UNEP, Nairobi.

Promochem (1993) *Reference materials for pesticide residue analysis.* Herts, England.

RIVM (1992) *The environment in Europe: a global perspective.* Report 481505001, RIVM, Bilthoven, The Netherlands.

Roberts, T M, Skeffington, R A and Blank, L W (1989) Causes of type I spruce decline in Europe. *Forestry* **62**(3), 179–222.

Secrétariat d'Etat auprès du Premier Ministre, Chargé de l'Environnement (1988) *L'observatoire de la qualité des sols.* Neuilly-sur-Seine, France.

Soane, B D (1982) Soil degradation attributable to compaction under wheels and its control. In: Boels, D, Davies, D B and Johnston, A E (Eds) *Soil Degradation,* pp 27–45. A A Balkema, Rotterdam.

Stigliani, W M, Doelman, P, Salomons, W, Schulin, R, Smidt, G R B and van der Zee, S E A T M (1991) Chemical time bombs: predicting the unpredictable. *Environment* **33**, 4–30.

Szabolcs, I (1991) Salinisation potential of European soils. In: Brouwer, F M, Thomas, A J and Chadwick, M J (Eds) *Land use changes in Europe: processes of change, environmental transformations and future patterns*, Warsaw, 5–9 Sept 1988, pp 293–315. Kluwer Academic Publishers, Dordrecht, The Netherlands.

Thornton, I and Abrahams, P (1983) Soil ingestion – a major pathway of heavy metals into livestock grazing contaminated land. *Sci Total Envir* **28**, 282–94.

Tjell, J C, Christensen, T H and Bro-Rasmussen, F (1983) Cadmium in soil and terrestrial biota, with emphasis on the Danish situation. *Ecotoxicology and Environmental Safety* **7**, 122–40.

UK Ministry of Agriculture, Fisheries and Food (1993) *News Release 3 March 1993*. Whitehall Place, London.

UNCED (1992) Agenda 21. UNCED Secretariat, Conches, Switzerland.

UNECE (1992a) Prevention and Control of Water Pollution from fertilizers and pesticides. *United Nations/Economic and Social Council Final Report*, ENVWA/WP.3/R.29. United Nations Economic Commission for Europe, Geneva.

UNECE (1992b) *The environment in Europe and North America. Annotated Statistics.* United Nations Economic Commission for Europe, New York.

UNEP (1987) Promotion of soil protection as an essential component of environmental protection in Mediterranean coastal zones. *MAP Technical Reports Series* **16.** United Nations Environment Programme. United Nations, Regional Activity Centre, Split.

UNEP (1992) *World atlas of desertification*. Edward Arnold, London.

Van Breemen N and Verstraten, J M (1991) Soil Acidification and N cycling. In: Schneider, T and Heij, G J (Eds) *Acidification research in The Netherlands*, pp 289–352. Final report of the Dutch priority programme on acidification. Studies in the Environmental Sciences 46. Elsevier Science Publishers, Amsterdam.

Van Lynden, G W J (1994) *The European soil resource: current status of soil degradation in Europe: causes, impacts and need for action*. ISRIC, Wageningen. Council of Europe, Strasbourg.

North Yorkshire, UK
Source: Spectrum Colour
Library

8 Landscapes

THE SIGNIFICANCE OF LANDSCAPES

The concept of landscape

The richness and diversity of rural landscapes in Europe is a distinctive feature of the continent. There is probably nowhere else where the signs of human interaction with nature in landscape are so varied, contrasting and localised. Despite the immense scale of socio-economic changes that have accompanied this century's wave of industrialisation and urbanisation in many parts of Europe, much of this diversity remains, giving distinctive character to countries, regions and local areas.

Landscapes can be divided into natural and cultural types. Chapter 9 is dedicated to the former by presenting a view on Europe's relatively small but valuable percentage of natural and semi-natural ecosystems. However, there are practically no areas in Europe that can be considered 'natural' in the sense that there is no human influence whatsoever, and few where there is no human presence. Even the Nordic subalpine birchwoods, or the tundra and taiga of Russia, which are often thought of as 'untouched', have been subject to some human impact. Other areas, such as the openfields or bocage landscapes have sometimes replaced former forest landscapes and are not only influenced by but are in fact the very result of centuries-old human landuse. The term 'cultural landscapes' characterises this distinctive interrelationship between nature and people and encompasses a group of mostly rural landscapes. By prevailing over the remaining natural types of land-cover, cultural landscapes play a significant role for the state of Europe's environment. The interrelationship between nature and people varies from place to place, due to differences in physical conditions, such as

topography, climate, geology, soils and biotic factors, and the type of human use or occupancy that can range from minimal to intensive. Landuse patterns have evolved around two significant factors: the type and accessibility of natural resources and the dynamics of demographic processes. Both factors are closely interlinked through a network of economic, ecological, social and cultural components. By acting as visual documents for the complex nature of these linkages, landscapes often represent aesthetic values in the perception of our environment.

The complexity of factors that contribute to the shaping of Europe's cultural landscapes is reflected in the diversity of values that are attached to them. Since these differences of perception set natural limits to any generalised approach of landscape evaluation, this chapter will only highlight some of the most obvious environmental aspects of cultural landscapes. Despite the many close links between this chapter and that on 'nature and wildlife' (Chapter 9), the values and problems of cultural landscapes are so much more based on economic and social aspects that they need to be set distinctively apart. However, conserving landscapes also helps protect the species and habitats within them and, taking action to protect species and habitats, contributes to safeguarding the richness and diversity of the landscape.

This chapter reviews the values which are attached to cultural landscapes, presents a typology of European landscapes, explains why landscapes are currently under stress, and describes landscape conservation measures underway at national and international levels. Given the existing discrepancy between the large variety of European landscapes and the lack of internationally harmonised approaches to describe and classify them, this chapter cannot be more than a first and incomplete attempt to tackle the subject. In particular, the large-scale map of European landscape types presented below must be considered as an indicative representation of the distribution of the main categories of landscape across Europe which has been developed for descriptive rather than analytical purposes. A prevailing theme throughout the chapter, however, is the importance of landscapes to the future of Europe's environment and their due place and importance in international efforts to safeguard the environment of this continent.

Values and functions of landscapes

Landscapes can be valued for a variety of reasons and they also provide a series of important functions. Five such values and functions are identified below: the role of landscapes in the sustainable use of natural resources, as wildlife habitats, providing economic benefits, scenery and open spaces, and possessing cultural heritage.

Sustainable use of natural resources

The character of many landscapes is often the cumulative result of human activities over many centuries. In parts of the Mediterranean region, for example, fire and overgrazing have led to the appearance of the *maquis*, now a very characteristic type of scrub and grassland ecosystem. However, one of the values which is to be found in some traditional landscapes is the presence of a sustainable pattern of landuse. The Iberian agro-silvo-pastoral landscapes of the montado and dehesa, the Scandinavian grazed deciduous woodlands, the puszta of Hungary and the sheep-grass downlands of Southern Britain are all examples of landuses which have also created environments that are rich in wildlife and nature. The sustainable methods of land management that are reflected in these landscapes have survived for many hundreds of years and therefore may provide information as to how similar areas could be better managed in future.

Wildlife habitats

Agri- and silvicultural landscapes can form transition zones between the controlled human environment of cities and the wilderness areas of flora and fauna. As such, landscapes have important functions for both humans and wildlife since they provide not only resources for human consumption but also important habitats for many plants and animals. The need to preserve endangered species and natural habitats or biotopes, or more generally maintain biological diversity (see Chapter 29), is thus intimately linked to the existence of natural and diverse landscapes. Given the absence of true wilderness areas in many parts of Europe, extensively used agri- and silvicultural landscapes have increasingly become the only remaining refuge for wildlife. Proper landscape management is therefore essential to give nature a place of its own.

Economic benefits

Cultural landscapes are the expression of past and present economic activity. In particular, farming is often the architect of landscapes. The importance of landscapes for recreation and tourism is much influenced by the very result of this impact: intensive, mono-cultured, large-scale agro-industrial complexes are clearly less attractive to people than natural and diverse landscapes. Across Europe, countryside with a varied pattern of fields, farms and woods attracts many millions of visitors from nearby cities and further afield. A wide range of activities are carried out in such areas (eg, bicycling, hiking, swimming, nature-excursions and use of off-road vehicles). The construction of purpose-built large-scale sport or recreational facilities (eg, golf courses, ski-ramps, amusement and safari parks) often involves major physical changes. For rural communities these developments can be economically very important. However, they can often also be the cause of problems from both direct and indirect impacts not only

affecting landscapes, but also through pollution, pressures they put on natural resources, urban development and congestion.

Open spaces and scenery

Landscapes and open spaces are often associated with harmony, stability and naturalness, and are generally viewed as pleasing contrasts to the quality of life in cities. 'Scenery' is a cultural/aesthetic expression of the land as it is seen and is primarily related to cultural landscapes. While human settlements represent a largely controlled environment, landscapes are appreciated for being the contrary: open, less controlled and seasonally changing. However, the role of human activities controlling, for example, the degree of openness or enclosure of a landscape, combined with the natural topography, is of primary importance in the creation of landscapes in Europe.

There is evidence that many people feel more comfortable in what they regard as the harmonious countryside of cultural landscapes rather than in the potentially threatening natural world of wilderness. Although difficult to quantify, the evidence indicates that people enjoy landscapes because they provide:

- scope for various forms of recreational and sporting activities;
- opportunities for inspiration and vision;
- a source of mental, physical and spiritual renewal; and
- a place of understanding and learning.

In this way, landscapes can provide many non-material benefits of life which all people seek. Sustainability is partly about people finding a new relationship with the natural world, and thus a greater valuation of the beauty and other intangible qualities of landscapes are relevant here.

Cultural heritage

European landscapes are important for the cultural elements that they contain; for example, ancient field systems of terraces, or vernacular architecture in farm buildings. This rich record of past landuse is to be found in many landscapes and has values which are comparable in many ways with the historic towns and cities of the continent (see also 'Cityscapes' discussed in Chapter 10). Other values exist in the association between landscape and art in its many forms. The celebration of landscape has achieved international recognition: in great paintings (such as those of the Flemish School, French impressionists or German Romanticism), in the music of Sibelius and in the poetry of Wordsworth. Of further significance is the part landscapes play in national and local consciousness. Since the landscapes found in Europe today are so often the outward expression of people's link to the land, they are often imbued with rich cultural associations – an expression of people's own identity.

Landscape as a European concern

Whereas the international dimension of natural landscapes has been readily recognised for many years through nature conservation, cultural landscapes have tended to be thought of almost exclusively as a national concern. Certainly the protection of landscapes, and the management of change within them, is primarily a matter for national and local action, but there is clearly a European scale of concern too. The particular richness and diversity of European landscapes, the visitors this attracts from within and outside Europe, combined with their many cultural associations, makes landscapes a matter of interest and concern to all. Ultimately the regional diversity and uniqueness of landscapes form collectively a common European heritage.

Table 8.1 (opposite)
The main characteristics of 30 European landscapes
Source: J Meeus

Since most landscapes are a by-product of human activities they are particularly exposed to change. This is an important characteristic of cultural landscapes which is not, *per se*, detrimental to the environment. Nevertheless, it is important to define an optimal potential where both the economic *and* ecological values of landscapes are balanced.

DESCRIPTION OF EUROPEAN LANDSCAPES

The approach

Identifying important landscape types of concern at European level needs to include numerous factors since, locally, landscapes possess many diverging connotations. This description focuses on the following factors: human activities, ecology, sustainability and scenery.

Landscapes of European concern need to represent:

- the main landforms that characterise the geological and climatic zones;
- areas where ecologically sound processes and sustainable use of natural resources are combined;
- areas which are extensively managed as semi-natural habitats for fauna and flora;
- regionally specific settlement patterns, ancient field systems, old trees, terraces and vernacular architecture; and
- examples of scenic quality and the visual characteristics of the continent.

Assessing such landscapes thus needs to take into account vegetational, geomorphological, agricultural, silvicultural and cultural influences.

A prime consideration for identifying cultural landscapes of European concern is the question of whether the ecological and anthropogenic elements of a landscape form a stable (or sustainable), functional and harmonious unit. To answer this question on an international level, and to set appropriate criteria to do so, needs a detailed study with a comprehensive and systematic approach. This was beyond the scope of the current assessment.

Figure 8.1
Holdridge diagram of vegetation types associated with different combinations of evaporation and precipitation
Source: WCMC/Meeus

The aim of this chapter is to provide a basis for an environmental appraisal of European landscapes from which future analyses of the priorities and concerns at European level can be developed. With the above considerations, an elementary typology of European landscapes was developed and applied across Europe at a scale of 1:6 000 000. The typology of Meeus et al (1990) is suitable here because it contains the appropriate elements and has been extended to incorporate additional landscape definitions to cover the whole of Europe. Attention is paid to the interaction of the following five factors.

Actual and potential vegetation

Noirfalise (1989) used soil and vegetation as the starting-point for making a distinction between different types of landscapes such as: bocage, openfields, agro-pastoral and agro-forestry landscapes, vineyards, Mediterranean landscapes and mountains. The map of the 'potential natural vegetation' of Europe (Map 9.1, Chapter 9, p 191), the landscape map of the USSR (Anon, 1988) and the Atlas of the Environment from the former Czechoslovakia (ČSAV, 1992) give an indication of potential natural vegetation based on interpretation of soil and climate. In the European part of the former USSR the following landscapes are distinguished: tundra; northern, middle and southern taiga; subtaiga; forest; mountain; steppe; and arid or desert landscapes.

Soil, landform and geomorphology

The soil map of the EC (CEC, 1985), the soil map of the world (FAO, 1982) and the descriptions of the major soils of the world (Driessen and Dudal, 1991) give an indication of the geography of soil types, and the formation and the use of land. The geomorphological map of the USSR (Anon, 1989) and the CORINE soil erosion maps of the southern part of Europe (CEC, 1992a) give information about the major processes of change in landscapes.

The agricultural use of the land

The landuse maps of Europe (Csati et al, 1980), CORINE land-cover (CEC, 1992b), the agricultural landuse map of the former USSR (Anon, 1991) and the agriculture types of Europe (Kostrowicki et al, 1984) give information concerning the distribution of crops, trees and forests grown in different regions of Europe.

Rural landscapes

Lebeau (1969) gives an extensive survey of the formation, morphology and geography of rural landscapes made and transformed by humans; Europe (excluding the former USSR) is divided into: bocage (hedgerows), openfields (three types: western, eastern and Mediterranean), linear villages in polders or forests, huertas, coltura promiscua and montados.

Forest landscapes

Forest landscapes have a history and scenery of their own. Bernes (1993) gives an overall view of the landscapes of the Nordic countries (Denmark, Iceland, Norway, Sweden and Finland). Different forests, tundra, mountains and agricultural landscapes are distinguished. Pisarenko (1993) divides the Russian forests into climatic zones: tundra, pre-tundra forest, northern taiga, middle taiga, southern taiga and a zone of broadleaved forests.

Vegetation, one of the criteria mentioned above, can be illustrated with the help of the so called 'Holdridge' diagram. The field in the simplified Holdridge diagram (Figure 8.1) represents different conditions of water availability controlled by the rates and levels of evaporation and precipitation. From this, ten types of potential natural vegetation are located in a field with the five climatic zones of Europe. Natural growing

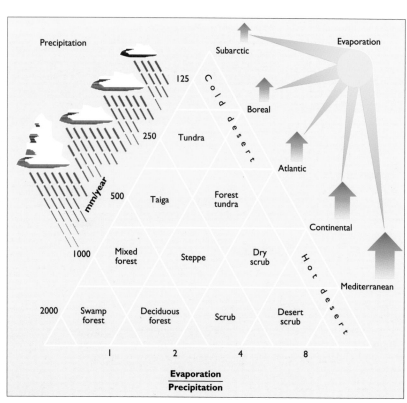

Landscape type	Landform	Vegetation	Character	Trends
Tundras				
1 Arctic tundra	Lowlands covered by ice and snow	Permafrost keeps vegetation scattered and low: moss/lichen	Desolate treeless wilderness	Wetland preservation
2 Forest tundra	Hills and lowlands bogs and fens	Dwarf shrubs (birch, alder) cover the valleys: heath everywhere	Wild, empty scattered forests	Fires and overgrazing
Taigas				
3 Boreal swamp	Peatland, mires, bogs and fens	Mixed, thin forests without production, only preservation	Inaccessible uncultivated wetlands	Drainage, extraction of peat
4 Northern taiga	Hills and plains, lowlands and lakes	Coniferous forest (pine, spruce) relics of grazed woodlands	Homogeneous forests	Clear felling, spruce becoming dominant
5 Middle taiga	Plains with sandy soils, moraines and mires	Mixed coniferous forests and some pastures or fodder crops	Some spaces within forests	
6 Southern taiga	Hills and plains of sand, silt and loamy soils	Mixed coniferous forests; pastures are in the minority	Semi-open forests	
7 Subtaiga	Outwash-plains and plateaus of loam/loess	Mixed broadleaved and coniferous forests and arable land	Silvi- and agricultural domination	Drainage and deforestation
Uplands				
8 Northern highlands	Hills and mountains, lakes, bogs and fens	Heath, grassland, rocks and relics of overgrazed woods	Desolate, rough and very open	Afforestation
9 Mountains	High mountains, glaciers, steep slopes and valleys	Moss, heath, grass and forest on slopes; intensive crops in valleys	Wild, rough, enclosed versus cultivated, open	Abandoning, afforestation and tourism
Bocages				
10 Atlantic bocage	Gentle slopes and plateaus of loam on rocks	Pastures and arable land surrounded by hedges, walls or trees	Enclosed, heterogeneous and cultivated	Plot enlargement and removing of hedges
11 Semi-bocage	Hills and middle mountains, wet conditions	Extensive grassland and crops, mixed forests and hedges	Hybrid open–enclosed	Abandoning and afforestation
12 Mediterranean semi-bocage	Hills and middle mountains, dry conditions	Extensive grass, arable and permanent crops, walls/forests	Relatively open	Extensification
Openfields				
13 Atlantic openfields	Loamy and clayey soils on undulating plains	Intensive arable land; trees only in valleys	Large-scale openness; monoculture	Intensification and set-aside
14 Continental openfields	Loess and loam on flat and hilly land	Mixed arable, grass and permanent crops; forests on hilltops	Diverse in scale	Diversification
15 Aquitaine openfields	Outwash plains and slopes of lime, loam, loess	Arable land on plateaus; forests on slopes; horticulture in valleys	Open and intensively cultivated	
16 Former openfields	Undulating plains with loamy and clayey soils	Land suited for arable crops, used for cereal, root crops, grass	Intensively cultivated; large-scale openness	Removing of trees on plots
17 Central collective openfields	Undulating plains with loess and loamy soils	Arable land without any other vegetation; pastures on lowlands	Large-scale, open and homogeneous	Water and wind erosion
18 Eastern collective openfields	Flat to undulating plains covered with 'black earth'	Treeless arable land and grass on moist lowlands	Extremely open, large and dry	
19 Mediterranean open land	Hills, plateaus, valleys, water-limited conditions	Forests and scrub v. cultivation: extensive crops and transhumance	Contrasting patterns between hills and valleys	In- and extensification, abandoning
Steppic and arid landscapes				
20 Puszta	Salt-affected soils on the Hungarian plain	Grassland and arable land	Treeless open space; extensive breeding	Salinisation, water and wind erosion
21 Steppe	Plains with brown earth, valleys and saltmarshes	Grassland and arable land, extensively cultivated	Treeless, dry, endless, windy, extremely open	Overgrazing, salinisation
22 Semi-desert	Lowland plains and saltmarshes	Grass, ephemeral and halophytic plants	Salt, dry, open and extensively cultivated	Changing levels of groundwater and sea water
23 Sandy desert	Mobile dunes, dry rivers and shifting sands	Absence of vegetation, only some ephemeral plants, sedges	Uncultivated	Pastoral husbandry protection
Regional landscapes				
24 Kampen	Undulating plains with brooks and sandy soils	Mixed crops, grassland; forests, heath, swamps and trees	Enclosed fields, mosaic, small-scale patchwork	Increase of scale and intensification
25 Poland's strip fields	Small elongated field systems	Mixed crops, horticulture, orchards and forests	Labour-intensive, small-scale diverse	
26 Coltura promiscua	Fertile valleys; loamy soils, remnants of traditional use	Three layers of permanent crops, cereals, horticulture, fodder crop	Heterogeneous and small-scale diverse	Homogenisation by in/extensification
27 Dehesa/montado	Poor, dry and stony soils on erosive, gentle slopes	Open evergreen forest (oak, olive), grazed and cultivated	Agro-silvo-pastoral parkland	Degradation; growth of shrubs
Artificial landscapes				
28 Polder	Estuaries on North Sea coast below sea level (clay, peat)	Intensive arable and grassland; trees along roads, canals, dikes	Flat, open, fertile, artificial and uniform	Intensification, set-aside
29 Delta (artificial forms)	Estuaries in coastal lowlands; deltas of large rivers	Intensive arable, fodder and permanent crops on irrigated fields	Intensive, flat, open, fertile and uniform	Salinisation and intensification
30 Huerta	Irrigated, fertile valleys on Mediterranean coast	Intensive horticulture and permanent crops (eg, fruits)	Irrigation, terraces, orchards	Expansion

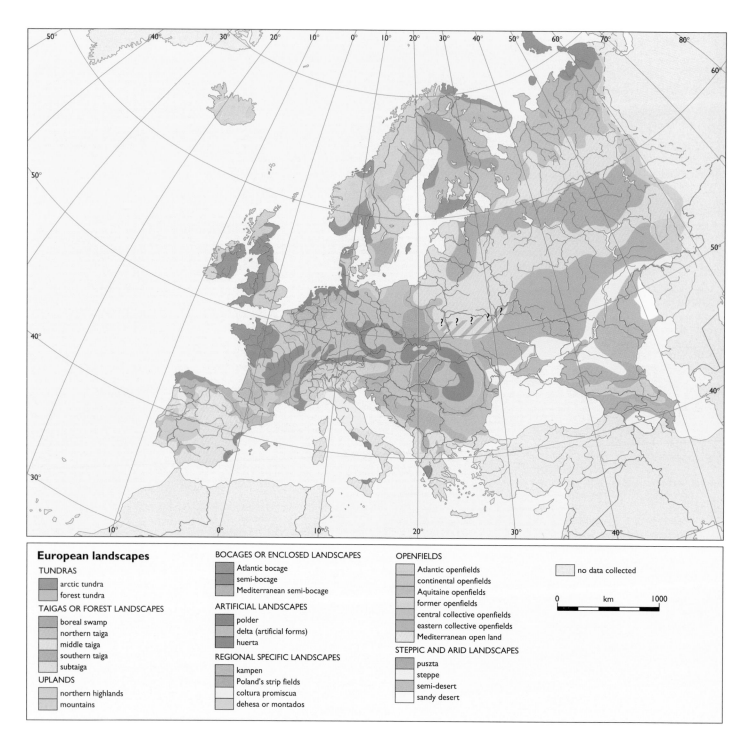

European landscapes

TUNDRAS
- arctic tundra
- forest tundra

TAIGAS OR FOREST LANDSCAPES
- boreal swamp
- northern taiga
- middle taiga
- southern taiga
- subtaiga

UPLANDS
- northern highlands
- mountains

BOCAGES OR ENCLOSED LANDSCAPES
- Atlantic bocage
- semi-bocage
- Mediterranean semi-bocage

ARTIFICIAL LANDSCAPES
- polder
- delta (artificial forms)
- huerta

REGIONAL SPECIFIC LANDSCAPES
- kampen
- Poland's strip fields
- coltura promiscua
- dehesa or montados

OPENFIELDS
- Atlantic openfields
- continental openfields
- Aquitaine openfields
- former openfields
- central collective openfields
- eastern collective openfields
- Mediterranean open land

STEPPIC AND ARID LANDSCAPES
- puszta
- steppe
- semi-desert
- sandy desert

- no data collected

0   km   1000

Map 8.1
European landscapes
Source: Compiled from various sources by Meeus

conditions will follow these climatological and biogeographic zones, but not all the combinations implied by this diagram may be found as such 'in the field'.

Combining climate and vegetation with landscape scenery (the openness or degree of closure of the scenery) allows the distinguishing characteristics of different landscapes in Europe to be defined. On this basis, 30 landscape types of European interest can be identified. Their main characteristics are listed in Table 8.1.

Comparison of 30 European landscapes

The geographical distribution of the 30 European landscapes identified in Table 8.1 are illustrated in Map 8.1. The landscapes vary from open (field) landscapes such as deserts, openfields and arctic tundra, to completely enclosed (forest) landscapes, such as bocage (hedgerows), forests, taiga and forest tundra. In total, eight broad categories are defined:

- tundras;
- taigas or forest landscapes;
- uplands;
- bocages or enclosed landscapes;
- openfields;
- steppes and arid landscapes;
- regional landscapes; and
- artificial landscapes.

Depending on the need, the application, the scale and the amount of detail, the 30 European landscape types identified could be divided into many subtypes – not attempted here. At this European level forests are only identified as landscape units in Eastern Europe. Indeed, some regional and artificial landscapes are too small or too unevenly distributed to be illustrated at the scale of the map; coastal and river linear landscapes are not included here either.

Box 8A Tundra and taiga

Located in the arctic and boreal zones, these landscapes resemble potential natural conditions. The fragile ecosystems of the arctic tundra are vulnerable to physical impacts which can produce irreversible changes; protection of these areas is therefore much warranted. The tundra and taiga are sparsely vegetated and open. Permafrost, low temperatures and water shortages make the vegetation period short and intense. While boreal swamps have been much drained during the last century, the subtaiga is a semi-open landscape with a combination of arable crops and forests. The northern taiga is illustrated in further detail below.

Northern taiga

The northern taiga is a semi-open coniferous forest in the hills and plains of the Nordic and the northern Russian region. The thickest forests are to be found in the river valleys and on the river terraces. The dominant trees are spruce, pine and birch, which give the northern taiga its appearance of uniformity and homogeneity (National Atlas of Sweden, 1990–92; Anon, 1988). Lichens, mosses and berry-bearing shrubs cover the ground in between a fine-meshed network of forest roads. Severe frost makes the growing season short (no more than 100 days), while the mean annual temperature is low (+1° C). There is an excess of water during the year, but because of a low water-retaining capacity of the sandy *podzols*, there can be severe drought problems during the growing season. Only some small-scale forestry and extensive grazing is possible. In the Nordic regions, forest fires are successfully controlled. The forest trees are cut mainly to provide the local population with wood. Reindeer pastures can be found as well. In Finland and Sweden in particular, many old wooded pastures and extensively used stands of birch and other deciduous trees have had to make way for conifer plantations (Bernes, 1993). Currently, the northern margin of the northern taiga is receding in a southerly direction, a shift which is being accompanied by an advance of the tundra. While this is most probably due to the establishment of industrial settlements and the construction of roads and railways (Symons, 1990), climatic changes might be another contributing factor.

Northern taiga, Hovfjallet, central Sweden
Source: redrawn from J Meeus

Box 8B Enclosed landscapes with hedges/bocage

Human influence controls the degree of enclosure or openness of the landscape. The landscapes in Europe vary from enclosed (bocages or hedges) to open landscapes (openfields). In the western parts of Europe, Atlantic and continental climatic influences, a high population density and a dispersed settlement pattern render the almost entirely human-made landscapes subject to swift changes in time and space.

In the French language the word 'bocage' refers both to the hedge itself and to a landscape consisting of hedges. The 'bocage' landscape of Brittany, central England and Ireland, with some 'offshoots' in Scandinavia, is the classic example of an enclosed agricultural landscape with a long history. Mixed cultivation has been pursued for many generations, with both agriculture and cattle farming side by side. Bocage landscapes usually have a slightly rolling landform, and are found mainly in maritime climates. Soil texture is usually sandy, loamy to clayey. Being a small-scale, enclosed landscape, the bocage offers much variation in biotopes, with habitats for birds, small mammals, amphibians, reptiles and butterflies. There is a close infrastructure network, and small plots, with some farmhouses surrounded by hedges or low walls. Plot fragmentation makes the majority of bocage landscapes increasingly less suitable for arable farming, contributing to the rise in popularity of cattle farming in these areas. In Ireland, for example, the bocage landscape with its patterns of hedges suits to the intensification of cattle farming.

Atlantic bocage, Brittany, France
Source: redrawn from J Meeus

There is a remarkable difference between the landscape types of most parts of Western and Eastern Europe. Natural soil and climate conditions are more extreme in the East but overall are also more homogeneous than in the West. Population is less dense and the landscape types follow climate zones more closely. In the Nordic countries and Russia, forest landscapes, or taiga, dominate, and human influence appears less pronounced. By way of contrast, in Central Europe, the so called 'collective openfields' have been completely cleared of trees and shrubs for intensive

Box 8C Openfields

The extremely open landscape of wide undulating plains with regular plots of arable land, extending from France to Germany, is called 'openfield'. Several subtypes of openfield are recognised: Atlantic, continental, Aquitaine, former and collective openfields. The complex and often pendulum-like changes which have led to the formation of landscapes found in Europe today, particularly the tension between open and enclosed forms, are well illustrated by the example of the former openfields in Denmark.

Former openfields in Denmark

In the sixteenth and seventeenth centuries the original forests of central Jutland (Denmark) were overutilised. Continuous felling, slash-and-burn agriculture, and overgrazing gradually turned the mixed oak woodland into an open landscape with local arable openfields, enclosed pastures, some left-over wooded land, heaths, bogs and fens. In the eighteenth and nineteenth centuries, heaths were ploughed up to feed the growing population, bogs and fens were drained, and roads and farmyards were built in uneven and scattered patterns. Since the land was suitable both for cattle breeding and for arable crops, the landscape was sometimes dominated by meadows and sometimes by cropland. After 1945, central Jutland began losing a large number of hedges; farms grew in size, plots extended, wet grasslands were drained. As a direct result, the landscape came to be characterised by greater uniformity over larger areas. Large numbers of pigs are now reared, making it necessary, for example, to grow more barley. To enable this, permanent grasslands have been ploughed up, while arable land on poor soils is abandoned, reforested or becoming covered in weed (Bernes, 1993). This has resulted in a 'hybrid', semi-open landscape, termed 'former openfields'. Nowadays, wind and water erosion threatens the soil and landscape. To prevent this erosion, each year many kilometres of new hedges are planted and thousands of hectares of evergreen crops are sown. These new hedges and crops are unlikely to restore the former rich wildlife but may enhance the sustainability of the landscape.

Former openfields, Slagelse, Sjælland, eastern Denmark
Source: redrawn from J Meeus

Box 8D Steppe and arid landscapes

East of the well-known Hungarian puszta landscape (see below), an east–west oriented belt of steppe/grasslands stretches from the central Russian uplands to the Caspian and Black Seas. These 'steppes' are vast treeless plains with low rainfall. In the lowlands, northwest of the Caspian Sea, the steppes form a transitional zone with deserts – areas known as semi-desert (see below) (Anon, 1988). Only further to the east of the lower Volga, on the continental divide with Asia (eg, in Kazakhstan), can true desert be found. Arid landscapes also exist in the western extremities of Europe, in the Iberian Peninsula and the Canary Islands (see Chapter 9).

Puszta

On the loess plateaus of the Hungarian plain, climate and vegetation are of a steppe nature. The landscape is open with grassy vegetation, and trees remain only in valleys. Farm plots are large and concentrated, and some are collective (Kostrowicki et al, 1984; Csati et al, 1980). Because of dryness during the summer, the fertile *chernozems* and *phaeozems* (reddish prairie soils) are used mainly for livestock breeding, but arable crops may also be grown. Highly mechanised and productive agricultural systems are separated from the less intensive ones. The most serious hazards are wind and water erosion.

Semi-desert

The sparse, open vegetation of steppe grasses is mixed with drought-resistant dwarf shrubs and ephemeral plants, which grow during the extremely short rainy season. Especially in depressions with *solonetz* and *solonchaks* soils, saltmarsh and halophytic vegetation can be found. Aridity, dryness and deep water tables are responsible for salinisation and alkalisation of the soil. Apart from the salinated *solonchaks* there are the *calcisols*, with their topsoil concentrations of calcium carbonate. The organic matter content of the surface of *calcisols* is low because of a sparse vegetation and rapid decomposition of vegetable debris (Driessen and Dudal, 1991). Sheep and cattle herds may graze in spring and autumn, but only where grazing remains extensive. The increase of anthropogenic influence and the lowering of the water table are the factors responsible for desertification of the meadow-steppe. The degree of salinisation and swamping but also the length of the flood periods are indicators of the productivity of the landscape (Zalibikov, 1992).

Semi-desert, Prikaspiyskaya, Russia
Source: redrawn from J Meeus

Box 8E Regional landscapes

In all parts of Europe, in addition to the main landscape types, specific regional landscapes have emerged. Examples are the open forest landscapes of the Iberian Peninsula (montado or dehesa), the coltura promiscua in Italy and Portugal, and Poland's strip fields, all of which are described below. The depiction of these regional landscapes on the map is particularly dependent on their size: only those areas greater than 100 by 100 km are included on Map 8.1.

Kampen

Found in Flanders (Belgium), the southern and eastern parts of The Netherlands, in North-Rhine-Westphalia (Germany) and also in Les Landes (France), the 'kampen' landscapes are generally enclosed, with a patchwork layout of woods, heath, swamps, mixed crops, scattered farmsteads and roads (Wijermans and Meeus, 1991). There is a great diversity of trees on plots and the poor sandy soils (generally *podzols*) are cut by stream valleys. This rich diversity of the kampen landscape makes it highly flexible for growing crops. There are some interesting ecological differences between cultivated land on the one hand and heath and wet pasture land on the other. The poorest soils are covered with woods. Intensification of agriculture, abundant use of fertilisers and manure and fragmentation of wildlife habitats have almost eliminated the contrast between open areas and enclosed farmland in the Dutch kampen. The quality of the kampen in this area is under pressure because of the vulnerability of the ecological system. Landscapes are split up, vegetation is being removed and there is a threat of soils drying out and groundwater becoming polluted. Increasing density of livestock (cows and pigs) results in large quantities of manure for disposal. Use of fertilisers and pesticides is abundant. The carrying capacity of the sandy soils is low and groundwater pollution has become a particular environmental problem of the Dutch kampen landscape. Drainage changes wet grasslands, heath and marshlands, depriving them of their natural character (Atlas van Nederland, 1984–90).

Coltura promiscua

The coltura promiscua is an enclosed landscape, characterised by intensive traditional mixed farming. The landscape displays a classic 'upright' pattern of trees, bushes and ground cover. Examples can be found in central Italy and in valleys of the Iberian Peninsula. In the foothills of the Apennines, coltura promiscua is combined with terraces (Lebeau, 1969), which both retain the scarce rainwater and keep the thin layer of soil in place. Scattered farmsteads can be found, and the population is concentrated in villages. In the expanse of the Mediterranean open land in Spain, there are several small areas with a mosaic of traditional landuse forms, but these areas are too small to be shown on Map 8.1. The traditional mixed cultivation of the coltura promiscua in central Italy and northern Portugal is strongly affected by the present tendencies in agricultural management. These traditional agricultural landscapes are therefore probably destined to disappear when the choice is made between intensification or marginalisation (Meeus et al, 1990).

Poland's strip fields

In the eastern part of Poland, traditional labour-intensive agriculture outlived the period of collectivisation. Since the 1980s, private agriculture on extremely small plots developed to a modern market-oriented agriculture and part-time farming (Kostrowicki et al, 1984). This landscape has a long historical connotation. The parcels are split up into a pattern of strips, buildings are concentrated in compact villages and there is a small-scale network of rural roads (Verhoeve and Vervloet, 1992). Mixed crops, orchards, horticulture and the presence of many people and horses on the small plots characterise this landscape.

Kampen in Kempenland, Flanders, Belgium
Source: redrawn from J Meeus

Coltura promiscua, northern Portugal
Source: redrawn from J Meeus

cultivation while steep slopes and valleys have been reafforested. In western parts of Europe, under continental or Atlantic climatic influences, a greater population density and a more dispersed settlement pattern has resulted in a considerable area of the land being characterised by small-scale landscapes as well as large-scale openfields. Thus, in general, in the North and East of Europe, landscapes are generally more continuous and more closely resemble potential natural conditions, while to the West, landscapes are fragmented, change more often over shorter distances and

reflect more strongly the impact of human activities.

The distribution of landforms following ground relief is a major factor differentiating landscapes. Uplands, comprising mountain ranges and highlands, are characterised by their relief, active processes of weathering and erosion and the variability of climate conditioned by both elevation and geology. These form largely natural landscapes and are discussed further as mountain ecosystems in Chapter 9.

In Southern Europe, under a Mediterranean climate, the landscapes reflect the ingenuity and past struggle of people

Box 8F Agriculture in dehesa/montado landscapes (Portugal, Spain)

The montados of southern Portugal and the dehesas of southwestern Spain are two very similar types of traditional agro-silvo-pastoral landscapes. Characterised by an unsystematic dispersion and density of cork and holm oaks (*Quercus suber, Quercus rotundifolia, Quercus ilex,* and *Quercus pyrenaica*), these montados represent some of the few 'small-scale' landscapes of the Mediterranean. In what was originally *maquis* ('charneca'), the trees became important as a source of cork and also as food (masts and acorns) for pigs and later for goats and cattle. Owing to a rotational system, goats are no longer than six to eight years in the same area before being moved to adjacent locations, while the soil is cultivated with cereal for two years. Subsequent grazing prevents the growth of a shrub-type vegetation. Besides being a landuse system which is based on sustainable principles and without the need for additional energy-input, the montado provides varied habitats for a high diversity of fauna and flora, including a number of rare and threatened species.

Three independent processes have, however, resulted in a substantial degeneration of the montados and dehesas in Portugal and Spain. One of them has been triggered by the 'wheat campaign' of the 1930s. Because of the emphasis on crop cultivation, the trees on the montados were reduced to a minimum, and mechanisation plus fertilisation was increased. While the oaks were severely affected by mechanised work and use of chemicals, the poor soils did not allow crop yields to be economically successful. Consequently, large portions of the land were abandoned in favour of establishing large-scale monoculture of fast-growing eucalyptus, designated for the cellulose industry. This development led to irreversible transformations and deteriorations of the landscape (Pinto Correia, 1991), accompanied by a loss of biological diversity that has affected regional socio-economic structures. Today, there is an increasing awareness that traditional forms of land management need to be re-established by increasing goat production, improving pastures and protecting and planting more cork oaks. The dehesa/montado is the typical candidate for a Red Book of threatened landscapes (Naveh, 1993). See also Chapter 23 (Box 23B).

Cork oaks (Quercus suber) in Portugal
Source: D Wascher

over many centuries to earn a living from the land in poor natural growing conditions; terracing, irrigation and low-input agriculture have been the methods used, altering the landscape accordingly.

In southeastern Europe, low average rainfall (below 400 mm/year) and dry, warm continental climate favour treeless steppes, semi-deserts and sand deserts with 'sagebrush' vegetation and low landuse intensity.

In all parts of Europe, in addition to the main landscape types, specific regional landscapes have developed (or remained) such as the open forest landscapes on the Iberian Peninsula (the montados or dehesas), the coltura promiscua in Italy and Portugal, Poland's strip fields, and the polders along the North Sea coast. These and other typical regional landscapes have their counterparts in diverse and traditional terraced landscapes, which do not have a fixed place 'in the field' and thus cannot be located on a map of Europe.

Other than polders, some forms of deltas and the huerta, there are some types of 'artificial landscapes' which are also not represented on the map. By corresponding to intensive agro-industrial landscapes, including such features as modern topographic terracing (monocultures, eg, vineyards), extensive glasshouses and concentrated animal husbandry, these artificial landscapes are relatively common and widespread components of many landscape types. Although artificial landscapes may have become characteristic for the type and intensity of a region's productivity, they are often uniform, unspecific and disintegrated in environmental and aesthetic qualities. Consequently, these are not considered further.

Some further details and examples of local characteristics, development and trends relating to the identified European landscape types are described in Boxes 8A to 8E

LANDSCAPES UNDER STRESS

The close interaction between people and nature is reflected in the way in which each European landscape has been shaped in some way by human activities. When the character of this interaction changes, the landscape is inevitably altered. Agriculture can be specially singled out as having a significant effect on landscapes, but there are a number of forces at work in Europe today which are disrupting the relationships of the past:

- agricultural intensification;
- agricultural abandonment;
- urban expansion;
- standardisation of building materials, designs, etc;
- infrastructure development, especially roads;
- tourism and recreation;
- mining/landfills; and
- loss of wildlife habitats.

Broader environmental problems, such as air pollution or the overexploitation of groundwater, may also have an impact on landscapes through, for example, the changes they bring about by damaging woods and trees (see, eg, Box 34A) or loss of wetlands.

An international European-wide 'landscape inventory' does not exist which could allow a systematic approach for assessing the dimension, rate and trends of landscape changes. In the absence of a comprehensive picture, the impact on landscapes in Europe from the forces mentioned above can only be briefly described and illustrated with the help of case studies. Most of the broader environmental problems associated with these changes are described in other chapters of this report, cross-referenced below.

Box 8G Urban expansion and polder landscapes (The Netherlands)

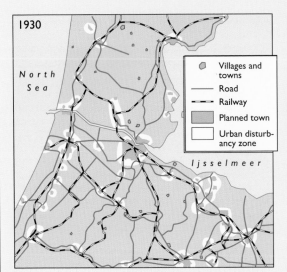

1930

North
Sea

Ijsselmeer

	Villages and towns
	Road
	Railway
	Planned town
	Urban disturbancy zone

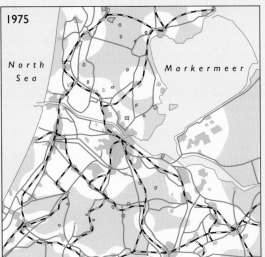

1975

North
Sea

Markermeer

Map 8.2 The urbanisation of the Amsterdam region between 1930 and 1975
Source: de Jong and Wiggers, 1982
Note: Urban disturbancy zone arbitarily set at 1.5 km in 1930 and 4 km in 1995 to reflect the increased mobility of people

land were often wetter and less intensively used. These old polders, with their peaty soils, are characterised by a backward agricultural structure and a rich natural potential. Along the coast, rivers and estuaries, and in the low-moorlands, there are the wetlands which are particularly important to nature. Wet grasslands provide habitats for wild species of plants and animals, such as meadow- and migratory birds. These areas are ideal for nature development and also for water recreation. Where the turf has been partially cut, the resulting lakes have become new nature areas. Through the relatively closed structure of willow and alder woodlands, these nature-like landscapes are attractive for small-scale water sports.

Urbanisation continues to put pressure on the traditional extensive landuse in the old polders (Map 8.2). The strong growth of population during this century (rising from 6 to 15 million between 1900 and 1990), combined with an increase in the use of available space per person, has resulted in a growth in urban areas from around 80 000 ha in 1900 to about 450 000 ha in 1980. Currently the annual rate of agricultural loss is 13 000 ha per year. This process affects mainly polder areas since the majority of the Dutch population lives in the western part of the country. If this trend continues, 16 per cent of The Netherlands (600 000 ha) will be covered by roads and urban areas by the year 2000 (CBS, 1992). Due to the increased mobility of people, there is also an increase in the zones of disturbance around the towns and villages of the area.

Together with the wet 'hay-lands', once common birds such as the corncrake (*Crex crex*), the Ballion's crake (*Porzana pusilla*) and white stork (*Ciconia ciconia*), or mammals such as the otter, have disappeared from many polder areas (de Jong and Wiggers, 1982). Despite massive government initiatives to protect wetlands, the habitats of meadow- and migratory birds are declining (Bink et al, 1994; RIVM, 1992).

Polder landscapes near Utrecht, Holland
Source: D Wascher

The flat and open 'polder' landscapes in the lowlands of the western parts of The Netherlands, originally estuaries of the rivers Rhine, Meuse and Scheldt, are now used by intensive agriculture, urbanisation and recreation.

'Old polders' in low fenland areas are perhaps the most characteristic landscapes of Holland. The broad landscape with numerous ditches and regular subdivisions and ribbon development are 'typically Dutch'. More than four centuries ago wetlands and waterlogged peat soils were drained. The results are long stretched parcelling of land in peat-reclaimed areas. The more remote parts of the reclaimed parcels of

Agricultural intensification

The modernisation of agriculture brings about changes in the landscape. The larger-scale impacts of collectivisation are known to result in an opening up of the landscape. In bocage landscapes, larger machinery favours bigger fields, demanding the removal of hedgerows. Everywhere, modern machinery needs bigger roads and buildings from which to operate. When monoculture replaces mixed farming regimes, the

landscape becomes more uniform and sterile. When land is drained for agriculture, or pockets of woodland are felled, the landscape is greatly changed (Park, 1989). On the other hand, irrigation can also have considerable impacts on a landscape's naturalness. In addition to these technology and production-related changes, there are also those resulting from changes in land ownership or management responsibility. Collectivisation in the East and land consolidation in many Western European countries, for example, have led to major changes in landuse

Box 8H Fragmentation of landscapes in Germany

Based on data of the location of different categories of motorways and railways, the German federal agency for nature conservation and landscape ecology studied the changes between 1977 and 1987 in the number of undissected areas larger than 100 km^2 (containing no tracks or roads or only those with low traffic frequency) considered as a resource for low-intensity recreation (nature-walks, relaxation, etc) (Lassen, 1990). The exercise showed that there had been an 18 per cent decline in such areas in this period in West Germany. In 1977 there were 349 undissected areas (56 184 km^2), while in 1987 these had dropped to 296, covering 45 876 km^2. Map 8.3 illustrates the findings for a section of southern Germany, where the decrease for the Federal State of Bavaria was 15 per cent (4198 km^2) and for the Federal State of Baden-Württemberg 20 per cent (1098 km^2).

A comparison of these findings with the distribution of landscape protection areas in both federal states shows that Baden-Württemberg's protected areas are located in regions where no undissected areas larger than 100 km^2 exist. Bavaria, on the other hand, has not only more such undissected areas, but also a higher congruence between these areas and landscape protection zones. Such an analysis could be one method of assessing the quality of protected areas for recreational purposes. A follow-up study, however, has demonstrated the difficulties of such analyses and the need for standard assessment criteria (Netz, 1990). Using criteria such as scenic value of a landscape and the quality of hiking-routes, a deeper evaluation was made of the 296 remaining undissected areas identified in 1987. This demonstrated that more than 75 per cent of Baden-Württemberg's undissected areas have relatively high qualities for recreation compared with only 50 per cent for Bavaria.

Map 8.3
Undissected areas (larger than 100 km^2) of low traffic frequency as a resource for low-intensity recreation, 1977 and 1987, in southern Germany.
Source: Federal agency for nature conservation, Bonn

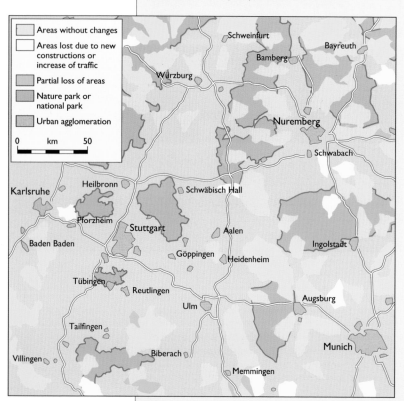

Areas without changes

Areas lost due to new constructions or increase of traffic

Partial loss of areas

Nature park or national park

Urban agglomeration

0   km   50

practices, leading in turn to different landscapes. The threats to the dehesa/montado of Portugal and Spain described in Box 8F illustrate how these pressures from agricultural intensification can affect a landscape (see also Chapter 22).

Agricultural abandonment

The opposite process has also been at work. Land that has been farmed for many years has been abandoned, or put to other uses, such as large-scale afforestation in several of the Atlantic seaboard countries. In some cases the effect on the landscape has been relatively slow and subtle, with the decline of farm holdings in marginal land and the encroachment of trees; many Alpine and other mountain areas have thus been affected, with the consequent disappearance of the former farmed landscape. But there are parts of Europe, and especially in poorer areas of the Mediterranean, where entire rural villages have been predominantly abandoned as the traditional farm economy has largely broken down (WRR, 1992).

Urban expansion and uncontrolled settlement

When towns expand, they incorporate the countryside and entirely destroy the landscape. The continually increasing transport network divides up the land (see below) and encourages further development. Beyond the physical expansion of towns, there is also the more subtle impact of an expanding urban presence felt within the countryside from increased mobility and, for example, the presence of second-home owners. Such developments can lead to rural communities having less influence in how the surrounding countryside is managed. The pressure from urban expansion is illustrated in Box 8G for the polder areas of The Netherlands.

Standardisation of building materials and design

The built features in the landscapes of Europe, such as farm buildings and walls, were traditionally constructed with materials such as local stone, timber and thatch. Styles of building evolved appropriately, and were immensely varied. Today, widely available modern standardised materials are displacing traditional ones (corrugated iron in place of tiles, or concrete breeze-blocks in place of local stone or brick, for example) due to cost-effectiveness or loss of traditional craft skills such as dry-stone walling or thatching. As a result, new buildings in different parts of Europe have increasingly come to resemble one another. The local vernacular is often lost along with local skills and sources of employment.

Infrastructure development and transport

In recent years the expansion and modernisation of national and even international infrastructure, such as roads, canals and power lines, have affected landscapes across Europe. Often this leads to ecological disruption and sometimes disrupts the local economy as well. Major roads, in particular, are alien elements in many landscapes; however, cumulatively, fewer improvements in the road network can also have a negative effect on the landscape. The way in which landscapes are becoming increasingly fragmented is illustrated with an example from Germany (Box 8H). See also Chapter 21.

Box 8 I Effects of tourism on Alpine landscapes

Mountain tourism is a leading recreational activity worldwide. The Alpine nations of Europe (Austria, France, Germany, Italy, Liechtenstein, Slovenia, Switzerland) share a yearly product of 52 billion dollars (25 per cent of the world production in tourism), and 70 per cent of the 12 million people in the Alps live directly or indirectly from the revenues of the tourist industry (Partsch and Zaunberger, 1990). Tourism thus forms an important base of the Alpine economy. However, tourism activities are also accompanied by a number of serious environmental repercussions affecting the very quality of the recreational values being exploited, having impacts on the state of Alpine landscapes and mountain ecosystems (see Chapter 9). On the whole, landscape damage from tourism derives from the construction of facilities required for sports activities, increased traffic and its associated infrastructure, as well as from the indirect effect of agricultural abandonment (see below).

Sports activities and facilities
According to official estimates, about 3000 cable-lifts move approximately 1.3 million individuals each year in the Alps (Weizäcker, 1990). The development of mountain climbing and skiing into mass-tourism activities puts tremendous pressures on Alpine resources. Heavy damage is done to soils and vegetation. Due to physical adaptations of the regional topography by scraping and earth construction works, the morphological structure of soils is altered, making them extremely vulnerable to (water) erosion. The resulting artificial landscapes are mono-functional (skiing) and lack natural vegetation cover such as various forms of snow-patch communities, acidophilous and calciphilous grasslands, and scree communities. During the summer months, the runoff from the hillsides, which often have heavily compressed soils and reduced vegetation cover, can be extremely high, increasing risks of flood and erosion. The growing number of cable-lift facilities and the necessary deforestation along the cable-tracks have similar effects and have become dominating features of the mountain landscape. The operation of snow cannons to provide artificial snow can also affect seasonal changes of the natural vegetation and is accompanied by high water consumption (Partsch and Zaunberger, 1990).

Traffic and infrastructure
The increase in tourist activities is enhanced by the infrastructure network of most Alpine regions. In fact, the accessibility of mountains at certain elevations is often the pre-requisite for tourist developments. Noise and air pollution follow an increasingly dense network of roads, serving mainly individual car traffic, into once-remote regions. Parking areas, petrol and repair stations additionally impact the landscape.

Agricultural abandonment
Two-thirds of Alpine landscapes are a result of agricultural landuse practices. Hence, mountain farmers have an important function in maintaining and caring for these landscapes. However, as a consequence of being forced into competition with the growing tourist industry, mountain farming nowadays concentrates substantially on the intensification and mechanisation of production. This has led to an abandonment of remote and less accessible Alpine regions where a high degree of manual work is required (MAB, 1984). The result of these factors is a loss of ecological stability and of traditional habitats for especially adapted flora and fauna.

In 1992, after more than 40 years of work, the Commission for the Protection of the Alps (CIPRA) succeeded in presenting a framework for the implementation of an Alpine Convention (CIPRA, 1993). CIPRA lists 15 possible actions to reduce the impacts of tourism on the Alpine environment, and comments critically on the official protocols of the convention.

Stilfser Joch, Italy
Source: D Wascher

Tourism and recreation

Tourist and recreational qualities of landscapes directly result, in general, from their scenic attractiveness or distinctiveness. The resulting effects of increased income and employment in areas which have often few other resources can provide the means for the maintenance and enhancement of these very landscape values. Unfortunately, the scale of some tourist developments, and the social and physical disruption associated with poorly planned tourism, is often completely alien to the landscape and wholly out of scale with the needs of the local economy. The problems of Alpine areas, which attract large and increasing numbers of tourists, and the effects on Alpine landscapes are discussed Box 8I. See also Chapter 25.

Mining, landfill and military areas

In many regions of Europe, mining, landfill and military activities have – sometimes dramatically – changed the topography and landuse patterns of landscapes. The shapes of mountains and lakes have been altered, river courses have been redirected, new visual perspectives have been formed by large and deep excavations, and very often the former vegetation has been replaced by a different subsequent landuse.

Some of the most obvious impacts result from large-scale coal and iron mining (see Box 8J and Chapter 20). Examples can be found in Europe's industrial regions, such as the Donbass coalfields (Ukraine), the Black Triangle Region (Czech Republic, Germany, Poland) and the Ruhrgebiet (Germany) as well as the coal and steel region around

Box 8J Mining in openfield/forest landscapes (Czech Republic)

A first phase of intense, deep mining of gold and silver in the Central European region during the thirteenth century was followed by a second boom with copper mining in Slovakia and silver/tin extractions in Bohemia. After the exploitation of these minerals, it was the opening of large coal mines and hundreds of small iron ore mines that gave Bohemia its reputation for being the blacksmith's shop of the Austro-Hungarian Empire. Over time, large parts of the country's forests disappeared, shaping a new type of open landscape. The surface mining (depths sometimes exceeding 300 m) and burning of brown coal in northern Bohemia are associated with a number of environmental problems affecting the landscape:

- changes in the configuration of the terrain and the extinction of centuries-old settlement structures located in the areas of gigantic open coal mines. In the foothills of the ore mountains an area of several hundred square

kilometres is affected in this way and includes devastated agricultural and other lands;
- degradation of agricultural soils, higher washout of heavy metals, such as aluminium;
- sulphur dioxide emissions destroying forests above the elevation of 800 m, and also affecting the health of forests in the whole of Central Europe and further afield (see Chapters 31 and 34);
- negative impacts of the transport of coal; dumping of valuable by-products such as sand, kaolin and clays for landfills.

The rehabilitation and recultivation of such areas is a slow process and often not environmentally sound. In some cases, the creation of artificial lakes had at least some positive impacts on the recreational qualities of the affected regions (CAS, 1992; FCE, 1991).

Map 8.4
Military and mining areas in Eastern Germany and Poland

Source: Statistisches Bundesamt, Weisbaden

Charleroi (Belgium). In many of these and other regions, large quantities of excavated materials form new landmarks, while deeply excavated mining areas often serve as landfills for waste or as artificial lakes. A multitude of relatively small gravel and sand pits have changed the character of landscapes by leaving on their surface a pattern of newly created open habitats (without vegetation) and waterbodies. When such features are created in areas where none have been before, they can sometimes form attractive points of recreation, or can be designed to function as second-hand biotopes.

National and international military exercise fields are relatively widespread throughout Europe. An example of concentrated military and mining activities at the border between Germany and Poland is illustrated in Map 8.4. The often extensive areas are located predominantly in regions where sandy soils permit large-scale machinery operations and which are far from human settlements. The presence of woodland provides natural shelter against outside observations as well as barriers to noise. In Central and Northern Europe, natural pine forests and heathlands often qualify for this function and, due to their relatively low economic productivity, their poor sandy soils are in little competition with other landuse interests. The destruction of vegetation and soil, the detonation of explosives and the release and storage of pollutants such as gas, oil, chemicals or radioactivity may pose serious environmental threats to the overall quality of the areas and their groundwater. However, in contrast to mining and landfill activities, military operations are often carried out in a fraction of the area designated and are inaccessible to the public as well as generally exempt from any other landuse activities.

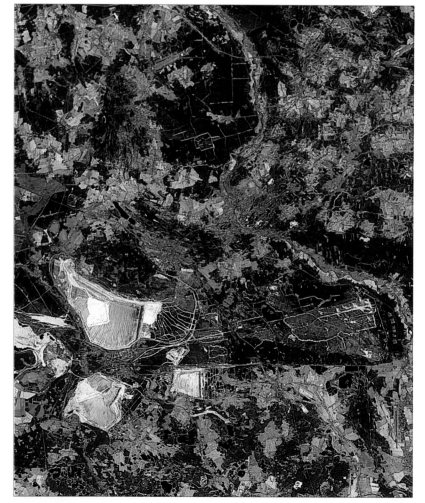

White stork (Ciconia ciconia) in extensively used landscape
Source: D Wascher

Box 8K Loss of habitat structures in open plains and semi-open mixed landscapes, Sweden

Besides forest landscapes, like the northern taiga, there are two main types of agricultural landscapes in Sweden: semi-open, mixed landscapes and 'open plains'.

The semi-open, mixed landscape of southeastern and central Sweden contains medium-scale fields of great variety which exist only in a belt along the forested landscape. Here, agricultural landuse includes animal husbandry, a wide variety of crops and some ecological farming. This has resulted in a small-scale mosaic of semi-natural habitats such as grasslands, meadows, pastures, wetlands and deciduous woodlands, connected by hedges, stone walls or creeks. There are some similarities with the scenery of the bocage landscape elsewhere in Europe, but the trend in the development is the opposite. Due to changing economic conditions the open cultivated lands are decreasing. From 1944 to 1988 this decrease amounted to about 10 per cent in southern semi-open mixed landscapes (Statistics Sweden, 1988). Marginalisation of agriculture continues unabated; many small farms are being abandoned, becoming overgrown, transforming the mixed semi-open landscape into one which is closed.

In the 'open plains' of southwestern parts of Sweden large agricultural fields dominate the flat or gently undulating landscape. There are only small plots of woodland and trees around the scattered farms. It is a productive landscape, where intensification of agriculture is the main trend. Here the scenery is becoming more and more open. Map 8.5 illustrates the process of fragmentation and decline of grasslands in open-plain landscapes as a result of changing landuse patterns through specialisation, intensification and increasing use of fertilisers. Intensification and marginalisation are both reducing the landscape heterogeneity and the diversity of biotopes and have negative effects on cultural landscape and wildlife. The remaining grasslands become fragmented. In the open plains most grasslands have been ploughed and the remaining ones fertilised, whereas in the semi-open, mixed landscape they have been overgrown by trees and bushes or planted with coniferous trees (Ihse, in press). Grassland species have decreased from 50 per cent to 10 per cent while mesophile and weed species have increased from 10 per cent to 50 per cent. Today, nearly 300 of the 700 species of flora and fauna in the agricultural landscape of Sweden are considered to be nationally threatened (Ingelög et al, in press). Drainage has caused a loss of 67 per cent of wetlands and ponds during the last 50 years, while the disappearance of ecologically valuable linear elements (rows of trees and hedges) affects the overall mobility of species because of larger distances between increasingly isolated habitats.

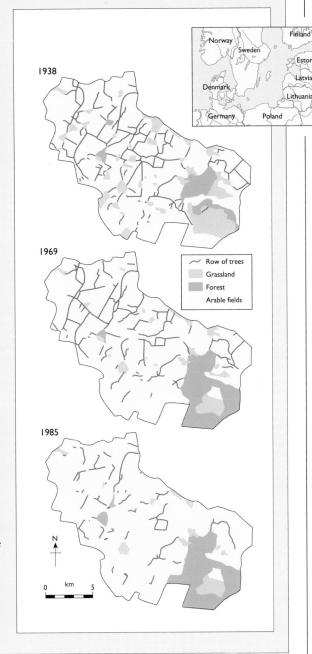

Map 8.5
Landscape changes from 1939 to 1985 in the 'open plain', Beden, southern Sweden
Source: Ihse (in press)

Furthermore, military exercises usually take place over relatively short periods of time and sometimes with many years interruption. Such areas, being extremely well sheltered against outside disturbances and in many ways less affected by human landuse than many other 'open' landscapes, can contain significant natural habitats and rare or endangered wildlife. Especially, abandoned military territories thus constitute an important source of natural landscapes to be managed and restored in an environmentally sound way.

Loss of wildlife habitats

Since it is characteristic for landscapes to change over time and be the result of human rather than of nature's influence, their function as habitats for fauna and flora might appear relatively unspecific and arbitrary. Thus, for example, the disappearance of certain landscape types does not automatically equal a loss in species numbers. There are however, some, general trends which are seriously affecting the role of landscapes as

wildlife habitats. These can be summarised as follows:

- small, diverse structures are decreasing;
- linear and point elements are disappearing;
- wetlands and waterbodies are decreasing as groundwater tables are lowered;
- substitute landscapes are often more uniform in physical and biological character;
- remaining habitats are often smaller, more fragmented and isolated; and
- the number of landscape components with evidence of pollution (eg, forest damage or eutrophication) are increasing.

All these types of change can have important effects on habitat qualities and the species compositions within them. In general, the diversity and the number of species in a given area decrease, or, where this is not the case, less specialised species replace highly specialised ones. Such changes affect not only nature and wildlife, but also the quality of

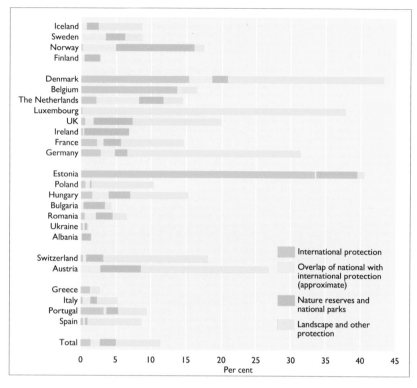

Figure 8.2
Protected areas as percentages of total country area
Source: WCMC, 1992, the Council of Europe, national sources and EEA-TF

agricultural landscapes themselves and their products. The disappearance of many natural, 'biologically active' organisms and the increase of inappropriate landuse activities (eg, crop farming on poor soils) has led to an overutilisation of fertilisers and pesticides, often with negative effects on soil, species and food-products. (See also Chapters 9, 22 and 29.) An example of the influence of landscape changes on the ecological value of an area is given in Box 8K for Sweden.

LANDSCAPE CONSERVATION AND STRATEGIES

Many of the pressures described above are inducing changes, both subtle and obvious, sometimes occurring with great speed. The combined effect of these changes often results in:

- the degradation of distinctive landscape features ;
- the diminishing of natural and cultural values; and,
- the weakening, and even breaking, of the links between people and the land.

The overall result is that the diversity, distinctiveness, and value of many landscapes in Europe are declining rapidly.

Landscape conservation faces the dilemma that, if applied in the sense of preservation, it will attempt to maintain a status quo of a scenery which is often the result of a specific type of landuse activity reflecting a certain historically or economically limited time-phase of a region. Preserving landscapes can therefore involve the unrealistic and unacceptable aim of preserving past economic patterns. Although there may be a limited role for 'museum landscapes' (that is, small areas of landscapes which are preserved in their entirety primarily for educational purposes), for the most part landscape conservation needs to focus on landscape management. An approach of this type would be designed instead to manage the process of change; this would aim to reduce the damaging effects on the landscape, and the natural and cultural values which it contains, of activities which would eventually modify the landscape, and encourage the creation of new landscape values. Several national and international instruments exist or are under preparation which can assist in such landscape conservation and management. These are discussed below.

Box 8L Proposed management objectives for the IUCN Category V sites

- To maintain the harmonious interaction of nature and culture through the protection of landscape/seascape quality and the continuation of traditional landuses, building practices and social and cultural manifestations.
- To support lifestyles and economic activities which are in harmony with nature and the preservation of the social and cultural fabric of the communities.
- To eliminate where necessary, and thereafter prevent, landuses and activities which are inappropriate in scale and/or in character.
- To provide opportunities for public enjoyment through recreation and tourism appropriate in type and scale to the essential qualities of the areas.
- To encourage scientific and educational activities which will contribute to the long-term well-being of resident populations and to the development of public support for the environmental protection of the survival of species (incorporating, as appropriate, breeding areas, wetlands, estuaries, grasslands, forests, spawning and feeding grounds of marine ecosystems).
- To identify areas where nationally or locally important flora, as well as resident or migratory fauna, are in need of improved habitat protection.
- Conservation of these habitats and species should depend upon active intervention by the management authority and, if necessary, through habitat manipulation.
- The size of the area should depend on the habitat to be protected and may range from relatively small to very extensive.

National protection of landscapes

The IUCN World Conservation Union has developed a system of categories of protected areas (see Table 9.14). Several of these are relevant to the conservation of landscapes, but Category V (protected landscapes and seascapes) is particularly so. Protected landscapes include areas in which people live and work, and in which there are farms, working forests and small settlements, as well as natural landscapes. The key criterion for their establishment is that they should be areas of distinctive landscapes in 'aesthetic, cultural and/or ecological' terms.

The IUCN definition of Category V Areas is as follows:

Areas of land, with coast and sea as appropriate, where the interaction of people and nature over time has produced an area of distinct character with significant aesthetic, cultural and/or ecological value. Safeguarding the integrity of this traditional interaction is vital to the protection, maintenance and evolution of such an area.

(IUCN, 1990)

Despite the relatively small number of protected landscape sites compared with national parks (264 sites) and nature reserves (more than 40 000), the area of land granted landscape protection status in Europe is greater than both of these put together. This is confirmed by an analysis of 25 countries, presented in Figure 8.2. There is, however, a wide variation between individual countries and some clearly make less use of this protection category.

For sites which satisfy the IUCN Category V definition for protected landscapes and seascapes, the management objectives listed in Box 8L have been proposed.

Properly used, protected landscapes can be effective instruments of landscape conservation. In some countries, protected landscapes provide powers and resources to:

- control undesirable forms of development;
- support traditional landuse practices;
- support nature conservation and the protection of the built heritage;
- encourage craft industries, sustainable tourism;
- manage visitors so that they do not damage the environment;
- develop sustainable landuse models, eg, for biosphere reserves; and
- support programmes of public education and community involvement, to increase awareness and support for landscape protection among residents and visitors.

There has, however, been a tendency in some countries to designate protected landscapes without providing the powers and resources to ensure the effective conservation of such areas.

World Heritage Convention

The purpose of this convention is discussed in Chapter 9, since it is an important instrument for international cooperation for the conservation of natural sites of great international importance. However, the criteria for the convention have recently been altered and it is now possible for cultural landscapes of outstanding universal significance to be inscribed on the World Heritage List. Since the guidance to be based on these criteria is still being formulated, it is not yet possible to know the full relevance of this development to the protection of Europe's landscapes. However, it is already clear that certain outstanding landscapes would merit inclusion on the list. Though this would be of limited benefit geographically, it would help to establish more clearly the international importance of certain landscapes in Europe. Moreover, within such areas, a major national effort will be required to protect the existing quality of the landscape.

A possible convention for the protection of Europe's rural landscapes

Discussions have taken place in regional meetings of the World Conservation Union (IUCN), the Federation of Nature and National Parks of Europe and elsewhere concerning the development of a European convention to address some of the problems of landscape protection covered in this chapter. The broad aim of such a convention (which might eventually be developed under the auspices of the Council of Europe) would be to strengthen the conservation of the rural landscapes of Europe, and its objectives might include the following: to encourage states to record their landscapes and to put in place measures to protect or enhance them; to develop a network of landscapes of European significance; and to support this with training, information exchange and perhaps a centre of European landscapes expertise.

Environmentally sensitive areas (ESAs) and related measures

Under EC Regulation 92/2078/EEC, governments may define certain areas as being important for conservation of biodiversity, landscapes or cultural features, where those qualities depend upon the survival of traditional forms of farming. Such areas are then eligible for grants, to which the EU makes a contribution to ensure that such traditional farming is maintained, with appropriate safeguards for environmental protection. In effect, a contract is made with the farmer to protect the landscape and other environmental qualities on the land in return for financial support for the farming system which sustains those qualities.

Article 19 of this Regulation has been widely used in several EU member states – within France, for example, nearly 170 000 ha have been so designated within the French system of regional nature parks which have been recognised as protected landscapes and seascapes. Similar schemes exist in some non-EU countries, such as in Sweden and Switzerland. Also, complementary schemes have been developed at national level in EU countries, including the UK, Germany and The Netherlands, some of which involve local authorities in the administration of the funds, or which provide in areas not given ESA status (Jongman, 1993).

Since Europe as a whole faces the likelihood of spare agricultural capacity for many years, this means of bringing new sources of income to farmers to maintain the landscape could have potentially wider applications.

Ecological networks and the landscapes

Within Europe, national and international ecological networks are considered to be of future importance for landscape conservation. They are of three kinds:

1 Networks of important sites – protected area systems (see Chapter 9), some of which are linked together to form European networks (such as the Council of Europe's network of biogenetic reserves or the evolving Natura 2000 network).
2 Ecological networks as part of nature conservation – a coherent framework for nature conservation, involving core areas, buffer zones, corridors and rehabilitation areas. Article 10 of the EU Directive 92/43/EEC on the conservation of natural habitats and of wild flora and fauna suggests the establishment of corridors as natural linkages between habitats. The Dutch ecological network (EECONET) and 'ecological bricks' proposal for the establishment of protected areas along the line of the former Iron Curtain are other examples. (These are discussed further in Chapter 9.)
3 Ecological networks as part of integrated planning – networks which form part of a nation's physical planning. There are countries, particularly in Central and Eastern Europe (eg, Estonia, the Czech Republic, Lithuania, Poland and the Slovak Republic), but also in Denmark and to a lesser extent elsewhere in Western Europe, where green corridors, 'belts' and 'lungs' are used as frameworks for physical planning. In the case of the 'Green Lungs' project (see Box 9NN), Poland is discussing with its neighbours how such frameworks can be extended across national boundaries.

All three approaches, but especially the last two, offer the potential for landscape conservation and management on a large scale. However, this potential cannot be realised solely by keeping urban development away from such green zones. Protective policies need complementing by others which address the factors of landscape change described in this chapter. This covers policies and instruments, such as ESAs, which offer the opportunity to restore damaged landscapes, and others (eg, support for sustainable forms of tourism) which generate income in the rural areas concerned.

In landscape terms, therefore, the ecological network approach is important because:

- it is a powerful tool for guiding landscape conservation and management;

- it links landscape conservation and management to the conservation of biodiversity; and
- it provides a framework for integrating landscape and nature conservation objectives into national planning for sustainable development.

Landscape issues cannot, however, be addressed solely through the network approach. Many landscapes will not be within the green corridors, lungs or belts. Therefore, networks need to be complemented by nationwide policies for landscape planning and management.

Planning and management

In order to avoid irreversible processes of decay, temporary collapse or radical deformation, integral landscape planning can help guide human landuse activities and provide compensation where necessary. To achieve this goal, landscape ecology needs to operate as a cross-sectional discipline by addressing cultural, economic and ecological issues in a scientifically and conceptually creative manner. As a special heritage of the community living within them, each landscape also requires national responsibility.

Numerous countries and local administrations with planning responsibilities have adopted aspects of landscape planning and management in their policy and management. The principles that seem to be most effective include:

- professional landscape planning and management addressing the processes of change rather than the changes themselves (environmentally sensitive areas are a good example);
- working with local people to produce better results; and
- an acceptance that landscapes *will* change, because they are a function of human landuse depending on social-economic and cultural factors (world markets, advances in technology, changes in society, etc).

CONCLUSIONS

Europe's landscapes are immensely diverse and rich in natural and cultural values. A series of factors are bringing pressures to bear on these landscapes causing changes that are both subtle and dramatic, and many of which are occurring at great speed. Often the changes being invoked are unintentional consequences of other activities where the roles and values of landscapes are not taken fully into consideration. This is resulting in a general replacement of natural and regional

diversity by artificial diversity or homogeneity, and in some cases is accompanied by more specific environmental degradation. In the past, the approach to conservation has been species- or site-specific. Now, however, the fragility of whole landscapes is an issue. Approaches to halt the loss of biodiversity and cultural identity in Europe's landscapes can be successful only if they encompass the economic viability of rural communities. An understanding of the important links that exist between cultural landscapes and the people who live within them is essential for promoting both environmentally sensitive changes and social-economic integration. This in turn requires a framework of national and international support which recognises that there is a European interest in the future of Europe's landscapes and rural communities.

To ensure the success of landscape planning and management, the following approaches are considered important:

- to study, record and monitor European landscapes for their ecological, social, cultural and economic values;
- to understand the processes of change, their causes and consequences;
- to take action to protect outstanding landscapes, using such recognised tools as designated landscapes and seascapes;
- to put in place effective landuse planning mechanisms for use in all areas (for making plans and for regulating what happens on the land);
- to make landscape considerations an important factor in shaping national and regional strategies for sustainable development – that is, treat landscape as an environmental resource in the planning process;
- to recognise the landscape scale as one that is important for strategies addressing the conservation of biodiversity, complementing species- and site-specific approaches;
- to develop support systems for rural communities, and farmers especially, to acquire greater prosperity without the need to destroy landscapes;
- to introduce measures to encourage farmers not only to protect existing landscape features, but also to create new elements in the landscapes; and
- to encourage greater public awareness of the value of landscapes, locally, nationally and internationally.

Landscape planning, management and protection are facing the challenge to embrace changes while at the same time understanding their implications and ensuring that they do not violate basic environmental and cultural values. All actions proposed should take into account the special dynamic aspects of landscape evolution over time.

REFERENCES

Anon (1988) *Landscape map of the USSR (1:4 000 000)*. Moscow.

Anon (1989) *Geomorphological map of the USSR (1:4 000 000)*. Moscow.

Anon (1991) *Agricultural land use map of the USSR (1:4 000 000)*. Moscow.

Atlas van Nederland (1984–1990). Stichting Wetenschappelijke atlas, The Hague.

Bernes, C (Ed) (1993) *The Nordic environment: present state, trends and threats*. Nord 1993:12, Nordic Council of Ministers, Copenhagen.

Bink, R G, Bal, D, Van den Berk, V M and Draaijer, L G (1994) *Toestand van de natuur 2*. IKC-NBLF-4, Ministry of Agriculture, Nature and Fisheries, Wageningen.

CAS (1992) *National report of the Czech and Slovak Federal Republic*, pp 40–42. Prepared for: UNCED Conference in Brazil, June 1992, by Czechoslovak Academy of Sciences (CAS) and the Federal Committee for the Environment. Prague.

CBS (1992) (Central Bureau for Statistics) *Netherlands national report to UNCED 1992*. Ministry of Housing, Physical Planning and Environment, Department for Information and International Relations, The Hague.

CEC (1985) *Soil map of the European Communities (1:1 000 000)* (Tavernier cs). Commission of the European Communities, Luxembourg.

CEC (1992a) *CORINE soil erosion risk and important land resources*. Commission of the European Communities, Luxembourg.

CEC (1992b) *CORINE land cover*. Commission of the European Communities, Luxembourg.

CIPRA (1993) *Die Alpenkonvention – eine Zwischenbilanz. Ergebnisse der Jahresfachkonferenz. 1.-3.10.1992. Schwangau, Bayern*. Walter Danz und Stephan Ortner, CIPRA, Munich.

Csati, E et al (1980) *Land use maps of Europe (1:2 500 000)*. Carthographia, Budapest.

ČSAV (1992) *Atlas of the environment and health of the population of the ČSFR*. Institute of Geography, Czechoslovak Academy of Sciences (ČSAV), Brno, and Federal Committee for the Environment, Prague.

Driessen, P and Dudal, R (Eds) (1991) *The major soils of the world*. Agricultural University/Katholieke Universiteit, Wageningen/Leuven, Belgium.

FAO (1982) *Soil map of the world (1:5 000 000)* Vol I–X 1971–1981, UNESCO, Paris.

FCE (1991) *State of the environment in Czechoslovakia.* Federal Committee for the Environment, Prague.

Ihse, M (in press) Swedish agricultural landscapes – patterns and changes during the last 50 years, studied by aerial photos. In: Burel and Baudry (Eds) *Agricultural landscapes in Europe. Landscape and Urban Planning* (special issue).

Ingelög, T, Thor, G, Hallingbäck, T, Andersson, R and Aronsson, M (in press) *Flora conservation in the agricultural landscape – threatened and rare species.* Sweden.

IUCN (1990) *United Nations list of national parks and protected areas.* IUCN, Gland, Switzerland and Cambridge, UK.

de Jong, J and Wiggers, A J (1982) *Polders and their environment in the Netherlands.* Keynotes international symposium 'Polders of the World'. Ijsselmeer Polders Development Authority, Lelystad, pp 223-43.

Jongman, R H G (1993) unpublished. *Developing ecological networks in Europe.* University of Wageningen, Wageningen.

Kostrowicki, J et al (1984) *Types of agriculture of Europe (1:2 500 000).* Polish Academy of Sciences, Warsaw.

Lassen, D (1990) Unzerschnittene verkehrsarme Räume uber 100 km2 - eine Ressource für die ruhige Erholung. *Natur und Landschaft* **65**(6), Nr 6: 326–7.

Lebeau, R (1969) *Les grands types de structures agraires dans le monde.* Initiations aux études de géographie. Masson, Paris.

MAB (1984) Landwirtschaftliche Forschungskommission: Die Berglandwirtschaft in Spannungsfeld zwischen Ökonomie und Ökologie, Kolloquium 23.11.84, Man and Biosphere programme (MAB), Berne.

Meeus, J, Wijermans, M and Vroom, M (1990) Agricultural landscapes in Europe and their transformation. *Landscape and Urban Planning* **18** (3/4), 289–352.

National Atlas of Sweden 1990–92: 2 (The Forests), 5 (Agriculture), 3 (Landscape and Settlement). Sna publishing, Stockholm University, Stockholm.

Naveh, Z (1993) Red Books for threatened Mediterranean landscapes as an innovative tool for holistic landscape conservation. Introduction to the Western Crete Red Book case study. *Landscape Urban Planning* **24**, 241–7

Netz, B (1990) Landschaftsbewertung der unzerschnittenen verkehrsarmen Räume – eine rechnergestützte Methode zur Ermittlung der Erholungsqualität von Landschaftsräumen auf Bundesebene. *Natur und Landschaft* **65**(6).

Noirfalise, A (1989) *Paysages. L'Europe de la diversité.* Commission of the European Communities, Luxembourg.

Park, J (Ed) (1989) *Environmental management in agriculture: European perspectives.* CEC, Belhaven Press, London.

Partsch, K and Zaunberger, K (1990) *Alpenbericht.* Sonthofen, Germany.

Pinto Correia, T (1991) Landscape ecology in EEC less favoured areas – problematic evolution in Southern Portugal. *Proceedings of the European IALE seminar on practical landscape ecology* (pp 115–24). Vol 1, Roskilde, 2–4 May 1991. Roskilde University Centre, Roskilde.

Pisarenko, A (1993) Russian boreal forests and global biosphere stability. Report of meeting held in Geneva, 15–18 March 1993, UNEP Geneva.

RIVM (1992) *National Environmental Outlook 2; 1990–2010;* National Institute of Public Health and Environmental Protection. Bilthoven.

Statistics Sweden (1988) *Jordbruksstatistisk Årsbok* (in Swedish), Solna.

Symons, L (Ed) (1990) *The Soviet Union: a systematic geography.* Routledge, New York.

Verhoeve, A and Vervloet, J (Eds) (1992) *The transformation of the European rural landscape: papers of the 1990 European conference for the study of the rural landscape.* NFWO, Brussels.

WCMC (1992) *Global biodiversity: status of the earth's living resources.* World Conservation Monitoring Centre. Cambridge.

Weizäcker, E U (1990) *Alpenpolitik als Erdpolitik* Institut f. Europäische Umweltpolitik e.V.

Wijermans, M and Meeus, J (1991) *Karakteristieke landschappen van Europa.* WRR (Netherlands Advisory Council on Government Policy), W 58, The Hague.

WRR (1992) *Ground for choices: four perspectives for the rural areas in the European Community.* Netherlands Advisory Council on Government Policy (WRR), 42, Sdu, The Hague.

Zalibikov, Z (1992) Classification on land ecosystems: problems of desert development. (*Problemy Osvoeniya Pustyn*) 4, 1–5.

Wetlands in Iceland
Source: Mercay, WWF

9 Nature and wildlife

INTRODUCTION

From the Atlantic plains to the steppes of the Caspian area and from the Lapland tundra to the Mediterranean maquis, Europe comprises diverse natural and semi-natural habitats. This diversity provides habitats suitable for a large number of animal and plant species some of which are endemic in certain regions. Map 9.1 illustrates how Europe would look without human intervention or environmental changes. Thus, 80 to 90 per cent of the land would be covered by forests. According to an analysis (Ellenberg, 1986), largely based on existing soils, topography and climate, Central and Western Europe would be dominated by deciduous trees, mainly beech (*Fagus sylvatica*), and oaks (*Quercus* spp).

Human intervention has resulted in a profound modification of the original landscape, through deforestation, agriculture, drainage of wetlands, coastline and river course modifications, mining, road construction, urbanisation, and so on (see Chapter 8). As a result, many animals and plants have had to find refuge in relatively small enclaves, sometimes only secure in legally designated protection areas. Lowland forests as well as wetlands such as peatlands and reed-swamps have disappeared in particular due to human landuse activities. Although a number of large pristine or quasi-natural areas persist in Nordic and Eastern European countries, human impacts can be felt everywhere to some degree.

In return, humans have created landscapes serving their own needs: grasslands, pastures and croplands, often intermixed with remaining woodlands and lined with hedges and waterways. Large mammals (such as bear, *Ursus arctos*; wolf, *Canis lupus*; lynx, *Lynx lynx*; and bison, *Bison bison bonasus*) have retreated to remote remnants of their original habitat; others (such as the tarpan, *Equus caballus* and the saiga, *Saiga tatarica*) have become extinct. Many species from distant biogeographic regions of Europe have become established in the newly created environment: singing-birds such as the lark (*Alauda arvensis*), or open-space species like the grey partridge (*Perdix perdix*) and the hare (*Lepus europaeus*), are directly associated with agricultural landscapes. Today, the invasion of species as a result of human activities is still continuing and can have negative impacts on native populations, for example, the increase of seagull (*Larus* spp) and black kite (*Milvus migrans*) near urban waste deposits, or the spread of exotic species such as Canadian pondweed (*Elodea canadensis*).

Up until the last century, biological diversity, in terms of habitat types as well as number of species had in general been on the increase in Europe (Kaule, 1986; Cox and Moore, 1973). Now, the trend is reversed: natural habitats are becoming smaller, more fragmented and less able to support wildlife. One crucial phenomenon is the isolation of small populations which are unable to maintain the biologically necessary links to larger gene-pools of the original ecosystem. Hence, the number of endangered species of flora and fauna has increased in many European regions (see below).

This chapter first presents a review of the state of Europe's natural (or quasi-natural) ecosystems, their geographic distribution, trends in the habitats' ecological functions and the main threats. This is followed by an assessment of European fauna and flora as well as of existing and potential nature conservation measures and strategies. An important constraint on this assessment, however, has been the unsatisfactory availability and quality of data concerning the various ecological parameters. Improvement will require internationally harmonised inventories and new scientific approaches.

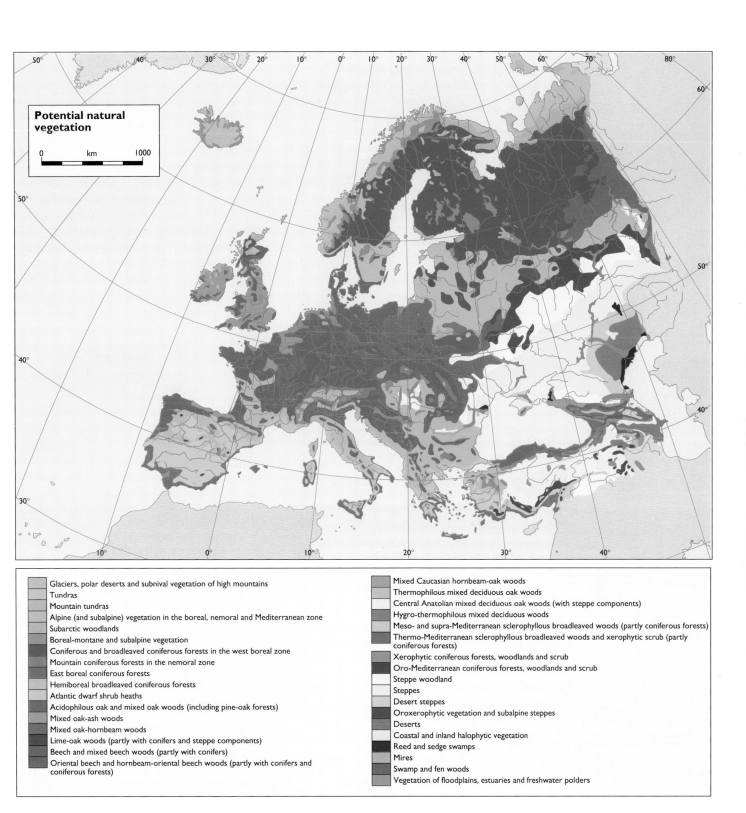

Potential natural vegetation

	Glaciers, polar deserts and subnival vegetation of high mountains
	Tundras
	Mountain tundras
	Alpine (and subalpine) vegetation in the boreal, nemoral and Mediterranean zone
	Subarctic woodlands
	Boreal-montane and subalpine vegetation
	Coniferous and broadleaved coniferous forests in the west boreal zone
	Mountain coniferous forests in the nemoral zone
	East boreal coniferous forests
	Hemiboreal broadleaved coniferous forests
	Atlantic dwarf shrub heaths
	Acidophilous oak and mixed oak woods (including pine-oak forests)
	Mixed oak-ash woods
	Mixed oak-hornbeam woods
	Lime-oak woods (partly with conifers and steppe components)
	Beech and mixed beech woods (partly with conifers)
	Oriental beech and hornbeam-oriental beech woods (partly with conifers and coniferous forests)
	Mixed Caucasian hornbeam-oak woods
	Thermophilous mixed deciduous oak woods
	Central Anatolian mixed deciduous oak woods (with steppe components)
	Hygro-thermophilous mixed deciduous woods
	Meso- and supra-Mediterranean sclerophyllous broadleaved woods (partly coniferous forests)
	Thermo-Mediterranean sclerophyllous broadleaved woods and xerophytic scrub (partly coniferous forests)
	Xerophytic coniferous forests, woodlands and scrub
	Oro-Mediterranean coniferous forests, woodlands and scrub
	Steppe woodland
	Steppes
	Desert steppes
	Oroxerophytic vegetation and subalpine steppes
	Deserts
	Coastal and inland halophytic vegetation
	Reed and sedge swamps
	Mires
	Swamp and fen woods
	Vegetation of floodplains, estuaries and freshwater polders

ECOSYSTEMS

According to Tansley (1935), 'an ecosystem can be defined as a spatially explicit unit of the Earth that includes all of the organisms along with all components of the abiotic (non-living) environment within its boundaries'. By itself, the term 'ecosystem' does not connote any specific dimensions (IGBP, 1986). Determining the often gradual boundaries can be more difficult for some terrestrial than for aquatic systems where, for example, the presence of water or wet soils helps to identify lateral boundaries of lakes and rivers.

In this report, ecosystems and habitats are classified according to the system developed for the EC CORINE biotopes project (CEC, 1991a), as explained in Box 9A. The following ecosystem groups are recognised:

- forests;
- scrub and grasslands;
- nland waters: rivers and lakes;
- wetlands: bogs, fens and marshes;
- coastal and marine ecosystems;
- mountains: rocks and screes;
- deserts and tundras;
- agricultural and urban ecosystems.

Comprehensive data on the current state and distribution of these ecosystems do not exist for the whole of Europe. Although

Map 9.1
Potential natural vegetation of Europe
Source: Federal Agency for Nature Conservation, Bonn

most European countries have a long history of gathering information and reporting on their native flora and fauna, the existing data are not always easy to access and can vary greatly in their age and reliability. Existing international statistics of nature conservation focus largely on protected areas only and often data do not always cover parameters such as habitat types, stresses and type of management. Hence, a complete and systematic assessment of Europe's ecosystems is presently not possible. In order to make the best use of information that exists on some very important examples of European ecosystems, the report attempts to identify and analyse a number of them as *representative* sites for the first seven ecosystem groups.

Representative sites

Candidate *representative* sites were identified and selected against criteria (see below) through a network of international experts under the guidance of the IUCN/EEA-TF ecosystem project group. From these, shorter lists of *representative* sites were established. Given the large differences of data

Box 9B Definition of status categories and their correspondence to the map legend for representative sites

Map legend	Status categories
◉ protected	● adequately protected (to maintain ecosystem integrity)
○ unprotected	● not protected
	● not protected, insufficient information
◔ protected, problems and/or stresses recorded	● inadequately protected (to maintain ecosystem in terms of size/percentage of protected areas or quality of regulations) vulnerable to threats
	● inadequately protected, predominantly due to management problems and internal threats;
	● inadequately protected, predominantly due to threats from outside the site
◉ unprotected, stresses recorded	● not protected and under threat.

availability between countries, many more sites could indeed have been included in some cases. Sites were chosen to picture the overall situation of the various ecosystem groups giving particular attention to their importance for nature conservation. From a purely national point of view the outcome may be incomplete, however, whenever possible, comments by the national focal points on these site lists were taken into account. The resulting maps and figures only illustrate the situation for the identified sites and cannot be representative for the whole ecosystem as such.

Selection criteria

The selection criteria for *representative* sites were:

- **representativeness:** habitats that are typical for a region, but have become rare or have degraded;
- **naturalness:** non-altered, species relatively identical with natural potential vegetation;
- **diversity:** number of habitat types and species within one area;
- **threat:** types of habitats that are suffering severely under environmental stresses; and
- **size:** those covering the largest areas fulfilling these criteria.

These selection criteria were adapted to the character of each individual ecosystem group. Often, scientific standards or specific priorities were needed to facilitate the selection, which was influenced by the quality and availability of information. Sites of regional or national importance were included when they were considered to be also of European interest. An explanation of the main criteria used for selecting sites accompanies the map for each ecosystem group.

Independent of their legal status, protected as well as unprotected sites have been identified and described (see below). Although protection status is considered to be an important part of the description of each site, the attributions given in this report are not meant to pre-judge how to respond to the threats identified. Current strategies and options for action to safeguard nature and wildlife are summarised at the end of the chapter.

Information on representative sites

For each ecosystem group, information has been compiled on the current state of the selected sites, including: their size, protection status, ecological conditions (habitat types, species, etc), and environmental stresses. The information collected was used to assign each site to one of seven status categories, which have been further merged into four classes in the maps in this chapter (see Box 9B). Further information on these sites, when available, can be found in the *Statistical Compendium*. The names of the sites indicated on the maps are listed at the end of the chapter, grouped by ecosystem and country.

Interpretation

As a result of the data analysis for representative sites it has been possible to:

- illustrate the spatial distribution, coverage and qualitative aspects of the representative sites;
- provide an indication of the present role of legal protection and resulting management practices (see Box 9B) in the maintenance of the integrity of the habitats and the species they support;
- highlight the significance of undesirable and damaging uses of resources and attempt to identify examples of conflict resolutions; and
- demonstrate, with the help of case studies, specific or very typical issues directly affecting the condition of an ecosystem.

Lists of representative sites are not claimed to be comprehensive. The large data gaps in Eastern and Central Europe point to the need for a systematic approach to meet long-term management objectives, something which the current assessment was unable to achieve with the data available.

Geographic regions

Throughout this chapter, the analytical presentations in graphs and tables frequently refer to the following seven 'geographic regions': boreal, Baltic, central, Atlantic, east, Alpine, and Mediterranean (Map 9.2). Although this delineation is mainly based on biogeographic factors such as climate, soils and vegetation, it is very schematic in nature.

Forests

Introduction

Around 500 BC, when the Mediterranean regions were already going through a phase of large-scale deforestation, Central and Northern Europe were still densely covered by forests. Starting at the time of the Roman Empire, intensive cultivation of the land led to a decline of the forests which continued until the fourteenth century (Bosch, 1983). Today, the average forest cover of Europe is 33 per cent, varying considerably between countries – from 6 per cent in Ireland to 66 per cent in Finland. Forest management practices during the last centuries resulted in substantial changes to the remaining forests, as far as their species composition and structure were concerned. In Central and Western Europe, deforestation has drastically reduced native deciduous forests, while coniferous trees dominated afforestation. The types of forest vary considerably depending on the major climatic regions. The main habitat types, ecosystem functions and threats of natural forest systems are explained below, followed by an interpretation of data deriving from 156 representative sites of natural and semi-natural forests, illustrated in Map 9.3.

Boreal region

In the northern hemisphere, boreal coniferous woodlands form a continuous belt around the whole Earth, covering much of Scandinavia and Northern Russia with trees such as Norway spruce (*Picea abies*), Scots pine (*Pinus sylvestris*) and downy birch (*Betula pubescens*) (see Box 9C). Boreal forests are very poor in vascular plants and numbers hardly exceed 250 different species. Besides being rich in mosses and lichens, the boreal region is home of the only forest ecosystem where some species of the original fauna (eg, large predatory animals) can be found in extensive habitats. However, in western parts the wolf and the brown bear have become very rare.

Atlantic region

While a trend from deciduous to coniferous trees has been recorded for many European forests, silviculture management has brought about large changes in the forests especially in this region. Once dominated by broadleaved deciduous forests of oaks (*Quercus* spp) and beech (*Fagus sylvatica*) with several tree layers and sometimes shrub as well as herbaceous layers, the Atlantic region shows a substantial increase of coniferous forests (spruce, *Picea abies*, Scots pine, *Pinus sylvestris*, etc), originally limited to mountains and some coastal areas in the south of Europe. The substitute coniferous forests lack the rich original herbaceous layer of deciduous forests, which consists mostly of hemicryptophytes (buds quite close to the ground) and geophytes (species with tubers, bulbs and rhizomes), growing predominantly in spring.

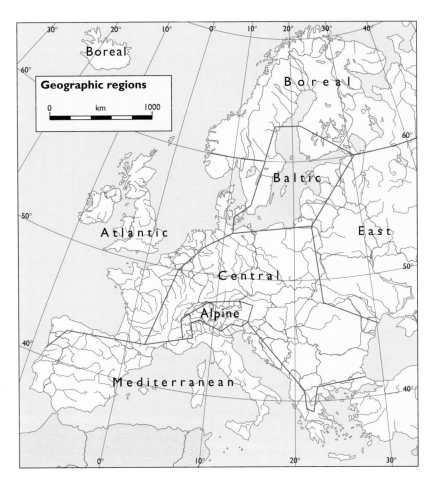

Map 9.2
Geographic regions of Europe used in this chapter

Mediterranean region

The Mediterranean climate is characterised by long, hot and dry summers with major rainfalls during the winter season. Early periods of exploitation and fires have left only small remnants of the original evergreen oakwood forests with holm oak (*Quercus ilex*), as well as cork oak (*Q suber*), turkey oak (*Q cerris*) and hungarian oak (*Q frainetto*). Deciduous forests prevail in regions with more rainfall. Coniferous forests (*Pinus* spp, *Abies* spp) and laurel forests once dominated the islands of the southern North Atlantic (the Canary Islands, Maldives and Azores), but have now disappeared from higher elevations.

Central European region

The whole of the Atlantic region (high rainfall) and the western transition zone of the central continental region (cold winters, moderately warm summers) favour temperate forests of deciduous trees such as oak (*Quercus robur*) and especially beech (*Fagus sylvatica*) with herbaceous and geophytic groundcover. Shorter periods of annual vegetation in the higher elevations of middle European mountain ranges and the decreasing influence of the oceanic climate towards the east result in an increasing dominance of coniferous trees such as spruce and pine associated with heather (*Erica*) on calcicolous or bilberry (*Vaccinium*) on acidophilous soils.

Alluvial forests

Azonal alluvial forests occur in all climatic zones in a very similar fashion. According to Ellenberg (1986), two main types prevail: hardwood forests with oak (*Quercus* spp), alder (*Alnus* spp), ash (*Fraxinus excelsior*), or elm (*Ulmus* spp), and softwood forests with willow (*Salix* spp) and poplar (*Populus* spp). Alluvial forests once fringed all European rivers as widely as inundations defined the floodplains. However, since alluvial plains of rivers have always been areas of preference for human settlements (transport, energy, food), many of the originally existing alluvial forests, especially of the large river systems in central Europe, have disappeared. The few remnants (for example along the Rhine and Danube) represent important wildlife habitats.

Mountain forests

In central European mountains, the beech (*Fagus*) is intermixed with fir (*Abies* spp) and larch (*Larix*), followed by spruce (*Picea* spp) at timber-line. In Britain, Scots pine (*Pinus sylvestris*) and birch (*Betula*) dominate, and in the Caucasus, deciduous oaks mix with oriental hornbeam (*Carpinus orientalis*), sycamore (*Acer pseudoplatanus*), Norwegian maple (*Acer platanoides*) and smooth-leaved elm (*Ulmus minor*). Above 1350 m, oriental beech (*F orientalis*) dominates, intermixed with Caucasian fir (*A nordmanniana*) and oriental spruce (*P orientalis*) followed by different birch (*Betula*) species at high altitude.

Functions and values

Forests provide a great range of different benefits. Due to the large amount of biomass that characterises this ecosystem, forests are able to create their own microclimate, influence general climatic conditions, improve air and water quality and temper the impacts of urban or industrial pollution on the environment. Not only do alluvial forests act as 'nutrient traps' which accumulate nitrogen, carbon and phosphorus, they also provide natural protection for river banks against erosion. Forests are also of substantial value for recreation and the sustainable maintenance of groundwater reserves, ensuring long-term water supply. Furthermore, forests play a crucial role providing natural protection against erosion and avalanches in mountain terrains. All these functions may be considered as important as, and in some cases to exceed, the provision of timber and cellulose.

In terms of nature conservation, forests – especially those in a pristine or semi-natural condition – represent very closely the potential natural vegetation, offering adequate habitats for invertebrates and birds, as well as natural refuges for many wildlife species, including the larger mammals which were characteristic of Europe's wooded landscapes (see Box 9D).

Habitat threats

Given past and present landuse practices, it is not surprising that mature natural forest ecosystems with highly diverse fauna and flora are very rare in Europe. In Northern and Central Europe this has generally been caused by intensive logging, resulting in a significant decrease in the average age of trees in stands, a tendency for forests to be divided into areas of uniform, even-aged, stands and a drastic reduction in the amount of dead timber. In other regions, particularly in the Mediterranean, forests have lost important habitat functions from excessive logging, fire and overgrazing by livestock, especially goats. Due to the extensive use of non-native species such as Sitka spruce (*Picea sitchensis*) in the north and eucalyptus (*Eucalyptus*) in the south, extensive reforestation and afforestation in European countries (Norway, UK, Ireland, France, Portugal, Spain) have resulted in the replacement of native, usually deciduous, broadleaved trees together with the original, ecologically much richer, composition of plants and animals. The modification of forests has recently become particularly severe with the increasing importance of pulp and fibre production. Consequently, a far wider range of tree species can be exploited and rotation times of plantations shortened.

Forests are especially threatened at the boundaries of their natural distribution such as in the forest-steppe regions of southern Ukraine and Russia or in the Mediterranean basin (see Box 9E). Most of Central Europe's alluvial forests along all major river courses have virtually disappeared. Remaining habitats are in need of protection and riparian corridors require regeneration to allow seasonal inundations for the re-establishment of alluvial forests.

Höga Kusten forest, Sweden
Source: G Heiss

Box 9C Coniferous forest: Muddus and Höga Kusten (Sweden)

Muddus National Park (Site 135 on Map 9.3) in the north of Sweden, is typical of Lapland's landscape. It is characterised by a mosaic of coniferous forests (pine and Norway spruce), lakes and bogs. Due to the inaccessibility of its swampy grounds, human exploitation has played only a minor role in the life cycle of this forest ecosystem. The unique quality of its extensive virgin forests (about 30 000 ha) and its representative landscape character led to the establishment of a national park in 1942. This large park, covering 50 350 hectares, has also been designated as a Biogenetic Reserve and awarded with the European Diploma.

Unlike the Mediterranean region, Muddus shows that forest fires do not always have negative effects on ecosystems. Here, forest fires have an indispensable role for ensuring the balance of pine and spruce in the boreal zone. Thus, fire is considered to be a natural part of the forest's life cycle. It results in periodic new growth and provides improved feeding grounds for local animal populations, especially mammals. Old pine trees which have died provide the substrate for the growth of yellow wolf lichen (*Letharia vulpina*), the only poisonous lichen species in Europe.

Another, quite different, example of the boreal forests of Northern Europe is the Swedish coastal forest of **Höga Kusten** (Site 134 on Map 9.3). Such coastal forests are rare in Europe. The location and the favourable climatic conditions of Höga Kusten make it particularly interesting from a botanical and faunistic standpoint. Here, the lichen old man's beard (*Usnea longissima*), a virgin forest indicator growing on Norway spruce, is still present. This species has disappeared from about 95 per cent of 122 locations in Sweden where it was found before 1970 (Esseen and Ericson, 1982). For the survival of this species it is essential that forest stands are able to develop naturally in unaltered conditions. It has been observed that managed forests are less likely to provide these habitat functions than non-managed forests. A part of the area (3250 ha) was established as a national park in 1984 and is surrounded by several nature reserves.

Airborne pollution (mostly in the form of acid deposition and photochemical smog) is causing severe damage to forests, in particular in Central and Eastern Europe. Forest decline was first spotted in the Black and Bavarian forests of Germany in the early 1970s when 8 per cent of mostly coniferous forests turned out to be damaged. A report (CEC, 1993) of the international cooperative

Box 9D Temperate forests: Bieszczady Region (Poland, Ukraine, Slovak Republic)

Where the borders of Poland, the Slovak Republic and Ukraine meet, the Bieszczady Region (Site 80 on Map 9.3) forms Europe's largest natural mountain beech forest ecosystem. The area's history and remoteness have made the site a priority target for European nature conservation.

The Polish part provides habitat for 70 per cent of the country's brown bear and wolf population; the European bison (*Bison bison*) has been reintroduced, and wild cats (*Felis silvestris*), otters (*Lutra lutra*) and about 100 lynx (*Lynx lynx*) find optimal habitat conditions within the territory. The most threatened species among the birds are eagle (eg, *Aquila* spp), Ural owl (*Asio* spp), hazel hen (*Bonasia bonasia*), black stork (*Ciconia nigra*), white-backed woodpecker (*Dendrocopos leucotus*), booted eagle (*Hieraaetus pennatus*) and lesser spotted eagle (*Aquila pomarina*). The entire area encompasses about 100 000 hectares of beech forests, mixed woods with fir (*Abies* spp), spruce (*Picea* spp), sycamore (*Acer pseudoplatanus*) and alder woods (*Alnus* spp) in damp valleys. The Wolosate valley forms the core area of the Bieszczady national park on Polish territory. This strictly protected nature reserve, covering 4400 ha, is a beech-dominated forest ecosystem with intermixed single sycamores and interspersed alder woods in the moist lowlands. Some years ago, the national parks in this region of Poland were tripled in size to 15 337 ha, protecting peat and rock formations in some small zones.

On the Slovakian side the protected landscape area of Vychodne Karpady with about 67 000 ha plus a buffer zone of about 31 000 ha is linked with the Polish national park and includes the famous Stuzica Reserve. This is believed to be the largest remaining area of virgin forest in Central Europe, covering about 700 ha. Additionally, on the Ukrainian side, a nature reserve of 2000 ha exists, planned to be increased to 5000 ha.

The lack of sufficient protection for most of the

Bieszczady Region forest Source: G Heiss

Bieszczady region allows human activities (such as tourism projects and forest exploitation in the immediate vicinity as well as inside the reserves) to affect the area unchecked. Recent efforts have been made to designate a large trilateral reserve with a high protection status. Nature conservation efforts might help improve the ecological quality of the region, but cannot halt effects from external pressures such as those from air pollution.

Box 9E Mediterranean forest: the Aoos Valley (Greece)

The south of Europe has a wide variety of forest communities, from the laurel forests on the Macaronesian islands, the Spanish fir forests of the Ronda range in southeast Spain, the Italian fir forests in Calabria, to the Grecian fir forests in southern Greece. The Aoos Valley (Site 58 on Map 9.3) in the Pindos range represents a region where a large number of those southern forest communities are grouped together. The nearby Vicos gorge is one of the most impressive canyons of Europe: vertical rock formations frame the river bed for about 20 km and their tops rise up to 1000 m above the valley bottom. One side-canyon links to the karst 'mesas' (high plateau) of the Astraka (2436 m) and Tymphi range (2496 m). Cliff faces are dissected by deep ravines and display the complete range of Greek forest communities: from the lowland oak forests up to the Bosnian pine (*Pinus heldreichii*) forests at the tree-line. The lowland forests of the Aoos gorge are composed of a rich variety of about 20 tree species forming small-scaled patterns.

Remoteness and inaccessibility of the region provide natural protection for a fauna of unique variety, comparable only with that of the Rhodopi mountains stretching from Greece to Bulgaria. Despite poaching, some brown bears survive in the region. Of special interest also are the wolf and wild cat populations as well as the presence of the otter. At least 100 bird species can be found, among them the black stork (*Ciconia nigra*), the griffon vulture (*Gyps fulvus*), the Egyptian vulture (*Neophron percnopterus*), the bearded vulture (*Gypaetus barbatus*), short-toed eagle (*Circaetus gallicus*) and

Aoos Valley Source: G Heiss

seven woodpecker species (*Dendrocopos* spp). Also noticeable is the high number of amphibians (9 species) and reptiles (16 species).

This wildlife paradise is endangered by tourist developments (foreign mountain clubs) and large-scale energy projects. Plans for hydroelectric power stations threaten the very existence of the Calda valley (a side valley) and the Aoos valley itself. Although two national parks (Vikos-Aoos with 12 600 ha and Pindos with 6927 ha) do already exist in the area, the level of protection and management still needs improvement.

programme on assessment and monitoring of air pollution on forests among 34 European countries (excluding Russia) found that 24 per cent of all sample trees taken from 184 million hectares of coniferous and deciduous forests must be classified as damaged. The problem of forest degradation due to atmospheric pollution is comprehensively covered in Chapter 34.

The main causes of biodiversity decline in natural forests are:

- decrease of non-managed forest ecosystems;
- drainage and alteration of floodplains of major river systems;
- occurrence of fire, especially in the Mediterranean region;
- introduction of non-native species (often especially vulnerable to fire, insects and wind throw);
- increased fragmentation and isolation of forests;
- increased tourism;
- inadequate wildlife and lifestock management (overgrazing, browsing, hunting); and
- increase of airborne pollution.

In order to avoid conflicts with already existing recreational functions, the establishment of large natural, strictly protected, forest reserves needs to go hand in hand with a development of alternative solutions for traditional local or seasonal tourism.

Representative sites: forests

A total number of 250 representative sites were selected, of which 156 have been evaluated and are presented on Map 9.3 according to the four classes defined in the legend. This evaluation is based on site-specific data concerning present status of protection, habitat management and prevailing pressures.

Criteria for site selection

Criteria as described in the introduction to this chapter were applied with some flexibility, except for the minimum size of 1000 hectares. The selection was easier in countries where natural forest inventories have already been implemented (Sweden, Norway) or are presently under preparation (Finland). Data are incomplete for Eastern Europe where even more forest sites of high quality can be expected to be identified.

Interpretation of the data

The 156 selected representative sites cover approximately 20 million hectares, or less than 4 per cent of Europe's forested areas. Figure 9.1 shows that two thirds of the total area (about 14 million hectares) are located in the boreal region, of which most belongs to the Finnish-Russian Woodland area. Only less than 20 per cent of semi-natural forests larger than 10 000 hectares remain in central and Atlantic regions.

Forest exploitation for human use has been identified as the major threat to most of the areas. Scandinavian forests are well monitored and documented: management problems derive mainly from grazing (reindeer), hunting and tourism. Intact and pristine forests are small in number and size (fragmentation) throughout most of the Atlantic region. A large number of forest sites with important functions for the landscape and for recreational needs have not been reported here due either to current management practices (lacking components of the natural native vegetation) or to their limited size. Most of the approximately 2 million hectares of Mediterranean sites are insufficiently protected and face various stresses (tourism, energy, etc). Many of the sites in Poland, the Czech Republic and Russia are known for being

Map 9.3
Forest ecosystems: representative sites (see end of chapter for site names)
Source: Compiled from multiple sources by IUCN/EEA-TF, GAF, Munich

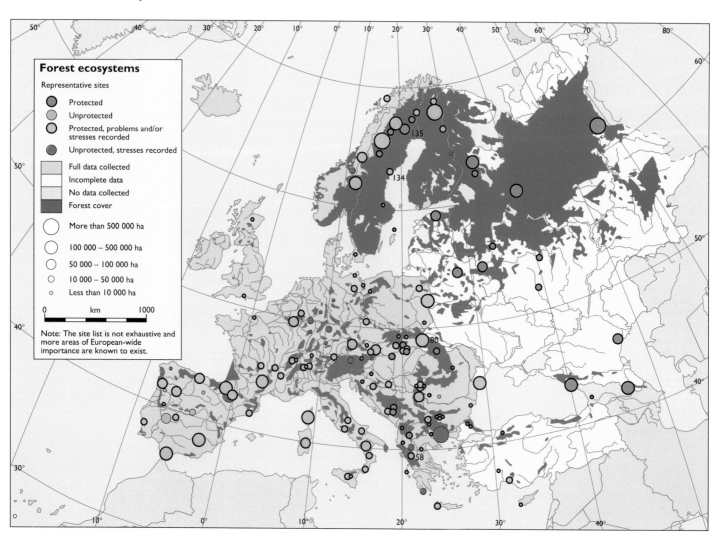

Forest ecosystems

Representative sites

- Protected
- Unprotected
- Protected, problems and/or stresses recorded
- Unprotected, stresses recorded

- Full data collected
- Incomplete data
- No data collected
- Forest cover

- More than 500 000 ha
- 100 000 – 500 000 ha
- 50 000 – 100 000 ha
- 10 000 – 50 000 ha
- Less than 10 000 ha

0   km   1000

Note: The site list is not exhaustive and more areas of European-wide importance are known to exist.

heavily affected by environmental pollution; the same can be expected for some sites in other Eastern European countries.

Conclusions

Natural forests require adequate protection and management. Many forest types are either not represented in national networks of protected areas or are of insufficient size for proper protection, especially with respect to wildlife such as large predatory mammals and birds. Even where forests are protected, excessive intervention by reserve managers seems to be the general rule. Most important for threatened species are those forest ecosystems which have remained (almost) untouched until now or which have retained the possibility to regain their natural complexity. While the maintenance of such ancient natural forest has top priority, it needs to be stressed that all remaining forests need to be managed according to silvicultural methods that integrate ecological principles. The relatively small average size of Central and Atlantic forests, as well as their higher degree of fragmentation, deserves special attention. The establishment of links and corridors could support valuable, but highly fragile, core areas, and facilitate the reintroduction of species.

The lack of national surveys in most European countries makes the identification of priority areas for forest protection extremely difficult. This demonstrates the need for more, and technically improved, forest surveys such as that in Sweden.

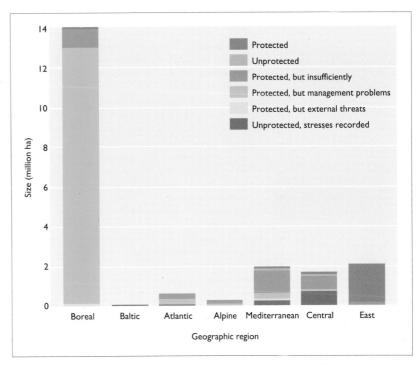

Figure 9.1
Forests: representative sites by region
Source: IUCN/EEA

Scrub and grasslands

Introduction

Scrub and grassland ecosystems provide a large number of habitat types that are important for both ecological and – given their strong links to agricultural landscapes – economical reasons. Because of the high diversity of European scrub and grasslands, the CORINE Habitat Classification recognises more types of vegetation in this category than for any other, covering a range of woody shrub communities (including maquis and dwarf shrub heath), tall herb stands and many types of what is commonly recognised as grassland. They occur from montane to lowland situations and can range from very sparse to very dense cover, and from a few centimetres to two or three metres in height. Seasonal variation of temperature and rainfall can play an important part in the overall appearance and composition of the vegetation (especially for certain types of grasslands). The main habitat types, ecosystem functions and threats for scrub and grassland systems are explained below, followed by an interpretation of data, deriving from 186 representative sites, illustrated in Map 9.6.

The scrub and grassland ecosystem includes the following habitat types:

Country	Permanent pasture a	Cropland c	Landuse ratio1	Dry grassland of potential conservation interest2,b	Number of sites	Percentage of permanent pasture
Albania	4030	7000	0.38	247	4	6.1
Austria	19 930	15 260	0.42	3	2	0.0
Belgiume	5564	8233	0.45	6	>35	0.1
Bulgariad	19 990	41 620	0.56	>3000	>10	15.0
former Czechoslovakia	16 730	50 600	0.53	10 377	6	62.0
Denmarkd	2079	25 484	0.64	<100	nd	4.8
Francee	111 980	192 340	0.55	490	>100	0.4
Germany	53 290	120 020	0.49	1603	>300	3.0
Hungary	11 730	52 870	0.69	2500	57	21.3
Ireland	46 940	9330	0.80	2300	nd	5.0
Italy	48 800	119 750	0.56	2053	21	4.2
The Netherlandse	10 800	9110	0.53	6	10	0.1
Poland	40 380	147 150	0.60	6057	55	15.0
Portugal	8380	31 730	0.43	540	9	6.4
Romania	47 780	100 200	0.62	352	5	0.7
Spain	103 000	200 890	0.60	23 000	63	22.3
Switzerlandd	16 990	4120	0.51	8	nd	0.0
UK (England only)	111 800	66 000	0.73	3815	184	3.4
former Yugoslavia	63 520	77 300	0.55	2794	21	4.4
Totals/average	743 713	1 279 007	0.56	>60 000	>850	9.2

Notes (size of areas in square kilometres):
1 Calculated as the sum of pasture plus cropland, divided by total land area.
2 Figures are very approximate and generally represent the maximum likely areas currently holding important populations of grassland species.
nd: no data.
Sources: a – FAO; b – Nature Conservation Bureau; c – Eurostat; d – national inventories; e – CORINE Biotopes Data

Table 9.1
Lowland dry grasslands in 19 European countries

- Heath and scrub, in temperate shrubby areas: Atlantic and alpine heaths, subalpine bush and tall herb communities, deciduous forests recolonisation, hedgerows, dwarf conifers;
- Sclerophyllous scrub: Mediterranean and sub-Mediterranean evergreen sclerophyllous bush and scrub (maquis, garrigue, mattoral), recolonisation and degradation of broadleaved evergreen forests, Macaronesian xerophytic communities;
- Phrygana: cushion-forming thermo-Mediterranean sclerophyllous formations, often thorny and summer deciduous;
- Dry calcareous grasslands and steppes: dry thermophilous grasslands of the lowlands, hills and montane zone, on mostly calcareous soils, sands, decomposed rock surfaces;
- Dry siliceous grasslands: poor Atlantic and sub-Atlantic mat-grasslands of strongly acid soils; grasslands of decalcified sands, Mediterranean siliceous grasslands;
- Humid grasslands and tall herb communities: unaltered or lightly altered wet meadows, tall herb communities; and
- Mesophile grasslands: lowland and montane mesophile pastures and hay meadows.

Other, very closely related, habitat types are the salt and gypsum continental steppes, included later in this chapter in the section on deserts and tundras. Scrub and grassland plant communities away from mountain tops or on very poor soils are generally highly modified by, if not entirely derived from and maintained through, human activities. In particular, wetland drainage or forest clearance for timber, arable farming and animal husbandry represent major components of a dynamic cycle that establishes scrub and grasslands, maintains them through low intensity (rotational) cereal growing, regimes of grazing and burning, to abandonment and eventual forest or wetland regeneration. Accordingly, their species turnover can be quite rapid, and the composition at any one time greatly reflects their previous history of human use, as much as environmental factors such as soil and aspect.

About half the land surface of Europe is either permanent pasture or under arable cultivation, ranging from 83 per cent in Ireland to 29 per cent in the former Yugoslavia. There are some 67.7 million hectares of permanent pasture, over 30 per cent of which occurs in just France, Spain and the former Yugoslavia. However, only about 5.7 million hectares (8 per cent) of this pasture are lowland dry grasslands, offering habitat for birds

such as quail (*Coturnix coturnix*), stone curlew (*Burhinus oedicnemus*) and little bustard (*Tetrax tetrax*). Throughout Europe, more than 700 sites of lowland dry grasslands have been identified (Tucker, 1991).

In Europe, lowland dry grasslands are represented in plains or gently undulating ground (ie, areas particularly vulnerable to arable intensification) with vegetation averaging less than one metre in height, and not dominated by ericaceous species. Lowland dry grassland can include a more or less substantial cover of low shrubby vegetation, especially in semi-arid areas, where grass species may be mainly annuals. They support distinct plant and animal communities; in birds, for example, species tend to be ground-nesting, have cryptic plumage, conduct conspicuous courtship and territorial displays, form flocks in the non-breeding season, and range freely over large areas (Goriup and Batten, 1990).

Very little useful statistical information is available concerning the extent of lowland dry grasslands in Europe. Hopefully, the data will improve at least in the EU as the provisions of the Flora, Fauna and Habitats Directive take effect. A compilation of available data from various sources is presented below; however, Greece and the former republics of the USSR have been excluded owing to lack of comparable data.

It is surprising that Austria, at the border of the central European plain, has hardly 300 ha of native lowland dry grassland left (see Table 9.1). On the other hand, it appears that the UK is much more important for lowland dry grasslands than had previously been appreciated. The main concentrations of habitats occur in Wiltshire, East Anglia and North Yorkshire. This realisation recently led English Nature, a statutory agency, to draw up a grassland conservation strategy for the English lowlands directed at identifying all remaining dry grassland sites and, where possible, reforming them into large contiguous areas through incentive schemes to encourage arable reversion in their vicinity.

Functions and values

Many animals and birds prefer unimproved or lightly managed pasture as their habitat, a type that resembles in structural terms the native grasslands, especially in the Iberian peninsula and Eastern Europe. Similarly, crops that are grown organically, or at least without using pesticides, often harbour important communities of plants and animals. Some birds are positively attracted to pesticide-free crops because of the better cover for breeding and abundance of insects for food. The careful management of these pseudo-steppes in the context of an overall regional landuse plan should form an important part of future conservation policies for scrub and grassland habitats.

Other than croplands, grasslands with low or no input of fertilisers and pesticides represent a type of agricultural landuse where ecological quality and economic benefits appear to be compatible with each other. The functional qualities of scrub and grasslands are:

- high biological diversity of species;
- presence of traditional, even historic (grazing) agricultural landscapes; and
- high aesthetic/recreational value.

In Austria, 1041 species of insects, of which 85 per cent are on the national Red List, depend on dry grasslands (Holzner, 1986). On less than one hectare of the Kaiserstuhl, for example, calcareous grasslands are home to 56 butterfly species and 131 bee species (Willems, 1990). Korneck and Sukopp (1988) list 588 species of higher plants for dry grasslands and 297 for wet grasslands in Germany.

Due to their close affinity with humans, scrub and grassland ecosystems in Europe (with the possible exception of heaths) have tended not to attract the same attention from ecologists as wetlands, forests or mountains. It is only comparatively recently that the biological value of scrub and

Map 9.4
Change in the distribution of the great bustard (Otis tarda) *in East-Central Europe: a) 1900, b) 1985*
Source: Nature Conservation Bureau, UK; P Goriup, personal communication

a) **Great bustard 1900**

b) **Great bustard 1990**

Box 9F Decline of the great bustard

Female great bustard, Estremoz, Portugal
Source: P Goriup, Nature Conservation Bureau, UK

It is no longer possible (if ever it was) for any one country alone to guarantee the future survival of many lowland dry grassland birds. In Hungary, for example, where great bustards are afforded strong protection and a large area of suitable habitat is maintained, the population has declined from over 3200 birds in the late 1970s to 1200 in 1992, because of emigration during hard winters. The population is forecast to decline to 800 by the late 1990s. The problem of the great bustard, and similar species is of such a scale that it can be tackled only on a continental basis, for example by an international agreement under the Bonn Convention on Migratory Species. This could eventually result in reintroducing the great bustard to environmentally sensitive areas holding lowland dry grasslands, with support grants paid to local farmers to manage the habitat.   An analysis of satellite data (CORINE Land-cover project) identified natural grasslands as potential habitat areas for the great bustard in Spain. Map 9.5 shows these areas as well as sites where the species has actually been recorded.

Map 9.5
Distribution of the great bustard
(Otis tarda)*in Spain*
Source: EEA-TF Bird Life

The total world population of the great bustard (*Otis tarda*) is estimated to be under 25 000 birds, of which about 60 per cent occur in Iberia alone. The bulk of the remainder are located in Russia and Ukraine (16 per cent) and Hungary (5 per cent). Over the last 50 years, the great bustard has retreated from western Germany, Poland and Yugoslavia, and just lingers (if at all) in Austria, the Czech Republic, the Slovak Republic, Bulgaria and Romania. By far the main causes for the overall fragmentation and decline of the great bustard since the end of World War 2 are the widespread ploughing of grasslands, and the application of agrochemicals as well as the more recent changes in cropping patterns (from forage to winter cereal crops). Map 9.4 illustrates the changes in distribution of this species in Central and Eastern Europe.

The population levels in Romania are fairly typical of what happened in other parts of Europe. As shown by the following figures, after World War 2, populations increased somewhat with greater arable farming, and then declined dramatically from the 1960s when intensification began:

1955	1957	1961	1970	1975	1976	1992
1110	1440	2200	550	300	29	100(?)

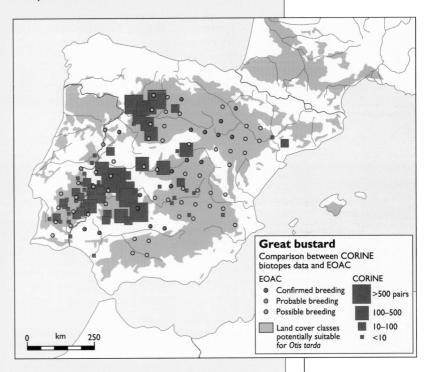

grassland ecosystems was recognised, but by then they had largely been lost to agricultural intensification.

In Europe as a whole the absence of herds of native ungulates like antelopes, gazelles and equids, grazing and browsing the vegetation, is a notable contrast to similar biotopes in Africa, Asia and North and South America. Native ungulates comprise mainly bovids (eg, ibex, *Capra hircus*) occupying Alpine areas or cervids (eg, roe deer, *Capreolus capreolus*) occupying forest margins and adapted to both browsing and grazing.

Habitat threats

The principal cause for the loss of scrub and grassland habitats over the last 50 years has been the advent of widespread agricultural mechanisation, which has allowed vast areas of natural and semi-natural grassland to come under the plough. As a consequence, much of the remaining area of scrub and grassland habitat (especially in the lowlands of northwest Europe – see Box 9G) has become highly fragmented, and restricted to steep slopes and ground with

thin soils. More recently, agricultural intensification has resulted in the loss of fallow land and stubbles, and the direct attrition of wildlife due to use of fertilisers and biocides. These two impacts, increasingly compounded by intensive tourism in protected areas, now threaten the viability of species that require large contiguous undisturbed areas of high quality habitat for their survival.

Today, the situation of many indigenous scrub and grassland species has become critical, and can no longer be neglected. Examples of threatened species include the grass Spanish Gaudinia (*Gaudinia hispanica*), Sardinian thistle (*Lamyropsis microcephala*), Alcon large blue butterfly (*Maculinea alcon*), Lilford's wall lizard (*Podarcis wilfordi*), meadow viper (*Vipera rakosiensis*), little bustard (*Tetrax tetrax*), great bustard (*Otis tarda*) and pardel lynx (*Lynx pardina*). The great bustard has suffered a rapid decline in Central and Eastern Europe (see Box 9F; and Map 9.4) and, if present trends continue, could be on the verge of a similar collapse in its Iberian stronghold (Map 9.5).

Since the early 1980s, grasslands and heaths have seemed to benefit from a shift in agricultural policies. In Germany and

Italy, the reform of EU policy has already released more than 300 000 ha of arable land for possible conversion to grassland in the lowlands ('set-aside' programme). While initially welcomed as an opportunity to increase the ecological values of such areas, these measures may also have negative impacts. The potentially most damaging developments identified by Baldock and Long (1987) are abandonment of traditional systems and inappropriate forms of forestry/afforestation (see Chapter 8). While 'less favoured areas' cover 17 million hectares (over 60 per cent of the agricultural area) in Spain, the regional administration of Castilla y Leon is developing agro-environment zonal plans that would result in the conservation of at least 1.5 million ha of dry farmland and steppe habitats. However, the overall future of grasslands affected by the 'set-aside' programme remains unclear.

In Eastern Europe, steppe ecosystems have become extremely rare in all of Ukraine and southeastern Russia. Some large-scale steppe areas remain only in the southern transitional zone towards the semi-deserts (north and west of the Caspian Sea: the Volga Delta and Terek region).

Representative sites: scrub and grasslands

A total of 615 representative sites were selected, of which 186 are presented on Map 9.6. The sites have been evaluated according to the four classes defined in the legend. This evaluation is based on site-specific data concerning present status of protection, habitat management and prevailing pressures.

Criteria for site selection

Criteria as they are described in the introduction to this chapter were applied in a flexible way, except for the minimum size, set at 1000 hectares. The most important and authoritative source was the inventory of Important Bird Areas (Grimmett and Jones, 1989). Accordingly, the site data presented here are biased toward sites of ornithological importance and no doubt a large number of sites of importance for endemic plants and insects have not been included. Furthermore, the dynamic nature of grassland maintenance and succession has to be taken into account: long-term fallow, fire management and grazing regimes can radically change the relative composition of habitat types on a site within a few months.

Interpretation of the data

The 186 representative sites that have been selected for this study cover a total of approximately 14 million hectares, most of which are located in the Mediterranean region (see Figure 9.2). More than half of these areas are of high ecological importance, but still without any protection and facing various forms of stress (hunting, road construction, drainage, etc). Many important sites which could not be selected (often less than 1000 ha) for this study, especially the large number of extensively used grasslands as part of the agricultural landscapes, are located in France, Germany and the Alpine region. A number of scrub and grasslands are fairly well protected in the UK and Denmark.

Data availability and assessment needs to be improved for Eastern and Central European countries. However, large and valuable grassland habitats are known to exist, especially in the steppe regions.

Map 9.6
Scrub and grassland
ecosystems:
representative sites
(see end of chapter
for site names)
Source: Compiled from
multiple sources by
IUCN/EEA-TF

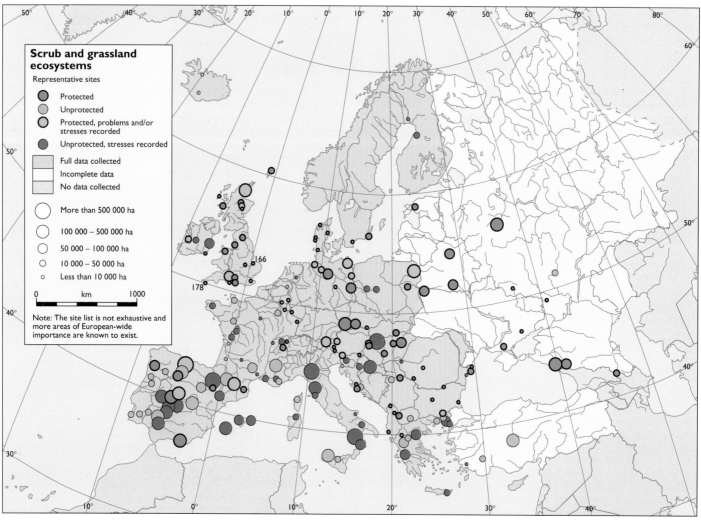

Box 9G Lowland heathland loss in eastern and southern Britain

Lowland heathlands in Britain generally occur on free-draining, nutrient-poor, acid soils. In historical times they were regarded as barren 'wastes', because they supported low woodland cover and were impossible to cultivate. Hence, many were given over to use as common grazing lands – and it is often these areas which survive today. Some existing heathlands are also derived from reversion of agricultural land and from felled woodland.

British lowland heathlands form part of the unique western European 'Atlantic' heaths. Although they are widespread in England and Wales, and also occur in Scotland, the most important sites are found in the coastal counties of eastern and southern England. In East Anglia (counties of Suffolk and Norfolk), heathlands are of a *Calluna vulgaris–Festuca ovina* type, and include the Brecklands (Site 166 on Map 9.6) and the Suffolk Sandlings. These are one of the main breeding grounds for the stone-curlew (*Burhinus oedicnemus*) in Britain. In southeastern England, including the Thames and Hampshire basins, the *Calluna vulgaris–Ulex minor* heath (Anglo-Norman heath) is dominant. These are important habitats for invertebrates, eg, Odonata, reptiles, including *Coronella austriaca*, and support the British populations of Dartford warbler *Sylvia undata*. Further west, *Ulex galili* replaces *Ulex minor* in heathland communities. In Dorset and on the Lizard peninsula (Cornwall, Site 178 on Map 9.6) associations of *Erica ciliaris* and *Erica vagnas* also occur. British heathland communities have been classified recently by Rodwell (1991).

A great many studies in the UK, particularly using data derived from old maps, have provided a picture of heathland change over the last century. Data held by English Nature (the statutory nature conservation body) indicate that the total amount of heathland in Britain (with greater than 10 per cent cover of *Calluna*) is about 60 000 ha (Farell, 1993). Past losses have been mainly due to agricultural reclamation and afforestation. In many areas these and other factors have led to

South Atlantic wet heath with furze, New Forest, UK
Source: D Wascher

severe fragmentation, so that much of the heathland which remains is found on small, isolated sites. This, in turn, has led to the impoverishment of fauna and flora. Current threats are urban encroachment and road building, pipeline installation, mineral extraction, fires and military use. Recreational pressure can be severe locally, causing erosion by trampling or from the illegal use of motorcycles or other vehicles. However, bracken (*Pteridium aquilinum*) and woodland encroachment due to lack of management remains one of the most important factors affecting the viability of the surviving British heathlands.

In response to these problems, actions have been taken to provide better protection of heathlands in the UK: in the framework of Biogenetic Reserves (four designations in 1992 by the Council of Europe), by the EC Bird Directive (Dorset Heathlands), and through a National Heathland Programme launched in 1993 by English Nature.

*Figure 9.2
Scrub and grassland ecosystems: representative sites by region*
Source: IUCN/EEA-TF

Conclusions

Very little accurate information is available on the current distribution of scrub and grassland in Europe, even less on the status and particular threats to individual sites. The countries with the greatest extent of such habitat are (in descending order) Spain, Italy, France, Greece, Hungary, the Czech and Slovak republics combined, and the UK.

Spain alone holds almost half of the identified area of scrub and grassland habitat in Europe, and in addition has the highest average site size (ie, the least degree of habitat fragmentation). Most of the dry grasslands of Europe result from human landuse activities, namely extensive grazing of areas cleared of wood. With the decline of grazing as a profitable economic activity, many grasslands have been altered and lost important habitat qualities. The remaining dry grasslands are facing the same threats. The special scientific, biogenetic, recreational, educational and aesthetic values, require an integration of grassland conservation and proper management into agricultural landuse practices.

The processes of land reform and investment in agricultural production in Eastern Europe are bound to lead to substantial decline in the extent of scrub and grassland habitats before proper surveys can be carried out and an adequate network of protected areas established. Nevertheless, efforts are being made in Hungary and Poland to introduce the concept of environmentally sensitive areas where farming should be carried out in a non-intensive manner.

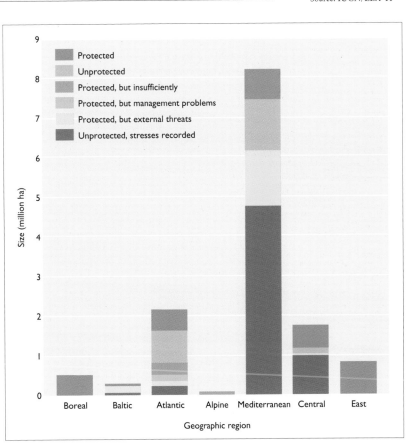

Inland waters: rivers and lakes

Introduction

Freshwater ecosystems like rivers and lakes form essential life support systems for a wide range of wetland habitats within their catchment areas. Being valuable habitats themselves, lakes and especially rivers perform a unique ecological and environmental function within the landscape by linking very different ecosystem types. Riparian corridors can extend over considerable distances and provide habitats for many species of plants and animals, while also allowing for the necessary movement and survival of populations of wildlife. Lakes, on the other hand, have important transitional and seasonal functions as resting areas for species during long-term migrations. Freshwater ecosystems are vulnerable to external pressures deriving from human landuse activities, namely water pollution and hydrologic modifications. Lakes as well as rivers do not end at their shores and banks and cannot be seen isolated from the land around them. An especially close link exists between freshwater systems and wetland habitats such as bogs, fens and marshes, which are discussed later in this chapter. All wetlands are profoundly affected by their local environment and by changes taking place on land, even at great distances (Klink, 1989).

The main habitat types, ecosystem functions and threats of natural river and lake systems are explained below, followed by an interpretation of data deriving from 184 representative river and 224 representative lake sites, illustrated in Map 9.7 (p 205).

Ecological functions: rivers

Rivers support a complex range of habitats which, under natural conditions, exhibit variability as well as both seasonal and long-term changes. These changes are the result of differences in climate and other environmental factors which control the amount and distribution of available runoff, geomorphology (which governs the pattern of flows and the ways channels and floodwater interact), and geology (which determines the character of sediment and substrate conditions).

The habitats of rivers and streams include the river channel with its mud, sand and gravel banks, and vegetation. Of equal importance, however, are the habitats which occur in the riparian corridor and especially the floodplain (Starkel, 1987). Some of these ecosystems, such as alluvial forests (see above), are intimately connected with and dependent on the floodwater regime and groundwater conditions along the river (see Chapter 5); others, such as marshes and fens, may occur because of the independent movement of regional groundwater or runoff down valley slopes.

Rivers and lakes are home for fish species, many of which are dependent in some way on adjacent wetlands – for food, spawning, nursery or other habitat requirements. The movement of fish is often closely linked to the hydrological cycle of the river basin and floodplain. The implications are clear:

- there are often complex and essential links between wetlands, rivers, lakes and the open sea;
- alteration of the water regime of a wetland or flooding cycle within the river basin, for example by drainage, irrigation or river regulation, can have considerable knock-on effects on aquatic species and fisheries production (see Chapter 5); and
- the environmental or economic benefits of associated habitats such as wetlands may be obtained at some considerable distance from their location and in a different region or country.

Pont de Chadron, Serre de la Fare, Haute-Loire, France
Source: M Rautkari, WWF

Box 9H The Loire (France)

The Loire basin drains more than 20 per cent of the whole of France. As one of the most untamed larger rivers of Western Europe, the Loire (Sites 39 to 44 on Map 9.7) and its largest tributary, the Allier, have their headwaters in the Massif Central. Their upper sections run through deep mountain gorges before entering the plains downstream of Roanne and Vichy. In the area of the confluence there are well-preserved alluvial floodplains with many braided channels, dead arms, river islands and open sand and gravel banks. Numerous bird, plant and invertebrate species, many of which are rare or endangered, occupy these floodplain habitats. The Loire is better known for its migratory fish species. Once very common, the salmon (*Salmo salar*) has suffered much from pollution and dams. At present it can only swim upstream along the Allier and the stock has been reduced to 10 per cent of the original amount.

The construction of dams is still a threat to the natural habitats of the Loire. The site of the first dam alone, planned to be built at Serre de la Fare on the upper Loire, but now cancelled, is home for 98 nationally protected species. Besides development schemes for dam constructions, the Loire ecosystems face other dangers: gravel and sand exploitation in the river bed, intensive agriculture in the floodplain, irrigation and water pollution through a lack of sewage treatment.

A NGO campaign (supported by the WWF, Berne Convention and Ramsar Bureau) under the guidance of a 'Loire Vivante Committee' led to the cancellation of two dam projects and to a review of two others. The Loire river serves as one of the representative sites for a functional analysis of European Wetland Ecosystems (EC STEP Programme) and, under the EC LIFE Programme, eight conservation projects along the Loire and the Allier have been approved and are currently being carried out. The projects aim at conserving the river's dynamics in order to maintain its biological richness and the water quality. Parallel to these activities, environmental education is being developed to lead to sustainable use of the river ecosystems and natural resources.

Ecological functions: lakes

The lakes and upland headwaters of northern and northwestern Europe are, in general, naturally poor in nutrients (oligotrophic) with rainwater as the main source. These areas are characterised by hard acid rock types, often overlain by extensive areas of peat. Waters have high oxygen levels, making for good fish-spawning conditions. They are frequently found in sparsely inhabited areas. However, oligotrophic lakes are very vulnerable to acidification from atmospheric sources or upland drainage because of limited natural buffering capacity.

Eutrophic lakes, conversely, are well supplied with

nutrients and are characteristic of lowland, soft-rock catchments with a base-rich geology and fertile soils. These are frequently enriched in nutrients due to a range of human activities within the catchment. An intermediate mesotrophic category is also recognised. Aquatic vegetation, from microscopic plankton to submerged plants rooted in the lake sediments, serves as a food base for the lake ecosystem. The type of vegetation and its productivity are controlled not only by light intensity and temperature, but also by the availability of nutrients, in particular nitrogen and phosphorus. Supply of these nutrients under natural conditions depends upon the climatic, geological, soil and vegetational characteristics of the catchments. In most areas these have been variously modified by landuse practices and other human-induced, environmental changes, which can seriously threaten or damage the wetland ecosystems.

Wetlands are vital for maintaining the integrity of lake ecosystems, not only in terms of supporting wildlife habitats and species, but also for a range of environmental benefits, including the capacity to improve water quality through such processes as denitrification, which reduces nitrate levels, and removal of phosphorus by incorporation into plant biomass. There are, of course, large benefits for recreational and economic use (Burgis and Morris, 1987).

Habitat threats: rivers and lakes

With a few exceptions (see Box 9H), most of Europe's rivers, especially the large ones, have undergone major physical changes due to stream corrections (disconnecting them from oxbows, wetlands and former floodplains) and flood control. Navigation and dam constructions can drastically change the environmental condition of rivers, often slowing down the water velocity and hence affecting the river's ecology. As a consequence, food-chains as well as the migratory patterns of fish species can be seriously affected (see Box 9I). Species which are especially adapted to particular river habitats are often replaced by a more general and less diverse fauna and flora. The widespread modification of river banks also includes the construction of concrete embankments and dikes. Artificial separation of many rivers from their adjacent riparian corridors and lowlands has caused a severe decline of alluvial forests, making them now one of the most endangered habitat types in Europe (Vivian, 1989). Both lake and river ecosystems are particularly vulnerable to changes in water quality (see Box 9J).

The phenomenon of accelerated nutrient enrichment (eutrophication) occurs widely where lakes have farmed or densely populated catchments. It is a consequence of intensive agricultural or fish farming operations and sewage effluent discharge. The results are an increase in phytoplankton production and microbial decay with consequent deoxygenation of water, production of toxins and a general decline in wildlife conservation value. The example of the Swedish Lake Hornborgasjön (see Box 9K) illustrates possibilities of undertaking the restoration of a damaged lake ecosystem. Non-impacted oligotrophic lakes are found in sparsely populated areas with relatively little human activity such as the northern Scandinavian peatlands (see Box 9L) or in mountain areas. However, some areas – especially in southern Scandinavia – have been the subject of acid rain deposition (see Chapter 5), derived partly from distant industrial sources, which has led to acidification of lakes and loss of natural populations of animals and plants.

The dramatic increase of acidification of European rivers and lakes is one of the best-documented environmental problems, especially in the southern parts of Scandinavia. During the 1950s and 1960s, at the peak of sulphur deposition, the rate of acidification was several hundred times that of the natural process. With pH-values

Box 9I The Volga (Russia, Ukraine)

Europe's largest river system, the Volga, is over 3500 km in length and drains an area of 1 360 000 km². It rises in the Valdai Hills, northwest of Moscow, and discharges into the northern Caspian Sea through a huge delta comprising more than 800 channels and with a seaward boundary over 200 km long. The Volga Delta (Site 154 on Map 9.7) is of international importance for wildlife conservation, its 652 000 ha area is designated a Ramsar site and contains a Biosphere Reserve. In the lower Volga extensive wetlands are found on the floodplain downstream of Volgograd. However, the natural character and functioning of the Volga ecosystems have been radically altered by the regulation of river flows through the construction of dams for hydroelectric power generation, flood protection and the creation of storage reservoirs for irrigation. Set against these economic benefits has been a serious decline in fishing yields and reduced soil fertility (Findlayson, 1992).

The fishing industry of the Caspian Sea and its major rivers, notably the Volga, accounts for 90 per cent of Russian sturgeon and 50 per cent of the world's beluga supply. During the period 1959–90 fisheries losses were estimated at almost 40 billion roubles (Findlayson, 1992). Dams at Volgograd and Saratov have prevented the important migration of many fish to spawning grounds, which had been formerly as much as 1500 km upstream. Now large spring floods are reduced by 25 per cent which has had an adverse effect on wetland ecosystem functioning.

Opportunities for fish to feed and spawn in the delta and floodplain have dramatically dwindled and food-chain support has declined through a reduction in water exchange and the sediment inputs which maintain fertility. Now some 7 million tonnes of sediment are trapped every year in reservoirs of the Volga. The delta and floodplain of the

Damtchik Canal, Volga Delta Source: W Fischer, WWF

lower Volga are exceptionally rich and diverse wetland habitats, containing over 250 bird, 400 vertebrate and some 430 vascular plant species. However, agriculture (rice paddies and reclamation of land into polders for irrigated cropping) has produced salinisation and contamination from fertilisers and pesticides. Major contributions to the pollution load are the extensive industrial, urban (eg, sewage waste from Volgograd) and agricultural developments within the wider catchment.

In order to address the concerns over wetland loss in the lower Volga a strategy and action plan for wetland conservation was developed by the World Wide Fund for Nature (WWF) in cooperation with local and national conservation bodies and directed at all parties with an interest in the wetlands concerned (Findlayson, 1992). The themes were: establishment and management of protected areas, sustainable use of biological resources, landuse and water pollution, hydrological regulation and water management, and integrated management structures. Currently, WWF-International is preparing the designation of three large-scale protection areas.

Box 9J The Danube

The Danube traverses or borders parts of nine countries and is fed by several thousand tributaries, including some of the most industrialised and highly polluted parts of Europe. Yet the Danube Delta and Valley represent one of Europe's last and most extensive wetlands in a natural state.

The Danube Valley

The Danube is a major international transport route with over 3500 ships passing annually through the delta. The river discharges 450 000 tonnes of nutrients per year, mostly agricultural chemicals, into the Black Sea, resulting in the loss of fisheries through algal bloom production. Untreated urban sewage discharges cause excessive pollution. Bratislava releases over half of its waste into the Danube (IUCN, 1990b).

Many internationally important wetlands are identified by Grimmett and Jones (1989) along the course of the Danube. In Bavaria, 38 000 ha of riverine forest and riparian lowlands are threatened by water abstraction, tourism, agricultural intensification and lately by construction plans in the framework of the Rhine–Main–Danube shipping route (see also Box 21B).

Some of the largest remaining near-natural tracts of floodplain forest in Central Europe are found between Vienna and Hainburg (see Site 1 on Map 9.8), where the possibility of hydroelectric power generation poses a major threat. In Slovakia the Danube Water Scheme, involving a major diversion of the Danube between Bratislava and Gabčikovo, with the construction of a canal and reservoir, is supposed to be developed as the Danube Protected Landscape Area, within which careful management will be required to protect wetlands and their functional integrity. In this respect, the most important Hungarian sites are at Szigetköz, Hansági and Gemenci.

Within the area of the former Yugoslavia over 6000 hectares of flooding river marginal wetland are identified in Croatia and in Serbia (Vojvodina) comprising forests, marshes, pools and streams. Since 1993 'Kopački rit', also

listed as a Ramsar site, is protected by law as a special reserve. A number of islands in the Danube in Bulgaria, bordering Romania, carry seasonally flooded forest and marsh habitats. The freshwater Lake Srebarna (a Ramsar Biosphere Reserve and World Heritage site) with fringing reedbeds is connected to the river and is surrounded by forests, vineyards, arable land and steppe.

The Danube Delta (Romania/Ukraine)

Its extensive reedbeds, maze of tributaries, canals and lakes with their great abundance of aquatic plants, its white willows and poplars, and not least its dunes with their mosaic of forests and semi-arid grasslands form a unique and important habitat complex.

The Danube Delta (Site 153 on Map 9.7) ecosystem holds the majority of the world's population of two endangered species – the pygmy cormorant (*Phalacrocorax pygmaeus*) and the red-breasted goose (*Branta ruficollis*). There are also significant populations of several water-bird species, whose populations have recently declined drastically, especially in Europe (Munteanu and Toniuc, 1992).

The ecological and economic (fishery) system as a whole is under threat, and continued decline can be expected unless urgent measures are taken. The main causes of decline are:

- interference in the natural hydrological cycle through the construction of canals, dikes and polders;
- the overexploitation of the biological resources by overfishing and overgrazing;
- ill-conceived attempts at intensive agriculture, fish farming and forestry development;
- pollution, including eutrophication as the result of the use of fertilisers, and toxicity from the use of pesticides.

In 1990, the Romanian government recognised the importance of the area by declaring an international 'Biosphere Reserve' over 442 000 ha and by establishing a Danube Delta Biosphere Reserve Authority to address the complex set of problems facing the area. A recent application for World Heritage listing was successful.

In June 1994, the convention for cooperation on the protection and sustainable use of the Danube river basin was signed (see Chapters 6 and 33).

Danube lowland near Hainburg, Austria
Source: K Pahlich

Box 9K Restoration: Lake Hornborgasjön (Sweden)

Substantial changes to the hydrological and ecological management of Lake Hornborgasjön (Site 384 on Map 9.7) have been initiated in order to restore the area to its former importance for bird conservation. Degradation of the site began in the early nineteenth century with drainage to increase the area of arable land. The level of the lake was lowered some 2 m between 1803 and 1993, as a result of canalisation for conducting water away from the site. Regularly flooded areas were greatly reduced, and in summer open water was found only in the canal and a few lagoons. The basin was filled with decomposing plant material and the former lake became invaded by reed and scrub. Original ecological functions became substantially altered and there was widespread concern for the decline in breeding birds in an area which had formerly been of considerable ornithological interest. Restoration of the lake required a raising of the water level,

implemented in two stages, by a total of 1.5 metres (Bjork, 1973), with the creation of dikes to protect surrounding arable land. Ten years later the plans were modified, reducing the water-level rise to 0.85 m and abandoning the diking proposals in favour of creating more marginal habitats, natural vegetation patterns and shallow water areas to encourage submerged vegetation. Flood meadows were also favoured to provide a source of seeds for ducks. More shallow water would also be of benefit to birds in restricting fish populations and thereby reducing competition for food resources. Reed clearance was a priority for management. The final government sanction for the plans was approved in 1990. The rise in water level is due to be completed by 1995. In 1991 it was already reported (Herzeman and Larsson, 1991) that some 7 km² of reeds had been cleared and open water areas re-established. Some 2200 hectares of surrounding land have been acquired by the Swedish Environmental Protection Agency in order to regulate management activities and preclude damaging operations. Already improvements have been recorded in breeding bird populations. Dunlin (*Calidris alpina*), ruff (*Philomachus pugnax*), little gull (*Larus minutus*) and black tern (*Chlidonias niger*) have begun nesting since the start of the restoration programme (Grimmett and Jones, 1989).

Lake Hornborgasjön, Sweden
Source: M Moser, IWRB

Box 9L Finnish lakes

Finland has more than 55 000 lakes with a surface area greater than one hectare, many of which are set in catchments with little human impact and dominated by vast areas of peatland and forest. There are some 2600 mostly natural lakes in excess of 1 km² in area, of which 500 are pristine, lacking measurable human impact other than atmospheric deposition (Sites 235 to 237 and 244 on Map 9.7). In total an area of 606 km² is legally protected on account of high ecological quality, most of which (555 km²) is found in the Vätsäri wilderness area. A typical example of an oligotrophic lake occupying a near-natural basin is Lake Inari in northeastern Finland. The area is sparsely populated and used for reindeer herding and contains extensive pine forests. At 1050 km² it is the second largest lake in Finland. Water quality is generally good and there are no indications of acidification, though there is localised wastewater discharge from lakeside settlements.

The main impact on the lake has been regulation of water levels from the hydroelectric power plant constructed in 1934 on River Paatsjoki within the former USSR. Regulation has led to a decline in some fish species such as lake trout (*Salmo trutta*) and arctic char (*Salvelinus alpinus*, see also Box 9GG). Vendace (*Coregonus vandesius*) has developed into an important commercial fisheries species since its introduction in the 1960s. Severely polluted lakes in extensively developed or exploited catchments and with inadequate treatment of wastewater or agricultural runoff account for only one per cent of the total number of lakes in Finland.

Untouched lake forest, Eastern Karelia, Kuhmo, Finland
Source: M Rautkari, WWF

significantly below 5.0, thousands of Scandinavian lakes have become too acid to function as fish habitats. In Norway over 2000 local fish populations have been affected or eliminated; in Sweden 21 500 lakes (larger than 1 hectare) have low alkalinity, covering more than 3200 km²; and in southern and central Finland 42 per cent of lakes larger than 1 hectare have a pH below 5.0 (Bernes, 1993). Despite a decline in atmospheric pollution during the last two decades and partial mitigation through liming activities, acidification continues to have detrimental effects on lake ecosystems (see also Chapters 5 and 31).

The use of lake and river water for industrial purposes often affects water quantity (when abstracting water) and water quality (when reintroducing it as wastewater, sometimes polluted or of higher temperature). Many industries, such as pulping and mining, wash large quantities of particulate matter into lakes and rivers.

Map 9.7 River and lake ecosystems: representative sites (see end of chapter for site names)
Source: Compiled from multiple sources by IUCN/EEA-TF

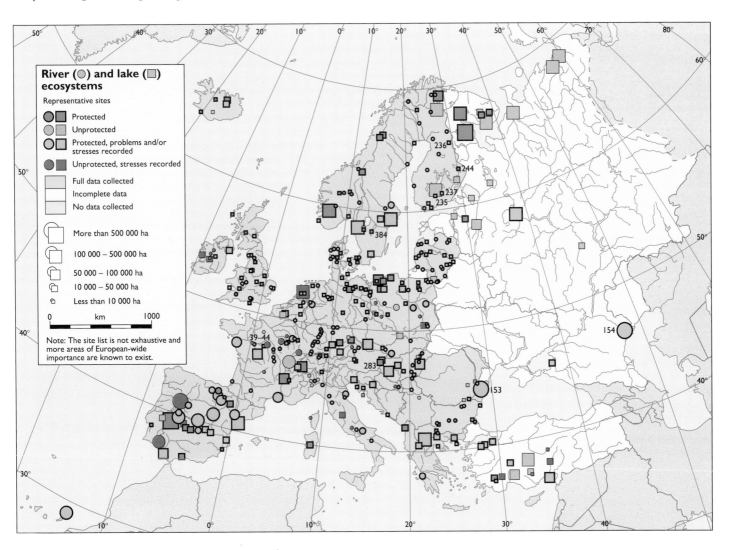

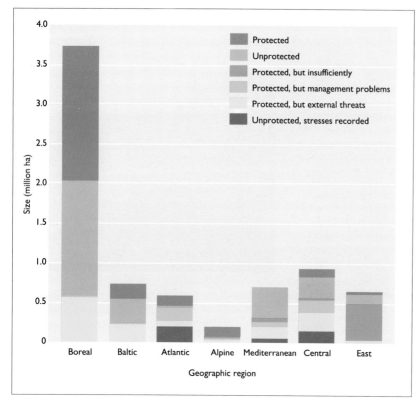

Figure 9.3
Lakes: representative sites by region
Source: IUCN/EEA-TF

Lake Balaton, near Czopak
Source: Spectrum Colour Library

Box 9M Lake Balaton (Hungary)

The 596 km² area of Lake Balaton (Site 283 on Map 9.7) in Hungary makes it the largest lake in Central Europe. It acts as a major tourist attraction and provides important facilities in summer, whilst forming a winter refuge for wetland birds – 70 000 geese, 30 000 ducks and 10 000 coots. In 1989 the lake was designated a part-time Ramsar site for October to April. Former water level fluctuations have now been more or less eliminated by regulation at the outflow canal leading from the lake, as an aid to navigation, lakeside recreation and agriculture.

The annual catch amounts to 1200 tonnes, mostly comprising bream (75 per cent) and pike-perch (15 per cent) and with important catches of eels. The lake is also used extensively by anglers, with 115 000 permits issued in 1986 (Hollis and Jones, 1990).

In summer the resident population of the lakeside of 250 000 swells to some 860 000 with visitors and tourists. This pattern became established during the 1950–75 period and eutrophication of the lake became a serious problem as a result of the growing use of fertilisers and inputs of phosphorus from sewage discharges from lakeside towns and tourist developments. Algal blooms became particularly frequent in the most polluted western parts of the lake. There is now a pollution control programme in place to reduce environmental impacts. Sewage treatment with phosphorus removal facilities has improved. Agricultural pollutants in the Zala River at the southwest end of the lake are now retained in a newly constructed series of reservoirs comprising a wetland complex of almost 15 000 ha of open water, reedbeds and *Salix* scrub, 1400 ha of which are designated a Ramsar site. Other provisions for improving water quality in the lake include the establishment of reservoirs on other major tributaries feeding the lake, diversion away from the lake of treated effluent produced in recreational areas, selected removal of phosphorus-rich lake sediment and the control of wastes from large stock-rearing units.

Representative sites: rivers and lakes

For inland waters, a total number of 184 representative sites for rivers and 224 sites for lakes were selected; all are shown on Map 9.7. The sites have been evaluated according to the four classes defined in the legend. This evaluation is based on site-specific data concerning present status of protection, habitat management and prevailing pressures. Difficulties arose from a paucity of data for river ecosystems. As a result, only a small fraction of representative sites could be fully evaluated. In the case of rivers, the question of defining site boundaries is of prime consideration.

Information on lakes was mainly retrieved from the ICBP/IWRB publication, *Important Bird Areas* (Grimmett and Jones, 1989), from Hollis and Jones (1990) and the CORINE Biotopes Programme (CEC, 1991b), as well as proposals from scientific experts and from a questionnaire on hydrological details for lakes of importance for nature conservation among European countries.

Table 9.2
Lake chemistry and impacts of 147 representative sites
Source: E Maltby, personal communication

Chemistry	Total	Threats					
		A	B	C	D	E	F
Freshwater	131	56	7	38	38	8	10
Oligotrophic	9	7	–	1	1	–	–
Mesotrophic	5	1	–	3	3	–	–
Eutrophic	34	12	2	18	10	3	2
Undifferentiated*	83	36	5	17	24	5	8
Brackish	4	1	3	2	1	1	–
Saline	12	5	2	6	4	3	1
TOTAL	147	62	12	47	43	12	11

* freshwater type unable to be identified owing to insufficient data.
A = non-significant or described
B = hydrological change
C = water quality deterioration
D = tourism/recreation
E = on-site impacts, excavations, developments
F = drainage of marginal vegetation

Criteria for site selection

The paucity of data on rivers made it necessary to be less restrictive in applying selection criteria as for other ecosystem types. For the selection of lakes the criteria as described in the introduction have been applied. In order to give a balanced picture of the present status of European lakes, examples of seriously impacted or degraded sites have been included.

Interpretation of the data

Map 9.7 shows that representative sites could be identified for only a few larger European rivers. Hence, the resulting portrayal of ecologically important riparian habitats understates the significance of this ecosystem type within Europe's natural environment. More than 50 per cent of all representative river sites cover less than 1000 hectares, due to relatively short river sections of ecological values or the lack of adjacent lowland habitats. Despite the obvious data gaps, the map mirrors the insufficient attention and protection for rivers throughout all geographic regions.

Besides entrophication, pollution, hydrological change and industrial use, problems derive also from intensive recreational use such as swimming, wind-surfing, motorboats, sailing, water-skiing and wildlife watching, with often negative impacts on the peripheral vegetation and threats to local or migratory wildlife populations (see Box 9M). Table 9.2 provides an overview of impacts that affect 147 of the representative lake sites. A more detailed interpretation is given in Figure 9.3. In the boreal region, lakes appear in large sizes and with relatively good environmental qualities. Half of these sites are well protected (Table 9.3). External threats play an important role in most geographic regions since lakes are often exposed to pollution and tourism. Table 9.4 illustrates the situation of the Ramsar sites in Europe. A high number of small lakes with important ecological functions do not show up on Figure 9.3.

Conclusions

Rivers often suffer from ecological changes and damages that can originate from very distant locations; this is the case, for example, with the Danube, Rhine and Volga. Maintenance of marginal river habitats in Europe is a particular problem, because many rivers cross national boundaries between states which have differing conservation and environmental protection policies and resources. With a few exceptions (such as northern Russia's Karelian and Kola region), riverine ecosystems have suffered from major impacts and losses. In the lowlands, water pollution is often considerable, though even in thinly populated uplands and northern latitudes acid impacts can be significant from either direct aerial deposition or through soil acidification. Relatively few data are available on the precise mechanisms by which river regulation and pollution levels stress the dynamics of associated ecosystems.

An EU initiative involving also the European Bank for Reconstruction and Development and the World Bank (Global Environment Facility Fund), is underway to improve knowledge of biological resources of the Danube, one of Europe's major rivers. The project can be viewed as a first step to maintain or rehabilitate environmental quality for the next century.

The fact that the list of representative river ecosystems includes only 184 river sites, still excluding inter alia some 30 major rivers, illustrates that river ecosystems have been severely neglected within existing inventories (including those related to the Ramsar wetlands Convention) and where they have been included it has been on the basis of small individual areas usually chosen for purposes of protection of a completely different habitat type (forest and grassland being the most common). In the creation of biological networks, however, rivers ought to play a crucial role as natural corridors and linkages between other ecosystems.

Lakes are extremely varied in morphometry, water chemistry, hydrological regime and in their associated aquatic and marginal ecosystems. They differ considerably in their buffering capacity against potentially damaging chemical inputs and in their vulnerability to a range of other deleterious impacts. In northern and northwestern Europe the many naturally oligotrophic lakes have low buffering capacity. Many in Scandinavia have suffered acidification from atmospheric inputs of acid rain from often distant industrial sources and/or the effects of acid runoff from adjacent forested land with acid peat or mineral soils. Elsewhere, lakes are predominantly eutrophic, fed by groundwater or surface flow frequently loaded with nutrients. The effects of eutrophication are damaging to ecosystem functioning, as are many of the human-induced hydrological changes (Table 9.4). Lowering of water levels by abstraction is common in the continental or semi-arid climates of much of Central and Southern Europe. In some places new waterbodies have been created by the

Total number of sites	International designation1	National designation2	No designation but international recognition3	No recognition4
224	58	109	47	10

1 All or part of site designated by one or more international measures for conservation, eg Ramsar or EU Special Protection Area.
2 All or part of site designated by one or more national measures for conservation, eg Nature Reserve or National Park.
3 Unprotected but recognised as being of international importance, eg Important Bird Area in Europe (Grimmett and Jones, 1989).
4 Unprotected and with no recognition of international conservation importance.

damming of major rivers, as in Spain, in order to meet rapidly increasing demands for drinking water and irrigation (see Chapter 5). Though these have had some recreational and ornithological benefits, it has been at the expense of destroyed marginal natural river ecosystems through flooding, regulation of flow and channelisation. Direct losses of lake marginal ecosystems have also resulted from tourism, urban and industrial expansion, landfill and activities of the extractive industries, though here final abandonment of sites can lead to the creation of valuable artificial lakes (as in the Thames Valley, UK) following gravel extraction.

Of major concern has been the continued loss and degradation of freshwater habitats in the Mediterranean basin which has included many inland lakes in addition to valuable coastal lagoons and estuaries. In Spain alone, more than 60 per cent of all inland freshwater wetlands have disappeared during the last 25 years (Casado et al, 1992). Causes include land reclamation for agriculture or tourist and urban development, groundwater overexploitation or water pollution.

Table 9.3
Legislative status of representative sites for lakes
Source: E Maltby and EEA-TF

Table 9.4
Threats and impacts to some Ramsar lake sites in Europe
Source: E Maltby and IWRB, personal communication

Number of sites	Unimpacted	Hydrological changes	Decline in water quality	Peat/mineral extraction	Tourism/ recreation pressures
54	16	13	25	1	18

Note: Sites can be subject to multiple threats and impacts.

Bogs, fens and marshes

Introduction

Europe's bogs, fens and marshes are important reminders of the primeval landscapes which once dominated extensive tracts of the land surface. Cool, continuously wet regions with mainly gentle relief, especially those in Northern Europe, are suitable for the extensive formation of peatlands with layers sometimes exceeding 50 metres (Polieski National Park, Poland). Peat, consisting of decomposed organic matter (plant material), is the key component of mires. The following habitats can be differentiated:

- Bogs: peat-based ecosystems which are maintained by precipitation, or where accumulation of peat occurs in water with exceptionally low nutrient content. They occur as raised bogs and blanket bogs (flat surface), both oligotrophic and dominated by bog mosses (*Sphagnum* spp) species. A third type, mainly existing in the pre-Alps, are the forested bogs with mountain pine (*Pinus montana*).
- Fens: often resulting from vegetational invasion of open water habitats and waterlogging by mineral-enriched groundwaters, fens are largely based on peat. Herbaceous vegetation such as sedges (*Carex* spp), reeds (*Phragmites communis*, *Typha* spp) and related communities is succeeded by scrub and woodland of alder or willow unless

management intervenes. Sedge-dominated fens extend well into the Mediterranean forest-steppe and the desert-steppe region of Europe.

- Marshes: herbaceous wetlands which are not peat-forming and which may be temporarily, seasonally or permanently waterlogged or flooded. They are often found fringing lakes, ponds or river channels.(Saltmarshes are discussed on p 214.)

The Nordic landscape is still rich in bogs, fens and marshes, forming a distinctive wetland landscape analogous to the so-called muskegs of North America. In Europe's mid-latitude lowlands, however, a dramatic loss of habitat has been a long and progressive feature of landscape development since early medieval times. Originally, some 178 million ha of peatland existed in Europe and the former USSR (Immirzi and Maltby, 1992). While some extensive areas still exist in the boreal and eastern regions, vast amounts of original existing peatlands of Scandinavia, UK, The Netherlands, Belgium and Germany have disappeared. In Finland, the once existing area of intact bogs and fens was about 11 million hectares – about 30 per cent of the land surface. Today, they are thought to cover less than 6.5 million hectares. Their loss is due to natural drying, drainage for agriculture and peat extraction. The situation is even more drastic in Germany. From the originally existing 330 000 hectares of raised bog ecosystems in Lower Saxony, less than 1 per cent (2500 hectares) remain intact, while 3800 hectares are under regeneration and a further 13 000 hectares are strongly affected by drainage and other impacts. Although local authorities have launched an ambitious programme targeting the re-establishment and conservation of 50 000 hectares, there is a definite 85 per cent loss of the original area (Drachenfels et al, 1984).

The main habitat types, ecosystem functions and threats of important wetland ecosystems are explained below, followed by an interpretation of data deriving from 182 representative bogs, fens and marshes illustrated in Map 9.8 (p 210).

Functions and values

- The European blanket bogs and raised bogs represent some of the best and scientifically most important examples, worldwide, of these ecosystems and habitat types.
- Fens and freshwater marshes form essential linkages with other ecosystems:
  – water flows from fens and marshes to rivers and lakes;
  – migrating birds use fens and marshes as important feeding, nesting and overwintering sites.
- Bogs and fens play a key part in carbon dynamics. When undisturbed they act as a sink for CO_2 and a source for CH_4, but when drained they provide a major source of CO_2 (Immirzi and Maltby, 1992).
- Fens serve as pools for the accumulation of nutrients such as phosphorus and nitrogen, giving them important functions as 'nitrogen traps'.
- Peat-forming systems are unique archives of environmental information available from the pollen, macrofossils and fall-out stored in sequential layers throughout the peat depth. They may include archaeological remains and even animal and human bodies.
- Bogs, fens and marshes provide essential habitats for sometimes highly specialised plant and animal species, some of which are threatened with extinction because of habitat destruction.
- Like other natural landscapes, bogs, fens and marshes provide important wilderness that is attractive for humans.

Habitat threats

Bogs, fens and marshes are particularly sensitive habitats. Any alteration of the quality or the quantity of the water supply will change the character of the habitat and affect the chances of

Clara Bog region, Ireland
Source: D Hogan

Box 9N Irish bogs

Almost all raised bogs in Ireland have been damaged to some extent, usually by a combination of hand cutting and burning. Of the original area of about 310 000 ha only 7.4 per cent have retained an intact surface and only 6.2 per cent can be classified in the top three categories for Nature Conservation value.

In Northern Ireland only 2270 ha of lowland peatland remained intact in 1987 from an original area exceeding 25 000 ha. By 1990 the figure was estimated to be less than 2000 ha. At the rate of decline of about 3500 ha per year it was estimated in 1990 that all Irish raised bogs would probably be seriously damaged by 1995.

Clara Bog (Site 94 on Map 9.8) is one of the largest remaining reasonably intact raised bogs in Ireland. It is a declared Nature Reserve (1987), a Ramsar site and a Biogenetic Reserve of the Council of Europe. The bog contains all the characteristic features of a raised bog: a hummock-hollow pattern, *Sphagnum* lawns, pools and a typical flora. An outstanding feature is the series of flushes where mineral rich water affects the surface of the bog resulting in the presence of species more typical of poor fens and including birch (*Betula pubescens*). The site has been acquired from Bórd na Mona (the peat board) by the Irish Department of fisheries and forestry and is now protected from future peat mining. However, survival can only be guaranteed if the surrounding hydrology is not substantially altered.

survival of important species. Alteration may take place as a result of physical damage to the site such as excavation of peat, diversion of surface water flow denying marshes or fens regular floodwater, or lowering of groundwater by overexploitation of aquifers resulting in a drop of the water table beyond the rooting depth of the marsh vegetation (Hutchinson, 1980; see Boxes 9N and 9O).

As Table 9.5 illustrates for Ramsar sites, peat bogs, fens and marshes continue to be threatened as much by indirect as by direct threats – lowered water tables due to aquifer exploitation, exploitation of particular species (eg, *Drosera* spp for pharmaceutical needs), reduced or regulated runoff, acidification and pollution of rainfall and surface waters, as well as by peat mining for energy and horticulture, afforestation and drainage for land development.

Atmospheric pollution, particularly through sulphur dioxide and associated acidification, affects the growth of *Sphagnum* moss which plays an essential part in the build-up of peat deposits, thus preventing active bog growth. This may pose a major problem in peatland regeneration strategies. A relatively large portion of Central and Eastern Europe's wetlands tends to be intact (see Box 9P) compared to many, also much smaller, examples found in Western Europe (IUCN, 1990b)

Wild boars in Doñana National Park, Spain
Source: E Kemf, WWF

Box 9O
Doñana National Park (Spain)

The marshes at the mouth of the River Guadalquivir in southern Spain comprise some 180 000 ha of diverse wetland ecosystems. That part protected within the Doñana National Park (Site 154 on Map 9.8), a designated Biosphere Reserve and holder of a European Diploma (Council of Europe), includes a Ramsar site of almost 80 000 ha, forming the largest natural wetland ecosystem in southwest Europe. The marshes are fed by the Madre des Marismas and contain old branches of the Guadalquivir River which form canal-like depressions. There is extensive spring flooding and also several shallow lakes which persist even during the dry summer period.

Outside the National Park to the north and east considerable areas have been converted to rice fields or reclaimed by drainage and used for irrigated cultivation. Fish farms and saltworks have also been established, while other activities involve cattle grazing, hunting, fishing and tourism.

Agricultural use of pesticides has caused serious contamination of waters leading to the poisoning of birds. There has been an increasing demand for groundwater for horticultural irrigation and adjacent tourist developments. This already has seriously depleted the freshwater aquifer supporting the marshes, thereby increasing the risk of saline incursion from the sea. Managing within the National Park includes the digging of wells for freshwater to attract birds away from those shallow contaminated pools which have been sources of botulism. Facilities have also been provided for bird-watchers and access has been regulated in order to minimise visitor disturbance.

Together with the adjacent habitats of scrub, woodland and dune systems of the Doñana National Park, the area is a major wildlife habitat of international importance.

Representative sites: bogs, fens and marshes

A total of 182 representative sites were selected and are presented on Map 9.8. The sites have been evaluated according to the four classes defined in the legend. This evaluation is based on site-specific data concerning present status of protection, habitat management and prevailing pressures.

Criteria for site selection

The main criterion for selection of representative sites is that they should be recognised as having international importance for the particular ecosystem in question. For illustrative purposes, only the larger well-known wetland locations are examined, as well as some sites in degraded ecological condition. A major reference source has been the important bird areas of Europe identified by Grimmett and Jones (1989) and the Council of Europe reports on peatlands of Western Europe (17 countries). For the EU, all sites have been compared and evaluated with the use of the CORINE Biotopes data (CEC, 1991b).

Total	Unimpacted	Impacted			
		Water level	Water quality	Peat/mineral extraction	Tourism/ recreation
25	9	12	9	2	4

Note: Sites can be impacted by multiple causes.

Interpretation of the data

Bogs, fens and marshes are distributed mostly throughout four geographic regions of Europe: the boreal, Atlantic, central and eastern regions feature each between 600 000 and 1 million ha of these habitat types. Protection appears relatively successful for most of the boreal and Atlantic sites.

However, Figure 9.4 conceals the fact that vast former peatland areas of the Atlantic region have vanished due to land cultivation and/or because they are extremely fragmented so that they were not selected for this study. Limitations of scale and other selection criteria have precluded illustration of many fen and bog habitats, such as the small but important types found in the Alpine region. Although Eastern Europe shows the largest amount of selected wetlands (approximately 1 million ha), reflecting very well the dominance of this habitat type in the eastern and northern landscapes, there are many more to include in future inventories. Data gaps must also be expected for the Mediterranean region, for which the MedWet Project (MedWet 1992) is expected to improve the situation. The sites that have been selected face stresses based on external threats (eg, pollution and eutrophication). Generally the map confirms the expectation that bog/fen/marsh wetlands are predominantly distributed in Northern Europe, mainly due to the supply of precipitation.

Box 9P Biebrza Valley (Poland)

Some 80 000 ha of wetlands have been identified (Grimmett and Jones, 1989) in the Biebrza Valley (Site 129 on Map 9.8) of northeast Poland. In the north, on the upper river course, sedge and moss fens are interspersed with intensively managed grassland on land which has been drained. Further drainage proposals pose an acute threat to some of the remaining fens. Scrubby fen and grassland is found in the central basin, which was partially drained in the nineteenth century. Forested areas of birch, alder, spruce and pine are important in this area. The lower valley basin of the south carries some well-preserved fens in an area some 5 to 12 km wide along a 30 km length of river. The river floodplain also includes many oxbow lakes. The fens are dominated by tall sedge with extensive reedbeds along the river channel. Scrub invasion is posing a threat to nesting water habitats. Many species of birds breed along the Biebrza Valley, including considerable numbers of waterfowl. Most of the land is at present unprotected though there are proposals for inclusion within National Park boundaries.

Table 9.5
Status of bogs, fens and marshes designated as Ramsar sites
Source: E Maltby and IWRB, personal communication

Biebrza Marshes, Poland
Source: Robert Harding Picture Library

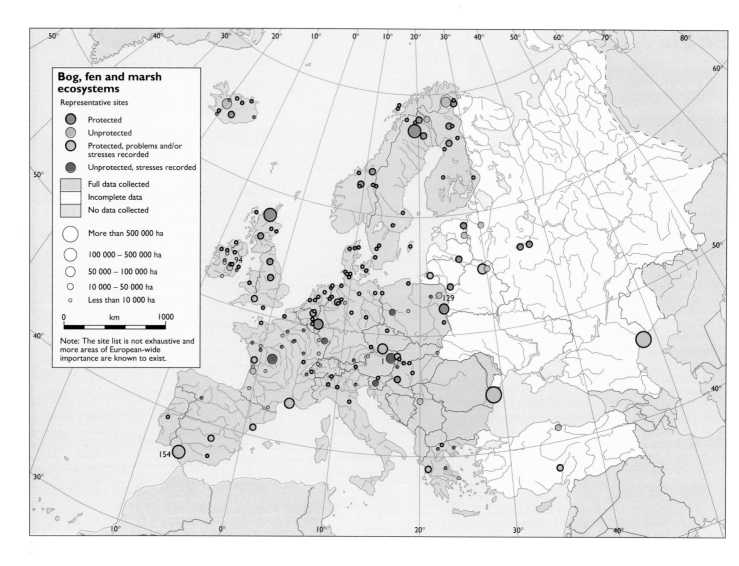

Map 9.8 *Bog, fen and marsh ecosystems: representative sites (see end of chapter for site names)*
Source: Compiled from multiple sources by IUCN/EEA-TF

Conclusions

Only a small fraction of the continent's wetlands is protected, although most countries are committed to increase the extent of these areas. Van Eck et al (1984) indicated that the protected area of virgin peatland in nine European countries gave a range of 0.1 per cent to 3.1 per cent with a mean of 1.3 per cent. Finland had 70 000 ha of peatland conserved (0.7 per cent of its peatland area), Norway's over 200 mire reserves extending to more than 2 per cent of the total mire area. In Estonia the figure is 13 per cent, but this is only 122 200 ha. With specific reference to raised bogs, the Council of Europe (Goodwillie, 1980) reported that protected areas in 18 countries averaged 3.2 per cent, ranging from 0.8 to 10.2 per cent. The International Mire Conservation Group in Vienna (IMCG) is targeting the protection of peatlands in Europe. Because of human landuse activities and pollution, oligotrophic or calcium-rich bogs and fens belong to the most endangered habitat types in this category. Their protection and regeneration is of great importance for the stabilisation of landscapes. With the exception of the national peat-bog inventories of Ireland, Finland and Sweden, there is presently no definitive policy for wetland protection in Europe as a whole and no comprehensive

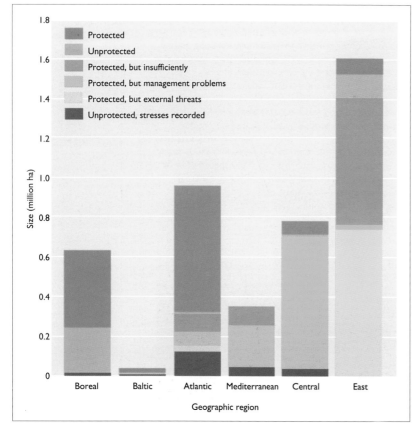

Figure 9.4 *Bogs, fens and marshes: representative sites by region*
Source: IUCN/EEA-TF

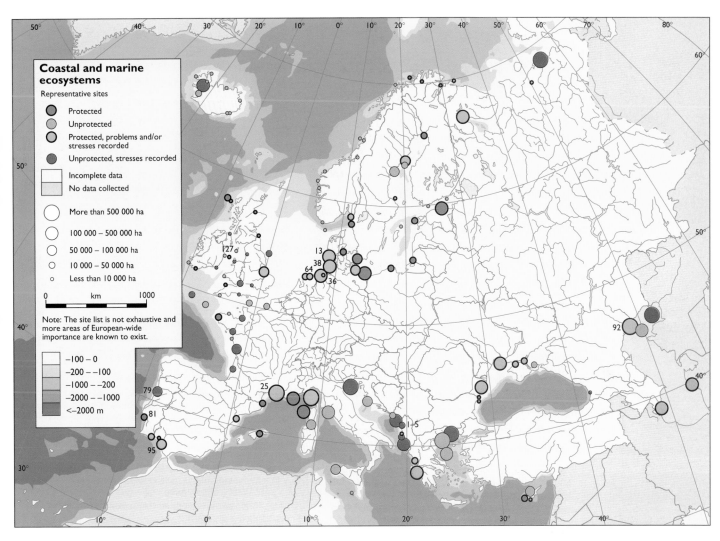

Coastal and marine ecosystems

Representative sites

- Protected
- Unprotected
- Protected, problems and/or stresses recorded
- Unprotected, stresses recorded

Incomplete data
No data collected

- More than 500 000 ha
- 100 000 – 500 000 ha
- 50 000 – 100 000 ha
- 10 000 – 50 000 ha
- Less than 10 000 ha

0   km   1000

Note: The site list is not exhaustive and more areas of European-wide importance are known to exist.

- −100 – 0
- −200 – −100
- −1000 – −200
- −2000 – −1000
- <−2000 m

Map 9.9
Coastal and marine ecosystems: representative sites (see end of chapter for site names)
Source: Compiled from multiple sources by IUCN/EEA-TF

strategy by which enhanced conservation and rehabilitation of degraded habitats can be achieved. Optimum functioning of these ecosystems requires the protection of their intrinsic hydrological systems as much as appropriate site management.

In a summary of present national inventories incorporating wetlands in Europe, Jones and Hughes (1992) report initiatives in various states of completion in 13 countries: the Czech Republic, Finland, France, Greece, Iceland, Ireland, Italy, the Slovak Republic, Spain, Sweden, Switzerland, Turkey and the UK. In addition the International Waterfowl and Wetlands Research Bureau (IWRB) is coordinating the project development phase of a wetland inventory for the Baltic Republics, the Commonwealth of Independent States (CIS) and Georgia as well as for the whole Mediterranean basin (MedWet, 1992). To date sub- and pan-European projects have culminated in the publication of Important Bird Areas in Europe (Grimmett and Jones, 1989) with an emphasis on water-bird habitats. In order to develop efficient habitat management and protection schemes more specific information on the precise status and functional character of bogs and marshes is vitally necessary (Maltby et al, 1993).

Coastal and marine ecosystems

Introduction

More than most terrestrial systems, coastal and marine ecosystems are often exposed to dynamic physical processes. Having evolved over millions of years, the European coastline is continuing to change today. Besides ongoing isostatic adjustments following deglaciation (uplift in the north, subsidence in the south), large-scale processes of

sedimentation and erosion are driven by the action of waves and currents, by geomorphological changes of the inland through incoming rivers, by meteorological factors like winds and storms, by alterations of the sea level and by geological processes such as seismic and volcanic activity. These natural processes are increasingly augmented by human alterations to the coast and to river regimes, from the building of groynes and breakwaters to the damming of major rivers and estuaries.

The main habitat types, ecosystem functions and threats of important coastal and marine systems are explained below, followed by an interpretation of data deriving from 135 representative sites, illustrated in Map 9.9.

The coast

Coastal areas are among the most attractive for settlement, and some 60 per cent of the world's population lives within 60 km of the shore. The use of coastal land for industrial and urban development is expected to intensify with most growth in less developed areas such as the eastern Mediterranean. Other more recent uses of coastal regions, such as tourism and development of leisure facilities, are also increasing, often in some of the better-preserved stretches of coastline. Based on the classification systems of the Ramsar Convention and the CORINE Biotopes project, this report focuses on the following coastal habitat types: rocky coasts, sand dunes and mud flats, delta areas, estuaries and lagoons.

Rocky coasts

A large part of Europe's coast is made up of rocks: cliffs, fjords, rock plateaus or islands steeply rising out of the sea.

Box 9Q Albanian lagoons

The Albanian coastline consists of a wide range of important habitat types in often exceptionally natural conditions and with a relatively high number of internationally important sites (Sites 1 to 5 on Map 9.9). Rocky shores, sandy beaches, estuaries and lagoons as well as phrygana and maquis vegetation host a flora and fauna of high biodiversity. Due to Albania's long period of political and socio-economic isolation, the still relatively unspoiled coastline is home to a number of rare and endangered species. In the 1950s and 1960s, Albania was known to be of world importance for breeding pygmy cormorants (*Phalacrocorax pygmeus*) and Dalmatian pelicans (*Pelecanus crispus*). Despite a substantial decline of these bird populations during the 1970s and 1980s (mainly due to hunting and overfishing), the recent re-establishment of Dalmatian pelicans makes the lagoons of the Karavastas/Divjaka complex an internationally important site (IUCN, 1991). In 1993, more than 200 Dalmatian pelicans and 400 pygmy cormorants were reported, and since 1991 at least 50 breeding pelicans have been recorded (Vangeluwe et al, 1993). Other equally important sites include Lag e Nartes, Kune-Vain Patok, Gjiri i Lalzit and Gjiri i Vlores, the latter being significant for the endangered monk seal (*Monachus monachus*; see Box 9BB).

Since the political changes of the early 1990s, Albanian coastal ecosystems now face new potential threats by becoming the target of large-scale tourist developments mainly from foreign investors (Vangeluwe et al, 1993). For example, the Karavastas/Divjaka site is currently being directly affected by the construction of recreational facilities in the direct proximity of the lagoons. Furthermore, uncontrolled expansion of hunting and fishing activities (attracting foreign interest) can have strong impacts on local wildlife. Although Albanian authorities are aware of the environmental situation (an 'Environmental Protection and Preservation Committee' has been created) there is still a lack of experience in operating with private investors and applying environmental policies.

Dalmatian pelicans on the Albanian lagoons
Source: IRSNB

Sand dunes and mudflats

As a result of the dynamics of sand and water, coastal dunes belong to the most richly varied landscapes of the world. In a continuous process, new drift-sand dunes may emerge, alternating with valleys, lakes, moors, bushes or forests. Often they protect mudflats (wadden) or lagoons. These, together with saltmarshes, are often the biologically richest natural areas.

Delta areas

Being situated at the interactive borderline of salt- and freshwater, river deltas form complex and varied landscapes, sometimes encompassing many of the habitat types described in this section.

Estuaries

Through storms and tidal forces, the mouth of a river can extend into vast estuaries fringed by mudflats and saltmarshes. The resulting differences in salinity levels affect waters many kilometres inland and produce great diversity of life forms.

Lagoons

If a bay or estuary gets cut off from the sea by silting or by dune formation, a lagoon may result. Size and depth can vary, and the water can range from saline to almost fresh.

The marine environment

Marine systems may consist of distinct and moving water masses, often identifiable by characteristic planktonic assemblages. The sea bottom (benthic systems) can be defined on the basis of sediment type, again associated with characteristic faunal and floral groups, though in some areas other features such as intermittent or permanent deoxygenation, turbidity, light or salinity are the essential determinants of community development (Mitchell, 1987). Based in Dublin, the BIOMAR project is presently undertaking the development of a classification system for northeast Atlantic marine habitats based on the hierarchical methodology of CORINE (CEC, 1991a), but also including faunistic communities.

Human impacts on open water systems and deeper benthic systems are often not habitat-specific, but there is an increasing number of incidents which may be symptomatic of widespread imbalance in European marine ecosystems, from algal blooms and epizootics to dramatic changes in community composition. Many of these issues are discussed on a regional basis in Chapter 6.

Functions and values

Ecosystem functions for marine and coastal areas can be exemplified in the context of the saltmarshes. Saltmarshes have a high gross and net primary productivity – almost as high as in subsidised agriculture. This high productivity is a result of subsidies in the form of tides, nutrient import and abundance of water, that offset the stresses of salinity, widely fluctuating temperatures and alternate flooding and drying. Detritus export and shelter found along marsh edges make saltmarshes important as nursery areas for many commercially important fish and shellfish. Saltmarshes have been shown at times to be both sources and sinks of nutrients, particularly nitrogen (Mitsch, 1986).

Being a saltmarsh habitat of its own, the Wadden Sea is also known as the largest continuous stretch of intertidal mudflats in the world. In terms of biomass (food supply, oxygen reserves) these areas belong, with rainforests, to the most productive systems of the world (Common Wadden Sea Secretariat, 1992).

Together with other habitat types, such as dunes, sandy beaches, rocky shores and lagoons, coastal plains provide the largest and often most undisturbed wetland habitats for millions of breeding and resting birds. By also being very attractive places for recreational activities as well as a leading resource of human food supply, coastal and marine ecosystems are often exposed to strongly competing interests (see Boxes 9Q and 9R).

Habitat threats

Algal blooms and pollution of the seas

Land-based pollution and diffuse sources of pollution are the principal agents in a general deterioration in water quality, from PCBs in the Baltic (WWF, 1991) to oil and pesticides in the Caspian (see Chapter 6). Pollution was identified as presenting a threat for 23 (18 per cent) of the representative sites. The most pristine sites are those of the less densely inhabited North Atlantic coast, particularly the fjordal habitats of Iceland and

Box 9R Sand dunes

A survey carried out by the European Union for Coastal Conservation in 1992 has produced some alarming statistics concerning the loss of European sand dunes (see Table 9.6). There are between 400 000 and 428 000 ha of dunelands remaining on Atlantic coasts, of which approximately 40 per cent are afforested. Around a third of the 40 per cent decline in coverage since 1900 has occurred in the last three decades, the principal cause being deliberate afforestation using exotic species. A further 5 per cent loss is predicted over the next two decades, primarily through development of recreational facilities such as golf courses. Other specific problems include lowering of the water table through pumping, and construction of buildings (Salman and Strating, 1992).

A further decline of 10 to 15 per cent through recreational developments is expected over the next two decades. Sandy beaches are the main attraction for many of the 100 million tourists who visit the Mediterranean each year. Unfortunately the peak of the tourist season coincides with the nesting season for the vulnerable loggerhead turtle (*Caretta caretta*) and the endangered green turtle (*Chelonia mydas*). Lara-Toxeftas in Greece is one of the most important beaches for turtles, providing nesting for some 200 loggerhead and 100 green turtles. Over 4000 hatchlings emerge each year from the nests of this monitored site.

Trampling by visitors may have significant local impacts, but restoration is possible by enclosure and fixation, where damage is moderate. The decline in Mediterranean dunes has been severe, with the present coverage of 30 000 to 36 000 ha representing an estimated 71 per cent loss since 1900, resulting mainly from construction of tourist facilities.

Sand dunes, Doñana National Park, Spain
Source: H Jungius, WWF

Table 9.6 *Loss of sand dune habitats between 1900 and 1990*
Source: EUCC, 1993

	Estimated area (ha) in 1900	Loss by 1990 (%)
Belgium	5000	46
Denmark	80 000	35
France	250 000	40–50 (Atl); 75 (Med)
Germany	10 000–12 000	15–20
Greece	15 000–20 000	40–50
Ireland	20 000	40–60
Italy	35 000–45,000	80
The Netherlands	45 000	32
Portugal	100 000	45–50
Spain	70 000	30 (Atl); 75 (Med)
UK	80 000	40

Mediterranean marine seagrass (Posidonia oceanica).
Source: E G Verriopoulos

Box 9S Marine seagrass

Rich and highly productive communities of attached macrophytes are found in many of Europe's coastal waters. Due to its high productivity, seagrass contributes to the process of water oxygenation: 1 m^2 of *Posidonia* bed produces between 4 and 20 litres of oxygen per day. More than 400 algae and 1000 animal species live in *Posidonia* beds, also providing shelter and nursery habitat for many crustaceans and fish.

As the only flowering group of marine plants, seagrasses form meadows which play an important role for stabilising sedimentary deposits and trapping nutrients. In fact, waves and currents are damped by the matte and the transient sediment is accumulated by the filtering mechanisms of their leaves and the rhizomes. The effectiveness of seagrass for coastal protection becomes apparent in areas where substantial changes have followed seagrass regression.

Several factors can contribute in synergistic ways to disrupt the equilibrium of seagrass beds: marine coastal construction, resulting alterations of the hydrological patterns, eutrophication and pollution of coastal waters, increased water turbidity, fishing activities and anchorages are among the possible factors responsible for the destruction and decline of seagrass beds.

Though often associated with tropical waters, there are a number of temperate seagrass species such as *Zostera marina*, and important meadows are found as far north as Breidafjördur in Iceland. Many Atlantic areas of *Zostera* have yet to recover from the deteriorating disease in the 1930s which was associated with raised water temperatures.

Of particular importance for the Mediterranean sea is *Posidonia oceanica* (see illustration) with an average biomass growth of 38 tonnes/year/hectare of dry weight. In the case of the Hyères archipelago in southern France, studies have shown a decline of *Posidonia* meadows from 30 000 hectares in 1900 to 22 500 hectares – a regression that is only expected to level off by the year 2015 if coastal development and pollution are reduced.

northern Norway. Threats in the Barents and White seas were generally of a specific nature, related to oil exploitation or military activity. However, even in scarcely inhabited areas, no site can be considered immune to the effects of pollutants owing to their dispersal in the marine environment or through the food web. Organic wastes include: sewage, (a seasonal problem in areas such as the Mediterranean – see Box 9S – where the sewerage infrastructure is insufficient to deal with the summer influx of tourists); by-products from industries (pulp and paper mills or tanneries); and fertilisers contained in runoff from agricultural areas (a problem particularly identified for Caspian sites). Although organic wastes are rapidly degraded, nutrient enrichment from this source is almost certainly a contributory factor to the increased frequency of algal blooms

Box 9T The Wadden Sea (Denmark, Germany, The Netherlands)

The Wadden Sea is one of the largest tidal wetland areas in the world, providing habitats for waterfowl and shorebirds and critical stopover points for coastal birds which migrate along the East Atlantic flyway with up to 12 million individuals of 50 species each year. The rich and varied ecosystem is also a critical spawning and feeding area for 102 species of North Sea fish as well as for the harbour seal (*Phoca vitulina*), the grey seal (*Halichoerus grypus*) and the bottlenose dolphin (*Tursiops truncatus*).

The representative sites of the Wadden Sea (Sites 36, 38 and 64 on Map 9.9) have been selected as examples of a number of habitat types: dune systems, saltmarshes, estuaries and mudflats, barrier islands, mussel beds and seagrasses. However, the listing of these three sites is not designed to single them out as requiring particular management efforts – indeed this would be at variance with the regional approach to conservation which has been developed over more than a decade.

Due to the long history of human intervention, people are an integral part of this system, but the extent and diversity of human uses is now threatening to undermine the integrity of the entire system. Today, the Wadden Sea is fringed by one of the most industrialised regions of Europe and the construction of port facilities and embankments has resulted in the substantial loss of habitats. During the last five decades, for example, more than 33 per cent of all saltmarshes were lost to embankments. Many remaining areas are exposed to other stresses such as dumping,

excavation, infilling, aquaculture, tourism, hunting, and changes to the water regime through civil engineering projects.

Mechanical damage to the benthos is a widespread problem, a result of both direct exploitation and incidental damage. Concern has been expressed over the extent of mussel and cockle harvesting from the Dutch Wadden Sea, and an important decision to limit negative ecological effects of these fisheries was taken at the last Wadden Sea trilateral conference (1991). Pollution through urban, industrial and agricultural runoff and sewage are also triggering community changes, with algal growth smothering the rooted halophytes in nutrient-rich waters. Unfortunately, the threats to the Wadden Sea are not local – and local efforts alone will not be sufficient to protect it indefinitely. Generic problems such as pollution from land-based sources and alteration to marine communities through non-sustainable practices will need to be addressed on the European or perhaps even global scale to ensure long-term conservation of this exceptional coastal marine area.

Jurisdiction over the Wadden Sea is shared between Denmark (10 per cent), Germany (60 per cent) and The Netherlands (30 per cent). The countries maintain sovereignty over those sections of the Wadden Sea for which they have jurisdiction, but in 1982 they signed the Joint Declaration on the Protection of the Wadden Sea. In addition, a Guiding Principle on the Trilateral Wadden Sea Policy was adopted at the Sixth Trilateral Meeting of Ministers in 1991, with the aim of wise use and the appreciation of precautionary principles in regional planning (Common Wadden Sea Secretariat, 1992). There are also proposals to make a joint nomination of the Wadden Sea to the World Heritage Convention, and to look at the possibility of a common Ramsar site and a common Natura 2000 site (see Nature Conservation below).

East-Friesian Wadden Sea, Germany

Source: D Wascher

in recent years. Examples of such blooms include the diatom slimes and dinoflagellates in the Adriatic, which principally cause nuisance problems such as fouling of fishing gears and beaches, to the more dangerous toxic *Chrysochromulina polyepsis* blooms of 1988 which caused massive mortalities to farmed salmonids and necessitated towing Norwegian salmon farm facilities to sheltered waters.

Overfishing

While it is difficult to establish a causal relationship, overfishing may well be responsible for major changes in fish communities subject to intensive exploitation in areas such as the North Sea. However, there may be longer-term ecological mechanisms involved in the 'biomass flips' observed. Competition between fishermen and species is also a growing problem: food shortages may have increased the vulnerability of the striped dolphin (*Stenella coeruleoalba*) to viral epizooics which led to hundreds of carcasses being washed up on Mediterranean beaches. Lack of food has also had impacts on other populations, from the Mediterranean monk seal (see Box 9BB, below) to sea-bird communities at Sem'ostrovov in the Barents Sea.

Some fishing techniques have direct detrimental effects on non-target populations. Trawling or dragging of fishing gear can have devastating local impacts on benthic systems such as the rare Norwegian coral communities (*Paragorgia arborea*) or the diverse communities associated with the horse mussel (*Modiolus modiolus*) at Strangford Loch (Site 127), threatened by trawling for the queen scallop, which has removed large areas of mussel beds. Drift-netting is particularly notorious for the by-catch of species such as sea turtles, cetaceans and seals. (See also Chapter 24.)

Saltmarshes: overgrazing

The historical decline in saltmarshes has been dramatic, the principal cause being drainage – or reclamation – to agricultural polders or pasture. Reclamation continues to threaten many saltmarshes and an estimated 50 per cent of the invertebrate fauna of marshes is now threatened by coastal protection measurements. Overgrazing has been identified as a problem at the Baie du Mont St Michel, one of the few European sites where extensive saltmarsh areas could be managed within a mosaic of coastal habitats. Other important sites in this respect include the entire Wadden Sea area (see Box 9T), Ria de Aveiro (Site 79), Doñana Gualdaquivir Odiel (Site 95), Camargue (Site 25) and Orbetello (Dijkema, 1984).

A wide variety of human uses present specific threats to saltmarsh communities, including dumping, excavation, infilling, aquaculture, tourism, hunting, and changes in the water regime through civil engineering projects and land reclamation (see Table 9.7). These stresses are particularly damaging to marsh habitats in seas such as the Caspian and Black Sea where the spatial distribution of different species is limited by water salinity (Balkas et al, 1990). Such intrusive activities often have a local focus, and are planned on a local basis. Their regional effect or interaction with other activities is all too often overlooked.

Estuaries and deltas: contamination

Estuaries, deltas and their associated habitats are among the most threatened of Europe's coastal habitats. Historically an expedient focus for transportation, estuaries have long been preferred sites for the construction of harbours and ports, and embankments and land reclamation facilities, often for waste disposal. Estuarine sites such as Varde å in the Danish Wadden Sea (Site 13), the Volga (Site 92) and the Tejo Estuary (Site 81) are particularly liable to become contaminated as materials from upstream industrial and urban effluent settle out under the changing salinity and energy regimes where freshwater flows into sea water. Such wastes often contain high quantities of inorganic wastes which may accumulate or disappear only very slowly in the natural environment. Metallic compounds and long-lived contaminants, such as organochloride pesticides residues, may reach dangerous concentrations in more enclosed seas like the Baltic (WWF/HELCOM, 1991).

Representative sites: coastal and marine

A total number of 135 representative sites were selected and are presented on Map 9.9. The sites have been evaluated according to the four classes defined in the legend. This evaluation is based on site-specific data concerning present status of protection, habitat management and prevailing pressures.

Criteria for site selection

A number of international designations (World Heritage Sites, Biosphere Reserves, Barcelona Convention, European Diploma Sites, Biogenetic Reserves and Ramsar sites) served as an important pool for the selection of sites. In addition, a vast database of information on European habitats gathered by the World Conservation Monitoring Centre (WCMC, 1990) and by the CORINE project (CEC, 1991b), and over 1241 of the sites described in the CORINE database including areas of coastal habitat, were used. No size limitations were applied.

Interpretation of the data

Figure 9.5 illustrates that the arctic seas (Iceland, Norway and the Barents Sea) and the Mediterranean are the locations of the largest marine and coastal representative sites, most of which are still without any protection. The largest site on Iceland is Breidafjördur (305 000 hectares), consisting of rocky islands and kelp forests/seagrass habitats. One of the environmental stresses is the harvesting of seaweed. For the site at the eastern coast of the Barents Sea (2.56 million hectares) oil pollution and erosion are current and potential risks. Another large North European site, the Franz Josef Land (6.4 million hectares) is outside the area of the map, but nevertheless is a very important feeding ground for polar bears, walrus and white whale, and as a habitat for bird colonies. The Wadden Sea of Denmark, Germany and The Netherlands comprises about 900 000 hectares of highly valuable coastal habitats – one of the richest biological resources, even on a global scale (see Box 9T). In the South there are numerous large estuaries of high ecological significance, many of them exposed to external threats (eg, river pollution). Generally, the percentage of sites that are protected and in good condition is relatively small compared to some other ecosystem types.

Conclusions

The impacts of human activities on marine and coastal areas fall into three main categories:

- habitat destruction and degradation through direct activities such as reclamation, coastal constructions and dredging, and damaging fishing techniques;
- deterioration in quality of habitat and of the water medium through pollution; and
- damage to biological communities and living resources through non-sustainable harvesting practices.

Such activities are presenting threats to many of the important areas presented on Map 9.9. Of these, it is direct intervention in coastal areas which has played the greatest and most immediate role in shaping Europe's coastal

environment. Almost half of the sites identified are considered to be threatened by one or more direct activities.

In open waters, pollution and resource abuse are generic problems, but there are no universal solutions – these are issues that need to be addressed on a number of fronts, including legislation and policy. Pre-emptive action to forestall a problem is obviously preferential to restoration, although this certainly now has an important role to play in Europe for the conservation of coastal environments and associated species.

The gradual degradation and loss of coastal systems is evidently rooted in a wide range of human activities, from bait digging and bird-watching to harbour development, mineral extraction and toxic waste disposal. In some areas there may be a few principal issues which can be addressed on a local level; in others, the essential inter-linkedness of marine and coastal systems requires that changes be made across the whole spectrum of human interactions. For these and other reasons discussed in Chapter 6, the problems of coastal zones have been identified as a prominent environmental problem of concern for Europe, and are examined further in Chapter 35.

While the selection of representative sites in terms of habitat provides a focus for identification of conservation priorities, such sites cannot be managed in isolation but need to be considered within the marine and coastal system as a whole. Integrated coastal zone management techniques provide a means to do this, to identify areas of particular conflict and to approach solutions through optimum planning (Salathe, 1992).

	Area (km2)	Protected area(%)	Area threatened by reclamation (%)
Denmark	81	30	0
France (excluding Camargue)	148	3	35
France (Camargue)	650	reserve	nd
Germany	188	55	60
Italy	740	30	nd
Ireland	175	4	5
The Netherlands	129	100	47
Portugal	92	28	80
Spain, (Doñana)	400	reserve	nd
UK	314	50	40

nd: no data

Table 9.7
Protection and threats for European coastal saltmarshes
Source: Dijkema, 1987

Figure 9.5
Coastal and marine ecosystems: representative sites by region
Source: IUCN/EEA-TF

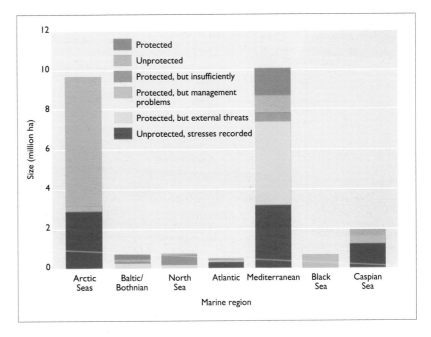

Mountains: rocks, screes, inland dunes and caves

Introduction

This section mainly covers rock and scree habitats above the tree-line; inland sand dunes and caves are covered to a lesser extent owing to lack of data. All of these habitats are summarised here as mountain ecosystems. Rock and scree communities are dominant features of Europe's large mountain regions, for which a brief summary is given below.

In the north, the Scandinavian Highlands and Ural mountains fringe the lowlands of the boreal region. In Central Europe, the Alps form a massive east-west spine, extending to the northeast with the Carpathians, and in the southeast to the Caucasus. In central southeast Europe, the mountains of the Balkan peninsula include extensive karst areas. Important mountain areas can also be found to the west

in the Iberian highlands. These are fringed by the high Pyrenees which also possess major karstic landscapes. The Apennines, which form the backbone of the Italian peninsula, represent one of the most active volcanic zones in Europe, next to Iceland.

The main habitat types and threats of important mountain systems are explained below, followed by an interpretation of data deriving from 97 representative sites, illustrated in Map 9.10.

Habitat types

Plant communities of inland rocks and screes are mainly composed of hemicryptophytes (plants having terminal buds close to the ground) and partially of chamaephytes (dwarf shrubs with buds up to 50 cm above the ground). Annual species are very rare due to the short vegetation period and low temperatures (Walter, 1970). Frequently, more than half the weight of mountain plants is located under the surface. Wind is the dominant factor for the distribution of seeds. Scree communities may develop only on steep and dynamic slopes. The vegetation cover of active screes is low (10 per cent), but root intensity and regeneration capacity is high (Wilmanns, 1973). If slopes become stabilised, Alpine grasslands will eventually replace the former scree communities.

The special feature of rock-inhabiting (rupicolous) communities is their high rate of endemic species. In the Alps, 35 to 40 per cent of all endemic species can be found on rocks and screes; half of the 40 species endemic to the Maritime Alps (see Box 9U) are rupicolous (Pawlowski in Ozenda, 1983). Among rupicolous communities, a remarkably diverse but little investigated flora has developed, for example in the genus navelwort (*Umbilicus*).

Because of the harsh conditions of the high mountain zone, fauna are limited in species and numbers of individuals. Nevertheless, rocks play a significant role as habitats for many especially adapted mammals, such as the endangered ibex (*Capra ibex*).

In 1986, 65 existing national parks (36 per cent) were completely or predominantly located above the tree-line in mountainous areas. The size of this percentage indicates that European nature conservation efforts have been particularly successful in establishing reserves in mountain regions where there is generally less conflict with human landuse activities (especially cultivation and construction) compared with other ecosystems.

Habitat threats

Recreational activities are the main sources of threats for mountain ecosystems (see Chapter 25). In general, for reasons of access, such threats are more intense at lower elevations, but they are also noticeable in higher and more remote regions. While in Scandinavia the impacts on mountains from recreation are rather low (see Box 9V), there are a number of regions in the Alps which appear to have reached their ecological carrying capacity (UNCED, 1992). The still increasing numbers of skiing facilities, including their large support systems, are damaging and altering many valuable mountain habitats. Uncontrolled hunting and selective plant gathering are still having negative impacts on the local fauna and flora, but seem to be on the decline.

Industrial activities in valley corridors and long-range transport of emissions from distant sources can affect the soil and vegetation (especially some sensitive lichen flora) even at the highest elevations. Human landuse activities such as agriculture, forestry and tourism often affect mountain habitats indirectly through soil erosion (see Box 9U).

During the last century, lowland habitats such as sand dunes were often stabilised by pine plantations which made

---

Box 9U The Maritime Alps

With about 40 species, the Maritime and Ligurian Alps have the highest concentration of endemic plants of the Alps, where they grow at exceptionally low altitudes (600 m and 1200 m respectively) owing to favourable climatic conditions. This number of species makes up 10 per cent of the approximately 400 vascular endemic plants (Ozenda, 1983), found in the whole Alpine mountain range.

In the Mercantour region (Site 14 on Map 9.10), a French national park was established in 1979 near the Italian border, protecting an exceptional variety of 2000 vascular plants on 68 500 hectares. Adjacent to this, Italy followed in 1980 by establishing a protected landscape area of 26 000 ha (Parco Regionale Argentera). Despite these designations, some of the most interesting sites on the French side, such as the Roya valley and the extraordinary fir forests of the Col de Turini, lie outside these areas and are without protection.

The whole area represents a typical example of the most serious threat to be found in the Mediterranean region: soil erosion. In mountain regions sun-exposed slopes are traditionally used for agriculture, whereas the shady valley slopes serve as timber reservoirs. Nowadays, the view from the 2800 m high Col de la Bonnette shows almost 'steppe-like' mountain ranges where tree growth has become impossible. Whole mountain slopes are being eroded and debris has buried the ancient road to St Etienne de Tinée. In order to prevent further degradation and to keep the site's function as a water reservoir for the Côte d'Azur coastal region, active management is needed to stop the destructive processes. Reforestation projects would best employ indigenous species, rather than exotics, such as black and Corsican pines (Heiss, 1987).

Erosion in the Maritime Alps, France
Source: G Heiss

Box 9V Jotunheimen – 'home of the giants'

Jotunheimen (Site 55 on Map 9.10) is located in the centre of the Scandinavian Highlands which stretch from Finmark to Stavanger, with its highest summit of 2469 m (Galdkøppigen). The Hurrungane, a mountain range within the region, has become a well-known symbol for the whole of Scandinavia, due to its pinnacles and bizarre mountain-shapes.

A large part of the area is covered by glaciers (200 km²) and numerous lakes contribute to its habitat diversity. One of the most beautiful lakes in Scandinavia is the 18 km long and 149 m deep fjord-like Gjende which receives its magnificent turquoise colour from soil particles mostly washed in from glacial rivers. From west to east the region undergoes an extreme climatic gradient. While in the western part close to the coast annual precipitation amounts to 2500 to 3000 mm, the eastern part receives only 250 to 300 mm (250 mm represents the upper limit of extreme aridity).

A large number of rare and endangered mountain flowers cover the mountain ridges. The area of Lake Gjende is home to a small moose population which feeds in birch forests. The Utladalen was well known in the past for its numerous bears. Currently, brown bear (*Ursus arctos*) and wolverine (*Gulo gulo*) can be observed only sporadically. Lynx (*Lynx lynx*) appears with a small but permanent population within the region. Arctic fox (*Alopex lagopus*) and otter (*Lutra lutra*) are rare, but present. The otter population is thought to have increased in recent years. In the western part, a population of wild reindeer (*Rangifer tarendus*) occurs, a species that has been

Jotunheimen National Park, Norway
Source: C Bowman, Robert Harding Picture Library

replaced by domestic reindeer in the eastern part. The area is frequented by one of the most typical mountain birds of Scandinavia, the purple sandpiper (*Calidris maritima*), characteristic for its delicate whistle which can be heard mainly at high elevations.

The struggle for legal protection status of the region dates back to the beginning of the century. In 1980 a large (1142 km²) area was finally protected as a national park covering a part of the region. Presently, tourism, road construction and power lines diminish the full value of the region. Above all, long-range air pollution is by far the most serious threat to this and other valuable ecosystems in southern and central Scandinavia.

Map 9.10 Mountain ecosystems: representative sites (see end of chapter for site names)
Source: Compiled from multiple sources by IUCN/EEA-TF

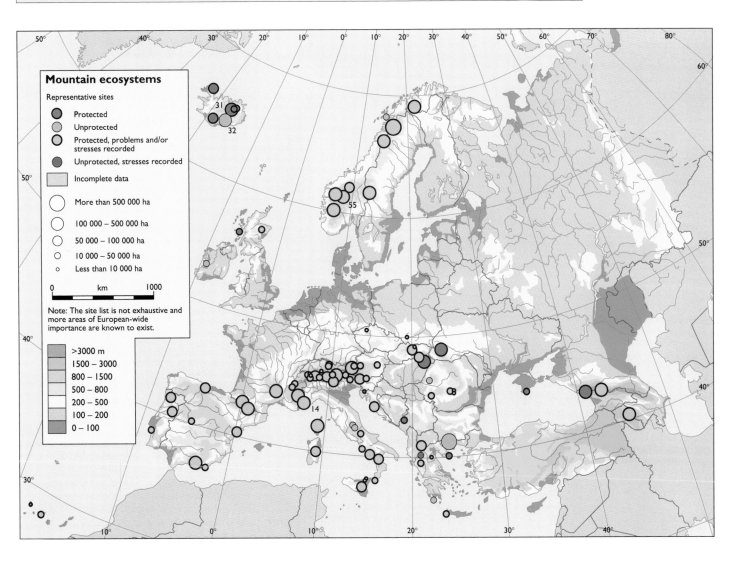

wandering inland sand dunes very rare. Caves are fairly well protected against human activities (although damage is frequently caused by pot-holers), and the effects of pollution may cause only minor damage, due to a very limited species composition of fauna and flora adapted to conditions under exclusion of sunlight.

Representative sites: mountains

A total number of 150 representative sites were selected, of which 97 are shown on Map 9.10. The sites have been evaluated according to the four classes defined in the legend. This evaluation is based on site-specific data concerning present status of protection, habitat management and prevailing pressures.

Criteria for site selection

Different information levels have influenced site selection: good information in Western Europe allowed a more accurate determination of specific sites, whereas paucity of information due to the far east of Europe (especially the Caucasus ranges) made the evaluation more difficult. Map 9.10 shows habitat types that include screes, rocks, and also some volcanoes and caves.

Interpretation of the data

As shown in Figure 9.6, the boreal region clearly has the largest area of selected sites – this is especially due to the sites in Iceland, of which Myvatn-Laxa and Skaftafjell-Grimsvötn (Sites 31 and 32) cover alone 1 million ha. The

Figure 9.6
Mountain ecosystems: representative sites by region
Source: IUCN/EEA-TF

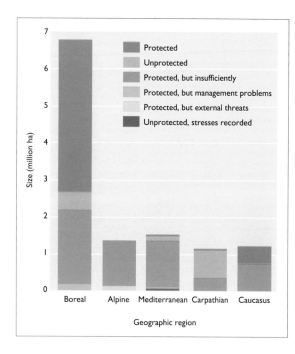

large mountain areas of Scandinavia are already now the best-protected mountain systems in Europe, and in the near future approximately 50 per cent will enjoy a high protection status. The construction of roads, however, generates stresses in many areas. In the Alps, flora and fauna of all selected sites are affected by noise and air pollution due to heavy road traffic, and mass-tourism in almost all of the Alpine provinces (see also Box 8I). The resulting environmental damage (from cross-country and downhill skiing, ski-lift constructions and serious soil erosion) has been well portrayed (FNNPE, 1992).

Map 9.10 shows that most of the 'mountain' areas are protected sites such as national parks. Their remoteness and inaccessibility limited other landuses and allowed the realisation of conservation goals. However, the increasing exposure of mountainous areas to tourism has affected

many of the original habitat functions, making the survival of often endemic plants and animals extremely difficult.

Conclusions

Mountain farming has been the primary cause for a lowering of the upper tree-line by several hundred metres in most mountain ranges outside the boreal zone. Widespread overgrazing has resulted in erosion within alpine grasslands and mountain forest belts, and this, in turn, has increased the exposure of screes and rocky grounds. Erosion has thus become a serious threat in Central and Southern European mountain ranges. Today, the impact of mountain farming has been replaced by an invasion of tourists.

Although the detrimental effects of tourism must be considered more severe on Alpine grasslands and montane forest belts, installation of large ski resorts in high mountain areas affects rupicolous and scree communities (UNCED, 1992).

Overgrazing by domestic reindeer in northern Scandinavian mountain ecosystems deserves special attention. Modern techniques of management (helicopter, snowscooter, four-wheel vehicles) and a lack of surveillance measures have increased population numbers enough to disrupt the ecosystem. Damaged lichen flora, the basic food for reindeer, negatively affects the hydrology and triggers erosion processes.

While overgrazing still plays an important role in Southern Europe, mountain farming has declined in Central Europe in recent decades. Considerable impacts may also result from construction of dams – especially in Scandinavia and the Alps, where habitats are destroyed by flooding.

Deserts and tundras

Deserts

Deserts are not normally thought to be included among European ecosystems. However, very arid conditions (often linked to high soil salinity) do prevail in parts of southern Italy and Spain, in the Canary Islands (particularly the eastern isles of Fuerteventura and Lanzarote) as well as regions around the northern Caspian Sea. Furthermore, deserts need not be hot and/or saline places: low air humidity can also occur in the Arctic zone, where available water is locked up for much of the year in ice and snow.

In deserts, much of the biomass is represented by soil organisms and the extensive root systems of shrubs. In biogeographical terms, the arid landscapes of Europe belong to the Indo-Saharan zone that extends from India to Mauritania. The vegetation is dominated by species of *Euphorbiaceae*, *Compositae*, *Caryophyllaceae*, *Crassulaceae* and *Leguminosae*. The Canary Islands represent a Macaronesian extension of this zone, with several rare endemic indicator species or subspecies that have evolved from more widespread North African species; for example the Purpurarian lizard (*Lacerta atlantica*) and Canary Islands chat (*Saxicola dacotiae*). The latter is found only on Fuerteventura, and is threatened by human disturbance and habitat loss. The most threatened bird in the arid region is the Houbara bustard (*Chlamydotis undulata*), which has an endemic race, *Chlamydotis undulata fuertaventurae*, confined to Fuerteventura and Lanzarote (see Box 9W).

Almost all of Europe's arid regions have arisen as a consequence of land degradation after centuries of human pressure on scarce resources. The original scrub or open forest vegetation was cleared, leaving very light soils exposed to sun and wind. In some places (such as the Canary Islands) the problem was compounded by water abstraction from wells or tapping springs at the source. The soils soon dried up and eroded through wind action, and the occasional rainfall drew salts to the surface as surface water rapidly evaporated. As a result a characteristic halophytic vegetation developed, usually heavily grazed by livestock (especially goats) so that much of the ground surface remained bare. Fuerteventura, for example, is no longer the lush island of woods with laurel pigeons (*Columba junoniae*) and permanent streams encountered by the French conquerors of the late 14th century. Perhaps the only natural desert to be found in Europe is the Medano plain of Tenerife, which lies in the rain shadow of El Teide – Spain's highest mountain (Bramwell and Bramwell, 1974).

Functions and values

As might be expected, the harsh conditions prevailing in arid landscapes means that biodiversity is rather low. While Europe's deserts are not significant on the global scale in terms of extent, arid landscapes are important for carbon fixation and storage. It has been estimated that dryland vegetation may hold as much as 10 per cent of total global carbon (through below-ground biomass, eg, extensive root systems).

Habitat threats and representative sites

Desert areas are generally threatened. They are easily accessible and thus subject to intensive disturbance by tourists and hunters using off-road vehicles, which further damages the soil and reduces vegetation cover (see Box 9W). Some large areas in Spain are coming under cloche-covered irrigation to produce market crops such as vegetables, fruit and flowers, as well as to establish almond groves. Paradoxically, exposed areas of the Canary Islands have been threatened by the installation of wind-farms to generate electricity.

Canarian Houbara bustard, Fuerteventura, Spain
Source: H Hinz

Box 9W Arid plains: the Canarian Houbara bustard

The endemic Canarian Houbara bustard (*Chlamydotis undulata fuertaventurae*) has a population of fewer than 200 birds. The Houbara bustards inhabit arid gravelly plains and areas with low sand dunes where shrubby vegetation can provide cover and food. Within such habitats, abandoned rain-fed fields (*gavias*) with annual weeds or irrigated crops of alfalfa are often used for feeding (Collar and Goriup, 1983).

A recovery plan for this species was drafted in 1985, but much of it remains to be implemented. However, some 2754 ha of the Matas Blancas (8 per cent of the total available habitat on Fuerteventura) has been designated as a Special Protection Area under the 1979 EC Bird Directive. A census made in 1988 revealed only five areas holding ten or more bustards; it is unlikely that other sites with fewer birds would support breeding groups exhibiting the normal lekking social system and achieving good recruitment to the population (Emmerson, 1989). Maintaining these larger breeding groups is therefore vital for populating less favourable areas in Fuerteventura. Such areas (and other potential sites) can be encouraged to attain higher breeding success by designating Environmentally Sensitive Areas and supporting local farmers to reduce rangeland grazing densities (by erecting stock housing units), and to restore *gavias*, growing crops like rape and alfalfa (to feed the housed livestock) which are also valuable for Houbara bustards (Dominguez and Diaz,1985). In 1991, the maximum extent of suitable habitat on Fuerteventura was about 34 300 ha, mainly in the north of the island, but with a significant tract of 4560 ha (13 per cent) covering the Matas Blancas (Jandia isthmus) in the south. The extent of habitat available is steadily eroding from desiccation, overgrazing, road building and tourist development (particularly vegetation destruction by off-road recreational vehicles). A few birds are poached each year by determined hunters, and any eggs found by goatherds are generally removed. A plan to erect a wind power-generation farm in the Matas Blancas with EC grant support was partly implemented during 1992, but efforts were being made to halt this development.

Site	Area (hectares)
Spain:	
Sierra Alhamilla - Campo de Njar - Desierto de Tabernas	105 000
Hoya de Baza	36 500
Depresion de Guadix	36 000
Bajo Alcanadre	25 000
Spain: Canary Islands	
Jable Istmo de Janda	2754
Jable de Lajares	2700
El Medano - Los Cristianos, Tenerife	2700
Jable de Corralejo	2300
Costa de Arinaga, Gran Canaria	1000
Llanos de La Corona - Las Honduras, Lanzarote	1000
Llanos de la Mareta - Hoya de La Yegua, Lanzarote	1000
Total area	**215 954**

Table 9.8
Arid landscapes: representative sites
Source: Nature Conservation Bureau, UK, personal communication

Without being able to take into consideration the arid regions of Eastern Europe, because of data problems, the very preliminary data collected for this report indicate the existence of more than 200 000 hectares of desert habitats (11 sites) with importance for nature conservation in Spain alone (see Table 9.8). With respect to site designations, ten desert sites covering about 213 000 ha are included in the BirdLife International list of Important Bird Areas (IBAs), while five sites covering up to 207 550 ha (96 per cent of the total area) receive some form of statutory recognition, varying from 'important landscape zone' to National Park.

Tundras

In contrast to arid landscapes, arctic habitats are largely natural, and generally undisturbed by humans. The importance of these areas lies more in their role in biophysical cycles: permafrost and ice-sheets in tundras hold enormous quantities of the world's freshwater and any change in global climate could have a very significant impact on the hydrological cycle of Europe. This habitat type can be found only where harsh temperatures do not permit trees and larger shrubs to grow: in the northern boreal region and in high Alpine altitudes. Tundra occurs in a narrow band in northern Norway, expanding further east into Russia with a total of 1.8 million km^2 (including Siberia). Iceland and the Svalbard

Reindeer antlers on permafrost, Siberian tundra
Source: J S Grove, WWF

Islands also support extensive areas of tundra which are very important as breeding areas for migratory species of auks, geese, ducks and waders. (A branch of tundra-like habitat also occurs on the highest plateaus of the Cairngorm Mountains of Scotland, but this is more related to Alpine grassland and has been included in the section of this chapter on scrub and grassland.) While the wild reindeer (*Rangifer tarandus*) is a rare and threatened species (occurring as tundra reindeer in Norway and as forest reindeer in a few Finnish sites) there is an overpopulation of domesticated reindeer which has caused severe management problems (eg, overgrazing).

Habitat threats and representative sites

The European tundra zone is generally well protected (if only by lack of interest) although locally the extraction of mineral resources and excess tourism eroding the vegetation can cause conservation problems. From the very preliminary data collected during the preparation of this report, it appears that Europe has at least 3.3 million hectares of tundra habitat (6 sites) of nature conservation importance (Table 9.9). With respect to site designations, five tundra sites covering about 3 million ha are included in the BirdLife International list of IBAs, all of which receive some form of statutory recognition.

Table 9.9
Tundra landscapes: representative sites
Source: Nature Conservation Bureau, UK, personal communication

Site	Area (hectares)
Norway: Svalbard	
Kong Karls Land - Nordaustlandet - Kvitya	1 555 000
Barentsya - Edgeya	645 000
South Spitsbergen	467 300
Northwest Spitsbergen	328 300
Prins Karls Forland	6700
Russian Federation	
Vaygach Island	270 000
Total area:	3 272 300

Conclusions

The above statistics clearly show that there is considerable under-recording of desert and tundra ecosystems in Europe, particularly so in arid regions in Eastern Europe, and for tundra in the European part of the Russia Federation. More examples of these habitat types are known to exist in other countries. For example, some sites in Italy, such as parts of Foggia in the southeast, and areas in Sardinia, as well as regions in Eastern Europe, are clearly arid in character, but have not been specifically identified in this report. Lack of data and lack of attention to such habitats has precluded both a reliable identification of representative sites and adequate assessment of threats.

With the exception of scrub and grasslands, this chapter has mainly focused on natural or quasi-natural ecosystems. Agricultural and urban landscapes, though closely related to human landuse, comprise a number of habitats that are also very important for many adapted species. By being strongly exposed to and often a product of the human cultural environment, these habitat structures form a link between the artificial and the natural worlds and provide a large number of social and ecological benefits. As such, they are important components of sustainable land management.

Over the last decade, there has been increasing controversy concerning the problems associated with the intensification of farming practices, such as the destruction of hedgerows, the drainage of wetlands, the use of pesticides and fertilisers, and the spread of monoculture. Their negative impacts on the natural environment and on human health are well documented.

Urban ecosystems appear in many forms: open lots, meadows, green fingers, parks, gardens, river banks, ponds, lakes, forests and even 'wasteland' are all urban open spaces and can serve as a refuge for specific plants and animals, having important functions for what is commonly known as *urban ecology*. The statement 'More nature in cities makes cities more human' (McHarg, 1969) unveils in a surprisingly simple way the natural bond that exists between people and their 'habitat'. However, for a number of reasons (including the price of property) the amount of open spaces in urban areas is continually decreasing and often replaced by shopping centres, offices and car parks.

Because of their entirely different character, it was not possible to integrate urban and agricultural landscapes into the evaluation process that has been applied for the other ecosystems included in this chapter. However, Chapters 8, 10 and 22 combined provide an overview of Europe's agricultural and urban environments, describing their functional aspects, as well as their environmental values and problems.

Conclusions

The incomplete availability, coverage and consistency of data on the distribution and condition of Europe's ecosystems makes it difficult to provide a systematic and comparative evaluation of the present situation. However, the selection and analysis of approximately 2000 representative sites for seven ecosystem groups, together with information from the most recent literature and reports, as well as a peer review by international experts and their associated networks, have allowed the following tentative conclusions to be framed:

1 Europe's ecosystems consist of a large variety of habitats that reflect the continents's high diversity in terms of geological, topographic, climatic, biogeographical and cultural characteristics. This complex ecological structure forms a pool of valuable natural resources, including a rich flora and fauna.

2 The states of the seven ecosystem types described (natural and prevailing distribution, characteristics and identified threats) differ from one another, with the consequence that any proposals concerning their future should be based on a consideration of individual situations, concerning not only the type of ecosystem but also their geographic basis.

3 All the evidence points to the inescapable fact that the extent and quality of remaining natural ecosystems is in decline and that in certain cases this decline is accelerating. The largest and most coherent portions of natural ecosystems are located on the extremities of Europe: in the far north, in the Mediterranean basin and in a sweep of territory across Eastern Europe. In West and Central

Europe, those remaining sites with semi-natural ecosystems are small and separated from each other in large numbers of fragmented 'islands'. Due to biogeographic conditions, the Mediterranean arc contains the highest diversity in the continent in terms of species numbers, but in habitats that are largely influenced by people and not afforded protection. This contrasts with the high species endemism, large expanses and strong viable populations of species contained in Eastern areas which, until now and the recent socio-economic changes, have been relatively well protected and researched.

4 Europe's ecosystems and their components (habitats and species) are facing a variety of human-inspired and controlled stresses. The maintenance, protection and restoration of degraded ecosystems will depend to a significant extent on the capacity and willingness to reduce and control these mechanisms. Those singled out by the experts include:
   – mass tourism, affecting mountains and coasts;
   – intensive agriculture replacing traditional farming and accelerating rural de-population and abandonment of marginal high-value lands; this, combined with the subsidisation of industrial farming, has had an enormous effect on western rural landscapes and continues to be the major threat to the remaining natural grasslands, bogs, fens and marshes;
   – the decline in water quality, affecting many lake, river and coastal ecosystems;
   – the strong focus of forestry management on economic targets, causing decline in biodiversity, soil erosion and other related effects;
   – policies pursued in the industry, transport and energy sectors, having a direct and damaging impact on the coasts (Atlantic and North Sea), major rivers (dam construction and associated canal building) and mountain landscapes (trunk routes).

5 The analysis of the expert opinion on the question of management effectiveness and the types of threats and impacts suggests that, overall in Europe, management of protected areas is inadequate to retain the integrity of representative areas of ecosystems in Europe (such as the representative sites identified in this report). Furthermore, it is concluded that even if management were effective in eliminating internally originating threats, external uncontrolled threats would continue to affect ecosystems and degrade their resources. Legislative designation of sites accompanied by the enforcement of regulations for the attainment of ecosystem conservation is likely to be a reasonable first step to begin to tackle these problems. The work revealed that formal protection is not followed by a consequent implementation of regulations. In many cases this is due to lack of staff and financial resources – in itself a reflection of the low priority given to the subject by governments and by European citizens.

6 An approach which might solve these problems is to aim for a self-regulating ecosystem network, based on the integration of wildlife and agricultural, industrial and domestic use of resources. Although the primary concern would be the maintenance of natural ecosystems of high quality, the spectrum of the biological and landscape heritage of Europe includes existing and re-created semi-natural habitats, including those maintained through human activities. Thus a pragmatic approach to future biological conservation would better encompass the whole countryside in a rural landuse framework.

7 The information on the distribution and conditions of ecosystems is in need of urgent and extensive improvement. Current national and international initiatives for developing habitat classifications and evaluation methodologies need further coordination. This is a prerequisite for efficient management and effective protection.

FAUNA AND FLORA

Introduction

Habitat requirements of plants and animals vary considerably among species. The cross-sectional character of their life cycles, feeding habits and migrational patterns make animal species frequent the components of many ecosystems. The previous section portrays only species which could be considered as parameters for assessing the condition of individual ecosystems. However, the state of major species classes, groups, families and some selected individual species across all European ecosystems deserves special attention.

Throughout evolution, plant and animal species have appeared and disappeared; some remained without ever changing and others divided up into subspecies. The climatic impacts in particular of several ice ages were probably responsible for developing endemic European subspecies. With the exception of major geological or cosmic impacts (volcanic eruptions, collisions with meteors) the decline and appearance of species usually takes place in geological periods of time (Olschowy, 1985).

During the last 10 000 years, however, the most dramatic impacts on the environment have been caused by the relatively rapid and omnipresent changes due to human activities. Today it is arguable that there is no place in Europe below 2000 metres that has not been altered by humans in one way or another. Human impacts often create new ecological conditions. Furthermore, many changes occur too quickly to allow species to adapt (such as in agriculture, where a succession of human modified systems may replace each other every few years). The net result is a progressive diminution of the number of plant and animal species in many parts of Europe. Since a number of them are endemic, their disappearance also means their total extinction. Even where species are not becoming overall extinct, 'gaps' appear in the ecological networks of many European regions and show detrimental effects on ecosystems and their component habitats.

The specific functions of all plants and animals in the natural processes of ecosystems make species monitoring an integral component of successful ecosystem management (Blab, 1986). Ultimately, the quality of an ecosystem is

Box 9X Use of the term 'threatened species' in this report

'Threatened species' comprise the categories *endangered* and *vulnerable* according to IUCN standards. Endangered species are those whose survival is unlikely if the causal factors, continue operating. Vulnerable species are candidate species, likely to move into the endangered category in the near future. Other IUCN categories, such as *rare, insufficiently known,* or *extinct* species have not been included because the causal factors of threat are likely to be more difficult to compare and to interpret.

The following sections of this chapter present in a series of maps the situation of threatened species by country as a percentage of the total number of each species group. Other species are threatened but could not be included because information is not complete. Because of this and the reasons given above, maps in the following section are likely to understate the situation. Due to different approaches by individual countries, direct comparisons between national data can be misleading, indicating the need for international harmonisation of species inventories.

Table 9.10
*Relative number of
threatened species
(according to IUCN
categories)*
Sources: WCMC, 1992;
Council of Europe, 1993;
EEA-TF

	WORLD		EUROPE	
	Total number	Threatened (%)*	Total number	Threatened (%)
Mammals	4327	16	250	42
Birds	9672	11	520	15
Reptiles	6550*	3	199	45
Amphibians	4000	2	71	30
Fish (freshwater)	8400*	4	227	52
Invertebrates	>1 000 000*	?	200 000*	?
Higher plants	250 000*	7	12 500*	21

* estimates

described by the presence of individual species as well as by the 'completeness' of the living communities within it. While these communities have been identified and classified relatively comprehensively within the field of plant sociology, there are still no satisfying concepts to define faunistic communities: the life-cycles of most animals are far more complex and more interlinked with multiple ecosystems, making classification very difficult (Riecken, 1990). After the description (in the previous section) of the state of ecosystems, mainly on the basis of physical and

phytosociological qualities, this section attempts to examine the main species groups (especially classses). Again, data on fauna and flora are far from complete. Inventories are available for some national or regional areas, but not for others. Those which do exist vary considerably in their reliability, due, for example, to the data being significantly out of date.

For the largest species group, the invertebrates, not all the species have been identified and classified, let alone inventories made of them. Nevertheless, long-term monitoring of species, undertaken by scientific organisations (universities, museums, NGOs, etc) and individual experts, provides the basis for describing the state of flora and fauna. The focus is on occurrences and the reasons for species decline, endemism, introduction, migration and wildlife trade.

In Table 9.10, the percentages of species threatened by extinction are given for Europe and for the world. Several groups of species are not included in this overview because of a lack of data; these are microorganisms (algae, bacteria, fungi, protozoa, viruses), nematodes, mites, lower plants (bryophytes, lichens) and soil macrofauna (Remmert, 1980). In a global comparison, relatively more species are threatened or have become extinct in Europe. Habitat loss and modification (fragmentation and pollution), overexploitation and the introduction of exotic species have been identified as the most important threats to biodiversity (McNeely et al,

Figure 9.7
*Changes of species
populations and
ecosystems by
geographic region
(1900–50)*
Source: RIVM, 1992

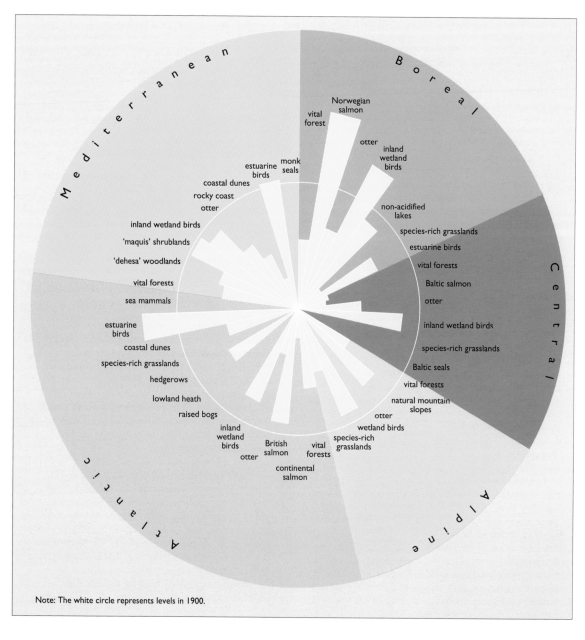

Note: The white circle represents levels in 1900.

1990). The correlation between the decline of individual species or species groups and their ecosystems and habitats, against the background of environmental stresses, can be expressed with the help of a radial plot (see Figure 9.7). This figure presents data on selected species and habitats in the reference period 1900 to 1950

The figure shows that in all parts of Europe, particularly in the central and western regions, parameters such as area size, species numbers, habitat diversity and water quality – identified to describe the quality of ecosystems – have declined severely. Species related to traditional agriculture practices are affected throughout Europe, as are organisms associated with primeval/natural forests. Several studies have shown substantial decline in areas covered by wetlands; only the boreal region has partly improved.

Mammals

Europe is home to nearly 250 species of mammals belonging to nine different orders, making up 5 per cent of the world's mammals. The great majority of species are indigenous (223 species, or 91 per cent of the total number of species found in Europe), while others (21 species, or 9 per cent of Europe's fauna) are not native to Europe but were introduced by humans, often on a very local scale. Such introductions have been particularly important in Britain, France, Germany and Italy. Certain species (44 species or 18 per cent of indigenous species) are endemic, that is, they are found only in Europe. Endemic species are found especially in Southern Europe, and in the Alps and the Caucasus.

No mammal species has become extinct throughout Europe in the last century. However, the smaller the geographical unit of reference (national, regional, local), the more evidence there is of species becoming extinct or endangered. About 60 per cent of Europe's indigenous mammal fauna (137 species) can be considered common, while nearly 40 per cent (86 species) are rare or found only in very localised populations, live in scattered populations, are on the decline or are endangered. Twenty-four of the 698 species of mammals which are considered endangered on a world scale (IUCN, 1990a) are found in Europe; this is 11 per cent of indigenous mammals (Beaufort, 1993). Map 9.11 presents the situation of mammals as the percentage of nationally threatened species (see Box 9X).

Terrestrial mammals

Not only are mammal species under threat, but today's remaining European populations of, for example, red deer (*Cervus elaphus*), bear little relationship to their indigenous ancestors. The genetic uniqueness of all autochthonous red deer populations has been lost due to extensive introductions of, and subsequent hybridisation with, other red deer subspecies, wapiti (*Cervus elaphus canadensis*) and sika deer (*Cervus nippon*). The critically endangered chamois subspecies, *Rupicapra rupicapra cartusiana* is threatened by genetic swamping from the introduced alpine chamois *Rupicapra rupicapra rupicapra*. Competition or hybridisation with introduced non-native species or feral domestic animals (eg, house cats) is a major (and usually ignored) factor in the decline of European mammals (IUCN-PSU, 1993).

The most threatened species include large and medium-sized land-dwelling carnivores and omnivores such as the brown bear (*Ursus arctos*, see Box 9Y), the wolf (*Canis lupus*), the Spanish lynx (*Lynx pardina*), European mink (*Mustela lutreola*) and the otter (*Lutra lutra*, see Box 9Z), which are considered a 'nuisance' in all intensively used areas because of their habitat requirements, their predatory habits or even their economic impact. Today, they are very unevenly distributed: they have become extinct in many countries and 60 to 85 per cent of their populations are found in the least intensively used areas of Europe, mainly in the east. Certain

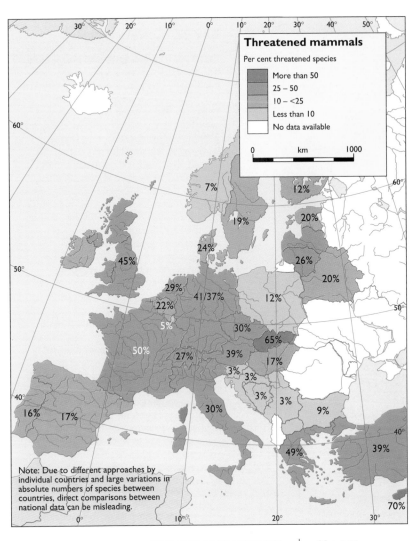

Threatened mammals

Per cent threatened species

More than 50
25 – 50
10 – <25
Less than 10
No data available

0   km   1000

Note: Due to different approaches by individual countries and large variations in absolute numbers of species between countries, direct comparisons between national data can be misleading.

Map 9.11
Percentage of nationally threatened mammal species
Sources: Multiple sources, compiled by EEA-TF

Box 9Y Brown bear

Brown bears live in deciduous and coniferous woods in mountain areas and in plain taiga. In mountain areas they have a seasonal vertical migration up to altitudes of 3000 m. Despite their reputation as predatory carnivores, bears feed mainly on plants, berries, insects, small vertebrates and eggs. They are generally active at night and live in relatively small areas with a home range of 500 to 2500 hectares. Brown bears used to live in all parts of Europe, from Britain and Spain in the west to the Urals in the east. Today, they have completely disappeared from most Western and Central European countries. In the Pyrenees, the Alps and the north of Greece, only very small isolated populations are left. Brown bears were generally considered a danger to domesticated livestock and were persecuted everywhere. However, loss of habitat (large and coherent forest areas, free of disturbance) is another main reason for their decline (Council of Europe, 1989).

Brown bear
(Ursus arctos)
Source: P Heineman, WWF

Common otter
(Lutra lutra)
Source: H Ausloos, WWF

Map 9.12
Distribution of the
common otter
(Lutra lutra)
Source: Council of
Europe, 1992a

Box 9Z Common otter

Otters live in fresh, brackish, and even salt water habitats of rivers, streams, lakes, canals, open marshes and coasts. In mountain areas they have been observed up to 2800 m. Otters are carnivores, feeding mainly on fish, amphibians and crustaceans, but also, to a lesser extent, on waterfowl, mammals and reptiles. They are mostly active at night which makes them difficult to study. Over the last three decades, their populations have declined dramatically in many Western and Central European countries. In the past otters were killed for their fur and also for perceived damage to fisheries. Current threats include habitat destruction by drainage of wetlands, river regulation, clearance of vegetation cover from river banks, and dam building. In poor habitats, disturbance by recreational activities may be detrimental, but the most important factor is water pollution. Research results suggest that contamination of fish prey by bioaccumulating organochlorine pollutants – both pesticides and polychlorinated biphenyls (PCBs) – was the major cause of decline (Mason and O'Sullivan, 1992). Because of their sensitivity to environmental stresses, otters can be considered as biological indicators for the quality of a wetland habitat, otter presence indicating good environmental conditions. Hence, the disappearance of the otter in many European regions reflects quite precisely the scale of environmental damage to which many wetlands have been exposed during the last decades. Today, there are many attempts to reintroduce the otter – an undertaking that can be successful only if the quality of the habitat has been improved accordingly.

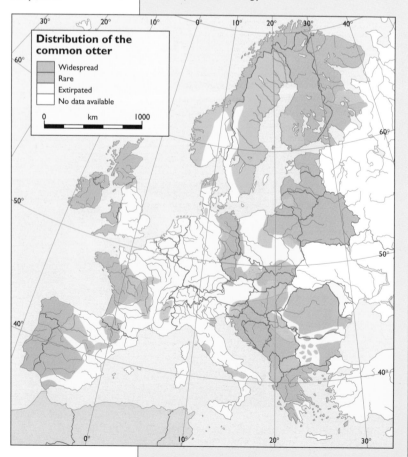

Distribution of the common otter

- Widespread
- Rare
- Extirpated
- No data available

0   km   1000

species are naturally rare and localised, such as the Pyrenean desman (*Galemys pyrenaicus*) or the Bavarian vole (*Microtus bavaricus*); some are marginal to Europe, such as the genet (*Genetta genetta*), mongoose (*Herpestinae*) and Asian wild dog (*Cuon javanicus*); while others are confined to very specific habitats, such as wolverine (*Gulo gulo*), arctic fox (*Alopex lagopus*) and polar bear (*Thalarctus maritimus*) in the Arctic zone. The main introduced species are deer (*Cervidae*), which have either escaped from parks or were introduced for hunting, carnivores such as the raccoon-dog (*Nyctereutes procyonoides*), the raccoon (*Procyon lotor*), and large rodents such as muskrat (*Ondata zibethicus*), coypu (*Myocastor coypus*), North American beaver (*Castor canadensis*), or the grey squirrel (*Sciurus griseus*) (Council of Europe, 1976b).

Existing nature reserves are mostly too small to accommodate viable populations of Europe's larger mammals so there are few possibilities for re-introduction of these species to parts of their range where they have become extinct. Loss of large predatory species has led to a need for repeated culling of deer by humans. Loss of beaver has led to the disappearance of certain types of wetlands created by their river-damming activities. The soil-upturning role of the wild boar in various types of forest cannot easily be replicated, leading to complications in tree regeneration in woodland reserves where the boar is absent.

Attempts to secure the future of Europe's wild fauna are by and large focused today upon sites from which the large mammal fauna has already been lost and in which the small mammal fauna is severely modified. Without a network of larger sites which can accommodate such species there is little hope for long-term survival of either the mammals or the habitats in which they evolved to play their part in ecosystem dynamics.

Aquatic mammals

Among aquatic mammals, the fate of whales has attracted much public attention. Whales and whaling have become a symbol for the dramatic decline of the world's large animal species, leading to one of the most controversial issues in today's debate on environmental ethics. The technological innovations in hunting techniques at the beginning of this century resulted in massive depletion of successive cetacean stocks, including blue (*Balaenoptera musculus*), fin (*Balaenoptera physalus*), sei (*Balaenoptera borealis*), minke (*Balaenoptera acutorostrata*) and humpback (*Megaptera novaeangliae*) whales. Cetaceans are among the most vulnerable to overfishing because of their slow growth, low fecundity and, possibly also through the link between social behaviour and reproductive success. During the earlier part of the twentieth century, some of these species came very close to extinction, and for some species, even with protected status, the situation is still precarious (eg, blue and humpback whales).

Though already founded in 1946 to 'protect all species of whales from further overfishing', the International Whaling Commission (IWC) continued to be driven by the economic interests of its 36 member states. Today this situation has clearly been improved, making whale conservation one of the prime targets of the Commission. However, the IWC's 1985 moratorium on commercial whaling did not succeed in preventing further exploitation. Since the moratorium came into effect in 1986, over 15 000 whales have been killed, 90 per cent of which have been minke whales (*Balaenoptera acutorostrata*) (Lean, 1990). Box 9AA provides some details of another threatened aquatic mammal, the common porpoise (*Phocaena phocaena*).

Seals have been hunted as intensively as whales in the Nordic region. Several arctic pinniped species, including the bearded seal (*Erignathus barbatus*) and the walrus (*Odobenus rosmarus*) are protected, following heavy overexploitation. In recent years, restrictions and slackening demand have also reduced the scale of hunting for species of seals that are still comparatively common. The catching of seals in Norway

Common porpoise (Phocaena phocaena)

Box 9AA Common porpoise

Common porpoises appear in shallow coastal waters and inland bays and sometimes move upstream into rivers. They feed on fish, squid and crustaceans, and used to be widely distributed in the coastal waters of the North Atlantic and the Baltic, White, Greenland and Black seas. Porpoises, like dolphins, are seriously threatened by sea pollution, especially by chlorinated hydrocarbons and heavy metals. Since these animals are situated at the top of the food-chain, they tend to accumulate relatively large quantities of these substances. Another threat is oil which can enter the breathing hole. The species has also been reduced by excessive commercial fishing.

mainly includes the harp seal (*Pagophilus groenlandicus*) and the hooded seal (*Cystophora cristata*). Currently, pups are no longer killed in either the Jan Mayen area or the White Sea, and the greater part of the catch has consisted of harp seals that are one year old or more (Central Bureau of Statistics of Norway, 1993). In Iceland the catch of seals consists mainly of the young of both the common seal (*Pocha vitulina*) and young grey seals (*Halichoerus grypus*). The number of common seals taken in Iceland has declined sharply in recent years (765 common seals in 1990; Icelandic Ministry for the Environment, 1992). Even if the pressure from hunting has eased, the populations of seals in northern Europe are still threatened by toxic pollutants (eg, PCBs), and virus epidemics, possibly related to human activities (eg, the 1988 phocine distemper virus in the Kattegat, Skagerrak and the Wadden Sea): both are cause for concern. The Mediterranean monk seal (*Monachus monachus*, see Box 9BB) continues to be one of the most endangered species, being driven close to extinction by direct killing.

Conclusions

With more than 40 per cent of species under threat, European mammals are one of the most critical groups. While the number of hoof mammals seems to be increasing, predatory species like the brown bear (*Ursus arctos*), the wolf (*Canis lupus*), the lynx (*Lynx lynx*) and others have become restricted to only a few, and often remote, regions. All large mammals have in common that they require large, coherent habitats which are not dissected or fragmented and where human landuse activities are not in conflict with the animals' demands for undisturbed space and shelter. Such habitat types have become especially rare in Central and Western Europe. Other threatened carnivores include wetland species such as the otter (*Lutra lutra*) and the European mink (*Mustela lutreola*), indicating the dramatic degeneration of many river systems. Despite the existing moratorium on commercial whaling (by the International Whaling Commission), large sea mammals continue to be threatened and require utmost international

Monk seal, Isla Sporadi Settentrionali, Greece
Source: F Di Domenico, WWF

Box 9BB Monk seal

The Mediterranean monk seal (*Monachus monachus*) has, since 1984, been listed by the World Conservation Union as one of the world's ten most endangered mammal species (IUCN, 1990a). Usually known for being an abundant species, the monk seal's range once extended from the Black Sea through the whole Mediterranean to the Atlantic Ocean around Madeira, the Canary Islands, and the northwest coasts of Africa as far south as Senegal. Habitats are the rocky coasts with sea-caves, used for both resting and reproduction. The most favourable caves have an underwater entrance and a pebble or sandy beach out of the reach of the waves.

Today, only fragmentary and small populations still remain, and there are only two viable Mediterranean populations of the monk seal, in the Aegean and Adriatic seas, and it is predicted that by the beginning of the next century only the former will remain. One other viable population of about 120 animals is found on the Atlantic coast of Morocco and Mauritania (Council of Europe, 1992b; IRSNB, 1993). The major reason for this decline is killing – deliberate and accidental – mostly by fishermen who perceive the species as a competitor. Other factors are the loss of undisturbed habitats due to coastal development, and lack of food resources linked to overfishing and pollution of the seas. The threat by fishermen appears to have played a dominant role in the decline of the species, except perhaps along the southern shores of the Mediterranean and the Atlantic. The world population is now probably fewer than 500 individuals and numbers throughout the Mediterranean are still declining. The EU has recently funded a major programme of research and conservation on this species which aims to identify and establish a network of protected areas throughout the Member States and to improve public awareness, and the feasibility of breeding in captivity.

attention. The same applies to the Mediterranean monk seal (*Monachus monachus*) as one of the most endangered mammal species. However, recovery programmes have been initiated.

Birds

General public interest in wildlife, birds and bird protection is playing an especially significant role (not least because of bird mobility) in fostering an international awareness and responsibility for seeking to remove specific threats to birds and their populations. Evidence of this can be seen in the growing membership of a number of bird protection organisations (in the UK for example, adult and junior membership of the Royal Society for the Protection of Birds now totals over 800 000), in

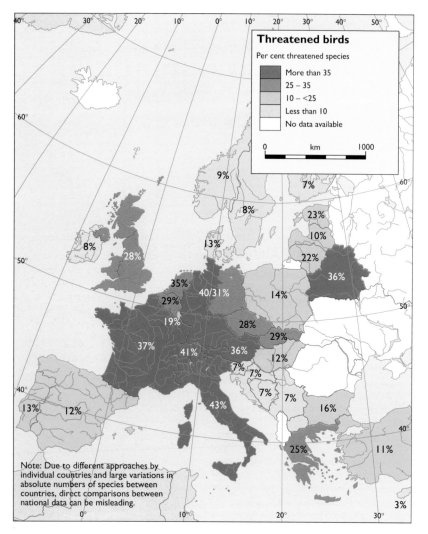

Map 9.13
Percentage of nationally threatened bird species
Source: Multiple sources compiled by EEA-TF

called Important Bird Areas. Presently, 29 out of 520 recorded species (6 per cent) of the European avifauna are believed to be threatened with global extinction (Collar and Andrew, 1988), while 78 (15 per cent) are threatened in Europe. Map 9.13 presents the situation of birds as the percentage of nationally threatened species (See Box 9X).

Data presented in Figures 9.8 and 9.9 are taken from the BirdLife International/European Bird Census Council European Bird Database 1993 which was established as part of BirdLife's Dispersed Species Project and the European Ornithological Atlas Project. The BirdLife Dispersed Species Project aims to establish wide-scale habitat conservation measures for European birds that are in need of adequate action and cannot be conserved by a network of protected areas alone (BirdLife, 1993). Preliminary analysis of these data reveals that at least 280 European bird species (over 50 per cent) are of conservation concern, and are referred to as Species of European Conservation Concern (SpECs). These species are divided into four categories: the top three categories, which are depicted in Figures 9.8 and 9.9, encompass those species which are declining, rare or very localised, including those that are globally threatened (Collar and Andrew, 1988); the fourth category includes species with a large part of their global population in Europe but which are currently considered to deserve attention.

Over one third of the bird species occurring in Europe are rare, declining or very localised at a few sites, and these occur in roughly the same proportions in different biogeographic regions where they constitute a significant percentage of the breeding avifauna (Figure 9.8). These species occur in a wide variety of habitats, although agriculture and grasslands, wetlands and forests support particularly large numbers. (Figure 9.9).

Habitat threats

The intensification of agriculture, fishing and forestry, together with continued urbanisation and industrialisation, have reduced the diversity and degraded the quality of Europe's natural and semi-natural bird habitats, and continue to do so. Significant problems arise from the drainage and reclamation of wetlands with which birds such as herons (*Ardeidae*), ducks (*Anas* spp), geese (*Anser* spp), swans (*Cygnus* spp), and shorebirds, are intimately linked for their survival. These important groups contain over 150 species, among them some of the most striking and colourful in the region, such as the elegant yet bizarre greater flamingo (*Phoenicopterus ruber*).

The importance of individual wetlands often varies greatly, reflecting seasonal movements of huge numbers of migratory birds. The need to maintain a network of wetland 'service stations' throughout their flyways to provide food and shelter is clearly vital (Hollis and Jones, 1991). Despite the growing awareness of the importance of wetlands for wildlife, economic pressures for more land from agriculture, tourist developments, industry, water supply, and so forth, continue to lead to the loss of important ornithological wetland habitats each year.

Birds of prey constitute a group of species that is especially exposed to environmental threats like pollution and contamination. Due to their position at the top of a food-chain they are exposed more than other birds to the danger of accumulating larger amounts of poisonous substances in their organisms. Hunting, disturbances and fragmentation of habitats also have negative impacts (see Boxes 9CC and 9DD).

the establishment of a regular system of communication between the different European bird protection societies, and the increasing liaison on particular issues within international agencies such as BirdLife International (formerly the International Council for Bird Preservation), the International Waterfowl and Wetlands Research Bureau and the European Bird Census Council (EBCC). These organisations have undertaken substantial research and international collaboration resulting in the identification of more than 2400 sites of European importance for the conservation of birds, the so-

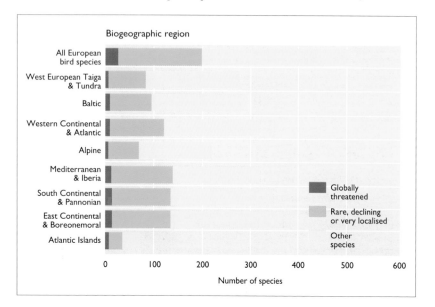

Figure 9.8
The number of European bird species occurring in broad geographic regions and the proportion of rare, declining or very localised European bird species
Source: BirdLife, 1993

Conclusions

Birds enjoy more attention from both conservation groups and the larger public than any other species group, a fact that is mirrored in the existence of more and better legislative action. Especially the EC Bird Directive, but also the Bonn Convention on Migratory Species and the Ramsar Convention on the Protection of Wetlands, are international responses to necessary cross-boundary cooperations and strategies. The progressive implementation of the EC Bird Directive has significantly contributed to improving the biological situation of most species. Several breeding species are no longer threatened with extinction in the EC, although the status of many of these species still gives cause for concern. There has been an increase in the population of almost all species of ducks *(Anatidae)*. Nevertheless, 15 per cent of Europe's avifauna are considered to be threatened, of which especially the wetland species and larger predators are suffering from loss of habitats, tourism and of course hunting and trapping, especially of migratory birds. Coupled to this, the depletion of food resources, through increasing use of pesticides, is an important factor in the decline of some bird species, particularly insectivorous birds with specific food requirements inhabiting agricultural areas.

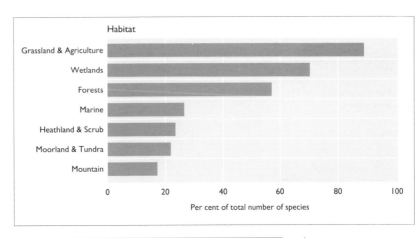

Amphibians and reptiles

Although amphibians and reptiles constitute two different vertebrate classes with different habitats and different conservation needs, they are traditionally taken together under the name of herpetofauna. Moving south in Europe, there is a clear increase in the number of species; the highest numbers of species are found around the Mediterranean and in southeastern Europe. A number of endemic species occur, and

Figure 9.9
Habitats important for rare, declining and very localised European bird species
Source: BirdLife, 1993

Map 9.14
Distribution of the common crane (Grus grus)
Source: EBCC/EOAWG, personal communication

Box 9CC Common crane

The common crane (*Grus grus*) has declined considerably in many of its original European habitats during the last 300 years.

Due to protection measures, the central European population has stabilised since the 1960s, but species numbers are probably still declining in the east (see Map 9.14). As a wetland species – roosting sites are located at shallow lakes, lagoons, rivers and bogs – the common crane is vulnerable to land drainage and has suffered from substantial destruction of its habitat. Other threats include human disturbances, collision with power-lines and hunting (mainly in southeastern Europe). The reintroduction and maintenance of low-intensity agriculture such as in Spain's dehesas (see Chapter 8) will play a crucial role for the species' long-term survival (Tucker and Heath, 1994).

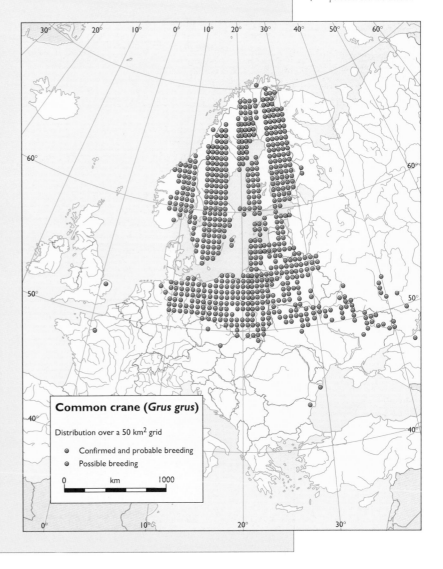

Common crane (*Grus grus*)

Distribution over a 50 km² grid

⬤ Confirmed and probable breeding
⬤ Possible breeding

0   km   1000

Common Crane, Lake Hornborgasjön, Sweden
Source: S Karlsson, WWF

Box 9DD Peregrine falcon

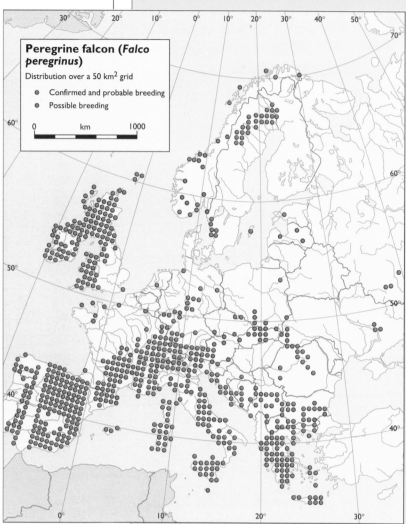

Peregrine falcon (*Falco peregrinus*)

Distribution over a 50 km² grid

- Confirmed and probable breeding
- Possible breeding

The peregrine falcon (*Falco peregrinus*) declined severely over much of its European range north of the Mediterranean from around 1955–65 onwards. The reason was widespread contamination by persistent toxic chemical residues, mainly from insecticides introduced between the late 1940s and mid-1950s. Typical for species at the top of the food-chain, the falcon was exposed to accumulating toxic residues already stored in the organs of its prey, such as the feral pigeon (*Columbia livia*). The lack of recovery in coastal zones of northern Scotland may reflect contamination by marine pollutants (including PCBs) through sea-bird prey. It has since shown variable population recovery due to improved environmental control measures. Map 9.15 shows that Greenland, the UK, France, Spain and Italy all have strong breeding populations. Recently, an increase of hunting and egg-collecting has been recorded (Tucker and Heath, 1994).

Peregrine falcon Source: W Moeller, WWF

Map 9.15
Distribution of the peregrine falcon (Falco peregrinus)
Source: EBCC/EOAWG, personal communication

Box 9EE Italian agile frog

The Italian agile frog (*Rana latastei*) is a rather small, brown frog, living in the plains of the Po river, including its northern tributaries up to southern Switzerland and to Istria (Croatia) in the East. The habitat of this frog typically consists of small streams, rivers or lakes with moist deciduous forests, some swampy ground and a lush herbaceous groundcover. Once widespread over the region, *Rana latastei* is now restricted to only a few sites of about 90 relict populations (80 in Italy, 6

in Switzerland, 4 in Croatia), often isolated from each other. The main causes for this dramatic decline have been the clearing of riparian forests and drainage of wetlands. Although several populations are living within protected reserves, current management practices are not concerned with efficiently safeguarding the species. A recovery programme based on habitat creation and restoration might improve the situation for this endangered species.

Map 9.16 Changes in the distribution of the Italian agile frog (Rana latastei)
Source: A Stumpel, personal communication

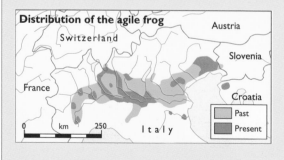

Distribution of the agile frog

Austria
Switzerland
Slovenia
France
Croatia
Italy

- Past
- Present

Italian agile frog
Source: K Grossenbacher

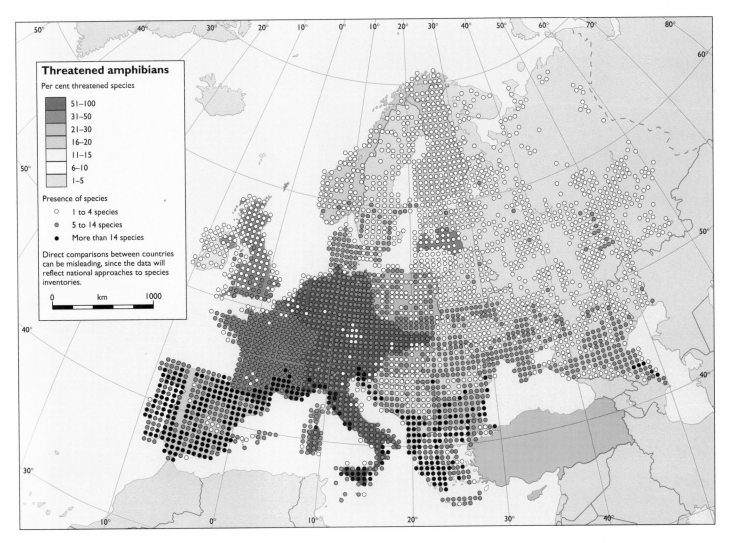

Threatened amphibians

Per cent threatened species

- 51–100
- 31–50
- 21–30
- 16–20
- 11–15
- 6–10
- 1–5

Presence of species

- ○ 1 to 4 species
- ◔ 5 to 14 species
- ● More than 14 species

Direct comparisons between countries can be misleading, since the data will reflect national approaches to species inventories.

0   km   1000

many others have restricted geographical ranges within Europe, or reach the borders of their natural areas of distribution. Currently in Europe there are 71 species of amphibians and 199 species of reptiles, including the marine turtles.

Threats

Amphibians and reptiles suffer on a large scale from habitat destruction, which is the main cause for their decline in Europe. Near-'sedentary' habits combined with hibernation needs make temperate amphibians and reptiles especially vulnerable to human-made changes to, and disturbance of, their habitats. Even short-term and modest changes in habitats can be catastrophic in their permanent effects. The speed of habitat change is often too great to allow survival or adaptation. Accordingly, many species are now in decline. Apart from this, many other human activities change and adversely affect herpetofauna habitats. Some modern forestry and agriculture practices destroy natural ecosystems on a large scale, and reforestation can prevent the development of habitats (see Boxes 9EE and 9FF). Reptile habitats are particularly vulnerable to fires, which both destroy the habitat and kill animals. Many of the best herpetofauna habitats are regarded as 'waste land' and are still being lost to agriculture. Large-scale monocultures replace the old-fashioned small-scale agriculture that has such value as herpetofauna habitat. In areas with low-level grazing, often rich herpetocoenoses were found because of the diverse vegetation structure. Now there is a tendency towards over grazing, destroying this structure. On the other hand, when grazing has been abandoned, the important half-open vegetation mosaic becomes too closed and loses its function as a habitat, particularly for reptiles.

Modern long-line fishery techniques are estimated to cause yearly at least 12 000 mortalities among marine turtles in the Mediterranean (Council of Europe, 1990). The very large increase in tourism during the last decade (see Chapter 25) has had large impacts on the herpetofauna. For example, seaside resorts attract people to the threatened marine turtle nesting beaches and to the sand dunes with important reptile habitats. Urbanisation and industry often lead to total habitat destruction. Motorways, other major roads, and canals constitute barriers in the landscape and they isolate habitats. Human settlements radiate negative effects on many neighbouring landscapes, frequently resulting in the decline or disappearance of amphibian or reptile populations. Trade and capture for pet use is reported to have increased. Rare species attract high prices, and declining numbers may seriously threaten remaining populations. Reptiles still suffer from persecution for various social and cultural reasons. Snakes particularly are often killed, sometimes systematically encouraged by bounty systems (Corbett, 1989).

Maps 9.17 and 9.18 present the situation of amphibians and reptiles as the percentage of nationally threatened species, and overall species distribution.

Most endangered species

The actual status of many European amphibians and reptiles is insufficiently known. However, the most endangered species living in the area of the Council of Europe member states have been identified and habitat conservation proposals have been made. Nevertheless, there are also species known to be threatened, but where no comprehensive habitat proposals can be made without further field assessment and survey. A few examples from the different groups of the most threatened taxa in Europe are given in Table 9.11.

Map 9.17
Distribution pattern of threatened European amphibian species and percentage of nationally threatened species
Source: Multiple sources, compiled by EEA-TF and Muséum National d'Histoire Naturelle, personal communication

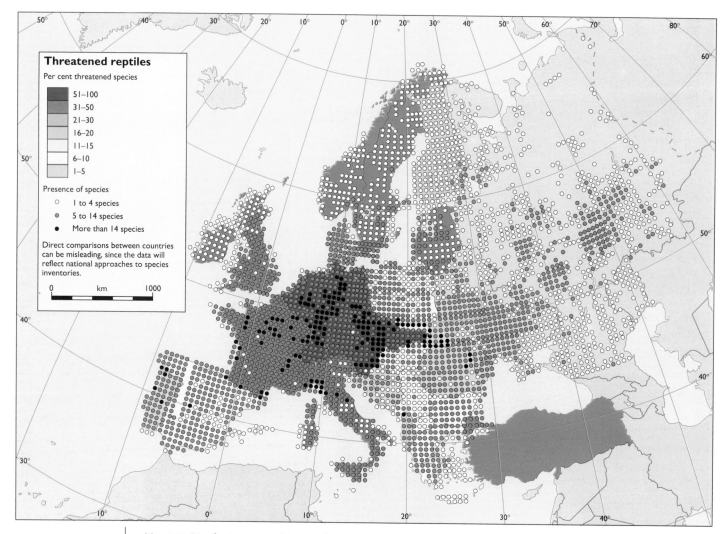

Map 9.18 Distribution pattern of threatened European reptile species and percentage of nationally threatened species
Source: Multiple sources, compiled by EEA-TF and Muséum National d'Histoire Naturelle, personal communication

Box 9FF Meadow viper

This viper *(Vipera ursinii rakosiensis)* is a particularly distinct subspecies and has become one of the most endangered reptiles in Europe, now restricted to only two areas in Hungary and Slovakia (Map 9.19) (Stumpel et al, 1992). Recently it has become extinct in Romania, and probably in Austria. This small, harmless viper has a two-year reproduction cycle and the females produce 2–20 young. It feeds predominantly on large invertebrates, especially crickets and grasshoppers, but lizards, frogs and nestling rodents are also taken. Being entirely restricted to lowland with sandy habitats of alluvial meadows and pusztas in Central Europe, the species requires wet and dry habitats in close proximity. Such habitats, which also support diverse

communities of invertebrates, lizards and frogs, are threatened by drainage and subsequent use for arable farming, vineyards, forestry and tourism. The last remaining sites are exposed to the application of fertilisers and pesticides, too-frequent mechanical mowing, collection for museums and private use, rearing of game birds, grazing and grubbing by geese and pigs. The relict meadow habitats of the Great Plain pusztas to the south of Budapest support the last remaining individuals of this species found in the wild. Their survival depends upon protecting this habitat against agricultural reclamation. The restoration of degraded habitat sites and the development of new ones is required to ensure the long-term survival of the species.

*Map 9.19
Changes in the distribution of the meadow viper* (Vipera ursinii rakosiensis)
Source: A Stumpel, personal communication

Meadow viper (male)   Source: R Podloucky

Box 9GG Arctic charr

The Arctic charr *(Salvinus alpinus)*, a member of the salmon family *(Salmonidae)*, has a holarctic distribution and is Europe's most northerly freshwater fish. It is characteristic of clean cold northern waters and is a very flexible species with seagoing populations in the northern part of its range and landlocked alpine stocks to the south in the Alps (Map 9.20). The species is important commercially in many countries. It is very susceptible to pollution – many populations have disappeared because of acidification and from other stresses.

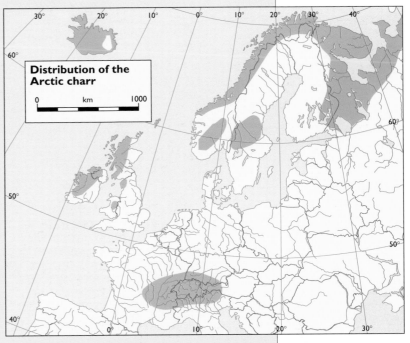

Arctic charr Source: P S Maitland

Map 9.20 Distribution of the Arctic charr (Salvelinus alpinus)
Source: P S Maitland, personal communication

Conclusions

Amphibians and reptiles are now heavily threatened animal groups in Europe and require urgent protection measures. The lack of critical site protection and of adequate habitat management have a direct bearing on the inability to control any or all of the above problems or to lessen the threats facing so many species. Amphibians and reptiles have not yet received a proportionate degree of conservation action or resource allocation in Europe compared to other species groups, such as birds and mammals. Their endangered status as well as their value as indicators for habitats with rich and diverse biocenoeses ought to elevate their significance in European nature conservation.

Fish

Due in part to the recent glaciations, the freshwater fish fauna of Europe are much less diverse than those of the tropics, and even within the continent there is, as for amphibians and reptiles, a marked reduction in species from south to north. Human impact has destroyed fish habitats on a wide scale and many populations of rare species have disappeared over the last two centuries. In addition, numerous distinct stocks of common species and many important fish populations have become extinct.

Of the 227 species of freshwater fish found in Europe, 200 are regarded as native and 27 as introduced (mostly from North America). Altogether, 122 of the native species are now included in the Berne Convention, 4 in Appendix II and 118 in Appendix III. Only a few of these species are also protected under the Convention on the International Trade in Endangered Species of Wild Flora and Fauna (CITES). This leaves 78 unlisted species which are assumed to require no protection at all. In fact, little information is available on

Table 9.11 Most endangered species of amphibians and reptiles and an indication of their habitat
Source: Andrén et al, 1991

Group/Taxon	Habitat
SALAMANDERS	
Sardinian brook salamander (*Euproctus platycephalus*)	Mountain streams and lakes
Golden salamander (*Salamandra atra aurorae*)	Mixed forests between 1300 and 1600 m
Gold-striped salamander (*Chioglossa lusitnaica*)	Borders of shaded streams in forested ravines
Olm (*Proteus anguinus*).	Subterranean caves in karstic regions
ANURANS	
Mallorcan midwife toad (*Alytes muletensis*)	Steep-sided limestone gorges close to torrents
Italian spadefoot toad (*Pelobates fuscus insubricus*)	Sandy areas, heathlands, deciduous woodlands in floodplains
Italian agile frog (*Rana latastei*)	Semi-moist deciduous riverine forests
MARINE TURTLES	
Loggerhead turtle (*Caretta caretta*)	Mediterranean
Green turtle (*Chelonia mydas*)	Eastern Mediterranean
TORTOISES AND TERRAPINS	
Hermann's tortoise (*Testudo hermanni*)	Open evergreen oak forests, maquis, garrigue, sand dunes
Nile soft-shelled turtle (*Trionyx triunguis*)	Estuaries of undisturbed rivers, coastal lagoons, sand dunes
LIZARDS	
European leaf-toed gecko (*Phyllodactylus europaeus*)	Rocky outcrops in deep vegetation and shrubs, trees
Lilford's wall lizard (*Podarcis lilfordi* subspp)	Small Mediterranean islands with a variety of natural habitats
Clark's lizard (*Lacerta clarkorum*)	Damp, subtropical forests
Iberian rock lizard (*Archeolacerta monticola*)	Mosaics of rocks, meadows and bushes above 1400 m
Hierro giant lizard (*Gallotia simonyi*)	Steep vegetated cliffs with crevices and small ledges
SNAKES	
Hungarian meadow viper (*Vipera ursinii rakosiensis*)	Sandy habitats of alluvial meadows and pusztas
Milos viper (*Vipera schweizeri*)	Sheltered, densely vegetated valleys with rough rocky slopes
Kaznakov's viper (*Vipera kaznakovi*)	Small glades and river valleys in damp sub-tropical forests
Large-headed water snake (*Natrix megalocephala*)	Damp subtropical forests

Box 9HH Ship sturgeon

Distribution of the ship sturgeon

*Map 9.21 Distribution
of the ship sturgeon*
(Acipenser
nudiventris)
Source: P S Maitland,
personal communication

*Map 9.22 Percentage
of nationally
threatened fish species*
Source: Multiple sources
compiled by EEA-TF,
personal communication

The ship sturgeon (Acipenser nudiventris) is one of several species of sturgeons found in Europe. All sturgeons have a similar life history, all are important economically (both for caviar and for their flesh) and all are severely threatened. The ship sturgeon is found in the Black, Azov and Caspian seas and formerly migrated into most of the larger, northern

Fishing for ship sturgeon, Volga
Source: W Fisher, WWF

rivers associated with these seas. Now, due to overfishing, pollution and the construction of river barriers, its numbers and distribution are much reduced.

the status and distribution of freshwater fish in many countries and substantial work remains to be done if conservation strategies for the fish fauna of Europe as a whole are to be properly developed. Map 9.22 presents the situation of fish as the percentage of nationally threatened species.

Certain characteristics of freshwater fish are especially

relevant to the structure of their communities and to their conservation. Their habitats are discrete and thus fish are contained within particular bounds. This leads to the differentiation of many independent populations with individual stock characteristics developed since their isolation. This is true even of migratory species: here, even though there is substantial mixing of stocks in the sea, the homing instinct has meant that there is a strong tendency for genetic isolation.

Because each population is confined to a single waterbody, within which there is usually significant circulation, the entire population is vulnerable to the effects of waterborne threats such as pollution and disease. Thus for each species the number of separate populations is usually of far greater importance than the number of individual fish. Migrations are a feature of the life-cycles of many fish and at these times they may be particularly vulnerable. In particular, in diadromous riverine species (which migrate seasonally between upstream sections of the river and oceanic waters), the whole population has to pass through the lower section of its river to and from the sea. If this section is polluted, obstructed or subject to heavy predation, populations of several species may disappear, leaving the upstream community permanently impoverished (Maitland, 1977).

Threats

Humans have been interacting with fish populations for many thousands of years, and it is often difficult to separate the effects of human impact from changes which have taken place due to more natural processes. Over the last 200 years, and particularly during the last few decades, various new and intense pressures have been applied to freshwaters and very many species have declined in range and in numbers. Many of the pressures are interlinked, the final combination often resulting in a complex and sometimes unpredictable situation.

The pollution of freshwaters is probably the single most significant factor in causing major decline in the populations of many fish species in Europe. In 1990 about 6300 of the 83 000 lakes in Sweden which are larger than 1 hectare had a pH below 5 (Bernes, 1991; see also Chapters 5 and 31).

Most pollution comes from domestic, agricultural or industrial wastes and can be so toxic that all the fish species

Threatened fish

Per cent threatened species

- More than 35
- 25 – 35
- 10 – <25
- Less than 10
- No data available

Note: Due to different approaches by individual countries and large variations in absolute numbers of species between countries, direct comparisons between national data can be misleading.

present are killed, or that a few sensitive species are selectively removed altering the environment in such a way that some species are favoured and others not. Considerable research has been carried out in this area and sensible water-quality criteria for fish are now available.

The impact of landuse changes (land drainage schemes, monoculture forests, filling in of ponds) often results in problems such as silting, increased acidification and alterations of the hydrology, affecting formerly important sites for fish. River and lake engineering works (see Chapter 5) have been responsible for the elimination of fish species, eg, different species of salmon, in freshwaters all over Europe (see Box 9GG, p 231).

Migratory species are particularly threatened by dams and other obstructions, and, if such fish are unable to reach their spawning grounds, they may become extinct in a few years. The combination of obstacles (weirs, locks, etc), severe pollution in lower river reaches and commercial fishing in estuaries has undoubtedly been a major cause in the decline of sturgeon (*Acipenseridae*, see Box 9HH), shad (*Alosa* spp), smelt (*Osmerus eperlanus*) and other migratory fish across the whole of their European distribution (Maitland, 1986).

The impact of fisheries (both sport and commercial) on the stocks which they exploit can cause extinction or lead to an ideal stable relationship of 'recruitment' and 'cropping' (see Chapter 24). The essence of success in fisheries management is to have a well-regulated fishery where statistics on the catch are consistently monitored and used as a basis for future management of the stock.

Conclusions

There is still substantial work to be done in the field of fish conservation. In addition to establishing the status of fish across Europe, the specific conservation goals of the most endangered species need to be identified and appropriate measures implemented. As well as habitat restoration, one of the most positive areas of management lies in the establishment of new populations, either to replace those which have become extinct or to provide an additional safeguard for isolated populations. Any species which is found in only a few waters is believed to be in potential danger, and the creation of additional independent stocks is an urgent and worthwhile conservation activity.

The general conclusion reached from this brief review is that there is insufficient information available on the status and distribution of freshwater fish in many parts of Europe. Although there has previously been some legislation and management in relation to both fish and various general aspects of conservation (such as the establishment of nature reserves), little of this has been aimed directly at the protection of fish species. If this situation is not improved it is very likely that many endemic European fish species and further valuable stocks of other native species will be lost. For commercial fish species (eg, sturgeons and salmonids), the eventual long-term value of restored stocks will be worth many times the actual cost of restoring them.

Invertebrates

Introduction

Apart from the vertebrates, 18 animal phyla are recognised as occurring in Europe. These 'invertebrates' (animals without an internal skeleton) are represented in Europe by

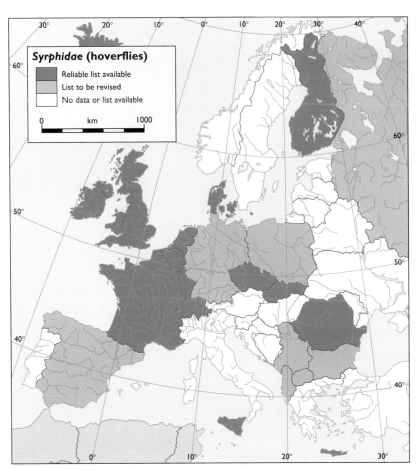

Map 9.23 Existing national lists for the insect family of hoverflies (Syrphidae)

Source: M Speight, personal communication

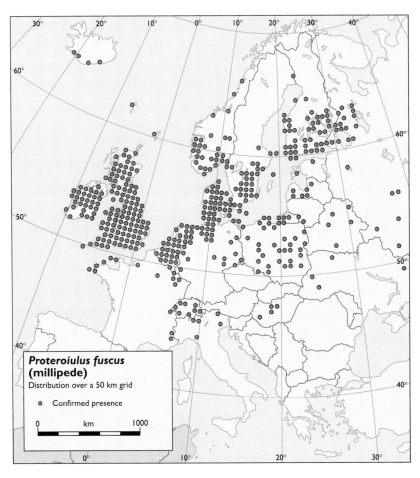

Map 9.24 Distribution of the millipede (Proteroiulus fuscus)

Source: Kime, 1993

Box 9 II The beetle *Rhysodes sulcatus*

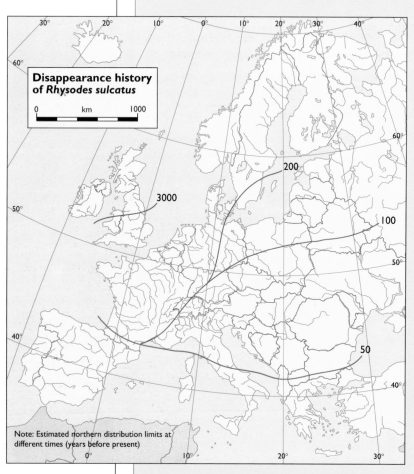

Disappearance history of *Rhysodes sulcatus*

0   km   1000

200

3000

100

50

Note: Estimated northern distribution limits at different times (years before present)

Map 9.25
Disappearance history of Rhysodes sulcatus
Source: M Speight, personal communication

A case of a species occupying a very particular habitat, and whose history is known to some extent, is offered by the beetle *Rhysodes sulcatus*. This beetle lives beneath the bark of large, standing, dead trees that are well-rotted, in circumstances where an appropriate microclimate is provided by surrounding forest. This habitat is now very

scarce in today's Europe, and over much of the continent potential sites for *R. sulcatus* do not exist. However, it has been observed in most places where the habitat does exist that the beetle does not. Is this because it has disappeared from these parts of Europe, or because it never occurred there in the first place? It is now known from fragments collected in dated archaeological contexts (Buckland and Dinnin, 1993) that 3000 years ago this beetle ranged as far west in Europe as southern England, but that it had disappeared there by the time entomologists became active. At the beginning of the nineteenth century *R. sulcatus* was still found in southern Scandinavia, but has subsequently disappeared there also. At the beginning of the 20th century this beetle was recorded in Central Europe, but since then has not been found. Since 1950, the species has only been found in Europe in a few isolated colonies in the Pyrenees and mountainous parts of northern Greece. This history is illustrated in Map 9.25. The implication is that this beetle has disappeared along with Europe's natural forest cover. Its European range has shrunk from what was probably a pan-European coverage in prehistory to a few colonies today. These colonies are now so remote from one another that genetically and in every other respect they are now entirely out of contact, being separated by vast tracts of land where habitats for the species do not exist.

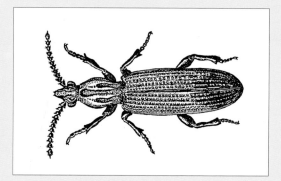

Rhysodes sulcatcus

---

some 200 000 species; 100 000 of these are either freshwater or terrestrial, the rest are marine. The four largest taxonomic groups are the insect orders *Coleoptera* (beetles), *Diptera* (flies), *Hymenoptera* (ants, bees, wasps, parasitic wasps and sawflies) and *Lepidoptera* (butterflies and moths).

A consistent data coverage on the distribution of European invertebrates does not exist. To consider the state of one major taxonomic group at national level, reasonably complete national lists exist for the *Diptera* of former Czechoslovakia, Finland and the UK, but not for other European countries. It might be expected that programmes for making inventories of the species occurring in national parks, nature reserves and other protected areas would by now be well advanced. But in reality, few such programmes exist and the sort of computerised databases needed to handle inventory information effectively have been set up for only a handful of national parks, (eg, the Berchtesgaden National Park in Germany). The same situation exists for nature reserves. Only one or two European countries have national systems up and running, such as the 'Recorder' system now in use for nature reserves in the UK.

One of the best-known families of *Diptera* are the *Syrphidae*, or hoverflies, many of whose larvae are recognised as beneficial because they consume aphids and other plant bugs, and whose adults are important as pollinators. Map 9.23 shows the availability of national lists for this insect family in

various European countries. Essentially, the map shows that for less than half of the surface of Europe usable regional lists of these insects exist; for the rest of the continent none are available. Similar pictures could be presented for other invertebrate taxonomic groups. For Ireland, which has a relatively small fauna, it was recently calculated (Ashe et al, 1988) that, in the case of the *Diptera* alone, in excess of 1500 species remain to be added to the national list and that, with the current level of effort and available resources, the task of listing the Irish *Diptera* will not be completed within less than 30 years.

Functions and values

Invertebrates can be seen as the glue which holds the biosphere together. One of the more important functions of invertebrates is pollination, mostly carried out by various insects, whose importance as pollinators has been responsible for a major development in plant evolution. But the role of invertebrates in recycling organic debris through the ecosystem is at least as important, if not more so. Between 30 and 50 per cent of European invertebrate species are involved in breakdown of the remains of plants or animals or their waste products. These breaking down processes are usually complex and involve joint action by invertebrates, fungi and microorganisms. An important aspect of the invertebrate

Box 9JJ Moonwort fern

The moonworts are small ferns possessing only one leaf up to 15 cm long, consisting of a sterile lamina and a fertile spike. Because of their complicated biology they are a good example of naturally rare plants. *Botrychium matricariifolium* is a species found in relatively dry, often acid grassland and open woodland habitats. Moonworts are saprophytic, associated with fungi, with which they develop a symbiosis. Only in favourable years does the single leaf appear; in bad years the plant survives below ground. Threats to the species are twofold: by acceleration of vegetation succession through nutrient input, and by loss of potential growth sites due to human urbanisation and settlement. A rapid loss of sites can be expected as a consequence. Map 9.26 shows the difference between long-term mapping by the Flora European project (Jalas and Souminen, 1972) and the estimation of the degree

of threat in the national Red Data Books. Effective protection of the species is difficult; creation of special nature reserves often takes too long, by which time the fern often disappears. Only larger areas, which are less polluted and with sustainable agriculture, can provide enough potential growth sites for the survival of the species.

Moonwort fern Source: M Schnittler

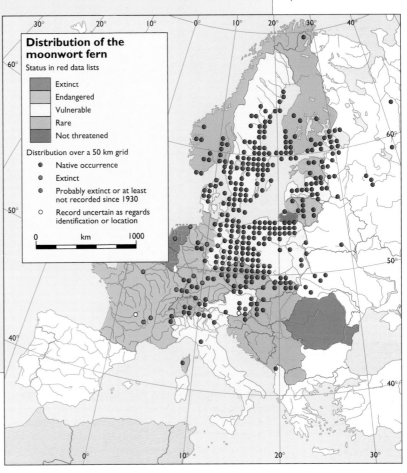

Map 9.26
Distribution of the moonwort fern (Botrychium matricariifolium)
Source: M Schnittler, personal communication

activities is rendering materials quickly accessible to the other decomposers, either by drilling holes through otherwise impermeable materials, such as bark, or by reducing them to small fragments through comminution – the chewing or rasping action of the animals breaking through cell walls. The role of certain types of invertebrates, such as earthworms (*Lumbricidae* and related families) can be seen to be of direct benefit for humans: in agricultural soils earthworm activity helps both to aerate and to maintain drainage capacity and is also important in soil mixing. For the most part, however, knowledge of the precise role of individual species is not available and for many invertebrates only very general functions can be identified, like helping in the regulation of populations of potential pests, or providing food for various vertebrates. With the exception of the land tortoises (*Testudo* spp), the majority of Europe's reptiles and amphibians are dependent upon invertebrates for food. Furthermore, 90 per cent of Europe's fish and bird species and many mammals, from the blue whale (*Sibbaldus musculus*) to the pygmy shrew (*Sorex minutus*) rely on invertebrates for their food supply.

Status

Few groups of invertebrates have received sufficient attention for firm statements to be made concerning the European status of their species. A rare exception are the millipedes (*Myriapoda*), for which data, like those shown in Map 9.24, for the species *Proteroiulus fuscus*, are beginning to appear (Kime, 1990). Rather more taxonomic groups have been the subject of comprehensive study in one part of Europe, or one European state, but not in others. Very localised endemic species form a category which has received close attention. In parts of Southern Europe endemism among invertebrates is a matter of some considerable significance. More than 900

endemic invertebrate species are already known from the Canary Islands alone.

Although the available information concerning the status of the vast majority of invertebrate species in Europe is far from adequate, the data from both well-studied and poorly studied taxonomic groups all point in the same direction: invertebrates are demonstrably as much affected by loss of habitat and general environmental deterioration in Europe as are other animals and plants. In habitat terms, one of the greatest concentrations of threatened invertebrates occurs in ancient forests. Economic exploitation of Europe's remaining forests has led to total eradication of over-mature trees from vast areas causing the virtual elimination of dead woody parts of trees from the ecosystem, a particularly rich habitat for many invertebrates.

Threats

In some cases, levels of human exploitation have been such as to lead either directly, or indirectly, to the inclusion of particular species, eg white-clawed crayfish (*Austropota-mobius pallipes*), escargot (*Helix pomatia*), on lists of invertebrates requiring protection at the European level (eg in the annexes of the Berne Convention). In general, and in contrast to the situation for vertebrates and plants, the collection of specimens by people cannot be identified as a factor causing significant reduction in populations of European invertebrates. Known instances relate to a few

Box 9KK Wintergreen family

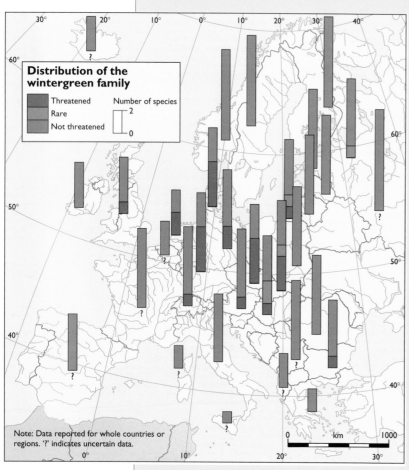

Distribution of the wintergreen family

Threatened
Rare
Not threatened

Number of species
2
0

Note: Data reported for whole countries or regions. '?' indicates uncertain data.

Map 9.27
Distribution of the wintergreen family (Pyrolaceae)
Source: M Schnittler, personal communication

The wintergreens *(Pyrolaceae)* are a family with ten species of small perennial herbs (including a few dwarf shrubs) occurring in Europe. They are typical and often abundant plants of the natural nutrient-poor woodlands and heaths in boreal and Central Europe. All species live in symbiosis with fungi (mycorrhiza), possessing long creeping rhizomes, buried in ground litter. Nutrient input (especially nitrogen) has negative effects on

mycorrhiza and favours other species such as grasses. Hence, the disappearance of this group may be regarded as a good indicator for the degree of nutrient input through air pollution. Map 9.27 illustrates the threatened state of family members especially in the highly industrialised regions of Central Europe. The wintergreens still occur frequently only in more natural regions such as the Alps. In Switzerland, only one species is threatened in the whole country, but all species are threatened in the Alpine 'Mittelland' region. Because of the indirect influence of nutrient inputs into ecosystems, effective conservation for species such as those of the *Pyrolaceae* cannot be achieved by the creation of nature reserves alone. Almost all species of the group are widely distributed in Central and Northern Europe and therefore are not threatened with extinction. However, the wintergreens are a good example of the loss of functional components of ecosystems within Central Europe.

Wintergreen Source: D Wascher

showy dragonflies, butterflies and shells.

The mechanism recommended by IUCN for establishing the threat category into which a species should be put, is thus to a significant extent inapplicable to invertebrates, because invertebrate specialists have been forced to develop a methodology for assessing threat status which is largely restricted to the use of present-day distribution data. In consequence, the number of invertebrate species recognised as under threat is probably under-estimated, and possibly quite

Map. 9.28
Percentage of endemic plants in South European mountain ranges
Source: Favarger, 1972

Endemic plants

France
Germany
Slovak Republic
Switzerland
Austria
18%
12%
Romania
13%
Spain
14%
Italy
38%
13%
36%
Greece
37%

dramatically so. Nonetheless, disjunct distribution patterns like those now exhibited by *Rhysodes sulcatus* (see Box 9II) imply that, on a regional scale, hundreds, if not thousands of species have disappeared over wide areas.

Particular effort has been put into establishing the status of the species of a small number of taxonomic or ecological groups of invertebrates at international level: butterflies (Heath, 1981), dragonflies (van Tol and Verdonk, 1988), saproxylic invertebrates (Speight, 1989), non-marine molluscs (Wells and Chatfield, 1992). The most recent of these studies, on non-marine molluscs, provides evidence that more than 15 per cent of the European species are under threat, most of them endemic to the continent. This confirms other studies, which demonstrate that between 15 and 40 per cent of the species reviewed are under threat at European level. Extrapolating to the situation of the entire invertebrate fauna of both freshwater and terrestrial systems, suggests that a minimum of 10 000 invertebrate species may now be under threat in Europe. As mentioned above, there are grounds for believing that this must represent an underestimate of the actual situation. No studies of marine invertebrates have been conducted that would enable a similar estimate to be made of the proportion of invertebrate sea-life under threat.

The year 1988 saw incorporation of a list of close to 100 invertebrates into the appendices of the Berne Convention,

Lavanda, Balkan Peninsula
Source: W Leonardi, Frank Spooner Pictures

Box 9LL Endemic plants

A plant is called endemic if it is native to the region and its distribution within that region is restricted to a narrow geographical area. The term 'endemic' is only useful in connection with the region for which it is valid, eg, 'endemic to the Canary Islands' or 'endemic to Europe'.

The main force in the evolution of endemic species is the geographical isolation of populations. But evolution needs time, and hence a strong negative correlation can be recognised between the numbers of endemic species per country with the southern frontier of ice during glaciation. Areas especially rich in endemics include the southeastern Alps, the Balkan and Iberian peninsulas, as well as islands like the Canary Islands.

as species requiring special measures to be taken for their protection in the territory of states ratifying the Convention. Recognition of the need to take special measures to protect an invertebrate implies a knowledge of the requirements of the species. All too frequently adequate information on the autecology of invertebrates is lacking and there are few taxonomic groups for which the available data have been brought together in a form usable for habitat management purposes. An exception is provided by the species accounts compiled for dragonflies by Dommanget (1987). More general prescriptions for habitat management to conserve invertebrates are also required, as attempted by Kirby (1992). But, without detailed information on the autecology of invertebrate species, site management remains an uncertain tool, and as yet insufficient research has been carried out to establish the particular needs of most threatened European species.

Developing parallel with work on invertebrate conservation has been the use of invertebrates as tools in environmental interpretation. Increasingly, there is feedback between these two fields of endeavour. Sampling the freshwater invertebrate fauna now provides a standard tool in water quality assessment, and application of the same techniques to site evaluation processes, for 'environmental impact survey' purposes, has heightened interest in lists of internationally or nationally threatened invertebrate species as aids in gauging site quality.

Conclusions

Although estimates suggest that more than 10 000 species are currently under threat, existing nature reserves and other protected areas in Europe have been selected almost entirely without any reference to the invertebrate fauna. Furthermore, management plans have been developed without reference to invertebrates. One pressing need is to establish what role these protected areas can play in the protection of Europe's invertebrates and how their management can be modified to accommodate the requirements of the invertebrate fauna. This objective requires making inventories of invertebrate faunas of protected sites, which in turn requires field surveys by specialists, and invertebrate distribution information.

While it is true that many invertebrates are linked directly (by, for instance, food plants) or indirectly (by ecological parameters) to particular phytosociological associations, other organisms live at the boundaries of different associations, such as some flying insects along the sides of woods and hedges. The importance of such ecotones should not be overlooked. Also, the larval stages of many invertebrates have ecological requirements very different from those of the adults so that the juxtaposition of different types of habitats becomes important. Biodiversity is related to spatial variability of the terrain, and particular combinations of biotopes (Kime, 1993).

Unless and until a more comprehensive, coordinated effort is made, at the international level, to identify species requiring protection and sites appropriate for protection of their habitats, many invertebrate species and sites will be lost before the significance of the situation is recognised. Better autecological data pertaining to invertebrates require research and funding for that research.

Map 9.29
Percentage of nationally threatened plants species
Source: Multiple sources compiled by M Schnittler/EEA-TF

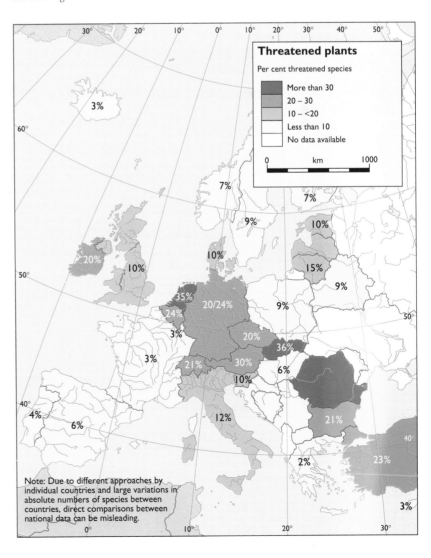

Note: Due to different approaches by individual countries and large variations in absolute numbers of species between countries, direct comparisons between national data can be misleading.

Higher plants

Introduction

Compared with other parts of the world, Europe is relatively poor in plant species. As a result of several ice ages during the last million years, the richness of flowering plants (some 12 500 species) shows an unequal distribution across Europe (Tutin et al, 1964/1980). A considerable proportion of Europe's endemic plants (see Box 9LL, p 237) grow in mountain regions such as the Alps, Apennines, Pyrenees, Sierra Nevada, Carpathians and those of the Baltic region and Balkan peninsula. These areas are recognised as major locations of diversity and endemism by the World Conservation Union (IUCN) and the World Wide Fund for Nature (WWF). Some of these mountain areas have a very high concentration of endemic species (Map 9.28, p 236). The Sierra Nevada, for example, houses about two-thirds of Spanish (mainland) plant endemics and Mount Olympos in Greece has 26 endemic plant species. Although more than 2200 species are recorded as endangered, vulnerable, rare or indeterminate according to IUCN categories of threat, only 27 higher plants have been recorded to have become extinct in Europe (Heywood, 1993) – Map 9.29 (p 237) presents the situation of plants as the percentage of nationally threatened species.

Because of differences in natural conditions and human impact, the main plant geographical areas of Europe are discussed separately, as follows.

Boreal and Arctic Europe

This area comprises Northern Scandinavia, the Northern Baltic region and the North-European part of the Russian Federation. In analysing the causes of threat, the Northern part of Britain and Ireland can also be included here, but the highly industrialised areas of Southern Sweden and Denmark should be excluded.

Only a small number of threatened species have their distribution centres in boreal or arctic zones and causes of threat are mainly indirect:

- acidification (see Chapter 31), especially damaging aquatic ecosystems; and
- potential global warming (climatic models indicate that this might especially affect the northern flora).

A system of rather large national parks and nature reserves, coupled with a well-developed monitoring system, allows effective nature conservation. The St Petersburg region and some areas of Northern Russia (eg, the Kola Peninsula) have problems caused by heavily polluting industries.

Central and Atlantic Europe

Here, the flora is relatively poor in species, with only a few endemics concentrated on the southeastern part, especially the Eastern Alps. The main problem is not extinction of species but the dramatic loss of species diversity in almost all ecosystems. This also concerns the naturally rare species (see Box 9JJ) that are typical elements for ecosystems, and the naturally not rare (see Box 9KK).

Central Europe has the highest percentage (25 to 35 per cent) of threatened plant species. The main causes of threats are:

- direct impacts by construction (especially dramatic is the loss of aquatic species associated with watercourse regulation and eutrophication);
- nitrogen input into terrestrial ecosystems;
- tourism (in particular in the Alps, shores and coastal zones); and
- intensification of agriculture and forestry.

A well-developed system of nature reserves and monitoring can often avoid species extinction, but is increasingly only able to keep species in the garden-like situation of small nature reserves. Most of the existing nature reserves are too small for effective species conservation and there is no effective protection against indirect influence such as through the input of nutrients.

Southern Europe

The Mediterranean and Macaronesian regions have the largest number of species in Europe, often with a high percentage of endemics (up to 48 per cent endemics for the flora of the Canary Islands). Like the Mediterranean region, large parts of the Balkan Peninsula, central France and the southern slopes of the Alps are influenced by a long history of human cultivation and landuse activities. Due to difficulties in regenerating woodland ecosystems, a number of species have become very rare.

Although species extinction and decline are key problems of southern regions, the percentage of threatened species as part of the whole flora appears to be relatively low (1 to 10 per cent), simply due to the overall species richness (see Map 9.28, page 236). However, because of the small size of their original habitats (sometimes only a few square kilometres) endemics continue to be endangered.

The main causes of threat are:

- tourism and settlement, mainly in coastal areas and islands;
- ineffective forestry management (fire, promotion of pine and eucalyptus); and
- agriculture (especially overgrazing).

Aquatic vegetation

Based on replies to questionnaires presented in Box 5E, a survey of the past (period 1900–50) and present distribution of 108 aquatic plant species in Europe has been undertaken. The species were divided into four categories of threat, as proposed by the IUCN in their Red Data Book Category Definition: common, rare but not threatened, vulnerable or endangered, and extinct.

Due to the homogeneity of the aquatic medium most aquatic plants are spread all over Europe, some even over the world. The most species-rich aquatic flora is presently found in countries with a great diversity of aquatic habitats (eg, UK, Sweden and the Russian Federation) where 60 to 70 aquatic plant species are considered to be common and only small distributional changes have occurred over the years. On the contrary, countries in Central and Eastern Europe (eg, Austria, the Czech Republic, Poland, Latvia, Moldova and Ukraine) have many fewer common species, ranging from seven in the Czech Republic to 28 in Latvia. For the same countries a dramatic change in species occurrence has taken place, implying that many species that were common in the past have now become rarer or even extinct. For example, the Czech Republic and Moldova have lost five aquatic species in this century.

For Europe as a whole results from the questionnaires received from 18 countries (expressed as the average of distributional changes within countries) show that 74 per cent of the aquatic plant species which were common in the past are still common, 9 per cent of formerly common species have become rare but not threatened, 10 per cent vulnerable or endangered and 1 per cent have become extinct. For the remaining 6 per cent, information about their present distribution is not available. For those species which were rare but not threatened or vulnerable or endangered in the past, the present situation is of more concern. For each of these groups, 34 and 32 per cent of the species respectively have moved towards the endangered group or have become extinct. It should be noted, however, that these figures illustrate country averages and it should be stressed that no aquatic plant species has disappeared from Europe in this century.

Due to differences in the location of their leaves in the water column, not all aquatic species are equally threatened. In particular, rooted species with their photosynthetic organs developed as floating leaves (Nymphaeids) or the free floating Lemnids, are less affected by water pollution than are the species with their leaves in rosettes on the bottom (Isoetids) or on long shoots in the water (Elodeids). The latter two groups are the first to disappear when light penetration through the water column becomes reduced due, for example, to eutrophication.

Conclusion

The number of threatened plant species in Europe, especially in Central Europe, is high and will continue to increase. The traditional strategy of conservation by the establishment of nature reserves can stop the extinction of species only by keeping them as rarities; this approach cannot avoid the loss of species diversity. For Central and Northern Europe the main task is the conservation of species diversity. A considerable number of species have almost lost their original Central European range of distribution, now with higher occurrences being restricted to Eastern Europe. However, economic growth and intensification of agricultural production in Eastern Europe might result in further extinctions, if adequate action for safeguarding existing valuable plant resources is not taken in time.

The unique diversity of habitats and the large number of endemic plants in some eastern and southern regions, such as the Carpathian, Transcaucasian and Balkan mountains, the Maritime Alps, Cyprus, the Greek mountains, Crete, Sierra Nevada and the Canary Islands, make their decline or loss a matter of global importance.

The establishment and promotion of a network of large nature reserves such as Natura 2000 as part of the EC's Fauna-Flora-Habitat Directive (92/42/EEC), is an important contribution to the successful, long-term protection of rare and endangered plant species. Further steps in support of species diversity include:

- reduction of environmental pollution and contamination in general;
- establishment of an international system of large nature reserves (as planned in the EC Habitat Directive) primarily in still comparatively unaffected regions such as the Alps, Central and Eastern Poland, northeastern France or the Slovak Republic; and
- avoiding further loss of the most threatened plant habitats: open, muddy ground and damp meadows of large watercourses, dry grasslands, coastal dunes, oligotrophic bogs and lakes, inland salt springs.

Introduced species

Introduction of new species has been a widespread phenomenon in Europe for thousands of years. This process – sometimes also referred to as 'biological invasion' – results from events such as climatic changes, and from a large variety of human activities: migration, trade, colonisation, agriculture, introduction of domestic and wild species. The release of genetically modified organisms (GMOs) is discussed in Chapter 17. Many of these activities have influenced and continue to influence the distribution of species in Europe and the world.

In some cases the introduction of new species has been devastating. Thus a disease introduced to Europe by North American crayfish (*Procambarus clarkii*) has contributed to the near disappearance of Europe's native crayfishes, *Astacus astacus* and *Austrapotamobius pallipes*.

The introduction of species has not been limited to individual continents but occurs everywhere and on a global scale. This is well documented for species such as the European rabbit (*Oryctolagus cuniculus*) in Australia and Chile, the rat (*Rattus rattus*) and the house mouse (*Mus musculus*), the blackberry (*Rubus*) in Chile, and the maritime pine (*Pinus pinaster*) in South Africa (Drake et al, 1989). On the other hand, whole plant communities have also been altered as a result of introductions. For example, the native species of the grazing lands of California, Chile, Argentina, South Africa and Australia have been partially replaced by Old World species, particularly Mediterranean basin herbs.

The percentage of alien species in Europe and the Mediterranean basin, especially among vascular plants, is much lower (5 per cent) than in most other regions of the world. On the other hand, 80 per cent of all aliens in these regions are of Eurasian origin.

With the Old World 'discovery' of the American and Australian continents, species introduction into European ecosystems began. Among them are very conspicuous and peculiar examples such as *Agave americana*, *Yucca* and particularly *Opuntia* – succulents that have become characteristic in Mediterranean landscapes. In addition, some notable shrubs or trees have been successful invaders, including: *Nicotiana glauca* from South America, *Robinia pseudoacacia* from North America, *Acacia* and *Eucalyptus* from Australia. Most new plant invaders (also cereals) became established in disturbed and open sites and arable fields. Only a very few introduced plants such as Canadian pondweed (*Elodea canadensis*), early goldenrod (*Solidago gigantea*), Indian balsam (*Impatiens glandulifera*), a grape-vine (*Vitis riparia*), or Canadian fleabean (*Erigeron canadensis*) have proved able to colonise natural ecosystems. However, nowhere in Europe is there a scale of natural ecosystem invasions that can match that of South Africa by *Hakea* (from Australia) and by *Pinus pinaster* (from Europe).

Invasions have taken place in most European rivers: fish such as *Gambusia affinis*, *Salmo gairdneri* (rainbow trout), *Ameiurus nebulosus*, all from North America, and several molluscs and parasites. Mammals, like the North American muskrat (*Ondatra zibethica*), raccoon (*Procyon lotor*) and nutria (*Moyocaster coypus*) from South America, and even the water hyacinth (*Eichhornia crassipes*), have also invaded European riverine ecosystems.

The environmental impacts of these species on Europe's ecosystems, especially in correlation with environmental stresses such as forest decline, eutrophication of waterbodies and their repercussions on the extinction rate of native species are not yet known.

Wildlife trade

Many of the products obtained from wildlife are exploited commercially. Worldwide, the commercial trade in wild plants and animals has been valued at $5 billion a year (Lean, 1990). Most of it is entirely legal, controlled by national laws and the Convention on the International Trade in Endangered Species of Wild Flora and Fauna (CITES, see below). However, about a third to a quarter of the trade – worth around $1.5 billion a year – is unlawful commerce in rare and endangered species, usually poached and smuggled across frontiers.

Millions of live animals are shipped around the world

Confiscation of ivory and rhino horns at Schipol airport, The Netherlands
Source: R Webster, WWF

every year to supply the pet trade. Furs, leather and even ivory are also traded in vast quantities. Lean (1990) estimates that typically a single year sees 50 000 live primates, tusk ivory from 70 000 African elephants, 4 million live birds, 10 million reptile skins, 15 million pelts, about 350 million tropical fish and about 1 million orchids bought and sold around the world. Meanwhile the African elephant has been included in CITES Appendix I (1989) and since 1992 all commercial trade in ivory is banned under CITES; enforcement and control, however, are still insufficient.

Europe is a major marketplace for reptile skins, monkeys and small cats, and trade in small exotic birds is still taking place in many European cities.

Conclusions

Although relatively few species of Europe's fauna and flora have actually become extinct during this century, the continent's biodiversity is affected by decreasing species numbers and the loss of habitats in many regions: approximately 30 per cent of the existing 1200 vertebrates and 21 per cent of the 12 500 higher plants are classified as 'threatened' according to IUCN categories. Threats are directly linked to the loss of habitats due to destruction, modification and fragmentation of ecosystems, as well as from overuse of pesticides and herbicides, intensive farming methods, hunting and general human disturbances. The deterioration of air and water quality adds to the detrimental influence. Insufficient knowledge on most of the 200 000 invertebrates and on lower plants makes it very difficult to give an accurate assessment of their current situation. Some data show, however, that also lower species of fauna and flora suffer from the environmental problems mentioned above.

The analysis of existing data and expert opinion on the major taxonomic groups of fauna and flora can be summarised as follows:

1 Mammals are a very threatened (42 per cent) species group. Large and predatory species have, in particular, disappeared due to landuse changes, habitat modifications and hunting. Long-term recovery of populations such as otters (*Lutra lutra*), monk seals (*Monachus monachus*), bears (*Ursus arctos*) or wolves (*Canis lupus*) can only be achieved by improving and re-establishing habitats in large coherent areas. Worldwide, threats to whales continue to be of major concern.

2 More than 50 per cent of all threatened bird species are associated with inland and coastal wetlands, as well as with natural grasslands. Drainage, floodplain management and intensification of agriculture have caused a substantial loss of habitats. Remaining important sites require adequate land-management and protection. Water pollution is a major threat, often originating from external sources.

3 Until now, amphibians, reptiles, fish and invertebrates have rarely been priority objectives of nature conservation. Inadequate knowledge, their often highly specialised habitat requirements – sometimes involving several different biotope-types – and their relative inconspicuousness have placed them outside most nature conservation schemes. Fish represent the most endangered group (52 per cent) and suffer from a lack of species-related conservation/management activities, while water pollution (eutrophication, acidification and contaminants) can have substantial impacts on populations. For all of them, however, especially invertebrates, there is a need for systematic inventories and management plans, both within and outside protected areas.

4 Although 21 per cent of Europe's higher plants are threatened, Europe is very fortunate to possess a large number of human and institutional resources devoted to the conservation of plants. For many countries, comprehensive plant inventories and detailed information on their distribution exist. Among plants, aquatic, calcareous and acidic species have especially been declining for many years and in many regions. Conservation strategies need to include the whole range of land management activities, giving special attention to relatives of crop plants (progenitors).

The unique diversity of habitats and the large number of endemic plants in some eastern and southern regions, such as the Carpathian, Transcaucasian, and Balkan mountains, the Maritime Alps, Cyprus, the Greek mountains, Crete, the Pyrenees, Sierra Nevada and the Canary Islands, make their decline or loss a matter of global importance, deserving special attention at national and international level.

5 While it is true that many species are closely linked to particular habitats, other organisms live at boundaries of different associations. These 'ecotones', for example, between woods and lakes, or hedges and grasslands ('edge-effect'), and the various patterns in which they appear form important functional components of the natural environment. Biodiversity is related to spatial variability of the terrain, and particular combinations of biotopes. Adequate management of natural and semi-natural ecosystems involves the integration of species that require particular ecotones and a combination of sometimes very specific habitat structures.

Available data on species are often patchy and incomplete, affecting the quality of international status reports and the compilation of 'red lists'. The identification, monitoring and active management of endangered, introduced and migrating species is a prime objective for national and international nature conservation. The role and function of indicator species for assessing environmental conditions needs to be further explored. All these activities require reliable information on the distribution, condition and vulnerability of plants and animals, especially for the species listed in the EC Habitats and Bird Directives, CITES, the Bonn Convention and other international agreements discussed below.

Box 9MM Protected areas: six management categories

Category I: Strict Nature Reserve/Wilderness Area: managed mainly for science or wilderness protection

Category II: National Park: managed mainly for ecosystem protection and recreation

Category III: Natural Monument/Natural Landmark: managed mainly for conservation of a specific natural feature

Category IV: Habitat and Species Management Area: mainly for conservation through management intervention

Category V: Protected Landscape/Seascape: managed mainly for landscape/seascape protection and recreation (see also Chapter 8)

Category VI: Managed Resource Protected Area: managed mainly for the sustainable use of natural resources

Note: the first five categories have been in use since 1978; the sixth was added in 1992

Source: Commission on National Parks and Protected Areas (CNPPA)

NATURE CONSERVATION

Although action to conserve nature has historic roots in Europe, for many centuries the motivation was the preservation of game for hunting. It is only in the past 100 years or so that countries have begun to develop policies for the protection of nature (national parks were first set up in Sweden in 1909 and in Spain in 1918). And it is only in the past few decades that international action for the protection of nature has developed within Europe.

A number of trends can be identified in the development of nature conservation policies in recent years in Europe, the most important of which are trends away from:

- the protection of species, towards the protection of habitats;
- the protection of species and habitats, towards the protection of the natural processes upon which these depend;
- nature protection as an isolated exercise, towards the integration of nature conservation into the planning and management of the terrestrial and marine environment as a whole, and into each economic sector;
- isolated local or national initiatives, towards coordinated programmes of international cooperation, in which standards and criteria are agreed internationally; and
- the conservation of nature for its scientific and aesthetic qualities towards a recognition of the importance of ecosystems, species and the variety within species for the process of sustainable development (expressed in the phrase 'the conservation of biological diversity').

The present position in Europe as regards the conservation of nature and wildlife is very complex. Each country has its own set of policies for conservation; these can vary greatly and are often developing fast. There is, however, increasing harmonisation of some aspects of national policy through international agreements.

In order to provide an overview of this situation, this section:

- gives a summary, with national action for conservation;
- describes existing international, especially European initiatives for nature conservation; and
- considers new initiatives currently emerging.

This section thus attempts to integrate the assessment of ecosystems and fauna and flora, presented above, into the context of nature conservation as a human response. The state of 'Nature and Wildlife' is closely linked to and influenced by policies and programmes that exist at national and international levels. There is in fact only a fine division between the assessment of ecosystems, species and nature conservation policies discussed in this chapter. Nature conservation is discussed here separately since the various legal and strategic concepts encompass different sets of habitat types or species lists and thus a full description of them in the assessments were not possible.

Since nature conservation is an issue that goes beyond the borders of protected areas, this section can also be seen as a bridge to those parts of the report that deal with human activities such as agriculture, forestry, fisheries, tourism, transport and others.

National action for nature conservation

National action for the protection of nature and wildlife varies greatly from country to country. In some, this subject excites a high level of public and political interest and numerous laws have been passed. In others, there has been far less action. Some countries have adopted a centralised approach; others, politically decentralised, have to delegate responsibility for nature and landscape protection to provincial or regional governments ('Länder' or 'cantons', for example). In many countries, local governments right down to the commune level have an important part to play. In some countries, non-governmental bodies are significant owners of specially protected areas.

This variety cannot be discussed fully here, but instead some broad generalisations are made, illustrated with some

	World	Europe	European Union	Germany	Lower Saxony 1	Osnabrück 2
Total number of plant species	250 000	12 500	6500	2728	2135	993
Threatened plant species (IUCN, local governments)	18 694	2200	1200	664	767	404
Berne Convention plant species	–	523	499	(25)	(7)	(2)
FFH-directive plant species (Annex IV)	–	–	173	(?)	(?)	(?)
CORINE handbook plant species	-	–	725	(40?)	(?)	(?)
Rural wild-flower programme	–	–	–	–	262	(?)

1 Federal state of Lower Saxony
2 Community of Osnabrück in Lower Saxony
( ) estimate
(?) not known

specific examples of action at the national level. Three kinds of action taken to conserve wildlife and nature can be identified:

1. action for the protection of species (such as laws that control hunting);
2. action to protect special areas (eg, as national parks, nature reserves or protected landscapes);
3. action designed to integrate nature conservation into national and regional planning, and into policies for different economic sectors.

The protection of species

The development of the 'Red Data Books' concept by Sir Peter Scott during the 1960s was the starting point for many initiatives to categorise species at risk according to the severity of the threats facing them and the estimated imminence of their extinction. While originally compiled exclusively on a global level, the concept was soon adopted at the national and supranational level in several countries. Today, almost all European countries (and even federal states or 'Länder' and 'cantons') have produced their individual lists of nationally threatened species groups, often differentiating the following categories, based on a former approach of IUCN:

- Extinct (Ex)
- Endangered (En)
- Vulnerable (V)
- Rare (R)
- Indeterminate (I)
- Insufficiently known (K).

(Although these IUCN categories were altered in 1993, this report as well as most 'Red Data' lists are still based on the original concept.)

In recent years attention has spread from terrestrial vertebrates to invertebrates and plants. The situation of plant protection based on lists for endangered species is illustrated in Table 9.12 for international and national levels and two regional levels in Germany. While the total number of identified plant species decreases as the size of the reference area becomes smaller, the number of 'threatened plants' follows the same principle only down to the national level. As the table shows, the federal state of Lower Saxony has monitored more higher plants that are threatened (767 or 36 per cent) within its boundaries than the federal agency in the list for the whole of Germany (664 or 24 per cent). This difference illustrates the fact that species which are not threatened for a whole country might still be declining dramatically in some smaller regions. The same applies for different listings between the federal state and a community (in this case between the federal state of Lower Saxony and its community of Osnabrück). This indicates the basic need for identifying biogeographical regions on the reference areas for endangered species. Such a system would allow specific regional aspects to be taken into account.

Table 9.12
International, national and regional threatened plant lists and programmes
Source: EEA-TF

When establishing threatened species lists on the regional level, it is important to keep in mind that species can have their natural distribution boundaries outside a region – making its rareness a natural feature rather than a matter of concern. However, regional governments should be encouraged to make inventories and monitor threatened species and to respond with appropriate protection programmes. In the case of Lower Saxony, a special programme for the protection of rare and endangered farmland wild flowers has been designed. This programme lists 262 species that have previously been typical of farming landscapes, and for which special protection measures are proposed and financial compensations for participating farmers and communities are offered. Only 91 plants (35 per cent) of these are also included in the regional threatened plants list of Lower Saxony – thereby extending possibilities for various farmlands to participate in the programme throughout the region.

Since international lists do not respond to the national level of species decline (eg, the Berne Convention covers only 18 plant species in Central Europe), national and regional governments face more responsibility to ensure that biodiversity is maintained and enhanced on *each* level. The results of these activities, listings and programmes could be the basis of an hierarchical model for an extended list that includes Europe's nationally and regionally threatened species.

Red Data Books and species protection programmes can be successful only if they are flanked by initiatives that regulate activities which affect flora and fauna, such as the use of pesticides, methods and seasons of hunting, and physical constructions etc.

The protection of sites

Every country in Europe has passed legislation for the protection of certain sites of importance to the conservation of nature as 'protected areas'. IUCN (1993) defines a protected area as:

an area of land and/or sea especially dedicated to the protection of biological diversity, and of natural and associated cultural resources, and managed through legal or other effective means.

The titles given to such protected areas at the national level in Europe vary greatly, however. As well as national parks and nature reserves, there is national legislation to establish biogenetic reserves, forest reserves, marine reserves, nature parks and regional nature parks, protected landscapes, wildlife reserves and many more besides. To complicate matters further, the same name means different things in different countries, even sometimes in the same country (eg, 'nature reserve' in Sweden); and the same kind of protected area will often be classified differently at the national level.

To help clarify this confused situation, IUCN has developed a system of protected area categories. These have recently been revised, following the IVth World Congress on Protected Areas in Caracas in 1992 (see Box 9MM, page 240). The logic behind such a system is as follows:

- the basis of categorisation should be the primary management objective;
- assignment to a category is not a commentary on management effectiveness;
- the categories system is to be used at the international level; national names will continue to vary;
- all categories are equally important, but their interest is different; and
- they imply a gradation of human intervention.

The application of the categories system in Europe enables comparisons to be made between countries, and between Europe and the rest of the world. The World Conservation Monitoring Centre (WCMC) has collected data under these categories from across the world (though with a minimum cut-off of 1000 hectares, many smaller protected areas are excluded). Table 9.13 compares the distribution of the first five types of protected areas between the world and Europe (excluding the former USSR).

Category V, Protected Landscape areas (see also Chapter 8) in the European region, accounts for nearly two thirds of the 42.45 million hectares which are recorded by WCMC as being under protection. In many such protected landscapes, the protection of landscapes and of nature and wildlife are achieved through integrated policies: by protecting the landscape, the setting for wildlife is secured, and by protecting wildlife, essential elements in the landscape are safeguarded.

Occurrence of the different categories varies greatly between countries. Thus some Scandinavian countries have much of their protected land in categories I and II (over 62 per cent of Finland), whilst several of the most intensively settled countries (eg, Denmark and the UK) have little or no land in these categories.

The total size of protected areas in Europe has expanded rapidly in recent years. Two thirds of all protected land in Europe has been so designated in the 20 years after the 1972 Stockholm Conference on the Human Environment; and 10 million hectares have been added (an area larger than Hungary) since 1982.

However, the designation of protected areas is not, unfortunately, a guarantee of their success. A recent review undertaken by the Commission on National Parks and Protected Areas of IUCN and the Federation of Nature and National Parks of Europe for the Caracas Congress concluded as follows:

- Although Europe has a complex and extensive system of more than 40 000 protected areas (most of which are too small to figure in the above statistics), there are dramatic differences between countries. In general more investment in management and extension is needed in the Mediterranean countries, and in some countries of Eastern and Central Europe.
- Whilst there has been progress in the creation of marine protected areas in the Mediterranean and the Baltic, marine protected areas in general lag behind those on land.
- Despite the rapid expansion in the number of protected areas, large tracts of natural and semi-natural vegetation (particularly in Eastern and Central Europe) deserve protection, but do not currently receive it.
- Most protected areas in Europe are under heavy pressure and in some their qualities are being destroyed, due to both external threats and a lack of management resources. Threats range from agricultural intensification to infrastructure development, and from airborne pollution to war and civil unrest.
- Pressure from visitors is causing a mounting strain on many protected areas; much needs to be done to make sure that park-based tourism is sustainable.
- The changed political outlook in Europe gives rise to novel challenges, for example, the pressure to privatise state-owned land in protected areas in former Communist countries.

Table 9.13
Protected areas – five management categories as a percentage of the total protected areas (overlap between the categories is possible)
Source: WCMC, 1992

Category*	World	Europe (exc. former USSR)
I	6.6	6.9
II	49.9	11.8
III	2.6	0.9
IV	30.3	14.9
V	17.9	65.5

* see Box 9MM for definitions

Despite these conditions, the prospects for protected areas in Europe are generally good, and certainly better than in many other parts of the world. There are few countries in Europe where population pressures and rural poverty create demands on wild areas as they do in many developing countries. The existence of surplus agricultural capacity in Europe (and to some extent surplus military land) means that it is possible to think about restoring the nature conservation values of land where this has been lost or damaged in the past (eg, restoring the wildlife value of farmland that has been subject to intensive farming).

International action for nature conservation

International action has mainly addressed two issues: the protection of species, and the protection of sites. There are three ways in which this has been done: through international conventions (treaties); through international programmes; and within the EU European legislation.

In terms of geographical areas covered there are initiatives of five distinct types:

- global, which affect Europe;
- Europe-wide;
- legislation affecting the EU countries only;
- others, affecting parts of Europe only; and
- transfrontier parks.

Table 9.14 presents an overview for six types of internationally designated areas as well as 'national parks' – a national category of protection which differs in terms of status and regulation from country to country. Table 9.15 lists European strategic initiatives for conservation of biological, landscape and marine diversity by type, responsible body and objectives.

Global initiatives affecting Europe

The Ramsar Convention
The Convention on Wetlands of International Importance was signed in 1971 in the Iranian city of Ramsar and came into force in 1975. The Convention's Bureau is backed by the International Waterfowl and Wetlands Research Bureau (IWRB) and the IUCN. It is a worldwide agreement, bringing together 70 nations which have expressed a common interest in resolving the environmental problems and opportunities of

Table 9.14
International designated areas and national parks in Europe
Source: Multiple sources compiled by EEA-TF

Nation	Barcelona Convention Total area (ha)	Number	Biogenetic Reserves Total area (ha)	Number	Biosphere Reserves Total area (ha)	Number	Bird Directive Total area (ha)	Number	European Diploma Total area (ha)	Number	Ramsar Convention Total area (ha)	Number	National parks Total area (ha)	Number
Albania	1800	2											9600	6
Austria			192 496	54	27 600	4			46 300	2	102 541	7	197 183	3
Belarus					253 301	2							87 607	1
Belgium			7242	17			431 306	36	3988	1	7945	6		
Bulgaria					32 116	17					2097	4	294 349	11
Croatia					150 000	1					80 455	4	69 420	7
Cyprus	2668	2	3550	2										
Czech Republic					299 921	5					18 109	4	111 120	3
Denmark							960 092	111			734 468	27		
Estonia					1 560 000	1					48 634	1	176 922	4
Finland					350 000	1					101 343	11	699 590	24
France	265 532	68	43 395	35	646 583	8	660 940	91	274 075	6	422 585	8	352 598	6
Germany			677	1	1 158 149	12	841 588	214	60 934	8	672 852	31	699 902	10
Greece	24 151	8	22 272	16	8 840	2	191 637	26	4850	1	104 400	11	68 742	10
Hungary					128 884	5					114 862	13	159 139	5
Iceland											57 500	2	180 100	3
Ireland			6906	14	8808	2	5991	21			13 035	21	36 798	5
Italy	26 040	10	33 568	37	3807	3	310 378	74	77 478	5	56 950	46	852 248	14
Latvia											43 200	3	92 048	1
Lithuania													132 900	5
Luxembourg			75	1			709	5	38 880	1	1837	4		
FYROM													85 588	2
Malta	5	2	11	2							11	1		
The Netherlands			270 379	18	260 000	1	306 935	13	4400	1	314 928	15	73 690	11
Norway			1 568 523	11	1 555 000	1					16 256	14	2 328 110	20
Poland					359 277	7					7141	5	164 758	17
Portugal			45 642	8	395	1	332 344	36	269	1	30 563	2	71 422	1
Romania					614 268	3					647 000	1	408 300	12
Russia					1 594 726	10			84 996	2	858 000	2	1 968 000	18
Slovak Republic					197 166	4					1855	4	199 724	5
Slovenia											650	1	84 805	1
Spain	21 045	6			859 827	13	2 398 910	140	79 899	3	102 205	17	132 478	10
Sweden			1 152 491	43	96 500	1			460 540	4	382 750	30	632 418	16
Switzerland			11 446	9	16 870	1			16 870	1	7049	8	16 887	1
Turkey	50 685	3	12 624	2						1			108 133	9
Ukraine					159 585	3					211 051	3	169 803	3
UK			7990	18	44 258	13	260 475	76	216 753	5	215 277	57	1 393 067	11
former Yugoslavia	68 282	10			200 000	1					18 094	2	237 673	9
Europe Total	460 208	111	3 379 287	288	10 585 881	122	7 444 596	843	1 370 232	41	5 395 643	365	12 268 122	264
World Total	657 000	132	3 379 287	288	169 633 411	311	7 444 596	843	1 370 232	41	36 000 000	632	358 868 900	1483

Note: The figures are totalled directly from the data supplied by the organisations responsible for the relevant conventions based on national data.

Table 9.15
Global, European and regional initiatives for nature conservation
Source: IUCN

Title	Secretariat/focal point	Aim/comment
Pan-European		
Berne Convention (1979)	Council of Europe	To provide international obligations for the conservation of European flora and fauna and their natural habitats
The European Network of Biogenetic Reserves	Council of Europe	Chosen as representative examples of Europe's natural heritage
European Diploma	Council of Europe	Awards for good examples of protected areas
Environment Programme for Europe	UN/ECE, EC, UNEP, IUCN et al	To prepare the intergovernmental environment programme for Europe
EuroMAB	EuroMAB, c/o UNESCO	To provide a European network for international scientific cooperation, especially on biosphere reserves
European Nature Conservation Year (ENCY) 1995	Council of Europe	To develop a campaign in 1995 for conservation
European Community		
Fifth Environmental Action Programme (1992)	European Commission DG XI	The European Community's programme of policy and action on environment and sustainable development
Bird Directive (79/409/EEC)	European Commission DG XI	To protect wild birds and their habitats, including through Special Protection Areas
CORINE	European Commission DG XI/ European Environment Agency	To develop a database for nature conservation in the EC, now being extended in part to the rest of Europe
Habitats Directive (92/43/EEC)	European Commission DG XI	To conserve fauna, flora and natural habitats of Community importance
Natura 2000	European Commission DG XI	The network of protected areas set up under the Bird and Habitats Directives
Central and Eastern Europe		
Ecological Bricks for our Common House of Europe	Initiative for Ecological Bricks, c/o WWF-Austria	To promote the establishment of 24 transboundary protected areas
Green Lungs of Europe, 1993	Institute of Sustainable Development, Warsaw	Based on Poland's experience, to create sustainable development zones (Belarus, Estonia, Latvia, Lithuania, Poland, Russia, Ukraine)
Arctic		
Arctic Initiative	Working Group on the Conservation of Arctic Flora and Fauna	To prepare a common Arctic Environmental Protection Strategy
Nordic Arctic Conference	c/o Nordic Council of Ministers	Environmental protection
Baltic Sea		
Helsinki Convention, 1974, 1992 The Baltic Sea Joint Comprehensive Environmental Action Programme	HELCOM, Helsinki	To improve the quality of the Baltic environment, including through marine and coastal protected areas
Black Sea		
Black Sea Action Plan (BSAP)	UNEP, UNDP, GEF and others	Environmental management programme for the Black Sea
Bucharest Convention on the Protection of the Black Sea against Pollution, 1992	Interim Secretariat, Istanbul	
North Sea		
Convention for the Protection of the NE Atlantic	Oslo–Paris Commission, London	To prevent pollution of the northeast Atlantic
Ministerial Conference on the North Sea	Secretariat in Ministry of Environment, Denmark	
Mediterranean		
Mediterranean Action Plan	UNEP-Europe, Geneva (Regional centres in Athens, Valbonne, Tunis, Malta and Split)	To improve the quality of the Mediterranean environment under the Barcelona Convention, 1976, including through a protocol on Specially Protected Areas
MedSPA	European Commission DG XI	To protect the Mediterranean environment, including protection of biotopes

Table 9.15
continued

Title	Secretariat/focal point	Aim/comment
MED-PAN	EIB/World Bank	To strengthen links between managers of protected areas
MedWet	Various	To conserve Mediterranean wetlands
Mediterranean Technical Assistance Programme (METAP)	World Bank/EIB	2nd Phase of European Programme for the Mediterranean (EPM), to reverse present environmental degradation
Nicosia Charter (1990)	European Community	To provide closer cooperation on sustainable development in the Euro-Mediterranean region, including on nature conservation
Others		
Alpine Convention 1991	c/o CIPRA, Vaduz	Conservation of the Alps
Black Triangle Regional Programme, 1992	Programme Coordination Unit, Ústi nad Labem, Czech Republic	Environmental protection for the very polluted Czech–German–Polish border zone
Danube River Basin Programme, 1991	European Commission DG XI	Environmental protection for the Danube
International Commission for the Protection of the Rhine against Pollution (ICPRP)		
Joint Declaration on the Protection of the Wadden Sea	Common Secretariat for the Cooperation on the Protection of the Wadden Sea, Wilhelmshaven	Trilateral cooperation between The Netherlands, Denmark and Germany
Conservation of Migratory Wild Animal Species, including West Palaearctic Flyway Agreement	Convention Bureau, Bonn	

wetland areas in their countries. The Convention works through countries pooling their technical and financial resources. Sharing a common belief in the value of wetlands as valuable and irreplaceable economic, cultural, scientific and recreational resources, these countries commit themselves to proper management of wetlands for the present and future benefit of their people.

When signing the Ramsar Convention a country gives four main undertakings:

1 to designate at least one wetland of international importance;
2 to promote the sustainable use of wetlands;
3 to consult one another on the implementation of the convention – especially where water systems affect other countries;
4 to establish nature reserves on wetlands in their country.

So far 362 wetlands have been designated in Europe (from a total of 632 sites), covering over 5 million hectares. Designated wetlands vary enormously in size and nature. For example in Europe the Danube Delta with its vast, varied but fragile waterfowl population on the Romanian Black Sea coast covers 650 000 hectares, whilst Llyn Idwal in the Snowdonia National Park, North Wales, is only 14 hectares, though still possessing vital and fascinating natural communities for protection and study. Worldwide, a Ramsar site can be a coral reef, a river valley, or a tropical mangrove swamp.

Article 1 of the Ramsar Convention describes wetlands as areas of marsh, peatland or water, whether natural or artificial, permanent or temporary, with water that is static or flowing, fresh, brackish or salt, including areas of marine water the depth of which at low tide does not exceed six metres.

The Ramsar Convention has three main ways of achieving action at local level. It monitors the situation of wetland sites in the contracting countries and helps actions to prevent deterioration or destruction of wetlands, particularly those on the Ramsar list. It coordinates the 'Rational Use of Wetland Project', which gives governments detailed guidance, based on current project work, on how to make rational use of natural resources which local communities derive from wetlands, taking into account the natural, institutional, legal and technical factors involved. Lastly, in 1990, the Convention set up the Ramsar Fund, a wetland conservation fund for developing countries.

Thus, with its established network of wetlands and wide local experience, Ramsar forms one important element in the environmental protection of specific sites in Europe and throughout the world.

The World Heritage Convention
Adopted in Paris in 1972, the Convention concerning the Protection of the World Cultural and Natural Heritage came into force in December 1975. The Secretariat is provided by UNESCO. The convention aims to designate sites of 'outstanding universal value' as World Heritage Sites, and promotes international cooperation in safeguarding these important areas. Sites may be considered for their cultural or natural qualities (Chapter 8 describes how cultural sites may include certain cultural landscapes – here, only natural sites are considered). Criteria for inclusion in the World Heritage List are agreed by the World Heritage Committee, and the advice of IUCN is sought on sites nominated by states party to the convention. There are at present 20 natural and 120 cultural sites in Europe on the World Heritage List. All European countries except Albania, Belgium, Iceland and Luxembourg have joined the World Heritage Convention. Conservation bodies have repeatedly called on those states which have not done so to join, and asked all members to nominate suitable sites.

CITES

The Convention on International Trade in Endangered Species of Wild Fauna and Flora (CITES) was signed in 1973. Its Secretariat is provided by the United Nations Environment Programme and is based in Geneva, Switzerland. The convention attempts to prevent commercial trade in species which are in danger of extinction or might become so if their trade was allowed to continue unchecked. The Convention covers not only live plants and animals, but also their derivatives. Most European States have joined CITES, but the following are not yet members: Albania, Greece, Ireland, Romania, San Marino, Vatican City and the states of former Yugoslavia. CITES has been ratified by over 110 states worldwide. The Convention has financed studies of particularly endangered species in order to establish the degree and nature of the threat presented by such trade. The trade in species covered and listed in the three appendices mentioned above is subject to progressive levels of restriction.

The Bonn Convention

The Convention on Migratory Species was signed in Bonn and came into force in 1979. The Convention requires that 'range States', that is those which share certain endangered migratory species, cooperate in joint measures for conservation, research, management, etc. Few countries in Europe are members of the Convention, which also suffers from a lack of finance.

Biosphere Reserves

Developed under UNESCO's Man and the Biosphere programme (MAB), Biosphere Reserves differ from other types of site in that they are not designated purely to protect the site *per se*. They are chosen rather as representative international examples of biogeographical habitats and ecosystems where practical management and research can be practised, and where exchanges of information with managers, policy makers, researchers and local inhabitants and interests can take place. They combine the preservation of ecological and genetic diversity with consideration of the local economy, research, environmental monitoring, education and training.

The intention is that each site should consist of a central core which is managed primarily for its ecological value, surrounded by a larger buffer zone where experimental research and management can take place side by side with normal human activities and traditional landuses. One aim is to demonstrate both to local people and to the international community that sustainable modes of living in close touch with the natural environment, and in harmony with it, can help to maintain a viable local economy in the long term.

The Biosphere Reserve concept first emerged from the UNESCO's Man and Biosphere (MAB) programme in the early 1970s. It was the 'Biosphere Conference' convened by UNESCO in 1968 that recommended the setting up of an intergovernmental and interdisciplinary programme of research. Since 1971 when the MAB programme officially began, a total of 122 reserves have been established in Europe. The demand from governments was for an emphasis on finding practical, low-cost, sustainable solutions to problems by putting long-discussed ideas on ecological approaches to land management to the test at field level.

Convention on Biological Diversity

A recent development is the signing at the 1992 Earth Summit of the Convention on Biological Diversity. The impact of the convention is yet to be felt (the convention entered into force on 29 December 1993). Its significance is discussed below.

European initiatives

The Berne Convention

The Convention on the Conservation of European Wildlife and Natural Habitats (Berne Convention) came into force in 1982. The Convention was promoted and developed by the Council of Europe and now includes most Western European countries, the European Community, some African states and, more recently, certain Eastern European countries in negotiation with, or now with membership of, the Council of Europe.

The principal aim is to protect flora and fauna and their habitats, and to promote international cooperation amongst the contracting parties in resolving transfrontier issues, with particular emphasis on the protection of endangered and vulnerable species and their habitats, particularly migratory species. The convention includes three annexes, listing these species. There is a governmental 'standing committee' which monitors the enforcement of the Convention. This committee can organise international seminars and special studies to promote understanding of the issues, and adopts recommendations to the governments of contracting parties designed to further the protection of specific habitats and species.

Biogenetic Reserves

These areas are representative samples of various natural habitat types and have been promoted by the Council of Europe since 1975. Under Resolution (76)17 of the Committee of Ministers of the Council of Europe, a Biogenetic Reserve is one which can be shown to contribute to the maintenance of the biological balance and the conservation of representative samples of Europe's natural heritage, and act as a living laboratory for research into the operation and evolution of natural ecosystems. The intention is that the scientific knowledge gained from this can be put to use in campaigns to generate public interest in environmental issues, and to provide information, education and instruction material.

By 1993 over 3 million hectares within 288 reserves had been designated. The areas concerned are natural or semi-natural. A Biogenetic Reserve could consist of anything from a tiny island with rare plants or animals to a large area of heathland or moor. Priority is given to reserves according to the degree to which the species found there are unique, endangered, rare or typical for a particular habitat. The criteria for selection are the value of the site for nature conservation and the effectiveness of the protection being given to them.

Biogenetic Reserves enable member states of the Council of Europe to cooperate and coordinate elements of their nature protection policies with reference to the biogenetic reserves in their partner countries and the exchanges of information which designation stimulates. The Council of Europe sees its biogenetic reserve programme as giving substantial support to national undertakings to conserve the habitats of wild flora and fauna as agreed under Article 4 of the Berne Convention.

The European Diploma

In 1965 the Committee of Ministers of the Council of Europe inaugurated the European Diploma. Up to 1993, a total of 41 awards had been given for natural areas, sites or features of international value and European interest which can be shown to be adequately protected.

There are three categories:

1 for the protection of the European heritage of flora and fauna, its environment and ecosystem (Category 'A', currently 991 342 hectares);
2 for the protection of the character of the landscape and natural features of special aesthetic or cultural value (Category 'B', currently 124 580 hectares);
3 for the combination of social and recreational activities with the maintenance of biological and aesthetic characteristics which have produced a first-class environment (Category 'C', currently 296 920 hectares).

In practice a government will suggest to the Council of Europe that they wish to enter a site or defined area for the Diploma. Not all applications succeed. In any case a local examination is undertaken and systematically repeated every five years, when the Diploma is due to be renewed, or – when the area is threatened – with the possibility to withdraw it. This control over Diploma sites instils a pride and sense of achievement in those involved, often key to the sites' protection and maintenance. Those awarded can include the sponsoring government, the site managers and their employers, wardens, local councillors or the NGO which manages the site. Experience of the resulting benefits is twofold:

1 the sponsoring government is more likely to refuse permissions for developments ecologically detrimental to the site, since they have acknowledged its international importance and know that the Diploma can be withdrawn if the habitat deteriorates;
2 government funds may be more readily available for internationally recognised areas.

At local level the prestige which results from the Diploma status can help all those involved in the area participate in its conservation, encouraging in turn greater cooperation from all concerned – the farming and local residential communities, local firms, potential tourist or mine concerns, or military neighbours. Knowing that a review of this status takes place every five years can also help to encourage all concerned to keep on top of the environmental status at the site into the future.

The Bird Directive

The EC Directive on the Conservation of Wild Birds (79/409/EEC) is legally binding on all Member States. All states are duty bound to 'preserve, maintain or re-establish a sufficient diversity and area of habitats' for all wild birds. They must ensure that a sufficient habitat is maintained to enable the survival of all migratory species and a list of 175 specifically mentioned species. Sites set aside for this must be classified as Special Protection Areas (SPAs). The European Court of Justice has ruled that EU Member States have no power to modify or reduce the extent of such areas once they have been defined. So far more than 800 sites have actually been designated, covering a total of more than 7 million hectares. However, the ICBP and IWRB claim that twice as many sites qualify as SPAs, covering an area of 14 million hectares, and that their protection is mandatory if the Directive is to succeed in its aim. These organisations are of the opinion that the sites designated so far fail to form an effective network for internal movement or vital international migration, and that Member States are being slow to designate sites and comply fully with the Directive.

The Habitats Directive

The EC Habitats Directive (92/43/EEC) is a recent EC legislative instrument, following on from the Bird Directive. It establishes a common framework for 'the conservation of natural and semi-natural habitats and of wild fauna and flora' and allows the development of a network of special areas of conservation called Natura 2000, described below. The Habitats Directive aims to combine the concern to protect endangered species with a wider concern to protect and enhance habitats of interest in their own right. It is further planned that measures should be taken to allow species to move between sites (identification of habitat corridors) and increase their range or regain lost territory.

For the purpose of the Directive, 'Habitat' means a natural or semi-natural area with particular, unique biogeographical characteristics. Annex I of the Directive lists 200 habitat types including rare and small habitats, such as Alpine lakes and shifting dunes, as well as habitats known for their high biological diversity, such as calcareous grasslands. Other examples are estuaries (important for migratory species), traditionally managed agricultural lands (eg, 'dehesas', see Chapter 8) and certain continental broadleaved forests.

The Directive establishes a list (Annex II) of 134 vertebrates (no birds), 59 invertebrates and 278 species of plants whose habitats must be protected for their survival. Annex IV provides strict protection for another 173 species of plants, 71 species of invertebrates and 160 species of vertebrates. A ban is placed on the deliberate capture and killing of the animal species concerned and on the disturbance of these species, particularly during critical stages of their life-cycle (breeding, rearing, hibernation and migration); deliberate picking, collecting, uprooting and destruction of the plants are prohibited.

Concerning the implementation of the Habitats Directive it is foreseen that within six years of the adoption of the Directive (by 1998), the European Commission should have

---

Box 9NN
The 'Green Lungs' of Europe

The 'Green Lungs' is an area of about 760 000 km^2 covering seven states, and has a population density of about 45 people per km^2. The main functions of this area are agriculture, forestry, industry, marine economy and tourism. As a follow-up to UNCED, the seven states involved agreed to an outline concept for the 'Green Lungs of Europe' in February 1993. Much of the land already has protected status (14 national parks, 18 state nature reserves – all in the former USSR – and 5 biosphere reserves). The area includes the Mazuria Lakes (north of Warsaw). Management of the area will rely on extensive consultation, and the participation of municipalities and international agencies. Legislative and landuse planning tools will be used to promote the appropriate development of different zones to integrate landscape and nature conservation goals into sustainable forms of: productive agriculture, development of tourist infrastructure, health resorts and ecological zones (Polish Ministry of Environmental Protection, Natural Resources and Forestry et al, 1993).

Map 9.30
The 'Green Lungs' of Europe area
Source: Polish Ministry of Environmental Protection, Natural Resources and Forestry et al, 1993

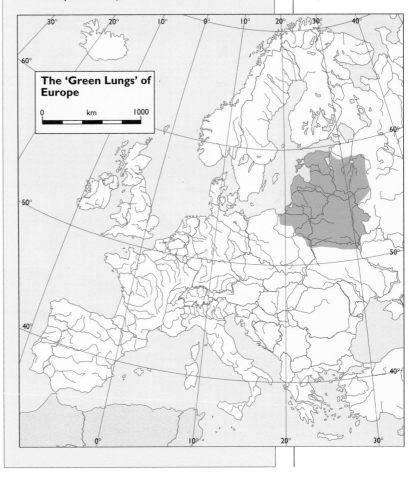

drawn up a list of sites of Community importance on the basis of national information. Within a further six years (ie by 2004) Member States should have designated the sites, to be called Special Areas of Conservation (SACs), which are intended to form the network Natura 2000. Adequate protection measures are to be taken by Member States, who will also be obliged to strengthen the ecological coherence of the sites.

Regional initiatives

There are a number of other initiatives (including conventions) which affect certain parts of Europe only. Some of these, (eg, those for the Arctic and the Mediterranean), extend to cover adjoining areas outside Europe. The most important are:

● The Alpine Convention 1991: to promote international cooperation for the protection of the Alps.
● The Arctic Initiative: to promote cooperation among Arctic countries. There is also a treaty for the protection of polar bears among these countries.
● The Baltic Sea: under the Helsinki Convention (1992) to improve the quality of the Baltic environment, including marine and coastal protected areas.
● Mediterranean Action Plan: developed under the Barcelona Convention (UNEP, 1975, see eg UNEP/IUCN, 1990): covering (1) a series of legally-binding treaties; (2) the creation of a pollution monitoring and research network, and (3) a socio-economic

programme to reconcile development with a healthy environment. The Convention for the Protection of the Mediterranean Sea against Pollution (Barcelona Convention) was adopted in 1976; there are four related protocols on: dumping from ships and aircraft; cooperation in pollution emergencies; pollution from land-based sources; and endangered Mediterranean species of animals and plants, and the marine and coastal areas essential to their survival. Related activities include the Mediterranean Pollution Monitoring and Research Programme (MEDPOL); and the Blue Plan which is designed to help the Mediterranean governments protect their coastal environments.
● Northeast Atlantic: under the Paris Convention, 1992, the Governments concerned cooperate for the protection of the northeast Atlantic marine environment.

In addition, the Danube Convention – developed to encourage international action to protect the river's environment – was signed in June 1994.

Transfrontier parks

Transfrontier parks are needed where important biogeographical areas are artificially divided by national boundaries. Parks of this kind not only provide joint protection and management of resources and environments, but can also help promote the cause of peace, and preserve and enhance cultural values and diversity among human communities living in or near such areas. Many of Europe's nations share important areas with their neighbours, notably mountain ranges (eg, Alps, Pyrenees, Balkans, Carpathians, and in Scandinavia), regional seas (eg, Mediterranean, Baltic,

Map 9.31
Internationally designated areas
Source: EEA-TF

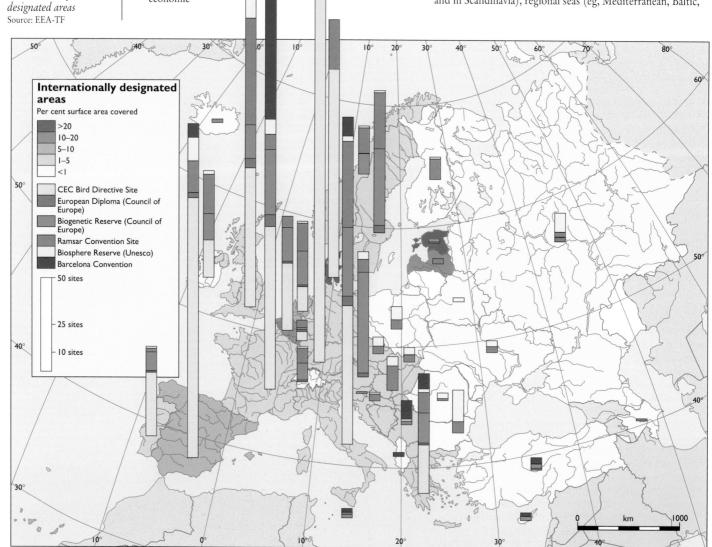

Box 9 OO 'Ecological Bricks' as transfrontier parks

Following the political changes that affected Central and Eastern Europe in 1989, the plan *Ecological Bricks for Our Common House in Europe* (Langer, 1990) identified and documented the state of 24 transboundary areas of major importance for nature conservation. The intention was to improve the information on these areas thus preventing potential damages or threats and helping to preserve identified values.

1 Finnish-Russian Woodland Area
2 Biebrza Marshes
3 Białowieza Virgin Forest
4 Schorfheide/Chorin
5 Spreewald
6 Sächsische Schweiz
7 Krkonoše Area
8 Tatra Area
9 Pieniny Area
10 Bieszczady Region
11 Slovakian Karst
12 Danube: Thaya and March Floodplain
13 Thaya Valley
14 Trebonsko Pond Region
15 Bavarian Forest/Sumava Forest
16 Lake of Neusiedel
17 Mur Floodplain
18 International Karst Park
19 Lower Tributaries of the Drau and Kopački-Rit
20 Sava Floodplain
21 Danube Delta
22 Lake Scutari
23 Prespa Area
24 Rhodope Mountain, Neslos Delta

The initiative has raised funds and is developing management plans for these areas.

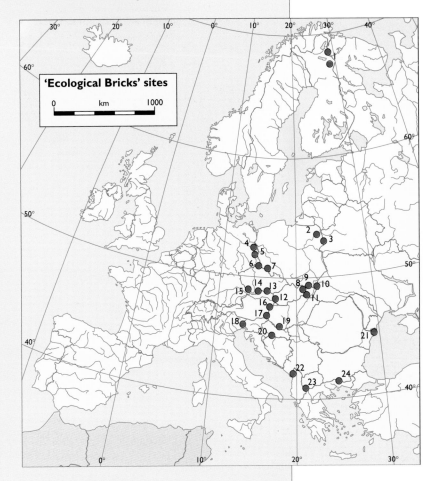

'Ecological Bricks' sites

0 km 1000

Map 9.32
Ecological Bricks sites
Source: Initiative for Ecological Bricks

North, Black and Wadden seas), and river basins (eg, Danube, Rhine and Elbe). With its many small countries, Europe has particular need of this approach. Examples of international cooperation of this kind include the Green Lungs of Europe (Box 9NN), the Ecological Bricks initiative (Box 9OO) and the cooperation between the Argentera Nature Park in Italy and its trans-Alpine neighbour the Mercantour National Park in France (they were twinned in 1987, and cooperate in the administration, promotion and scientific work of the parks).

New initiatives

The current preoccupation of those concerned with the environment in Europe is that in spite of all the measures taken over the last century or so, ecological diversity is still declining on a regional scale. Areas have been protected, anti-pollution laws, controls and campaigns have been instituted; positive management programmes to protect threatened species have been initiated. Furthermore, today every European country has some type of environmental agency or administrational unit to carry out this work. Considerable energy goes into international cooperation on the treaties and other positive measures, examples of which have been mentioned above. Nevertheless the decline of species and the loss of habitats continues in many regions.

For this reason a number of organisations feel that it is time to look at the way national measures for nature conservation and environmental protection are viewed in relation to the international scene as a whole. Hence, several new initiatives, both governmental and non-governmental, are currently being developed at the international level which are

designed to exploit these opportunities in an imaginative way and which could become powerful mechanisms for carrying forward the intentions of European governments as, for example, set forth in their decisions at the European conference of environmental ministers held in Lucerne in April 1993. These include:

- the follow-up to the ratification of the Convention on Biological Diversity;
- the development of the Natura 2000 network, through the implementation of the Birds and Habitats Directives;
- the development of a European Ecological Network;
- the publication of IUCN's Action Plan for Protected Areas in Europe.

Convention on biological diversity

The signing of the Convention on Biological Diversity was one of the principal achievements of the Earth Summit. Virtually all European States signed the convention at Rio de Janeiro and many have now started the process of ratification. Under Article 6, each signatory has to prepare national plans for the conservation and sustainable use of biodiversity, and to integrate these into other sectors. To help achieve one of its main objectives – the conservation of biodiversity – Article 8 (a) requires States to 'establish a system of protected areas or areas where special measures need to be taken to conserve biological diversity'. In conserving biodiversity, the Convention gives priority to *in situ* measures, including protected areas, and sees the main purpose of *ex situ* efforts (botanic gardens, zoos etc) as to complement *in situ* conservation.

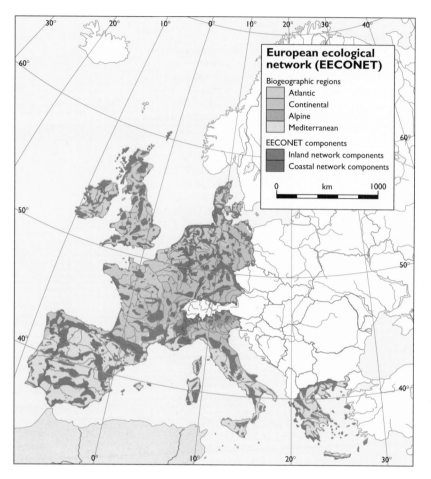

Map 9.33
European ecological
network
(EECONET)
Source: Agriculture
University, Wageningen

The real value of the Convention lies not so much in its measures, which are in general rather imprecise, but in the achievement of putting this matter onto the international agenda and the follow-up action required at the national level. Properly used, the Convention could become the foundation for efforts in each country of Europe for the conservation of endangered species and ecosystems (see also Chapter 29).

Natura 2000

Developed for the EU, the term and concept of Natura 2000 is a component of the directive on the Conservation of Natural Habitats and Wild Fauna and Flora (92/43/EEC) which states under Article 3(1):

> *A coherent European ecological network of special areas of conservation shall be set up under the title Natura 2000. This network, composed of sites hosting the natural habitat types listed in Annex I and habitats of species listed in Annex II, shall enable the natural habitat types and the species' habitats concerned to be maintained or, where appropriate, restored as a favourable conservation status in their natural range. The Natura 2000 network shall include the special protection areas classified by the Member States pursuant to Directive 79/409/EEC.*

Based on the Bird and Habitats Directives, the network sets the minimum standard for biodiversity conservation in the 12 Member States, encompassing a wide range of issues and containing a number of concrete obligations. This concept is strengthened by the Treaty of Rome (Article 139r), amended by the single European Act and later by the Maastricht Treaty, according to which all Community policies and instruments must comply with the Community's environmental statutes, including the Habitats and Bird Directives.

European Ecological Network (EECONET)

In 1991 the Institute for European Policy, in collaboration with several European research institutes, proposed the creation of a European Ecological Network under the name EECONET (IEEP, 1991). The objective of EECONET is to conserve the ecosystems, habitats and species which are of European importance and to enhance the ecological coherence of the continent. The initiative includes a proposal to develop an international cooperative framework through which necessary action can be followed.

The physical ecological network comprises four main elements:

1 *core areas* representing the habitats of European importance;
2 *corridors* to facilitate, where appropriate, dispersal and migration;
3 *buffer zones* to protect the network from adverse external influences; and
4 *restoration areas* to rehabilitate damaged habitats.

The design of the network was the subject of a seperate study which considered future landuse in the EU (Bischoff and Jongman, 1993). An international conference on EECONET was held in Maastricht in November 1993. The EECONET Declaration, adopted by the Conference (Bennett, 1994), urged the Council of Europe to coordinate the preparation of a European Biological and Landscape Diversity Strategy, for submission at the October 1995 Sofia conference of European Environment Ministers.

IUCN Action Plan for Protected Areas

The IUCN Action Plan for Protected Areas in Europe was developed and launched during 1994. It has the aim of developing an adequate, effective, well-managed network of protected areas in Europe by proposing a set of recommendations for actions by governments and others, and by identifying some 20–30 priority international projects of a catalytic nature. The plan gives much attention to placing Europe's protected areas in their wider context, to addressing the needs of priority sub-regions and countries, to improving management, and to creating the climate of public and political support needed to ensure the success of protected areas. By taking into account existing initiatives such as EC Directives, Natura 2000 and EECONET, the plan has been drawn up through a continent-wide process of consultation among those who deal with protected areas on a day-to-day basis (IUCN, 1993).

Integration of nature conservation into planning and other sectors

The protection of species and of sites has traditionally formed the cornerstone of national efforts to conserve nature, but it has become increasingly apparent that this approach alone will fail. Just as species cannot be safeguarded unless their habitats are also protected equally, a reliance on a network of protected areas, however good that network may be, will not suffice. If protected areas are treated as 'islands' of conservation, then:

● the protected areas themselves will be subject to all manner of externally driven pressures (eg, all the different forms of pollution by air, surface and groundwater, landuse demands, unregulated tourism and so forth) which do not take the needs of conservation into account;
● no protection will be afforded other areas outside the formally designated protected sites, and yet there is no part of the environment where conservation should not be considered, (in the farmed countryside and even in towns) as a means of bringing nature closer to where people live;
● public understanding and support will depend on whether people will be able see how protected areas can contribute to sustainable development, for example by creating income

through tourism, by providing places for economically important fish and wildlife to breed and feed, and as a 'treasure house' of biological knowledge and in other ways.

For these reasons, it is now accepted that the conservation of nature requires that it be integrated into national and regional planning at every level. In particular, the following are the key areas for action:

- the integration of biodiversity conservation with national planning for sustainable development countries have the opportunity to implement this as they prepare, for example, national plans for sustainability in response to Agenda 21, and for biodiversity conservation as required by the Convention on Biological Diversity;
- the integration of nature conservation within national systems for the planning and management of the use of land. Landuse planning is widely practised in Europe and provides the appropriate framework for the control of potentially damaging developments in sensitive environments; in particular, the use of environmental impact assessment will greatly assist nature conservation;
- the integration of conservation and the planning and management of the marine environment, especially in the 'Exclusive Economic Zone' and through Coastal Zone Management (see also Chapter 35),
- pollution control, since the conservation of species and ecosystems depends on the reduction of pollution of air, water and soils;
- the involvement of local communities in the management of protected areas, through devices such as incentive payments to farmers for conservation, the encouragement of volunteer groups, and the establishment of local consultative fora.

Equally important, conservation needs to be integrated into sectoral policies, such as agriculture, forestry, tourism, transport and industry.

CONCLUSIONS

Europe has a rich and varied nature and wildlife, despite the extent of modification to which the natural environment has been subjected. The trends, however, give cause for concern: valued ecosystems are everywhere facing stresses and decline, and the list of endangered species continues to lengthen, and for many species there is a lack of information. Much action has been taken, at the international and national levels, to address these problems, especially in the past 20 years.

Global initiatives like the protection of 362 wetlands under the Ramsar Convention, the designation of about 120 Biosphere Reserves and 20 World Heritage Sites are supplemented by European approaches like the Network of Biogenetic Reserves (288 sites) or the 41 European Diploma awards, both established by the Council of Europe. For the European Union, more than 800 Special Protection Areas have been classified in application of the Directive on the Conservation of Wild Birds, while the Habitats Directive was adopted in 1992, encompassing more than 200 habitat types and 500 plant and animal species. Although several of these initiatives could do with strengthening, a bigger problem is that habitat protection is progressing too slowly in many countries.

However, on the national level, a large variety of site designations (130 different types of protection areas have been identified alone within the European Union) has led to an estimated 40 000 protected sites in Europe. Less than 900 of these sites are included in the United Nations list of protected areas requiring a minimum size of 1000 hectares. Much has been done during the last ten years: the area of national parks increased 50 per cent with over 50 000 ha created in each of Austria, Bulgaria, former Czechoslovakia, Finland, Germany, Norway, Romania and former Yugoslavia. A large number of actions have been initiated, financed or backed by non-governmental organisations, making them one of the driving forces in national and international nature conservation.

One important component for the preparation as well as the implementation of future strategies, regulations and actions is the availability of reliable information on the state of ecosystems, species and protection for the whole of Europe. Although far from being complete, data gathered for this report clearly indicate that all existing initiatives have not yet succeeded in stabilising Europe's ecosystems and stopping the decline of fauna and flora. Activities in favour of nature are still clearly outnumbered in time and space by those human activities which have the reverse effect. There are a host of threats to nature and wildlife from, among other things, pollution, the destruction of habitats, poorly planned agricultural and forestry operations, short-sighted tourism, urbanisation of formerly rural areas, damaging infrastructure schemes, and unregulated or poorly regulated taking of species.

Conservation of biodiversity has taken on a new urgency since UNCED in 1992. Due to the high level of public awareness and concern in Europe, there is realisation of the need to complement an essentially defensive approach, based on protection, with new policies designed to create or restore nature and wildlife in Europe. Concepts like Natura 2000 are promoting more actions to be taken outside the strictly 'protected areas' and to introduce the principles of sustainability for all forms of natural resource management. In this context, priorities need to be identified to ensure that the most valuable areas are preserved or regenerated first. Parallel to the systematic reduction of pervasive environmental stresses, linkages between important, but fragmented, conservation areas need to be established. In response, new network strategies and new approaches towards sustainable landuse outside protected areas are being discussed and are also the theme of the 1995 European Nature Conservation Year (ENCY), organised by the Council of Europe. Biological diversity is considered further in this report as a prominent environmental problem of concern to Europe, in Chapter 29.

ECOSYSTEM SITE LIST

Below are listed the names of the representative sites of Maps 9.3, 9.6, 9.7, 9.8, 9.9 and 9.10, grouped by ecosystem and country. Details on the identification and location of the sites may be found in the *Statistical Compendium.*

FORESTS (MAP 9.3)

ALBANIA
1 Lura
2 Tomori

AUSTRIA
3 Dachstein-Waldgebiet/ Kemetgebirge
4 Dürrenstein
5 Leitha-Gebirge
6 Podyji-Thayatal
7 Thaya-March-Donauauen
8 Wienerwald

BELGIUM
9 Hautes Fagnes/Eifel
10 Haute-Sûre/Ardenne méridionale (G D de Luxembourg)

BULGARIA
11 Alibotouch a Líbotouch
12 Bayuvi Doupki-Djindjiritza
13 Boatin
14 Djendema
15 Ouzounboudjak
16 Parangalitza
17 Rhodopi
18 Steneto
19 Tzaritchina
20 Vitosha
21 Zlatni Pyassatsi

CROATIA
22 Crni Lug
23 Šume donjeg toka Drave & Kopački rit
24 Poplavna Savska nizina

CYPRUS
26 Troodos

CZECH REPUBLIC
114 Beskydy
25 Karkonoski NP/Krkonoše

ESTONIA
27 Lahemaa NP

FINLAND
28 Elimyssalo
29 Finnish-Russian Woodland Area (partly Russia)
30 Lemmenjoki/Øvre Anarjokka (Norway)
31 Oulanka
32 Inaria/Øvre Pasvik (partly Norway)
33 Pallas-Ounastunturi

FRANCE
34 Cévennes
35 Parc Regional Corse
37 Forêt de Cérisy
38 Haut Sommet du Pilat
39 Le Vercors
40 Pyrénées Occidentales
42 Volcans d'Auvergne

GERMANY
43 Ammergebirge
44 Bayerischer Wald/Sumava (Czech Republic)
45 Bienwald
46 Ederbergland
47 Elm
48 Jasmund
49 Müritz/Serrahn
50 Nördlicher Steigerwald
51 Rheingau-Gebirge
52 Unterliezheimer Forst

GREECE
53 Ainos
54 Grammos
55 Lefka Ori
56 Olympos
57 Taygetos
58 Vikos-Aoos-Valia Calda

HUNGARY
59 Aggtelek
60 Bükki
61 Pilis
62 Zemplén

ITALY
63 Abruzzi
65 Calabria
66 Gennargentu
67 Le Madonie
68 Monti Nebrodi
69 Pollino
70 Foresta Umbra
71 Valle dell Orfonto
72 Val Grande

LUXEMBOURG
157 Grünewald/Müllerthal

NORWAY
73 Ånderdalen
74 Femundsmarka-Rogen-Langfjället (partly Sweden)
75 Gressåmoen
76 Roltdalen

POLAND
77 Babia Góra NP
78 Puszcza Bialowieska (partly Belarus)
79 Biebrza river valley
80 Bieszczady Region
81 Drawieński NP
82 Roztoczański NP
83 Woliński NP

PORTUGAL
84 Peneda-Gerês
85 Serra de Arrabida
86 Serra de Marao
87 Serra de Monique
88 Serra de Nogueira
89 Serra de Montezinho
90 Vallée de Loriga
91 Vallées du Angueira et de Macas

ROMANIA
92 Ceahlau-Politele cu crini
93 Cerna-Mehedinti
94 Danube delta
95 Muntii-Almajului
96 Retezat
97 Seaca-Optasani
98 Southern Semenic Range with Izvoarele Nerei
99 Tismana

RUSSIAN FEDERATION
100 Adzhametskiy
101 Berezinskiy
102 Karpatskiy
103 Kavkazskiy
104 Kivach
105 Lagodekhskiy
106 Oksky
107 Panayarvy/Karelia
108 Pechoro-Ilychsky
109 Russky Sever/Vologda
110 Smolensk Pooserie
111 Tsentral'nolesnoy
112 Voronezhskiy
113 Volga Delta

SLOVAK REPUBLIC
115 Pieninsý národný park
116 Polana
117 Slovenský raj

SLOVENIA
153 Krakovski gozd

SPAIN
118 Alt Aneu-San Mauricio-Bohí-Berit
119 Cazorla y Segura
120 Covadonga/Picos de Europa/Saja
121 Garajonay
122 Los Alcornocales
123 Monfragüe
124 Monte de El Pardo
125 Montseny
126 Rio Eume
127 Sierra de Grazalema
128 Sierra de San Pedro
129 Valle del Río Arlanza
130 Valles de Hecho-Ansó-Roncal-Salazar

SWEDEN
131 Båtforsområdet
132 Blaikfjället
133 Gotska Sandön
134 Höga Kusten
135 Muddus
136 Pärlaelven
137 Pessinki
138 Sjaunja
139 Söderåsen
140 Vindelfjällen

SWITZERLAND
36 Creux du Van et Gorge de l'Areuse
41 Vallée de Joux
141 Albiskette – Reppischtal
142 Pfynwald-Illgraben
143 Maggia

TURKEY
145 Kasnak Meresij (Isparta/Eğridir)
146 Koprulu Canyon NP
147 Sakagolu Longuzu (Istanbul/Demirkoy)
148 Yedigoeller

UK
149 Abernethy
150 New Forest

YUGOSLAVIA (former)
151 Djerdap
152 Durmitor and Tara Gorge NP
154 Pelister
155 Sutjeska
156 Tara

SCRUB AND GRASSLANDS (MAP 9.6)

ALBANIA
1 Ilogara (Vlorë)
2 Lura (Peshkopi)
3 Thethi (Shkodër)

AUSTRIA
4 Hohe Tauern and Rauristal
5 Karwendel
6 Mussen
7 Parndorfer Platte
8 Seewinkel

BELARUS
9 Zakataly and Belokany
10 Belovezhskaya pushcha, Brest, and Grodno
11 Berezina Floodplains (Berezinsky Reserve)
12 Pripyat, Zhitkovichi, Lelchitsy and Petrikov, Gomel

BELGIUM
13 Côte Bajocienne
14 Lesse-et-Lomme
15 Spa - Malchamps

BULGARIA
16 Atanasovosko Ezero
17 Rusenski Lom
18 Steneto
19 Studen Kladenetz

CROATIA
20 Pokupski bazen
21 Poplavno Podrucje Rijeke Sava
22 Nacionalni Park Paklenica
23 Nacionalni Park Krka
24 Planina Dinara

CZECH REPUBLIC
25 Šumava Mountains
26 Třeboňsko
27 Krkonoše Mountains
28 Pálava

DENMARK
29 Hanstholm Reservatet
30 Værnengene
31 Tøndermarsken
32 Kallesmærsk Hede
33 Nørre Ådal

ESTONIA
34 Matsalu

FINLAND
35 Islands of Kainuunkylä
36 Liminganlahti- Lumijoenselkä

FRANCE
37 Plaines de Pons-Rouffiac
38 Plaines de Niort Sud-est/ouest
39 Plaines de Saint-Jean-de-Sauves
40 Plaines de Saint-Jouin-de-Marne
41 Plaines de Mirebeau et de Nauville-du-Poitou
43 Les Monts d'Arrée
44 Prairie humide de Carentan
45 Landes des trois Pignons
46 Camp militaire de Suippes
47 Prairie humide de la Sarthe
48 Val de Saône
49 Prairies du Jura
50 Causse de Quercy
51 Les Causses
52 Camp militaire de Captieux
53 Pelouses alpines des Pyrénées occidentales
54 Pelouses alpines des Pyrénées centrales
55 Pelouses alpines des Alpes
56 Plateau de Valensole
57 Camp militaire de Canjuers
58 Garrigues de Pic Saint Loup
59 La Crau
60 Pelouses alpines de Corse

GERMANY
62 Lüneburger Heide
63 Wolferskopf
64 Badberg/Haselschacher Buck
65 Berchtesgaden NP
66 Diepholzer Moorniederung
67 Elbtalaue
68 Unteres Odertal (Odra)
69 Spreewald
70 Greifswalder Bodden
71 Steckby - Lödderitzer Forst und Zerbster Ackerland

GREECE
72 Oros Olympos
73 Kato Olympos, Tembi, Ossa, Delta Piniou
74 Ori Athamanon
75 Delta Tou Evrou
76 Vouna Tou Evrou
77 Lakes in Thrace - Vistonis, Porto Lagos, Lafri, Lafroud, Xirolimni, Karatza, Alyki, Ptelea, Elos, Mana
78 Potamos/Kilada Filiouri
79 Antikhassia Ori/Meteora
80 Limnothalassa Mesolongiou/Aetolikou/ Ekvoles Acheloou Kai Evinou (Mesolongi wetlands)
81 Ori Thryptis

HUNGARY
82 Aggtelek
83 Hortobágy
84 Kiskunsági NP
85 Dévaványa
86 Pusztakócsi mocsarak
87 Pusztaszer

ICELAND
88 Veidivötn
89 Eylendid

IRELAND
90 River Shannon/River Little Brosna/River Suck Callows
91 The Burren

92 Wexford Slobs
93 Wicklow Hills

ITALY
94 Bacino del fiume Cecina ed area di Volterra
95 Carso Triestino
96 Coastal areas south of Naples to Reggio Calabria
97 Altipiani delle Alpi Italiane
98 Livorno to Grosseto Hills
99 Steppe di Manfredonia
100 Monte Pecodaro and vicinity
101 Le Madonie
102 Le Murge di Bari
103 Logudoro – Ozieri
104 Sila

THE NETHERLANDS
105 Holterberg and Haarlerberg
106 Oirschotse Heide

POLAND
107 Bagna Biebrzańskie
108 Dolina Środkowej Warty
109 Wielki Łeg Obrzański
110 Dolina dolnego Bugu

PORTUGAL
111 Arada and Freita Mountains
112 Castro Verde Plains
113 Estrela Mountain
114 Gerês Mountain
115 Montesinho and Nogueira Mountains
116 Mourão - Barrancos
117 Southwestern coast of Portugal

ROMANIA
118 Domogled Mountain
119 Iezerele Cindrelului
120 Lacul Istria, L. Nuntasi and southern L. Sinoie
121 Lunca Muresului: Ceala, Pecica and Bezdin
122 Padurea Letea
123 Perisor-Zatoane-Sacalin nature reserve

RUSSIAN FEDERATION
124 Caucasus Biosphere Reserve
125 Teberda, Karachaevo
126 Tsentralno-Chernozemny (Central Black Earth) Biosphere Reserve
127 Zavidovo Reserve

SLOVAK REPUBLIC
128 Podunají (Danube flood plain)
129 Slovenský kras

SLOVENIA
130 Slovenske goric (partly)
131 Ljubljansko Barje
132 Triglavski narodni park

SPAIN
133 Altos de Barahona
134 Bardenas Reales
135 Belchite-Mediana
136 Cáceres
137 Cordillera Cantábrica
138 Ibiza
139 Fresser-Setcases
140 La Alcarria-Aranjuez
141 La Serena
142 Llenera-Peraleda de Zaucejo-Guareña
143 Los Monegros
144 Menorca
145 Monfragüe-Sierras de las Villuercas
146 Montes de Toledo-Cíjara
147 Pirineos
148 Sierra de Tramontana
149 Sierra de Gredos
150 Sierra de Pela - Embalse de Orellana - Zorita
151 Sierra Morena Occidental y Central
152 Sierra Nevada
153 Tierra de Campos-Los Oteros
154 Villafáfila - Villalpando - Villardefrades

SWEDEN
155 Alvaret
156 Helgeån

SWITZERLAND
61 Creux du Van et George de la Reuse
157 Haut Jura Vaudois

TURKEY
158 Acigöl and Calti Gölü
159 Bafa Gölü
160 Tuz Gölü

UKRAINE
161 Askaniya-Nova Reserve
162 Borisoglibivka
163 Luganskk Reserve
164 Ukrainian Steppe Reserve

UK
165 Berwyn Mountains
166 Breckland Heaths
167 Caenlochan and Clova glens
168 Cairngorms
169 Hartland Moor, Arne Heaths and Studland
170 Na Meadhoinean Iar (including Loch Druidibeg)
171 New Forest Heaths
172 Salisbury Plain
173 Scolt Head, Holkham, Blakeney Point and Cley-Sal
174 Somerset Levels
175 Speyside Pinewoods and Moorlands
176 Sutherland and Wester Ross Moorlands
177 Swale and South Sheppey
178 The Lizard
179 Tiree and Coll
180 Unst, Shetland
181 Upper Teesdale and Moor House
182 Y Wyddfa, Glyder and Carneddau Mountains

YUGOSLAVIA (former)
183 Deliblatska Pescara
184 Klisura Crna Reka
185 Obedska Bara
186 Planina Korab i Klisura Radike

RIVERS (MAP 9.7)

AUSTRIA
1 Inn: Innstausee

BELGIUM
2 Lesse: Furfooz

BULGARIA
3 Beli Lom
4 Cherni Lom
5 Cherni Osam
6 Demyanitza
7 Yantra
8 Ropotamo
9 Vladaiska

CROATIA
10 Danube: Kopački rit
11 Rijeka Drava/Dravske šume

CZECH REPUBLIC
12 Morava: Soutok
13 Vltava: floodplain

DENMARK
14 Esrum å
15 Højen Bæk
16 Idom å
17 Karup å
18 Linding å
19 Simested å
20 Skjern å
21 Suså

FINLAND
22 Ivalonjoki
23 Kiskonjoki - Perniönjoki
24 Lapväärtinjoki - Isojoki
25 Lestijoki
26 Näätämönjoki
27 Oulankajoki
28 Kiiminkijoki
29 Simonjoki
30 Tenojoki
31 Tornionjoki - Muonionjoki

FRANCE
32 Allier: Mars-sur-Allier (Bourgogne)
33 Allier: Val d'Allier
34 Bois du Parc
35 Courant d'Huchet
36 Dordogne: Gorges de la Dordogne
37 Etang de la Horre
38 Garonne: Palayre
39 Loire: Angers - Gennes
40 Loire: Angers - Ingrandes
41 Loire: Briare - Varennes-Vauzelles
42 Loire: Digoin - Marcigny
43 Loire: Feures - Nantes
44 Loire: La Charité-sur-Loire
45 Marne: Vitry-Lefrançois - Epernay
46 Meuse: St-Mihiel - Lacroix-sur-Meuse
47 Rhône: la Camargue
48 Saône: Bresse and flood zones
49 Seine: Monterau - Bray-sur-Seine
50 Vallée de la Lanterne et de la Saône
51 Vallée de Fango, Corse
52 Vallée de l'Eure de Chartres à Cherisi

GERMANY
53 Dollart
54 Danube: Neu-Ulm - Lauingen
55 Danube: Donau-Auen/Donau Ried
56 Danube: Regensburg-Vilshofen
57 Elbe: Kirnitzsch/NP Sächsische Schweiz
58 Elbe: Schnackenburg - Lauenburg
59 Ems-Außendeichsflächen: Leer - Emden
60 Inn: Haiming - Neuhaus
61 Isar: Gottfieding - Plattling
62 Main: Eltmann
63 Main: Fahr - Dettebach
64 Main: Schweinfurt
65 Oder: Malczyce - Kaczawa
66 Oder: Schwedt
67 Rhine: Avenhein/Kehl-Greffen
68 Rhine: Bibermühle
69 Rhine: Gimsheim - Eicher Altrhein und Fischsee
70 Rhine: Greffen-Murgmündung
71 Rhine: Haltingen - Neuenburg
72 Rhine: Hordtes Rheinaue
73 Rhine: Insultheimer Hof
74 Rhine: Kuhkoff - Knoblauchsaue
75 Rhine: Lampertheimer Altrhein
76 Rhine: Neuenburg - Breisach
77 Rhine: Nonnenweiher - Goldscheuer
78 Rhine: Wagbauchniederung
79 Rhine: Weisweil - Nonnenweiher
80 Rhine: Wesel - Emmerlich
81 Rhine: Eltville - Bingen
82 Weser: Weserstaustufe Schlüsselburg

GREECE
83 Acheloos: delta (Mesolonghi Lagoons)

HUNGARY
84 Danube: Hansági
85 Danube: Gemenci
86 Danube: Pilisi
87 Tisza: Tiszacsegei hullámtér
88 Tisza: Közép - Tisza
89 Tisza: Labodár és Sasér
90 Tisza: Mártély
91 Tisza: Tiszaalpári
92 Tisza: Tiszadobi ártér
93 Tisza: Tiszaluc - Kesznyéteni puszta
94 Tisza: Tiszetelek - Tiszaberceli ártér

IRELAND
95 Shannon: Portumna - Athlone
96 River Little Brosna Callows
97 River Suck Callows
98 River Camlin

ITALY
99 Grotticelle
100 Lago di Nazano
101 Monte Velino
102 Montefalcone
103 Pania de Corfino
104 Po: delta
105 Po: Garzaia della Cascina e Bosco Basso
106 Po: Garzaia di Somaglia
107 Po: Garzaia Cascina/San Alessandro
108 Po: Garzaia della Tenuta Baraccone
109 Po: Garzaia di Valenza
110 Po: Isola di Penedo
111 Po: Lago di Sartirana Lomellina
112 Po: Punte Alberete e Valle della Canna
113 Po: Viadana - Ostiglia
114 Tevere: Gola del Tevere

LATVIA
115 Abava Valley, Kroja - Abava mouth
116 Daugava Primeval Hollow
117 Gauja Primeval Hollow/Vireši
118 Salaca valley
119 Škervelis valley
120 Venta valley, Padure - Abava

LITHUANIA
121 Bartuva
122 Bilsinycia-Ratnycia
123 Dubysa
124 Grabuosta
125 Jura
126 Karkle
127 Merkys
128 Minija
129 Virinta
130 Zeimena

MOLDOVA
131 Prut

THE NETHERLANDS
132 Rijn: Waal
133 Rijn: Waal/Maas

NORWAY
134 Atnaelva
135 Forra
136 Glomma
137 Lågen
138 Reisaelva
139 Sjoa

POLAND
140 Bug: lower river valley
141 Czarna Hańcza
142 Drwęca
143 Dunajec/Pieniński NP
144 Tanew
145 Warta: Krajkowo reserve
146 Warta: Dolina Środkowej Warty
147 Warta: Santok region
148 Wieprz/Roztoczański NP
149 Wisła: Dolina Środkowej Wisły
150 Wisła: Rzeka Wisła

PORTUGAL
151 Rio Sabor, Douro, Águeda
152 Val de Rio Guadiana

ROMANIA
153 Danube: delta

RUSSIAN FEDERATION
154 Volga: delta

SLOVAK REPUBLIC
155 Morava: Záhorie

SPAIN
156 Delta del Ebro
157 Marismas del Guadalquivir
158 Alto tajo
159 Arca y Buitrera
160 Arribes del Duero
161 Carrizales y Sotos de Aranjuez
162 Embalse de las Canas
163 La Geria
164 Laguna de Pitillas
165 Las Hoces del Rio Duraton
166 Monfrague
167 Orbarenes - Sierra de Cantabria
168 Penas de Iregua, Leza y Jubera
169 Riberas de Castronuno
170 Sierra de Alcarama y Rio Alhama
171 Tablas de Daimiel

SWEDEN
172 Baftors
173 Hovran area
174 River Ume älv delta

UK
175 Avon (Hampshire)
176 Barle (Somerset)
177 Derwent/Cocker (Cumbria)
178 Eden (Cumbria)
179 Itchen (Hampshire)
180 Teme (Hereford & Worcester)
181 Wensum (Norfolk)
182 Wye (Hereford & Worcester and Wales)

UKRAINE
153 Danube: delta

YUGOSLAVIA (former)
183 Danube: Donje Podunavlje
184 Danube: Kovilijski Rit

LAKES (MAP 9.7)

AUSTRIA
201 Almsee
202 Altaussener See
203 Grabensee
204 Grundlsee
205 Neusiedler See
206 Seewinkel

BELGIUM
207 Bassin de la Haine
   (ponds & marshes)
208 Bokrijk, Limburg
   (lakes & marshes)
209 De Maten
   (ponds & wetlands)

BOSNIA HERZEGOVINA
210 Hutovo Blato
211 Skadarské (partly Albania)

BULGARIA
212 Atanasovsko Ezero
213 Dourankoulak
214 Lake Mandra
215 Lake Srebarna
216 Pirin Mountain lakes
   (glacial pristine)
217 Rila Mountain lakes
   (glacial pristine)
218 Studen Kladenetz

CROATIA
219 Botoniga
220 Jezero Njivice
221 Vransko jezero/Cres
222 Pokupski bazen

CZECH REPUBLIC
223 Nove Mlyny Middle
   Reservoir
224 Trebonsko Area

DENMARK
225 Arresø
226 Esrum Sø
227 Furesø and Farum Sø
228 Grane Langsø
229 Hald Sø
230 Hampen Sø
231 Maribo søerne
232 Mossø
233 Rørbæk Sø
234 Tissø, Lille Åmose and
   Hallenslev Mose

FINLAND
235 Evo Lake District
236 Hossa Lakes
237 Kuijärvi/Sonnanen
238 Lake Inari
239 Lake Orivesi and others
240 Melkuttimet Lake District
241 Nuuksio Lake Plateau
242 Lake Päijänne
243 Pöyrisjärvi
244 Suomunjärvi
245 Vätsäri Wilderness Area

FRANCE
246 La Brenne
247 La Dombes
248 Lac de Grand - Lieu
249 Lac de la Forêt d'Orient
250 Lac de Madine
251 Lac du Der-Chantecoq et
   Etangs Latéraux
252 Lac et Marais du Bourget
253 Sologne des Étangs
254 Vallée de la Scarpe

GERMANY
255 Bayerische Seen
   (Ammersee, Starnberger See)
256 Bodensee
257 Charlottenhofer
   Weihergebiet
258 Dambecker Seen
259 Dümmer
260 Federsee
261 Galenbecker See und
   Putzarer See
262 Großer Plöner See
263 Koblentzer See und
   Latzig See
264 Krakower Obersee
265 Ostufer Müritz, Großer
   Schweriner See
266 Prierowsee
267 Rietzer See
268 Stausee Berga-Kelbra
269 Steinhuder Meer
270 Teichgebiet Peitz
271 Weserstaufstufe
   Schlüsselburg

GREECE
272 Kerkini
273 Mitrikou
274 Prespa
275 Vistonis
276 Vegoritis and Petron
277 Volvi-Langada

HUNGARY
279 Fertö-tó
280 Hortobágy Nagyhalastó
281 Kardoskut - Fehér-tó
282 Kiskunsági Szikes-Tavak
283 Lake Balaton
284 Velence - Dinnyés

ICELAND
285 Eyjavatn
286 Myvatn Laxá
287 Skumsstadavatn
288 Veidivötn
289 Vestmannsvatn

IRELAND
290 Lough Corrib
291 Loughs Ree and Derg

ITALY
292 Lago di Mezzola
293 Lago Trasimeno
294 Laguna di Caorle
295 Lago Maggiore,
   di Como e d' Iseo

LATVIA
296 Babītes
297 Ciriss
298 Drīdzis
299 Engures
300 Klaucēnu
301 Papes
302 Priekulānu
303 Raipala
304 Sventes
305 Zebrus

LITHUANIA
306 Lake Zuvintas

MOLDOVA
307 Beleu (centre of the lake)

THE NETHERLANDS
308 Ijsselmeer
309 Naardermeer
310 Oostvaardersplassen and
   Lepelaarasplasse
311 Veluwemeer, Harderbroek
   and Kievitslan
312 Vennen van Oisterwijk and
   Kampina
313 Zuidlaardermeer

NORWAY
314 Atnsjøen
315 Bessvatn
316 Borrevatn
317 Gjende
318 Femunden
319 Lakes of Jæren
320 Nordre Øyeren

POLAND
321 Lake Dobskie
322 Lake Karaś
323 Lake Łuknajno
324 Lake Óswin
325 Lake Świdwie
326 Przemków fish ponds
327 Drawski Landscape Park
328 Goczałkowice Reservoir
329 Hańcza
330 Iławski Landscape Park
331 Koszalin and Słupsk (coast)
332 Lake Drużno
333 Lake Miedwie
334 Lipa fish ponds
335 Liwia Łuża
336 Milicz fish ponds
337 Nidzkie
338 Sławskie lakes
339 Słowiński NP
340 Szczecinek Region lakes
341 Tuknajno
342 Wełtyń
343 Żydowo - Biały Bór Region

PORTUGAL
344 Murta Dam

ROMANIA
345 Balta Alba, Amara and
   Jirlan lakes
346 Mostistea lakes

RUSSIAN FEDERATION
347 Chalmny-Varre, Lovozero
348 Iokanga/Ponoy watershed
349 Kanin Peninsula
350 Lake Ilmen
351 Lake Ladoga
352 Lake Onega
353 Lake Vyalye
354 Lakes of northern Karelia
355 Lakes Vashutkiny and
   Khargeyskiye ozer
356 Lapland, Monchegorsk
357 Narva Reservoir
358 Primorsko -
   Akhtarsk/Grivinskaya Salt
359 River Chernaya
360 River Ponoy middle reaches
361 River Sura floodplain
362 Rybinsk Reservoir
363 Strelna/Varzuga watershed
364 Varandeyskaya Lapto
   Peninsula

SPAIN
365 Albufera de Valencia
366 Embalse de Borbollón
367 Complejo lagunar de Pedro
   Muñoz
368 Complejo lagunar de Alcázar
   de San Juan
369 Embalse de Alcántara
370 Embalse de Cedillo
371 Embalse de Cíjara
372 Embalse de Gabriel y Galán
373 Embalse de Puerto Peña -
   Valdecaballeros
374 Embalse de Valdecañas
375 Embalses de Rosarito y
   Navalcán
376 Lagunas de Fuente de Piedra
377 Laguna de Medina
378 Ondon in Alacant
379 Pantano de El Hondo
380 Embalse de Sant Llorenç
   Montagai

SWEDEN
381 Lake Ånnsjon
382 Lake Åsnen
383 Lake Hjälmaren
384 Lake Hornborgasjön
385 Lake Mälaren
386 Lake Östen
387 Lake Persöfjärden
388 Lake Tåkern
389 Lake Tärnasjön
390 Lake Tjålmejaure
391 Lake Vänern

SWITZERLAND
392 Lac de Neuchâtel
393 Lac Léman
394 Bodensee (Untersee)

TURKEY
395 Acigöl and Calti Gölü
396 Akşehir and Eber Gölü
397 Apolyont Gölü
398 Beyşehir Gölü
399 Burdur Gölü
400 Cukurova deltas
401 Eğirdir and Hoyran Gölü
402 Ereğli Sazligi
403 Iznik Gölü
404 Karapinar Ovasi
405 Karataş Gölü
406 Manyas Gölü
407 Tuz Gölü
408 Tuzla Gölü

UK
409 Bosherston Lake (Dyfed)
410 Cheshire/Shropshire meres
411 Cumbrian lakes
412 Hatchet Pond (Hampshire)
413 Llangors Lake (Powys)
414 Llyn Tegid (Gwynedd)
415 Loch Leven (Fife)
416 Loch Lomond (Strathclyde)
417 Lochs Druidibeg, a'Machair
   and Stillig
418 Loughs Neagh and Beg
419 Malham Tarn
420 Norfolk Broads
421 Oak Mere
   (Cheshire/Shropshire)
422 Rutland Water
423 Swanholme Lakes
   (Lincolnshire)
424 Upton Broad (Norfolk)

BOGS, FENS, AND MARSHES (MAP 9.8)

AUSTRIA
1 Donau östlich v. Wien
2 Hanság
3 Kaltbrunner Ried
4 Sablatnigmoor

BELARUS
5 Karachevskoye Marsh
6 Obol Marsh, Polotsk and Shumilino

BELGIUM
7 Hautes Fagnes
8 Leopoldsburg - Zwarte Beek
9 Mechelse Heide/Vallei Ziepbeek
10 Les Tailles

CROATIA
191 Ušće Neretva

CZECH REPUBLIC
11 Pančavská and Labska Louká and Úpská Rašelina
12 Třeboňsko, protected landscape
13 Modrava Peatbogs

DENMARK
14 Holmegårds Mose and Porsemose
15 Hostrup Sø, Assenholm Mose and Felsted Veste
16 Jerup Hede, Råbjerg Mose and Tolshave Mose
17 Lille Vildmose
18 Sølsted Mose
19 Store Vildmose
20 The River Valley of Sønderå
21 Tissø, Lille Åmose and Hallenslev Mose
22 Vejlerne, eastern & western part

ESTONIA
23 Muraka Marsh
24 River Emajogi Mouth

FINLAND
26 Koitelaiskaira
28 Lätäseno
29 Luiro
30 Martimoaapa
31 Patvinsuo
32 Riisitunturi
33 Salamajärvi
34 Sammuttijänkä

FRANCE
35 Anse de l'Aiguillon, Marais de Charron
36 Basse Vallée du Doubs; Ile Girard
37 Camargue
38 Complexe des Etangs du Charnier et de Scamand
39 Etangs de la Brenne
40 Etangs de Saint-Hubert - Pourras - Corbet
41 Haute Vallée de l'Aisne de Mouron à Rilly Sûre
42 L'Etang Noir et la Zone d'arrière-Dune du Mar
43 Lac de Grand-Lieu
44 Marais de Balacon
45 Marais de Grande-Brière
46 Marais de l'Erdre, Tourbière de Logne
47 Marais de la Sangsurière et des Gorges
48 Marais de Lavours et bords du Seran
49 Marais de Sacy
50 Marais de Saint-Gond
51 Marais du Blayots et Banc de Saint-Seurin
52 Marais et Etangs d'arrière - Dune du Littoral
53 Massif du Grand Ventron, Complexe de Tourbier
54 Massif du Roc Blanc et Vallée de l'Oriège
55 Pays de Bitche
56 Plaine et Etang de Bischwald
57 Pointe de Grave Entre-deux-Mers
58 Sologne du Loiret (Vannes-sur-Cosson)
59 Vallée de la Mouche et Vallon de Senance
60 Vallée de la Seine

GERMANY
61 Ahlener Moor
62 Bastau-Niederung
63 Diepholzer Moorniederung
64 Elb-Marschen (Teichgebiet Lewit)
65 Ewiges Meer
66 Fehntjer Tief/Flumm Niederung
67 Hochharz
68 Neustädter Moor
69 Ostenholzer Moor und Meißendorfer Teiche
70 Osterseen
71 Peenetalmoor und Anklamer Stadtbruch
72 Presseler Heide- und Moorgebiet
73 Tinner und Staverner Dose
74 Treene zwischen Treia u.Tressee, Bollinstedt

GREECE
75 Amvrakikos Kolpos (Gulf of Arta)
76 Artzan Marshes
77 Ekvoli Potamou Strymona
78 Limni Agra
79 Limni Distòs (Evvia)
80 Ormos Sourpis/Stomion

HUNGARY
81 Kis Balaton
82 Ócsa

ICELAND
83 Arnarvatnsheidi-Tvídaegra
84 Eylendid
85 Hjaltastadablá
86 Oxarfjördur
87 Pollengi
88 Safamyri
89 Skogar
90 Svarfadardalur
91 Thjórsárver

IRELAND
92 Ballyeighter Loughs
93 Blackwater River Callows
94 Clara Bog
95 Donegal Moors
96 Errisbog and Bogland North
97 Errisbog, Raheenmor Bog
98 Knockmoyle/Sheskin Bog Complex
99 Lough Barra Bog
100 Lugnaquillia and Glenmalur
101 Owenduff Bog
102 Owenduff catchment
103 River Shannon Callows, Athlone-Portumna
104 Slieve Bloom Mountains

ITALY
105 Bolle di Magadino
106 Lago di Busche/Vincheto di Cellarda
107 Palude Brabbia
108 Torbiere d'Iseo
109 Valle Santa, Valle Campotto e Bassarone

LATVIA
110 Teiču bog, Jēkabpils and Madona

ESTONIA
111 River Cepkeliai Marshes

THE NETHERLANDS
112 Bargenbeen
113 Botshol and Ronde Hoep
114 De Peel and Strabrechtse Heide
115 De Weeribben and de Wieden
116 Fochtelörveen, Esmeer and Huis ter Heide
117 Oostelijke Vechtplassen

NORWAY
118 Astujaeggi, Bardu
119 Dovrefjell
120 Dvergberg, Andoya
121 Ferdesmyra
122 Fokstumyra
123 Øvre Forra
124 Havmyran
125 Øvre Pasvik
126 Skogvoll
127 Sølendet

POLAND
128 Białowieża Forest
129 Biebrza Valley
130 Bielawskie Bog
131 Carbonate Marshes near Chełm
132 Great Obra Marshes
133 Kramskie Marshes
134 Nietlickie Marshes
135 Rozwarowo Marshes
136 Skoszewo Meadows

PORTUGAL
137 Paúl de Boquilibo

ROMANIA
138 Danube Delta

RUSSIAN FEDERATION
139 Dubna Marshes
140 Moskovskoye Morye
141 Nemunas Delta
150 Sources of river Oredezh
142 Volga Delta

SLOVAK REPUBLIC
143 Čičovské mŕtve rameno
144 Latorica
145 Parížske Močiare
146 Senné rybníky
147 Šúr
148 Niva Moravy

SLOVENIA
149 Ljubljansko Barje

SPAIN
151 Delta del Ebro
152 Laguna Dulce de Campillos
153 Las Tablas de Daimiel
154 Marismas del Guadalquivir
155 Salinas de Villafáfila

SWEDEN
156 Florarna
157 Former Lake Kvismaren
158 Hamrafjället
159 Jordbäsmuren
160 Komosse
161 Muddus National Park
162 Rocknenområdet Årshultsmyren - Tönnersjöheden
163 Sjaunja
164 Store mosse
165 Svarvarnåset - Fleringe - Rute - Träskmyr
166 Tavvavuoma

SWITZERLAND
167 Baie de Fanel et du Chablais
168 Les Grangettes

TURKEY
170 Kizihrmak Delta
171 Sultan Marshes

UK
172 Cors Fochno/Cors Caron
173 Inchs Marshes
174 Lewis Peatlands
175 Muir of Dinnet
176 New Forest Valley Mires
177 North Pennine Moors
178 Peak District Moors
179 Peatlands of Caithness and Sutherland
180 Rannoch Moor
181 Somerset Levels and Moors

YUGOSLAVIA (former)
182 Obdeska Bara

COASTAL AND MARINE SITES (MAP 9.9)

ALBANIA
1 Gjiri i Vlores
2 Karavastas/Divjaka
3 Kune-Vain
4 Lag e Nartes
5 Skadar Lake

BULGARIA
6 Dourankoulak Lake
7 Kaliakra

CROATIA
8 Kornati - Krka

CYPRUS
9 Akamas Peninsula
10 Limassol
11 Northern sandy beaches

DENMARK
12 Stavns Fjord & adjacent waters
13 Varde å estuary and Skallingen

ESTONIA
14 Lahemaa National Park - marine extension
15 Vilsandi bay

FINLAND
16 Saaristomeri National Park
17 Perämeri National Park
18 Merenkurkku Archipelago
19 Itäinen Suomenlahti National Park
20 Oura Archipelago
21 Pohjanpitäjänlahti Coast

FRANCE
22 Baie de Bourneuf/Loire Estuary
23 Baie du Mont St Michel
24 Bonifacio Strait
25 Camargue
26 Falaises de Pays de Caux
27 Girolata Porto Fango Scandola
28 Gironde to Charente Estuary
29 Hyères Islands archipelago
30 Landes Dunes
31 Massive Dunaire et Estuaires Picards
32 Mer d'Iroise
33 Vendéenne (Sand)
34 Sept Isles (representing N Britanny rocky)

GERMANY
36 Niedersächsisches Wattenmeer
37 Norderhever & Süderaue
38 Schleswig-Holsteinisches Wattenmeer
39 Spiekeroog
40 Süd-Ost-Rügen
41 Vorpommersche Boddenlandschaft NP

GREECE
42 Amvrakikos Gulf
43 Mont Athos/Thessaloniki coast
44 Nestos-Evros delta
45 Sporades Islands
46 Zakinthos and Kephalonia islands

ICELAND
47 Breidafjördur
48 Hornbjarg and Haelavikurbjarg
49 Latrabjarg and Keflavikurbjarg
50 Myrar-Löngufjördur
51 Nypslon/Skogalon
52 Olafsfjardarvatn
53 Papafjördur and Lonsfjördur
54 Reykjanes
55 Skardsfjördur
56 Vestmannaeyjar

IRELAND
112 Lough Hyne
120 Shannon Estuary
121 Skelligs
123 South Wexford

ITALY
57 Ligurian Sea
58 Northern part of the Adriatic
59 Orbetello, Feniglia ed Arcipelago Toscano
60 Western tip of Sicily

LITHUANIA
61 Kursio Nerija NP - Cuonian Spit

MALTA
62 Malta Gozo Strait

THE NETHERLANDS
63 Bosplaat, Ter Schelling Island
64 Rottumerplaat-Rottumeroog
65 Schiermonnikoog

NORWAY
66 Framvaren
67 Indre Porsangerfjord
68 Lindåspollene
69 Neiden-Munkefjord
70 Nøtterøy-Tjøme
71 Risøy-Flatvær
72 Skarnsundet
73 Skorpo-Nærlandsøy
74 Sør-og Nordsandfjord
75 Utvær-Indrevær
76 Vega-Lovunden
77 Vistenfjorden

POLAND
78 Słowiński NP (marine extension)

PORTUGAL
79 Ria de Aveiro
80 Ria Formosa
81 Tejo Estuary

ROMANIA
82 Danube Delta

RUSSIAN FEDERATION
83 Ainovy Islands
84 Eastern Coast of Barents Sea (Pechora River to Yugorsky Shar Strait)
85 Franz Josef Land
86 Kandalakshsky Strict Nature Reserve
87 Krasnovodsky and Severo-Chelekensky Bays
88 Kyzyl-Agachsky Bay
89 Sem'ostrovov
90 Shallow waters of northeastern Caspian Sea
91 Shallow waters of northern Caspian Sea
92 Volga Delta

SPAIN
93 Islas Baleares Mallorca-Menorca
94 Cerbere-Cabo de Creus-Islas Medas
95 Doñana Guadalquivir Odiel
96 Delta del Ebro
97 Ría de Huelva

SWEDEN
98 Haparanda archipelago
99 Falsterbo peninsula with Makläppen
100 Gullmar fjord
101 Kopparstenæna/Gotska Sandön/Salvorev
102 Landsort/ Askoe/Hartsoe
103 Storö/Bockö/Stora Nassa/SvenskaHögarna/ Svenska Björn
104 Trysunda area/Ulvöarna/Ullanger/Ulv ödjupet
105 Väderöarna/Tjärnö archipelago/Koster archipelago

UK
106 Calf of Man
107 Fleet & Chesil
108 Isles of Scilly
109 Lambay islands
110 Loch Maddy
111 Loch Sween
113 Lundy Island
114 Menai Straits
115 Monach Isles
116 Moray Firth
117 Morecambe Bay
118 Rathlin Island
119 Severn Estuary
122 Skomer
124 St Kilda
125 St Mawes
126 St Abbs
127 Strangford Lough
128 Thames
129 The Wash

UKRAINE
130 Karkinits'ka Zatoka with Lebedyni Ostrovy
131 Kazantip Peninsula and Bay
132 Obitochna Kosa
133 Sivash Zatoka
134 Yagorlits'ka and Tendrvs'ka Zatoky

YUGOSLAVIA (former)
135 Kotor Bay

MOUNTAINS
(MAP 9.10)

AUSTRIA
1 Dachstein
2 Hohe Tauern
3 Kalkhochalpen
4 Östliche Karawanken

BULGARIA
5 Rhodopi

CROATIA
6 Velebit

FRANCE
9 Cévennes
10 Inner Corse and Gulf of Porto
11 La Grande Chartreuse
12 Le Vercors
13 Les Ecrins
14 Mercantour - Upper Pesio valley - Col de Turini - Argentera (partly Italy)
15 Pyrénées-Occidentales

GERMANY
18 Berchtesgaden- Hagengebirge
19 Sächsische Schweiz

GREECE
20 Athos
21 Grammos
22 Lefka Ori
23 Olympos
24 Taygetos
25 Vicos-Aoos-Valia Calda

HUNGARY
26 Zemplén

ICELAND
27 Askja i Dyngjufjollum
28 Hekla-Fjallabaki
29 Hornstrandir
30 Jökulsargljufur
31 Myvatn-Laxa
32 Skaftafjell-Grimsvötn-Kverkfjoli

IRELAND
33 Burren

ITALY
34 Adamello - Brenta
35 Aspromonte
36 Cilento
37 Dolomiti Bellunesi e Feltrine
38 Etna
39 Gennargentu
40 Maiella
41 Monti della Laga
42 Monti Sibillini
43 Pollino
44 Puez Geisler - Sextner Dolomiten -Fanes - Sennes - Prags
45 Stelvio
46 Stromboli
47 Triglav - Tarvisio
48 Vesuvio
49 Vulcano

NORWAY
50 Dovrefjell
51 Femundsmarka-Rogen-Langfjellet
52 Hardangervidda
53 Indrefjord-Øksfjord
54 Jostedalsbrœen-Breheimen-Mørkrisdalen
55 Jotunheimen

56 Reisadalen
57 Spitsbergen
58 Svartisen-Saltfjellet
78 Tysfjord-Hellemo

POLAND
59 Ojcowski NP/Cracow-Częstochowa Upland
60 Pieniny Mountain

PORTUGAL
61 Peneda-Gerês
62 Pico
63 Serra de Estrela

ROMANIA
64 Bucegi
65 Codru Moma
66 Fagaras
67 Piatra Craiului
68 Retezat

SLOVAK REPUBLIC
97 Malá Fatra
7 Východní Karpaty
8 Vysoké Tatry (partly Poland)

SPAIN
69 Caldera de Taburiente
70 Cañadas del Teide
71 Cuenca Alta del Manzanares
72 Ordesa-Bielsa
73 Puertos de Beceite y Tortosa
74 Saja-Picos de Europa-Covadonga
75 Sierra del Cabo de Gata
76 Sierra de Gredos
77 Sierra Nevada

SWEDEN
78 Sarek-Padjelanta-Stora Sjöfallet
79 Stora Alvaret

SWITZERLAND
80 Berner Hochalpen & Aletsch-Bietschhorngebiet
81 Dent Blanche-Matterhorn-Monte Rosa
82 Diablerets-Vallon de Nant-Derborence
83 Gelten-Iffigen
84 Karstlandschaft Bödmeren-Silberen-Charetalp-Glattalp
85 Oberengadiner Seenlandschaft und Bernina-Gruppe
86 Säntisgebiet
87 Silvretta-Vereina
88 Speer-Churfirsten-Alvier
89 Val Verzasca

UKRAINE
91 Carpathian
92 Kryms' ke

UK
93 Cairngorms
94 Rhum

YUGOSLAVIA (former)
95 Durmitor & Tara Gorge
96 Galicica-Pelister-Prespa

REFERENCES

Andrén, C, Balletto, E, Bea, A, Corbett, K, Dolmen, D, Grossenbacher, K, Nilson, G, Oliveira, M, Podloucky, R and Stumpel A (1991) *Threatened amphibians in Europe requiring special habitat protection measures*. Report Societas Europeae Herpetologica, Conservation Commission. Bern.

Ashe, P, Speight, M C D and Williams, M de Courcy (1988) A review of the state of taxonomy of the Diptera (Insecta) in Ireland. In: Moriarty, C (Ed) *Taxonomy: Putting Plants and Animals in their Place*, pp 141–53. Royal Irish Academy, Dublin.

Baldock, D and Long, A (1987) *The Mediterranean Environment under Pressure: the influence of the CAP on Spain and Portugal and the 'IMPs' in France, Greece and Italy*. Report to WWF.

Balkas, T, Dechev, G Mihnea, R, Serbanescu, O and Unluata, U (1990) *State of the Marine Environment in the Black Sea Region*. UNEP Regional Seas Reports and Studies No 124. UNEP

Beaufort, F (1993) *Status and Population of European Mammals*. Report to the EEA-TF, Muséum National d'Histoire Naturelle, Paris, July 1993.

Bennett, G (Ed) (1994) *Conserving Europe's Natural Heritage: Towards a European Ecological Network*, Graham and Trotman, London.

Bernes, C (1991) Acidification and liming of Swedish freshwaters. *Monitor 12*. Swedish Environmental Protection Agencies, Solna.

Bernes, C (1993) *The Nordic environment – present state, trends and threats*. Nord 1993:12, 142. Nordic Council of Ministers. The Nordic Council, Stockholm.

Biebelriether, H and Schreiber, R L (1989) *Die Nationalparke Europas*. E. Pro-Natura Buch, Munich.

BIOMAR (1993) Marine Biotopes Classification Work Shop. *Marine Coastal Zone Management: Identification, Description and Mapping of Biotopes* (under preparation). Joint Nature Conservation, Peterborough.

BirdLife (1993) *Report to the EEA-TF by BirdLife International*, June 1993. Cambridge.

Bischoff, N T and Jongman, R H G (1993) *Development of Rural Areas in Europe: The Claim for Nature. Preliminary and Background Studies*. Netherlands Scientific Council for Government Policy. The Hague.

Bjork, S (1973) *Hornborgasjoutredningen. Del 2. Den limnologiska delutredningen*. Statens naturvardsverk PM 280. Solna.

Blab, J (1986). *Grundlagen des Biotopschutzes für Tiere. Ein Leitfaden zum praktischen Schutz der Lebensräume unserer Tiere*. Bonn – Bad Godesberg.

Bosch, C (1983) *Die sterbenden Wälder. Fakten, Ursachen, Gegenmasnahmen*. Beck, Munich.

Bramwell, D and Bramwell, Z.I (1974) *Wild Flowers of the Canary Islands*. Excmo. Cabildo Insular de Tenerife (Aula de Cultura) in association with Stanley Thornes (Publishers) Ltd, London.

Buckland, P C and Dinnin, M H (1993). Holocene woodlands, the fossil insect evidence. In: Kirby, K J and Drake, C M (Eds) *Dead Wood Matters: the Ecology and Conservation of Saproxylic Invertebrates in Britain*, pp 6–20. English Nature Science, no.7, English Nature, Peterborough.

Burgis M J and Morris, P (1987). *The Natural History of Lakes*. Cambridge University Press, Cambridge.

Casado, S, Florin, M, Molla, S and Montes, C (1992) Current status of Spanish wetlands. In: *Managing Mediterranean Wetlands and their Birds. Proceedings of an IWRB Symposium*, Grado, Italy, 2/1991. IWRB Special Publication 20, Slimbridge.

CEC (1991a) CORINE Biotopes Programme. *Manual: Habitats of the European Community. Data Specifications – part 2* EUR 12587. Commission of the European Communities, Luxembourg.

CEC (1991b) CORINE Biotopes Programme. *Biotopes Database of Sites.* Commission of the European Communities, Luxembourg.

CEC (1993) *Forest Condition in Europe. Results of the 1992 Survey. 1993 Executive Report.* Commission of the European Communities/United Nations Economic Commission for Europe. Brussels, Geneva.

Central Bureau of Statistics of Norway (1993) Natural Resources and the Environment 1992. Resources Accounts and Analysis. *Rapporter 93/1A*, Oslo.

Collar, N J and Goriup, P D (Eds) (1983) Report of the ICBP Fuerteventura Houbara Expedition, 1979. *Bustard Studies* 1, 1–92

Collar, N J and Andrew, P (1988) *Birds to watch: the ICBP world checklist of threatened birds.* International Council for Bird Preservation (Techn Publ 8) Cambridge.

Common Wadden Sea Secretariat (1990) *The International Wadden Sea. World wide importance, Worth-while protecting.* The Common Wadden Sea Secretariat, Wilhelmshaven.

Common Wadden Sea Secretariat (1992) *Sixth Trileteral Governmental Wadden Sea Conference,* 1991. The Common Waddensea Secretariat, Wilhelmshaven.

Corbett, K (Ed) (1989) *Conservation of European Reptiles and Amphibians.* Christopher Helm, London.

Council of Europe (1976) *Threatened Mammals in Europe.* Nature and Environment Series, No 10. Council of Europe, Strasbourg.

Council of Europe (1981) *Birds in Need of Special Protection in Europe.* Nature and Environment Series, No 24. Council of Europe, Strasbourg.

Council of Europe (1989) *Workshop on the Situation and Protection of the Brown Bear (Ursus arctos) in Europe.* Environmental encounters series, No.6, Council of Europe, Strasbourg.

Council of Europe (1990). *Marine Turtles in the Mediterranean: Distribution, Population, Population Status, Conservation.* Nature and Environment Series, No 48, Council of Europe, Strasbourg.

Council of Europe (1991) *Inventory and Cartography of Natural Biotopes in Europe: The Two Networks of Protected Areas of the Council of Europe.* Council of Europe PE-SEM/COR (91)2, Strasbourg.

Council of Europe (1992a) *Conserving and Managing Wetlands for Invertebrates.* Environmental Encounters Series, No 14, Council of Europe, Strasbourg.

Council of Europe (1992b) *Conservation of the Mediterranean Monk Seal – Technical and scientific aspects.* Environmental Encounters Series, No 13, Council of Europe, Strasbourg.

Council of Europe (1993) *The state of the Environment in Europe: the Scientists take Stock of the Situation.* International Conference, 12–14 December 1991, Council of Europe, Milan.

Cox, C B and Moore, P D (1973) *Biogeography: An Ecological and Evolutionary Approach.* Blackwell Scientific Publications, Oxford.

Dijkema, K S (1984) *Saltmarshes in Europe.* Nature and Environment Series, No 30. Council of Europe, Strasbourg

Dijkema, K S (1987) *Selection of Saltmarsh Sites for the European Network of Biogenetic Reserves.* RIN, Texel.

Dominguez, F and Diaz, G (1985) *Plan de Récuperacion de la Hubara Canaria.* Servicio Provincial del ICONA Santa Cruz de Tenerife.

Dommanget, J-L (1987) Etude faunistique et bibliographique des Odonates de France. *Inventaires de Faune et de Flore* 36, 1–283. Sécretariat de la Faune et de la Flore, Paris.

Drake, J A, Mooney, H A, di Castri, F, Groves, R H, Kruger, F J, Reijmanek, M and Williamson, M (Eds) Biological Invasions: A Global Perspective. *SCOPE 37.* John Wiley and Sons, Chichester.

Drachenfels, O V, May, H and Miotk, P (1984) Naturschutzatlas Niedersachen. Erfassung für den Naturschutz wertvoller Bereiche. *Naturschtz Landschaftspflege Niedersachsen,* **13.**

Dugan, P (1993) *Wetlands in Danger: a Mitchell Beazley World Conservation Atlas.* IUCN - The World Conservation Union, London.

Ellenberg, H (1986) *Vegetation Mitteleuropas mit den Alpen.* Ulmer Verlag, Stuttgart.

Emmerson, K (Ed) (1989) *Censo de la Poblacion de Hubara Canaria en la Isla de Fuerteventura, Diciembre* 1988. Ornistudio S L, Santa Cruz de Tenerife.

Esseen, P-A and Ericson, L (1982) *Granskogar med langskagglav i Sverige.* Naturvardsverket, Rapport PM 1513.

EUCC (1993) *European Coastal Conservation Conference,* 1991: Proceedings. The European Union for Coastal Conservation. The Hague/Leiden.

Farrell, L (1993) *Lowland Heathland: the Extent of Habitat Change.* English Nature, Peterborough.

Favarger, C (1972). Endemism in the Mountain Flora of Europe. In Valentine, D H (Ed) *Taxnomy, Phytogeography and Evolution.* Academic Press, London.

Findlayson, C M (Ed) (1992) *A Strategy and Action Plan to Conserve the Wetlands of the Lower Volga.* IWRB, Slimbridge, UK.

FNNPE (1992). *Loving Them to Death? The Need for Sustainable Tourism in Europe's Nature and National Parks.* Federation of Nature and National Parks Europes, Sustainable Tourism Working Group, October 1992, Grafenau.

GLE (1993) *Green Lungs of Europe, Outline Concept.* Document adopted by representatives of governmental agencies of environmental protection from Belarus, Estonia, Latvia, Lithuania, Poland, Russia and Ukraine, 10–11 February 1993 in a meeting in Warsaw.

Goodwillie, R (1980) *European Peatlands.* European Committee for the Conservation of Nature and Natural Resources. Nature and Environment Series 19, Council of Europe, Strasbourg.

Goriup, P D (1988) The avifauna and conservation of steppic habitats in Western Europe, North Africa, and the Middle East. In: Goriup, P D (Ed) *Ecology and Conservation of Grassland Birds.* ICBP Technical Publication No 7. ICBP, Cambridge.

Goriup, P D and Batten, L (1990) The conservation of steppic birds - a European perspective. *Oryx* 24, 215 - 23.

Gorodkov, K B (1982) *Maps 126–78: Provisional Atlas of the Insects of the European part of USSR,* Academy of Sciences of the USSR, Moscow.

Grimmett, R F A and Jones, T A (1989) *Important Bird Areas in Europe.* ICBP Technical Publication No 9, International Council for Bird Preservation, Cambridge.

Heath, J (1981) *Threatened Rhopalocera (butterflies) in Europe.* Nature and Environment Series No 23. Council of Europe, Strasbourg.

Heiss, G (1987) *Situation of Natural, Ancient, Semi-Natural Woodlands within the Council of Europe Member States and Finland.* Council of Europe, Environmental Encounters Series No 3.

Herzeman, T and Larsson, T (1991) Hornborgasjön - genmale och Kommentarer – Skaraborgsnatur 1986

Heywood, V H (1993) Flora conservation. In: *Naturopa,* No 71–1993, 24–5. Council of Europe, Strasbourg.

Hollis, G E and Jones, T A (1990) Europe and the Mediterranean Basin. In Findlayson, C M and Moser, M (Eds) *Wetlands,* pp 27–56. IWRB and Facts on File, Oxford.

Holnzer, W, Horvatic, E, Kollner, E, Köppl, W, Pokorny, M, Scharfetter, E, Schramayr, G and Strudi, M (1986) *Österreichischer Trockenrasenkatalog. Grüne Reihe des Bundesministeriums für Gesundheit und Umweltschutz,* Band 6.

Hutchinson, J N (1980) The record of peat wastage in East Anglian fenlands at Holme Post 1848–1978 AD. *Journal of Ecology* **68**, 229–49.

Icelandic Ministry for the Environment (1992) *Iceland – National Report to UNCED,* Reykyavík.

IEEP (1991) *Towards a European Ecological Network. EECONET.* Report by the Institute for European Environmental Policy. Arnhem, The Netherlands.

IGBP (1986) SCOPE/MAB Workshop, *Definition and Designation of Biosphere Observatories for Studying Global Change.* International Geosphere–Biosphere Programme, UNESCO.

Immirzi, C P and Maltby, E (1992) *The Global Status of Peatlands and their Role in Carbon Cycling.* Friends of the Earth, London.

IRSNB (1993) *IRSNB and the Community Programme for the conservation of the Monk Seal.* L'Institut Royal des Sciences Naturelles de Belgique, Brussels.

IUCN (1990a) *1990 IUCN Red List of Threatened Animals.* IUCN, Gland.

IUCN (1990b) *The Wetlands of East and Central Europe.* IUCN, Gland and Cambridge.

IUCN (1991) *Environmental Status Reports: 1990, Volume Two: Albania, Bulgaria, Romania, Yugoslavia.* The World Conservation Union, East European Programme, IUCN, Oxford.

IUCN (1993) *Action Plan for Protected Areas in Europe.* Second Draft. November 1993. The World Conservation Union, IUCN, Gland.

IUCN-PSU (1993) Communication with Roland Wirth, Mustelid, Viverrid and Procyonid Specialist Group of IUCN, October 1993.

Jalas, J and Suominen, J (Eds) (1972) *Atlas Florae Europaeae. Distribution of Vascular Plants in Europe. Volume I. Pteridophyta/Gymnospermae.* Cambridge University Press, Cambridge.

Johns, T A (1993) *A Directory of Wetlands of International Importance, Part Three: Europe.* Ramsar Convention Bureau, Gland.

Jones, T A and Hughes, J M R (1992) *Wetland Inventories and Wetland Loss Studies – a European Perspective*. Workshop E. IWRB Executive Board Meeting XXXV. St Petersburg Beach, Florida, USA.

Kaule, G (1986) *Arten und Biotopschutz, UTB Große Reihe*. Ulmer Verlag, Stuttgart.

Kime, R D (1990) *A Provisional Atlas of European Myriapods, Pt 1. Fauna Europaea Invertebrata*, 1. European Invertebrate Survey, Luxembourg.

Kime, R D (1993) Written Contribution to the EEA-TF, 2 December 1993.

Kirby, P (1992) *Habitat Management for Invertebrates: a Practical Handbook*. RSPB/JNCC, Sandy, Bedfordshire.

Klink, A (1989) The Lower Rhine: palaeoecological analysis. In: Petts, G E, with Moller, H and Roux, A L (Eds) *Historical Change of Large Alluvial Rivers: Western Europe*, pp 183–201. Wiley, Chichester.

Korneck, D and Sukopp, H (1988) Rote Liste der in der Bundesrepublik Deutschland ausgestorbenen, verschollenen und gefährdeten Farn- und Blutenpflanzen und ihre Auswertung fur den Arten- und Biotopschutz. *Schriftenreihe Vegetationskunde 19*. Bundesforschungsanstalt für Naturschutz und Landschaftsökologie, Bonn–Bad Godesberg.

Langer, H (1990) *Ecological Bricks for Our Common House in Europe*. Global Challenges Network e V Verlag für Politische Ökologie. Sonderheft 2 (October 1990). Verlag für Politische Ökologie, Munich.

Lean, G (1990) *World Wildlife Fund Atlas of the Environment*. Prentice Hall Press, New York.

Macdonald, S M and Mason, C F (1992) Status and conservation needs of the otter (Lutra lutra) in the Western Palearctic. In: *Report of the Council of Europe T-PV3(92)43*. Strasbourg.

Maitland, P S (1977) *Freshwater Fishes of Britain and Europe*. Hamlyn, London.

Maitland, P S (1986) *Conservation of Threatened Freshwater Fsh in Europe*. Council of Europe, Strasbourg.

Maltby, E, Hogan, D V, Immirzi, C P, Tellam, J H and Van der Peijl, M (1993) *Building a New Approach to the Investigation and Assessment of Wetland Ecosystem Functioning*. Proceeding of INTECOL IV, Columbus, Ohio.

Mason, C F and O'Sullivan, W M (1992) Organochlorine pesiticide residues and PCBs in otters (Lutra lutra) from Ireland. *Bulletin of Environmental Contamination and Toxicology* 48, 387–93.

McHarg, I (1969) *Design with Nature*. Doubleday/Natural History Press, New York.

McNeely, J A, Miller, K R, Reid, W R, Mittermeier, R A and Werner, T B (1990) *Conserving the World's Biological Diversity*. IUCN, WRI, WWF-US, World Bank. IUCN, Gland.

MedWet (1992) *An Integrated effort for the Conservation of Mediterranean Wetlands* (in press). Cooperation between CEC, IWRB, WWF, ICONA, Portugal, Spain, Italy, France, Greece. MedWet Secretariat, Roma.

Mitchell, R (1987) *Conservation of Marine Benthic Biocenoses in the North Sea and the Baltic*. Nature and Environment Series No 37. Council of Europe, Strasbourg.

Mitsch, W (1986) *Wetlands*. Van Nostrand Reinhold Company Inc, New York.

Munteanu, D and Toniuc, N (1992) The present and future state of the Danube delta. In: Findlayson, C M et al (Eds) *Managing Mediterranean Wetlands and their Birds*, pp 43–46. Proc Symp, Grado, Italy, 1991, IWRB Spec Publ No 20, Slimbridge, UK.

OECD (1991) *OECD Environmental Data Compendium 1991*, Paris.

Olschowy, G (1985) Warum Artenschutz? Gutachtliche Stellungnahme und Ergebnisse eines Kolloquiums des Deutschen Rates für Landespflege. In: *Der Schriftenreihe des Deutschen Rates für Landespflege*, pp 537–651, Heft 46. Bonn.

Ozenda, P (1983) *The Vegetation of the Alps*. Council of Europe, Nature and Environment Series No 29.

Polish Ministry of Environmental Protection, Natural Resources and Forestry; Belarus State Committee for Ecology; Estonian Ministry of Environment; Latvian Environmental Protection Committee; Lithuanian Environmental Protection Department; Russian Ministry of Ecology and Natural Resources; Ukrainian Ministry of Environmental Protection (1993) *Green Lungs of Poland: Outline Concept*. Document adopted on 10–11 February 1993 at the Meeting in Warsaw, published by Polish Ministry of Environmental Protection, Natural Resources and Forestry, Warsaw.

Remmert, H (1980) *Ökologie. Ein Lehrbuch*. Third Edition. Berlin.

Riecken. U (1990) Möglichkeiten und Grenzen der Bioindikation durch Tierarten und Tiergruppen im Rahmen Raumrelevanter Planungen. In: *Schriftenreihe f. Landschaftspflege und Naturschutz*, Heft 32. Bonn.

RIVM (1992) *The Environment of Europe: a Global Perspective*. National Institute of Public Health and Environment Protection, Bilthoven.

Rodwell, J S (1991) *British Plant Communities. Vol 2: Mires and Heaths*. Cambridge University Press, Cambridge.

Salathe, T (1992) *Towards Integrated Managment of Coastal Wetland Types*. Commission of the European Communities, XI/669/92.

Salman, A H P M and Strating, K M (1992) European Coastal Dunes and their Decline since 1900: A Pilot Study. In: EUCC, INCC *Sand Dune Inventory of Europe*.

SOVON (1993) Report to the EEA-TF by Samenwerkende Organisaties Vogelonderzoek Nederland (SOVON), May 1993, Beek-Ubbergen.

Speight, M C D (1987) *Saproxylic Invertebtrates: their Role in Natural Forests and their Potential as Indicators of site Quality in Selection of European Forests to be used as Nature Reserves*. Environmental Encounters Series No 3, Council of Europe, Strasbourg.

Speight, M C D (1989) *Saproxylic Invertebrates and their Conservation*. Nature and Environment Series No.42 Council of Europe, Strasbourg.

Starkel, L (1987) The evolution of European rivers – a complex response. In: Gregory, K J, Lewis, J and Thornes, J B (Eds) *Palaeohydrology in Practice*. Wiley, Chichester.

Stumpel, A, Podloucky, R, Corbett, K, Andrén, C, Bea, A, Nilson, G and Oliveira, M E (1992) Threatened reptiles in Europe requiring special conservation measures. In Korsos, Z and Kiss, I (Eds) *Proceedings of the 6th Ordinary General Meeting of the Societas Europaea Herpetologica, 19–23 August 1991*, Budapest, pp 25–434. Hungarian Natural History Museum, Budapest.

Tansley, A, G (1935) The use and abuse of vegetational concepts. *Ecology* 47, 733–45.

Tucker, G (1991) The status of lowland dry grassland birds in Europe. In: Goriup, P D, Batten, L A and Norton, J A (Eds) *The Conservation of Lowland Dry Grassland Birds in Europe*. JNCC, Peterborough.

Tucker, G and Heath, M (1994) *The birds in Europe – their conservation status*. Birdlife International, Cambridge.

Tutin, T G, Heywood, V H, Burges, N A, Valentine, D. H, Walters, S M and Webb, D A (1964/1980). *Flora europaea, Volumes 1–5*. Cambridge University Press, Cambridge.

UNCED (1992) *An Appeal for the Mountains*. Prepared by the initiating group of Mountain Agenda-UNCED 1992 for the Occasion of the United Nations Conference on the Environment and Development, University of Berne.

UNEP (1988) *The Blue Plan: Futures of the Mediterranean Basin: Executive Summary and Suggestions for Action*. United Nations Environment Programme, Athens.

UNEP/IUCN (1990) *Répertoire des Aires Marines et Cotièrs Protégées de la Méditerranée (Barcelona Convention)*. MAP Technical Reports Series No 36, UNEP, Athens.

van Eck, J, Govers, A, Lemaire, A. and Schaminee, J (1984) *Irish Bogs: a Case for Planning*. Catholic University of Nijmegen, The Netherlands.

Vangeluwe, D, Beudels, M-O and Lamani, F (1993) *Conservation Status of Albanian Coastal Wetlands and their Colonial Waterbird Populations (Pelicaniformes and Ciconiiformes)*. Institut Royal des Sciences Naturelles de Belgique, Brussels.

van Tol, J and Verdonk, M J (1988) The protection of dragonflies (Odonata) and their biotopes. *Nature and Environment Series* 38, pp 1–181. Council of Europe, Strasbourg.

Vivian, H (1989) Hydrological changes of the Rhone River. In: Petts, G E, with Moller, H and Roux, A L (Eds) *Historical Change of Large Alluvial Rivers: Western Europe*, pp 57–77. Wiley, Chichester.

Walter, H (1970) *Vegetationszonen und Klima*. Eugen Ulmer Verlag, Stuttgart.

WCMC (1990) *The IUCN Red List of Threatened Animal Species*. World Conservation Monitoring Centre, Cambridge.

WCMC (1992) *Global Biodiversity. Status of the Earth's Living Resources*. World Conservation Monitoring Centre. Chapman and Hall, London.

Wells, S M and Chatfield, J E (1992) *Threatened Non-Marine Molluscs of Europe*. Nature and Environment Series No 64, Council of Europe, Strasbourg.

Wells, S M, Pyle, R .M and Collins, N M (1983) *The IUCN Invertebrate Red Data Book*. IUCN, Gland, Switzerland.

Willems, J H (1990) Calcareous grasslands in continental Europe. In: Hillier, S H, Walteon, D H W and Wells, D A (Eds) *Calcareous grasslands – Ecology and Management*. BES/NCC symposium, Sheffield, 1987. Bluntisham Books, Bluntisham, Huntingdon.

Wilmanns, O (1973) *Ökologische Pflanzensoziologie*. Eugen Ulmer Verlag, Stuttgart.

Wolkinger, F and Plank, S (1981) *Dry Grasslands of Europe*. Nature and Environment Series No. 21. European Committee for the Conservation of Nature and Natural Resources, Strasbourg.

WWF/HELCOM (1991) Nature Conservation and Biodiversity in the Baltic Sea Region. Working Documents for a WWF/HELCOM-Seminar, May 1991, Stockholm.

Pedestrian zone
Source: Frank Spooner
Pictures

10 The urban environment

THE URBAN (ECO)SYSTEM

Many concerns for the state of the environment first developed in urban areas where changes in environmental conditions began to affect human health. Today, virtually all cities share concern for the quality of their environment. It is also in cities that many regional and global environmental problems originate. Concerns for the sustainability of cities have increased rapidly, as it has become evident that the environmental challenges of the future will be confronted in an increasingly urbanised world (Box 10A).

As part of ecological systems, cities affect and are affected by natural cycles. Cities depend on the availability of natural resources. They import water, energy and materials which are transformed into goods and services and ultimately returned to the environment in the form of emissions and waste. Their high concentration of people and activities make cities major contributors to local, regional and global environmental change. On the other hand it is the same concentration of people that provides unique opportunities for economies of scale and resource conservation. Thus, it is in cities that many environmental problems can be effectively addressed and resolved.

This chapter analyses the quality of the urban environment in Europe. The flows of natural resources that sustain cities are described to illustrate the interdependence between urban systems and the regional and global environment. Current patterns of urban development are examined in relation to the degree of pollution and exploitation of natural resources. An experimental set of

urban environmental indicators are used to identify major urban environmental problems in a selected number of European cities and to assess regional differences and priorities (Map 10.1). Finally, the options for improving the urban environment and successful examples in several European cities are examined.

European urban areas

In Europe more than two thirds of the total population live in urban areas. The net result of urbanisation processes between 1950 and 1990 has been an overall increase of urban population. However, different patterns of urbanisation have occurred across Europe as a result of differences in economic and social development (Figure 10.2). In Southern and Eastern European cities, rural-to-urban migration combined with high birth rate continued to produce rapid population growth throughout the 1960s and the 1970s. During the same period, Western European cities have experienced an alternate cycle of growth and decline in the form of population concentration followed by geographic expansion and suburbanisation and finally de-urbanisation. Parallel to this trend has been the faster growth of smaller and medium cities. During the 1980s the pattern of urban decline spread from the largest to the smaller cities and from northwestern to Southern and Eastern Europe (CEC, 1992).

In contrast with past trends of urban decline are the most recent signs of re-urbanisation. European cities have responded to the crisis of the 1980s with programmes for urban renewal and restructuring. The common purpose of

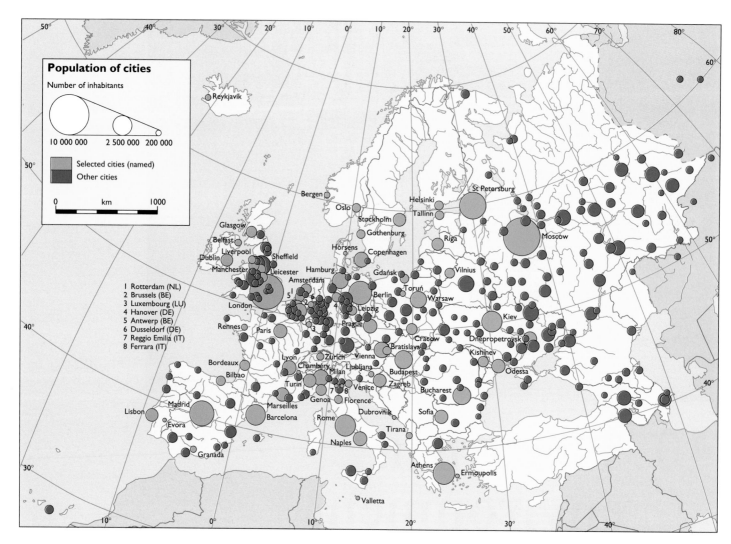

Map 10.1
Population of selected European cities
Data source: OECD/Eurostat

Figure 10.1
Current and projected urban population by region, 1950–2000
Source: United Nations, 1991

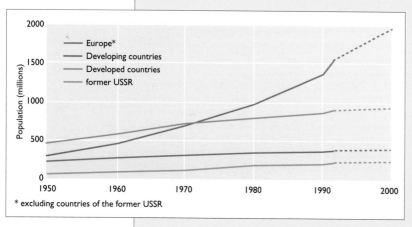

these programmes is economic revitalisation and the improvement of the quality of life of the urban population that both the rapid urbanisation and urban decline have caused to deteriorate. The quality of the urban environment is increasingly recognised to be a key ingredient of the economic regeneration of European cities. Understanding the common causes of degradation of Europe's urban environment is one crucial condition for their success.

Urban systems and the environment

Cities can be described according to their geographical position, climate and morphology. Alternatively, they can be described according to their population, social relationships and economic activities. The urban form, structure and landuse are other important variables which affect the quality of the urban environment and the impact of urban

Box 10A Urban growth

By the turn of the century, almost half of the world population will live in urban areas. The United Nations estimates that by the year 2000, 45 per cent of the population of developing countries (1972 million) and 75 per cent of that of developed countries (945 million) will live in cities. The urban population has grown rapidly, increasing from 14 per cent at the beginning of this century to 25 per cent in 1950. Between 1950 and 1990, the number of people living in cities has more than tripled, increasing from 733 million to 2.26 billion (United Nations, 1991). Urban growth patterns have not been consistent over time and across the various geographic areas (Figure 10.1). Of the developed world's population, 54 per cent was already urban in 1950, reaching 73 per cent by 1990. The developing world's urban population was only 16.9 per cent in 1950, increasing to 34 per cent in 1990. The bulk of urban growth is now taking place in the developing world where the urban population has quadrupled in the last 35 years, while in the developed regions it has doubled. In the developing countries urban growth rates in the 1980s reached 3.6 per cent a year compared to a rate of 0.8 per cent a year in the developed countries.

settlements on the regional and global environment. In terms of systems ecology, the city can be compared to an ecosystem with its own structure, functions and metabolism. Describing the interactions between the urban system and the environment is, however, a difficult task because of the complex relationships that take place in cities between physical, social and economic variables.

An analysis of the urban dimension of environmental change in Europe should consider both the state of the environment in cities and the impacts that cities have on the regional and global environment. Changes in the quality of the urban environment are rooted in the patterns of urban activities. Urban density, mobility and lifestyles are reflected in the demand for space and the flow of resources. Concentration of people and activities in a limited space places a heavy demand on local natural habitats. However, the impact of cities on the global environment is not an attribute of the city itself, but depends on the activities cities accommodate, their spatial arrangement and the management of natural resources that support these activities. From a global perspective, cities can thus play an important role in achieving global sustainability.

Urban sustainability

Ecological sustainability requires that development should allow response to present needs of the human population without impairing the capacity of future generations to satisfy their own needs. Translating this concept to the urban scale requires that inhabitants' needs are met without imposing unsustainable demands on local, as well as on global, natural systems. Indeed no city sustains itself within its own boundaries. Although cities are generally described as discrete geographical entities, the total area required to sustain an urban region is much greater than that contained within urban boundaries (Rees, 1992). Inputs of energy, water and materials are made into the urban system, where resources are processed, used and converted into residuals. Based on available data in selected European cities and national data, it is estimated that on average a city of 1 million inhabitants in Europe every day requires 11 500 tonnes of fossil fuels, 320 000 tonnes of water and 2000 tonnes of food. It also produces 300 000 tonnes of wastewater, 25 000 tonnes of CO_2 and 1600 tonnes of solid waste.

The interdependence between the local and the global environment is crucial to understand the relationships between global and local sustainability. Since cities sustain themselves by drawing on the natural capital of other regions, achieving sustainability at the local level does not ensure ecological sustainability at the global level. Cities may achieve good local environmental conditions in the short term (eg, reducing local air pollution) by placing unsustainable demands on natural resources elsewhere (eg, importing large quantities of energy and exporting large amounts of waste). It is only when seen from a long-term perspective that the link between local quality and global sustainability becomes evident. Indeed these same cities will also suffer from the effects of global environmental problems such as climate change, acidification and stratospheric ozone depletion.

Selected urban indicators

In an attempt to assess the state of Europe's urban environment, 55 urban environmental indicators grouped into 16 urban attributes have been selected, focusing on urban patterns, urban flows and urban environmental quality (Table 10.1). The experimental set of 55 indicators has been selected to identify major urban environmental problems on the basis of the best available information and is not claimed to provide a complete picture of the complex relationships between cities and the environment. A systematic analysis of these relationships would have required a far more complete and detailed list of indicators.

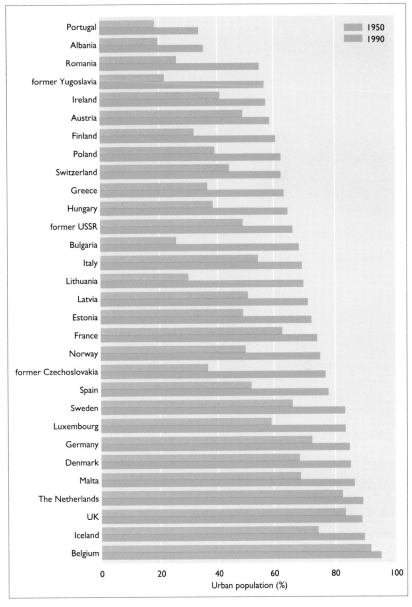

Figure 10.2
Urban population in Europe, 1950–90
Source: UNEP, 1993

The indicators applied were selected on the basis of their relevance for a preliminary assessment in 72 European cities, acknowledging the constraints imposed by the limited comparable information that is available or could be obtained from existing sources. Cities were selected in order to account for the wide variety of city size, geographic location and predominant economic activities. The list and regional classification of cities as referred to in this chapter are given in Table 10.2. The range of city size in the sample finally included in the report (Table 10.3) was between 12 000 (Valletta) and 8 818 000 (Moscow) inhabitants. Half of the sample cities had over 500 000 inhabitants, one third over 1 000 000 and one third less than 300 000. Three cities (Moscow, London and St Petersburg) had more than 4 500 000 inhabitants. City area ranged between 2 (Valletta) and 1578 (London) km².

Because of the lack of data in several cities, only a limited number of indicators (20) proposed for the assessment of the urban environment are presented in this report. Data reported are also limited to 51 cities out of the 72 originally selected. A great deal of attention has been given to ensure the quality of the data. However, much caution should be exercised in comparing and analysing the data, since different methods of measurement and variations in accuracy limit comparability. All figures summarised in the indicator table can be found in the *Statistical Compendium*. Other sources of data cited in this

Attributes		Indicators
A Indicators of urban patterns		
1 Urban population	a) Population	● number of inhabitants in city (1) in conurbation (2)
	b) Population density	● population per km² (3)
		● area by density classes (4)
2 Urban land-cover	a) Total area	● area in km² (5)
	b) Total built-up area	● area in km² (6)
		● by landuse (7)
	c) Open area	● area in km² (8)
		● % green areas (9)
		● % water (10)
	d) Transportation network	● motorway length (km) (11)
		● railway length (km) (12)
		● % of total urban area (13)
3 Derelict areas	Total area	● area in km² (14)
		● % of total urban area (15)
4 Urban renewal areas	Total area	● area in km² (16)
		● % of total urban area (17)
5 Urban mobility	a) Modal split	● number (18) and average length (19) of trips in km per inhabitant per mode of transportation per day
	b) Commuting patterns	● number of commuters into and out of conurbation (20)
		● as % of the urban population (21)
	c) Traffic volumes	● total (22) and inflow/outflow (23) in vehicle-kms
		● (24) number of vehicles on main routes
B Indicators of urban flows		
6 Water	a) Water consumption	● consumption per inhabitant in litres per day (25)
		● % of groundwater resources in total water supply (26)
	b) Wastewater	● % of dwellings connected to a sewage system (27)
		● number (28) and capacity (29) of treatment plants by type of treatment
7 Energy	a) Energy consumption	● electricity use in GWh per year (30)
		● energy use by fuel type and sector (31)
	b) Energy production plants	● number (32) and type (33) of power and heating plants in the conurbation
8 Materials and products	Transportation of goods	● quantity of goods moved into and out of the city in kg per capita per year (34)
9 Waste	a) Waste production	● amount of solid waste collected in tonnes per inhabitant per year (35)
		● composition of waste (36)
	b) Recycling	● % of waste recycled per fraction (37)
	c) Waste treatment and disposal	● number of incinerators (38) and volume (39) incinerated
		● number of landfills (40) and volume (41) received by waste type
C Indicators of urban environmental quality		
10 Quality of water	a) Drinking water	● number days per year that the WHO drinking water standards are exceeded (42)
	b) Surface water	● O₂ concentration of urban surface water in mg per litre (43)
		● number of days pH is > 9 or < 6 (44)
11 Quality of air	a) Long term: SO₂+TSP	● annual mean concentrations (45)
	b) Short-term concentration: O₃, SO₂, TSP	● exceedances of AQGs : O₃ (46) SO₂ (47), TSP (48)
12 Acoustic quality	Exposure to noise (inhabitant per time period)	● exposure to noise above 65 dB (49) and above 75 dB (50)
13 Traffic safety	Fatalities and casualties from traffic accidents	● number of people killed (51) and injured (52) in traffic accidents per 10 000 inhabitants
14 Housing quality	Average floor area per person	● m² per person (53)
15 Accessibility of green space	Proximity to urban green areas	● percentage of people within 15 minutes' walking distance of urban green areas (54)
16 Quality of urban wildlife	Number of bird species	● number of bird species (55)

*Indicator number in parentheses

Table 10.1 Selected indicators for the urban environment

chapter derive from international, national and municipal reports as well as from a number of case studies that compensate for the paucity of available data on historical trends at the urban scale. Additional information is obtained from recent studies on urban energy and on urban carbon dioxide emissions in a number of European cities (ICLEI, 1993; LRC, 1993; IBGE, 1993). The results of a recent survey on urban travel in OECD cities were also used (OECD/ECMT, 1993).

The need for better urban environmental information

The current analysis shows that knowledge of the European urban environment is incomplete. Even when data exist, it is extremely difficult to obtain them. Environmental data at the urban scale are often dispersed among various local agencies and levels of government dealing with different urban environmental aspects. The attempt made for this report to apply an experimental set of indicators shows the areas of environmental information that are lacking or inadequate for a comprehensive assessment of urban environmental problems. An overview of the state of environmental information at the urban scale in Europe is given in Table 10.4.

Monitoring activities at the urban scale are well established in certain areas, such as for example air pollution, where environmental quality standards are available. Such activities are generally less developed for acoustic quality and green space. Data on electricity consumption, water use and waste production can be easily obtained at the municipal level, although their comparability is uncertain due to different measurement methods. Statistics on fossil fuels and consumption of materials are not collected at the urban scale and only a few cities have inventories of energy consumption by fuels, activities or districts. Data on landuse, mobility and infrastructure, when available, are difficult to compare due to different definitions and classification systems. Data availability and reliability vary also from city to city and between regions. In most large cities in Europe, environmental monitoring activities are well established. However, in many others, environmental monitoring has begun only recently.

It is clear that the capacity to address urban environmental problems requires improved information. The importance of different problem areas and contributing causes will only be understood when adequate information becomes available on the urban scale. Increased attention to urban patterns and their relationships to the use of natural resources will be crucial to measuring local and global environmental impacts and exploring alternatives for urban development. The success of important efforts undertaken by local authorities to improve information on the state of the urban environment is essential for a better understanding of priority problems in different urban areas. It is clear, however, that the desired quality and comparability of information across European cities will not be available in the short term. Making use of the best available information now is required.

URBAN ENVIRONMENTAL QUALITY

The quality of the urban environment depends on both physical elements and conditions for community life. Essential to urban life are clean air, water and soil, but also adequate housing, green areas and open space. Other important elements are safety, accessibility and opportunities for economic activities, social interaction and recreation. In cities these factors are interdependent. An attempt has been made to a certain extent to underline the relationships between the state of the urban environment and socio-economic change. However, an analysis of social and

Cities (country code)	City population**	Total city area** (km2)	Total built-up area (km2)
Amsterdam (NL)	713 493	202	187
Barcelona (ES)	1 635 067	99	59
Belfast (UK)	279 237	150	125
Bergen (NO)	215 967	445[1]	
Berlin (DE)	3 456 891	889	380
Bilbao (ES)	385 000	41.3	16
Bordeaux (FR)	213 274	49.4	
Bratislava (SK)	445 730	368	119
Brussels (BE)	951 217	161	105
Budapest (HU)	2 434 064	525	
Chambéry (FR)	55 603	21	15.6
Copenhagen (DK)	465 000	88	53
Cracow (PL)	750 500	327	90.3
Dnepropetrovsk (UA)	1 185 500	397	222
Dubrovnik (HR)	34 521	6.7	5.57
Ermoupolis (GR)	12 386	3	
Evora (PT)	39 229	16	11
Ferrara (IT)	94 106	414	23
Gdańsk (PL)	465 100	257.5	69
Glasgow (UK)	688 600	202.7	145
Gothenburg (SE)	432 112	449	100
Hannover (DE)	522 354	204	102
Helsinki (FI)	501 518	185[2]	81
Kiev (UA)	2 651 300	825	345
Liverpool (UK)	454 400	113	107
Ljubljana (SI)	224 817	230	
London (UK)	6 679 699	1578	
Madrid (ES)	3 010 492	607	
Milan (IT)	1 406 818	182	77
Moscow (RU)	8 818 000	996	531
Odessa (UA)	1 086 700	333	
Oslo (NO)	4 67 090	427[1]	135
Paris (FR)	2 152 423	105.4	81.3
Prague (CZ)	1 212 010	495	200
Reggio Emilia (IT)	131 880	231.5	25
Rennes(FR)	203 533	50.4	34
Reykjavík (IS)	100 850	114	33
Riga (LV)	900 000	307	89.2
Rotterdam (NL)	589 678	273	138
Sheffield (UK)	529 300	363	109
Sofia (BG)	1 148 550	183	108
Stockholm (SE)	684 574	188	64
St Petersburg (RU)	4 948 000	1400	363
Tirana (AL)	270 000	19	16
Toruń (PL)	202 200	116	26.4
Valletta (MT)	12 403	1.6	0.6
Venice (IT)	77 000	17	13
Vienna (AT)	1 500 000	415	134
Warsaw (PL)	1 653 300	485.3	202
Zurich (CH)	365 043	92	30.36
Zagreb (HR)	706 770	297	74

Notes:
* Only 51 cities out of 72 selected were finally included in the report for reasons of data availability and comparability.
** Figures relate to administrative boundaries of municipalities.
1 The total city area includes the land area only (there are large forest areas within the municipality).
2 Total city area of Helsinki is 588 km^2, of which 185 km^2 is covered with land and 403 km^2 with water.

Table 10.3 Selected European cities: size
Source: Compiled from multiple soucres, see *Statistical Compendium*

Northern Europe	Western Europe	Southern Europe	Central Europe	Eastern Europe
Bergen (NO)	Amsterdam (NL)	Athens (GR)	Bratislava (SK)	Dnepropetrovsk (UA)
Gothenburg (SE)	Antwerp (BE)	Barcelona (ES)	Bucharest (RO)	Kiev (UA)
Helsinki (FI)	Berlin (DE)	Bilbao (ES)	Budapest (HU)	Kishinev (MD)
Oslo (NO)	Bordeaux (FR)	Ermoupolis (GR)	Cracow (PL)	Moscow (RU)
Reykjavík (IS)	Brussels (BE)	Evora (PT)	Dubrovnik (HR)	Odessa (UA)
Stockholm (SE)	Chambéry (FR)	Ferrara (IT)	Gdańsk (PL)	Riga (LV)
	Copenhagen (DK)	Florence (IT)	Ljubljana (SL)	St Petersburg (RU)
	Dublin (IR)	Genoa (IT)	Prague (CZ)	Vilnius (LT)
	Düsseldorf (DE)	Granada (ES)	Sofia (BG)	Tallinn (EE)
	Glasgow (UK)	Lisbon (PT)	Tirana (AL)	
	Hamburg (DE)	Madrid (ES)	Toruń (PL)	
	Hanover (DE)	Milan (IT)	Warsaw (PL)	
	Horsens (DK)	Naples (IT)	Zagreb (HR)	
	Leicester (UK)	Reggio Emilia (IT)		
	Leipzig (DE)	Rome (IT)		
	Liverpool (UK)	Turin (IT)		
	Lyons (FR)	Valletta (MT)		
	London (UK)	Venice (IT)		
	Luxembourg (LU)			
	Marseilles (FR)			
	Paris (FR)			
	Rennes (FR)			
	Rotterdam (NL)			
	Sheffield (UK)			
	Vienna (AT)			
	Zurich (CH)			

economic trends of European cities is beyond the scope of this report. This chapter focuses on the quality of the physical urban environment.

Table 10.5 provides a summary of different environmental problem areas in the 51 European cities examined in this report. While the poor quality of the urban environment emerges as a major concern for Europe, the importance and severity of various problems varies widely between European cities. Air quality is a major concern in most cities. In cities where data are available, unacceptable levels of noise affect between 10 and 50 per cent of urban residents.

Table 10.2
Selected European cities: regional classification

Table 10.4
Assessment of urban environmental information

Indicators	Quality of data				
	Availability	Accuracy	Reliability	Disaggregation	Comparability
Urban patterns					
Population	◔	◔	◔	◔	◔
Area	◔	◔	◔	◔	◔
Landuse	◔	◔	◔	◔	●
Mobility	◔	◔	◔	◔	◔
Infrastructure	◔	◔	◔	◔	●
Urban flows					
Energy consumption	●	◔	◔	●	●
Water consumption	◔	◔	◔	◔	◔
Waste water	◔	◔	◔	◔	◔
Materials	●	◔	◔	◔	◔
Waste	◔	◔	◔	◔	◔
Urban environmental quality					
Air quality	◔	◔	◔	◔	◔
Water quality	●	◔	◔	◔	◔
Green areas	◔	◔	◔	◔	●
Acoustic quality	●	◔	◔	◔	◔
Housing quality	◔	◔	◔	◔	◔
Traffic safety	◔	◔	◔	◔	◔

◔ Good
◔ Poor
● Very poor

City (country code)	Green areas: accessibility of green space (% of persons 15 min walking)	Green areas (% of city area)	Outdoor noise: population exposure (%)		Housing space per capita (m²)	Traffic safety per 10 000 inhabitants		Wastewater (% of dwellings connected to sewer)	Air pollution Long-term SO₂ + TSP (annual mean)	Short-term O₃ (max 1 hr)	Short-term SO₂ (max 24 hrs)	Short-term TSP
			> 65 dB	> 70 dB		People injured	People killed					
Amsterdam (NL)		○	●	●	●	●	●	●	●	●		●
Barcelona (ES)	●	○		●	●	●	●	●	○			
Belfast (UK)	●			●		●	●	●				
Bergen (NO)	●		●	●	●	●	●	●				
Berlin (DE)						●						
Bilbao (ES)	●	●	●	○	●	●	●	○				
Bordeaux (FR)						●	●					
Bratislava (SK)	●	●		●	●	●	●	●	●		●	●
Brussels (BE)	●	○	●	●	●	●	●	●	●	●		●
Budapest (HU)	●	○	●	●	●	●	●	●	○	●	●	●
Chambéry (FR)		●			●	●	●	●				
Copenhagen (DK)	●			●	●	●	●	●	○	●	●	●
Cracow (PL)		○	●	●	●	●	●	●	○			
Dnepropetrovsk (UA)	●	○	○	●	●	●		●				
Dubrovnik (HR)	●			●	●	●	●	●				
Ermoupolis (GR)	●			●		●	●	●				
Evora (PT)	●	●				●	●	●				
Ferrara (IT)	●				●	●	●	●				
Gdańsk (PL)	●	●			●	●	●	●				
Glasgow (UK)	●					●						
Gothenburg (SE)	●		●	●		●	●	●	●	●	●	●
Hanover (DE)		○			●	●	●	●				
Helsinki (FI)		○			●	●	●	●	●	●	●	●
Kiev (UA)	●	○		●	●	●	●	●	●		●	●
Liverpool (UK)	●					●	●	●	●			●
Ljubljana (SI)					●	●	●	●	●		●	●
London (UK)												
Madrid (ES)		●			●	●	●	●				
Milan (IT)					●	●	●	●	●		●	●
Moscow (RU)					●	●	●	●	●			
Odessa (UA)		○		●	●	●	●	●	●		●	
Oslo (NO)	●		●	●	●	●	●	●	●	●	●	●
Paris (FR)	●		●	●	●	●	●	●	●	●	●	●
Prague (CZ)	●	○	●		●	●	●	●	●		●	●
Reggio Emilia (IT)	●			●	●	●	●	●				
Rennes (FR)		○				●	●	●				
Reykjavík (IS)	●				●	●	●	●	●		●	
Riga (LV)					●	●	●	●	●			
Rotterdam (NL)		○			●	●	●	●	●	●	●	●
Sheffield (UK)												
Sofia (BG)	●			●	●	●	●	●	●		●	
Stockholm (SE)	●		●	●	●			●	●	●	●	
St Petersburg (RU)				●				●	●	●	●	
Tirana (AL)	●				●	●	●	●	●		●	●
Toruń (PL)		●			●	●	●	●				
Valletta (MT)	●			●		●	●	●				
Venice (IT)	●			●		●	●	●				
Vienna (AT)					●	●	●	●	●		●	
Warsaw (PL)	●				●	●	●	●	●	●	●	●
Zagreb (HR)	●	●		●	●	●	●	●	●	●	●	●
Zurich (CH)	●	○	●		●	●	●	●	●	●	●	

Green areas (% of population)
- ● >50
- ● ≤50

Green areas (% of city)
- ● >30
- ○ 5–30
- ● <5

Outdoor noise (%)
> 65 dB   > 70 dB
- ● <25 | ● <5
- ○ 25–50 | ● 5–10
- ● >50 | ● >10

Housing space per capita (m²)
- ○ >40
- ● 20–40
- ● <20

Traffic safety
people injured   people killed
- ○ <25 | ● <1
- ● 25–50 | ● 1–3
- ● >50 | ● >3

Wastewater (%)
- ○ 100
- ● 90–99
- ● <90

Long-term air pollution
Annual mean concentrations of air pollutants at city background stations (µg/m³)
SO₂ + TSP
- ○ <100
- ● 100–200
- ● >200

Short-term air pollution
Exceedances of AQGs (µg/m³)
AQGs for: O₃ = 150 (max 1 hr), SO₂ = 125 (max 24 hrs), TSP = 120 (max 24 hrs)
O₃, SO₂, TSP
- ● <1 AQG
- ○ 1–3 AQG
- ● >3 AQG

Table 10.5
Urban environmental quality
Source: compiled from multiple sources, see *Statistical Compendium*

Traffic accidents are another dominant feature of most European urban centres: in 25 per cent of cases more than 50 people out of every 10 000 inhabitants are injured every year, and one in every 10 000 inhabitants is killed. Housing quality is a particular concern in Eastern European cities, where floor area per capita is half as much as in Western European cities. Access to green areas within a 15-minute walking distance is possible for more than 50 per cent of urban residents in the majority of European cities, and for more than 90 per cent in two thirds of cases where data are available. However, the size and quality of green areas and open spaces vary widely between cities. Although land degradation is a shared concern in most European cities, few data are available on derelict land.

Urban air quality

Major air pollutants in urban areas are sulphur dioxide (SO_2), particulate matter, nitrogen oxides (NO_x), carbon monoxide (CO), ozone (O_3), lead (Pb), other heavy metals and organic compounds arising from various activities. The results of the air quality survey made for this report show that in 70 to 80 per cent of the European cities with more than 500 000 inhabitants, air pollution levels of one or more pollutants exceed WHO Air Quality Guidelines (AQGs) at least once in a typical year (Chapter 4). Major sources are space heating, electricity generation, industrial activities and road traffic. Concentrations of air pollutants vary considerably across European cities depending on the density of activities, fuel

---

Box 10B Air quality in London

London is a good example of how air quality standards have reduced traditional air pollution in Western European cities over the last 30 years. Annual mean sulphur dioxide concentrations have dropped dramatically from the 300 to 400

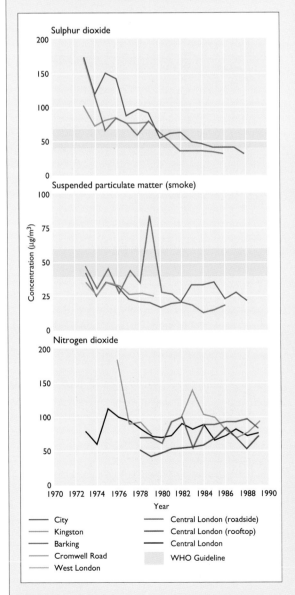

$\mu g/m^3$ of the 1960s to 20 to 30 $\mu g/m^3$. Concentrations of suspended particulate matter (smoke) levelled to 30 $\mu g/m^3$. These levels are well below WHO annual guidelines. Both are the result of the switch in fuel type from fuel oil and coal to smokeless coal and natural gas as well as of the introduction of pollution control technologies in the manufacturing industry. A factor which has contributed to the drastic reduction of these pollutants in the city centre is also the relocation of these activities out of the city. However, high pollution episodes are still a problem during the winter. Exceedances of sulphur dioxide short-term guidelines (500 $\mu g/m^3$ 10-minute means, and 350 $\mu g/m^3$ one-hour mean) are still a regular occurrence.

A drastic reduction in lead-in-air concentrations also occurred in the second half of the 1980s as a result of the phase-out of permissible lead levels in petrol which entered into force in 1986. Annual concentrations are now below WHO guidelines at background sites, and roadside concentrations do not exceed the upper limit of WHO guidelines at the central London site (Figure 10.3).

Carbon monoxide concentrations show mixed patterns across different monitoring sites. The number of episodes of the 8-hour WHO guideline exceedances in central London are of major concern. Emissions from road traffic, which account for 95 per cent of total carbon monoxide emissions, are trapped by ground-based temperature inversions, building up high localised concentrations. High traffic volumes are also the cause of increased concentrations of nitrogen oxides. While a number of sites show stable trends in nitrogen oxides concentrations, other sites indicate marked increases. During 1989, 16 sites in central London exceeded the EC 98 percentile limit of 200 $\mu g/m^3$, and all 64 sites throughout London but two had an estimated 50 or 98 percentile greater than the EC guide values of 50 and 135 $\mu g/m^3$. Measurements of ambient tropospheric ozone concentrations at a number of sites in central London show no clear trends, but they indicate the frequent occurrence of exceeding the hourly mean value of 214 $\mu g/m^3$ (100 ppb), the upper limit of the WHO guideline range.

Emissions from motor vehicles are the most significant source of carbon monoxide, nitrogen oxides and secondary pollutants such as ozone. These are expected to increase further due to the increased number of cars. Greater London had 2.7 million motor vehicles in 1988 (one car per three inhabitants) and together with the metropolitan area accounts for over 6 million cars (35 per cent of all UK cars). According to OECD estimates, road traffic in London has increased by 70 per cent during the last 20 years. A 7 per cent increase has occurred since 1989 alone.

Figure 10.3 Annual mean concentration of SO_2, smoke and NO_x, London, 1973–89

Source: UNEP/WHO, 1992

mix and technologies employed. Relevant differences in the occurrence and frequency of high concentration episodes can be associated with different local meteorological and topographical conditions. (An extensive assessment of local air pollution is presented in Chapter 4.)

Important changes in the level and composition of urban air pollution in Europe have occurred during the last 20 years. Air quality trends in London and Moscow, two of the ten world 'megacities' recently studied by UNEP and WHO, illustrate well the different patterns occurring in Western and Eastern European cities (UNEP/WHO, 1992) (Boxes 10B and 10C). A major improvement in Northern and Western European cities, and to a lesser extent in Southern European cities, is reflected by the decreasing trend in SO_2 concentrations achieved by imposing strict emission standards. In most cities, concentrations of SO_2 are less than their level in the late 1970s. Helsinki, Stockholm, Gothenburg, Oslo and Copenhagen, for example, show a decrease of 80 per cent (Bernes, 1993). In the majority of cities surveyed, annual means of SO_2 concentrations detected at individual sites are generally less than 50 $\mu g/m^3$. However, there are still a large number of cities in all parts of Europe where concentrations exceed a short-term WHO Air Quality Guideline (AQG) level (24-hour) for SO_2 and/or total suspended particulates (TSP)/black smoke (winter smog) (Table 10.5).

During the last decade, SO_2 concentrations in Central and Eastern European urban areas were lowered due to the adoption of several technical measures. Recently, because of economic stagnation and restructuring of industries, air quality has considerably improved. For example, in Prague, Bratislava and Warsaw, SO_2 annual means have declined by half between 1985 and 1990 (OECD, 1993d). However, the exceedences of WHO guidelines for combined exposure to SO_2 and particulate matter of 50 $\mu g/m^3$ (annual average) and 125 $\mu g/m^3$ (24-hour average) show that both short-term and long term SO_2 exposure levels are still unacceptably high. Several Central, Eastern and Southern European cities, regularly exceed WHO guidelines for SO_2 many times over. In Toruń SO_2 concentrations above 1000 $\mu g/m^3$ (1-hour/average) are not rare. This is often due to the burning of high-sulphur fuels for heating and industrial purposes, and to outdated technologies.

A downward trend with relative variations between cities can be detected for particulate concentrations (OECD, 1993c). In the majority of cities, suspended particulate concentrations range between 50 $\mu g/m^3$ and 100 $\mu g/m^3$. However, TSP concentrations are generally higher than SO_2 concentrations. In several cities, especially in Central and Eastern Europe, but also in Southern Europe, particulate concentrations show upward trends. High TSP and black smoke concentration levels in Athens originate from the combined effects of various emission sources (Box 10D).

In contrast to SO_2 and TSP trends, is the stability or upward trend in NO_x and CO concentrations in all European cities (OECD, 1993c), which is mainly due to the increase in urban traffic. In Western cities, road traffic is a significant contributor to emissions, ranging from 30 to 50 per cent of total NO_x and 90 per cent of CO emissions. The share of road traffic in urban air pollution in Eastern European cities is expected to rise substantially in the next few years.

Nitrogen oxides, together with VOCs, are also precursors of ozone, which is a major constituent of photochemical smog (see Chapter 32). Photochemical smog occurs frequently in the majority of European cities. WHO computation of maximum potential for excess ozone exposure in urban areas indicates that up to 93 per cent of Europe's total urban population of 278 million was exposed in 1989 to levels above the 150 $\mu g/m^3$ (or 75 ppb) hourly limit value and up to 57 per cent to levels above 200 $\mu g/m^3$ (or 100 ppb).

The urban population in Europe is also exposed to varying concentrations of heavy metals and persistent organic pollutants. Although concentrations of heavy metals

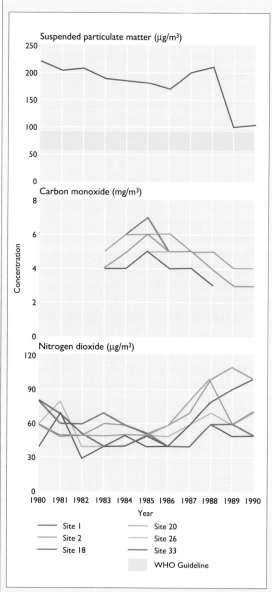

Box 10C Air quality in Moscow

Moscow is a good example of the changing air pollution patterns resulting from current transition in fuel consumption away from oil and coal to natural gas. Emissions of sulphur dioxide and suspended particulate matter (smoke) have dropped drastically in the last few years. However, measurements of ambient sulphur dioxide concentrations are not consistent with emission figures, which may be related to the sensitivity of the measurement methods. Smoke concentrations have decreased considerably in recent years, although mean daily concentrations of smoke still exceed WHO guidelines.

Lead concentrations in air do not exceed WHO guidelines and have decreased since control measures prohibited leaded petrol sales in Moscow. A decrease in carbon monoxide concentrations has been observed between 1975 and 1990. However, carbon monoxide maximum concentrations frequently exceed WHO guidelines. In addition, concentrations of nitrogen oxides have increased significantly between 1986 and 1990 as a result of increased vehicle traffic and power generation.

Source: UNEP/WHO, 1992

Figure 10.4 Annual mean concentration of smoke, CO and NO$_2$, Moscow, 1980–90

Box 10D Particulate pollution in urban areas: the case of Athens

The air pollution phenomenon in Athens is known as 'the *nefos*' (the cloud), a name which underlines its visible character. Visibility impairment during pollution episodes is due to high burdens of aerosol particles, and the yellowish-brown colour of the cloud is due to nitrogen dioxide.

The local monitoring stations measure concentrations of suspended particles by two methods: the black-smoke method, capturing carbonaceous particles mainly emitted by diesel vehicles and heating, and the 'total suspended particulates' (TSP) method taking into account both black and non-black (ie non-carbonaceous) particles. The latter are generated by industrial activities as primary aerosols, including dust, or formed in the atmosphere as secondary aerosols by chemical conversion of gaseous pollutants. These secondary non-black aerosols have the greatest effect on visibility reduction through scattering mechanisms.

Satellite multispectral images provide a general view of how the *nefos* spreads over the Athens basin and help explain the spatial distribution of particulate pollution at a single point in time. Such images illustrate that, during a typical autumn episode in the city, the main burden of non-black suspended particles is confined to the industrial district west of the city centre.

Figure 10.5 shows the relative concentration levels of the different particles monitored at two representative stations: in the city centre (Station V) and in the industrial area (Station R). Black-smoke concentrations are three times higher in the city centre, while TSP concentrations are slightly higher in the industrial area. This is because non-black and/or dust concentrations are much higher in the industrial area. The trends show that there has been a decline in black-smoke levels in recent years thanks to the control measures applied to vehicles and domestic heating. Such measures have little effect, however, on non-black particulate concentrations, which is why TSP levels are relatively stable.

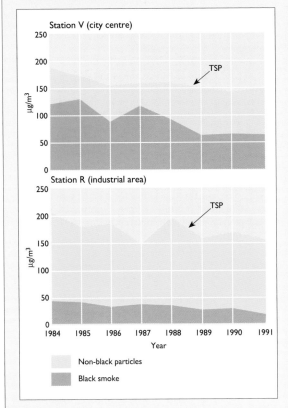

Figure 10.5 Athens: the nefos

Source: N Sifakis
Data sources: EURIMAGE, PERPA-Athens

(eg, lead, mercury, cadmium) are monitored in several European cities, measurements are incomplete. Even fewer data are available for trace organics (eg, benzene, benzo(a)pyrene (BaP), dioxins and furans, formaldehyde) that are measured in only a few cities. Existing data show contrasting patterns across Europe. Lead concentrations have substantially declined in urban areas in those countries (mostly in Northern and Western Europe) which have reduced the content of lead in petrol. Exposure to high ambient levels of lead, however, is still a problem in several Central and Eastern European cities, especially in mining towns. Estimates from model calculations (RIVM, 1992) suggest that benzene is likely to exceed acceptable limits in most European cities and BaP particularly in Eastern European cities (see also Chapter 4).

Urban air pollution can be exacerbated by local weather conditions and wind patterns. Some towns are situated in valleys where polluted air accumulates and directly affects the urban population. Prague is one example (see Box 10I, p278). Other examples are Sofia, which is situated in a basin attracting heavily polluted air from industries around the city, and Ljubljana, where, due to its specific meteorological situation, air pollution can reach levels three times above WHO standards. In Cracow (Poland), polluted air from the Nowa-Huta metallurgy complex located to the east of the city is drawn into the lower lying city. Levels of SO_2 are relatively high in Cracow particularly in winter, due to individual domestic heating installations fuelled with high sulphur-content coal.

Particularly relevant for its combined effect with air pollution is the phenomenon of the urban heat island – the air being warmer above cities than above the surrounding countryside. This urban heat island is generated by the urban structure with its dense built-up area and paved surfaces together with building and traffic heat losses. As a consequence, precipitation and evaporation patterns are heavily influenced above large cities (Box 10E).

Urban water

An assessment of the quantity and quality of Europe's water resources is presented in Chapter 5. Urban management of water supply and wastewater is treated in the 'Urban flows' section of this chapter. Threats to water resources become visible in cities, where the supply for drinking water, and for recreational purposes, is endangered. Cities affect and are affected by modifications of the hydrological regime, and changes in the groundwater table may affect the urban structure. Coastal cities are at an increased risk from coastal erosion and flooding. New threats for coastal cities derive from the potential impact of a possible sea-level rise related to climate change (see Chapter 27).

Surface water is an important visual feature, determining the character of most European cities and towns. But its importance is more evident when considering the many functions that water serves in the cities. Most European cities have grown by important waterbodies such as lakes, rivers, estuaries, and coastal waters. In all cities water is a vital economic resource in addition to its role for commercial traffic, transport and recreation. Surface water is also important as a habitat for wildlife and for its influence on the urban climate, helping to cool the air and stimulate air circulation.

Urban waterbodies in European cities are under pressure due to intensified expansion of built-up areas, uncontrolled use of land and water, and releases of pollutants. Sealing off land greatly accelerates runoff; extensive surface sealing causes cities to drain off up to 90 per cent of rainwater as opposed to 10 to 15 per cent in unbuilt areas (Miess, 1979). Following heavy rain episodes, severe contamination of the recipient waterbody can occur due to the combination of sewer overflows or by direct rain runoff through separate sewer systems. Effects can include: increasing bacterial concentrations, oxygen deficiency from the breakdown of excess organic matter, increased concentration of nutrients, heavy metals and polyaromatic hydrocarbons (PAHs), as well as general loss of aesthetic value. In spite of important progress achieved in reducing water pollution in several Western European cities, in most of Europe, especially in Eastern and Southern European countries, rivers crossing urban areas are in bad or poor condition.

The water quality of a river running through a city reflects the geological conditions and human activities taking place in the catchment upstream from the point of measurement. Depending on the location of the measurement point, the water quality reflects the pollution loads in the city itself in addition to those from activities taking place hundreds or more kilometres away from the point of measurement. Figure 10.6 summarises the water quality of 16

rivers running through some of the larger European cities. As far as possible each river monitoring site has been selected to represent the water quality of the river entering the city. Data are not available to assess the impact of cities on the river quality by, for example, comparing upstream and downstream concentrations. The quality indicators reported and the classification used have been chosen to describe river water quality in cities with respect to two widespread sources of river pollution in Europe: sewage discharge (eg, BOD, ammonium and total phosphorus) and pollution by agriculture (eg, nitrates). According to the classification adopted for Figure 10.6, it is evident that sewage pollution represents a bigger problem than does agricultural pollution. Organic matter content (BOD), ammonium and total phosphorus concentrations generally range from fair to poor, with the French river Rhone being the only notable exception having good quality. Nitrate conditions range from good to fair in the majority of rivers and concentrations are in all rivers below the maximum admissible concentration (50 mg/l) in water used for drinking water abstraction, but often exceed the threshold value for eutrophication.

Aquifers below many cities are threatened due to overexploitation and contamination of groundwater. In cities,

Box 10E The urban heat island

The process of urbanisation produces radical changes in the nature of the surface and atmospheric properties of a region. It involves the transformation of the radiative, thermal, moisture and aerodynamic characteristics, and thereby dislocates the natural solar and hydrologic balances. For instance, the dense urban construction materials make the system store heat and waterproof the surface; the block-like geometry creates the possibility of radiation trapping, air stagnation or undesirable increased wind speed at pedestrian level depending on the height-to-width ratio of buildings. The seemingly inevitable increase of air pollution affects the radiation balance and supplies extra nuclei around which cloud droplets may form.

It has been estimated that the air above cities such as Berlin or Brussels is on the average 2 per cent drier than in the surroundings in winter, and 8 to 10 per cent drier in

summer (Table 10.6). Compared with rural areas, clouds above these cities occur between 5 and 10 per cent more frequently, fog in winter twice as often, and in summer one third more. Rainfall is increased by 5 to 10 per cent, and there is 5 per cent less snow (Horbert et al, 1982).

The urban heat island, by which the air in the urban canopy is warmer than that in the surrounding countryside, is probably both the clearest and the best-documented example of inadvertent climate modification. The intensity of the urban heat island measured as the maximum temperature difference between city centre and the surrounding rural area is found to be proportional to the logarithmic of the urban population. In fact there is a better correspondence with the urban density expressed as the height-to-width ratio of the street canyons in the city centre. The commonly hypothesised causes of the urban heat island are illustrated in Table 10.7.

Table 10.6
Average changes of climate parameters in built-up areas
Source: Horbert et al, 1982

Climate parameters	Characteristics	In comparison to the surrounding area
Air pollution	Condensation	10 times more
	Gaseous pollution	5–15 times more
Solar radiation	Global solar radiation	15–20% less
	Ultraviolet radiation (winter)	30% less
	Ultraviolet radiation (summer)	5% less
	Duration of sunshine	5–15% less
Air temperature	Annual mean average	0.5–1.5°C higher
	On clear days	2–6°C higher
Wind speed	Annual mean average	10–20% less
	Calm winds	5–20% more
Relative humidity	Winter	2% less
	Summer	8–10% less
Clouds	Overcast	5–10% more
	Fog (winter)	100% more
	Fog (summer)	30% more
Precipitation	Total rainfall	5–10% more
	Less than 5 mm rainfall	10% more
	Daily snowfall	5% less

Altered energy balance terms leading to positive thermal anomaly	Features of urbanisation underlying energy balance changes
1 Increased absorption of solar radiation	Canyon geometry – increased surface area and multiple reflection
2 Increased long-wave radiation from the sky	Air pollution – greater absorption and re-emission
3 Decreased long-wave radiation loss	Canyon geometry – reduction of sky view factor
4 Anthropogenic heat source	Building and traffic heat losses
5 Increased sensible heat storage	Construction materials – increased thermal admittance
6 Decreased evapo-transpiration	Construction materials – increased 'water-proofing'
7 Decreased total turbulent heat transport	Canyon geometry – reduction of wind speed

Table 10.7 Possible causes of the urban heat island
Source: Oke, 1987

groundwater can help to conserve the urban fabric, particularly in areas of clay and peat where buildings need stilts (piles) with foundations in deep sand layers in the soil. Most of the buildings in some historic town centres in The Netherlands, Belgium, Denmark and Germany, but also in Italy, have wooden stilt foundations conserved in groundwater. In many places drainage systems are installed in order to stabilise groundwater levels (equally important for green spaces as well as for building foundations).

In cities with high groundwater levels, sewage pipes lie below the water table. When the sewage system is old or not well maintained, sewage pipes can serve as drains, lowering the groundwater table. In other cities intensive abstraction of groundwater for industry, and for the production of piped water for households, is responsible for the lowering of the water table. This can affect the stability of urban soils. It can also lead to the deterioration of wooden foundations, which require water for their conservation. Water table lowering may affect the stability of historic city centres, as it is occurring in Amsterdam and Venice. The complex relationships between cities and water are well illustrated by the case of Venice (Box 10F).

Acoustic quality

Many Europeans experience noise, caused by road-, rail- and air-traffic, by industry and recreational activities, as the main local environmental problem (see Chapter 16). The methods that are generally used to assess the exposure to noise in European urban areas are described in Box 10G. The findings of many studies undertaken in European countries on the effects of noise point out that, to ensure desirable indoor comfort, the outdoor level should not exceed a daytime L_{eq} of 65 dB(A) (OECD, 1991). In the case of new residential areas, the outdoor level should not exceed 55 dB(A). Urban areas with noise levels between 55 and 65 dB(A) are referred to as 'grey areas', while areas with levels in excess of 65 or 70 dB(A), which are perceived as annoying, are referred to as 'black spots'.

In large cities, the proportion of people exposed to unacceptable levels of noise is two to three times higher than the national average (OECD, 1992). It is recognised in many European countries that the percentage of population living in the 'grey areas' with noise levels between 55 and 65 dB(A) is increasing, and therefore noise has become a more significant problem than was predicted some years ago. The percentage of the urban population suffering from noise increases with the number of inhabitants. In West Germany, regular public opinion polls within the last ten years show that, in towns with up to 5000 inhabitants, 14 to 16 per cent of the population are strongly/severely annoyed by street noise. In towns with between 5000 and 20 000 inhabitants this percentage is 17 to 19 per cent, in cities with between 20 000 and 100 000 inhabitants, 19 to 25 per cent, and in cities of 100 000 and more inhabitants, 22 to 33 per cent (see, eg, UBA 1988).

Exceedances of the maximum acceptable level of 65 dB(A) occur in most cities – affecting between 10 and 20 per cent of inhabitants in Western Europe and up to 50 per cent in some cases in Central and Eastern Europe. The 70 dB(A) threshold, which is well beyond an acceptable level of noise, is also exceeded in 40 per cent of the cases where data were available. In Sofia 47 per cent of the urban population is exposed to more than 70 dB(A), as are 25 per cent in Budapest, 20 per cent in St Petersburg, 19 per cent in Cracow, 17 per cent in Dnepropetrovsk and 12 per cent in Bratislava. Some cities are 'better off': only 4 per cent in Copenhagen, 3.2 per cent in Kiev and 2 per cent in Amsterdam of the population are exposed to noise levels of 70 dB(A) (Figure 10.7). A larger number of people live in unacceptable acoustic conditions, if this is considered as a noise level above 65 dB(A). In Budapest this proportion is 50 per cent, in Prague 45 per cent and in

Dnepropetrovsk 33 per cent. Other cities where data are available are Amsterdam (19 per cent), Oslo (16 per cent), Bergen (12 per cent) and Bilbao (10 per cent).

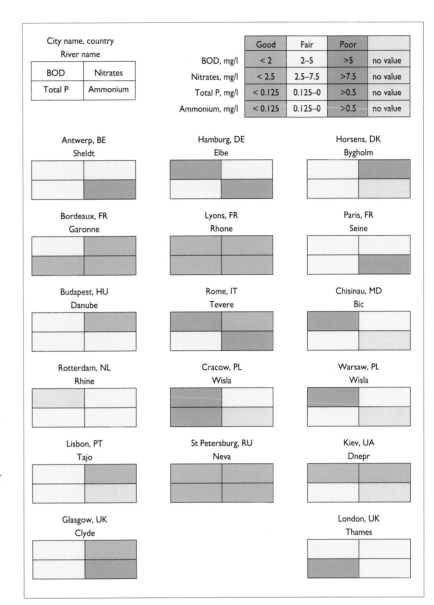

Figure 10.6
Water quality of large rivers in selected European cities
Source: Compiled from multiple sources, see *Statistical Compendium*

Figure 10.7 Acoustic quality in selected European cities
Source: Compiled from multiple sources, see *Statistical Compendium*

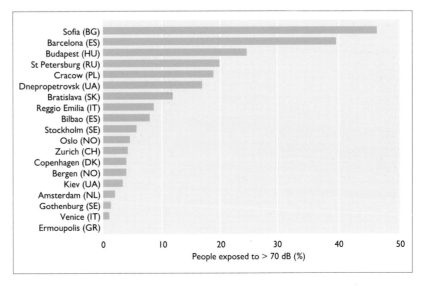

Box 10 F Venice, a city by the water

Venice is exposed to natural and artificial modifications of the environment at the interface between the Adriatic Sea and the large alluvial basin extending from the Alps to the coast. The combination of a fragile environment with its extraordinary architecture and historical value make Venice a unique case in some respects. However, Venice's hydrological problems are common to many other European cities near coastal waters. Venice is exposed to increasing erosion and flooding which affect the urban fabric. The delicate equilibrium of the lagoon is threatened by increased pollution. The impact of sea-level rise related to global climate change could be enormous.

From the beginning of this century the difference in level between land surface and water surface in Venice has changed by 23 cm. Sea level has risen approximately 11 cm while the land has sunk by 12 cm. On average, flooding now occurs 40 times a year, which is six times the average at the beginning of the century. In Venice, groundwater overexploitation has accelerated the process of salt water intrusion and subsidence, and successive human interventions are also modifying the lagoon bed. According to recent estimates, the lagoon is losing one million cubic metres of sediment a year. At this rate, there is strong concern that the lagoon could, in the future, gradually become a branch of the sea, unprotected from marine action.

Water pollution has also affected the biology of the lagoon. During the last decades, pollution of the water and sediments in the lagoon have reached unacceptable levels. The intense growth of algae that explodes nearly every summer and the decreasing biodiversity of the ecosystem are all symptoms of eutrophication caused by the enormous quantities of nutrients discharged into the lagoon. Since the beginning of the century organic pollution has tripled, and the nutrient load is five times higher. The lagoon receives each year between 10 000 and 12 000 tonnes of phosphorus and nitrogen. Urban pollution accounts for 15 to 20 per cent of the total. The industrial sector accounts for 20 to 25 per cent. However, the largest proportion (40 to 60 per cent) of the nutrient input comes from agricultural, industrial and urban activities taking place in the catchment of the lagoon.

These modifications affect Venice's urban environment and may compromise its future. In Venice, the degradation of historical buildings has increased as an effect of exposure to more frequent and severe storms. Subsidence, eustacy (change in sea level) and storm floods threaten the stability of the old city structure and cause the continuous deterioration of buildings. Pollution of water has substantially constrained its uses. Socio-economic factors, such as the significant change in activities and depopulation that occurred during the last decades, have contributed to the process of degradation. Indeed, accelerated degradation is also a result of inadequate maintenance of buildings and infrastructure.

During the last 30 years, alternative solutions have been studied and proposed for physical defence from flooding as well as for environmental clean-up and restoration. Plans for a project of defence by means of tide gates at three openings of the land barrier separating the lagoon from the sea were completed in October 1989 by the Consorzio Venezia Nuova. In addition, other important measures are under way, including: rehabilitation of lagoon morphology; environmental clean-up; restoration of the shore-line; building of a sewage and rainwater drainage system; and restriction of port traffic (eg, oil tankers). However, concrete measures have still to be taken. The mobilisation of the international community is also important to help protect one of Europe's most valuable historical cities.

Venice
Source: G Arici/Grazia Neri/Milano

Box 10 G Measuring noise in the urban environment

Quantitative method		Qualitative method:	Acoustical quality
Typical noise levels		landuse designation	
L_{95}/dB(A)	L_{eq}/dB(A)		

day	25	35	Areas designated for	Only natural sounds being
night	25	35	nature conservation	heard, like rustling of leaves
day	40	50	Areas designated for hospitals,	Very quiet rural environment
night	30	40	sanatoria, homes for the aged, 'soft' recreation	
day	45	55	Purely residential areas	Quiet suburban neighbourhood
night	30	40		
day	55	65	Mixed residential areas with	Typical urban noise situation
night	40	50	some light industry and civic centres	
day	65	75	Predominantly industrial areas	Very noisy, unfit for permanent
night	65	75		human habitation

Table 10.8 Two methods of assessing the acoustic environment
Source: A von Meier, 1993, and K M Müller (CEC DGXI)

The decibel scale of sound measurement is logarithmic and ranges from 0 (the threshold of normal human audibility) to 130 (the threshold of pain). A 10 dB increase in acoustic power corresponds to an approximate twofold increase in apparent loudness. The A-weighted sound level, indicated with dB(A), is the sound level approximating the frequency sensitivity of the human ear.

One method of measuring sound quality uses a quantitative descriptor, the background noise level. In addition, L_{eq} measures the average of different noise intensities over a defined period of time. The second method is based on a qualitative descriptor of the landuse designation of a specific area. While the use of a quantitative descriptor which can be objectively measured is preferable, it is not always practicable and is often substituted by a qualitative evaluation. Table 10.8 describes two assessment criteria in a sequence of degrading acoustic quality ranging from nature conservation areas to industrial areas.

For further details on the measurement of noise see Box 16B.

Source: von Meier, 1993

Box 10H Urban wildlife: fauna and flora

The heterogeneous structure of the city gives rise to a highly diversified mosaic of dispersed biotopes (Council of Europe, 1982). Cities support a relatively wide variety of plants and animals in comparison to the surrounding countryside. Some are specifically adapted to the urban climate, water and soils, and are an integral part of the urban landscape. More plant species can be encountered in urban spaces than on equally large surfaces of the surrounding areas. Of the approximately 80 to100 breeding species of birds found in any one area of Britain, for example, about 70 to 80 can be found in cities.

The variety of the urban wildlife is due to two main factors:

1 the greater diversity of the urban landscape, combining extreme ecological conditions;
2 the human introduction of non-native species, directly by domestication of plants and animals, and indirectly by transportation of seeds and insects by cargo and cars.

The number of native and alien species is related to the size of human settlements (Table 10.10). Birds, occupying a special position in urban ecological food-chains, require detailed and differentiated habitat conditions. The vegetation structure is undoubtedly the major factor affecting birds in towns, but pollutants such as heavy metals and PCBs also influence bird life. The danger is greater around industrial areas and high traffic lanes. Birds are exposed to direct and indirect effects of air and aquatic pollution through the food-chain. In this way therefore, the number of bird species living in towns and cities can act as an indicator of urban environmental quality.

Source: Falinski, 1971

Type of settlement	Native (%)	Alien (%)
Forest settlements	70–80	20–30
Villages	70	30
Small towns	60–65	35–40
Medium towns	50–60	40–50
Cities	30–50	50–70

Table 10.10 Proportion of native and alien species in villages and cities in Poland

	UK	France	The Netherlands	Switzerland
	m^2/inhabitant			
Sports	4	4	5	4
Playgrounds	1.5	1.5	3	1.5
Parks and public gardens	4	4.5	5	6
Private gardens	7		5	3

Table 10.9 Green space requirements for urban development projects
Source: Data from International Union of Architects

barriers and insulation, pedestrian zones, speed limits of 30 km/h, traffic flow control and noise-absorbent road surfaces.

Green areas

City planning has increasingly recognised the importance and varied functions of green space in urban areas. Standards have been established in several European countries to determine the requirements of green space in new neighbourhoods (Table 10.9). One of the most important functions of green areas in cities is recreation for the urban population. Another is optical dispersion and urban design. But green areas perform other important functions:

- improvement of the urban climate (air circulation, balancing of humidity, capture of dust and gases);
- recycling of organic waste (compost) and upgrading of wastewater (marshlands, fishponds, clarification ponds);
- physical exercise of citizens (walking, jogging, cycling, allotment gardening);
- nature conservation, education and research;
- a higher experience value of the urban environment (seasonal change, colours, odours).

Green space in European cities varies considerably in size and type as well as in distribution across the city spatial structure (Figure 10.8). Recent studies on the mental and social aspects of urban life show that the quality and accessibility of available green spaces in cities is far more important than their sizes. A walking distance of 15 minutes or less from homes to green space in urban areas is suggested as an indicator of urban environmental quality. In Brussels, Copenhagen, Glasgow, Gothenburg, Madrid, Milan and Paris, all the citizens live within 15 minutes walk from public green space. This is also the case in most smaller cities, such as Evora, Ermoupolis, Ferrara, Reggio Emilia and Valletta. In Prague and Zurich the corresponding figure is 90 per cent, in Sofia 85 per cent, in Bratislava 63 per cent, in Venice 50 per cent and in Kiev 47 per cent. In the majority of European cities,

Figure 10.8 Green space in selected European cities
Source: Compiled from multiple sources, see *Statistical Compendium*

Data required for a systematic analysis of noise exposure are poor in most cities. However, a recent OECD/ECMT survey on the perceived severity of urban problems, shows that 25 per cent of the cities ranked noise as a severe or very severe problem, and half indicated that noise levels were increasing (OECD/ECMT, 1993). In Europe it is estimated that 113 million people are affected by noise levels of over 65 dB(A) and 450 million people by noise levels of over 55 dB(A). Road traffic is the major offending source of noise in terms of the number of people disturbed. Second and third come neighbourhood and aircraft noise (see also Chapter 16).

In urban areas, tackling the problem of ambient noise is difficult because of the combination of different sources. A number of European countries including The Netherlands, Germany and Switzerland have adopted regulations and noise zoning which establish quality standards for different landuse designation. In Switzerland cities are required to apply specific standards including planning limits, ambient limits and alarm limits for four different zones. For example, a number of measures have been adopted by the city of Zurich to achieve targets. These include: rerouting traffic, quieting vehicles, noise

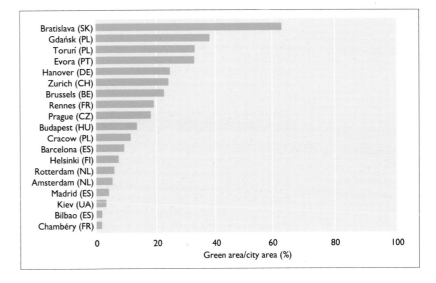

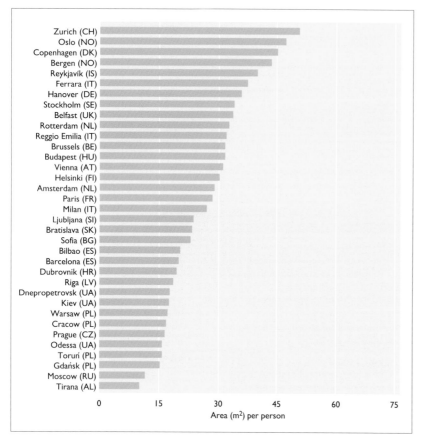

Figure 10.9
Housing floor space in selected European cities
Source: Compiled from multiple sources, see *Statistical Compendium*

most European cities. Housing conditions of marginalised groups – and often entire districts – in many cities are poor, and particularly so during periods of economic recession. However, such groups in Western Europe are still relatively well off when compared with those in Central and Eastern Europe.

Differences in housing standards are clearly evident when comparing Western European with Central and Eastern European cities. Housing quality, measured as average available area per person, is twice as high in Western Europe as in Central and Eastern Europe (Figure 10.9). Moscow has 11.6 m² net living space per person available, Bratislava 15.4, Prague 16.4, Kiev 17.2, Riga 18.3, Sofia 22.6 and Ljubljana 23.7 m². Paris has 28.2, Rotterdam 32.8, Copenhagen 45.1, Oslo 47.2 and Zurich 50.6 m² per inhabitant. The emergence of housing problems in Central and Eastern European cities is also evident when examining the number of housing units without basic services. Lack of basic sanitary requirements such as indoor piped water, flush toilet and fixed bath or shower is reported in many Eastern European countries (WHO, in press). Although rural dwellings are more likely to lack these requirements, the proportion of these dwellings in cities is still high. In Russian cities, for example, every fifth house is without running water, sewage and central heating. Even when the basic infrastructure is present, the state of housing stock in Eastern European countries is often reported to be in need of significant repair (Kosareva, 1993).

The design, layout and building materials are important elements affecting the quality of housing. Design and layout of buildings affect urban microclimate, natural elements and the urban landscape. Building technologies and materials are particularly relevant for the indoor environment. All these factors also determine buildings' environmental performance (eg, energy requirements). This latter aspect is treated further on in this chapter.

The quality of the indoor environment in cities is increasingly of great concern, as its link with human health has become evident (see Chapter 11). Public attention to indoor pollution is relatively recent, and little is known regarding the state of the indoor environment in European cities. However, it is increasingly recognised in most European countries that indoor pollution is an important area of inquiry, since the majority of the population spends most of its time indoors. Among the many pollutants found indoors are nitrogen dioxide, carbon monoxide, radon and asbestos, formaldehyde (from insulation), volatile organics, tobacco smoke and other numerous substances contained in consumer products. The concentrations of indoor pollutants are generally higher indoors than outdoors (Nero, 1988). Pollution from traffic may also have important effects on the quality of indoor air in urban areas (WHO/UNEP, 1990).

more than half of the population meet this criterion.

Size and location of green spaces in the city, together with type and frequency of their use, are important indicators of the quality of these spaces as biotopes. Large parks in the inner cities are habitats for a wide and diverse wildlife (Box 10H). The diversity of species between the countryside and the city has been attributed to the transition character of the biotopes which occur in these areas. Railway tracks and green lanes allow immigration and emigration of species across urban areas. Vacant land in the inner city is relevant for its typically urban wasteland ecosystems which include a surprisingly high diversity of species adapted to urban environmental conditions and pressures.

Biotope mapping has been carried out in a number of European cities. The municipality of Lidingö in Greater Stockholm has carried out a comprehensive project to economise use of land and water in the built-up area and achieve a better integration of natural elements in landuse planning. In addition to mapping all vegetation, 'ecological corridors' were marked to connect the various biotopes in the town with reproduction areas outside the town. Other examples in Sweden emphasise that biotope protection in cities can be achieved only if the urban infrastructure is integrated with the green infrastructure. Several other examples exist in European cities. However, no systematic monitoring is carried out to assess natural biotope loss due to increased demand for built-up areas or caused by disturbances such as air pollution and noise.

Housing

Social and economic changes which have accompanied the urbanisation process in Europe are reflected in the housing quality within European cities. During the last century, large sectors of the Western European population have experienced a general improvement in living standards reflected for example in the increase in the average living space per capita.

Simultaneously with this trend, rapid demographic changes in European cities were a cause of increasing marginalisation of urban population groups and housing problems. High standard districts together with very poor housing conditions coexist in

Road traffic

Cars dominate the public spaces of European cities. According to a recent OECD/ECMT study on urban travel in 132 cities, there are at least three reinforcing trends which make urban mobility one of the major environmental problems facing cities. These are: increasing levels of car

Road traffic, Milan Source: Michael St Maur Sheil

ownership; modal transfer from public transport to car; and the decentralisation of population and employment from inner to outer areas (OECD/ECMT, 1993).

In addition to traffic-related air pollution and noise problems, most large European cities share growing concern for traffic congestion and accidents which are increasingly deteriorating the quality of life of urban communities. Comparable data on traffic congestion are difficult to obtain and have not been included among the indicators selected here. Data from the OECD/ECMT study provide an overview of the general trends. According to OECD estimates vehicle speeds declined by 10 per cent over last 20 years in major OECD cities. Congestion in Paris in the same period increased by 10 per cent per annum. Since 1971, inner London transport speeds have fallen to less than 18 km/h, and on all London roads the average speed is less than 26 km/h. Speeds, and spatial and time distribution in OECD cities suggest that congestion is particularly significant in the morning peak compared to the inter-peak period. Likewise, speeds are lower in the central business district (CBD) area than in the built-up area (Table 10.11) (OECD/ECMT, 1993).

While important progress in the implementation of traffic safety measures has been achieved in European cities, traffic-related accidents remain a prominent urban concern. Over the past 20 years in EU countries alone more than a million people died in traffic accidents and more than 30 million were injured and/or permanently handicapped. This corresponds to an annual average of 55 000 people killed, 1.7 million injured and 150 000 permanently disabled due to traffic accidents. Expenditure for related health care figures prominently in public budgets.

In European countries, between a quarter and a third of casualties in built-up areas arise from accidents on residential streets. Pedestrians account for 20 to 60 per cent of victims, depending on cities and countries. In London in 1991, for example, 59 per cent of the victims were pedestrians (LRC, 1993). Children are particularly vulnerable. In Brussels, traffic accidents cause three times as many victims among the 5 to 9 age group than among the population as a whole (Région de Bruxelles Capitale, 1993).The number of people killed or injured in traffic accidents in selected cities is given in Figures 10.10 and 10.11. High figures are reported in Paris, Bordeaux, Milan and Reggio Emilia.

Improvement of public transport plays a role in reducing car use and accidents. Many Western European cities have tried to improve road safety through a variety of measures. Speed limits, improved design of road environments and traffic calming have helped considerably. In a limited number of cities in Western Europe, public transport systems are promoted. Examples are: integrated networks of rapid trams, subways, buses and trains as introduced in urban regions in Germany, Switzerland and Scandinavia. Several instances of good public transport can be found in Central and Eastern European cities; one is Prague.

In Denmark, France, Germany, Sweden and The Netherlands bicycling in cities has been promoted since the early 1970s. In relatively flat towns, over 40 per cent of journeys to work can be made by bicycle (IIUE, 1992). With appropriate safety measures, bicycling in cities can contribute to a substantial drop in the number of traffic accidents. In Sweden, for example, bicycle travel increased by one third in ten years, and at the same time the fatal casualty rate was reduced by more than 50 per cent, and the number of deaths by 40 per cent. However, bicycling still remains dangerous in cities with a large number of cars. In The Netherlands, where every citizen has one or more bicycles, a bicyclist is still five times more likely to be involved in a traffic accident than a car driver. Public transport is relatively safe. Public transport passengers have only a 10 per cent chance of being involved in an accident, compared to car drivers. The automobile causes most accidents. Compared to a bicycle, a car causes 20 times more accidents per passenger, while public transport causes

km/h	Built-up area		Central business district	
	am peak (%)	inter-peak (%)	am peak (%)	inter-peak (%)
<19	13	2	38	19
20–29	29	18	41	35
30–39	24	20	17	17
40–49	25	20	2	17
50+	9	40	2	12

Table 10.11
Number of cities in different speed bands, 1990
Source: OECD, 1993

only 5 times more (IIUE, 1992).

Due to the increasing numbers of cars, the effect of measures to promote traffic safety remains limited. Some cities in Europe are anticipating more drastic solutions, for instance by promoting 'Cities without cars'. A 'car-free cities' network promoted by the European Commission has been created recently in Europe to exchange experience and undertake common actions to reduce the use of cars in cities (Car Free Cities Charter, 1994).

Cityscapes

Cityscapes are the visual scenery of urban life. They consist of the architecture and archaeology of built-up areas, but also of the building design, layout and the presence and integration of natural elements into built-up and open areas. They are shaped by the cultural and historical development of individual cities and by their social and economic developments. Planning plays a critical role. Cityscapes are affected by urban density,

Figure 10.10
People injured in traffic accidents in selected European cities
Source: Compiled from multiple sources, see *Statistical Compendium*

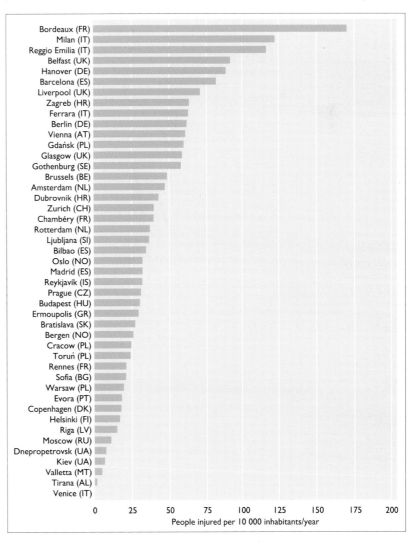

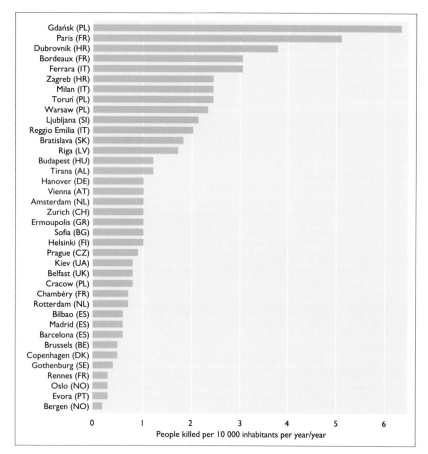

Figure 10.11
People killed in traffic accidents in selected European cities
Source: Compiled from multiple sources, see *Statistical Compendium*

Battersea and Albert Bridges, London
Source: H Girardet

protecting skylines in several European cities (such as London, Liverpool, Paris and Munich). In London, for example, ten strategic views of the skyline of central London (including St Paul's Cathedral and the Palace of Westminster) are safeguarded: this may be increased to 12 views in the near future (LPAC, 1993a, 1993b).

URBAN FLOWS

The impact of cities on the global environment can be described by analysing the flows of natural resources that support their activities. Cities import vast quantities of oxygen, water and organic matter (in the form of fossil fuels and food), and release corresponding quantities of emissions and waste. The land surface required to support any given city and to absorb its waste is many times greater than the city itself. The concept of a city's 'ecological footprint' on the Earth has been proposed to measure the aggregated land area required to support various urban communities (Rees, 1992). The total energy, water and materials which enter a city and the amount of emissions and waste released to the environment depend on city size and average living standards of its inhabitants, but also on the capacity of cities to exploit economies of scale provided by urban concentrations.

Three indicators of urban flows are given in Table 10.13. Particular caution should be exercised when analysing this information due to comparability problems (Table 10.4). Several case studies have been used, as indicated in the following sections, to compensate for the lack of data and to confirm the interpretation of available information. Energy, water and material flows are treated separately in the following sections by examining their impacts and options for an efficient management. As an example, a more comprehensive picture of urban flows in the city of Prague is provided in Box 10I.

Few comparable data on energy consumption are available at the urban scale. It is even more difficult to obtain disaggregation by fuel type or end use. Electricity consumption per capita in cities where data are available average 5700 kWh per year, ranging from 500 to 23 000 (see *Statistical Compendium*). The data show that electricity consumption per capita is clearly influenced by climate patterns, but provide some indication of better performance (eg, for Helsinki). The lack of data on wastewater, waste composition and waste disposal facilities have led to the omission of these indicators from Table 10.13. Except in a few cases (Bergen, Milan, Oslo, and Reykjavík), water consumption per capita is clearly higher in Central and Eastern European cities. The large variation from the range of 100 to 400 litres per capita per day may be due to differences in measurements, rather than reflecting major differences in consumption levels. One third of European cities produce annually more than 600 kg of waste per capita.

land-use and infrastructure, and by the degree of integration or separation of urban activities in the various districts. These elements have created unique environments in European cities with their own image and identity. Urban change has affected the visual scene of European cities during rapid growth and decline, suburbanisation and reurbanisation. Loss of historical and natural values, as a result of uncontrolled urbanisation and redevelopment in European cities, is an increasing concern. Although many European countries have systems for listing historical buildings, quantification of the loss of the historical fabric in cities at the European scale is difficult due to different listing systems and criteria.

Perhaps the key elements that contribute to cityscape are the impact of buildings beyond the 'human-scale' and the impact of infrastructure, especially that brought about by the private car. In many cases, there are management responses to preserve certain components of cityscapes: traffic management schemes are probably the most widespread. Reducing car traffic in towns through traffic calming, pedestrian zones and encouraging public transport, have brought tremendous benefits to town centres (eg, improved environmental quality, reduced social stress, improved commercial turnover and so on) (Friends of the Earth, 1992). Another example has been the development of strategic planning guidelines for

Table 10.12
End-use energy consumption in selected European cities
Source of data: ICLEI, 1993; LRC, 1993; IBGE, 1993

City	Residential (%)	Commercial (%)	Industrial (%)	Transport (%)	Total (GJ/capita)
Berlin	33	29	15	23	78.10
Bologna	36	21	11	32	67.30
Brussels	43	29	5	23	94.70
Copenhagen	48	26	6	20	78.16
Hanover	28	25	26	21	112.43
Helsinki	34	23	9	34	89.50
London	36	24	11	29	89.10

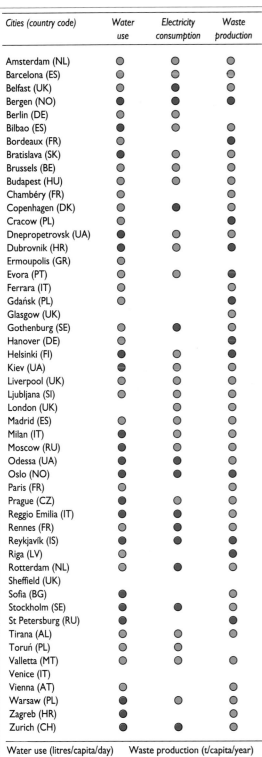

Cities (country code)	Water use	Electricity consumption	Waste production
Amsterdam (NL)	◐	◐	◐
Barcelona (ES)	◐	◐	◐
Belfast (UK)	◐	●	◐
Bergen (NO)	●	●	●
Berlin (DE)	◐	◐	
Bilbao (ES)	●	◐	○
Bordeaux (FR)	◐		●
Bratislava (SK)	●	◐	◐
Brussels (BE)	◐	◐	◐
Budapest (HU)	◐	◐	◐
Chambéry (FR)	◐	●	◐
Copenhagen (DK)	◐	●	◐
Cracow (PL)	◐		◐
Dnepropetrovsk (UA)	●	◐	◐
Dubrovnik (HR)	●	◐	●
Ermoupolis (GR)	◐		
Evora (PT)	◐	◐	●
Ferrara (IT)	◐		◐
Gdańsk (PL)	◐		◐
Glasgow (UK)			◐
Gothenburg (SE)	◐	●	◐
Hanover (DE)	◐		◐
Helsinki (FI)	●	◐	◐
Kiev (UA)	●	◐	◐
Liverpool (UK)	◐	◐	◐
Ljubljana (SI)	◐	◐	◐
London (UK)		◐	◐
Madrid (ES)	◐	◐	◐
Milan (IT)	●	◐	◐
Moscow (RU)	●	◐	◐
Odessa (UA)	●	●	◐
Oslo (NO)	●	●	●
Paris (FR)	◐	◐	◐
Prague (CZ)	●	◐	◐
Reggio Emilia (IT)	●	◐	◐
Rennes (FR)	◐	●	◐
Reykjavík (IS)	●	●	●
Riga (LV)	◐		●
Rotterdam (NL)	◐	●	◐
Sheffield (UK)			
Sofia (BG)	●		◐
Stockholm (SE)	●	●	◐
St Petersburg (RU)	●		●
Tirana (AL)	◐	◐	◐
Toruń (PL)	◐	◐	
Valletta (MT)	◐	◐	◐
Venice (IT)			
Vienna (AT)	◐		◐
Warsaw (PL)	●	◐	◐
Zagreb (HR)	●		◐
Zurich (CH)	●	●	◐

Water use (litres/capita/day)	Waste production (t/capita/year)
◐ <150	◐ <0.4
◐ 150–300	◐ 0.4–0.6
● >300	● >0.6

Electricity consumption (kWh/capita/year)

◐ <3000
◐ 3000–6000
● >6000

Note: Greater electricity consumption is not necessarily related to greater environmental impact. Cities ranked as high electricity consumers may indicate high electricity consumption or a greater use of electricity in the share of energy sources.

Table 10.13 Urban flow indicators
Source: Compiled from multiple sources, see *Statistical Compendium*

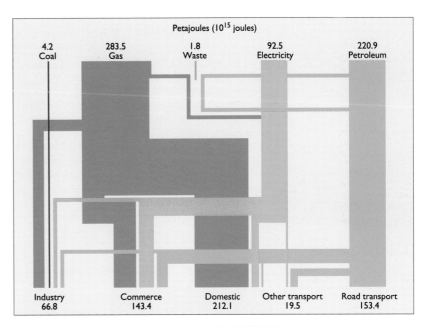

Energy flows

The main source of energy entering the city is the sun, but cities have still very limited technological capability to capture and transform it so that people and machinery can use it.
A large part of the solar energy is immediately radiated out of cities. Only a small part is captured by urban vegetation or absorbed in soils and green spaces. Taking Barcelona as an illustration, the relative proportions of energy flows into the city can be measured (Pares et al, 1985). In Barcelona one third (31.6×10^{12} kcal/year) of the total solar energy entering the city is absorbed by vegetation. About two thirds falls on buildings (71.4×10^{12} kcal/year), where it is absorbed in cool periods and/or radiated out into the air in warmer times. In addition to natural energy radiation, 15×10^{12} kcal of combusted fuels and 3×10^{12} kcal of energy in the form of food are imported each year into Barcelona. Human-induced energy in Barcelona accounts for about one fifth of the total quantity of the solar energy flux.

Primary sources of energy used in European cities are fossil fuels, including petroleum products, gas and coal.

Figure 10.12
Energy flow in London, 1991
Source: LRC, 1993

Figure 10.13
End-use energy trends in London, 1965–91
Source: LRC, 1993

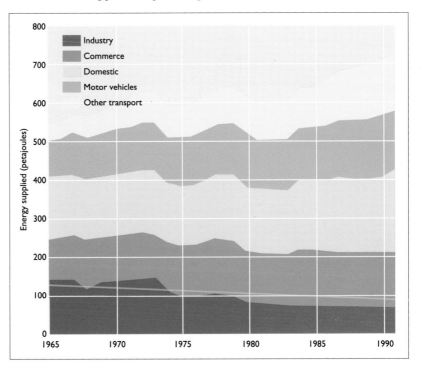

Box 101 Metabolism of a city: Prague, Czech Republic

Prague, the capital of the Czech Republic, is one of Europe's most beautiful historical cities and part of national and European cultural heritage. With its 500 km² and a population of 1.2 million, Prague is the largest urban area in the former Czechoslovakia. Unfortunately, it is also a city suffering from serious environmental problems.

Prague's urban metabolism is summarised in Table 10.14. Besides features common to all cities, Prague has some specificities which concern especially the structure of sources of energy and which have important environmental consequences. The total input of anthropogenic energy contained in delivered fuels and electricity to the territory of

Map 10.2
Landuse in Prague
Source: City of Prague

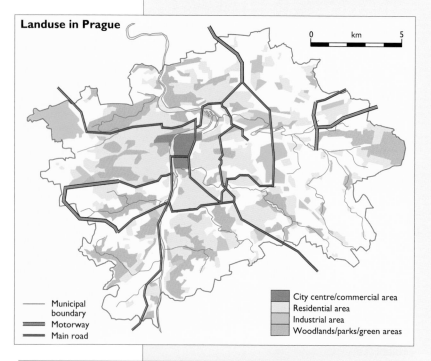

Landuse in Prague

0   km   5

— Municipal boundary
═══ Motorway
━━ Main road

■ City centre/commercial area
□ Residential area
▨ Industrial area
▨ Woodlands/parks/green areas

Area	494.8 km²
Population	1 212 010 (1991)

Consumption of materials (tonnes per year)
(approximate data for 1985)

Excavated rock and soil	11 000 000
Sand, gravel, building stone	5 900 000
Asphalted gravel for roads	1 800 000
Bricks	260 000
Other construction materials	8 600 000
Iron, steel	4 200 000
Other metals	35 000
Paper	300 000
Chemicals	160 000
Paints	70 000
Plastics	55 000
Food including packaging	4 100 000
Beverages	500 000

Consumption of energy (TJ per year)

Type of fuel or energy	Delivery on the territory of the city	Consumption after transformations
Coke	10 665 (11.3%)	10 665 (13.3%)
Other solid fuels	18 360 (19.4%)	8 110 (10.1%)
Liquid fuels	13 087 (13.8%)	8270 (10.3%)
Natural gas	37 735 (40%)	19 122 (23.9%)
Electricity	14 578 (15.4%)	13 394 (16.7%)
District heating (centralised heating)	–	20 566 (25.7%)
Total consumption	94 425 (100%)	80 127 (100%)

Transport

Number of all motor-vehicles	430 000
Number of passenger cars	340 000
Number of people per car	3.6

Transportation mode – percentage of journeys done by public transport in an average working day:
- for all intra-urban journeys	75%
- for all extra-urban Journeys	58%

Percentage of transport means in the public transport:
- underground railway	36%
- tram	33%
- bus	31%

Road network:
Total length in the city	2 700 km
Density	5.4 km roads per km² urban territory

Landuse (data for 1980)
(% of the city)

Residential	17.3
Industrial	4.5
Agricultural	39.6
Green spaces	18.7
Sport and recreation	1.4
Public amenties (culture, education, commerce, services, health and social care)	3.3
Transport and technical facilities	7.9
Water reservoirs, rivers, ponds	1.7
Other	5.6

Charles Bridge, Prague
Source: H Girardet

Table 10.14 Selected data, City of Prague, 1991
Source: City of Prague, personal communication

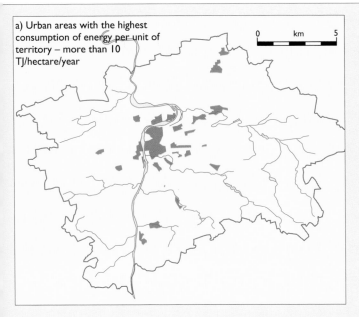

a) Urban areas with the highest consumption of energy per unit of territory – more than 10 TJ/hectare/year

0 km 5

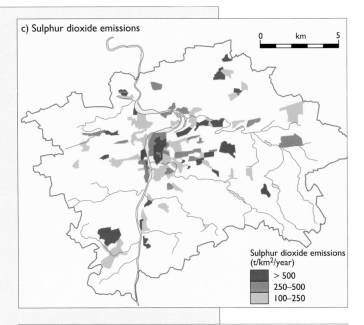

c) Sulphur dioxide emissions

0 km 5

Sulphur dioxide emissions (t/km²/year)

> 500

250–500

100–250

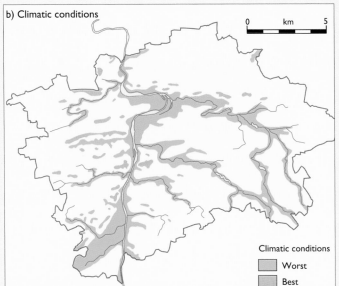

b) Climatic conditions

0 km 5

Climatic conditions

Worst

Best

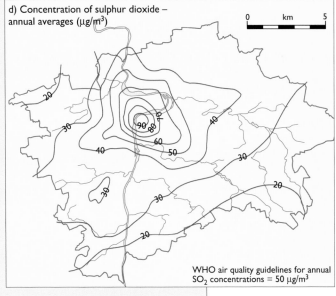

d) Concentration of sulphur dioxide – annual averages (µg/m³)

0 km 5

WHO air quality guidelines for annual SO₂ concentrations = 50 µg/m³

Prague is about 94 500 TJ (10¹²J) per year. The final consumption of energy after various transformations (for instance burning of solid and liquid fuels in the heat-and-power plants and boiler stations for district heating systems) is about 80 200 TJ per year. The differences are losses during transformations.

There are more than 200 large emission sources in Prague (large heat-and-power plants and boiler stations with capacity over 5 MW), nearly 10 000 medium sources (boilers in about 5000 small heat plants and boiler stations with 0.2 to 5.0 MW capacity supplying one or several buildings) and more than 200 000 local heating devices (stoves). In addition to these stationary sources, there are about 430 000 motor vehicles in Prague including 350 000 private cars.

The delivery of fuels into the city and their subsequent conversion to heat and energy using rather outdated technical infrastructure have very undesirable effects on the quality of the environment. Total annual emissions from all stationary and mobile sources of pollution in Prague are approximately 56 200 tonnes of sulphur dioxide, 24 000 tonnes of nitrogen oxides, 24 000 tonnes of particulate emissions (dust, fly-ash), 46 200 tonnes of carbon monoxide and 12 000 tonnes of hydrocarbons. The density of emissions in Prague (in tonnes per km² per year) is several times higher than in the districts of heavily polluted North Bohemia.

The geomorphology of Prague – characterised by the valleys of the Vltava (Moldau) and its tributaries and a central kettle-like depression surrounded by hills with total altitude differences of more than 200 metres – provides specific microclimatic conditions not favourable to the ventilation and dispersion of air pollution. Thermal inversions occur very often in the lower layers of the atmosphere, and, together with typically low wind velocities, provide the basic conditions for smog episodes, especially during winter. Critical situations occur in the central part of the city where there are thousands of small and medium ground-level emission sources burning solid fuels (very often brown coal of low quality). The ambient air quality is generally poor and health standards are very often breached. Sulphur dioxide concentrations reach the highest levels in the historical core of Prague during the heating season, while nitrogen oxides and airborne aerosol (dust) are more equally distributed over the year.

The characteristics which control the spatial distribution o annual average concentrations of sulphur dioxide in ambient air in Prague (Map 10.3d) are illustrated in Maps 10.3a, b and c. The highest concentrations are, as a rule, in the historical core of Prague, the 'Old Town'. This is a result of the combination of geomorphology, climate, energy consumption patterns and density of emission sources.

Map 10.3a – d Spatial distribution of annual average concentrations of SO₂ in ambient air in Prague
Source: City of Prague, personal communication

The urban areas with the highest consumption of anthropogenic energy per unit of territory are shown in Map 10.3a. The red areas, mostly in the centre of Prague, show energy consumption above 10 TJ per hectare per year (the map may also be seen as one of 'energy consumption density' or 'intensity of energy metabolism'). Since the annual input of solar radiation in Prague is about 30 TJ per hectare per year, the marked areas have an annual consumption of anthropogenic energy equal to or larger than one third of annual solar energy input. During winter months the input of anthropogenic energy in such areas is several times higher than the solar one.

The production of waste heat together with accumulation of solar radiation energy in buildings and other urban structures creates the 'urban heat island' effect (see Box 10E) which is typical of the urban climate. The red areas on the map can be called 'the heart' of the urban heat island in Prague.

Map 10.3b shows the natural climatic conditions in Prague classified as best or worst zones for insolation, ventilation and occurrence of thermal inversions. The worst climatic conditions are at the bottom of the Vltava valley and include the lower parts of the historical core of Prague (protected by law as the Prague urban reserve).

The map of sulphur dioxide (SO_2) emission density, expressed in tonnes per square kilometre per year (Map 10.3c) shows very clearly that the centre of the city is heavily burdened, because of the great number of small and medium sources located there. Other areas of the city with high density of sulphur dioxide emissions are those where large industrial plants are located.

The annual average sulphur dioxide concentrations in Prague in 1990 varied between 31 and 110 $\mu g/m^3$ and the maximum concentrations between 184 and 681 $\mu g/m^3$. The relative number of measurements exceeding the daily health standard of 150 μ/m^3 was between 0.3 and 27.1 per cent. The results for nitrogen oxides were even worse, with average annual concentrations of 37 to 134 $\mu g/m^3$ and the relative number of measurements above the daily recommended limit of 100 $\mu g/m^3$ from 5.2 to 61 per cent. Although nitrogen oxides as well as airborne dust have their maximum concentrations also in the inner city, they are more influenced by other sources. For nitrogen oxides, automobile traffic is an important source.

Air pollution in Prague causes high economic losses. The annual losses due to accelerated corrosion, increased occurrence of diseases in Prague inhabitants and damage to vegetation were estimated to be almost 2000 million Czech crowns ($70 million).

The solution of the problems caused by energy conversion and use in Prague faces no technical obstacles. The efforts fail because of lack of finance and because of serious shortcomings in organisation at all levels. Moreover, the necessary economic instruments for pollution abatement are still missing.

The major goal is to switch energy sources in the city from solid and liquid fuels to electricity, natural gas (with proper low-emission combustion technologies) and district-wide central heating. The priority is being given to the historical core of Prague. New legislation (Air Protection Act from 1991) should help state and municipal administration in these efforts.

Source: City of Prague

Electricity is mainly derived from fossil fuels or produced by nuclear installations and hydropower plants; in rare cases it is generated also from geothermal power. Some small settlements use biomass from natural systems in the immediate vicinity. In some cities, advanced systems are in use to derive energy from waste: gas from waste dumps or through fermentation in waste installations; heat from incineration. The use of wind and sun energy (active through solar panels and photovoltaic cells or passive through appropriate architecture and urban design) is still very limited.

Energy consumption in cities varies with differences in climate, degree of urbanisation and industrial structure. Recent studies have explored the relationships between per capita energy consumption and patterns of landuse, transportation modes, building technologies and lifestyle. Compared with North American cities, European and Japanese cities are much more energy-efficient (Lowe, 1991), often explained by their more compact urban system and density. However, important differences can be also observed between European cities. Although comparable data on energy consumption by sectors and by fuels are not generally available, a number of cities, such as Berlin, Bologna, Brussels, Copenhagen, London, Helsinki and Hanover, have charted urban energy flows and their environmental impact (Table 10.12, p 276). As an example, the energy flow of London is illustrated in Figure 10.12. From the examples of these cities, consistent patterns and trends emerge. The domestic sector is the largest consumer of energy (from 48 per cent in Copenhagen to 28 per cent in Hanover – see Table 10.12). This is in large part due to heating and cooling of residential buildings. A main area of growth in domestic energy consumption is appliances, such as driers and dishwashers. Transport is the second most important sector in terms of urban energy consumption, even if considerable variations can be found between European cities. Commercial and industrial energy consumption vary considerably according to the predominance of economic activities.

Total energy consumption in cities has increased following the general trend in European countries (see Chapter 19). However, different patterns can be analysed according to different sectors. Industrial use of energy in Western European cities has declined considerably in the last 30 years, due to urban de-industrialisation and an increase in efficiency. In London, for example, the industrial sector accounts for 11 per cent of total energy consumption, which is 15 points lower than it was in 1965 (26 per cent) (Figure 10.13). In contrast to this trend is the increasing importance of transport; in the group of cities mentioned above this accounts for 20 to 33 per cent of total energy consumption. Transport in London accounts for 29 per cent of energy use, which represents 11 points growth from 1965 (18 per cent). This is particularly due to greater mobility of citizens and the increased use of cars (LRC, 1993).

Energy efficiency in cities requires the proper allocation of energy supply to the various end uses. District heating or cogeneration (combined heat and power) are important means to improve the urban environmental energy balance, but this potential for energy saving is not yet fully exploited in Europe. Rapid progress is being made in Belgium, Denmark, Finland, Germany and The Netherlands. Cogeneration contributes significantly to the improved energy efficiency in Helsinki (Box 10J). In Copenhagen, the district heating network covers two thirds of the heating requirements in the municipality and it is planned to achieve 95 per cent by the year 2002. However, the proportion of district heating in European cities varies considerably (Nijkamp and Perrels, 1989).

Box 10J Energy balance in Helsinki

Cities can implement measures to control energy flows; increasing efficiency, and thus conserving energy sources and reducing environmental impacts. In Helsinki (Figure 10.14) the total net efficiency (excluding motor vehicles) provided by the Helsinki Energy Board in 1993 was 82 per cent. Including motor vehicles, it becomes lowered to 73 per cent, which shows the low energy efficiency of motor vehicles compared with heating and electricity supplies. Of the total process losses (4061 GWh per year), more than half (2016 GWh per year) are due to motor vehicles which account for a quarter of total energy used in traffic. The remaining losses are from electricity generation and heating production which totals 1630 GWh per year and accounts for only one sixth of total production.

Cogeneration of heat and electricity at an urban scale is economically advantageous and saves raw energy. By means of cogeneration Helsinki has been able to save 25 to 35 per cent of fuel compared with the earlier separate production of the same amount of heat and electricity (in condensing power plants and private boilers). An additional advantage is reduced dependence on oil. The fuel in thermal power plants in Helsinki is now mainly coal, in heat supply stations it is mainly heavy fuel oil, and in boilers for single houses it is light fuel oil. In the near future more natural gas will be used. The level of emitted carbon dioxide decreases considerably with combined energy production and with increased use of natural gas.

Source: Helsinki Energy Board, 1993

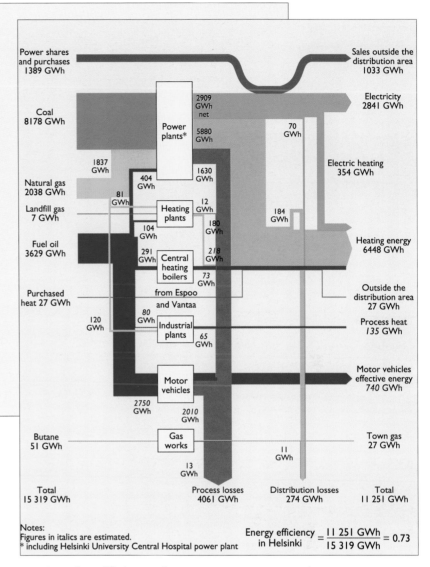

Notes:
Figures in italics are estimated.
* including Helsinki University Central Hospital power plant

$$\text{Energy efficiency in Helsinki} = \frac{11\,251\ \text{GWh}}{15\,319\ \text{GWh}} = 0.73$$

Figure 10.14
Helsinki energy flow
Source: Helsinki Energy Board, 1993

Environmental impacts

The impact of urban energy flows on the local, regional and global environment depends on the total energy demand from various activities, and also on fuel mix and end uses. This can be seen by examining the energy flows in London according to different emissions of air pollutants and end uses (Figure 10.15). Different fuels emit very different amounts of pollutants, so that the fuel mix may offset lower levels of energy consumption when comparing energy flow impacts between different cities. Lower levels of energy consumption per capita in some cities do not necessarily mean that these cities release fewer pollutants into the air. Likewise, the environmental impacts of various urban activities are dependent on the allocation of energy supply.

Local impacts of urban air emissions from urban energy production and consumption for domestic heating have been significantly reduced in most European cities as a result of switching fuels and the introduction of more efficient technologies. In most Central and Eastern European cities, however, power plants and domestic heating are still major causes of local problems. In addition, increasing production of nitrogen oxides as a consequence of increased car use in both Western and Eastern countries is now a major concern because of local and regional impacts (see Chapters 4, 31 and 32).

Cities produce a considerable amount of greenhouse gases which contribute to climate change. An overview of carbon dioxide (CO_2) emissions by sectors in a number of European cities is given in Figure 10.16. Major contributions derive from road transport. Second and third rank the residential and commercial sectors. Emission of CO_2 per capita in European cities is relatively low compared to North American cities. This is explained by the higher population densities and the increasing use of district heating in Europe. However, the difference in the level of CO_2 emissions per capita in European and North American cities is not as large as one would expect from the difference in energy use per capita. This could be explained by the higher CO_2 intensity of the fuels used in European cities (ICLEI, 1993). This suggests that considerable room for reduction in CO_2 emissions exists even at the same level of energy consumption.

Options for efficient urban energy management

Cities provide a wide range of opportunities for improving energy efficiency and reducing the impacts of their energy consumption. Some energy management measures apply specifically to cities because of their high concentration of activities. They include:

- energy conservation (minimising end use by energy-saving devices, insulation, and by information to the public);
- use of renewable energy sources (passive solar energy in urban design, wind and water, biogas);
- increased energy efficiency (district heating, cogeneration, public transport instead of individual passenger cars, use of waste heat from industry);
- fuel switching and technological innovation (natural gas instead of coal and oil, burning of municipal waste).

Water flows

Urban water flows consist of natural flows (surface water, groundwater, precipitation and evaporation) and artificial flows (piped water and drained water). Cities rely on piped water for their water supply. The average per capita water supply in European cities is 320 litres per day, ranging in most cities between 100 and 400 litres per day (Figure 10.17). However, in several cities, especially in Central and Eastern Europe, including Prague, Moscow, Zagreb and Warsaw, reported

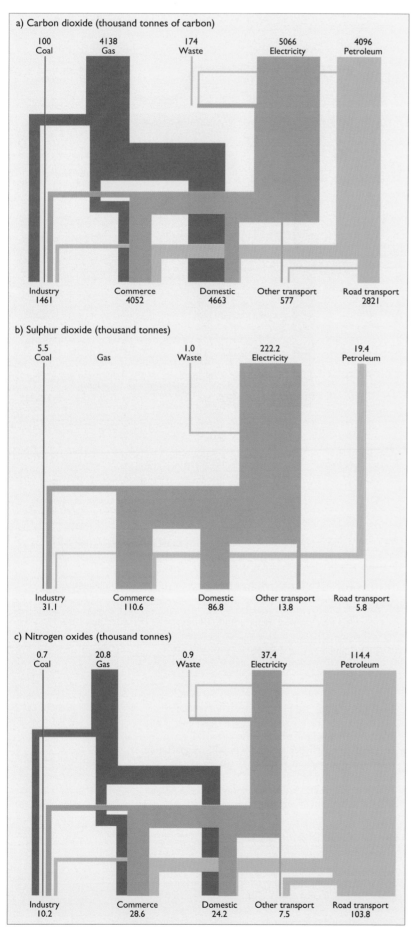

Figure 10.15
Emissions from energy consumption in London, 1991
Source: LRC, 1993

figures are much higher. Higher values are also reported in Nordic cities such as Oslo (663), Reykjavík (580), and Bergen (584). However, high figures in several cities suggest uses for purposes other than domestic supply (eg, for industry).

The use of groundwater for drinking water purposes generally exceeds other sectoral uses. For Europe as a whole, groundwater supplies on average 65 per cent of drinking water. In recent years a tendency towards more intensive use of surface water for drinking water supplies has been observed. This is partly due to the easier accessibility of surface water and partly to the widespread deterioration of groundwater reservoirs (see Chapter 5).

During the International Drinking Water and Sanitation Decade (the 1980s) an analysis was made of the public water supply in 32 European countries. The finding revealed that more than 16 million people in urban areas, and 27 million in rural areas, still rely on public fountains for their daily water supply. In the slum districts of a few urban agglomerations, 1.6 million people do not yet have adequate water supplies. Twenty-two of the countries involved offer a full network service to their urban citizens. Full network supply to homes does not, however, guarantee that water will be available when it is needed. There are many reports about shortages, either seasonally (mainly due to low precipitation) or during the day (mainly due to variation in consumption or technical insufficiency).

On a European scale it seems that drinking water resources now and in the foreseeable future are sufficient. However, as shown in Chapter 5, the resources may not always be present where and when they are needed.

Lack of drinking water resources in many places is also due to lack of proper resource and supply management or to insufficient maintenance of the distribution system. In many cities there are no individual water meters, which leads to reduced awareness of water consumption. In Moscow and Bratislava, but also in several Western and Southern European cities, a large amount of water is lost because of leaking pipes before it reaches the user. Leakages as high as 30 to 50 per cent are frequently reported.

Comparable data on the quality of drinking water are not available, but many countries report failure of compliance with standards; in some countries as many as 40 to 50 per cent of samples do not comply. Two pollutants, nitrates and pesticides, are repeatedly mentioned as causing problems. In many places special treatments to remove nitrates have to be implemented to achieve compliance. Other contaminants occasionally causing problems are heavy metals (particularly lead), aluminium, manganese, sulphate, microbiological pollutants and organic matter. (For further information see WHO, in press.)

Environmental impacts

Urban water flows are controlled to a certain degree, and losses of water occur during transportation and use. Figure 10.18 depicts the flows of water to, through and out of Barcelona. Between the points of abstraction and delivery, about one fifth of the water (50 million m³ per year) is lost in leakage and inadequate processing. After use in households (105 million m³ per year), industries (65 million m³ per year) and other public utilities (15 million m³ per year), the water evaporates (19.6 million m³ per year), runs in drains (30.4 million m³ per year), or vanishes in groundwater (4.7 million m³ per year). Urban water flows catch atmospheric deposits (wet and dry), pollutants in runoff from roofs, streets, roads, parks and gardens, and pollutants from traffic, households and industries.

Urban water use

Until a few decades ago, in many cities, requirements for piped water could be met from natural resources such as rivers, lakes and aquifers. Today, water resources used to prepare drinking

Figure 10.16 CO$_2$ *emissions in selected European cities, 1988*
Source: ICLEI, 1993

water are collected in large reservoirs outside the city. The need to look for clean water sources has led to the extraction of deep groundwater around cities. In Germany, this has caused irreversible changes of heathland around cities, and in The Netherlands to drying out of nature conservation areas. Water consumption in Europe in the domestic sector generally increases with living standards. In most European countries the consumption of piped water in households has increased over the past 15 years from 30 to 45 per cent, with the exception of Sweden and Switzerland (countries where policies to limit end use of water have been implemented). Urban industries have reduced their use of water.

The provision of drinking water for household purposes is often not the most efficient way to allocate a valuable resource. For example, it has been estimated in both Switzerland and The Netherlands that only 5 per cent of all piped water used in houses is for drinking and cooking, nearly two thirds is for bathing and washing clothes and dishes, while nearly one third is to flush toilets and clean houses and streets. It has been demonstrated that more than 60 per cent of the drinkable water could be substituted by 'grey' water (upgraded wastewater to be used in toilets), and rainwater (for bathing and washing). Pilot projects in Berlin and other German cities have tested new systems for re-use of wastewater and use of rainwater.

Urban water supply systems require basins for storage in order to meet consumption peaks. In such basins water can evaporate or leak away. In long, hot, dry periods the water levels in basins can drop dramatically, as was the case in Genoa, Italy in 1991, resulting in rationing of the available piped water. Systems can become inoperative; for instance, the water supply of cities along the Rhine is frequently threatened by pollution from upstream industries, and in certain periods no water is taken for the preparation of drinking water. For many cities piped water has to be transported over great distances, which reduces the efficiency of piped water systems. As distances increase, more transformation stations and pumps are needed, requiring more energy and materials. Longer distances imply more risks and losses, especially when the water transportation infrastructure is not well maintained.

Sewage systems and wastewater treatment plants

In several European cities the wastewater from users is collected together with rain water in sewers and discharged to seas, lakes and rivers without any purification. A very few cities in Europe have installed costly 'separate systems', with one system of channels combining domestic and industrial wastewater and one system for rainwater. Since the rainwater drain is not usually connected to purification installations (in contrast to the channel for wastewater), street dirt (including pollutants from traffic) flows with rainwater into lakes, seas and rivers.

In the majority of cities more than 95 per cent of city dwellings are connected to a sewage system. In Amsterdam, Barcelona, Bergen, Copenhagen, Madrid, Moscow, Paris, Rotterdam, Oslo and Valletta all city dwellings are now connected. In Evora, Helsinki, Liverpool, Ljubljana, Rennes and Reykjavík less than 1 per cent of all houses are excluded. However, in Southern and Eastern European cities, and also in several cities in Western Europe, lack of wastewater treatment is still a major problem (eg, Brussels and Venice have no treatment plant).

Wastewater plants were rapidly built in the 1960s and

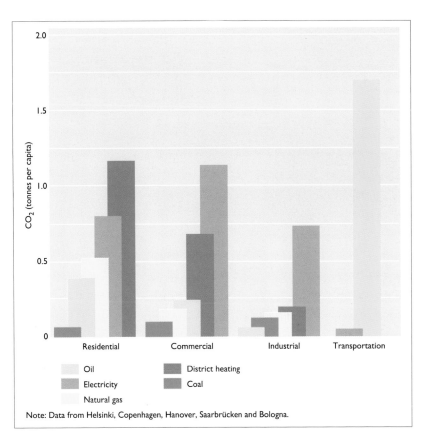

Note: Data from Helsinki, Copenhagen, Hanover, Saarbrücken and Bologna.

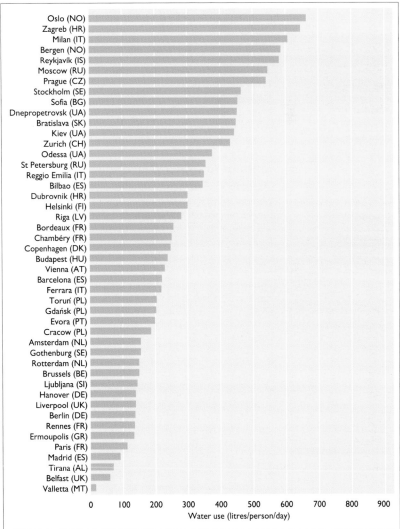

Figure 10.17 Water use in selected European cities
Source: Compiled from multiple sources, see *Statistical Compendium*

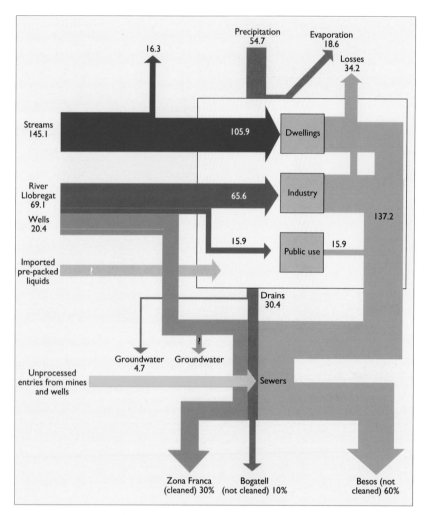

Precipitation 54.7
Evaporation 18.6
Losses 34.2
16.3
Streams 145.1
105.9
Dwellings
River Llobregat 69.1
65.6
Industry
Wells 20.4
137.2
15.9
Public use 15.9
Imported pre-packed liquids ?
Drains 30.4
Groundwater 4.7
Groundwater ?
Unprocessed entries from mines and wells
Sewers

Zona Franca (cleaned) 30%
Bogatell (not cleaned) 10%
Besos (not cleaned) 60%

Figure 10.18
Water cycle of Barcelona (million m³)
Source: Pares et al, 1985

1970s, particularly in Northern and Western European cities. Between 70 and 100 per cent of the wastewater of Denmark, Finland, Germany, Luxembourg, Sweden, The Netherlands and the UK is now treated in sewage plants and up to 90 per cent with chemical or biological treatment (see Chapter 14).

Options for water management

The flows of water in cities are influenced by urban landuse patterns and by the design and management of infrastructures for piped water and drained sewage water. Spatial planning and management of water flows in relation to infrastructure development have become necessities to reduce adverse environmental impacts. A few cities in Europe have integrated water management systems based on careful analysis of the various possible options, their environmental costs and benefits. However, most cities suffer from poor water management and old infrastructures.

The most advanced examples for the management of water flows include:

● the design of a diversified provision system of water related to use;
● efficient processing and transport of piped water (careful selection of the site where water is extracted), reduction of end use (through information to users and installation of individual water meters), reduction of leakage from pipes (through technical improvements), and quality control (to avoid health risks, etc);
● re-utilisation of wastewater (for instance by filtering or infiltration of wastewater from kitchens, sinks and showers in marshes or soil systems, using this water for toilets); the design of a differentiated sewage system to separate rainwater from wastewater from toilets, industries, parking lots, etc.

Material flows

Cities import a wide range of raw and elaborated materials, semi-finished products and end products. Materials brought into cities are transformed into other materials or products, distributed, used and re-used and exported as products or in the form of wastes. All the activities involved, from extraction to production, transport into cities, use and disposal, are potential causes of environmental impacts. The amount and quality of materials imported and exported are therefore a measure of the burden that cities place on the local, regional and global environment.

Building materials are one major component of this flow. Sands, stone and gravel are prime constituents of buildings and road construction. Other building materials used in cities are wood, aluminium, steel, iron, glass and rubble. Enormous quantities of these materials end up as demolition waste which contributes approximately 20 per cent of the total waste produced in cities (where data are available). Cities also import vast quantities of food. In London, for example, 2.4 million tonnes of food were imported in 1990, including milk products (83 000 tonnes), meat products (353 000 tonnes), fish (52 000 tonnes), fruits and vegetables (1 150 000 tonnes), grains (536 000 tonnes), eggs (24 000 tonnes) and beverages (26 000 tonnes) (UK MAFF, 1990). In Europe, the balance of imports and exports varies considerably among cities according to local economic structure and consumption patterns. At European level, few comparable data are available in cities which allow a comparison of imported and exported goods. The assessment of environmental impacts of material flows can be partially made on the basis of urban waste.

Environmental impacts

Materials and products that enter cities are exported as products or transformed into waste. Municipal waste consists of organic substances, paper, metals, textiles, glass, synthetic materials and a large variety of small quantities of toxic substances. In Europe, between 150 and 600 kg of municipal waste are produced per person each year. On average each European produces more than 500 kg of waste per annum or 1.5 kg of waste each day. An assessment of municipal waste in Europe is presented in Chapter 15.

Waste production in selected cities is presented in Figure 10.19. The majority of Western European cities, including Amsterdam, Brussels, Copenhagen, Glasgow, London, Paris and Rotterdam, produce between 300 and 600 kg per capita per year. The same level is reported also in Southern European cities, including Barcelona, Ferrara, Madrid, Milan and Reggio Emilia. In general, Eastern European cities produce annually less than 400 kg of solid waste per capita. These include Budapest, Kiev, Ljubljana, Odessa, Prague, Sofia, Tirana and Warsaw. Some exceptions such as Gdańsk, Cracow, Riga and St Petersburg, but also Bratislava and Moscow, suggest that reported quantities include waste other than generally classified as municipal. Greater amounts of waste per capita are reported in most Northern European cities such as Bergen, Helsinki, Oslo and Reykjavík.

Although historical data on municipal waste production at the European scale are incomplete, estimates provided by the OECD for Western Europe indicate an increase in production of municipal waste at a rate of 3 per cent per annum between 1985 and 1990 (OECD, 1993e). In addition, a major shift is occurring in the composition of municipal waste with the increase in plastic and packaging materials. The introduction of Western consumer products in to Eastern European markets will extend this effect to these countries.

Municipal waste is generally collected in most European cities, although in deteriorating neighbourhoods removal systems do not always work adequately, due to lack of public funding. A large proportion of municipal waste (60 per cent in OECD Europe) from cities is taken to landfills. Tipping, which is the most common method of disposing of urban wastes in landfills in Europe, is not always controlled. Building and demolition waste are often re-used. Demolition waste is sometimes used for landscaping of recreational areas outside cities. Toxic substances which may be found in those materials are an increasing concern (see Chapter 15).

Incineration of municipal waste in Western Europe is used on average for 20 per cent of waste produced (OECD, 1993c). A few cities compost and ferment their household waste (only 4 per cent of all municipal waste in OECD countries). This requires selective collection of waste to ensure that toxic or non-biodegradable substances do not contaminate the resulting compost. Special handling procedures for toxic and hazardous waste from industrial waste and hospitals do not apply to the increasing amount of the so-called small quantities of hazardous waste produced by households, including paints and inks, batteries, medicinal products and pesticides. This has raised concern for the potential risk they pose in both waste disposal and recycling.

Options for waste reduction and recycling

Efforts are undertaken now in many cities in Europe to set examples of good practice. The flows of materials in several European cities are being remodelled, according to the waste avoidance concept. The aim is to reduce the unnecessary import of materials and to reduce the volume of wastes leaving the city. Recycling in cities is an important way to achieve a better balance between the import and export of materials. This requires selective collection, which means a more differential organisation and management of flows of materials. A number of cities have started differentiated collection programmes and have achieved up to 50 per cent recycling of municipal waste. In Copenhagen, for example, a waste management plan adopted in 1990 has increased the share of recycled waste from 17 per cent in 1988 to 48 per cent in 1993.

Reported levels of recycling achieved in different cities are often not comparable, since they refer to different components of waste materials. Out of the total volume of urban waste, Reggio Emilia now recycles 5 per cent, Moscow recycles 15 per cent of its municipal waste and 54 per cent of its industrial waste, St Petersburg recycles 15 per cent, Oslo 21 per cent and Bratislava 23 per cent.

URBAN PATTERNS

The changes in the quality of the environment in European cities are correlated with changes in production and consumption processes which have occurred in the last decades. The decline in sulphur dioxide air concentrations, for example, is explained by the fuel switch in domestic heating and the increases achieved in energy efficiency. The increased importance of nitrogen oxides in urban air pollution is clearly linked to increased urban traffic. However, the way these patterns are linked together with the changes in the urban structure is often not fully appreciated. Density and location of urban activities influence mobility. Public transportation infrastructure affects people's choice of mode of transport. Changes in the quality of the environment and in the environmental performance of European cities can be explained by examining the changes in urban patterns as a result of socio-economic developments.

Table 10.15 gives an overview of urban size and density patterns in the sample of cities examined. Cross-comparisons should be made with extreme caution since measurements of urban density are particularly sensitive to urban boundary definitions. Administrative city boundaries usually do not

Table 10.15 Urban patterns
Source: Compiled from multiple sources, see *Statistical Compendium*

Cities (country code)	Size — City population	Size — Total city area	Structure — Population density	Structure — Landuse intensity built-up/total area	Infrastructure — Transport network/total city area
Amsterdam (NL)					
Barcelona (ES)					
Belfast (UK)					
Bergen (NO)					
Berlin (DE)					
Bilbao (ES)					
Bordeaux (FR)					
Bratislava (SK)					
Brussels (BE)					
Budapest (HU)					
Chambéry (FR)					
Copenhagen (DK)					
Cracow (PL)					
Dnepropetrovsk (UA)					
Dubrovnik (HR)					
Ermoupolis (GR)					
Evora (PT)					
Ferrara (IT)					
Gdańsk (PL)					
Glasgow (UK)					
Gothenburg (SE)					
Hanover (DE)					
Helsinki (FI)**					
Kiev (UA)					
Liverpool (UK)					
Ljubljana (SI)					
London (UK)					
Madrid (ES)					
Milan (IT)					
Moscow (RU)					
Odessa (UA)					
Oslo (NO)*					
Paris (FR)					
Prague (CZ)					
Reggio Emilia (IT)					
Rennes (FR)					
Reykjavík (IS)					
Riga (LV)					
Rotterdam (NL)					
Sheffield (UK)					
Sofia (BG)					
Stockholm (SE)					
St Petersburg (RU)					
Tirana (AL)					
Toruń (PL)					
Valletta (MT)					
Venice (IT)					
Vienna (AT)					
Warsaw (PL)					
Zagreb (HR)					
Zurich (CH)					

City population
- <500 000
- 500 000–1 500 000
- 1 500 000–3 000 000
- 3 000 000–4 500 000
- >4 500 000

Total city area (km²)
- <100
- 100–500
- >500

Population density (inhabitants per km²)
- <1000
- 1000–5000
- >5000

Landuse intensity (% of total area)
- <30
- 30–50
- 50–70
- >70

Transport network/city area (%)
- <5
- 5–15
- >15

* The total city area includes the land area only (there are large forest areas within the municipality).
** Total city area of Helsinki is 588 km², of which 185 km² is covered with land and 403 km² with water.
† See Table 5.9 for country codes.

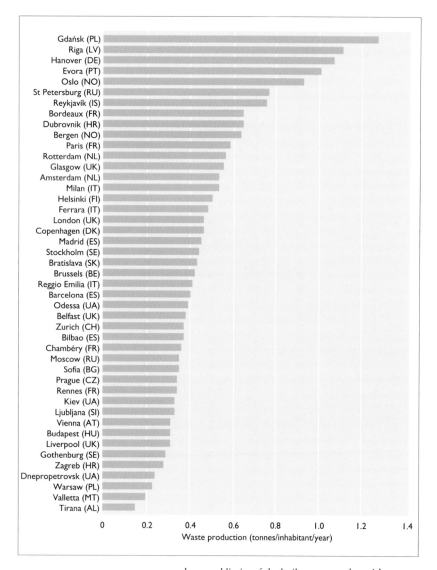

Figure 10.19
Municipal waste
production in selected
European cities
SSource: Compiled from
multiple sources, see
Statistical Compendium

respect the actual limits of the built-up areas; they either go beyond them or exclude significant portions. Consequently, estimated densities will often misrepresent the true urban densities of particular cities. Differences in boundary definitions are a major constraint in the attempt to establish correlations between density patterns and urban quality or urban flows. Other important indicators selected to describe European urban patterns such as landuse and mobility patterns (number of trips, trip length and modal share) are not available or comparable for the majority of the cities and are not included in the table.

The majority of cities examined have a density between 2000 and 5000 inhabitants per km². Population densities greater than 5000 inhabitants per km² occur in 20 per cent of the cities selected. Paris ranks the highest with more than 20 000 inhabitants/km². The intensity of built-up area averages 50 per cent of total city areas, ranging from 11 per cent in Dubrovnik and Reggio Emilia up to more than 90 per cent in Amsterdam. Except for Paris and Barcelona, where high built-up density as well as high density of transportation network, correspond to high density of the city population, in the majority of other cases these variables show mixed patterns. Built-up density is influenced by the city form and location. More compact cities such as Amsterdam have developed according to scarce availability of space. In other cases, the spread of the city and lower population densities play a role in the greater density of the transportation network.

The influences of urban patterns on the state of the environment are still poorly understood. The lack of

comparable environmental information at the urban scale does not allow general conclusions to be derived. Notwithstanding, the findings of the current analysis together with several thematic studies provide new evidence of the impact of alternative urban patterns on environmental problems. The importance of variables such as population density, landuse, structure and infrastructure, but also mobility and lifestyles, shows that urban environmental problems will be resolved only if serious attention is given to the patterns of urban development.

Urban change

Urban population change in European cities shows mixed patterns. While Southern and Eastern European cities are growing faster, most Western European cities have reached stabilisation and a number of them are experiencing a phase of decline (Figure 10.20). Nevertheless, the proportion of people living in urban areas in Southern and Eastern European countries remains in most cases below that in Northern countries. Recent trends suggest that European urban systems have become more balanced in population size: larger cities have stopped growing in favour of smaller and medium-sized cities. However, increased imbalance exists between cities as to the levels of economic development and quality of life.

Different urban patterns in Europe reflect different stages in the economic transition of various European countries, but are also dependent on the economic and institutional capacity of each individual city to respond to these transformations. Accordingly, different priorities of environmental problems between European cities reflect regional differences in urban development. Northern and Western European cities are facing issues related to shifts in activities and specialisation. These changes are expected to bring increased congestion and stress if environmental implications are not fully understood and addressed. Disregard for environmental implications during rapid urban growth has placed heavy demands on local ecosystems. More recently, urban decline and the loss of functions in urban areas as a result of a decline in traditional industrial sectors have left derelict and contaminated sites. The process of suburbanisation in the last decades has raised major concerns for the local and global impacts of increased demand for land and the increased commuting distance to work of the urban population.

Most Southern and Eastern European cities are facing the pressure of increased demand of housing and infrastructure to support increased economic activities and migration of population from rural areas. Several Southern European cities and particularly Eastern European cities find it difficult to meet acceptable environmental standards for large sectors of their populations. Those cities are also now facing problems of air, water and soil contamination as a result of poor environmental practices in the past.

Density and landuse

Urban growth has direct impacts on the use of land. According to recent estimates, in Europe 2 per cent of agricultural land is lost to urbanisation every ten years. In addition to urban population growth, the expansion of urban areas is the result of increased demand for urban land per capita. During the last century, the surface of urban land per capita in Europe has increased tenfold (Hahn and Simonis, 1991). Even when the city population has not increased, suburbanisation and urban sprawl have induced increased demand for land.

A recent OECD/ECMT survey of 132 cities shows that whether a city is growing or declining, there is a consistent trend towards decentralisation (from inner to outer areas) of both people and jobs in the majority of cities (OECD/ECMT,

1993). This trend is particularly characteristic of the largest cities in Northern and Western European countries, such as Brussels, Copenhagen and Paris. Decentralisation of the urban structure is accompanied by increased separation of urban landuses as a result of urban economic restructuring. Important factors affecting the location of activities (eg, communication and transportation) have changed, reinforcing the tendency to separate home from work and both (home and work) from commercial and recreation areas.

Density patterns vary considerably between European cities, as illustrated in Maps 10.4 to 10.11. Low densities in city centres occur in cities where depopulation is linked to the rise in the Central Business District (Vandermotten, 1994). Examples are Brussels and Copenhagen, where the core districts have less then 5000 inhabitants per km². In contrast, high densities in the central city core are characteristic of most Mediterranean cities such as Rome and Madrid, with densities higher than 10 000 inhabitants per km². However, high densities may also be found in several Northern and Western European cities as a result of particular land-forms and very strict landuse regulation and planning, for example, Amsterdam and Rotterdam. Central and Eastern European countries have the highest population densities in the core built-up areas, reflecting the centralised spatial planning of the former centrally planned economies. Examples are Bratislava, Cracow, Warsaw and Moscow.

Different settlement and landuse patterns have been analysed in relation to the pressure on the local and global environment generated by selected cities worldwide (ICLEI, 1993; Lowe, 1991; Newman and Kenworthy, 1989). A city's form and landuse may determine the efficient use of energy, materials, water and space. Dense settlement patterns together with mixed landuses which are generally associated with European cities are considered more efficient in the use of natural resources. The density and location of urban activities as well as the provision of infrastructure also affect travel patterns and petrol consumption and hence the level of emissions from urban transport. Research conducted by the UK government (ECOTEC, 1993) suggests that landuse policies could reduce projected transport emissions by 16 per cent over 20 years. More compact cities, when appropriately planned, reduce travel needs and provide many opportunities for efficient public transport and other forms of energy saving.

However, density alone is not a measure of better environmental performance. Other factors such as centralisation, location of activities and degree of landuse mix play an important role. For example, when combined with the separation of activities in different urban districts, high concentration of activities generates increased mobility and hence increased energy consumption and traffic congestion. In addition, the infrastructure is also crucial. In high density districts, such as the historical cores of several European cities, increased pressure of urban activities combined with the old structure and infrastructure inevitably produces congestion and environmental degradation. The structure of these districts cannot accommodate the increased volume of motorised traffic. The infrastructure for water and energy can be adapted to contemporary demands only at great cost.

Mobility and transport

The increases in mobility and in car ownership and use, as indicated in a recent study of 132 cities in OECD countries (OECD/ECMT, 1993) are consistent with the trend of declining population living in the city core in favour of the suburbs. This study includes 93 cities located in 15 European countries and thus does not provide a complete overview of urban travel trends in all cities selected for this report. However, it does provide the most up-to-date base of comparative information on urban travel in a wide spectrum of cities. Travel distances have increased over time with a

Construction site for the Olympic Games, Barcelona, 1992
Source: Frank Spooner Pictures

Figure 10.20 Population growth in selected European cities
Source: UNEP/WHO, 1992

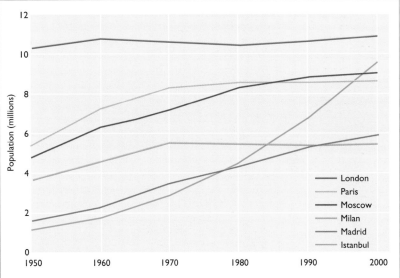

shift in trip length from the 5 to 10 km to the 10 to 15 km band. Modal split in the majority of the cities is increasingly dominated by the private car. Table 10.16 shows a marked increase in car ownership, with 45 per cent of the cities having almost one car for every two people in 1990, compared to only 9 per cent in 1970.

Travel demand and modal split are influenced by settlement size, urban density and structure. A recent joint study by the UK Departments of Environment and Transport (UK DoE/DoT, 1993) indicates that higher residential densities are likely to be associated with a relatively lower demand for travel and a higher efficiency in the mode of transport. Travel demand in UK cities is seen to rise markedly as density falls below 15 inhabitants per hectare, and falls as density reaches the threshold of 50 inhabitants per hectare. The relationship between settlement size and emissions-efficient transport is more complex. It is however evident that both the dispersal of population from major city centres and smaller settlement sizes are typically accompanied by higher car use. More centralised urban structures have the effect of increasing journey length, but,

Map 10.4
Population density in Brussels, 1990
Source: Vandermotten, 1994

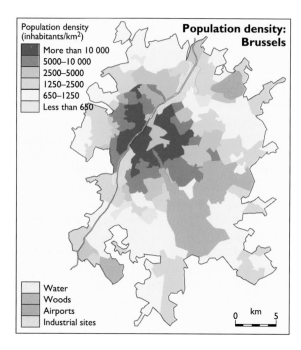

Cars/inhabitant	1970 (%)	1980 (%)	1990 (%)
<0.20	57	12	3
0.20–0.29	23	31	19
0.30–0.39	10	39	33
0.40–0.49	6	13	26
0.50–0.59	3	3	15
0.60+	–	1	4

Table 10.16 Car ownership in OECD cities
Source: OECD/ECMT, 1993

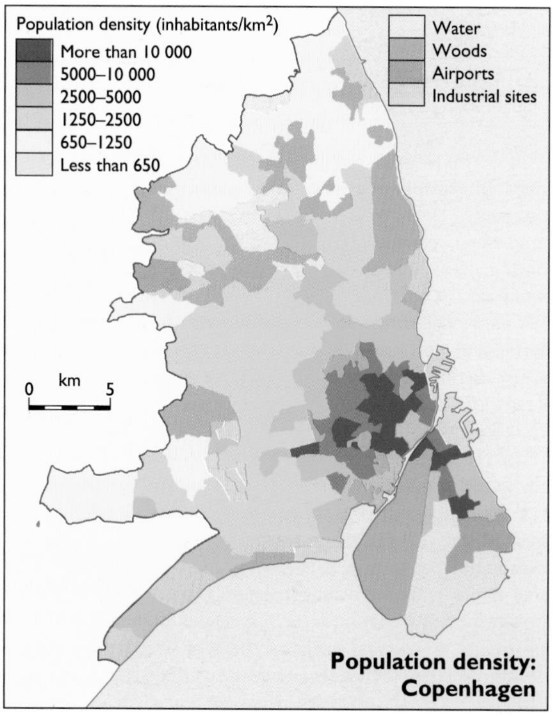

Buildings and infrastructure

The urban structure influences also the efficiency of resources use and the opportunities for implementing ecologically efficient solutions through its buildings and infrastructure. Recent studies by the European Commission (DG XVIII PERU programme) and by OECD (1993a, b, c) have examined the influence of urban structure on energy requirements. In European cities, different levels of efficiency can be generally associated with the typology of buildings in the various city districts. Building design (eg, layout and orientation), density and materials are important factors influencing, for example, the capacity to save energy for space heating and cooling. Detached houses have a greater heat loss compared with flats because of their greater surface/volume ratio. Smaller housing units, both terraced housing and low-rise flats, have the advantage of maximising passive solar gains. In addition, attention to microclimate in building design can save at least 5 per cent in energy requirements (OECD, 1993b).

Other important factors influencing the environmental efficiency of buildings are age, maintenance and tenure of building stocks. The structure of old buildings in the historical centres of many European cities imposes severe constraints in the implementation of energy-saving measures. Most of these buildings were built before or at the beginning of the century and hence do not respond to the energy-related building standards (eg, insulation) progressively introduced in most countries. However, the new system-built apartment buildings erected during the 1960s and 1970s, as well as most of the most recent high-rise buildings erected to accommodate business offices or housing, often disregard the minimum microclimate and other environment-related considerations.

Incorporating environmental considerations into the design of the infrastructure which supplies a city with water, energy, food and transportation can be even more critical to prevent and minimise environmental problems. The quality and maintenance of urban supply systems influence the volume of flows of resources and services necessary to the functioning of the city as well as the level of environmental impacts generated. Urban renewal projects implemented in a number of European cities (eg, Berlin, Copenhagen, and Helsinki) have proved that upgrading the urban infrastructure may save energy and water. The provision of a diversified transportation infrastructure and efficient public transportation system is even more important. In Copenhagen, Zurich, Bordeaux, Oslo, Bergen and Stockholm, measures adopted to improve access to public transportation together with alternative infrastructures (eg, bicycling pathways) have been proved to encourage energy-efficient choices in the mode of transport used, and hence to reduce pollution and congestion. In the majority of Western European countries, up to 70 per cent of national investments in the building sector are being channelled into urban renewal and restoration. Urban redevelopment projects are being implemented not only to restore old areas of major

according to the UK study, they also increase the use of public transport.

The location of urban activities, especially the location of workplaces, as well as the provision of infrastructure, is crucial to the choice of mode of transport. In Copenhagen, a study of seven districts (Jorgensen, 1993) shows that only 7 per cent of workers actually work in their home district and 18 per cent in the neighbouring district. Consequently, only one quarter of the employed residents work in a range of 5 km from their homes. Another study in Copenhagen shows that the relocation of a large office from the city centre to a suburban location increased the proportion of car-use by the employees from 26 to 54 per cent, while the average travel distance remained the same. A similar study extended to four large offices shows that car use is between 10 and 15 per cent higher in offices located more than 500 metres from an underground station.

The growth in urban mobility and the predominance of road traffic are among the principal causes of urban air pollution, congestion and noise (see Chapters 21 and 37).

Map 10.5
Population density in Copenhagen, 1990
Source: Vandermotten, 1994

urban centres, but are also being increasingly carried out in the districts of more recent suburban development or former industrial zones. These projects provide important opportunities for achieving more livable urban districts while increasing the efficiency of resource use.

Urban lifestyles

The root cause of many urban environmental problems is ultimately unsustainable production and consumption patterns. Consumption of water, energy and materials and production of waste are general concerns of the sustainability of national economies, and not exclusively an urban concern (see Chapter 12). They become an urban concern since the majority of population and economic activities are concentrated in cities. Urbanisation and increased economic affluence undoubtedly have an influence on lifestyles. Cities attract people because of their increased opportunities for economic activities and social interactions. Increase in income has an important influence on the ability of people to choose. The combination of these factors therefore gives urban communities a critical role in achieving better environmental performance.

Changes in lifestyles are examined in detail in Chapter 26, which describes current trends in households and their environmental implications. In cities, mixed patterns emerge. On the one hand, change in the household structure (average household size) and the number of households together with increased household income and the availability of goods, have resulted in increased consumption. On the other hand, recent years have seen a raised awareness of the problems involved, largely as a result of concerned pressure groups influencing consumer behaviour.

While plans and policies may be framed at the national, regional and local levels, the actual test of their effectiveness is whether they are taken up in modified urban lifestyles. Opportunities in cities are greater. In some cases the possibility of personal choice is controlled, or at least guided, by encouraging environmentally friendly behaviour (eg, traffic management schemes) or by providing the necessary infrastructure and information to make better choices. However, increasingly, the role of the individual in promoting environmental improvement has become vital given the increased reliance on voluntary actions (eg, for recycling). This suggests that the implementation of measures to improve the quality of the urban environment and cities' environmental performance can be achieved only by increasing individual and collective awareness, together with promoting citizens' participation in the design of such measures.

TOWARDS SUSTAINABLE EUROPEAN CITIES

Urban environmental problems

Europe is a highly urbanised continent with more than two thirds of the total population living in urban areas. The state of the environment in cities is consequently of great importance to most Europeans. On the other hand, environmental problems, from local to global, are often rooted in increasing urban activities and their pressure on natural resources. The level of ecological awareness in designing and managing cities is thus a crucial test for achieving sustainable development in Europe.

The urban environmental indicators which have been selected to describe European cities provide a basis for identifying major urban environmental problems and assessing regional differences and priorities. From the assessment of 51 European cities, the environmental quality of urban areas emerges as a major concern for Europe (see

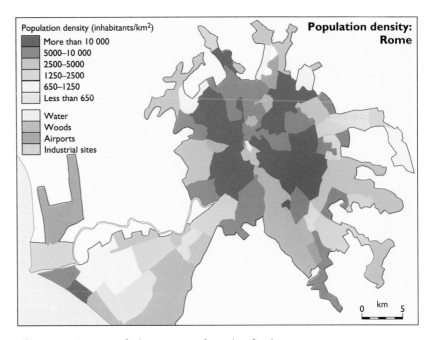

Population density (inhabitants/km²)
More than 10 000
5000–10 000
2500–5000
1250–2500
650–1250
Less than 650

Water
Woods
Airports
Industrial sites

Population density: Rome

0   km   5

Chapter 37). However, the importance and severity of various problems vary widely between European cities and regions. Three interconnected areas of concerns are: air quality, noise and road traffic. Despite the achievements in the reduction of traditional emissions (SO_2 and particulate matter) the majority of cities still exceed short-term WHO AQGs. Compared with Eastern and Southern European cities, Northern and Western cities are better off in terms of long-term exposure to SO_2, but are now faced with other types of threats (eg, NO_x and VOCs). Even for strictly regulated air pollutants, there is an enormous variation in the level of implementation of standards and guidelines between European cities. Poor acoustic quality is also of increasing concern, particularly in urban areas. Between 30 and 40 per cent of cities indicate that more than 10 per cent of the residents are exposed to unacceptable levels of noise (above 70 dB). Furthermore, increased traffic congestion and accidents are other dominant characteristics of most European urban centres.

Greater variations can be detected when examining the quality of urban space and housing. Land availability varies according to geographic location and land form. However, economic and political factors play a decisive role in resolving the tension between different landuses in cities. Thus, landuse regulations in Europe have been vital to balancing urban development with the need to satisfy urban quality criteria and the protection of natural areas and historical centres. More recently, under increased economic pressure, most municipalities have found it increasingly difficult to protect and maintain open space and green areas, as well as to prevent the replacement of historical buildings and sites. Different patterns of housing quality and capacity are clearly linked to regional differences. Housing problems are more relevant in Eastern European countries, where basic standards for housing are not met for a considerable proportion of urban residents. However, high-standard districts together with very poor housing conditions coexist in most European cities.

While all European cities are faced with increasing environmental pressures and deterioration, the causes are often diverse. One important example is given by the uneven contribution of various activities to urban air pollution problems across Western and Eastern Europe. Since the early 1970s, sulphur dioxide air concentrations in Western European cities have considerably decreased as a result of reduced coal burning for domestic heating. Instead, increased concentrations of nitrogen oxides, carbon monoxide and carbon dioxide show that increased traffic volumes are a major cause of urban air pollution. Different patterns of air pollutant

Map 10.6
Population density in Rome, 1981
Source: Vandermotten, 1994

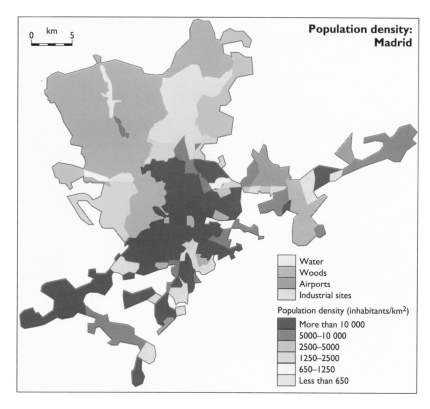

Map 10.7
Population density in Madrid, 1986
Source: Vandermotten, 1994

Root causes of urban degradation and poor environmental performance of cities are often the result of inadequate environmental consideration in landuse planning and urban management. The deterioration of the urban environment accelerates particularly in periods of rapid change. Both urban growth and decline in Europe have caused increased pressure on the urban environment. Current trends towards the decentralisation of urban structure and increased mobility dependent on the private car are likely to exacerbate current environmental problems in the major European cities.

The need for an integrated approach

Urban environmental problems can be tackled by reducing the pressure of urban activities through landuse planning, and the efficient management of the flows of resources as well as through measures directed towards protecting the quality of the urban environment. Addressing the causes of urban environmental problems instead of focusing on their symptoms is a necessary condition for success. Moreover, the interrelated nature of urban environmental problems requires that the actions undertaken at the various levels should be part of an integrated approach. The need for integration was acknowledged in an OECD report (OECD, 1990) which examined the organisational and institutional implications. At the EU level this was recognised in the Green Paper on the Urban Environment, 1990 (CEC, 1990). Both reports emphasised the importance of a specifically urban approach if environmental problems are to be effectively addressed and resolved. The options and examples illustrated below show the wide scope for change that an adequate consideration of environmental concerns in urban planning and management can provide.

Changing urban patterns

Landuse planning and space management are powerful tools to improve the condition of the urban environment. Through ecological management of the urban space, the demand for mobility can be reduced and access to essential public services increased. Urban renewal and re-use of old industrial sites in urban areas provide the opportunity for creating open spaces and ecological restructuring of urban centres (Box 10K). Also the replacement or rehabilitation of inadequate urban infrastructure can improve access to urban areas and services and reduce the pressure of urban travel.

The shift from private to public transportation in urban areas is highly dependent on the urban structure and integration of urban activities. It also depends on the level of accessibility of public facilities. While landuse planning is one of the most efficient tools for reducing the need of people to move, an integrated public transportation system together with proper traffic management can influence the choice of transportation mode (Boxes 10L and 10M). In particular the provision of differentiated infrastructure for public transportation, bicycles and pedestrians in urban centres is important to satisfy a differentiated demand for urban mobility (Box 10N).

Managing urban flows

The flows of energy, materials and water in the urban system can be reduced by improved management and up-to-date technologies. Considerable opportunities for energy and water savings as well as material recycling are available at the urban level. However, if urban ecological cycles are to be made more efficient, there will have to be active integration of the technical and green infrastructures. As one of the first national reports on the state of the urban environment produced by Sweden, Ecocycles (SEAC, 1992), pointed out, in order to improve the urban environment, supply systems need to be adapted to environmental cycles, and the cycles

concentrations in Central and Eastern European cities show that heavy industrial activities and energy production plants are relatively more important as sources of air pollution in the city.

A measure of the contribution of European cities to regional and global environmental problems is the amount of energy and water consumed as well as the per capita production of emissions and waste. Most Western European cities have achieved important results in energy efficiency as well as reducing the amount of emitted pollutants per unit of product and service provided. However, the total amount of natural resources used by urban activities has increased as a consequence of changes in lifestyle and living standards. Furthermore the total amounts of waste have increased. On the other hand, in many Southern and Eastern European cities, limited economic resources are a major constraint in the process of replacing highly polluting and inefficient technologies which still place enormous burdens on the local and global environment.

Map 10.8
Population density in Amsterdam, 1989
Source: Vandermotten, 1994

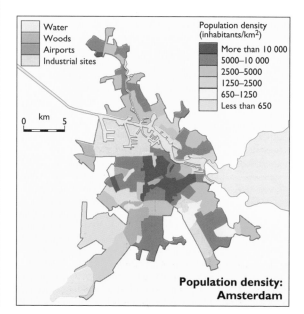

Box 10K Ecological urban renewal in Copenhagen

The Agency of Environmental Protection, City of Copenhagen has developed a Planning Model for Ecological Renewal of old European city areas, using as a basis the ecological renewal experience in the Vesterbro district of Copenhagen. The Copenhagen Model describes the ecological planning process, a study of the potential for minimising the consumption of resources and the environmental impact of energy use, water and waste and an analysis of the barriers to, and tools for, the implementation of this potential. A number of guidelines for ecological urban renewal have been developed in cooperation with four other European cities: Madrid, Amsterdam, Berlin and Genoa.

Copenhagen Source: Michel St Maur Sheil

Ecological planning in Copenhagen is meant to introduce as many ecological solutions as possible connected to an up-to-date definition of urban renewal standards. In 1990 a working group was established at the Copenhagen Municipality's 5th department (Energy, Water Supply and the Environment) in order to integrate ecological elements in the urban renewal action plan for Vesterbro.

The district of Vesterbro was urbanised in the period between 1852 and 1990 and almost 90 per cent of the buildings are pre-1990. It contains 23 blocks and about 4000 apartments for 6500 residents. Old buildings together with low housing quality and social problems in inner Vesterbro are characteristics of many old areas in European cities. According to the working group, the ecological and social dimensions need to be combined into an urban renewal programme if the improvement of these areas is to succeed.

A study carried out to assess the potential for conservation of resources in Inner Vesterbro indicated:

- Electricity saving: 50 per cent
- Saving on heating: 30 per cent
- Water saving: 50 per cent
- Re-use of waste: 48 per cent
- Use of energy content of waste: 38 per cent

A number of ecological solutions have been studied and implemented in three testing blocks:

- Buildings:
  - double-glazed windows
  - insulation
- Installations:
  - low temperature district heating
  - water-saving fittings
  - water-saving toilets
  - electricity-saving on light
  - installations for sorting waste
- Green yards:
  - green plants
  - permeable cover which allows rainwater to seep into the ground
  - rainwater for watering the yard
  - integrated systems for waste sorting
  - compost for the garden

The realisation of the first urban renewal phase has produced a reduction of approximately 32 000 GJ of end-use energy per year which corresponds to a 2500 t (14 per cent) reduction in carbon dioxide emissions.

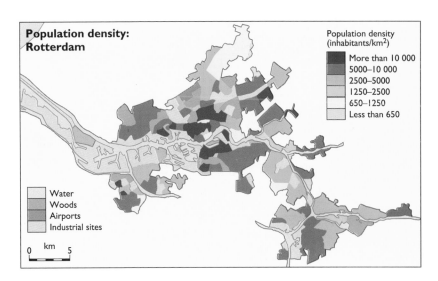

Population density: Rotterdam

Population density (inhabitants/km²)

- More than 10 000
- 5000–10 000
- 2500–5000
- 1250–2500
- 650–1250
- Less than 650

- Water
- Woods
- Airports
- Industrial sites

0 km 5

Map 10.9
Population density in Rotterdam, 1990
Source: Vandermotten, 1994

must be made visible in the cities. Successful examples of eco-restructuring and implementation of small-scale ecological infrastructures have been developed in Northern and Western European cities (Box 10 O). Important efforts are also being made in Central and Eastern European cities. These experiences constitute an important base of information for exploring new and creative solutions.

Improving urban environmental quality

Although important progress has been achieved in establishing air pollution standards for cities, their implementation is still a major problem in several European countries. Improved monitoring of air quality is a priority in most Eastern European cities. New mechanisms, such as local taxation and permitting systems need to be developed to achieve effective implementation of established standards. Comparable standards for water resources, soil and noise still remain to be established to protect the health of people living in urban areas.

Setting urban sustainability targets

The World Conference on Environment and Development (UNCED, 1992), recognised the crucial role of cities and local authorities in achieving sustainable development. Chapter 28 of Agenda 21 highlights that:

local authorities construct, operate and maintain economic, social and environmental infrastructure, oversee planning processes, establish local environmental policies and regulations, and assist in implementing national and subnational environmental policies. As the level of governance closest to the people, they play a vital role in educating, mobilising and responding to the public to promote sustainable development.

Among the other objectives, Chapter 28 establishes that by 1996:

local authorities in each country should have undertaken a consultative process with their population and achieved a consensus on a Local Agenda 21 for the community.

A starting point for developing local Agenda 21 in European cities is to achieve a better understanding of what a sustainable city looks like. Ideally, local Agenda 21 should be designed to ensure compliance with sustainability limits. In practical terms, there is vast uncertainty on the exact limits of local and global ecosystems to accommodate increasing demands of environmental goods and services. Notwithstanding, the evidence that current urban patterns have reached critical

Map 10.10
Population density in
Moscow, 1989
Source: Vandermotten,
1994

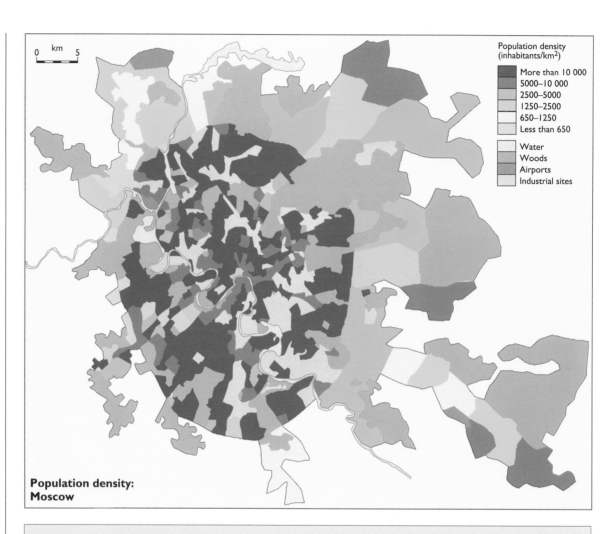

Population density
(inhabitants/km²)

- More than 10 000
- 5000–10 000
- 2500–5000
- 1250–2500
- 650–1250
- Less than 650

- Water
- Woods
- Airports
- Industrial sites

Population density:
Moscow

Box 10 L Traffic management in Bergen and Oslo, Norway

Bergen

Bergen, in Norway, has about 310 000 inhabitants in the region; about 216 000 people live in the city itself. Bergen was the first European city to introduce road pricing as a solution to serious traffic problems. Since the 1970s the city has been faced with congestion, accidents and noise due to local use of the motor car, but also because regional traffic was forced to cross through the city to reach other destinations. The number of victims from car accidents were continually increasing: the number of people killed per year doubled in the second half of the 1970s. Discussions took place on how to combat the problems of traffic. One of the major proposals was to build a serviceable road network to divert through-traffic from the city. However, due to necessary investments this would take 30 years to achieve. It was decided therefore to introduce a toll ring system, the proceeds from which could cover about half of the funds needed for road construction. The Bergen toll ring was opened in 1986. There are now toll gates for all the main access roads to the city. This has reduced the traffic entering Bergen. At the same time, the number of people killed in car accidents has been reduced to the level of the 1970s and continues to drop; the number of injuries has stabilised.

In connection with the introduction of the toll system, Bergen adopted a new traffic strategy in which environmental quality and safety were main objectives. Measures include: tighter restrictions on parking, restrictions on access for private cars and reductions in transport requirements in general by improved public transport and landuse planning. An important element is the building of tunnels, which will divert 25 000 vehicles per day of the total 65 000 now entering through the toll ring into the city. Noise measurements in 1991 indicate that about 20 000 dwellings were exposed to a noise level of 60 dB(A) or more; 40 000 persons were disturbed by noise, of which 18 000 were seriously disturbed. Due to the traffic strategy, it is planned that the number of persons exposed to a noise level of 60 dB(A) will be reduced by 50 per cent before the year 2010, when it is planned that there will be no dwellings with an interior noise level of more than 35 dB(A).

Oslo

The Oslo conurbation has approximately 680 000 inhabitants, and 467 000 live in the city. In 1990, Oslo followed the Bergen example to introduce road pricing, after ten years of public debate between professionals and politicians over the advantages and disadvantages of a toll system. The discussions focused on whether it was right to make people pay for the toll and whether money could be raised in a more cost-effective way through alternative systems (for instance by increasing the yearly tax for all car owners or by increasing taxes on petrol). In 1990, 19 stations were installed, 3 to 8 km from the city centre, so that it was possible to drive into the inner city by paying a fee. Approximately 210 000 cars pass every day (towards the city) which amounts to 40 per cent of all car trips in the Oslo region. Since the introduction of the toll ring the number of cars entering the city has been reduced by 3.5 per cent. The toll ring has led to a slight increase in the share of public transport: the increase in trips crossing the toll ring is estimated to be 10 per cent. People also travel less in general because the new toll system.

Box 10M Public transportation in Zurich

The city of Zurich with its 360 000 inhabitants is the centre of an agglomeration of about 1 million people. Since 1973, all traffic technical measures taken by the city have been aimed at controlling car use, and with good results.

Around 80 per cent of all displacements into and out of the inner city, and 50 per cent of the displacements within the municipality take place on public transport. The citizens of Zurich average 490 trips a year by public transport, compared with 320 for Amsterdam, 150 for Cologne and 131 for Manchester. Zurich has carefully planned its investments in transport infrastructure. Cost-benefit analysis has shown that speed-up programmes for trains and buses were 20 times more successful than the construction of an express motorway and four times better than improvement of the S-Bahn (underground). It was therefore decided to concentrate on developing the bus and tram system in Zurich.

Investment in the area of public transport includes the laying on of free bus and tram lanes, minimising the waiting times at traffic lights, promoting regularity and punctuality, and in addition extending the S-Bahn network. The city has now the highest rate of public transport use in Western Europe (500 trips per person per year). The result has been a stabilisation of city traffic since 1990, and a decline in car ownership since 1982 (Friends of the Earth, 1992, p 25). Noticeable is the high percentage of season-ticket holders among the travellers, a consequence of companies offering free or reduced tariffs to their employees. Also noticeable is the extremely good image enjoyed by public transport. Eighty-seven per cent of the population supports further promotion of public transport.

Parking policy is an important accompanying measure. There are strict norms for laying out parking facilities. The city of Zurich has drawn up standards to determine the number of new parking places allowed based on the quality of public transport at the given points and on the protection of the city's appearance. These standards comprise minimum and maximum requirements with regard to the number of parking places.

The minimum requirement is used to prevent the expected number of cars putting pressure on the other parking capacity. The maximum requirement is used to avoid generating excessive motor traffic in a given area. The canton of Zurich has drawn corresponding guidelines, including additional reducing factors based on the air quality and expected pollution, with respect to the air quality management plan of 1990.

Despite the measures taken, the number of private parking places has doubled in the last two decades. Formerly, there was more concentration on the minimum requirement, now it is more the maximum requirement. Furthermore, there has been a shift in the type of jobs to a larger share for services. The growth in the number of private parking places can ultimately be explained by the labour market. As a kind of fringe benefit, employers sometimes want to ensure a sufficient number of parking places. Around 85 per cent of the car commuters in Zurich have their own parking space. Thus, to reduce car traffic further it is the aim to try and push down this percentage especially, by raising effective parking taxes. Apart from this the focus will be on visitors to the shops and businesses. The supply of parking places for this category will be reduced.

In the late 1970s, a bicycle route network of 200 km was designed and adopted in 1984. To date, the share of the bicycle in the total amount of traffic is about 7 per cent .

The main objective of the traffic policy in Zurich is to guarantee the accessibility for employees and visitors in an attractive and inexpensive way with positive side-effects such as a reduced energy level, lower noise levels, less air pollution, more space for cyclists and pedestrians and a safer urban environment. The results of the transport policy in Zurich show that, contrary to the belief of some entrepreneurs, economy and ecology are indeed compatible.

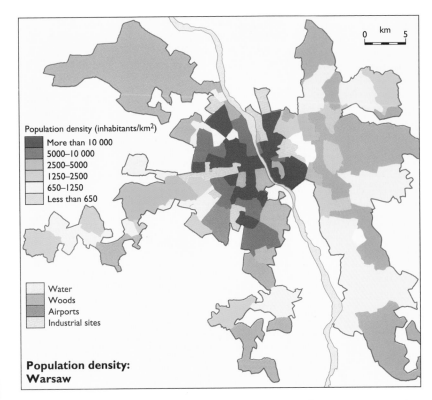

Population density (inhabitants/km²)
- More than 10 000
- 5000–10 000
- 2500–5000
- 1250–2500
- 650–1250
- Less than 650

- Water
- Woods
- Airports
- Industrial sites

Population density: Warsaw

Map 10.11
Population density in Warsaw, 1988
Source: Vandermotten, 1994

thresholds in such areas as air pollution and energy consumption is already an important basis for setting targets to move in the direction of sustainability. Progressive improvements of these targets and the means to achieve them will be possible only with the improvement of our understanding of what sustainability entails and of different priorities to achieve sustainable targets in each local context.

The design of local Agenda 21 should also help in balancing the local and global perspective. Considering the various scales of urban environmental problems, including district, city and metropolitan, as well as their interactions with the regional and global scales, will help clarify problem areas where different perspectives (eg, local and global) may conflict. Cross-national cooperation of local authorities and international programmes is being pursued to accomplish this task (Box 10P). This will help improve the institutional framework to address urban environmental problems.

Building institutional capacity

Important progress achieved in understanding the nature of urban environmental problems has led to relevant changes in urban planning and management in Europe.

Local authorities in several European cities are rearranging their functions and setting new mechanisms to achieve the necessary integration between departments and across different sectors of activities. In these attempts, cities have discovered that the scope for integrating environmental considerations into urban decision making is vast and needs to be fully exploited if measurable improvements in the quality of European cities are to be achieved. Eco-auditing and environmental accounting at the municipal level could help to establish feedback mechanisms in order to ensure that progress is being made in all sectors of urban activities that affect environmental quality and performance. Increased attention is placed on economic instruments to achieve better internalisation of environmental costs and to encourage less polluting behaviours. In addition, full compliance with ambient quality and emission standards for the outdoor and indoor environment is to be ensured. The need for better integration and coordination across the various levels of government is also required.

New forms of cooperation with all sectors of the urban community have proved crucial to the success of programmes of urban renewal and environmental improvement in several European cities. Amsterdam, Berlin, Copenhagen, London and Manchester have successfully established partnerships between private and public sectors in urban redevelopment projects. A similar approach is now being considered in a larger number of cities.

Monitoring urban environmental change

One major challenge for the orientation of future urban environmental programmes is building the necessary technical and institutional capacity for monitoring urban environmental change. As anticipated in the introduction to this chapter, the improvement of environmental information at the urban scale is a priority for sound environmental action. Important progress is being made at the municipal level in most European countries to establish systematic monitoring and reporting on the state of the urban environment. For the present analysis, outstanding efforts have been made by the selected cities to provide the information presented in this chapter. This is an effort which needs building upon if future monitoring and reporting on the urban environment is to be improved.

An important distinction must be drawn between the comparability and availability of information. Obtaining comparable information on the state of the urban environment at the European scale has gained increased importance as countries have recognised the European dimension of most urban environmental problems. However, environmental information is often not available or adequate in the first place. Indeed, in many cities environmental monitoring activities are still limited in scope. Furthermore, most environmental statistics are not collected at the municipal level. The sectoral organisation of data collection across various agencies usually does not allow for cross-sectoral analyses and comparisons because of different classification systems and accuracy.

The influence of different urban patterns on the quality of the urban environment can be understood only when quantitative links between landuses, transport, energy use and emissions can be established. The improvement of environmental monitoring at the urban scale and its integration with other urban statistics is crucial to interpret changes in environmental quality and assess the effects of urban policies and programmes. There are several advanced examples of cities in European countries such as France, Finland, Germany, Switzerland and Sweden where systematic monitoring of urban environmental quality is effectively integrated with emission inventories and dispersion modelling. These experiences could provide a useful framework for integrating and harmonising urban monitoring across European cities.

Obtaining objective and comparable information on the state of the urban environment also requires comparable urban environmental indicators to monitor progress and assess the effectiveness of programmes and measures designed to improve the state of European urban environments. The experimental indicators used in this chapter demonstrate the benefits of this approach, but considerable improvements are required to obtain a more accurate picture. Existing networks of cities may play an important role in this process.

Box 10N Bicycling network in Delft

In the 1970s the municipality of Delft created a bicycle network plan. It is made up of three networks, each having its own functional and design characteristics:

- The city level network consists of a grid of cycle paths situated approximately 500 metres apart. The paths run directly through the city, and are connected with the regional bicycle-path system. The network is designed for the purpose of linking intensive flows of cyclists with important urban activity centres: schools, university, stations, offices and industrial areas, sport and recreation areas. Physical barriers (such as canals and railways) call for expensive engineering structures to be built to avoid detours.
- The district level network has two major functions. It connects the various facilities within the district (schools, shops, etc) and collects and distributes bicycle traffic to and from the city level network. The links at this level are spaced 200 to 300 metres apart. The bicycle traffic flows on this network are assumed to be less heavy than at the city level and used for shorter distances. The facilities necessary at this level are relatively simple: separated bicycles lanes, small bridges and so on.
- The sub-district level network connects housing areas to local amenities, catering for short trips. This particular network is often used by children. The sub-district level network is a fine-grained system with links at 100 metre intervals, with a simple structure and provisions which can also be used by pedestrians.

Implementation of the network has had the following results:

- *Increased bicycle use:* the project's primary goal is to encourage cycling. Investigations demonstrate that bicycle use increased in terms of length and number of trips made, thanks to the implementation of the bicycle network plan. The distance travelled by bicycles increased from 6 per cent to 8 per cent (depending on type of trip). This figure does not include increases caused by factors other than the bicycle network plan.
- *Improved comfort and safety:* cycling comfort and safety improved as a result of the implementation of the bicycle network plan. Analysis of the use of the network by cyclists shows that the distance travelled by bicycling on separate and independent cycle paths increased, as well as the total number of bicycle trips on these specific cycle lanes. Bicycle volumes at linkages with mixed traffic decreased.
- *Reduced car use:* one of the objectives of the bicycle network plan is to reduce car traffic by encouraging cycling. Recent studies have indicated that the number of car trips in the Delft area have not increased as in other Dutch towns. The number of car trips in the centre decreased.
- *Better competition between modes:* a combination of all the aforementioned improvements and changes to the advantage of the cyclist, resulted in a shift in modal split: the share of the bicycle rose from 40 per cent to 43 per cent. The car and walking mode both remained stable at 26 per cent. The public transport share declined from 6 per cent to 4 per cent (although the total number of passengers remained the same).
- *Applicability:* Delft is a typical medium-sized Dutch town. Bicycle use varies distinctly between medium-sized towns in The Netherlands (population 50 000 to 200 000), with the proportion of trips made by bicycle from 20 per cent to 50 per cent. Based on the Delft project it can be assumed that 55 per cent is the maximum attainable share of internal bicycle trips in The Netherlands. This means that possibilities are open in many cities and towns to increase bicycle use.

Box 10O Urban environmental technology in Berlin

In Berlin, some 75 environmentally sound renewal projects of housing quarters and of inner-city districts have been realised. Many of these projects started at the grass-roots level: small groups of citizens took the initiative to improve their immediate surroundings. These small-scale initiatives had a great multiplier effect in the city and created the basis for developing larger projects of ecologically sound urban development.

An example is a building complex with 106 dwellings at Dessauer Straße, focusing on urban water management. This was the first in a range of projects aimed at developing a decentralised inner-city water management system. The housing units are equipped with water-saving devices widely available on the market. These include toilets that can be flushed with 4 to 6 litres instead of the usual 9; flow rate limiting devices (with a saving of 20 to 40 per cent); one-tap systems to mix warm and cold water which enable quick regulation of water temperature (with water savings of 5 to 10 per cent); water meters in every house for individual payment; separate piping systems for clean and recycled wastewater. Each apartment in the block can be easily connected to or disconnected from the water recycling system, according to user requirements.

All the wastewater leaving the apartments connected to the recycling system is collected in a sedimentation pond in the courtyard of the housing block. It is then led through a small area of vegetated wetlands and ends up in a purification pond. At this stage it is purified to such an extent that it can be used to flush toilets and to water green spaces and gardens.

Most of the rainwater is kept in the area: it is not disposed of through the municipal sewerage system, but is collected in a filtration basin with plants, from which it is drained in a pond. All measures brought together in this decentralised water management system led to a 50 per cent saving of the amount of piped drinking water compared to the average elsewhere.

Although this project focused on water conservation, this is not the only environmental aspect. When integrated with other decentralised infrastructures and supply systems, various other environmental benefits can be realised. An example is a special installation to derive heat from wastewater; this heat is used to warm up clean water to be used in homes.

Another element is green spaces: 60 per cent of the project area is covered by plants, on roofs and gardens. This helps to catch rainwater and keep the area humid, and to filter the air and trap dust particles. It also contributes to insulation and energy conservation in buildings. In a 'regular' project of this size only 10 to 30 per cent of the area would be covered by plants.

A third element is waste reduction. In this project, information is given to residents on how to prevent waste, and to separate waste for collection and recycling. The various waste fractions are collected and processed separately. The project was the first social housing project with facilities designed to store glass, paper, plastics, metal, textiles, organic waste and chemicals. The separate collection of organic waste yields 7 m^3 of compost annually: sufficient for the maintenance of the whole green area of the project.

Another important element is visibility: local people can observe the immediate impacts of their behaviour on the management of green spaces and on the volume and composition of solid wastes and the possibilities for re-use. It is only a small step from environmental technology and design to environmentally sound behaviour.

Box 10P International initiatives for sustainable cities

The **Sustainable Cities Project** is an initiative launched by the European Commission Expert Group on the Urban Environment in 1993 as a follow-up to the 1990 Green Paper on the Urban Environment. The first outcome of this project is a report which will formulate recommendations for integrating urban environmental considerations in the European Union, and national and local policies. Three main areas of integration are considered: economy, mobility and landuse planning. The project also includes the preparation of a Best Practice Guide and the exchange of information and experience through the establishment of a Sustainable City Network.

The **Sustainable Cities Programme** was established by the **UNCHS** (United Nations Centre for Human Settlements – Habitat) in 1990 to provide municipal authorities and their partners in the public and private sectors with an improved environmental planning and management capacity. The Programme focuses on developing countries and brings expertise from different regions of the world to help strengthen ability to define the most critical environmental issues and to identify available instruments and mechanisms to address them.

The **Ecological City Project** was launched by the **OECD** Environment Group on Urban Affairs in 1993 to identify strategies for integrated and coordinated policies at the national level that can help cities to address environmental problems more effectively. The project focuses on urban issues such as transport and infrastructures, energy production and consumption, landuse patterns and the potential for renewal and redevelopment in suburban and inner-city districts.

The **Healthy Cities Project**, established by **WHO** in 1991, seeks to commit municipalities to a concerted programme for improving the quality of the urban environment and health in cities. A selected list of health-related environmental indicators is being collected through the Healthy Cities network. It involves more than 500 municipalities from various world regions including Europe, North America, Latin America and Africa.

The **Urban Environmental Management Program** is funded by the **UNDP** and jointly executed by **UNCHS** (Habitat) and the **World Bank** to strengthen the capacity of cities to address urban problems. The programme seeks to define urban environmental strategies for setting and implementing local action plans in selected cities.

The **Local Agenda 21 Initiative** is an international project by the International Council of Local Environmental Initiative (**ICLEI**) aimed at developing a planning framework for local sustainable development. The project is a direct follow-up of UNCED and responds to the need to assist local authorities in the implementation of local Agenda 21 as established in Chapter 28 of Agenda 21. The initiative by ICLEI in partnership with the International Union of Local Authorities (IULA) places specific attention upon the application and redesign of environmental planning mechanisms such as consultation, audits, target-setting, monitoring and feedback to identify options and assess conflicting interests and values implicit to sustainability.

REFERENCES

Bernes, C (ed) (1993) The Nordic Environment – present state, trends and threats. *Nord 1993: 12*, Nordic Council of Ministers, Copenhagen.

Car Free Cities Charter (1994) Adopted at the Car Free Conference, 24 – 25 March. Eurocities, Amsterdam.

CEC (1990) Green Paper on the Urban Environment. COM(90) 218. Commission of the European Communities, Brussels.

CEC (1991) Europe 2000, outlook for the development of the Community Territory. Office for Official Publications of the European Communities, Luxembourg.

CEC (1992) *The Future of European Cities.* Drewett R, Knight R and Shubert U (Eds). FAST, Commission of the European Communities, Brussels.

City of Copenhagen (1993) Ecological Urban Renewal in Copenhagen: A European Model. Report prepared for the Commission of the European Communities, DGXI, Agency of Environmental Protection, Copenhagen.

Consorzio Venezia Nuova (1993) Il recupero morfologico della laguna di Venezia. Supplemento *Quaderni Trimestrali.* Consorzio Venezia Nuova. Anno I (I), Gennaio-Marzo.

Council of Europe (1982) Nature in cities. Nature and Environment Series No 28, Strasbourg.

ECOTEC (1993) Reducing Transport Emissions through Planning. Report to the Department of the Environment. HMSO, London.

Falinski, J B (1971) Synanthropisation of plant cover. II. Synanthropic flora and vegetation of towns connected with their natural conditions, history and function. Mater Zakl Fitosoc Stos U W Warszawa-Bialowieza **27**, 1–317.

Friends of the Earth (1992), *Less Traffic, Better Towns, Friends of the Earth's Illustrated Guide to Traffic Reduction.* Friends of the Earth Trust Limited, London.

Hahn, E and Simonis, U E (1991) Ecological urban restructuring. *Ekistics* **58** (348/349), May/June–July/August.

Helsinki Energy Board (1993) Energy Supply in Helsinki. A solution saving energy and environment. Pirvola, I and V Hokkanen (Eds). Paper submitted at the Convention of European Municipal Leaders on Climate Change, Amsterdam, 29–31 March.

Horbert, M, Blume, H P, Elvers, H and Sukopp H (1982) Ecological contribution to urban planning. In: Bornkamm, R, Lee, J A and Seaward, M R D (Eds). *Urban Ecology.* 2nd European Ecological Symposium, pp 255–75. Blackwell Scientific Publications, Oxford.

IBGE (1993) Constitution d'un système d'éco-géo-information urbain pour la région Bruxelloise intégrant l'énergie et l'air. Convention Institut Wallon et Mens en Ruimte. Institut Bruxellois pour la Gestion de l'Environnement, Brussels.

ICLEI (1993) Draft Local Action Plans of the Municipalities in the Urban CO$_2$ Reduction Project. International Council for Local Environmental Initiatives, Toronto, Ontario.

IIUE (1992) Towards Sustainable Urban Traffic and Transport: A Preliminary Survey of Approaches in Western Europe. Report prepared for the Commission of the European Communities. International Institute of the Urban Environment, Delft.

Jorgensen, G (1993) Ecological Land-Use Patterns: Which strategies for redevelopment? OECD Workshop on the Ecological City. OECD, Paris.

Kosareva, N B (1993) Housing reforms in Russia. First steps and future potential. *Cities* **10** (3), 198-207.

Llewelyn, D (1993) Urban Capacity: A study of housing capacity in urban areas. Interim Report. The Joseph Rowntree Foundation.

Lowe, D M (1991) Shaping Cities: The Environmental and Human Dimensions. Worldwatch Paper 105, Worldwatch Institute, Washington DC.

LPAC (1993a) Draft 1993 advice on strategic planning guidance for London, for consultation. London Planning Advisory Committee, London.

LPAC (1993b) *London's Skylines and High Buildings.* Report by Greater London Consultants and London Research Centre for the London Planning Advisory Committee, London.

LRC (1993) *London Energy Study Report.* London Research Centre, London.

von Meier, A (1993) Noise Pollution. Report prepared for the Commission of the European Communities. M + P Raadgevende ingenieurs bv, Aalsmeer, The Netherlands.

Miess, M (1979) The climate of cities. In: Laurie, I C (Ed). *Nature in Cities,* pp 9–114. Wiley, Chichester.

Nero, A V (1988) Controlling indoor air pollution. *Scientific American* 258(5), 42–9.

Newman, P and Kenworthy, J (1989) *Cities and Automobile Dependence: An International Sourcebook.* Gower, Aldershot, UK.

Nijkamp, P and Perrels, A (1989) Exploration in Urban Energy Planning. In: Juul, K and Nijkamp, P (Eds) *Urban Energy Planning and Environmental Management.* Commission of the European Communities.

OECD (1990) *L'environnement Urbain: quelles Politiques pour les Années 1990?* Organisation for Economic Cooperation and Development, Paris.

OECD (1991) *The state of the environment.* Organisation for Economic Cooperation and Development, Paris.

OECD (1992) Cities and the Environment. Background note for the International Conference on the economic, social and environmental problems of cities, November 18–20, 1992. Organisation for Economic Cooperation and Development, Environment Directorate's Urban Affairs Division, Paris.

OECD (1993a) Cogeneration, District Heating and Urban Environment. Group on Urban Affairs, Environment Directorate, ENV/UA/E(93)2, Organisation for Economic Cooperation and Development, Paris.

OECD (1993b) Energy and Land Use Planning. Group on Urban Affairs, Environment Directorate, OECD/GD (93)2, Organisation for Economic Cooperation and Development, Paris.

OECD (1993c) *Environmental Data Compendium 1993.* Organisation for Economic Cooperation and Development, Paris.

OECD (1993d) *Environmental Information Systems and Indicators: A Review of Selected Central and Eastern European Countries.* Organisation for Economic Cooperation and Development, Paris.

OECD (1993e) The Role of Local Energy Advisors and Energy Services. Group on Urban Affairs, Environment Directorate, ENV/UA/E(93)4, Organisation for Economic Cooperation and Development, Paris.

OECD/EECT (1993) Urban Travel and Sustainable Development: An Analysis of 132 OECD Cities. Urban Affairs Division, Organisation for Economic Cooperation and Development, Paris.

Oke, T R (1987) *Boundary Layer Climates.* Methuen, London and New York.

Pares, M et al (1985) Descobrir el medi urba. Ecologia d'una ciutat: Barcelona. Centre del Medi Urba, Barcelona.

Région de Bruxelles Capitale (1993) Changeons ensemble la circulation avant qu'elle ne nous change. Ministère des Travaux Publics et des Communications de la Région de Bruxelles Capitale, Brussels.

Rees, W (1992) Ecological footprints and appropriated carrying capacity: what urban economics leaves out. *Environment and Urbanisation* **4** (2), October, 121–30.

RIVM (1992) The environment in Europe: a global perspective. Report 461505001, National Institute of Public Health and Environmental Protection (RIVM), Bilthoven.

SEAC (1992) Ecocycles: The Basis of Sustainable Urban Development. Sweden Environmental Advisory Council Ministry of the Urban Environment and Natural Resources, SOU 1992:43, Stockholm.

Torrie Ralph (1993) Findings and Policy Implications from the Urban CO$_2$ Reduction Project. Paper, International Council for Local Environmental Initiatives (ICLEI), Toronto, Ontario.

UBA (1988), *Lärmbekämpfung 88, Tendenzen, Probleme, Lösungen.* Umweltbundesamt, Berlin.

UK DoE/DoT (1993) *Reducing Transportation Emissions Through Planning.* UK Department of Environment and Department of Transport. HMSO, London.

UK MAFF (1993) *Annual Report of the National Food Survey Committee.* UK Ministry of Agriculture, Fisheries and Food, London.

UNCED (1992) *Agenda 21.* UNCED Secretariat, Conches, Switzerland.

UNEP (1993) Environmental Data Report 1993–94. United Nations Environment Programme. Blackwell Publishers, Oxford.

UNEP/WHO (1992) *Urban Air Pollution in Megacities of the World.* Blackwell Publishers, Oxford.

United Nations (1991) *World Population Prospect.* United Nations, New York.

Vandermotten, C (1994) *Comparative Atlas of Major European Cities.* Université Libre de Bruxelles (ULB), Brussels.

WHO (in press) *Concern for Europe's tomorrow.* World Health Organisation, Copenhagen.

WHO/UNEP (1990) Indoor environment: health aspects of air quality, thermal environment, light and noise. Geneva.

Whitelegg, J (1993) *Transport for a Sustainable Future: The Case for Europe.* Belhaven Press, London.

Zanda, L (1989) The case of Venice. In: *Cities on Water: Impact of Sea Level Rise on Cities and Region.* Marsilio, Venice.

Cyclist in London
Source: David Townend,
The Environmental
Picture Library

11 Human health

INTRODUCTION

In 1989, the first Ministerial European Conference on Environment and Health adopted The European Charter on Environment and Health, stating that:

Good health and well-being require a clean and harmonious environment in which physical, physiological, social and aesthetic factors are all given their due importance. The environment should be regarded as a resource for improving living conditions and increasing well-being.

(WHO, 1990)

This basic statement results from a self-evident but often neglected fact that human health depends on the availability and quality of food, water, air and shelter. It is a basic requirement of health that the global cycles and systems on which all life depends are sustained. The potential of the environment to have adverse effects on health has been realised for centuries. However, in recent years, public awareness of environmental health hazards has increased. This has been due, in part, to the rapid development of industry and new, potentially hazardous, technologies. The progress of scientific research revealing the existence of previously undetected hazards, which may have been present for some time, has also increased public concern.

This chapter summarises the information on the main issues related to the health status of the European population and on known links of health with environmental conditions in Europe. It is derived from material presented in the report *Concern for Europe's Tomorrow* (CET) (WHO, in press). The main objectives of the CET report are the evaluation of the environmental issues of clear health significance, the assessment of European population exposure to the environmental factors possibly affecting health, and the indication of health impacts of these factors.

The scope of the CET report is determined by the definition of health adopted by WHO, which states that *'Health is a state of complete physical, mental and social well-being and not merely the absence of disease or infirmity.'* In the description of health status of the European population presented in the CET and summarised here, a major limitation is the difficulty in providing a proper measurement of health. Mild deficits of health are very subjective; small physiological changes can be measured but their health significance, for example, as predictors of a disease, is not always clear. More serious health deficiencies, even those requiring treatment, are rarely recorded in a way enabling inference about the population health status. Epidemiological studies are necessary to establish prevalence of certain diseases, but the availability of such studies in Europe is very limited due to the costs and resources needed to conduct this type of research. For this reason most of the information on health status is based upon the registration of death – the most severe health deficiency. Mortality data have been collected in all European countries for many years and recorded according to a uniform classification of causes of death. Total and age-specific mortality data, combined with information on age structure of a population, serve to calculate that widely used demographic indicator, 'life expectancy'.

Life expectancy and cause of death constitute the basis for the descriptive analysis of the main aspects of population health status in Europe presented below. The latest data available relate to 1990, and to the countries existing in 1990. For the indices of health status available for all, or most, countries, the results of national data analysis are presented for three groups of countries: countries of Central and Eastern Europe (Poland, East Germany, former Czechoslovakia, Hungary, Romania, Bulgaria, former

Yugoslavia), former USSR (including its parts of Asia), and the remaining countries hereafter referred to in this chapter as 'Western Europe' or 'European OECD countries'. This grouping is dictated largely by the available statistics which, for example, do not separate data related to European and non-European parts of the former USSR. However, the relative homogeneity of the mortality patterns within these three groups allows for this kind of classification. Whenever possible, the descriptions based on mortality data are supplemented by information from epidemiological studies evaluating less severe health deficiencies. Further, the known health hazards of specific environmental media and factors are reviewed, and their possible impact on health in Europe is evaluated. These descriptions are preceded by a short summary of issues to be considered before the links between health and environment are assessed.

ENVIRONMENTAL HEALTH HAZARDS

Humans are exposed to a variety of environmental factors through air, water and other beverages, food, and materials that contact the skin. Exposure occurs in a variety of settings: residential, industrial, occupational, during transport, and both indoors and outdoors. Exposure occurs when there is contact between a person and an environmental factor. The degree of exposure depends on the specific concentration or intensity of the factor and the time interval. Consideration of exposure to environmental factors is crucial to assessing their potential health effects. Occasional contact with a contaminant may have negligible effect, while prolonged contact with the same factor may have serious, though in many cases non-specific, health impacts. If a factor is present in various environmental media, the exposure of an individual occurring from each of these media accumulates (determining *total exposure*). This must be considered when preventative measures are designed, to evaluate both costs and the effectiveness of reducing total exposure when the factor is eliminated from one medium only. Exposure, together with the capacity of the organism to absorb, distribute and metabolise the factor, determine the individual's *internal dose*. This influences the likelihood, type and intensity of health effects.

Adverse health reactions to an exposure may take a whole range of forms, extending from a disturbance of aesthetic values or psychological and physical discomfort, through physiological changes of uncertain health significance, to varying intensity of clinical disease from mild to severe or, in an extreme, uncommon case, to a disease leading to death. A number of health effects become apparent after a delay (latency period for some cancers is as long as 20 years or more) and, in some cases, an exposure accumulated only over several years produces a disease. An individual's response to environmental exposure may be related to personal (including genetic) susceptibility. For example, the effects of air pollution on the respiratory system are different in people with a pre-existing disease (for example, with asthma) than in healthy subjects. Personal behaviour may modify the extent of exposure or the dose of an environmental factor: exercise increases respiration rate and the amount of an air pollutant absorbed; children playing in a sandpit contaminated with lead, or licking walls covered by lead-containing paint, increase their exposure to lead. Synergistic effects may also occur. For example, the risk of lung cancer related to asbestos in air increases ten-fold in people smoking cigarettes as compared with non-smokers. Nutritional status may also modify the response to an environmental agent.

The assessment of the health hazard of an exposure is based upon toxicological evaluation of the factor, which can involve the assessment of health effects of a controlled exposure in a group of animals or humans, usually to

relatively high levels of the factor. An assessment of the actual health risk due to the presence of the hazard in the environment is based on the information concerning the population exposure to the factor and on the knowledge of the exposure–response relationship. Epidemiological studies are an important source of this information. In these studies, the relationship between an environmental factor and a disease, or other health parameter, is assessed taking into account possible influences of other factors on the relationship. This approach is extremely important since, in most cases, the diseases in question are of multifactor aetiology, that is, they may be associated with a variety of factors and their interactions. In spite of a development of epidemiological methodology in the last decades, the quantitative assessments of many health–environment relationships are still incomplete, and the uncertainty of the risk estimates is rather substantial. In part, this is due to incomplete knowledge concerning the health effects of exposures of low intensities occurring in the environment. Also, sufficiently precise assessments of environmental exposure are lacking for most of the European population. Improvements in both require substantially increased resources and expenditure. The estimates of health effects of the environment in the European population summarised below thus have a large margin of uncertainty.

It is important to stress that the environmental health policy should focus on prevention of exposure to the environmental hazards and on the avoidance of health effects in the population. Air, water and food quality standards and guidelines have been prepared by national and international bodies with the purpose of protecting the health of the population. Guidelines are set according to the agent, mode of exposure and health effect, and the consequence of the exceedance of one is difficult to compare with another. Furthermore, exceedance of a guideline does not directly imply the immediate onset of a health effect. In populations this relationship is usually better described as a continuum, where the guideline figure represents a value above which the probability of the onset of a health effect is regarded as undesirable given the nature of the exposure and uncertainties related to the estimation of this figure. Taking into account the inherent uncertainties of the risk assessments leading to guideline setting, the adoption of the precautionary principle of preventative action is advisable.

In most European countries, the public health services have succeeded in identifying, regulating and controlling many of the factors potentially affecting health. Analysis of population exposure to environmental hazards and of population health status may indicate areas where the risk to health is still significant, and where more intensive action of the involved economic sectors, environmental health authorities and society is needed. A factor-by-factor approach should be supplemented by a comprehensive assessment of the combination of environmental, occupational, lifestyle, social and personal factors. It is quite common that selected parts of society, frequently the poorest or most underprivileged members, are exposed to multiple hazardous environmental conditions (see, for example, Chapter 10). The health impact of such combinations may significantly exceed the estimates based on evaluations of individual factors.

THE HEALTH OF THE EUROPEAN POPULATION

Perhaps the simplest indicator of population health is based on self-assessment of health evaluated in population surveys. These data, available from 14 European countries (WHO, 1991), show that the residents of Sweden and Norway are the most satisfied with their health, with 95 per cent of people assessing their health to be fair or average, or better. The worst scores were reported from Poland and Hungary, where the above proportion does not exceed 80 per cent. In all countries, except Finland, the proportion of males rating their health as good is higher than that of females.

The life expectancy at birth in European countries was 74.9 years in 1988. It ranged from 69.3 to 78.2 years, with a clear tendency for the lowest values to occur in Central and Eastern Europe (Figure 11.1). Over the 1980s, there was a steady increase of average life expectancy (from 73.2 years in 1980, or 0.2 years annually). This was a continuation of the average trend observed over the previous decade. However, most of the countries with low life expectancy at the end of the 1980s have shown very little improvement in the past ten years.

In contrast to male perceptions in assessing their health, (above) life expectancy in females is actually greater than that in males: the average difference is about 6.8 years (ranging from 3.7 to 8.8). Large differences are found in countries with overall life expectancies which are relatively high (France) and low (Poland, Hungary). Only about one third of countries in Europe have shown small reductions in this difference after 1980. For the others, the difference is stable or still increasing.

Infant mortality, that is, deaths of children before one year of age, has an important impact on the life-expectancy-at-birth indicator. This indicator improved in all countries of Europe, declining from an average of 16.0 per 1000 live births in 1980 to 10.8 in 1989. Twenty-one per cent of the total increase of average life expectancy at birth between 1980 and 1989 can be attributed to the reductions in infant mortality. The rates of decline in infant mortality were similar in the Western and Eastern parts of Europe, with the result that the pattern of substantial differences between countries was maintained. In 1988–90, the highest rates (22 to 27 per 1000) were noted in Romania, Albania, former Yugoslavia and the former USSR, and the lowest (below 6 per 1000) in Iceland, Finland and Sweden. The most significant contribution to the infant mortality difference observed between various parts of Europe is the number of deaths due to communicable diseases. Their frequency ranged from 0.4 per 1000 live births in Western Europe, through 4.7 in Central and Eastern European countries, to 9.2 per 1000 in the former USSR. This can be attributed, at least in part, to poorer sanitary conditions of human settlements (eg, lack of piped water supply at home) in the Eastern part of the region. A study conducted in the Czech Republic has detected an increase of infant mortality, and in particular of the post-neonatal mortality due to respiratory diseases, in regions with elevated concentrations of suspended particulates in the air (Bobak and Leon, 1992). Though the observed association has yet to be confirmed by other studies, the possibility of a causal link between infant health and air pollution cannot be excluded.

Similarly to life expectancy at birth, life expectancy of people at ages after infancy, on average, increased in Europe during the 1980s. However, a clear difference in trends can be seen when countries are divided into groups. Whereas in Western European countries the life expectancy at ages 1, 15, 45 and 65 displayed a steady increase over the decade, there was a stagnation of the indicators in Central and Eastern Europe, and even a decline in the former USSR.

The above difference in trends of life expectancy between the groups of countries is due to differences in trends of age-specific mortality. In Western Europe, mortality has declined in all age groups, with approximately constant rates in the 1970s and 1980s. Mortality in the remaining countries of Europe followed this pattern until the mid-1970s (though on a higher absolute level). However, in the following years, in Central and Eastern Europe, the decline of mortality was much smaller or remained static, and in males aged 15 to 64 years there was an increase in the 1980s (Figure 11.2).

When the mortality at all ages is considered, the most common causes of death at the end of the 1980s were diseases of the circulatory system (42 per cent of all deaths in Western

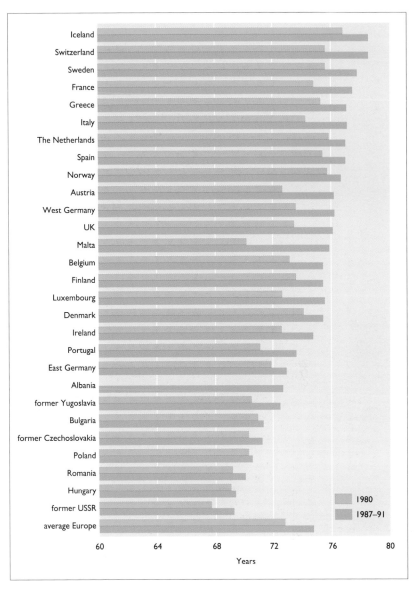

Europe, 55 per cent in Central and Eastern European countries, and 58 per cent in the former USSR), followed by cancers (25, 17 and 16 per cent, respectively) (Figure 11.3). The third leading group of causes of death was injuries and poisonings (6 to 9 per cent), followed by respiratory diseases (6 to 7 per cent). Communicable diseases were a relatively infrequent cause of death, responsible for approximately 1 per cent of all deaths, although the rate was twice as high in the former USSR as in Western Europe. In all parts of Europe, mortality due to communicable diseases halved over the last two decades.

The structure of causes in total mortality reflects, to a large extent, the relative frequency of diseases in the oldest age groups. In age groups below 45 years, the structure of causes of deaths is different. Among young males, close to 50 per cent of all deaths are due to injury and poisoning. Also among young females, fatal accidents are a very common cause of death (24 to 33 per cent of mortality) but in the 15 to 44 age bracket, in all countries except the former USSR, neoplasms are the leading cause of female death (29 to 33 per cent).

Circulatory system diseases

The main group of causes of death responsible for the difference in trends between the Western and Eastern parts of Europe were the circulatory system diseases. A steady decline of this mortality was observed in Western countries throughout the 1970s and 1980s, while an increase or, at best,

Figure 11.1
Life expectancy at birth in 1980, 1987–91, by country
Source: WHO (in press)

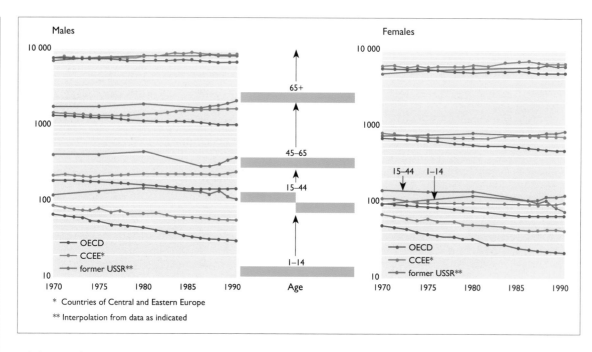

Figure 11.2
Trends in age-specific mortality by group of countries and sex per 100 000 people (1970–90)
Source: WHO (in press)

stabilisation of rates was seen in Central and Eastern Europe. At the end of the 1980s, there was a more than two-fold difference in rates between the groups of countries in people aged 15 to 64 years, and a 70 per cent excess of mortality in Central and Eastern Europe compared with Western Europe in the oldest age group. The prevalence of cardiovascular diseases, assessed in population studies in some European countries, is also higher in the Eastern countries than in the West. The main recognised risk factors of cardiovascular diseases, and coronary heart disease in particular, are hypertension, high cholesterol levels and smoking. Less clear are the roles of obesity and inadequate physical activity. Diet, especially the intake of saturated fats and salt, can have an impact on the risk factors and, indirectly, on disease. There seems to be a rather limited relationship between cardiovascular diseases and environmental factors, although adequate environmental conditions (housing, work, recreational facilities) can influence behaviour and healthy lifestyles, and, indirectly, can modify the risk of cardiovascular diseases. Potentially, increased levels of carbon monoxide in air, due to traffic emissions (eg, in traffic tunnels) or fossil fuel combustion indoors, can aggravate arrhythmia or angina symptoms in patients with cardiovascular disease.

Cancers

The differences of mortality due to malignant neoplasms (cancers) in various parts of Europe are less dramatic than those for cardiovascular diseases. In the late 1980s, in age groups below 65 years, this mortality was higher in Central and Eastern Europe by up to 25 per cent, or 50 per cent in the former USSR, than in the rest of the continent. This was the result of a combined effect of an increasing trend in Central and Eastern countries accompanied by a decrease of the rates in Western Europe. In people 65 years of age and over, cancers were 25 per cent more commonly diagnosed as a cause of death in Western Europe than in Central and Eastern parts, and the rates changed very little during the last two decades. A detailed analysis of spatial patterns of cancer mortality at sub-national level was conducted recently for EU countries (Smans et al, 1992). The known differences in diagnostic and coding practices may bias the comparisons. However, some spatial trends, for example lower mortality rates for many cancer sites in the south of Italy than in the centre or north, may be at least partly real. At present, it is

difficult to find a convincing explanation for these differences.

Although the aetiology of most cancers is still obscure, several risk factors have been identified. Tobacco smoking undoubtedly increases the risk of lung cancer (85 per cent of all cases in men are attributable to smoking) and also of several other cancers (oesophagus, oral cavity, pharynx, larynx, bladder, kidney). Dietary factors are suspected to be related to digestive tract cancers. For stomach cancer, a diet rich in starchy food as well as smoked, salted and fried foods is suspected to increase the risk. High intake of fat is suspected to increase risk of colon and rectum cancers. Diet rich in vegetables, citrus fruit and fibre may reduce the risk. Various hormonal factors may increase the risk of breast, corpus uteri and prostate cancers. A sexually transmitted human papilloma and other viruses are suspected as risk factors for cervix uteri cancer, and chronic hepatitis B virus infection increases liver cancer risk. Various chemicals present in the environment are known to have carcinogenic properties, for example: benzene, soots, arsenic and its compounds, asbestos, nickel compounds, radon and its decay products. In most cases, the population exposure to these substances is rather limited and a quantitative assessment of the contribution of environmental exposures to total cancer incidence is difficult. However, as discussed further in this chapter, some components of urban air pollution, selected occupational exposures and ionising radiation have been shown to increase incidence of, and mortality from, lung and other cancers in the European population.

Respiratory system diseases

The main inter-country differences in mortality due to respiratory system diseases are as a result of mortality in children. In the former USSR, death rates (mainly from pneumonia and acute respiratory infections) exceeded those in Western Europe 20-fold, and those in Central and Eastern Europe, three-fold. In people 15 to 64 years of age, mortality due to respiratory diseases steadily decreased in Western Europe over the last two decades but was stable, or decreased much less, in Central and Eastern Europe. The differences in trend were greater in males than in females. As a result, by the end of the 1980s, the rates differed two-fold between Eastern and Western Europe, although they were similar 20 years earlier. In people 65 years of age and older, where chronic respiratory diseases are dominant, the rates and trends in mortality were similar in all parts of the region, and decreasing. Data from population studies indicate that close to 10 per cent of the adult population may suffer from chronic respiratory diseases. Since these diseases may severely restrict normal activity of people for decades of their life, the public health significance of chronic respiratory diseases is greater than that indicated by the proportional mortality rates. Besides tobacco smoking, which is probably the main risk factor of chronic obstructive airways disease, air pollution (outdoor and indoor, both occupational and residential) may play a significant role in their aetiology. Host factors (for example, bronchial hypersensitivity) are important risk factors for the asthmatic form of the disease.

Communicable diseases

Declining mortality from tuberculosis of the respiratory system is a main cause of the diminishing mortality brought about by communicable diseases in Europe. The incidence of these diseases was markedly reduced in recent decades in most European countries. However, epidemics of some communicable diseases are still observed in selected regions. An example is the dramatic increase of diphtheria incidence (eliminated from a number of European countries) in some parts of the former USSR during 1991–92. High incidence of hepatitis B and tuberculosis is also observed in some countries (Romania, Bulgaria, the former USSR, Poland). The aetiology of most of the communicable diseases is well understood and many of them are related to basic environmental conditions, for example: proper water sanitation, communal waste disposal, hygienic housing conditions, food processing and storage. These factors may have the most pronounced impact on the occurrence of the communicable diseases in infants, and on infant mortality, as mentioned above. Lifestyles, as well as the availability and effectiveness of preventative measures (eg, vaccination) are important factors related to the incidence of some of these diseases.

The AIDS epidemic increased rapidly in Europe up to 1990. The rate of increase in the number of new cases diagnosed was slower at the beginning of the 1990s than in the previous decade, but the incidence is still growing. Over 95 per cent of European cases have been registered in Western Europe.

Injury and poisoning

Mortality due to injury and poisoning has, on average, declined in Europe over the last two decades but the rates in Eastern Europe remain double those in Western countries. Substantial variability of the rates is seen between countries, also within particular parts of Europe. The main cause of accidental deaths was motor vehicle traffic accidents (in various countries, from 21 to 52 per cent of male, and from 11 to 50 per cent of female, deaths in accidents) and suicide (from 7 to 71 per cent). The number of casualties from road traffic accidents (over 350 people killed and 6000 injured *daily* in Europe) exceeds by several orders of magnitude the number of victims of other types of accidents or emergencies. Accidental poisonings are not a common cause of death in most countries, accounting for 3 per cent of male and 5 per cent of female accidental deaths in Western Europe, 8 and 6 per cent respectively in Central and Eastern Europe, but as much as 21 per cent in the former USSR. The important characteristics of mortality due to accidents, and mortality due to traffic accidents in particular, are their similar intensity in all ages over 15 years; this is in contrast to deaths due to the remaining causes, where the rates in the oldest age groups exceed those in young adults by two orders of magnitude. As a result, the proportion of deaths due to accidents in young people is relatively high (reaching 19 per

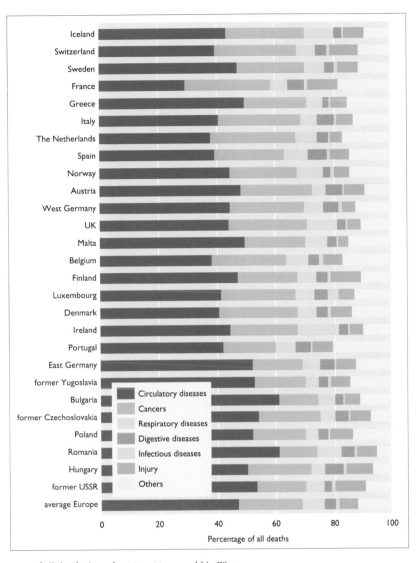

cent of all deaths in males 15 to 44 years old in Western Europe, and 12 to 14 per cent in the remaining parts of Europe), and the social costs ('years of life lost') are very significant. Even higher are the costs of a larger number of non-fatal injuries resulting from traffic accidents, involving temporal or permanent restrictions of human activity and disability. Traffic accidents are, at least in part, due to inappropriate development of the transport networks and their associated infrastructure, an integral part of today's urban environment (see Chapters 10 and 21).

ENVIRONMENT-RELATED HEALTH PROBLEMS IN EUROPE

The CET analysis (WHO, in press) of exposure of the European population to environmental factors, and of the health significance of the exposure, has been carried out in the context of a lack of detailed and comprehensive studies of health effects due to environmental factors in Europe. The CET was compiled in order to obtain an idea of the situation in Europe, recognising from the start that few reliable data are available on which to base extrapolations, and that there will be errors in the assessment of numbers of people allocated to a particular exposure level. The unsatisfactory nature of the available information arises not only from problems of data collection but also from the complexity of health–environment relationships *per se*, and the number of confounders to understanding the true causes and effects.

Notwithstanding these important caveats, the results of the CET confirm that a considerable proportion of the

Figure 11.3
Proportional distribution of main causes of death, 1987–91, by country
Source: WHO (in press)

European population is exposed to environmental factors to an extent which is considered to pose a threat to some aspect of health. The precision of the estimates is limited, and the large errors which may be associated with them will be due mainly to the incompleteness of the data on population exposure to particular factors. The results presented below represent the best estimates currently available; nevertheless, the figures should be interpreted with caution.

Air pollution

The most common environment-related health problems concern exposure to excessive levels of air pollutants, possibly affecting hundreds of millions of people in Europe. In most cases, the exposure relates to urban populations and occurs in short episodes. In the past decade, many epidemiological studies have observed health effects related to this type of exposure, for example, in Athens, Barcelona, Cracow and several French and German cities. Some of the cities where the concentrations of the most commonly monitored air pollutants exceed WHO air quality guidelines are listed in Chapter 4. Among the air pollutants considered, particulate matter (PM), measured as 'total suspended particulates' (TSP) or black smoke, is estimated to pose the greatest potential burden to health. An example of the results supporting this conclusion is the proportion of chronic respiratory diseases expected to be related to PM: as many as 15 per cent of asthma cases and 7 per cent of obstructive airways disease (OAD) occurring in the urban European population (at least 18 000 cases a year of each disease) are estimated to be possibly related to prolonged exposure to high concentrations of particulates. In cities with the highest dust concentrations, the proportion of cases attributable to pollution reaches 23 per cent of asthma and 11 per cent of OAD cases. Among other health problems expected to be caused by excess PM and affecting a considerable number of Europe's residents are various short-term respiratory symptoms and diseases, chronic and transient decrements of lung function, and even precipitation of mortality. Various respiratory problems, some requiring hospitalisation, are estimated to be related to ambient air pollution characterised by elevated levels of sulphur dioxide (SO_2) and nitrogen oxides (NO_x). Up to 30 per cent of the European population may experience these pollutants in concentrations exceeding daily WHO air quality guideline levels, though for most of these people the exposure is limited to less than 30 days per year. Exposure to elevated ozone levels, though affecting a very large part of the European population, seems to produce less severe health effects, such as transient decrements in lung function (especially in children or in exercising people), cough or eye irritation episodes.

Indoor air pollution was also considered as an important factor. Examples of the most common types of indoor air pollutants are carbon monoxide (CO) and nitrogen dioxide (NO_2) generated by indoor combustion sources (eg, unvented gas stoves), tobacco smoke, volatile organic compounds emitted from building materials or consumer products (eg, paints and cleaning agents), radon and its decay products (see below), and asbestos fibres. Population exposure to these pollutants is very heterogeneous and depends on a variety of factors, including the geological structure of the soil under the building (radon), the construction of the building (ventilation rates, building materials) and personal behaviour (use of pollutant emitting agents, presence of smokers, time spent in various indoor spaces). The assessment of such exposures is thus an extremely difficult task and is not available for most of the European population. Consequently, the possibility of a quantification of the health impact of indoor air pollution is very limited. However, the risk due to the exposure to several indoor air pollutants is established relatively well. For

example, the risk of lower respiratory tract disease in infants exposed to tobacco smoke due to a mother smoking at home is increased by some 50 to 100 per cent, according to a recent comprehensive review (USEPA, 1992). The risk of respiratory disease may be increased by some 20 per cent due to nitrogen dioxide exposure associated with the use of gas stoves at home, as indicated by combined evidence from several studies (Hasselblad et al, 1992).

Besides the acute or chronic respiratory system diseases, which are the most common health effects of air pollution, some components of urban or indoor air pollution may increase the risk of cancers. For instance, asbestos, benzene and soots are classified by the International Agency for Research on Cancer as 'carcinogenic', and benzo[a]pyrene and diesel engine exhausts as 'probably carcinogenic'. The limited database, however, does not allow an accurate quantification of the potential impact of present ambient air pollution on cancer incidence in Europe; only site- and case-specific conclusions can currently be drawn. An epidemiological study conducted in Cracow, Poland, has shown an increase in lung cancer risk in residents of the city centre which, especially in the past, was heavily polluted (Jedrychowski et al, 1990). The main source of pollution was combustion of coal for heating. Also, there is evidence for increased lung cancer risk in the populations surrounding some types of industry, in particular non-ferrous smelters, where arsenic emissions may be of importance (Pershagen and Simonato, 1993). Better established than the impact of ambient air pollution on cancer is the relationship between the risk of cancer and indoor air pollution. In particular, there is sufficient evidence to expect a 20 to 30 per cent increase in risk of lung cancer in non-smokers married to smokers, related to the environmental tobacco smoke exposure (Pershagen, 1993).

Contaminated food and water

Contamination of drinking water and food with microbiological agents can be a source of a variety of communicable diseases, such as hepatitis A, salmonellosis or shigellosis. Though the contamination sources, routes of transmission and preventative measures are, in general, known, there is evidence that the protection of the population is not sufficient in several areas of Europe (WHO, in press). Outbreaks of diseases due to microbiological contamination of drinking water have been reported in the UK, France, the Russian Federation, former Yugoslavia, Romania and Scandinavian countries. Microbiological contamination of bathing water, mainly in the Mediterranean, is estimated to result in over two million cases of gastrointestinal diseases annually. Potential health hazards of chemical water pollution are also significant. Predicted nitrate concentrations of groundwaters in several areas across Europe with intensive agriculture were found to exceed guideline levels designed to protect infants from methaemoglobinaemia, a serious, life-threatening disease (see Chapter 5). Concentrations of arsenic found in waters of selected regions of Bulgaria, Hungary and Romania can also, potentially, lead to health problems (skin cancer). In Estonia, high levels of fluoride have been observed and were found to cause endemic fluorosis. However, most of the hazardous chemicals are effectively regulated and controlled, and the exposure to chemicals in drinking water seems, generally, to have less health impact on the total population of Europe than does the microbiological contamination. The data on the local situations, where the exposure may indeed pose a health risk, are very scarce and this hampers a more precise estimation of the potential problem.

Lead

From a variety of pollutants to which an exposure may occur through various media, lead is identified as a substance posing a health risk in several regions of Europe. The available data indicate a substantial decrease in exposure to lead in a number of countries in Western Europe over the last decade, due mainly to the reduced quantity of lead in petrol. However, in selected locations in Central and Eastern Europe, mainly around lead-emitting industries (eg, Hungary, Romania), the measured population exposure to lead is still high, possibly resulting in impaired mental development of children and in behavioural problems. It is estimated that at least 400 000 children in Eastern parts of Europe may be affected.

Radon

The main source of population exposure to ionising radiation is naturally occurring radon and its decay products (Chapter 16). The populations most at risk are miners and residents of particular areas where radon is emitted naturally from the soil. According to the existing data (WHO, in press), approximately 2 million people in Europe are estimated to be at elevated health risk from this cause. The main health effect of ionising radiation is neoplasms. Recent studies from Finland, Norway and Sweden indicate that as many as 10 to 20 per cent of all lung cancer cases in these countries can be attributed to residential radon exposure (Simonato and Pershagen, 1993). For the EU it is estimated that up to 10 000 cancer deaths (ie, 1 per cent of all cancer deaths) are caused annually by radon (Trichopoulos and Karakatsani, 1993).

Ultraviolet radiation

Experimental evidence indicates that the ultraviolet part of solar radiation (and its shorter-wavelength part, UV-B, in particular) is a risk factor of skin cancer (IARC, 1992) (see Chapter 16). Also, UV-B is a recognised risk factor for cataract, and it is suspected that UV-B may suppress normal immune responses. It is estimated that the increase in terrestrial levels of UV-B by 1 per cent, related to stratospheric ozone depletion, may result in 1 to 2 per cent excess in the health effects. However, the assessment of a possible increase of cancer or cataract cases in Europe due to UV-B exposure is difficult to evaluate because of the insufficient precision of the disease registers, lack of exposure data and the possible influence of personal, or genetic, characteristics on the dose–effect relationship (see Chapter 28). Furthermore, many of the potential hazards of UV-B can be eliminated by individual conduct, for example by avoidance of excessive exposure to sunlight.

High intensity electromagnetic fields

The potential health risks related to the relatively widespread exposure to low-frequency (50 to 60 Hz) high intensity electromagnetic fields is still being investigated and no conclusive evidence is available (Draper, 1993) (see Chapter 16).

Wastes

Waste collection and disposal and treatment procedures are a potential health hazard because wastes can, in principle, contain any hazardous chemical, biological and physical agent. Through an emission to ambient air or leakage to surface and groundwater, the environment surrounding a waste site can be biologically or chemically contaminated (see Chapters 7 and 15). Several technical and organisational means can (and do) minimise or even eliminate the waste-related threats to human health. Studies conducted to evaluate health impact of waste sites have failed to provide conclusive evidence of a causative role of waste sites in the incidence of more serious health effects (such as cancers, reproductive outcomes). In part, this may be due to considerable methodological difficulties faced by the studies, mainly in obtaining adequate data on exposure to relevant pathogenic agents. However, increased anxiety, diarrhoea and reports of disturbing odours have been registered more often in populations living in the vicinity of these installations than in the control groups (WHO, in press). Notwithstanding lack of epidemiological evidence – potential, present and future – more serious health impacts of the poorly managed or uncontrolled waste sites should not be understated.

Housing

For human health, housing conditions are a very important element of the environment. To a large extent, they depend on the type and status of urban development, described in detail in Chapter 10. Among the main factors affecting health, deficiencies in sanitary equipment of houses were identified by the CET project as the most significant. Poor sanitary conditions (eg, lack of piped water supply at home) increase the risk of communicable diseases. Improper housing conditions may result in exposure of the residents to hypo- or hyperthermia and, in the case of insufficient ventilation, to indoor air pollutants, with some of the health consequences summarised above. An example of the effect of insufficient ventilation in newly built, air-tight buildings is the growth of mite allergen, possibly promoting allergic sensitivity among residents, observed in cold parts of Sweden (Wickman et al, 1992). Approximately 40 000 fatal accidents are estimated to occur annually in houses in Europe, although it is difficult to assess the proportion of these accidents related to unsafe housing conditions or faulty design, construction or maintenance. The number of less severe accidents, resulting in an injury, is expected to be a hundred times greater. Physical or psychological discomfort or stress (for example, caused by excessive noise – see Chapter 16 – estimated to affect a quarter of the European population) are also adverse health effects related to deficient housing conditions and, sometimes, to improper urban planning and development.

Workplace hazards

Health effects of occupational factors, in contrast to other environmental exposures, can usually be anticipated in certain types of industries or workplaces. Appropriate preventative measures can reduce health impacts of inherently harmful types of manufacturing or other work-related human activities, though it is rarely possible to eliminate some of these hazards completely. As a consequence, the impact of unhealthy working conditions is still significant in the European population. Probably the most acute are accidental injuries, experienced by 25 per cent of workers in high risk occupations (construction, mining, fishing), resulting in approximately 25 000 deaths annually and causing permanent disability with a rate of between 20 to 30 per 100 000 people (ILO, 1992). It is estimated that between 1 and 20 per cent of all cancers are work-related. The rates of chemically induced occupational diseases decreased in recent years due to increased awareness of the hazards and as a result of preventative measures. However, differing degrees of sophistication of occupational health and safety legislation result in large differences between countries.

Accidents

The significant figures from road traffic accidents dominate accident casualty statistics, for which the road traffic environment is certainly partly responsible. Next to this, the health effects of industrial and nuclear accidents, or biological contamination of food and drinking water, can be documented by hundreds of fatalities and thousands of injured or sick registered annually in Europe and ascribed to a known, though unexpected, event. Between 1980 and 1991, over 1100 people were killed in major chemical accidents which occurred in Europe and were registered by the OECD. Many of the most serious accidents were related to the transport of gas, oil or chemicals, including transport by pipelines. The spectrum of health problems of the accidents covers a wide range of acute effects – injuries, burns, poisonings – as well as long-term, or delayed, outcomes, including increased risk of neoplastic diseases or of congenital malformations in children of exposed parents. Mental health can also be affected by the event and the emergency situation (see also Chapters 18 and 30).

CONCLUSIONS

Although the dominating risk factors of the most frequent severe diseases are related to various host characteristics (eg, genetic predisposition or other individual susceptibility) or behavioural and lifestyle factors (eg, tobacco smoking, diet, excessive alcohol drinking), a number of environmental factors may, and probably do, adversely influence the health of the European population. The analysis conducted for the CET report (WHO, in press) focuses on the observed or observable effects, that is, changes in health occurring immediately or shortly after an exposure to an environmental agent. Small physiological changes, possibly related to prolonged exposures of low intensities and especially to combination of exposures, may accumulate and result in an emergence of adverse health effects in the long term. At present, a number of the basic health indicators (life expectancy, mortality, incidence of communicable diseases) show positive changes. One can assume that these positive changes are due largely to alterations in lifestyle and improvements in medical care, but they may also result partly from improvement in living and environmental conditions (improved sanitation, safer working conditions, reduced air pollution, and better housing preventing overcrowding and the influence of adverse climatic conditions). Deviations of health indicators from these positive trends observed in some parts of the European population should be of significant concern and a reason for comprehensive responses. Besides the promotion of healthy lifestyles, these should aim at creating an environment which promotes health. Effective prevention or reduction of population exposures to the recognised health hazards should be of the highest priority. An important condition of effective preventative action and management of the environmental health risk is the ability to assess the risk properly. This requires adequate support by environmental health research and monitoring and, in particular, European-wide epidemiological studies.

REFERENCES

Bobak, M and Leon, D A (1992) Air pollution and infant mortality in the Czech Republic, 1986–88. *Lancet* **340**, 1010–14.

Draper, G (1993) Electromagnetic fields and childhood cancers. *British Medical Journal* **307**, 884–5.

Hasselblad, V, Eddy, D M and Kotchmar, D J (1992) Synthesis of environmental evidence: nitrogen dioxide epidemiology studies. *Journal of Air and Waste Management Association* **42**, 662–71.

IARC (1992) Monographs on the evaluation of carcinogenic risk to humans. Vol 55: *Solar and ultraviolet radiation*. International Agency for Research on Cancer, Lyons.

ILO (1992) *International labour statistics 1980–1991*. International Labour Office, Geneva.

Jedrychowski, W, Becher, H, Wahrendorf, J and Basa-Cierpialek, Z (1990) Case-control study of lung cancer with special reference to the effect of air pollution in Poland. *Journal of Epidemiology and Community Health* **44**, 114–20.

Pershagen, G (1993) Passive smoking and lung cancer. In: Samet, J M (Ed) *Epidemiology of lung cancer*. Marcel Dekker, New York.

Pershagen, G and Simonato, L (1993) Epidemiological evidence on outdoor air pollution and cancer. In: Tomatis, L (Ed) *Indoor and outdoor air pollution and cancer*, pp 135–48. Springer-Verlag, Berlin, Heidelberg, New York.

Simonato, L and Pershagen, G (1993) Epidemiological evidence on indoor air pollution and cancer. In: Tomatis, L (Ed) *Indoor and outdoor air pollution and cancer*, pp 119–34. Springer-Verlag, Berlin, Heidelberg, New York.

Smans, M, Muir, C S and Boyle, P (Eds) (1992) *Atlas of cancer mortality in the European Economic Community*. IARC Scientific Publications 107. International Agency for Research on Cancer, Lyons.

Trichopoulos, D and Karakatsani, A (1993) Environment and human health. In: *Environment and health*. Project paper, working document PE 164.709, part A, Brussels, 25 October.STOA (Scientific and Technological Options Assessment), European Parliament and the Commission of the European Communities Directorate General for Research.

USEPA (1992) *Respiratory health effects of passive smoking: lung cancer and other diseases*. Publication EPA/600/6-90/006FF, US Environmental Protection Agency, Washington DC.

WHO (1990) *Environment and health: the European charter and commentary*. WHO Regional Publications, European Series No 35. World Health Organisation, Copenhagen.

WHO (1991) *Second evaluation of the strategy health for all by the year 2000*. World Health Organisation, Regional Office for Europe. EUR/RC41/Inf.Doc./3., Copenhagen.

WHO (in press) *Concern for Europe's tomorrow*. World Health Organisation, Copenhagen.

Wickman, M, Nordvall, S L and Pershagen, G (1992) Risk factors in early childhood for sensitisation to airborne allergens. *Pediatric Allergy and Immunology* **3**, 128–33.

Part III
Pressures

Introduction

Many environmental problems that have come to light in the assessment of Europe's environment are rooted in the increased demand and transformation of natural resources and in the increased emissions and waste associated with current patterns of economic development. While contributing to improved living standards, the unprecedent growth of economic activities during the last few decades, together with the development and spread of new technologies, have significantly affected the state of the environment.

This section examines the causes of changes in Europe's environment by focusing on the pressures induced by human activities – the agents and means by which the environment is actually impacted. The ways in which human activities interact with environmental processes are not always obvious. While, for instance, energy production and transportation are clearly related to carbon dioxide emissions and the greenhouse effect, the relationships between other activities and the environment are not straightforward. To understand these interactions, the relationships between the state of the environment and human activities need to be systematically identified and assessed through the pressures they exert on the environment. Causes of environmental change are diverse, but are linked together through production and consumption processes.

People and activities contribute to the alteration of environmental processes in a number of ways:

- exploitation and degradation of natural resources (renewable and non renewable);
- direct transformation of physical structure and habitats;
- release of emissions and wastes;
- removal and manipulation of species from the biotic system; and
- disaster events and accidents which lead to one of the above.

Chapter 12 examines the links between population trends, production and consumption patterns and environmental change. Chapter 13 addresses the exploitation of natural resources. Emissions to air and water are analysed in Chapter 14, and waste in Chapter 15. Physical fields, such as noise and radiation, are examined in Chapter 16, chemicals and genetically modified organisms in Chapter 17, and natural and technological hazards in Chapter 18.

Each category of pressure described above requires different considerations for its adequate control. Concentrating on individual effects or activities has been shown to be unsatisfactory for identifying adequate solutions. Action programmes dealing with different environmental media or human activities separately have often resulted in environmental problems being moved from one part of the environment to another. The development of an integrated approach has emerged as crucial to understanding the underlying patterns which link environmental changes and human activities. New concepts, such as pollution prevention, clean production and integrated pollution control, have emerged as integrated strategies to reduce the overall environmental impact of production and consumption.

Kurfürstendamm,
Berlin
Source: Eye Ubiquitous

12 Population, production and consumption

INTRODUCTION

Population is a key element in the human impact on the environment. Population growth is often held to be the root cause of environmental problems, but the interactions between population, economic development and environmental change are too complex to support this conclusion. Other important factors, such as distributional patterns, migration and living standards, have a major role affecting the impact of population on natural resources and the environment.

Population growth

World

In 1992 the world population was 5.5 billion and, on the basis of UN projections, is expected to increase by 1 billion by the year 2000. World population growth between 1950 and 1992 added to the planet more people (3 billion) than the total number of people who inhabited the planet in 1950 (2.5 billion). This exponential increase is still occurring, although the rate of population growth has fallen since 1970 from 2.1 to 1.7 per cent. In fact the population growth rate has not fallen as fast as the population base has grown. Thus, the global population increases by 240 000 a day and by over 90 million every year. This increase is mainly in Asia, Africa and Latin America (Figure 12.1). The very young age structure in these countries contrasts with the increasing old-age dependency ratio in developed countries, and will maintain continued vast population growth even if fertility rates drop significantly (Figure 12.2).

The absolute increase in population during the last few decades has placed a greater demand on natural resources worldwide. However, to understand the impact of population on the environment it is essential to consider the distribution of population growth as well as its absolute numbers. The impact of people on the environment depends in fact on the regional concentration and population density, as well as on the level of economic development and per capita consumption of natural resources in the different world regions.

Europe

The European population reached about 680 million people in 1990, which represented 12.8 per cent of the world's population. Comparable population estimates for Europe, as defined in this report, are not available for the preceding years. Historical trends can be examined by using UN statistics which provide historical data for Europe (excluding the European part of the former USSR) and for all the former

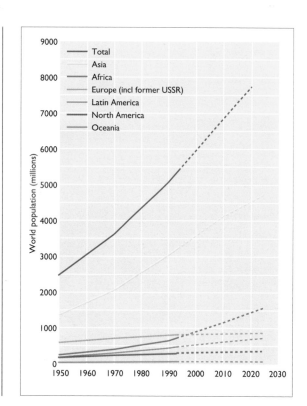

Figure 12.1
World population
growth by regions,
1950–2025
Source: UN, 1991

Figure 12.2
Population age
pyramids, 1985
and 2025
Source: UN, 1990

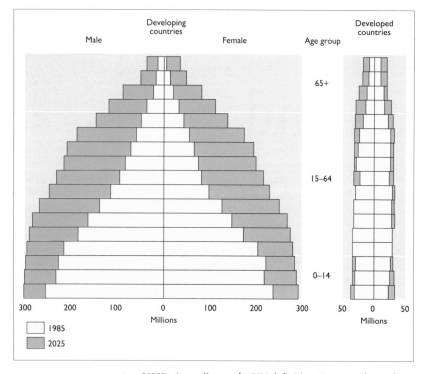

Table 12.1
Vital rates within
European regions
Source: UN, 1993

USSR's share of world population declined from 7 to 5 per cent and is expected to decline further to 4 per cent by 2025.

After a rapid increase in the nineteenth century and a slower growth rate in the first part of the twentieth century, Europe's own population has remained relatively steady. European population, excluding the European part of the former USSR, reached about 509 million in 1990 (398 million in 1970). Including all of the former USSR, this figure was 779 million people in 1990 (692 million in 1970). The total population for Europe and the former USSR is projected to reach 871.6 million in 1995 and 966.8 million in 2025. However, there are considerable variations in population growth within Europe. Population in EU countries is projected to rise from 351.6 million in 1995 to 368.0 million in 2025 (an increase of 4.45 per cent), whereas for the non-EU part of Europe (including the whole former USSR) population is projected to rise from 519.9 million to 598.8 million in the same period (an increase of 13.17 per cent).

Distribution and density

Regional diversities in European population trends are shown in Table 12.1, which compares vital rates across European regions, defined according to the UN classification, between 1980 and 1985 and between 1985 and 1990. Growth rates differed markedly between 1980 and 1985, with low growth rates in Northern and Western Europe (respectively 0.2 and 0.1 per cent) and high growth rates in Eastern and Southern Europe (both at 0.5 per cent). A major change in regional patterns is represented by the convergence of regional growth rates between 1985 and 1990, ranging between 0.3 and 0.5 per cent: a combined effect of increased international migration towards Western and Northern European countries and decreased birth rate in Eastern and Southern European countries.

Compared with other world regions, Europe (with the exception of the former USSR) is very densely populated. However, density patterns vary across European countries, as illustrated in Map 12.1.

The impact of population growth on the environment can be better understood in combination with such trends as mobility and urbanisation. Movements of people can be both cause and effect of environmental change. Degraded environmental conditions can prompt emigration, while the influx of new people can strain a local carrying capacity. In Europe, the most manifest example of how migration has changed the environment was the rapid urbanisation which started in the eighteenth century, as illustrated in Chapter 10.

Migration

Migration inside Europe has historically been characterised by long-term movements from less developed to more prosperous regions. In the EU there was, however, a virtual standstill in inter-regional migration in the 1970s and early 1980s, and there are as yet no signs of a revival. Amongst EU countries, with the exception of Ireland, net migration from weaker regions has been close to zero and in some cases even slightly negative over the second half of the 1980s (CEC, 1991). Eastern Europe has experienced considerable migration in the post-war period. Until 1989, Latvia had the highest population growth in all of Europe, due to movements from Russia, Belarus and Ukraine. Most immigration in Latvia has been to urban areas – which is leading to environmental conflicts, while the natural population growth rate continually decreased from the late 1950s to the 1980s (Latvia Environmental Protection Committee, 1992).

A new wave of migration between Eastern and Western Europe is now taking place as a result of political change, the main flow entering West Germany. Flows from Eastern to Western Germany are expected to continue at an estimated rate of 200 000 people a year, implying an annual loss to the Eastern German population of 1.25 per cent up to the middle of the 1990s (CEC, 1991).

USSR. According to the UN definition, Europe's share of world population declined from 16 per cent in 1950 to 9 per cent in 1990. UN projections show a further decline to 6 per cent by 2025 (UN, 1993). Between 1950 and 1990 the former

Region	1980-85			1985–90		
	Growth (%)	Birth(‰)	Death (‰)	Growth (%)	Birth(‰)	Death (‰)
Eastern	0.5	16.4	10.7	0.3	14.9	11.1
Northern	0.2	13.1	11.3	0.3	13.6	11.2
Southern	0.5	13.2	8.9	0.3	11.6	9.1
Western	0.1	12.3	11.0	0.5	12.2	10.6

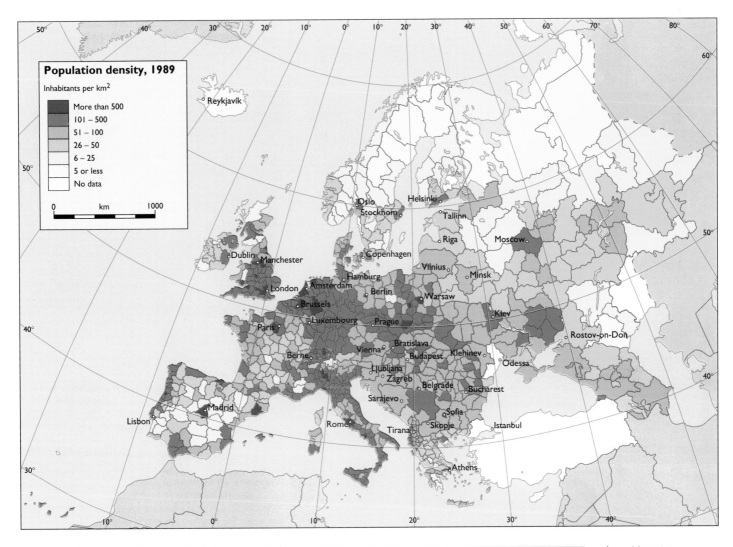

Population density, 1989

Inhabitants per km²

- More than 500
- 101 – 500
- 51 – 100
- 26 – 50
- 6 – 25
- 5 or less
- No data

0   km   1000

Map 12.1
Population density
in Europe, 1989
Source: WHO/Eurostat

Internal migration in Europe is also being progressively replaced by immigration from southern countries such as Algeria, Morocco, Tunisia and Turkey. In Belgium, France, Germany, The Netherlands, Sweden and Switzerland in the 1980s, non-European foreign residents increased from 30 to 50 per cent of all foreign residents.

Urbanisation

Rapid urbanisation is often the cause of enormous pressure from urban areas on rural and natural environments. In 1920, the urban population was 14 per cent of the world's population. It reached 25 per cent in 1950. Between 1950 and 1985 the number of people living in cities almost tripled, increasing by 1.25 billions worldwide (UN, 1991). As already illustrated in Chapter 10, the share of urban population is expected to grow. According to the UN medium scenario, almost half of the world population will live in urban areas in the year 2000 (UN, 1991). However, patterns of urbanisation differ from region to region. In 1950, more than half of the population in developed countries lived in cities, and urban population has grown steadily up to approximately 67 per cent in 1970 and 73 per cent in 1990. This percentage is expected to reach 80 by the year 2025. The rate of urban population growth between 1950 and 1990 has been much faster in developing countries, from 17 per cent to 34 per cent.

In Europe more than two thirds of the population now lives in urban areas. High rates of rural-to-urban migration characterised the 1960s and 1970s. Differences in the stage of the urbanisation process can be detected between Western European countries, which have now reached stabilisation, and Southern and Eastern European countries, where rural-to-urban migration is still important (Maps 12.2 and 12.3).

The complex links between population, the environment and economic development are still not fully understood. Both poverty and affluence may cause significant environmental degradation. Poverty in the developing regions of the world is often linked to high birth rate and low life expectancy. Poverty and population pressure force people to cultivate more marginal land. On the other hand, affluence in the more developed countries has led to enormous and wasteful consumption of natural resources. Appreciating the ways both poverty and affluence cause environmental degradation may help in understanding what can reverse it. Although no clear relationships can be established between human development and environmental change, there is no doubt that improvements in education, health and nutrition, together with advances in science and technology, broaden the choice to use natural resources sustainably. Equal rights for women play an important role in population issues. Recent trends in three important indicators – fertility, life expectancy and living standards – across the various regions of the world may provide important insights for assessing future trends.

Fertility

The total fertility rate worldwide during the 1950s and 1960s was around five children per woman. A major decline in the total fertility rate started in the 1970s: it went down from 5 to 4.5 between 1970 and 1975, and to 3.6 in 1980–85. Between 1985 and 1990, it fell to 3.45. The UN medium scenario projects total fertility rates of 3.0 and 2.5 for the periods 2000–2005 and 2020–25, respectively.

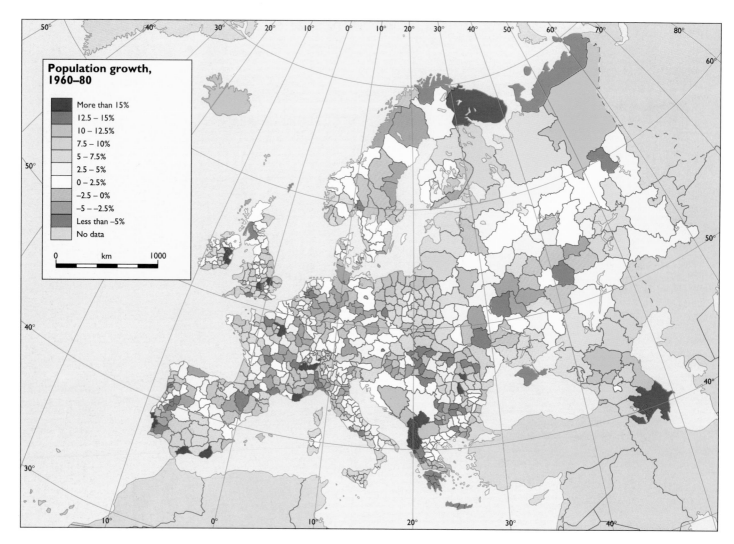

Map 12.2
Population growth in Europe, 1960–80
Source: Decroly and Vanlaer, 1991

Marked differences in fertility trends can be noted between the more and less developed regions of the world. Fertility rates in the most developed regions have declined from 2.8 in 1950–55 to 1.9 in 1985–90. In these regions (excluding the former USSR), the current fertility rate is below the replacement level. A substantial decline in the fertility rate has also occurred in less developed regions, with a decrease from 6.19 children per woman in the period 1950–55 to 4.54 in the period 1975–80 and 3.94 in the period 1985–90.

In Europe, where fertility rate is lower than the average level of developed regions, the decline of fertility rate has occurred earlier in Northern and Western parts. The later and slower decline in Southern Europe has been characterised by very steep trends reaching very low levels during the 1980s. By this time, Northern and Western European countries had already reached stabilisation with, in some cases (such as in Denmark, Sweden and the UK), even small increases in fertility rates beginning to occur. In Eastern Europe, fertility declined more slowly than in Southern Europe, while in the former USSR it has increased.

Life expectancy

Life expectancy is considered as one of the most representative indicators of health- and environment-related quality of life (see Chapter 11) as well as an important indicator of human development. Improved health and adequate nutrition of the population are fundamental to sustainable development. Large discrepancies in life expectancy between world regions show, however, that such conditions are still to be met for most of the world population.

In 1985–90 life expectancy at birth for the world was 63.9 years. Life expectancy is 61.4 years in the less developed

regions and 75.0 years in the industrial countries. The weighted average life expectancy at birth for European countries was 74.9 years in 1988 (based on 27 countries), and ranged from 69.3 to 78.2 years. The lowest life expectancies in Europe occurred in Central and Eastern Europe. Over the 1980s there was a steady increase of average life expectancy (from 73.2 years in 1980, ie, 0.2 years annually). This is a continuation of the average trend observed over the previous decade. However, most of the countries with low life expectancy at the beginning of the 1980s have shown very little increase in the past ten years.

The average difference in life expectancy between females and males in Europe is about 6.8 years (range from 3.7 to 8.8). Large gender differences are found both in countries with relatively high (France) and low (Poland, Hungary) overall life expectancy. Only about one third of countries in Europe have shown a small reduction in this difference after 1980. For the others, the difference is stable or still increasing.

Different trends in life expectancy and fertility rate have changed the distribution of age structure of the population across world regions. The effects of the necessary social and economic adjustment to the different age structures (younger in the developing countries and older in the developed countries) on the patterns of production and consumption are still unfolding.

Living standards

National income is one measure of living standards across different countries, although it does not reflect marked regional differences. Expressed in GNP (Gross National Product) per capita, the average incomes in developed versus

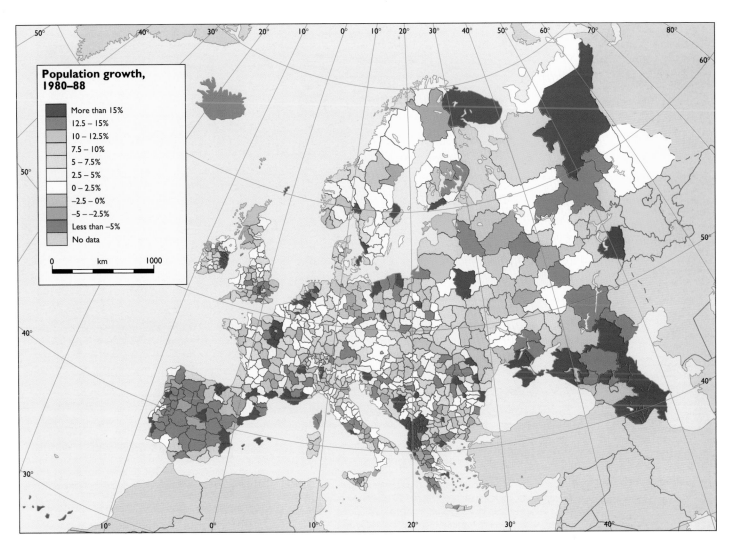

Population growth, 1980–88

- More than 15%
- 12.5 – 15%
- 10 – 12.5%
- 7.5 – 10%
- 5 – 7.5%
- 2.5 – 5%
- 0 – 2.5%
- –2.5 – 0%
- –5 – –2.5%
- Less than –5%
- No data

0   km   1000

Map 12.3 Population growth in Europe, 1980–88
Source: Decroly and Vanlaer, 1991

developing countries was $12 510 versus $710 in 1990. On average, the income of one person from the North is 18 times greater than that of one person in the South. This means that 77 per cent of the world's population earn only 15 per cent of the world's income (UNDP, 1991).

Regional differences in income and living standards between European countries are responsible for different patterns of production and consumption between Western, Southern and Central and Eastern Europe. Comparing real GDP (Gross Domestic Product)per capita in selected European countries, adjusted with the UN Purchasing Power Parity (PPP, which accounts for the differences in the cost of living), reveals that important discrepancies exist across European regions (Figure 12.3).

PRODUCTION AND CONSUMPTION

The patterns

Production and consumption have grown worldwide more rapidly than population over the last 40 years. In 1950 the world produced only about a sixth of the goods it does today (Figure 12.5). However, marked differences can be observed across world regions. Increases in production and consumption are particularly concentrated in a limited number of countries which total only one quarter of the world's population. Disparities in income between the rich and poor countries, measured in volume of GDP, have widened. The income gap between developed and developing countries is reflected in the per capita consumption of the

Figure 12.3
Living standards in selected countries
Source: UNEP, 1993

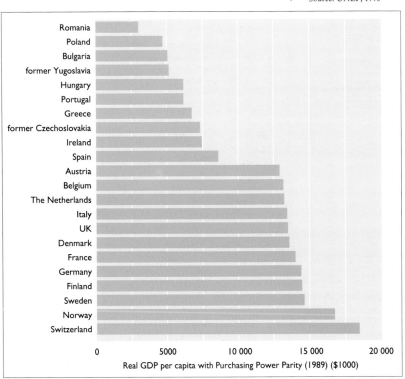

Romania
Poland
Bulgaria
former Yugoslavia
Hungary
Portugal
Greece
former Czechoslovakia
Ireland
Spain
Austria
Belgium
The Netherlands
Italy
UK
Denmark
France
Germany
Finland
Sweden
Norway
Switzerland

0   5000   10 000   15 000   20 000
Real GDP per capita with Purchasing Power Parity (1989) ($1000)

Table 12.2
Consumption patterns for selected commodities
Source: Parikh and Parikh, 1991, cited in UNEP, 1992

Category	Products	World total (Mt)	Share (%)		Per capita (kg or m2)		Proportion
			Developed	Developing	Developed	Developing	(developed/developing)
Food	Cereals	1801	48	52	717	247	3
	Milk	533	72	28	320	39	8
	Meat	114	64	36	61	11	6
Forest	Roundwood	2410	46	54	388	339	1
	Sawnwood	338	78	22	213	19	11
	Paperboards	224	81	19	148	11	14
Industrial	Fertilisers	141	60	40	70	15	5
	Cement	1036	52	48	451	130	3
	Cotton and wool fabrics*	30	47	53	15.6	5.8	3
Metals	Copper	10	86	14	7	0.4	19
	Iron and steel	699	80	20	469	36	13
	Aluminium	22	86	14	16	1	16
Chemicals	Inorganic chemicals	226	87	13	163	8	20
	Organic chemicals	391	85	15	274	16	17
Transport	Cars**	370	92	8	0.283	0.012	24
vehicles	Commercial vehicles**	105	85	15	0.075	0.0006	125

Notes:
Cereal, milk and meat data refer to 1987. Roundwood, sawnwood and paperboards data refer to 1988. Fertiliser consumption includes nitrogen, phosphate and potash fertilisers. Cotton and wool fabric excludes synthetics as well as data from the former USSR and China. Data sources: Statistical Year Book 1987; Handbook of industrial statistics 1989; International Trade Statistics Yearbook; UN FAO Book of Production 1989; and Handbook of Industrial Statistics 1988 UNIDO. Per capita data are calculated.

* Cotton and wool fabric world total consumption in billion square metres and per capita consumption in square metres
** Cars and commercial vehicles in millions

world's natural resources and in the generation of emissions and waste. Industrialised countries, including Eastern and Central European countries, account for 82 per cent of world commercial energy consumption. On average, per capita consumption of energy from fossil fuels is 10 times higher in OECD countries than in developing countries. OECD countries, which account for only 16 per cent of the world population, also generate 68 per cent of the world's industrial waste. Relevant disparities in consumption patterns among developed and developing countries emerge from recent estimates on a number of commodities (Table 12.2).

Regional differences in production patterns in Europe are evident when examining and comparing changes in macro-economic indicators over the last 20 years. Significant growth in GDP per capita and industrial production occurred in Western European countries (Figure 12.4). The growth of economic activities has been accompanied by structural changes in the production system with a shift from material- and energy-intensive sectors to services. Particularly relevant from an environmental perspective is the development in industrial production, shifting from traditional industries (such as iron and steel and petroleum refining) towards

Figure 12.4
Trends in GDP in selected Western countries
Source: OECD, 1991a

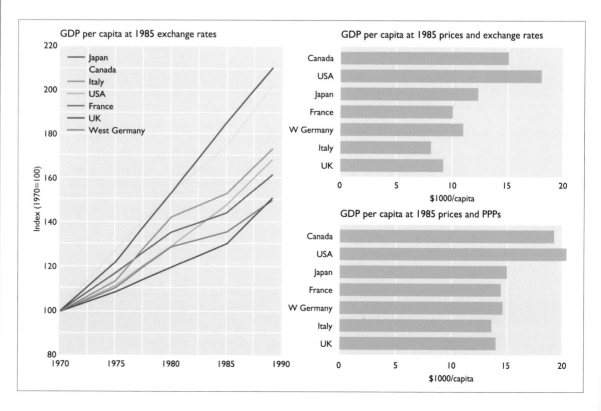

electronic and electrical industries, telecommunication, data processing and fine chemicals. This trend, coupled with some effects of energy efficiency improvements, has led to a significant reduction of energy intensity (or total primary energy requirements per unit of GDP) with an improvement of more than 25 per cent in the last 20 years (Figure 12.6). Structural changes with relevant environmental implications have occurred also in the agriculture system with the increasing intensity and specialisation of food production, as the result of, among other factors, the increasing vertical integration of the food sector.

Trend analysis of the Gross Material Product in Central and Eastern European countries shows a general growth of economic activity between 1970 and 1990 (Figure 12.7). Except for the last few years, during which production has considerably dropped, the increase of economic activities ranged between 45 and 75 per cent compared with 1971. GDP per capita in Central and Eastern European countries remains, however, much lower than average GDP per capita in Western European countries (Figure 12.8). However, energy intensity of production is still high despite a decrease of 20 per cent achieved between 1970 and 1990. While per capita energy consumption in these countries is often lower than the Western European average, relative consumption of energy per unit of product is three or more times higher than in Western European countries, due to the high share of heavy industries and obsolete technologies. The transition from centrally planned economies to market economies in Central and Eastern Europe is expected to bring substantial changes towards the restructuring of industrial activities. However, the current economic recession suggests that these changes will not have an effect as fast as was envisaged.

Processes and technologies

The nature of production and consumption must be taken into account when considering the links between environmental impacts and economic development and how they vary over time. Environmental impacts depend on processes and technologies as well as on the volume of

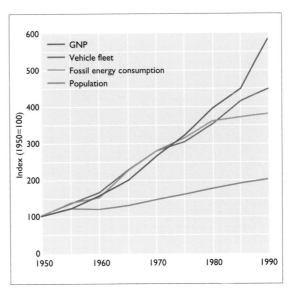

Figure 12.5
Trends in world population, GNP, fossil energy consumption and vehicle fleet, 1950–90
Source: RIVM, 1992; UNEP, 1989, 1993

consumer goods and services provided. The amount of energy and materials involved and emissions and waste generated may vary to a great extent depending on product design, methods of production and technologies employed. Technology is a key factor influencing environmental transformation processes due to human activities.

Increases in oil prices in the early 1970s, together with growing environmental concerns and environmental regulation, have led to economic restructuring in industrialised countries worldwide. Technological change in industrialised countries has occurred particularly in the field of energy conservation, material substitution and pollution control. This is especially evident when comparing energy and raw material consumption to production trends. It is also evident when examining the considerable growth of the specific industrial sector concerned with environmental protection, which accounted already at the end of the 1980s to between $70 and $100 billion and a workforce of approximately 1.5 million (OECD, 1991b).

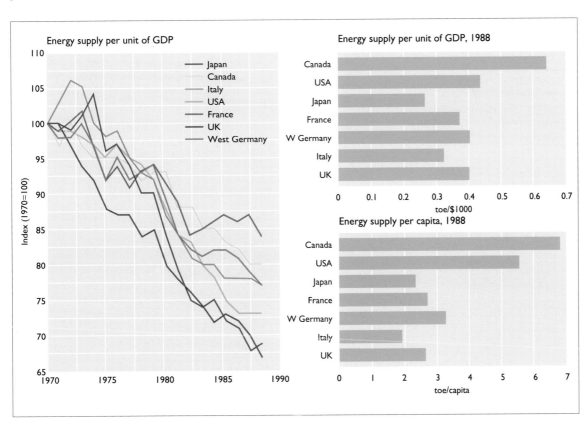

Figure 12.6
Trends in energy consumption in selected Western countries
Source: OECD, 1991a

Figure 12.7
Trends in Net Material Product in selected Eastern European countries
Source: OECD, 1993

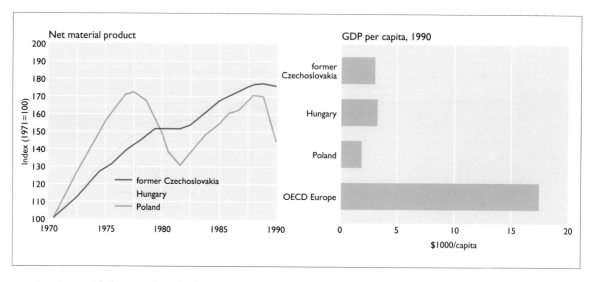

Consistent with these trends, technological transformation and substitution of industrial processes in Western European countries have increased substantially the energy and material productivity of production. Compared with production trends, the growth rate of energy and raw material use is still elevated in Eastern Europe. The substitution of production processes in Eastern European countries is a more recent trend associated with political change and adjustment of the prices of natural resources.

In the medium to long term, it is becoming clear that the transition to a 'new technological system' will affect the economy in both Western and Eastern European countries. The main areas of this transition are biotechnology, new materials and new energy sources. In addition, the development of information technology and artificial intelligence may affect both the way resources are managed and the capability for environmental restoration and pollution control. However, the potential risks and new constraints imposed by the wide application of these new technologies are still poorly understood.

While improving the efficiency of production processes, technological innovation worldwide has raised new environmental concerns. Product substitution has led to an increase in the type and amount of new substances, the effects of which on the environment and human health are not always well known (see Chapter 17). Increased environmental concerns have emerged for the growing amount and toxicity of waste and the environmental hazards in the recycling and disposal of new materials (see Chapter 15). In addition, development and widespread use of more complex technologies also increase risk of accidents (see Chapter 18).

Environmental impacts

Production and consumption activities have both direct and indirect effects on environmental systems, from the extraction of raw materials, through intermediate processing, manufacturing of final products, consumption and disposal. At all stages, these activities involve the use of materials and energy as well as the emission of pollutants and waste (Figure 12.9). From an environmental standpoint, production and consumption can be described as the sum of total materials and energy transformation to convert raw materials into products, to use these products to satisfy human needs and to dispose of wastes.

Industrialised countries worldwide have increasingly recognised the need to address the full impacts of their production and consumption processes. New concepts such as clean production and integrated pollution control have emerged as strategies to minimise the environmental impact of production and consumption activities in the various stages of the product life-cycle. Life-cycle analysis (LCA) methodologies have been implemented and are currently being proposed as a tool to improve product design and manufacturing processes (Box 12A). Increasingly, however, concerns are emerging as to whether reductions of the volume of certain activities will be required to achieve sustainable levels of production and consumption.

Figure 12.8
Trends in energy consumption in selected Eastern European countries
Source: OECD, 1993

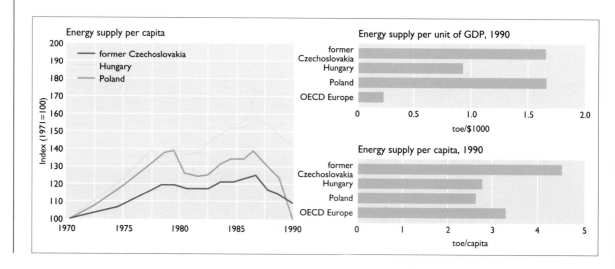

Box 12A Life-cycle analysis

Life-cycle analysis (LCA) is a technique which allows the environmental implications of various products to be assessed. The impact of products and services is the sum of the resources required plus the amount of emissions and waste. The material and energy intensity determines the level of efficiency of the process employed to deliver a given product or service. LCAs are inventories of the materials and energy requirements as well as of the waste released at each stage of a product's life. The methodology of compiling these inventories is a powerful tool to assist companies to improve products and reduce energy and material consumption in production processes. It can also assist governments to formulate product policies and implement ecolabelling schemes. Dissemination of this information can guide consumers towards more environmentally responsive choices.

Several studies have been developed in various European countries attempting to assess the impact of products. Major difficulties in compiling environmental inventories have been encountered because of the detailed information required. More difficulties emerge in the assessment of the results of LCAs. In some cases different approaches and assumptions have led to contradictory conclusions when trying to compare different products on the basis of their overall environmental impacts. Among other examples are the controversial conclusions of three studies on packaging alternatives for milk. Because of different approaches and categories of emissions considered, the results of the studies were considerably divergent. A Swedish study concluded that milk cartons were preferable to refillable glass, while a Swiss study found in favour of refillable glass. In a German study the milk carton and the glass bottle were found to be equally acceptable: the milk carton ranked best as regards atmospheric emissions, while the glass bottle ranked best considering emissions into water.

Contradictory conclusions of product LCAs can be explained by differences in the definition of processes and data inventory, which affect the whole accounting procedure. Notwithstanding, these studies have shown that the number of times a beverage package is re-used, and the transportation involved, as well as technology, play a major role. While LCA provides a tool to improve product design and manufacturing processes, comparing products on this basis alone could be misleading until standardised LCA methods are established.

ECONOMIC GROWTH VERSUS DEVELOPMENT

The notion of sustainable development has raised new questions about the physical limits to population, production and consumption and the extent to which successful application of new technologies may expand these limits. Although there is a wide range of opinions on what sustainability entails, a general principle can be derived from the definition provided in the Report of the World Commission on Environment and Development (WCED, 1987). Sustainability is a relationship between human and ecological systems which allows the improvement and development of the quality of life while maintaining the structure, functions and diversity of life-supporting systems.

To achieve a sustainable pattern of resource use and population it is crucial to understand the interactions of population and per capita consumption as mediated by technology, culture and value (Arizpe et al, 1992). The relationship between people, resources and development is

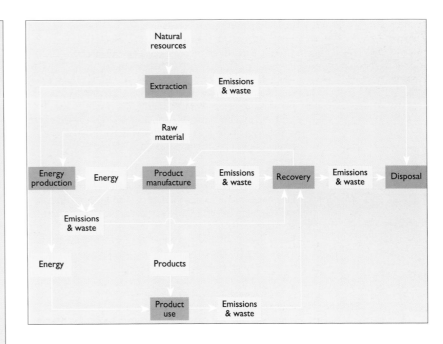

Figure 12.9
Product life-cycle
Source: EEA-TF

Box 12B Strategic environmental assessment

Strategic environmental assessment (SEA) is the term used to describe the environmental assessment process for policies, plans and programmes. The concept of SEA evolved from two decades of experience and debate on the implementation of environmental impact assessment (EIA) in the authorisation process of individual projects. The recent growth of interest in the use of environmental assessment at earlier stages of the planning process stems from two sources (Lee and Walsh, 1992):

1 a growing recognition that some important aspects of environmental assessment cannot be satisfactorily undertaken at the project evaluation stage (EIA) and must, therefore, be carried out earlier in the planning process;
2 increased appreciation that the implementation of sustainable development strategies will require the use of environmental assessment procedures and methods in the formulation of policies, plans and programmes for the principal sectors of national economies.

SEA and EIA share the same objectives, should relate to each other closely within the same policy and planning process, and are intended to be complementary. However, while formal provisions for EIA as an integral element of project approval and decision-making now exist in many countries and are included in the procedures of many development agencies and financial institutions, formal provisions for SEA, as an integral element of appraisal and decision-making for policies, plans and programmes, are much less developed.

Without a formal provision, SEA is already used in both sector and landuse planning in several countries. Others are in the process of adopting procedures for incorporating some form of environmental assessment into the early stages of the planning process. More are carrying out studies into the desirability and practicability of doing so. The need to articulate such principles in specific procedures and methodologies has emerged as countries have started to put this approach into practice.

highly complex and still poorly understood. Several studies have indicated that there is no linear relationship between growing population and density and environmental degradation (Caldwell, 1984). Recent studies on the relationships between population, use of natural resources and environmental impact show that no single factor dominates the changing patterns of total impact across time (Clark, 1991). Clark (1991) has also demonstrated that the same level of pollution loading can come from radically different combinations of population size, consumption and production.

One important distinction that has emerged in the last decades is between economic development and growth. Economic growth implies increase in quantity of production and consumption, while economic development is increase in well-being without a necessary increase in consumption. Technological advances, no doubt, have increased the efficient use of materials and energy. Different opinions exist, however, regarding whether a sustainable economy can be achieved by pushing further the energy efficiency of production systems through technological change without requiring changes in lifestyle, population and wealth distribution. The degree to which technological change can eliminate resource constraints is uncertain. Limited also is the knowledge of the potential impact that new technologies can pose.

Because of the large uncertainty, increasing attention at scientific and institutional levels is being placed on precautionary principles and more integrated approaches to environmental protection and economic development. This requires an increased capacity to anticipate and assess potential environmental consequences of economic and development programmes for the long term. Important steps have been made for extending the concept and procedures of environmental impact assessment (EIA) from individual projects to economic decisions and policies. The notion of strategic environmental assessment (SEA) has emerged to provide policy makers with a critical tool to assess alternative scenarios and policy options (Box 12B). A number of European countries such The Netherlands, Denmark and

Norway have launched the concept of Green Plans (Box 12C) to better integrate environmental considerations into economic decisions.

These considerations also influence the framework for environmental monitoring, assessment and reporting as discussed in Chapter 1.

Box 12C Green Plans

Green Plans are long-term, comprehensive strategies for addressing environmental problems at source by integration of environmental considerations into national policies and programmes of all sectors of activity. Several examples exist in Europe and more are currently being developed. Denmark, France, Ireland, Norway, The Netherlands and the UK have already adopted a Green Plan which establishes targets up to the year 2000. Targets are specified for themes (eg, climate change, acidification, eutrophication) and sectors and groups (eg, industry, energy, agriculture, transport, consumers). Partnership with all sectors of society through new forms of participation is pursued for ensuring implementation, and monitoring mechanisms and state-of-the-environment reporting are generally developed. A few examples are:

- Denmark: *Our Common Future*, the Danish Government's Action Plan;
- France: *Le Plan Vert;*
- Ireland: *An Environmental Action Programme;*
- The Netherlands: *National Environmental Policy Plan and NEPP Plus;*
- Norway: Report to the Storting No 46 and *The Future is Now;*
- UK: *This Common Inheritance;*
- EU: *Towards Sustainability*, the fifth Environmental Action Programme.

REFERENCES

Arizpe, L, Costanza, R and Lutz, W (1992) Population and natural resource use. *An agenda of science for environment and development into the 21st century.* In: Dooge, I C I, Goodman, G T, LaRivière, J W M, Marton-Lefèvre, J, O'Riordan, T and Praderie, F (Eds). Cambridge University Press, Cambridge.

CEC (1991) *Europe 2000: outlook for the development of the community territory.* Commission of the European Communities DGXVI, Brussels.

CEC (1994) *Alternative scenarios for Central Europe,* Hans Van Zon (Ed). FAST Programme, Commission of the European Communities DGXII, Brussels.

Caldwell, J (1984) *Desertification: demographic evidence, 1973–83.* Australian National University. Occasional Paper No 37.

Clark, W (1991) Paper presented at the Annual meeting of the American Association for the Advancement of Science, Washington DC.

Decroly, J and Vanlaer, J (1991) *Atlas de la Population Européenne.* Editions de l'Université Libre de Bruxelles.

Latvia Environmental Protection Committee (1992) *National Report of Latvia to UNCED 1992,* Republic of Latvia Environmental Protection Committee, Riga.

Lee, N and Walsh, F (1992) Strategic environmental assessment: an overview. *Project Appraisal,* 7 (3), September, 126–36.

OECD (1991a) *Environmental indicators: a preliminary set.* Organisation for Economic Cooperation and Development, Paris.

OECD (1991b) *The state of the environment.* Organisation for Economic Cooperation and Development, Paris.

OECD (1993) *Environmental information systems and indicators: a review of selected Central and Eastern European countries.* Organisation for Economic Cooperation and Development, Paris.

RIVM (1992) *National environmental outlook 1990–2010.* RIVM, Bilthoven.

UN (1990) 1990 Revision of world population prospects: computerised data base and summary tables. United Nations, Department of International Economics and Social Affairs, New York.

UN (1991) *World population prospects: 1990.* United Nations, New York.

UN (1993) *World population prospects: 1992.* United Nations, New York.

UNDP (1991) *Human development report 1991.* United Nations Development Programme, Oxford University Press, Oxford.

UNEP (1989) *Environmental data report, Second Edition 1988/89.* United Nations Environment Programme. Blackwell Publishers, Oxford.

UNEP (1991) *Environmental data report, Third Edition 1990/91.* United Nations Environment Programme. Blackwell Publishers, Oxford.

UNEP (1992) *The world environment 1972–92: two decades of challenge.* Tolba, M K, El-Kholy, O A, El-Hinnawi, E, Holdgate, M W, McMichael, D F and Munn, R E (Eds.) United Nations Environment Programme. Chapman and Hall, London.

UNEP (1993) *Environmental data report, Fourth Edition 1993/94.* United Nations Environment Programme. Blackwell Publishers, Oxford.

WCED (1987) *Our common future.* World Commission on Environment and Development. Oxford University Press, Oxford.

North Sea oil rig
Source: Eye Ubiquitous

13 Exploitation of natural resources

INTRODUCTION

The Earth's natural resources are vital to the survival and development of the human population. However, these resources are limited by the Earth's capability to renew them. Freshwater, forests and harvesting products are renewable, provided that exploitation does not exceed regeneration. Fossil fuels and metal ores are non-renewable. Although many effects of overexploitation are felt locally, the growing interdependence of nations, and international trade in natural resources, make their demand and sustainable management a global issue. This chapter focuses on major developments in the use of renewable and non-renewable resources in Europe in the context of global trends. Available statistics to monitor changes in the use of natural resources at the global and European levels are described in Box 13A. The use of natural resources by sectors of activity are detailed in Part IV of this report (Chapters 19 to 26).

RENEWABLE RESOURCES

Food, water, forests and wildlife are all renewable resources. For resource use to be sustainable, the consumption rate should be maintained within the capacity of the natural systems to regenerate themselves. Current rates of depletion of the Earth's stocks of renewable resources and levels of pressure imposed on their regenerative capacity by means of production and consumption might already be, in some cases, beyond this threshold.

Food

Since 1950 global food production has increased substantially. Increases in grain outputs, seafood and livestock have outpaced population growth. This has been achieved by raising the production base and by substantially rising productivity. World grain production more than tripled between 1950 and 1990, rising from approximately 700 million tonnes to more than 1900 million tonnes. The annual average increase of 2.7 per cent was greater than that of population. Over the same period, world fish catch increased from 22 million tonnes to 100 million tonnes, more than doubling seafood availability per person from 9 to 19 kg. Today, however, there is increasing evidence of important constraints to expanding production of grain and of major sources of protein as fast as global population. The global area of available agricultural land is decreasing as a result of desertification and urbanisation. Between 1981 and 1991, global harvested area of grain dropped from 735 million hectares to 693 million hectares. After 1989, fish catch started to decline both in absolute terms and per capita. According to FAO, the ocean may not be able to sustain a level of catch greater than that reached in 1989 (FAO, 1992). The environmental impacts of modern production methods (see Chapters 22 and 24) may impose additional threats to food production in the long term.

Global figures of increasing food production also mask relevant regional differences. In the industrialised countries with relatively stable population and higher levels of income, food security is no longer a problem, although poverty in

	Total cereals production (1000 t)		Cereals and cereal products			
			Exports (1000 t)		Imports (1000 t)	
	1980	1990	1980	1990	1980	1990
World	1 565 208	1 952 175	220 542	221 228	217 312	221 903
Western Europe	177 560	192 133	34 614	59 078	45 500	34 001
Eastern Europe & former USSR	283 297	314 449	6036	3453	47 306	37 764

	Total meat production (1000 t)		Meat and meat products			
			Exports (1000 t)		Imports (1000 t)	
	1980	1990	1980	1990	1980	1990
World	134 626	175 102	8084	11 596	7909	11 407
Western Europe	29 515	33 350	3673	5635	3761	5190
Eastern Europe & former USSR	25 130	30 095	738	768	956	966

	Marine and freshwater fish and shellfish production (1000 t)		Fish and fish products			
			Exports (1000 t)		Imports (1000 t)	
	1980	1990	1980	1990	1980	1990
World	71 874	95 148	10 308	16 934	9776	16 606
Western Europe	11 354	11 083	3758	4861	4478	6634
Eastern Europe & former USSR	10 751	11 122	702	1 077	671	1268

Table 13.1
Food production, exports and imports (thousand tonnes)
Source: FAO, 1992

these countries still exists. In developing countries and former central-planned economies relying on imports, purchasing power is often not enough to meet food needs. In addition, emerging difficulties in expanding food outputs pose increasing concerns about whether food production increases can continue indefinitely and about the ecological and economic constraints to its growth. If predictions of global population growth between now and 2025 prove accurate and the patterns of consumption in the developed world remain unchanged, food production should increase at an annual rate of 1.4 per cent to maintain current per capita production levels; and greater increases in production are required to eliminate current undernourishment.

Statistics on the production of food compiled by the Food and Agriculture Organization of the United Nations (FAO) show an increase in production of cereals and meat between 1980 and 1990 in both Western and Eastern Europe, including the former USSR (Table 13.1). A slight decline in the production of fish and shellfish occurred in the same period in Western Europe, while an increase occurred in Eastern Europe and the former USSR. However, the capability of these countries to meet their food requirements has been uneven. Western Europe has continued to produce surpluses of cereals and meat, though it imports animal feed, while most Eastern European countries have relied on imports.

Food surpluses and deficits across world regions are compensated for by international trade. However, lack of ecological considerations in trade policies may cause increased pressure on natural resources by encouraging unsustainable levels of production in some countries and increasing dependency of countries which rely on imports. These imbalances derive from economic mechanisms and national policy interventions intended to support the income of farmers and upgrade production. International trade is driven primarily by the demand and supply of certain food products, but it is also affected by national subsidies and incentives systems. Increasing food surplus in Western Europe is a result of direct and indirect subsidies which stimulate overproduction even in the absence of demand.

Box 13A
Statistics on the use of natural resources

The statistics presented in this chapter for food and energy utilise the UN regional definition of Western and Eastern Europe.

Food
The responsibilities of the Food and Agriculture Organisation of the United Nations (FAO) include compiling global information on food and agriculture, forestry and fisheries activities. Data on crop production and trade are compiled on annual bases from questionnaires by the relevant statistical authorities in individual countries. Indices of agriculture and food production are calculated by FAO to assess the trends in disposable output of individual countries' agriculture sector relative to a given base period (FAO, 1992).

Water
Information on the availability and use of freshwater at the global level is not widely reported, but estimates exist in most countries. UNEP has attempted to estimate both supply and uses by individual countries (UNEP, 1992). Estimates of renewable water resources and water abstractions for European countries are also made available by UNECE and Eurostat.

Forests
FAO routinely compiles global data on the production and prices of trees and wood products. Since 1980, FAO and UNEP have conducted assessments of forests and rate of deforestation and reforestation based on published data and results of national surveys. More recent assessments also make use of available satellite data from AVHRR-based normalised differentiated vegetation index (NDVI) imagery and high resolution LANDSAT and SPOT imagery. In Europe, data on forests and forest products are produced by UNECE with FAO and the CEC.

Fossil energy
Global data on energy resources, energy production and consumption are regularly compiled and published by the World Energy Council (WEC) and the International Energy Agency (IEA). Energy statistics are also published regularly by the United Nations Statistical Office (UNSTATO) in the *Energy Statistical Yearbook*. Data on nuclear energy are produced by the International Atomic Energy Agency (IAEA). Eurostat has compiled data on energy from various sources including the IEA and national statistics.

Materials
World reserves of major metals are published on a regular basis by the US Bureau of Mines. The World Bureau of Metal Statistics publishes production and consumption data for aluminium, cadmium, copper, lead, nickel, tin and zinc. Consumption of iron ore and crude steel is published by the United Nations Commission on Trade and Development (UNCTAD). Iron and steel consumption data are also published by the International Iron and Steel Institute. In Europe, metal production and consumption data are published by UNECE and Eurostat.

This leads to distortion in the price of products on the international markets, affecting the capability of emerging economies to enhance food production.

Global export of cereals as a percentage of total production declined worldwide except for Western Europe, where cereal and cereal product exports increased significantly (Table 13.2). In 1990 in Western Europe the equivalent of 30 per cent of the total production of cereals was exported. Conversely, the import of cereals and cereal products by Western European countries declined between 1980 and 1990. By contrast, the former USSR and Eastern Europe in 1990 imported over ten times more cereal products than they exported.

Global trends in the export of meat, fish and fish products increased. In Western Europe, export of fish and meat as a percentage of total fish and meat production increased respectively from 33.1 and 12.4 per cent in 1980 to 43.9 and 16.9 per cent in 1990. Export of fish and fish products as the percentage of total production increased slightly also in Eastern Europe, while export of meat declined slightly.

During the 1980s in Europe there was a general decline in the demand for cereal and cereal product imports. Imports of cereals and cereal products to both Western and Eastern Europe declined by about 20 per cent. Compared with other world regions, Western Europe is still a major importer of some food products such as meat, fish, and animal feed. While only accounting for about 15 per cent of the world imports of cereals and cereal products, Western Europe accounts for about 45 per cent of world imports of total meat and about 40 per cent of world imports of fish and fish products (FAO, 1992). By 1990 Western Europe was importing almost as much meat as it was exporting, and importing considerably more fish and fish products than it was exporting.

Trends in food production and trade reflect regional differences in environmental conditions. However, methods of production together with economic incentives may increase the imbalance between exporter and importer countries, causing adverse environmental effects. One example is the parallel nutrient accumulation and exhaustion in soil in different regions as a result of current methods of crop production (Hoogervorst, 1992). Availability of agricultural land, fertility of soils and climatic regime determine which crops can be grown, what inputs are required and what outputs are possible. International trade causes a large-scale redistribution of the worldwide stock of nutrients. Chemical fertilisers and imported fodder add large quantities of nutrients (nitrogen, phosphorus and potassium) to the mass balance of European countries, causing build-up of these substances in the soil. In other regions, where nutrients are subtracted and not replenished, due to lack of finances, the quality and productivity of agricultural land is affected.

Water

Globally, water use is estimated to have increased more than 35-fold over the past three centuries (L'vovich and White, 1990). Current water abstractions have been increasing at a rate of 4–8 per cent per year. Globally, water withdrawal is of the order of 3240 km^3 year. Of this, agriculture accounts for 69 per cent, industry for 23 per cent and the domestic sector for 8 per cent. Total and per capita water use also varies markedly between world regions. In Europe, 700 m^3 per capita is the average annual water use, which is slightly higher than the global average (Table 13.3), much lower than the 1692 m^3 in North and Central America, but higher than the 526 m^3 in Asia and 244 m^3 in Africa. Current assessments predict a doubling of global water use by the year 2000 and a slight shift in the share that each continent demands (Figure 13.1).

Europe has only 8 per cent of the world's renewable freshwater resources but it accounts for about 15 per cent of the total world withdrawals (see Chapter 5). By coincidence, compared with the world average, Europe as a whole uses a greater proportion (15 per cent) of its renewable freshwater resources than the world average (8 per cent). These figures suggest that Europe is currently exploiting only a small portion of its renewable water resources. However, in practice the situation is rather different because the

	Cereals and cereal products		Total meat		Fish and fish products	
	1980	1990	1980	1990	1980	1990
Exports as a percentage of total production						
World	14.1	11.3	6.0	6.6	14.3	17.8
Western Europe	19.5	30.7	12.4	16.9	33.1	43.9
Eastern Europe & former USSR	2.3	1.1	2.9	2.6	6.5	9.7
Percentage of world imports						
Western Europe	20.9	15.3	47.6	45.5	45.8	39.9
Eastern Europe & former USSR	21.8	17.0	12.1	8.5	6.9	7.6

Table 13.2
Food exports and imports
Source: FAO, 1992

availability of good quality water can vary significantly from country to country and even within different regions in the same country. National statistics show that some European countries are already using a significant proportion of their water resources, such as Malta (130 per cent), Belgium (72 per cent), Cyprus (42 per cent), Spain (32 per cent), Italy (32 per cent), Poland (26 per cent).

A comparative analysis of the principal uses of water indicate that Europe as a whole uses over half (53 per cent) of its freshwater withdrawals for industrial purposes and 19 per cent for domestic purposes. These proportions are higher than those of North America and other industrialised regions of the world. In contrast, agricultural withdrawals stand at 26 per cent, lower than the average world use of about 70 per cent (Table 13.4).

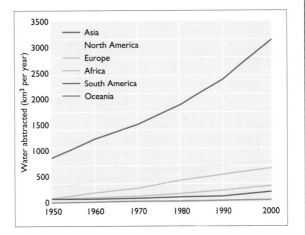

Figure 13.1
Trends in water abstraction by continent
Source: Meybeck et al, 1989

Forests

Exploitation of forest resources such as roundwood and other products has increased worldwide during the last two decades. Global softwood production has risen by 28 per cent and hardwood production by 54 per cent. The total world consumption of roundwood in 1988 was 2972 million m^3. Consumption data show that 52 per cent of roundwood was

Table 13.3
Global and European freshwater resources and withdrawals
Sources: EEA-TF; WRI, 1992 (world figures)

		Freshwater withdrawals		
	Renewable freshwater resources (km3/year) (1992)	Total (km3/year)	Per capita (m3/year)	Proportion of renewable resources (%)
World	40 000	3240	656	8
Europe	3100	480	700	15

	Freshwater withdrawn (%)		
	Domestic	Industry	Agriculture
World	8	23	69
Europe	19	53	26

used as industrial roundwood and 48 per cent as fuelwood. However, the share of industrial wood and fuelwood in developed countries is 82 and 18 per cent respectively, while in developing countries these proportions are reversed (UNEP, 1993).

According to the recent FAO assessment (FAO, 1992), global forest area has declined by about 6 per cent in the last 20 years. Between 1981 and 1990 the world's average tropical deforestation was 16.9 million hectares per year, which is a 50 per cent increase in deforestation on the aggregated figure for 1980. In contrast, it is estimated that over the same period there has been a 5 per cent increase of temperate and boreal forest cover (mainly due to afforestation and reforestation).

Between 1981 and 1990, the European region (excluding Iceland and the former USSR) showed a net increase in forest and wooded land of 1.9 million ha (see Chapter 23). The former USSR showed an increase of 22.6 million ha over the same period (UNEP, 1993). While deforestation is the major cause of loss of forest area in the tropical regions, in some parts of Europe forest loss and damage is caused mainly by atmospheric deposition of air pollutants and consequent increase in vulnerability to stress. However, in Russia it is now due to uncontrolled logging (see Chapter 34).

Tropical deforestation is caused by landuse changes, shifting cultivation and various forms of exploitation of forest products. Population pressure and poverty are major causes of deforestation in these regions. In developing countries, forests are often the only hope of subsistence for the increasing human population (Mather, 1990). Together with landuse shifts, another important cause of deforestation is improper land management and development of infrastructure. Large-scale projects such as hydropower plants or mining activities open up the forests by construction of roads. Exploitation for fuelwood and timber together with increasing worldwide trade of forest products are important additional factors threatening these ecological reserves.

Tropical deforestation does not directly affect the state of Europe's environment in the short term, although it does pose increasing threats to the global environment in the longer term. Forests perform a crucial ecological function for the Earth: they support a large variety of species and supply a wide range of resources; they play a crucial role in regulating the atmosphere and the climate and act as a major store of carbon; and they also contribute to soil formation and watershed protection. Loss of tropical forests is both a local and a global concern.

Because of increasing interdependence of world economies, to a certain extent, tropical deforestation is also linked with Europe's demand and supply of natural resources through economic processes. European countries contribute

indirectly to deforestation in tropical regions through the use of wood and wood products from these regions. Imports of tropical hardwoods decreased from 6.8 to 3.5 million m³ between 1970 and 1990. In 1990, this constituted about 20 per cent of the total imports into Europe, North America and Japan. FAO estimates that only 6 per cent of wood harvested in developing countries enters international trade; most is used domestically (FAO, 1992).

NON-RENEWABLE RESOURCES

Minerals, oil, gas and coal are non-renewable resources: their use as materials and energy sources leads to depletion of the Earth's reserves. However, the time period during which reserves can be available can be extended by recycling or improving the efficiency of use. Eventually, limitations to the extent to which more efficient processes may expand the use of non-renewable resources stocks will be reached, requiring substitution with renewable resources and restrictions on the volume of activities that can be sustained by existing stocks.

Fossil energy

It took roughly one million years for nature to produce the present annual worldwide consumption of fossil fuels (Gibbons et al, 1989). Worldwide consumption of energy has grown rapidly over the last decades: from 1650 million tonnes of oil equivalent (toe) in 1950, to 8100 million toe in 1990 (UNEP, 1992). The annual growth rate in energy consumption of 2.2 per cent at the beginning of the century has more than doubled between 1950 and 1970. After the increase of oil prices in the early 1970s, this rate has stabilised at 2.3 per cent between 1970 and 1990.

Although estimates of proved recoverable reserves of fossil fuels have increased in the last decade, these resources are non-renewable and cannot be relied upon for sustainable development. In addition, the use of fossil fuels is a major cause of environmental impacts (see Chapter 19). Combustion of fossil fuels contributes a large proportion of global anthropogenic emissions of carbon dioxide (55–80 per cent), sulphur dioxide (90 per cent) and nitrogen dioxides (85 per cent) (IEA, 1990).

The link between increase in economic activity and energy consumption is evident if the distribution of energy consumption is examined across more and less developed countries. In 1990, OECD countries, together with Eastern European countries and the former USSR (which in total represent only 22 per cent of the world population), accounted for 82 per cent of world commercial energy consumption. Of this, approximately half (40 per cent) is consumed in Europe (including the former USSR).

Energy consumption trends vary across European countries. Between 1980 and 1990, net consumption of energy decreased in Europe, excluding the former USSR. During the same period, energy consumption in the former USSR increased by one quarter of the 1980 figure. Europe and the former USSR together have 36 per cent of the world's proved recoverable reserves of coal and anthracite, and 47 per cent of the world's proved recoverable reserves of sub-bituminous coal and lignite (with the former USSR having the greatest proportions, at 23 per cent and 27 per cent respectively) (UNEP, 1993). Consumption of energy in Europe, excluding the former USSR, is, however, greater than production, as shown in Table 13.5. This implies that Europe is contributing to the depletion of energy resources in other regions of the world.

In Western European countries, the total amount of energy consumed increased despite the considerable decrease in the energy intensity achieved in these same countries; and despite the lower share of energy consumption in most Eastern European countries, the energy intensity of production is still elevated (see Chapter 20). Current trends in

	World	Europe	former USSR
Production (10¹⁵ J)			
1980	269 244	37 125 (14)	56 287 (21)
1990	318 735	38 703 (12)	69 072 (22)
Consumption (10¹⁵ J)			
1980	251 844	64 148 (25)	44 181 (18)
1990	301 442	62 832 (21)	56 602 (19)
Note: ( )% World total.			

the energy mix show also that Eastern European countries are still highly dependent on the most polluting fossil fuels. Indeed (as described in Chapter 19), among fossil fuels there are wide variations in environmental impacts, with coal ranking first as to its content of carbon per unit of energy compared with oil (which contains 17 per cent less carbon per unit) and natural gas (43 per cent less).

Worldwide, the increase in energy consumption has been accompanied by a major shift in the energy mix. Coal, which accounted for about 80 per cent of the total world energy consumption in the 1920s, dropped to 32 per cent in 1970, and to 29 per cent in 1990. The share of oil decreased from 47 per cent in 1970 to 41 per cent in 1990, while the other sources increased: natural gas from 19 to 22 per cent and nuclear energy from 1 to 6 per cent. Parallel to this trend, European countries including the former USSR have achieved a significant reduction in the share of coal and oil. However, the share of coal in Eastern European countries in 1990 still accounted for about half of total energy use (Table 13.6).

Prime raw materials

Most materials generally used in intermediate and consumer products are substances obtained by physical and chemical methods using raw materials as input. Worldwide, the use of metals and ores increases (UNEP, 1992), although marked differences exist across countries and world regions. During the last decades, important changes in the use of basic materials such as steel, aluminium, copper, lead, zinc and nickel have occurred in most industrialised countries as a result of changes in production and products, material substitution and recycling. The material intensity (expressed in kg per GNP unit) has declined due to higher energy prices and availability of new technologies (eg, glass fibre). Replacing one mineral resource with another or with non-mineral materials can reduce the energy requirements during extraction and processing, as well as use, hence reducing environmental impacts. However, substitution does not always result in less pressure on the environment if substituted materials introduce new environmental problems.

In spite of these changes, the replacement of traditional materials and increased recycling have not reduced the pressure on mineral reserves. Current proved viable reserves of some raw materials could be exhausted in a few years at the current extraction rate. This does not imply that in a few years depletion of raw materials contained in the Earth's crust will be reached. Instead it means that exploitation of future reserves will require increased energy consumption and higher prices. Indeed, the availability of raw materials is often assessed on the basis of the reserves/production ratio (De Vries, 1992). Figure 13.2 shows the world reserves and production of raw materials. It is important to notice that reserve estimates increase as production levels increase, due to increased motivation for exploration.

Since Europe has already depleted its high grade mineral reserves it relies mostly on imports (principally from Africa). The assessment for 1991 given in Table 13.7 shows that Europe has only a small proportion of the world's reserves of some important minerals. On the contrary, unlike the rest of Europe, the former USSR is still a major producer and consumer of many minerals (UNEP, 1993). In addition, Eastern Europe and the former USSR are responsible for a larger share of world steel consumption (21 per cent in 1991) than all other world regions (the 12 EU countries consumed 16.1 per cent), and they also consume the largest share (31 per cent) of scrap metal for steel-making (UNIDO, 1992).

PHYSICAL INTRUSION

Environmental change is also caused by modifications to the land surface. Human activities cause alterations to the local and global environment by physical intrusion (eg, mining and quarrying, dam-building, river channelisation) or by altering the nature of the land surface (eg, by urbanisation, infrastructure-building, waste tipping), affecting landscapes, landuses and productivity (eg, deforestation, desertification, erosion).

Detailed surveys of landuse are carried out in several European countries. Such surveys are essential to help planners allocate land for different purposes, such as development, recreation, protected areas, waste disposal or reclamation. However, the categories of landuse vary from country to country and, at present, there is not a standardised international collection of such detailed information.

The OECD collects landuse statistics for member

		Coal	Oil	Gas	Nuclear	Hydro	Others
World	1970	31.7	46.8	18.8	0.9	1.7	–
	1990	28.5	40.5	22.2	6.3	2.5	–
Western	1970	27.6	60.3	6.3	1.2	3.0	1.6
Europe	1990	19.3	44.0	16.6	14.4	3.0	2.7
Eastern	1970	60.1	23.3	13.1	–	1.0	2.4
Europe	1990	48.4	24.2	20.1	4.5	1.1	1.6
former	1970	36.8	34.8	22.3	0.2	1.6	4.3
USSR	1990	21.0	30.0	42.0	4.0	1.6	1.3

Table 13.6
World and European primary energy consumption, 1970–90 (toe, %)
Source: UNEP and Eurostat

Figure 13.2
Reserves and production of raw materials
Source: US Bureau of Mines, 1985, cited in De Vries, 1992

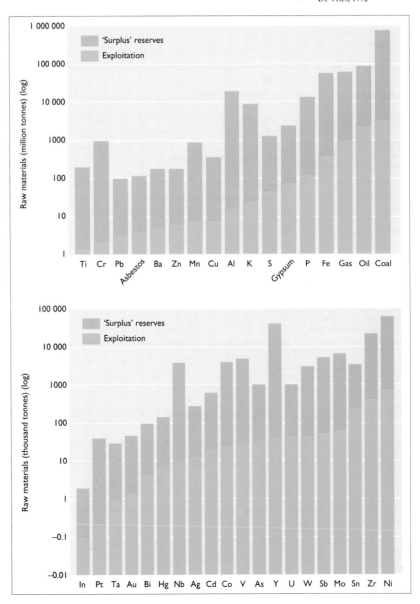

Reserves (million tonnes metal content)	Copper	Lead	Iron ore	Bauxite
World	321.0	70.4	64 648	21 559
Europe	23.0	9.7	3214	1342
former USSR	37.0	9.0	23 500	300

Table 13.7
World and European reserves of selected minerals, 1991
Source: UNEP, 1993

countries under the categories of: arable and cropland, grassland, forest and woodlands, and other uses. This broad classification does not highlight those landuses which physically change the natural landscape and its ecosystems. On the other hand, differences in the categories of more detailed landuse statistics collected by European countries make international comparisons, or the estimation of regional totals, difficult. As part of the EC CORINE programme a nomenclature of 44 land-cover classes grouped in a three-level hierarchy has been proposed and gradually applied to EU and some Central and Eastern European countries to produce comparable information across Europe (Box 13B).

A basic set of European landuse statistics used for this report has been presented in Chapter 3. Statistics available for individual European countries show relevant variations between countries, but confirm that only a small proportion of the total land area is used by, or for, the categories which cause the most severe land degradation, such as mining, industry, urbanisation and transport. Despite the small size of such areas compared with other landuses, their impacts could be large and affect the global environment. The importance of these impacts depends on the sensitivity of various land areas. Particularly relevant is the impact of overuse of critical areas such as, for example, the fragile environment of coastal zones (see Chapter 35).

Landuse conflicts can result from the construction of new infrastructure such as plants related to energy production. The construction of new dams affects farmland and forests and can displace communities. The impact of transportation infrastructures on the environment, but also of telecommunication links and electricity pylons, merits special attention; they consume substantial areas, virtually to the exclusion of other uses. In some cases the impacts of such links can be reduced through engineering solutions, such as road tunnels and underground electricity cables. However, this is not always possible for practical or cost reasons. The building of major new transport links is not without controversy for most means of transport (eg, public protest against bypasses, sea tunnels and bridges, canals and high-

speed train links). However, landuse requirements which differ considerably between transport modes, form only one of the associated problems (see Chapter 21). Furthermore, the building of infrastructure, particularly for transport, uses vast quantities of natural resources and energy during construction, and also generates substantial waste products which imply additional impacts and physical intrusion by quarrying and tipping of construction waste.

Box 13B CORINE Land-Cover

CORINE Land-Cover (CEC, 1993) is one of a number of inventories undertaken in the framework of the EC CORINE programme. Its aim is to provide the European Commission and Member States with reliable quantitative information on land-cover which is consistent and comparable across Europe.

The basic characteristics of the information gathered by the CORINE Land-Cover inventory are :

- the mapping scale used: 1:100 000;
- the nomenclature, with 44 classes grouped in a three-level hierarchy;
- the size of the smallest unit mapped (unit area), that is, 25 ha;
- the methodology: 'computer-assisted photointerpretation of Earth observation satellite images, with simultaneous consultation of ancillary data';
- the digitisation of the cartographic results and their integration into the CORINE information system.

The CORINE Land-Cover database can generate simple cartographic representations or statistical overviews and also provides one of the inputs for the production of more complex information on other environmental themes (eg, soil erosion, or emissions into the air by vegetation). The inventory has recently been applied to large transfrontier portions of the European Union, and already offers the potential for a wide array of uses. The applications are wide-ranging in both the variety of the themes (eg, management of protected areas, rational exploitation of natural resources, layout of road and railway networks, urban planning) and in the different levels of geographical area/decision making (ie, EU, national and regional).

References

CEC (1991) *Europe 2000: outlook for the development of the Community territory.* Commission of the European Communities, Luxembourg.

CEC (1993) *CORINE Land-Cover: Guide technique.* EUR 12 585 FR Commission of the European Communities, Luxembourg.

De Vries, H J M (1992)) The supply of raw materials. *National Environmental Outlook 1990–2010,* pp 90–5, RIVM. Bilthoven.

FAO (1992) *The state of food and agriculture 1991.* FAO Agriculture Series No 24, Food and Agriculture Organisation of the United Nations, Rome.

Gibbons, H J, Blair, D P, and Gwein, H L (1989) Strategies for energy use. *Scientific American Special Issue: Managing Planet Earth* September 1989, 136–43.

Hoogervorst, N J P (1992) Nutrient management and food production. *National Environmental Outlook 1990–2010,* pp 100–14. RIVM, Bilthoven.

IEA(1990) *Energy and the environment: policy overview.* International Energy Agency, Organisation for Economic Cooperation and Development, Paris.

L'vovich, M I and White, G F (1990) Use and transformation of terrestrial water systems: In :Turner, B L (Ed) *The earth transformed by human action: global and regional changes in the biosphere over the past 300 years.* Cambridge University Press, Cambridge.

Mather, A S (1990) *Global forest resources.* Belhaven Press, London.

Meybeck, M, Chapman, D and Helmer, R (Eds) (1989) *Global freshwater quality: a first assessment.* GEMS: Global Environment Monitoring System. Published on behalf of the World Health Organization and the United Nations Environmental Programme by Blackwell Publishers, Oxford.

UNEP (1991) United Nations Environment Programme, *Environmental data report 1991/92.* Blackwell Publishers, Oxford.

UNEP (1992) *The world environment 1972–1992: two decades of challenge.* Tolba, M K, El-Kholy, O A, El-Hinnawi, E, Holdgate, M W, McMichael, D F and Munn, R E (Eds), United Nations Environment Programme. Chapman and Hall, London.

UNEP (1993) United Nations Environment Programme, *Environmental data report 1993/94.* Blackwell Publishers, Oxford.

UNIDO (1992) *Industry and development global report 1992/93.* United Nations Industrial Development Organisation, Vienna.

WRI (1992) *World Resources 1992–93.* World Resources Institute, Oxford University Press, New York.

Young, J E (1992) *Mining the Earth.* Worldwatch Paper 109, Worldwatch Institute, Washington DC.

Emissions from steelworks, Gijon, Spain
Source: Simon Fraser/Science Photo Library

14 Emissions

INTRODUCTION

Many of the environmental problems currently facing Europe result from the discharge of pollutants into the environment. Information on the characteristics, quantities and locations of such emissions are therefore essential to the assessment, prediction and understanding of their fate and potential environmental impact. A systematic analysis of emissions into the various environmental media is also crucial for establishing quantitative and qualitative relationships between these problems and the pressures from specific sectors of activities, production processes and technologies. Strategies for controlling and reducing discharges will become more effective as the quality and quantity of information on these emissions develop.

The term 'emissions' can encompass a wide range of agents, from chemical susbstances to noise and radiation. In this report, 'emissions' refer specifically to substances which are of no further use for the purposes of production, transformation or consumption and which are released to the environment – air, water or land – rather than recycled or re-used. These are usually referred to as atmospheric emissions when directly released to air, as wastewater when discharged to waterbodies and as waste when dumped to land or handled further before final dumping. Wastes are a particular form of emissions; throughout their extended life, wastes may change nature and location as they are handled and processed before disposal to land, causing emissions to air, water and soil.

This chapter presents an overview of emissions to air and water in Europe – their physico-chemical characteristics, magnitude, pathways and sinks in the different environmental media. Wastes are treated separately in Chapter 15. The impacts of emissions are examined in detail in Chapters 4 to 11 and further developed in Chapters 27 to 38 as part of the presentation of prominent European environmental problems.

Emissions are treated by sources in Chapters 19 to 26. Radioactive emissions are addressed in Chapters 16 and 18.

EMISSION INVENTORIES

Information on pollutant emissions is usually compiled in emission inventories. Direct measurements are still infrequent and therefore emissions are usually estimated with the help of emission factors applied to statistics on human activities. Emission inventory methodologies have evolved over the past 20 years and there is now general appreciation of methodologies and of the types of data to be included in an emission inventory to ensure its completeness, consistency and transparency and hence its quality and usefulness (Box 14A).

Emission inventories serve several purposes – they help:

- assess the fate of pollutants in the environment and their potential impacts;
- identify the activities responsible for different emissions and their relative contribution;
- assess potential impacts and implications of different development strategies;
- identify priorities for emission reductions;
- set explicit sustainability objectives and constraints;
- improve the information available to policy makers, industry and the public;
- evaluate the environmental costs and benefits of different policies;
- ensure that policies, agreements and regulations are being complied with; and
- assess the effectiveness of policy actions in achieving desired objectives.

Box 14A Emission inventory methodology developments

To provide a useful and informative emissions database, a clear explanation of the assumptions made to derive the estimates of emissions should be included in an emission inventory. Such an inventory should include the following types of information:

- emission measurements;
- emission factors;
- individual emission source specifications (location, activity, processes, controls, emission heights, conditions, etc);
- activity statistics (energy use, production, population, employment, transport, etc);
- references to the sources of the data included; and
- comments on the assumption made in deriving the emissions;

as well as emission estimates derived from the above data.

Emission inventories usually provide :

- the distribution of emissions across relevant technologies and socio-economic sectors;
- the spatial distribution of emissions; and
- trends in emissions over time.

National, and hence international, emission inventories have included very few direct measurements of emissions from individual sources but have derived emission estimates from activity statistics (for example fuel consumption, vehicle kilometres and production statistics) and representative scaling factors or **emission factors**. Large, localised sources of pollution, for example power-plants, refineries and factories, are often included separately in the inventory as **point sources**. Increasingly, point source emissions are based on plant-specific information, including on-site measurements, provided to the regulatory authorities by the plant operators in compliance with legal requirements (for example EU Large Combustion Plant Directive or various schemes for toxic release inventories). In contrast, smaller or more diffuse sources, such as domestic housing, motor vehicles and agriculture, are treated as **area sources** and accounted for on an administrative area (for example county, *département* or *Land*) or regular grid basis (for example 50 x 50 km) based on activity statistics and appropriate emission factors. In some inventories, where activity statistics are available on individual roads, motor vehicles can be treated as **line sources**.

The methodology of emission inventory developed steadily throughout the 1980s and certain fundamental requirements have now been recognised. These are **completeness, consistency** and **transparency**. An emission inventory should cover all known sources (completeness). To enable users to understand how the estimates were derived, inventories should include activity data and emission factors plus references to the source of these data as well as the emission estimates or measurements themselves (transparency). Finally, emission inventories need to use the same definitions and subdivisions (consistency) to allow comparisons to be made between national totals and contributions from sub-sectors. However, these requirements have seldom if ever been fully achieved in emission inventory systems established to date.

A wide range of emission inventories already exists at the national level in several European countries. Attempts to produce international integrated inventories of emissions have been made more recently in response to increased concerns for their transboundary environmental impacts (see Box 14B). Nevertheless, most of the existing inventories cover only emissions to the atmosphere. Emissions of pollutants into the air have been quantified most systematically and in greatest detail in recent years because of the increasing evidence of their impact on the local, regional and global levels. Emissions into waterbodies have long been considered to have more restricted dispersion and hence more local (or regional) impacts. Consequently, these have not been quantified in such detail to date. Instead, more attention has been given to measuring concentrations of water pollutants and monitoring water quality. Releases into or onto land, including the various types of waste, have been addressed with different approaches in European countries and accordingly information is incomplete and fragmentary. As yet, no agreement has been reached on a consistent classification system for developing inventories of such releases.

In addition to the limited coverage of existing inventories, another major obstacle to a European-wide assessment of emissions is their limited comparability. Many of the national inventories which do exist differ fundamentally in their scope and structure as well as in their method of data collection, geographic scale and range of emissions considered. The harmonisation of emission inventory methodologies at the European level is increasingly recognised as a priority (Box 14C).

Furthermore, the real value of information on emissions occurs not when data are available as disparate estimates from different sources into different media, but rather when data can be readily compared and integrated (Box 14B). A study on the feasibility of a European integrated emissions inventory (Briggs, 1993), carried out for the European Commission, has identified the links between emissions to air, water and land from a range of human activities, and the major environmental problems and targets of the EU Fifth Environment Action Programme. Many substances are released into more than one medium (or potentially transferred from one medium to another), are relevant to more than one environmental problem and are dependent on the same types of activities.

Consequently, data collected on activities relevant to one substance, one medium or one problem are likely to be relevant to other substances, media and problems. However, to date, most national and international programmes to quantify emissions have addressed the various policy areas

Species	European emissions (Mtonnes/year)	Share of world total (%)
CO_2	8070	30
CH_4	55	16
$CFCs$[2]	0.5	36
N_2O	0.5	7
NO_x	22	21[3]
CO	125	11[3]
VOC (NMHC)	25	25[3]
SO_2	39	25[3]

Notes:
1 This table does not represent the large differences in per capita, unit area and GDP emission rates which exist between European countries of the OECD area and former planned economy countries.
2 Indicated as CFC-11 equivalents.
3 Global estimates are highly uncertain.

Table 14.1
European emissions of selected compounds[1]
Sources: Compiled from multiple sources (see Table 4.7)

Box 14B European atmospheric emissions inventories

The first recent attempt to produce a consistent European inventory was the Emission Inventory of Major Air Pollutants in OECD European Countries in 1980 (OECD, 1990), which was designed to assess pollution by large-scale photochemical oxidant episodes in Western Europe and to evaluate the impact of various emission control strategies on such episodes.

This inventory covered three pollutants – SO_2, NO_x and VOCs – categorised into nine main source sectors, and provided national totals, sectoral, large-point sources (utility power plants >200 MW electric) and 50 x 50 km gridded data for 17 European OECD countries based on official submissions from these countries. The final OECD inventory was reasonably complete (except for VOCs, for which problems of completeness and consistency remain) but was not fully consistent or transparent, despite the efforts of the OECD project team to refine the official submissions.

The experience of the OECD project was used in the development stage of the CORINAIR 1985 inventory. CORINE – CO-oRdination of INformation on the Environment – was established by a Decision of the European Communities (85/338/EEC) to develop an experimental project for gathering, coordinating and ensuring the consistency of information on the state of the environment in the European Community. One of the CORINE projects, CORINAIR, aimed to compile a coordinated inventory of atmospheric emissions from the 12 Member States of the Community in 1985.

CORINAIR85 covered the same three pollutants as the OECD project and recognised eight main source sectors. It was a major step forward in international emission inventory methodology, although it is now recognised (McInnes and Pacyna, 1992) that the requirements of completeness, consistency and transparency were not fully achieved. The VOC inventory was not complete, there were inconsistencies in the use of some emission factors which were not fully resolved, and restrictions on the number of fuels which could be specifically included reduced transparency.

Protocols on the reduction of emissions or transboundary fluxes of sulphur, oxides of nitrogen and volatile organic compounds (the SO_2, NO_x and VOC Protocols), adopted under the UNECE Convention on Long Range Transboundary Air Pollution (LRTAP), commit Signatories to reductions in emissions or transboundary fluxes within agreed timetables and to submit national emission data annually.

During the 1980s it was recognised that the data submitted by the Signatories did not fully meet the requirements of completeness, consistency and transparency. Consequently, the Cooperative Programme for Monitoring and Evaluation of the Long Range Transmission of Air Pollutants in Europe (EMEP) established guidelines for the estimation and reporting of emission data (Pacyna and Joerss, 1991) and set up a Task Force on Emission Inventories (McInnes and Pacyna, 1992) to improve guidance on international atmospheric emission inventory methodology.

Recognising the case for consistency and harmonisation, this activity collaborated with the CORINAIR project team to develop a common CORINAIR/EMEP source sector split with 11 main source sectors, which was adopted by the Convention Executive Body in 1991.

The European Commission has maintained the CORINAIR Expert Group to produce an emissions inventory for 1990 – CORINAIR90 – which from now on will be coordinated by the European Environment Agency.

The CORINAIR90 project has developed the original CORINAIR source nomenclature to cover about 250 activities and eight pollutants within the main 11 source sectors agreed with EMEP. A more detailed data processing system has also been developed to improve the completeness, consistency and

transparency for the 1990 inventory.

The system has been made available to about 30 countries in Europe to help them develop their national methodologies and their submissions to both the LRTAP Convention and, where appropriate, the European Commission. The first results from the CORINAIR90 project only became available in the closing stages of preparing this report and they were therefore unable to be fully integrated into the current assessment. An example of the type of data that will be available from CORINAIR90, is illustrated in Map 14.1 where national emissions of SO_2 and the contribution of 11 main sectors to the national totals are presented.

Map 14.1
European SO_2
emissions by country
and main sector,
reported by
CORINAIR, 1990
Source: CORINAIR90

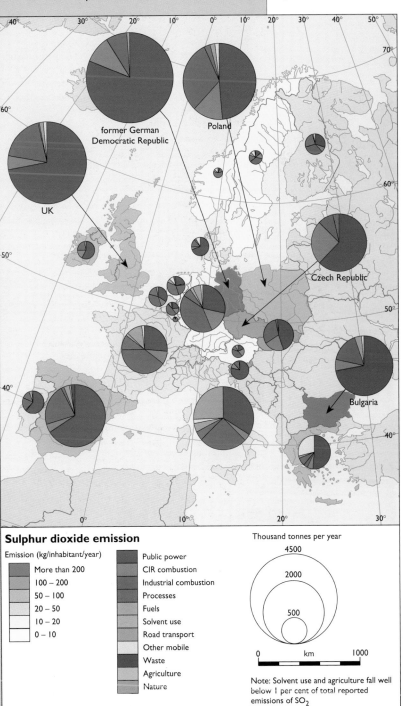

Sulphur dioxide emission

Emission (kg/inhabitant/year)

- More than 200
- 100 – 200
- 50 – 100
- 20 – 50
- 10 – 20
- 0 – 10

- Public power
- CIR combustion
- Industrial combustion
- Processes
- Fuels
- Solvent use
- Road transport
- Other mobile
- Waste
- Agriculture
- Nature

Thousand tonnes per year

4500
2000
500

0   km   1000

Note: Solvent use and agriculture fall well below 1 per cent of total reported emissions of SO_2

former German Democratic Republic

Poland

UK

Czech Republic

Bulgaria

separately. As a result, these programmes have produced non-comparable information – a major obstacle to a more integrated approach to emission control. This implies the need not only to extend existing emission inventories to other areas, sources and media, but also to make the data within them more consistent and comparable (Briggs, 1993).

Figure 14.1
SO₂ emissions by sector for 20 European countries
Source: CORINAIR90

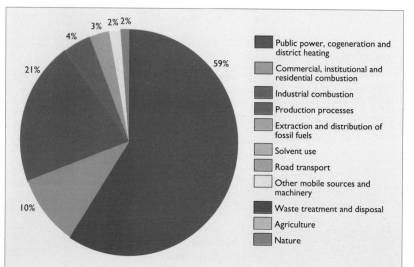

- 59% Public power, cogeneration and district heating
- 10% Commercial, institutional and residential combustion
- 21% Industrial combustion
- 4% Production processes
- 3% Extraction and distribution of fossil fuels
- 2% Solvent use
- 2% Road transport
- Other mobile sources and machinery
- Waste treatment and disposal
- Agriculture
- Nature

ATMOSPHERIC EMISSIONS

Emissions to air arise from human activities and from natural processes. The share between global anthropogenic and natural emissions for a selected number of atmospheric compounds is given in Chapter 4 (see Table 4.7). Anthropogenic emissions occur during the extraction, distribution and combustion of fossil fuels, from various industrial processes, from waste treatment and disposal, from agriculture, and from a range of consumer products. Natural sources – such as volcanic eruptions, forests, grasslands and wild animals – are not further considered in this chapter.

Atmospheric emission inventories have developed from rough estimates of national emissions of sulphur dioxide, derived from national energy statistics, to highly detailed databases containing spatial information on human activities and their resulting emissions of a wide range of pollutants. Most countries have developed their own systems for compiling inventories of atmospheric emissions and there have been several initiatives to produce multinational inventories. However, national inventories are often not comparable in terms of, *inter alia*, source category splits and pollutants covered. On the other hand, international inventories have been limited either in the number of countries covered or in their source category splits (in some cases differentiating only between mobile and stationary sources).

In many inventory systems, attempts are made to differentiate between anthropogenic (or man-made), biogenic and natural emissions. However, the boundaries between these categories are not always well defined or definable and hence differences arise between inventories. For example, emissions from forests, including those which have been developed by humans for logging or other purposes, tend to be categorised as natural, whereas domesticated animals tend not to be considered as natural. The discussion below will avoid these ambiguous terms and focus on the different categories of human activities relevant to emissions, including agriculture, domesticated animals and managed land. Where inconsistencies arise or natural emissions not dependent on human activity are known to be included, these will be noted.

It is difficult to give a complete and consistent overview of the contribution of different sources of the various pollutants for all European countries; the completion of the CORINAIR90 inventory (Box 14C) will facilitate this task. Nevertheless, using existing inventories, it is possible to put European emissions of a number of pollutants into a global context, and to determine the main contributors and some spatial and temporal trends. The European contribution to global emissions of SO₂, NO$_x$, N₂O, CO, CO₂, CH₄, CFCs and VOCs is summarised in Table 14.1. Temporal trends and spatial patterns are analysed below for sulphur dioxide, nitrogen oxides, volatile organic compounds, ammonia, carbon monoxide, carbon dioxide and methane. Except for heavy metals and metalloids, emissions of the various components of aerosols (both primary and secondary) and particulate matter have not been addressed due to the lack of consistent data from the different countries of Europe. The data used to produce maps and graphs presented in the following sections can be found in the *Statistical Compendium*.

Sulphur dioxide

Sulphur dioxide (SO₂) is released during the combustion of sulphur-containing fossil fuels and from certain industrial processes. It has effects on human health (see Chapters 4 and 11) and as one of the principal contributors to acidification has an impact on ecosystems and buildings (see Chapters 5, 9 and 31). As such, SO₂ emissions have been quantified in the

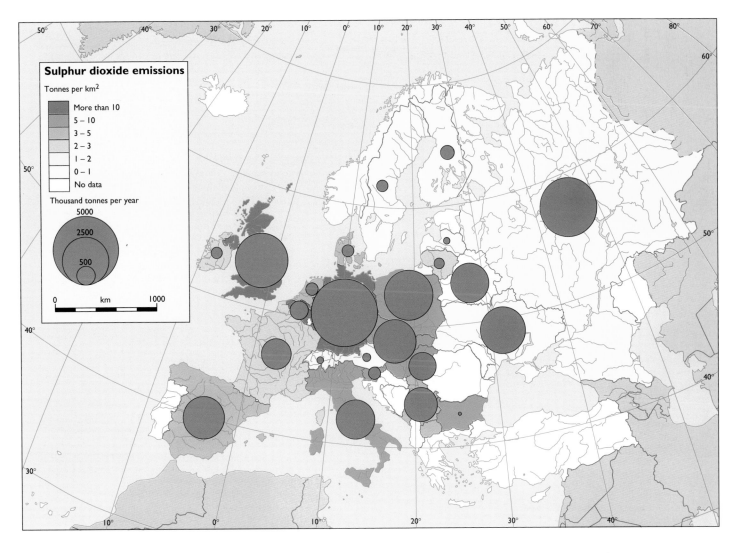

Sulphur dioxide emissions

Tonnes per km²

- More than 10
- 5 – 10
- 3 – 5
- 2 – 3
- 1 – 2
- 0 – 1
- No data

Thousand tonnes per year

5000
2500
500

0   km   1000

greatest detail for the longest period of time.

Table 14.1 indicates that Europe accounts for about one quarter of global emissions. The distribution in national total SO_2 emissions from human activities across Europe in 1990 (Map 14.2) shows that Germany, the Russian Federation and the UK were the main contributors to SO_2 emissions – each accounting for over 10 per cent of the European total in 1990 (see also Map 4.1a).

In most countries, the production of electricity accounts for at least half of total SO_2 emissions, although this sector is not always separated from other stationary combustion source emissions in national inventories. The contribution from public power, cogeneration and the district heating sector (of which public power is by far the main contributor) is available for 20 European countries (EU12, Austria, Finland, Norway, Sweden, Bulgaria, Czech Republic, Poland and Slovakia) in 1990 from the first results of CORINAIR90. This sector accounted for 58 per cent of the total of 25 million tonnes of SO_2 (Figure 14.1). National contributions from this sector ranged from 1 per cent in Luxembourg and Norway to 72 per cent in the UK and Bulgaria.

The other major source sectors were industrial and non-industrial (stationary) fossil-fuel combustion accounting for 21 and 10 per cent respectively of total European emissions of SO_2 in 1990 and 28 and 15 per cent respectively of the Central European total in 1988 (Austria, Croatia, former Czechoslovakia, Hungary, Italy, Poland and Slovenia). By contrast, transport (mainly diesel-powered) accounted for about 3 per cent of these SO_2 emissions across Europe as a whole.

There have been reductions in SO_2 emissions in most

countries over the last ten years as countries have addressed the issues of air quality and acidification and the requirements of the SO_2 Protocol of the LRTAP Convention and related legislation on emission reductions. Figure 14.2 shows the downward trend in national total emissions from a selection of countries in more detail between 1980 and 1990. While a flat rate cut of 30 per cent in national emissions was the basis of the first sulphur protocol, critical loads now form an integral part of the new 1994 protocol (see Chapter 31).

Nitrogen oxides

Nitrogen oxides (NO_x) consist of the two gases, nitrogen dioxide (NO_2) and nitric oxide (NO), formed in combustion processes from both the nitrogen present in the fuel and from the oxidation of nitrogen in air. NO_x can reduce plant growth and cause visible damage to plant crops, contribute to acidification and the formation of ground-level ozone. Acute exposure to high concentrations of NO_2 can cause breathing problems in humans.

Europe accounts for just under one quarter of global human-made emissions (Table 14.1). The distribution in national total NO_x emissions across Europe in 1990 (Map 14.3) shows that the UK, the Russian Federation and (Western) Germany were the main contributors, each accounting for more than 10 per cent of the total NO_x emissions in 1990 (see also Map 4.1b).

In most countries, road transport and the production of electricity are the main sources of NO_x emissions. These sectors accounted for 46 and 22 per cent of European (20

Map 14.2
European SO_2
emissions by country,
1990
Source: UNECE

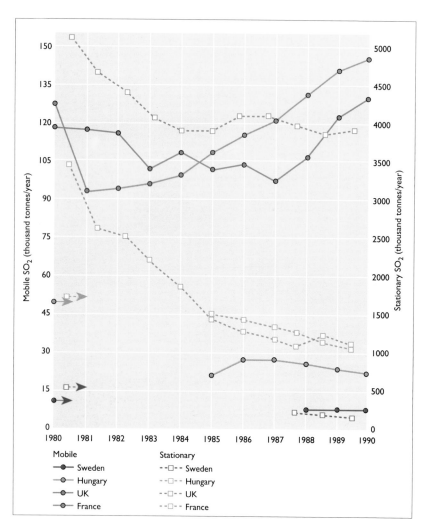

Figure 14.2
SO₂ emission trends,
1980–90
Sources: UK: WSL, 1992;
France: CITEPA, 1992;
Sweden: Statens
Naturvårdsverk, 1992;
Hungary: J Kutas,
personal communication

Legend:
Mobile
— Sweden
— Hungary
— UK
— France

Stationary
- -□- - Sweden
- -□- - Hungary
- -□- - UK
- -□- - France

In effect the VOC Protocol addresses NMVOC, and the two definitions cover the same species. Nevertheless, inconsistencies continue and care needs to be taken when using VOC inventories, including CORINAIR85, to confirm whether or not methane (CH4) and other species are included in any VOC (or even NMVOC) emission estimate. Methane is a greenhouse gas and should now be quantified separately from NMVOC.

Map 14.4 presents estimates of national total VOC emissions for 1990 as reported to the UNECE (see also Map 4.1d). Due to the above complications, data for VOCs are less complete and consistent than for SO₂ and NOₓ. Hence, the spatial distribution in national total VOC emissions across Europe should be viewed with some caution.

The combustion of fossil fuels, responsible for almost all SO₂ and NOₓ emissions, accounts for only one third of NMVOC emissions in the CORINAIR90 inventory. The use of solvents, extraction and distribution of fossil fuels (solid fuel and gas) and forests (both coniferous and deciduous) accounted for about 50 per cent of the 1990 NMVOC total (Figure 14.5).

It is not yet possible to detect clear trends in VOC (or NMVOC) emissions; some countries appear to have reduced their emissions while others appear to show some increase over the past ten years. It will be necessary to await the production of more detailed, consistent time series of VOC emissions before confirming whether these trends are real or simply a reflection of the uncertainties in past methodologies.

The VOC Protocol under the LRTAP Convention adopted in 1991 introduces a stepped approach to controlling VOC emissions. The first step, which Parties may opt for on signature of, or accession to, the Protocol (although alternative options are available), is to take effective measures to reduce national annual emissions of VOC by at least 30 per cent by the year 1999 using 1988 (or another specified year between 1984 and 1990) as the baseline.

Ammonia

Interest in ammonia (NH₃) developed in the late 1980s to explain its contribution to acidification. The distribution in emissions across Europe shows that the traditional farming countries made the largest contributions to European emissions (see eg, Map 4.1c). The main sources of atmospheric ammonia are animal manure and the use of fertiliser. Hence ammonia emissions are dependent on the numbers of farm animals and the intensity of land-cultivation. Preliminary estimates made in the UK (WSL, 1992) indicate that about two-thirds of UK emissions in 1990 were from animal manure (in particular from cattle) and 22 per cent from the use of fertiliser. These sources accounted for 98 per cent of ammonia emissions in both Norway in 1991 (CBS, 1993) and The Netherlands in 1988 (VROM, 1992).

Carbon monoxide

Carbon monoxide (CO) is derived from incomplete combustion of fossil fuels. It is toxic at high concentrations and contributes indirectly to global warming as a precursor of ozone. Emissions arise mainly from road transport, and Europe emits about 125 Mtonnes, or 11 per cent of the world total.

Carbon dioxide

Carbon dioxide (CO₂) is a key natural component of the carbon cycle. It is also the main contributor to enhanced global warming. Emissions from human activities in Europe are almost exclusively dependent on the combustion of fossil fuels. The production of electricity, combustion in industry,

countries) NOₓ emissions respectively in 1990 (Figure 14.3).

In contrast with the trend in SO₂, there has been little change in total NOₓ emissions across Europe between 1980 and 1990. Some countries show a downward trend produced by the introduction of emission controls, while others show an upward trend where increased fuel consumption, particularly in the mobile sector, has not been compensated for by emission controls. Figure 14.4 shows the trends in annual NOₓ emissions in a selection of countries in more detail between 1980 and 1990. The NOₓ Protocol under the LRTAP Convention committed parties to a freeze on national emissions at 1987 levels by 1994, together with a package of abatement measures.

Volatile organic compounds (VOCs)

The term VOC potentially covers thousands of chemical species, some of which are toxic to health or are precursors of photochemical pollution, greenhouse gases or stratospheric ozone depletion. In contrast to SO₂ and NOₓ, few sources have been measured in any detail. Furthermore, different measurement campaigns have quantified different subsets of VOC species. The LRTAP Convention VOC Protocol has defined VOC as 'unless otherwise specified, all organic compounds of anthropogenic nature other than methane that are capable of producing photochemical oxidants by reaction with nitrogen oxides in the presence of sunlight'. The current EMEP Guidelines for Estimation and Reporting of Emission Data, however, define non-methane VOC (NMVOC) as 'all hydrocarbons and hydrocarbons where hydrogen atoms are partly or fully replaced by other atoms (for example sulphur, nitrogen, oxygen or halogens) which are volatile under ambient conditions, excluding CO, CO₂, CH₄, CFCs and halons'.

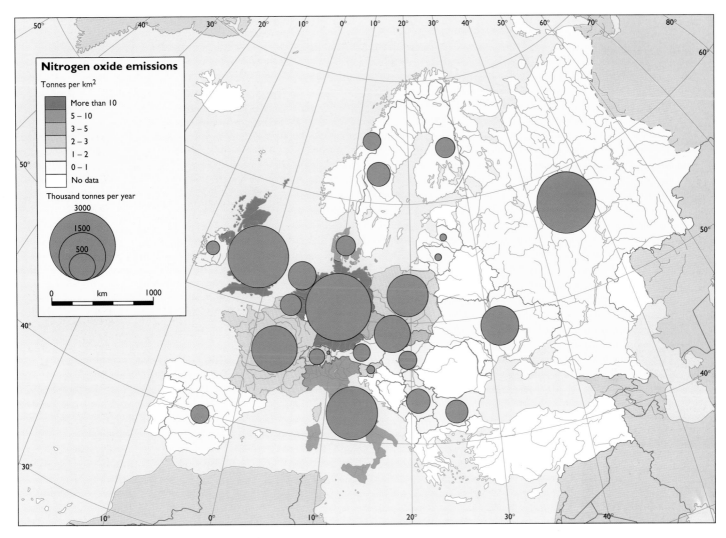

road transport and other non-industrial combustion (in approximately that order) are the main source sectors. Europe is responsible for about 30 per cent of global CO_2 emissions. The distribution in contributions from individual European countries in 1990 (Map 14.5) identifies Germany and the UK as the major emitters. Trends in emissions from France and the UK over the period 1980 to 1990 are presented in Figure 14.6.

A major project led by the scientific assessment working group (WGI) of IPCC and OECD (OECD, 1991) is under way to develop guidelines on emission inventory methodology for greenhouse gas sources (and sinks) and help countries across the world produce comparable greenhouse gas inventories for the Framework Convention on Climate Change (FCCC). This project is working closely with the CORINAIR90 project to ensure compatibility for submissions to the FCCC from European countries. The FCCC was adopted in Rio de Janeiro in 1992, and entered into force in March 1994. Several countries, including collectively the 12 EU Member States, have adopted the objective of stabilising CO_2 emissions by 2000 at 1990 levels.

Methane

Methane (CH_4) is a greenhouse gas and plays an important role in the production of tropospheric ozone. It has been estimated that European emissions are about 55 Mt (Table 14.1). However, methane emissions have not yet been quantified widely or systematically for many individual countries of Europe. Estimates will become

available through the CORINAIR90 and the IPCC/OECD projects. Global emissions of methane are estimated to range from 240 to 590 million tonnes per year, with rice production, enteric fermentation in animals and coal-mining making the largest contributions. In the UK, it has been estimated (WSL, 1992) that animals, landfill, offshore oil/gas production, coalmining and gas distribution contributed 26, 23, 21, 19 and 8 per cent respectively to the total of nearly 4.4 million tonnes of CH_4 emitted.

Map 14.3
European NO_x emissions by country, 1990
Source: UNECE/LRTAP

Figure 14.3
NO_x emissions by sector for 20 European countries
Source: CORINAIR90

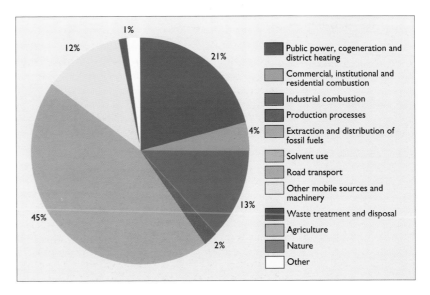

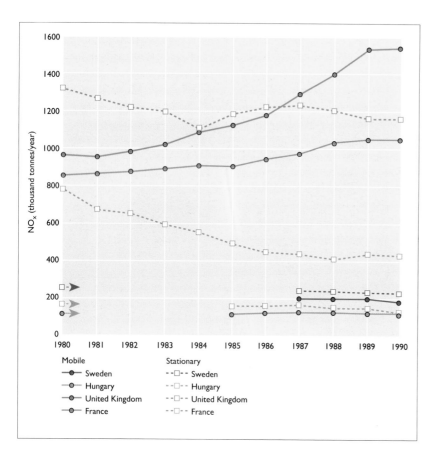

Figure 14.4 *NO$_x$ emission trends, 1980–90*
Source: See Figure 14.2

Heavy metals and metalloids

Heavy metals are released during fuel combustion or from a range of industrial processes. However, emission inventories for these pollutants are less well developed than for the above gaseous pollutants. The most systematic and comprehensive work carried out for Europe to date has produced an unofficial set of national emission estimates for arsenic (As), cadmium (Cd), mercury (Hg), lead (Pb) and zinc (Zn) for the year 1982 (Axenfeld et al, 1992, which also included emission estimates for lead in 1985, not presented here).

Axenfeld et al's study (1992) indicates that the non-ferrous metal industry was the main source of As, Cd and Zn in 1982, accounting for 74, 65 and 57 per cent respectively. Power plant and other fuel combustion each accounted for 38 per cent of mercury emissions.

Road traffic (that is, the combustion of leaded petrol) accounted for three quarters of lead emissions in 1982. These emissions from road traffic have decreased markedly since 1982 although this sector probably remains the main source of lead in the atmosphere. Downward trends in national lead emissions from road traffic over the period 1980 to 1990 were due initially to the introduction of low-leaded fuel, then

Map 14.4 *European total VOC emissions by country, 1990*
Source: UNECE/LRTAP

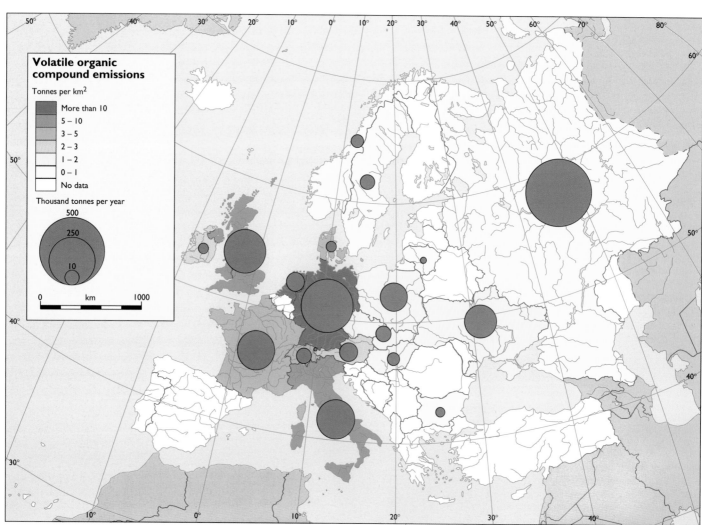

subsequently to the introduction of unleaded fuel and, in the case of Germany, the early introduction of vehicles equipped with catalysts which could not use leaded fuel.

Increasing attention is being given to heavy metal emissions, and more detailed and officially produced up-to-date heavy metal inventories are likely to become available within the next few years. The Executive Body of the LRTAP Convention has established a Task Force on Heavy Metals which will elaborate a state-of-the-art report on these pollutants, including emission inventory, atmospheric dispersion and deposition, analytical problems, technologies for the control of heavy metals, and economic problems, to provide the basis for elements of a possible protocol. The Paris Convention for the Prevention of Marine Pollution from Land-Based Sources (which covers the North Atlantic including the North Sea) and the Helsinki Convention on Protection of the Marine Environment of the Baltic Sea Area are also preparing atmospheric emission inventories for several heavy metals and other pollutants.

EMISSIONS TO WATER

Emissions to water consist of a multitude of substances having different adverse effects on the environment. They can occur either as point or non-point (diffuse) source pollution arising from a variety of human activities. The fate and effect of emissions of polluting substances in a particular waterbody will depend not only on the amount of substances emitted, but also upon the hydrological, physical, chemical and biological conditions characterising the receiving body. Major point sources are:

- sewage networks (with or without sewage treatment);
- industry;
- accidental spills;
- mining;
- storm runoff.

Major non-point sources are:

- agricultural practices;

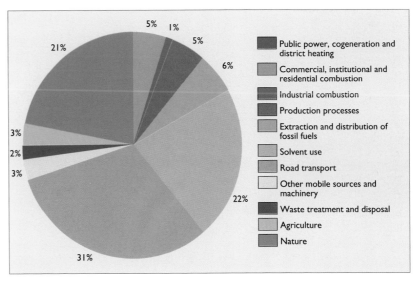

Figure 14.5
Non-methane VOC emissions by sector for 20 European countries
Source: CORINAIR90

Map 14.5
European CO$_2$ emissions by country, 1990
Source: UNECE/LRTAP

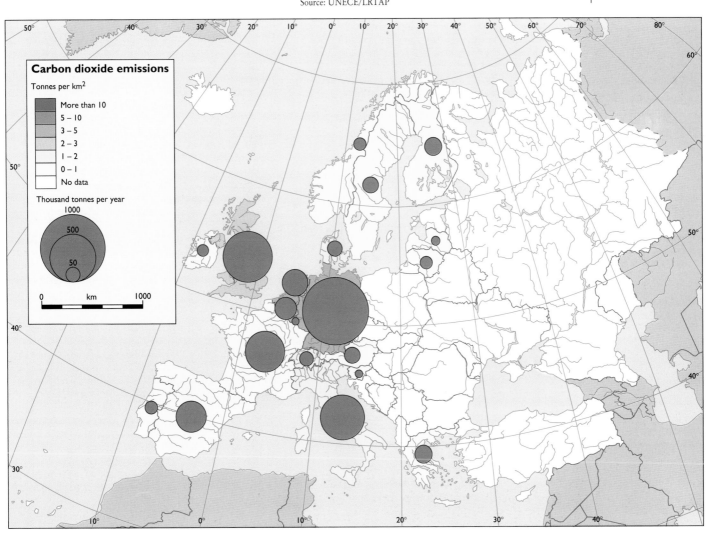

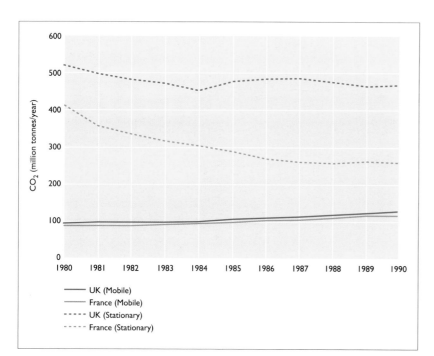

Figure 14.6
CO$_2$ emission trends,
1980–90 UK and
France
Sources: UK: WSL, 1992;
France: CITEPA, 1992

- forestry practices;
- scattered dwellings;
- construction works;
- atmospheric deposition;
- transportation on river barges or sea-going ships.

Over the last few decades, there has been an increasing tendency in several European countries to reduce point source emissions. Pollution control measures have been adopted in the domestic sector and industry. This has not been followed by a simultaneous action for non-point source pollution and, consequently, the relative importance of diffuse source emissions in total emissions has increased.

The origins of emissions are treated in Chapters 19 to 26 and the relationships between emission of important pollutants (eg, nutrients and organic matter) and their aquatic effects have been treated in detail in Chapter 5. Very little quantitative information on emissions to water is available from international data banks, and, where available, the scale (national or regional statistics) is of little relevance for an assessment of river and lake pollution. Emission data on the relevant scale for these waterbodies (the catchment) is almost completely lacking. An important exception is the River Rhine, for which the International Commission for the Protection of the Rhine against Pollution has produced an emissions inventory. Two approaches to improve estimates of emissions to water are presented in Box 14D. Data on emissions of important polluting substances for some of the major European seas (eg, nutrients and heavy metals) have been gathered and they are treated in Chapter 6 describing the state of the European seas.

The discussion below expands upon the information referred to above and establishes links between the human activities and the state of the European waterbodies. The almost complete lack of emission data on the catchment scale makes this analysis difficult and consequently the information presented here is given by way of illustration. Special focus has been given to emissions from agriculture and from wastewater, for which more information is available. In many parts of Europe, these emissions are known to seriously restrict usage of water resources. A few examples, to illustrate the contribution of·industry as a source of emissions into the aquatic environment, are also given.

The concern attached to nutrients discharged to waterbodies arises from the problem of eutrophication which they cause (Chapters 5 and 33). Nutrients in rivers and lakes originate from numerous sources such as the natural runoff from soil and bedrock weathering, and the enhanced runoff related to various human activities in catchment areas such as deforestation and agricultural activities, fertiliser application and irrigation. Many rivers are used to remove human waste products as either sewage discharge, stormwater runoff, or industrial effluent. The increase in nutrient concentrations above the natural background level depends on the demographic, industrial and agricultural development in the catchment. In sparsely populated areas with a low proportion of farmland and industrial development, the increase is low, in contrast to densely populated areas with an intensive agricultural production and many industries. Another important factor is the wastewater treatment process: an efficient high technology sewage plant may reduce the phosphorus discharge by more than 90 per cent compared with low technology sewage plant discharges.

Phosphorus

There is a positive relationship between annual mean phosphorus concentrations in the water and the population density in river catchments (Figure 14.7). In catchments with fewer than 15 inhabitants per square kilometre, the phosphorus concentration is generally lower than 20 μg P/l, and in catchments with more than 100 inhabitants per square kilometre, phosphorus concentrations are usually higher than 200 μg P/l.

In densely populated areas such as Denmark, Germany, Latvia, and the River Po catchment in which one third of the Italian population lives, 43 to 64 per cent of the phosphorus discharge to inland surface waters is related to sewage discharge from the various municipalities (Figure 14.8). Industrial effluent is in general less important and makes up only 3 to 12 per cent of the total discharge. In these areas 22 to 41 per cent of the phosphorus discharge stems from agricultural activities. If there was no pollution, phosphorus levels would be only 5 to 10 per cent of the present level.

Sweden is sparsely populated, with many major towns located along the coast. The country has one of the highest levels of sewage water treatment in Europe, and river phosphorus concentrations are consequently low (see Figure 5.19 and Map 5.16). About 27 per cent of the phosphorus discharge to inland surface waters in Swedish rivers can be related to sewage and industrial discharges (Figure 14.8) and about 10 per cent to agricultural activities, while the remaining discharges originate from diffuse runoff from land areas.

Nitrogen

Elevated nitrate concentrations in European surface waters have been strongly connected with modern agricultural practices, particularly the usage of nitrogen fertiliser (Stibe and Fleischer, 1991; Wright et al, 1991; Edwards et al, 1990; Pierre and Prat, 1989; Neill, 1989). There is a significant relationship between nitrogen concentrations and the percentage of farmland in river catchments (Figure 14.9). In rivers located in catchments with less than 10 per cent agricultural land, nitrogen levels are generally below 0.3 mg N/l, whereas nitrogen levels lie between 0.5 and 2.5 mg N/l in rivers where agricultural land constitutes 10 to 50 per cent of the catchment, and in rivers with the highest proportion of cultivated land in the catchment area, the nitrate and total nitrogen levels are generally above 1.5 and 2 mg N/l, respectively.

In Denmark, where 65 per cent of the total land area is farmland, approximately 80 per cent of the nitrogen discharge

Box 14D How to obtain better estimates of emissions from agriculture to water

In order to improve our knowledge about the emissions of nutrients from agriculture and from other sectors to the aquatic environment, two approaches seem appropriate: the National Network Approach and the Type Catchment Approach. Common to both approaches (exemplified below using nitrogen) is the establishment of a permanent river-monitoring network or the adaptation of existing networks to include continuous measurements and estimates of the discharge of flow and frequent water sampling for subsequent analysis of nitrogen species and phosphorus.

The National Network Approach

This mass-balance approach relies on setting up a monitoring network with the aim at determining the total flux of nitrogen (N_{tot}) through major rivers from country to country and to the sea. Knowing the nitrogen contributions from point sources, N_{point}, either from measurements or by applying emission factors; the contribution, N_{scat}, from scattered houses and dwellings (if important); the retention of N in rivers and lakes, N_{ret}, by sedimentation and denitrification; the area coefficient for N from uncultivated land, N_{back}, and the atmospheric deposition, N_{dep}, directly on freshwater surfaces, the contribution from agriculture, N_{agri}, can be obtained by solving the mass-balance equation:

$$N_{agri} = N_{tot} - N_{point} - N_{back} - N_{scat} - N_{dep} - N_{ret}$$

The reliability of this estimate is heavily dependent on the accuracy with which it is possible to measure or estimate the other important sources or sinks. If these are properly estimated, this approach applied to all major rivers of a country allows the calculation of (i) the total nitrogen load to the sea, and (ii) the partition of this load among the major nitrogen contributors (eg, agriculture, industry, domestic sewage). With minor extensions of the list of determinants to include, for example, other nutrients and heavy metals, the National Network Approach may serve multiple purposes with only limited additional costs.

The Type Catchment Approach

The agricultural nitrogen input to surface waters can also be estimated by comparing fluxes in watercourses draining well-defined small catchments on the same soil type but with different degrees of utilisation. Heath or forest covered catchments may serve as baseline catchments and provide estimates of the background concentration of nitrogen. The nitrogen flux in watercourses draining catchments dominated by agriculture is composed of the background and the agricultural inputs and subtraction of these contributions provides an estimate of the agricultural input. For the purpose of catchment selection 'baseline' and 'type' catchments might be defined respectively as catchments with more than 90 per cent natural areas and 90 per cent agriculture.

These approaches are recommended for use in UNEP/WHO's Global Environment Monitoring System, GEMS/WATER programme, and have, for example, been successfully applied for source partitioning of nutrient emissions to water in several countries bordering the North and Baltic seas.

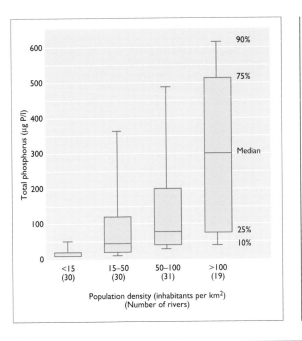

Figure 14.7
Relationship between annual mean total phosphorus concentrations and population density in river catchments
Source: Compiled by NERI and EEA-TF from multiple sources listed in Box 5E

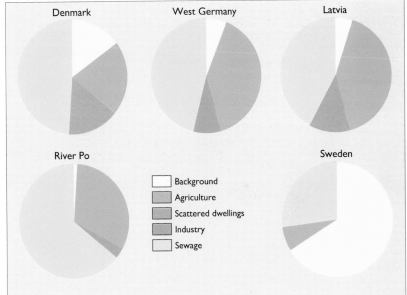

Figure 14.8
Sources of phosphorus discharge to rivers and lakes
Sources: Revised from Kronvang et al, 1993; Umweltbundesamt, 1992; Carl Bro Group, 1992; Italian Ministry of Environment, 1992; Löfgren and Olsson, 1990

to inland waters is a consequence of agricultural activities (Figure 14.10). In Latvia a high proportion of the nitrogen discharge (70 per cent) can be related to agricultural activities. In Germany and the River Po catchment, agriculture is the single most important nitrogen source, representing about 50 per cent of the total amount discharged. Here, point source nitrogen discharges play an important role, with about 45 per cent of the total discharge, three quarters being attributable to municipality wastewater and the remaining quarter to industrial effluents. If there was no pollution, nitrogen levels in the above-mentioned areas would be only 10 per cent of the present levels.

In Sweden, only 6 per cent of the land area is cultivated, and river nitrogen concentrations are usually low (see Figure 5.20 and Map 5.18). Here approximately 40 per cent of the nitrogen discharge into inland surface waters can be related to human activities, the remaining 60 per cent being diffuse discharge from forested and uncultivated areas (background in Figure 14.10). Agricultural activities are responsible for more than two thirds of the nitrogen discharge related to human activities, the remaining third being attributable to point sources. However, in southern Sweden, with its relatively higher proportion of farmland, nitrogen concentrations increase, and a higher percentage of nitrogen discharge originates from agricultural activities (Fleischer et al, 1987). Additional indication of the importance of European agriculture for the pollution of surface waters with nutrients is presented in Table 14.2.

The average agricultural contribution from eight countries/regions with intensive agriculture (excluding Norway, Sweden and Finland) amounts to 64 per cent of total nitrogen loads, ranging from 50 per cent in Germany to 81 per cent in Denmark. The average agricultural contribution to the total phosphorus load of the aquatic environment is 31 per cent, ranging from 21 per cent in The Netherlands up to 41 per cent in Latvia. These are considered below in further detail.

Although these estimates of the contributions of nitrogen and phosphorus from agriculture to the total load of the aquatic environment have been obtained using different, and not always comparable, methods, Table 14.2 clearly indicates that agriculture plays a substantial role in the pollution of the aquatic environment all across Europe, not only with nitrates, but maybe more surprisingly also with phosphorus.

Nitrogen emissions from agriculture

Many studies have linked increasing nitrate concentrations in rivers to intensification of agriculture and associated changes in soil structure, the extent of land drainage and especially fertiliser use. A substantial surplus of nitrogen is yearly applied to European croplands as artificial fertilisers or as manure. For two Western European countries (The Netherlands and Denmark) with very intensive agriculture, the inputs of nitrogen to agricultural land greatly exceed the amount removed with harvested crops. In The Netherlands (RIVM, 1992) the total yearly input to agricultural land was

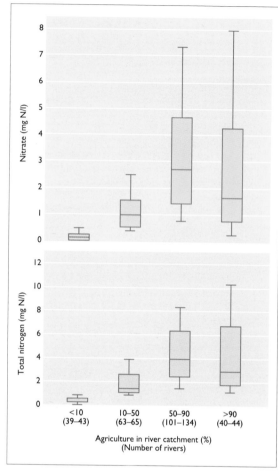

Figure 14.9 *Relationship between annual mean nitrate and total nitrogen concentration and percentage of agricultural land in river catchment*
Source: Compiled by NERI and EEA-TF from multiple sources listed in Box 5E

1.25 million tonnes N in 1986 compared with 0.46 million removed with crops, and in Denmark (National Agency of Environmental Protection, 1991) the figures in the late 1980s were 0.72 and 0.36 million tonnes N, respectively. This surplus nitrogen is potentially lost as nitrates dissolved in the water leaching out of the root zone, or as ammonia emissions, or denitrified to the atmosphere. Other nitrogen species, for example ammonia and organic nitrogen, can also leach, but compared with nitrate these leaching losses are normally much lower (Table 14.3).

Nitrate leaching losses

The relationship between nitrate leaving the plant's root zone (leaching) and nitrogen applied as fertiliser is not straightforward. Leaching losses show large variations in space and time depending on several factors, of which climate, soil and crop type, quantity of nitrogen present in the soil (from either natural sources or fertiliser inputs) and the temporal and spatial distribution of nitrogen applications are among the most important. For a discussion of these relationships the reader is referred to Chapter 5.

Figure 14.10
Sources of nitrogen discharge to rivers and lakes
Sources: Revised from Kronvang et al, 1993; Umweltbundesamt, 1992; Carl Bro Group, 1992; Italian Ministry of Environment, 1992; Löfgren and Olsson, 1990

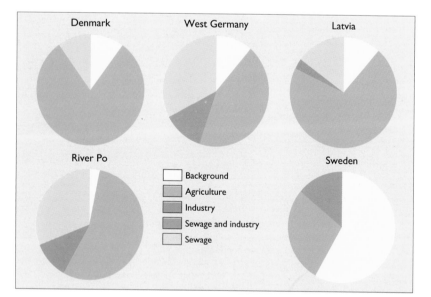

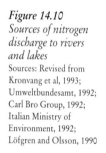

Country/region	N (% of total)	P (% of total)	Reference year
Germany (West)[1]	50	40	1989
The Netherlands[2]	68	21	1989
Italy[3]	62	33	1986
Switzerland[3]	61	no data	1985
Norway: North Sea[4]	45	23	1991
North Sea catchment[3]	60	25	1987
Denmark[5]	81	22	1991
Latvia[6]	71	41	1990
Sweden[6]	28	7	1990
Finland[7]	24	40	1989
Poland[8]	62	34	1990
River Po[6]	56	32	1990

Table 14.2
Agriculture's share of total emissions of nitrogen and phosphorus to the aquatic environment in several countries/regions of Europe
Sources:
1 Umweltbundesamt, 1992; 2 RIVM, 1992; 3 Isermann, 1990;
4 State Pollution Control Authority, 1993; 5 Christensen et al, 1993;
6 see Figures 14.8 and 14.10; 7 Wahlström et al, 1993; 8 MEPNRF, 1991

Table 14.3 shows that nitrate leaching may be 10 to 50 times higher in areas with intensive agriculture compared with leaching from natural areas with similar soil type. However, in many places denitrification processes in the soil (eg, microbiological reduction of nitrate by ferrous sulphide, ferrous ions, methane or organic matter under anaerobic conditions) are able to transform considerable amounts (up to 100 per cent (Postma and Boesen, 1991)) of the nitrates leached either to harmless nitrogen gas (N_2), or to nitrous oxide (N_2O), a gas that contributes to greenhouse warming and to destruction of the ozone layer (see Chapters 27 and 28). The impact of denitrification on groundwater quality, however, is not always beneficial. For example, nitrate reduction by ferrous sulphide may lead to increased concentrations of sulphate and ferrous ions which, especially in areas with low buffer capacities, increase the risk of acidification of groundwater and surface waters fed by groundwater (see Chapter 31).

Nitrate losses to surface waters

Areal estimates of nitrate leaching give an indication of the potential emission to the aquatic environment. The nitrate loads that actually reach surface waterbodies from agricultural land are carried to rivers and lakes by seeping groundwater, lateral subsurface flow in the soil or in drainage tiles, and these loads are in general considerably lower than the amount of nitrate leached. This is not only due to denitrification in the soil and subsoil: important denitrification also takes place in meadows and alluvial forests bordering rivers and lakes (Christensen et al, 1992). These waterlogged habitats (for a detailed description of the function of these riparian zones the reader is referred to Chapter 8) have unfortunately been drained in many places and turned into arable land with less capacity to remove nitrates. The re-establishment of riparian zones on the floodplains would almost certainly help solve the nitrate problem of surface waters. This would be especially beneficial along headwater tributaries of the river network, since it is within these zones that the background water quality of the main river is determined. These small streams also comprise a significant proportion of the total channel length in any basin.

Reliable estimates of the input of nitrogen to surface waters from agriculture (often referred to as area coefficients, as they represent an integrated estimate of nitrogen losses from a river catchment) can best be obtained using the catchment approaches described in Box 14D. The figures presented in Table 14.4 clearly indicate that such inputs are significant, but considerably lower than the leaching losses shown in Table 14.3 for the same type of soils and landuse. Many studies have been carried out across Europe with the aim of establishing area coefficients for nitrate and to establish the relationships between nitrate losses and the factors controlling it.

Table 14.5 summarises the results from selected studies for nitrate losses to surface waters from areas with and without agriculture in different parts of Europe. These data

Crop	Manure application	N-leaching (kg N/ha/year)		
		Nitrate		Total N
		Range	Average	Average
Natural areas	no	0.5–3.3	2.5 (6)	
Green fallow	no	2–5	4 (4)	
Permanent grass	no	8–35	16 (4)	
Mixed clover and grass	no	25	25 (1)	48 (1)
Winter crops and rape	no	39–44	41 (3)	49 (2)
Winter crops and rape	yes	29–88	62 (6)	78 (6)
Spring barley	no	19–166	72 (12)	94 (11)
Spring barley	yes	48–147	86 (5)	106 (5)
Root crops	yes	28–154	88 (6)	111 (6)

Note: Numbers in parentheses indicate sample size

Table 14.3
Measured nitrogen leaching rates from Danish agricultural and non-agricultural areas
Source: Miljøstyrelsen, 1991

	Area coefficient (kg N/ha/year)		
	Nitrate		Total N
	Range	Average	Average
Natural areas	0.2–3.3	1.4 (7)	2.2 (7)
Agriculture	4.6–49	15.0 (42)	15.7 (58)
Agriculture and point sources	4.2–48	17.2 (167)	20.0 (173)

Note: Numbers in parentheses indicate sample size.

Table 14.4
Nitrogen contribution to surface waters from different landuse types in Denmark
Source: Miljøministeriet, 1992

Table 14.5
Area coefficients for nitrate losses to surface waters from river catchments with different landuse characteristics

Country/region	Characteristics of catchment area	Area coefficient, kg N/ha/year	Reference
11 EU/EFTA countries (average of 30 catchments)	Forest, scrub, heath or grassland	1.6**	Hornung et al (1990)
France	Forest	2–7	Pierre and Prat (1989)
Nordic countries	<10 per cent agriculture	0.6	EEA-TF questionnaire
Finland	Forest	1.3–3.1*	Rekolainen (1989)
Sweden	Forest	1.3	Ryding and Forsberg (1979)
Sweden	Mixed agriculture	16	Ryding and Forsberg (1979)
Switzerland	Mixed agriculture	16–21	Gächter and Furrer (1972)
Finland	Mixed agriculture	7.6–20*	Rekolainen (1989)
UK (Devon)	Grassland	8–51	Webb and Walling (1985)
Northern Ireland	Mainly grassland	3.4–17	Smith (1977)
Poland (Pomerania)	55–75 per cent arable land	1–1.3	Taylor et al (1986)
Denmark	Averages for various types of agriculture on different soil types	13–20	Christensen et al (1993)
France	Intensive cereal production	15–18	Pierre and Prat (1989)
Central and Eastern Europe (Poland, Czech Republic, Latvia, Lithuania)	>50 per cent agriculture	3.0	EEA-TF questionnaire
former USSR (different regions)	Various types of arable land	31–104*	Kudeyarov and Bashkin (1984)
UK (Shenley Brook)	Clay soils, 23 per cent arable land	17–74	Roberts (1987)
Wales	Agricultural grazing	5–25	Brooker and Johnson (1984)

Notes: *Total nitrogen. **A few of these baseline catchments, however, show relatively large output of nitrate (>10 kg N/ha/year); this is either because of changed N dynamics due to forest damage or site preparation for afforestation including ploughing and drainage.

were taken from a variety of conventional surveys using different methodologies. Although not comprehensive, they are indicative of the increase of nitrogen loading of the environment as landuse progresses from forest through grassland to arable agriculture. Thus, for example, Neill (1989) estimated that the average area coefficient from land to rivers in the southeastern part of Ireland was 2.0 kg N/ha/year for unploughed land compared with 76 kg N/ha/year for ploughed land. In areas with intensive agriculture (eg, northwestern Europe) the great majority of area coefficients for nitrate falls between 15 and 25 kg N/ha/year compared with natural areas with nitrate losses of about one tenth of those found for agricultural soils. For Central and Eastern Europe (Czech Republic, Latvia, Lithuania and Poland) area coefficients are in general lower than in northwestern Europe. There are two major reasons for this: both runoff and application rates of nitrogen fertilisers are lower in these countries than in northwestern Europe (see Map 5.7).

Precipitation variability within and between years is another factor heavily influencing areal losses. The largest nitrate losses occur in wet periods during the autumn and winter as the nitrogen demand of the crops is reduced and mineralisation of soil organic matter increases the nitrate content of the soil water. Particularly high loss rates have been observed during the first significant rainfall following drought periods (Roberts, 1987; Burt et al, 1988) and storm events (Webb and Walling, 1985).

Phosphorus emissions from agriculture

Agricultural phosphorus balances for seven Western European countries in the temperate zone (Sibbesen, 1989) clearly indicate that more phosphorus in fertiliser and manure is added to soils than removed with harvested crops. By far the greatest proportion of this surplus, ranging from 5 kg P/ha/year in the UK to 59 kg P/ha/year in The Netherlands is, however, accumulated in the soil. Losses of phosphorus from agricultural soils to the aquatic environment are very small compared with the amount of phosphorus fertiliser applied and, until recently, these have been considered to be insignificant in relation to other anthropogenic inputs contaminating the aquatic environment. The loss of phosphorus from agriculture can be divided into a point source contribution (farm contribution) and a non-point contribution (area coefficient) carried to the recipient by either water or wind, the latter being of minor importance.

*Table 14.6
Area coefficients for losses of total phosphorus from agricultural catchments in the Northern temperate zone*

Phosphorus loss from point sources

If storage facilities for excreta and silage of feed lots are either too small or not suitably constructed, phosphorus will sometimes be lost from these sources and find its way to the aquatic environment via surface runoff and drainpipes. Such losses are very difficult to monitor, but following episodes of heavy rain they can locally be large and very harmful to the environment.

Loss of phosphorus to surface waters

The phosphorus area coefficient (a measure of phosphorus input to freshwater via groundwater, drainage water and surface runoff, including particulates) varies in the Nordic countries between 0.2 and 0.4 kg P/ha/year in Denmark and 0.7 to 1.8 kg P/ha/year in Finland and Norway, due to differences in soil type, topography, climate, fertiliser application and tillage conditions (Nordic Council of Ministers, 1991). As shown in Table 14.6, similar loss rates have been found elsewhere in countries in the Northern temperate zone.

The majority of measured phosphorus losses from fields are fairly similar for the temperate part of Europe and fall between 0.3 and 1.0 kg P/ha/year. Although these values seem low, recent Danish research (National Agency of Environmental Protection, 1991) has shown that such agricultural losses result in river and lake concentrations too high to prevent or reverse eutrophication even if all point source contributions were stopped.

Phosphorus losses are lower from areas without agriculture. The majority of reported loss rates for catchments with natural vegetation (mostly forested catchments) in the Northern temperate zone range from 0.05 to 0.1 kg P/ha/year (Nordic Council of Ministers, 1991), which is 30 per cent or less of loss rates from agricultural areas.

Wind erosion

Substantial amounts of soil and phosphorus can be lost due to wind erosion on susceptible soils. Particularly sensitive are fine sandy soils in open areas and without crop cover during springtime. The wind is able to carry soil particles smaller than 0.1 mm diameter far away, whereas larger soil particles (0.1 to 0.5 mm) tend to be deposited in ditches, in streams and by hedges. Losses of about 2 to 3 cm soil sometimes occur in Denmark from land with a large proportion of fine sand (Sibbesen, 1989). Two to three centimetres of soil is equivalent to about 200 to 300 kg P/ha. Actual measurements of phosphorus loss by wind erosion are sparse, but phosphorus losses up to 18 kg/ha have been recorded by Sdobnikova (1989) from Russian fields subject to severe wind erosion. For further details on the environmental effects of wind erosion the reader is referred to Chapter 7.

Emissions of pesticides and their residues

Very little (if any) systematic and comparable information is available on emissions of pesticides and their residues to the aquatic environment. Based on model calculations, a European overview showing the extent of the problem has been presented in Chapter 5. A few countries, however (see *Statistical Compendium*), have published results of their surveys of groundwater concentrations of some of the most widespread pesticides, of which the herbicides atrazine and simazine are among the most common substances to breach the EU drinking water standard of 0.5 μg/l for total pesticides and 0.1 μg/l for individual pesticides. Although the acute toxicity of these substances to animals and humans is only slight (LD_{50} < 5g/kg (Swanson and Lloyd, 1992), high concentrations of atrazine (>1 μg/l) are commonly found in groundwater under fields where this herbicide has been applied to kill dicotyledonous weeds. For example, high atrazine concentrations, detected in groundwater and surface waters of the Po Valley, Italy, in the mid-1980s, raised public awareness mainly because the long-term carcinogenic effects of atrazine are uncertain. As a consequence, sale and use of atrazine was banned by the Italian government in 1990. At the same time Germany also banned atrazine.

Region/country	Area coefficient, (kg P/ha/year)	Reference
Northern temperate zone	0.46	Dillon and Kirchner (1975)
The Netherlands	0.72	Berbee (1987)
West Germany	0.36	Schulte-Wülwer-Leidig (1983)
Ireland	0.78	Phillips-Howard (1985)
France	1.4	Seux et al (1985)
Lithuania	0.2–0.8	Pauliukevicius et al (1989)
Poland	0.3–1.0	Ryszkowski et al (1989)
Northern Ireland	0.58–1.4	Smith (1977)
East Germany	0.92	Harenz (1989a; 1989b)

Treatment and emissions of wastewater

Wastewater is an important source of emissions to waterbodies. Wastewater comprises mainly industrial effluents (with or without separate treatment), household connections and stormwater runoff. Agricultural waste, leachate, and contaminated groundwater may also occasionally contribute to wastewater when sources are connected to sewer systems. Infiltration water into sewers below the groundwater table also acts as a source, driven by hydraulic forces. National statistics on wastewater emissions by volume and weight of polluting substances (eg, BOD, N, P and heavy metals) exist for several European countries, but such national figures are, from an environmental point of view, of only marginal interest because the water quality of a particular recipient (river, lake or coastal area) reflects human activities taking place in the catchment draining to the waterbody regardless of national boundaries. In addition, these statistics are often difficult to compare due to different interpretations between countries of what to include in them. For example, to obtain a national figure for volume discharge of wastewater, some countries add up emissions of untreated and treated effluents; some countries include stormwater runoff while others do not. The need to extend the inventory approach to water emissions has been stressed in Box 14C. Due to the lack of comparable wastewater statistics the following section concentrates on a brief description of the most widespread sewage treatment technologies and summarises their distribution across Europe.

Wastewater treatment technologies

Sewer systems conduct wastewater to the treatment plant. Combined sewers are constructed for transport of both wastewater and stormwater runoff, whereas separate sewers have one network for wastewater (dry weather sources) and another for stormwater runoff (wet weather sources). There are numerous different designs of wastewater treatment plants, involving combinations of physical, chemical and biological processes for separation and degradation of the pollutants.

Primary or physical treatment is of limited value in sewage treatment (Table 14.7) and involves the screening of the wastewater to remove large solids and the production of sludge by settling of particles on the bottom of large sedimentation lagoons or tanks. Recognising the inadequacies of primary treatment alone, biological or secondary treatment was introduced in the 1950s primarily to treat sewage from large population centres. In addition to physical treatment of the sewage, secondary treatment involves efficient bacteria-mediated degradation of organic matter. In the 1970s and 1980s low-technology methods were introduced in Germany and the Scandinavian countries for treating domestic sewage from very small villages and individual houses or farms which could be connected to the sewerage system only with difficulty. These methods (biological sand-filters and constructed reedbeds) apply microbiological degradation of organic matter and are also able to remove parts of the nutrients from the effluent.

Although giving rise to substantial environmental improvements (eg, improving oxygen conditions and decreasing the ammonia content of the receiving waters) secondary treatment was only able to remove a minor proportion of the nutrients creating eutrophication problems (see Chapter 5). Despite introduction of secondary treatment at many treatment plants in the 1960s, eutrophication problems continued to grow, particularly due to increased use of phosphorus-containing detergents. Chemical treatment was introduced in some European countries in the late 1960s and 1970s as a third step in the sewage treatment process, mainly to overcome eutrophication problems. Very efficient removal of phosphorus (Table 14.7) can be achieved by adding a chemical – most often lime, aluminium or iron salts – to the biologically treated sewage. These chemicals react

with phosphorus to form compounds which precipitate and can be removed from the treatment plant with the organic sewage sludge (see Chapter 15). Removal of nitrogen is a further biological process. First, ammonia and organic nitrogen in the wastewater are transformed to nitrate in the process of nitrification. This process requires oxygen whereas the subsequent transformation of nitrate to gaseous nitrogen (denitrification) occurs only under oxygen-free conditions. Treatment plants with nitrification/denitrification technologies may remove as much as 85–95 per cent of wastewater nitrogen.

Distribution of sewage treatment technologies across Europe

In 1990 about 60 per cent of the European population, in 24 countries for which data were available, was served by sewage treatment of some nature. This average figure covers large variations between countries (Map 14.6) and it is based on OECD statistics, supplemented by data from some Eastern countries obtained from national environment reports and other published sources. However, comparisons and interpretations of sewage treatment statistics should be carried out with caution because it is not always clear how countries define treatment technologies. For example, tertiary treatment may in some countries consist of phosphorus removal only, while other countries in their definition, in addition to phosphorus removal, require nitrification/ denitrification to be included as well. For this report, it has not been possible to obtain data from all Eastern countries and therefore the estimated percentage of population served by sewage treatment plants may be too high. In Albania there are no sewage treatment plants at all and untreated domestic and industrial sewage is discharged directly into rivers and the Mediterranean. Sewage treatment facilities are also poorly developed (with less than 50 per cent of population served) in several Western and Southern countries (eg, Iceland, Ireland, Portugal, Belgium and Greece). On the other hand, in The Netherlands, Luxembourg, Switzerland and the Scandinavian countries, more than 90 per cent of the population was served by some sort of wastewater treatment plants, with Denmark serving 98 per cent of its population.

As illustrated in Map 14.6, there is also large variability between European countries concerning the percentage of population connected to any of the three most widespread sewage treatment technologies. The importance of inefficient primary treatment decreased in most countries over the period 1970–90 during which time, the construction of treatment plants applying more efficient secondary treatment increased significantly in almost all countries. However, in 1990 only a small percentage of Europe's population was served by the most efficient technology, tertiary treatment. The highest percentages are found in the Scandinavian countries, ranging from 21 per cent in Denmark to 84 per

Treatment process	Suspended solids	BOD	P	N	E. coli	Viruses	Heavy metals Cd, Zn	Heavy metals Cu, Pb, Cr
Primary	60–70	20–40	10–30	10–20	60–90	30–70	5–20	40–60
Constructed reedbeds[1]	–	60–80	20–40	25–50	–	–	–	–
Secondary	80–95	70–90	20–40	20–40	90–99	90–99	20–40	70–90
Tertiary	90–95	>95	85–97	20–40[2]	>99	>99	40–60	80–95

Notes:
1 Miljøstyrelsen, 1990.
2 N-reduction increases to 85–95% with nitrification and denitrification processes included.

Table 14.7
Reduction efficiencies of different wastewater treatment processes (%)
Sources: Modified from Winther et al, 1978; Miljøstyrelsen, 1990

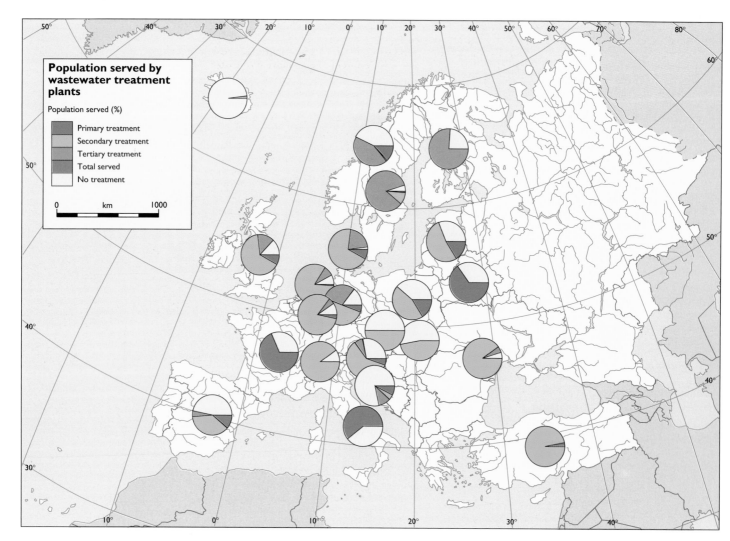

Map 14.6
Population served by different levels of wastewater treatment, 1990
Source: Multiple sources – see *Statistical Compendium*

cent in Sweden. Also in Germany (30 per cent of population served) this technology is relatively widespread. Although efficient wastewater treatment technologies are able to improve water quality of receiving waters considerably by reducing inputs of dissolved or suspended pollutants, this happens at the expense of the production of large volumes of sewage sludge. As environmental and health pressures intensify to reduce discharges of untreated sewage into surface waterbodies, a substantial increase in production of sewage sludge is likely to occur. The disposal and environmental impacts of sewage sludge is dealt with in Chapter 15.

For a part of Eastern Europe, the Danube basin, WASH (1992) showed that all of the older and larger cities in five countries – Czech Republic, Slovak Republic, Hungary, Romania and Bulgaria – have combined sewer systems serving a majority of the population. Interceptors and pump stations have been built to receive flows from sewer systems and carry wastewater to treatment sites, but rarely do the interceptors serve all of a city. Secondary wastewater treatment plants have generally been designed for the larger cities, and have been built to various stages of completion. However, effective operation and maintenance of wastewater treatment plants has been achieved in only a handful of cities in the five countries. Critical pieces of equipment in treatment plants are often defective or inefficient, and many municipal plants are overloaded. Sludge treatment and disposal is a major problem at essentially all the plants. None of the plants provided disinfection of the treated effluent, and none was designed or equipped for removal of the nutrients, nitrogen and phosphorus.

Industrial emissions

Industry and mining are the principal sources of synthetic organic chemicals and heavy metals in freshwater (WRI/UNEP/UNDP, 1992). Other important emissions from industry include organic matter and nutrients. There are few industries which do not make use of water, either directly as raw material as part of the manufactured product or indirectly for cooling, steam source, cleaning and circulation. A recent study (Meybeck et al, 1989) suggested that most synthetic organic chemical pollution is from chemical and petrochemical refineries, pharmaceutical manufacturing, iron and steel plants, wood pulp and paper processing, and food processing. Industrial sources of heavy-metals include: discharge of heavy-metal solutions from smelting and metal processing; use of metals and metal compounds in manufacture of paints, plastics and batteries; and tanning. Many of these activities generate liquid effluents which may contain many different chemicals, as well as organic matter, depending on the nature of the industrial processes involved (see Table 20.2). These discharges are a source of water pollution and have, especially during the 1960s and 1970s, caused considerable concern worldwide.

Examples from Western Europe – The Netherlands and the River Rhine – indicate that aquatic metal pollution has declined since its peak in the 1970s, partly as a result of improved wastewater treatment and reduced emissions from industry. Similarly, in Sweden discharges from most iron and steel works have now been considerably reduced. However, the flux of metals from many mines is continuing as a result of seepage from tailings and other mining wastes and these discharges now account for the bulk of inputs of cadmium, copper, lead and zinc to inland waters in Sweden. In the Nordic countries producing pulp and paper, the industrial emissions of organic matter have been substantially reduced in recent years. In Sweden, for example, the yearly emission of organic matter (taken as BOD) has decreased from some 700 000 tonnes at the beginning of the 1960s to 130 000 tonnes despite a 50 per cent increase in production (Nordic Council of Ministers, 1993).

In addition to heavy-metal pollution, mining activities can also result in severe acidification and salinisation of surface water (and groundwater). Coal and potash mining in, for example, Poland, East Germany and Alsace are but a few examples where such problems still occur. Further details are given in Chapter 5.

St Petersburg, with its five million inhabitants, numerous factories and inadequate sewage treatment, is one of the biggest polluters of the Baltic (Lothigius, 1993). There are approximately 300 metal-finishing firms here discharging, for example, large quantities of metals into the city's sewers. Apart from polluting the Baltic, these discharges disturb treatment processes at municipal sewage works and contaminate the sludge produced there.

In a study of industrial emissions (WASH, 1992) from plants producing, among other things, chemicals, steel, processed food and automobiles or refining petroleum products in five Danube countries (Czech Republic, Slovak Republic, Hungary, Romania and Bulgaria), it was shown that industrial wastewater treatment or pre-treatment was primitive and operated ineffectively. In addition, large industrial complexes have been developed at locations where available water resources cannot dilute or assimilate the emissions.

Generally, only few data on discharge of wastewater and polluting substances by sectors of industry are available. Systems of wastewater management, emissions inventories and reporting differ from one country to another, which makes comparisons almost impossible. However, there are a few examples of environmental agencies in some countries measuring discharges to water by industrial sectors and the tonnage and concentration of inputs from industry to rivers – some even publish lists of the most polluting 'companies'. Some data for Eastern European countries have been published in national reports. Countries which publish relevant information of this kind at the national level include: Belgium (Flanders), Bulgaria, France, The Netherlands, the Scandinavian countries and the UK.

TOWARDS INTEGRATED POLLUTION PREVENTION AND CONTROL

During the past two decades, European countries have achieved important progress in reducing emissions of traditional pollutants to air and water as a result of a number of concurrent factors, including:

- the establishment and implementation of emission standards and economic instruments;
- changes in the economic structure and particularly in energy supply; and
- widespread use of less-polluting technologies and pollution control measures.

These factors have not been equally important for all pollutants, across all economic sectors or across all European countries. Reduction has been achieved primarily in the emissions of sulphur dioxide associated with high-sulphur-containing fossil fuels and certain industrial processes; sulphur dioxide emissions have been cut on average about 40 per cent in Western European countries, but remain a concern at the local and regional level particularly in Eastern European countries. Emissions of NO_x and VOCs have not decreased, but have either remained stable or increased. This is a cause of increasing concern for both their local and global impacts. In the next few years, European countries, which are responsible for about 30 per cent of global CO_2, will have to stabilise these emissions to comply with the global Framework Climate Change Convention adopted at the UN Conference on Environment and Development (Rio, 1992).

Concerning emissions to water, the reduction of point source emissions of traditional water pollutants has been achieved in most Western European countries as a result of implementing advanced treatment technologies and improved industrial processes. Nevertheless, increased use of hazardous substances in production processes and consumer products has changed the composition of wastewater. In addition, intensive agriculture and livestock farming have combined with increasing use of chemicals to add an enormous pressure on the aquatic environment from non-point source emissions.

As policies to reduce emissions have been implemented, some important lessons have been learned. Emerging air and water pollution problems in the last two decades have been addressed separately. Most severe local pollution problems have been successfully dealt with by national programmes and regulations focusing on specific media. However, this approach has proved inadequate to face the cross-media nature of environmental problems. Pollution control technologies have shifted pollution problems from the air and surface water to soil and groundwater. The most visible effect is the increasing amount of waste in the form of ashes or sludges with high concentrations of hazardous substances from pollution control processes and the treatment of emission flows.

Criticism of the single-medium approach to environmental regulations emerged during the 1970s and 1980s. With the recognition of complex environmental problems, such as acidification, soil contamination and climate change, the need to move towards an integrated approach has been more fully appreciated. Several Northern and Western European countries have now adopted new regulations aiming at integrated pollution control. Adopting this approach should help prevent pollution problems instead of transferring them from one part of the environment to another. Following this line is also expected to improve the ability to set priorities and choose more efficient solutions. The European Commission (CEC, 1993) has submitted to Council a proposal for a Directive on Integrated Pollution Prevention and Control with the aim 'to prevent, wherever practicable, or minimise emissions from installations within the Community so as to achieve a high level of protection for the environment as a whole'.

REFERENCES

Axenfeld, F, Munch, J, Pacyna, J M, Duiser, J and Veldt, C (1992) *Test-Emissionsdatenbasis der Spurenelemente As, Cd, Hg, Zn und der speziellen organischen Verbindungen Lindan, HCB, PCB und PAK für Modellrechnungen in Europa*. Dornier, Friedrichshafen.

Berbee, R P M (1987) Pollution of surface water in the Netherlands by non-point sources. *5th International Symposium of CIEC*, Balatonfured (Hungary). Symposium Document Vol 3, Section 2, 3–7.

Briggs, D J (1993) An integrated emissions inventory for europe. *Final Report of a feasibility study on behalf of the European Environment Agency Task Force*. Institute of Environmental and Policy Analysis Research Report No 93/1, Huddersfield.

Brooker, M P and Johnson, P C (1984) The behaviour of phosphate, nitrate, chloride and hardness in twelve Welsh rivers. *Water Research* 18, 1155–64.

Burt, T P, Arkell, B P, Trudgill, S T and Walling, D E (1988) Stream nitrate levels in a small catchment in south west England over a period of 15 years (1970–1985). *Hydrological Processes* 2, 267–84.

Carl Bro Group (1992) *Pre-feasibility study of the Gulf of Riga and the Daugava river basin*. Technical report, prepared for the Baltic Sea Environment Programme & Nordic Investment Bank, Helsinki.

CBS (1993) *Natural resources and the Environment 1992*. Central Bureau of Statistics, Oslo.

CEC (1993) Proposal for a Council Directive on Integrated Pollution Prevention and Control. Commision of the European Communities, OJ No C 311.

CEC (1994) *Corinair-Inventaire des émissions de dioxyde de soufre, d'oxydes d'azote et de composés organiques volatils dans la Communauté européenne en 1985*. Report EUR 13232 FR. Commission of the European Communities, Luxembourg.

Christensen, P B, Thyssen, N and Kristensen, P (1992) Nitrogen removal in freshwater ecosystems and coastal areas. In: Baltic Sea Environment Proceedings No 44. Nitrogen and Agriculture International Workshop, Schleswig, Germany, pp 165–76. Baltic Marine Environment Protection Commission, Helsinki Commission.

Christensen, N, Paaby, H and Holten-Andersen, J (Eds) (1993) *Miljø og samfund – en status over udviklingen i miljøtilstanden i Danmark*. Faglig rapport nr 93. Danmarks Miljøundersøgelser, Roskilde.

CITEPA (1992) *Evolution des émissions de pollutants atmosphérique en France 1980–1990*. Paris.

Dillon, P J and Kirchner, W B (1975) The effect of geology and land use on the export of phosphorus from watersheds. *Water Research* 9, 135–48.

Edwards, A C, Pugh, K, Wright, G, Sinclair, A H and Reaves, G A (1990) Nitrate status of two major rivers in NE Scotland with respect to land use and fertilisers additions. *Chemistry and Ecology* 4, 97–107.

Fleischer, S, Hamrin, S, Kindt, T, Rydberg, L and Stibe, L (1987) Coastal eutrophication in Sweden: reducing nitrogen land runoff. *Ambio* 16, 246–51.

Gächter, R and Furrer, O J (1972) Der Beitrag der Landwirtschaft zur Eutrophierung der Gewässer in der Schweiz. *Schweizerische Zeit für Hydrologie* 34, 41–70.

Harenz, H (1989a) Probleme der Schliessung des Phosphorkreislaufs. *Nachrichten Mensch-Umwelt* 3, 23–52.

Harenz, H (1989b) Phosphor-Kreislauf und Verluste. *Spectrum Chemie* 20, 14–16.

Hornung, M, Roda, F, and Langan, S J (Eds) (1990) *A review of small catchment studies in Western Europe producing hydrochemical budgets*. CEC Air Pollution Research Report 28, Brussels.

Isermann, K (1990) Share of agriculture in nitrogen and phosphorus emissions into the surface waters of Western Europe against the background of their eutrophication. *Fertilizer Research* 26, 253–69.

Italian Ministry of Environment (1992) *Report on the state of the environment*.

Kronvang, B, Grant, R, Kristensen, P, Ærtebjerg, G, Hovmand, M and Kierkegaard, J (1993) Nationwide monitoring of nutrients and their ecological effects. *Ambio* 22, 176–87.

Kudeyarov, V N and Bashkin, V N (1984) Study of landscape-agrogeochemical balance of nutrients in agricultural regions: Part III, nitrogen. *Water, Air, and Soil Pollution* 23, 141–53.

Löfgren, S and Olsson, H (1990) Tillförsel av kväve och fosfor till vattendrag i Sveriges inland. Report No 3692 from Naturvårdsverket. Stockholm.

Kutas, J (1992) Personal communication, Budapest.

Lothigius, J (1993) Industrial inputs keep Baltic on the sick-list. *Enviro* 15, 27.

McInnes, G and Pacyna, J M (1992) *Proceedings of the First Meeting of the Task Force on Emission Inventories*, London, UK, 5–7 May 1992. EMEP/CCC-Report 4/92, Lillestrøm.

MEPNRF (1991) *The state of the environment in Poland: damage and remedy*. Ministry of Environmental Protection, National Resources and Forestry. Warsaw.

Meybeck, M, Chapman, D V and Helmer, R (Eds) (1989) *Global freshwater quality: a first assessment*. WHO/UNEP Global Environmental Monitoring System. Blackwell, Cambridge and Oxford.

Miljøministeriet (1992) Ferske vandområder; Vandløb og kilder. Vandmiljøplanens Overvågningsprogram 1991. Faglig rapport nr 62. Danmarks Miljøundersøgelser, Miljøministeriet, Copenhagen.

Miljøstyrelsen (1990) Lavteknologisk spildevandsrensning i danske landsbyer. Spildevandsforskning fra Miljøstyrelsen 5. Miljøstyrelsen, Copenhagen.

Miljøstyrelsen (1991) Kvælstof og fosfor i jord og vand. Samlerapport. NPo-forskning fra Miljostyrelsen, Miljøstyrelsen, Copenhagen.

National Agency of Environmental Protection (1991) *Environmental impacts of nutrient emissions in Denmark*. Ministry for the Environment, Copenhagen.

Neill, M (1989) Nitrate concentrations in river waters in the south-east of Ireland and their relationship with agricultural practice. *Water Research* 23, 1339–55.

Nordic Council of Ministers (1991) *Phosphorus in the Nordic countries – methods, bio-availability, effects and measures*. Svendsen, L M and Kronvang, B (Eds). Nord 47. Copenhagen.

Nordic Council of Ministers (1993) *The Nordic environment – present state, trends and threats*. Nord 1993: 12. Copenhagen.

OECD (1990) *Inventaires d'émissions des principaux polluants atmosphériques dans les pays européens de l' OCDE*. Monographies sur l'environnement No 21. Organisation for Economic Cooperation and Development, Paris.

OECD (1991) *Estimation of greenhouse gas emissions and sinks*. Final report from OECD Experts Meeting, 18–21 February 1991. Organisation for Economic Cooperation and Development, Paris.

Pacyna, J M and Joerss, K E (1991) *Proceedings of the EMEP Workshop on Emission Inventory Techniques*, Regensburg, Germany, 2–5 July, 1991. EMEP/CCC Report 1–91, Lillestrøm.

Pauliukevicius, G, Gulbinas, Z, Dilys, A, Baubinas, R and Grabauskiene, L (1989) Phosphorus migration in lake landscapes of Lithuania. In: Holm, Tiessen (Ed) *Phosphorus cycles in terrestrial and aquatic ecosystems*, pp 173–7. Proc SCOPE/UNEP Regional workshop 1: Europe, Czerniejewo.

Phillips-Howard, K D (1985) The anthropic catchment-ecosystem concept: an Irish example. *Human Ecology* 13, 209–40.

Pierre, D and Prat, M (1989) The use of catchment basins in measuring the impact of agriculture on the quality of drainage water. Methodological problems. Examples of application. Cases of nitrates. *Toxic Environmental Chemistry* 20–21, 149–61.

Postma, D and Boesen, C T (1991) Nitratreduktionsprocesser i Rabis hedesletteaquifer. NPo-forskningsprojekt B8. Danmarks Tekniske Højskole, Copenhagen.

Rekolainen, S (1989) Phosphorus and nitrogen load from forest and agricultural areas in Finland. *Aqua Fennica* 19, 95–107.

RIVM (1992) *National environmental outlook 2 1990–2010*. National Institute of Public Health and Environmental Protection. Bilthoven.

Roberts, G (1987) Nitrogen inputs and outputs in a small agricultural catchment in the eastern part of the United Kingdom. *Soil Use and Management* 3, 148–55.

Ryding, S-O and Forsberg, C (1979) Nitrogen, phosphorus and organic matter in running waters. Studies from six drainage basins. *Vatten* 1, 46–58.

Ryszkowski, L, Karg, J, Szpakowska, B and Zyczynska-Baloniak, I (1989) Distribution of phosphorus in meadow and cultivated field ecosystems. In: Holm, Tiessen (Ed) *Phosphorus cycles in terrestrial and aquatic ecosystems*, pp 178–92. Proc SCOPE/UNEP Regional workshop 1: Europe, Czerniejewo.

Schulte-Wülwer-Leidig, A (1983) Gelöstes, organisch gebundenes und partikuläres Phosphat in Wasserläufen unterschiedlich genützter Einzugsgebiete. *Mitteilungen der Deutschen Bodenkundlichen Gesellschaft* 38, 709–14.

Sdobnikova, O V (1989) Phosphorus in agroecosystems. In: Holm, Tiessen (Ed) *Phosphorus cycles in terrestrial and aquatic ecosystems*, pp 163–7. Proc SCOPE/UNEP Regional workshop 1: Europe, Czerniejewo.

Seux, R, Soulard, B, Boutin, P and Bechac, J P (1985) Azote et phosphore dans un cours d'eau du nord de la Bretagne: origine des apports, estimation des flux. *Trib Cebedeau* **38**, 11–24.

Sibbesen, E (1989) Phosphorus cycling in intensive agriculture with special reference to countries in the temperate zone of Western Europe. In: Holm, Tiessen (Ed) *Phosphorus cycles in terrestrial and aquatic ecosystems*, pp 112–22. Proc SCOPE/UNEP Regional workshop I: Europe, Czerniejewo.

Smith, R V (1977) Domestic and agricultural contributions to the inputs of phosphorus and nitrogen to Lough Neagh. *Water Research* **11**, 453–9.

State Pollution Control Authority (1993) *Pollution in Norway*. State Polution Control Authority, Oslo.

Statens Naturvårdsverk (1992) *Utsläpp till luft av forsürande ämnen 1990*. Rapport 3995, Statens Naturvårdsverk, Stockholm.

Stibe, L and Fleischer, S (1991) Agricultural production methods – impact on drainage water nitrogen. – *Verhandwagen des Internationalen Vereins für Limnologie* **24**, 1749–52.

Swanson, T M and Lloyd R (1992) The regulation of chemicals in agricultural production: a joint economic toxicological framework. Paper presented at the European Science Foundation's Symposium on Environmental Economic Analysis and Toxicology: *Aims, accomplishment and status of a pilot study on Atrazine*. 14–15 October 1992, Dublin.

Taylor, R, Florczyk, H and Jakubowska, L (1986) Run-off of nutrients from river watersheds used for agricultural purposes. *Environmental Protection Engineering* **12**, 51–65.

Umweltbundesamt (1992) *Daten zur Umwelt 1990/91*. E Schmidt Verlag, Berlin.

UNECE (1992) *The Environment in Europe and North America: Annotated statistics 1992*. Conference of European Statisticians. Statistical Standards and Studies No 42, United Nations, New York.

UNCED (1992) *Climate Change Convention*. Rio de Janeiro

Vlaamse Milieumaatschappij (1992) *Emissiejaarverslag 1992: Structuur en resultaten van de Emissie-Inventaris Vlaamse Regio*, Vlaamse Milieumaatschappij. Erembodegem.

VROM (1992) *Emission inventory in the Netherlands: industrial emissions for 1988: summary*. Ministerie van Volkshuisvesting Ruimtelijke Ordening en Milieubeheer. Report Nr 6, The Hagne.

Wahlström, E, Hallanaro, E-L and Reinikainen, T (1993) *The state of the Finnish environment*. Environment Data Centre and Ministry of the Environment, Helsinki.

WASH (1992) *Point source pollution in the Danube Basin – summary*. Water and Sanitation for Health Project, Field Report No 374. Prepared for the Europe Bureau, US Agency for International Devélopment under WASH Task No 271. USAID, Washington DC.

Webb, B W and Walling, D E (1985) Nitrate behaviour in streamflow from a grassland catchment in Devon, UK. *Water Research* **19**, 1005–16.

Winther, L (Ed), Linde Jensen, J J, Mikkelsen, I, Jensen, H T and Henze, M (1978) *Teknisk Hygiejne – Spildevandsteknik*. Polyteknisk Forlag, Copenhagen.

WRI (1992) *World resources 1992–93: a guide to the global environment*. World Resources Institute, Oxford University Press, Oxford.

Wright, G G, Edwards, A C, Morrice, J G and Pugh, K (1991) North east Scotland river catchment loading in relation to agricultural intensity. *Chemistry and Ecology* **5**, 263–81.

WSL (1992) *UK Emissions of Air Pollutants 1970–1990*. Warren Spring Laboratory report LR 887 (AP), Stevenage.

Household Waste
Source: Oscar Poss,
Spectrum Colour Library

15 Waste

INTRODUCTION

Economic development involving increased production and consumption has caused increased production of waste worldwide. According to the OECD, the total amount of waste produced in 1990 in OECD countries was 9 billion (thousand million) tonnes (OECD, 1991). Of those, 420 million tonnes were municipal wastes and 1.5 billion tonnes were industrial waste, including more than 300 million tonnes of hazardous waste. About 7 billion tonnes of waste were residues from energy production, agriculture, mining, demolition and sewage sludge.

Concerns over the environmental impacts of the increasing volume and toxicity of waste have emerged dramatically in the last two decades. Improper management of waste has caused numerous cases of contamination of soil and groundwater, and threats to the health of the exposed population. Existing disposal facilities are reaching saturation, and difficulties emerge in the siting of new facilities. Increased movement of waste to less developed countries is seen to threaten these countries. The World Conference on the Environment and Development has pointed out that the increasing production of waste poses significant threats for the environment and is no longer acceptable (UNCED, 1992).

To date, there is no realistic inventory of waste production, composition and disposal paths for Europe as a whole. Systematic data collection on waste is recent and usually coincides with the enactment and implementation of waste regulations. Waste statistics across countries are often not comparable due to diverging definitions, classification systems and scope (Box 15A). The need to harmonise classification systems at the international level is gaining major attention since countries have recognised the importance of coordinated actions to control waste movements and to reduce the potential environmental threats

of improper waste management practices (Box 15B).

This chapter analyses present trends in waste production and management in Europe on the basis of the most up-to-date information provided by national authorities through a joint OECD/Eurostat survey (1993) and/or reported in national state-of-the-environment reports. It also assesses the potential threats to human health and the environment resulting from current waste management practice. In addition, current progress achieved by European countries in reducing waste and recycling materials through integrated processes is examined. The information gap and the quality of existing statistics are briefly discussed in order to assess the state of knowledge and to identify future information needs.

MATERIAL LIFE-CYCLE

The nature of waste issues can better be understood by examining the material life-cycle from extraction to final disposal (Figure 15.1). Materials are transformed into waste as a result of a broad range of production and consumption processes (see Chapter 20). Residuals from transformation processes that are discharged directly into air or into water have been referred to in this report as *emissions* (see Chapter 14). Residuals which are further handled before being discarded are referred to as *waste*. Once generated, waste may be re-used (through in-plant processes), recycled (after being treated) or transferred to a treatment plant (to reduce its toxicity) or to an incineration plant (to reduce its volume). Non-recoverable materials are generally sent to a disposal facility. At each stage in the material life-cycle, various control options can be considered to reduce the amount and toxicity of waste materials. Each step in waste management has potential environmental impacts, as the various management methods imply various releases of pollutants into different environmental media (Table 15.1).

Box 15A Information on waste

Data availability

Statistics on waste are undeveloped in most European countries. Data collection on waste production, movements and management is often limited in its scope and coverage. Few data are available on waste production before 1985. Most recent data are based on *ad hoc* surveys or estimates. In particular, national statistics on hazardous waste are seldom available or accurate. Thus far, only a few European countries have adopted regulatory requirements for industries to report on the amount of hazardous and other wastes generated. Because of the enormous variability in production processes and technologies, estimates based on production and consumption trends have a wide range of error. In addition, a considerable portion of waste still escapes control and therefore is often not recorded at all in official statistics.

Data quality

The most accurate data are generally recognised as available through properly regulated systems of notification and reporting. Since such 'control systems' in Europe are still not fully and uniformly implemented, statistics on waste are particularly unreliable. In most European countries, the quantities and types of waste produced are often reported only on a voluntary basis and not as part of a systematic monitoring activity. Confidentiality is often the cause of problems in data collection and disclosure. An additional factor which hampers the reliability of waste statistics is that materials destined for recovery operations are often not recorded as waste.

Data comparability

In Europe, there are striking differences in the scope, detail, accuracy and comparability of national waste statistics. Consequently, great caution should be exercised when comparing such data across European countries. National figures on the amount and type of waste produced reflect differences in definitions, classifications and methods of data collection. The generic terms 'municipal waste', 'industrial waste', and 'hazardous waste' may include different waste streams in each national context. Caution should also be used when deriving historical trends; often the definition and classification systems have changed over time. For hazardous waste, in particular, historical data are seldom consistent with more recent information.

Box 15B Harmonising waste classification systems

The harmonisation of waste classification systems is crucial for developing comparable waste control systems across European countries. In particular there is a need to harmonise and cross-reference the existing lists of waste which require control under international and regional conventions. Attempts to harmonise definitions and establish an international waste classification system are currently being undertaken by international organisations as summarised in this box.

The International Waste Identification Code

Developed by the OECD in 1988, the International Waste Identification Code (IWIC) (OECD Council Decision C(88)90 Final) provides a uniform classification system to describe waste considered to be hazardous. Six tables are used, each describing waste from a different approach, as follows:

1 the reasons why the material is intended for disposal;
2 disposal operations;
3 generic types of potentially hazardous wastes;
4 their constituents;
5 the hazardous characteristics;
6 the activities generating them.

The IWIC provides a coded list for each of these tables which together allow a 'cradle-to-grave' dossier to be produced. In this way, any batch of wastes undergoing transfrontier movement can be described.

The core list of hazardous waste

During the last decade, the emergence of international agreements on controlling the transfrontier movement of hazardous waste prompted national governments to recognise the need for a common list of waste, specifying the characteristics which make them hazardous. A core list of 17 generic waste types and 27 constituents was adopted by OECD countries in 1988. The 45 categories of waste to be controlled by the Basle Convention (1989) on the Control of Transboundary Movement of Hazardous Wastes and their Disposal correspond to a large extent to the OECD core list.

Green, amber and red lists

OECD countries have also adopted a classification system for waste destined for recovery operations. This distinguishes wastes in terms of the 'green', 'amber' and 'red' lists, depending on the level of control which applies for their transfrontier movement. These lists have been introduced into EC Council Regulation 93/259/EEC on the supervision and control of shipments of waste.

The European waste catalogue

The initiative to create a general nomenclature for waste has been undertaken by the European Commission to develop a European Waste Catalogue (EWC). The EWC has been developed and approved (EC Decision 94/3/EC) in compliance with EC Council Directive on waste 75/442/EEC as amended by EC Directive 91/156/EEC (Framework Directive), which requires a common reference list of waste to be established across the EU Member States. Its implementation will provide a common basis for cross-referencing national lists, and facilitate the implementation of EU waste management policies. The catalogue is not meant to replace national classification schemes nor to follow strictly a single approach. Instead it should allow EU Member States to harmonise waste control systems and reporting on waste. The system classifies waste according to sources, processes and waste streams, providing the basis for producing comparable and compatible statistics on waste in EU Member States. The EU and UNECE are cooperating towards extending the catalogue to all European countries.

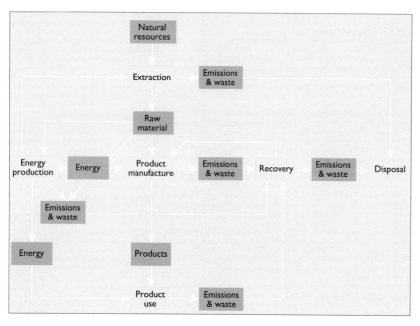

Figure 15.1
Material life-cycle
Source: EEA-TF

Sources of waste

All human activities are potential sources of waste. Wastes can be classified according to their major sources, which include: municipal, industrial, agricultural, extraction, energy production, sewage sludge and dredged spoils. Figure 15.2 shows estimated totals of waste arising in five of these major source categories in the OECD European area (OECD, 1993a). Agricultural, mining, industrial and municipal activities are, in order of importance according to quantity, the major sources of waste in Europe. Industrial and mining wastes are generally more important when wastes are compared according to their potential environmental impacts.

The relative importance of the various sources of wastes varies between countries and according to their economic structure. This is evident from comparing the major sources of waste in European countries. Data from the 1993 Eurostat/OECD survey show that the relative contribution of industrial and municipal wastes to total waste arising in Western European countries is greater than that in Central and Eastern Europe. In most countries of Central and Eastern Europe, mining activities rank first, with an extremely high

share of waste compared with other sources. Mining activities in Bulgaria and the former Czechoslovakia, for example, produce, respectively, five and ten times the amount of waste generated by the industrial sector (see *Statistical Compendium*). However, differences between countries should be regarded with caution since these may simply reflect differences in national definitions and classification systems.

Waste streams

Waste can also be classified according to different waste streams. Figure 15.3 shows the share of paper, plastics, glass, metals and organic matter in municipal waste in selected European countries. Although the composition of municipal waste varies widely from country to country, some general patterns can be detected. Organic waste, for example, accounts for a large share of municipal waste in most European countries. Paper is still a major waste stream despite recycling efforts. Waste plastic is especially found in increased proportions in Western European countries.

Wastes from industrial processes include a wide range of materials which may have varied chemical composition and physical state. Depending on the different industrial sectors, they may contain varying proportions of organic and inorganic compounds. It is their heterogeneity which makes their treatment and disposal difficult. Major categories of industrial wastes which are considered hazardous include solvents, waste paint, waste containing heavy metals, acids, and oily waste (Figure 15.4). Waste from mining activities includes topsoil, rock and dirt and may occur as inert or mine tailings which are contaminated by metals and chemicals from the mining process. Large amounts of ash are often the product of energy generation processes.

Other waste streams of concern for their increased amount and toxicity are sewage sludge, agricultural, demolition and hospital waste. Sewage sludge from the treatment of domestic and industrial wastewater and dredged spoils from harbours and rivers generally concentrate heavy metals and synthetic organic compounds. Sewage sludge and agricultural manure are organic wastes with high contents of nutrients. In addition, waste from agricultural activity may contain high concentrations of pesticide residues. Demolition and construction waste, including asphalt road planings and concrete, steel, timber and cement, may also contain relevant concentrations of toxic substances such as asbestos. Hospital wastes contain contaminated materials and are generally required to be segregated from other municipal waste.

Specific regulations are generally adopted for radioactive wastes, and these are considered later in this chapter.

Management options

During the last two decades, European countries have established various control systems for the management of waste, giving increased attention to waste prevention strategies. Since 1976, OECD countries have adopted a hierarchy of preferred options for waste management (see Chapter 36). Waste prevention is preferred to recycling; recycling is preferred to incineration; and disposal onto and into land is the least preferred option of the accepted methods of disposing of waste. More recently the EU has adopted a strategy for waste management in which primary emphasis is laid on waste prevention, recovery of materials and optimisation of final disposal (CEC, 1990) (Box 15C). However, despite the increasing emphasis on waste prevention, wastes have increased. Landfill and incineration, rather than recycling, are still the predominant practices in the management of waste.

On average, more than 60 per cent of municipal waste generated in Western Europe is currently disposed of into or onto land, about 19 per cent is incinerated, 4 per cent is composted and 3 per cent is subjected to mechanical sorting

Figure 15.2
Waste production by sources in OECD Europe, 1990
Source: OECD, 1993

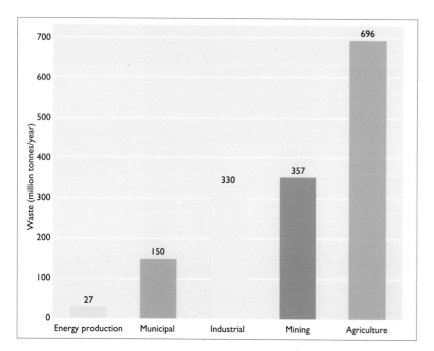

	Air	Water	Soil	Landscape	Ecosystems	Urban areas
Landfill	Emissions of CH_4, CO_2; odours.	Leaching of salts, heavy metals, bio-degradable and persistent organics to groundwater	Accumulation of hazardous substances in soil	Soil occupancy; restrictions on other landuses	Contamination and accumulation of toxic substances in the food-chain	Exposure to hazardous substances
Composting	Emissions of CH_4, CO_2; odours			Soil occupancy; restrictions on other landuses	Contamination and accumulation of toxic substances in the food-chain	
Incineration	Emissions: SO_2, NO_x, HCl, HF, NMVOC, CO, CO_2, N_2O, dioxins, dibenzofurans, heavy metals (Zn, Pb, Cu, As)	Deposition of hazardous substances on surface water	Landfilling of slags, fly ash and scrap	Visual intrusion; restrictions on other landuses	Contamination and accumulation of toxic substances in the food-chain	Exposure to hazardous substances
Recycling	Emissions of dust	Wastewater discharges	Landfilling of final residues	Visual intrusion		Noise
Transportation	Emissions of dust, NO_x, SO_2; release of hazardous substances from accidental spills	Risk of surface water and groundwater contamination from accidental spills	Risk of soil contamination from accidental spills	Traffic	Risk of contamination from accidental spills	Risk of exposure to hazardous substances from accidental spills; traffic

Table 15.1 Potential environmental impacts of waste management practices
Source: EEA-TF

Figure 15.3 Composition of municipal waste in selected European countries, 1990
Sources: OECD/Eurostat; EEA-TF (see *Statistical Compendium*)

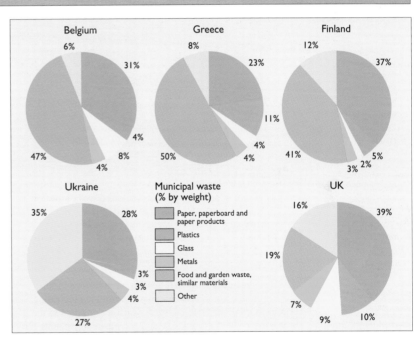

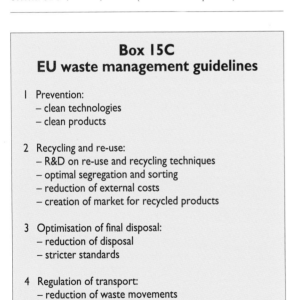

Box 15C
EU waste management guidelines

1 Prevention:
 – clean technologies
 – clean products

2 Recycling and re-use:
 – R&D on re-use and recycling techniques
 – optimal segregation and sorting
 – reduction of external costs
 – creation of market for recycled products

3 Optimisation of final disposal:
 – reduction of disposal
 – stricter standards

4 Regulation of transport:
 – reduction of waste movements
 – ensure safe disposal
 – monitoring

5 Remedial action:
 – R&D on techniques for site mapping and clean-up
 – development of financial instruments

Source: CEC, 1990a

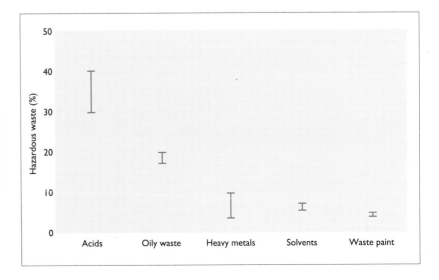

Figure 15.4 Major categories of industrial waste considered hazardous in OECD Europe, 1990 (% range of total hazardous waste, by weight)
Source: OECD, 1991

Figure 15.5 (left)
Management of
municipal waste in
OECD Europe,
1990
Source: Yakowitz, 1992

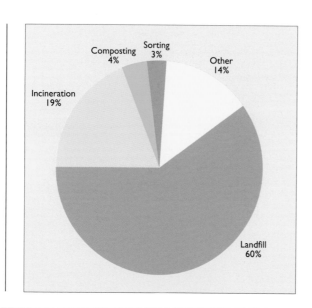

Figure 15.6 (right)
Management of
hazardous waste in
OECD Europe,
1990
Source: Yakowitz, 1992

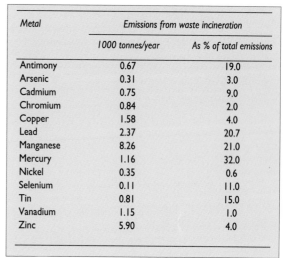

	Unit	Leachate average	Leachate maximum	Groundwater average	Groundwater maximum	Target
Chlorine	mg/l	743	7122	214	9600	100
Sodium	mg/l	2988	4335	195	1760	
Total Nitrogen	mg/l	438	2250	23	413	2–10
Arsenic	µg/l	51	499	184	2350	10
Cadmium	µg/l	4	140	0.2	10	1.5
Chromium	µg/l	67	1750	7	40	1
Copper	µg/l	30	830	14	884	15
Mercury	µg/l	1	26	0.4	53	0.05
Nickel	µg/l	92	1050	13	180	15
Lead	µg/l	394	30 300	7	300	15
Zinc	µg/l	720	30 000	122	10 000	150
Mineral oil	mg/l	1386	30 200	26	330	50

Metal	Emissions from waste incineration	
	1000 tonnes/year	As % of total emissions
Antimony	0.67	19.0
Arsenic	0.31	3.0
Cadmium	0.75	9.0
Chromium	0.84	2.0
Copper	1.58	4.0
Lead	2.37	20.7
Manganese	8.26	21.0
Mercury	1.16	32.0
Nickel	0.35	0.6
Selenium	0.11	11.0
Tin	0.81	15.0
Vanadium	1.15	1.0
Zinc	5.90	4.0

Table 15.3
Leachate and
groundwater
composition at
landfill sites in The
Netherlands
Source: RIVM, 1992

Table 15.2 Worldwide atmospheric emissions of trace metals
from waste incineration
Source: Nriagu and Pacyna, 1988

for recovery (Figure 15.5). From industrial wastes, more than 70 per cent of hazardous waste is still disposed of into or onto land, 8 per cent is incinerated and only about 10 per cent is recovered as secondary materials (Figure 15.6).

Environmental impacts

The increased rate of waste generation in Europe means that production and consumption activities use increasing quantities of materials each year which are returned to the environment in a degraded form, potentially threatening the integrity of renewable and non-renewable resources (see Chapter 13). Furthermore, managing waste has a wide range of potential environmental impacts, since natural processes act to disperse pollutants and toxic substances throughout all environmental media (see Table 15.1). The nature and dimension of these impacts depend upon the amount and composition of waste streams as well as on the method adopted for disposal.

At disposal sites, waste materials may contaminate the soil and groundwater by the leaching of toxic substances, such as heavy metals and metalloids, nitrogen compounds, chlorinated compounds and other organics such as hydrocarbons from households and industrial residues. From organic materials, the leachate may be enriched by inorganic materials such as nitrates. These may cause severe pollution of drinking water sources and eutrophication of surface water in surrounding areas. Biodegradation of organic matter in landfills also generates toxic and hazardous gases. Methane, one of the main components of landfill gases, is explosive at concentrations of 5 to 15 per cent in air. Other gases, such as hydrogen sulphide, are toxic. Two important landfill gases (CO_2 and methane) contribute to the greenhouse effect.

Incineration of solid waste contributes to the emission of heavy metals, such as mercury, cadmium and lead, contained in consumer products (Table 15.2). In addition, organic compounds produced by the combustion process and known as products of incomplete combustion (PICs) are released into the atmosphere at incineration plants. Among these products are dioxins and furans known to be highly toxic.

Figure 15.7
Contribution of
incineration plants to
emissions in The
Netherlands
Source: VROM, 1991

Chlorinated dibenzodioxins and -furans	
Hydrochloric acid	
Mercury	
Cadmium	
Zinc	

Percentage of total emission into the air of each substance from incineration

Polyaromatic hydrocarbons (PAHs), such as benzo(a)pyrene (BaP), are other highly toxic substances which can be formed during combustion. Filter residues are constituted by ashes and slags with high concentrations of hazardous substances.

Although the concentration of contaminants in waste from different sources is unknown, some European countries have attempted to estimate the contribution of emissions to the air and water from the disposal and processing of waste. Figure 15.7 illustrates that a high percentage of the total emissions of dioxins, hydrochloric acid and mercury into the atmosphere in The Netherlands is caused by incineration plants. The average composition of leachate at a number of disposal sites in The Netherlands is illustrated in Table 15.3. It is clear, however, that specific considerations of potential emission pathways of waste management facilities should be incorporated into monitoring activities if environmental impacts of waste management practice in Europe are to be monitored and their adverse effects prevented.

WASTE PRODUCTION

Given current national statistics on waste (See Box 15A), it is not possible to provide a figure of total waste produced in Europe. In most European countries, waste statistics are incomplete or inadequate. When available, data on waste are often not comparable due to different definition and classification systems. An estimate for total waste generation in 1990 in the OECD European area, excluding radioactive waste, is over 2000 million tonnes (Yakowitz, 1992). More recently, disaggregated figures for each sector were estimated (OECD, 1993). Of those, there were: 150 million tonnes of municipal wastes; 330 million tonnes of industrial wastes; 700 million tonnes of agricultural wastes; 360 million tonnes of mining wastes; and 27 million tonnes of energy production waste. In addition, there were 620 million tonnes of sewage sludge, 259 million tonnes of construction waste and 46 million tonnes of dredge spoil.

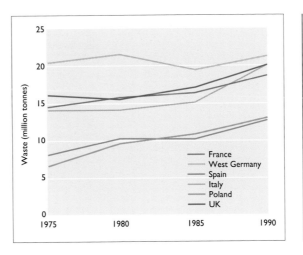

Figure 15.9
Municipal waste production trends in selected European countries
Sources: OECD/Eurostat; EEA-TF (see *Statistical Compendium*)

Figure 15.10
Municipal waste production in European countries, late 1980s
Sources: OECD/Eurostat; EEA-TF (see *Statistical Compendium*)

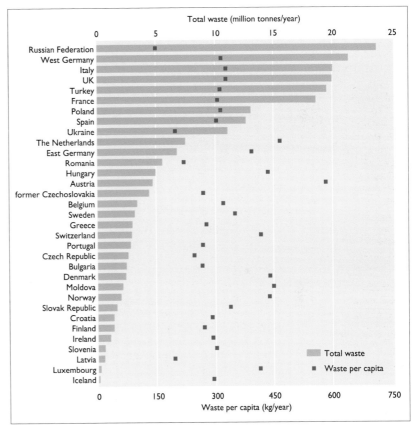

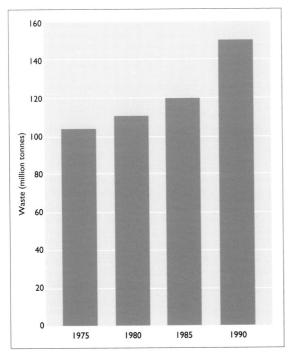

Figure 15.8 Municipal waste production trends in OECD Europe
Sources: OECD, 1993; OECD/Eurostat; EEA-TF (see *Statistical Compendium*)

Figure 15.11 Municipal waste production as a function of GDP
Sources: OECD/Eurostat; EEA-TF (see *Statistical Compendium*)

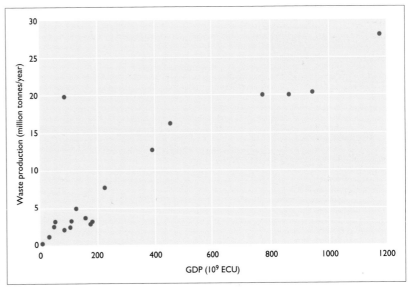

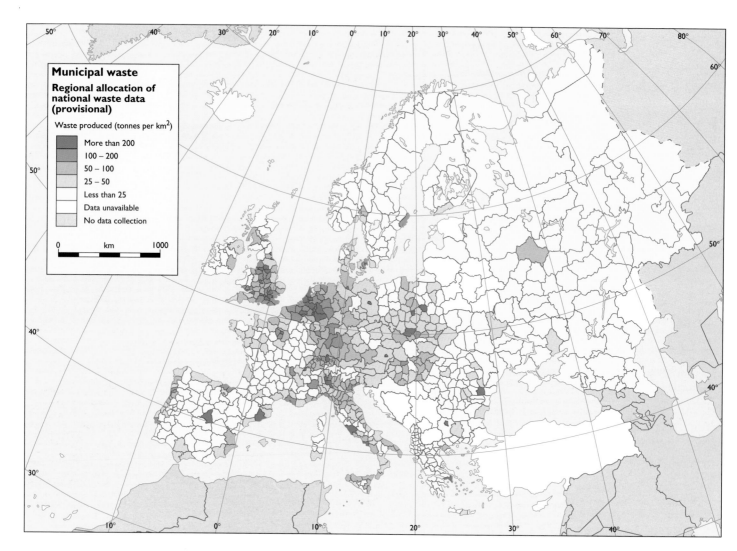

Map 15.1
Municipal waste
production in
Europe, late 1980s
Sources: EEA-TF
estimates based on
OECD/Eurostat data

Municipal waste

Municipal wastes in Europe have increased markedly over recent decades. According to a preliminary estimate based on data from 33 countries, the total amount of municipal waste produced in European countries is of the order of 250 million

tonnes. In the OECD European area, the production of municipal waste between 1975 and 1990 increased by about 30 per cent (Figure 15.8), with differences in growth rates between countries (Figure 15.9). The growth rate increased particularly in the five years up to 1990. The average rate of increase of municipal waste was 3 per cent per year between 1985 and 1990, compared with about 1 per cent per year between 1980 and 1985.

Municipal waste production per capita in European countries ranges between 150 and 600 kg per year (Figure 15.10). Western European countries generate more than 1 kg of municipal waste per day per person, which is higher than the amount of waste produced per capita in most Central and Eastern European countries. Comparing waste production in OECD countries, the level of municipal waste production appears to be correlated to the level of industrialisation and the level of income (Figure 15.11).

Map 15.1 shows the regional allocation of municipal waste produced per square kilometre. The highest density levels of waste generation occur in the most densely populated areas in Western Europe. In these areas it is increasingly difficult to provide the necessary treatment and disposal capacity for meeting good practices and safe standards of waste management.

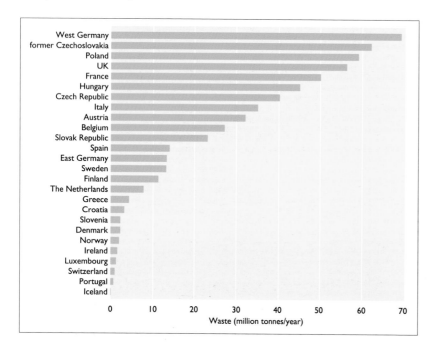

Figure 15.12 Industrial waste production in European countries, late 1980s
Sources: OECD/Eurostat; EEA-TF (see *Statistical Compendium*)

Hazardous liquid chemical waste
Source: Adam Hart-Davis/Science Photo Library

Industrial waste

OECD Europe is estimated to have generated, in 1990, nearly 330 million tonnes of industrial waste. Between 1985 and 1990, industrial wastes in these countries increased on average at an annual rate of 3 per cent, that is at a rate of about 9 million tonnes per year. In Central and Eastern Europe in the late 1980s, it is estimated that 520 million tonnes of industrial wastes were generated per year. This could increase rapidly in the next few years if these countries increase their industrial output. The most up-to-date figures of industrial waste production are given in Figure 15.12.

Manufacturing is, however, only one part of a wide range of activities involved in industrial production. The extraction of raw materials and energy production are also important sources of waste and, particularly in Central and Eastern European countries, the cause of major problems.

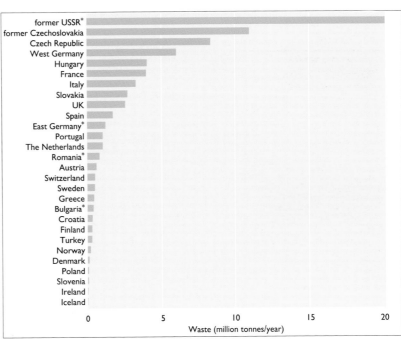

A significant proportion of industrial waste and extraction waste is regarded as hazardous waste.

Hazardous waste

Statistics on hazardous waste are still incomplete, and existing data are particularly unreliable (see Box 15A). Furthermore, the definition of hazardous waste varies between countries, which implies that cross-comparisons are misleading. In general, waste containing metallic compounds, halogenated solvents, acids, asbestos, organo-halogen compounds, organophosphate compounds, cyanides or phenols is regarded as hazardous waste.

On the basis of official national statistics, OECD estimates that worldwide annual production of hazardous waste is about 16 per cent of the total industrial waste (338 million tonnes out of 2100 million tonnes). In the OECD European area, the estimate for hazardous waste ranges between 30 and 45 million tonnes per year. In addition, Central and Eastern Europe are estimated to generate annually about six million tonnes of hazardous waste. A further 25 to 30 million tonnes of hazardous waste are estimated to be produced each year in the former USSR.

Figure 15.13 gives an overview of total hazardous waste production in selected European countries. Hazardous waste production measured according to national definitions shows a large variation between countries, which reflects, in most cases, the relevant difference in these definitions. Indeed, countries that report high levels of hazardous waste are also those countries with

Figure 15.13
Hazardous waste production in European countries, late 1980s
Sources: OECD/Eurostat; EEA-TF (see *Statistical Compendium*)

Note: *Estimated values.

Figure 15.14
Hazardous waste per unit GDP in Europe
Source: Yakowitz, OECD, personal communication

Note: For definition of core and IWIC lists, see Box 15B.

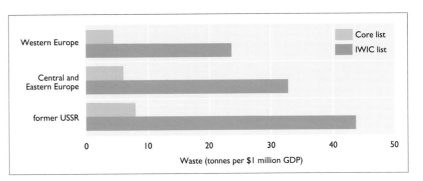

more inclusive lists of waste regarded as hazardous. Differences across countries can be examined by using the core list of hazardous waste. The difference is even greater when comparing Western and Central and Eastern European production of hazardous waste per unit of GDP (Figure 15.14).

Household wastes commonly include hazardous items, such as batteries, oils, paints, resins, and out-of-date medicines, sometimes defined as small chemical waste. Consumer products containing hazardous substances are summarised in Box 15D. Although the importance of this source is increasing, reliable data are extremely difficult to obtain. In The Netherlands, household chemical wastes are estimated to be 70 per cent (41 thousand tonnes) of the small chemical waste produced in one year. Household hazardous wastes are especially important because they are ultimately disposed of as municipal waste without the same precautions required for hazardous waste. The operation and regulation of facilities managing such waste consequently need to take this fully into account. The presence of hazardous items in municipal waste indeed limits the opportunities for municipal waste recycling since they may contaminate recoverable materials with toxic substances. Some countries are segregating such waste at source.

Agricultural waste

Agricultural waste consists of crop residues, animal manure, animal carcasses, agrochemical residues and containers. Such waste accounts for more than one third of the waste produced in Western European countries. Data available for Central and Eastern European countries show that residuals from agricultural activities constitute a significant share of waste generated in these countries as well. The importance and composition of agricultural wastes vary across countries according to agricultural systems (see Chapter 22).

Most agricultural residues are organic and biodegradable and hence should be suitable to conversion by biological, chemical and physical processes into energy, animal feed or organic fertiliser. However, problems for their management arise from their occurrence in large volumes and high concentrations, and as a result of the increased use of chemicals in agriculture. In general, farm crops produce significant amounts of residues. Intensification of agriculture and livestock farming is the cause of increased pressure on the environment due to increased quantities and concentration of such residues. Opportunities for recycling are still not fully exploited.

Figure 15.15 Glass recycling in selected European countries, late 1980s
Source: FEVE (Fédération Européenne du Verre d'Emballage)

WASTE MANAGEMENT

In Europe most waste is disposed of in landfills, despite the accepted principle that waste disposal in or on land should be considered as one of the least desirable options. However, the extent of the use of landfill varies between countries (see Chapter 36). In some countries, which have reached saturation of landfill capacity or have imposed restrictions on the landfill of certain waste (Austria, Denmark, Germany and The Netherlands), an increasing proportion is now incinerated or submitted to other treatment processes. In addition, increasing concerns for the emissions of toxic substances from incineration plants have led countries such as Germany and The Netherlands to adopt new programmes for waste prevention and recycling.

Recycling

Most waste streams contain significant amounts of valuable materials which can be recovered and re-used in production processes or other useful applications. Re-use and recycling

Plastic bottles awaiting recycling
Source: Peter Ryan/Science Photo Library

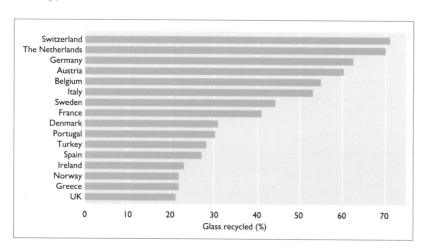

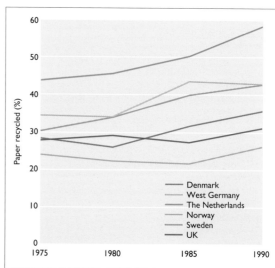

Figure 15.16 Paper recycling in selected European countries, 1975–90
Source: OECD, 1993

	Aluminium	Steel	Paper	Glass
Energy use	90–97	47–74	23–74	4–32
Air pollution	95	85	74	20
Water pollution	97	76	35	–
Mining waste	–	97	–	80
Water use	–	40	58	50

Table 15.4 *Potential savings (%) in material recycling*
Source: Bartone, 1990

activities have the advantage of reducing the demand for raw materials and for energy while minimising the impact of waste disposal. An attempt to estimate the potential saving in energy, water use, mining waste and emissions to air and water that can be achieved when specific materials are recycled is presented in Table 15.4.

At present, the total amount of waste material recovered in Europe is not known. Based on OECD data, it is possible to estimate that between 30 and 40 per cent of municipal waste produced in OECD European countries is submitted to operations leading to recovery of material or energy. Recovery of materials from municipal waste substantially increased between 1975 and 1985, but remained constant between 1985 and 1990 (Yakowitz, 1992).

The range of different recycling efforts varies enormously between countries (Figures 15.15 and 15.16). Among the major flows of recovered materials are paper and cardboard (20 to 60 per cent), aluminium (about 30 per cent) and glass (8 to 63 per cent). In addition, between 20 and 25 per cent of the organic fraction of municipal waste is incinerated to recover energy (Yakowitz, 1992).

Incineration

In OECD Europe, the average rate of incineration of municipal waste (19 per cent) and of hazardous waste (8 per cent) remained constant between 1985 and 1990. In Central and Eastern European countries the range of incineration rates is much lower.

A few Northern (Norway and Sweden) and Western (Denmark, France, Germany and Switzerland) European countries already rely on a significant incineration capacity. Several countries such as The Netherlands,

Country	Year	Incineration capacity/year (1000 tonnes)	Energy reclaimed (%)
Austria	1990	370	100
Croatia	1990	5	40
Finland	1990	50	100
France	1990	8700	68
Germany	1993	9500	–
Italy	1991	1912	–
Luxembourg	1990	150	100
The Netherlands	1990	2850	94
Romania	1990	757	–
Spain	1990	606	61
Slovak Republic	1992	398	–
Sweden	1990	1800	97
Switzerland	1990	2300	91
Ukraine	1990	880	–

Table 15.5 *Incineration capacity in selected European countries*
Sources: OECD/Eurostat; EEA-TF (see *Statistical Compendium*)

Country	Year	Number of landfill sites	Total capacity (1000 tonnes)
Austria	1990	160	3216
Belgium	1990	30	–
Croatia	1990	350	1342
Czech Republic	1990	844	–
Estonia	1991	250	–
Finland	1990	750	25 000
France	1990	484	16 000
Germany	1987	10 400	–
Italy	1991	1463	33 681
Latvia	1990	504	–
Luxembourg	1990	4	150
Moldova	1990	1353	–
The Netherlands	1990	373	–
Norway	1990	500	–
Poland	1990	13 813	79 298
Portugal	1990	303	821
Slovak Republic	1992	7200	–
Slovenia	1991	56	3502
Spain	1990	94	9376
Sweden	1990	282	7500
Switzerland	1990	60	20 000
Ukraine	1990	2760	2 500 000
UK	1990	4193	–

Table 15.6 *Number of landfills in selected European countries*
Sources: OECD/Eurostat; EEA-TF (see *Statistical Compendium*)

Belgium, the UK, Italy and Switzerland have planned to increase their capacity to meet waste treatment needs. However, increasing difficulties have been encountered by public agencies in siting new incineration plants, since the significant contribution of these plants to the total emissions of toxic substances to the atmosphere has become evident. Public opposition to the siting of new facilities, and increased capital and operating costs due to higher environmental standards, have prevented the expansion of incineration capacity. On the other hand, the current capacity of European countries for incineration is expected to reach saturation in the next few years (Table 15.5).

Composting

In a number of European countries, some of the organic fraction of municipal waste is composted. The share of composting is extremely limited except for a few countries such as Spain, Portugal, Denmark and France, where composting reaches respectively 21, 10, 9 and 6 per cent of municipal waste collected. A small contribution is also found in The Netherlands, Germany and Italy. Although composting played an important role in the past, difficulties in finding a market for compost products have kept this option at a fairly modest level during the last two decades. Other ways of utilising organic waste include the production of biogas to recover energy from agriculture residue. This is, however, extremely limited in Europe.

Landfill

Disposal of waste on land is still the major disposal route in Europe. On average, more than 60 per cent of municipal waste and about 70 per cent of hazardous waste was delivered to landfill sites in OECD European countries in 1989 (Table 15.6). However, there are major variations between countries. In the UK, 75 per cent of municipal waste is landfilled; in Norway it is 70 per cent. The percentages of waste sent to landfill in Sweden and Switzerland are, respectively, 30 per cent and 10 per cent.

WASTE MOVEMENTS

Transfrontier movements of hazardous waste have also increased. More than 2 million tonnes of hazardous waste move each year across national frontiers in the OECD European area (Table 15.7). The patterns of hazardous waste movements show two clear directions: from North to South and from West to East. In addition, transfrontier movements of hazardous waste across Western European countries are seen to match the degree of stringency of regulations adopted by those countries for the management of hazardous waste. Map 15.2 gives an overview of major flows of hazardous waste between European countries. These include waste destined both to final disposal and to recovery. A number of examples of disputed transfrontier movements of wastes are given in Box 15E (p 354). In all these cases, the types of wastes exported were highly hazardous.

North–South

Europe exports legally about 120 000 tonnes of hazardous waste to developing countries. Increasing evidence that the magnitude of these transfrontier movements is far larger than recorded has emerged in recent years from the detection of illegal flows of hazardous waste from northern to southern countries. The improper management of these wastes in countries with less adequate technologies and fewer measures for their control poses risks for health and the environment and threaten development in southern countries.

West–East

Another major route of hazardous waste movement is from Western to Eastern European countries. Although the records of these movements are incomplete and rarely consistent, a few examples are sufficient to understand the magnitude of such a flow.

In 1988 more than 1 million tonnes of hazardous waste were exported from Western Europe to East Germany (the former DDR). According to the German Federal Ministry for the Environment, 685 000 tonnes of hazardous waste (65 per cent of the total waste exported) were exported in 1988 from West to East Germany. Other exporter countries included Austria (50 000 tonnes), Italy (50 000 tonnes), The Netherlands (35 000 tonnes) and Switzerland (4000 tonnes). Other examples of export from Western to Eastern European countries are polychlorinated dibenzo-furan exported from Austria to the former Czechoslovakia (4000 tonnes) and to the former USSR (1500 tonnes) to be recovered.

Within Western Europe

Movements of hazardous wastes between Western European countries are seen to parallel the patterns of regulations. Waste moves from countries which have adopted and enforced stringent standards, or impose higher costs for the management of such waste, to less regulated countries or those which have lower costs. The total amount of waste exported from West Germany in 1988 reached 1 058 100 tonnes, that is, more than 20 per cent of the annual hazardous waste produced. In addition to East Germany, destination countries included Belgium (128 400 tonnes), France (197 500 tonnes), the UK (36 200 tonnes), The Netherlands (11 400 tonnes) and Switzerland (300 tonnes).

The large increase of waste imports into the UK from 1980 to 1988 (Figure 15.17) was due to the increasing stringency of regulations in continental Europe and North America. Waste imports increased from 5000 tonnes in 1983 to 183 000 tonnes in 1986–87. Her Majesty's Inspectorate of Pollution (HMIP) of the UK estimates that 80 000 tonnes of 'special waste' were imported in 1987. This figure dropped to 40 000 tonnes of imported hazardous waste in 1988 after the implementation of the EC Directive on transfrontier movement of hazardous waste.

The amount of waste imported into France (250 000 tonnes per year) is extremely high; and according to the Belgian Ministry of Health and Environment the amount of waste imported into Belgium per month in 1989 was 54 700 tonnes.

CLOSING MATERIAL CYCLES

Waste prevention has been recognised worldwide as a priority objective to reduce the pressure on the environment from production and consumption processes. Reducing the amount of emissions and waste can be achieved by closing the cycle of materials in the production and consumption processes. Closing material cycles has the advantage of both reducing the pressure on the stock of natural resources and reducing the environmental impact of waste and emissions.

In principle, product substitution and modification of manufacturing processes could help avoid or minimise the production of waste. Furthermore, re-use and recycling of materials contributes to waste avoidance. In practice, achieving waste prevention is a complex task which requires different strategies depending on each specific waste stream, including the redesigning of processes.

Although the magnitude of the various waste streams varies across European countries, it is possible to identify waste streams that require specific consideration for reduction based on their increasing importance and potential impact on the environment. Four examples are plastic waste, packaging waste, chlorinated solvents and used tyres.

Plastic waste

Sources

The production and consumption of plastics have increased enormously in the last 30 years. Plastic materials have replaced others such as metal, wood and glass in many sectors. Different types of plastics and their major uses are described in Box 15F (p 355). Although plastic wastes have increased in all sectors of economic activity, the increased share of plastic in household waste is the most evident example of this trend. In most Western European countries, the plastic component of household waste has increased eight-fold since the 1960s. In Central and Eastern Europe, plastic is now increasingly replacing other materials, raising the share of plastic waste, and this trend is expected to continue in the near future.

In Western Europe, more than one third of all plastic is used for packaging (Figure 15.18, p 385). PET bottles, PVC packaging and the PE carrier bags account for the most

Table 15.7
Import/export of hazardous waste in selected European countries, 1990
Sources: OECD/Eurostat; EEA-TF (see *Statistical Compendium*)

Country	Total production	Imports		Exports	
	1000 tonnes	1000 tonnes	per cent	1000 tonnes	per cent
West Germany	6000	62.6	1	522.1	9
France	3958	458.1	12	16.0	0.4
Italy	3246			20.0	1
UK	2540	45.0	2	1.0	< 0.1
Spain	1708			20.2	1
Portugal	1043	82.3	8	2.0	0.2
The Netherlands	1040	199.0	19	195.4	19
Switzerland	736	12.0	1	132.0	17
Austria	616	20.2	3	68.2	11
Sweden	500	47.2	9	42.6	9
Finland	250	20.0	6	12.0	4
Norway	200			22.0	11
Denmark	106	2.0	2	13.2	12
Iceland	5			0.2	4

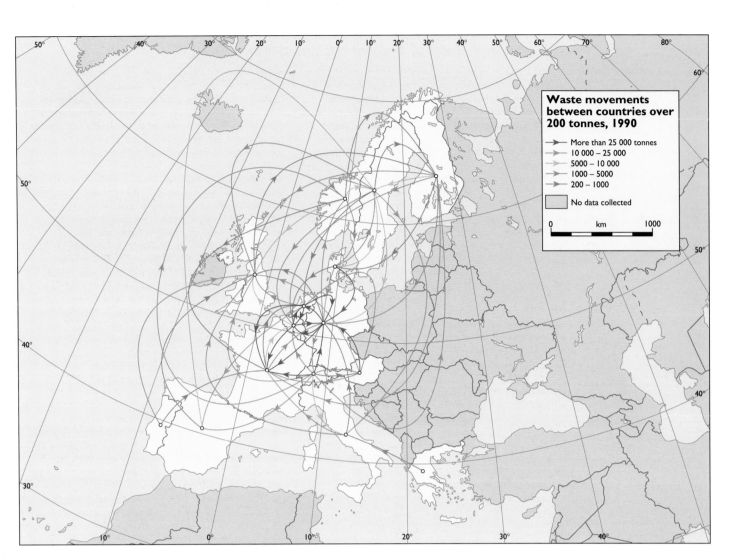

Note: Includes waste destined for recovery and disposal.
The map is included to illustrate the numerous movements of waste across European countries and not to compare waste flows. Figures of waste exports cannot be compared because of the differences in the definitions of hazardous waste from country to country.

significant part of plastic consumption. These products have a shorter life than consumer goods, which explains the exponential growth in waste plastic in a short time. Other important sources of plastic waste are building construction (19 per cent), transport (7 per cent) and the electrical engineering industry (7 per cent).

Impacts

The environmental impacts of disposing of plastic waste concern both landfill and incineration processes. Two major problems of landfilling plastic waste are their non-degradability and the space requirements. Chlorinated polymers, particularly PVC, are one of the principal sources of hydrochloric acid emissions from incineration plants and are considered responsible for the formation of dioxins. The contamination of fly ash and slag from incineration with heavy metals is also ascribed to the metal content of plastic waste.

Options

Several options for reducing the impact of plastic waste are available. On a qualitative level, this would imply cutting down the content of toxic substances in plastics as well as replacing PVC with other plastics. For product design, the options include reducing the total amount of packaging, the replacement of plastics with other materials and the design of products with an extended life-cycle.

Packaging waste

Sources

The increasing portion of plastic and cardboard in municipal waste is due to the packaging of consumer products. In Northern and Western European countries, it has been estimated that between 30 and 35 per cent of the mass of total municipal waste and 50 per cent of its volume is due to packaging material. OECD Europe produces nearly 45 million tonnes of packaging waste per year with a recovery rate ranging between 10 and 15 per cent. The proportion of packaging material in municipal waste in Central and Eastern Europe is lower, but is expected to rise as a consequence of introducing Western consumer products (OECD, 1992).

Map 15.2
Transfrontier movement of hazardous waste in Europe, 1990
Source: Based on waste export data provided by individual countries to OECD (Lieben, OECD, personal communication)

Figure 15.17
Hazardous waste imports in the UK, 1981–90
Source: UK DoE, 1992

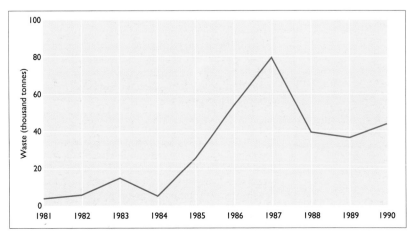

Box 15E Cases of disputed transfrontier movement of hazardous waste

The international non-governmental organisation Greenpeace has been monitoring transfrontier movement of hazardous waste worldwide since 1986. This initiative is aimed at raising environmental awareness about the threat posed by waste shipments. An inventory of waste trade operations was compiled in 1990 (Greenpeace, 1990). Up-to-date information on the status of these operations and additional illicit movement of waste are published in the *Toxic Trade Update* newsletter and made available on the computer network Econet (named haz.net) (Greenpeace, 1993a and b). A large number of examples involving the movement of highly hazardous waste across European countries and between Europe and other world regions are provided. Some of these shipments disregard national regulations and international conventions on the transboundary movement of hazardous waste. A large part of the shipments find their way through regulations by the promise of recycling. In fact, what happens is that wastes move from countries with stricter regulations and higher disposal costs to countries with fewer regulations and where disposal of such waste is cheaper. This is a situation that the Basle Convention should now alleviate. A few examples reported by Greenpeace (1993a and b) are given below.

German waste in the UK and Poland
In November 1989, some 470 tonnes of hazardous waste containing organic solvents, acid, alkalis and heavy metals were sent from Germany to the UK and stored in two warehouses. Some were sent directly to landfill sites in Essex. Three loads of 20 tonnes were sent to Poland. The wastes were exported as dangerous goods and described as non-ferrous metal residues. Therefore no transfrontier shipping documentation accompanied the waste. This waste importation was discovered in August 1990 by the Essex County Council. The shipment was made by a German waste broker and a private UK company in Wales. The case ended up in court and the waste was returned to Germany.

Mercury waste at Almaden, Spain
In June 1990, thousands of tonnes of mercury-contaminated waste were found in a dump site of the district of Almaden, 300 km south of Madrid. The case involved the dumping, since 1980, of about 11 900 tonnes of hazardous waste from chlor-alkali, battery and pesticide manufacture in Germany, Italy,

France, UK, Norway, Switzerland, Sweden and The Netherlands. Part of the waste also came from the USA and Australia. In 1988 and 1989, 2000 tonnes of waste were transferred and buried in a landfill at San Fernando de Henares, east of Madrid. In February 1991, under order from the Spanish government, the waste was buried in a new landfill.

German waste in Albania
More than 500 tonnes of pesticides either banned or expired were exported from Germany to Albania between 1991 and 1992. According to the German Environment Ministry, the pesticides included formulations of lindane and other organochlorine compounds and herbicides with high concentrations of dioxins. Disposing of such waste in Germany would have cost at least DM 8000 per tonne. According to Greenpeace, the wastes were still sitting in barrels deposited on six sites in August 1994: the port of Durres; in Bazja railway station in northern Albania; in a warehouse at Milot; and in the towns of Lushnja, Fier and Skodri in western Albania. In January 1993 the Albanian authorities asked Germany to help resolve the problem. The German Environmental Ministry has recently agreed to take back the waste and has allocated a budget of DM 9.6 million for remediation.

UK waste in Mexico and Bolivia
Between 1992 and 1993, a UK company dismantling the Capper Pass metal smelter in Humberside, northeast England, was reported to be sending tin slags to Bolivia and Mexico. The tin smelter closed in 1991 after operating since 1937. The clean-up of the site resulted in 3500 tonnes of dusts originating from the electrostatic precipitator of the plant's secondary furnace. The shipment of 600 tonnes of this waste went to a disposal plant in the area of Oruro in the high Andean Plateau. This plant had closed several months before the arrival of the wastes. A consignment of 500 tonnes of waste was also sent from Felixstowe, UK, destined for Mexico and, after a protest from Greenpeace, was returned to the original site. However, a shipment of 39 tonnes of toxic waste had already been sent to the Mexican industrial site at San Luis Potosi. Over 3000 tonnes of the waste are now stored at the Capper Pass site.

Sources: Greenpeace, 1990; Greenpeace, 1993a; Greenpeace, 1993b

Impacts

The pressure of packaging waste on the environment is due to both the environmental impacts of disposing of a significant proportion of plastic materials and the consumption of natural resources. The environmental impacts of disposing of packaging materials are similar to those mentioned above for plastic waste, due to the increased component of plastic in the packaging waste. The relative importance attributed to packaging as a source of environmental impacts is due to the short life of packaging products, their potential for exponential increase in the short term and their frequent dumping in improper sites.

Even if in Europe there is no systematic inventory of packaging waste, some countries have started to monitor the amount of various forms of packaging (eg, one-way beverage containers) and composition (eg, plastic content). In Germany, for example, between 1970 and 1988, the overall use of one-way packaging containers increased significantly, while the proportion of returnable containers fell from 90 per cent to 74 per cent. In The Netherlands it has been estimated that a large portion of paper and

cardboard for packaging has been replaced with plastics. Packaging waste is also considered to contribute more than 10 per cent to the total household small chemical waste produced each year.

Options

Several options are being considered to reduce the amount and potential threat of packaging waste. The most obvious is to reduce the use of packaging and packaging materials. A second option is the replacement of certain materials with others which have less environmental impact. Recycling and recovery of the various packaging components is the focus of most of the current programmes aimed at reducing the total amount of packaging waste.

Several European countries have undertaken actions to reduce the quantity of packaging and packaging waste and to recover packaging material. In The Netherlands, for example, a 'packaging covenant' between the Dutch government and the packaging industry sets a reduction target for the year 2000 for packaging newly put on the market to 10 per cent below the level of packaging in 1986.

One of the most significant initiatives for recycling packaging waste has been taken by Germany to cut substantially the amount of packaging waste per unit of product, and requires industries to collect packaging materials for recycling. The three-stage scheme concerns three categories of packaging: transport packaging (eg, pallets for loading lorries), secondary packaging (eg, cardboard boxes) and sales packaging (eg, plastic layers around the products). Since April 1992, German industries are required to accept secondary packaging at the point of sale. In order to cover the cost of the scheme, a consortium of more than 3000 companies has been created to collect packaging materials from consumers. The eligible packaging products for the system are identified by a green dot which entails a fee of up to 0.1 ECU per package. A total of one billion ECU per year is expected to be generated by this fee in order to operate the scheme. The use of packaging in Germany has dropped by 3.1 per cent between 1992 and 1993 as a result of implementing this policy. However, the system has collected more material than can be re-used or recycled in the country, causing impacts on the markets for these materials.

At the EU level, a Directive on packaging for liquid food was adopted in 1985 (85/339/EEC). More recently, a proposal for a new Directive has been advanced by the European Commission aiming at harmonising national measures concerning the management of all packaging and packaging waste. The Directive's first priority is prevention of packaging-waste generation. Additional principles are re-use, recycling and recovering packaging waste and reducing its disposal.

Chlorinated solvents

Sources

Production and consumption of chlorinated organic compounds have increased in the last two decades because of their attractive properties in several industrial applications, notably as solvents. Chlorinated solvents are used in metal cleaning and degreasing, dry-cleaning, paint stripping and as an intermediate in chemical production. The many chlorinated compounds used as solvents include:

- methylene chloride (MC);
- 1,1,1 trichloroethane (1,1,1,-T);

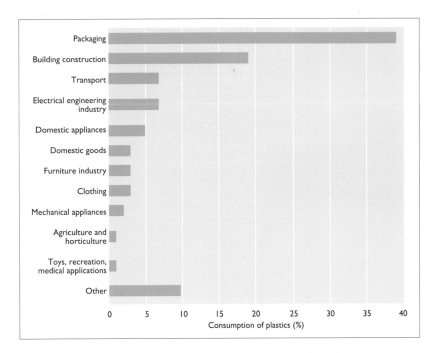

Figure 15.18
Consumption of plastics in Western Europe
Source: APME

- trichloroethylene (TRI);
- perchloroethylene (PER);
- CFC-11;
- CFC-113.

In EU countries, it is estimated that between 600 000 and 800 000 tonnes of chlorinated solvents were used in 1989 (CEC, 1992). Of those, MC accounted for the largest proportion (160 000 tonnes), followed by TRI, PER, 1,1,1-T (each 130 000 tonnes), CFC-113 (44 500 tonnes), and CFC-11 (5500 tonnes).

Although the largest portion of chlorinated solvents emissions is released into air during their use, a part of their residue is left as waste. Existing statistics in Europe are not comparable, due to different definitions of chlorinated/halogenated solvents, and therefore the total amount of chlorinated waste at the European level is unknown. In EU countries, 700 000 tonnes per year of waste containing chlorinated solvents are estimated to be produced. The chlorinated solvents content of such waste is estimated to be approximately 200 000 tonnes. It is also estimated that approximately 90 000 tonnes out of 150 000 tonnes of chlorinated solvents in waste streams are recycled. However, recycling and useful applications of such waste are still considered too limited compared with the potentials that they offer.

Impacts

Chlorinated solvents pose threats to human health and the environment in several ways. Exposure to chlorinated solvents in high concentrations and over long periods of time is toxic and is suspected as a cause of carcinogenic, mutagenic and teratogenic effects.

Exposure to chlorinated solvents may occur through their emission into the air (the main pathway for chlorinated products) but also through supplies of drinking water. In addition, migration of solvents through environmental media may affect the environment at various levels of the biological system. A number of these substances also have global impacts. For example, 1,1,1,-T, CFC-11 and CFC-13 are particularly important because they play an important role in stratospheric ozone depletion and as greenhouse gases. The use of these substances will be phased out by the Montreal Protocol by 2005 and in EU countries by the year 1995.

Box 15F Major thermoplastics

Polymer name		Main applications
PE	(Polyethylene)	Containers, toys, hosewares, films, bags, pipes
HDPE	(High density polyethylene)	
LDPE	(Low density polyethylene)	
PET	(Polyethylene terephthalate)	Bottles, film, food packaging
PP	(Polypropylene)	Film, battery cases, microwave containers, automotive industry, electrical components
PS	(Polystyrene)	Pharmaceutical packaging, electrical appliances, thermal insulation, tape cassettes, cups, plates
PVC	(Polyvinyl chloride)	Window frames, pipes, flooring, wallpaper, bottles, cling film, toys, credit cards, cable insulation, medical products

Options

The reduction of waste containing chlorinated solvents may be achieved in several ways, including changing production processes or substituting certain solvents with other substances. Such waste may also be reduced by better use and in-house recovery of chlorinated solvents. There are several examples in the metal and printing industries where chemical solvent-based processes have been replaced with water-based ones. Examples of successful substitution of chlorinated solvents by alkaline solutions for degreasing in the metal industry are reported in several European countries, such as Denmark, Sweden and The Netherlands, where industries were able to cut costs at the same time (UNEP/IEO Cleaner Production Working Group on Halogenated Solvents).

Used tyres

Sources

Waste tyres are the most significant source of waste rubber products in terms of volume and importance. They are generated both when car tyres are replaced by new tyres, and when vehicles are scrapped. Total volume of used tyres released in the EU in tonnes per annum has been estimated at 1 955 000 tonnes on the basis of available figures for Member States. In the EU, almost half of the total waste tyres (46 per cent) is disposed of by landfill. One third (31 per cent) is submitted to processes leading to recovery of materials or energy, and about 23 per cent is re-used as retreaded tyres.

The most important factors affecting future development of car tyres are car ownership, car use, and the average life of car tyres. According to current trends in car ownership and use (see Chapter 21), the number of waste tyres is likely to increase in the next decade. The world market for tyres is forecast to grow at 1.4 per cent for cars and 2.2 per cent for commercial vehicles. On the other hand, the average life of passenger car tyres, as far as wear is concerned, has increased by 5 per cent over the last ten years.

Impacts

The disposal of waste tyres poses increasing environmental concerns. Tyres are made from natural rubber (truck tyres) or artificial rubber (car tyres). Depending on their design and application, tyres vary in size and total weight. On average, the weight of a car tyre is 7 kg. Average composition includes rubber hydrocarbon (48 per cent), carbon black (22 per cent), steel (15 per cent), textile (5 per cent), zinc oxide (1.2 per cent) and sulphur (1 per cent). Once treated with sulphur, they become immune to breakdown from bacteria so that, when disposed of in landfills, they remain intact for decades. If they are not stored in proper conditions they can pose a fire risk. Major fire episodes may last for a long time, affecting the quality of air, surface waters and wildlife.

When tyres are burned to recover energy for steam production, or as supplementary fuel in cement kilns, air pollutants are emitted. Although scrubbers remove zinc oxide and 90 per cent of SO_2, incineration still produces 24 grams of CO_2 per million joules. This is the same amount emitted by a coal-fired plant. The energy content is estimated at approximately 63 kJ per tyre. After incineration in a grate furnace fly ash (mostly zinc oxide) makes up approximately 4.5 per cent of the total residuals, and slag (mostly iron) between 13 and 18 per cent.

Options

Opportunities for prevention are linked mainly to the reduction of road transport and particularly to the use of the car. Indeed, extending the life of car tyres, which is the other preventative option, is limited for safety reasons. Retreading is a method of re-using tyres for their original application. It has the advantage of reducing the quantity of tyres being disposed of, and lowering the demand for new tyres. Useful applications for whole tyres or rubber range from engineering works (coastal protection, highway structures, erosion barriers) to insulation of building foundations and sound-absorbing walls.

Processing options include incineration and pyrolysis (chemical breakdown on heating). Burning waste tyres as a fuel in cement kilns or for energy or steam production are considered relatively clean options compared with other recycling processes. However, there are differing opinions as to the efficiency of energy production processes when compared with fossil fuel power plants. Useful products can also be obtained with pyrolysis, such as steel, carbon black, oil and gas. However, the variability of product quality and high capital costs are major constraints for the application of pyrolysis technology.

CONTAMINATED WASTE SITES

Increasing concern has emerged during the last decade over the threat to the environment posed by contaminated sites originating from the poor management of hazardous waste in the past, including poorly designed disposal sites. Several thousand contaminated sites are reported by national governments in Europe. Old hazardous waste sites are a potential threat to soil and groundwater and may pose health risks for the exposed population. Although it is extremely difficult to quantify the problem, the cost of damage from improper waste management practices in Europe is estimated to be extremely high.

Contaminated sites

At present, the number of contaminated sites from old waste disposal practices in Europe is unknown. The type, amount and potential risks of toxic substances present at these sites are even more difficult to determine. Attempts to estimate the number of potentially contaminated sites have been made in some European countries. A tentative list of registered contaminated sites is presented in Chapter 7. National environmental agencies are aware that reported figures underestimate the actual level of contamination from past hazardous waste practices.

Cost of clean-up

Estimating the cost of clean-up is extremely difficult since it varies enormously from site to site and depends upon the substances that are found at each site. A very rough estimate indicates that about ECU 1 to 1.5 billion per year are currently spent in Europe for the clean-up of such sites. National estimates include: DM 22 billion for Germany, more than HFL 3 billion for The Netherlands, and at least DKR 400 million for Denmark. In the USA, where legislation has been enacted since 1986, the estimate ranges from $20 to $100 billion.

Clean-up priority sites

Since costs of clean-up are extremely high, attempts at developing national lists of priority clean-up actions are being undertaken by several European countries on the basis of a preliminary assessment of identified sites. However, only a few countries have established programmes to produce systematic inventories of contaminated sites and their potential risks. Because of the enormous variability of conditions, estimates are very inaccurate. Potential risks of contamination from old hazardous waste sites vary according to the amount and type of substances present at different sites and according to the level of vulnerability of the soil and groundwater.

Building inventories of sites and establishing monitoring networks is expensive, sometimes prohibitively so. This is particularly the case in Central and Eastern Europe, where there is a need and requirement for cooperative efforts to provide technical and economic resources to assess the current threats of past waste management practices.

RADIOACTIVE WASTE

The impact of radioactive waste on the environment and human health depends on the type of radiation emitted, the period of radioactive decay, the waste form, and the radioactive waste management activity being considered.

Classifications therefore relate to the factors mentioned above:

- Low-level waste is low in radioactivity and 'short lived' (decaying to negligible levels after several centuries at most). It is produced by all nuclear related activities, and principally by the nuclear industry.
- Medium-level waste contains higher concentrations of radioactive materials, is also 'short lived' and is produced by the nuclear industry.
- Alpha waste contains significant quantities of alpha emitters which are 'long lived', ie, these wastes remain dangerous for periods of time exceeding centuries. They represent some per cents of the total waste volume.
- High-level waste is highly radioactive, heat emitting, and contains significant quantities of long-lived alpha emitters. These wastes include the vitrified waste resulting from spent fuel reprocessing and, when not reprocessed, the spent fuel itself. They contain some 99 per cent of the radioactivity contained in all radioactive waste as a whole but represent only 1 per cent of their total volume.

Each country has its own classification(s) for radioactive waste set up by the regulatory authorities. However, these classifications are very similar and generally aim at answering acceptance criteria for future disposal. For safety reasons, long-lived and high-level waste are not disposed of on the surface or near surface facilities; according to the present state of technology they are planned to be disposed of in deep underground repositories (so called 'geological disposal'). All radioactive waste must be solid or solidified in an appropriate long-lasting matrix before disposal.

Detailed data are available in Western European countries on most types of radioactive waste produced by civil activities. The amount of waste handled at military sites is usually not available.

A recent report (CEC, 1993) gives an overview of radioactive waste management in the Member States of the European Union. The report provides data on past and present waste production and a forecast for the year 2010. For an operational electronuclear capacity of nearly 110 GWe, the total production rate of conditioned low-level, medium-level and alpha waste is estimated at present to be about 80 000 m^3/year (corresponding weight is roughly 160 000 tonnes of heavy metal (HM)) for the EU as whole. In addition some 3400 tonnes (HM) of spent nuclear fuel and 150 m^3 of highly radioactive vitrified waste from spent fuel reprocessing are produced annually.

Under conservative assumptions, concerning the future of the electronuclear programmes, it is predicted that annual waste production will remain approximately the same until the year 2000. Production will increase significantly after 2000 due to new waste arisings resulting from the decommissioning of obsolete nuclear plants.

The production of radioactive waste outside of the nuclear industry is linked with the level of development of the country. Within the EU, a figure of some 10 m^3 per million inhabitants per annum may be deduced.

Data from Central and Eastern Europe are very limited, due notably to the fact that information on civil nuclear operations was until recently restricted under military secrecy rules. In the former USSR wastes were classified as:

- research, industry, medical waste;
- power plant waste; and
- reprocessing waste.

The first category has been managed by a network of 35 regional disposal centres. The most important, RADON, near Moscow receives some 3000 m^3/year, and 60 000 m^3 have already been disposed of. Part of the reactor wastes are stored on site (after concentration of the liquid waste); 1991 estimates for the former USSR were of some 150 000 m^3 of liquid waste and 100 000 m^3 of untreated solid waste under storage. A major part of the liquid waste has apparently been released into lakes or rivers, or injected under pressure into the ground (see Chapters 16 and 18).

In addition, Ukraine has to face the special problem of the waste resulting from the Chernobyl accident (see Box 18E Chapter 18), the overall volume of which might be evaluated at 1 million m^3.

A special mention should be made of the waste materials produced by uranium milling activities associated with uranium mining. These waste materials, called tailings, result from the processing of uranium ores. Tailings contain natural radioisotopes resulting from the decay chain of uranium-238 and notably radon gas. As in any other mining activities, the disposal and conditioning of mill tailings should be part of an integrated process of the rehabilitation of the site. Uranium mining and milling has been relatively modest in the EU (and limited to France, Germany, Portugal and Spain); however, they produce large quantities of waste materials. In Central and Eastern Europe these activities are situated in Ukraine and the Russian Federation.

SUMMARY AND CONCLUSIONS

A rigorous pan-European assessment of environmental impacts caused by waste generation and disposal requires much more complete and reliable information on the various aspects of waste (including the amount of waste by types, and the nature and toxicity of substances, as well as the methods used for disposal). In addition, if waste minimisation policies are to succeed, it is necessary to know the various activities and processes through which wastes are generated. However, in most European countries, waste statistics are incomplete or inadequate. In addition, comparing waste statistics across countries is made extremely difficult because of different definitions and classification systems.

With the data that are available, the following conclusions can be drawn:

1   In spite of the increased attention to waste prevention strategies, waste has increased in all European countries. Major sources of waste are (in order of importance per tonnage): agricultural, industrial, municipal and mining activities. However, the importance of various waste streams varies across Europe depending upon the economic structure and development of each country.

2   Waste has also changed in composition. Increasing portions of industrial waste are considered hazardous. A major shift is visible in the composition of municipal waste in Western European countries, with the increase of

plastics and packaging materials. The increased introduction of Western consumer products in Eastern European markets will probably produce a similar effect in these countries.

3 In Europe, most waste is disposed of in landfills which have, in cases of improper management, caused the emission of pollutants into soil and groundwater, and the formation of methane, CO_2 and other toxic gases. As a result of more stringent regulations which have imposed restrictions on the landfill of certain wastes, a significant proportion is now incinerated or submitted to other treatment processes. In spite of increased control, however, incineration is still responsible for the major part of the emissions of dioxins, hydrochloric acid and mercury.

4 Important progress in the minimisation of selected waste streams has been achieved in a number of European countries through integrated waste management, separate collection and recycling programmes. However, waste recycling efforts are still too limited and offset by high waste production rates.

5 In spite of the increased control introduced through international conventions, a significant amount of hazardous waste is still transferred across European countries and from Europe to developing countries. Transfrontier shipments of hazardous wastes still escape control. Data show that hazardous wastes move from countries with more stringent regulation to less regulated countries.

6 The current and potential threats for the environment and public health posed by contaminated sites due to improper waste management is unknown. However, drawing upon the preliminary results of ongoing assessments at a number of sites, the size of this problem and its potential consequences could be enormous.

Current patterns in the production and management of waste are not likely to change in the short term. The growth of economic activities as a result of European economic integration and economic restructuring in Central and Eastern Europe is likely to lead to a significant increase in the overall generation of waste. Waste issues will be a dominant concern for Europe's environment in the next decades, if reduction targets and safe management standards are not achieved in all countries.

REFERENCES

Bartone, C (1990) Economic and policy issues in resource recovery from municipal solid wastes. *Resources Conservation and Recycling* 4, 7–23.

CEC (1990) *A Community strategy for waste management.* Council Resolution of 7 May 1990 OJ No C122.

CEC (1992) Analysis of Priority Waste Streams: Chlorinated Solvents. Information Document. CEC, DG XI, Brussels, October 1992.

CEC (1993) *Communication and Third Report from the Commission on the Present Situation and Prospects for Radioactive Waste Management in the European Community.* COM(93) 88. Brussels, 1 April 1993.

Greenpeace (1990) *The international trade in waste. A Greenpeace Inventory.* Greenpeace, Washington DC.

Greenpeace (1993a) *Toxic trade update No 6.1.* First Quarter 1993. Greenpeace, Washington DC.

Greenpeace (1993b) *Toxic trade update No 6.2.* Second Quarter 1993. Greenpeace, Washington DC.

Nriagu, J O and Pacyna, J M (1988) Quantitative assessment of world wide contamination of air, water and soil with trace metals. *Nature* 333, 134–9.

OECD (1991) *Environmental Data Compendium 1991.* Organisation for Economic Cooperation and Development, Paris.

OECD (1992) *Reduction and Recycling of Packaging Waste.* Environment Monographs No 62, Organisation for Economic Cooperation and Development, Paris.

OECD (1993) *Environmental Data Compendium 1993.* Organisation for Economic Cooperation and Development, Paris.

RIVM (1992) *National Environmental Outlook 1990–2010.* National Institute for Public Health and Environmental Protection (RIVM), Bilthoven.

UK DoE (1992) *Digest of environmental protection and water statistics.* UK Department of the Environment. HMSO No 15, London.

UNCED (1992) *Agenda 21.* United Nations Conference on Environment and Development, Conches, Switzerland.

VROM (1991) *Essential Environmental Information: The Netherlands.* Ministry of Housing, Physical Planning and Environment, The Hague.

WRI (1992) *World resources 1992–93: a guide to the global environment.* Oxford University Press, Oxford.

Yakowitz, H (1992) Waste management in Europe. Paper presented at Globe Conference, Strasbourg May 17–20, 1992. OECD, Paris.

Traffic noise barriers
Source: Michael St Maur
Sheil

16 Noise and radiation

INTRODUCTION

Pressures on the environment may occur through chemical, physical and biological agents. The subject of this chapter is physical agents, where *physical fields* are examined. The term physical fields is used here to encompass acoustic fields (noise and vibrations) and radiation and electromagnetic fields (ionising and non-ionising). These are considered below under three headings: noise, non-ionising radiation and ionising radiation.

Physical fields are part of the natural environment. Depending upon their properties and the surrounding conditions, physical fields are transmitted through the spaces in which we live, potentially affecting human health and nature. Human activity can add to, modify, and enhance and reduce the intensity of these fields. Furthermore, changes in physical fields can be linked to global environmental problems such as stratospheric ozone depletion (causing a potential increase in ultraviolet-B radiation) and climate change.

NOISE

The ability to hear sounds is a sensory function vital for human survival and communication. However, not all sounds are wanted. 'Unwanted sounds', for which the term 'noise' is normally used, often have their origin in human activities (mainly transport, industry and households).

This chapter addresses *environmental noise* – also referred to in the scientific literature as community, or residential, noise. Environmental noise is the noise experienced by people generated outside households. Noise experienced by people in the workplace or occupational environment is not considered in this chapter.

The results of an acoustic quality survey in European cities is presented in Chapter 10. The current section first concentrates on the concepts involved with the measurement of noise and vibration. This is followed by an assessment of the effects of noise on human health and nature, and then by a review of noise sources in Europe.

The availability and comparability of data on noise pollution in Europe are generally poor (Box 16A). Available data published by the OECD show that exposure to noise, which was fairly stable at the beginning of the 1980s, had increased by the end of the decade in some Western European countries (eg, France, Germany, The Netherlands, Switzerland) (OECD, 1991a). For sound levels L_{eq} greater than 65 dB(A) (see Box 16B for a definition of L_{eq}), exposure appears to have stabilised in some cases and increased in others. However, within the range 55 to 65 dB(A), exposure has significantly increased, apparently as a result of the fast-growing volume of road traffic . In the highly industrialised European countries such as Belgium, France, Germany, Italy, The Netherlands and the UK, but also Austria, the Slovak Republic, Spain and Switzerland, more than 50 per cent of the population is exposed to noise levels from road transport which are above L_{eq} 55 dB(A), which is the level at which people become seriously annoyed during the daytime (WHO, 1993a).

The effects of noise

The nuisance effects of noise are difficult to quantify, as people's tolerance to noise levels and different types of noise vary considerably. Distinct variations in noise intensity and noise levels can occur from place to place (even within the same general area), and from one moment to the next. Similarly there can be large variations during each day, week or year. The main effects on people from noise occur along roads in both cities and rural areas, around airports and in residential areas.

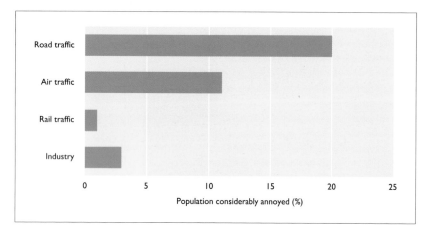

Figure 16.1
Annoyance caused by noise in The Netherlands, 1987
Source: Van den Berg, 1992, in von Meier, 1993

The impact of noise on sensitive groups deserves particular attention (eg, school children, the elderly, the sick). The reaction of these groups may be considered as warning signals as to what may happen to other groups on exposure at higher noise levels. This group could constitute as much as 30 per cent of the population (K Andersson, Swedish Environmental Protection Agency, personal communication, 1994).

Annoyance

This is perhaps the most common adverse effect of noise on people and many complaints are made every year about many different types of noise. The feeling of annoyance results not only from interference with communication and sleep disturbance, but also from less well-defined feelings of being disturbed and affected during all kinds of activities as well as during periods of rest. There is, for example, evidence of a clear relationship between degrees of individual annoyance and noise levels; for example, it has been demonstrated that less use is made of private gardens and public parks when there is too much noise (OECD, 1991b).

If the percentage of the population which feels 'considerably annoyed' by the noise is determined through sociological surveys, then a picture like that in Figure 16.1 emerges. It shows how the various classes of noise sources, namely road traffic, air traffic, rail traffic and industry, contribute significantly to annoyance.

Whether and to what extent such exposure is ultimately harmful to human health and well-being has not yet been fully and conclusively explored, except at very high sound levels, when it causes hearing loss and *tinnitus* (ringing in the ears). The present state of knowledge, however, clearly indicates that long-term health effects due to environmental noise exposure cannot be excluded. In addition, a number of well-defined harmful effects on the quality of sleep, communication and psycho-physiological behaviour can be identified. There is a lack of evidence to indicate that such reactions to noise diminish with time, although within certain limits tolerances may be built up. However, it seems that complete physiological habituation to sleep-disturbing noise does not occur, not even after several years of exposure (Suter, 1992, in WHO, in press).

Sleep disturbance

Sleep disturbance is probably the most apparent effect of environmental noise. It can also be interpreted as a reduced quality of sleep, and may even occur when the people affected are not aware of it (H Ising, BGA, Berlin, personal communication). To ensure undisturbed sleep, single noise events (such as a passing aeroplane) should not exceed a maximum sound pressure level of approximately 55 dB(A) (H M Müller, DG XI, CEC, personal communication).

Interference with communication

The degree of interference of noise with speech or music depends on the noise level in relation to the level that conveys the desired information. An increasing noise level requires speakers to raise their voice and/or to get closer to the listener in order to be understood. Noise levels from about 35 dB(A) and above are seen to interfere with speech communication until, at noise levels of about 70 dB(A), normal speech communication becomes virtually impossible (Verein Deutscher Ingenieure, 1988, in von Meier, 1993). In classrooms where teaching has to take place over long distances, noise levels should not exceed 25 dB(A) (K Andersson, Solna, personal communication).

Extra-auditory effects

A great number of psycho-physiological effects of noise have been reported in the literature (WHO, 1993a). The most common responses are physiological stress, and at higher noise levels, cardiovascular reactions. Mental health effects and influences on performance and productivity have

Box 16B Measurement of noise

There are two main ways of assessing the influence of noise: by physically measuring sound pressure levels, and by recording the discomfort or annoyance caused by noise. Physically noise is treated as an acoustic phenomenon called sound. A sound event as a physical phenomenon can be fully described by four parameters:

1 the *strength* or *sound pressure*, mostly expressed in terms of the amplitude of the sound pressure waves, and is usually measured as sound pressure levels in decibels (dB);
2 the *frequency* or *pitch*, measured in Hertz (most noises consist of a mixture of sounds with various pitches and frequencies, and hence do not have a recognisable pitch in any musical sense);
3 the fluctuation of sound with time (also known as the *time history*), measured as sound pressure level as a fluctuation of time;
4 *sound character*, which describes the particular features of a sound (eg, tonal and harmonic qualities).

Decibels are measured on a logarithmic scale, so that an increase of the sound pressure level (SPL) by a factor of 10 would result in a 10 dB increase in sound level, an increase of SPL by a factor of 100 would result in a 20 dB sound level increase, a factor of 1000 would result in a 30 dB increase, and so on. The decibel scale ranges from 0 (the threshold of normal human audibility) to 130 (the threshold of pain). For most purposes the frequency scale is weighted by the frequency sensitivities of the human ear, known as A-weighting, transforming the (unweighted) sound pressure (dB) into A weighting sound pressure level ('dB(A)'). The dB(A) scale strongly attenuates the low frequencies and moderately attenuates the very high frequencies. The way in which the dB(A) scale corresponds to everyday noises is shown in Figure 16.2, where it can seen that the range of everyday noises varies from roughly 45 dB(A) to 115 dB(A).

The main descriptors used in the assessment of environmental noise are shown in Table 16.1. For describing the impact of noise on humans, the so-called Equivalent Sound Pressure Level (L_{eq}) needs to be calculated, that is, the mean value of sound intensity over time expressed in decibels. The significance of the L_{eq} is to be seen in the hypothesis that a noise that varies through time is equivalent in its disturbance on humans to a steady constant source of noise, over the same interval of time, if the noise level (sound pressure level) of the constant source of noise is equal to the L_{eq} value of the noise that varies through time. However, L_{eq} is not enough for the characterisation of environmental noise. It is equally important to measure and display the maximum values of the noise fluctuations (L_{max}), preferably combined with a measure of the number of noise events (L_n) (WHO, 1993a). For most people, noise pulses are more annoying than a steady pulse of noise.

Apart from the descriptors listed in Table 16.1, there are a number of other descriptors used in European countries, notably with regard to aircraft noise (eg, in the UK, The Netherlands and Norway). The use of such a variety of descriptors makes international comparisons of noise exposure difficult, if not impossible. However, as far as can be judged from the serious efforts being made by the scientific community, through various international bodies, such as the International Organization for Standardisation (ISO) and the European Committee for Standardisation (CEN), it appears there is a preference towards those descriptors based around the concept of L_{eq}.

The actual annoyance to people caused by noise is normally measured by sociological investigations, in the laboratory, or by analysing complaints by type of noise source to local authorities. Such figures should, however, be treated with caution, as people tend to be disturbed by, and react differently to, different types of noise. It is possible to relate annoyance to noise measured in terms of L_{eq} to obtain 'dose-effect curves'. These have assumed a prominent role in quantifying the impact of environmental noise on people. The general conclusion from dose-effect curves is that people are becoming 'considerably annoyed' if environmental noise levels are in excess of about 40 dB(A) at night and 50 dB(A) during the day.

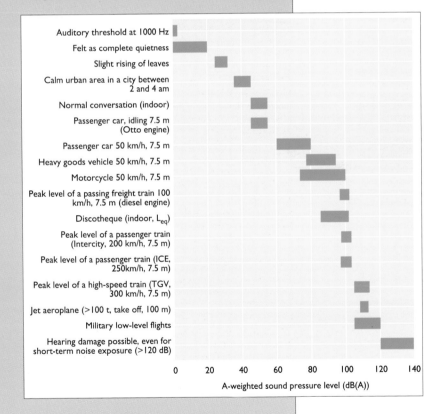

Figure 16.2 Maximum A-weighted sound pressure level (dB)
Source: Müller, DG X1, CEC, personal communication based on US EPA, 1979

Note: For modes of transport, the speed, distance from source and weight of carrier are indicated.

also been observed and documented (von Meier, 1993).

Intensive research on these subjects is ongoing (eg, Miedema, 1993), but it can be generally concluded from the present state of knowledge that exposure to environmental noise acts as a stressor to health, as it leads to measurable changes in, for example, blood pressure and heart rate. But there is not sufficient evidence yet to relate the exposure to environmental noise levels directly to specific health effects, although such relationships can by no means be excluded. It is presumably the total load of stressors, of which environmental noise is only one, that has a harmful and lasting effect on physical and mental health, seen as a statistical entity that allows great individual variations (see WHO, in press, for more details).

Low frequency noise and vibrations

While noise is pressure variations in air, vibration is wave motion in solid bodies, normally of low frequency. Low frequency noise and vibration are hence closely related. Vibrations will intensify the noise annoyance. Both have been linked with causing disturbance in residential areas. Vibration has not been well analysed.

Descriptor	Symbol	Definition	Field of application
Maximum sound pressure level	L_{max}	The maximum sound pressure level occurring during a certain period of time or during a single noise event.	To characterise noises of short duration. Mostly used with regard to sleep disturbances.
Equivalent sound pressure level	L_{eq}	The level of the average sound intensity taken over a certain period of time eg, 24 hours, daytime, night-time.	To measure average environmental noise levels to which people are exposed.
Day/night sound pressure level	L_{dn}	Equivalent sound pressure level over a 24-hour period with all sounds occurring between 22.00 and 06.00 raised by 10 dB(A).	To describe average environmental noise exposure by taking into account the increased sensitivity of humans during night-time.
Sound exposure level	SEL	Level of the sound intensity summed up over the period of time of a noise event related to 1 second.	To describe the noise of a specific, usually short, noise event.
Statistical sound pressure level	L_n	The sound pressure level of a time-varying noise that is exceeded during a certain N per cent of the time during the period measured.	To describe time-related properties of the noise eg, the background sound level L_{95} is the level which is exceeded during 95% of the time. Used to describe the acoustical quality of a residential area.

The effects of noise on wildlife

The effects of noise on the natural environment have not yet been fully explored. Research results available point to extra-auditory effects, mainly unspecific stress reactions, on animals with an acute sense of hearing, under extremely high noise exposures from low-flying aircraft (Umweltbundesamt, 1987). However, noise effects of major proportions or of lasting harmful consequences on nature have not been reported. The evidence available so far is too sketchy and inconclusive fully to exclude such effects.

Economic costs

Environmental noise also has implications for costs and benefits in the economy more generally. For example, cost effects are known to exist as a result of the general degradation of residential areas by exposure to environmental noise (eg, lowering of rents and property prices). Noise-induced illnesses, losses of productivity, and a higher rate of accidents caused by sleep disturbance have obvious cost effects, but these are extremely difficult to quantify. There are recent indications that certain producers of low-noise equipment may gain some economic advantage over their

competitors as public awareness of environmental noise disturbances increases.

The sources of noise

All European countries have adopted similar classifications of the sources of environmental noise. This classification relates the sources to specific human activities, thus making them easily identifiable. The major sources are road traffic, air traffic, rail traffic, industry and recreational activities. As a matter of expediency sometimes one or two additional classes are also included, such as noise from construction sites and from traffic on waterways. In the following paragraphs a brief source-specific analysis of the noise exposure data for Europe is provided.

Road-traffic noise

Road traffic is responsible for the highest percentage of the European population exposed to noise levels greater than 55 dB(A) (L_{eq}, 24-hour). In particular, the high proportions of L_{eq} between 55 and 65 dB(A) can be attributed to the growth of traffic volume during recent years (Figure 16.3). This more than offsets any possible results of tightening noise limits from road vehicles through engine/exhaust noise controls as exercised for example by EC legislation. In addition, tyre/road noise is now the dominant source of road-traffic noise, particularly in free-flowing traffic at speeds above 40 to 50 km/hour (von Meier, 1992). Roughly speaking, each increase of vehicle numbers by 25 per cent will result in a noise level increase of 1 dB, whereby one heavy goods vehicle is equivalent to about eight passenger cars.

Airport and air-traffic noise

Noise levels around airports have been reported for 27 airports (for nine European countries) by the OECD (OECD, 1993). Most exposure data are site-specific, depending on geographic location. It is difficult to deduce trends with poor time-series data, and to make comparisons between countries, owing to different limit values, measurement techniques and the diversity of descriptors being used. Average data for all German airports indicates, however, the number of people exposed to outdoor noise levels of daytime L_{eq}>67 dB(A) has increased from 500 000 to 610 000 between 1980 and 1990, and for outdoor noise levels of daytime L_{eq}>75 dB(A) from 100 000 to 122 000 (OECD, 1993). If such trends exist in other countries (which seems likely, due to increased air traffic over the last decade) it is a cause for concern.

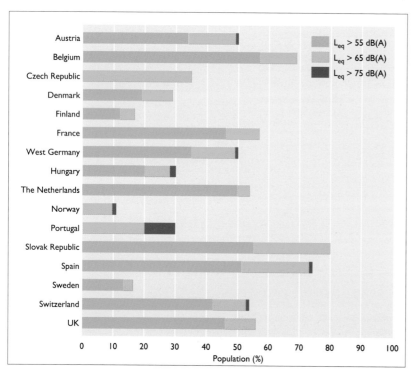

The OECD has presented data for air traffic for 11 European countries (OECD, 1993). With the exception of The Netherlands, the extent of noise exposure due to air traffic is seen to be considerably less than to road traffic. Two opposing trends determine the present state and future levels of environmental noise from air traffic in Europe: on the one hand the increasing use of quieter aeroplanes, and on the other the rapid growth of air traffic (commercial, general aviation – eg, from light propeller aircraft used for recreation, executive jets and helicopters – and military aircraft). This later trend is compounded by an increase in shifting freight transport from day- to night-time.

Rail-traffic noise

Nine European countries supplied railway noise data to the OECD survey (OECD, 1993). Somewhat surprising is the high percentage of exposed people in the Slovak Republic. Switzerland is known to have a railway noise problem because the topographical constraints dictate that railway lines run through the same valleys as those where human habitation is concentrated.

Trends in rail-traffic noise are not clearly detectable. Great efforts are being made in Western European countries to limit rail-traffic noise by curative measures, mainly by screens along railway lines, but the introduction of high-speed trains in France, Germany and Spain – and shortly also in The Netherlands, Belgium and the UK – and planned extensions of railway networks, will increase the number of people exposed to noise.

On the whole, however, rail-traffic noise appears to be of much less importance than road-traffic noise. In dose–effect studies, railway noise is generally judged to be less annoying to people than other noises, in particular during the night (von Meier, 1993).

Noise from industry

Noise originates from all sorts of industrial activities, ranging from small workshops to livestock activities (abattoirs), outdoor factories (eg, for stone crushing), and heavy industry. Because of this great variety of industrial noise sources and of the different characters of the noises they emit, with regard to frequency content and time history, it is very difficult to compile an inventory of people non-occupationally exposed to industrial noise which is both representative for a certain country and comparable with other countries.

Exposure levels of more than 75 dB(A) rarely exist for industrial noise in Europe. Indeed, most exposure is below 55 dB(A). This does not mean, however, that industrial noise is a negligible environmental problem. On the contrary, people exposed to it can be very seriously annoyed and affected in their well-being. This depends on the character of the noise and on the attitude people have towards the source (eg, whether or not it is seen as economically important).

Recreational activities

Few data are available for noise from recreational activities in Europe. There is often an overlap with industrial noise for certain activities, such as catering. Loud noise from recreation tends to be confined to locally concentrated sources such as sports arenas, theme and leisure parks, shooting ranges, discotheques and rock venues. There is also some evidence of an increase in complaints from street noise (eg, from dogs barking, radios and recorded music) (UK DoE, 1992). New forms of tourism and recreation, such as motor sports, off-road racing and use of speedboats and water scooters, create noise which disturbs other (less noisy) holiday-makers. On the other hand, some forms of recreation require almost complete silence, such as bird-watching.

Overall noise environment in Europe

From the still somewhat limited noise exposure data available (for example, for road traffic, see Figure 16.3) it can, with some caution, be concluded that, at least at the national level, there are no systematic differences between groups of countries in Europe (eg, north–south or east–west). This is not necessarily the case for individual cities where Central and Eastern European cities appear to be noisier than Western European cities (see Chapter 10). But the possibility of such differences existing between countries cannot be ruled out, as they could be brought to light as more accurate data with wider coverage become available. In particular the dearth of data for Central and Eastern European countries accounts for the still somewhat fragmentary character of this inventory. However, a tentative analysis relating to the exposure to noise of the total population of Europe can be attempted (given a total population of Europe of 680 million). The result is shown in Figure 16.4.

Attention should be paid to interpreting Figure 16.4, as it does not take account of possible overlap between exposures from different sources (ie, some of the exposure attributed to road traffic could be attributed to rail traffic and vice versa). Nevertheless, Figure 16.4 is very clear in identifying road traffic as the predominant source of noise pollution. By relating Figure 16.4 to a dose–effect relationship for road traffic (which after all accounts for most of the cumulative exposure of the European population), it is possible to conclude the following:

- About 450 million people (about 65 per cent of the European population) are exposed to environmental noise levels above L_{eq} 24h 55 dB(A), causing annoyance and disturbance of sleep.
- About 113 million people (about 17 per cent of the European population) are exposed to environmental noise levels above L_{eq} 24h 65 dB(A), the level at which serious impacts of noise are noticeable.
- About 9.7 million people (about 1.4 per cent of the European population) are exposed to such high environmental noise levels (above L_{eq} 24h 75 dB(A)) as to be considered unacceptable.

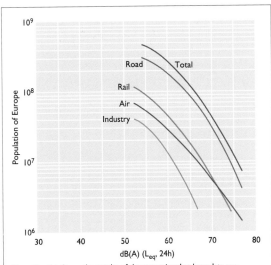

Figure 16.4
Cumulative exposure of the European population to noise
Source: von Meier, 1993

Note: For this figure the results of the countries that have data are extrapolated to the rest of Europe and summed for each band of noise exposure. The states in the European part of the former USSR have been accorded a weighting factor of 0.5 (to account for the low degree of urbanisation in most of these countries).

Policies and strategies for noise abatement

There is a general consensus in OECD countries that the outdoor level of noise should not exceed 65 dB (daytime L_{eq}). In the case of new residential areas the outdoor levels should not exceed 55 dB (daytime L_{eq}). These levels now frequently serve as the reference for defining noise 'black spots', where the levels of exposure to noise exceed 65 or 70 dB(A) (L_{eq} from 6 am to 10 pm), and 'grey areas' where the noise level is between 55 and 65 dB(A) (OECD, 1991b).

The European Commission stated in its 5th Environmental Action Programme that, 'no person should be exposed to noise levels which endanger health and quality of life'. This translates into the following targets for night-time exposure:

- exposure of the population to noise levels in excess of 65 dB(A) should be phased out; at no time should a level of 85 dB(A) be exceeded;
- the population at present exposed to levels between 55 and 65 dB(A) should not suffer any increase (ie, in an area with 56 dB(A), a new industrial plant causing an increase in noise level to a total of 58 dB(A) is not desirable);
- the population at present exposed to levels of 55 dB(A) or less should not suffer any increase above 55 dB(A) (based on CEC, 1993a).

Extrapolating the first goal for the whole of Europe would mean that the number of highly exposed people (above 65 dB(A) during night-time) would be reduced by 9.7 million. The other goals would not reduce the number of people disturbed by noise, but in any case their number should not be allowed to increase. Noise pollution impinges also on other policy areas (particularly transport).

There are three main options available to achieve these goals: engineering (including landuse planning), legal, and education and information. Both engineering and legal measures need constant feeding through research and can succeed only if supported by the public. Hence, education and information of the public is required. Such measures should be seen as complementary actions which cannot yield any substantial results without the help of others. Some examples of such measures are provided in Box 16C. Noise protection measures should always be introduced at a very early stage, in order to improve acoustical quality and increase cost-effectiveness.

At the same time as the CEC's 5th Environmental Action Programme, other initiatives have been taken by different countries and organisations. The WHO recently issued an external review draft on an Environmental Health Criteria Document aimed at community noise (WHO, 1993a). Sweden at the same time published an Action Plan against noise (Kihlman, 1993). Several other countries, inter alia, The Netherlands, Denmark and Switzerland, are already implementing comprehensive programmes for improving noise, emphasising urban areas.

Outlook and conclusions

The estimate of 113 million people, or 17 per cent of the total European population, exposed to environmental noise levels that have serious negative impacts is an important cause for concern. This situation has arisen out of a gradually deteriorating situation over about the past 10 years. If this is extrapolated into the future, then the outlook can only be of continuing and increasing concern. Economic development in Central and Eastern European countries is likely to exert additional pressures in the future, particularly as a result of

---

Box 16C Noise abatement measures

Engineering measures

- Landuse planning and orientation of buildings (eg, separating out incompatible functions; using bypass roads for heavy traffic; establishing areas of silence or noise abatement zones, using buildings for screening purposes in architectural design; tunnels, and so on).
- Emission reductions by source modification (eg, low-noise product development).
- Measures to obstruct the path of noise (eg, noise screens; enclosures around machinery).
- Measures to protect the unintended listener (eg, insulation of dwellings; double or triple glazing; facade design for restoring old houses and for new houses).
- Reducing tyre noise, and developing new low-noise road surfaces.
- New engine technology (eg, for road vehicles and aircraft).
- Traffic management schemes.

Legal measures

- Defining limit or guide values for exposure levels for day- and night-time, and agreeing criteria for measuring noise.
- Provision of noise zones along traffic routes and around airports and industrial installations (eg, in The Netherlands).
- Developing regulations for composite noise situations (for where noise comes from more than one source).
- Controlling noise emissions at source for transport

vehicles (eg, EC Directives on road vehicles, subsonic jet airliners, motorcycles).
- Legal specification of maximum noise emissions from products (eg, EC Directives on construction plants and equipment, lawn mowers); establishing appropriate frameworks for measuring noise (eg, labelling products such as spare car components with noise levels, and so on).
- Minimum requirements for acoustic properties of dwellings, normally part of national building codes (eg, specifications of sound insulation of facades, roofs etc); sound-classified dwellings.
- Vesting power in local authorities to control recreational noise through licensing (eg, for rock venues).
- Speed limits (eg, for freight transit across Austria).
- Quick and efficient enforcement of existing regulations.

Education and Information

- Improved monitoring of noise in urban areas.
- Increasing the number of qualified noise experts.
- nitiating appropriate research and development (eg, low-noise road surfaces and tyres).
- Sharpening noise awareness of the public by providing factual information, for example, on the number of noise enforcement actions taken and the number of complaints made.
- Influencing behaviour to avoid noise generation (eg, reducing speed when driving a motor vehicle).

Box 16D Measurement and concepts of physical fields

The electromagnetic spectrum is presented in Table 16.2. Electromagnetic waves propagate at the speed of light with an approximate velocity, c, of 3×10^8 metres/second. The wavelengths of these waves, λ, depend upon their frequency, f, according to the relation $\lambda = c/f$. The wavelength therefore varies inversely with frequency, as shown in Table 16.2.

Physicists use the term 'field' to refer to a region in which objects separated in space exert a force on each other.

The strength of an **electric field** at any given point is determined by the voltage of the conductor and the distance of the point from the conductor. The **electric field strength**, E, is thus measured in volts per metre (V/m).

Magnetic fields are those in which magnetic objects (ones in which there are electric currents) exert force on other magnetic objects. The strength of a magnetic field at any given point is determined by the power of the source of electric current and the distance of the point from the source. The **magnetic field**, H, is thus measured in amperes per metre (A/m). Other special units of measurement are often used instead of A/m, for example: 10 milligauss (mG) = 1 microtesla (μT) = 0.80 (A/m); some of these are used in this chapter.

In high-frequency electromagnetic fields, such as radiowaves or visible light, electric and magnetic fields interact and their effects are not separable. At very low frequencies (less than 1000 Hz), however, electric and magnetic fields have different effects and can be characterised separately (for further information, see WHO, 1993b).

Static fields (not to be confused with static electricity) are created by objects that have a constant charge or current flow (DC or direct current). Objects whose charge or current periodically fluctuates create fields that also fluctuate, as in the case of radar (AC or alternating current).

Wavelength (metres)	Frequency range (in cycles/sec and Hertz)	Typical sources/applications/types
–	0	DC power lines
10^6–10^7	10–100	Power line frequency 50/60 Hz
		Telephone
10^5	1000	Telephone
10^4	10^4	
10^3	10^5	VLF communication systems
10^2–10^3	10^6 (1MHz)	RF induction heaters
		Medium-wave radio (535–1605 kHz)
1–10^2	10^7 (10MHz)	Short-wave radio (3–30 MHz)
		Radio frequency sealant, short-wave diathermy, hyperthermia for cancer therapy
10^{-1}–1	10^8 (100MHz)	VHF television
		FM radio (88–108 MHz)
		UHF television
10^{-2}	10^9 (1GHz)	Radar for sea and air navigation
10^{-3}	10^9 (1GHz)	Microwaves
		Microwave oven (2.45 GHz)
		Satellite communications; electrical power lines
10^{-3}	10^{10} (10 GHz)	Microwave links for long-distance telephone and TV communications
		Cellular telephones
10^{-6}–10^{-3}	10^{12} to 10^{14}	Infrared
10^{-6}	10^{15}	Visible light
10^{-8}–10^{-7}	10^{15} to 10^{16}	Ultraviolet light
10^{-11}–10^{-8}	10^{16}	X-rays
10^{-14}–10^{-11}	10^{17}	Gamma (γ) rays
shorter than 10^{-14}	above 10^{17}	Cosmic rays

Table 16.2 Electromagnetic spectrum and some typical examples of applications
Source: Gandhi, 1990 and others

growing road traffic, as the first data from the 'New Bundesländer' (East Germany) have indicated (Umweltbundesamt, 1992b). To meet the objectives of the CEC's 5th Environmental Action Programme this trend will have to be reversed, requiring that all available noise control options be applied, with road-traffic noise being identified as the top priority.

Environmental noise pollution is thus a European problem of growing extent and concern. Only concerted actions on a technological as well as on a political level are able to improve the present situation and to reverse existing trends. The elements required to set up a successful noise control strategy are known and, as far as technology is concerned, readily available, but great efforts still need to be made on both national and European levels to set up and enforce noise legislation and to support this by education and information programmes and by promoting research and development.

NON-IONISING RADIATION

Physical fields arising from non-ionising and ionising radiation are additional agents exerting pressure on organisms in the environment. Box 16D explains the concepts and terminology associated with such fields. Their effects on human health are examined briefly in Chapter 11. Some parts of the electromagnetic spectrum, such as gamma rays and X-rays from nuclear sources and electron beams, which have wavelengths less than 10^{-8} metres, are said to be *ionising* (ie, radiations and fields that have enough energy to produce ionisation of matter such as gases and biological matter). These are discussed below in the next section. Radiations

with a wavelength longer than 10^{-8} metres are said to be *non-ionising*. These include electromagnetic fields and non-ionising ultraviolet (UV) radiation, which are discussed in this section. The effects of low-frequency electric and magnetic fields are included (especially below power transmission lines) as well as exposure to UV-B radiation. Microwaves are not covered in this overview.

Human-induced sources of electromagnetic fields are generally small when compared with natural sources. However, suspicion has been directed towards certain human-made frequencies and wavelengths which happen to be far greater than those occurring naturally, although evidence of health effects from EMF is insubstantial.

With respect to ultraviolet light, the evidence of biological damage from exposure is clear. Some epidemiological evidence suggests that certain frequencies and/or wavelengths may be associated with human health effects. However, much of this exposure for humans appears to be related more to choice of lifestyle (such as sunbathing) than to environmental factors. This may yet be exacerbated by the depletion of the stratospheric ozone layer, whose equilibrium has been disrupted by emissions of CFCs and halons (see Chapter 28).

There are a few other types of non-ionising radiation which have demonstrated human health effects, including high-frequency fields, ultrasound and laser light. Medical applications of all three of these have demonstrated clear diagnostic or therapeutic benefits. In human terms, exposure is presently significant only in the occupational context, which is beyond the general scope of this overview.

Electromagnetic Fields

Sources

Natural

Electromagnetic fields (EMFs) are present naturally. The Earth has a 'static' (steady) magnetic field strong enough to turn a compass needle. EMF of various strengths also occur in thunderstorms, in the static electricity associated sometimes with clothes and furnishings, and in the electrical activity of the human body itself. The Earth's core is thought to be responsible for massive electric currents inside the Earth, which in turn are thought to be the source of most of the Earth's static *magnetic* field. This field has a strength of about 40 A/m (50 μT). A natural *electrical* field is also normally present at the Earth's surface due to the electrical charges in the upper atmosphere and from solar activity. At ground level during good weather this field is about 100 V/m, but during severe weather it may rise to many thousands of volts per metre.

Artificial

EMF are produced by power transmission lines and by most equipment that uses a mains supply, including everyday home appliances. Table 16.3 provides a basis for comparison of the field strengths associated with various sources in the environment. It is clear from this that the human-induced sources of EMF generally occur at much higher intensities than the naturally occurring fields, especially at the power frequencies of home appliances (50 Hz). Some of the fields to which humans are occasionally subjected are extremely strong. There is a general but unproven feeling, however, that DC (also referred to as 'static') fields, such as the Earth's magnetic field and that involved in magnetic resonance imaging (MRI), are less hazardous or problematic than AC (or 'non-static') fields. Human exposure to non-static fields generally comes through electricity use.

Effects of EMF

EMF may have direct or indirect effects on living organisms. Direct effects arise when electric fields induce a surface charge on an exposed body. This results in a distribution of electric current in the body, depending on exposure conditions, size, shape and position of the exposed body in the field. Observable biological effects may include vibration of body hair, stimulation of sensory receptors and cellular interactions. Indirect effects of EMF are defined as those that may occur when an electric current passes completely through the body when in contact with a conducting object. Observable effects may include the occurrence of visible discharges, sometimes known as 'sparks', or interference with medical implants such as cardiac pacemakers.

EMF act in a very different manner from most other environmental agents. A straightforward measure of field strength, analogous to a measure of the concentration of a chemical in the air or water, may not provide an accurate measure of exposure or risk. The exposure to EMF can be characterised by several different parameters (field strength, field direction, field orientation in relation to the body exposed, field complexity, and so on), though it is not known which of these parameters are associated with risk to human health and the organisms.

The WHO recently summarised the results of research studies on the impact of EMF in the laboratory on animals, on humans, at the cellular level, and on the biological effects of Extremely Low Frequency (ELF) fields (WHO, 1993b). Evidence for thermal effects of exposure to EMF is identified. There are also some epidemiological studies which claim increased incidence of cancer in children and adults, as well as workers occupationally exposed to magnetic fields of 50 to 60 Hz over long time periods (see reviews in UK NRPB, 1992). The greatest concern relates to children reportedly exposed to field strengths of 0.1 to 0.4 μT living in proximity to power lines.

Table 16.3
Typical exposure levels from some electric and magnetic field sources
Source: WHO (in press)

Field sources	Frequency range (Hz)	Magnetic flux density		
		Average level	Peak level	Electric field strength
Earth	0	50 μT		130 V/m
	50 and 60	10–6 μT		3–20 kV/m (thunderstorms)
	5–1000			10–4–0.5 V/m (atmospherics)
Offices and households	50 and 60	0.03–1 μT	1–40 μT (at 30 cm	2–500 V/m (at 30 cm distance
		0.6–12 μT (residences	distance from various	from various appliances)
		with electric heating)	appliances)	
30 cm from a VDU	50	0.8 μT	5–18 μT	30–1500 V/m
High-voltage AC power transmission				
Right under overhead transmission line (380 and 756 kV)	50 and 60	10–30 μT	400 μT (during failure)	3–10 kV/m
25 m from midspan (380 kV)	50	6–8 μT		1–2 kV/m
Railway transmission line (110 kV)	16 2/3	12–25 μT		2–5 kV/m
Inside generating stations	50 and 60	20–40 μT	270 μT	5–16 kV/m
Industry				
Electrolytic processes	0 and 50	1–10 mT	50 mT	
Welding machines	0 and 50		130 mT	
Electric furnaces	1–1000		25 mT	
Induction heating	50–10 000	1–6 mT	25 mT	
Medicine				
Nuclear magnetic resonance imager	0	1–5 mT (operator)	2T, 6T/s (patient)	
			30–250 mT (operator, outside magnet)	
Therapeutic equipment	12, 15, 20, 50, 75	1–16 mT (patient)		

Most studies have been criticised as no biological mechanism has been suggested to explain these effects and because of the low statistical significance of the results due to the low case numbers and inadequate controls. Consequently, there is still great uncertainty surrounding the possible carcinogenic effects of magnetic fields. Two recently published studies, from Sweden and Finland, of children living close to power lines (Verkasalo et al, 1993; Feychting and Ahlbom, 1993) concluded that electric and magnetic fields of transmission lines do not constitute a major public health problem regarding childhood cancer. The increased risk of childhood leukaemia noted in the Swedish study (one extra case per year) was indicated to be insignificant from a public health point of view.

At its meeting in May 1993, the International Commission on Non-Ionising Radiation Protection (ICNIRP) reviewed all available data on the possible carcinogenicity of 50 to 60 Hz magnetic fields. They concluded that there is no firm evidence of the existence of a carcinogenic hazard from exposure to ELF fields, but that the findings justify the proposal of a specific programme for further research (WHO, in press).

Policy and research

Present public safety standards for electromagnetic fields are based primarily upon laboratory experiments and modelling studies. These focus on exposure conditions known to cause physiological or chemical effects associated with the development of cancerous cells.

Some international standards have been set by, for example, the European Committee for Standardisation (CEN) and its subsidiary the European Committee for Electrotechnical Standardisation (CENELEC). Limits for electric and magnetic fields at power frequencies of 50 to 60 Hz in some European countries are shown in Table 16.4, together with the guide limits of the International Radiation Protection Association. Some individual countries such as Belgium, the former Czechoslovakia and the former USSR have set limit values for overhead power lines in different settings (general purpose, road crossings, accessible or inhabited areas and edge of right of way). However, there are no internationally agreed limits for overhead power lines (Maddock, 1992). See WHO (1993b) for details on other frequency ranges.

There are currently no proposals to legislate on the exposure of the general public to EMF in the EU. Attention in the EU is currently centred on concerns about occupational exposure of workers to the risks arising from physical agents (CEC, 1993b) – in the case of EMF, this includes primarily the exposure of those working on high-tension lines – and is further restricted to the thermal effects of such exposure.

The majority of European countries have protection standards which regulate occupational exposure to EMF, but the standards differ from each other in the permitted exposure levels, coverage of EM spectrum, permitted exposure time, and technical standards for EMF sources. These differences arise from disparate research findings consequential from questions on what constitutes a health hazard, and methodological differences in collection, processing and interpretation of results. An international research project at European level, in the framework of the European Commission's COST programme (COST 244 – 'Biomedical Effects of Electromagnetic Fields') aims to coordinate European research in this field. Its objective is to ensure the scientific background for developing common European protection standards in the whole spectrum of EMF, oriented both to the general population and to occupationally exposed personnel. The project entered the research phase in 1994 and is planned to end in 1997.

Outlook and conclusions

Public concern about EMF continues. The WHO predicts increases in EMF exposure in coming years as increased electrification accompanies socio-economic development (WHO, in press). This general scenario might perhaps be tempered in some respects by the following considerations.

1 There are differences between Western Europe and Central and Eastern Europe in the trends for electricity consumption, and in the use of electrical appliances in the home. While trends put forward in the WHO socio-economic development scenario may well apply to Central and Eastern Europe for the next two decades, it is doubtful that anything similar will apply in Western Europe.

2 In Central and Eastern Europe, most future electricity expansion could be accompanied by more modern equipment and appliances and more efficient generation, distribution and consumption practices, which may reduce EMF exposure.

3 The spread of information and concerns about EMF in Western Europe will lead to prudent avoidance of EMF where possible, particularly through burying cables, as well as appliance designs that seek to limit EMF, improved energy efficiency, and so on. Such activities place in doubt an expected increasing trend in EMF as witnessed in Western Europe over the past decades, and may result in exposures beginning to decrease overall.

Table 16.4
Electric field and magnetic field limits at power frequencies of 50 to 60 Hz
Source: Maddock, 1992

	Occupational	Public	Status	Basis
Electric field limits (kV/m)(rms)				
former Czechoslovakia	15	–	S	P, H
West Germany	20.7	20.7	S	J
Poland	20[a], 15	10, 1[e]	S	P, H
UK	12.3[b]	12.3[b]	G	I
former USSR	25 to 5[c]		S	J
IRPA	30 to 10[d]	10[f], 5[g]	G	P, H
Magnetic-field limits (mT)(rms)				
West Germany	5	5	S	J
UK	2[h]	2[h]	G	I
former USSR	7.5 to 1.8[i]		G	W
IRPA	25[j], 5[k], 0.5[l]	1[f], 0.1[g]	G	J

Notes: IRPA: International Radiation Protection Association
G Guideline; H Health-concern for possible effects; I Basic restriction is induced current; J Basic restriction is of induced current density; P Perception of spark discharges or tingling sensations; S Standard, order or rule, usually with legal force; W these values seem to have been developed primarily for electric-arc welding

a 2 hours maximum
b reference levels may be exceeded if the basic restriction is observed (a limit on the continuous induced current of 1.03 mA at 50 Hz in any arm, hand, leg, ankle or foot)
c depends on duration (t, hours per work day) of exposure, t = 50/E-2 for E between 5 and 20 kV/m; between 20 and 25 kV/m, only 10 minutes' exposure is permitted
d depends on duration (t, hours per work day) of exposure, t ≤ 80/E for E between 10 and 30 kV/m
e 1 kV/m applies where there are homes, hospitals, schools and the like
f for up to a few hours per day and can be exceeded for a few minutes per day provided precautions are taken to prevent indirect coupling effects
g for up to 24 hours per day – this restriction applies to open spaces in which members of the general public might reasonably be expected to spend a substantial part of the day, such as recreational areas, meeting grounds, and the like
h reference levels may be exceeded if the basic restriction is observed (see note b)
i varies with duration of exposure from 1 to 8 hours per work day
j for limbs
k maximum exposure duration is 2 hours per work day
l for whole working day

Table 16.5
*Effective UV-B
radiation doses*
Source: Reported by CEC
1993c

Note: Figures represent
percentage increases in the
biologically effective
radiation dose from UV-B
in Antarctica assessed
according to the action
spectra of four biological
targets.

Box 16E Effects of enhanced UV-B radiation

According to CEC (1993c), recent measurements in Antarctica by American scientists demonstrated that a 58 per cent decrease in stratospheric ozone resulted in ground-level increases of 300 per cent for UV-B, 31 per cent for UV-A, and 32 per cent for visible light. In order to determine how important such increases were in biological terms, the scientists referred to 'action spectra': the results of a complicated series of scientific experiments on a given type of organism which attempt to determine which wavelengths of light or radiation are most important to the organism, ie, which wavelengths elicit certain biological reactions. If the increases in radiation in Antarctica are assessed according to the action spectra of different biological targets, the calculated biologically effective radiation doses are those shown in Table 16.5.

Action spectrum for:	Method used	Increase in biologically effective dose (%)
DNA damage	Setlow	638
Erythema	Robertson-Berger	415
Generalised plant damage	Caldwell	689
Fish	Hunter	267

The problem of understanding EMF effects on health might be more rapidly resolved through epidemiological studies of electrical workers than through studies of the general public, who generally have much lower exposure.

The existing situation in Europe could be considered adequate, if the measures and limits internationally agreed for the protection of workers are properly implemented. Until further research results suggest otherwise, the IRPA (International Radiation Protection Association) limit for continuous exposure of the general public of 5 kV/m for electrical field strength (or a magnetic flux density of 0.1 mT at 50/60 Hertz) may be considered to provide substantial protection from possible health effects.

Ultraviolet radiation

Sources

Natural

Ultraviolet (UV) radiation is emitted by the sun with wavelengths between 10^{-7} and 10^{-8} metres as well as by various human-induced sources. The UV radiation band can be divided into three smaller bands: UV-A (320 to 400 nm), UV-B (280 to 320 nm) and UV-C (<280 nm). About 5 per cent of the total solar radiation that reaches the Earth's surface is ultraviolet. The amount and intensity (of UV-B especially) varies with the angle of the sun (time of day, season and latitude), the weather (clouds, mist, fog, etc) and the altitude. At sea level, about 95 per cent of the UV radiation is UV-A, 5 per cent is UV-B.

Most solar UV radiation which would otherwise reach the Earth's surface is absorbed by the atmosphere. However, human activities have been recognised to be changing the composition of the atmosphere, including the ozone layer (see Chapter 28). Such a change could permit a small but significant increase in the amount of UV radiation reaching the Earth's surface, with repercussions for human health and ecosystems.

Although ozone layer measurements have indicated a decrease in levels of stratospheric ozone, there is little evidence of increases in UV flux at the Earth's surface, with the exception of a large increase recorded in the South Pole region (see Box 16E). It has been suggested that the most common monitors (Robertson-Berger (RB) meters), designed before there was a perceived need for detailed knowledge of the UV spectrum, are not sensitive enough to detect the changes, which are occurring most importantly in the shorter wavelengths recorded by the meter (CEC, 1993c).

A recent Canadian study (Kerr and McElroy, 1993, cited in WHO, in press) showed that levels of UV-B in Toronto during the winter have increased by more than 5 per cent per year from 1989 to 1993, as ozone levels have decreased. At the time it was produced, this study was the most reliable of its kind with regard to the situation in the northern hemisphere.

Artificial

Technology has also introduced new sources of UV radiation and visible light which can have deleterious effects on human health. Much of the working population, and more and more of the general population, of Europe is now exposed to the UV radiation emanating from fluorescent and high-intensity quartz tungsten-halogen lamps over long periods. Likewise, industrial and medical uses of UV radiation are widespread, including such diverse applications as photo-polymerisation, disinfection and sterilisation, welding, and the widespread use of lasers.

Changing fashion and human behaviour patterns have also led to increased exposure to the sun and UV radiation, which has already demonstrated profound effects on human health, in particular skin cancers. The use of sun-lamps and sun-beds, most frequently for cosmetic reasons, is a common source of additional UV exposure. Prior to the mid-1970s, mercury lamps were generally used as sun-lamps, sources of high levels of both UV-B and UV-C. While some of these devices are surely still around, since the late 1970s fluorescent lamps, generating almost exclusively UV-A (the most important for tanning) and visible light, have mostly been produced for this application. A clear link has not been proven between the use of sun-lamps or sun-beds and the incidence of skin cancer. However, the use of artificial tanning equipment is widespread and largely uncontrolled. In addition, it should be noted that a 'safe' level of exposure has not been defined. It is estimated that 10 per cent of people in the Western world use sun-lamps or sun-beds (WHO, in press).

There is also an increasing use of UV light in working environments (Moseley, 1988, and WHO, in press):

- Fluorescent lamps consist of low pressure mercury vapour lamps generating a strong emission at a wavelength of 254 nm. Depending upon the nature of the phosphor coating which is excited by this wavelength, there may be emissions of UV radiation.
- Other low pressure mercury vapour lamps are used for disinfection and sterilisation of drinking water, swimming-pool water, sewage, and so on. No conclusive studies have linked these UV emissions with adverse effects, but further studies have been called for (IARC, 1992).
- High-intensity (several kilowatts) quartz-halogen lamps may be used to induce polymerisation during the production of protective coatings or printed circuit boards, or the curing of special inks.
- Large xenon arc lamps are used to promote weathering of test samples, or to illuminate broad areas such as football fields. These efficient lamps generate significant amounts of UV radiation, which is usually designed to be attenuated by the glass housing of the lamps; however, some UV-B usually escapes.
- Electrical arc welding also gives rise to UV-B, which has been associated with specific health effects (Mariutti and Matzue, 1987).

- Lasers and other high-power sources of UV and visible radiation are used in the chemical and electronic industries, and in clinical medicine. Such high intensities can cause not only immediate burns, but also risks of stray or scatered light.

Effects

Effects on humans

UV light is sufficiently energetic to rupture chemical bonds or to energise molecules into excited states which can initiate a variety of chemical and biological processes. UV-C is especially lethal to many living organisms since it interacts particularly with proteins and DNA. However, UV-C is not an environmental problem from natural sources, because it is almost completely absorbed by oxygen and ozone (even at reduced levels) in the atmosphere before it reaches the Earth's surface (CEC, 1993c). Artificial sources, on the other hand, deserve attention. The longer wavelengths represented by UV-A are little absorbed by the ozone or the atmosphere, but, compared with UV-C and much of UV-B, they are also relatively harmless to most living things.

UV-B, by contrast, which is partially absorbed by ozone in the atmosphere, is of special concern because it may have damaging effects on the biosphere. Unfortunately, slight changes in the ozone layer, which is especially suited to blocking UV-B wavelengths around 300 nm, may have a significant effect on the amount of UV-B that reaches the surface of the Earth. In fact, a reduction in ozone permits a very specific increase in UV-B radiation between 290 and 315 nm – precisely that waveband where solar irradiance is normally reduced by 10 000 times due to absorption by ozone. Extensive studies on humans have demonstrated that UV-B wavelengths of 290 nm are 1000 to 10 000 times as effective in producing damage to DNA, killing cells, and causing skin erythema and skin cancer than are visible light wavelengths longer than 300 nm (Moan et al, 1989, and McKinley and Diffey, 1987).

Many biological reactions are particularly sensitive to the 290 to 315 nm waveband, which may result from evolutionary adaptation because of the protection provided by the Earth's ozone layer. An instantaneous (in geological terms) change in this protection will not permit time for a similar modification to take place. Even the low levels of UV-B radiation that normally reach the Earth's surface can damage DNA, and cause sunburn, eye cataracts and some skin cancers (IARC, 1992), not to mention suppressing the normal human immune responses (Morrison, 1989). It is possible, therefore, that if ozone depletion or any other mechanism leads to significant increases in UV-B fluxes at the Earth's surface, this could have an important effect on the human immune response to certain allergens or infectious agents, as well as on the development of some cancers. The 'International Programme on Health, Solar UV Radiation and Environmental Change' (ITERSUN), a joint IARC/UNEP/WHO study which involves measurement of UV radiation at the surface, cancer incidence, ocular cataracts and immune suppression, may help resolve some of these uncertainties.

Effects on non-human targets

UV radiation can penetrate sea water to a depth of up to 50 metres, which is sufficient to cause various ecological effects. UV-B damage to phytoplankton in the laboratory has been reported, which goes beyond DNA effects to include impairment of cellular metabolism and motility (Smith, 1989). Increased levels of UV-B may bring irreversible damage or death to zooplankton, which are an essential element of the food-chain for all higher marine life forms. Other marine organisms in the upper levels of the sea include the eggs and larvae of fish, which would certainly be susceptible to damage from raised levels of UV-B (Smith, 1989).

The effects of enhanced levels of UV-B on plants varies between species, as well as between different populations of the same species, indicating the additional influence of reflections from the soil or other local variables (Tevini and Teramura, 1989). Most plants respond to some extent through adaptation, including photo-repair, accumulation of UV-absorbing pigments and delay in growth. Various food crops such as soya beans and cereals are especially sensitive to UV radiation. The causes are not well understood, but have been shown to involve an impairment of the photosynthetic mechanism. In general, increased levels of UV radiation are expected to lead to an increase in the production of less economically valuable plants, and a decrease in the production of the more valuable plants.

Most animals have developed a certain coat or pigmentation which protects them from the harmful effects of UV radiation, but many (cattle and sheep are the most widely studied) seem susceptible to ocular cancers at a rate which is directly proportional to UV exposure (Kopecky, 1978). Any significant increase in UV-B is expected to have significant effects on food-producing animals, and these effects may be assumed to be greater at higher latitudes. A question which now arises, however, is what impact ozone depletion, through increasing levels of UV-B radiation, has already had on the welfare of the biosphere (SCOPE, 1993).

Policies and strategies

Policies for reducing UV exposure have been determined, until now, through strategies for protecting the ozone layer (see Chapter 28). These have focused on two areas: reduction and eventual elimination of emissions of ozone-depleting substances, and 'active adaptation' to an environment of greater exposure to UV radiation. The Montreal Protocol (concluded in September 1987) and the London and Copenhagen amendments (June 1990 and November 1992, respectively) address the first of these. In many cases, controls already in force or being considered within the EU and some other European countries are even stricter than those agreed at Copenhagen (CEC, 1993c). Even if all countries comply with these agreements, however, ozone depletion may exceed any natural variations in the ozone layer that humans have had to cope with until now.

Outlook and conclusions

There is clear evidence from research involving both animals and humans of a direct causal relationship between UV radiation exposure and skin cancers. The rapid increase in European skin cancer cases in recent years is undoubtedly connected to social and recreational behaviour – the key factors – as well as to other less clear causes. Increased public awareness of the hazards of UV exposure is necessary, especially for individuals with sensitive skin types and for young children.

To date, there is little direct evidence in Europe that depletion of the stratospheric ozone layer has led to increases in terrestrial UV-B levels. It is widely suspected, however, that the consequences of ozone depletion may be substantial, including a broad range of effects on human health. The need to counter further depletion of the ozone layer is clear.

Efforts are continuing to ensure that the Montreal Protocol and subsequent agreements are respected. Meanwhile, active adaptation would involve behavioural change to avoid excessive exposure to the sun. Over the medium to long term, active adaptation could imply, for example, that biotechnology might contribute to the development of food crops with greater resistance to UV radiation.

Table 16.6
Worldwide annual average effective dose to adults from natural sources of ionising radiation
Source: UNSCEAR, 1993

Component of exposure	Annual effective dose (mSv)	
	In areas of normal background	In areas of elevated exposures
Cosmic rays	0.38	2.0
Cosmogenic radionuclides	0.01	0.01
Terrestrial radiation: external exposure	0.46	4.3
Terrestrial radiation: internal exposure (excluding radon)	0.23	0.6
Terrestrial radiation: internal exposure from radon and its decay products:		
inhalation of Rn-222	1.2	10
inhalation of Rn-220	0.07	0.1
ingestion of Rn-222	0.005	0.1
Total	2.4	–

Note: Other sources not included in the worldwide averages include: the increase in cosmic radiation exposure during air travel, the exposure of radiological workers, exposure due to the Chernobyl accident and fall-out from nuclear weapons testing. Radiation doses are expressed in sieverts (Sv), which are a measure of the amount of energy absorbed by unit mass of tissue (see Effects and risks, p 372.

Figure 16.5
World average dose from natural radiation sources: total 2.4 mSv/year
Source: UNSCEAR, 1993

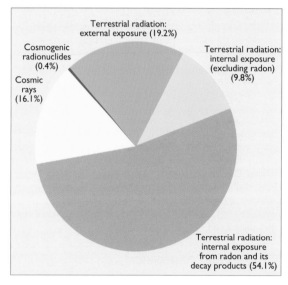

A broad range of measurements, programmes and research is already being carried out in Europe. This interest is motivated in large part by the scientific interest and broad policy implications of ozone layer depletion. However, this research requires increasing coordination, or at least harmonisation, so that definitions, measurements, procedures and data will be comparable between European countries.

IONISING RADIATION

The energy level of some types of radiation is high enough so that when they interact with matter they cause the formation of electrically charged particles or ion-pairs and break molecular bonds. These so-called ionising radiations are by their nature potentially harmful to life; at high doses they can be lethal and at lower doses can cause genetic damage. Ionising radiation occurs naturally, and life on Earth has always been exposed to it. Human activities can, however, enhance exposure, and new sources have been created. The toxicity of radionuclides derives almost exclusively from the effects of the radiation which emanates from them, although

some radionuclides, such as the isotopes of plutonium, are also highly chemically toxic.

Sources

Doses can be received from external radiation (the radiation source is outside the body in air, soil, etc) and from internal radiation (from radionuclides which enter the body by food, by drinking water and through inhalation). Terrestrial radionuclides are almost entirely responsible for internal exposures.

Natural

Natural sources include: primary and secondary cosmic radiation; radon and thoron in the air; and diverse radionuclides in the soil which may enter water and food.

The worldwide average annual effective dose from ionising radiations from natural sources to the average person is calculated to be about 2.4 mSv (UNSCEAR, 1993) (Table 16.6 and Figure 16.5). Corresponding values for the countries of Western Europe show somewhat higher levels, ranging 2 to 7 mSv (Figure 16.6).

Cosmic rays originate both in outer space and from the sun. Those coming from space remain fairly constant, but those from the sun are produced in bursts during solar flares. The intensity and number of cosmic rays which enter the Earth's atmosphere are affected by the earth's magnetic field; more cosmic rays enter near the poles than near the equator, so exposure decreases with decreasing latitude. Exposure to cosmic radiation increases with altitude due to the shielding effect of the atmosphere. Furthermore, some differences in indoor exposure to cosmic radiation arise due to variations in the shielding factor of different building designs. Other naturally occurring radionuclides acting as sources of ionising radiation are those continuously being produced by the interaction of cosmic rays and the atmosphere, eg, tritium (H-3).

All materials in the Earth's crust contain trace amounts of uranium, thorium and other naturally radioactive elements (terrestrial radiation). These nuclides include the long-lived isotopes potassium-40, uranium-238 and thorium-232, and, by radioactive decay, the medium to short-lived radionuclides radium-226 and radon-222, as well as their short-lived decay products. The activity level depends on the type of rock or soil – there is relatively high activity in granites, low activity in some sedimentary rocks and intermediate activity in soils. The gamma rays produced by these materials irradiate the whole body more or less uniformly. Since most building materials are extracted from the earth, they also contain natural radioactivity (particularly potassium-40, uranium-238 and thorium-232). Exposure to terrestrial radiation is therefore experienced both indoors as well as outdoors as a function of geology and the structure and type of the buildings.

Apart from exposure of the lungs to radon, thoron and their daughter products discussed below, the human body is internally exposed from natural radionuclides such as beryllium-7, potassium-40 and members of the uranium-238 decay series (especially lead-210 and polonium-210) and the thorium-232 decay series. These radionuclides enter the body by air, food and water. Exposure due to potassium-40 is essentially constant over a given population, but lead-210 and polonium-210 exposure are more variable with diet. Overall doses from all of these sources together do not vary much between individuals, but of course the cumulative dose increases with age.

Radon-222 is a radioactive gas which emits alpha particles. It is the direct daughter product of radium-226 formed during the radioactive decay of naturally occurring uranium-238. As this occurs in the Earth, radon eventually escapes to the open air, where it is quickly dispersed. When the gas enters a dwelling, however, either through the floor or from the walls, dispersal is much slower and the concentration may become higher than outdoors. As the gaseous radon itself decays (half-life 3.8 days), alpha- and beta-emitting daughter radionuclides are formed. These normally attach themselves to small particles or aerosols in the air and, when inhaled, deposit in the respiratory tract and lungs, irradiating these internal tissues. Concentrations of radon may vary appreciably from one locality to the next (frequently by a factor of ten or more), and they may vary dramatically from one house to another (sometimes by a factor of 100), particularly in relation to the ventilation rate which may have been reduced to conserve energy. Thoron (another name for radon-220) is produced by the decay of thorium-232, but levels and doses are generally much lower than those of radon-222 (about 10 per cent). The relative magnitude of the most important natural sources of radiation exposure in Western European countries is illustrated in Figure 16.6. Comparable data are not currently available for Central and Eastern Europe.

Artificial

Humans have increased their exposure to radiation from the environment through a number of activities. These include industrial activities where raw materials containing naturally occurring radionuclides at low concentrations are processed on a large scale (eg, the production of phosphate fertilisers, the extraction of oil and gas and the burning of fossil fuels), mining, the extraction of water from deep wells, the testing of nuclear weapons, the generation of nuclear power, and the use of radiation or products with high radioactivity in medicine, other industry and research. Ionising radiations can be made artificially either by the generation of appropriate electromagnetic fields, as in X-ray apparatus, or by the creation of artificial radionuclides produced in accelerators or by nuclear fission (atomic weapons or nuclear reactors). The general population is exposed to these ionising radiations either directly (eg, from medical uses) or through their presence in the environment (eg, discharges from the nuclear industry). Table 16.7 compares doses resulting from such artificial human-induced sources to an equivalent period of exposure to natural radiation sources (UNSCEAR, 1993). This perspective indicates that the increased exposure from most of these activities is, on the whole, not a great cause for concern. However, artificial radiation may in fact contribute a significant fraction of the dose over which there is some control. Particular attention is best devoted therefore to such sources of exposure, some of the most important of which are discussed below.

Medical exposure

With respect to the various artificial sources of exposure to ionising radiation, medical exposure is by far the greatest for the average individual and probably represents the area in which reductions in exposure would be most cost-effective. In any given year, probably a third of the Western European population is exposed to X-rays, with a dose up to 42 mSv/year (Vaas et al, 1991). In 1986, close to 50 per cent of the Russian population received X-rays for medical reasons (Nikitin et al, 1991); the increased exposure compared with Western Europe was caused largely by technical problems.

Nuclear medicine (diagnostic examination with radio-pharmaceuticals) contributes additional exposure of an estimated 1 to 2 per cent of the Western European population, with doses up to some 100 mSv/year (Vaas et al, 1991). Comparable figures are not available for Central and Eastern Europe.

Radiation therapy gives rise to very high (though usually localised) exposure of patients, but in this case such high

Source	Basis	Equivalent period of exposure to natural sources
Medical exposures	One year of practice at the current rate	90 days
Nuclear weapons tests	Completed practice	2.3 years
Nuclear power	Total practice to date	10 days
	One year of practice at the current rate*	1 day
Severe accidents	Events to date	20 days
Occupational exposures	One year of practice at the current rate	8 hours

*Note: This is not the dose received in one year but the cumulative effects of discharges taking place in one year.

Figure 16.7
Exposures to human-induced sources expressed as equivalent periods of exposure to natural sources
Source: UNSCEAR, 1993

levels of exposure are necessary to attain the objective of destroying cancerous cells.

Radioactive emissions and waste disposal

Most radioactive wastes are, or are planned to be, disposed of by concentration and confinement in surface, near surface or deep underground licensed facilities (see Chapter 15). Disposal by discharge to and subsequent dilution and dispersion in the environment is severely regulated at national and international level; its use is limited to some low-active liquid or gaseous effluents resulting from the operation of nuclear facilities (electronuclear, medical, etc). The authorised and controlled releases and discharges to water (lakes, rivers, and seas) and atmosphere from such facilities contributes directly to the dispersion of the radionuclides to the environment. In Western Europe, the spent fuel reprocessing plants (Sellafield, UK, and La Hague, France, and, to a lesser degree, Marcoule, also in France) are the most significant points of controlled release of liquid and gaseous effluents, the contribution from electronuclear power plants being individually much less. As a result of improved working

Figure 16.6
Average doses from natural radiation sources
Source: Based on Green et al, 1993

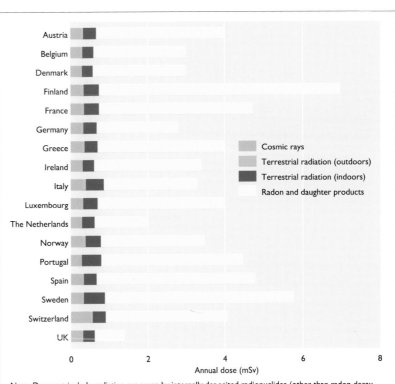

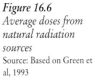

Cosmic rays
Terrestrial radiation (outdoors)
Terrestrial radiation (indoors)
Radon and daughter products

Note: Does not include radiation exposure by internally deposited radionuclides (other than radon decay products), which would typically add 0.2 to 0.3 mSv per year.

practices, the level of radiologically significant discharges to the Irish Sea from Sellafield is now about 1 per cent of peak levels in the 1970s, when they were by far the most important from any single site in Western Europe (CEC, in press, see also Chapter 18).

The practice of dumping radioactive waste at sea has been regulated since 1975 by the London Convention (more correctly referred to as the International Convention on the Prevention of Marine Pollution) by dumping of waste and other matter. Up to 1983 dumping of low-level radioactive waste packages complying with specific criteria has been authorised and practised by some Western Europe countries in the northeast Atlantic under the aegis of the Nuclear Energy Agency of the OECD. No environmental contamination has been detected up to now. This practice was suspended in 1983, following a voluntary moratorium. Since 1994 the prohibition of all sea dumping has been accepted by all signatories to the convention other than the Russian Federation, which has not as yet been able to implement alternative approaches. Potential unquantified threats of environmental contamination are found in the Barents and Kara seas around the Novaya Zemlya archipelago. They result from the dumping of radioactive objects, including highly radioactive material, by the former USSR (Yablokov et al, 1993). For further information see Chapters 6 and 18.

Also in the former USSR, emissions into the Techa river (near Chelyabinsk and Sverdlovsk in the southern Ural mountains) in the 1940s and 1950s and associated operations of the Mayak nuclear facility (also the site of the Kyshtym radiation accident – see below) has resulted in severe contamination of the area. Furthermore, a small lake (Lake Karachay) continues to be used to store high-activity waste material and is probably the worst site of radioactive waste accumulation on earth (Burkart and Kellerer, in press).

Radiation accidents

Uncontrolled release of radionuclides to the environment can occur as a result of accidents during the handling, transport, use and disposal of radioactive materials and during the operation of nuclear facilities. From an environmental point of view a distinction should be made between abnormal nuclear events having consequences limited to a nuclear plant or site, and abnormal nuclear events causing off-site radioactive pollution. According to the International Nuclear Events Scale (INES), first introduced on a trial basis by the IAEA in 1989 and now formally adopted with some revisions (IAEA, 1992), only these latter should be called accidents, the former being anomalies or incidents. Accidents have been few, the most prominent being the Windscale reactor fire accident (UK, 1957), and, considerably worse, the severe accidents at the Kyshtym military reprocessing plant (USSR, 1957) and the Chernobyl power plant (Ukraine, 1986). The Chernobyl accident had the most widespread effect in Europe for increasing exposure to ionising radiations (see Box 18E). No accident has been reported since 1987 in Western Europe. Even relatively minor abnormal nuclear events have to be reported to the national safety authorities according to the provisions of national law. A number of international reporting systems are also in use on a voluntary basis, such as that directly associated with INES under the aegis of the IAEA.

Further information on accidents at nuclear installations can be found in Chapter 18, where the origin and impact of radiation accidents are examined. Radiation accidents are treated as a problem of environmental concern for Europe in Chapter 30.

Weapons testing

Atmospheric testing of nuclear weapons has been a major source of artificial radioactivity in the environment. The extensive testing in the 1950s and 1960s released enormous quantities of radionuclides into the upper atmosphere which have now dispersed throughout the whole environment. In terms of radiation exposure, the most significant radionuclides from this source are strontium-90 and caesium-137 with half-lives of approximately 30 years. Since the signing of the Test Ban Treaty in 1963, atmospheric tests have almost been eliminated (the last atmospheric test was conducted by China on 16 October 1980). However, fall-out continues to provide a certain component of routine radiation exposure, although annual doses have become very small compared with the variations in natural background radiation.

Effects and risks

Whether natural or artificial, the effects of ionising radiation on matter are the same – transferring energy which can break and rearrange the bonds holding molecules together. In biological materials, such chemical changes can have various types of consequences: effects can be deterministic (with a threshold dose above which the effect can be anticipated) or stochastic (with a statistical probability of occurrence decreasing with the dose but with no lower limit). Radiation sickness or organ and skin damage are deterministic effects. They occur only at high radiation doses at dose rates above a threshold, and the severity is related to the dose. Cancer and hereditary disorders are stochastic effects. Their probability, but not their severity, is assumed to be proportional to the dose even at very low doses (see below; ICRP, 1991).

Different types of radiation vary in their ability to penetrate matter and thus in the efficiency with which they transfer energy and induce biological effects. The conversion of absorbed doses to a common 'effective dose' permits different types of radiation exposure to be aggregated on a common basis. This effective dose is expressed in sievert (Sv) or sub-units thereof.

Because of its high rate of energy transfer, alpha radiation is one of the most damaging types of radiation. Internal exposure, primarily through the inhalation or ingestion of alpha emitting radionuclides, is of most concern for alpha radiation. Gamma radiation looses its energy over greater distances, causing less biological damage but being more penetrating.

In 1977, the United Nations Scientific Committee on the Effects of Atomic Radiation (UNSCEAR) assessed the risk of a fatal cancer from ionising radiation at 2.5 per cent per dose of 1 Sv (UNSCEAR, 1977). Since then, with the availability of additional data from Japan and improved methods of risk evaluation, the estimated risk has been increased to 5 per cent per Sv for the general population, ie, a cumulative dose of 1Sv gives a one in twenty chance of a fatal cancer occurring (ICRP, 1991). To apply this coefficient to the ordinary circumstances of exposure in order to develop a risk factor for protection purposes, allowing for the overall potential detriment to health, the risks of non-fatal cancers and hereditary defects suitably weighted for lethality and severity are also taken into account. This results in an all-inclusive risk factor of about 7.3 per cent per Sv. Applied to the worldwide (somewhat higher in Europe) average annual dose rate of 2.4 mSv per year, this implies a theoretical lifetime risk to the average individual of premature death from natural radiation-induced malignancy of about 1 per cent. In order to put this risk in perspective, it can be compared to an overall probability of death by cancer. For the population of the USA it has been estimated that 30 per cent of fatal cancers are induced primarily by smoking, 35 per cent by substances ingested in food, and 3 per cent by alcohol consumption (Doll and Peto, 1981). These effects have meant that even large-scale epidemiological studies have not been able to identify an influence on the incidence of cancer attributable to natural radiation.

Policies and strategies

The main areas of action have been for occupational safety (particularly in the medical and nuclear industries) and exposure to the general public. The non-occupational standard of 1 mSv/year applies to populations which may be exposed to artificial ionising radiation. ICRP (1991) recommends lifetime dose for the public from planned activities should not exceed 70 mSv, with a maximum of 5 mSv in any one year. The dose averaged over 5 years should be less than 1 mSv/year. This does not apply to natural background or medical exposure.

Natural exposure

Little can be done to prevent exposure to terrestrial and cosmic radiation. With regard to radon exposure, however, some authorities, including the European Commission, have recommended that high exposure to radon in the home should be avoided. There are also legal requirements in some European countries to limit exposure to radon in the workplace (Green et al, 1993). Public understanding of radon risk must be improved, and home owners should be encouraged to undertake the necessary measures and remedies.

Medical exposure

Radiation exposure of patients is assumed to provide direct benefits to these patients which more than offset the associated risks, and hence it would be counter-productive to impose general limits, especially in radiation therapy applications. In diagnostic applications, however, there is certainly scope for using more efficient X-ray techniques giving a reduction in exposure without affecting the benefits.

Outlook and conclusions

Exposure to radiation in general, and especially the role of the nuclear industry, attracts a great deal of consideration in politics and society as a whole, disproportionate to scientific claims or evidence. Therefore, there is a significant challenge to weigh these considerations properly together with the valid scientific and socio-economic evidence.

Little can be done about the risks of exposure from natural sources of ionising radiation; indeed, humans have always had to cope with such risks. Therefore, much of the potential future problem lies in the proliferation of different kinds of artificial radiation sources, and in particular from medical applications and the nuclear industry.

In medical sciences in particular, changes in tissue weighting factors proposed by ICRP (Clarke, 1991) may make for important changes in exposure assessments, due to the proposed increase in specific values for abdominal organs. This would render average medical exposures nearly as significant as natural exposures for humans, and create even more pressure for appropriate monitoring and control of these exposures. New recommendations have also recently been published for radon (ICRP, 1993).

Finally, estimates of average exposure of humans cannot reflect accurately the wide-ranging exposure of individuals. A margin of safety must be incorporated to ensure that particular groups, such as miners, airline staff, and those living in high-radon areas, are not exposed beyond the recommended control levels (see also WHO, in press).

REFERENCES

Burkart, W and Kellerer, A M, (Eds) (in press) A first assessment of radiation doses and health effects in workers and communities exposed since 1948 in the southern Urals. *Science of the Total Environment.*

CEC (1993a) *Towards sustainability, a European Community programme of policy and action in relation to the environment and sustainable development.* OJEC 17 May 1993, No C 138, Commission of the European Communities, Luxembourg.

CEC (1993b) *Proposal for a Council Directive on the minimum health and safety requirements regarding the exposure of workers to the risks arising from physical agents.* DG V, CEC, COM(92) 560 final – SYN 449, draft 8 February 1993, Commission of the European Communities, Luxembourg.

CEC (1993c) *Environmental UV radiation: causes–effects–consequences,* Acevedo, J and Nolan, C (Eds), DG XII, Commission of the European Communities, Brussels.

CEC (in press) *Radioactive effluents from nuclear power stations and nuclear fuel reprocessing plants, discharge data (1977–86).* Commission of the European Communities, Luxembourg.

Clarke, R H (1991) The causes and consequences of human exposure to ionising radiation. *Radiation Protection Dosimetry* 36 (2–4) [Statistics of Human Exposure to Ionising Radiation], Sinnaeve, J and O'Riordan, M C (Eds), Proceedings of a workshop held in Oxford, 2–4 April 1990. Nuclear Technology Publishing, Ashford, Kent, UK.

Doll, R and Peto, R (1981) The causes of cancer: quantitative estimates of avoidable risks of cancer in the United States today. *Journal of the National Cancer Institute* 66(6), 1193–308.

Feychting, M and Ahlbom, A (1993) Magnetic fields and cancer in children residing near Swedish high-voltage power lines. *American Journal of Epidemiology* 138, 467–81.

Gandhi, O (1990) *Biological effects and medical applications of electromagnetic energy.* Prentice Hall.

Green, B M R, Hughes, J S and Lomas, P R (1993) *Radiation atlas: natural sources of ionizing radiation in Europe.* UK National Radiological Protection Board, Commission of the European Communities, DG XI, Brussels (EUR 14470).

IAEA (1992) *INES: The international nuclear event scale user's manual.* International Atomic Energy Agency/Nuclear Energy Agency, IAEA, Vienna.

IARC (1992) Solar and ultraviolet radiation. *IARC Monographs on the evaluation of carcinogenic risks to humans,* 55. International Agency for Research on Cancer (IARC), WHO, Geneva.

ICRP (1991) Recommendations of the International Commission on Radiological Protection (ICRP). Radiation Protection, *ICRP Publication* 60. International Commission on Radiological Protection, Pergamon Press, Oxford.

ICRP (1993) Protection against radon-222 at home and at work. Radiation Protection, *ICRP Publication* 65. International Commission on Radiological Protection, Pergamon Press, Oxford.

Kerr, J B and McElroy, C T (1993) Evidence for large upward trends of ultraviolet-B radiation linked to ozone depletion, *Science* 262, 1032–35, American Association for the Advancement of Science, Washington DC.

Kihlman, T (1993) Sweden's action plan against noise, *Noise/News International,* December 1993, 194–208. Institute of Noise Control Engineering of the USA Inc, Poughkeepsie, NY.

Kopecky, K E (1978) Ozone depletion: implications for the veterinarian. *Journal of the American Medical Association* 173, 729–33. Chicago.

Maddock, B J (1992) Guidelines and standards for exposure to electric and magnetic fields at power frequencies. Paper prepared for the *International Commission on Non-Ionising Radiation Protection (ICNIRP),* Panel Session on Electric and Magnetic Fields, 31 August 1992. National Grid Technology and Science Laboratories, Leatherhead, Surrey, UK.

Mariutti, G and Matzue, M (1987), Measurement of ultraviolet radiation in welding processes and hazard evaluation. In: Passchier, W F and Bosnjakovic, B F M (Eds), *Human exposure to ultraviolet radiation: risks and regulations,* pp 387–90. Elsevier, Amsterdam.

McKinley, A F and Diffey, B J (1987) A reference action spectrum for ultraviolet-induced erythema in human skin. In: Passchier, W F and Bosnjakovic, B F M (Eds), *Human exposure to ultraviolet radiation: risks and regulations,* pp 83–6. Elsevier, Amsterdam.

Miedema, H M E (1993) Response functions for environmental noise in residential areas. Nederlands Instituut voor Praeventieve Gezondheidszorg, Applied Scientific Research (TNO), NIPG-publikatienummer 92.021, June 1993, Leiden, The Netherlands.

Moan, J et al (1989) Ozone depletion and its consequences for the influence of carcinogenic sunlight. *Cancer Research* **49**, 4247–50. Cancer Research Institute, Baltimore.

Morrison, W L (1989) Effects of ultraviolet radiation on the immune system in humans. *Journal of Photochemistry and Photobiology* **50**. 515–24.

Moseley, H (1988) Non-ionizing radiation: microwaves, ultraviolet and laser radiation. *Medical Physics Handbook 18*. Adam Hilger, Bristol, UK.

Nikitin, V V, Yakubovskij-Lipskij, Y O, Kalnitskij, S A and Bazjukin, A B (1991) Statistics of medical exposure in the USSR. In: *Radiation protection dosimetr*, Vol **36** (2–4) [Statistics of human exposure to ionising radiation], Sinnaeve, J and O'Riordan, M C (Eds), Proceedings of a workshop held in Oxford, 2–4 April 1990, Nuclear Technology Publishing, Ashford, Kent, UK.

OECD (1991a) Fighting noise in the 1990s. Organisation for Economic Cooperation and Development, Paris.

OECD (1991b) The state of the environment. Organisation for Economic Cooperation and Development, Paris.

OECD (1993) Environmental data compendium. Organisation for Economic Cooperation and Development, Paris.

SCOPE (1993) Open letter from J W B Stewart, President, and E C De Fabo, Chairman, *Scientific Committee on Problems of the Environment (SCOPE)*, 1 October 1993. Paris.

Smith, R C (1989) Ozone, middle ultraviolet radiation and the aquatic environment. *Journal of Photochemistry and Photobiology* **50**, 459–68.

Tevini, M and Teramura, A H (1989) UV-B effects on terrestrial plants. *Journal of Photochemistry abd Photobiology* **50**, 479–87.

UK DoE (1992) *The UK Environment*. Publication of the Government Statistical Service, UK Department of the Environment. HMSO, London.

UK NRPB (1992) Electromagnetic fields and the risk of cancer. *Report of an Advisory Group on Non-Ionising Radiation*, Doc. NRPB 3(1). National Radiological Protection Board (NRPB), London.

Umweltbundesamt (1987) Rat von Sachverständigen für Umweltfragen: Umweltgutachten 1987, Auswirkungen von Lärm auf Tiere.

Umweltbundesamt (1989) *Lärmbekämpfung '88 Tendenzen – Probleme-Lösungen*. Erich Schmidt Verlag, Berlin.

Umweltbundesamt (1992a) *Daten zur Umwelt*. Erich Schmidt Verlag, Berlin.

Umweltbundesamt (1992b) *Jahresbericht 1991*, Umweltbundesamt, Berlin.

UNSCEAR (1977) *Sources and effects of ionising radiation*. United Nations Scientific Committee on the Effects of Atomic Radiation (UNSCEAR), United Nations, New York.

UNSCEAR (1993) *Sources and effects of ionising radiation*. United Nations Scientific Committee on the Effects of Atomic Radiation (UNSCEAR), United Nations, New York.

Vaas, L H, Blabber, R O and Leenhouts, H P (1991) Radiation sources, doses and dose distributions in the Netherlands. In: *Radiation Protection Dosimetry* Vol **36** (2–4) [Statistics of Human Exposure to Ionising Radiation], Sinnaeve, J and O'Riordan, M C (Eds), Proceedings of a workshop held in Oxford, 2–4 April 1990. Nuclear Technology Publishing, Ashford, Kent, UK.

Van den Berg, M (1992) Het Voorspellen van Hinder Door Geluid van Wegverkeer. *Geluid* **4** (December), pp 147–51. Samson H D, Tjeenk Willink bv, Alphen aan den Rijn, The Netherlands.

Verein Deutscher Ingenieure (1988) Wirkungen von Verkehrsgeräuschen. Verein Deutscher Ingenieure (VDI)-Richtlinie 3722, Blatt 1, August. Beuth Verlag, Berlin.

Verkasalo, P K, Pukkala, E, Hongisto, M Y, Valjus, J E, Järvinen, P J, Heikkilä, K V and Koskenvuo, M (1993) Risk of cancer in Finnish children living close to power lines. *British Medical Journal* **307**, 9 October 1993, 895–9. British Medical Association, London.

von Meier, A (1992) Thin porous surface layers – design principles and results obtained. *Paper presented at Euro-Symposium The Mitigation of Traffic Noise in Urban Areas*, 12–15 May 1992, Laboratoire Central des Ponts et Chaussées, (LCPC)-Centre de Nantes, Bouguenais, France.

von Meier, A (1993) *Noise pollution*. Report prepared for the EEA-TF, DG XI, European Commission. M+P Raadgevende Ingenieurs bv, Aalsmeer, The Netherlands.

WHO (1993a) Community noise. *Environmental Health Criteria Document, External Review Draft*, 28 June 1993, World Health Organisation, Geneva.

WHO (1993b) Electromagnetic fields (300 Hz to 300 Ghz). *Environmental Health Criteria* **137**. World Health Organisation, Geneva.

WHO (in press) *Concern for Europe's tomorrow (CET)*. World Health Organisation, Copenhagen.

Yablokov, A V, Karasev, V K, Rumyantsev, V M, Kokeyev, M Ye, Petrov, O I, Lysysov, V N, Yemelyanenkov, A F and Rubtsov, P M (1993) *Facts and problems related to radioactive waste disposal in seas adjacent to the territory of the Russian Federation*. Office of the President of the Russian Federation, Moscow. Translated by Gallager, P and Bloomstein, E, Small World Publishers, Inc, Albuquerque, USA.

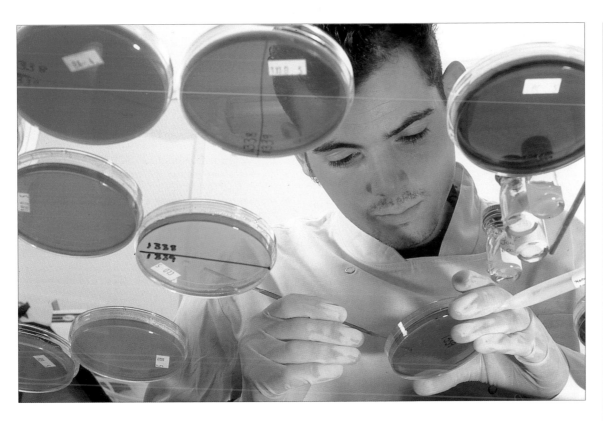

*Lab technician
preparing cultures*
Source: Science Photo
Library

17 Chemicals and genetically modified organisms

INTRODUCTION

This chapter considers the pressures on the environment and potential risks arising from the use and marketing of chemical substances and the development, production, field testing, and release through placing on the market of living, genetically modified organisms. Such substances or organisms may not necessarily be released directly into the environment, but may eventually end up there either during their production and use, or at the end of the product's useful lifetime.

CHEMICALS AND CHEMICAL PRODUCTS

Natural processes are continually transforming chemicals and redistributing them in the environment. Background ('natural') concentrations can be described by media or location, but there are large variations. Many natural chemical compounds are, at specific concentrations, necessary for sustaining life, and most have existed for billions of years. Human activities are increasingly changing the way in which chemicals are distributed in the environment and, with the introduction of new compounds (xenobiotics), are causing new pressures on the environment, both actual and potential.

Technology has provided new materials, methods and processes which have given rise to a whole range of novel industries making use of the ever-increasing number of new compounds in the manufacture of new products. In the service of this need the production of chemicals has increased enormously over the past few decades. It is estimated that more than 13 million synthetic chemical compounds have been produced in laboratories. These either reproduce the structure of compounds found in nature or are made of completely new structures. They can be organic or inorganic, chemically active or inert, and are used in different sectors of the economy. According to EINECS (the European Inventory of Existing (commercial) Chemical Substances), about 100 000 chemicals are marketed in the EU (Box 17A) and it is commonly understood that the chemicals industry markets 200 to 300 new chemicals each year.

The speed and complexity of the development and use of chemical products over the past few decades has not allowed knowledge of environmental effects to keep pace with development. This has already led to a number of unexpected environmental impacts. The state of knowledge of the toxicity and ecotoxicity of chemicals in use and circulation, particularly about their potential impacts on humans and the environment, is in fact very unsatisfactory (see Box 17A). This state of affairs is one of the main reasons why, so often,

Box 17A Current knowledge on the toxicity of existing chemicals

The European Inventory of Existing (commercial) Chemical Substances (EINECS) lists more than 100 000 individual substances. Around 1000 of these are used for 95 per cent of the total production of chemicals in Europe.

The EC Existing Chemicals Directive (93/793/EEC) is expected to lead to a significant improvement in basic information about chemicals over a three-stage long-term programme.

In 1984 the US National Research Council (USNRC) published statistics on the availability of toxicity data for a selected choice of around 50 000 chemicals which were used in the USA at that time (USNRC, 1984). Particularly noteworthy from these statistics are the large number of chemicals involved and the general lack of toxicity data for them – both the more basic, high-volume chemicals, as well as those produced in smaller quantities. A separate more recent study (USEPA, 1991) shows that the main substances released to water are industrial chemicals (about 60 per cent of the total in 1989). However, the USNRC statistics show that little toxicological information is available for these chemicals, such information being confined mainly to pesticides, cosmetics, drugs and food additives. The general lack of toxicological information for chemicals should be seen against the fact that almost 80 per cent of hazardous waste comes from the chemical industries (UNEP, 1992a) (see also Chapter 15).

A thorough and systematic risk assessment and risk management of chemicals can be performed only with complete information of all existing, commercially available chemicals. The size of this task is such that international programmes have been initiated, which collect data and coordinate the results. These programmes are the main sources of information concerning chemicals. They include:

- the EC Existing Chemicals Directive and notification scheme;
- the chemical programme of the OECD (voluntary);
- the International Register of Potentially Toxic Chemicals (IRPTC from UNEP);
- the International Programme of Chemical Safety (IPCS from WHO, UNEP and ILO);
- the ECETOC programme (European Centre for Ecotoxicology and Toxicology) sponsored by the chemical industries in Europe;
- ECDIN (Environmental Chemicals Data and Information Network) of the European Commission's Joint Research Centre;
- IARC (International Agency for Research in Cancer); and
- those of the FAO (Food and Agriculture Organisation of the UN) for evaluating chemicals in food.

Currently IRPTC has a data profile for about 800 chemicals. The profiles include 17 different categories of general and dangerous qualities of each chemical. The legal file records data from 12 different countries and 6 international organisations, and in 1989 it contained 42 000 records on more than 8000 chemicals (Lönngren, 1992).

unexpected, undesirable effects on human health and the environment occur. Rather than just reacting to problems as they arise, more strategic, pro-active approaches are being developed with the aim to control chemical risk (see Chapter 38). The backlog of work to be done testing the numerous existing marketed chemicals, about which little is known concerning their health and environmental impacts, is such that a great deal of time and resources will be necessary to complete the work. The EC Existing Chemicals Directive (93/793/EEC) and notification scheme designed to address this problem and improve basic knowledge about chemicals is already being implemented.

The control of chemical risk, including the goals and strategies being taken to tackle this problem, is considered in Chapter 38.

Impacts of chemicals

Chemical substances vary greatly in their potential to cause undesirable impacts on the environment. Over the last three decades, many chemical substances have been shown to induce severe impacts on natural processes, and organisms including humans. In particular, the release of highly toxic chemicals causes serious problems. A series of events, in the 1950s and 1960s, provided clear evidence that some manufactured chemicals could be a threat to the environment and to human health (Chapter 11) and many questions arose concerning the properties of new chemicals and the risks they pose.

The main sources of risks from chemicals arise from the regular accumulation of contaminants in environmental compartments, leading to long-term exposure, and from accidents. Worldwide, over the period 1977–87, accidents involving chemicals caused about 5000 deaths (not including poisoning by pesticides), 100 000 injuries and evacuation of 620 000 persons. The most serious accident remains that at Bhopal in India in 1984, causing about half of the total number of casualties reported above. Accidents as a cause of environmental pressure are discussed fully in Chapter 18.

Toxic chemicals can be found in air, water, soil and biota (see Chapters 4 to 11), and, as a consequence, they can also be present in food for human consumption. Thus, humans and other living species can be exposed to such compounds through the air they breathe, the food they eat, the water they drink, or by direct contact.

Chemicals display many different properties. They can be highly reactive, flammable, explosive and corrosive. Toxic chemicals that reach sensitive tissues in the human body can cause discomfort and injuries, loss of function and change of structure, and can lead to disease and mortality. Health impacts can include acute poisoning, carcinogenicity and genetic effects (mutations) in humans and other organisms (see Chapter 11). Chemicals can also affect reproduction and the nervous and immune systems, and can cause sensitisation. Other less serious but still highly bothersome effects are also possible, such as irritability and allergic reactions. Impacts can be restricted to isolated individuals, or can affect a whole population in a large area or even future generations.

Toxic chemicals may also have ecological impacts by affecting the biochemical reactions of organisms, altering the balance of processes and their equilibria, as well as having direct effects on fauna and flora. Many unforeseen impacts of chemicals on wildlife have been reported. Chemicals may be lethal to some species, reduce the survival ability of others, or accumulate in the food-chain until toxic levels are reached. The determination of causes and effects in wildlife is usually even more complex than examples in the human health sector.

Certain chemicals can also cause global physico-chemical effects, eg: CFCs and the depletion of the stratospheric ozone layer; NO_x in their role of increasing tropospheric ozone; and the contribution of increased levels of carbon dioxide and methane in the atmosphere to the greenhouse effect (see Chapters 27, 28 and 32).

The extent to which a chemical is dangerous is dependent on its physico-chemical properties, its toxicity and ecotoxicity, the use to which it is put, the manner in which it is used and, most importantly, the quantities employed.

Unexpected impacts of chemicals usually arise from lack of knowledge related, for example, to the toxicity or degradability of the substances, or to the presence of contaminants. Furthermore, human behaviour can affect chemical concentrations in different compartments of the environment, greatly influencing their environmental impacts (see for example Chapters 22 and 26).

Despite the increasing number of chemical compounds being produced, exposure of the general population and non-human targets is restricted mainly to a low number of so-called high-volume chemicals. Dangerous practices and problematic uses of chemicals can nevertheless lead to exposures to compounds used in lesser amounts (see also Chapter 18). A greater awareness in the use and effects of chemicals is clearly required. In tandem, this needs the assessment of total intake or exposure, based on biological modelling or on environmental monitoring, a knowledge of dose–effect relationships, and innovative approaches to the control of risks associated with new products being put on the market. In order to produce a hazard assessment for a potentially harmful chemical, information is needed about its emission patterns, physico-chemical characteristics, distribution in environmental compartments, bioaccumulation, effects on living organisms, persistence and mobility (Vighi and Calamari, 1992). Predictive instruments such as the so-called QSAR models (Quantitative Structure-Activity Relationships), have been developed for the evaluation of chemicals, and of their environmental distribution and fate (evaluative models) (Karcher and Devillers, 1991). Although these techniques are promising, the complete set of data needed to apply the models is only rarely available for the substance of interest although the physico-chemical properties of most substances are well known. Once obtained this information can be used to calculate the environmental distribution of the substance and to predict its concentration in the different non-living compartments: air, water and soil. Effects on living organisms, bioaccumulation and environmental persistence are much more difficult to deal with due mainly to lack of experimental evidence.

The sources of chemicals

The chemicals industry is well developed throughout Europe. In the EU and EFTA total production was equivalent to about 250 million tonnes for all chemicals, and 15 per cent of the total was exported outside Western Europe. Here, the chemicals industry grew by an average of 2.2 per cent in 1992 and sales totalled ECU 332.9 billion in 1991. In the EU chemicals represent 20 per cent of all exports in financial terms (CEFIC, 1993). In countries of the former USSR the heavy chemicals sector, comprising organic and inorganic chemicals, fertilisers and rubber, is relatively well established. In 1991 production declined by nearly 8 per cent, due, among other reasons, to reduced access to energy supplies from the Russian Federation, the loss of the former CMEA (Council for Mutual Economic Aid) partners, and local military conflicts (CEFIC, 1993). In Central and Eastern Europe, declines in the production of chemicals were recorded in almost all of the former CMEA countries in 1992. Only in Poland did production rise (by 3 per cent) (CEFIC, 1993). (See Chapter 20 for further relevant details of the chemicals industry in Europe.)

In EU and EFTA countries, the main volumes of chemicals produced are basic chemicals, pesticides and pharmaceuticals (see Table 17.1). In the EU, the markets where most chemicals go are consumer goods, service (hospital, research etc) and agriculture (see Table 17.2).

The release of chemical substances to the environment is an everyday occurrence. How chemicals are employed in agriculture, in industry or in households greatly affects what enters the environment. Cultural influences, social behaviour and economic reasons strongly affect the extent

Group of chemicals	Per cent
Basic organics	19.7
Pharmaceuticals	19.7
Plastics	11.0
Inorganics	7.2
Perfumes and cosmetics	6.4
Paints	4.7
Detergents	3.7
Fibres	3.7
Dyes	1.6
Others	22.3

Table 17.1
Chemicals produced in the EU and EFTA countries (by volume)
Source: CEFIC, 1991

Industry	Per cent
Consumer goods	27.4
Service (hospitals, research, etc)	19.2
Agriculture	10.3
Textile	6.6
Metal industry	6.5
Construction	5.7
Paper	3.9
Electronics	3.8
Automotive industries	3.6
Food industry	3.1
Mechanical engineering	2.4
Others	7.5

Table 17.2
End-user markets of chemicals in EU and EFTA countries
Source: CEFIC, 1991

and manner of use, particularly those compounds in the hands of the normal consumer. Releases can occur at household level as well as on the larger industrial scale. They can come about as the result of routine actions (the general use and discarding of products or the emission of gaseous and liquid effluents), by the more formal disposal of waste, or as a consequence of accidents (see Chapters 14, 15 and 18). In heavily populated areas, chemicals are often released into waterbodies in exceptionally large volumes through industrial discharge pipes and municipal sewage.

Selected chemicals of concern

A comprehensive discussion of the different types of chemical substances which adversely affect the environment and human health is beyond the scope of this report. A number of chemical substances have attracted a great deal of attention over the past few decades and for these environmental monitoring is carried out.

Some of the chemicals given the most treatment, by, among others, WHO and OECD, are discussed below. These include in particular: eutrophying compounds (phosphorus from washing powders, and nitrates), heavy metals and metalloids, polychlorinated biphenyls (PCBs), dioxins, benzene, polyaromatic hydrocarbons, polyvinylchloride (PVC), chlorofluorocarbons (CFCs) and asbestos. According to WHO, chemicals which are creating most concern for human health are arsenic, mercury, cadmium, lead, nickel, nitrate, benzene, polycyclic hydrocarbons and PCBs. Many other compounds, however, are of potential concern. For example, recent effort is being put into assessing (mainly in conditions of occupational exposure) the carcinogenicity of styrene, the possible infertility effects of glycol ethers, chronic neurotoxicity effects of solvents, respiratory allergens, the effect of polypropylenes, perchlorethylene, polyalkylene glycols, acrylates and metacrylates (ECETOC, 1992). Recently the occurrence of pesticides and their residues in groundwater has caused great concern in several countries. Unless otherwise stated the following discussion is based on information from: Axenfeld et al (1992), CEC (1992), CEFIC (1993), UNEP (1992a, b) and WHO (1990).

Eutrophying substances (nitrogen and phosphorus)

The survival of all living organisms is dependent on a certain level of essential chemical elements in their surroundings. Exceedance of this level can lead to a series of undesirable environmental effects called eutrophication. The process of eutrophication is caused by nutrient imbalances which can disturb the natural biochemical balance of ecosystems. In particular, eutrophication restricts the intentional uses of waterbodies over large areas (see Chapters 5 and 6). The nutrients of particular concern for the environment are nitrogen (from ammonia and nitrate) and phosphorus. These originate from a wide range of uses of which agriculture dominates nitrogen emissions, and the major parts of phosphorus emissions come from domestic (eg, washing powders, see below) and industrial sources (see Chapter 14). The widespread and serious groundwater contamination with nitrates and its health aspects is an area of particular concern (see Chapter 11).

Heavy metals and metalloids

Heavy metals and metalloids of particular environmental concern are lead, mercury, cadmium, arsenic and nickel. They have a wide field of use and are often found as contaminants in other high-volume chemicals and in emissions. The problems related to copper and zinc contamination of inland waters are discussed in Chapter 5.

Lead is a potentially toxic substance with no known biological function. The world production from mining in 1989 was 2.7 million tonnes and from secondary production (recycling) 2.1 million tonnes. Consumption in the EU in 1992 was about 1.5 million tonnes out of a world total of about 5 million tonnes. Lead is used as an additive in petrol and paints, and can be found in some children's toys. Alkyl-lead used as a petrol additive is highly toxic. Because of the introduction of lead-free petrol, lead consumption is decreasing. There is also concern over lead in drinking water. Exposure leads to increased concentrations in blood. One of the most worrying effects of lead is the decline of cognitive ability in children, which occurs at very low blood concentrations.

Mercury has no known biological function. The world production in 1986 was 7000 tonnes. The consumption in the EU in 1986 was about 3100 tonnes and the recovery rate 1500 tonnes. Mercury is used primarily in chloralkali plants, electrical and paint industries and fungicides. Mercury is also present in dental amalgams and in measuring instruments, and is used as a catalyst. It is also emitted during the combustion

of fossil fuels, especially coal. Its most toxic derivative is methylmercury, which affects the nervous system and can enter the food-chain. The worst disaster caused by this derivative occurred in Japan in the late 1950s (the Minamata disease), causing the death of more than 200 people and serious illness to many more.

Cadmium is used mostly in electroplating and as a stabiliser in plastics. It is also employed in nickel-cadmium batteries. Cadmium is a natural contaminant in phosphate fertilisers. Because cadmium is strongly adsorbed to soil particles, long-term fertiliser application may lead to enrichment in topsoil. In 1992, EU consumption was about 5400 tonnes out of a world total of about 19 000 tonnes. Cadmium is a category 1 carcinogen (IARC), and a correlation has been noted between cadmium concentration in the human body (especially in the kidneys) and renal damage (WHO, 1990). This element also has undesirable effects on the microbial flora and invertebrates of aquatic ecosystems and soils (see Chapters 5 and 7).

Arsenic is used mainly in pesticides, plant desiccants and wood preservatives. Arsenic emissions come from metal smelters and through the combustion of coal. World production in 1986 was about 5600 tonnes. Inorganic arsenic is an established human carcinogen. Ecotoxicologically, arsenic is a phytotoxic agent, which is especially accumulated in aquatic organisms. A current concern is its content in drinking water as a result of groundwater contamination.

Nickel is used for alloys, man-made fibres, glass, ceramics, electronics and in the fashion jewellery industry. Emissions additionally come from the burning of heating fuels and vehicle exhaust. World production in 1986 was 800 000 tonnes. Consumption in the EU (in 1992) was about 210 000 tonnes, and in the world about 780 000 tonnes. Nickel can cause allergic skin reactions and cancer.

Polychlorinated biphenyls (PCBs)

PCBs are a group of completely artificial compounds; 209 are theoretically possible and 60 have been identified. They have been used in closed, semi-enclosed and open systems and in this way become released to the environment. The desirability of these compounds stemmed from their great stability under different conditions. Up to the end of the 1980s a total amount of about 1.1 million tonnes had been produced worldwide. Main fields of use have included capacitors, transformers, hydraulic and heat exchange systems and pumps, as well as in plasticisers, surface coatings, paints and adhesives. Because of their stability (and non-degradability) PCBs have now found their way into all parts of the environment including the Arctic and Antarctic. The potential primary routes of human exposure are ingestion, inhalation and skin contact. The toxicity of PCBs has been recognised in two major accidents in Japan which led to the Yusho (oil) disease and Yu-cheng disease (WHO, 1990). There is also evidence for carcinogenic properties of PCBs. Restrictions on the use of PCBs and limits to concentrations in food have been established in many countries. Although PCBs are known to accumulate in the food-chain, acute ecotoxicological effects are unknown, but there are effects on reproduction in some mammalian species (UNEP, 1992b).

Dioxins

Dioxin compounds (75 different chemicals or 210 including furans) form a dangerous group of contaminants (not produced for their own sake). Dioxins have recently been extensively reported on and discussed, raising serious public concern. Their main environmental pathways are shown in Figure 17.1.

Dioxins caused serious health and environmental effects after the Seveso accident in 1976 (see Chapter 18). The pesticide 2,4,5-T, for instance, contains small amounts of

Figure 17.1
Sources and pathways of dioxins
Source: UK DoE, 1992

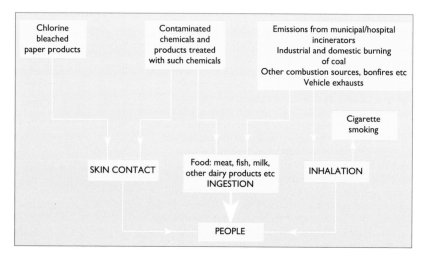

dioxins (mainly 2,3,7,8-tetrachlorodibenzo-*p*-dioxin) after synthesis. This impurity has a high acute toxicity to some animals but less to others. For example, for male rats the LD_{50} (a given dosage necessary to kill 50 per cent of a group of test animals) is as low as 0.022 mg/kg body weight, whereas this dioxin is 200-fold less toxic to a related organism – the hamster – with an LD_{50} of 5 mg/kg body weight. Because 2,4,5-T is a widespread herbicide, dioxin-like compounds are found in soils, sediments and biota throughout the world. Dioxins are possible carcinogens according to IARC.

Oil

Of total world oil production of approximately 3.2 billion tonnes per year, an estimated one thousandth of this (3.2 million tonnes) entered the marine environment in 1981 from all sources. This amount decreased to 2.35 million tonnes in 1990 (GESAMP, 1993) but shows large variations between years depending on number and size of shipping accidents (Chapter 18) as well as the effects of war. Seepage of oil from natural sources is estimated to account for at least 15 per cent of the total entering the environment. The environmental effects of oil spills are more conspicuous in shallow coastal waters with poor circulation and long residence times than in open marine areas, in particular, because coastal waters offer spawning ground for many marine animals (fish species and their prey) and because young animals are generally more sensitive than adults to oil pollution. For marine mammals, sea-birds and turtles the physical effect of oil pollution (oiling) is of greater concern than the toxic effect of dissolved hydrocarbons. A detailed discussion of oil-related pollution problems in Europe's seas is contained in Chapters 6 and 18.

Benzene is a natural component of crude oil which is found in petrol and is emitted by motor vehicles and by the chemicals industry. The world production in 1989 was about 15 million tonnes of which the EU had a share of about 5.5 million tonnes. Acute exposure to benzene depresses the central nervous system, and benzene is also a well-known human carcinogen.

Polyaromatic hydrocarbons (PAHs)

The major sources for PAHs are incomplete combustion of organic compounds, industrial solvents and wood stoves. Several hundred compounds exist, the best known of which is benzo(a)pyrene (BaP), which represents around 5 per cent of all the emitted PAHs. BaP, in particular, is considered to be a probable human carcinogen by IARC.

Polyvinylchloride (PVC)

PVC is used for a great range of purposes (eg, for pipes, films, bottles, cable installations, floors, walls). World production in 1991 was 18 million tonnes, with 29 per cent from EU and EFTA countries. Among plastics PVC has attracted attention because it contains chlorine, and the monomer vinylchloride can cause cancer. Burning PVC produces chlorine gas which, with water, forms hydrochloric acid, contributing to the acidification problem (see Chapter 31). Burning PVC can also lead to the formation of dioxins and related substances.

Chlorofluorocarbons (CFCs) and halons

When CFCs became available to the general public, at the beginning of the 1930s, they were considered excellent substitutes for ammonia in refrigeration appliances due to their inert properties. This view was sustained until the 1970s when they were the first substances suspected of depleting ozone at high elevations in the atmosphere (see Chapter 28). Their very stability at low elevations, where CFCs and halons are still considered to be basically inert gases, allowed them over time to reach high into the atmosphere where exposure to UV-B radiation breaks them down into chemically active components.

CFCs are especially used as propellants in aerosols, solvents, for foam blowing, refrigeration and air conditioning. Halons are used primarily as fire extinguishers. The EU production of CFCs in 1990 was 284 000 tonnes and of halons 11 600 tonnes. Due to their stratospheric ozone depleting properties, they may be responsible for increased UV-B radiation reaching the Earth's surface, with potential health and ecological effects. CFCs and halons are now being phased out and substituted by other less harmful products. (eg, HCFCs – see Chapter 28).

Asbestos

The term asbestos includes several fibrous silicate minerals which occur naturally as serpentines and amphiboles. It is a widely used material (over 1000 uses are described) which can be woven into incombustible fabric, used as asbestos cement, roof sheets, wall boards and pipes. It is also used as a filler for plastics, as a heat insulant, and in vehicles in brake and clutch linings. World production in 1986 was 4 million tonnes. Major health effects of concern involve the respiratory system, lung cancer and mesothelioma. Exposure to the general population may occur where asbestos is used in buildings. Mesothelioma, whose only known cause is asbestos exposure, has, among other things, been related to the deterioration of insulation in office buildings (Stein et al, 1989).

The Directive on the prevention and reduction of environmental pollution by asbestos (87/217/EEC) has been in force since 1991 and lays down stringent measures and provisions for the working of products containing asbestos, the demolition of buildings as well as the transport and disposal of waste containing asbestos fibres. The aim is to encourage the replacement of asbestos fibres by other materials less harmful to human health. EU countries may go beyond these standards (asbestos has thus been banned in Germany) or may set up more detailed provisions if environmental conditions require it.

Washing powders and detergents

The world production of washing powders and detergents is approximately 15 million tonnes per year. They are used in washing and dyeing of textiles and fabrics, degreasing and tanning of leather, degreasing of metals, floatating processes in ore beneficiation, additives and anti-corrosives, etc. Laundry powders, which represent the largest quantity of detergents, are formulated from five different groups of substances: surfactants, builders, bleaching agents, additives and fillers. The major constituents are phosphates (mainly sodium tripolyphosphate), sodium carbonate, sodium silicate and zeolites. Phosphate is a problematic ingredient because of its eutrophying effect on aquatic ecosystems (see eutrophying substances, above). Zeolites are thought to have undesirable effects as they may release strongly bound heavy metals from organometal complexes.

Pesticides

According to WHO up to 20 000 fatalities per year at world level may result from acute poisoning by pesticides. UNEP (1992b) reports that 3 million tonnes of pesticides are produced worldwide; 30 per cent of this is used in the EU.

A wide range of different chemical compounds are currently in use as insecticides, fungicides and herbicides. About 600 single, active compounds are in use in the EU. Insecticides include, among others, organochlorines, organophosphates, carbamates and pyrethroids. Fungicides include dithiocarbamates, organomercury compounds, organochlorines, phthalimides and a series of other compounds. Herbicides include chlorophenoxy acids, carbamates, urones, sulphonylureas, triazines and many other compounds, eg, glyphosate.

In addition to the danger to workers and users who handle pesticides, these compounds have a direct or indirect impact on flora and fauna, and they may pollute groundwater, rivers and lakes (see Chapter 5). Some pesticides may remain in soils for prolonged periods due to their high persistency. Because most pesticides are applied directly in the environment, direct lethal impact on non-target organisms (eg, honey bees and other pollinators) in agricultural areas has been observed. The major threat to the biodiversity of arable land by pesticide application is the reduction of the amount and quality of food and habitats for the fauna, and the reduction of the seed pool. (See also Chapters 7, 9 and 22.)

To prevent rotting and insect attacks, wood is treated with a wide range of compounds: creosote, mineral oil, arsenic, copper, zinc, boron, mercury and chromium salts, chlorophenols and other insecticides and fungicides. Many of these chemicals can cause cancer (eg, chlorophenol, chromium salts, naphthenates, arsenic), or are highly toxic (eg, mercury salts, organomercury compounds) or can cause allergic reactions.

Conclusions

The manufacture and use of chemical products results in a wide variety of chemical compounds being released to the environment. Once in the environment these compounds can have undesirable impacts on human health and welfare and ecosystems. This can lead to loss of biodiversity and detriment to human health and welfare from loss of living space by contamination, sickness, and change of human behaviour.

The total exposure of an individual to a particular chemical is affected by social factors covering industrial development, lifestyle, cultural habits and personal activities. To estimate the total uptake from all media and the target tissue dose, environmental measurements of chemicals at point locations in individual media are needed with detailed information on individual exposure patterns and toxicity. Much of this information, however, is not available for the many chemicals in use and circulating in Europe today. In particular, insufficient information is available on the toxicity, ecotoxicity and exposure to chemicals concerning environment and health impacts.

Although international initiatives are beginning to improve the situation through increased research, monitoring and notification procedures, much still has to be done before unexpected effects of those chemicals already in circulation and present in the environment are minimised. Environmental auditing, legislation, education and research can contribute to an improvement of risk assessment and risk management procedures required to avoid dangerous and unexpected effects.

The importance of problems related to chemicals and the way in which they potentially affect all parts of the environment and human activities are the reasons for including the issue of chemical risk as one of the prominent environmental problems of concern to the whole of Europe (Chapter 38).

GENETICALLY MODIFIED ORGANISMS

In recent years, the development and use of new techniques of genetic engineering have profoundly changed the traditional methods and scope of biotechnology. These sophisticated techniques enable the identification of many genes which confer desirable characteristics, and the transfer of these genes to organisms which did not possess them before. The bacteria, fungi, viruses, plants, insects, fish and mammals which are designer-made in this way are referred to as genetically modified organisms (GMOs), and can be defined as organisms in which the genetic material has been altered in a way that does not occur naturally by mating and/or natural recombination.

Traditional selection and breeding techniques have been used for a long time in industry and agriculture to produce organisms with more desirable characteristics. However, the new, more powerful tools of molecular biology allow biological barriers to be bypassed and novel organisms with new or enhanced properties to be created. For example, a gene from a microbe responsible for expressing a certain natural toxin against insects can be introduced into the genetic material (DNA) of crop plants, which can then protect themselves from insect attack without the need for application of pesticides. In principle, genes from any species could be inserted into any other species.

The development of a growing number and range of GMOs has opened up significant potential for many useful applications in agriculture and food-processing, pharmaceuticals and diagnostics, environmental clean-up, chemicals production and the development of new materials and energy sources. At the same time, however, there are some concerns about the potential risks to human health and the environment associated with the use and the release of these novel organisms into the environment, and in particular in relation to longer-term effects which are very difficult to predict. The need to undertake environmental risk assessments and to implement risk management measures as required has been recognised by the EU, and also by the USA and the other industrialised countries who are members of the OECD. Many non-OECD countries are also realising the need to take a preventative approach, especially as a result of the discussions at the UNCED in Rio de Janeiro in 1992 and the implications for biotechnology of the UN Biological Diversity Convention.

Uses and applications

The potential for application of the techniques of molecular biology to products and processes has increased dramatically in the last few years. Advances in transformation techniques have resulted in a greater number of different GMOs being developed for a broad range of applications.

By far the most extensive use of GMOs at present is the use of microorganisms in laboratories and production facilities (contained use). Many products can be made by fermentation processes using modified microorganisms with novel characteristics. For instance, human genes encoding for the expression of substances such as insulin, growth hormone or the blood clotting factor have been successfully transferred into the DNA of microorganisms. This allows large-scale production of these substances for medical use.

However, in recent years, new applications, in particular in agriculture, require that new organisms are created for release to the environment. The development and field testing of modified organisms, and plants in particular, has been a major growth area in terms of research, but only now is there a gradual movement to place GMOs on the market as products.

Plants are being genetically modified and released with resistance to herbicides, viruses, fungi and insects, for improved storage of protein, yield or ripening characteristics, and for differentiated flower colour. More basic research is being carried out to develop plants producing pharmaceutical substances or petroleum product substitutes.

Genetically modified viruses are being released as biopesticides (eg, modified *Baculovirus*) or as live viral animal vaccines (eg, anti-rabies vaccine for foxes and raccoons). Nitrogen-fixing bacteria (*Rhizobium*) have been modified to improve incorporation of atmospheric nitrogen into plants, while other bacteria have been released as bio-pesticides (Bt-toxin containing *Pseudomonas*).

Other interesting applications for genetically modified bacteria will be forthcoming, as research progresses in the development of bacteria with the ability to degrade toxic chemicals, to leach metals from low grade ore deposits, to enhance the flavour and the characteristics of processed or fermented food products.

Fish (eg, rainbow trout, *Salmo gairdneri*) are being modified to alter their growth characteristics, tolerance to temperature changes and resistance to disease. Larger mammals (eg, pigs, sheep, goats, cattle) are being modified to enable them to produce specific chemicals or to alter their growth characteristics, while smaller mammals (eg, mice) are being modified to make them more suitable indicators for certain types of testing (eg, carcinogenicity).

Potential undesirable effects

Concerns have been expressed that the release into the environment of novel organisms which are not a result of natural selection may entail potential risks related to human health and the environment. Examples of such concerns are that organisms with novel properties could cause adverse effects in the environment if they survive and establish themselves, out-competing existing species or transferring their novel traits to other organisms. The introduction of co-competition could have evolutionary effects.

The migration of inserted genes (in particular resistance genes) from cultivated plants to wild species could cause weediness, while an artificially created selection pressure could lead to a dominance of GMOs. Once a GMO is released into the environment, it could be impossible to recall or prevent its spread and therefore adverse effects must be avoided as they might be irreversible.

While the potential environmental risks are more wide-ranging where the release of genetically modified microorganisms to the environment is involved, concern is also expressed with relation to the contained use of genetically modified microorganisms (GMMs), since organisms which are intended to be kept under containment could affect human health or the environment if they are released. Modified pathogenic or potentially pathogenic organisms can be used for certain contained operations, and organisms not designed and approved for release to the environment may find their way there.

Risk assessment and risk management

The introduction and use of GMOs is a recent development, where any associated risks are potential, but also where any longer-term effects are not known. Appropriate risk assessment and risk management are necessary to avoid upsetting the natural ecological processes and, if possible, to avoid undesirable evolutionary effects.

Potential adverse effects to the environment depend not only on the characteristics of the new organism, but also on the interaction with the environment and with the different organisms present in a certain ecosystem. Effective risk management means that a careful assessment must be made of each case, and the necessary measures taken, in relation to the potential risks posed.

The EU has taken a clear preventative approach in adopting legislation to establish a common set of environmental risk assessment requirements and safety measures. This is intended to protect the health of citizens and the environment, as well as ensuring a single unified market for biotechnology.

The EC Directive (90/219/EEC) on the contained use of genetically modified microorganisms covers all activities with GMMs, whether in research facilities or industrial installations. It provides for an appropriate system of risk management, recognising that not all GMMs, and their uses, entail the same risk.

The EC Directive (90/220/EEC) on the deliberate release of genetically modified organisms to the environment covers the environmental assessment and release approval of all GMOs through all the stages of release to the environment. Both small-scale and large-scale experimental introductions, as well as releases through product marketing, fall within the scope of this legislation. The focus of the Directive is risk management. As with other technologies, potential risks associated with the release of GMOs need to be identified and assessed and the appropriate measures taken. Risk management is about the effective assessment and control of any potential adverse consequences that could occur as a result of releasing new GMOs to the environment. The Directive introduces an approach of risk management based on the concept of step-by-step development and testing of novel organisms, whose risks and impacts are analysed on a

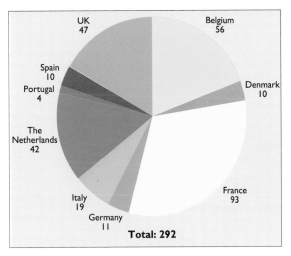

Figure 17.2
Field releases of GMOs notified in the EU in 1991–94 by country
Source: European Commission

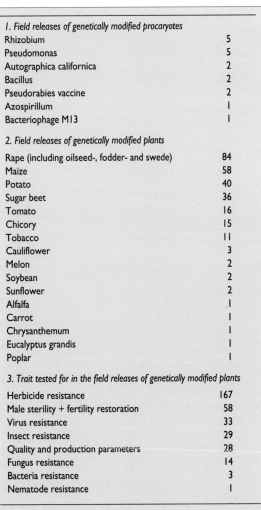

Table 17.3
Field releases of GMOs notified in the EU in 1991–94, by groups of organisms and traits
Source: European Commission

1. Field releases of genetically modified procaryotes	
Rhizobium	5
Pseudomonas	5
Autographica californica	2
Bacillus	2
Pseudorabies vaccine	2
Azospirillum	1
Bacteriophage M13	1

2. Field releases of genetically modified plants	
Rape (including oilseed-, fodder- and swede)	84
Maize	58
Potato	40
Sugar beet	36
Tomato	16
Chicory	15
Tobacco	11
Cauliflower	3
Melon	2
Soybean	2
Sunflower	2
Alfalfa	1
Carrot	1
Chrysanthemum	1
Eucalyptus grandis	1
Poplar	1

3. Trait tested for in the field releases of genetically modified plants	
Herbicide resistance	167
Male sterility + fertility restoration	58
Virus resistance	33
Insect resistance	29
Quality and production parameters	28
Fungus resistance	14
Bacteria resistance	3
Nematode resistance	1

case-by-case basis.

The Directive on GMO releases contains provisions for circulation of information on notifications for deliberate release of GMOs for research purposes between the EU Member States. The Directive also contains provisions for a common EU marketing approval procedure for GMOs placed on the market as products.

The provisions for circulation of information on notifications provides the EU Member States with a unique opportunity to follow the development of deliberate release of GMOs. Table 17.3 and Figure 17.2 give information on the release notifications sent to the European Commission from 23 October 1991 until 1 July 1994. The figure shows that most of the EU countries have had notifications for approval of releases. However, the number of releases varies greatly between countries.

Of the 292 notifications received so far only 18 have been for releases of genetically modified microorganisms. The rest have been for releases of genetically modified plants, as no notifications for releases of genetically modified animals have yet been circulated. Of the 16 different plant species released in the EU, the most frequently field tested are rape (*Brassica napus* ssp *napus*), maize (*Zea mays*), potato (*Solanum tuberosum*) and sugar beet (*Beta vulgaris* ssp *vulgaris*).

The trait which has been inserted in most organisms is herbicide resistance, and it has been introduced in over 50 per cent of the notified releases. In some cases herbicide resistance is inserted together with other main traits, as for instance male sterility or fertility restoration.

Notifications for the placing on the market of GMOs are now also being submitted for approval. Two vaccine products have been given environmental clearance and a bromoxmynil resistant tobacco has been approved. Herbicide (Basta) resistant rape seed is currently being evaluated.

CONCLUSIONS

Genetic engineering – the ability to transfer genetic material in new ways and radically alter the intricate genetic structure of individual living cells – has developed rapidly since it first became feasible in the mid-1970s, and promises to have enormous impacts on many sectors of the economy. These new techniques will, within a few years, result in many GMOs being placed on the market as products. This involves the use of new medical diagnostic tests, treatments and vaccines, improved foods, and agricultural and industrial products. However, the use of GMOs needs to be assessed and managed, in order to ensure that damage to human health and the environment is avoided. An environmentally sound management of biotechnology is a prerequisite for its safe and sustainable development.

REFERENCES

Axenfeld, F, Muench, J, Pacyna, J M, Duiser J A and Veldt, C (1992) *Test-Emissionsdatenbasis der Spurenelemente As, Cd, Hg, Pb, Zn, und der speziellen organischen Verbindungen Lindan, HCB, PCB und PAK für die Modellrechnung in Europa*. Umweltbundesamt 104-02-588, Berlin.

CEC (1992) The state of the environment in the European Community. Overview Vol III Com (92) 23. Brussels.

CEFIC (1991) *The chemical industry in 1991, annual review*. New York.

CEFIC (1993) *The chemical industry in 1992, annual review*. New York.

ECETOC (1992) *Annual report 1992*. CEFIC, Brussels.

GESAMP (1993) Impact of oil and related chemicals and wastes on the marine environment. IMO/FAO/UNESCO/WMO/WHO/IAEA/UN/UNEP Joint group of experts on the scientific aspects of marine pollution Rep. Stud. GESAMP 50, London.

Karcher, K and Devillers, J (Eds) (1991) Practical application of quantitative structure-activity relationships (QSAR) in environmental chemistry and toxicology. *Proceedings of Eurocourse, JRC Ispra, Italy, June 11–15, 1990*. Kluwer Academic Publishers, Dordrecht, EUR 12831.

Lönngren, R (1992) *International approaches to chemicals control*. KEMI, Stockholm.

Stein, R C et al (1989) Pleural mesotheliomia resulting from exposure to amosite asbestos in a building. *Respiratory Medicine* **83**, 232–9.

UK DoE (1992) *The UK environment*. UK Department of the Environment. HMSO, London.

UNEP (1992a) *The world environment 1972–1992: two decades of challenge*. Tolba, M K, El-Kohly, O A, El-Hinnawi, E, Holdgate, M W, McMichael, D F and Munn, R E (Eds). United Nations Environment Programme. Chapman and Hall, London.

UNEP (1992b) *Chemical pollution: a global overview*. United Nations Environment Programme. Geneva.

USEPA (1991) *Environmental progress and challenges: EPA's update*. EPA-230-07-88-033, Washington DC.

USNRC (1984) *Toxicity testing*. National Research Council, Washington DC.

Vighi, M and Calamari, D (1992) Ecotoxicological risk indicators for environmental chemicals. In: Colombo, A G (Ed), *Environmental impact assessment*, pp 261–75. ECSC, EEC, EAEC, Brussels and Luxembourg.

WHO (1990) *Cancer: causes, occurrences and control*. IARC Scientific Publication 100. Tomatis, L (Ed), International Agency for Research on Cancer, Lyons.

Flooding in Vaison-LaRomaine (Vaucluse)
Source: Chardon-Martin, Gamma, Frank Spooner Pictures

18 Natural and technological hazards

INTRODUCTION

Naturally occurring hazardous events and technological accidents are separate causes of environmental impacts. The information available on their occurrence and consequences is far from comprehensive (Box 18A) and, in particular, there are significant gaps in our understanding of the long-term environmental impacts of accidents. However, on the basis of data which are available from a range of sources, it is possible to indicate how these two kinds of events are different and why they need to be treated separately from other pressures. This chapter examines the characteristics and importance of accidents and natural hazards as causes of environmental impacts. It summarises their causes and consequences in Europe, identifying different types of damage which can result from different sources.

HAZARDOUS EVENTS: A SEPARATE CAUSE OF IMPACTS

In the six preceding chapters, different causes of environmental impacts have been discussed. These are the pressures which come about through the routine performance of human activities, particularly in the different economic sectors (see Chapters 19 to 26). However, these activities can also exert pressure on the environment through the consequences of accidents. In this case the medium by which the environment is impacted is less important as a distinguishing factor determining the scope for potential

impacts than the nature of the accident itself. The examples in Boxes 18B, 18C and 18D illustrate this well. Generally, it is important to consider accidents separately from other causes of impacts because of two distinguishing characteristics, discussed below: the uncertainty in magnitude, nature and timing, and the non-continuous character of the events; and the impact mechanisms and types of damage which result. Similar characteristics distinguish naturally occurring hazardous events as a cause of environmental impacts, but there are also important differences, as discussed below.

Uncertainty and duration

Accidents and natural hazards manifest themselves in largely unpredictable, singular events. Accidents arise when routine operations go wrong, resulting, for example, in the unexpected release of toxic substances. The forces of nature cause hazardous events such as earthquakes and flash floods, mass movements, mud and snow avalanches, cold- and heat-waves, tidal waves, drought, volcanic eruptions, storms and tornados. Sometimes, natural hazardous events have a longer duration than accidents (eg, heatwaves, droughts) and in many cases it is easier to foresee their consequences.

Usually, it is unlike with accidents, little can be done to prevent or reduce the magnitude of a naturally occurring hazardous event. However, much can be done to reduce potential impacts; landuse and emergency planning are the keys to this. Landuse planning is particularly important for understanding and predicting potential interactions between natural hazards and human activities as a cause of environmental impact (eg, soil erosion from floods). This is a common element in the range of tools available for mitigating the impacts of hazardous events and accidents from all sources (see Chapter 30). In practice, planning for natural hazards relies on a type of quantitative risk assessment. Because it is often uneconomic to protect against the most severe foreseeable event, designs are chosen to plan, for example, for the 10-year or 100-year storm, based on the historical record and the design life of the structure.

In contrast to natural hazards, much can be done to prevent accidents. The first step in accident prevention is good planning, management and control of the routine activities in question. This, together with a thorough implementation of codes of good practice, goes a long way to ensuring that emission levels and other interactions with the surrounding environment are contained within limits considered to be acceptable. These limits can be specified in site licence conditions, in principle, allowing local environmental impacts due to any single source to be anticipated and accounted for (through auditing and monitoring programmes) and, if necessary, mitigated. It also

Box 18A Data availability

Industrial accidents

Data on accidents provide valuable lessons not only on how an accident happened but also about the clean-up or remediation process. They are important as inputs to predictive analyses for risk management and emergency planning. A better knowledge of the behaviour of various contaminants in the environment and more comprehensive data on small spills and accidental discharges are necessary to establish a reliable picture of the impacts of accidents on the environment. Until this information is available, it is difficult to be sure of the appropriateness of the priorities adopted for risk management (see Chapter 30).

Most data are currently collected on a national basis. Few international databases exist. There is a paucity of information for Eastern Europe, while Northern Europe is covered best of all. The main sources of data on major accidents worldwide are Lloyd's Weekly, the Loss Prevention Bulletin, the Rhine Commission Reports, the French Survey of Accidents, the MHIDAS database at UKAEA, the AHE database at USEPA, the ENVIDAS database, the FACTS database at TNO in The Netherlands and MARS of the European Commission's Joint Research Centre. The information compiled usually includes a description of the location and nature of the accident, as well as any known causes. Information on the type of release as well as its size is also usually given, although this information can be very approximate.

The impacts of individual accidents on human health are usually well reported, and especially so if the accident has caused fatalities. However, there are often surprising discrepancies between sources concerning consequences even in major accident reporting, and varying opinions on causes and effects, not only as part of legitimate scientific debate, can cloud a clear understanding of impacts. Numbers of human fatalities are widely used to assess the significance or scale of an accident (see Chapter 30). However, information on environmental damage is usually restricted to observations of the more obvious short-term impacts, such as immediate fish kills. More detailed information on long-term impacts is often not given or not known. (This has been well researched recently by Lindgaard-Jorgensen and Bender (1992) for the Community Documentation Centre on Industrial Risk.) There is a particular lack of information on the impacts of accidents affecting the terrestrial environment. For this purpose the technical journals and periodicals are a more useful source of information, and they tend to report the results of long-term post-accident monitoring of the affected environment in more detail.

Often accidents are reported incompletely and much important information is missing. A standard system of reporting would help to alleviate this problem and enable accidents to be compared. Overall, there are significant gaps in the data on accidents that should be filled, the most important of which are ecological effects, long-term recovery of the environment and clean-up actions.

Within the context of ecological effects and environment recovery there is a significant paucity of information on the ecotoxicology of chemicals (see Chapters 17 and 38). While the chemical and physical characteristics of most chemicals are well known, their complex environmental behaviour is poorly understood. Furthermore, baseline information is usually insufficient to be able to assess the environmental effects of accidents. Data from long-term environmental monitoring is very scarce. Clean-up actions are often reported incompletely, if at all. In some cases no clean-up is undertaken following an accident, while in other cases details are simply not recorded.

Natural hazards

In general, data on natural hazards are widely available from a number of sources. However, the type of information varies considerably according to the reporting purpose, and information on environmental impacts is seriously lacking.

Data are collected at the regional, national and international levels. Most reliable data come from developing countries, as well as certain parts of Eastern Europe and the USA. This is because of the propensity for natural disasters to occur in these regions. Some of the main sources include: the major re-insurance companies (such as Swiss Re), the International Red Cross and Red Crescent Societies, the UN Department of Humanitarian Affairs (DHA), the UN International Decade for Natural Disaster Reduction (IDNDR), the Centre for Research on the Epidemiology of Disasters (CRED), the European Programme on Climatology and Natural Hazards (EPOCH), the Bradford University Disaster Prevention and Limitation Unit, the Disaster Research Centre at the University of Delaware and the Natural Hazards Centre at the University of Colorado.

The classification of the impact of a natural hazard depends on the source of data. International sources such as the UN have traditionally classified 'impact' as the number of deaths and/or economic loss, whereas re-insurance companies focus on the number and value of insurance claims. The information records usually list the affected country and region together with the type of disaster. The level of 'damage' given is based on the chosen definition of impact.

It has not been routine to report upon the environmental impacts of natural hazards. Although reports on the link between natural hazards and environmental impact are occasionally released, no systematic reporting structure exists. There is a particular lack of information on the potential interaction between natural hazards and human activities as a cause of environmental impact. A significant issue which monitoring should take into account is the transfer of impact geographically and often transnationally (eg: drought, irrigation and water resource depletion elsewhere; deforestation, soil erosion, floods; coastal erosion, coastal defences, loss of sediment supply elsewhere, erosion elsewhere).

Box 18B Examples of accidents – industrial installations

Flixborough incident, UK, 1974

The Flixborough Works of Nypro (UK) Limited were virtually demolished by a huge explosion in 1974. Of those working on the site at the time, 28 were killed and 36 others suffered injuries. Outside the works, injuries and damage were widespread but no one was killed. There were 53 people recorded as casualties, while hundreds more suffered relatively minor injuries. Property damage extended over a wide area. The cause of the accident was the rupture of a piece of temporary pipework which allowed cyclohexane to be released. The cyclohexane formed a cloud of vapour (mixed with air) which then exploded.

Accidental release of tetrachlorodibenzo-p-dioxin (TCDD), Seveso, Italy, 1976

On 10 July 1976 a chemical reactor safety valve lifted at the ICMESA plant at Seveso, due to the development of an uncontrollable exothermic reaction within a process vessel during the production of trichlorophenol. The fluid mixture burst through the valve orifice and was dispersed over a wide area by wind. Following this, vapours escaped from the plant and approximately 1800 ha of densely populated land was contaminated. Plants and animals were severely affected, and people who had been exposed to the toxic cloud developed dermal lesions (chloracne). Contamination of the soil with dioxin was found up to ten years later.

Explosion leading to release of arsenic, Manfredonia, Italy, 1976

Ten tonnes of arsenic trioxide, 18 tonnes of potassium oxide and 60 tonnes of water were released from an ammonia cooling column at a petrochemical factory. Following the incident, 1000 people were evacuated from an area of 2 km^2 around the plant. A number of sheep were poisoned by arsenic and tests on sea water and fish showed abnormally high levels of arsenic. An area of 15 km^2, consisting mainly of olive and almond plantations, was contaminated. Arsenic levels in the soil ranged from 200 mg/m^2 at the edge of the factory area to more than 2000 mg/m^2 in the factory area.

Fire at the Sandoz agrochemical warehouse, Schweizerhalle, Switzerland, 1986

A fire broke out on 1 November 1986 at an agrochemical warehouse at Sandoz's Schweizerhalle works, near Basle in Switzerland. Extensive pollution of the river Rhine resulted when 30 tonnes of substances, including the fluorescent dye Rhodamine B, two organophosphorus pesticides (disulfoton and thiometon), 150 kg of mercury and a fungicide (ethoxyethylmercuryhydroxide) were washed into the river during fire-fighting operations. The city of Basle was also affected by a cloud of mercaptans associated with the warehouse fire. The accident resulted in massive fish kills in the river downstream of Basle. An estimated 500 000 fish were killed, including 150 000 eels (Anguilla anguilla). Benthic organisms were virtually eliminated over a 400 km stretch of the Rhine. The wider and longer-term impacts of the accident were not as great as originally predicted. The suspended solids of the river water appear to have reduced the effects of the contaminated runoff by adsorbing a large quantity of the pesticides, thus limiting their effective concentrations.

Fire at chemical plant, Auzouer-en-Touraine, France, 1988

A fire started after an explosion at a chemical plant which was producing a variety of different chemicals. Runoff water from the fire-fighting operation flowed into the river Brenne carrying with it a mixture of highly toxic chemicals, including phenols, arsenic and cyanides; 19 km of the river Brenne and up to 50 km of the Loire were polluted and more than 200 000 residents in the surrounding towns and villages had their water supplies cut off for a number of days.

---

means that the cumulative impact of numerous sources in a region can be estimated for the purpose of strategic planning and policy development; this is becoming increasingly important.

Accidents, however, such as the spilling of toxic chemicals into a river, are neither routine nor planned. This does not mean that they cannot be avoided or mitigated. Estimates of the most probable 'worst cases' can usually be made to assist planning. The cause of any particular accident can be traced back to a combination of operational failures relating to one or more aspects of the technical and management systems which control the activity. However, this uncertainty means that pollutant discharges and other sources of potential impact resulting from major accidents can be neither anticipated easily nor accounted for in the same way as normal sources from routine operations. In this respect, they represent an additional and largely unquantified source of impacts on the environment.

The cumulative effect of minor spills and discharges can be considered differently, since the precise time and location of spills are less important in determining their impacts and potential contribution to long-term environmental damage.

Impact mechanisms and types of damage

Accidents

Accidents can result in pollutant discharges and physical effects on the environment (eg, fire and explosions) which would be neither expected nor allowed during the course of normal industrial operations. The basic differences between accidents and routine operations, in terms of their potential pressures on the environment and human populations, relate to the following general parameters: the toxicity of discharges, volume and rate of release, and flammability and explosiveness.

The damage pathways are often very complex, involving direct and indirect effects to more than one environmental medium. The resulting damage can differ from that associated with routine activities in a number of ways. For example, the development of unexpected, uncontrollable conditions in an industrial plant could result in the production of large quantities of toxic compounds. If the emergency facilities fail, this could result in release to the environment of large quantities of toxic gases and/or fire and explosion. Depending on the prevailing weather conditions and the sensitivity of the surrounding environment, a single event such as this could cause considerable short- and long-term damage to natural resources (groundwater, rivers, soils), to terrestrial and aquatic ecosystems and to humans. The Seveso accident in 1976 provides an example of this (see Box 18B).

In the case of a failure of a storage vessel or a rupture of a road tanker container in an accident, the sudden and rapid release may be caused of a large quantity of a substance which in normal circumstances might not be released at all, or only in small quantities and under controlled conditions. This can result in off-site concentrations of a pollutant which exceed the damage threshold level of the surrounding environment and/or cause serious human health effects, including fatalities. In some cases the consequences may be dramatic, for example in the form of fireball, as occurred tragically at Los Alfaques in Spain in 1978 (see Box 18C).

Natural hazards

Unlike accidents, natural hazards are an important 'dynamic' of the environmental change process. To try to mitigate these events would indirectly constitute a form of environmental impact which could result in even greater, unforeseen impacts in the future.

Like an industrial accident, the nature and extent of the impact of a natural hazard on human society (number of deaths, structural damage, insurance costs) do not depend exclusively on characteristics of the event itself, eg, a storm path. Other factors are just as significant; for example, proximity to populations, construction of earthquake- and/or flood-resistant buildings and infrastructure, disaster prevention and emergency planning.

Similarly, the potential for environmental impacts resulting from natural hazards is to a large extent linked with human activities. These may imbalance natural systems, such as forest cover or groundwater level, and thereby make a natural disaster more likely. They may also result in the location of polluting activities in areas where natural hazards are likely to occur, and thereby enhance the impact potential of the activities. These linkages between human activities, natural hazards and environmental impact can become very complex and cyclic, the cause–effect relationship sometimes being unclear. Nevertheless, it is an area which is likely to become more important as both the pressure on natural resources and the intensity of polluting human activities increase in future.

SOURCES AND CONSEQUENCES

The sources of hazards which cause damage are discussed in this section under the following headings:

1 accidents at industrial installations;
2 inland transport accidents;
3 marine transport and offshore installation accidents;
4 accidents at nuclear installations and during transport of radioactive material;
5 natural hazards.

The distinguishing characteristics of hazards arising from these sources and the important aspects of resulting impacts are outlined below. The order in which they are treated does not imply a ranking of importance, which is not possible given the complexity and variety of causes and effects between different sources of hazards.

Industrial installations

Accidents at industrial installations can result in a wide range of possible damage routes (see Box18B). This is because of the vast range of toxic, flammable and explosive substances which are now used, produced or processed at high temperatures or pressures in large quantities. These substances may be released accidentally, frequently as a result of, or resulting in, fires and explosions.

By the end of 1991 a total of 121 accidents had been included in the Major Accident Reporting System (MARS) database reported to have occurred in the EU under Council Directive 82/501/EEC. Table 18.1 summarises the consequences of the accidents notified. An analysis of these accidents leads to the identification of the following key points:

● Most of the accidents occurred in petroleum refineries and in the petroleum industry, whereas the ceramic, cement, surface coating and dyes industries were least accident prone.

Box 18C Examples of accidents – inland transport and distribution

Road tanker accident, Los Alfaques, Spain, 1978

The pressurised cargo tank of an articulated road tanker carrying liquefied propylene disintegrated into three parts which were dispersed widely as it passed a seaside campsite. The tank contained 23 tonnes of liquid propylene, although its permitted load was only 19 tonnes. Furthermore, it was not fitted with a pressure relief device. A total of 217 people, mostly campers, were killed and a further 67 were injured. A discotheque and smaller buildings attached to the camp were demolished and some apartment houses suffered damage. Altogether 74 motor vehicles were damaged, of which 23 were totally destroyed. This Spanish campsite disaster is the worst-ever road accident on record involving the transport of dangerous goods .

The accident still raises many questions about what exactly happened and the precise cause of the human fatalities. Various investigators have put forward different opinions: it could have been a vapour cloud explosion, it could have been a boiling liquid expanding vapour explosion (BLEVE), or it could have been a flashfire.

River Roding insecticide spill, Essex, UK, 1985

A road accident involving a tanker transporting insecticides resulted in about 500 litres of an organophosphorus insecticide (Dursban 4E) being spilt into a tributary of the river Roding in the UK. Despite a prompt clean-up emergency response, the insecticide polluted a 20 km stretch of the river (and within 48 hours elevated concentrations of

the contaminant were recorded in the river Thames). Surveys conducted soon after the accident indicated a higher than 90 per cent mortality of fish and aquatic arthropods in the affected stretch of the river. Some birds were also killed, although no data are available on this. It took about two years to restore the river Roding to its pre-spill condition. Long-term impacts, such as the persistence of chlorpyrifos (the active ingredient of this particular insecticide) in sediments, and potential bioconcentration effects have not been investigated.

Pipeline explosion leading to rail disaster, Siberia, Russia, 1989

The gas pipeline causing the explosion lay to the west of Chelyabinsk and was the main line carrying natural gas liquids – a mixture of propane, butane and other components. The explosion occurred just as two trains carrying 1200 people were passing each other on the Trans-Siberian route close to the pipeline. It is thought that gas had leaked from the pipeline, accumulated in pools and started to evaporate; a spark from the wheels of one of the trains ignited it. Consequently, 645 people died in this disaster and about the same number were injured. The then Soviet prosecutors alleged that gross violations of construction and operations rules contributed to the accident. The pipeline had been damaged by an excavator in 1985 and the damage ignored; the pipe was buried and later leaked.

- Highly flammable gases were the substances most often involved in notified accidents; also, chlorine was commonly released.
- Most accidents happened during normal operations, although the frequency of accidents is probably higher during maintenance, start-up, shut-down and other non-standard plant conditions.

Usually, public attention is given to major accidents which result in the most obvious and dramatic impacts. It was the major accident at Seveso (Italy, 1976) which stimulated action by the European Community to improve plant safety and minimise the likelihood of similar events in the future. Similarly, well-publicised accidents at Flixborough (UK, 1974), Bhopal (India, 1984) and elsewhere were instrumental in raising public awareness of, and concern about, the reality of industrial accidents.

However, relatively frequent minor accidents can also contribute to environmental damage in a different, but potentially significant, way. This is because minor spillages of liquid chemicals can accumulate to considerable amounts in some areas without being reported, resulting in potentially serious contamination of soils and groundwater. It is likely that this represents a significant source of uncertainty in regional and national inventories of total pollutant loadings to soils and the aquatic environment. The growing requirement for monitoring and reporting of such spills should ensure that current levels and future trends can be estimated with more confidence. (This is discussed further below when marine transport and offshore installations are considered).

Inland transport and distribution

Box 18C provides examples of transport and distribution accidents which illustrate their considerable potential for impacts with regard to flammability, toxicity and persistence in the environment.

The transport of hazardous chemicals by road and rail (and to some extent by barge), is a major potential cause of environmental and human health damage whose importance is growing as both the amount of traffic and the quantities of chemicals requiring transportation increase. Releases of toxic, flammable and explosive substances can occur following collisions and accidents resulting from mechanical and operational failures.

The possible accident and release scenarios associated with the transportation of one substance by road, or even several substances by rail, are more limited compared with industrial installations, where process complexity and bulk storage mean that the range of possible events and releases can be considerable. The volumes transported are relatively small compared with site storage and marine transport (from about 20 m³ for road tankers, 50 to 100 m³ for rail tankers and several thousand cubic metres for barges).

However, there are many more possible impact scenarios resulting from transport accidents, since the receiving environment and the surrounding population can vary considerably according to the chosen route of the vehicle and the time of the accident. This means that the potential for limitation of damage following an accident is significantly less than for fixed installations, where there is usually an emergency response plan to take account of the likely impacts on the surrounding environment and population. This is the most important factor which distinguishes inland transport accidents from others and necessitates special consideration for the purpose of risk management.

Even buried pipelines can, if not maintained properly, be the cause of catastrophic damage and loss of life, as illustrated by the 1989 disaster in Siberia (see Box 18C). However, the potential for environmental damage in such cases is low, except where liquid fuel is being piped close to a sensitive waterbody.

Consequences	Numbers of accidents
No or negligible damage	24
Fatalities	22
– on-site	21
– off-premises	2
Injuries	55
– on-site	53
– off-premises	10
Material damage	82
– within the establishment	79
– off-premises	19
Traffic disruption	16
Environmental damage	14
Public evacuation	9
Plant evacuation	9
Public annoyance	9
Public deprived of potable water	2

Table 18.1 Consequences of accidents notified on the MARS database, situation end 1991
Source: Drogaris, 1993

Marine transport and offshore installations

Nearly 85 per cent of accidents at sea involve tankers (Table 18.2), whereas blowouts and pipeline incidents each account for less than 5 per cent. Clearly, these less frequent events can be serious, such as the Piper Alpha explosion and Ixtoc I blowout (see Box 18D); however, the potential environmental damage associated with tanker accidents is greatest, not only because of their frequency but also because of their tendency to take place near coastlines.

Figure 18.1 shows the worldwide trend in oil spills resulting from accidents involving tankers, combination

Type of accident	Frequency (%)
Exploration blowout	0.7
Other blowouts	3.9
Tankers	83.6
Pipelines	4.6
Others	7.1

Table 18.2 Frequency of accidents at sea
Source: Sharples et al, 1990

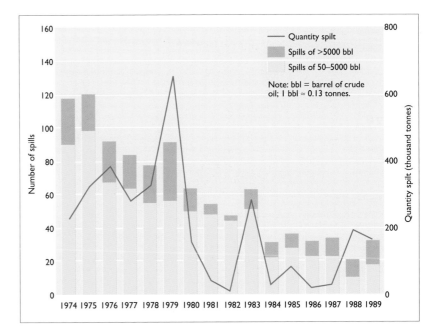

Figure 18.1 Oil spills resulting from accidents involving tankers, combination carriers and barges, 1974–89
Source: International Tanker Owners Pollution Federation Limited, London (unpublished data)

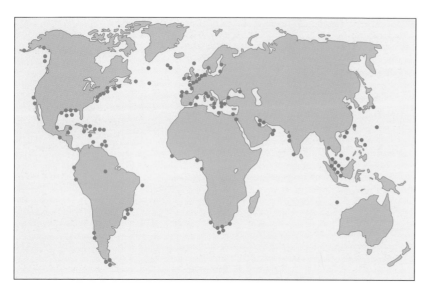

Map 18.1
Distribution of tanker spills greater than 10 000 bbl since 1974
Source: International Tanker Owners Pollution Federation Limited, London

Table 18.3
Number of main nuclear plants in different regions of Europe (1992)
Source: IAEA, 1993
Notes: 1 Pierrelatte (FR), Capenhurst (UK), Almelo (NL), Sverdlovsk (RU)
2 Chelyabinsk (RU), Cap de la Hague (FR), Sellafield (UK)
See Table 5.9 for country codes.
(Russian plants situated east of the Urals are not reported here)

carriers and barges between 1974 and 1989. Between 1975 and 1985 there were 74 reported spills of between 10 000 and 100 000 barrels of oil, but only 16 spills in the range 100 000 to 200 000 barrels and 24 spills of more than 200 000 barrels (Cooper, 1990). Both the number of oil spills and the quantity of oil spilled declined during this period. In addition, it appears that a growing percentage of the reported oil spilled has been due to a relatively small number of large spills. However, most unreported spills are likely to be at the smaller end of this range. It is likely, therefore, that the proportion of the total volume spilled coming from relatively small-scale incidents is higher than the official statistics would suggest. Nevertheless, the carrying capacity of the marine environment is less likely to be exceeded following individual small spills and so they have not been considered as a major cause of concern.

Map 18.1 shows the global distribution of large tanker spills since 1974. Not surprisingly they are concentrated in the areas of heaviest shipping traffic, which inevitably is found in coastal areas. Hence the high potential for ecological damage from this source.

As with industrial installations, it is the short-term impacts of large, dramatic marine oil spills which gain the public's attention. However, the environmental damage caused by accidents at sea can vary considerably, depending primarily on where the spill occurs – especially whether or not the pollutant is released in coastal waters which are often very sensitive ecologically – on the nature of prevailing weather conditions and on the type of oil spilled. Even small spills can cause significant damage under adverse conditions in a sensitive area.

For example, the world's biggest oil spill (550 000 tonnes over 9 months), from the Ixtoc I oil well in the Gulf of Mexico in 1979, apparently did remarkably little damage, the source being well away from any coasts (see Box 18D). In contrast, the *Amoco Cadiz* accident in 1978 (also Box 18D) caused considerably more ecological damage along the French coast: although the amount of oil spilled was less than half of that discharged from the Ixtoc blowout (220 000 tonnes), it was discharged in only a few days.

Prevailing weather conditions determine not only the overall direction in which an oil slick will move, but also how quickly it will break up. For example, the short-term impacts resulting from the *Braer* oil spill in 1993 (see Box 18D) were not as severe as had been expected from such a large spill in ecologically very sensitive coastal waters. This was because of the dispersive action of hurricane-force winds.

There are other aspects of marine spills which could also be significant in the longer term: the impacts of smaller spills (eg, on bottom fauna, flora and sediments); and the largely unknown impacts of spilled toxic chemicals, including heavy metals and chlorinated hydrocarbons.

The cumulative impacts of the many smaller accidents and spills which take place, both reported and unreported, are potentially significant in the longer term, depending on the persistence of the substance. However, very little long-term monitoring of impacts has occurred, even following major spills. Much more extensive monitoring and research will be required before the potential chronic effects of accumulated toxic spills will be known.

Nuclear installations and transport of radioactive material

In Europe, all sections of the civil nuclear industry are operational, from the production and use of radioisotopes to the production of electricity, involving all types of nuclear industrial facilities, such as: uranium enrichment plants, reactors and electronuclear power plants, and spent fuel reprocessing plants (Table 18.3). The nuclear power industry is discussed in Chapter 19, where the capacities and locations of nuclear power plants in Europe are presented (Map 19.2).

Nuclear accidents can potentially occur at a range of installations, including all those belonging to the nuclear industry as well as at military and medical facilities and research institutions. The transport of radioactive materials (eg, nuclear fuels, radioisotope sources and waste products) is also a potential source of accidents.

Accidents in such contexts are distinguished from all others by their unique potential to release radionuclides into the environment. The subsequent radioactive contamination can be extremely damaging (see Chapter 16) and some radionuclides are very persistent in the environment. The Chernobyl accident in 1986 (see below) demonstrated how nuclear accidents can have major impacts on the environment and human health.

Transportation

The transport of radioactive materials during the last 30 years has not resulted in accidents with significant radioactive releases or other consequences for the environment (CEC, in press (a)). Some 2 million shipments are made yearly within the European Union, 90 per cent of which are medical isotopes and less than 5 per cent comprise nuclear fuel cycle material, including radioactive waste. Chemical hazards are also taken into account for shipments of uranium hexafluoride (a chemical compound used in industrial processes to enrich uranium). A hazard evaluation undertaken by the International Atomic Energy Agency (IAEA), following the sinking off the Belgian coast in 1984 of the freighter *Mont-Louis*, which was carrying this compound, concluded that there had been no significant radioactive or chemical consequences to the public or the environment.

Operation of nuclear installations

Due to accumulated experience and technological progress, particularly in the field of radioactive waste management, public exposure to radiation and controlled radioactive releases to the environment, under normal conditions of plant operation, have been continuously decreasing during the last

Type of plant	EU	EFTA	C & E Europe	Total Europe
Nuclear power reactors	132	17	68	217
Uranium enrichment plants[1]	3	0	1	4
Reprocessing plants[2]	2	0	1	3

Box 18D
Examples of accidents – marine transport and offshore installations

Ixtoc I oil well blowout, Gulf of Mexico, 1979
The Ixtoc I blowout is probably the world's biggest oil spill in terms of the amount of oil released into the environment. More than 500 000 tonnes of oil were released into the sea over a period of nine months. However, since the blowout occurred well away from any coasts, the environmental damage it caused was remarkably small.

Piper Alpha oil and gas platform explosion, North Sea, 1988
On 6 July at approximately 10.00 hrs a large explosion occurred at Occidental's Piper Alpha platform, located approximately 190 km northeast of Aberdeen. The cause of the accident is unclear, but it is thought that it may have been a maintenance error. Within seconds a major crude-oil fire developed and all but the lower parts of the platform and wellhead area were engulfed in a choking black smoke.

The majority of the 226 people on board were trapped in the accommodation areas of the platform, unable to make their way to the survival craft because of the smoke and the initial fires. Approximately 20 minutes later, a major failure of a connecting pipeline occurred and a fireball more than 50 m in diameter engulfed and rose above the platform. The subsequent fires caused the platform to collapse, and 167 people lost their lives. The environmental impact of the accident was not examined in view of these human fatalities.

Amoco Cadiz oil spill, Brittany, France, 1978
After the Amoco Cadiz was wrecked off the coast of Brittany in March 1978, a total of around 220 000 tonnes of oil were spilt. Following the accident, over 4500 oiled birds of 33 species were recovered from beaches, and it was suggested that many more had actually died but had not been recovered. The most severely affected species were puffins (Fratercula arctica), razorbills (Alca torda) and guillemots (Cepphus grylle). Extensive areas of saltmarsh were damaged and some animals, such as the heart urchin (Echinocardium cordatum), were almost completely wiped out. The affected coastline was also being intensively used for oyster culture, which was severely disrupted by the spill. A high initial mortality was accompanied by heavy contamination that resulted in shellfish being unsuitable for consumption.

Braer oil tanker, Shetlands, UK, 1993
The Liberian-registered tanker Braer ran aground at Garths Ness on the Shetland coast on 3 January 1993, resulting in a spill of 80 000 tonnes of crude oil. The cause of the accident was a complete power failure which left the tanker adrift, and eventually led it to run aground. Investigations of the extent of the damage are still under way, but it seems that the environmental impact is not as bad as had been expected immediately following the accident. Fortunately there were strong winds blowing offshore which helped disperse the oil slick. Numerous salmon farms have, however, experienced large losses and one month after the accident it was estimated that 2.5 million farmed salmon worth £35 million had been contaminated. The long-term environmental effects of the damage cannot yet be assessed.

Source of contamination	Location and date	Suspected total activity (TBq)*
LRW from Navy and Murmansk Maritime Shipping Line	North Atlantic Ocean Northwest Pacific Ocean 1959–1991	About 925 Over 450
SRW from Navy and Murmansk Maritime Shipping Line, including sunken reactors	Same	About 11 100 (experts estimate not over 92 500)
Sunken nuclear submarines	Atlantic and Pacific oceans	Under 24 000
Lost NWHs, RTGs, satellites, etc.	Atlantic, Indian and Pacific oceans	Many tens of thousands
Discharge of radioactive waste from Yenisey and Ob Rivers	Arctic seas	Many tens of thousands
Total	All the world's oceans	Not over 370 000

decade and kept below regulatory levels, at least in Western European countries. The environmental impact appears reasonably controlled and radiological exposure of the public well below the authorised limits in these countries (CEC, in press (b); Mayall et al, in press).

The situation is different in the countries of the former USSR, and especially in the Russian Federation. Here, waste disposal practices have been limited in many cases to direct releases of radioactive liquids to the environment (rivers, lakes, seas, soil – see Chapters 5, 6 and 7), to ground storage of untreated solid waste, and to the sea dumping of waste and even marine nuclear reactors in violation of the Convention on the Prevention of Marine Pollution by dumping of waste and other matter (usually known as the 'London Convention'). This convention was signed in 1972, taking effect in the former USSR in January 1976. Most of the high-level waste was dumped by the former USSR prior to that date; low-level waste, however, continues to be dumped in shallow waters. Information about dumping nuclear reactors by the former USSR has been released by the Russian Federation Governmental Commission on matters related to radioactive disposal at sea (Table 18.4) (IMO, 1993; Yablokov et al, 1993). A comprehensive evaluation of the potential impact on the environment of these practices is not available and remains to be made, but initial international monitoring surveys indicate there is no immediate risk to health. See also Chapter 6.

Abnormal events
The International Scale for Nuclear Events of Safety Significance (INES), adopted on a trial basis in 1989 under the aegis of the International Atomic Energy Agency and now formalised (IAEA, 1992), grades the impact of abnormal events which occur during the operation of nuclear installations on the environment (Table 18.5) as follows:

- 'Anomalies' (level 1) and 'incidents' (levels 2 and 3) do not require off-site protective measures and are therefore of no concern from an environmental point of view.
- 'Accidents' (levels 4 to 7) imply off-site contamination and if above level 4 are likely to require off-site protective actions, ranging from possible local food controls to (level 7) measures aimed at meeting extremely severe long-term environmental consequences.

The INES system, having been designed to characterise abnormal events in the nuclear industry, was not intended initially for events resulting from activities involving the use of radionuclides outside this industry. Available information shows that most events are accidental irradiation of personnel due to the mishandling of radioactive sources in industry and medicine. However, uncontrolled release or disposal of spent sources caused some pollution (a severe accident of this type was reported in 1987, at Goiána, Brazil).

Available information concerning the last decade

Table 18.4
Summary data on the scale of contamination of world oceans by radioactive wastes from former USSR territory, 1961–90
Source: Yablokov et al, 1993
Notes: *Terabecquerels (10^{12} Bq) at time of disposal
LRW – Liquid radioactive waste
SRW – Solid radioactive waste
NWH – Nuclear 'warhead'
RTG – Radioisotope thermoelectric generator.

Table 18.5
The international
nuclear event scale
for prompt
communication of
safety significance
Source: OECD, 1993

Level	Description	Criteria	Example
Accidents			
7	Major accident	● External release of a large fraction of the radioactive material in a large facility (eg, the core of a power reactor). This would typically involve a mixture of short- and long-lived radioactive fission products (in quantities radiologically equivalent to more than tens of thousands of terabecquerels of iodine-131). Such a release would result in the possibility of acute health effects, delayed health effects over a wide area, possibly involving more than one country; and long-term environmental consequences.	Chernobyl nuclear power plant, former USSR (now in Ukraine), 1986
6	Serious accident	● External release of radioactive material (in quantities radiologically equivalent to the order of thousands to tens of thousands of terabecquerels of iodine-131). Such a release would be likely to result in full implementation of countermeasures covered by local emergency plans to limit serious health effects.	Kyshtym reprocessing plant, former USSR (now in Russian Federation) 1957
5	Accident with off-site risk	● External release of radioactive material (in quantities radiologically equivalent to the order of hundreds to thousands of terabecquerels of iodine-131). Such a release would be likely to result in partial implementation of countermeasures covered by emergency plans to lessen the likelihood of health effects.	Windscale pile, UK, 1957
		● Severe damage to the nuclear facility. This may involve severe damage to a large fraction of the core of a power reactor, a major criticality accident or a major fire or explosion releasing large quantities of radioactivity within the installation.	Three Mile Island, USA, 1979
4	Accident without significant off-site risk	● External release of radioactivity resulting in a dose to the most exposed individual off-site of the order of a few millisieverts.* With such a release the need for off-site protective actions would be generally unlikely except possibly for local food control.	
		● Significant damage of the nuclear facility. Such an accident might include damage to nuclear plant leading to major on-site recovery problems such as partial core melt in a power reactor and comparable events at non-reactor installations.	Windscale Reprocessing Plant, UK, 1973; Saint-Laurent nuclear power plant, France, 1980; Buenos Aires Critical Assembly, Argentina, 1983
		● Irradiation of one or more workers which results in an overexposure where a high probability of early death occurs.	
Incidents			
3	Serious incident	● External release of radioactivity above authorised limits, resulting in a dose to the most exposed individual off-site of the order of tenths of millisieverts.* With such a release, off-site protective measures may not be needed.	
		● On-site events resulting in doses to workers sufficient to cause acute health effects and/or an event resulting in a severe spread of contamination, for example a few thousand terabecquerels of activity released in a secondary containment where the material can be returned to a satisfactory storage area.	
		● Incidents in which a further failure of safety systems could lead to accident conditions, or a situation in which safety systems would be unable to prevent an accident if certain initiators were to occur.	Vandellos nuclear power plant, Spain, 1989
2	Incident	● Incident with significant failure in safety provisions but with sufficient defence in depth remaining to cope with additional failures.	
		● An event resulting in a dose to a worker exceeding a statutory annual dose limit and/or an event which leads to the presence of significant quantities of radioactivity in the installation in areas not expected by design and which require corrective action.	
1	Anomaly	● Anomaly beyond the authorised operating regime. This may be due to equipment failure, human error or procedural inadequacies. (Such anomalies should be distinguished from situations where operational limits and conditions are not exceeded and which are properly managed in accordance with adequate procedures. These are typically 'below scale'.)	
Below scale/zero	Deviation	No safety significance	

Note:

* The doses are expressed in terms of effective dose equivalent (whole body dose). Those criteria where appropriate can also be expressed in terms of corresponding annual effluent discharge limits authorised by national authorities.

Year	INES rating			
	1	2	3	4 and above
1987	0	1	0	0
1988	1	0	0	0
1989	0	0	0	0
1990	11	11	0	0
1991	28	25	6	0
1992	28	15	2	0
1993	11	7	3	0

Table 18.6 *Nuclear events reported to the INES service of IAEA by year of occurrence (worldwide)*
Source : IAEA, INES information service, personal communication, 1993

Location	Kyshtym (Eastern Urals)	Chernobyl (Ukraine)
Date	September 1957	April 1986
Estimated radioactivity release (TBq)	740 000	2 million
Nature of plant	Chemical processing	Electricity generation
Main radionuclides of concern	Ce-144, Pr-144 (short term) Sr-90 (long term)	I-131, Cs-134, Cs-137 and rare earths
Mode of release	Instantaneous (chemical explosion)	Over some days (criticality explosion and subsequent phenomena)
Significantly contaminated area	Some 20 000 km²	Some 20 000 km²
Affected area	–	Northern hemisphere
People evacuated	11 000	116 000 (first year after accident)

Table 18.8
Severe nuclear accidents (level 6 to 7 of INES scale)
Source: IAEA, 1991; SCOPE, 1993

(Table 18.6) shows that nearly all nuclear events have been 'anomalies', and a few of them 'incidents'. The most important 'incidents' in Western Europe are reported in Table 18.7 in relation to the type of facility and the probable cause of the incident. The most significant was probably the Sellafield incident (UK, 1983) where a total of 5.9×10^{13} becquerels, mostly of ruthenium-106, was released to the Irish Sea, leading to temporary advice being given not to access the beaches in the vicinity unnecessarily, as a precautionary measure. 'Accidents', that is, nuclear events leading to environmental contamination, have been exceptional and none have occurred since 1980 in Western Europe (Windscale pile, UK, 1957; Windscale reprocessing plant, UK, 1973; and Saint-Laurent power plant, France, 1980). A reservation should be made, however, as far as the safety record of nuclear plants or ships of the former USSR is concerned: because many of them have a military character, at least partially, available information has been limited. Two extremely severe accidents have been recorded (Table 18.8) (IAEA, 1991; SCOPE, 1993): the Kyshtym accident (Southern Urals, former USSR, 1957) in a military processing plant and the Chernobyl accident at an electronuclear power plant (level 7 of INES scale, Ukraine, former USSR, 1986). According to the more open policy of the Russian Federation, information about 'anomalies' and 'incidents' (levels 1 to 3 of INES scale) is also released, in particular through the INES channels (Chernobyl nuclear power plant, 1982; St Petersburg nuclear power plant, 1991 and 1992; Tomsk military reprocessing plant, 1993).

Environmental effects

The few nuclear accidents listed above are individual in cause and consequence and do not easily allow general conclusions to be drawn about the effects of important accidental releases of radioactivity into the environment. These are discussed under five headings below and related to the particular experience of the Chernobyl and Kyshtym accidents. The Chernobyl case is particularly noteworthy here, and a description of the accident and its consequences is detailed in Box 18E.

Important radionuclides

During an abnormal event, the radionuclides released depend on the nature of the accident and of the type of nuclear plant; these determine the main sources of exposure to radiation and changes with time. In the Chernobyl case, after the decay of short-lived radionuclides, radioiodine was predominant for some weeks after the accident, present in milk and fresh vegetables. Since 1987, the radiobiological significance of contamination has been defined mainly by caesium-137 and -134, and, more locally, by strontium and plutonium isotopes.

Radionuclide distribution

The distribution of the radioactive fall-out depends strongly on the conditions of radioactive release and on the meteorological conditions at the time of the accident. The continuous release of radioactivity during the first ten days of the Chernobyl accident, with changing wind directions and the intermittent presence of rain along the path of the airborne activity, explains the broad spread and somewhat

Table 18.7
Important incidents in nuclear fuel cycle facilities (Western Europe)
Source: OECD, 1993

Nuclear fuel cycle facility	Location	Date	Type of abnormal event	Scale	Probable cause
Spent fuel reprocessing plant	Windscale, UK	1970	Nuclear criticality	No injuries; no release into the environment	Insufficient monitoring during transfer of liquors from one tank to another
Uranium enrichment plant	Pierrelatte, FR	1977	Release of hazardous chemical products	No injuries; small chemical hazards outside the plant	Accidental atmospheric release of 7016 kg of U hexafluoride
Fuel element fabrication facility	Hanau, DE	1990	Chemical explosion	No release to environment; 2 workers injured	Explosion of an off gas scrubber due to abnormal heating of a pump
Spent fuel reprocessing plant	Cap de la Hague, FR	1980	Fire	No significant release to environment; contamination limited to site	Exothermic reaction during storage of combustible and pyrophoric materials
Spent fuel reprocessing plant	Sellafield, UK	1983	Release of radioactive liquid	Unauthorised release to the environment; temporary off-site protective measures needed	Mishandling; release of radioactive liquid to the sea
Spent fuel reprocessing plant	Cap de la Hague, FR	1980	Loss of power supply	No injuries; no external release of radioactivity	Short circuit in an electric cable and subsequent fire

Note: See Table 5.9 for country codes.

Box 18E The Chernobyl nuclear accident

Chernobyl site after the accident
Source: Frank Spooner Pictures

Description of the accident

At the end of 1985 the USSR was operating 14 RBMK pressure tube reactors, each generating 1000 MW(e) of electricity. Four reactors were at Chernobyl on the Pripyat river, about 100 km north of Kiev (Ukraine) and 140 km south of Gomel (Belarus). Unit 4 came into operation in December 1983 and was destroyed accidentally on 26 April 1986. A combination of factors and circumstances contributed to this catastrophe. First, there were design drawbacks with reactors of this type and an imperfect control and protection system. Secondly, some erroneous actions taken by the operators guided by poor technical specifications and provisions for safe operation resulted in an unstable operation mode of the reactor.

The combination of these circumstances culminated in an instantaneous increase in the thermal power at 01.23 Moscow time on 26 April 1986 (Lakey, 1993). Two non-nuclear (steam/hydrogen) explosions in quick succession penetrated the reactor pressure vessel and blew the roof off the Unit 4 reactor building. Concrete, graphite and other debris heated to very high temperatures then escaped to the atmosphere. Flames and fragments spread 30 secondary fires around the Unit 3 and the turbogenerator building (IAEA, 1991).

The radionuclides, normally accumulated inside the reactor core, immediately began to be released into the environment, and this continued for ten days. Widespread distribution of airborne radioactivity thereby resulted across Europe, and eventually, at appreciably lower levels, throughout the northern hemisphere. The deposition of airborne radioactivity resulted in the contamination of soil, plants and animals, leading to foodstuff contamination.

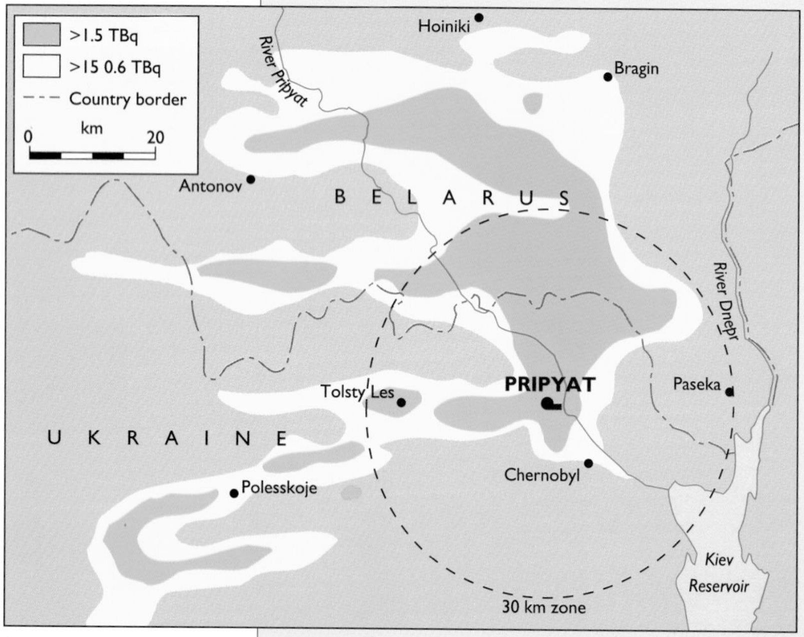

Consequences and countermeasures

On the site

The first response to the fire alarm was the power station fire team of 15 men followed by the Pripyat Fire Brigade. Eventually 15 men were fighting the spot fires close to the damaged reactor, under very difficult circumstances. Later on, reinforcements from Kiev arrived. All the first firemen received serious radiation doses. To achieve thyroid blocking, potassium iodide tablets were issued to plant personnel around 03.00. By 5.00 the fire brigade had extinguished the main fire sites, except for the fire in the central reactor hall, where graphite continued burning. In order to extinguish the burning graphite and to suppress the radioactive release, more than 5000 tonnes of material (mostly lead, sand and clay) was dumped by helicopters onto the reactor during the subsequent days. This process continued until June 1986.

There were 31 deaths and, of the 301 people who were initially diagnosed, 203 people were confirmed to show clinical effects due to radiation exposure or burns. In order to isolate the destroyed reactor, it was decided to build a containment, better known as the 'sarcophagus', around the destroyed reactor building. This work was finished by mid-November 1986. The gaps in the structure, due to the difficult conditions under which it was built, as well as the need to ventilate it, are continuously monitored for radioactive emissions (Lakey, 1993).

The near field

The town of Pripyat, 3 km from the reactor, was closed in the first few hours and people were instructed to seek shelter indoors. Later, iodine tablets were distributed to households by volunteers. Despite the social impact and disadvantages associated with relocation and resettlement, on 27 April the authorities decided to evacuate the inhabitants of the town using some 1100 buses and 200 lorries. On 2 May the evacuation zone was expanded to an area with a radius of 30 km from the plant – the exclusion zone (Map 18.2). A real exodus then started, and thousands of inhabitants and thousands of farm animals were transported. This was completed on 6 May. New buildings were constructed in Kiev, Zhitomir and Chernigov to shelter the evacuees.

After the people in the 30 km zone were moved, decontamination work started. Over a period of several months efforts were made to reduce external exposure due to radioactive materials that were deposited on surfaces: removal of soil to a depth of 15 cm, asphalting and covering of soil and cleaning of buildings. Later on, these efforts were found to be only moderately effective and natural decontamination, ie, natural decay of radionuclides and migration through the soil, seemed to be more efficient.

To prevent the highly contaminated ground from being washed away, or the radionuclides from migrating through the soil and thus contaminating the rivers and the Kiev reservoir, 140 dams and dikes were built. Initially a deep concrete wall was constructed around the plant to prevent the passage of contamination into the groundwater. Subsequently a long concrete dike was constructed along the bank of the Dnepr to separate it from ground liable to flooding, which would in turn allow activity from the heavily contaminated soil to be transferred downstream and thus appear in drinking water.

Map 18.2
Caesium contamination around Chernobyl after the accident
Source: Gagarmski, 1990

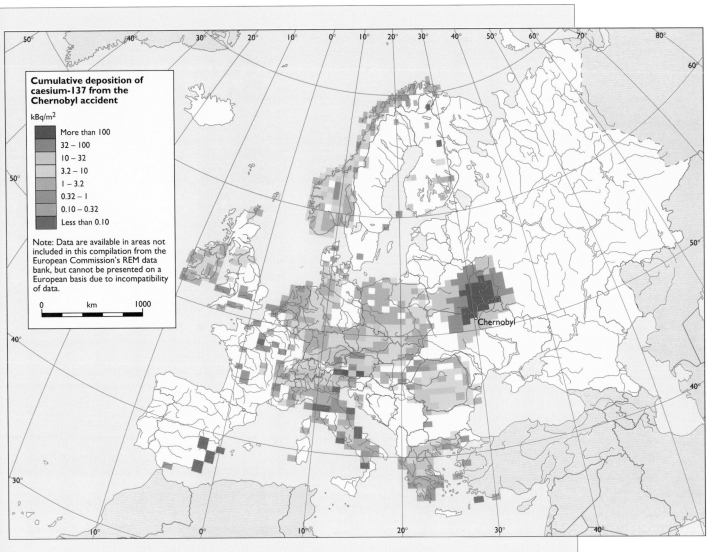

Cumulative deposition of caesium-137 from the Chernobyl accident

kBq/m^2

More than 100
32 – 100
10 – 32
3.2 – 10
1 – 3.2
0.32 – 1
0.10 – 0.32
Less than 0.10

Note: Data are available in areas not included in this compilation from the European Commission's REM data bank, but cannot be presented on a European basis due to incompatibility of data.

0 km 1000

Chernobyl

Map 18.3
Cumulative deposition in Europe of Cs-137 from the Chernobyl accident
Source: De Cort et al, 1990

The far field
The amount of caesium-137 deposited in Europe, outside the former Soviet Union (Map 18.3), varied from less than 1 kBq/m^2 (in the southern part of the UK, western France, the Iberian peninsula, and some parts of Italy and southern Greece) to 100 kBq/m^2 (in some 'hot spots' in Austria, Greece and Scandinavia, this value was even exceeded) (De Cort et al, 1990). This patchy deposition pattern led to important variations in foodstuff contamination, which induced differences in individual doses. The estimated doses range from 200 μSv (UK) to 960 μmSv (Greece) (UNSCEAR, 1988).

The European Commission and the Council of Ministers had to act urgently to set common intervention levels for food control in order to avoid large trade conflicts within the EC since, according to the Treaty of Rome, Member States are allowed to override all normal requirements for the free access of imports if no common rules exist concerning health on a particular issue. On 30 May 1986 a Regulation was adopted with the following permitted caesium levels applicable to the import of food into the EC: 370 Bq/kg for milk and foodstuffs for infants during the first four to six months of life; and 600 Bq/kg for all other products. In the meantime some 20 countries outside the EC have also adopted these levels (Luykx, 1993).

Health effects
Although precise knowledge of the full consequences of the accident is unlikely, some unexpected effects have emerged. Five years after the accident, a sharp increase in the reported incidence of thyroid cancer among children has been observed in Belarus. To date there is no proof of an increase in the incidence of any other type of radiation inducible cancer. However, the indirect adverse effects on health may be more important than the direct effects due to radiation (WHO, in press).

Follow-up
In 1992 the CEC/CIS joint programme on the consequences of the Chernobyl accident was initiated. The programme was initially conceived for implementation within the framework of CHECIR (Chernobyl Centre for International Research) which was established by an agreement between the former USSR and the IAEA in 1990. However, because of the political changes that occurred in the former USSR in 1991, a different procedure had to be followed, which culminated in an agreement being signed in June 1992 between the Commission and the three new states of Belarus, the Russian Federation and Ukraine to establish the joint programme of work. The main purpose of the CEC/CIS collaborative programme is to complement and to assist those in the three republics responsible for evaluating and mitigating the consequences of the accident, and to gain an improved understanding and knowledge of the health and environmental impact of radioactive contamination and of how it can be reduced. (Kelly and Cecille, 1994).

patchy distribution of the resulting radioactive deposition (see Maps in Box 18E). Biogeochemical processes such as radionuclide migration through soil, plant-root uptake and leaching processes subsequently modify the distribution of deposited material.

Radiation damage to vegetation

For both the Chernobyl and Kyshtym accidents, various degrees of radiation damage to vegetation were detected; during the first and second years in particular after the incidents, trees and herbaceous vegetation were affected, and especially the tree species most sensitive to radiation such as conifers (*Pinus* spp). The so-called 'Red Forest', to the north of the Chernobyl plant, died, was cut down and buried where it stood. Other forests in the vicinity of the plant continue to be a significant source of contamination for the biosphere, although wild animals are prospering in the zone. Since 1987, there has been evidence that forests have stopped showing the effects of radiation exposure and contamination in terms of mutations.

Radiation damage to fauna

Radiation damage to fauna appears to have been limited to the first years after both accidents. In the Kyshtym case, diminished reproduction was apparent in the vulnerable herbivorous fish species a few years after doses to eggs exceeded lethal levels. By 1960 no ecological effects were observable; effects on plankton, invertebrates or aquatic plants have not been detected. After the Chernobyl accident, most animals, including cattle and horses, that were not removed from the 30 km exclusion zone died; surviving cattle were all hypothyroidic and of stunted growth; the second cattle generation seems to be normal. There were no clinical signs of health effects recorded in cattle from outside the exclusion zone. A fact-finding mission, led by the Food and Agriculture Organization of the United Nations (FAO), investigated evidence from reports that had been widely published in newspapers around the world concerning birth anomalies in domestic and wild animals in the affected areas. The team concluded that none of the abnormalities reported were different from others observed elsewhere in the world.

Lessons learned and needs

The consequences of both accidents were mitigated, first, through evacuation of the population at risk, then through countermeasures related to food production and monitoring of agricultural products, and finally by reorganisation and modification of agriculture and forestry techniques. When applied to large areas, available techniques for land remediation are unsatisfactory and need improving in terms of both cost and efficiency.

Besides the more obvious questions of reviewing the major design and operation changes to the RBMK-type reactor and the training of operators and fire brigades, the Chernobyl accident certainly highlighted the need for a better emergency preparedness nationally as well as internationally in case of a large nuclear accident (see Chapter 30).

In the first instance it became clear that there was an urgent need for sufficient coverage of radioactivity monitoring stations and the necessity for rapid automatic transmission and processing of the collected data. EU countries have already updated or are in the process of updating their national early detection systems using gamma-ray detectors, and their continuous monitoring systems by means of on-line aerosol detectors. However, differences in objectives of these systems, reflected in the detection limits and rapidity of response of the systems as well as the spatial density of the network, still exist between countries. Also more attention is being given to the role of mobile monitoring equipment, in order to assess the radiological situation adequately, to examine the need for countermeasures and to provide reliable information to the public (Janssens et al, 1993).

The risks associated with natural disasters worldwide, in particular in developing countries, have increased over recent decades. This has been partly due to demographic changes which have increased population density in vulnerable areas such as coastlines and river basins. This has been coupled with a general increase in polluting human activities (mainly industrial), which have not been exclusive to areas with low natural hazard potential. Although these trends have been most acute in the developing world, certain parts of Europe (particularly in the south and east) have been undergoing similar changes. This interaction between human activities and natural hazards has increased the potential for natural hazards to contribute to environmental impacts. This is an important issue which is highlighted in this section and provides a further reason for landuse planning to play a vital role in mitigating or even avoiding such impacts.

Natural hazards are grouped below into three main sources of impact for the purpose of this review:

- storms and floods;
- heatwaves, fires and droughts;
- earthquakes and volcanoes.

Storms and floods

In Europe, as worldwide, storms and floods are the most common natural disaster and also, in terms of economic and insured losses, the most costly. The threat of coastal and estuary flooding as a result of storm surges has particularly been recognised and planned against since the North Sea floods of 1953. These resulted in serious human and infrastructural impacts to low-lying areas in the UK and The Netherlands in particular, and stimulated the development of coastal defences and more sophisticated preventative measures such as The Netherlands delta plan and the Thames barrage. Coastal defence works need a regional and transboundry approach so that impact is not just transferred elsewhere (see also Chapter 35).

Although some areas are more vulnerable than others, the transboundary nature of inland flood potential can be appreciated by considering that the river Danube runs through nine different countries. The catastrophic flooding in southern France and Switzerland in 1992 and 1993, caused by continuous thunderstorms, provides an example of the susceptibility of regional populations to inland flooding. In addition to the local damage caused by flooding, there can be unforeseen events such as landslips, mudflows or river surges caused by dam failure. This happened in northeastern Romania in 1992 when the Tazlu dam gave way, killing 107 people and resulting in economic losses of $50 million.

Environmental impacts typically include soil erosion, sedimentation and tree levelling. However, when such events interact with the pollution resulting from human activities there is the potential for more catastrophic environmental damage. This kind of potential interaction between pollution from human activities and natural hazards introduces a new dimension to international risk management and emergency planning which urgently needs to be addressed.

Heatwaves, fires and droughts

Heatwaves periodically have serious effects in Europe, including crop damage, reduced water supplies, drying up of rivers and lakes (with associated ecological impacts), human health effects and even deaths. Structural damage in urban areas has also been observed. As with floods, however, the most important potential for environmental impacts is in conjunction with human activities:

- inevitably, heatwaves increase the potential for forest fires, whether started from natural causes or human activities;

- health effects are particularly likely when the heatwave coincides with or causes urban air pollution.

In addition to direct effects on vegetation, droughts can lead to soil erosion, lowering of water tables, excessive demand on surface waters and ecological effects resulting from saline intrusion. Though a climatological drought (lack of precipitation) can be enhanced by a heatwave, the most serious consequences are often a result of a long-term decrease in precipitation in conjunction with excessive demand on water resources and poor distribution and planning. Intensive farming techniques and inappropriate forestry practices (eg, eucalyptus plantations in Portugal – see Chapter 23) can exacerbate a drought situation. Again, this indicates the important role to be played by environmentally aware landuse planning in preventing environmental disasters.

Earthquakes and volcanoes

Direct impacts of earthquakes on the environment are very small, the most serious effects on society usually being due to infrastructural damage and consequent fires, disease and lack of shelter. Clearly, however, there is the potential for earthquakes and volcanoes to cause the release of pollutants and to have serious effects (eg, by causing damage to industrial facilities). This is another aspect of risk management which needs consideration as pressure increases for industrial expansion in areas subject to seismic activity. In fact, design codes for industrial installations and nuclear plants against seismic effects are well established in earthquake-prone areas, and even in Northern Europe, including for offshore installations.

The main environmental consequence of volcanic eruptions relates to air quality. Apart from the considerable local impacts of particulate emissions, there is now evidence that emissions into the stratosphere can have other impacts, particularly on climate.

Over the past century, serious earthquakes have occurred in Italy, Greece, Turkey, Yugoslavia, Armenia and Romania, resulting in catastrophic damage and loss of life. However, there are few active fault lines in Europe and such events are infrequent compared with other regions of the world. Volcanic activity is even less frequent.

HAZARDOUS EVENTS AS A CAUSE OF DAMAGE

The identification of trends in accidents over the years has to be undertaken with some caution, since only major accidents are reported and there are different reporting limits for different chemicals. The reporting of accidents has improved over the past 20 years, but inconsistencies still occur (Box 18A). Some geographical areas receive less attention (eg, Eastern Europe) and certain types of accidents are often not reported at all.

Coincidentally there has been an apparent increase in impacts from natural hazards in recent decades (Swiss Re, 1993). Whether this is due to an increase in the frequency or intensity of events, an increased vulnerability to them, or perhaps a combination of these, is not totally clear. However, as mentioned above, it is in any case increasingly relevant to consider the importance of natural hazardous events, in conjunction with human activities, as a cause of environmental impacts.

Short- and long-term impacts

Accidents have a unique potential to cause massive short-term impacts and sometimes irreparable damage to local populations and ecosystems. This results from the acute effects of highly toxic chemicals or the impacts of explosions or large-volume releases of pollutants over a short length of time. Natural hazards have the potential to precipitate or enhance the effects of such events. In such cases, the risks to the surrounding environment presented by the hazardous activity will be significant, if not dominant, compared with contributions from more chronic sources of pollution.

In general terms, accidents are likely to cause greatest *ecological* damage through aquatic impacts, whereas atmospheric emissions are the greater cause of *human health* effects and fatalities. It is also likely that transport accidents present the greatest potential for ecological damage, taking account of oil tanker spills.

It is not realistic, therefore, to categorise accidents according to their ecological or human health impacts. Increasingly, it may be more useful for impacts to be expressed in an integrated way, accounting for both types of impact, as well as potential risks to natural resources such as groundwater (see Chapter 30).

Serious environmental impacts of single accidents are generally perceived in terms of local, immediate and short-term effects. To appreciate better the long-term environmental effects of accidents, extensive post-accident monitoring is required. Long-term monitoring is still, however, very rarely undertaken following an accident, so the long-term effects of accidents on the environment are still very poorly understood. The investigation of long-term impacts of the Basle accident (Box 18B) has revealed less damage than originally feared, whereas the river Roding accident in the UK (Box 18C) provides an indication of the potential long-term nature of resulting damage. The long-term monitoring of the *Braer* oil-spill impacts (Box 18D) will also provide useful information in this respect.

Relationship with other causes of impacts

The direct effects of a single accident on the surrounding population and environment can be very serious and warrant special consideration, as discussed above. Nevertheless, overall, the general health of the population and the quality of the environment are more affected by chronic, ongoing pollution than by the acute effects of singular events. To a large extent, therefore, the significance of the environmental and human health impact of accidents needs to be seen in the context of environmental stresses (pollution loadings, etc) from routine sources (as discussed in Chapters 12 to 17).

Only major accidents are routinely reported and smaller accidents often receive no attention and do not become part of accident statistics. There is concern, however, that the discharges from these smaller accidents can accumulate and contribute significantly to overall pollutant loads to soils and waterbodies. This source of impact is currently largely unquantified in Europe because of the lack of past reporting. As an indication, however, in North America the total amount of polychlorinated biphenyls (PCBs) released in spills between 1981 and 1985 has been estimated to be nearly twice the amount of PCBs routinely discharged into the Great Lakes (Peakall, 1989).

In addition, growing evidence of accumulated soil and groundwater contamination is resulting from site investigations, indicating that this general problem could be very serious in some areas of Europe with a long history of industrial activity (see also Chapter 15).

There is a similar concern in relation to marine oil spills. They attract a large amount of public attention and can have serious short-term impacts on the environment. The importance of tanker spills derives from their relatively rapid release of large volumes of oil, which can easily overwhelm the local receiving waters in coastal environments. However, when the possible longer-term effects of cumulative discharges are considered, other sources become important.

Table 18.9
Estimated world input of petroleum hydrocarbons to the sea
Source: Clark, 1989

Source	Input (million t/year)
Transportation	
Tanker operations	0.70
Tanker accidents	0.40
Bilge and fuel oils	0.30
Dry docking	0.03
Non-tanker accidents	0.02
TOTAL	1.45
Fixed installations	
Coastal refineries	0.10
Offshore production	0.05
Marine terminals	0.02
TOTAL	0.17
Other sources	
Municipal wastes	0.70
Industrial waste	0.20
Urban runoff	0.12
River runoff	0.04
Atmospheric fall-out	0.30
Ocean dumping	0.02
TOTAL	1.38
Natural inputs	0.25
SUM TOTAL	3.25

Petroleum hydrocarbons reach the sea by many other routes and tanker accidents are not of major significance alongside other inputs, such as operational discharge of oil during tanker operations, fuel oil discharges, and municipal and industrial wastes. An indication of the contributory sources is given in Table 18.9. It can be seen that reported tanker accidents are likely to account for about 25 to 30 per cent of the total input from transportation and only about 10 to 15 per cent of the total input due to human activities.

REFERENCES

CEC (in press (a)) Third report from the standing Working Group to the Commission on the safe transport of radioactive materials in the European Community. Commission of the European Communities, Luxembourg.

CEC (in press (b)) Radioactive effluents from nuclear power stations and nuclear fuel reprocessing plants, discharge data 1977–1986. Commission of the European Communities, Luxembourg.

Clark, R B (1989) *Marine pollution* (2nd edn). Clarendon Press, Oxford.

Cooper, A D (Ed) (1990) *The Times atlas and encyclopaedia of the sea* (2nd edn). Times Books, England.

De Cort, M, Graziani, G, Raes, F, Stanners, D, Grippa, G and Ricapito, I (1990) Radioactivity measurements in Europe after the Chernobyl accident. Part II: fallout and deposition. Commission of the European Communities. Ispra, Italy, EUR 12800.

Drogaris, G (1993) Major accidents reporting system – lessons learned from accidents notified. Community Documentation Centre on Industrial Risk, Joint Research Centre, Commission of the European Communities, EUR 15060 EN. Elsevier Science Publishers, Amsterdam.

Gagarmski, A (1990) Nuclear power and the environment. *Jornadas sobre la generación de electricidad y el medio ambiente*, 20–21 March 1990, Instituto de la Ingeniería de España, Madrid.

IAEA (1991) The International Chernobyl Project – an overview. Assessment of radiological consequences and evaluation of protective measures. Report by an International Advisory Committee for the International Atomic Energy Agency. IAEA, Vienna.

IAEA (1992) *INES: The international nuclear event scale user's manual*. International Atomic Energy Agency and the Nuclear Energy Agency. OECD, IAEA, Vienna.

IAEA (1993) International Atomic Energy Agency Bulletin, 1/1993, p 52. International Atomic Energy Agency, Vienna.

IMO (1993) International Maritime Organisation. Document No LC/IGPRAD6/UNF 4, 30 June 1993.

Janssens, A, Raes, F and De Cort, M (1993) Environmental monitoring in accidental conditions. In: Lakey, J R A (Ed) *Off-Site Emergency Response to Nuclear Accidents. Textbook based on training courses of the SCK/MOL, Belgium, 1991 and 1992*, pp 191–207. SCK/MOL and Commission of the European Communities, Luxembourg.

Kelly, G N, and Cecille, L (1994) Assessing the radiological impact of past nuclear activities and events. IAEA Technical Document 755, 115–123. International Atomic Energy Agency, Vienna.

Lakey, J R A (1993) Review of radiation accidents. In: Lakey, J R A (Ed) *Off-Site Emergency Response to Nuclear Accidents. Textbook based on training courses of the SCK/MOL, Belgium, 1991 and 1992*, pp 41–53. SCK/MOL and Commission of the European Communities, Luxembourg.

Lindgaard-Jorgensen, P and Bender, K (1992) Review of environmental accidents and incidents. Community Documentation Centre on Industrial Risk, Commission of the European Communities, Luxembourg.

Luykx, F (1993) Foodstuff intervention levels. In: Lakey, J R A (Ed) *Off-Site Emergency Response to Nuclear Accidents. Textbook based on training courses of the SCK/MOL, Belgium, 1991 and 1992*, pp 141–58. SCK/MOL and Commission of the European Communities, Luxembourg.

Mayall, A, Thomas, S, and Cooper, J R (in press) The radiological impact on the population of the EC of discharges from EC sites between 1977 and 1986. In: CEC (Eds) *Radioactive Effluents from Nuclear Power Stations and Nuclear Fuel Reprocessing Plants, Discharge Data 1977–1986*, Volume II. Commission of the European Communities, Luxembourg.

OECD (1993) The safety of the nuclear fuel cycle. Nuclear Energy Agency, Organisation for Economic Cooperation and Development, Paris.

Peakall, D B (1989) Ecotoxicological considerations of chemical accidents. In: Bourdeau, B and Green, G (Eds) *Methods for Assessing and Reducing Injury from Chemical Accidents*, SCOPE 40 IPCS Joint Symposia 11 SGOMSEC 6, pp 189–98. John Wiley and Sons, England.

SCOPE (1993) *Radioecology after Chernobyl*. SCOPE 50. F Warner and R M Harrison (Eds). John Wiley and Sons, London.

Sharples, B P M et al (1990) Oil pollution by the offshore industry contrasted with tankers – an examination of the facts. Paper presented at the Institution of Mechanical Engineers, London, 25 January 1990.

Swiss Re (1993) Natural catastrophes and major losses in 1992: insured damage reaches new record level. In: Rudolph, E (Ed) Sigma Economic Studies.

UNSCEAR (1988) *Exposure from Chernobyl: Annex D*. United Nations Scientific Committee on the Effects of Atomic Radiation. United Nations, Vienna.

WHO (in press) *Concern for Europe's tomorrow*. World Health Organisation, Copenhagen.

Yablokov, A V, Karasev, V K, Rumyantsev, V M, Kokeyev, M Ye, Petrov, O I, Lysysov, V N, Yemelyanenkov, A F And Rubtsov, P M (1993) *Facts and problems related to radioactive waste disposal in seas adjacent to the territory of the Russian Federation*. Office of the President of the Russian Federation, Moscow. Translated by Gallager, P and Bloomstein, E, Small World Publishers, Inc, Albuquerque, USA.

Part IV
Human Activities

Introduction

The importance of sustainable development and its relation to economic growth are now widely recognised. The environment and natural resources are the basic foundation of all socio-economic activity. The satisfactory guardianship of the environment is a precondition for sustainable development. Human–environment interactions are complex.

Human activities are the underlying cause and driving forces of environmental degradation. Environmental policy-makers now realise that the most effective way of improving environmental quality is to integrate environmental considerations in the management of human activities.

The purpose of this part of the report, covering Chapters 19 to 26, is to present the basic facts on energy, industry, transport, agriculture, forestry, fishing and aquaculture, tourism and recreation and households relevant to a better understanding of their environmental impacts. Some of these sectors have already been singled out in the EC Fifth Environmental Action Programme for priority attention (industry, energy, transport, agriculture and tourism).

Photo overleaf shows a mosaic of satellite images, showing the night-time city lights of Europe.
Source: National Snow and Ice Data Center/ Science Photo Library

Electricity pylons
Source: Atomic Energy
Authority Culham/
Brooks, Krikler Research

19 Energy

INTRODUCTION

The activities related to energy use may be analysed in three stages; the production of primary energy, its conversion to derived energy, and the sector in which fuels are finally consumed, or end use (see Box 19A for energy related definitions). Primary energy resources are unevenly distributed across the countries of Europe and this leads to different levels of production activity. The main factors which determine the quantity of energy consumed in any particular country include the number of people, their income level, the level and structure of production in the economy, the technology in place, energy efficiency, and energy prices. High levels of energy consumption are particularly associated with the countries of Central and Eastern Europe where energy prices have been very low in the past (see Box 19B for an explanation of data sources and country groupings used in this chapter).

Big gains in energy efficiency since the early 1970s have resulted in a weakening of the links between growth in population, GDP and energy consumption. This demonstrates that, through greater energy efficiency, it is possible to obtain the same amount of energy services (that is, the practical end use to which energy is put) using less energy input. Improvements in energy efficiency can reduce all the main environmental impacts from energy use, and are especially important for reducing carbon dioxide emissions (for which no cost-effective control technologies are currently available). This can be economically profitable at the same time by avoiding the investment in new capacity (eg, building new power stations and energy distribution systems). Improvements in efficiency of a few per cent per year can go a considerable

way towards reducing the demand for energy when an economy is growing at, say, 2 to 3 per cent per year. Further improvements in energy efficiency are possible for existing generating installations in Europe, and also for specific end-use sectors, for instance, tighter control over heating levels and/or improved insulation in European homes.

The mix of fuels consumed (ie, energy derived from solid fuels, gas, oil, nuclear, renewables and derived energy sources such as electricity) is subject to many influences, including energy and environmental regulation and policy, prices of various fuels (influenced either by the market or government intervention), technological developments, and the need for security of supply. Depending on the relative influence of these factors in the various parts of Europe, some quite different patterns of fuel mix have emerged, which in turn lead to varying contributions to environmental impacts from the generation of energy from different sources.

In 1990, the world gross energy consumption was about 8250 million tonnes of oil equivalent (Mtoe), an increase of 2.2 per cent per year since 1985, of which the EU accounted for some 15 per cent, Central Europe approximately 4 per cent, the former USSR around 16.5 per cent and EFTA countries 1.8 per cent (CEC, 1993b).

Each region in Europe consumes a mix of primary energy containing all three fossil fuels: oil, gas and solid fuels (coal, lignite), in different proportions. In the EU oil is the main fuel consumed, in Central Europe it is solid fuels, while in the former USSR gas and oil are the most important. In EFTA the main fuel is also oil, but significant proportions of hydropower and biomass are used (CEC, 1993b). In all country groups gas is likely to take an increasing share of energy consumption in the future,

Box 19A Energy definitions

The official SI units for energy measurement are Joules (J) and Watt hours (Wh) and their metric multiples (1kWh = 3600 kJ). However, the tonne of oil equivalent (toe) is widely used to help visualise the quantities involved (1 Mtoe = 41.86 PJ, where P (peta) is 10^{15}), 1 GWh = 86 Mtoe, and 1 TWh=1000 GWh. This chapter uses mostly million toe (Mtoe).

Energy may be broadly divided into two categories:

1 *Primary* energy sources include non-renewable fossil fuels (mainly solid fuels, crude oil, natural gas), nuclear power and renewables such as hydropower, geothermal, biomass and solar energy. Combined together, they provide a measure of *primary energy production*. Primary sources may be divided into two further categories in respect of their impact on global warming: carbon-intensive (solid fuels, oil, gas) and low- or zero-carbon (wind, solar, biomass, hydropower, geothermal and nuclear).

2 *Derived* energy sources are produced from the primary energy sources by converting them into other forms of energy for end use consumption. Examples are electricity, petroleum products and heat.

The main convention used for energy data is to refer to the *actual* energy content of each energy source at each stage. In particular, this means that the amount of electricity generated by hydropower and by wind power is considered directly as primary energy production.

For nuclear power, the amount of heat supplied for the generation of electricity is the primary energy input. Given the average thermal efficiency of a nuclear power station, the amount of heat supplied is about three times the amount of electricity generated. The amount of electricity produced in a conventional thermal power station may also be as little as one third of the amount of fossil fuel inputs, correctly reflecting the efficiency of traditional power stations (CEC, 1988).

- *Gross energy consumption* corresponds to the total primary energy consumed, including quantities delivered to marine bunkers.
- *Gross inland consumption* (or Total Primary Energy Supply (TPES)) is indigenous primary production, plus imports, minus exports and international marine bunkers, and plus/minus stock changes of primary energy.
- *Final energy consumption* is the consumption of primary and derived energy by the end-use sectors: mainly industry, transport, and households and services/commerce. Final energy consumption is always lower than gross inland consumption since it does not include the energy losses in conversion and distribution. Where 'consumption' is referred to in this chapter, this is gross inland consumption, unless otherwise specified, since this is more representative of the actual energy consumed in an economy. Where energy use is subdivided by end use sector or by fuel type, these data relate to final energy consumption.
- *Energy intensity* is a statistical measure which relates energy consumption (eg, gross inland consumption) to the level of economic activity (eg, GDP). Thus trends in energy intensity reflect changes in the amount of energy needed to produce a unit of economic output. This indicator is dependent on the efficiency of using energy for the various energy services required (eg, light, heat, power) and the structure of economic and social activities (eg, a high proportion of heavy industries consuming large amounts of fuel being used at comparably low efficiency, versus a service-oriented society).
- *Energy efficiency* is a measure of the overall efficiency of providing energy services, ie, the efficiency with which energy is produced from primary resources, transformed into useful forms, delivered to end users and consumers.
- *Energy conservation* is usually taken to refer just to the energy saving on the demand side.

because of both increased availability and lower environmental impact relative to other fossil fuels.

ENVIRONMENTAL IMPACTS

All energy types have potential environmental impacts, to varying degrees, at all stages in their cycle of use from extraction through processing to end use. The environmental impacts caused by energy sources that have attracted most attention from policy makers in recent years are atmospheric – acid rain and global warming – both of which stem largely from the combustion of fossil fuels. Impacts on water, soil and land may also occur but are more localised.

Specific impacts vary considerably depending on the type of energy, but also on the efficiency of the technologies employed (eg, in power stations) and the use of pollution control technologies (eg, emissions control). The construction of any energy installation (including those based on renewable sources) involves energy consumption, and full life-cycle analysis is required to assess the overall environmental effects of generating energy using a particular source. A summary of the environmental impacts caused by the main energy sources is given in Table 19.1.

Impacts from fossil fuels

Producing energy from fossil fuels is the most obvious source of environmental pressure, because several or all of the following steps are required, each having their own impacts: mining or extraction, processing, transportation, conversion, further transportation, combustion and waste disposal.

Combustion of fossil fuels results in emissions to the atmosphere of carbon dioxide (CO_2), sulphur dioxide (SO_2), nitrogen oxides (NO_x) and particulate matter (and dust), as well as metals and radionuclides. CO_2 is the major contributor to global warming, while SO_2 and NO_x cause acid rain, and together with particulate matter contribute to bad air quality. NO_x emissions also contribute to the formation of tropospheric photochemical oxidants (see Chapter 32). Methane (CH_4), another potent greenhouse gas is emitted during the extraction and processing of all fossil fuels (including coal), from leaks in the gas transportation system, and during combustion. The amount of CO_2 produced per unit of energy from coal, oil and gas is in the approximate proportion of 2 to 1.5 to 1 (IEA, 1991).

Combustion of fossil fuels (including that for transport and industry) contributes around 80 per cent to total worldwide anthropogenic CO_2 emissions. Emissions in Europe are about 30 per cent of this total. In Europe, solid fuels and oil each contributed similar amounts to total CO_2 emissions in 1987 (IEA, 1991). Coal production and gas

Box 19B Energy sources, availability and country groups

Key energy data sources

There are three main international organisations which gather energy data from questionnaires and studies, according to the scope of their responsibilities. The main data sources are Eurostat (CEC), the International Energy Agency (IEA) and the United Nations (UN). Results from questionnaire exercises are published regularly by these organisations, and there is substantial coordination between them in terms of classification, methodology and definitions.

- *Eurostat:* publications such as *Energy Balance Sheets* (eg, CEC, 1991) and the *Energy Yearly Statistics* (eg, CEC, 1993a) series give full and complete coverage for the EU but do not include forecasts, or data for Central and Eastern Europe. The Directorate General for Energy (DG XVII) of the CEC uses Eurostat data for the EU, supplemented by data from other sources, to provide up-to-date commentary on energy such as in its *Energy in Europe* series (eg, CEC, 1993b). These include complete energy balance data, with forecasts as well as historical data. Data are presented by individual EU countries and grouped into world regions.
- *International Energy Agency (IEA):* the IEA compiles annual statistics for OECD countries on energy production and consumption by fuels and sectors, such as in the publications *Energy Balances of OECD Countries* (eg, IEA, 1993a) and *Energy Statistics of OECD countries* (eg, IEA, 1993b). The IEA also publishes energy balances for non-OECD countries, including Central and Eastern European countries (eg, IEA, 1993).
- Until 1993 UNECE collected data which were published each year (eg, UNECE, 1993c). Data cover basic production and consumption of primary energy and reserves, and include renewable energy sources – by type and by individual country. No forecasts are included.

From these sources energy data on total production and consumption are available for most countries except the newly independent states of the former USSR. However, sectoral fuel consumption data, on which to base analyses of energy efficiency prospects, is limited for all countries. Substantial errors are likely to exist in particular for Central and Eastern Europe where the use of energy was not measured in the past. For example, the IEA reports differences between IEA's own estimates and those made by the former Czechoslovakia for 1990 energy balance data, using IEA estimates as the base – see Table (IEA, 1992).

	IEA method	former Czecho-slovakia method	Difference in data	
	Mtoe	Mtoe	Mtoe	per cent
Total final consumption of solid fuels	10.8	21.6	10.8	+100.0
Industry petroleum	2.7	6.3	3.6	+133.0
District heating and combined heat and power(CHP)	12.7	2.4	–10.3	–81.1

Even larger errors are to be expected in emissions data, and in some countries estimates do not yet exist for all major gases. In addition to better information on fuel type and energy production, improvements in emissions reporting and forecasts will require the establishment of detailed inventories of major energy-using plant and equipment (ie, boilers, power stations, vehicle stock), their age, size, type of technology and current use of pollution control equipment. This task is being tackled in the CORINAIR project (see Chapter 14, Box 14C).

Country groupings

Given the form, described above, in which existing energy statistics are published, the following country groups have been used to present the data in this chapter:

- European Union (EU): Twelve Member States for both current and historical data. Up to and including 1990 this excludes East Germany, which is included in the Central and Eastern Europe group. Forecasts of data presented refer to Germany beyond 1990, and therefore include East Germany as part of the EU or Western Europe.
- EFTA: Austria, Finland, Iceland, Norway, Sweden, Switzerland, Liechtenstein.
- Western Europe: EU and EFTA combined.
- Central Europe: Albania, Bulgaria, Croatia, Czech Republic, Hungary, Poland, Romania, Slovak Republic, Slovenia, and the remaining territory of the former Yugoslavia.
- Former USSR: the whole of the former Soviet Union, including the Baltic Republics, Belarus, Ukraine, Moldova and the Russian Federation.
- Central and Eastern Europe: Central Europe and the European part of the former USSR.

Where Europe as a whole is referred to this includes the EU, EFTA, Central and Eastern Europe, Cyprus and Malta; the degree to which the former USSR is included or excluded is always stated.

---

leakage contribute significant amounts to methane emissions. (See Chapters 4 and 14.)

Consumption of fossil fuels provides the major contribution to anthropogenic atmospheric emissions of SO_2 and NO_x. In 20 European countries, for 1990, around 95 per cent of SO_2 emissions and 97 per cent of nitrogen oxides arose from the combustion of fossil fuels. The highest emissions come from those sectors, electricity generation and industry, which use the most energy and the highest sulphur-content fuels, that is, solid fuels and high sulphur heavy fuel oil. Energy efficiency has increased in Western countries, meaning fewer pollutants are emitted per unit of energy consumed. In Central and Eastern Europe, however, emissions are still very high per unit energy consumed (see also Chapter 20). This is caused by the high share of solid fuels in the total fuel mix, the poorer quality and high sulphur content of these (lignites), the low

quality of power generation installations, and the lack of effective emissions control technology. For instance, pollution from the combustion of coal, in terms of kilogrammes of pollutants per toe used in the energy sector, is consistently worse in Poland than in OECD nations. Sulphur dioxide emissions in 1989 were estimated at 32 kg per toe of coal in Poland, well ahead of most OECD countries, where typical values are in the range 2 to 4 kg per toe (*Financial Times*, 1993).

Solid fuels are the most polluting fossil fuels locally and globally. Solid fuels range from hard, 'clean-burning' bituminous coals to soft brown coals and lignites, which have high proportions of combustion waste and pollutants such as sulphur, heavy metals, moisture and ash content. The mining of solid fuels may have major local impacts on water quality and the landscape (for example, in Poland: see Box 19C). Impacts from coal mining differ substantially between deep

Environmental media	Air	Water	Soil/Land	Nature and wildlife/Landscapes
Energy source				
Fossil fuels	• combustion → increasing levels of greenhouse gases → climatic effects • combustion → emission of particulates and dust, SO_2, H_2S, NO_x, CO, CO_2, CH_4, small amounts of heavy metals and radionuclides (coal) → poor air quality, acid rain, and global warming • storage and transportation of gas → leaks, explosions and fires • cooling towers → microclimatic impacts	• electricity generation → thermal releases from cooling raising water temperature • mining activities → pressure on available water resources • coal mining → liquid mine waste (containing acids and salts) → water pollution • coal washing → water pollution • NO_x and SO_2 emissions → deposition → acidification and eutrophication of waterbodies • stockpiles of coal → water pollution • oil spills and leaks → water pollution • combustion of liquefied petroleum gas (LPG) → problems of liquid residual disposal	• mining activities, plants, pipelines → physical intrusions → pressure on land resources • deposition of acidifying compounds → increasing soil acidity • solid wastes and ash disposal (spoil tips) from coal mines → contamination of water percolating through slag heaps → groundwater and soil pollution • coal mining activities → land subsidence	• procuring fossil fuels → disturbance of natural habitats, exploitation of wilderness or natural areas for surface mining • deposition of acidifying compounds → acidification → impacts on flora and fauna • accidental oil → threats to wildlife and coastal zones • mining, electricity generation, waste disposal and transportation (eg, pipelines, pylons) → visual impacts on the landscape • cooling towers and transmission cables → visual impacts
Nuclear power	• nuclear accidents → potential release of radionuclides • cooling towers → microclimatic impacts	• uranium mine operation → groundwater contamination • mine tailing water → release of toxic metals and radiological wastes • plant operation → thermal releases from cooling water and emissions of radionuclides • accidental releases → contamination risk of water resources	• nuclear accidents → risk of soil contamination due to deposition from air • improper storage of radioactive wastes → potential contamination • decontamination and decommissioning of nuclear power plants → potential contamination • risk of accidents during transport ofspent fuels → potential contamination	• nuclear accidents → risk of impacts through air, soil and water contamination → secondary impacts on organisms transmission • cooling towers and transmission cables → and visual impacts
Hydropower	• dams → micro-climatic effects (from flooding of large areas and physical intrusion) • inundation of land to create reservoirs → CH_4 → global warming	• dams → impacts on hydrological cycle (re-routing of rivers), water quality and reliability • dam rupture → risk of major flooding	• dams → land irreversibly flooded (loss of fertile land and forests) • dam construction → landslide risks • dams → increased erosion and sedimentation	• flooding → habitat changes, fish migration • dams and associated infrastructure → visual intrusion and landscape changes
Other renewables (biomass, geothermal, solar, wind)	• biomass combustion → emissions of CO, VOCs, NO_x, particulate matter • geothermal energy → emissions to air	• biomass conversion → water pollution • geothermal energy → emissions to water	• energy plantations (biomass), wind farms, and solar parks → pressure on landuse • decommissioning of photovoltaics → toxic pollution	• renewable energies → some habitat and wildlife disturbance (eg tidal barrage schemes) • wind farms and solar parks → visual intrusion

Table 19.1 Overview of significant and potential environmental impacts from energy sources
Sources: EEA-TF and ERM; WHO, 1992

Eula Saxe open air coal mine near Leipzig
Source: Patrick Piel, Frank Spooner Pictures

Box 19C Impacts of coal mining in Poland

In Poland, coal accounts for over 90 per cent of energy production and 75 per cent of gross inland consumption, as well as being the country's main export. The abundance of domestic coal reserves and years of subsidised pricing have led to widespread inefficiency, in both the coal sector and the energy sector in general. Before the recent political changes, other sources of energy were virtually ignored. As a result of energy policies which neglected environmental concerns, coal mining became a major polluter, with serious health consequences for the communities living and working in the mining regions.

The environmental effects of this high level of mining include air emissions, liquid effluents and solid wastes. The global air emissions caused by mining are not significant in comparison with those produced during combustion, but include significant local impacts such as dust originating from drilling and blasting, and methane emissions (estimated at an average rate of 5 cubic metres per tonne of coal mined).

The main environmental problem resulting from mining is acid waters from mine drainage. The continuous acid discharges from Polish mines seriously affected aquatic ecosystems, since acid waters containing heavy concentrations of dissolved heavy metals will support only limited water flora, and will not sustain fish. Acid waters are not suitable for human consumption and are too corrosive for most industrial uses.

The Polish government has set medium-term goals for the next eight to ten years, aiming to cut dust and acid water from mines, to repair physical damage caused by mining, to restore mined lands, and to consume the methane cleared from mines (in addition to cutting emissions of SO_2 and NO_x from coal combustion).

Sources: *Financial Times*, 1993; World Bank, 1991

underground mining (which is visually less intrusive but expensive) and surface or opencast mining (which is cheaper but results in major local impacts, both visually and on air quality).

Compared with solid fossil fuels, gas and oil are less polluting. Gas is the least polluting fossil fuel. The major impacts from oil are associated with accidental spillages during transportation both at sea and on land. The resultant damage to coastal areas and marine life can be dramatic in the short term and may also have long-term consequences; clean-up is very expensive. The majority of major marine environmental accidents in the last ten years have involved oil tankers. Data stretching back over 20 years are available on oil spills from pipelines in Western Europe. Out of a total of 50 spillage incidents between 1987 and 1991, only three caused no environmental pollution (CONCAWE, 1992). (See also Chapters 18 and 30.)

Nuclear power

Normal operation of nuclear power stations does not result in serious environmental impacts, although nuclear power does present the problem of storage and disposal of radioactive waste, particularly from spent fuel or waste processing (see Chapter 15), and the risks of major environmental impacts from nuclear accidents such as Chernobyl (see Chapter 18).

Long-term storage of radioactive waste is a major source of concern for which there is not yet a clear solution. The earlier interest in the reprocessing of spent fuel has diminished in favour of the search for safe methods of long-term dry storage of spent fuel or waste (with the notable exception of THORP in the UK). There is also concern about the risks and problems of waste storage associated with decommissioning of reactors at the end of their lifetime (see Chapters 16, 18 and 30 for further discussion of this topic).

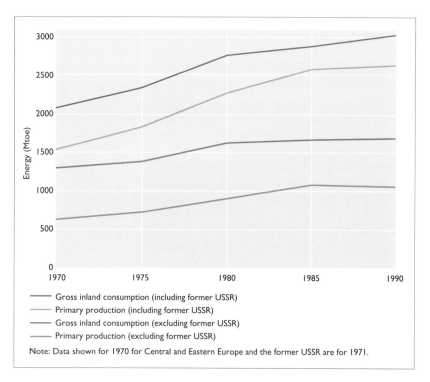

Note: Data shown for 1970 for Central and Eastern Europe and the former USSR are for 1971.

Map 19.1 Oil and
gas distribution and
production in Europe
Source: GISCO, CEC,
based on various sources

Figure 19.1 *Trends in primary production and gross inland*
consumption in Europe, 1970–90
Source: Special data set compiled for this chapter based on International Energy
Agency and Eurostat data (see *Statistical Compendium*)

Renewable energies

Generating energy from renewable sources also has some
environmental impacts. With the exception of large hydropower
schemes, these are generally small and localised. The
environmental impacts of hydro schemes are closely related to
their scale and siting. The social and environmental impacts in
upstream areas include resettlement of people living in the areas
to be submerged, loss of valuable fertile land and forest,
alteration or destruction of the landscape by the infrastructure
required to support such schemes, effects on wildlife and fish,
and loss of livelihood for people relying on fishing, farming and
related activities. These impacts are frequently found to be
unacceptably high, and prevent the development of large-scale
hydropower schemes. In addition, the re-routing of rivers to
supply dams may lead to conflicts between competing uses for
water (such as those over the construction of the Gabcikovo
dam on the Danube on the Slovak/Hungarian border). The
ecological consequences of such schemes are difficult to predict
and may be very complex (see Chapter 5).

In a few countries renewables provide a high proportion of
total energy requirements (for instance, Iceland has pioneered use of
geothermal energy). Exploitation of sources of renewable energy in
the future would require commitment of significant areas of land
(which, for biomass, may become possible with the reforms under
way in agricultural policy – see Chapter 22) and would inevitably
have some environmental consequences: visual impacts (from wind,
and solar), noise impacts (wind), and small quantities of air emissions
(geothermal, forest residues, biomass conversion). However, these
alternative sources appear to have at least the potential to ease
emissions from the energy sector of greenhouse gases and other
pollutants (such as SO_2, NO_x and particulate matter).

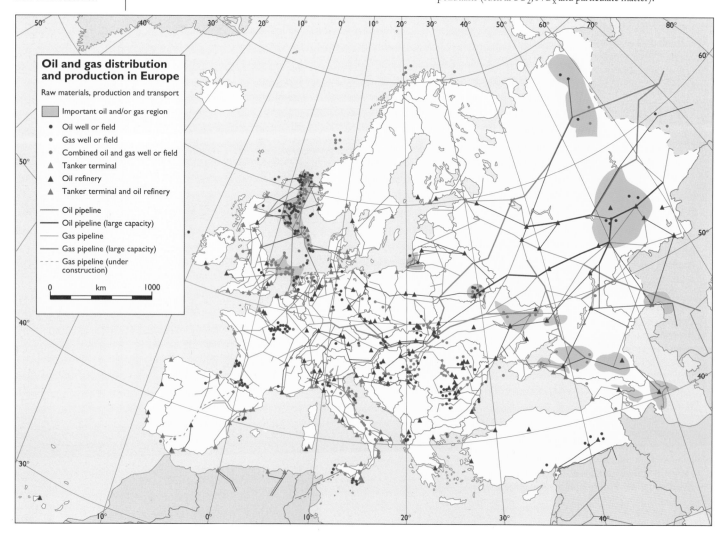

TRENDS IN ENERGY PRODUCTION AND CONSUMPTION

There is no direct link between consumption and production of energy in a region or country, since production depends on the availability of primary (principally fossil fuel) resources and export markets, while consumption is linked to the energy intensity and structure of economic activity. In this section the trends in the production of primary fuels and derived fuels, gross inland consumption and final energy consumption are presented.

Since 1970, the overall trends in energy production and consumption in Europe have moved in parallel, but with production growing slightly faster. In 1970, primary production of energy in Europe (excluding the former USSR) was half of gross inland consumption; by 1990 it was 62 per cent, as the demand for energy stimulated greater production after the oil shocks of the 1970s.

From 1970 to 1990 the total gross inland consumption in Europe (excluding the former USSR) grew from almost 1300 Mtoe to 1670 Mtoe, an average annual growth rate of 1.3 per cent, while primary production grew from just over 635 Mtoe to 1040 Mtoe, 2.5 per cent per year on average (excluding the former USSR). Both production and consumption grew much faster in the 1970s than during the 1980s. From 1970 to 1980, consumption growth averaged 2.2 per cent per year (production grew by 3.6 per cent per year) while in the 1980s the annual growth rate averaged 0.4 per cent (production 1.4 per cent), in all cases excluding the former USSR. There, total gross inland consumption

increased by 74 per cent from 770 Mtoe in 1971 to about 1340 Mtoe in 1990. Over the same period primary production increased by 82 per cent, from 884 Mtoe in 1971 to around 1600 Mtoe in 1990. Figure 19.1 shows the markedly different trends between the two decades.

Production trends

Primary energy

Primary energy refers mainly to oil, gas, solid fuels and nuclear energy. The distribution and production network in Europe of oil and gas is shown in Map 19.1. The sites and capacities of nuclear power plants are illustrated in Map 19.2.

The 1970s was a period of very rapid growth of total primary energy production after the oil price shocks, followed by a period of much more limited growth in the 1980s. Nuclear power production continued to grow throughout the two decades while solid fuels production declined; the trends for oil and gas were more mixed.

Changes in total primary production reflect changes in demand for energy both domestically and worldwide (since fossil fuels are internationally traded commodities), and new discoveries and exploitation of reserves. Between 1970 and 1980, in the EU, total primary energy production rose by over 40 per cent, reflecting mainly the exploitation of North Sea oil and gas by the UK and The Netherlands. Nuclear power also increased significantly. The increase of primary energy production over the same period in the former USSR was higher, at over 50 per cent (due to a greater exploitation

Map 19.2
Location and maximum net capacity of European nuclear power plants by site, 1992

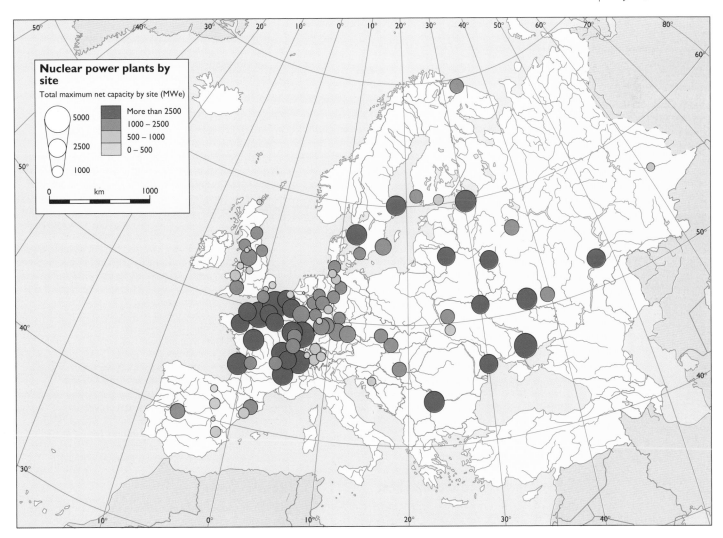

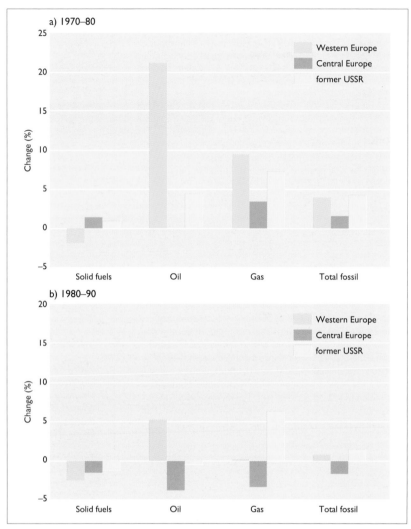

a) 1970–80

b) 1980–90

Figure 19.2
Changes in primary production of fossil fuels in Europe,
a) 1970–80;
b) 1980–90
Source: Special data set compiled for this chapter based on International Energy Agency and Eurostat data

consumption was 18 per cent in the EU, 29 per cent in EFTA, 14 per cent in Central Europe, and 12 per cent in the former USSR (see *Statistical Compendium*). Electricity intensity of an economy tends to increase as it develops, since electricity is the principal fuel used in commercial and service activities. This explains the high and growing proportion of electricity consumption in Western Europe. Electricity consumption in Central Europe and the former USSR, while lower than in Western Europe, has been encouraged by very low electricity prices.

The use of district heating is highest in the Central and Eastern European countries, but also has significant use in EFTA countries. In 1990, the percentage contributed by heat to final energy consumption in the EU was negligible at 0.6 per cent, while it was 17 per cent in Central Europe, 12 per cent in the former USSR and 4 per cent in EFTA (see *Statistical Compendium*). Outside the EU, district heating, sometimes also involving the cogeneration of electricity(ie, combined heat and power), has long been established as a principal means of delivering heat to the urban domestic sector. The energy saving potential of this approach has, in Nordic countries, been shown to be substantial. However, the very poor standards and maintenance of heat distribution systems sometimes found in Central and Eastern Europe means that it is possible, in some cases, that as much as 50 per cent of the heat sent out of district heating boilers is lost (ERM Energy, personal communication, 1993).

Figure 19.3 shows the relative contributions of different sources to electricity production in 1990. Some key features are:

- currently, the EU uses a higher proportion of nuclear power – at 35 per cent – than any of the other country groups;
- EFTA relies on fossil fuels for only 11 per cent of its electricity, while the lion's share is generated using renewable sources, particularly hydropower;
- in Central Europe solid fuels provide nearly two thirds of the electricity produced;
- in the former USSR over one third of electricity is generated using gas.

In the future, the share of gas in power generation is expected to grow in all country groups, and this will help to reduce emissions from the sector.

The contribution of nuclear power in meeting Europe's demand for electricity has expanded rapidly in recent years, but the rate of expansion is slowing down. Not all countries have nuclear power programmes and several countries have moratoria on the development of future nuclear power stations. Only six EU countries currently operate nuclear power stations, and the share of nuclear power in electricity generation varies considerably between European countries. For example, there is a very high proportion of electricity generated by nuclear power in France (84 per cent), Belgium (62 per cent) and Switzerland (66 per cent), but in other countries where nuclear power is used the share is less than half.

The use of nuclear power to generate electricity has increased in all country groups during the 1970s and through much of the 1980s, with big increases, particularly in the EU and former USSR, between 1980 and 1985 (see Figure 19.4). This shows the energy inputs from nuclear power, which is about three times the output in terms of electricity generation, due to inefficiencies in the energy transformation process. The EU has by far the highest growth rate in electricity generation from nuclear power. After 1985 the growth rate slowed in all country groups, but in the EU, use of nuclear power to generate electricity has still grown faster than in the former USSR, Central Europe, or EFTA.

of large natural gas reserves). A different pattern emerged in the period 1980 to 1990, when, in the EU and the former USSR, total primary energy production increased by around 20 per cent. In EFTA countries, primary production increased during the 1970s more than threefold (230 per cent), but in the 1980s this fell to around 90 per cent. Central Europe shows the biggest contrast between decades, with a 22 per cent increase between 1970 and 1980, but a 13 per cent *fall* in primary production between 1980 and 1990.

Figures 19.2a and b show the changes in production of different fossil fuels in Europe by country group since 1970. From 1970 to 1980, the major increase in oil production in Europe (outside the former USSR) was due to rapid increases in North Sea oil production. There were also increases in the production of gas, but an overall stagnation in the production of solid fuels. Increases in gas production were due to a desire to diversify the supply of energy away from oil (to limit the impact of any future oil price rises), to shift production away from solid fuel for environmental reasons, and to enable cheap gas to penetrate new markets. In the former USSR, production rose for every fuel, particularly gas, of which the country has vast reserves.

In the period 1980 to 1990 production of solid fuels continued to stagnate in both groups and, while gas production passed its peak and fell in Western Europe, it continued to rise in the former USSR. Liquid fuels production grew in Western Europe but fell in the former USSR due to production difficulties.

Derived fuels: electricity and heat

The use of electricity in Europe is highest in Western Europe. In 1990, the proportion of electricity in final energy

Renewable sources

Both fossil fuels and the uranium used in nuclear plants are, ultimately, finite resources, and this has contributed to interest in the potential of exploiting renewable sources of energy. These include hydropower, solar, wind, biogas, biomass and geothermal energy.

Renewable energies in the EU accounted for about 3 per cent of gross inland consumption, of which 61 per cent was accounted for by biomass, and 31 per cent by hydropower (EP, 1992). The European Commission has developed programmes to encourage the penetration of renewable energy sources in the EU, and increase trade in products, equipment and services within and outside the EU (see 'Policy' below).

The Nordic countries have been leaders in exploiting renewable sources of energy in Europe, with Sweden, Finland, Iceland and, particularly, Norway producing large quantities of hydropower (in Norway, hydropower provides nearly all the electricity); Bernes (1993) reports that Denmark produced some 2 per cent of its total electricity output from wind power – over half of all Europe's wind-power capacity; Iceland generated 31 per cent of its gross inland consumption from geothermal (the main countries exploiting geothermal energy in the EU are Italy and, to a lesser extent, France). Iceland additionally generated about 37 per cent of its gross inland consumption in 1990 from hydropower, so that a total of 68 per cent of its gross inland consumption (and 99.9 per cent of its electricity generation) is generated from renewable sources (Icelandic Ministry for the Environment, 1992).

The main problem with wind power is that it requires considerable space to set up the generating capacity. For instance, it would be technically possible to build a 1000 MW wind farm near Pori on the west coast of Finland, but this would require 2500 generators in a 100 km[2] area (EDC-MEF, 1993). However, conventional sources of energy also need large areas of land, which is, if all stages in the energy cycle from mining through processing to final waste disposal are counted, at least as much as the area required for wind power. For countries with large areas of open, sparsely populated land, therefore, wind power may be a realistic option for meeting a part of the demand for energy.

There has been increasing interest, from both environmental and agricultural perspectives, for greater use to be made of biomass to generate electricity and heat, usually in combined heat and power (CHP) plants, or in district heating plants. Biomass fuels include ethanol from agricultural products, vegetable oils as diesel fuel, wood and wood waste, and other wastes. Wood is a common fuel for heating in many European households, although well-established sources of data on the amounts involved are not generally available (see Chapter 26). Use of biomass is CO_2 neutral, and is also viewed in some quarters as a possible solution to the problem of agricultural surpluses (see Chapter 22).

Biogas recovered from sewage, landfill sites and agricultural wastes is a potential energy source at the local level. As these gases are mostly methane (a potent greenhouse gas – see Chapter 27), capturing and burning them is in any case environmentally worthwhile, and deriving energy is an added bonus. Biogas contributed over 50 ktoe to primary energy production in 1989 in the UK, The Netherlands and Italy (EP, 1992).

Over the last decade significant progress has been made in all basic applications of active solar energy: production of domestic energy, space heating, heating of swimming pools, and heat for commercial, agricultural and craft end uses. For the EU, the annual production of solar energy is about 200 ktoe, mostly in Greece, France, Spain and Portugal, although the growing markets are in more northerly countries. There is also increasing interest in photovoltaic systems, but the generation of electricity on this basis is not at present competitive, being about ten times more expensive than other forms of generation (EP, 1992), except in remote locations.

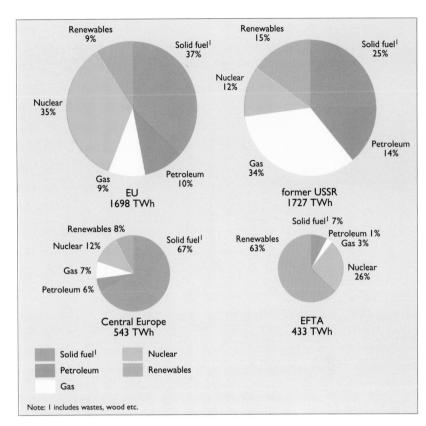

Consumption trends – by fuel type and by end-use sectors

The gross inland consumption shown in Figure 19.1 masks significantly different trends between the European country groups. Figure 19.5 compares the trends between these groups.

In the EU, EFTA, Central Europe and the former USSR, the average annual increases in consumption between 1970 and 1980 were 1.8, 2.2, 3.5 and 3.8 per cent respectively. In the EU, the increase in consumption in the 1970s was much lower than the rise in production, resulting in an increase in exports and a reduction in imports of oil. This was a deliberate response to the oil price shocks of the 1970s and the desire to decrease import dependency. In the former USSR, consumption rose by an amount roughly matching its own production. The former USSR, being self-sufficient in energy and a net exporter of oil and gas, was insulated from fluctuations in world energy prices and therefore did not

Figure 19.3
Electricity generation by source by country group, 1990
Source: Special data set compiled for this chapter based on International Energy Agency and Eurostat data

Figure 19.4
Nuclear input to electricity generation in Europe, 1970–90
Source: Special data set compiled for this chapter based on International Energy Agency and Eurostat data

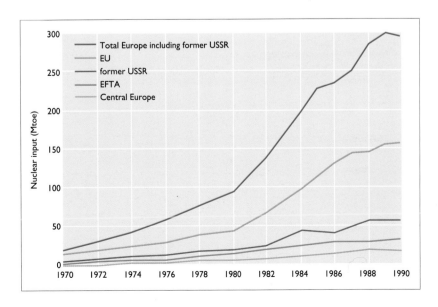

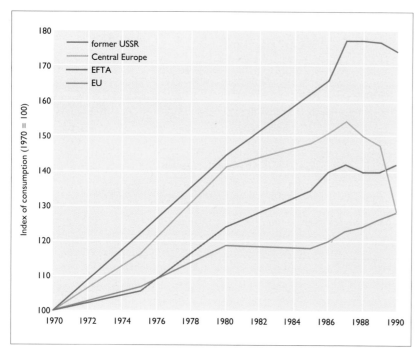

Figure 19.5
Trends in gross inland consumption by country group, 1970–90
Source: International Energy Agency and Eurostat (see *Statistical Compendium*)

reduce their import dependency by developing their own resources, diversifying supply and improving energy efficiency. In Western Europe, this trend began in the 1970s. Central and Eastern Europe were not exposed to world energy prices until the late 1980s with the start of the economic reform process and independence from the former USSR. The latter has more recently moved from a barter trade system to one in which the aim is to sell its energy at or near to world prices for hard currency. Before these reforms, Central and Eastern European countries became accustomed to plentiful energy supplies at heavily subsidised prices. This led to large inefficiencies in the use of energy, since there was no incentive to invest in energy efficiency. Also, a concentration of manufacturing production in the energy intensive industrial sectors had developed, since low energy prices gave these sectors a comparative economic advantage in world markets.

Figure 19.6 compares energy intensity and gross inland consumption per capita for selected European countries. Consumption per capita is similar in Western and Central European countries, but higher in the Russian Federation. Higher energy intensity illustrates greater inefficiency in the use of energy and higher energy demand for a given level of output. Due to the distorting effect of unstable exchange rates, it is not possible to calculate energy intensity with GDP for Central and Eastern Europe and thus make an accurate quantitative comparision of the high energy intensity of the countries of the former USSR with the countries of the rest of Europe.

Low energy prices in Central and Eastern Europe have been a major influence on energy consumption. In Hungary, for example, where reform started relatively early, energy prices rose between two and four times between 1987 and 1991, but were still well below world prices and below their costs of production. In Lithuania, energy prices in 1991 at current exchange rates were about ten times too low in relation to their economic costs (and 15 to 20 times too low in the case of diesel and fuel oil) (ERM energy estimates, personal communication, 1993).

The problem of low energy prices in Central and Eastern Europe was compounded by the absence of effective price signals, and of any means to control energy consumption. Consumers are charged a flat monthly amount for their gas and heat (and sometimes also electricity) consumption, irrespective of how much energy they use. The lack of metering or effective pricing perpetuates a very poor understanding of the quantities, and costs, of energy consumed. This problem can extend right up to the national level; for example, in Lithuania even the bulk delivery of gas by pipeline from Russia is not currently metered at the border.

However, the non-metering and absence of control over consumption of energy is not unique to Central and Eastern Europe, since in many European countries, consumers in domestic (households) and commercial (small business/service) sectors – unlike most large industrial consumers – have little control over their heating systems, or are not aware of how much energy they use.

Trends in final energy consumption by fuel type are shown in Figure 19.7. In the EU, about half of (1990) final energy consumption is of oil, and almost all the remaining consumption is of gas, electricity and solid fuels. In Central Europe in 1990, oil and solid fuels each accounted for around one quarter of final energy consumption, with heat (here meaning CHP), electricity and gas accounting for the remainder. Overall there is small use of renewables (about 2 per cent) (DG XVII, CEC, personal communication). In EFTA countries oil still provides some 45 per cent of final energy consumption. In the former USSR, oil and gas each account for about 30 per cent of final energy consumption, with solid fuels, electricity and heat together accounting for the remaining 40

respond to the 1970s oil price increases in the same way as Western European countries. In the 1980s consumption growth rates fell substantially in all groups, but particularly in Central Europe. Between 1980 and 1990, the average annual growth rates for energy consumption in the EU, EFTA, Central Europe and the former USSR were respectively 0.8, 1.4, –0.9 and 1.9 per cent per year.

The divergence of trends was even more marked in the second half of the decade. From 1985 to 1990, energy consumption in the EU rose by 8 per cent, in EFTA by 6 per cent and in the former USSR by 7 per cent, while in Central Europe it fell by 13 per cent. The markedly different trends in the former USSR and Central Europe reflect the changing economic and political situations in the two areas. By 1990 economic reform had already started to have a major impact on Central Europe, but not on the former USSR.

Western as well as Central and Eastern Europe have tried to substitute their consumption away from imported fuels and

Figure 19.6
Energy intensity, and gross inland consumption per capita for selected countries, 1990
Source: Special data set compiled for this chapter based on International Energy Agency and Eurostat data

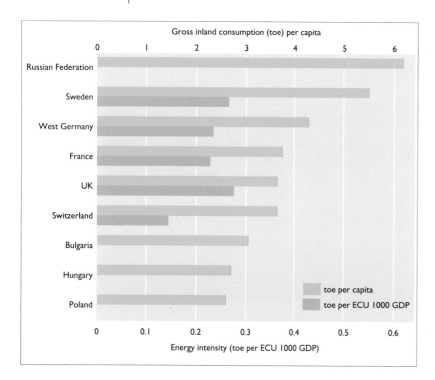

per cent. Renewable energy is not shown in Figure 19.7, but the EFTA countries have the greatest contribution to final energy consumption (mainly biomass and hydropower).

The mix of fuels is influenced by the levels of activity in different sectors of an economy. The use of energy by economic sectors (mainly industry, transport, and domestic/commerce) is examined in the other chapters of this report covering those activities. The advanced economies of the EU have a relatively high share of activity in the commercial/services sector (where electricity is the dominant fuel) and the transport sector (where fuel consumption is almost entirely of petroleum products). The economies of Central and Eastern Europe have a much greater dependence on the industrial sector, where the demand for boiler fuels (solid fuels, oil or gas, depending on supply and cost) is a large proportion of total demand. In Central and Eastern Europe there is also a considerable dependence on solid fuels as household heating fuel. This is particularly the case in those countries with a large indigenous solid fuel resource, such as the Czech Republic, Poland and the Slovak Republic, and it is often the poorer qualities of fuel which are supplied to the domestic market. These low-level sources of air pollution are, in many urban areas, a greater contributory factor to poor urban air quality than the emissions from the high stacks of solid fuel (coal)-burning power stations (see Chapters 4 and 10).

The variable percentage breakdown of energy use between sectors in different country groups is illustrated in Figure 19.8 for 1970–90. A higher proportion of energy is consumed by industry in the former USSR and Central Europe, and in the EU the share on transport is higher.

There has been little incentive for increasing energy efficiency or implementing conservation measures in energy-intensive industries in Central and Eastern Europe because of the subsidised energy made available to the industrial sector. The use of energy by the industrial sector in these countries is therefore expected to fall as prices increase under market influences (see Chapter 20). The transport sector accounts for a relatively high percentage of final energy use in the EU. This is related to large increases in road transportation (see Chapter 21).

FUTURE ENERGY USE: UNDERLYING DRIVING FORCES

The main factors influencing future growth of gross inland consumption and final energy consumption will be the rate of growth of the economy and its structural development, the scope for energy efficiency improvements and conservation measures, technological improvements and energy prices. The major change in fuel mix is expected to be a significantly growing share for gas in all areas of Europe.

Recent projections suggest that final energy consumption will grow across Europe between 1990 and 2005, with growth fastest in the EU (1.2 per cent) and slowest in Central Europe (0.6 per cent) (CEC, 1992). However, future trends in Central and Eastern Europe are highly dependent on the pace of economic recovery in those areas; final energy consumption is expected to fall over the next few years as economic output declines, before starting to grow again. The assumption underlying the projections given below is that the economies will grow at around 2 per cent per year, averaged over the whole period, in Western and Central Europe, but at less than 1 per cent in the former USSR (these CEC projections differ from the more recent UNESC projections given in Chapter 2). This implies an improving trend in energy intensity in most of Europe but a worsening situation in the former USSR, where final energy consumption is expected to grow by about 1 per cent per year, somewhat faster than GDP (CEC, 1992).

Factors which will influence the future mix of fuels consumed include the end use sectoral share of demand,

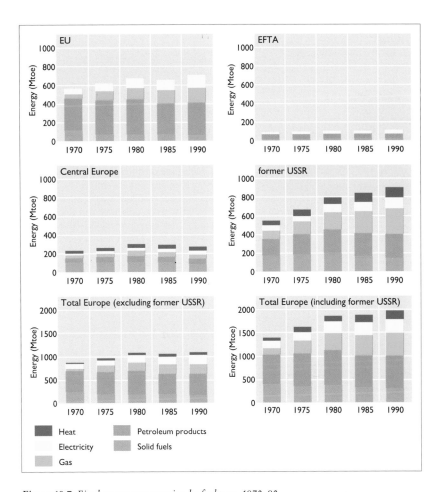

Figure 19.7 *Final energy consumption by fuel type, 1970–90*
Source: Special data set compiled for this chapter based on International Energy Agency and Eurostat data (see *Statistical Compendium*)

Figure 19.8 *Final energy consumption by end-use sector, 1970–90*
Source: Special data set compiled for this chapter based on International Energy Agency and Eurostat data (see *Statistical Compendium*)

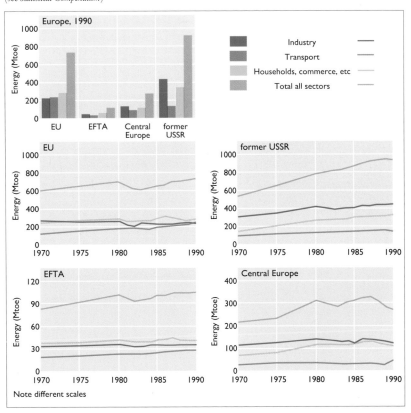

Region	Percentage shares 1990				Percentage shares 2005			
	Oil	Coal	Gas	Nuclear	Oil	Coal	Gas	Nuclear
EU	44	24	17	13	42	19	24	12
former USSR	31	20	42	4	26	13	55	3
Central Europe	24	49	20	4	21	35	37	3

Note: Percentages do not sum to 100 since hydropower and other renewables are not included.

Table 19.2
Current and future shares in gross inland consumption, by country group
Source: ERM Energy, based on CEC, 1992

relative fuel prices, technological developments, security of supply, environmental pressures and government commitment to supporting certain fuel industries (see the next sub-section). In the scenario presented in Table 19.2 the projected changes in fuel mix are considerable; these changes are clearly shown in Figure 19.9. Consumption of solid fuels is projected to decrease in all areas. Oil's share will fall a little everywhere, but fastest in the former USSR. In all areas gas is expected to increase its share substantially.

Resources and technological developments

Tighter environmental standards in EFTA and the EU, together with increasing attempts by Central and Eastern Europe to attain similar standards eventually, will lead to wider use of pollution control technology, in energy generation and consumption. The heavy use of solid fuels in Central and Eastern Europe, without the use of desulphurisation, has led to significant air pollution and acid rain problems in the region (see Chapters 4 and 31). With increasing use of flue-gas desulphurisation (FGD), filtering and various other technologies, and a trend towards the use of cleaner solid fuels and gas, these problems will be alleviated. Typical performance data and costs of control technologies for reducing SO_2 and NO_x emissions from stationary combustion sources are given in Chapter 31. The former USSR is showing an increasing tendency to use gas and this will significantly lessen the emissions of greenhouse and acidic gases (provided the gas leaks from gas processing and pipeline systems are kept low), relative to the heavier use of solid fuels and oil in previous years. Western Europe will tend to turn more to gas and this will also help diminish emissions of greenhouse and acidic gases relative to the previous emissions from an energy sector dominated by oil and solid fuel.

Figure 19.9
Projected changes of gross inland consumption by fuel type, 1990–2005
Source: ERM Energy, based on CEC, 1992

The choice of fuel type and conversion technologies is strongly influenced by cost, and any new technologies which had a significant impact on relative fuel costs would also alter the fuel mix and overall demand. The exploitation of pollution control technology and scope for efficiency improvements on the demand side varies considerably between end-use sectors, with, for instance, households and small business activities sluggish to adapt compared with the more vigorous transport sector (see Chapters 21 and 26). Interest in new technologies on the supply side has focused on the so-called advanced or clean coal technologies, improvements in emission control, renewable energy technologies and fuel cells. Research may also bring environmental benefits in the long term:

- Clean coal technologies such as integrated gasification combined cycle plant offer improved overall combustion efficiency and considerably lower emissions of pollutants per unit of energy produced. They are likely to find wider application in Central and Eastern Europe (where solid fuels are the main indigenous resource and new coal stations may be required to replace old stations, including nuclear power stations) than in Western Europe.
- Improvements in emission control will include reductions in the sulphur content of fuels, improved combustion efficiencies, and the further use of emission control technologies such as FGD and low NO_x burners.
- Interest in renewables continues to grow and their costs are gradually falling, although they still require significant financial and other support in most applications. The European Commission has proposed setting a target of doubling the share of renewables in primary energy by 2005, although this target is not reflected in the scenario shown in Table 19.2. Principal contributing technologies would be wind power, waste use, landfill gas and biomass. The target is technically achievable but will require vigorous policy backing.
- Fuel cells are unlikely to make an impact in the next few years, but offer good prospects for a high efficiency and low environmental impact conversion technology after the turn of the century.
- Research is continuing into 'inherently safe' nuclear reactors but the outlook for their application is uncertain. Technologies for the control, capture and disposal of CO_2 are also being researched and developed. Long-term research is still ongoing into possibilities for electricity generation from fusion power.

The current trend is to increase the exploitation of gas and its penetration into new markets. Significant resources exist in the non-European part of the former USSR but their pace of exploitation will be dependent on the establishment of a stable political framework and legal and commercial basis for the encouragement of overseas investment. At current production rates the proven gas resources in the Russian Federation will last about 60 years (BP, 1992). Norwegian gas reserves have an expected life of 44 years from 1990, and oil reserves of 13 years. Proven reserves probably underestimate the ultimately recoverable resource and continuing exploration activity will possibly extend the lifetime of the reserves by one or more decades (*Statistical Compendium*).

The limitation of an energy source and depletion is, however, only a matter of concern if there are no alternatives to depleting energy sources, or where knowledge does not improve to harness the alternatives, or to use existing resources more efficiently. Present limitations to energy use are therefore linked more to the tolerating capacity of the atmosphere (ie, with respect to greenhouse gases – see Chapters 4 and 27) and ecosystems (with respect to pollutants such as NO_x and SO_2 – see Chapter 31).

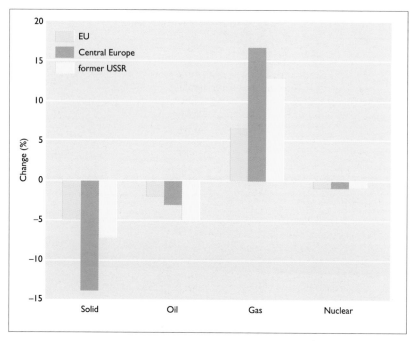

Policy

The role of governments in guiding the development of supply sources, technologies and demand is very important. Driving forces in policy issues include the need for security of supply and low energy prices (in Western Europe), the establishment of prices reflecting market realities (in Central and Eastern Europe), and the reduction of the environmental impacts of energy production and use. In respect of the last goal, it is necessary to progress steadily to an environmentally sustainable energy future.

Security of supply has been tackled largely by diversification of both energy sources and geographic origin of deliveries, and by improvements in energy efficiency. Security of supply can also be enhanced by improving the relationship and mutual understanding of consumers and producers, who respectively seek security of supply and demand. The European Energy Charter is a political declaration which aims to advance this understanding, and that of environmental aspects of the production, transport and consumption of energy, by committing its signatories to cooperate in trade, investment and policies in energy sectors at a pan-European level.

Key trends in energy are also linked to broad factors in economic development, environmental policy and resource development. For example, in Central and Eastern Europe, a major decline in gaseous emissions is occurring from their previously high levels through ongoing efforts to modernise industrial production and in reducing the dependence on solid fuels in the region. Efficiency improvements will also be required in other sectors, especially households, services and transport.

Energy and environment policies may also influence demand by affecting energy prices, which should take full account of all costs in the energy cycle. For instance, the aim of a tax on carbon/energy would be to reduce emissions of greenhouse gases by improving energy efficiency and fuel switching. Policies may also take the form of subsidies, accelerated depreciation for energy/environment investments, and cross-compliance measures. For example, the European Commission has a number of programmes designed to encourage a shift in the pattern of energy use to achieve environmental objectives. These include measures targeting promotion, research and development of 'cleaner' and renewable energy technologies, and improving energy efficiency (see Box 19D). The OPET network (Organisations for the Promotion of Energy Technology) was created by the European Commission to ensure effective application of available energy technologies, and to encourage implementation of new technologies. There are 12 European Commission Energy Centres in Central and Eastern Europe which carry out these roles.

Another policy direction is driven by the increasing efforts to bring private sector finance and competition into energy markets. This has resulted in some moves to privatisation (eg, in the UK), breaking up vertically integrated monopolies and allowing new entrants to compete with former monopolies. New market structures are leading towards different investment choices; for example, the move towards smaller power stations (typically 350 MW, rather than the 1000 plus MW power stations built in the past), gas-fired power stations, and CHP for large users. The integration of European gas and electricity networks is bringing opportunities for growing trade in energy products.

Prices, demand and energy efficiency

Price changes affect overall demand, relative fuel shares and energy efficiency; the most obvious example of this was the impact of the two oil price rises in the 1970s. Prices of natural gas and solid fuels were also affected and a period of rising real prices stimulated changes in energy intensity in the short term, and long-term investment in research and development for improved technologies. In the last few years some energy prices have been falling in real terms and, furthermore, inflation has eroded at least some of the initial impact of efficiency gains.

Energy intensity trends reflect changes not only to efficiency improvements and technological developments, but also to changes in the structure of the economy, international trade, the fuel mix, and sectoral energy policies. Reducing the energy intensity of an economy, through any of these changes, will reduce the demand for energy and improve security of energy supply. In addition, without the decline in energy intensity since 1973 in Western Europe, CO_2 emissions would have been much higher.

Establishing prices which reflect market realities in Central and Eastern Europe will play a crucial role in improving energy efficiency in those markets. Energy at competitive prices is essential for economic development. Any economy which has subsidised or distorted energy prices, as is the case in many countries of Central and Eastern Europe, is likely to be able to achieve substantial energy savings (and hence CO_2 and other emission reductions) by raising their prices towards international energy price levels. Table 19.3, for example, shows the changes in emissions that are estimated could occur in Central and Eastern Europe if the price subsidies were to be removed. No such dramatic changes are expected in Western Europe, where energy prices already broadly reflect their costs of production, but throughout Europe the increasing interest in internalising the external environmental costs of the energy cycle into energy prices is likely to become a key influence on the demand for energy and in underlining the need for further improvements in energy efficiency in the future.

Box 19D
Key European Union policy initiatives

- The CEC's energy support programmes, JOULE, Thermie, SAVE, Altener – research and development, demonstration and funding programmes for new energy technologies, information dissemination and promotion of innovation. Actions have focused on energy efficiency and renewables, in support of the CEC's carbon dioxide stabilisation target.
- Fifth Environmental Action Programme – *Towards Sustainability* – sets the EU approach for achieving sustainable development and requires a move towards a less carbon-intensive economic structure.
- Internal energy market – aims to allow free movement of energy, improve security of supply and improve competitiveness. Will lead to increased trade in gas and electricity, with expected slightly lower costs.
- CO_2/energy tax – fiscal mechanism aimed at stabilising carbon dioxide emissions in 2000 at 1990 levels. The proposed tax on fuels of an equivalent of $3 per barrel of oil in the first year would rise to the equivalent of $10 in the final year of introduction (around 2000).
- Energy Cooperation Programmes, PHARE, TACIS, European Energy Charter, Regional Energy Programmes: – cooperation programmes between the EU and countries in Central and Eastern Europe, and the rest of the former USSR. Establishment of a pan-Europe energy agreement between the EU and Central and Eastern Europe to exchange technologies for energy development and emissions control and facilitate the exploitation of resources.

	Estimated change in Central and Eastern Europe
Estimated reduction in CO₂ emissions, 1995	
Amount (millions of tonnes of carbon)	446MtC
As share of projected Central and Eastern European emissions	29%
As share of projected global emissions	7%
Cumulative CO₂ reduction, 1991–2000	
Amount (millions of tonnes of carbon)	3796MtC
As share of projected Central and Eastern European emissions	24%
As share of projected global emissions	6%

Note: The base case is derived from World Bank projections of energy demand. In this scenario, worldwide CO_2 emissions increase by about 20 per cent between 1990 and 2000.

Table 19.3 Effects on carbon dioxide emissions of eliminating subsidies on commercial energy in Central and Eastern Europe
Source: Adapted from World Bank, 1992

SUMMARY AND CONCLUSIONS

- All sources of energy generation have some environmental impacts, whether on the atmospheric environment or terrestrial and aquatic habitats, or through generating waste that is difficult to deal with. Therefore, generating energy from any source involves making choices between impacts and about how far those impacts can be tolerated – at the local and global scales. The construction of any energy installation (including those based on renewable sources) involves energy consumption, and full life-cycle analysis is required to assess the overall environmental effects of generating energy using a particular source.

- The fossil fuel cycle (extraction, processing, combustion, disposal of waste) has significant environmental impacts, mainly on the atmosphere. Combustion of fossil fuels leads to emissions to the atmosphere of CO_2, SO_2, NO_x and particulate matter, as well as metals and radionuclides. Carbon dioxide is the major contributor to global warming, while SO_2 and NO_x cause acid rain. NO_x emissions also contribute to the tropospheric ozone problem. Methane (CH_4), another potent greenhouse gas, is emitted during the extraction and processing of all fossil fuels (including coal), from leaks in the gas transportation system, and during combustion.

- Since 1970, the overall trends in energy production and consumption in Europe have moved in parallel, but with production growing slightly faster. In 1970, primary production of energy in Europe (excluding the former USSR) was half of gross inland consumption; by 1990 it was 62 per cent. Both production and gross inland consumption grew much faster in the 1970s than during the 1980s. From 1970 to 1990 the total gross inland consumption in Europe (excluding the former USSR) grew at an average rate of 1.3 per cent per year, while primary production grew 2.5 per cent per year on average. From 1970 to 1980 the growth in consumption averaged

2.2 per cent per year (production grew by 3.6 per cent per year) while in the 1980s the annual growth rate averaged 0.4 per cent (production 1.4 per cent).

- In Europe as a whole, the mix of fuels used for energy generation has shifted away from solid fuels and nuclear power towards increased use of gas. The share of solid fuels has been falling substantially except in parts of Central and Eastern Europe, where they are the main indigenous energy resource. In these cases, where changing the mix of fuels will be difficult in the short term, 'clean coal technologies' offer a more realistic solution to pollution problems. For renewables there is little use at present, but increasing interest; generation from these sources is particularly well established in EFTA countries.

- Improvements in energy efficiency are one sure way to reduce emissions per unit of useful energy from the energy sector, and, furthermore, save money. There have been big gains in energy efficiency, particularly in Western Europe, since the early 1970s, and this has demonstrated the possibility of breaking the link between economic growth and energy consumption. However, efficiency gains have been eroded by falls in real oil prices, and on the supply side there remain differences between the operating efficiency of power generation installations and energy distribution networks across Europe. On the demand side, scope for efficiency improvements in energy use varies between sectors and countries, with, for example, the transport and household sectors capable of big efficiency gains.

- Major driving forces for future patterns of energy use and the future mix of fuels consumed include the share of demand by end-use sectors, sectoral energy policies, relative fuel prices, technological developments, security of supply, environmental pressures and government commitment to supporting certain fuel industries. Consumption of solid fuels is projected to decrease in all areas. Oil's share will fall a little everywhere, but fastest in the former USSR. In all areas, gas is expected to increase its share substantially.

- Competitive energy prices are essential for economic development. The energy intensity of an economy is increased or kept high if energy prices are held below market levels. Increases in real prices of energy encourage more efficient energy use and implementation of conservation measures. The full costs of the energy cycle should be included in energy prices.

- Improvements in emission control and investments in technology can reduce emissions per unit of useful energy and improve combustion efficiencies of fuels. Options being researched and developed for the future include the possibility of 'safe' nuclear reactors, fusion power, fuel cells and CO_2 control, capture, and disposal. Further investment in renewable energy sources is also likely to bring dividends and encourage a shift away from fossil fuel consumption, although many renewable technologies are currently unable to compete with the lower economic costs of conventional power generation.

- Although there is no established methodology for calculation and integration of external costs of the energy cycle, energy extraction, production and consumption are inherently subsidised because market prices for various forms of fossil fuel energy do not take account of the cost of environmental damage resulting from atmospheric emissions. The increasing interest in incorporating these costs is likely to become a key influence on the demand for energy and in underlining the need for further improvements in energy efficiency in future.

REFERENCES

Bernes, C (Ed) (1993) *The Nordic environment: present state, trends and threats.* Nord 1993 : 12 Nordic Council of Ministers, Copenhagen.

BP (1992) *Statistical review of world energy.* June 1992, British Petroleum plc, London.

CEC (1988) *Principles and methods of the energy balance sheets.* Commission of the European Communities, Luxembourg.

CEC (1991) *Energy balance sheets, 1988–1989.* Commisson of the European Communities, Luxembourg.

CEC (1992) *Energy in Europe: A view to the future.* September 1992, DG XVII Energy, Commission of the European Communities, Luxembourg.

CEC (1993a) *Energy yearly statistics 1991*, Commission of the European Communities, Luxembourg.

CEC (1993b) *Energy in Europe: annual energy review.* Special issue April 1993, DG XVII Energy, Commission of the European Communities, Luxembourg.

CONCAWE (1992) *Performance of oil industry cross-country pipelines in Western Europe: statistical summary of reported spillages, 1991.* Conservation of Clean Air and Water in Europe (CONCAWE), Report 4.92, Brussels.

EDC-MEF (1993) *The state of the Finnish environment.* Environment Data Centre and Ministry of the Environment, Finland, Helsinki.

EP (1992) *Possibilities and limitations of alternative energy sources in the context of EC policy.* Energy and Research series W4, EN-10-92, European Parliament, Directorate General for Research, Luxembourg.

Financial Times (1993) Financial Times Profile: Poland. *East European Report 17*, Financial Times Newsletters, London.

Icelandic Ministry for the Environment (1992) *Iceland: National Report to UNCED.* Icelandic Ministry for the Enviroment, Reykjavik.

IEA (1991) *Greenhouse gas emissions: the energy dimension*, International Energy Agency, OECD, Paris.

IEA (1992) *Energy policies, Czech and Slovak Republics: 1992 survey.* International Energy Agency, OECD, Paris.

IEA (1993a) *Energy balances of OECD countries, 1990–91.* International Energy Agency, OECD, Paris.

IEA (1993b) *Energy statistics of OECD countries, 1990–91.* International Energy Agency, OECD, Paris.

IEA (1993c) *Energy statistics and balances of non-OECD countries, 1990-91.* International Energy Agency, OECD, Paris.

UNECE (1993) *Annual Bulletin of General Energy Statistics for Europe 1993.* United Nations , New York.

WHO (1992) *Report of the Panel of Energy.* World Health Organisation Commission on Environment and Health, WHO/EHE/92.3, Geneva.

World Bank (1991) *Environmental assessment source book volume 3: guidelines for environmental assessment of energy and industry projects.* Technical Paper 154, World Bank, Washington DC.

World Bank (1992) *World Development Report 1992: Development and Environment.* Oxford University Press, Oxford.

Paper plant at St Gaudens, France
Source: Michael St Maur Sheil

20 Industry

INTRODUCTION

Industry is for most countries one of the main contributors to generating income. In 1991, it accounted for up to 50 per cent of Gross Domestic Product (GDP) in countries such as Bulgaria and Romania, and in the Russian Federation about 40 per cent of GDP (UNECE, 1993a); in the EU manufacturing industry accounted for almost 25 per cent of GDP in 1991 (CEC, 1993), while in Nordic countries in 1990, the proportion of manufacturing in GDP was one quarter or less (Nordic Council of Ministers, 1992).

The purpose of industrial activities is primarily to manufacture goods for final consumption, and for the manufacture of other products (intermediate consumption). In meeting these demands, manufacturing industries have an impact on the environment, through processing of raw materials and their subsequent manufacture into finished products. Even if the processes used are designed to minimise emissions (that is, if they are environmentally 'clean'), or employ emissions abatement technology (known as 'end-of-pipe'), any manufacturing industry will contribute to some extent to environmental impacts through the use of energy and raw materials. The environmental significance of a manufacturing activity can be greater if the raw materials used are non-renewable. The main impacts arise directly: as a result of emissions to air or water, or by their effects on the land and soil, mainly near the site of production, as well as from wastes generated and deposited.

In addition to manufacturing, industrial activities also include service industries such as catering, cleaning, or financial services, and although carrying out these activities can result in sometimes significant environmental impacts, these are not generally covered in this chapter.

Although manufacturing industry itself is a contributor to many and various types of contamination, pollution and use of non-renewable resources, industry also plays a major role in providing solutions to environmental problems. This includes the development of new processes and the machinery necessary for effective pollution abatement, and introducing new technologies and product modification. Industry in Europe is under continuous pressure to improve its productivity and product quality, but at the same time must adapt to strong and growing constraints – often legal – on pollution emissions and energy consumption. On the one hand, the use of pollution control devices is, in general, detrimental to the efficiency of the plant; on the other, improvement in the efficiency of processes and equipment will directly reduce the specific emissions per unit of output (Carvalho and Nogueira, 1993).

The approach in this chapter is first to look at the importance of 'industry' as a whole in terms of emissions and use of natural resources, and to give a general overview of some of the industrial data relevant to environmental questions. The aim of the second part of the chapter is to illustrate, using specific sectors, some examples of 'environmental performance data' from different industrial activities in parts of Europe. The chapter also considers the ways in which business practice has changed in response to environmental challenges faced by industry.

DEFINITIONS OF INDUSTRY

For the purposes of this report, 'industry' refers to manufacturing activity involving transformation of raw materials into products. This therefore includes processing of non-energy related materials, and manufacturing activity in general, under normal operating circumstances. The energy sector is considered in Chapter 19. The impacts related to the use, consumption and disposal of finished products are considered in Chapter 15 and partly in Chapter 26. Threats to the environment arising from emergency situations and accidents and the management of risk by industry are covered in Chapters 18 and 30.

The term 'industry' has to be interpreted flexibly, since definitions for some countries include mining and energy production, whereas for others forestry, and even agriculture, may be included. Data availability on industry and country groupings used in this chapter are described in Box 20A.

ENVIRONMENTAL IMPACTS

A great range of manufacturing and service industries give rise to environmental impacts. Table 20.1 provides a summary of the types of emissions from selected sectors of industry which may have notable impacts. There are many complex links in the way that resources are used by different sectors: some process resources which are used by others, and manufacturers make products which are used to manufacture other products. During this process, some emissions result, and once the good is produced it must be stored and

Table 20.1
Overview of significant and potential environmental impacts by industrial sectors
Source: OECD (1991), WHO (1992), and van der Most and Veldt (1992)

Sector	Air	Water	Soil/Land
Chemicals (industrial inorganic and organic compounds, excluding petroleum products)	● many and varied emissions – depending on processes used and chemicals manufactured ● emissions of particulate, matter, SO_2, NO_x, CO, CFCs, VOCs and other organic chemicals, odours ● risk of explosions and fires	● use of process water and cooling water ● emissions of organic chemicals, heavy metals (cadmium, mercury), suspended solids, organic matter, phenols, PCBs, cyanide → water quality effects ● risk of spills	● chemical process wastes → disposal problems ● sludges from air and water pollution treatment → disposal problems
Paper and pulp	● emissions of SO_2, NO_x, CH_4, CO_2, CO, hydrogen sulphide, mercaptans, chlorine compounds, dioxins	● use of process water ● emissions of suspended solids, organic matter, chlorinated organic substances, toxins (dioxins)	
Cement, glass, ceramics	● cement → emissions of dust, NO_x, CO_2, chromium, lead, CO ● glass → emissions of lead, arsenic, SO_2, vanadium, CO, hydrofluoric acid, soda ash, potash, speciality constituents (eg, chromium) ● ceramics → emissions of silica, SO_2, NO_x, fluorine compounds, speciality constituents	● emissions of process water contaminated by oils and heavy metals	● extraction of raw materials ● metals → soil contamination and waste disposal problems
Iron and steel	● SO_2, NO_x, emissions of CO, hydrogen sulphide, PAHs, lead, arsenic, cadmium, chromium, copper, mercury, nickel, selenium, zinc, organic compounds, PCDDs/PCDFs, PCBs, dust, particulate matter, HCs, acid mists ● exposure to ultraviolet and infra-red radiation, ionising radiation ● risks of explosions and fires	● use of process water ● emissions of organic matter, tars and oil, suspended solids, metals, benzene, phenols, acids, sulphides, sulphates, ammonia, cyanides, thiocyanates, thiosulphates, fluorides, lead, zinc (scrubber effluent) → water quality effects	● slag, sludges, oil and grease residues, HCs, salts, sulphur compounds, heavy metals → soil contamination and waste disposal problems
Non-ferrous metals	● emissions of particulate matter, SO_2, NO_x, CO, hydrogen sulphide, hydrogen chloride, hydrogen fluoride, chlorine, aluminium, arsenic, cadmium, chromium, copper, zinc, mercury, nickel, lead, magnesium, PAHs, fluorides, silica, manganese, carbon black, HCs, aerosols (local exposures depend on the particular material being processed)	● scrubber water containing metals ● gas scrubber effluents containing solids, fluorine, HCs	● sludges from effluent treatment, coatings from electrolysis cells (containing carbon and fluorine) → soil contamination and waste disposal problems
Refineries, petroleum products	● emissions of SO_2, NO_x, hydrogen sulphide, HCs, benzene, CO, CO_2, particulate matter, PAHs, mercaptans, toxic organic compounds, odours ● risk of explosions and fires	● use of cooling water ● emissions of HCs, mercaptans, caustics, oil, phenols, chromium, effluent from gas scrubbers	● hazardous waste, sludges from effluent treatment, spent catalysts, tars
Leather and tanning	● emissions including leather dust, hydrogen sulphide, CO_2, chromium compounds	● use of process water ● effluents from the many toxic solutions employed, containing suspended solids, sulphates, chromium	● chromium sludges

Notes: HCs: hydrocarbons; VOCs: volatile organic compounds; PAHs: polyaromatic hydrocarbons; PCBs: polychlorinated biphenyls; PCDDs: polychlorinated dibenzodioxins; PCDFs: polychlorinated dibenzofurans.

Box 20A
Data availability and country groups

Some general data are available for describing 'industry' or 'manufacturing industry' as a whole. There are many data on production, employment and trade flows in various industrial sectors, and some limited data are available which cover the types of resources (raw materials, energy and water) used by the sectors. Such data are published by, for example, the European Commission (CEC, 1991), the Nordic Council of Ministers (1992), and the UN system (eg, UNECE, 1992; UN, 1990). Sources of data for Central and Eastern European countries include national reports submitted to the 1992 UNCED conference, and the Eurostat/Statistisches Bundesamt (E/SB) report on Central and Eastern Europe, and other parts of the former USSR (E/SB, 1991).

It is also possible to describe trends in energy use by industry as a whole. Total energy consumption and consumption by fuel type are available for most European countries. However, disaggregated information on energy use by specific industrial sector is available for only a few countries.

Data on emissions (to air and water, generation of waste) which have a potential environmental impact are scarce (see Chapters 14 and 15). Information about the types of substances arising from various processes can be found, although this is often qualitative, and not generally monitored by national authorities at the sector level. A few countries – eg, The Netherlands (VROM, 1991) – have made progress towards development of integrated emissions inventories, which enable estimations to be made of contribution by industrial activities of emissions to air and water (see Chapter 14). Some examples of emissions trends in specific industrial sectors are available from company or industry reports. Results concerning improvements in environmental performance are available, but data showing that emissions to the environment are increasing are much harder to come by (see the section on industry response to environmental factors later in this chapter).

As explained in Chapter 15, statistics on waste are particularly unreliable and national estimates of the amount of industrial waste (especially hazardous waste) generated have a wide margin of error. Data on soil and groundwater pollution differ sharply between regions in detail and reliability (see Chapters 5 and 7).

The country groupings used in this chapter are as for Chapter 19: EU, EFTA, Western, Central, former USSR, Central and Eastern Europe – see Box 19B.

The spatial significance of impacts will also vary, from the local to the national, from groups of countries, to the European and global levels. Industry has some specific links to environmental quality in media, and at different scales; for instance, emissions from industrial activities can have both transboundary and global effects (eg, emissions of sulphur dioxide leading to acidification of soil and waters, and producing carbon dioxide emissions leading to climate change) as well as more localised health and water quality effects (see Chapters 5 and 11).

Industrial accidents which may have consequences for the environment and the general public are often the events which attract media interest (see Chapters 18 and 30). War damage to industrial plants can also endanger the environment. Although major acute events of this kind can affect many countries, the general environmental situation is often more accurately defined by a larger number of chronic problems. For instance, in Eastern Europe this includes numerous localised events such as poor air quality in many industrial centres, and related chronic problems resulting from industrial activities such as refuse tips, slag heaps, ash dumps, tailings dumps and spoil heaps which can cover large areas (Danilov-Danilyan and Arski, 1991).

The health of the general population or local community may be affected through the manufacture of potentially hazardous products or by-products, or through exposure to industrial emissions into air, water or soil. Acute (short-term) effects more often result from emergency conditions, while chronic (long-term) effects may be associated with years or decades of exposure to agents, often at levels not much above background concentrations (WHO, 1992). Industrial activities also have the potential to affect the health of workers employed in industrial facilities, although occupational health concerns are outside the scope of this chapter (see Chapter 11).

The links between emissions and environmental quality are explained in Chapter 14. The following sections of this chapter concentrate on the main sources of industrial emissions to air, water and soil.

The main determining factors for the type and significance of emissions from industry are:

- type of product manufactured, processes used;
- raw materials used, content of fuels and ores;
- resource use intensity (air, water, energy);
- size of plant and location;
- technology employed;
- surrounding environment;
- dispersion potential of emitted pollutants (including from accident occurrence);
- resource efficiency of the manufacturing process (particularly energy with respect to emissions to air); and
- quality and efficiency of abatement technology.

To a great extent it is the status and effectiveness of technology employed, environmental abatement equipment, and the effective training and management of staff which determine the level of emissions from an industrial plant of equal capacity in different countries. Gains in energy efficiency are also of key importance, and the weakening of the link between economic growth and energy consumption (see Chapter 19) has meant that, in Western Europe at least, increases in industrial production do not necessarily lead to increases in energy consumption and therefore emissions to the environment (see 'Energy use' below).

transported before it can be used. Production of many goods has evolved over the years to take environmental effects into account. The environmental impact varies at different stages in the life-cycle of a product, depending on the raw materials used, the product make-up, the technology and research used to enable its manufacture, the transformation/manufacturing processes employed, the nature of goods produced, the packaging of the product, the way that the product is distributed to consumers, the nature of consumption, and finally the eventual fate of a product – whether it is disposal of any waste residual, re-use, or recycling (see Chapter 12). Many goods made by industry do not, under normal operating circumstances, have environmental impacts of the same order at the manufacturing stage as compared, for example, with impacts associated with their use, consumption, or disposal.

Emissions to air

For many manufacturing sectors, it is the combustion of fuels to generate energy, heat, steam or other power which has the biggest impact on the environment, and therefore the main industries which contribute to atmospheric emissions tend to be the 'energy intensive' sectors, ie, those with high energy requirements such as base metal and chemical industries. Few data are available which show the relative contribution of different industrial sectors to atmospheric emissions. Some data show contributions in certain countries, but these cannot be used to make meaningful comparisons between countries, or to draw Europe-wide conclusions. Data available for Central and Eastern European countries from national environmental reports and UNCED national reports often do not separate the industry contributions for different types of pollutants. Industry is often described as contributing a certain percentage of total air pollution – but it is often not clear what pollutants or emissions are included. Industrial emissions data for these countries may also be lumped together with emissions from domestic sources, which again makes the industry contribution difficult to assess (see Chapter 14 for more details on what data are available, and the need for agreement on designating sources of pollution).

Information is available which gives a rough idea of which sectors contribute to certain emissions in other parts of Europe. For 20 European countries in 1990, industry contributed one quarter of total sulphur dioxide emissions and about 14 per cent of total nitrogen oxides emissions (see Chapter 14).

Some data based on the CORINAIR methodology have been developed by the International Institute for Applied Systems Analysis (IIASA) for the Central European Initiative (CEI) countries, which cover the Austria, Czech Republic, Hungary, Italy, Poland, Slovak Republic and Slovenia (Klimont et al, 1993). This provides information for 1988 on the contribution of emissions from industry (from combustion of fuel and processes) to total emissions. Pollutants covered were sulphur dioxide, nitrogen oxides, particulate matter (PM) and carbon dioxide. Figures 20.1 and 20.2 show summary results of the contribution from industry to total emissions, in absolute and relative terms, respectively.

Detailed data are available for industrial emissions of sulphur dioxide in Norway. Industrial activities were the source of about 60 per cent of total emissions of sulphur dioxide (1989) in Norway. About half this amount originated from the manufacture of metals, but substantial emissions also arose from manufacture of industrial chemicals and oil refining. Over the period 1973–91 there were large reductions in sulphur dioxide emissions from industrial sources (CBS, 1992).

Carbon dioxide emissions arise from the combustion of fuels in industry, particularly the cement industry (one of the main contributors to industrial emissions of carbon dioxide). Carbon dioxide is also produced from the decomposition of calcium carbonate in the cement. Sulphur dioxide is emitted mainly from combustion of sulphur-containing fuels in power generation and from some industrial processes (smelting of sulphide ores). Industrial emissions of nitrogen oxides arise from combustion of fuels and processes, particularly refining. The contribution of industrial activities to potential acid deposition, control measures taken and abatement technologies available are reported in Chapter 31.

Emissions from production processes can also have impacts, however (see Table 20.1). The main industrial source of VOC emissions, for example, is evaporation from the use of solvents. Distribution of oil products and processes relating to extraction of oil and gas are significant sources as well. Extraction of fossil fuels, including deep-mined coal,

also contributes to methane emissions. For the UK it is estimated that extraction of deep-mined coal contributed about 22 per cent to total methane emissions of 3.4 million tonnes in 1991 (UK DoE, 1993).

Industry also contributes to carbon monoxide emissions, although the amounts involved are usually small (for instance, in Norway, industry contributes just 10 per cent to carbon monoxide emissions). Most ammonia emissions come from agricultural activity (especially livestock farming), but some are also emitted during extraction and distribution of fuels.

Estimates of emissions of heavy metals (eg, cadmium, chromium, lead, mercury, nickel, manganese and vanadium) from various sources are given in Chapter 14. The main industrial sources are non-ferrous metals processing and iron and steel manufacture.

In Central and Eastern Europe, some of the most severe health problems have been caused by atmospheric emissions from non-ferrous metal smelting, coking and chemical plants, and steelworks located near communities or cities (see Chapter 11).

Where concentrations and densities of industries in a particular area are high, exposure to air pollutants can contribute to some acute health and environmental problems (see Chapters 4 and 11). The health impact is magnified in cities with a high use of coal by numerous small point sources (including heavy industries and domestic heating). The high concentration of industrial point sources (primarily metallurgical and energy generation) in the 'Black Triangle' (Upper Silesia in Poland, Northern Bohemia in the Czech Republic and Saxony in the eastern part of Germany) and the Donetsk basin in Ukraine is the cause of many of the main health problems (World Bank, 1993), as it is also in the Copsa

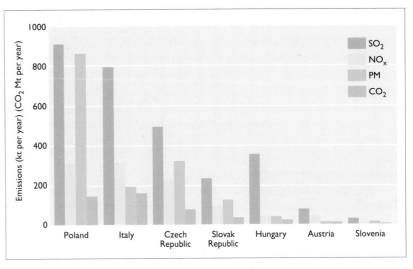

Figure 20.1
Industrial emissions from CEI countries, 1988
Source: Klimont et al, 1993

Figure 20.2
The relative contribution of industry to total emissions, CEI countries, 1988
Source: Klimont et al, 1993

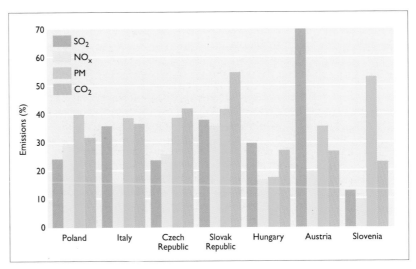

Mica area in Romania. In urban areas, emissions from industrial sources can lead to high concentrations of particulates. For example, in Athens, non-controlled industrial emissions of particulates have been shown to contribute to the *nefos* smog phenomenon (see Chapter 10).

Emissions of dust from industrial activities are sometimes a problem, particularly in Central and Eastern Europe, where many inefficient solid-fuel-burning power and heating plants and industrial plants – both small and large scale – are operating with or without poorly operating dust filters. The major exposure in the glass, ceramics, and especially cement industries for the general population, as well as to workers, is also dust; for example, large amounts are emitted from cement plants in Kunda, Estonia, and on the Latvian border in Lithuania. Metal dusts are commonly emitted by non-ferrous, iron and steel processing, and metallurgical plants which do not have effective pollution abatement equipment. Other emissions, such as sulphur oxides and fluorides, also occur from glass and ceramics works.

Emissions to water

Discharges from industry are a source of water pollution. Industry's impacts on the aquatic environment arise mainly from the discharge of process effluents to watercourses, estuaries and the sea. Most synthetic organic chemical pollution is from industrial sources, including chemical and petrochemical plants, refineries, pharmaceutical manufacturing, iron and steel plants, wood processing, pulp and paper manufacture, and food processing. Industrial sources of heavy metals include: discharge of heavy metal solutions (from smelting) and metal processing; use of metals and metal compounds in manufacture of paints, plastics and batteries; and tanning (Meybeck et al, 1989). Discharges of heavy metals also occur from existing or abandoned mines and leaching from contaminated ground.

The reduction of pollution from this source depends firstly upon what an industrial plant allows into the effluent stream, and secondly on the efficiency and effectiveness with which industrial effluent is treated. Large amounts are spent by industry on industrial wastewater treatment plant and operation (in the EU, about ECU 3000 million per year) (Williams and Musco, 1992). The special treatment requirements for industrial effluents differ from those of other types of water treatment. In some European countries water purification is poor because of the inefficiency of existing wastewater treatment plants (see Chapter 14).

Industry is also a source of phosphorus and nitrogen, but is much less a source of pollution than municipal sewage and agriculture (see Chapter 14). In Sweden and Finland, the forest products sector accounts for the bulk of industrial emissions of phosphorus and nitrogen to water. In the Nordic countries in particular, large industrial plants usually treat their own water, and are generally not connected to municipal sewage works. Discharges of nutrients and organic matter from many large industrial plants have fallen substantially in recent years, not only as a result of effluent treatment, but also thanks to changes in manufacturing processes.

Food and agro-product processing industries, including slaughterhouses, are large producers of waste organic matter. When released in untreated contaminated effluent to the water environment, this consumes oxygen and therefore reduces the dissolved oxygen content of the receiving waterbody. Slaughterhouses present numerous point sources spread throughout Europe, and the increases in demand for meat and livestock products add to pressure on the aquatic environment from this source (see Chapter 14).

Micropollutants are also commonly emitted from industrial sources. Polychlorinated biphenyls (PCBs) are present in different industrial effluents and also released from discarded electrical equipment containing PCBs (eg, transformer oil) when such equipment is unsatisfactorily stored or destroyed by incineration (NIVA, 1992).

Even though industrial activity has declined with the introduction of freer market conditions, privatisation and removal of subsidies, industrial water emissions remain a major source of pollution in the countries of the Danube basin, including Bulgaria, Hungary, Romania and parts of the Czech and Slovak republics. Large-scale industries developed during a time when environmental protection was not a concern, and the technologies used for industrial processing, manufacturing and treatment of industrial wastewater are generally outdated. Industrial wastewater treatment and pre-treatment plants are often primitive and ineffectively operated. Furthermore, in many cases, large industrial complexes have been developed at locations where there are insufficient water resources available to dilute or assimilate emissions.

In the past, many of the large enterprises in the countries of the Danube basin and elsewhere in Central and Eastern Europe traditionally pre-treated their wastewater before discharge into the municipal wastewater systems. But as firms are split up and privatised, and the costs of industrial pre-treatment are found to be high, wastewater treatment may not be seen as a priority, and industrial effluents containing, for example, heavy metals, chemicals and PCBs are discharged directly into municipal sewers which are not equipped to handle such wastes (World Bank, 1993). In some countries – for example, Lithuania – efforts have been made to collect all wastewater from industries into a common network and to purify it in municipal wastewater treatment plants. But these municipal facilities are overloaded, and there are not enough of them to treat all polluted water. In addition, it does not reduce the need for pre-treatment of wastewater.

Industrial cooling waters are returned to the aquatic environment in large quantities, but these are not generally heavily contaminated (Williams and Musco, 1992). Increased temperature of cooling waters discharged to a waterbody decreases solubility of oxygen and increases the rate of decomposition of organic matter, which in turn leads to further decreases in oxygen concentrations. The implications of changes in oxygen content in the aquatic environment are explained in Chapter 5.

Soil contamination

Soil contamination from industrial activities can occur in three ways (Barth and L'Hermite, 1987):

1 Emissions to air which are deposited nearby and at a distance after transport from the site of production. In heavily industrialised regions, this will result in a general spread of common pollutants deposited from the atmosphere, such as soot, polyaromatic hydrocarbons (PAHs) and metals (see Chapters 4, 7 and 14).

2 Discharges in wastewater – for example, where discharges are made via ditches which were not lined to prevent seepage to soils. Incidental pollution of river sediments from effluent discharges can also occur, and dredged material can contain elevated concentrations of heavy metals, hydrocarbons and other contaminants from industrial sources.

3 Site-specific effects from the location of the activity – a production site may have lasting effects on the soil or the land at a particular site long after production has ceased (contaminated sites are covered further in Chapters 7 and 15).

One of the most crucial, and most difficult, issues to be tackled, particularly in Central and Eastern Europe, is liability for clean-up of industrial sites contaminated in the past. The implications of this for investment in industry is briefly covered at the end of this chapter. Many industrial sites are contaminated by asbestos and lead: asbestos was widely used in industry as a structural or filler material and in lagging for insulation; and many industries also made extensive use of lead-based paints. Soil around old painted structures and in particular those repainted (stripping) often shows elevated lead concentrations (Barth and L'Hermite, 1987). Mining operations can also contribute to degradation in some mining areas, where spoil heaps of tailing from mining operations (and slag from smelting) contribute to toxic metal contamination of nearby soil (see also Chapters 7 and 15).

Contaminated soils can be found at numerous industrial sites where chemicals have leaked, or been buried or disposed of. Leakage has, for example, occurred on hundreds of sites in Finland and Sweden where timber used to be impregnated with arsenic, creosote or chlorophenols. The areas affected are generally small, but concentrations of contaminants can be very high. Some of these sites have been cleaned up, but the cost of such measures is always considerable (Bernes, 1993).

Soil degradation is attributable mainly to contamination from accumulation of industrial waste and, in major urban areas and heavily industrialised regions of Europe, from deposition of airborne pollutants emitted from local sources. In Scandinavia, transboundary air pollution from distant industrial regions is a major contributor to soil acidification. In Poland, the second most important threat to human health in the Katowice-Cracow area (after exposure to airborne pollutants) is the accumulation of metals (especially lead) in the soil, after deposition (World Bank, 1993; see Chapter 4).

Waste

There are different approaches to classifying industrial waste in European countries, and in the national context 'industrial waste' may include different waste streams. Nevertheless, some data on industrial waste arisings are available, and these have been presented in Chapter 15. Around 330 million tonnes of industrial waste were generated annually (late 1980s) in OECD European countries.

Procedures to destroy, store, or transport industrial wastes all carry some risks for the environment and human health. Toxic wastes from industrial processes are of particular concern (WHO, 1992). There is no single generally accepted definition of 'hazardous wastes'. Categories of industrial wastes which are considered hazardous include solvents, waste paint, waste containing heavy metals and acids, and oily waste.

In OECD Europe, over two thirds of the total weight of hazardous waste is disposed of to land, 8 per cent is incinerated and only about 10 per cent is recovered as secondary materials. In most Central and Eastern European countries, industries and factories are required to store hazardous waste on their grounds or in temporary storage sites until a hazardous waste management system has been built. Inadequate organisation and control of temporary storage has meant that some of this waste finds its way into the water supply and to the environment.

In Central and Eastern European countries in particular, reliable information on volumes and composition of industrial and hazardous wastes is scarce, and evidence of the fate of various wastes is often only qualitative. For instance, in the former USSR, there is very little processing of the vast quantities of solid wastes generated by industry and households. These are piled on dumps and at every possible storage site, occupying large

Cleansing contaminated soil, Holland
Source: Michael St Maur Sheil

and ever-increasing areas. More than 15 billion tonnes of mining wastes are brought to the surface each year, but only 7 to 10 per cent are utilised. As a result, at least 13.5 billion tonnes of solid waste are added to spoilbanks, slag-heaps, tailing dumps and other waste piles. (Danilov-Danilyan and Arski, 1991).

Modifications of many production processes are possible to minimise and better manage waste arisings. In many cases, reductions in waste are effective in minimising product loss, so waste reduction is not only environmentally, but economically, worthwhile. Product substitution, modification of manufacturing processes, and re-use and recycling of recoverable waste materials can help to minimise industrial waste arisings. An analysis of specific waste streams (plastic, packaging, car tyres, mining waste and chlorinated solvents) is given in Chapter 15. Methods of waste management and industrial waste minimisation are covered in Chapter 36.

Various examples of end use, or facilitating end-use or recycling of waste arisings, are available. In Denmark it is common to use slag and similar materials for infilling and for road construction, instead of disposing of it to landfill. The same is true for excavation spoil in many parts of the Nordic region (Bernes, 1993). The Czech Republic operates a waste recycling register to assist higher utilisation of industrial waste as a source of secondary raw materials and energy. The register includes some 700 organisations, and holds information on metal bearing, organic and inorganic wastes (MECR and CEI, 1992).

RESOURCE USE BY INDUSTRY

Industries can be described by the nature and intensity of their resource use (a concept elaborated and applied in Statistics Canada, 1991):

- Those which are primary transformers of raw natural resources can be described as 'resource intensive industries'; examples are metal-smelting, food production, pulp and paper and wood industries.

- 'Energy intensive industries' include non-ferrous smelting and refining, cement industry, utility industries, pulp and paper industries.
- 'Water intensive industries' are major consumers of water within the production process; they include non-metallic mineral industries, concrete products, ready-mix concrete, motor vehicle parts and pulp and paper industries.

Countries which are endowed with indigenous natural resources are able to build up manufacturing industries based on exploitation (and effective management) of those resources. For instance in 1991, Finland and Sweden together

Compared with Western European industrialised countries, the average efficiency of raw material and fuel consumption is generally very low in Central and Eastern European economies, which have often been geared to highly resource-consuming industrial activities, but exposure to more open market conditions will encourage greater efficiency of resource use, although this transformation will take quite some time.

Energy use

Consumption of energy is (in many cases) the main source of atmospheric emissions from industry. However, improvements in energy efficiency, and diversifying of industry, has meant that increases in industrial output no longer lead necessarily to increases in energy consumption (see also Chapter 19). Final energy consumption by industry in Europe is shown in Figures 20.3 and 20.4, as absolute values and as a proportion of total energy consumption.

Final energy consumption by industry in all Europe has increased from 711 Mtoe in 1970 to 815 Mtoe in 1990 (Figure 20.3). Particularly noticeable is the increase in total energy consumption in the former USSR, from 306 Mtoe in 1970 to 430 Mtoe in 1990. However, as a whole, energy consumption by industry in Europe has declined from 49 per cent to 41 per cent of total final energy consumption between 1970 and 1990 (Figure 20.4). In the EU, industry's energy consumption as a proportion of total energy consumption had declined to around 30 per cent of the total by 1989, although the energy intensive iron, steel and chemical industries combined still accounted for about 13 per cent of final energy consumption in the late 1980s (CEC, 1992b). The way in which these energy demands are met is described in Chapter 19.

Energy consumption for the production of basic materials can be reduced in several ways, including more efficient energy production and energy conservation, or more efficient use of materials (by improving the properties or by recycling). Emissions from the use of coal as a boiler fuel for process heat can be reduced by switching fuels (to natural gas, for instance); by electrifying the process; or by using one of the various pre-combustion, combustion and post-combustion emissions reduction technologies, such as devices for removing particulates, flue gas desulphurisation technologies ('scrubbers') to reduce sulphur dioxide emissions, and technologies to improve efficiency of fuel combustion. Better control of furnace temperatures also enables emissions of nitrogen oxides to be reduced significantly. However, the existence of better technologies does not guarantee that they will be adopted. For small firms in particular, the costs of emissions control in relation to output may be large (World Bank, 1992a). Costs of implementing technological options for reducing emissions of sulphur dioxide and nitrogen oxides are outlined in Chapter 31.

For many Central and Eastern European countries, the dependence on energy intensive industries is demonstrated by the high proportion of electricity consumed by industry. Industry accounts for over half of electricity consumption in a number of these countries, and as much as two thirds of the total electricity consumption in Romania (E/SB, 1991). Industrial activities in the former USSR have more recently been affected by reduced access to supplies of energy from the Russian Federation and producers in other parts of the former USSR in which there is falling fuel output (eg, 12 per cent decline of oil extraction between 1991 and 1992 in the former USSR) (UNECE, 1993b).

In the EU the proportion consumed by industry of final electricity consumption fell from around 48 per cent in 1980

Figure 20.3
Final energy consumption by industry, 1970–90
Source: Compilation based on IEA and Eurostat data

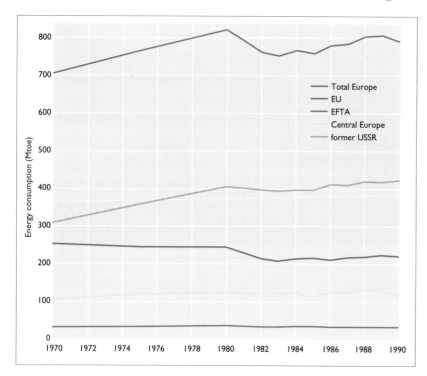

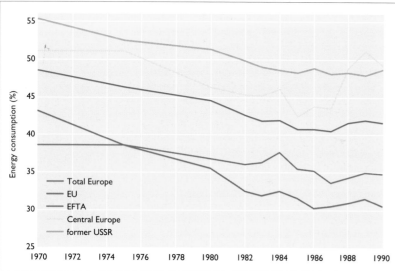

Figure 20.4
Industrial energy consumption: percentages of total final energy consumption
Source: Based on IEA and Eurostat data (see *Statistical Compendium*)

accounted for 26 per cent of total EU and EFTA production of paper, but for only 6 per cent of consumption (see *Statistical Compendium*). The former USSR exploited resources in the Asian part of the territory – east of the Urals – where over 70 per cent of energy reserves are to be found (Éditions Atlas, 1992). These fuel resources are transported by an extensive pipeline system to the European part of the former USSR, where the bulk of industrial activity takes place.

to 44 per cent in 1990, although there were considerable variations between countries, with industry in Portugal, Italy, Luxembourg, Spain and Belgium all consuming more than half of all electricity (1990), while in Denmark, Ireland, the UK and France, industry consumed less than 40 per cent of final electricity consumption (special data set compiled for this report based on Eurostat data).

The proportion of electricity consumed by industry may vary for a number of reasons. For instance, it may be relatively high because of low consumption in other sectors. In many Central and Eastern European countries, the proportion consumed by households is low, because the range of modern appliances used in Western Europe is not available. Where there is insufficient generating capacity to meet all users' needs, industry may have (historically) been favoured over households (as in Romania). Nuclear power produces a fairly fixed flow of electricity, irrespective of demand. Therefore industry is encouraged to use electricity, especially during off-peak periods.

Mining

Mining for resources is necessary to satisfy energy requirements, but also to satisfy other metals and other raw materials requirements of industry. Mining and quarrying can seriously alter the composition of a landscape, disrupting landuse and drainage patterns and removing habitats for wildlife (see Chapter 8). This is particularly so with opencast mining.

There are three major industrial regions in Eastern Europe (Donbass, Moscow, the Urals) which all import coal from Vorkuta in northern Russia. While the Donbass coalfield (Ukraine) produces more high-grade coal than the UK or Germany, it is also known for environmental problems that are only associated with Europe's old industrial regions (Cole and Cole, 1993).

In the area of the Black Triangle, most notably in and surrounding Krokonose, problems result from the use of the locally mined and extremely sulphur-rich lignite and coal. The open lignite mines in the former East Germany (near Leipzig) not only affect the landscape, but also contribute to high levels of air pollution (Knook, 1991).

In the Czech Republic (coal mines near the town of Most in Bohemia), the opencast mining of brown coal, which is concentrated in a few areas, has the most damaging effects on land reserves and landscape and local hydrology, but the mining of non-ore materials (kaolin, clay, sand, grit, rock, brick material, quartz and limestone) also has significant effects. The area used for mining amounts to almost 10 000 hectares (see Chapter 8, Box 8J). Reclamation of around 10 per cent of opencast areas is taking place (MECR and CEI, 1992).

Among the most extensive opencast mining operations in Western Europe are the lignite mines on the western border of Germany (North Rhine Westphalia), between Aachen, Cologne and Monchen Gladbach. Five mines produce about 120 million tonnes of burnable fuel per year. Much of this is for the power stations which supply the Rhine/Ruhr area with electricity. The lignite is extracted with huge machines, and by 1990 about 250 km^2 of land had been excavated. The usable fuel is burnt and the waste is dumped in heaps. To keep the pits dry for mining operations, about 1.2 billion m^3 of groundwater are pumped out each year (this is more than the amount used by the whole population of The Netherlands). Consequently, a rapid decline in the level of the water table is occurring (Harle, 1990).

Water use

Many industrial activities use large amounts of water in the manufacture of products. Industry accounts for just over half of total water withdrawals in Europe (see Chapter 5). The main use of water is for cooling, which accounts for about 70 to 80 per cent of industrial water use, and most of this is used in power generation (see Chapter 19).

Some industries, such as manufacture of pulp and paper, cement, motor vehicle parts, and processing of petroleum, can be described as 'water intensive' since they are major consumers of water in the production process (that is, if more than 15 per cent of the activity's total water intake is consumed).

Normally, surface water is used for industrial purposes, although the quality of water required depends on the product. For example the quality of cooling water may be low, but that required for paper and pulp manufacture must be higher.

Industrial use of water usually accounts for between 20 and 60 per cent of total water supplied for all purposes (see *Statistical Compendium*). There have been continuous efforts over the years by industries to reduce their consumption of cooling water, and UNECE data indicate that supply for manufacturing activities in many countries has decreased (UNECE, 1992). Consumption by firms can be reduced in a number of ways: by improving water supply and sewage systems, construction of closed recirculating water systems for cooling purposes, as opposed to 'once-through' systems, building of plants that do not discharge wastewater into watercourses, and use of processes that do not require water. One of the main trends in reducing wastewater discharge by large industrial enterprises has been the introduction of local recycling of wastewater and its purification into certain stages of the cycle of production (UNECE, 1991).

A combination of industrial and management practices, use of less water per unit of output across different sectors, and more efficient use of water is expected to lead to a fall in the industrial demand for water, especially process water, over the next decade, despite growth in industrial output. In the UK, for example, industrial abstractions have been decreasing for ten years, reflecting in part a contraction in some industries but also more efficient water usage, including recycling. This trend is likely to continue in the medium term. Recent estimates of water use by sectors of industry suggest that, in the EU at least, demand for water by all sectors of industry is likely to fall up to 1995, even though output of the sectors requiring high volumes of water – paper and pulp, and chemicals manufacture – is expected to grow faster to 1995 than other sectors of industry (Williams and Musco, 1992). The reductions is expected mainly as a result of wider use of more water-efficient technologies in the production process.

TRENDS IN MANUFACTURING ACTIVITIES

The importance of industry in the national economy is greater in Central and Eastern Europe compared with Western Europe. However, between 1989 and 1992, the industrial output for a group of Central European economies fell by around 40 per cent, even faster than the 30 per cent drop in national product over the same period (the grouping includes Bulgaria, the Czech Republic, Hungary, Poland, Romania, the Slovak Republic and the former Yugoslavia) (based on UNECE, 1993a). In the countries of the former USSR, industrial output declined substantially over the same period, with the average fall at nearly 8 per cent in 1991. The fall varied from around 11 per cent for Moldova to 1.5 for Belarus. Trends in industrial

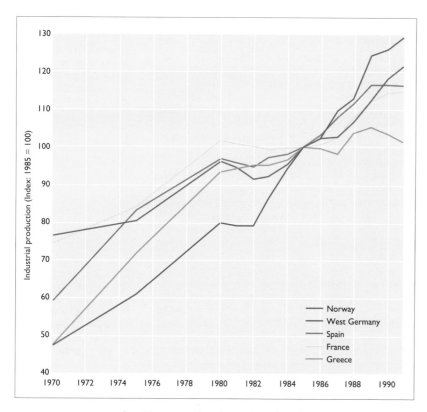

Figure 20.5 Trends in industrial production, selected Western European countries, 1970–91
Source: Multiple sources, see *Statistical Compendium*

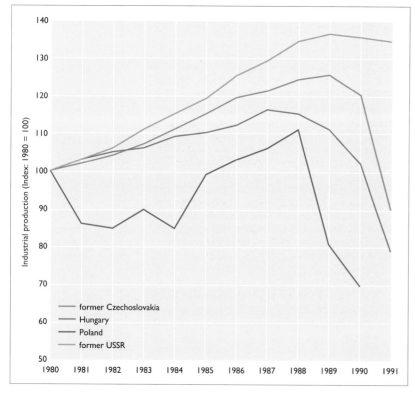

Figure 20.6
Trends in industrial production, selected Central and Eastern European countries, 1980–91
Source: Multiple sources, see *Statistical Compendium*

reports and are also, for example, among those sectors identified by the Central Bureau of Statistics in Sweden as exerting the highest environmental pressure (Bureau Central de Statistiques de la Suède, 1984, drawing on Canadian experience). The next most notable sectors which have not been overviewed in this section (and which were identified as exerting a high or moderate pressure on the environment by the Swedish study) include: refineries and petroleum processing, the glass and ceramics industries, leather and tanning, the food industry, pharmaceuticals, wood, rubber and detergent manufacture. The environmental impacts of some of these sectors are summarised in Table 20.1.

This section draws on the information available at the sector level from international statistics, from other state-of-the-environment reports, and from material produced by the sectors themselves. Sometimes more economic- or production-type data can also be useful to give a fuller picture of the activity.

Chemicals

The chemicals industry is one of Europe's – particularly Western Europe's – most important economic sectors, and its activities give rise to many emissions from combustion of fuels, and also from processing and development of chemicals by the sector (see Table 20.1). Further information about chemicals and the treatment of chemical risk is found in Chapters 17 and 38. In 1988, the EU accounted for some 28 per cent of world production, EFTA for 15 per cent, and Central and Eastern Europe and other parts of the former USSR for 16 per cent (CEC, 1991). Its products and technologies are used by virtually every other industry and activity. The sector is usually described on a subsectoral basis because it is so diverse in terms of products and processes. The relationship between the main sectors is shown in Figure 20.7. The outlets for the two bulk (upstream) sectors are almost

Figure 20.7 Relationship between the main chemicals industry subsectors
Source: March Consulting Group, 1992

Raw materials: oil, coal, gas, air, water, minerals

Chemical processing

Bulk inorganics and organics
Ammonia, gases, acids, salts, aromatics, olefins

Chemical processing

Main product groups
Fertilisers, industrial chemicals, plastics, resins, elastomers

Chemical process industry
Rubber and plastic goods
Fibres and textiles
Paints and dyestuffs
Soaps and detergents
Pharmaceuticals
Agrochemicals

Other industries
Metals, Glass
Cement
Motor vehicles
Paper, Textiles
Food products,
Agriculture

Consumer needs

production for some European economies up to 1991 are shown in Figures 20.5 and 20.6.

Given the large number of activities and sectors that industry can cover, it is useful to look beneath these general trends at a few sectors in greater detail. The activities considered here are: chemicals, paper and pulp, cement, the production of iron and steel, and non-ferrous metals. These five sectors usually feature in state-of-the-environment

exclusively the downstream sectors, which in turn supply either other industries (eg, fertilisers for agriculture) or the end user (eg, soaps and detergents).

Production characteristics, and hence emissions in the industry, are more dependent on subsector than geographical location. Within any subsector there are no significant differences between countries in Western Europe in terms of energy efficiency or technologies used, and, because of the multinational nature of the sector, new technologies spread rapidly across national boundaries (although competitive considerations do sometimes get in the way). Newer plants will generally be more efficient in the use of energy and raw materials than older ones, thanks to improvements in plant design and cleaner technologies.

The chemicals industry's share of Western Europe's GNP is about 3 per cent, and the industry accounts for some 12 per cent of manufacturing value added (CEFIC, 1991). Table 20.2 shows the percentage of total world production of various chemicals produced in Europe including the former USSR in 1988. Figures for individual countries can be found in the *Statistical Compendium*.

There are over 13 million known chemical substances; some 60 to 70 thousand are in regular use, but only about 3000 account for 90 per cent by mass of the total used (UNEP, 1992). The production of chemicals has increased exponentially, as have their many uses.

In general, the chemicals industry is also very energy intensive. Energy intensity (measured as the cost of energy per unit output) averages around 10 per cent across the industry in the EU, but varies between a fraction of 1 per cent in high value speciality chemicals, to 75 to 80 per cent in chlorine production. Other subsectors of the chemicals industry with high energy intensities include plastics (20 to 25 per cent), bulk organics, inorganics and fertilisers (around 15 per cent). There is also a wide variety of energy-consuming equipment. Because of the high energy cost and low value of bulk inorganics and fertilisers, there has been a tendency in recent years for the EU industry to move towards higher-value speciality chemicals production (March Consulting Group, 1992).

Energy consumption per unit of output by the chemicals industry in Western Europe (EU and EFTA) fell by about 30 per cent between 1980 and 1989 (see Figure 20.8). This reflects improvements in energy efficiency and reductions in energy consumption.

Total final energy use in 1989 was 1790 peta joules (PJ), which is almost 20 per cent of the EU industry total. France, Germany, Italy, The Netherlands, Spain and the UK together account for 92 per cent of energy use in the EU chemicals industry, with the sector being the largest industrial consumer of electricity. The most important subsectors in terms of energy use are the two bulk sectors, with organic around twice the size of inorganic. Together, these two subsectors account for nearly 60 per cent of energy use in EU industry (March Consulting Group, 1992).

Where energy costs are a large proportion of total costs, and where the final product is easy to transport, subsectors are often located or are likely to move to areas where there is plenty of cheap energy available. For instance, much of the world production of fertiliser is now located in Saudi Arabia and Canada (although, in the latter, energy is still plentiful but no longer at low cost), the production of ammonia being directly linked to the price and availability of energy.

The chemical industry as a whole is probably the most prosperous of all EU industry sectors, with a high propensity to invest in new technologies in order to remain competitive on a world scale. There are some 9000 companies in the EU chemical sector, although 40 per cent of production is shared between the top five companies (CEC,1991).

The Russian Federation has by far the largest chemicals industry in the former USSR, with a focus on bulk chemicals. The chemicals industries of the former USSR entered into

Product	Europe and the former USSR, %
Phosphate fertiliser[1]	42
Insecticides, fungicides, disinfectants	51
Washing powders, etc[2]	58
Sulphuric acid	44
Styrene	18
Ethylene	38
Xylene	35

Notes:
1 P_2O_5 content other than natural fertilisers.
2 Organic surface-active agents, surface-active preparations and washing preparations whether or not containing soap.

Table 20.2
Proportion of world production of selected chemical substances in Europe, 1988
Source: UNEP, 1992

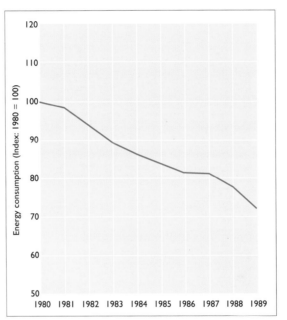

Figure 20.8
Energy consumption per unit output of the Western European chemicals industry – electricity included, 1980-89
Source: CEFIC, 1991

recession later than those of other Central and Eastern European countries. In the economies of Central and Eastern Europe and the former USSR, with the unique exception of Poland, chemical production has been falling, sometimes sharply, since 1989, although falls have been delayed in the countries of the former USSR. Even though some restructuring and modernising efforts are ongoing in the chemical industries of these economies, there are a large number of plants which are 20 years old or more still in operation. These are virtually obsolete and their continued operation adds to environmental pressures. Better environmental performance and increased safety of operations are two major issues now receiving more attention, including closing the most polluting plants. Direct investment by Western firms has so far been limited to low risk and high demand domestic areas, such as the Hungarian household chemicals sector.

The German chemicals industry is the biggest in Western Europe and the fourth largest in the world. A vast privatisation programme of chemical companies in East Germany is being carried out. Of 130 enterprises, about two thirds have been privatised and 28 closed down (1992). The restructuring programme is also aimed at improving environmental and quality performance, as well as taking steps to solve the environmental problems caused by previous activities (UNECE, 1993b).

High emissions from chemical factories in Bulgaria are mainly the result of physically worn-out systems for the production of sulphuric acid, the lack of gas desulphurisation installations in thermal power plants and the inappropriate treatment facilities for fluorine compounds, as well as unsatisfactory control of furnace gases (Bulgarian Ministry of Environment, 1992). The Bulgarian chemicals industry suffered huge drops in output during 1990 and 1991; one of the main reasons for this was the reduced access to crude oil previously supplied by the former USSR.

The chemicals industry in Belarus is perhaps the most diversified in the former USSR, and accounts for about 6 per cent of the former USSR output of chemicals (UNECE, 1993b). The Novopolotsk area has a high concentration of petroleum processing and petrochemical industries. In this area, atmospheric air pollution from chemicals, especially PAHs, has resulted in increased incidence of bronchial asthma cases registered by the public health authorities. A marked relationship between chemical pollution and various illnesses has also been established in other large industrial centres (Gomel, Minsk, Soligorsk etc) (SCERB, 1992).

There has so far been a lack of investment in the chemical industries of Central and Eastern European countries by Western European chemical companies for a number of reasons. The markets are undeveloped, with depressed demand in domestic markets, and little likelihood of their developing in even the medium term. Faced with overcapacity and low levels of demand and profitability, there are moves in the sector in Western Europe towards restructuring so that companies become more specialised in particular types of chemicals production, such as industrial or 'fine' chemicals, where innovation of the chemicals industry in the Western countries is highest. The preoccupation with tackling the structural problems of the sector in Western Europe in recent years has been an unfortunate factor for Central and Eastern European economies at a time when investment is most needed.

Cleaning up the chemicals industry

During the 1970s, most companies responded to increasing environmental pressure through investing in 'end-of-pipe' technology. Now they are moving towards waste minimisation in production processes. In Western Europe, about 10 per cent of total capital investment by the chemicals industry is spent on environment protection (CEFIC, 1991). This will include development of cleaner technologies, waste and wastewater treatment, recycling processes, biotechnology processes, catalysts, membranes, desulphurisation plants, noise reduction, and other products manufactured with an environment protection purpose.

An example of effective pollution abatement measures limiting wastewater discharges can be taken from a Danish manufacturer of pesticides, Cheminova, situated on the northwest coast of Jutland. Until 1989, effluent discharges of phosphorus from the production of pesticides was around 1100 tonnes per year, amounting to 10 per cent of total phosphorus emissions from all controllable measured sources (point sources) in Denmark, and one third of direct discharges by industry (Miljøstyrelsen, 1988). During the 1960s, discharges of effluent from the plant had damaged nearby lobster populations, and the effluent was still deadly to fish even after dilution by a factor of 50 000. In 1989, biological treatment of wastewater was initiated, and by 1993 phosphorus emissions had been reduced by more than 98 per cent of the pre-1989 values. Figure 20.9 shows the reduction in emissions of phosphorus achieved in the pesticide manufacturing plant as a result of wastewater treatment between 1987 and 1993. There was no clear trend in production over the period, although levels in 1992 were similar to those in 1987.

The chemicals industry has had a reputation for being a 'dirty' industry, but has been a leader in trying to establish guiding principles for integrating environment into management practice, to improve its public image and to gain public trust (see below in this chapter the section on industry response to environmental factors). The Responsible Care programme was initially developed by the US Chemical Manufacturers Association in 1988, and is now being adopted and implemented by the national chemical associations in 18 European countries under the general guidance of the European Confederation of Chemical Industries (CEFIC). The programme is a self-imposed, voluntary commitment by the industry to reduce waste, pollution and accidents, to improve energy efficiency, and to introduce clean technologies. An important aspect of the programme is measuring progress and communicating results to the public. Based on a set of guiding principles, similar to those of the International Chamber of Commerce (ICC) Charter (see Box 20B, p 430), the Responsible Care programme promotes codes of practice including:

- community awareness and emergency response;
- pollution prevention;
- process safety;
- employee health and safety;
- product stewardship.

However, many groups are still struggling with the amount of information to be provided at the level of the individual company. Different companies report their environmental results differently (eg, those on emissions), and results are mostly not comparable or consistent between companies or industry groups. Questionnaires and checklists for self-assessment of performance are provided, but, so far, continued participation in the Responsible Care programme, as with other guidelines of this kind, does not require external verification of the environmental performance reported.

Paper and pulp

The primary activity of the paper and pulp industry is the manufacture and processing of paper pulp, paper and cardboard based on raw pulp (see *Statistical Compendium*).

The main emissions of concern from pulp and paper mills are those to the aquatic environment, as a result of water used in the manufacturing process (see Table 20.1). It is in the manufacture of paper where the main potential environmental impacts lie, rather than in its consumption or disposal. Paper products are recyclable, and can also provide a source of non-fossil fuel energy after initial use. Specific exposures to emissions from pulp and paper mills have not

Figure 20.9
Emissions to water of total phosphorus from the Cheminova plant, Denmark, 1987–93
Source: County of Ringkøbing, NW Jutland, personal communication

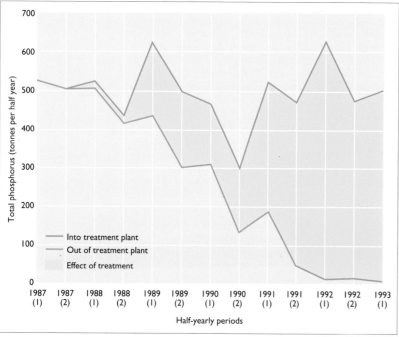

been well documented, although in the past mercury compounds were used as slimicides and caused environmental contamination (WHO, 1992).

The pulp and paper industry is a capital-, energy-, and resource-intensive industrial sector. Setting up a typical modern paper mill requires huge investment, and the lifetime of a paper machine may be 15 to 20 years, although plants in Central and Eastern European countries are often much older. The high use of energy in pulp and paper manufacture contributes to emissions of nitrogen oxides and sulphur dioxide, almost all of which are from combustion of fuels (rather than the production process). In Finland, the industry consumes almost one third of total electricity, and in Sweden about 44 per cent (63 TWh) of total industrial energy use (Komppa, 1993). However, modern pulp mills produce from wood more than enough thermal energy (steam) and electricity needed to run the pulp-making process. The surplus electricity and steam can be used by nearby industries, and to heat local homes. From an environmental point of view, what is important is the proportion of energy derived from renewable natural resources (wood) and the quantity which has to come from non-renewable sources.

Because of the nature and diversity of the industry, and the breadth of the capital base, technology is improved in steps, and there are many variations in environmental performance between individual paper mills. The large, modern mills are much more efficient than older and smaller mills in terms of specific consumption of energy, water and wood raw materials (that is, per tonne of paper produced), as well as in the use of labour. However, as a result of increased production capacity and improved quality of product, the total consumption of energy (electric power and heat) is increasing continuously, even in the most efficient mills (Komppa, 1993). Older paper mills use more energy because of inefficient processes and losses, and they also have large emissions to air and water.

Trends in paper and board production in Europe are illustrated in Figure 20.10, which shows that production is greatest in the EU. In terms of total consumption, the paper and board market of the EU is the largest market in the world. Consumption per capita, however, is lower in the EU than in Scandinavia or Japan (CEC, 1991). The prosperous development of publishing activities as well as the increasingly intensive use of packaging continue to sustain the development of the industry.

Paper originates from a renewable resource – wood – and in Europe it is produced mostly from forests managed on sustainable principles. Wood pulp is derived from the forestry industry, and 85 per cent of total roundwood production in Europe is used for industrial purposes (the rest is used for fuel and charcoal); 40 per cent of industrial roundwood produced is used in the production of pulp and paper (1989 data, see *Statistical Compendium*) (see also Chapter 23). Wood, including thinnings and sawmill waste, is converted into paper by means of pulping which can be either chemical or mechanical. Each pulping method has technical and environmental advantages and disadvantages, depending on the type of wood used and the quality of paper required, although as a general rule more energy is required to produce pulp of higher quality. Mechanical processes require relatively large amounts of energy but are more efficient in the use of wood than chemical processes, and otherwise have little impact on the environment. Chemical processes have lower energy requirements, but need more wood per tonne of paper produced. Chemical methods also generate sulphur and large amounts of organic matter, which can deprive aquatic organisms (fish) of oxygen. Different methods of chemical pulping can be used, and by far the most common is the sulphate (or 'kraft') process, which accounts for 90 per cent of the total world pulp production (see CEPI, 1993).

Chemicals (bleaches, homogenisers, dyes, fillers and sizers), together with heat, steam and large amounts of water,

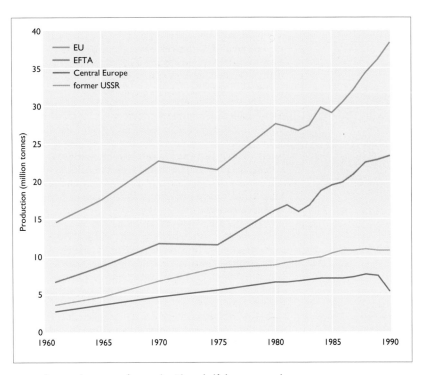

are used to produce paper from pulp. About half the water used by the manufacturing industry in Sweden is used in paper production, whereas in countries with smaller paper industries, such as Poland or Austria, less than 15 per cent of industrial water use goes into paper production (UNECE, 1992).

Bleaching is necessary to meet the standards of brightness, strength and quality preferred by consumers of paper, though growing awareness is increasingly leading consumers to choose unbleached products in cases where whiteness is unnecessary, such as toilet paper. Chlorine is by far the most commonly used chemical for bleaching pulp, although in much of Europe it is no longer used as a bleaching agent. Since the 1970s, use of chlorine gas in production processes in EU and EFTA countries has been gradually displaced by chlorine dioxide, oxygen, hydrogen peroxide or dithionite, largely as a result of technological advances and a recognition by the industry of the need to reduce emissions from pulp mills.

In some countries with a long tradition in paper making, particularly the Scandinavian countries, technological advances in the production of paper, including reductions in energy consumption per unit of final product and reductions in the amounts of water used and discharged (by recycling process waters), have reduced emissions of many harmful substances in wastewater. The Finnish paper industry can

Figure 20.10
Production of paper and board in Europe, 1961–90
Source: FAO (see *Statistical Compendium*)

Figure 20.11
Production rate and effluent load in the Finnish paper industry, 1970–90
Source: Komppa, 1993

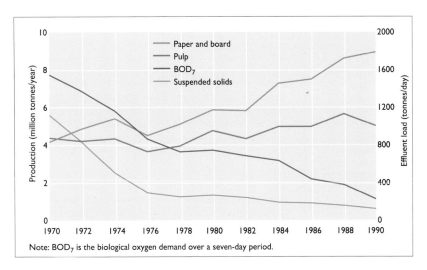

Note: BOD$_7$ is the biological oxygen demand over a seven-day period.

demonstrate considerable successes in reducing specific energy consumption and effluent emissions. During the last 20 years production of paper by the Finnish industry has more than doubled, but effluent load has decreased to one sixth of the previous level over the period (Figure 20.11) (see also Chapter 14).

The Western European paper industry is undergoing considerable restructuring and much of the production is being combined in a few large units. The Scandinavian countries have fewer paper mills than the EU, but their average capacity is twice as large. In contrast to paper making, the paper and cardboard processing industry still consists of many small and medium-sized companies, which enables more flexibility and adaptation to changing market conditions.

The EU paper industry has an increasing dependence on imports of commercial paper pulp, which in the past has led to suggestions that this dependency should be reduced. Some countries are moving in this direction, with intensification of collection of paper for recycling, the use of recycled paper on a large scale, and the use of existing forestry resources as well as the creation of new resources.

However, although increased recycling reduces the amount of wood required for pulp manufacture, it also

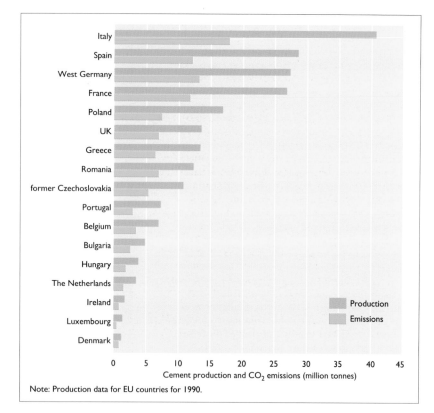

Note: Production data for EU countries for 1990.

Figure 20.12
Production of cement and emissions of CO_2, selected European countries, 1989
Sources: E/SB, 1991; CEC, 1992c; Carbon dioxide analysis information centre (cited in WRI, 1992)

leads to increases in sulphur dioxide emissions, and to a significant increase in the consumption of non-renewable resources, compared with that consumed in manufacture of recycled paper from pulp. Therefore reducing imports of 'virgin' pulp by moving towards recycled pulp can be only a partial solution to certain resource problems.

Cement

The emissions of most relevance from this sector are atmospheric: dust, carbon dioxide and nitrogen oxides are the most important. Other pollutants, such as sulphur dioxide, lead and fluorides, are also emitted. Emission of dust can be checked by installation of appropriate dust control systems. EU plants, at least, are investing increasing amounts on

environmental protection, and investments in dust removal equipment can reach 15 per cent of total investment (CEC, 1991). Other exposures include carbon monoxide (up to 50 ppm near limestone kilns), asbestos and chromium (WHO, 1992).

Cement is essential for the construction sector, either directly or mixed with sand or gravel to form concrete. In general, cement production matches the national consumption in all countries, mainly because of high transport costs. For this same reason cement plants are spread throughout Europe, near to raw materials and consumers. Growth in this sector is linked closely to general economic development, and to the building and construction sector in particular. In 1989, world production of cement reached 1143 Mt, 30 per cent more than in 1980. The EU as a whole produced around 170 Mt in 1989, the largest producer in the European region, contributing about 15 per cent of world production (CEC, 1991). The former USSR produced around 140 Mt in 1989, which accounted for one eighth of world production. Figure 20.12 shows cement production and carbon dioxide emissions for the main producing countries in 1989.

The sector is energy intensive, and the main atmospheric emissions from cement production result from combustion of fuels and the decomposition of limestone (producing carbon dioxide). Energy cost is a major concern of the sector, and companies focus on both saving energy and finding the cheapest fuels, which may include industrial waste and spent tyres. Coal is the most common fuel used (around 60 per cent of EU total energy consumption in the sector), with France, Denmark and The Netherlands using less than the average, but Portugal, for instance, using almost 90 per cent coal. Coke, petroleum and electricity (based on a wide range of fuels) each comprise more than 10 per cent of the total energy consumption. Other fuels are mainly lignite, heavy fuels, natural gas and industrial wastes. The Netherlands uses far more electricity than the average (which is explained by the nature of production in that country – mainly cement grinding, which consumes electricity only) (March Consulting Group et al, 1992). High temperatures are required to prepare raw materials for cement manufacture. In the EU, average consumption of energy per kg of cement clinker (intermediate cement product) produced is around 900 kcal/kg, and varies between 800 and 1400 kcal/kg per country, depending on the production method used (CEC, 1991).

Cement can be manufactured using wet or dry process technology. Wet processes are more energy consuming and may therefore lead to higher emissions per tonne of cement clinker produced. Denmark (one of the smallest producers in Europe) uses the wet process technology only and hence has a higher fuel consumption per kg of cement clinker (about 5.2 GJ/tonne) and higher carbon dioxide emissions relative to output (see Figure 20.12). The UK, Belgium, France and Italy also have some production based on this technology. Portugal, Germany and Spain have levels of fuel consumption closer to 3.5 GJ/tonne cement clinker. Germany has low electricity consumption per unit of output owing mainly to energy-efficient technology (March Consulting Group et al, 1992).

Choice of fuel in cement manufacture is dictated largely by economics; when oil and gas prices are high, coal and coke are often used as the fuel source in production of cement clinker. Unlike most other industrial combustion processes, high levels of sulphur and other impurities do not present a problem in cement manufacture because residual sulphur and ash are incorporated into the cement clinker without significant increases in emissions or extra requirements for abatement costs. Supplementary fuels such as used tyres, spent oils, lubricants and solvents are increasingly being used as partial replacements for the principal fuel. In addition, domestic refuse and hazardous industrial material can also be burnt in cement kilns, since combustion temperatures in the kiln are so high.

The Central European market is around 45 million tonnes of cement per year, and is expected to be the growth area for cement production up to the year 2000 and beyond. The overall picture of the sector is one where the leading companies are multinational, but concentrated and dependent on their home market. There are also many small companies competing on the home market (March Consulting Group et al, 1992).

The Akmene cement factory in Lithuania is the largest cement enterprise in the Baltics, producing 3.4 million tonnes of cement per year. Primitive and out-of-date technology means that large amounts of fuel are required, and considerable atmospheric emissions of particulates, mainly dust, occur (62 000 tonnes per year). With the plant being located near Lithuania's border with Latvia, transboundary dispersion is an issue. Reconstruction of this industry on the basis of new technology that will limit pollutant emissions is considered as essential (Lithuanian Environmental Protection Department, 1992).

In modern cement manufacturing facilities, dust levels of 15 to 20 mg/m^3 in the flue gases are usually the upper limits measured. However, with older processing techniques, dust levels may be up to 114 mg/m^3, with levels varying according to the various steps in manufacture. With appropriate use of modern air pollution control devices, dust levels near cement factories can be reduced to 5 to 10 per cent of levels seen with old processing techniques (Parmeggiani, 1983, in WHO, 1992). The source of dust within the production process is important because different types of dust are produced: some settle mainly within the confines of the plant, whereas other emissions, if dispersed from tall stacks, can contribute to long-range and transboundary air pollution. Most dusts are products of the various manufacturing processes and, since they are generally not wastes or by-products that need disposal, they can be returned to the process from which they came. Consequently, it is in the interest of manufacturers to design and operate plants with efficient dust capture machinery to help minimise product loss.

The possibility of EC legislation on carbon dioxide is clearly of direct relevance to this sector. If legislation is strict, competition on cement clinker production will increase, since this is the most energy intensive part of the production process.

Iron and steel

Iron and steel manufacture provides the most important material for a number of other industries, in particular for sectors manufacturing metal products, mechanical engineering equipment, electrical machinery, transport equipment and construction materials. The production of iron and steel contributes to various emissions to air and water, and some solid wastes (see Table 20.1). Emissions from these industries (especially sulphur dioxide and particulate matter), which tend to settle quickly from the atmosphere, can lead to rising concentrations in the soil around iron and steel mills and foundries, and have historically been the source of acute air pollution problems localised to specific regions – examples are the pollution episodes of the Meuse valley, Belgium, in December 1930, and Donora, Pennsylvania, USA, in October 1948 (Masters, 1971). Axenfeld et al (1992) have suggested that for the year 1982 the main emissions of heavy metals from this sector were of zinc, which contributed 9400 tonnes (23 per cent) to total emissions in Europe

(including the former USSR). Emissions of cadmium, arsenic and lead from this sector each contributed about 4 to 5 per cent of their respective total emissions in Europe (see Chapter 14 for further details). The significance, levels of exposure and numbers of people exposed in the general population are not known. Health effects of iron and steel manufacturing in the general population are not well documented.

The former USSR is the biggest producer of iron and steel in the world, accounting for over 154 Mt of crude steel in 1990. In the same year, EU countries produced 137 Mt, Central European countries produced about 49 Mt, and EFTA produced much less, around 13 Mt (see *Statistical Compendium*).

In the EU, energy consumption used by the steel industry has fallen as a proportion of total energy used by industry from the 1960s to the late 1980s. This is due largely to the introduction of more energy-efficient electric arc furnaces (see *Statistical Compendium* for data between 1980–90). However, steel plants of varying levels of production capacity are found throughout Europe, and this development of the industry, and therefore the potential for environmental impacts, is not spread uniformly. Figure 20.13 shows the production of steel in European countries in 1990.

Metal smelting is a resource-intensive process. The main raw material input to the production process is iron ore. Together with consumption of scrap, this provides the bulk of the input to smelting furnaces. In integrated plants, the proportion of scrap used varies between 5 and 35 per cent of the total input. For the EU, the proportion of recycled scrap used in steel mills is about half 'circulating' (ie, recycled from spent products) and half 'process' (ie, a by-product of steel manufacture) scrap. On this basis, recycling rates in the industry are among the highest for any industry (CEC, 1991, Section 3).

Iron and steel are produced by a number of different processes/furnaces, including electric arc, oxygen based, and flame combustion (open hearth). Electric arc is generally the cleanest and most energy efficient, while open hearth is dirtier and more inefficient. Plants using these low fuel-efficient and energy intensive processes emit more pollution per unit output than do comparable plants in industrial market economies. A higher proportion of iron and steel in Central and Eastern Europe is produced from open hearth and Siemens processes, but the proportion is falling fast. Figure 20.14 shows the decline in use of the open hearth procedure, but some remaining high use in Central and Eastern European countries. The proportion of raw steel produced by this method has declined in all European countries over this period, and will continue to fall further as old plant and equipment is replaced.

The state and effectiveness of pollution abatement

Figure 20.13
Total steel production, selected European countries, 1990
Source: Based on Eurostat and E/SB (1991) data (see *Statistical Compendium*)

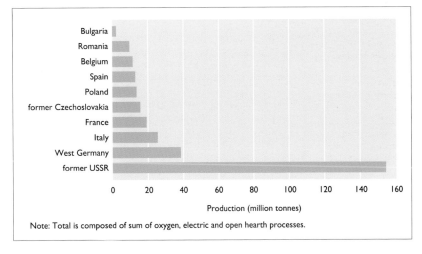

Note: Total is composed of sum of oxygen, electric and open hearth processes.

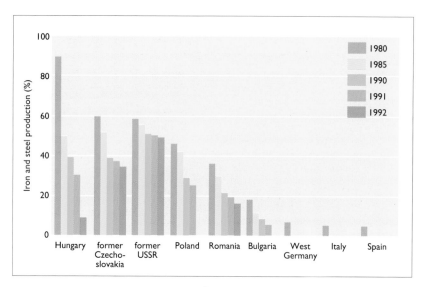

Figure 20.14
Percentage of iron and steel produced by open hearth method, selected countries, 1980–92
Source: based on Eurostat, UN and Statistisches Bundesamt data (see *Statistical Compendium*)

equipment in the metallurgical industry will differ tremendously between old and new plants, and this difference will be greater in Central and Eastern European countries, where out-of-date technology and obsolete equipment is common, than in Western Europe. Environmental control technology, if installed, has often been allowed to deteriorate over time so that dust-removing filters may be operating well below their design specification, and become ineffective. In other cases, the original environmental controls may have been primitive and they have not been upgraded over time (World Bank, 1993). However, some new plants developed are of an equivalent standard to those in Western Europe. Restructuring of the EU steel industry in the 1980s has resulted in the closure of many older installations and their replacement by modern plants fully equipped with the whole range of pollution equipment which upgraded these plants to the level of the best available technology in this field.

There has been a general contraction of the iron and steel industry in Europe, and despite modernisation and restructuring of the industry in the late 1970s and the first half of the 1980s it appears likely that the capacity will have to be reduced still further during the 1990s. There has also been a substantial increase in production by newly industrialised countries, and large quantities of iron and steel are often dumped at prices bearing no relation to production costs. This has placed pressure on the steel industry in Europe, particularly in the EU.

The image of steel-making as a 'dirty' industry is still prevalent even though up to one fifth of expenditure is now made on pollution control equipment (WHO, 1992). Over the years many of the processes used in manufacture have been modified and become more energy efficient as technology is developed and updated, such that many types of energy saving technology are now used in iron and steel processes. The motivation for developing countries to invest in control technology is that energy savings from use of more fuel efficient technology may offset the costs of control programmes. Most of these methods recover heat and gases which would normally be released to the atmosphere, for further use, or re-use, in the production process. Recovered dusts can also be recycled. Much technical effort in process research is also being directed at reducing the amount of coke consumed, both to reduce energy consumption and to minimise environmental impacts.

Treatment of scrap has also developed: for example the preheating of scrap for electric arc furnaces at high temperatures, using waste gas. Low temperature preheaters have caused environmental problems in Europe, particularly in Germany

with respect to incomplete combustion of hydrocarbons and the production of chlorinated hydrocarbon emissions. Low temperature scrap preheaters are no longer in use in the EU and are being replaced by high temperature preheaters.

Non-ferrous metals

Operation of non-ferrous metal plants results in various atmospheric emissions. These emissions arise from the various stages in the preparation of ore, and in the production, melting and refining of metals. The environmental and health hazards of this industry are relatively well documented (WHO, 1992). The main atmospheric emissions are of heavy metals, although other contaminants are also emitted, including arsenic. A 1982 study (reported in Axenfeld et al, 1992) indicated that the non-ferrous metals industry was the source of three quarters of total arsenic emissions, two thirds of cadmium emissions, and 57 per cent of zinc emissions in Europe, including the former USSR (see Chapter 14). Non-ferrous metals production does not contribute much to total emissions of nitrogen oxides and sulphur dioxide by industry, although emissions of carbon monoxide and carbon dioxide may be more significant. Heavy metals, except mercury, are generally not emitted as pure substances but as constituents of dust.

Deposition of these substances can sometimes lead to contamination of food crops near smelters and exposure of the local population, although the type of emission depends on the ores being processed. The volume of individual heavy metals or compounds emitted varies among the numerous processes and also depends very much on the composition of the raw material. Given the large number of different raw materials used and the various processes applied, emissions can include virtually all heavy metals and their compounds.

Broadly, emissions are of two types: stack gas emissions and fugitive emissions. Emissions to air can be transported away from source and eventually deposited or washed off to soil and waterbodies. Heavy metals can also reach water by leaching from soils after improper storage of materials, by surface runoff from solid waste, and by direct discharge of liquid effluent from scrubbers. Stack gas emissions occur at confined points and can usually be captured and disposed of using appropriate abatement technology. Fugitive emissions occur from a range of sources, or from parts of the preparation process, such as the handling and storage of raw materials or as dust released during the production process. Such emissions are difficult to capture but can be controlled by better practices. Usually there are no direct measuring methods for heavy metals emissions (although in some cases continuous monitoring can be carried out, and semi-quantitative results can be obtained). Since fugitive emissions cannot be absolutely evaluated, examples of heavy metals emissions data are difficult to find (see Chapter 14 for available data). Currently, the only approach to estimate such emissions consists of measuring contaminant concentrations in the environment (see Chapters 4, 5 and 7).

INDUSTRY RESPONSE TO ENVIRONMENTAL FACTORS

The earlier parts of this section have looked at the performance of industry in terms of its emissions to air, water and land. This part briefly examines changes in the way industry is managing the environmental aspects of its business. These changes take, and have taken, place in response to both internal and external pressures, including increasingly stringent and better enforced legislation, public awareness and perception, pressure groups, green consumerism and employee attitudes. For industries which supply raw materials or intermediate products, pressure to adapt practices to meet environmental objectives comes mostly from government action and activism at the local level. On the other hand, for industries where pollution is due more to end use or disposal of products, pressure to change practice tends to come from consumers (see Chapters 15 and 26).

As mentioned earlier, although industry is a major polluter, through innovation and response to the demands of its consumers, it is also contributing to the development of technical solutions to environmental problems.

Change has been led by the large national and multinational companies. Those with strong links with the USA recognised that the high industrial costs and liabilities associated with stricter environmental standards were likely to spread to Europe, while others became targets for environmental pressure groups and consumers. Among the first targets were those companies in industrial sectors perceived as 'dirty', with a high public profile, such as the chemical, oil and gas industries. Leading companies adopted stated policies setting out environmental principles and objectives. These then began to be implemented through the establishment of environmental management systems, which covered the monitoring and assessment of emissions, employee training, waste handling, spill prevention and response, as well as environmental auditing programmes for the periodic assessment of environmental performance.

An historical perspective

Environmental management practice has changed significantly over the last decade in Western Europe. Before the 1980s most companies tended to have an uncoordinated, technical approach to environmental problems. Plant and processes were designed with little regard to their environmental impact, with an 'end-of-pipe' technology appended aimed simply at meeting regulatory requirements. External pressures were often deflected by public relations departments. As a result of internal and external pressures, the need to manage environmental costs and risks increased, and the more proactive companies recognised that risks and liabilities associated with poor environmental management could translate into serious cost and image problems. Better informed consumers put pressure on industries to manufacture more environmentally neutral – natural, benign and cleaner – products.

During the 1980s, there was a general shift by the leading companies towards a more integrated approach based on prevention, and incorporating environmental factors at early stages into decision making and the design of equipment. This involves managing environmental aspects of business as a line management responsibility, and as an integral part of every function from operations to strategic planning. Environmental management systems with environmental policies, strategies and programmes, use of life-cycle analysis, (see Chapter 12) and environmental auditing have been introduced (Figure 20.15).

Current developments

Progressive companies in Western Europe are responding positively to the environmental challenge by seeking to play an active role in the formulation of new regional, national and international environmental policies rather than passively accepting them, and are taking a more positive lobbying stance, seeking to improve rather than minimise environmental legislation.

Some industry associations, for example, the International Chamber of Commerce (ICC), have developed environmental codes of behaviour ('Guiding Principles') in an effort to encourage wider adoption of good environmental management practices (see Box 20B). Members may be requested to sign up to these principles (as with the ICC), or their adoption may be voluntary. Specific industry sectors are also taking initiatives to implement self-imposed programmes relying on the voluntary participation of companies in the sector, for instance, the CEFIC sponsored 'Responsible Care' programme (see the section on the chemicals industry in this chapter).

However, most guidelines of this kind do not require signatories or companies to demonstrate, via some form of reporting or verification of actual environmental performance, that, in practice, they are actually adhering to the principles to which they publicly claim to be committed.

There is a move towards integrated and preventative management practices, and away from 'end-of-pipe' solutions. Companies are seeking competitive advantage in many ways, including the integration of the environment into strategic planning; the use of life-cycle analysis in the product development cycle; the use of clean process technology; and waste minimisation.

Other important developments include the incorporation of environmental auditing and of life-cycle analysis (see Chapter 12) into the overall environmental management cycle. Environmental audit is used mainly as a management tool for evaluating company environmental performance and to assess and prioritise environmental risks (eg, site investigations, environmental liability). The results of these analyses can be used in product conception and design so that ways of reducing the impacts can be identified (eg, through product alteration or process adaptation).

To date, however, the adoption of integrated environmental management systems has been very much the domain of the few, rather than the many, and is generally limited to a minority of large multinational companies. Wide variations still exist in the extent to which environmental management systems have been integrated into business management systems. For example, while the large chemical and oil companies are among the leaders, many other large companies, and smaller service companies with lower environmental impacts, have made little headway. Within sectors there is also variation, with the leaders often those either with progressive management or which have historically had major environmental problems. As a result, a number of schemes have been developed, at both national and international level, aimed at encouraging wider and more rapid adoption of good environmental management practice in industry.

An example is the EC Eco-Management and Audit Scheme (EMAS) regulation, which applies from April 1995 (EC Regulation 93/1836/EEC). Aimed specifically at industrial companies, participation is voluntary and requires the companies to: conduct an environmental review of its

	Pre-1970s	1970s–80s	1990s
General approach	Few regulations/ limited focus – air/water	Compliance/ reactive	Prevention/ proactive
	Hazardous waste not an issue	'End-of-pipe' control	Life-cycle approach
Management			Environmental audit
Organisational structure	Limited corporate environmental presence	Corporate environmental presence functionally isolated	Full environmental integration throughout business
Costs	Environmental costs low	Environment is a cost to be minimised	The environment is a strategic opportunity to be seized

Figure 20.15
Historical environmental management trends
Source: ERM

Box 20B The Business Charter for Sustainable Development

The Business Charter for Sustainable Development is a set of 16 principles of environmental management launched in April 1991 by the International Chamber of Commerce (ICC). The principles are to:

- give corporate priority to environmental management through the adoption of policies, programmes and practices;
- integrate the environment with overall management;
- continuously seek improvement with regulation as the starting point;
- train and educate employees on environmental issues;
- assess environmental impacts of new and past activities;
- provide products and services that have no undue environmental impact;
- advise customers on environmental impacts of products and services;
- operate facilities taking into account inputs, outputs and environmental effects;
- promote research on environmental impacts of activities and products (life-cycle analysis);
- prevent serious or irreversible environmental degradation caused by products, services and activities;
- promote these principles with contractors and suppliers;
- develop emergency preparedness plans with authorities;
- contribute to the transfer of environmentally sound technology;
- contribute to the development of public policy and education;
- improve internal and external communications with employees and the public;
- measure environmental performance and conduct audits.

operations, set a policy, establish a programme, implement comprehensive environmental management systems, and conduct regular environmental audits of its activities. Participants will have to assess their environmental performance and demonstrate that reasonable, continuous improvement is being achieved, and to make the results public by issuing a statement accredited by external verifiers.

The development of an environmental management systems standard (BS 7750) by the British Standards Institute (BSI) was the first initiative of its kind. Companies will be able to participate on a voluntary basis and must implement specific elements of an environmental management cycle including: setting a policy, implementing it via appropriate management systems, periodically assessing and reviewing progress and making modifications as necessary.

One issue high on both government and industry association agendas is to ensure participation in such programmes by small- and medium-sized enterprises (SMEs). Such enterprises are important in this context owing to their wide variety of geographical location and their flexibility of response to environmental concerns (CEC, 1992a). Methodologies applied to environmental management systems in larger companies, often reflected in government schemes and standards, are not always wholly appropriate for an SME. For example, documentation and decision-making mechanisms are likely to be different and more informal in SMEs, and if encouragement of SMEs to adopt better environmental management practices is to be successful, it

will need to take account of such factors.

Many Central and Eastern European countries have well-established and quite comprehensive environmental regulatory frameworks, with, for example, laws specifying emission limits. However, enforcement has often been lax. As a result, most Central and Eastern European companies or state industries have not yet faced the same pressures and problems as their Western counterparts and there has been little progress in implementing environmental management systems and practices.

Future developments

In Western Europe, the development of business environmental management practice is likely to continue at a rapid pace as many companies try to catch up with the leaders and as government schemes such as EMAS and BS 7750 come into operation. The European Committee for Standardisation (CEN) and the International Organisation for Standardisation (ISO) are jointly expected to come up with their own scheme by the end of the 1990s. Such schemes should reinforce the trend towards greater openness taking place in response to demands for companies to make more information on environmental performance available to their shareholders, insurers, banks and the public. Indeed, banks and insurers are trying to reduce their own environmental risks by putting pressure on their business clients to reduce theirs.

Adoption of good environmental management practice by industry leaders, particularly the larger multinationals, will be felt through the supply chain as they attempt to reduce their liability and exposure to poor practices of suppliers and even buyers. Certification to formal management systems such as BS 7750 in the UK and EMAS may become as common as certification to quality management standards. In Western Europe, the similarity between quality and environmental management is being recognised, and this may result in their integration.

Life-cycle analysis techniques are likely to become more fully developed and incorporated into product development and management cycles. Integrated pollution control will replace single medium pollution control (see Chapters 14 and 15).

As a result of greater awareness about environmental problems generally, and also because of greater public willingness to take action (see Chapter 26), assessment of the environmental impacts of proposed new plants is increasingly considered. Siting or extensions of industrial plants now often need careful environmental impact assessment (EIA) before authorisation for construction or operation is granted. EIAs require a company to take account of such factors as likely resource use, visual impacts, habitat destruction, emissions to the atmospheric and aquatic environments, waste and noise impacts.

As companies in Central and Eastern Europe try to adapt to freer market economies, significant and sometimes costly changes to company business practice to take account of environmental concerns will not always be of the highest priority. However, Western companies investing in Central and Eastern Europe are displaying concerns over environmental liability, and this is already starting to encourage changes. Combined with the transfer of know-how from West to East, this is likely to encourage a more positive attitude and approach, in the long term, to how companies in Central and Eastern Europe manage their environmental performance.

There is also a growing trend at government level to move away from relying solely on traditional regulatory approaches and adopt the use of other instruments such as pollution charges, tradable emission permits and the encouragement of voluntary schemes. Examples of this include the Dutch programme of 'covenants' which represent voluntary agreements between specific industry sectors and government, and which sets out various targets (VROM, 1993).

Economic instruments are used increasingly to encourage industry to internalise its environmental costs and make appropriate environmental investment decisions. An example of this might be use of charges or incentives to encourage the installation of an effluent treatment plant, and the development of cleaner processes or of substitute products whose manufacture is less polluting.

Various suggested guidelines are issued which emphasise techniques for preventing pollution and minimising waste, including cleaner processes and conservation of resources. For instance, the World Bank, in its lending programmes, gives considerable attention to the four 'Rs' – reduction, re-use, recycling and recovery (World Bank, 1992b).

POLLUTION ABATEMENT AND INVESTMENT DECISIONS

Technologies for emissions control often exist, but the costs of implementing them, particularly for small firms, can be high. New industries, or industries new to a particular country, have the advantage of being able to consider investing in the latest available industrial capital, with waste minimisation and emissions abatement technology fully integrated into processes. The flexibility of a process for improving abatement control varies between industrial sectors depending on the product being manufactured. In many companies, existing industrial capital may not allow much flexibility, and it can be difficult to accommodate even small changes in production processes. As a result, industrial companies have tended to control emissions by use of 'end-of-pipe' technology. Expenditure on end-of-pipe emissions control is generally the only abatement expenditure which can be separately identified, although up to the last few years most expenditure in industrialised countries will have been of this type. Assessments of pollution abatement expenditure often include estimates of expenditure by industry, but comparisons between countries are misleading because the relative structure of economies, in terms of the balance between public and private sector activities, can be quite different (see for example OECD, 1993a, for information about pollution abatement expenditure in OECD countries).

In all fields, many company groups and individual companies cannot be clear about levels of environmental investment. Defining what is related to the environment in a new plant, and what is best practice, is almost impossible. Some groups do not differentiate between spending on health, safety and the environment (Abrahams, 1993). Thus data are not comparable between firms or sectors.

Investments in many sectors (for instance cement) are very large, and finance required for construction of new plants is often committed over a short pay-back period (which may be as short as three years). This gives very little flexibility to companies when making investments. Furthermore, with economic climate and legislation changing fast, there are very strict requirements for investment performance. This can be a major check on new investments and – since devices for energy and water savings and environmental abatement/clean technology are often integrated in new technology – on mitigating emissions of environmental contaminants and efficiency of resource use.

For many products and industries, the Central and Eastern European market is expected to provide the growth area for production up to the year 2000 and beyond as a result of a growing economy following reconstruction. But the caution shown by firms before getting involved in Central and Eastern Europe stems largely from the uncertainty over the pay-back periods for investments, the continued political instability in some countries, and environmental factors.

Many Central and Eastern European companies will benefit from the transfer of technology, skills and responsibility to local management as investment by Western multinationals and smaller Western companies increases. However, the high liabilities associated with many sites, especially from soil contamination, is likely to limit the extent of this investment. This may encourage clean-up and the adoption of better environmental practices, although the costs involved in the former may be extremely high in the short term (see Chapter 7).

The perception of environmental conditions in Central and Eastern Europe may be more significant than an objective assessment when investment decisions are taken by companies. A 1992 survey of the major US and Western European companies (summarised in OECD, 1993b) found that, although environmental factors did not discourage most companies from considering industrial investing in Central Europe, environmental issues are an important concern, which can impede direct industrial investment in Central Europe. Over half the companies that had rejected potential investment sites reported that environmental problems had played a very important role in this decision.

Existing pollution is perceived to be widespread in the countries of Central Europe and, overall, investors are most concerned about costs associated with inheriting liability for past pollution since, in general, liability entails responsibility for environmental contamination of existing or former industrial sites caused by a previous owner. Other environmental issues were also important: liability arising from future practices; uncertainty about the rules for environmental liability; and the costs of bringing host country facilities acquired into compliance with environmental standards (that is, either host country requirements, or internal corporate standards, which 70 per cent of respondents claimed were stricter than those of the host country).

CONCLUSIONS

- All branches of manufacturing industry contribute to some extent to environmental impacts through use of energy and raw materials. The main impacts on the environment arise directly: as a result of emissions to air or water, or by effects on the land and soil. Such impacts can occur at local, transboundary and global levels, and have implications for human health. Industrial accidents deserve particular attention and are covered in Chapter 30. However, although manufacturing industry is a contributor to environmental pressures, it also has the capability to play a major role in providing solutions to environmental problems. It can do this by developing new processes and machinery necessary for effective pollution abatement, by introducing new technologies and modified products, through better product quality, and by improving industrial productivity.
- Industrial energy consumption in Europe as a whole has declined from 49 per cent to 41 per cent of total final energy consumption between 1970 and 1989. Industry's share of total energy use has fallen in all regions of Europe, but only in the former USSR has this been coupled with an increase in the actual amounts of energy consumed over the period. However, the weakening of the link between growth in industrial output and energy consumption (largely because of efficiency improvements, and diversifying in industry) has meant that increases in industrial output no longer lead necessarily to increases in energy consumption.
- Many industrial activities need to use large amounts of water in manufacturing products. Industrial use of water accounts for just over half of total water withdrawals in Europe, although abstraction for manufacturing purposes has decreased in many countries. In some water-intensive industrial sectors, consumption by firms can be reduced in a number of ways: by improving water supply and sewage

systems, by construction of closed recirculating water systems, by building of plants that do not discharge wastewater into watercourses, and by using processes that do not require water. Firms in sectors such as pulp and paper manufacture and some sectors of the chemicals industry have demonstrated that water use can be reduced, along with wastewater emissions, without sacrificing product quality.

- Many industrial firms have adapted their behaviour to take account of environmental factors in their business practice. These changes in practice take place in response to both internal and external pressures. For 'upstream' industries which produce raw materials or intermediate products and are big consumers of natural resources in the production process, pressure to adapt practice to meet environmental objectives comes mostly from government action and activism at the local level. On the other hand, for industries where pollution is due more to end use or disposal of products, pressure to change practice tends to come more from consumers. In any case, in many sectors there is an increasing recognition that efficiency gains and minimisation of waste and product loss have benefits which reach beyond the environmental sphere.

- Environmental factors have also led to cautious attitudes being adopted at a corporate level in investment decisions, particularly with respect to Western European companies considering investing in Central and Eastern Europe. For many products and industries, the Central and Eastern European market is expected to provide the growth area for production up to the year 2000 and beyond, but due largely to the uncertainty over the pay-back periods for investments, the continued political instability in some countries and environmental factors, new investment has been sluggish. The perception of environmental conditions in the region may be more significant than an objective assessment when investment decisions are taken by companies, but environmental concerns, particularly concern about costs associated with inheriting liability for past pollution, has impeded direct industrial investment by Western investors in Central and Eastern Europe.

- In the future, small- and medium-sized companies will have an important role to play in meeting environmental objectives. Given the growth in the service sector and the encouragement given to SMEs across Europe, special attention should be accorded to services and enterprises in future state-of-the-environment reports for the pressures they exert on the environment and the special opportunities they offer for flexible responses to environmental issues.

REFERENCES

Abrahams, P (1993) Survey on chemicals and the environment. Friday, 18 June 1993. *Financial Times*, London.

Axenfeld, F, Munch, J, Pacyna, J M, Duiser, J and Veldt, C (1992) *Test-Emissionsdatenbasis der Spurenelemente As, Cd, Hg, Zn und der speziellen organischen Verbindungen Lindan, HCB, PCB und PAK für Modellrechnungen in Europa*. Dornier, Friedrichshafen.

Barth, H and L'Hermite, P (Eds) (1987) *Scientific basis for soil protection in the European Community*. Elsevier Applied Science, London and New York.

Bernes, C (Ed) (1993) *The Nordic Environment – present state, trends and threats*, Nord 1993: 12. Nordic Council of Ministers, Copenhagen.

Bulgarian Ministry of Environment (1992) *Environment and development of the Republic of Bulgaria*. Report for UNCED. Bulgarian Ministry of Environment, Sofia.

Bureau Central de Statistiques de la Suède (1984) Classification des activités industrielles selon leur pression sur l'environnement. Conférence des Statisticiens Européens. Réunion sur les problèmes généraux de méthodologie des statistiques de l'environnement, Paris, 16-19 April 1984.

Carvalho, M G and Nogueira, M (1993) Energy efficiency and pollution abatement in glass furnaces, baking ovens and cement kilns. In: Pilavachi, P A (Ed) *Energy Efficiency in Process Technology*, pp 753–66. Elsevier Applied Science, London and New York.

CBS (1992) *Natural resources and the environment 1991*. Rapporter 92/1A, Central Bureau of Statistics of Norway, Oslo.

CEC (1991) *Panorama of EC Industries 1991–92*. Commission of the European Communities, Luxembourg.

CEC (1992a) Industrial competitiveness and protection of the environment. SEC (92) 1986 final, 4 November 1992. Commission of the European Communities, Luxembourg.

CEC (1992b) *Environment Statistics, 1991*. Eurostat, Commission of the European Communities, Luxembourg.

CEC (1992c) *Panorama of EC industry, statistical supplement 1992*. Commission of the European Communities, Luxembourg.

CEC (1993) *National Accounts ESA (European Systems of Integrated Economic Accounts)*, Eurostat. Commission of the European Communities, Luxembourg.

CEFIC (1991) *Facts and figures – West European chemicals industry*. European Confederation of Chemical Industries, European Chemicals Industry Council, Brussels.

CEPI (1993) *1992 Annual Review*. Confederation of European Paper Industries, Brussels.

Cole, J and Cole, F (1993) *The geography of the European Community*. Routledge, London.

Danilov-Danilyan, V I and Arski, J M (1991) *The environment in the USSR: economics and ecology*. UNECE, Senior Advisers to ECE Governments on Environmental and Water Problems, Joint Working Group on Environment and Economics (second session, 18–20 December 1991), Geneva.

Éditions Atlas (1992) *Europe Le Grand Atlas*, Éditions Atlas, Paris.

E/SB (1991) *Country Reports: Central and Eastern Europe 1991*, Eurostat/Statistisches Bundesamt, Commission of the European Communities, Luxembourg.

Harle, N (1990) The ecological impact of overdevelopment: a case study from the Limburg borderlands. In: *The Ecologist* **20**, (5), pp 182–9, September/October 1990.

Klimont, Z, Amann, M, Cofala, J, Gyarfas, F, Klaassen, G and Schöpp, W (1993) *Emission of air pollutants in the region of the Central European initiative – 1988*. IIASA, Laxenburg, Austria.

Knook, H D (1991) Coordination of EC assistance to the Black Triangle. In: *International Spectator*, November, xlv, pp 709–17, Dijkman, The Hague.

Komppa, A (1993) Paper and energy: a Finnish view. In: Pilavachi, P A (Ed) *Energy Efficiency in Process Technology*, pp 101–10. Elsevier Applied Science, London and New York.

Lithuanian Environmental Protection Department (1992) National Report to UNCED. Environmental Protection Department, Vilnius, Lithuania.

March Consulting Group (1992) *Energy technologies in the chemicals industry sector*. Report prepared for DG XVII Energy, Commission of the European Communities, Brussels.

March Consulting Group, MAIN, and Cowiconsult (1992) *Energy technology in the cement industry sector*. Report prepared for DG XVII Energy, Commission of the European Communities, Brussels.

Masters, R L (1971) Air pollution–human health effects. In: McCormac, B M (Ed) *Introduction to the Scientific Study of Atmospheric Pollution*, pp 97–130. D Reidel Publishing Company, Dordrecht, The Netherlands.

MECR and CEI (1992) *Environmental year-book of Czech Republic, 1991*. Ministry of the Environment of the Czech Republic (MECR) in collaboration with the Czech Ecological Institute (CEI), Prague.

Meybeck, M, Chapman, D V and Helmer, R (Eds) (1989) *Global environment monitoring system: global freshwater quality, a first assessment*. Blackwell, Oxford.

Miljøstyrelsen (1988) *Fosfor – kilder og virkninger*. Redegørelse fra Miljøstyrelsen Nr 2 1988. Copenhagen.

NIVA (1992) Paris Convention – annual report on direct and riverine inputs to Norwegian coastal waters during the year 1991. Norwegian Institute for Water Research, Oslo.

Nordic Council of Ministers (1992) Yearbook of Nordic Statistics 1992. Nord 1992: 1, Vol 30, Nordic Council of Ministers and the Nordic Statistical Secretariat, Copenhagen.

OECD (1991) The state of the environment. Organisation for Economic Cooperation and Development, Paris.

OECD (1993a) Pollution abatement and control expenditure in OECD countries. OECD Environment monographs No 75. OCDE / GD (93)91, Organisation for Economic Cooperation and Development, Paris.

OECD (1993b) Environmental policies and industrial competitiveness. Organisation for Economic Cooperation and Development, Paris.

Parmeggiani, L (1983) Encyclopedia of occupational health and safety. International Labour Organisation, Geneva.

SCERB (1992) Governmental report of the state of the environment in the Republic of Belarus. The State Committee for Ecology of the Republic of Belarus, Minsk, Belarus.

Statistics Canada (1991) Human activity and the environment. Ministry of Industry, Science and Technology, Ottawa.

UK DoE (1993) Digest of environmental protection and water statistics No 15, 1992. Department of the Environment. HMSO, London.

UN (1990) Industrial statistics yearbook 1988, Volume II, General industrial statistics. United Nations, New York.

UNECE (1991) Rational use of water and its treatment in the chemicals industry. United Nations, New York.

UNECE (1992) The environment in Europe and North America. Annotated Statistics 1992. United Nations, New York.

UNECE (1993a) The Steel Market in 1992. United Nations, New York.

UNECE (1993b) The chemicals industry in 1992. Annual review. Production and trade statistics 1989–91. United Nations, New York.

UNEP (1992) Chemical pollution: a global overview. United Nations Environment Programme, Earthwatch, Geneva.

van der Most, P F J and Veldt, C (1992) PARCOM-ATMOS: Emission factors for air pollutants 1992. TNO Institute of Environmental and Energy Technology, Apeldoorn, The Netherlands.

VROM (1991) Emission inventory in The Netherlands. Industrial emissions 1985–87, Summary. No 2, July 1991. Dutch Ministry of Housing, Physical Planning and Environment (VROM), Inspectorate General for Environmental Protection, Leidschendam, The Netherlands.

VROM (1993) Milieuprogramma 1994–97. Dutch Ministry of Housing, Physical Planning and Environment (VROM), The Hague.

WHO (1992) WHO Commission on Health and Environment: Report of the Panel on Industry. World Health Organisation, Geneva.

Williams, H and Musco, D (1992) Research and technological development for the supply and use of freshwater resources, strategic dossier. Sast Project 6, EUR-14723-EN, DG XII, Commission of the European Communities, Luxembourg.

World Bank (1992a) World development report 1992. Development and the Environment. World development indicators. Oxford University Press, Oxford.

World Bank (1992b). The World Bank and the environment. Fiscal 1992. World Bank, Washington DC.

World Bank (1993) Environmental Action Programme for Central and Eastern Europe. Document submitted to the Ministerial Conference, Lucerne, Switzerland, 28–30 April 1993.

WRI (1992) World Resources Institute. World Resources 1992–3, a guide to the global environment. Oxford University Press, Oxford.

Congestion on M25, UK
Source: Michael St Maur Sheil

21 Transport

INTRODUCTION

An efficient transport system is a crucial precondition for economic development and an asset in international competition. Personal mobility for work, study and leisure purposes is considered a key ingredient of modern life. With the integration of markets in Europe, economic growth and higher levels of income, transport is also a major growth sector. In the EU, the transport service industry accounts for about 7 to 8 per cent of GDP including 'own account' (transport by and for the same enterprise) and private transport (CEC, 1992a). To grow together, the continent needs the mobility of people and goods made possible by an efficient transport system.

The benefits of transport, however, come at a high price. The Task Force on the Environment and the EC Internal Market considered transport: '...the most important environmental impact of the Internal Market' (Task Force Environment and the Internal Market, 1990). Not only are the building and maintenance of transport infrastructure a significant item in government spending and accidents a heavy social cost (see Chapters 10 and 11), but nuisances from noise, air pollution and the consumption of energy and natural resources also represent considerable environmental liabilities. Carbon dioxide (CO_2) emissions from transport are a major contributor to the greenhouse effect. Road transport is currently the greatest offender, accounting for 80 per cent of CO_2 emissions from transport and 60 per cent of total nitrogen oxides (NO_x) emissions. Routine and accidental releases of oil or chemical substances into the environment by lorries and tankers contribute to the pollution of soils, rivers and the sea. Transport infrastructure covers an increasing amount of land to the virtual exclusion of other uses, cuts through ecosystems and spoils the view of natural scenery and historic monuments.

No mode of motorised transport is environmentally friendly. However, some modes of transport, notably rail and inland waterway, have lower environmental impacts than others, such as road and air. An analysis of transport developments and their impact on the environment must therefore distinguish between different modes of transport, and whether these modes are transporting passengers or freight.

The social and environmental costs in Germany of road transport alone have been put at 2.5 per cent of GDP (Kågeson, 1993). For OECD countries as a whole the figure for road transport has been put even higher at nearer 5 per cent (OECD, 1988). It is therefore questionable whether the

Figure 21.1 The contribution of road transport to environmental problems in The Netherlands
Source: VROM, 1991

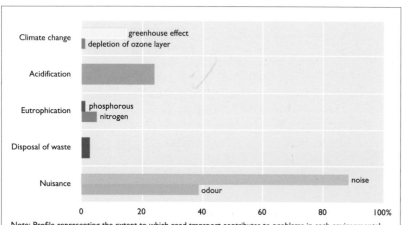

Note: Profile representing the extent to which road transport contributes to problems in each environmental theme. All target group contributions together are taken to be 100% per theme.

present organisation and level of transport in Europe are environmentally sustainable.

Figure 21.1 shows the contribution of road transport to a number of environmental problems in The Netherlands. For example, road transport is the main cause of noise nuisance in The Netherlands (over 80 per cent), while contributing little to waste disposal problems (under 5 per cent). Comparisons of the range of environmental impacts from transport would be possible with such 'theme profiles' for other countries. Unfortunately, theme profiles have so far only been developed for The Netherlands (Adriaanse, 1993).

ENVIRONMENTAL IMPACTS

Overview

Transport is a key sector for policy actions to improve the quality of the European environment and combat global environmental problems. Transport has impacts on both the natural and built environments, and on human health. Table 21.1 provides an overview of significant types of environmental impacts from different modes of transport. This section elaborates some of the principal impacts of transport in terms of emissions, energy consumption and transformation of land resources. Other environmental impacts are covered elsewhere in the report. Noise from transport is treated in Chapter 16, used tyres in Chapter 15 and traffic accidents and associated hazards in Chapters 10, 11

Table 21.1
Overview of significant transport impacts on the environment
Sources: EEA-TF, based on OECD, 1991; UNEP, 1992

Environmental media / Transport activity	Air	Water	Soil/Land	Nature and wildlife/Landscapes
Road transport	● combustion of petroleum products → emissions of NO_x, CO, CO_2, VOCs, particulate matter → local to global environmental impacts, health effects ● NO_x and VOCs emissions → tropospheric O_3, and PAN ● use and release of fuel and additives → emissions of lead and VOCs (eg, benzene) ● road transport → noise and air pollution (and human casualties)	● runoff containing oil, salts and solvents from road surfaces → pollution of surface and groundwater ● NO_x and SO_2 emissions → acidification ● roads → modification to hydrological systems	● road construction → land lost for infrastructure and service stations → pressure and fragmentation of land resources ● transport of hazardous substances → risk of accidents → soil contamination and human casualties ● discarded vehicles, waste oil, batteries, old tyres, old cars → disposal problem	● extraction of road-building materials and road construction → landscape degradation ● infrastructure → severance and fragmentation of habitats, possible hindrance to wildlife migration
Rail transport	● electricity power generation to run electric trains → air emissions ● diesel trains → emissions to air ● steam trains (coal powered) → emissions to air	● railways → modification to hydrological systems	● transport of hazardous substances → risk of accidents	● dereliction of obsolete facilities → landscape degradation ● railway infrastructure → possible hindrance to wildlife migration
Water transport (maritime and inland)	● concentrated harbour activities → emissions to air ● enclosed seas and crowded traffic routes → emissions to air ● bunkering of fuel and loading of fuel → emissions to air (VOCs)	● discharge of ballast water from ships → water pollution ● accidental and operational spills at seas (including oil) → water pollution ● sewage and waste from ships → water pollution ● antifouling paints → water pollution ● transport of hazardous substances → risk of potential accidents	● disposal of dredged material from canal construction and dredging → waste disposal problem	● construction of shipping berths and channels → landscape impacts ● terminal dereliction → landscape impacts ● canalisation of rivers → landscape impacts
Air transport	● aircraft → emissions of NO_x and CO_2 (high emissions especially during take-off, taxiing and landing) → ground level smog and acid rain ● contributions to stratospheric ozone depletion and global warming at higher levels; contrails ● associated road traffic at airports → increased emissions	● runoff from airports containing oil and anti-freeze → water pollution ● airport construction → modification of hydrological systems	● construction of airports → pressure on land resources	● extraction of airport building materials → landscape degradation ● airport construction → landscape changes ● airport construction → disruption of ecological areas
Pipelines	● → emissions to air (CH_4) → global warming	● leaks of oil → potential water pollution		● possible barrier to wildlife migration if above ground

	EU		EFTA		Central Europe		former USSR		Total Europe2	
	1970	1990	1970	1990	1970[1]	1990	1970	1990	1970	1990
Total energy for transport: all products (Mtoe)	111.4	229.6	16.8	28.3	30.5	34.1	89.6	139.5	248.3	433.1
Petroleum products (Mtoe)	107.2	226.0	16.0	27.2	21.9	31.8	76.7	121.8	221.8	406.8
Petroleum products: road transport (Mtoe)	86.7	191.7	12.8	22.5	19.6	28.5	29.2	58.1	148.3	300.8
% transport in final energy consumption	18.6	31.9	20.1	26.3	13.1	12.8	16.8	15.4	17.3	22.2

Notes:
1   1971 for Central countries.
2   including former USSR.

Table 21.2 *Energy consumption by transport in Europe, 1970 and 1990*
Source: Special data set compiled for this report, based on International Energy Agency and Eurostat data (see *Statistical Compendium*)

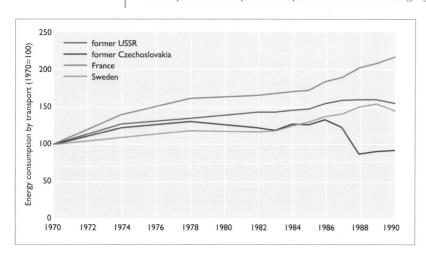

Figure 21.2 *Energy consumption by transport in selected European countries, 1970-90*
Source: Special data set compiled for this report, based on International Energy Agency and Eurostat
(see *Statistical Compendium*)

Figure 21.3 *Share of energy consumption by different transport modes in selected*
European countries, 1990 Source: Special data set compiled for this report, based on International Energy
Agency and Eurostat (see *Statistical Compendium*)

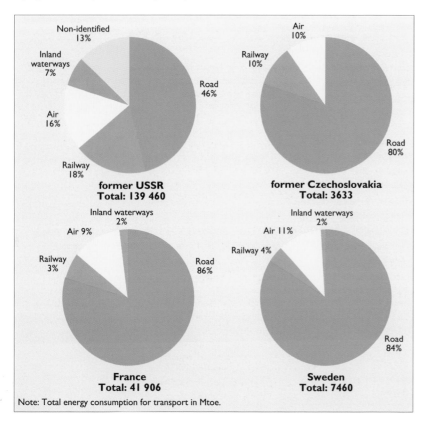

and 18. Transport in urban areas is covered in Chapter 10.
Box 21A provides an overview of definitions, data availability
and quality related to transport.

Energy consumption and climate change

Energy consumption

Since the beginning of the 1970s, transport has become a
major consumer of non-renewable energy resources. All
power-driven transport consumes energy. Oil currently fulfils
almost all transport energy needs. In the EU, road transport
currently accounts for over 80 per cent of oil consumption of
the transport sector (see Table 21.2). Thus, the demand for oil
products by transport is largely responsible for the depletion
of non-renewable resources, energy-related emissions, and
environmental impacts arising from the oil industry.

Energy consumption by transport can be assessed by country
or city, by mode, and by comparison with energy consumption by
other human activities (notably industry, agriculture and
households). Some data showing the share of transport in total
energy consumption by country and by city are presented earlier in
this report (see Chapters 19 and 10 respectively).

Table 21.2 shows trends in energy consumption by
transport for different parts of Europe over the last 20 years.
Energy consumption by transport doubled in the EU
between 1970 and 1990. In Central Europe the increase in
total energy used in transport was far more modest, though
consumption of petroleum for road transport increased by
nearly 50 per cent between 1970 and 1990. The trends in
energy consumption by transport for some selected European
countries are shown in Figure 21.2: the greatest increase
relative to 1970 is for France, and the greatest decrease for the
former Czechoslovakia.

Figure 21.3 shows the modal split of energy consumption
for the same countries in 1990. Road and air transport
accounted for 95 per cent of energy consumed by transport in
France, and in Sweden, compared with 90 per cent in the
former Czechoslovakia (on the other hand, the energy used in
rail transport was 7 per cent more in the former
Czechoslovakia than in France). In the former USSR, road
transport in 1990 accounted for only 46 per cent of energy
consumption by transport, with railways accounting for 18 per
cent, air for 16 per cent and inland waterways for 7 per cent.
The low figure for road transport reflects the low degree of
motorisation in the former USSR, while the high share of air
transport reflects the size of the country. Generally speaking,
in Western Europe road and air transport are both the biggest
energy consumers within the transport sector and have
experienced strong growth in the past. The situation in Central
and Eastern Europe is different due to another modal split.

Air transport, passenger cars and lorries consume
proportionally more energy per passenger-kilometre or
tonne-kilometre than railway and inland waterway transport.
(Worldwide, air transport accounts for about 15 per cent of

Transport mode	CO_2 emission estimates
Passenger transport	*Passenger transport [g/pass-km]*
Passenger car	133–200
Bus	35–62
Train	39–78
Air	160–465
Freight transport	*Freight transport [g/tonne-km]*
Lorry	207–280
Train	39–48
Inland waterway	40–66
Air	1160–2150

Table 21.3
Specific carbon dioxide emission estimates for different transport modes
Source: CEC, DG XI internal data (Samaras, personal communication, 1993)

energy use by transport.) Energy requirements for passenger and freight transport vary not only by mode of transport, but also according to the occupancy rate (ie, the number of passengers in a car, and the loading factor in lorries).

There have been improvements in road vehicle efficiency across Europe. During the period 1970 to 1990, the average fuel consumption of new cars in the EC decreased from 10 to 8.2 litres per 100 kilometres (Gwilliam and Geerlings, 1992). Currently available technologies, plus weight reduction, could have the potential to reduce fuel consumption by more than a third. However, such improvements in efficiency have been more than offset by increases in vehicle ownership. Another trend, leading to higher fuel consumption, is the increasing amount of larger and more powerful vehicles.

Climate change

Transport is a major contributor to emissions of greenhouse gases. The most important of these are emissions of CO_2, and to a lesser extent nitrous oxide (N_2O) and methane (CH_4) (Rypdal, 1993). In the EU, transport currently accounts for about one quarter of total energy-related CO_2 emissions and reaches a share of over 30 per cent in Luxembourg, France and Spain (ECMT, 1994). In some Central and Eastern

European countries in 1988 (the former Czechoslovakia, Hungary and Poland), the transport sector was responsible for between 7 and 11 per cent of total energy-related CO_2 emissions, while the share in Slovenia was some 21 per cent (Klimont et al, 1993). Thus transport is a less important contributor to CO_2 emissions in Central and Eastern Europe than in Western Europe.

Road traffic currently accounts for some 80 per cent of total transport emissions of CO_2 in the EU, followed by aircraft with a share of 15 per cent. Freight transport contributes about one third of total transport CO_2 emissions, and passenger cars about 45 per cent (DG XI, CEC, personal communication). Table 21.3 shows estimates of actual emissions by passenger and freight transport by taking into account different 'loading factors'. For example, for freight transport, the load factor for light- and heavy-duty road freight transport is low (30–40%) compared with trains and inland waterways (nearer 100%). At these assumed loading factors, lorries emit 5 to 6 times more CO_2 than trains per tonne/km.

Looking at the life-cycle of passenger cars, almost one third of greenhouse gas emissions arise from causes other than vehicle operation. The IEA have estimated that about 72 per cent of greenhouse gases from passenger cars are emitted from the exhaust during vehicle operation, 17 to 18 per cent of car life-cycle emissions arise from fuel extracting processing and distribution, and a further 10 per cent from vehicle manufacture (IEA, 1993).

Outlook

The consumption of energy by transport varies not only in absolute terms between different countries, but also by mode of transport. Trends indicate that in a 'business as usual' scenario, energy consumption and hence transport-related CO_2 emissions will increase by almost 25 per cent between 1990 and 2000 (CEC, 1992a). The main increase in consumption of energy and CO_2 emissions will be for road transport. Growth trends may even be greater in Central and Eastern Europe.

Conventional emissions

As described in earlier chapters, transport is a major factor contributing to atmospheric emissions (Chapters 4 and 14). Emissions arising from road transport are the most significant. However, emissions into the air from aircraft and shipping and operational discharges to water from shipping are also cause for concern. When emission figures are cited in this section, they refer to total emissions from all transport modes, unless otherwise qualified. The term 'mobile sources' is used when UNECE statistical data are cited, which in principle covers all modes of transport (UNECE, 1992a). The human health effects of emissions arising from transport are not covered in this chapter (see Chapter 11).

Vehicle type	Vehicle stocks (thousands)	NOx emissions (thousand tonnes/year)
Diesel engines: passenger cars and light-duty vehicles	20 496	414
Diesel engines: heavy-duty vehicles and buses >3.5t	3780	1926
Petrol engines: passenger cars and light vehicles	115 241	3228
Petrol engines: heavy-duty vehicles and buses >3.5t	785	77
Other road transport[1]	24 529	94
Total	164 831	5739

Note: For other transport-related activities (railways, inland waterways, marine activities, off-road vehicles, airports, service stations) not currently available from CORINAIR90.

1   Two-wheeled vehicles and liquefied petroleum gas.

Table 21.4
Nitrogen oxide emissions by type of road vehicle in the EU, 1990
Sources: B, DK, F, GR, IRL, L: CORINAIR90 (Samaras, personal communication 1993); D, I, NL, P, E, UK (Samaras and Zierock, 1992) (see Table 5.9 for country codes)

The most significant conventional emissions from road transport are of NO$_x$, CO, VOCs, lead and particulate matter. Some exhaust emissions from road transport give rise also to secondary pollution, such as photochemical oxidants (see Chapter 32).

Nitrogen oxides from road transport

Nitrogen oxides (NO$_x$) contribute indirectly to the 'greenhouse effect' and directly to acid rain and the build-up of tropospheric ozone. In 1990, about 45 per cent of total NO$_x$ emissions to the atmosphere in 20 European countries were attributable to road transport (Chapter 4 and 14). Emissions of NO$_x$ result from the combustion of fuels under high pressure and temperatures. As the energy efficiency of engines depends on high compression ratios, there is a conflict between the objectives of energy efficiency (and lower CO$_2$ emissions) and low NO$_x$ emissions. Consequently, for example, the inherently energy-efficient diesel engine has comparatively high NO$_x$ emissions. Within the EU for 1990, petrol-powered cars and light-duty vehicles, which accounted for 70 per cent of the vehicle fleet, contributed 56 per cent to NO$_x$ emissions from road transport, while diesel-powered heavy goods vehicles and buses, which accounted for only 2 per cent of the vehicle fleet, contributed 34 per cent to NO$_x$ emissions from road transport (Table 21.4).

Trends in emissions from mobile sources of NO$_x$ vary between different European countries . The share of NO$_x$ emissions from mobile sources, of total NO$_x$ emissions, varies between 25 per cent in the Czech Republic to 78 per cent in Norway (CORINAIR90, 19 countries – see Chapter 14). Even with the implementation of legislation to reduce NO$_x$ emissions from road vehicles, throughout the 1980s emissions of NO$_x$ have risen steadily in Western Europe. This is mainly because of the continuing increase in road transport. In addition, there is also a considerable difference between prescribed emission standards for new passenger cars and the actual emission performance of the existing fleet. This is due mainly to the time needed for emission control technologies to penetrate the vehicle fleet. In addition, the replacement rates for car fleets can differ considerably between countries. For example, for 1991, in Spain 22 per cent of passenger cars were 15 or more years old, while in the UK the corresponding figure is only 4 per cent (UNECE, personal communication). In Hungary, 42 per cent of the passenger fleet is over 10 years old, and 62 per cent over 7 years old (World Bank, 1993). Finally, in all parts of Europe, in-service emissions from vehicles are often very high due to poor maintenance.

Fewer data are available for NO$_x$ emissions from road transport for Central and Eastern Europe, and hence it is more difficult to detect trends. However, increased motorisation in Central and Eastern Europe is likely to lead to emissions rising

in the future. For example, NO$_x$ emissions for passenger cars from Poland have been projected to increase from around 55 thousand tonnes per year in 1989 to about 100 thousand tonnes per year by 2000 before declining as catalytic converters are introduced (Fergusson, 1991). Planned restrictions and the introduction of technological control devices are hoped to reduce total European NO$_x$ emissions by 2000 by about 20 to 30 per cent compared with 1990 levels, and transport must make a contribution (UNECE 1988 Sofia protocol on NO$_x$ emissions – see Chapter 31).

Some of the air pollution problems occur along major trunk roads linking towns. This is particularly true for NO$_x$ emissions, which increase with vehicle speed. Such vehicle emissions contribute to the formation of tropospheric photochemical oxidants. Map 21.1 provides an example of the emissions arising from journeys between Helsinki and Oulu in Finland.

Other emissions from road transport

The greatest volumes in absolute terms of toxic emissions from road transport are of carbon monoxide (CO) created through incomplete fuel combustion from petrol engines. Emission estimates from mobile sources vary between 30 and 90 per cent of total CO emissions (CORINAIR90, 19 countries – see Chapter 14). In Western Europe, emissions of CO are almost exclusively produced by petrol-powered passenger cars (particularly in urban areas), while in Central and Eastern Europe there are also considerable emissions from stationary sources. Carbon monoxide has significant human health impacts, particularly interfering with absorption of oxygen.

Volatile organic compounds (VOC) emissions contribute to the build-up of tropospheric photochemical oxidants. They arise from incomplete fuel combustion and the evaporation of fuel from petrol engines and service stations. Consistent data are not available on emissions of VOCs from mobile sources (see Chapter 14), but road transport is believed to have accounted in the late 1980s for about 35 to 40 per cent of VOC emissions in Europe (CEC, 1992b). Emissions of VOCs from mobile sources in most countries are quite stable or slightly increasing, which reflects the trend in total VOC emissions (UNECE, 1992a). Emissions related to service stations (resulting from stockage of petrol and its distribution from terminals to service stations) account for about 5 per cent of all anthropogenic VOC emissions (CEC, 1992b).

Particulate matter is associated primarily with diesel engines, with emissions 30 to 70 times greater than for petrol engines. Particulate matter can remain in the air for considerable periods of time and contribute to particulate smog. Particulates are also damaging to health, particularly those particles fine enough to remain within the lung (leading to respiratory diseases and cancer).

Off-road emissions

While most transport emissions originate from passenger and freight transport, emissions from off-road vehicles (eg, agricultural tractors, construction machines, industrial machines and forest machines) can also be significant. In Finland, it has been estimated that work machinery contributed 15 per cent of total NO$_x$, 9 per cent of total CO, 5 per cent of total VOCs, and 4 per cent of total particulate matter emissions in 1990. Similar studies in Sweden and The Netherlands have equally shown that emissions from work machinery are significant (Puranen and Mattila, 1992).

Aircraft emissions

The main environmental problem of aircraft emissions relates to NO$_x$ and CO$_2$. A particular concern here is that emissions are injected at such high altitudes. The suspected effects of air transport on stratospheric ozone depletion and global warming are already described in previous chapters (4 and 14). Although the share of the total volume of transport accounted for by air transport is small, its share of emissions is considerable because of high energy consumption per km travelled. The share of NO_x produced by aviation could actually increase, because newer aircraft have more efficient engines with higher combustion temperatures, and hence a tendency to produce more NO_x (IATA, 1992). Changing engine technology and the design of planes, combined with alterations to flight patterns, mean that the average altitude of flights is slowly increasing, raising the altitude at which CO_2 and NO_x gases are emitted. This in turn may enhance their effect on global warming (Gwilliam and Geerlings, 1992).

Railway emissions

There are few data on emissions arising from rail transport, which are low compared with road and air transport. Electrification of railway lines can reduce considerably the overall energy burden needed to operate a network. For example, in Germany the energy use has been reduced to about 25 per cent of that in 1953, while the volume of traffic has increased by a third (Ellwanger, 1993). Overall reduction in the energy burden in turn reduces energy-related emissions. The alternative to using electricity is normally diesel fuel. In the EU the degree of electrification varies considerably, from zero in Greece to 80 per cent in Luxembourg (CEC, 1993a). In Central and Eastern Europe, the extent of the electrified train network is more limited than in the EU, and hence there are more direct emissions arising from rail transport.

Maritime and inland waterway emissions

As with rail transportation, there are few data available on emissions from maritime and inland water transport per tonne-km travelled. Maritime transport contributes to some local environmental impacts, mainly around harbours. The bunkering and loading of fuels can give rise to evaporative VOC emissions. Sulphur dioxide emissions from sea-going vessels can be considerable due to the high sulphur content of bunker fuel used in maritime transport. This can be quite high in crowded shipping lanes (eg, emissions of SO_2 from ships in the Strait of Dover involved in international trade were 4.0 to 9.9 tonnes/km^2/year between 1985 and 1990 (CORINAIR data)). The emissions to air arising from freight transport by inland waterways are lower per tonne-km. However, both maritime and inland waterway transport give rise to significant emissions to water. While most attention is normally focused on periodic accidents (see Chapters 18 and 30), most discharges to water are of an operational nature.

Outlook

Conventional emissions from road transport are cause for concern. Emissions of NO_x, CO and VOCs per vehicle should reduce (at least temporarily) with the gradual penetration of passenger car fleets with catalytic converters in Western Europe over the next few years. Increasing road traffic, however, will cause increases in total emissions from road transport. In Central and Eastern Europe, however, trends in conventional emissions are likely to increase until new technologies are introduced.

Land intrusion

Transport infrastructure consists of roads, railways, waterways, harbours, airports, garages and depots and parking places. This covers land which could possibly be put to other purposes or left to nature. The use of land by different transport modes has been reviewed in Chapter 10. In general, natural habitats are irreversibly destroyed when transport facilities are built. Damage also arises at extraction sites of building materials and at dumps for rubble from

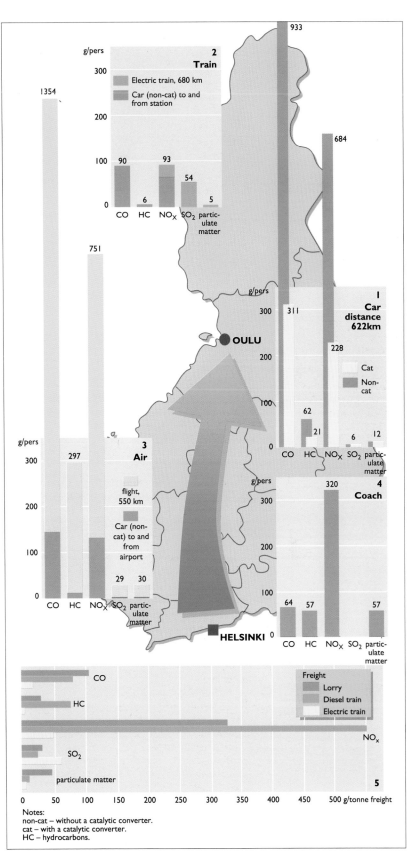

infrastructure works. In many cases, recultivation measures for quarries, spoil tips and abandoned lines are not carried out. In addition, transport infrastructure creates barriers within natural habitats where they impede the migration of animals. However, in certain cases, transport infrastructure can also create natural corridors in urban areas or areas of intensive agriculture. Box 21B provides an example of a recent infrastructure development of European significance.

Different modes of transport use varying amounts of

Map 21.1
The emissions resulting from four ways to travel from Helsinki to Oulu, Finland
Source: Translated from Wahlström et al, 1992

Box 21B The Rhine–Main–Danube Canal

The opening of the Rhine–Main–Danube Canal in September 1992 marked the completion of one of Europe's most ambitious transport projects (see Map 21.2). The scale of its technical details is indeed impressive: during a construction time of about 30 years, more than DM 6 billion have been spent to create the 178 km linkage between Bamberg and Kehlheim in Bavaria (Sturm, 1992). The 60 m wide canal required the excavation of 100 million cubic metres of soil and went hand in hand with alterations of the Main as well as the Danube to allow larger ships to take this long awaited 'short cut' between the Black Sea and the North Sea (Bryson, 1992). For canal users, the transport distance and prices are roughly reduced to one tenth compared with the alternative passage through the Mediterranean Sea (Sturm, 1992).

However, the environmental impacts of the enterprise seem to be of a similar scale to its technical accomplishments. By cutting through some of the most scenic and environmentally sensitive landscapes in Germany, the canal has left a series of scars on the landscape. For example, in the 300 000 hectare Altmühltal Naturpark, 38 per cent of wetlands, 21 per cent of floodplain areas and 16 per cent of reedbeds had to be destroyed to make way for the canal (Weiger, 1992). Although DM 280 million have been invested to establish new habitats and carry out ecological designs, making it the world's most expensive landscape development scheme, the canal's negative impacts on the environment are far from being mitigated (Bryson, 1992). Accompanying projects, like the creation of 55 hydroelectric power stations (Bryson, 1992), land reforms for agricultural development plans (intensification), improved flood control (Koch and Vahrenholt, 1983) and recreational projects for several large artificial lakes have affected the natural balance in regions far beyond the direct impacts of the place of construction.

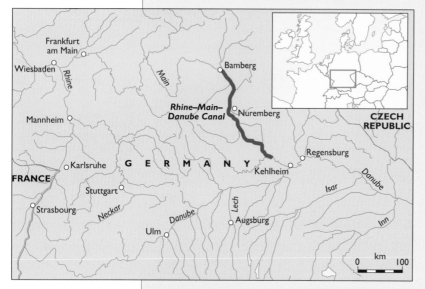

Map 21.2 *Location of the Rhine–Main–Danube Canal*
Source: EEA-TF

land. There are no comprehensive data available on the land taken for different modes of transport. However, it has been estimated that the road network consumes about 1.3 per cent of the total land area of the EU, as compared with about 0.03 per cent for the railway network (CEC, 1992c). An important measure of land-take is landuse in terms of infrastructure needed to move one transport unit, expressed in persons or tonnes of freight, over a given distance. Landuse is most efficient for air and sea transport due to relatively small land-take for a given journey, followed by inland navigation. It is less efficient for rail and least for road (CEC, 1992a).

In all European countries, the total road network is much longer than the total rail network. In addition, while the total length of the rail network diminished by 4.1 per cent in European OECD countries between 1970 and 1985, the road network has expanded (UNEP, 1992). For most analyses the road network is split into motorways and 'other roads'. Motorways account for a small fraction of the total length of the road network. In the EU the current length of the motorway network presently exceeds 35 000 km (CEC, 1993b). However, the greatest rates of increase in road construction in Europe have been for motorways, while the length of 'other roads' has increased only slightly. In the EU the construction of some 12 000 km of new motorway is planned by 2002 (CEC, 1993b). Also in Central and Eastern European countries, the motorway network has been expanded. Between 1980 and 1988, its length grew by 130.4 per cent in Bulgaria, 59.2 per cent in the former Yugoslavia, 48.8 per cent in Hungary and 17.7 per cent in Romania. However, the length of motorways in Central and Eastern Europe was very low in 1980 by Western European standards (absolute figures are contained in the *Statistical Compendium*).

An indicator which gives some impression of the extent to which land is taken by transport infrastructure is road density, defined as the total length of the road network in relation to land surface. Figure 21.4 shows the road density (the length of all roads, per unit of total land area) between 1970 and 1990 for selected countries, and illustrates the large differences between them.

Generally speaking, the environmental evaluation of transport infrastructure has to consider both direct and indirect effects as well as unavoidable trade-offs. For example, besides taking increasing amounts of land, the extension of old roads and the building of new roads and motorways indirectly tends to induce new road transport as bottlenecks are reduced and convenience and accessibility are increased. New motorways can also increase pressures for development, especially close to interchanges (eg the M25 ring around London). Consequently, the environmental effects described above can be worsened. Even though rail and waterways are environmentally preferable transport modes, they can also have deleterious environmental impacts. Transport infrastructure is never environmentally neutral.

Figure 21.4
Road density in selected European countries, 1970–90
Source: Eurostat

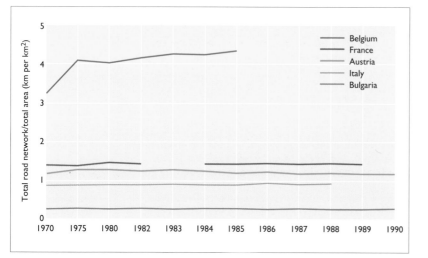

Mode	Transport volume		Growth (% per year)	Modal share (%)	
	1970	1990	1970–90	1970	1990
Inland freight (thousand million tonne-km)					
(15 countries)					
Road	430.98	952.36(E)	4.04	50.06	66.31
Waterways	105.83	109.58	0.17	12.29	7.63
Rail	252.76	234.09	–0.38	29.36	16.30
Pipelines	66.58	121.71	3.06	7.73	8.48
Air	4.83[2]	18.35	9.31	0.56	1.28
Total	861.43	1436.09		100	100
Passengers (thousand million pass-km)					
(14 countries)[3]					
Total road[4]	1799.80 (E)	3419.47 (E)	3.26		
Cars	1556.36 (E)	3071.41	3.46	72.66	75.69
Buses	243.44 (E)	348.06 (E)	1.80	11.37	8.58
Rail	202.90	263.35	1.31	9.47	6.49
Air	139.21[2]	374.93	6.83	6.50	9.24
Total	2141.91 (E)	4057.75 (E)		100	100

Note: It is not possible to indicate figures for the total annual average growth, as data refer to different years.
1   includes Turkey
2   1975
3   excluding Greece
4   total road = total of cars and buses
(E): estimate

Table 21.5
Trends and development of transport modal split, 1970–90, Western Europe[1]
Source: Based on ECMT and ICAO data (see *Statistical Compendium*)

Mode	Transport volume		Growth (% per year)	Modal share (%)	
	1970	1990	1970–90	1970	1990
Inland freight (thousand million tonne-km)					
(4 countries)					
Road[1]	39.62	97.09	4.58	15.29	30.18
Waterways	10.87	11.25	0.17	4.19	3.50
Rail	194.24	182.95	–0.30	74.95	56.88
Pipelines	14.38	30.15	3.77	5.55	9.37
Air	0.04[2]	0.21	11.70	0.02	0.07
Total	259.15	321.65		100	100
Passengers (thousand million pass-km)					
(4 countries)[3]					
Cars[4]	7.27	47.00	9.78	not possible	not possible
Buses	77.54	138.19	2.93	to calculate	to calculate
Rail	83.49	92.44	0.51		
Air	5.90[2]	15.83	6.80		

Table 21.6
Trends and development of transport modal split, 1970–90, former Czechoslovakia, Hungary, Poland and former Yugoslavia
Source: Based on ECMT and ICAO data see *Statistical Compendium*

Notes: Data on passenger and freight transport exist for other Central and Eastern European countries (eg, Romania and Lithuania) but originate from other data sources and are only partially complete. It is not possible to indicate figures for the total annual average growth, as data refer to different years.

1   former Yugoslavia: transport for hire and reward only
2   1975
3   as data are only available for Hungary for private cars it is impossible to give a complete picture of the passenger traffic situation for these countries
4   data on private cars only available for Hungary

TRENDS AND UNDERLYING FORCES IN THE TRANSPORT SYSTEM

The picture of transportation in Europe has changed considerably over the last few decades, albeit less so in Central and Eastern Europe than in Western Europe. It is the direction of these changes which turns transport into an increasingly important pressure on the environment and on the quality of life of European citizens. Developments are determined by two major trends: growing demand for transport, and the changing market share of transport modes. An overall picture is provided in Tables 21.5 (Western European countries) and 21.6 (Central European countries). While both these trends have been evident in Western Europe for many years, they are likely to materialise also in the Central (and Eastern) European countries as their economies are transformed.

Growing transport demand

The growing demand for transport is governed by a complex interaction between economic growth, changes in industrial structure for freight transport, and socio-economic factors such as higher levels of income and new landuse patterns for passenger transport. While the costs of private transport have declined in real terms for both passenger and freight transport, the same cannot be said for public transport.

Economic growth

Changes in the volume and structure of economic activity have immediate repercussions on the transport system. In the 20-year period between 1970 and 1990, an average annual GDP growth of 2.6 per cent was accompanied by an annual growth rate of 2.5 per cent for freight transport generally, and a 4.1 per cent increase for road freight in ECMT countries. Since 1985, the growth in freight transport even exceeded the increase in GDP in ECMT countries (Short, 1992).

Over the same period for passenger transport there was an annual growth rate of 3.1 per cent; for passenger cars the annual growth rate was 3.4 per cent (Short, 1992). The general pattern of growing demand for both freight and passenger transport is likely to persist if economic growth is maintained, and there are no significant changes in real costs and other factors. It is difficult to make such comparisons for passenger and freight transport for Central and Eastern

European countries due to difficulties in devising historic GNP figures for these countries.

Changes in industrial structure affecting freight transport

Changes in industrial structure have resulted in locational shifts from urban to new industrial sites and contributed towards a dispersal of economic activities. These changes have been amplified by continuing processes of economic integration within the EU (the completion of the Single Market, the European Economic Area and the movement into market economies of the Central and Eastern European countries).

The development of strongly service-based economies has made transport patterns more diffuse. The number of points of departure and arrival has increased, and flexibility and speed have become key parameters in transport decisions. At the same time, changes in production organisation and logistics have increased the demand for transport. In an effort to optimise production and save on storage costs and stock capital, for instance, business has introduced just-in-time (JIT) delivery systems. JIT is based on the splitting of loads into smaller batches which means a higher number of transport movements for a given load. Generally speaking, the nature of freight has shifted from heavy bulk goods to lighter high-value goods. There has therefore been a reduction in shipment size and an increase in shipment frequency.

A large part of the growth in freight transport is accounted for not by the increase in freight volumes, but by the fact that goods are shipped over longer distances. In Germany, for example, between 1970 and 1990, while total domestic freight transport on all modes grew roughly in line with GDP, long distance road freight transport rose by an average rate of 5.4 per cent (OECD, 1993b). The total international north–south transport of freight in Europe accounts for about 65 million tonnes a year, and according to some estimates will increase to 120 million tonnes in 2010. During the last decades the share of road goods in international road transport has increased most and now exceeds freight transport by rail (UNECE, 1991). For example, freight traffic over the Alps in France, Switzerland and Austria has grown by one half, from 50.4 to 74.6 million tonnes between 1979 and 1990: the share of freight carried by road transport has increased from 44.2 per cent to 55.6 per cent. The disturbance caused by road freight transport is not confined only to the level of traffic. The size of the freight vehicles is also important. As a result, the transit treaty between the EU and Switzerland contains certain provisions to limit the number of trips by 40 tonne trucks (BUWAL, 1992).

The opening up of Central and Eastern European countries and the development of corresponding trade links will enhance transport demand for both freight and passenger transport (UNECE, 1992b). Even under conservative assumptions concerning economic developments in the former socialist countries, west–east road transport is predicted to rise by a factor of around four, and transport flows on rail by a factor of around three in the next 15 to 20 years (Empirica, 1993, p 160). The European Commission and UNECE have proposed the strengthening of pan-European networks to cope with anticipated increases in traffic (CEC, 1992a, UNECE, 1992b).

Socio-economic factors affecting passenger transport

The increase of net income and demographic changes have led to higher rates of car ownership and increased holiday and leisure time for travelling. The bulk of the increase of passenger transport is due to the increased use of the private car, which has almost doubled in Western Europe, and

Figure 21.5
Rate of increase in car ownership in selected European countries, 1970–90
Source: See *Statistical Compendium*

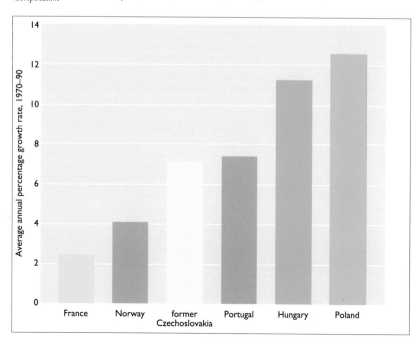

increased over sixfold in Hungary between 1970 and 1990 (see Tables 21.5 and 21.6). Current levels of car ownership are greater in Western Europe, with an average of 34.9 cars per 100 inhabitants in 1990, compared with an average of 17.6 cars per 100 inhabitants for former Czechoslovakia, Hungary and Poland. However, in Central and Eastern European countries, although fewer people have a car, the annual rate of increase in car ownership is greater than in Western Europe (Figure 21.5). In Poland, for example, the number of motor cars per 100 inhabitants has nearly doubled from 6.7 in 1980 to 13.8 in 1990. In Central and Eastern Europe, rising incomes are likely to lead to further increased demand for personal transport.

Passenger transport by air is growing rapidly in Europe. However, the number of aircraft movements has not increased as rapidly as the number of passengers. This is due to an increase in the size of aircraft. This growth is particularly associated with the movement of tourists, which tends to be concentrated in summer months (see Chapter 25).

At the local level, the functional segregation of landuse (work, leisure, shopping) has increased the need for commuting and for travelling in daily life. Many services can be obtained only by travelling over longer distances. Out-of-town shopping centres have become commonplace (see Chapter 10).

Changing market shares of different transport modes

The choice between different modes of transport is determined by user needs, on the one hand, and the services and costs offered by each mode, on the other. User needs change with developments in production and distribution systems for freight transport, and landuse and lifestyles for passenger transport. The services and costs of different transport modes are determined by the operators themselves and, partly, by government policies. In essence, changes which have taken place in recent years have favoured road over less polluting rail and waterway transport. The distribution of goods and passengers between these modes has changed accordingly.

Regarding freight transport, the greatest increase has been for air, road transport and pipelines in Western Europe (see Table 21.5). There has been a decline in the movement of goods transported by rail: only in Austria is there currently still more freight movement by rail than road. In Central European countries, more freight is moved by rail than road, though the share of rail in freight transport declined by almost 20 percentage points between 1970 and 1990 (Table 21.6).

For many purposes, road freight transport has a number of advantages over other modes of transport. It offers rapid, flexible and reliable door-to-door services. These correspond well to current trends in transport needs and logistics (eg, JIT). While heavy, low-value bulk goods can well be shipped by rail and waterways, these modes are presently not well suited to the transport of small batches of high-value freight. There has also been a proliferation of courier express companies which use the fastest means of transport (usually by road or air) at each stage in a journey to ensure rapid delivery.

While its superior service quality is the main trump card of road transport, even as far as costs are concerned, rail and combined transportation are less competitive, particularly over short distances. Sixty-six per cent of all goods in the EU are transported within 50 km, and a further 20 per cent within distances of between 50 and 150 km, leaving only 14 per cent for longer distances (CEC, 1992a). It is estimated that just under half of all national and international rail transport in the EU takes place over distances of more than 150 km and about 15 per cent over distances in excess of 500 km. The

transport of rail freight in the EU is mainly over medium distances (50–150 km).

Nevertheless, even on longer distances, road haulage is increasingly the preferred option of shippers. The phenomenon of 'trucked airfreight' illustrates the competitive power of road haulage over long hauls. The advantage of road transport also lies in the extensiveness of its infrastructure. Roads link up all places of production, distribution and consumption. At the same time, smaller railway lines are closed down. In many cases, therefore, road transport is the only option.

In the passenger transport sector, a large variety of factors determine modal choice. Income, car ownership, demographic characteristics, trip length, time of the trip and journey purpose are important in the decision to take the car or, instead, use some form of public or non-motorised transport. Crucial, of course, are the availability of public transport services as well as their quality (reliability, frequency, speed, comfort). Especially on shorter trips, public transport is often not a viable alternative to using the car. Access to many facilities (eg, out-of-town shopping centres, places of leisure) is possible virtually only by car. Although comprehensive data for passenger transport are patchy, data for two large EU Member States point to the predominance of short-distance travel (CEC, 1992a). For one country, for example, half the car passenger journeys were over a distance of less than 5 km, and only a quarter over more than 10 km. Data from the other country show that a private car is on average used for only 2.7 journeys over a distance of more than 200 km per year, and that half of all private cars never travel that distance. People may, however, be deterred from using their own vehicle by restrictive policies such as parking restrictions in urban areas. Positive measures also include the building of cycle lanes and walkways to make these non-motorised forms of mobility more attractive. This clearly highlights the significance of public infrastructure and landuse policies in influencing choice of means of transport.

Outlook

Growing demand for transportation and changes in the split between transport modes have characterised developments in the field in Western Europe. In the absence of new policy measures, growth will continue to concentrate on road transport of freight and passengers. To what extent these patterns will be repeated in the Central and Eastern European countries depends both on the progress of social and economic transformation and on public policies. In the past, there were strong differences in the transportation systems in Western and Central and Eastern Europe. Although the vehicle fleet in the former socialist countries is older and hence noisier and more polluting than in Western countries, there was little of the growing transport demand and the changing modal split that has been increasing environmental pressures in Western Europe. In fact, transport policies under socialism were often geared towards public transport services, although these were often not well maintained. The road's share of total freight transport (in tonne-kilometres) was much lower than in Western Europe, although it rose during the 1980s in Bulgaria, the former Czechoslovakia and Hungary. Even in Bulgaria, the country with the highest road share in Central and Eastern Europe, road transport accounted for only about one third of total freight transport (E/SB, 1991). If the transport system in the former socialist countries were to develop along Western lines and with similar growth rates, despite technical improvements in vehicles, the local environment would suffer to an increased extent. Furthermore, the pressure on the European and global environment in terms, for example, of energy consumption, global warming and air pollution, would worsen.

Box 21C Measures to control the environmental impact of transport

Technical measures:

- emission standards on CO, VOCs, NO_x and particulates for all kinds of motor vehicles (eg, by the EC and UNECE);
- fuel quality standards concerning, for example, lead, sulphur and benzene (eg, by the EC);
- emission standards on smoke, CO, VOCs and NO_x for aeroplanes (by ICAO);
- noise standards for motor vehicles (by the EC);
- noise standards for aeroplanes (eg, by the EC and ICAO)
- development of electric cars and fuel cells.

Construction measures:

- noise protection walls along major roads and motorways, low-noise asphalt;
- bridges and tunnels for animals crossing roads and railways;
- integration of transport infrastructure into landscape (eg, through impact assessment).

Transport planning and traffic management:

- provision and improvement of public transport facilities;
- provision of separate cycling tracks along roads and in cities;
- restriction of car use in inner cities and residential areas through pedestrian zones, speed limitations, parking restrictions, road safety measures, alternate odd and even number plate access;
- extension of rail, waterway and combined transport;
- bans on through traffic.

Economic instruments:

- internalisation of external costs for all transport modes through taxes and fees (eg, energy tax, fuel taxes, road pricing and parking fees);
- differentiated purchase taxes (eg, between leaded and unleaded petrol);
- scrappage benefits to encourage owners to replace older polluting vehicles with cleaner vehicles fitted with catalytic converters;
- differential circulation taxes (eg, in the EU).

Others:

- regular in-service emission tests for motor vehicles;
- time restrictions on transport movements, especially night-flying bans for airports and bans on night and weekend driving for lorries;
- lowering and enforcement of speed limits for motor vehicles;
- encouraging smooth driving behaviour;
- educational campaigns;
- car pooling;
- staggered working hours; encouraging working from home, now possible through better telecommunications techniques;
- carrying through existing resolutions (eg, ECMT resolutions on Transport and Environment (No 66) and Power and Speed (No 91/5) and conventions (eg, UNECE Sofia protocol on NO_x emissions, 1988).

(See Chapter 4 for further details on measures to improve air quality.)

OUTLOOK

Forecasts of growth in transport demand show, at least as far as Western Europe is concerned, that in a 'business as usual' scenario, with reasonably favourable economic conditions, the expansion of the road sector is likely to be buoyant. Under these conditions, a near doubling of road transport for both passengers and freight between 1990 and 2010 seems likely. Even if lower economic growth slows the rate of deterioration of the environment for a time, the risk of development of the transport sector being unsustainable in the medium to long term due to its broad environmental impact remains real (CEC, 1992a).

Action to alleviate the environmental consequences from transport has been taken in all European countries and by the international organisations concerned. It includes a variety of technical and non-technical measures (see Box 21C). Through these measures, the encroachment of transport on the environment and on the quality of life of European citizens

has been, and will further be, significantly reduced. Further improvements along these lines are possible through advanced technology and better maintenance. Mandatory emission standards combined with regular in-service controls have emerged as the universal strategy to combat air pollution from road transport. Emissions and noise from the different types of road vehicles have been cut in several steps over the last two decades. Carbon dioxide emissions cannot be controlled by any end-of-pipe technology, such as catalytic converters for NO_x emissions. However, CO_2 emissions from individual motor vehicles can be curbed by higher fuel efficiency (lean burn engines) and a shift to lighter vehicles. Efforts to develop recycleable and electric cars are progressing.

Nevertheless, especially in the long run, technical measures by themselves will not suffice to meet environmental targets. The growth trends in road and air transport in Western European countries are bound to cancel out the technical improvements achieved in the vehicle fleets. Due largely to its strong growth, and to the unfavourable distribution of transport demand across different modes at least in Western Europe, transport is becoming an increasingly important pressure on the environment in comparison to other human activities. Thus, a common problem for policy makers in Europe today is how to respond to the twofold issue of growing transport demand and how this demand can be best allocated to different transport modes. To supplement technological approaches, there is a need to consider the more wide-ranging use of different instruments (including economic instruments) as a way to curb growth in transport demand in order to meet environmental targets.

The price of transport services can be influenced through taxation or fees such as road tolls. In turn, transport prices influence overall transport demand and the relative use of different transport modes. From both an economic and an environmental point of view, transport

Table 21.7
External social costs arising from transport in Germany, 1993
Source: Kågeson, 1993

Mode	Air	CO_2	Noise	Accidents	Total
Freight (ECUs per 1000 tonnes)					
Lorry (long-distance)	5.6	2.5	0.6	3.5	12.2
Train (electric)	0.8	1.9	0.3	1.4	4.4
Ro-ro vessel[1]	6.0	0.6	0.0	0.1	6.7
Passenger (ECUs per 1000 pass-km)					
Car	14.6	4.5	1.2	13.7	34.0
Train (electric)	0.9	2.2	0.3	1.4	4.7
Aircraft	7.3	9.2	1.6	0.2	18.3

Notes: ECU rate of exchange of 29 April 1993.
1 shipping which accommodates roll-on, roll-off (ro-ro) lorries.

prices should not only cover the costs of providing and maintaining transport infrastructure and services. Other external costs such as pollution and nuisances, greenhouse gas emissions, landuse and accidents should also be reflected in transport prices. However, in Europe, up until now, this has generally not been the case. Hence, transportation is artificially cheap and demand for transport exaggerated. Moreover, the external costs of transport are internalised in transport prices to a varying extent for different modes. This creates distortions of competition between different modes. Table 21.7 provides some estimates of the external social and environmental costs arising from transport for Germany. Road transport is clearly the main cause of environmental problems. As mentioned in the introduction, the social and environmental costs of road transport in Germany have been translated from these estimates to approximately 2.5 per cent of GDP, split by accidents (1.16 per cent), air pollution (0.74 per cent), energy/CO_2 emissions (0.39 per cent) and noise (0.17 per cent) (Kågeson, 1993).

Central and Eastern European countries face both a special responsibility and a unique opportunity in deciding how to construct their future transport systems. If the same trajectory is followed as that taken so far by Western countries, environmental pressures will be further exacerbated. For example, it is expected that the large increases in road transport in East Germany after reunification in 1990 will lead to an increase in CO_2 emissions there by 140 per cent by 2005, compared with an increase of 24 per cent in CO_2 in West Germany (ECMT, 1993). Hence, a transport strategy which squares the need for economic development and high-quality transportation services with environmental concerns is urgently needed.

While concern about the environmental consequences of transport is rising, there are no easy solutions at hand. Some actions, besides technological improvements, have already been taken. The development of high-speed networks, already well established for example in France, makes the railways more competitive against road and air transport. However, it should also be added that high-speed train links are not the most important factor in the general state and development of the rail network, and conventional links and regularity of services are also important considerations. In the freight sector, efforts are under way to improve the quality and capacity of combined transport (transport of freight using more than one mode for a given journey), for example, for Alpine transport. However, combined transport can only replace pure road transport if the capacity and flexibility of European railways are significantly enhanced. The development of inland and coastal shipping and the use of pipelines for freight transport are also being encouraged by bodies such as the European Commission and UNECE. The use of cars has been severely curtailed in a growing number of European cities (see Chapter 10). The European Commission has set the ambitious objective of making the transport system as a whole environmentally sustainable through the concept of sustainable mobility (CEC, 1992a and 1992c), although this concept still lacks clear 'operationalisation'.

If transportation systems are to be brought in line with the imperative to protect and improve the environment, substantial changes are required in the organisation of economic activities as well as in private lifestyles. The key question is whether the same or higher levels of the quality of life can be achieved with lower transport demand. So far, the 'transport intensity' of the economic system and of private lifestyles has been growing, and

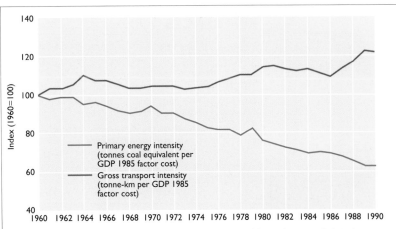

Note: Gross transport intensity is calculated by dividing a measure of the total amount of physical movement in the economy by GDP. This means that all movements of passenger, freight, road vehicles and other forms of transport modes (eg, passenger cars and trains) are included. Their masses are multiplied by distance travelled and summed together. The total is then divided by GDP. Transport intensity is for Great Britain (ie, excluding Northern Ireland), whereas energy intensity is for the UK. GDP for both transport and energy intensity is for the UK (see Peake, 1994).

higher levels of welfare have been bought with higher levels of transport. Figure 21.6 demonstrates, for Britain, that, in contrast to the continuing decline of energy intensity, transport intensity is still rising; whereas energy intensity has fallen by 40 per cent between 1960 and 1990, transport intensity has risen by 20 per cent.

For the future, sharp political choices are required mainly in relation to the pricing of transport, and to landuse and infrastructure planning. While the principle has been accepted that users should pay for the full social and environmental costs of transport, as called for, for example, in the resolutions of the ECMT, this principle has still to be put fully into practice. Transport and urban and regional planners, for their part, face the challenge of devising solutions which reduce transport demand and satisfy mobility requirements with a less environmentally harmful mix of transport means.

Figure 21.6
Gross transport intensity and primary energy intensity in Great Britain and the UK, 1960–90
Source: Peake, 1994

CONCLUSIONS

● Emissions of traditional pollutants per vehicle, such as particulate matter, CO, NO_x and VOCs will be substantially reduced in Western Europe in the next few years, as technology begins to take effect. However, this will be offset by expected increases in vehicle numbers. The emission of traditional pollutants is of concern in Central and Eastern Europe.

● Emissions of CO_2 from transport are increasing and are a serious cause for concern in the context of the greenhouse effect.

● The change in modal split towards road transport and the strong growth in air transport are exacerbating environmental impacts.

● Reductions in road transportation could be achieved by decoupling transport and economic growth, aimed at inhibiting increases in passenger and road freight transport. If transport does not pay its real social and environmental costs, then the benefits arising from technological improvements will be outweighed by disbenefits from increased transport growth.

● Technical solutions to reduce environmental impacts will need to be supplemented with actions to limit transport demand and improve modal split by economic instruments, appropriate transport infrastructure and landuse planning.

REFERENCES

Adriaanse, A (1993) *Environmental policy performance indicators*. Dutch Ministry of Housing, Physical Planning and Environment (VROM), The Hague.

Bryson, B (1992) Linking Europe's waterways: Main-Danube Canal, *National Geographic*, August 1992, 3-31. Boulder, USA.

BUWAL (1992) *The Swiss Report to UNCED*. Swiss Federal Office of Environment, Forests and Landscapes (BUWAL), Berne.

CEC (1992a) *The future development of the common transport policy: a global approach to the construction of a Community framework for sustainable mobility*. COM(92)494 final, Office for Official Publications of the European Communities, Luxembourg.

CEC (1992b) *Proposal for a common position of the Council on the proposal for a Council Directive on the control of volatile organic compound (VOC) emissions resulting from storage of petrol and its distribution from terminals to service stations (the so-called 'Stage 1' Directive)*. COM(92) 277 final, 30 July 1992, Office for Official Publications of the European Communities, Luxembourg.

CEC (1992c) *Green paper on the impact of transport on the environment: a Community strategy for 'sustainable mobility'*. COM(92)46 final, 20 February 1992, Office for Official Publications of the European Communities, Luxembourg.

CEC (1993a) *Projet de rapport sur un réseau transeuropéen des chemins de fer*. Groupe de Travail Rail Conventionnel, DG VII Transport, Office for Official Publications of the European Communities, Luxembourg.

CEC (1993b) *Trans-European networks: towards a master plan for the road network and road traffic*. Motorway Working Group, DG VII Transport, Office for Official Publications of the European Communities, Luxembourg.

Ellwanger, G (1993) Transport ferroviaire et environnement. *UNEP Industry and Environment*, (January–June) 16(1–2), 60–64. UNEP Industry and Environment Programme Activity Centre, Paris.

Empirica (1993) The spatial consequences of the integration of the new German Länder into the Community and the impact of the developments of the countries of Central and Eastern Europe on the Community Territory, Part I, the impact of the developments of the countries of Central and Eastern Europe on the Community Territory. *Draft Final Report to DG XVI Regional Policy*. Commission of the European Committee, Empirica, Bonn.

ECMT (1993) *Reducing transport's contribution to global warming, Conclusions of ECMT Seminar 30 September–1 October 1992*. European Conference of Ministers of Transport, European Conference of Ministers of Transport, Paris.

ECMT (1994) *Transport and the greenhouse effect in ECMT member countries: data and forecasts*. Committee of Deputies Ad Hoc Group on Transport and Environment, 3 March 1994, CEMT/CS/ENV(94)1, ECMT, Paris.

E/SB (1991) *Country Report: Central and Eastern Europe 1991*. Eurostat and Statistisches Bundesamt, Office for Official Publications of the European Communities, Luxembourg.

Fergusson, M (1991) *Road traffic pollution in Central and East Europe*. Worldwide Fund for Nature Discussion Paper, Gland, Switzerland.

Gwilliam, K M and Geerlings, H (1992) *Research and technology strategy to help overcome the environmental problems in relation to transport: overall strategic review*. Monitor-SAST Activity, DG XIII Information Technologies and Industries and Telecommunications, Brussels.

IATA (1992) *Air transport and the environment*. International Air Transport Association, Geneva.

IEA (1993) *Cars and climate change*. International Energy Agency. OECD, Paris.

Kågeson, P (1993) *Getting the prices right: a European scheme for making transport pay its true costs*. European Federation for Transport and Environment, Stockholm.

Klimont, Z, Amann, M, Cofala, J, Gyarfas, F, Klaassen, G and Schöpp, W (1993) *Emission of air pollutants in the region of the Central European initiative-1988*. IIASA, Laxenburg, Austria.

Koch, E R and Vahrenholt, F (1983) Die Lage der Nation. Umwelt-Atlas der Bundesrepublik: Daten, Analysen, Konsequenzen, p 423. GEO, Hamburg.

OECD (1988) *Transport and the environment*. Organisation for Economic Cooperation and Development, Paris.

OECD (1991) *The state of the environment*. Organisation for Economic Cooperation and Development, Paris.

OECD (1993a) Indicators for the integration of environmental concerns in transport policies. *Environment Monographs 80*. Organisation for Economic Cooperation and Development, Paris.

OECD (1993b) *Environmental performance reviews: Germany*. Organisation for Economic Cooperation and Development, Paris.

Peake, S (1994) *Transport in transition: lessons from the history of energy*. Royal Institute for International Affairs. Earthscan, London.

Puranen, A and Mattila, M (1992) Exhaust emissions from work machinery in Finland. *Environment International 18*, 467–76.

Rypdal, K (1993) *Anthropogenic emissions of the greenhouse gases, CO_2, CH_4 and N_2O in Norway*. Central Bureau of Statistics of Norway, Oslo.

Samaras, Z and Zierock, K-H (1992) *Assessment of the effect in EC Member States of the implementation of policy measures for CO2 reduction in the transport sector*. Office for Official Publications of the European Communities, Luxembourg.

Short, J (1992) Freight transport as an environmental problem. Paper presented at the *Workshop on Sustainable Mobility in the Internal Market: the Environmental Dimension of Freight Transport in Europe*, 21–23 May 1993, European University Institute, Florence.

Sturm, N (1992) Bayern hat den Donau-Kanal noch immer nicht voll, *Süddeutsche Zeitung 222*, 25 September, 1992, Stuttgart, p 24.

Task Force Environment and the Internal Market (1990) *'1992': the environmental dimension*. Economica, Bonn.

UNECE (1991) *Transport information 1991*. United Nations, New York.

UNECE (1992a) *The environment in Europe and North America: annotated statistics 1992*. United Nations, New York.

UNECE (1992b) *International transport in Europe, an analysis of major traffic flows in Europe*. United Nations, New York.

UNEP (1992) *The world environment 1972–1992: two decades of challenge*. Chapman & Hall, London.

VROM (1991) *Essential environmental information: The Netherlands 1991*. Dutch Ministry of Housing, Physical Planning and Environment (VROM), The Hague.

Wahlström, E, Reinikainen, T and Hallanaro, E-L (1992) *Ympäristön tila Suomessa*. Gaudeamus, Helsinki.

Weiger H (1992) *Le canal Rhin-Main-Danube-un cauchemar pour les protecteurs de l'environnement*, translated by Guicharrousse, H and Parreaux, P. Bund for Umwelt und Naturschutz Deutschland, Munich.

World Bank (1993) *Environmental Action Programme for Central and Eastern Europe*. Document submitted to the Ministerial Conference, Lucerne, Switzerland, 28–30 April 1993, (29 March 1993).

Crop spraying in Germany
Source: Michael St Maur Sheil

22 Agriculture

INTRODUCTION

Agriculture is essentially a manipulation of ecosystems to produce or raise organic matter (crop plants or livestock) from the use of land. By employing various technologies and techniques, production can be maximised (use of fertilisers, genetic developments, irrigation, mechanisation), while other methods are used to minimise loss of crops through pests and weeds (including mechanical weeding, biological control, use of pesticides such as insecticides, fungicides, herbicides).

The purpose of agriculture has traditionally and primarily been to meet the demand for agricultural products, mainly food, but also raw materials for fibre manufacture. Although the underlying purpose of agriculture has not changed, the nature, structure and ways in which these demands have been met have all changed greatly over the last few decades, and will continue to do so.

Changes have resulted from a variety of factors. These include: patterns of consumption of agricultural products; food distribution and processing; genetic development of agricultural

---

Box 22A
Data sources, availability and country groupings

This chapter has been prepared with data supplied mainly by Eurostat compiled from multiple sources. These sources include Eurostat databanks, FAO and UNECE data. Some examples have been drawn from national reports submitted to UNCED in 1992. Most data relate to the structure of production, the inputs to the land to maximise production and the resulting production itself. However, fewer data are available on the size and relative contribution to impacts on the environment of agricultural production and changes in agricultural systems. Details of the data can be found in the *Statistical Compendium*.

For EU and EFTA countries, plenty of data are available at national level on production, employment structure, fertiliser and pesticide use, farm sizes and livestock demographics. However, in other cases (for instance, irrigation, crop types grown) it is useful to consider geographical groupings of countries. Where possible, therefore, statistical data have been presented by specific country groupings based on physical and

geographical regions relevant to important landuse characteristics. These are referred to as: Nordic, Eastern, Southern and Western regions. These should not be confused with the common political association of countries which these may imply. The precise countries included in these regions are as follows (see also Figure 5.3):

- *Nordic:* Finland, Iceland, Norway, Sweden
- *Eastern:* Belarus, Bulgaria, Czech Republic, Estonia, Hungary, Latvia, Lithuania, Moldova, Poland, Romania, Russian Federation, Slovak Republic, Ukraine
- *Southern:* Albania, Bosnia-Herzegovina, Croatia, Cyprus, Greece, Italy, Malta, Portugal, Serbia-Montenegro, Slovenia, Spain, former Yugoslav Republic of Macedonia (FYROM)
- *Western:* Austria, Belgium, Denmark, France, Germany, Ireland, Liechtenstein, Luxembourg, The Netherlands, Switzerland, UK

production and other technological developments; the progressive globalisation of agricultural markets; and the influence of national and international agricultural policies.

The failure of European agriculture to meet demand during World War 2 and shortly after, made security of food production the main objective of agricultural policy from the late 1940s. Every country in Europe has encouraged its farmers to produce more food through a variety of mechanisms, including price support, other subsidies, and support for research and development. As a result, European agriculture has been able, on the whole, to meet the needs of a population of about 680 million people in Europe ever more abundantly, and with increasing cost efficiency.

The success achieved in agricultural production has, however, entailed increased impacts on the environment, ranging from pollution of groundwater to loss of habitats for plants and animals. The natural vegetation of Europe, which once consisted predominantly of woodlands, with a relatively small number of animal and plant species (see Chapters 8 and 9), has been greatly modified as a result of agricultural activities, and many habitats as well as landscapes are reliant on extensive agriculture. During the last three decades the introduction of modern, more intensive, agricultural techniques has accelerated these changes.

European agriculture is an extremely diverse and heterogeneous economic sector in terms of the products generated, the nature and structure of production units, and the variability of potential impacts on the environment, and it is difficult to make meaningful generalisations. Nevertheless, using the available data (see Box 22A), this chapter aims to show the trends in agricultural structure and practice which have evolved to meet demands, and to indicate the associated potential impacts of agriculture on the environment.

ENVIRONMENTAL IMPACTS

Agriculture is a human activity that, without doubt, affects the environment. However, it is important first to consider the extent to which agriculture itself is a victim of a degraded natural environment. Increased concentrations of sulphur dioxide and nitrogen dioxide in the atmosphere are harmful to plant growth. Tropospheric ozone can have a marked effect on yield, with yields diminishing as concentrations increase (UNECE, 1992) (see also Chapter 32). Increases in ultraviolet-B solar radiation may possibly inhibit the growth of certain plant crops and a whole range of organisms extending from bacteria to vertebrates (UNECE, 1992) (see Chapters 16 and 28). Air pollution can have effects on crop growth, and climate change may also have effects on some important crops grown in Europe. Chapter 27 examines the potential impacts of climate change and of increased carbon dioxide levels in the atmosphere on agriculture.

The extent and causes of the environmental impacts from agricultural practices vary significantly across Europe, particularly by farm type and by crop type. 'Industrial'-type farms, consuming large amounts of energy, raising livestock intensively and growing crops which demand high levels of fertilisers and pesticides, are commonly found in Europe. Often, on these types of productive units, farming has been carried out without due consideration for the environment. However, these intensive units are only part of the picture, and other, smaller, farms in Europe, particularly Southern Europe, bring a heterogeneity to the agricultural sector and its techniques, and therefore variations in agricultural practice, which determine the effect on the landscape and environmental media. Smaller farms where land is used intensively contribute to soil erosion, for instance, in Southern Europe. There are also many variations in the nature of the receiving environment (for example, different soil types or climatic and weather patterns) (eg, see Chapters 3 and 7). Table 22.1 summarises selected environmental impacts from agricultural practices.

Agriculture contributes to a variety of emissions to the atmosphere (including, in particular, ammonia and methane). About 90 per cent of European ammonia emissions (totalling around 8–9 million tonnes per year) originate from livestock farming and land application of manure (Hartung, 1992; Pretty and Conway, 1988). These ammonia emissions do not only contribute to damage to terrestrial and aquatic ecosystems (with nutrients in rainwater changing nutrient-poor habitats), but also represent an economic loss of valuable nitrogen fertiliser. This problem is particularly acute in agricultural areas where animal husbandry is practised intensively (eg, in parts of Belgium, The Netherlands, Denmark), although often these countries (eg, The Netherlands and Denmark) have specific programmes for abatement of ammonia emissions from agriculture.

Livestock ruminants are also a significant source of methane emissions. In the UK, for instance, cattle and sheep accounted for about one quarter (26 per cent) of total methane emissions (4.4 Mt per year) in 1990 (WSL, 1990). In Poland, animal husbandry accounts for about 35 per cent of total methane emissions (about 2.7 Mt per year) (MEPNRF, 1991). Methane is a potent greenhouse gas, and its role is covered further in Chapters 4, 14, and 27. Acidic deposition to the soil from agriculture as a result of ammonia emissions is covered in Chapters 7 and 31.

Use of fertiliser can contribute to the greenhouse gas nitrous oxide (N_2O) emissions (see Chapter 14). Denitrification by bacteria, especially in poorly drained soils, generates N_2O. Pesticides drift can result in long-distance atmospheric transport of pesticides, and pesticides residues in rainwater.

The main types of pollution from agricultural activities are from nitrates, pesticides, runoff of silage effluent, and slurry. Although the main source of phosphorus pollution is from sewage, agriculture, and more specifically animal manure, is also an important source (furthermore, once phosphorus is in groundwater or soil, it is more difficult to remove than nitrates). Rising nitrate levels threaten the quality of drinking water. In addition, levels of nitrates and phosphates, much lower than those resulting in groundwater contamination, can lead to eutrophication of fresh and coastal waters (see Chapters 5, 6 and 35). Losses of nitrate from agriculture depend on the type of farming system operated. The quantity of nitrate lost from a particular farming system is determined largely by the balance between inputs of nitrogen in the form of fertilisers and effluents of livestock, and the nitrogen output from the farm in terms of harvested crops or animal products. It is also dependent on whether the farming system protects the soil from leaching during winter, by ensuring vegetation cover.

The use of pesticides, particularly herbicides, which are also used in non-agricultural circumstances such as on roads and railways, can lead to contamination of waterbodies and residues in drinking water supplies. Accidental spills and leaks of materials high in organic matter (eg, slurry, silage effluent) into waterbodies can deprive aquatic organisms of oxygen and lead to serious loss of aquatic life. Non-accidental but steady releases of such materials which are also high in nutrients can lead to eutrophication. Erosion of agricultural soils and associated runoff can lead to sedimentation of waterbodies. Suspended sediments in surface waters clog waterways, increase the operating costs of water treatment facilities and silt up dams/reservoirs, reducing their flood control and/or

Table 22.1 Overview of significant environmental impacts from agricultural practices
Sources: EEA-TF, ERM

Environmental media / Agricultural practice		Air	Water	Soil	Nature and wildlife/ Landscapes
Specialisation and concentration	Increasing field sizes, land consolidation, removal of vegetation cover		• removal of vegetation cover → increased surface runoff and sediment load → sedimentation, contamination, eutrophication	• removal of vegetation cover → soil erosion • inadequate management → soil degradation	• loss of hedgerows, woodlands, small watercourses and ponds → decrease in landscape variety and reduction in species diversity • land degradation if activity not suited to site
	Intensive animal husbandry	• emissions of methane, ammonia	• silage effluent → organic matter and nutrients in waterbodies (see Fertilisation)	• spreading of manure high in heavy metal content → elevation of soil concentrations	• construction of storage silos → changed landscape
	Intensive cropping		• soil erosion → increased sediment runoff → water pollution (see Fertilisation)	• loss of organic matter in soil → deterioration of soil structure, soil biological activity → decline in soil fertility and adsorbtion capacity → increased erosion and runoff	• increasing field sizes and land consolidation possibly required → changed landscape
Fertilisation	Animal manure (slurry or solid)	• ammonia and nitrous oxide volatilisation • unpleasant odours	• spills of organic matter and nutrients to waterbodies → eutrophication → oxygen depletion → excess algae and water plants, fewer fish • leaching to groundwater → pollution of drinking water supply	• accumulation of heavy metals and phosphates in soil (may enter food-chain) • overapplication → local soil acidification possible	• potential loss of nutrient-poor habitats
	Commercial (nitrogen, phosphorus)	• ammonia and nitrous oxide release	• nitrate leaching and phosphate runoff → elevated nutrient levels → eutrophication of fresh and coastal waters, and contamination of aquifers	• accumulation of heavy metals → effects on soil microflora and entry into food-chain • overapplication → local acidification → deterioration of soil structure, imbalance in nutrients	
	Sewage sludge		• leaching of nutrients and other chemicals to aquifers	• accumulation of heavy metals and organic micro-pollutants (may enter food-chain)	• direct contamination of fauna and flora with microbial agents and chemicals
Pesticides application (insecticides, fungicides, herbicides)		• evaporation and pesticide drift → adverse effects in nearby ecosystems → long-range transport of pesticides in rainwater	• leaching of mobile residues and degradation products → groundwater → possible impacts on wildlife and fish and drinking water resources	• accumulation of persistent pesticides and degradation products → contamination and leaching to groundwater • use of broad spectrum pesticides → impacts on soil microflora and may affect or eradicate non-target organisms	• possible wildlife poisoning incidents (non-target organisms) • loss of habitat and food source for non-target species • resistance on some target organisms
Irrigation/water abstraction		• increase of nitrous oxide and methane emissions (greenhouse gases)	• lowering of the water table → soil salinisation/alkalinisation → impacts on surface and groundwater quality → drinking water • high abstraction required for some crops → strain on resources in some areas	• waterlogging → salinisation/ alkalinisation of soils • use of saline or brackish waters for irrigation in hot climates (high evaporation) → increased salt precipitation and carbonates → possible salinisation/ alkalinisation	• soil salinisation/ alkalinisation → losses of species, desertification • drying out of natural elements affecting river ecosystems
Drainage		• chemical changes in soil → greenhouse gases	• channelisation → hydrological changes → possible decrease in aquatic biodiversity • water abstraction → lowering of water table	• oxidation of organic soils → reduction in organic content, acidification and changes in soil structure	• potential loss of wetlands and changes in botanical composition of grassland, fens and other habitats
Mechanisation	Tillage, ploughing	• increase in dust and particulate matter in air	• increased surface water runoff, sediment load and associated particles → sedimentation, contamination, eutrophication	• ploughing up and down slopes → soil erosion (water and wind)	
	Use of heavy machinery		• soil compaction → increased surface runoff and sediment load → sedimentation, contamination, eutrophication	• compaction and erosion of topsoil	

Lowland wetland: Poitevin marshes, France
Source: Michael St Maur Sheil

energy producing capacity. These problems are covered in more detail in Chapters 5 and 7.

In some parts of Europe the agricultural demand for water is putting considerable strain on freshwater resources, particularly where irrigation is required. The extent of this problem is described in Chapter 33.

Soil

Many agricultural practices, if improperly carried out, can result in a decline in soil quality. Erosion of topsoil by surface runoff of water and, to a lesser extent, wind, can be exacerbated by a number of agricultural activities. Soil compaction occurs mostly in areas with highly productive soils and heavily mechanised agriculture (see Chapter 7). The degree of compaction depends on the type of soil, soil water content, the slope of land on which machinery is used, the size of machinery, axle load (weight), the timing and frequency of passage of machinery, and agricultural practices (eg, irrigation).

Overapplication of fertilisers can alter the chemical composition of the soil. Local soil acidification may result, although this can be mitigated by adequate liming. In some circumstances acidification is enhancing the leaching of nitrates and heavy metals to groundwater (see Chapters 7 and 31). Manure, artificial fertilisers and sewage sludge used on agricultural soils all contain heavy metals and other trace elements in addition to nutrients. Sewage sludge used on agricultural soils contains far more heavy metals than artificial fertilisers. However since less sewage sludge is applied to land than artificial fertilisers it is a less important source of heavy metals. Heavy metals can accumulate in the topsoil and may lead to reduced yields by reducing growth, or to elevated levels of these elements in agricultural products, which enter the food-chain if not adequately regulated (see Chapter 7). They can also be directly ingested with soil by grazing animals.

The factors which determine potential impacts of fertilisers include: the compound used; the level of application; the soil type (including the organic matter content of the soil) to which the application is made; the season when the application is made; the intensity of farming (whether cropping or livestock); and the mitigating actions taken (for instance, liming).

Improper soil management, resulting in a net loss of nutrients and organic matter from the soil, affects 3.2 million ha of land in Europe (Oldeman et al, 1991). Loss of organic matter will result in the degradation of soil structure, and this limits root penetration, soil moisture and permeability, which in turn increases the risk of erosion and runoff and reduces the biological activity of the soil.

In Europe (including the European part of the former USSR), it has been estimated that over the total land area about 7 per cent (64 million ha) has been degraded by improper management of agricultural land, and about 5 per cent (50 million ha) has been degraded by overgrazing (Oldeman et al, 1991). The problem is particularly severe in southern Mediterranean areas such as Portugal, where more than 20 per cent of the current agricultural area is highly erodible (Gardner, 1990). This could lead to problems of desertification. In Bavaria, erosion is a problem on up to half of agricultural land (Umweltbundesamt, 1987). In Romania, soil erosion is estimated to affect about one third of arable land, and this is aggravated in some areas by large irrigation projects (IUCN, 1991). Erosion of soil can result from a combination of agricultural practices, including mechanisation, concentration, overapplication of chemicals and irrigation (see Chapter 7).

In fragile, semi-arid environments, such as in the Mediterranean areas, land degradation has increased recently for a variety of reasons and is estimated to threaten over 60 per cent of the land in Southern Europe (UNEP, 1991). The renewed pressure on land resources through migration, the changes in agriculture both in terms of what is produced (cash-crops) and the mode of production (intensive agriculture), the increased demands on water through the development of irrigation schemes, together with the impact of land degradation on flooding, groundwater recharge, salt-water intrusion and soil salinisation has in many cases been responsible for land desertification (CEC, 1994a).

Salinisation and alkalinisation can result from irrigation with improper drainage, and from use of irrigation water with saline content. Excessive levels of salt on and in the soils will seriously affect crop productivity and change the soil structure. Drainage will lower the groundwater table, thus potentially creating conflicts with demands for water resources.

Waterlogging of the soil can result from inadequate drainage and from compaction combined with heavy rainfall or overwatering, causing problems for raising crops on agricultural land.

Most soils contain heavy metals, including copper, zinc, nickel, chromium and cadmium, but these natural contents can be increased by human activities (heavy metals originating, for instance, from road traffic, industry, or waste incineration). There will also be some input of heavy metals from fertiliser (natural phosphates of sedimentary origin contain, for example, cadmium), sewage sludge, and plant protection products, especially copper in viticulture. Some crops take up heavy metals from the soil to such levels that plants may become unfit for human consumption.

Landscape, habitats and biodiversity

Agriculture plays a significant role in the shaping of the landscape, and many natural landscapes and habitats have been moulded by agricultural use over time, such that the abandonment of agricultural activities would change considerably the appearance of the landscape. Issues of rural development are inextricably linked to agriculture and agricultural policy, and the restructuring of agriculture in Europe (under way in the EU) will lead to further decreases in the agricultural workforce, which in turn may mean less management of the land and soil. Particularly alarming are the loss of many wetlands and peatlands (see Chapter 9) and removal of hedgerows. Rural landscapes are also influenced by forestry practices, such as afforestation (see Chapter 23), tourism and recreation (Chapter 25), by development of infrastructures such as rural road networks, urban spread and industrial development. These factors have all played a role in the reduction of Europe's biodiversity, but it is likely that agricultural development has had the greatest influence (Umweltbundesamt, 1987).

The loss of natural habitats, combined with rising levels of toxic pesticide residues in the environment (which can enter the food-chain via groundwater and surface waters, the soil, the crop itself or through direct contact), has had significant effects on European wildlife and biodiversity. A further impact of pesticide use are effects on non-target species. The channelisation of streams and small waterways running through agricultural land to improve land drainage has often been associated with a decrease in aquatic biodiversity in these waterways (see Chapter 5). However, more environmentally sympathetic solutions such as multi-stage channels are now more often adopted for new works (eg, in the UK and Denmark).

Also of concern in Europe, particularly Western Europe, is loss of visual amenities by changes in the rural landscape, caused for example by removal of hedgerows, clearing of woods, the realignment of watercourses, the disappearance of meadows and riparian forest along waterways, and the abandonment of terraces, often carried out to make mechanised farming easier. Many countries have tried to protect such areas by establishing protected zones, but much of what remains is unprotected and significant changes to the countryside are still occurring. Some of the landscape, habitat and biodiversity questions are examined in more detail in Chapters 8 and 9.

THE EVOLUTION OF AGRICULTURAL PRODUCTION IN EUROPE

There have been many changes in the nature of the production structure of agricultural systems in Europe since World War 2. These include developments in international trade and agricultural policies (see below), technological developments and greater availability and affordability of many agricultural products, and of the inputs necessary to produce them. These have each influenced the yield, labour productivity and efficiency, and have led to some significant changes in the underlying structure of European agriculture.

Decreases in agricultural employment in all parts of Europe, decreases in the number of farms in most countries, and merging of land and livestock into increasingly larger units are all elements of agricultural 'concentration'.

Greater concentration of farms, land and livestock, and specialisation in a limited range of products, have often been accompanied by an intensification in the use of land, fertilisers, pesticides, feedstuffs and energy. Together, these factors have increased the pressure on the environment caused by agricultural activities.

Figure 22.1
Loss of biotopes between 1950 (left) and 1970 (right) on a testing area near Groenlo, The Netherlands
Source: Arnold and Villain, 1990

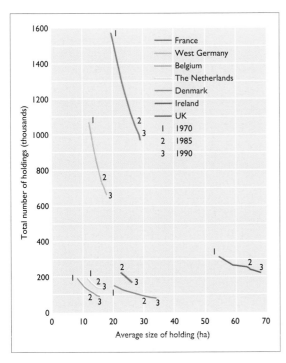

Figure 22.2 Agricultural holdings in selected countries of Europe, 1970–90
Source: Compiled by Eurostat from multiple sources
(see *Statistical Compendium*)

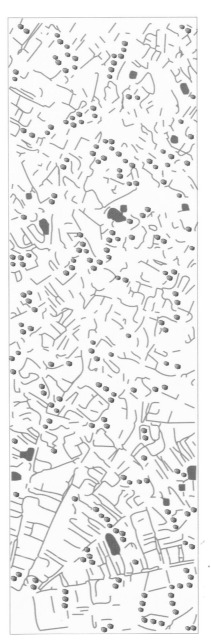

Hedgerows
Woodland
Trees

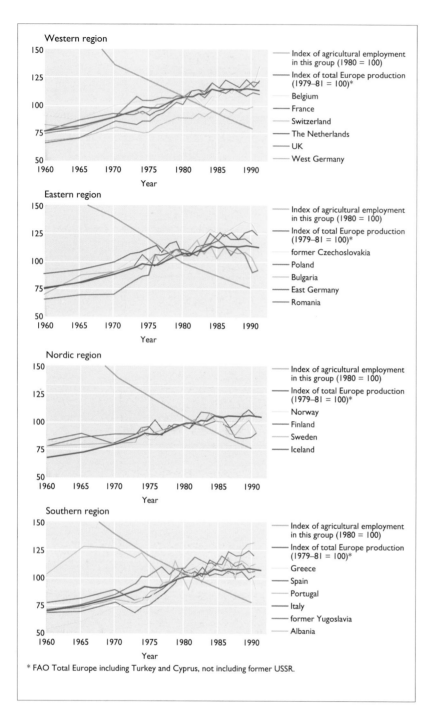

Figure 22.3
Agricultural
production and
employment in
selected European
countries, 1960–91
Source: Compiled by
Eurostat from multiple
sources, mainly FAO (see
Statistical Compendium)

landscape considerably (see Chapter 8).

The heterogeneity of European agriculture is marked by the wide diversity of the size of farm holdings. In the EU (1989–90, excluding East German Länder), about 8.2 million farms have been counted, of which 6 per cent are larger than 50 ha, but 60 per cent smaller than 5 ha. The average farm size in Greece was 4 ha in 1989–90, but 68 ha in the UK. Larger holding sizes are more efficient in terms of economic costs; fewer labour hours and less energy are required for fields unbroken by hedgerows, watercourses and other landscape features. For example, England and Wales lost around 20 per cent of hedgerows between 1947 and 1985. Figure 22.1 shows loss of biotopes between 1950 and 1970 near Groenlo, The Netherlands.

In the Eastern and Southern regions for which there are data, there is a large number of holdings of fewer than ten hectares, and in most countries in the western region the average size of holding is larger, with fewer farms. This second group includes countries like Denmark, the UK, The Netherlands, Belgium and France, where the trend over time has been an increase in average size of holding while the number of farms has been decreasing. Figure 22.2 illustrates this pattern. For instance, in France the average farm size increased from 19 ha in 1970 to 28 ha in 1989–90, and in the UK from 54 ha in 1970 to 68 ha in 1989–90.

In the Eastern region there has been a slower increase in the average size of holdings than in the countries shown in Figure 22.2. The development of agricultural structure has been heavily influenced by the policy stance of the former centrally planned economies. In East Germany, for instance, the three successive land reforms between 1945 and 1960 eradicated the dual structure of quasi-feudal properties and family farms, replacing it with an almost universal system of large-scale agricultural holdings.

The development of agricultural structure in Poland, with respect to average holding size, was different from other countries which had centrally planned economies. Farms tend to be much smaller because arrangements for land tenure (with 72 per cent of farms privately owned) were maintained under communism. The system was therefore not replaced by the industrial-type units common to a number of the countries in Central and Eastern Europe (MEPNRF, 1991).

Agricultural production and employment

Since the 1950s, the demand for more secure, cheaper and more plentiful supplies of food in Europe has been met by significant increases in average yields and efficiency of production in terms of inputs of labour and capital. Variations across Europe reflect the heterogeneous nature of agriculture as well as differences in local climate, soil quality and available water resources. The diversity of European agriculture is also expressed in the variety of production types used in different parts of Europe.

The importance of agriculture in a country's economy can be measured by its share of GDP. For the eastern region, up until recently, this has been measured by the gross material product (GMP), which is based on material product balances. In the EU, farming makes a small and declining contribution to GDP in each country: the average was 2.8 per cent in 1991. It was lowest in Germany (1.3 per cent), the UK (1.4 per cent) and Belgium (2.1 per cent); and highest in Greece (16.1 per cent) and Ireland (8.1 per cent) (CEC, 1994b). In the Eastern region of Europe, data for the combined contribution of agriculture and forestry to GMP at current prices are available. In 1989, these sectors contributed 13 per cent to GMP in Bulgaria, 15 per cent in Poland, 16 per cent in Romania, 10 per cent in the former Czechoslovakia, and 14 per cent in Hungary (1988) (E/SB, 1991). In most of these countries, the contribution of agriculture to national product has declined

Highly productive 'industrialised' systems in Western Europe fall into two broad categories. Firstly, there are areas of intensive field-crop farming, dominated by large holdings concentrated in France (the Loire valley to Calais region), eastern England, eastern Denmark, northern Germany and much of The Netherlands. Secondly, there are areas of very intensive agriculture specialising in animal production (dairy products and meat) and/or fruit and vegetable farming found in the coastal and southern areas of The Netherlands, northern Belgium (Flanders), western Denmark, parts of Germany, Provence (France), along the Mediterranean coast of Spain, and central and northern Italy.

Agricultural land accounts for more than 42 per cent of the total land area in Europe, although the proportion varies between less than 10 per cent in Finland, Sweden and Norway, to 70 per cent or more in Hungary, Ireland, Ukraine and the UK. Farming has shaped, and in many ways enhanced, the landscape to which most Europeans are accustomed, and if agricultural landuse changes significantly (for instance under 'set-aside' arrangements in the EU) this could alter the

since 1985 (there was no clear trend throughout the 1980s).

Figure 22.3 shows the index of agricultural production for selected countries in the different regions of Europe since 1960. The corresponding decline in agricultural employment is also shown. The gains in labour productivity in most parts of Europe are clear: fewer people are working on the land, but increases in production have nevertheless been achieved, explained mostly by higher yields from increased inputs, mechanisation and genetic development of products. The area of agricultural land used in Europe has hardly changed since 1960, and if anything it has decreased by a few per cent.

There have also been consistent gains in agricultural production and labour productivity in the Eastern region up until 1989. For cereals, an increase in production of 18 per cent was achieved during 1975–90. Despite these steady gains, much of the Eastern region remains a net importer of food. In fact, the gains in production and labour productivity for most countries in the Eastern region were not as significant as in other parts of Europe. Furthermore, dramatic reductions in overall agricultural production took place in some of these countries following the instabilities and changes of the late 1980s (see Figure 22.3).

In the Western region of Europe, total agricultural production has increased significantly since the 1950s. In the EU in particular, increases in production have been sustained to the extent that there is an excess supply of some products such as cereals, meats and dairy products. Wider availability of agricultural products, increased purchasing power of consumers enabling greater exercise of choice between products, and changes in consumer tastes resulted in increased supply of some products, such as red meat and dairy products. However, the same is not true for animal feed, and some EU countries import large amounts of this each year for the intensive feeding of livestock. This creates a nutrient surplus at the farm level in some countries (eg, in Northwest Europe), originating from the disposal of animal wastes, which contributes to problems of eutrophication in the freshwater and marine environment, and increased ammonia emissions to the air.

Employment in agriculture has shown a sharp decrease in all regions of Europe since 1960, but the average percentage of the total workforce in most of the Eastern region countries is still high when compared with the EU average of 6.2 per cent in 1991. For example, in East Germany it is estimated to be about 10 per cent; in Poland and Romania around one quarter of the working population is employed in agriculture and forestry combined, while in Bulgaria the proportion is about 20 per cent (E/SB, 1991). Although the average proportion employed in agriculture in the EU is low, dipping to less than 3 per cent in Belgium and the UK, in Greece it is 21.6 per cent and in Portugal 17.5 per cent (CEC, 1993).

Crop yields

Yields, measured in tonnes of harvested crop per hectare (t/ha) of agricultural land, have improved because of a number of factors: genetic improvements in crop types, intensification and better technology. For example, total European production of cereals alone increased from 199 million tonnes in 1970 to 283 million tonnes in 1990 (ie, by 42 per cent; world production increased by 62 per cent over the same period). Against this, the area harvested for cereals in Europe declined from 72 million ha in 1970 to 66 million ha in 1990. This increase in production has therefore been derived from a reduced area of land. This is significant because it is against productivity in terms of yield (t/ha) of agricultural product from an area that the quality of the (manipulated) ecosystem needs to be assessed. If agricultural activity is modifying the ecosystem in such a way that yields cannot be maintained in the longer term, then the activity can be defined as 'unsustainable'.

In the EU, there have been considerable increases in crop yields: average wheat yields increased from 3.2 t/ha in 1975 to 4.9 t/ha in 1991. In the Eastern region (including the former USSR) average wheat yield was 3.0 t/ha in 1990. In 1990, the greatest quantities of cereals were produced by France (55 Mt) and Ukraine (48 Mt). France also had one of the highest yields (6.1 t/ha), whereas yields in Ukraine were about half these (3.6). The highest average yields for cereals are achieved in Northwestern Europe. The Netherlands has the highest yields (7 t/ha), but it is also one of the smallest producers. Belgium, the UK, Denmark and Ireland all produce average yields of about 6 t/ha.

Within the EU there are many variations in average yields. For instance, the average yield for cereals in France is 6.1 t/ha compared with 1.9 t/ha in Portugal, due to the differing crop types, soil type and climatic conditions. For durum wheat (*Triticum durum*), grown in many Mediterranean countries, the highest yields will always be lower than the best yields achieved for different wheat varieties grown in more northerly countries. Variations are partly the result of different climatic and natural conditions, but they also reflect variations in other critical factors such as the level of modernisation and use of chemical inputs.

In the Eastern region, countries with the highest yields in 1990 were the former Czechoslovakia, East Germany and Hungary. But there are also large variations in yields between countries: in 1990 the former Czechoslovakia achieved 5.2 t/ha for cereals compared with only 1.9 t/ha in the Russian Federation, 2 t/ha in Belarus and 3.6 t/ha in Ukraine (see *Statistical Compendium* for further details on crop yields).

The rate of increase in cereal yields has begun to level off in the Western region. Projections suggest that yields will continue to increase, but at a lower rate of up to 1 per cent per year. In the Eastern region, increases in yield are likely to rise from 1.5 per cent to about 2 per cent per year.

AGRICULTURAL PRACTICES

The improvements in agricultural yields and productivity described above result to a great extent from the changes in agricultural practice and techniques over the last few decades in Europe. Few practices have remained unaltered by the modernisation of agriculture. Tilling, sowing and harvesting have become increasingly mechanised, and the methods of applying fertilisers and pesticides have become more sophisticated. Intensive agricultural practices, such as the shortening of traditional 3 to 4 year crop rotations, and repeated planting of the same crops year after year (which together deplete the nutrients in the soil) and irrigation, have become widespread.

This section shows trends in those agricultural practices which have the potential to cause environmental damage, and aims to highlight variations in these practices across Europe. The impacts of these practices are summarised in Table 22.1.

Many changes within the agricultural system can be summed up by 'intensification'. The result and aim of intensification has been to achieve increases in production, yields, and labour productivity in agriculture. Intensification of agricultural output is characterised by changes in a number of key factors, some of which can be readily quantified. They include: specialisation and concentration of crops and livestock; use of fertilisers and pesticides; drainage, irrigation and water abstraction; and mechanisation and physical practices (tillage and ploughing). If taken together, these factors broadly indicate that, as a whole up to the end of the 1980s, farming in Europe was becoming more intensive with most of these variables increasing consistently over recent decades in most countries. However, the overall area of agricultural land used in Europe has not seen large changes.

Intensification of farming has been particularly acute in northern EU countries, especially in The Netherlands,

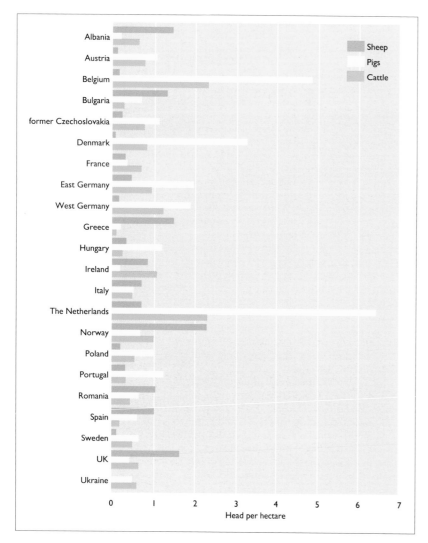

Figure 22.4
*Livestock intensity:
head of cattle, sheep
and pigs per hectare
of agricultural land,
1990*
Source: Compiled by
Eurostat from multiple
sources, mainly FAO (see
Statistical Compendium)

pesticides, energy). Specialisation is often regionalised, and can be characterised by a reduction in the number of 'mixed' farms which undertake a variety of agriculture. Above all, this has led to the separation of arable land and livestock farming. This has upset the nutrient cycles that were typical of traditional, integrated farming systems. Increasing amounts of fertilisers have been used to try to counteract this imbalance.

Over the last few decades, the range of products generated by European farms has reduced, and become ever more specialised. A loss of genetic potential could result if only a small number of varieties of a crop or livestock covers a large area, or if the number of varieties being used by farmers were to decline. In a specific study into wheat varieties used in UNECE countries, however, no dominant tendency was discernible over the period between 1950 and 1988; some countries increased the number of varieties they used, while others reduced them (UNECE, 1992).

Within the split between crop and livestock farming, there has also been a trend, again more pronounced in the Western region, towards a gradual intensification of livestock farming with bigger herds being grazed on less area of pasture per animal, or perhaps with animals being kept even more intensively in stalls rather than grazed on open land (see Figure 22.4). For example, in The Netherlands, concentrations of cattle and pigs averaged around 2.4 and 7 head per hectare of agricultural land respectively in 1990 (see *Statistical Compendium*).

Price guarantees for cereals and other crops grown for livestock feed have resulted in a shift of livestock farming from the traditional upland and grassland areas of the EU to the northwestern coastal areas of Europe. Here, because of the high internal grain price in the EU and the low import cost of so-called 'grain substitutes', livestock producers can gain a competitive edge by purchasing the same feedstuffs but at lower imported 'world' prices (von Meyer, 1989; 1993).

A consequence of this trend has been a significant increase in the quantities of excreta produced in some countries. Excreta produced can lead to increased nitrogen input to soils, and estimates of this effect for Europe are illustrated in Map 22.1. The estimates of nitrogen inputs from manure have been derived from landuse types and livestock populations at regional level in Europe. It is the excreta produced by area – that is, per hectare – which is important. In The Netherlands, for example, livestock excrete much more manure than the land can absorb. The average load of nitrogen from animal manure in The Netherlands has more than doubled since 1960, from 110 kg/ha per year to 241 kg/ha per year in 1988 (CBS, 1961; 1989).

A significantly higher portion of animal excreta is handled as slurry rather than straw manure in Western region countries. This has led to a rising number of pollution incidents related to inappropriate methods of storage and accidental spills of slurry into surface waters and contamination of soils and groundwater. This increases the load of nitrogen and organic matter to the soil, groundwater and surface waters (see Chapters 5 and 7).

Nitrate loads to watercourses and groundwater are also increased by seepage from inappropriately stored animal dung in fields, and field stocks of silage. In some countries (for example, Denmark), the extent of the problem has been recognised and the practice of storing manure and silage in field heaps has now been prohibited. The Netherlands is to address the problem of nitrates and organic loading to waterbodies from manure by having 'quotas' on production and storage of dung.

In some countries (eg, The Netherlands and Denmark) methods of treating excess excreta to produce energy as a source of fuel for heating (biogas) are encouraged, and ways to reduce it to more manageable volumes (eg, dehydration) for easier disposal are also being investigated.

The increased intensity of livestock rearing depends in

northern France, the southeast of the UK, Belgium, Denmark and parts of northern Germany. In the Southern, Nordic and Eastern regions it has been less intense but is nevertheless occurring. In some countries of the Eastern region, the problems of intensification are exacerbated by the inappropriate use of agricultural inputs (see fertiliser and pesticide use below).

Energy consumption in agriculture is a further factor which can illustrate the intensification of farming systems. Even though agriculture is not overall a large consumer of energy, perhaps accounting for no more than 5 per cent of total energy use, mechanisation does involve increases in energy consumption. Unfortunately, data on energy use in agriculture are rarely available, or separate from, those of other activities (for example, UNECE data combine energy use for forestry and fishing with that of agriculture). Calculations of energy efficiency within the farming sector show that some traditional subsistence farming systems use energy much more efficiently than modern, highly mechanised, farming systems (UNECE, 1992). One off-farm activity on which agriculture depends – the production of nitrogenous fertilisers by industry – has a high energy intensity, although this is usually considered consumption of energy by industry.

Specialisation and concentration of crops and livestock

Specialisation encompasses a tendency to simplify the countryside in order to enlarge plots of agricultural land, with the purpose of raising just a few products (concentration); and the increased use of agricultural inputs (fertilisers,

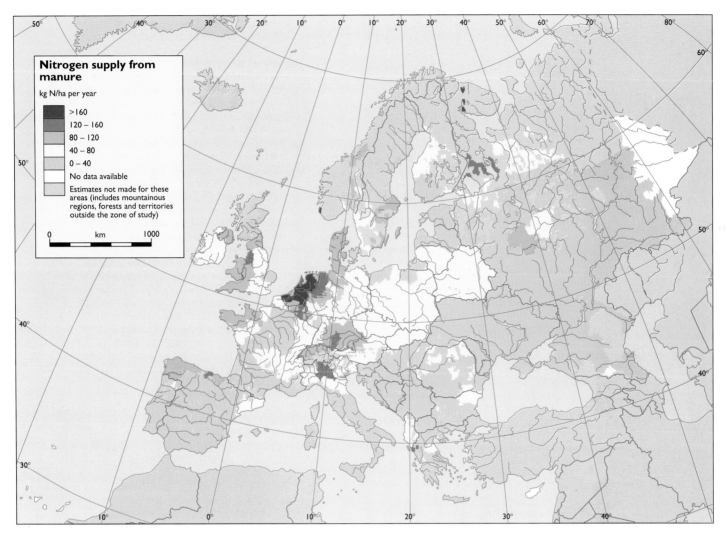

Nitrogen supply from manure

kg N/ha per year

- >160
- 120 – 160
- 80 – 120
- 40 – 80
- 0 – 40
- No data available
- Estimates not made for these areas (includes mountainous regions, forests and territories outside the zone of study)

0   km   1000

particular on large amounts of imports of protein-rich feedstuff from non-European countries. Over one third of the world's grain is now fed to livestock. Of this, the production of pork uses more grain worldwide than any other meat industry, and the EU alone produces one fifth of the world's pork. Much of the livestock feed imported to Europe comes from the USA, Thailand, Argentina and Brazil, where there have been huge increases in the area devoted to cultivating animal feed, most of which is exported once harvested. Although, for example, more than 60 per cent of the land surface in The Netherlands is farmland, another five times as much land in other countries, mostly in the developing countries, is used for the production of fodder for Dutch livestock.

Imports of protein-rich animal feed constitute a major addition of nutrients (nitrogen and phosphorus) to the agricultural system, and contribute to the disruption of the nutrient cycle on some farms in Northwestern Europe. In this way, therefore, specialisation and concentration towards animal husbandry in specific areas can lead to mineral surpluses, which contribute to problems such as eutrophication.

Fertiliser use

Increased use of agrochemicals, including fertilisers, pesticides and other chemicals, has been one of the most influential factors (alongside the genetic development of crop types) increasing yields from agricultural land in Europe. Practices such as short crop rotations and the use of high yield crop varieties demanding large amounts of fertilisers have contributed to nutrient imbalances on European agricultural soils.

Fertilisers can be split into two broad categories, commercial fertilisers – and organic fertilisers (manure) – the main elements in both being nitrogen, phosphorus and potassium. Leaching of nitrogen (as nitrates) and excess runoff of phosphorus (as phosphates) tend to occur because it is hard for the farmers to 'integrate' fully the fertiliser into the soil in just the right quantities needed and at the right time for the crop. Experiments appear to show that 10 to 60 per cent of fertiliser nitrogen is not taken up by the crop for which it was applied (UK MAFF, 1992). Direct losses of nitrogen occur when there is high rainfall soon after the fertiliser is applied, and application depends on other aspects of the weather during the months which follow. This uncertainty makes an assessment of future yield difficult and, as a result, farmers tend towards overapplication.

In some areas of Europe, agriculture is estimated to be responsible for up to 80 per cent of the nitrogen loading and 20 to 40 per cent of the phosphorus loading of surface waters (see Chapter 14). Furthermore, intensification of livestock farming has led to an animal manure (natural fertiliser) storage and disposal problem in Europe. Use of fertilisers also contributes to atmospheric emissions. For instance, fertilised soils produce 2 to 10 times as much nitrous oxide as unfertilised soils.

Average consumption of nitrogen fertiliser per hectare of agricultural land has increased in Europe (excluding the former USSR, but including Turkey) by 75 per cent between 1970 and 1989. The average consumption per hectare in 1989 was about 31 kg nitrogen/ha for Europe, including the former USSR and Turkey, but double that – at 63 – if the former USSR is excluded.

High inputs of nitrogen fertilisers occur mainly in arable

Map 22.1
Nitrogen supply from manure in agricultural soils in Europe
Source: RIVM estimates, based on Eurostat, FAO data (see *Statistical Compendium*)

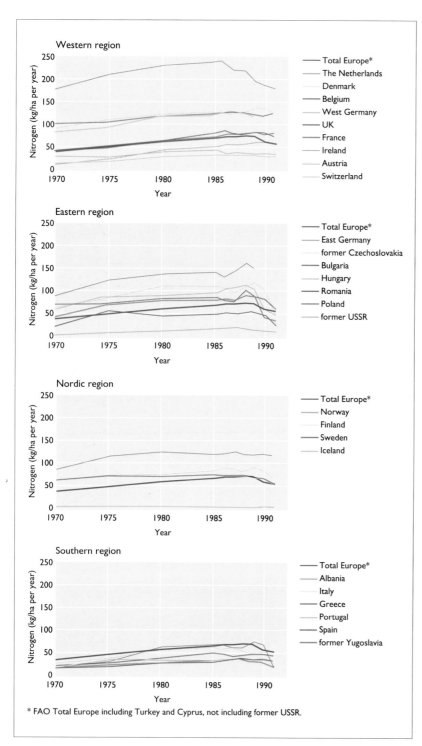

Figure 22.5
Consumption of nitrogen fertiliser per hectare of agricultural land in selected European countries, 1970–90
Source: Compiled by Eurostat from FAO (see *Statistical Compendium*)

land, with the exception of the intensive grasslands of Northwestern Europe. Figure 22.5 shows the consumption of nitrogen fertilisers per ha for the different regions of Europe during 1970–90. Nitrogen fertiliser consumption in most countries in the Western region is higher than for most other regions. In the Western region fertiliser use rose steadily until 1970. Since then, in most countries, consumption has not increased at the same rate as before, and towards the end of the 1980s consumption of nitrogen fertiliser began to level off or even fall slightly in some countries. There are significant variations between countries in consumption per hectare of agricultural area, with less than 50 kg nitrogen/ha in, for example, Spain, Austria, Portugal and Switzerland, and below 20 kg nitrogen/ha per year in the former USSR and Iceland, to over 190 kg nitrogen/ha in The Netherlands. In The Netherlands, the total amount of nitrogen provided by artificial and natural fertilisers has been well above 200 kg

nitrogen/ha per year for most of the 1980s, although since 1985 it appears to have fallen rapidly. Denmark alone accounts for over two thirds of the nitrogen leached from farmland in Scandinavia.

In the Eastern region the trend of steady increases in fertiliser use has been similar to that of the Western region. In some countries the problems of farm intensification have been exacerbated by the inefficient use of agricultural inputs.

By contrast, average application of phosphorus fertilisers in Europe has remained fairly constant over the last 20 years at just over 30 kg/ha (P_2O_5 unit) of agricultural land per year. In 1990 this ranged from over 50 kg phosphorus/ha in Belgium to 23 kg phosphorus/ha or less in Albania, Austria, Iceland, Romania, Sweden, Switzerland, Turkey, the UK and former Yugoslavia (Figure 22.6). In general phosphorus is strongly fixed by most soils. Only on soils with low adsorption capacity (such as sandy soils), where shallow water tables and phosphorus-demanding crops are grown (such as maize), is phosphorus released to groundwater (as in parts of The Netherlands).

Map 22.2 shows the nitrogen supply from fertiliser in agricultural soils in Europe (see also Chapters 5 and 7) estimated from fertiliser consumption and regional landuse data. (These same data form the basis of the estimates given in Chapter 5 of the risk of nitrate leaching to groundwater.) Map 22.3 shows the combined nitrogen supply to agricultural soils from fertiliser and manure (see 'livestock' above.) The highest nitrogen loads are clearly in Northwest Europe, and particularly Belgium, The Netherlands and northwestern Germany (atmospheric deposition, originating mostly from combustion of fossil fuels and volatilisation of ammonia, generally contributes about 10 to 20 kg/ha per year in most parts of Europe).

A further consideration when examining the use of fertilisers in agriculture is their contribution to air pollution. The upward trend in the production of manure and silage (Map 22.2), together with the overapplication of commercial fertilisers, has significantly increased emissions of nitrous oxide and ammonia from agriculture.

Agriculture's contribution to acidification is, however, much greater than just that deriving from deposited air pollutants. In some countries of Northwestern Europe there is a problem of potential acidification of soils and groundwater as a direct result of ammonia use as fertiliser. Ammonia fertilisers have been used on agricultural land since the 1960s when appropriate equipment became available to spread them (they are in liquid form). Even sufficient liming may not always prevent acidification: if nitrification takes place below the limed depth, then acidification continues, and in sandy areas may lead to groundwater and surface water becoming acidified (Rebsdorf et al, 1991). (See Chapters 5, 7 and 31.)

Nutrient replenishment in European agricultural soils will continue to be an important issue in the near future. However, increasing awareness among, and counselling of, farmers is likely to lead to changes in the practices of fertiliser use as rates of application are lowered and methods of application become more precise, together with alternative crop management techniques. These include: using organic agricultural practices (see Box 22B, p 459); letting fields lie fallow; periodically planting legumes to reduce excess nitrogen by fixation from the air and reducing the necessity for applying nitrogen fertilisers; ensuring vegetation cover in winter by growing wintergreen crops or 'catch crops' to provide soil cover and prevent nitrogen leaching.

However, it is difficult to determine the precise amounts to apply and types of compounds to use so as to minimise potential environmental impacts. This difficulty arises from the considerable variation in field crops, the soil type and groundwater depth, rotation profile and geomorphology. Furthermore, the expense of some of the relatively new technology required to identify at what exact time of the day and season applications should be made renders these determinations additionally complex.

Pesticide use

Pesticides are toxic chemicals used to control or kill pests (see Chapter 17), and include herbicides, insecticides and fungicides. 'Growth regulator' is also often included in pesticides data. Use of these products has played a significant role in raising the yields of crops from European agricultural land. Heavy use of fertilisers to increase yields intensifies the risk of fungal infections in some crops. The development and widespread application of pesticides has taken place during the post-war decades, with a succession of more sophisticated and effective pesticides being introduced. More modern pesticides generally break down more quickly and have fewer unintended side-effects as they act more selectively than earlier formulations. However, there are many different pesticide products available, containing almost a 1000 types of active ingredients (see Chapter 17). This makes it difficult to deduce environmental impacts.

Data on the consumption of pesticides are far from comprehensive, making assessments of their use difficult; accurate data for many European countries are often lacking altogether. Trends are hard to discern since compounds used vary enormously, and their application depends on seasonal, climatic, crop and geomorphological factors. Consumption is usually measured in terms of the 'active ingredient' (AI), the chemical which controls or kills the pests. The amount of AI is expressed as the weight of the ingredient in individual pesticides. Formulation weight is the actual weight of the product bought in the form that it is to be used. The relationship between these two measures is not constant, and this adds to the difficulties of deriving meaningful patterns from data on pesticide use.

Apparent declining application rates may be misleading because newer generations of pesticide have high potency and require relatively lower application rates (for example, 'sulphonyl urea' or 'imidazoline' and synthetic 'pyrethroid' insecticide formulations). As with the use of fertilisers, there are considerable variations in application rates between different countries; in 1990 application rates for pesticides (per hectare of arable land and land under permanent crops) in The Netherlands were over four times what they were in West Germany, which in turn was estimated to be over five times as much pesticide per hectare as Estonia. Figure 22.7 shows pesticide consumption in some European countries for the latest year available.

Map 22.4 shows the average load of pesticides per hectare of soil under arable and permanent crops. The average used has consistently been around 20 kg/ha per year in The Netherlands. Within the average figure for usage, there is a wide variation. It is estimated that bulb-growing consumes about 120 kg/ha, mushrooms consume 112 kg/ha and greenhouse production takes 106 kg/ha. Two thirds of pesticides are consumed in arable farming, but at the much lower rate of 19 kg/ha (WWF, 1992).

Various factors are forcing much of European agriculture to rethink practices of pesticide use. The trend towards growing large monoculture crops (often repeatedly) has provided ideal conditions for pest infection and many pests have developed resistance to the pesticides employed. Awareness of the dangers of pesticide pollution in drinking waters is also developing, and it is widely recognised that herbicides can reduce biodiversity of natural flora, as more aggressive, resistant weeds take over hedgerows. Pesticides also have an impact on non-targeted wildlife, and many poisonings of animals and birds are recorded each year in European countries (see Chapter 9). Pesticides can affect species directly, through build-up in the food-chain, and also by affecting their food supplies and by changing habitats. It has been estimated that less than 0.1 per cent of the pesticides applied to crops reaches the target pests. Thus, more than 99 per cent of applied pesticides has the potential to impact non-

target organisms and to become widely dispersed in the environment (Pimentel and Levitan, 1986).

Many new management approaches to pest control aim to optimise efficiency of pesticide use, and may result in reductions in amounts of active ingredient used. These often involve the integration of several pest control practices and are commonly referred to as integrated pest management (IPM). IPM is essentially a chemically based approach: monitoring pest levels on crops in order to determine the most efficient use of pesticides, while considering alternatives to chemical applications. This move towards more refined approaches, along with the development of genetically engineered pest resistant crops, the use of biological control, assessments of meteorological conditions to anticipate likely changes in pest populations, and more carefully timed application of different, lower-dose pesticides is likely to become more widespread, and may lead to reductions in

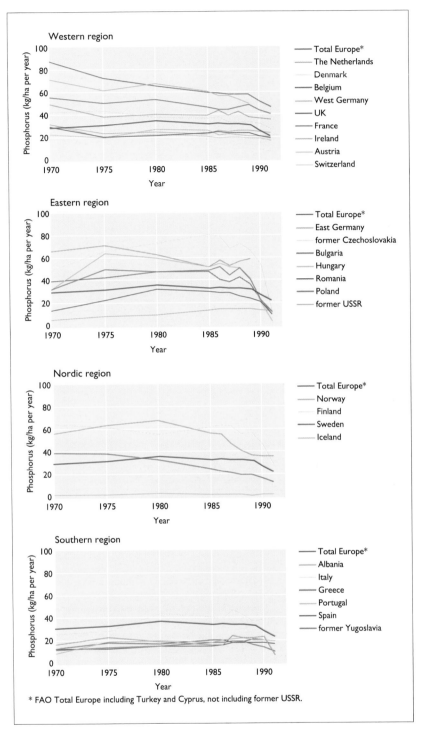

* FAO Total Europe including Turkey and Cyprus, not including former USSR.

Figure 22.6
Consumption of phosphorus fertiliser (P_2O_5 unit) per hectare of agricultural land in selected European countries, 1970–90
Source: Compiled by Eurostat from FAO (see *Statistical Compendium*)

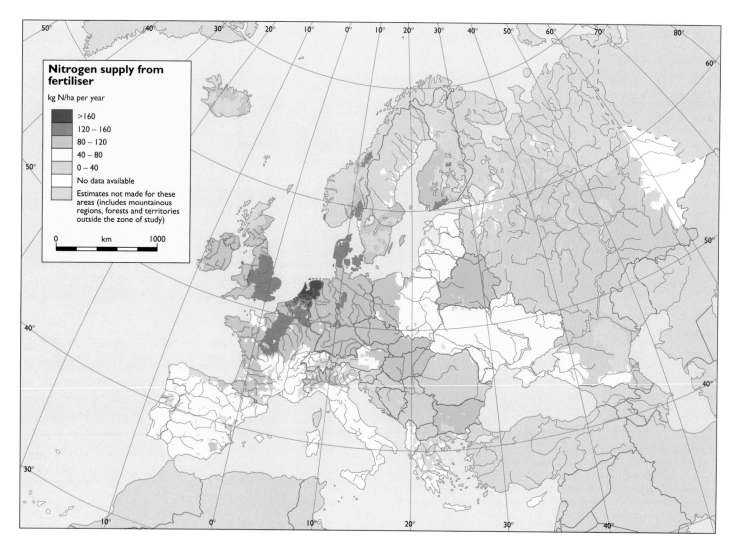

Map 22.2
Nitrogen supply from fertiliser in agricultural soils in Europe
Source: RIVM estimates

pesticide use. Varied crop rotations are also an important way of reducing dependency on pesticides. On the other hand, genetic engineering is also used to make crop plants more resistant to herbicides, which may result in such plants having increased dosages.

Pesticides also find their way into the food-chain. Residues of banned compounds can be detected in food long after the compound has been banned for agricultural use in Europe. These residues can also be introduced through direct imports of food from outside Europe, or (more commonly) through imports of animal feed from outside Europe, which is then fed to European-reared livestock. However, it should be noted that it is difficult to distinguish between banned residues from this source and those persistent in the environment from application before banning. Data for Finland of DDT (and PCB) residues in pigs and cattle have been published (UNECE, 1992), which show that, despite the long-standing ban on DDT, residues of this compound are still detectable.

A few European countries have reviewed pesticide use in agriculture and horticulture, and have set specific objectives to reduce use of these chemicals; examples include Denmark, The Netherlands and Sweden.

Drainage, irrigation and water abstraction

The purpose of drainage is to lower the groundwater table, mainly to remove excess water from clay soils and organic soils to facilitate the production of crops from agricultural land. Lowering the groundwater table leads to the oxidation and thus the loss of organic matter. It can also lead to erosion of agricultural soils and the loss of wetlands (see Chapters 5, 7 and 9). Intensive drainage is evident in many countries of Northwestern Europe, and has led to the loss of wetlands in, for example, Ireland (Baldock, 1990).

Water is abstracted for irrigation, and this can also lead to lowering of the water table. Excessive abstraction, in the same way as drainage, can in some places lead to a loss of wetlands. In some areas, particularly those subject to low rainfall combined with high average temperatures and many hours of sunshine, such as in Southeastern parts of Europe, irrigation is leading to the salinisation of significant tracts of arable land. The proportion of agricultural land irrigated (including land irrigated by controlled flooding) is about 5 per cent for Europe as a whole (including the former USSR and Turkey) but varies considerably between countries (Figure 22.8). Albania has irrigated some 38 per cent of its agricultural lands, The Netherlands 28 per cent, but in most countries in the Nordic region the percentages are much lower, due mainly to higher rainfall. In Romania, it has been estimated that, of 3.2 million ha irrigated, 200 000 are salinised (see also Chapters 7 and 33).

Box 22B
Organic agricultural practices

Modern agricultural practices have brought with them a number of serious on- and off-farm environmental impacts, and the recognition of these by consumers has put increasing pressure on agriculture to develop less damaging alternative practices. There is consequently an increasing interest in Western and Northern Europe in 'organic' farming, and an expanding, though still tiny, market for organically produced crops. Produce tends to be expensive because of the higher labour intensity of the farming methods used. Organic farming techniques are not yet practised widely in Europe, accounting for less than 0.5 per cent of farms and of agricultural area in the EU, and less than 1 per cent of each in EFTA countries. Such practices are almost non-existent in Central and Eastern Europe, and in only a few of these countries can organic farms be found; in 1992, Hungary had about 100 farms covering about 3500 hectares, and in the former Czechoslovakia there were about 130 organic farms covering about 15 000 ha.

The features that make organic farming distinct from traditional farming affect a wide range of practices. Organic farmers use new ecological methods as well as selected traditional techniques. Methods rely primarily on farm-derived, renewable resources in preference to external synthetic chemical inputs such as pesticides and fertilisers. As far as possible, these systems rely on extended and more diversified use of crop rotations, crop residues, animal manures, use of legumes, green manures, off-farm organic wastes, and aspects of biological pest control to maintain soil productivity, to supply plant nutrients, and to control pests. However, other elements of modern farming methods are also employed, including modern seeds, mechanisation, proven husbandry techniques and soil conservation. Land converted from non-organic to organic use takes two to three years before it is free of agrochemicals. Labour costs in organic farming tend to be high, since physical measures for removing weeds are commonly used in preference to pesticides, and production methods involve less intensive use of land. This increases the cost price of such products.

The EU has drawn up standards for the production and labelling of organic foodstuffs in a 1991 Council Regulation (91/2092/EEC), which came into effect at the beginning of 1993.

Organic farming may involve the replacement of chemical fertiliser by (organic) manure, which is often seen as more environmentally friendly; there is little evidence for or against this view, and more research is needed to compare the environmental costs and benefits of both systems. Such research is difficult because it is not easy to find farms where land conditions and factors influencing production are comparable. Experiments (at Rothamsted, UK) have, however, shown that, while farmyard manure can be at least as effective as a chemical fertiliser in improving yields, the potential for nitrate loss from these manures is greater than from inorganic fertilisers because these manures are often applied throughout the autumn or winter as soil conditions permit and when convenient (Addiscott, 1988). Since timely, accurate applications of nitrogen are difficult, substantial leaching of nitrate and heavy metals is likely to result (UK, MAFF, 1992).

So far, research into organic-type farming has given priority to work on soil fertility (especially chemical fertilisers) or on soil erosion. But more research is needed on biological processes, and soil compaction, before firm conclusions can be drawn on the benefits or otherwise of these farming practices.

Source: Partly from Greenpeace, 1992

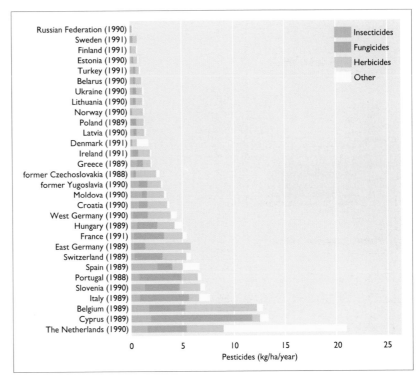

Figure 22.7 Total consumption of pesticides per hectare of arable and permanent cropland in selected European countries, latest year available
Source: Compiled by Eurostat from multiple sources, mainly OECD (see *Statistical Compendium*)

Mechanisation and physical practices (tillage and ploughing)

The level of mechanisation or degree of 'capitalisation' on farms has been following a steadily upward trend over the last two decades, although there are signs that in some regions of Europe saturation has been reached or neared. Despite its benefits in increasing yields, mechanisation has clearly had some adverse environmental effects. The potential impacts of the use of heavy machinery are described in Table 22.1.

Deep ploughing exposes more soil to wind and water erosion. Crop residues can be removed (as opposed to ploughing back into the soil) by, for instance, stubble burning and stubble mulching (removing the stubble and leaving it on the soil surface). Removal of residues can lead to a serious loss of organic content in the soil, which may increase the risk of soil erosion by limiting the ability of the soil to retain humidity. Stubble burning can create a major public nuisance, and in some countries, such as Denmark and the UK (England and Wales), it has been banned, and ploughing straw back into the soil has become much more common.

Although the level of mechanisation in Central and Eastern Europe is lower than that of the West, it has increased steadily since 1970. For example, in 1970 in West Germany, there were 10.6 tractors per km[2] of agricultural land. This had increased to 11.8 by 1990. The corresponding figures for East Germany were 2.4 for 1970 and 2.8 in 1990. In Poland in 1987 it was estimated that there were still approximately one million horses in use for a total of 2.8 million farm holdings. More than 1.1 million tractors were in use by 1990 (see *Statistical Compendium*).

It is easier and more efficient to work in large open fields when using tractors, but on some soils this can lead to soil compaction, particularly with large and heavy vehicles; and, even if it has improved the efficiency of agricultural capital, the practice of increasing field sizes has led to the loss of hedgerows, small streams, ponds and other wetland habitats (see Chapter 8).

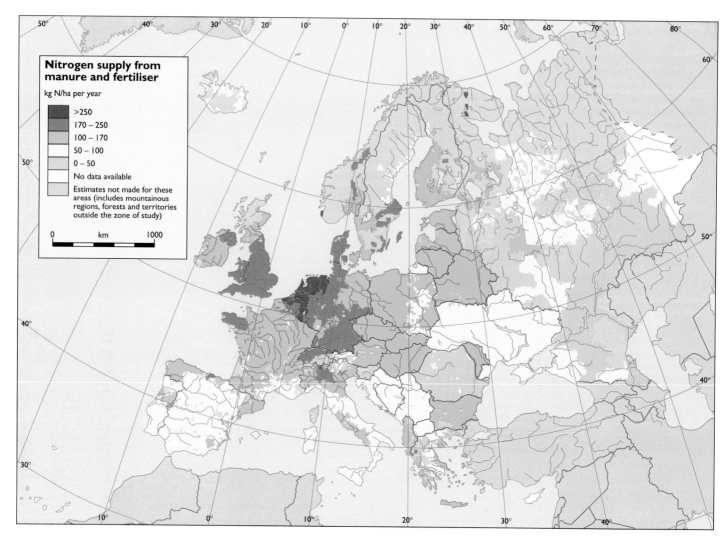

Map 22.3 *Nitrogen supply from manure and fertiliser in agricultural soils in Europe*
Source: RIVM estimates

AGRICULTURAL POLICIES

Since World War 2, the context within which agricultural policies have been formed has varied considerably across Europe. Whatever the circumstances, however, the main objectives have been to increase farm outputs and productivity. Objectives have also included support for farmers' incomes and price stability for consumers.

In the EU, the Common Agricultural Policy has played a primary role in the formation of agricultural policy at the national level: it had the goal of ensuring self-sufficiency in Europe (Box 22C). Policies developed in EFTA countries such as Norway, Sweden and Switzerland have been strongly influenced by national concerns to ensure agricultural self-sufficiency and the protection of rural communities; most EFTA countries provide a higher level of support to agriculture than does the EU. In most Central and Eastern European countries agricultural policies have been developed within the framework of central planning, with an emphasis on 'collective' farming and its growth as a public industry, rather than a collection of private farms as in the EU.

Despite the different policy regimes, many similar measures have been adopted to achieve greater and more efficient output from agriculture. These include guaranteed prices regardless of the levels of output and demand for certain products (dairy products, cereals, beef and wine); financial support through subsidies and low-interest loans for investment in certain types of capital; high border protection (with internal prices high and unrelated to world markets); land reclamation and farm rationalisation; support for farming in marginal areas (often with serious environmental consequences); and an expansion of research services. Policies

using these measures were particularly effective among the countries of Northwest Europe, where agricultural production and labour productivity have risen substantially since the 1960s (see also Figure 22.3). However, this success has led to excess supply of some products in the EU. Some of the most important effects on the environment that can be linked to agricultural policy have resulted from price support systems for agricultural products – one of the main features, for example, of the CAP.

The challenge therefore is for agricultural policy to recognise that there are three different basic needs to reconcile: the production of food and agricultural products, the protection of the environment, and the maintenance of the socio-economic fabric of rural areas. Specific measures (CEC, 1993) designed to help farmers reduce impacts on the environment and to protect the rural landscape include financial compensation or assistance for:

- using less intensive farming methods – eg, reducing the amounts of fertiliser and pesticide used, or reducing the number of livestock grazed per hectare;
- switching arable land to grasslands and meadows;
- preserving farming areas/methods that maintain high quality biotopes (hedges, mixed woodland/farmland) and biodiversity;
- taking some agricultural land out of productive use and ensuring long-term 'set-aside' for other purposes (eg, wetlands);
- afforestation projects;
- farmers wishing to convert to or to maintain organic production (using financial support, low interest loans).

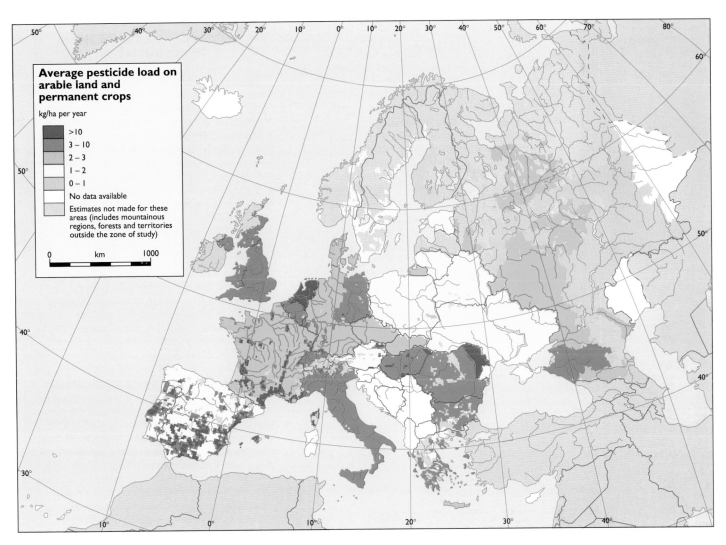

Average pesticide load on arable land and permanent crops

kg/ha per year

>10
3 – 10
2 – 3
1 – 2
0 – 1
No data available
Estimates not made for these areas (includes mountainous regions, forests and territories outside the zone of study)

0   km   1000

In some countries, for example The Netherlands, Germany, Denmark and the UK, other initiatives are also being taken by national governments which aim to complement the objectives of the CAP reforms, and also those of the EC Directive (91/676/EEC) concerning the protection of waters against pollution caused by nitrates from agricultural sources. These include setting limits for the amounts of chemical inputs that can be used and the quantities of manure produced (and therefore the numbers of livestock maintained) in defined nitrate-sensitive zones (the UK) or in the whole country (The Netherlands, Germany and Denmark).

In Central and Eastern Europe the highest priorities for agriculture, as with other economic sectors, have since 1989 been to increase production, efficiency and exports, particularly to Western Europe. Many Central and Eastern countries are at the same time trying to recast policy from that of the command-led economy to more market-based systems. This brings pressures to remove subsidies to allow prices, more similar to market levels, to prevail. A further aim is to maintain or increase revenues to avoid rising agricultural unemployment. The direction of agriculture in Central and Eastern Europe will be strongly influenced by the policies followed in EU and EFTA countries; these policies will also have a wider impact on socio-economic development of rural areas in Central and Eastern European countries, where environmental concerns have not yet been integrated into agricultural policy as much as in Western Europe.

CONCLUSIONS

European agriculture is an extremely diverse and heterogeneous economic sector in terms of the products generated, the nature and structure of production units, and the variability of potential impacts on the environment. Because of this, it is difficult to make meaningful generalisations. Nevertheless, some overall description of the main developments in European agriculture is possible from the information available.

● There are serious environmental impacts associated with European agricultural activities. A problem of particular concern in certain parts of Europe is the excessive enrichment of surface waters and groundwater by nutrients

Map 22.4
Average pesticide load on arable and permanent cropland in Europe
Source: RIVM estimates, based on various sources and years (see *Statistical Compendium*)

Irrigation in Peloponnisos, Greece
Source: A. Nychas

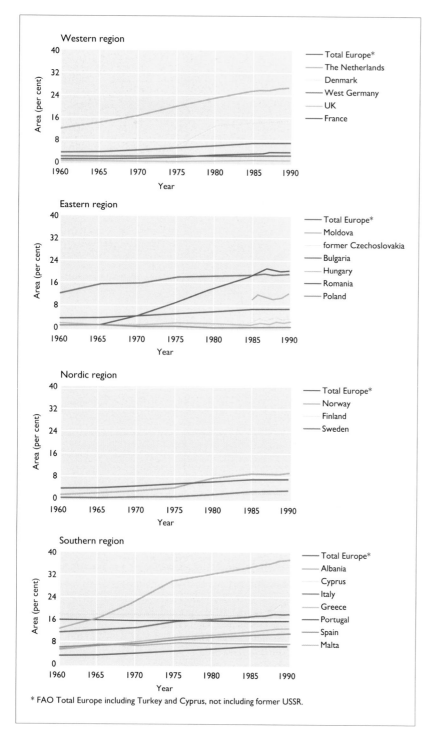

*Figure 22.8
Agricultural land
under irrigation in
selected European
countries, 1960–90
(%)*
Source: Compiled by
Eurostat, mainly from
FAO (see *Statistical
Compendium*)

(nitrogen and phosphorus). Impacts also include the acidification and erosion of soils (the latter particularly in parts of Southern Europe), and the reduction of natural habitat and consequent losses of biodiversity resulting from landscape changes (loss of hedgerows, wetlands, watercourses, ponds, etc). Many of these impacts are associated with the intensive, highly specialised and concentrated farming areas of Northwestern Europe (and the northern countries of Central and Eastern Europe).

● Agricultural productivity (measured in terms of both yields and labour productivity) in Europe has increased enormously since World War 2, and agriculture can be described as being a victim of its own success. This success in productivity is attributable largely to intensification: increased specialisation and concentration of crops and livestock, greater mechanisation of many farming activities, drainage and irrigation, and the development and increased use of fertilisers and pesticides. The use of fertilisers and

Box 22C The EU Common Agricultural Policy

The EU Common Agricultural Policy (CAP) is the single most influential agricultural policy in Europe. Before its 1992 reform, the heart of the policy, in most cases, was the system of guaranteed high prices for unlimited production. These guarantees encouraged surpluses of produce such as cereals, beef, dairy products (milk) and wine. Quotas on some products were introduced during the 1980s, but the purpose of these was to maintain guaranteed high prices, not to deliver (even indirectly) environmental benefits.

The CAP has evolved considerably since its inception in the late 1960s, but it was not until 1985 that the EU acknowledged that agriculture had a direct and significant impact on the environment (CEC, 1985). In 1987 a detailed measure was adopted within the CAP which permitted EU Member States to make payments to farmers in environmentally sensitive areas affected by agriculture by setting aside land. This was later incorporated into the CAP, requiring Member States to make such payments (under the agri-environment EC Regulation 92/2078/EEC).

The main driver to reform of the CAP in 1992 was the high cost of surplus production subsidised by EU consumers/taxpayers, and the increasing amounts of EU money being spent on storing and selling off surpluses at subsidised prices on world markets (coupled with the desire to tackle the trade problems arising out of these subsidies, and the need to improve the competitiveness of EU agriculture on both internal and external markets).

The main aims of the 1992 changes were to reduce surpluses, cut prices for consumers and, to a certain extent, decouple support for farmers from production. Farmers' incomes have been restructured, and are now much less linked to price support. Consequently, the incentive to unlimited production, and therefore to increased application of chemicals, is substantially reduced. Objectives to bring production down to levels more in line with market demand and to encourage farmers to remain on the land also have relevance to the way agriculture affects the environment. Environmental matters were also given fuller specific consideration in development of the 1992 CAP reform, in which one of the five main objectives was stated to be to protect the environment and develop the natural potential of the countryside. The 1992 CAP changes included a declaration by the European Council affirming its commitment to pursuing environmental protection as an integral part of the CAP and calling on the European Commission to make further proposals on this.

Under the pre-1992 CAP arrangements, technology and genetic engineering were encouraged to develop higher milk yields or more productive cereals strains, but the reform of the CAP aims to encourage studies into varieties requiring lower inputs and producing lower yields.

The dual environmental objectives of the 1992 CAP do create strains: one objective is to remove a proportion of land from productive use, and another is to reduce applications of chemicals to agricultural land in some parts of the EU. Taking land out of productive use may lead farmers to try to maintain or increase yield by farming more intensively on the remaining areas of productive land. However, the intention is to reduce chemical inputs, as farmers' incomes are no longer so closely linked to production.

pesticides has increased significantly over the last two decades. In addition to the problems of drinking water contamination and eutrophication of waterbodies, use of fertilisers and pesticides (in particular) has led to changes in habitats and loss of biodiversity.

- The increased intensity of livestock rearing, particularly in farms in Northwest Europe, depends in particular on large amounts of imports of nutrient-rich feedstuff from countries outside Europe. Imports of this kind, and the resulting amounts of livestock effluent, constitute a major input of nutrients (nitrogen and phosphorus) to the European environment in general and to the agricultural system in particular. In this way, specialisation and concentration towards animal husbandry in specific areas can lead to nutrient surpluses, which contribute to problems such as eutrophication.

- The structure of European agriculture has been strongly influenced by national and international agricultural policies, especially the EU's common agricultural policy (CAP). Traditionally these policies have tended to focus on providing financial, technical and training support to farmers and on encouraging increased production levels through guaranteed prices. However, agricultural policies, particularly in EFTA, and in the EU with the reform of the CAP, have adopted measures aimed at minimising environmental impacts as well as reducing food stocks. The main trigger for reform of the EU CAP was its ever-increasing costs; however, integration

Oilseed rape
Source: Michael St Maur Sheil

of environmental considerations was also an important element of the CAP reform.

There is an increasing interest in various forms of low input, extensive and traditional farming systems. In the EU, Member States are now required to offer incentives to encourage such systems where they can bring environmental benefits. One such system that may be encouraged to expand as a result is 'organic' farming.

REFERENCES

Addiscott, T (1988) Farmers, fertilisers and the nitrate flood. *New Scientist* **120** (1633), 50–54.

Arnold, R and Villain, C (1990) New directions for European agricultural policy. Centre for European Policy Studies (CEPS). Paper nr 49, Brussels, 3.

Baldock, D (1990) *Agriculture and habitat loss in Europe.* WWF International CAP discussion paper number 3.

CBS (1961) *Mestbalansen.* Dutch Central Bureau of Statistics (CBS), Voorburg, The Netherlands.

CBS (1989) *Mestbalansen.* Dutch Central Bureau of Statistics (CBS), Voorburg, The Netherlands.

CEC (1985) *Perspectives for the Common Agricultural Policy,* Green Paper. Commission of the European Communities, Brussels.

CEC (1993) *The situation of agriculture in the EC, 1992 report.* Commission of the European Communities, Luxembourg.

CEC (1994a) *Desertification and land degradation in the European Mediterranean.* Report EUR 14850, Commission of the European Communities, Directorate-General XII, Brussels.

CEC (1994b) *The agricultural situation in the Community.* 1993 Report. Commission of the European Communities, Luxembourg.

E/SB (1991) *Central and Eastern Europe 1991, Country Reports.* Eurostat and Statistisches Bundesamt, Luxembourg.

Gardner, B (1990) *European agriculture's environmental problems,* paper presented at the First Annual Conference of the Hudson Institute, Indianapolis, Indiana, April 1990.

Greenpeace (1992) *Green fields grey future: EC agriculture policy at the crossroads.* Greenpeace, Amsterdam.

Hartung, J (1992) An inventory of measures to reduce ammonia emissions from animal husbandry. In CEC Air Pollution Research Report 39, pp 355-64. Angeletti, G, Beilke, S and Slanina, J (Eds). Guyot, Brussels.

IUCN East European Programme (1991) *The environment in Eastern Europe: 1990.* Environmental Research Series 3, International Union for Conservation of Nature and Natural Resources, Cambridge.

MEPNRF (1991) *The state of the environment in Poland, damage and remedy.* Ministry of Environmental Protection, Natural Resources and Forestry, Warsaw.

Oldeman, L R, Hakkeling, R T A, Sombroek, W G (1991) *Global assessment of soil degradation. An explanatory note for the World Map of the status of human-induced soil degradation.* International Soil Reference and Information Centre, (ISRIC), UNEP. ISRIC, Wageningen, The Netherlands.

Pimentel, D and Levitan, L (1986) Pesticides: amounts applied and amounts reaching pests. *Bioscience* **36**, 86–91.

Pretty, J N, and Conway, G R (1988) *Agriculture as a global polluter,* pp 5–7. International Institute for Environment and Development, London.

Rebsdorf, A, Thyssen, N and Erlandsen, M (1991) Regional and temporal variation in pH, alkalinity and carbon dioxide in Danish streams, related to soil type and land use. *Freshwater Biology* **25**, 419–35.

UK MAFF (1992) *Solving the nitrate problem, progress on research and development.* UK Ministry of Agriculture, Fisheries and Food.

Umweltbundesamt (1987) *Daten zur Umwelt 1986/87.* Erich Schmidt Verlag. GmbH & Co, Berlin.

UNECE (1992) *The environment in Europe and North America: Annotated Statistics 1992.* United Nations, Geneva.

UNEP (1991) Status of desertification and implementation of the United Nations plan of action to combat desertification. United Nations Environment Programme, Nairobi.

von Meyer, H (1989) Wie Agrarpolitik Böden zerstören oder schützen kann. In: *Forum Wissenschaft, Studienhefte* **7**, 51-63.. Forschungs- und Informationsstelle des BdWi. Marburg, Germany.

von Meyer, H (1993) Agriculture and the environment in Europe. In: *The European Common Garden,* 69-85. Group of Sesimbra, Brussels.

WSL (1990) *UK atmospheric emissions inventory 1970–1990.* Warren Spring Laboratory, UK.

WWF (1992) *Pesticide reduction programmes in Denmark, The Netherlands, and Sweden.* World Wide Fund for Nature International Research Report. Gland, Switzerland.

*High forest,
Condroz, Belgium*
Source: Walphot

23 Forestry

INTRODUCTION

How forests are used and managed can significantly affect environmental quality. This chapter examines the changes in the nature of Europe's forests and in how they are used. It summarises how the associated activities and practices can result in impacts on the environment, both negatively and positively, and identifies the principal underlying driving forces influencing these changes, including such factors as policy, demand for wood and ownership patterns (see Box 23A).

There has been extensive loss of forest cover in Europe since the early post-glacial era 10 000 years ago, when an estimated 80 to 90 per cent (Delcourt and Delcourt, 1987) of Europe's land was forested. Deforestation was due partly to changing climatic conditions but was caused primarily by human activities such as land clearance for farming and the harvesting of forests for fuelwood and wood for building, shipping and mining material. The first important deforestation in Europe took place during the Roman period but the largest deforestation period ever in Central Europe occurred in the Middle Ages. In Central and Eastern Europe, however, vast primary forests were still present until the beginning of the twentieth century.

Today, forests cover about 312 million hectares, that is 33 per cent of Europe's land (see Chapter 3), from which 166 million hectares are located in the European part of the Russian Federation. This percentage ranges from 6 per cent in Ireland up to 66 per cent in Finland (see Map 23.1). Recent trends from different sources (UNECE/FAO, 1992a; see *Statistical Compendium*) indicate that there has been an increase of Europe's forest cover, in both area and volume, over the past 30 years (see Figure 23.1). In addition, the composition and use of many of Europe's forests have changed greatly.

In Europe, there are practically no forests that can be considered 'natural' or 'virgin' in the sense that there has been no human influence whatsoever (see Chapter 9). It is estimated that as little as 1 per cent of Europe's forests remain essentially untouched by humans and are still in their original, natural state (Dudley, 1992), most of it being in Russia. Even the Nordic subalpine birch woods, or the tundra or taiga of Russia, which are often thought of as untouched or virgin, have been subject to human impact. Therefore, most of the forests in Europe are the result of past use (or misuse) of the environment: in parts of the Mediterranean basin, for instance, fire and overgrazing have led to the appearance of a degraded brush vegetation. However, in the Iberian region (agro-forestry landscapes of the montados or dehesas (see Box 23B), or in Scandinavia, grazed deciduous woodlands are the result of sustainable landuses that have survived over the centuries, are rich in wildlife and nature, and are considered as harmonious landscapes which deserve protection (see Chapter 8). Although most European forest has been modified by human influence, changes are not necessarily for the worst. Forest is still one of the components of the European environment where human impact has been relatively slight and natural values dominate.

ENVIRONMENTAL IMPACTS

In many cases, the ways in which forests are managed and used in Europe result in overall benefits for the environment, helping to preserve, and sometimes even increase, biodiversity and landscape value, in addition to their role in soil quality protection and water regulation. However, some practices, especially those associated with the intensive use of forests for large-scale pulp production, can have detrimental environmental impacts.

Box 23A Data availability and country groupings

There are various definitions of what constitutes forest; in this report data on forest usually refer to any land area with a tree crown cover (a measure of tree density) of more than about 20 per cent of the ground area with trees growing to more than about 7 metres in height and able to produce wood (UNECE/FAO, 1992a); whenever no such data are available, 'forest and wooded land' is used – a broader term which also includes open woodland, scrub, shrub and brushland. The term 'forestry', as used in this chapter, is taken to include all human activities (not just those associated with wood production) relating to the management of forests for whatever purpose.

A reasonable amount of data available for Western Europe on forests (eg, UNECE/FAO, 1992 a; b; c; CEC/UNECE, 1993, and Eurostat). However, the vast majority relate to factors important to the management of forests for timber production. This includes such measurements as: area by broad forest type (coniferous versus broadleaved), rates of reforestation and afforestation, volumes of the standing stock of wood and annual growth, annual fellings and removals. Since the UNECE and CEC merged their datasets on forest condition in Europe in 1991, comparable information on 90 000 trees and 4400 plots has been made available every year on defoliation, discoloration and species (from 1993, additional data on soil and foliar analyses, deposition measurements and increment studies have been collected). Data are also available on economic aspects, but again these relate almost wholly to physical measures of production and trade of wood-derived products such as roundwood, pulp and charcoal. Accurate data are generally lacking on such factors as:

- the area of forest subject to harvesting practices such as large-scale clear felling and the use of heavy machinery, and the area of forest grown specifically for pulp production;
- the extent and rate of afforestation with monoculture stands;
- the extent and rate of planting of introduced tree species (exotic trees);
- the amount of ancient/natural forest remaining;
- the extent to which forests are used for non-wood production purposes such as recreation (visitor frequency), hunting and conservation;
- employment trends.

The availability and quality of data relating to forests in Eastern European countries vary considerably. Many countries have centralised forest inventories but there are very few available data for the Baltic States, Moldova and the European part of the Russian Federation. Economic data on production, trade and other factors, such as contribution to GDP/GNP, are either lacking or sometimes obsolete.

Few data are compiled on the specific environmental effects associated with forestry. This arises partly because parameters such as biodiversity, landscape value and conservation value are more difficult to measure than purely physical impacts such as soil quality and water pollution. In addition, efforts to assess such parameters on a comprehensive basis have yet to be mounted in most countries.

Although attempts have recently been made to collect data on factors relating to the non-wood-production forestry such as recreation, conservation, hunting, and so on, the material is generally restricted to questionnaire-based surveys (UNECE/FAO, 1992b) and thus very qualitative in nature. Furthermore, a bias may be introduced into the data depending on who is being surveyed.

Because of the diversity of conditions (for example as that between the Mediterranean basin, the mountainous areas of Europe, the North European plain and the boreal shield), it is impossible to generalise trends to the whole of Europe.

The country groupings adopted here are the same as those in Chapter 22 – Nordic, Eastern, Southern and Western European regions – and are based on physical and geographical areas which generally present similar behaviours (see Box 22A, for a precise description of the countries included in these regions). Since data are systematically, or often, lacking for some countries (Albania, Baltic States, Belarus, Cyprus, Malta, Moldova, Russian Federation, Ukraine), when the whole of Europe is mentioned, the figures given will always be underestimates. The Nordic region includes Finland, Norway and Sweden only, since there is hardly any forest in Iceland.

The environmental impacts reviewed below relate to ways in which forests are used and managed. However, forests are also vulnerable to impacts arising from other human activities not related to the direct use of forests. About 60 000 fires damage an average of 700 000 hectares of wooded land in Europe each year, and atmospheric pollution has caused extensive damage to forests in Germany and Scandinavia as well as in a number of Central and Eastern European countries, especially those of spruce (*Picea abies*) and fir (*Abies alba*) (see Chapter 34). In addition to obvious damage from gaseous emissions such as sulphur dioxide close to industrial concentrations, there is evidence to suggest that acidic deposition resulting mainly from SO_2 and NO_x emissions is harmful to tree growth, and actually stresses trees, making them more vulnerable to disease and other natural stresses such as adverse weather conditions, insects, fungi, nutrient deficiency and forest fire. In the next few decades forest ecosystems will potentially grow in steadily increasing atmospheric carbon dioxide concentrations and possibly changed climates. Consequent changes in the structure of these ecosystems and in their water and nutrient requirements are likely to influence the impact on forests from climate changes and their sensitivity to pollutant stress (see Chapter 27).

The nature and extent of human impacts on forests, both

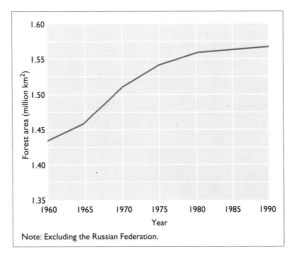

Figure 23.1
Evolution of wooded area in Europe
Source: Compiled by Eurostat from multiple sources, mainly: UNECE/FAO, 1992 (see *Statistical Compendium*)

Note: Excluding the Russian Federation.

desirable and undesirable, and the resulting various uses of forests, differ widely across Europe. For these reasons the data have been grouped in geographical regions showing similar behaviour (see Box 23A). The management of forests can, however, change with time (especially when forest uses

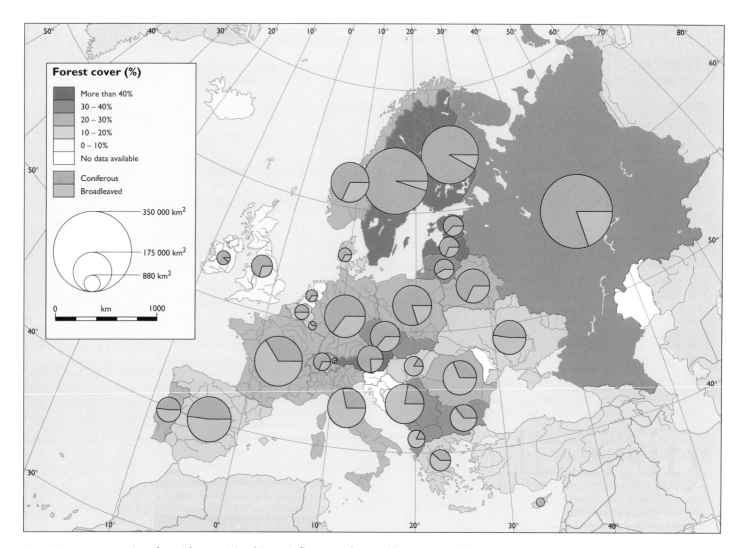

Map 23.1
Forest cover in Europe
Source: Eurostat, CEC/UNECE, 1993 (see *Statistical Compendium*)

change) because it is subject to influences such as world markets, technological progress and changes in society. Table 23.1 summarises the more significant environmental impacts associated with these different uses. Many of these impacts are linked to the modernisation practices associated with the management of forests for wood or pulp production. Such practices occur in most European countries but, except perhaps in the Nordic region, they are generally practised in only limited areas.

Nature and wildlife

As mentioned above, a managed forest is not necessarily a poor forest from an environmental perspective. Forests which are subject to non-intensive forms of management may have great value for biodiversity; for instance the impact of grazing and coppicing in mixed oak–beech forest enhances their evolution to the semi-natural oak–hornbeam forest which is much richer in species. Across much of Europe, forestry mostly contributes positively to preserving biodiversity and

Box 23B The abandonment of the dehesas and montados in Southern Europe

Dehesa, Portugal
Source: C Steenmans

In Spain and Portugal, multi-purpose ecosystems called dehesas and montados were traditionally used and formed a valuable landscape of pastures mixed with open woodland: the open tree stratum dominated by oak species (*Quercus suber*, *Quercus ilex*, *Quercus pyrenaica* and *Quercus rotundifolia*) was grown for cork, timber, fuel and charcoal, but also tannin and acorns, together with shifting cultivation of cereals, and grazing of pigs, sheep, goats and cattle on fallow land. This has led to an artificial system with a high degree of patchiness and

characterised by a tree cover of 10 to 80 trees per hectare at most (Joffre, 1992). These mixed systems were ecologically rational and economically productive, but required intensive recurrent management practices, including the use of fire (Catarino, 1992). Since the 1970s and the rapid development of intensive animal husbandry, these agroforestry systems were progressively abandoned. This led to the invasion of the open landscape by shrubs, therefore increasing fire risk. These systems still cover almost 5 million hectares in southwestern Spain and more than half a million hectares in Portugal, but are under tremendous pressure. Despite their ecological importance in the Mediterranean area (for maintaining greater water availability), the remaining dehesas and montados undergo steady erosion due to the combined pressures of livestock grazing, frequent fires and urban and rural development (Catarino, 1992). (See also Box 8F, Chapter 8.)

Forestry practices	Water	Soil	Landscapes	Nature and wildlife
Wood production:				
Planting	● Accumulation of litter from acidifying tree species ➤ soil acidification ➤ groundwater acidification ● Cultivation of productive water-demanding tree species reducing groundwater availability	● Accumulation of litter from acidifying tree species ➤ soil acidification	● Uniform planting ➤ major changes in form, colour and texture arising from sharp boundaries of evergreen forest stands	● Plantation of monoculture and introduced tree species ➤ uniformity, loss of biodiversity
Clear felling	● Bare land after clear felling ➤ water erosion ➤ increased sediment and organic matter loads ➤ eutrophication, sedimentation	● Bare land after clear felling ➤ wind and water erosion ● Use of vehicles ➤ compaction ● Sudden decrease in water demand following clear felling ➤ waterlogging	● Large clearances ➤ scarred landscapes	● If dead/decaying wood removed ➤ loss of plant and animal species which depend on this ➤ loss of biodiversity
Draining	● Lowering of groundwater level reducing water availability ● Oxidation of organic soils ➤ soil acidification ➤ groundwater acidification	● Oxidation of organic soils ➤ acid-sulphate formation ➤ soil acidification	● Drying of land causing changes in plant communities and hence landscape	● Lowering of water table ➤ loss of wet forests and wetlands high in biodiversity
Weeding, cleaning, thinning	● Use of herbicides ➤ groundwater pollution	● Increased frequency of vehicle use ➤ erosion, compaction	● Removal of greenery ➤ uniformity	● Removal of understorey, an important habitat for many animal species ➤ loss of biodiversity
Pesticide and fertiliser application	● Leaching of applied substances ➤ groundwater pollution	● Fertilisation in waterlogged conditions ➤ denitrification ➤ greenhouse gas emission ➤ contribution to climate change	● Changes in plant communities and hence landscapes	● Release of chemical pesticides ➤ poisoning of non-target species Fertiliser applications ➤ changes in plant communities
Heavy machinery use	● Soil erosion ➤ increased sediment load in surface waters ● Oil leakage/spills ➤ water pollution ● Soil sealing ➤ increased runoff, decreased infiltration to groundwater	● Increased frequency of vehicle use ➤ soil compaction and erosion ● Oil leakage/spills ➤ soil pollution		● Increased frequency of vehicle use ➤ disturbance of wildlife
Recreation:	● Increased use of water ➤ reduction of water availability and pollution through effluent from tourist centres, camping grounds, etc	● Trampling leading to erosion and compaction	● Infrastructure development (access roads, recreation centres, etc) ➤ changes in landscape	● Increased number of visitors in forests ➤ wildlife disturbance ● Recreational infrastructure development ➤ increased groundwater abstraction affecting tree growth
Hunting:		● Soil contamination from lead pellets	● Reduced access to forest at hunting periods	● Removal of some animal species (eg, wolf, bear, lynx) from original range ➤ loss of biodiversity ● Selection of hunted species to other species detriment ➤ loss of biodiversity ● Poisoning of avifauna from lead pellets ● Damage from overstocked game
Grazing and browsing:	● Overdensity of grazing/browsing animals ➤ soil erosion and compaction ➤ increased sediment load, decreased infiltration to groundwater	● Overdensity of grazing/browsing animals ➤ soil erosion and compaction	● Overdensity of grazing/browsing animals ➤ erosion and changes in landscape	● Overgrazing and browsing (overpopulation of game) ➤ damage to young plants, trees and habitats

habitats. In some regions, however, the implementation of certain forestry practices and techniques for wood production has created the potential for negative impacts.

The use of ground preparation techniques and large-scale clear felling can have a dramatic impact on ground flora and fauna by displacing a wide range of species from the bare site. Many return as the new generation of forest grows but others cannot re-establish themselves before the trees are felled again.

It is common practice to thin productive coniferous stands. Numerous insects, fungi and lichens which rely on the presence of dying or dead trees have therefore become rare or have disappeared altogether. In addition, some productive coniferous species such as spruce have a thick canopy which limits light in the forest and undergrowth. Of particular concern is the loss or replacement of pristine forests in wet habitats, particularly those growing alongside rivers. These are habitats with high biodiversity, often supporting unique species or assemblages of flora and fauna. Draining such forests greatly enhances their timber productivity but at the same time eliminates their wetland character and thus many

Table 23.1
Overview of significant and potentially negative environmental impacts of forestry
Source: EEA-TF, ERM

Box 23C European alluvial forests endangered

Alluvial forests are among the richest and most complex ecosystems of Europe. They present an exceptional diversity of tree species, such as European oak (*Quercus robur*), elm (*Ulmus campestris*), poplars (*Populus* spp) and ash (*Fraxinus* spp). The forest along the Rhine contains 24 different trees, 24 shrubs and 3 climbers belonging to 21 families and 34 genera (Schnitzler-Lenoble and Carbiener, 1993). The Shannon index (which takes into account the total number of species as well as the frequency of each species), a good indicator of biodiversity, ranges from 0.6 to 0.8 for the terrestrial temperate forest and from 3 to 4 in the alluvial temperate one. The presence of species here which have disappeared in other European forests is due to the richness of the soil in mineral particles, and the abundance of light in the understorey which allows a complex stratification to take place. Alluvial forests are also rich in animal species; in the Rhine forest, 38 bird species have been recorded with about 200 pairs per 10 hectares of forest, which is twice the number of pairs to be found in temperate oak–beech forest.

Since the middle of the twentieth century, these forests have been endangered through the diking of major European rivers and reduction of the natural flooding zone, deforestation, pollution of air, water and soil, bad management of water resources and non-adapted silvicultural practices. Most of the original riparian forest has disappeared along the four major European rivers (Danube, Po, Rhine, Rhone). Forest used to cover about 2000 km² along the Rhine. Nowadays, it is very much fragmented and covers a total of 150 km², of which less than 1.5 km² is still semi-natural. Only some parts of the Danube (in Hungary, Serbia-Montenegro and Romania) survive under a relatively natural regime, due to a delay in technological development. The perturbation of the flooding regime has altered the delivery of nutrients to the riparian forests. Flood water, rich in mineral particles (such as phosphates) and rainfall, is no longer filtered by soils and trees, causing eutrophication of groundwater.

of the fauna and flora dependent upon them (see Chapter 9 and Box 23C).

Tree species themselves can be endangered because of the evolution of forest management and related changes towards more productive species or gene transfers. This is generally the case for oak (*Quercus* spp) and more specifically for the cork oak (*Quercus suber*), which shows a general die-back in most of its natural range (Iberian peninsula); black poplar (*Populus nigra*) is endangered because of fast modifications of its natural riparian habitats and uncontrolled hybridisations with Euramerican hybrids; wild *Rosacea* appear also threatened by changes in silvicultural regimes, from mixed deciduous forests to mono-species pure deciduous forests or conifer plantations (Ministerial Conference on the Protection of Forests in Europe, 1993a).

Soil

In Europe, forestry practices generally have few known negative impacts on soil quality and, where they do, it is mostly restricted to certain soils in those parts of Europe practising large-scale, mechanised forestry management.

- Shallow soils, for example, with restricted rooting depths, are difficult to plough and may become degraded by the frequent passage of heavy vehicles.

- Afforestation with conifers, drainage and use of some fertilisers have the potential to increase soil acidification (UK Forestry Commission, 1989). However, in areas sensitive to atmospheric pollution, such as Scandinavia, the acidifying effect of deposition is estimated to be more than twice that related to forestry practices (Bernes, 1993).
- Problems of soil compaction and erosion can occur where heavy machinery is used for forestry operations.
- Ground that is suddenly exposed to direct rainfall and runoff, following ploughing or clear felling, is susceptible to soil erosion; however, forests protect soil against erosion: erosion on bare soil can be 50 times higher than on soils protected by a dense forest cover (Marchand, 1990).

Water pollution and resources

Forests play an important role in water regulation. The large-scale plantation or clear felling of large tracts of forest can dramatically alter the water regulation dynamics of the area, leading to excessive groundwater abstraction on the one hand or increased runoff and risk of floods on the other. Water resource problems have been reported (Adlard, 1987) in Portugal concerning certain fast-growing, water-demanding eucalyptus plantations where serious depletion of water reserves has occurred when tree roots reach the water table; this is the case in the dry zones of southern Portugal.

The use of fertilisers is practised to a significant degree only in Sweden, Portugal and the UK, and to a certain extent in Finland and Norway, and fertilised forests represent only 0.3 per cent of total forest area in Europe. The amounts of fertilisers and pesticides used in forestry have decreased appreciably in recent years in the Nordic region. In Finland and Sweden, for example, fertilising forest reached its peak in the 1970s, when for each country some 20 000 tonnes of nitrogen fertiliser was applied annually from the air, dropping to less than 2000 tonnes in both countries in the early 1990s (EFMA, 1992; Bernes, 1993). The quantities and application rates used are relatively small when compared to those of agriculture (Table 23.2).

Landscapes

When monoculture is introduced, forests become more uniform, affecting the landscape. When a forest is drained or felled, landscape is also greatly changed. Large clear felled sites, particularly in hilly areas, can be seen from long distances. To many people, such sites are unattractive and seen as scars or wounds on the landscape. Blankets of

Table 23.2
Fertiliser applications on forests
Source: EFMA, 1992

	Fertilised forest area (ha)	Total forest area (1000 ha)	Proportion of fertilised forest (%)	N (kg/ha)	P (kg/ha)	K (kg/ha)
Nordic Countries						
Finland	16 000	19 511	0.08	112	37	24
Norway	3 000	6 638	0.05	130	0	0
Sweden	60 000	22 048	0.27	150	0	0
Eastern Countries			No data			
Southern Countries						
Portugal	20 000	2346	0.85	30	60	30
Western Countries						
UK	56 000	2207	2.54	5	135	125
Total	155 000	52 750	0.29			
Mean application on fertilised forests (EU and EFTA)				2	1	1
Average application of fertilisers for all crops (EU and EFTA)				78	34	38

straight-lined evergreen monoculture plantations reduce the scenery value of landscape. Examples, including the replacement of heathlands with spruce plantations in some parts of Western Europe, and grazing land with eucalyptus plantations in areas of Southern Europe, are still common.

In addition to these changes brought about by technological developments and the pressure to increase fast production, there are also changes that follow from modifications in forest uses (abandonment of traditional practices such as coppicing for fuelwood or grazing) and in land ownership or management responsibility (the balance of public and private).

CHANGES IN EUROPEAN FORESTS AND FORESTRY

This section examines the changes in the nature of Europe's forests and how they are used and managed. The changes in the underlying driving forces which affect the way forests are used and managed, such as demand for wood, government policy and ownership patterns, are identified and reviewed. The way in which changes in underlying driving forces have translated into changes in forestry practices and techniques, are examined, and the implications for the environment assessed.

The nature of forestry in Europe

There have been important changes in the nature of Europe's forests and in the ways in which they have been used over the last few decades. Many of these changes are regional in nature, with broad differences evident between Nordic, Western, Eastern and Southern regions of Europe.

Overall, Europe's forest area has increased by over 10 per cent (see Fig 23.1) since the early 1960s. Most of this increase is accounted for by the Southern and Western regions, in spite of forest cover remaining stable or decreasing in many countries from the Eastern region and the former USSR due to overexploitation of forest resources. This increase is due partly to the afforestation of poor soils unsuitable for agriculture, traditionally used for rough grazing, and partly to the afforestation or spontaneous invasion of abandoned agricultural land. Large-scale afforestation and reforestation have occurred in several of the Atlantic seaboard countries (eg, 600 000 ha of eucalyptus plantations in Portugal) through encroachment of trees in marginal agricultural land and the abandonment of coppices for fuelwood. Many Alpine and other mountain areas have been affected, with the consequent disappearance of the former farmed landscape. In certain parts of Europe, especially in poorer areas of the Mediterranean, whole rural villages have been largely abandoned as the traditional farm economy has broken down and trees have invaded the whole landscape. Accompanying this overall increase in forest cover, however, has been a drastic change in the composition and use of some of Europe's forest area.

Much attention is focused on the split between

Map 23.2
Broadleaved and coniferous species
Source: CEC/UNECE, 1993

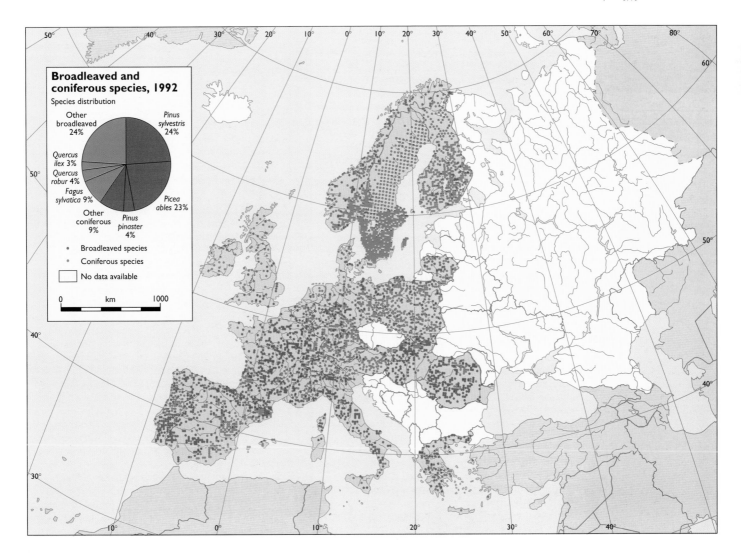

Figure 23.2
Public and private ownership of forest and other wooded land
Source: Compiled by Eurostat from UNECE/FAO, 1992 (see *Statistical Compendium*)

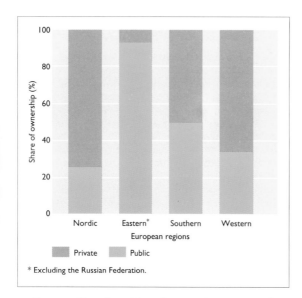

international trade and the demand for greater availability of cheaper wood, wood-based products and paper, but also in the cost of personnel, the development of new technology, and changes in uses. This has led to significant increases (see below) in the yield, labour productivity and profitability of wood-production forestry on the one hand, and to abandonment of coppices and other wooded land on the other.

At the same time, society's growing awareness of the other values and functions of forests, particularly in Northern Europe, has led to the reappraisal of many of the modern forestry practices associated with wood production, and is beginning to influence how forests are managed, what they are managed for, and thus the nature of forests and forestry itself (UK Forestry Commission, 1989). In many places, however, there is a tradition of a more multi-function silviculture, where cutting is selective and natural generation allowed.

The Nordic region

The Nordic region has witnessed a significant shift towards the industrialisation or 'intensification' of the management and forestry techniques used for wood-production forestry. This has involved the mechanisation of the planting and harvesting of trees, the use of fast-growing conifers but also broadleaved species such as birches (*Betula* spp) planted in monoculture crops, to be clear felled at regular intervals. Fertilisers are also used to increase productivity, and applications, mainly of nitrogen, occur every 8 to 10 years. However (as in Norway in 1992), the fertilised forest area has been drastically reduced recently. The application of lime is often practised in areas where atmospheric pollution is contributing to forest degradation. In some cases this had led to a change in the composition of forest (see below) and to a reduction in biodiversity and landscape value.

Intensive silviculture and forest improvement practices, however, tend to be limited to those areas of forests owned and controlled by the large forestry companies or to some public-owned forests. In Sweden approximately half the exploitable forest area is controlled by companies, with forest holdings averaging 313 000 ha for private companies and 145 500 ha for state companies. Much of the remaining area is managed by numerous private farmers, with average forest holdings of just 53 ha, who apply more mixed and less intensive silviculture practices. In Norway and Finland a much greater proportion of forest is managed by private farmers with holdings typically under 50 ha.

Western and Eastern regions

Following the extensive replacement of mixed forest by plantations (especially in German-speaking countries in the latter part of the nineteenth century) there has been a more cautious application of modern forestry methods in the Western and Eastern regions of Europe, combined with a recognition of the need for greater environmental concern. As a result mixed species forest is more common and in certain areas management is extensive because of poor access and low wood values. Nevertheless some areas, particularly in France, Germany, the former USSR, Poland, Romania and the Czech and Slovak republics, are still subject to intensive high wood production practices. The use of fertiliser is rare (mainly in Portugal and the UK). In the UK, applications of mainly phosphorus and potassium are made when planting young trees, and there is only one fertiliser application for the whole rotation to remediate soil deficiencies. Recently, applications of lime, phosphate, potash and also magnesium have been used in some parts of Germany to improve the vitality of the trees.

The diversity of ownership categories and of the resulting styles of management found in Western Europe may, on balance, have been an advantage for the environment, in contrast to the position in many of the countries which had centrally planned economies. Average size of holdings in the

coniferous and broadleaved forest because of concern over the replacement of broadleaved forest with monoculture, often of exotic, coniferous forest. The proportion of coniferous species in the total has remained almost constant, at about 60 per cent for Europe as a whole since the early 1950s (Kuusela, 1993). However, this proportion is due partly to needs of the timber industry and, without the intervention of human activities, deciduous tree cover would be considerably higher. Certain countries, however, have witnessed significant changes. Denmark, Ireland, the UK, The Netherlands, Belgium and Bulgaria have all increased by over 10 per cent the proportion of their exploitable forest, which is accounted for mainly by afforestation with coniferous trees and not necessarily by the replacement of existing broadleaved or mixed forests with evergreen species. In the pan-European survey on forest condition in Europe (CEC/UNECE, 1993), within the 184 million hectares of forest covered by the survey, 113 different species have been recorded and 6 species only accounted for two thirds of all trees: *Pinus sylvestris* (24 per cent), *Picea abies* (23 per cent), *Fagus sylvatica* (9 per cent), *Pinus pinaster* (4 per cent), *Quercus robur* (4 per cent), and *Quercus ilex* (3 per cent). This is illustrated in Map 23.2.

About half (49 per cent) of Europe's forest is privately owned, ranging from farmers with small holdings of woodland to large private companies, but there are large variations between the different regions of Europe (see Figure 23.2). In many Central and Eastern European countries all the forests fall into the public domain, while the share of private ownership exceeds 70 per cent in many countries, such as Austria, Denmark, France, Norway and Portugal. The division of responsibility and the sheer number of decision makers with differing objectives has profound implications for any government's ability to guide the practices adopted by owners. Not all private owners make use of professional guidance and technical assistance but they are usually under some kind of supervision by public authorities. The importance given to wood production in forest management varies considerably between public and private sectors in different countries. In some countries, such as Belgium, Greece, Luxembourg, Poland, the UK, and the former Yugoslavia, wood production (fuelwood and industrial roundwood) is given far more importance in public forests than in most private ones, while in other countries, such as Denmark, Finland, Italy and Sweden, it is the opposite. Priorities of public forest administrations are also very different from country to country. Only a few countries have financial goals and ensure that marketable goods and services are given special weight in forest management.

Many of the changes in forest cover and composition, particularly in Northern and Western regions, can be linked to modifications in the ways in which forests are managed (or not) for the production of wood. This in turn reflects trends in

latter vary between just under 2000 ha in Hungary to just over 300 000 ha in the former USSR (see *Statistical Compendium*).

The Southern region

With the exception of some parts of Spain and Portugal, much of Southern Europe has witnessed the abandonment of many traditional forestry practices. Unable to compete with cheap imports from the highly productive Nordic and Eastern regions, this area has witnessed a steady decline in the management and use of its forests for wood production, coupled with a trend towards the abandonment of traditional practices such as coppicing for firewood and the collection of barks, resins, acorns and tannin. On average, the productivity of Mediterranean forests is low because of the hot, dry summers. Many forests have become neglected or abandoned, increasing fire risk and fire damage; grazing may have been another contributing factor. In Italy, for example, the production of roundwood (that is all wood obtained from removals) fell by 21 per cent between 1965 and 1989. From a financial point of view, Mediterranean forests are costly to manage and the part that forestry plays in GNP is relatively small (Marchand, 1990). However, the role of forest and shrubland in these areas for soil protection and nature and landscape conservation, together with the incomes generated by non-forest functions, is of major importance in Southern Europe. Rural depopulation and related abandonment of agricultural land in the South is responsible for spontaneous tree encroachment taking place. Afforestation of abandoned land is difficult partly because the average size of the former fields is too small: 35 per cent of farm holdings are less than 1 hectare in Italy, 31 per cent in Portugal, 26 per cent in Greece and 14 per cent in Spain (see Chapter 22).

Wood and wood products – demand and supply

Much of the intensification of wood production practices can be linked to the need to meet the changing demand for wood, in terms of both volume and type. In Europe, the apparent consumption of roundwood, which gives an indication of the overall demand, has increased slowly (28 per cent) over the last 25 years (Figure 23.3). Consumption in the Western region has increased steadily since 1965 by a total of 30 per cent. In the Eastern region consumption peaked around 1985 but has since been declining.

Much of the wood supply needed to meet this increasing demand has been provided by increasing production (see below) in Europe itself. Some of it, however, has been met by imports (Figure 23.4). Net imports of roundwood for the whole of Europe, for example, increased by 25 per cent between 1965 and 1989; the Eastern region is the only part of Europe where imports have decreased during the same period (see *Statistical Compendium*).

The increased level of imports is not just a result of demand in Europe outstripping domestic supply. In recent years, the amount of wood being felled from Europe's forests has fallen significantly below the annual rate of increase in its wood stock (net annual increment), meaning there is a surplus availability of wood for harvesting (Figure 23.5). In Russia, however, the annual fellings are higher than the annual production of wood. This is due to several factors: from the management side, there is a tendency in traditional silviculture for precaution felling favouring the accumulation of growing stock; recently, there has been an excess of wood supply coming from the harvesting of trees affected, for example, by air pollution, insect attacks, or uprooted after the exceptional storms from 1987, 1989 and 1990 (in 1990, the production of roundwood in West Germany was of 73 million m³ or nearly twice as much as 1989 due to wind throws); or from the huge influx of wood from the Eastern

region, especially from the former USSR (Box 23D), to Western markets. These factors have driven down prices which in turn affect the way some areas of Europe's forests are managed; indeed, forest owners tend to delay the frequency of felling, increasing the stock and affecting future prices.

Of Europe's total roundwood production of about 368 million m³ in 1989 (data from 1990 were not taken into account because of the exceptional harvesting due to wind throw in Germany), 31 per cent is from the Nordic region, 21 per cent from the Eastern region, 16 per cent from the Southern region and 32 per cent from the Western region. The major uses of wood (see Figure 23.6) are as sawnwood and panel products used in the construction and furniture industries, as paper and board for use in packaging, printing and writing, and as firewood or charcoal largely for household heating and cooking. The Nordic region produces 45 per cent of the total wood for pulping while the Western region produces 36 per cent of the total industrial roundwood other than pulp and 35 per cent of the fuelwood. Of the many products concerned, only charcoal and firewood have experienced a sharp and consistent decline over recent decades (see *Statistical Compendium*), but in the Southern region they are still a major energy source.

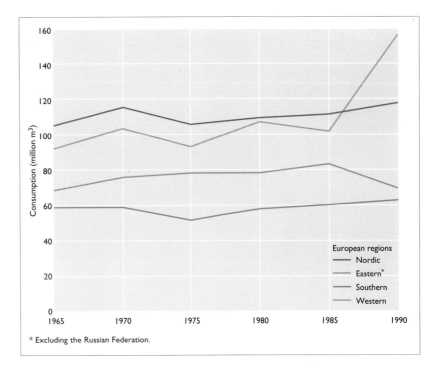

* Excluding the Russian Federation.

Figure 23.3
Trends in consumption of roundwood
Source: Compiled by Eurostat from FAO

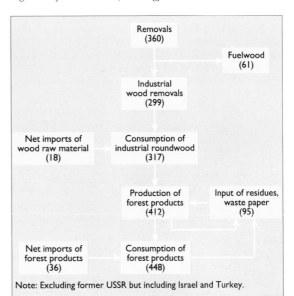

Note: Excluding former USSR but including Israel and Turkey.

Figure 23.4
Trade flows of roundwood in Europe, 1987 (million m³)
Source: UNECE/FAO, 1990

Figure 23.5
Annual fellings and increment in exploitable forest in Europe
Source: Compiled by Kuusela, 1993, from UNECE/FAO, 1992

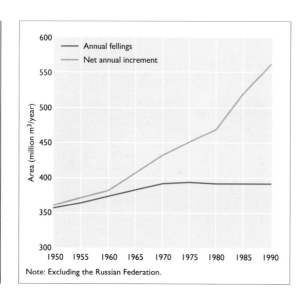

Note: Excluding the Russian Federation.

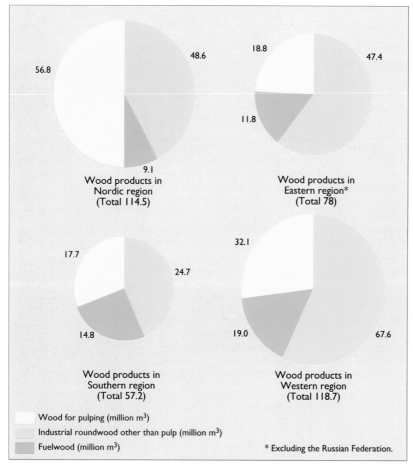

Wood for pulping (million m³)
Industrial roundwood other than pulp (million m³)
Fuelwood (million m³)

* Excluding the Russian Federation.

Figure 23.6
Share of different wood products in the totals produced in the Nordic, Eastern, Southern and Western regions of Europe, 1989
Source: Compiled by Eurostat from FAO (see *Statistical Compendium*)

One characteristic of wood industries' demand is the requirement for large quantities of products that do not vary in quality and are obtainable at the most competitive prices possible. The nature of this demand has basic effects on how forests are managed for wood production, the most advantageous forests being those providing vast areas of the same kinds of wood that can be easily harvested using mechanical means. Hence the trend towards clear felling practices accompanied by the planting of large monoculture stands and the introduction of fast-growing tree species.

Box 23D Destruction of the boreal forest

Boreal forests are mostly coniferous and dominated largely by Norway spruce (*Picea abies*) and, in drier areas, by Scots pine (*Pinus sylvestris*). From Europe, they stretch from Scandinavia to Siberia. They grow on shallow soils. Boreal forests take a long time to grow, and productivity is very low; for example, the net annual increment varies from 0.4 m³ per hectare in the Vologda region up to 2 m³ per hectare in the Murmansk area, both regions providing the major bulk of felling in the European part of Russia (72 per cent).

Boreal forests are very poor in species but in general have been little affected by human activity (see Chapter 9). One problem area is permafrost, where changes in vegetation (through logging for example) can cause melting. Until relatively recently, logging has been scattered and the forests are so huge that regeneration was achieved through natural regrowth. However, in recent years in the Russian Federation, where timber is one of the country's major exports, logging has increased enormously, generally by clear cutting, causing serious erosion problems and changes in plant and animal habitats. These forests are under increasing threat from internal mismanagement and the short-term opportunism of foreign companies (Dudley, 1992). Because of the low level of harvesting technologies, large amounts of wood are wasted (about 50 per cent) and left on the spot. Much of the logging is fostered by giant timber companies from South Korea, the USA and especially Japan. The amount logged (often clear felled) reached 40 000 km² each year, or as much as the annual deforestation in Brazil (FAO, 1992). In 1991, ten Japanese firms signed a five-year agreement with the Russian Federation to harvest in Siberia 6.5 million m³ of timber per year for export (MacKenzie, 1994). In addition to the destruction of the forest resource, forest clearing releases large amounts of carbon dioxide and methane to the atmosphere. With the forest destroyed, it can no longer absorb carbon dioxide from the atmosphere through tree growth (see Box 23F).

Siberia's forests are also being depleted by fires, affecting at least 10 000 km² each year. Being situated in high latitudes in the far north, where the greatest potential temperature rises in the northern hemisphere are expected from climate change (see Chapter 27), boreal forests could soon suffer marked desiccation, making them even more vulnerable to fires (Myers, 1993).

It is often pointed out (Dudley, 1992) that, although northern industrial countries have made tropical timber producers adopt a policy to sell timber only from sustainable managed forests in a new International Tropical Timber Agreement (see Box 23E), similar restrictions do not exist for the protection of their own forests.

Forestry productivity

Increased demand for cheaper and more plentiful supplies of wood in Europe over the last few decades has been met largely by increases in yields and the efficiency of production in terms of labour and capital inputs. Europe is the world's second highest industrial roundwood producing region. Total production of roundwood in Europe was about 368 million m³ in 1989, an increase of over 18 per cent since 1965. The biggest increase over the period was in the Western region (43 per cent), compared with 13 per cent in the Eastern region, 11 per cent in the Nordic region and 3 per cent in the Southern region (Figure 23.8). Some of this increased production has been achieved simply by expanding the total area of forest, but much is the result of increased productivity in most parts of Europe.

Box 23E Demand for tropical wood in Europe

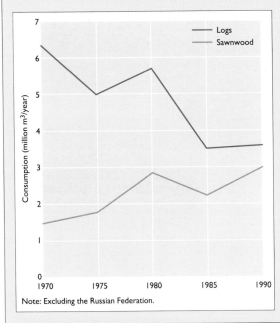

Note: Excluding the Russian Federation.

The EU is one of the three main consumers of tropical timber exports along with Japan and the USA. Apparent consumption of tropical logs in Europe decreased by about 50 per cent between 1970 and 1985 and has now levelled off at about 3 million m³ per year (see Figure 23.7).

The current price of raw logs and processed wood from tropical forests is very attractive since current exploitation of the tropical forest does not include renewing the resource base. A number of measures have been proposed or implemented by certain Member States. These include: a voluntary agreement in The Netherlands between the government, non-governmental organisations and tropical timber traders and industries to market only sustainably produced timber from 1995 onwards; Tropenwald, an initiative taken by the German tropical trade and industry to promote and certify sustainably produced timber; and the Austrian attempt to introduce an 'Ecolabel'. Unilateral actions might, however, have unintended negative impacts; if they succeed in lowering the demand for tropical timber, reducing associated revenues and profits, tropical forests may be converted more rapidly to higher-value purposes, such as export agriculture (cash crops) or cattle ranching (World Resources Institute, 1992).

Figure 23.7 Apparent consumption of tropical wood in Europe, 1970–90 Source: Compiled by Eurostat from UNECE/FAO (see *Statistical*

The net annual increment (NAI) in exploitable forest is an indicator of annual production and productivity. In 1990, NAI was 564 million m³ overbark for Europe and has steadily increased since 1950 (see Figure 23.5). There is a considerable range of NAI per hectare which can be related largely to climatic and soil conditions (Figure 23.9). Thus, Greece has a net increment per hectare of exploitable forest of 1.4, Norway 2.7 and Finland 3.6, while Austria shows 6.6, former Czechoslovakia 6.9 and Ireland 8.4. Species differences account for some of the range of values found; France with much broadleaved forest grows 5.3 m³/ha, while East Germany produces 6.7. Recent studies have attributed the important recorded increase of productivity in the long term to elevated levels of atmospheric CO_2 and nitrogen deposition from the atmosphere (Becker and Serre-Bachet, 1992; see Box 23F). The critical aspect is not so much the rate of volume production per unit as the efficiency with which wood-producing silviculture and harvesting are combined to keep down the costs of maintaining and exploiting the forest.

Besides site conditions, the increase in productivity has been achieved in a number of ways including the introduction of faster growing and more disease resistant trees, improved management techniques, including in some cases the use of fertilisers and pesticides, some mechanisation (for cutting, log removal, ground preparation, planting, etc), and in addition fertilisation through atmospheric pollution (see Chapter 34). A number of these practices, however, have the potential to impact on the environment (see above).

In Table 23.3, the participation of the different regions in the area and production of exploitable forest is summarised. The Nordic region (Finland, Sweden and Norway) has 35 per cent of Europe's exploitable forest but 25 per cent of the growing stock. Growth is relatively modest, averaging only about 3.5 m³/ha/year. The Western region (10 countries) has 21 per cent of the exploitable forest area (France and West Germany alone accounting for 70 per cent of the total), 30 per cent of the growing stock and an average growth of 6.7 m³/ha/year. The Eastern region accounts for 27 per cent of Europe's exploitable forest area and 31 per cent of the growing stock.

It is difficult to assess employment trends in wood-production forestry because forestry is often aggregated with agricultural employment statistics. For those few countries that do track segregated forestry employment levels, the

trends suggest a steady decline in the number of people employed, indicating a continuous increase in labour productivity and specialisation.

Technological progress

If technological progress has been a vital factor in the rapid productivity improvements over the last three to four decades, its extension has been rather limited compared with other sectors. Indeed, in many forests of Europe (Mediterranean forests, mountain forests, small woodlands), growing conditions are difficult, accessibility is often limited, and the use of relatively costly operating techniques is not profitable.

Technological progress has allowed considerable changes

Figure 23.8 Production of roundwood, 1965–90 Source: Compiled by Eurostat from FAO (see *Statistical Compendium*)

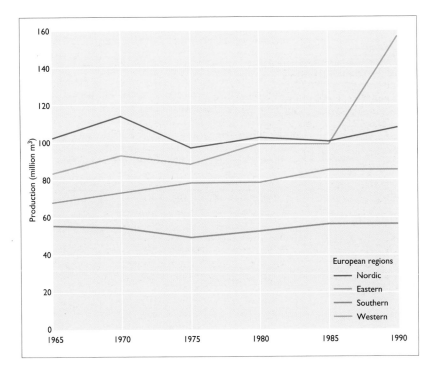

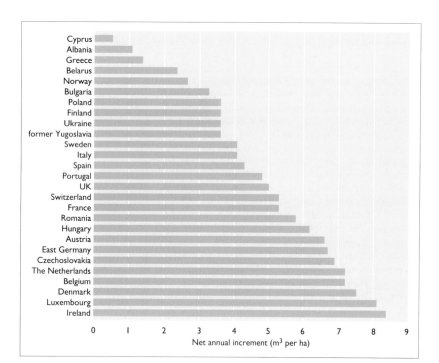

Figure 23.9
Net annual increment per hectare in exploitable forest
Source: Compiled by Eurostat from UNECE/FAO (see *Statistical Compendium*)

Box 23F Forest as a carbon sink?

Recently, the issue of climate change has started a debate about the possible role of forests for carbon dioxide sequestration (MacKenzie, 1994). When they are burned fossil fuels produce CO_2; part of the CO_2 is absorbed by the oceans but most remains in the atmosphere having the potential to influence world climate. However, the increase in CO_2 in the atmosphere is far less marked than expected. One reason for this could be due to northern forests (boreal and temperate) acting as a sink for the carbon by way of afforestation and atmospheric pollution, absorbing CO_2 and turning it into wood (IPCC, 1992). Indeed, because of the increase in atmospheric CO_2 and fertilising nitrogen deposited by pollutants such as nitrogen oxides, it has been observed that some trees are growing faster, thus absorbing more carbon (Kauppi and Tomppo, in press). In the next few decades, consequent changes in the structure of the forest ecosystems and in their water and nutrient requirements are likely to influence both the interaction of the forest with the climate and the sensitivity to pollutant stress.

European forests contain about 2.8 Gt C in trees and litter, plus 3 to 4 Gt C in forest soils; to give a comparison, 2.8 Gt C is the amount of carbon emitted from fossil fuels in the EU in about 4 years (Cannell et al, 1992). This carbon represents only a small fraction of the total carbon in world temperate forests (200 to 250 Gt C). Using forests to sequester carbon dioxide from the atmosphere has been discussed for some time. To store most carbon, most rapidly, short rotation, fast-growing trees should be planted with wood products that last longer than one rotation, and in soils with initially small organic carbon contents. However, such measures cannot be used as the main strategy to reduce CO_2 atmospheric levels, since the flux through forests is simply too small. One estimate is that forests currently sequester around 85 to 120 Mtonnes of carbon each year (Kauppi et al, 1992), that is about 5 per cent of emissions. At such rates, this would imply vast plantations of monoculture of fast-growing trees, causing important impacts on the environment (landscape degradation, loss of biodiversity, etc). Furthermore, what is acting as a sink now may become a source later. Before 1890, boreal forests were a source of CO_2, mainly because of forest fires and tree felling (deforestation). After 1920, a steep increase in tree growth outstripped carbon losses, turning the boreal forest into a carbon sink, and storing CO_2 from fossil fuel burning. If forest loss and tree damage degradation are not controlled, instead of forming a continuous and possibly increasing carbon storage, these forests might begin to revert to sources of carbon.

in how forests are managed and associated practices. The advent of the power saw (in itself the cause of the only improvement in labour productivity in forestry for centuries) and subsequently the introduction of machines carrying out the full set of harvesting operations (from felling, through pruning and cross-cutting, to extraction to roadside) all contributed to significant productivity improvements.

This trend towards increased mechanisation has greatly encouraged and facilitated many of the intensive wood-production practices witnessed in certain parts of Europe, giving rise to actual environmental problems as discussed earlier in this chapter. However, increasing awareness of such problems is leading to the design of machines more favourable to the environment and to changing the way in which the machines are used.

Policy and planning activities related to forest functions

Because of the long-term return of wood production, sustainable management of forests has traditionally been the primary objective of most foresters in Europe. For most of the non-wood functions of the forest, supply and demand are generally not regulated by the market mechanism, and, because forests are easily degraded and destroyed unless given appropriate protection from careless cutting, fire and grazing, most countries have established national forestry policies and legal controls on forest management.

At the national level strict laws usually apply to woodland clearance, and restocking is required after felling. These requirements are more stringently applied in montane forests, where poor management puts at risk not only the forest but

Table 23.3
Area, growing stock and net annual increment in exploitable forest
Source: Compiled by Eurostat from UNECE/FAO (see *Statistical Compendium*)

also lower-lying land which may be flooded and subject to deposition of eroded soil. Where afforestation (that is tree planting on non-forest land, which extends the area of the forest) is concerned, governments commonly determine the location and type of new planting. In addition to these blanket controls, more attention is being paid to the detail of operations conducted and the form of forest they promise to create (UK Forestry Commission, 1989). In the EU, attempts to promote tree planting by private owners on agricultural land (as a way to reduce agricultural surpluses) have not been very successful. Heavy subsidisation levels are required to attract private owners because of the long rotation periods traditionally followed in Europe (of 100 years or more).

A number of countries are also adopting specific measures to tackle issues of particular national concern. These include concerns about:

European regions	Area of exploitable forest		Growing stock on exploitable forest		Mean net annual increment on exploitable forest
	km²	(%)	million m³	(%)	m³/ha/year
Nordic	481 970	(35)	4 721	(25)	3.5
Eastern*	365 980	(27)	5 893	(31)	4.8
Southern	234 750	(17)	2 641	(14)	2.8
Western	283 590	(21)	5 702	(30)	6.7

* Excluding the Russian Federation.

- the means of ensuring soil protection (Austria, France, Germany, Italy, Iceland, Slovak Republic, Spain, Switzerland);
- reductions in the natural richness (biodiversity) of the forests (most countries, but especially the Nordic countries and Germany);
- landscape change (UK, Czech Republic and Ireland);
- the desirability of conserving wild places (Denmark and The Netherlands);
- the effects of air pollution (Germany, Czech Republic, Poland); and
- the potential for increasing indigenous wood production and supporting rural communities (Hungary, Ireland, Spain).

All countries have taken action in recent decades to limit or prohibit practices which have deleterious effects on the environment. Most of the changes made have concerned the conduct of forest operations currently used. In addition, increasing attention is being paid to the broad matter of forest design, namely the type of silviculture being adopted and the arrangement of the trees from both visual and ecological standpoints. Many countries operate strict measures through the local landuse planning authority before afforestation schemes may be approved.

At the international level, coordination of forestry policy has been going on for a long time. The International Union of Forestry Research Organisations (IUFRO) was established in 1890 with the main objective of promoting international cooperation in forestry research. More recently, in June 1993 the second Ministerial Conference on the Protection of Forests in Europe was held in Helsinki, where 45 participating countries signed-up to a set of guiding principles regarding the sustainable management of Europe's forests and conservation of their biodiversity (Ministerial Conference on the Protection of Forests in Europe, 1993b). Based on the recommendations from the Rio Conference, a list of six pan-European criteria and 27 quantitative indicators for the sustainable management of forests has recently been adopted by the signatory states.

Forestry practices

As already mentioned, most forests in Europe are the expression of past and present economic activity. But the character of this interaction – especially forest uses – is changing due to a number of forces:

- intensification of roundwood (all wood and wood-derived materials) production;
- non-productive forests abandonment;
- recreation and tourism (including hunting) expansion;
- environmental threats, such as atmospheric pollution and fires.

As forests are also part of the environment, they can be affected by broader environmental problems, such as air pollution and fires, which in turn affect their management (for example, liming and thinning). These aspects are discussed in more detail in Chapter 34.

Intensification of roundwood-production forestry practices

The exploitation of Europe's forests for the production of roundwood has traditionally been seen as one its most important functions. Indeed for a number of European countries the harvesting and production of wood-based products represents a major national industry. It accounts for more than 4 per cent of industry export earnings in Finland, Austria, Norway, Portugal, Sweden and former Yugoslavia (Dudley, 1992). In the Western and Eastern regions of Europe, the need for timber led to the conversion of coppice grown for fuelwood to high forests.

The industrial revolution strongly affected silviculture in Europe. Exotic tree species were planted to increase the production of industrial wood on poor soils, not suitable for agriculture. The sensitivity of forest ecosystems growing on poor soils has been enhanced during the last decades due to the effects of air pollutants. Recently, the shift towards modernisation of silviculture methods in some parts of Europe has brought with it a number of implications for the environment, including issues of biodiversity, landscape value, soil quality, water regulation and recreational value. Mechanisation favours infrastructure development, uniform plantations, clear felling, plantations in rows and removal of dead wood. When forests are drained or felled, landscape is greatly changed. When monoculture replaces mixed forest regimes, forests become more uniform.

Accurate data on the extent to which clear felling and replanting is practised in Europe is not generally available. However, in the Nordic regions where pulpwood and saw timber productions are important economic activities, large-scale clear felling and replanting are widely practised (Bernes, 1993). In southern Finland, the share of total forest area made up of birch woods (indigenous deciduous) fell from 15 per cent in the early 1950s to barely 7 per cent 20 years later. However, in recent years, steps have been taken to reverse this process and now almost one in every ten trees planted in Finland is a birch (Bernes, 1993). In Denmark, because of afforestation, just over half the country's total area of forest is now made up of introduced species, primarily the Norway spruce (*Picea abies*) and Sitka spruce (*Picea sitchensis*), the latter originating from North America. Of a total of some 70 000 km² of wet forest originally present in Finland at the beginning of this century, more than two-thirds (some 50 000 km²) have been drained (Bernes, 1993).

The situation in the Eastern region of Europe is very heterogeneous. In the Baltic States, forestry is still largely carried out in a way which Western European foresters regard as inefficient and outdated. Furthermore, the majority of wetlands in these countries are still undrained, thus preserving the natural diversity of species of these habitats from afforestation.

In Southern and Western regions, mainly Portugal and the UK fertilise forests, as mentioned above (Table 23.2), to increase production. The application rates are negligible in comparison with those of agriculture and are not likely to pose a serious threat to the environment.

In Europe as a whole, mono-specific plantations are often used for afforestation, and geographic transfers of forest reproductive materials are frequent. A few highly productive species, each represented by a few improved varieties, are commonly used for reforestation. Consequently, local races and ecotypes can disappear or hybridise with the introduced forest reproductive materials. Additional threats come from infrastructure developments (such as urbanisation, dams, roads and recreation areas) and changes of the atmosphere and climate.

Recreation

The importance of forests as a source of recreation has gained increasing attention in Europe over the last two decades. Indeed, forests provide many of the non-material benefits needed by people, especially those living with the stress of a modern industrialised economy. In general, however, comprehensive data on visits to forests are rare and compiled only at long intervals. Few data on possible indicators, such as visitor frequency to forests, or what proportion of forests are managed deliberately for recreation, have been collected. In the Nordic countries, where people have the right to visit forests regardless of who owns them, it is estimated that at least 400 million visits a year are made to their forests (Bernes, 1993).

The impacts of recreational activities act to both the benefit and the detriment of forests. In general they tend to favour, and thus preserve, forests with a high level of biodiversity (for instance mixed woodlands and semi-natural forests), encouraging the conservation and protection of

Beech coppice and bluebells, Forest of Halles, Belgium
Source: Mark Roekaerts

wildlife, and helping to limit the loss of forest to urban spread and road building. However, if visitor pressure becomes excessive, problems of soil erosion can occur along and adjacent to footpaths, wildlife can be disturbed (especially important during the breeding seasons), excessive damage to plants and tree saplings can occur from trampling underfoot and discarded waste can become a problem (see Chapter 25).

Hunting

Hunting is a popular activity in a number of European countries (Chapter 25) and much of it is practised in forests or on land adjacent to forests as these provide the breeding, roosting and feeding grounds of much of the traditional game. The use and management of forests for hunting, however, has a number of implications for the environment: small woodlands (not more than 10 hectares) and shrublands are favoured; and buffer zones of spontaneous vegetation not spread with chemicals of at least 6 metres wide are maintained between forests and crops (in the UK).

Hunting can provide an incentive – often in the form of financial rewards derived from the fees paid for access rights by hunters – for forest owners to maintain mixed species forest and to resist pressures to allow the forest to be cleared or subject to intensive monoculture wood-production practices. However, it can also restrict access to the forest to other visitors, thus reducing its recreational and amenity value, and can lead to serious disturbances of other wildlife in the forest, and to tree damage, especially in young plantations from excessive game density and shooting. In Scotland for instance, red deer (*Cervus elaphus*) is considered as the main threat to the remains of the Caledonian forest (Pearce, 1993). The number of red deer has grown continuously over the past 200 years, encouraged by landowners wanting to attract revenue from stalkers, and because of the disappearance of the deer's natural predators.

As with many of the other non-wood forest functions, few data currently exist to help assess the level of importance attached to hunting and the extent to which it is practised in the different regions of Europe. A questionnaire-based survey conducted by UNECE and FAO (UNECE/FAO, 1992b) in 1990 found that hunting was listed as the second most important forestry function in Europe by the national forestry authorities. Countries of Eastern (Hungary, Poland), Southern (Italy, Spain) and Western regions of Europe (France, Germany and Switzerland) attach a great deal of cultural importance to hunting, and for many small-scale private forest owners hunting represents an important, if not the biggest, source of income generated by their forests. In France and the UK (Scotland), hunting represents an important contribution to the local economy of some regions (Pinet, 1987).

Grazing

Historically, forests have long provided good grazing grounds for a number of domesticated animals including goats, pigs and deer. In all countries, however, the general trend over recent decades has been to abandon this practice. Nowadays,

many European countries attach only a low importance to it (UNECE/FAO, 1992b) and many have laws to regulate or prohibit grazing in forests in order to protect them. However, in some countries, including Albania, Spain and Romania, grazing is important in more than 20 per cent of the forests (in Finland up to 80 per cent of the Lapland state-owned forest is grazed by reindeer).

The implications for the environment of grazing in forests are varied. A certain level of grazing by deer, wild boar, bears, etc, was always a natural element in the forest ecosystem. The practice of forest grazing has led to the decline of browsing-sensitive species and has favoured light-demanding and fruit-bearing tree species like oak to feed the animals. In Southern Europe, it was also used as a means to control undergrowth, therefore reducing fire risk. But grazing, if not well controlled, can also cause serious damage to trees, prevent the growth of new tree sprouts and lead to the compaction and erosion of soil. Such problems have occurred in areas of Greece and Bulgaria because of unrestrained grazing of domestic animals, particularly goats and sheep.

Conservation

Forests are an important reservoir of biodiversity and thus play a central role in the conservation of Europe's wildlife heritage. The growing awareness of conservation issues in European society, coupled with a realisation that much of Europe's virgin and semi-natural forest has already been lost, has encouraged many countries to protect part of their forests by creating national parks (see Chapter 9) and protected areas. In addition, specific regulations have been introduced to save trees located in wetlands and around lakes where rare birds and other animals breed. Such areas are typically managed with minimum interference and careful control on visitor/tourist pressure, including limited access to sensitive areas of the forest.

The protection and management of forests for conservation purposes is not necessarily incompatible with exploiting other forest functions such as wood production, provided the latter is achieved using low impact techniques such as selective cutting, minimum or careful use of heavy machinery, and no chemical inputs.

Protection of soil and water resources

In regions with a fragile soil, the forest often plays a vital protective role. This is particularly the case with soil that is sensitive to:

- wind erosion, as in lowland countries (Denmark, UK and Hungary) where forest is often used as wind protection (windbreaks); or
- water erosion, as in Southern European countries, such as Cyprus, Greece and Spain, where protection against erosion and flooding is one of the priority roles of the

Pine forest of Les Landes, France
Source: C Steenmens

forest – indeed, soil erosion in Mediterranean forest catchments varies from 2 to 3 t/ha/year, while on bare soils it is 20 to 30 t/ha/year (Marchand, 1990).

Forests also contribute to protection in mountain areas against natural hazards such as avalanches, falling stones and floods. In France, a study of the mountain forests in the Alps (400 000 ha) has shown that stands are getting old and damage through wind throw is increasing.

Forests have a major effect on water flow in catchment areas. The improved infiltration of water into the forest soil helps to replenish the groundwater. The quality of the water provided by a forest catchment area is generally excellent since erosion in the area is usually low. However, water with a low pH may come from forests which themselves have a litter with an acidic humus and poor decomposition characteristics. Forests also evaporate and transpire a great deal; any disadvantages of this greater water consumption by trees are usually offset by the benefits provided by forests, but in Southern Europe forests can create conflicts with demand for water resources (the problem of trees taking too much water and parching the land is particularly well known with eucalyptus).

Other products

Forests also provide a number of products other than wood. Such products include berries of different kinds, pine nuts, mushrooms, cork, game meat, honey and aromatic plants as well as Christmas trees, greenery for decorative purposes, aromatic and medicinal plants, resin and fodder. The importance and extent of harvesting of these various products in the different regions of Europe is difficult to assess because of lack of data. Nevertheless, for certain parts of Europe the significance of this forest function, at a local level, should not be overlooked. In Portugal the harvesting of oak bark to produce cork is a significant industry and provides Europe with 60 per cent of its cork. The harvesting of berries to sell as fresh produce or jams and other similar products supports a small cottage industry in many parts of Scandinavia, as does the harvesting of wild mushrooms in France. Although the collection of non-wood products will never constitute a major income, since labour costs are high, it can be important locally.

SUMMARY AND CONCLUSIONS

- Forests provide humans with a wide variety of products and services and are an integral part of the natural environment. As a result, much of Europe's forest area is now managed, in some way, and is subject to various human practices and activities. Certain of these activities, particularly those relating to the production of wood, have the potential to affect the environment adversely. Forests themselves are affected by human activities, gene pollution, disease, air pollution, hunting, conservation, grazing animals and fires (see Chapter 34).
- Today, forests cover about 312 million hectares, that is 33 per cent of Europe's land. The coverage varies from country to country, ranging from 6 per cent in Ireland to 66 per cent in Finland. On average, half of the forest in Europe is privately owned, but there are also wide variations between countries.
- Recent trends indicate that there has been a 10 per cent increase of forest cover in Europe, in both area and volume, over the last 30 years. The increase in wooded areas is due to reforestation policies and spontaneous forest growth in marginal areas, helping to reverse the trend in the loss of forest area which occurred between the post-glacial and medieval times.

- Forest composition has changed under human influence, with about 60 per cent of forest being coniferous, and has led to a more homogeneous and artificial forest.
- Forest productivity has steadily increased since 1950 but there has been no such increase in the amount of wood being felled.
- Some areas of Europe, particularly in the Nordic and Western regions, have witnessed the intensification of wood-producing techniques and practices over the last few decades. This has involved to some extent the mechanisation of the planting and harvesting of trees, the draining of 'wet' forests, the use of fast-growing tree species (particularly conifers but also broadleaved species such as poplars) planted in monoculture crops, to be clear felled at regular intervals, and, in a few countries, the use of fertilisers and pesticides. Locally, the implementation of such practices has a number of environmental impacts particularly in relation to biodiversity, habitat destruction and amenity or landscape value.
- The process of intensification has led to greatly improved wood productivity which has allowed Europe to meet much of its increasing demand for wood through the sustainable harvesting of its forests. Furthermore, contrary to popular belief, conifers are not replacing broadleaved forests on a widespread basis, although this is an issue in some parts of Europe including Belgium, the UK, Denmark, Ireland and Bulgaria. In fact the total area of broadleaved forest has increased. More of an issue is the replacement of native, natural forests with monoculture plantations – a concern in some parts of the Nordic countries and some Central European countries.
- The production of wood using intensive practices and techniques is generally limited to those areas of Europe where large state and private enterprises manage big forest holdings. This includes parts of the Nordic countries, parts of the former USSR, Romania, Poland, Germany, France and Austria. However, a significant proportion of Europe's forests, particularly in the West, belongs to numerous farms and private owners who hold relatively small areas of forest (less than 50 ha) which are not necessarily managed for wood production, and, in some cases, are not managed at all.
- Much of Southern Europe has witnessed the abandonment of its forestry practices. As traditional uses such as grazing, coppicing for firewood, and the collection of litter, bark, resin and tannin have diminished, and rural depopulation has occurred, many forests have become neglected or abandoned, increasing fire risk. Furthermore, as cheap wood imports from the more productive regions of Europe or elsewhere have become easily available, the management of forests for wood production has also declined. However, forest management for soil protection is still one of the highest priorities in Southern countries.
- There has been increasing international coordination of forestry policy in recent years, including a set of resolutions aimed at ensuring the sustainable management of Europe's forest resources and the protection of its biodiversity at the Strasbourg Conference in 1990 and at the Helsinki Conference in June 1993. At the national level, forest policy has traditionally focused on issues relating to the management of forests for wood production. However, the growing awareness by society of forests' other functions and values – particularly for conservation and recreation, soil protection and water regulation – is forcing national forest policy makers to resolve the often conflicting demands and requirements of a multiple-function forest.

REFERENCES

Adlard, P (1987) *Review of the ecological effects of eucalyptus.* Report prepared for Wiggins Teape at the Oxford Forestry Institute, Oxford.

Becker, M and Serre-Bachet, F (1992) Modification de la productivité des peuplements forestiers. In: Landmann, G (Ed) *Les recherches en France sur les écosystèmes forestiers. Actualités et perspectives,* pp 93–4. Ministère de l'agriculture et de la forêt. Paris.

Bernes, C (Ed) (1993) The Nordic Environment – present state. *Nord 1993: 12.* Nordic Council of Ministers, Copenhagen.

Cannell, MGR, Dewar, R C and Thornley, J H M (1992) Carbon flux and storage in European forests. In: Teller, A, Mathy, P and Jeffers, J (Eds) *Responses of Forest Ecosystems to Environmental Changes,* pp 256–71. Elsevier Applied Science, London. Commission of the European Communities, Brussels, EUR 13902.

Catarino, F M (1992) Strategy of maintaining biodiversity in Mediterranean areas. In: McGloughlin, P and Schmitz, B (Eds) *Biological diversity: a Challenge to Science, the Economy and Society,* pp 401–16. Poplar sc, Brussels.

CEC/UNECE (1993) *Forest condition in Europe.* Report of the International Co-operative Programme on Assessment and Monitoring of Air Pollution Effects on Forests. Results of the 1993 survey. Commission of the European Communities, DG VI FII.2, Brussels.

Delcourt, P A and Delcourt, H R (1987) Long-term forest dynamics of the temperate zone. *Ecological Studies* **63**, Chapter 10.

Dudley, N (1992) *Forests in trouble: a review of the status of temperate forests worldwide.* WWF International (World Wide Fund For Nature), Gland, Switzerland.

EFMA (1992) *Agriculture and fertilizer consumption in EFMA countries.* European Fertilizer Manufacturers Association yearly review, Brussels.

FAO (1992) Forest resources assessment 1990, tropical countries. Food and Agriculture Organisation, Forestry paper, 61 pp.

IPCC (1992) *Climate change 1992.* Intergovernmental Panel on Climate Change. Cambridge University Press, Cambridge.

Joffre, R (1992) The dehesa: does this complex ecological system have a future? In: Teller, A, Mathy, P and Jeffers, J (Eds) *Responses of Forest Ecosystems to Environmental Changes,* pp 381–8. Elsevier Applied Science, London and New York.

Kauppi, P E, Mielikainen, K and Kuusela, K (1992). Biomass and carbon budget of European forest, 1971 to 1990. *Science* **256**, 70–4.

Kauppi, P E and Tomppo, E (in press) Impact of forests on net national emissions of carbon dioxide in west Europe. *Water, Air and Soil Pollution.*

Kuusela, K (1993) Trends of European forest resources on the basis of the FAO ECE Timber Committee resource assessments. *Silva Fennica.* Finland **27**(2), 159–66.

MacKenzie, D (1994) Where has all the carbon gone? *New Scientist,* **19,** 8 Jan, 30–3.

Marchand, H (1990) Les forêts méditerranéennes: enjeux et perspectives. *Les fascicules du Plan Bleu 2.* PNUE – CAR/PB 1990 Economica, Paris.

Ministerial Conference on the Protection of Forests in Europe (1993a) *Report on the follow-up of the Strasbourg Resolutions.* Helsinki, 16–17 June 1993. Ministry of Agriculture and Forestry, Helsinki.

Ministerial Conference on the Protection of Forests in Europe (1993) *Sound Forestry – Sustainable Development.* Helsinki, 16–17 June. Ministry of Agriculture and Forestry, Helsinki.

Myers, N (1993) Taiga taiga burning bright. *Guardian,* 5 Aug.

Pearce, F (1993) Last chances for the wasting wilderness? *New Scientist* **18** Sept, 30–3.

Pinet, J- M (1987) *L'économie de la Chasse.* Comité National d'Information Chasse-Nature, Paris.

Schnitzler-Lenoble, A and Carbiener, R (1993). Les forêts galeries d'Europe. *La Recherche* **255**, June, volume 24, 694–701.

UK Forestry Commission (1989) *Land capability for forestry.* South-West Scotland. Field Book 6. Produced by the Macaulay Land Use Research Institute, Aberdeen.

UNECE/FAO (1990) Survey of medium-term trends for wood raw material, notably pulpwood, and wood for energy. *Timber Bulletin,* XLII (2). United Nations, New York.

UNECE/FAO (1992a) *The Forest Resources of Temperate Zones: The UN-ECE/FAO 1990 Forest Resource Assessment,* Vol 1. United Nations, New York.

UNECE/FAO (1992b) *The Forest Resources of Temperate Zones: The UN-ECE/FAO 1990 Forest Resource Assessment,* Vol 2. United Nations, New York.

UNECE/FAO (1992c) *The Forest Resources of the Temperate Zones, main findings of the UN-ECE/FAO 1990 Forest Resource Assessment,* ECE/TIM/60. United Nations, New York.

World Resources Institute (1992) Tropical deforestation. In: *The World Resources Institute Report 1992–1993.* Oxford University Press, Oxford.

*Fishing in the
North Sea*
Source: Frank Spooner
Pictures

24 Fishing
and aquaculture

INTRODUCTION

The interrelationships between fishing and marine and
freshwater ecosystems are complex and often not well
understood. This lack of understanding is exacerbated by the
sheer complexity of food-chains in these ecosystems.
Most concern has been focused on three broad issues:

1   the issue of unsustainable exploitation of certain fish
    stocks;
2   the impacts of fishing activities on non-target species,
    particularly marine mammals, seabirds and benthic
    (bottom living) organisms; and
3   the effect of emissions from aquaculture on marine and
    freshwater ecosystems.

The interrelationships are complicated by the fact that fishing
activities both have an impact on and are impacted by the
environment.

This chapter reviews the nature and relative importance
of the impacts of fishing and aquaculture on marine and
freshwater ecosystems. Those which are of greatest concern
are then examined in greater detail, with an assessment of
environmental impacts and underlying driving forces. The
chapter is concluded with an evaluation of the effectiveness of
existing fishing policies.

Box 24A provides definitions of fishing statistics used in
this chapter and summarises the availability and quality of
fishing data.

AN OVERVIEW OF FISHERIES RESOURCES IN EUROPE

Marine provinces

The major European marine provinces are the continental
shelf of the northeast Atlantic and the Mediterranean basin,
in which several distinct biogeographical regions can be
identified (see Chapters 5 and 6). The northeast Atlantic is
one of the richest areas in the world, accounting for
approximately 10 per cent of the world's marine fish catch.
This is the result of a combination of factors including the
influence of the warm waters of the Gulf Stream and the
extensive continental shelf. The Mediterranean basin has a
long coastline, but the continental shelf is generally narrow.
This reduces the productivity of the area with the exception
of the northern Adriatic. The eastern Mediterranean is also
noteworthy in that the diversity of fish is increasing because
of fish migration through the Suez canal from the Red Sea
(Caddy, 1993).

Attempts have been made to collect existing maps into
atlases showing the pan-European extent of the major fish
species, but these are unfortunately not always up to date (eg,
Couper and Smith, 1989; FAO, 1981). The best coverage is
currently for the North Sea (ICES, 1993b). Presented here
instead in Map 24.1 are the total nominal catches of fish and
invertebrates (including crustaceans and molluscs) by
administrative zones.

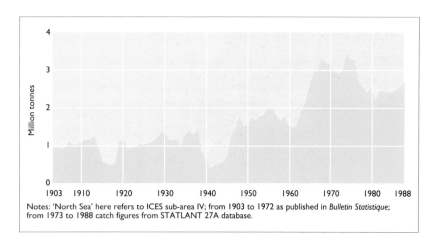

Notes: 'North Sea' here refers to ICES sub-area IV; from 1903 to 1972 as published in *Bulletin Statistique*; from 1973 to 1988 catch figures from STATLANT 27A database.

Figure 24.1 *Total international fish catch in the North Sea, 1903–88*
Source: ICES (1992b and personal communication)

Figure 24.1 shows total catches for the North Sea over this century. Total catches increased from 1 million tonnes just after 1900 to 2 million around 1960. During the 1960s, the catch increased steeply to almost 4 million tonnes, followed by a gradual decline to around 2.5 million in recent years.

In the northeast Atlantic and the Mediterranean, nominal catches showed a remarkable increase between 1960 and 1980. Since then there has been some decline (see *Statistical Compendium*). Until recently, the FAO used to produce estimates of the maximum sustainable yield (MSY) for its statistical areas. However, the FAO has recently decided that in the case of marine fisheries this is now a misleading concept, as fish resources are so severely depleted that the issue is now stock reconstruction (FAO, 1992a).

The risks of overexploitation are partly due to the common property status of fisheries resources and of the world need for protein. However, the potential for fisheries to meet the needs for increased world food supply is limited: fisheries currently account for only about 16 per cent of the world's total animal protein for human consumption (FAO, 1991). Further, although global harvests are presently below potential yields, many of the more valuable fish stocks are overfished, and the steady trend towards increased global harvests is partly through exploitation of new and less valuable species and of less accessible species which are costly to exploit.

Trends in fishing pressure vary between Western and Central and Eastern European countries. In Western Europe, rapid increases in production throughout the 1960s were stemmed after the establishment of Exclusive Economic Zones (EEZs) in 1977 (see Box 24B). Since then, nominal catches have remained more steady. For Eastern countries, catches have decreased steadily since the 1970s, following a rapid surge in the 1960s. This is due largely to their exclusion from waters of EU countries, through the establishment of

Map 24.1 *Total nominal catches of fish and invertebrates in European marine provinces, 1988*
Source: EEA-TF using FAO and ICES data

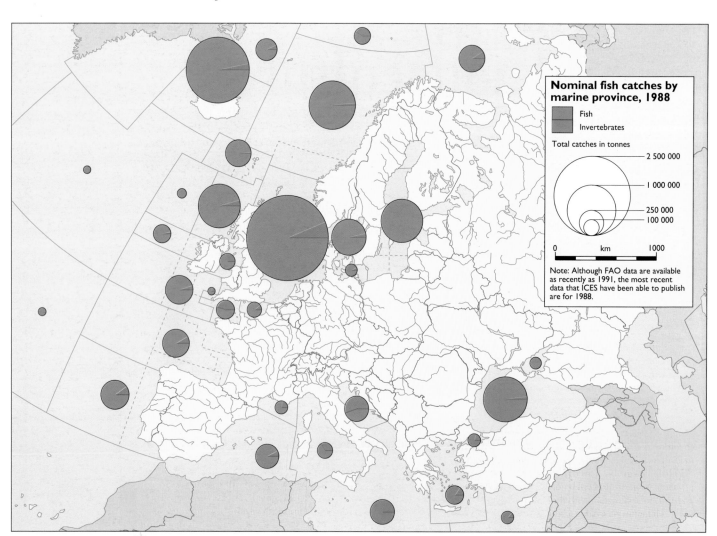

Box 24A Definitions, data availability and quality

The Food and Agricultural Organisation of the United Nations (FAO) explains clearly the various definitions used in fishing (FAO, 1993a). The gross catch is 'the total live weight of fish caught'. The concept of 'landings' refers to the 'quantities landed on a landed weight basis' (ie, what is actually recorded at shore). From this are derived the 'nominal catches' – 'the live weight equivalent of the landings' (ie, landings of each species adjusted for on-board processing such as gutting, filleting and drying). The data on nominal catches may or may not be the same as landings. Every year the FAO publishes, with a two-year delay, data on nominal catches of freshwater and marine animals (mainly fish species), by species, fishing area and country (FAO, 1993a). These data normally relate to commercial fishing rather than recreational fisheries. Eurostat also publishes a yearly report on fishing statistics (eg, CEC, 1992).

There are four broad groups of species which are of economic importance for fishing:

- pelagic species: these are fish living in the water column; pelagic species are normally split into two groups: small (eg, anchovies and mackerel) and large (eg, tuna and swordfish);
- benthic or demersal: fish living near the sea floor (eg, crustaceans, molluscs, flatfish, cod-like species);
- diadromous: these are species which can live in freshwater and marine environments (eg, sturgeon, salmon);
- freshwater species: those living in rivers, lakes and other freshwater bodies (eg, carp species).

For the marine environment, the FAO analyses the world's seas by statistical areas. In the European area these are the northeast Atlantic (area 27) and the Mediterranean and Black Sea (area 37). The International Council for the Exploration of the Sea (ICES), through its Cooperative Research Report series and Statistical Bulletins, provides more detailed data for the northeast Atlantic on fish catches and assessment of fish stocks (eg, ICES, 1992a; 1993a). ICES, for example, holds time-series data for the North Sea from the beginning of the century onwards (Figure 24.1). For the purposes of this report the principal sources of data for the marine environment are FAO and ICES. Data on inland fishing are available from the FAO for Europe, for two zones: Europe (area 04) and part of the former USSR (area 07) (FAO, 1992b). Aquaculture production is often included in marine and inland catch data.

The increasing adoption of production quotas as a means of managing fish stocks is a contributory factor in the growing decline in the quality of data on fish catches submitted to FAO (UNEP, 1989). Further, total catch data

are not reliable due to unreported catches or to the misreporting of fish species: for example, herring (*Clupea harengus*) are often declared as sprat (*Sprattus sprattus*), and mackerel (*Scomber* spp) as horse mackerel (*Trachurus mediterraneus*). The data available are also often not environmentally relevant, referring to administrative units rather than natural areas, and they monitor landings of commercially exploited species, which do not always include all species of importance in the food-chain.

While data are available on fish catches, data on fish stocks are harder to come by. Stock assessment involves the estimation of a variety of population parameters, in particular mortality due to fishing (from fish catches, accidental kills during fishing and natural mortality), numbers of fish in different age classes (differences between years reflecting rates of recruitment) and spawning stock biomass (SSB). Such analytical assessments are hindered by the reliability of the catch data and a lack of knowledge of the biology of the species concerned. Stock data are only meaningful in the context of historical trends and assessment of the long-term potential of stocks. Thus, in the long term, the SSB and recruitment are more useful indicators of a given species than the total stock. Stock estimation is further complicated by the fact that fishing targeted at one stock has an impact on other stocks in the same area, and ecological interactions between fish and shellfish stocks (ICES, 1992b, annex 3). Stock assessments from ICES are considered in the conservation measures adopted in policies (eg the EC Common Fisheries Policy) and by regional fisheries commissions in the northeast Atlantic area. The lack of reliable fishing statistics (stocks and catches) for the Mediterranean is a serious handicap in its management. Modern techniques for stock assessment are based on acoustic methods, and remote sensing may be particularly useful in the Mediterranean.

Few data are available on aspects such as trends in fishing techniques, employment levels and productivity. Those which are available are not particularly reliable. For example, productivity (measured, for example, as landings/fisherman/year) can be a misleading indicator because, although the total catch of a fleet may be increasing, productivity and catch per unit effort (CPUE) may be decreasing as fish stocks decline because of overexploitation. This is attributable partly to the lack of international standardisation on data collection and parameter definition (Cross, 1989).

EEZs (and the EC Common Fisheries Policy). Other possible factors include ageing fleets, polluted waters and more latterly the difficulties of embracing free market principles in the fishing industry.

Inland water provinces

Inland water systems, due to their geographic discontinuity, often develop specific fish populations. The most important freshwater systems of Europe in terms of fish resources are the Danube basin, the Volga basin and the ancient lakes of South Europe, such as Ohrid on the Yugoslav–Albanian border and Lake Balaton in Hungary (which alone harbours 512 species of fish). The numerous small lakes and ponds in areas such as Finland and the Mazurian system in Poland are also important habitats for fish, particularly carp (*Cyprinus* spp), which are the basis of traditional pond culture (Hollis and Jones, 1991).

ENVIRONMENTAL IMPACTS OF FISHING

Fishing activities have impacts on, and are impacted by, the environment in a number of ways (see Table 24.1). These can be classified loosely as direct and indirect impacts (ICES, 1992b). The most obvious direct impacts of fisheries arise from the commercial harvesting of fish at sea, impacts on non-target species and impacts arising from aquaculture. The indirect effects of fishing are far less easy to describe, but would include impacts on predator or prey species and disruption of ecological equilibria (Northridge, 1991).

However, it is often difficult to isolate the specific impacts of fishing alone, as the marine environment is also impacted by other human activities, as well as by natural fluctuations of, for example, ocean currents and climate. Further, the lack of knowledge about the impact of fishing on

Figure 24.2
*Norwegian spring-
spawning herring
(Clupea harengus),
1952–92*
Source: ICES 1993a

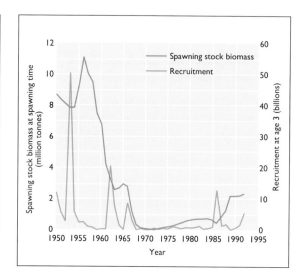

the environment, exacerbated by the sheer complexity of
marine ecosystem food-chains (Chapter 6), means that
cause–effect relationships between fishing and environmental
impact are not well defined.

Impacts associated with harvesting of fish at sea

Ecosystem effects of fishing may occur at all scales of space
and time. Changes in ecosystems may be described and
quantified in a variety of ways. In this chapter, the practice
used by the International Council for the Exploration of the
Sea (ICES) is followed by describing changes in the
abundance of individual stocks.

Fisheries exploit species against the background of a
variable natural environment. Water temperature is an

important influence on fish stocks. A temperature rise or fall
of one degree can have significant impacts on fish migration
and reproduction patterns. Water currents can have a major
effect by changing the drift pattern of fish eggs and larvae,
and hence their prospects for survival. The currents can also
have an impact on water temperature and salinity, which in
turn affect fish populations. Ecological factors such as
predation and food supply can influence fish populations,
which are known to fluctuate greatly in the long and medium
term (Gulland, 1983). Fish stocks can also be prone to
outbreaks of disease which can affect populations.

Anthropogenic pressures on the marine environment can
come from several sources, only some of which are directly or
indirectly related to fishing. It can, therefore, be difficult to
separate the long-term effects of fishing from changes due to
other human activities.

Being renewable resources, fisheries and aquaculture can
provide benefits to humans indefinitely, provided that they
are properly managed on a sustainable basis. However, due to
the commercial importance of fisheries resources and the
limited regulation of catches in the past, many of the marine
fish species of Europe are, or have been, overexploited
(Gulland, 1983). This is reflected in the estimates of fish
stocks for the northeast Atlantic (ICES, 1993a).

A classic example is the Atlantic herring (*Clupea
harengus*), which spawns in the spring in the Norwegian Sea.
The herring used to migrate away from these waters after
breeding, especially to the sea areas to the north and east of
Iceland. In the 1960s herring were fished so intensively in
Norwegian and Icelandic waters that the population was
virtually eradicated. In 1970, the fishery was almost totally
abandoned as it was no longer producing catches.
Recruitment remained at a low level until 1986, when a good
three-year age class was recorded (ie, spawned in 1983) (see
Figure 24.2). The reason for the substantial increase in the
spawning stock in 1988 was because the greater part of this

Table 24.1
*Overview of
significant impacts on
the environment
arising from fishing*

Fishing activity	Resources	Water	Nature and Wildlife
		Environmental media	
Marine fishing	● overexploitation ➔ depletion of certain fish stocks, reductions in genetic diversity, impacts on natural ecological dynamics	● processing plants and factory ships ➔ water pollution ● flushing of holds ➔ water pollution (including oil) ● anti-fouling preparations, such as tributyl tin from hulls of boats (now banned in EU countries) ➔ water pollution ● damage to oil/gas pipelines from trawling ➔ risk of rupture and threat to ecology ● discards of unwanted fish and offal ➔ water pollution and odour	● beam and otter trawls and dredges ➔ impacts on benthic organisms ● gill, drift nets and longlines ➔ impacts on vertebrates ● discarded fishing gear at sea ➔ impacts on birds and marine mammals
Freshwater fishing	● overexploitation ➔ depletion of certain fish stocks, reductions in genetic diversity, impacts on natural ecological dynamics (removal of top predators from freshwater bodies)		● lead from recreational fishing ➔ toxic effects on waterfowl ● discarded commercial fishing gear and anglers' tackle ➔ threat to waterfowl and aquatic mammals
Aquaculture		● discharge of surplus food and animal wastes from fish farms ➔ eutrophication ● use of anti-fouling paint on aquaculture facilities ➔ water pollution	● escape of farmed fish, exotic and transgenic fish ➔ impacts on gene pool of wild populations, increasing competition for same ecological niches ● increased demand for fishmeal (used for food in aquaculture) ● physical barriers established for making fish farms ➔ potential to hinder movement of migratory fish species

year class spawned for the first time in 1988. Fishing of herring was resumed on a small-scale basis in 1984. The estimated spawning stock of 2 million tonnes in 1992 is about three times as large as the spawning stock in 1987 (Central Bureau of Statistics of Norway, 1993). However, the catches are only 15 per cent of those recorded in the 1950s (Bernes, 1993). Nevertheless, this shows that even a severely overexploited fishery such as this one can restore its numbers over time.

In the Mediterranean, stock dynamics are less well understood due mainly to a lack of data, the large number of landing sites used by small vessels (which makes record-keeping difficult), and a wide diversity of species. Nearly all commercial stocks in the Mediterranean basins are considered fully- or overexploited, with the possible exception of lower-value small pelagics such as bogue (*Boops boops*) and horse mackerel (*Trachurus mediterraneus*) (FAO, 1993b). Catches of pelagic fish, such as swordfish (*Xiphias gladius*), bluefin tuna (*Thunnus thynnus*) and bonito (*Sarda sarda*), have all increased in the Mediterranean in recent years. The mean size of swordfish has declined, however, and there is concern that all stocks are being exploited too heavily. Despite increased

---

Box 24B Existing fishing policies and further options

The principal regulatory elements of fishing policies have been the establishment of Exclusive Economic Zones (EEZs) and fixing Total Allowable Catches (TACs). EEZs were established through international law during the 1970s to give countries the right to exclusive fishing rights within 200 nautical miles of their coasts. Within the European area, EEZs exist for the FAO northeast Atlantic area, but not for the Mediterranean (because countries are too close together and because of the geopolitical difficulties). TACs are determined for a commercial species on the basis of scientific and historical considerations. For each species, a quota is fixed for each country. In practice TACs and quotas are difficult to enforce, particularly in Europe, where the need to reduce TACs to rebuild stocks over the long term is opposed by the pressure to increase catches to maintain the profitability of fishing industries in the short term.

The major fisheries policies are listed below. It is also important to remember the relevance of more general policies such as the UNEP Mediterranean Action Plan, which considers fishing along with other human activities.

The European Union: the Common Fisheries Policy (CFP) began to be implemented in 1970 (CEC, 1991a). The two main measures used in the CFP to limit catches are the fixing of TACs and technical conservation measures to protect young fish, such as stipulating minimum mesh sizes for nets, minimum sizes of landed fish and levels of by-catch, and by restricting fishing activities in certain zones. Both have proved difficult to monitor and to enforce. As a result the main thrust of the CFP now consists of five-yearly multi-annual guidance programmes which establish fleet sizes for all Member States (in terms of tonnage and engine size), with the objective of reducing the fleet sizes by 2002 to a capacity which would not permit them to exceed optimum levels of exploitation on the stocks. Recent criticism has been levelled at the CFP for failing to reduce the size of its fishing fleet, and suggestions have been made that the efficiency of fishing vessels be taken into account when determining fleet sizes.

Other Western European countries: TACs, implemented in one form or another, are used for regulating fisheries elsewhere, including the Baltic Sea, Iceland, Norway and Greenland. In most cases TACs are set as part of multinational agreements, except in the case of Iceland which has virtual autonomy over exploitation of its fish resources and is relatively successful in enforcing legislation. Technical regulations are also enforced in most of these countries and individual transferable quotas (eg, by fisherman, or by boat) were recently implemented in Iceland.

Central and Eastern Europe: the fisheries policies of countries such as Poland, the former USSR and to a lesser extent East Germany were based upon the twin aims of supplying as much fish protein as possible to their populations (much of it coming from non-European waters) and to earn as much foreign exchange as possible. Few of these policies contained measures aimed at conserving fish stocks. The former USSR used TACs in order to manage its fisheries. Since independence, some of the former Soviet republics have implemented changes in fisheries policies. Estonia's reformed Fishing Law in 1991 and 1992 (Kangur, 1993) abolished limits on netting and catches except for internationally regulated species. The effect of such policies on the fisheries of the Baltic, which are generally declining, remains to be seen. It is difficult to assess the effectiveness of policies in Central and Eastern Europe due to lack of data. The enclosed Adriatic Sea is endangered by overfishing, in part because there is no agreed common policy between Italy and Croatia.

Regional policies: there are independent regional bodies in place in the northeast Atlantic (Northeast Atlantic Fisheries Commission (NEAFC)) and the Baltic (International Baltic Sea Fishery Commission (IBSFC)). Individual species stocks in these areas are managed by a series of TACs, set annually following scientific evidence from ICES. In the Mediterranean, the FAO has set up the General Fisheries Council for the Mediterranean (GFCM) and, in the Black Sea, the Mixed Commission for Black sea Fisheries (MCBSF). NEAFC has a regulatory role, but only in international waters as a result of the extension of national fishery limits to 200 miles in 1977. The Mediterranean and Black seas are almost totally unregulated except for national legislation, which applies only within fishery limits extending to a maximum of 12 miles in European waters (the so-called Territorial Sea), though less in certain countries (Greece's limit extends only six miles). Despite the CFP, the system of TACs does not currently apply to the Mediterranean. The Black Sea fisheries are exploited predominantly by the former USSR and Turkey. There are also regional commissions that focus on the management of certain species such as the International Commission for the Conservation of Atlantic Tunas (ICCAT).

Many fishery scientists now favour a switch to control of fishing effort, rather than fixing quotas. Possible additional measures include: limiting the number of boats in a fishery (through licensing), regulating the time spent at sea; reducing the number of nets per boat; use of area or seasonal closures or restrictions for certain shellfish and fishery areas. Sea ranching may lead to stock enhancement of threatened species. Encouraging small-scale local fishing activities rather than large-scale fishing may help reduce pressure on stocks. In the EU, more than 45 per cent of the fishing industry is employed in small-scale fisheries, particularly in the Mediterranean area (European Parliament, 1992b). The International Union for the Conservation of Nature and Natural Resources (IUCN) has recognised that small-scale, community-based fisheries account for almost half the world fish catch for human consumption, employ more than 95 per cent of people in fisheries, and use only 10 per cent of the energy of large-scale corporate fisheries (IUCN et al 1991).

catches in recent years, bluefin tuna populations in the east Atlantic and Mediterranean have diminished by more than 70 per cent since the 1970s (Lleonart, 1993). Among the small pelagics, anchovy (*Engraulis encrasicolus*) stocks and catches have declined, whereas the less valuable stocks of sardines (*Sardina pilchardus*) have increased. Hake (*Merluccius* spp) catches in the western Mediterranean have declined with decreasing recruitment.

However, in the Mediterranean, despite high exploitation rates, the overall landings for all species continued to rise in the 1970s and 1980s. It would normally be expected that landings would decline as effort further increased and stocks were depleted. Increases in fishing production in the 1970s and 1980s were particularly evident in the semi-enclosed basins such as the Black Sea and the Adriatic, where nutrient-enrichment is more pronounced, than in the more nutrient-poor regions of the eastern Mediterranean (FAO, 1992a). Total catches of small pelagics in the Black Sea are estimated to have risen steadily from a plateau of around 350 000 tonnes in the late 1970s to some 700 000 tonnes in the 1980s, before collapsing to a provisional estimate of close to 100 000 tonnes in 1991. This decline is related to the effect of mesotrophic conditions on food webs in enclosed and semi-enclosed seas (Caddy, 1993). In 1970 there were over 26 species of fish in the Black Sea that could be commercially exploited (Gulland, 1983). By 1990 this had dropped to five, principally as a result of overfishing and habitat degradation. The estuaries on the Black Sea have also undergone a decline in fisheries potential. For example, at the beginning of the century, over 1000 tonnes of sturgeon (*Acipenseridae*) were caught annually in the Romanian part of the Danube delta, whereas only 20 tonnes were landed in 1989 (Crivelli and Labat, 1992). The anchovy fishery has similarly collapsed in the Black Sea.

Significant declines have been noticed in catches of commercial fish species in the Caspian Sea (eg, sturgeon (*Huso huso*) and pike (*Esox lucius*)). This decrease has been attributed in part to overfishing, though habitat degradation is a contributing factor. The total fish catch from the Caspian Sea was 300 000 tonnes per year in the mid-1930s; this had dropped to 100 000 tonnes by 1970. Catches of sturgeon in the Caspian sea area of the former USSR fell from 10 830 tonnes in 1907 to only 850 tonnes in 1953 (Sokolov and Berdichevskii, 1989).

Overexploitation does not normally lead to the extinction of the species. The economics of fishing do not generally allow for a complete elimination of a target species. The only documented case of a marine fish population to have apparently disappeared in Europe due to overfishing is the skate (*Raja batis*) from the Irish Sea (Brander, 1981). In contrast, freshwater species are more vulnerable to extinction, as they often inhabit smaller ranges and generally have smaller stocks, but the few extinctions recorded (eg, the disappearance of the burbot (*Lota lota*) from Britain) are probably due to habitat loss.

Overexploitation, however, may reduce genetic diversity. For example, overfishing in the Baltic has led to large-scale salmon stocking schemes, for commercial and recreational fisheries. There are fears that this may cause a depletion of the gene pool of the natural salmon stock in the area. Overexploitation of certain fish species can potentially impact on the natural ecological dynamics of the aquatic environment and therefore can affect the abundance of other plant and animal species in the aquatic food-chain.

The Icelandic Ministry for the Environment has suggested that the fishing industry is being affected by air- and seaborne pollution from the rest of Europe and North America, which could endanger fish stocks and marine mammals by affecting their reproductive capacity (Icelandic Ministry for the Environment, 1992).

Technology has played a part in increased pressure on fishery resources. There has been a general increase in the size of nets used, as bigger and more powerful vessels have been built. Bottom trawls, used for fishing flatfish and benthic molluscs, have been modified to allow increasingly rougher ground to be worked. The development of sonar and echo-sounding equipment has increased the searching power of vessels. Synthetic twines have replaced natural fibres, potentially contributing to solid waste generation since they do not rot. The power block allows the use of larger purse seine nets, and helps in hauling in gill nets used in deeper waters. Sealing and whaling are covered in Box 24C.

Impacts on non-target species at sea

Fishing, by its nature, removes individuals of a 'target' species from the environment: most fishing methods, however, have also an impact on 'non-target' species. Fishing can impact on non-target species both directly through the entrapment of, or damage to, non-target species in nets or by hooks, or indirectly by possibly depleting stocks of certain fish which represent an important food source to other marine animals.

Non-target species taken by a fishery are of little importance to the fishery concerned. Certain species may be retained for sale (known as 'by-catch species'), even though they are not target species. Other species are discarded at sea. Discarded species are not normally recorded other than by dedicated scientific investigation. The discarding of dead fish and parts of fish are important food resources for scavenging organisms, notably sea-birds. In the North Sea, the populations of the fulmar (*Fulmarus glacialis*) have increased as they feed on discarded offal from fishing vessels (ICES, 1992b).

Obviously the fishing intensity, the season and area of fishing are important, but the type of impact is determined largely by the type of fishing gear being used. Two major gear types have the greatest impacts on non-target species: heavy dragged gear (such as beam trawls and dredges) used to catch demersal species, and fixed (especially gill nets) or drifting nets to catch pelagic species (Northridge, 1991).

Trawls and dredges generally impact on benthic organisms living on the sea floor. ICES has estimated the total swept areas for the North Sea for various types of trawl. However, it is difficult to relate such statistics to damage, as the fishing effort is very uneven, and thus certain areas will be fished many times while others are missed. Nevertheless, there have been some measures taken to prevent damage to organisms on the sea floor. For example, in the Mediterranean, the seagrasses (*Posidonia oceanica* and *Zostera marina*), which are important nursery areas, are destroyed by trawling. Consequently, trawling has been banned in the EU, to a depth of 50 metres or to a distance of 0.5 nautical miles from the coast (the limit of growth for seagrasses). Another secondary effect arising from fishing is that of 'ghost' fishing, where non-target species are caught or killed by lost fishing gear (mainly nets and traps).

Fixed and drifting nets have a greater impact on the vertebrate marine fauna (other than fish), especially those with low reproductive rates. Dolphins, for example, the bottlenose dolphin (*Tursiops truncatus*), and porpoises (*Phocoena phocoena*), loggerhead turtles (*Caretta caretta*) and monk seals (*Monachus monachus*) (see Chapter 9), are particularly vulnerable to entrapment and damage: furthermore, all are mentioned on the *IUCN Red List of Threatened Animals* (IUCN, 1990). For example, the number of porpoises taken in Danish fixed net fisheries in the Danish inshore waters may be between 500 and 1000 annually (Northridge, 1991). Despite existing international practice (and EC legislation) limiting the use of surface drift nets to lengths less than 2.5 km, their continuing use creates a major problem in the Mediterranean (legislation does not exist for

Box 24C Sealing and whaling in Europe

Seal hunting and whaling have been, and in some places still are, important parts of the fisheries of some Northern European countries (notably Iceland, Norway and Russia).

Cetaceans: cetaceans are normally classified as large or small. The first group is composed of baleen and sperm whales, while the second includes small toothed whales and dolphins. The International Whaling Commission (IWC) manages large cetaceans on a global basis, while populations of small cetaceans are not often considered by traditional fishery management bodies. For management purposes, the IWC divides the North Atlantic stocks into three sub-stocks: western stock, central stock (Iceland, Jan Mayen) and eastern stock (North Sea, Norwegian Sea and Barents Sea). The main whale species in the northeast Atlantic are: blue (*Balaenoptera musculus*), fin (*Balaenoptera physalus*), sei (*Balaenoptera borealis*), minke (*Balaenoptera acutorostrata*) and humpback (*Megaptera novaeangliae*) whales. Cetaceans are among the most vulnerable to overfishing because of their slow growth and low fecundity, and possibly also through the link between social behaviour and reproductive success. During the earlier part of the twentieth century, some of these species came very close to extinction, and for some species, even with protected status, the situation is still precarious (eg, blue and humpback whales).

In 1986 the IWC imposed a complete ban on all commercial whaling. Norway has now come to the conclusion that the stocks of minke whales are sufficient to withstand a certain level of hunting (Central Bureau of Statistics of Norway, 1993). Thus, Norway decided to resume commercial hunting of minke whales in 1993, although the IWC had not lifted its ban.

Sealing: in the past seals have been hunted as intensively as whales in the Nordic region. Several arctic pinniped species, including the bearded seal (*Erignathus barbatus*) and the walrus (*Odobenus rosmarus*), are protected, following heavy overexploitation. In recent years, restrictions and slackening demand have also reduced the scale of hunting for species of seals that are still comparatively common. Seals hunted in Norway are mainly the harp seal (*Pagophilus groenlandicus*) and the hooded seal (*Cystophora cristata*). Currently, pups are no longer killed in either the Jan Mayen area or the White Sea, and the greater part of the catch has consisted of harp seals that are one year old or more (Central Bureau of Statistics of Norway, 1993). In Iceland the catch of seals consists mainly of the young of both the common seal (*Pocha vitulina*) and the grey seal (*Halichoerus grypus*) (Icelandic Ministry for the Environment, 1992). Even if the pressure from hunting has eased, the populations of seals in Northern European waters are still threatened by toxic pollutants (eg, PCBs) and virus epidemics, possibly related to human activities (eg, the 1988 phocine distemper virus in the Kattegat and Skagerrak): both are cause for concern.

See also the case study on monk seals in the Mediterranean in Chapter 9.

---

gill nets). Although there is not much reliable information available, there is some evidence that significant, and for some cetacean species possibly unsustainable, numbers of marine mammals are being killed in the Mediterranean Sea due to unselective fishing methods (Northridge and Di Natale, 1991).

Sea-birds can become entangled in most types of fishing nets, particularly in gill and other fixed nets. It has been estimated that 25 000 sea-birds were killed by drowning in the southeast Kattegat between 1982 and 1988 – mainly guillemots (*Uria aalge*) (Oldén et al, 1988, in ICES, 1992b). In the Baltic, salmon gill net fisheries are known to take at least 20 000 birds annually (Northridge, 1991). However, further investigation is required to prove whether breeding sea-bird decline can be attributed to entanglement (ICES, 1992b).

Impacts of inland fisheries

Catches from inland waters by capture and aquaculture were about 16 per cent of the total world catch in 1991; of total European catches the figure was just under 6 per cent, and that of the former USSR about 7 per cent (FAO, 1993a) (see *Statistical Compendium*). In both Western and Central and Eastern Europe (excluding the former USSR) inland aquaculture accounts for the majority of inland catches. In Western Europe most capture is for recreational purposes. On the other hand there is more commercial capture in Central and Eastern Europe, especially in the important inland water resources based mainly on the Danube river system and delta and the Mazurian lakes. The former USSR had well-developed inland fishing resources and more than half of inland catches are through capture (rather than aquaculture). However, severe environmental damage in many river systems through pollution and impoundment led to a small fall in the total catch between 1984 and 1990. High levels of stocking are common in rivers in the former USSR in order to sustain commercially valuable fisheries (eg,

sturgeon-stocking schemes in the Volga and Ponto-Caspian rivers) (FAO, 1992b).

As inland catches are still not reported as a separate category in FAO statistics, analysis of data for inland fish catches is still at an early stage of development (FAO, 1992b). For some countries the reporting of fish catches extends back several decades (FAO, 1990; 1993c). Trend statistics for inland water are complicated by difficulties in reporting by countries due to differences between extensive inland aquaculture and intensive management of inland fisheries, and the reporting of diadromous species (even though statistical conventions exist for both of these difficulties) (FAO, 1992b). Recreational fisheries have been consistently difficult to quantify because of economic difficulties in collecting such statistics, and their localised nature.

Commercial fisheries

Freshwater fishing activities may have a considerable ecological impact due to the relatively confined nature of some freshwater bodies. This is caused primarily by habitat degradation brought about by pollution or riparian engineering works resulting in an impoverished fish fauna in many rivers (Backiel and Penczak, 1989; Haslam, 1990). However, in larger European rivers, fishing seems to have little effect.

Somewhat ironically, river pollution can reduce fishing pressure in certain situations. For example, one of the reasons the middle reaches of the Vistula river in Poland support an extensive fish fauna is that the pollution is not severe enough to kill the fish, but nevertheless can taint their flavour and make them unattractive to fishermen (Backiel and Penczak, 1989).

Many of the freshwater fish valued as food occupy the niche of top predator (eg, pike (*Esox lucius*), zander (*Stizostedion lucioperca*) and eels (*Anguilla anguilla*)). The removal of top predators from lakes and rivers may significantly affect the biological community. One of the most important induced effects can be increases in the amount of phytoplankton in lakes (Jeppesen et al, 1990).

Recreational fisheries

Generally, recreational fisheries have less impact on the aquatic environment than commercial operations, due to the small quantity of fish caught and the fact that many fish caught by anglers are returned alive to the water. Where recreational fishing can have a major impact on the environment is when a body of water with a limited population of a species considered good sport is fished out entirely. As mentioned above, such species are often top predators, such as pike, resulting in the previously described knock-on effects (Mann and Penczak, 1984). An example of this has occurred in the river Pilica in southern Poland (Backiel and Penczak, 1989). Abundant populations of barbel (*Barbus barbus*) or nase (*Chondrostoma nasus*) found by anglers have often been fished out entirely. Recreational fishing can also impact non-target species. For example, concern has been voiced over the toxic effects of anglers' lead weights on waterfowl, such as swans in the UK.

The impact of fisheries (both recreational and sport) on the populations which they exploit can range from the virtual extinction of populations, to a more or less stable relationship of recruitment and cropping which existed in many long-established fisheries (eg, that formerly true for brown trout (*Salmo trutta*) in Loch Leven, UK). The essence of success in management is to have a well-regulated fishery where statistics on the catch are consistently monitored and used as a basis for future management of the stock.

Any decline in freshwater fishery numbers may not be due to overfishing but to a variety of unrelated factors, especially pollution by domestic, agricultural and industrial wastewater, the introduction of exotic species, habitat destruction and river and lake engineering (Maitland, 1993). Acidification has been attributed as the cause of the loss of considerable fish stocks in large parts of Scandinavia. For example, it has been estimated that over the last 100 years between 92 000 and 300 000 adult salmon (*Salmo salar*) were lost each year in 25 southern Norwegian rivers (Hesthagen and Hansen, 1991).

ENVIRONMENTAL IMPACTS OF AQUACULTURE

Aquaculture is the farming of aquatic organisms, including fish, molluscs, crustaceans and aquatic plants (FAO, 1993d). It can be either freshwater or marine (mariculture), and is increasingly common throughout Europe. There are currently difficulties in separating out aquaculture from the marine and inland catch statistics, although it is generally recognised that they should not be mixed with traditional data. Current statistics indicate that aquaculture production in Europe and the former USSR accounted for 12 per cent of world aquaculture in 1991 (FAO, 1993d). Total aquaculture production in the EU increased from 755 000 tonnes in 1984 to 884 000 tonnes in 1990. The recent increase in aquaculture activity has been stimulated mainly by the high costs of raised species, which has in turn attracted entrepreneurial interest. Escalating costs of fishing (eg, higher fuel costs) and decreasing wild fish stocks are also important factors. Trends in future production are difficult to forecast due partly to concerns over environmental impacts. However, in the short to medium term, aquaculture production levels in Europe are likely to increase.

Salmonoid rearing – mainly Atlantic salmon (*Salmo salar*) and freshwater, mainly rainbow trout species (*Salmo gairdnerii*) – is the most widespread aquaculture activity in Europe, currently accounting for 35 per cent by weight of total European production. Almost 40 per cent of rearing of salmonoid species takes place in Norway. Non-salmonoid species involved with aquaculture are the eel (*Anguilla anguilla*), carp (*Cyprinus carpio*), sea bass (*Dicentrarchus labrax*), gilt-head sea bream (*Sparus auratus*), catfish (*Ictalurus*

spp and *Clarias* spp) and various molluscs.

Aquaculture has the potential to supplement fish catches and help offset the declining stocks of some fish species. Collection of comprehensive, reliable data on aquaculture production is often difficult as many operations are small in scale. In addition, data are not always comparable since some include seaweeds and others may include juvenile specimens which are subsequently released to enhance natural stocks (UNEP, 1989).

Aquaculture can lead to eutrophication caused by discharges of fish-food materials and fish excrements from fish farms. However, to date, evidence for fish farming having brought about changes in the nutrient status and eutrophication of coastal waters is limited and restricted to regions with distinct hydrographic characteristics (limited circulation and mixing due to the sea current regime). Organic fish farm waste has been shown to cause enrichment of the sediment ecosystem in the immediate vicinity of the operation. Enrichment also causes changes in the physical structure of the sediment, some aspects of sediment chemistry and the community structure of the benthic macrofauna. Most studies show that the eutrophication effects are localised to within 30 to 40 m of the fish farm. The few studies undertaken to date show that the sediment ecosystem can recover from the effects of organic fish farm waste, although this can take several years. Large-scale deoxygenation of coastal waters as a result of fish farming is taking place in deep, insolated fjords and inlets such as in Norway (Gowen, 1990). There is increasing concern of eutrophication in the Mediterranean (eg, in enclosed inlets with limited circulation), which can be attributed partly to aquaculture. However, other factors such as tourism and agriculture are also responsible for eutrophication in these waters.

In Denmark, in 60 per cent of aquaculture farms an unacceptable state of pollution was found, with higher degrees of pollution downstream than upstream. Separation of feed waste and fish excrement reduces watercourse pollution. Fodder quotients (kilograms of fodder per kilogram of produced fish) have been introduced and reductions have been observed in discharges of nitrogen and phosphorus to watercourses (Danish Ministry of the Environment, 1991).

The pellets used as feed for intensive salmonid culture are generally made of fishmeal. Organic matter produced by phytoplankton is transferred down the food-chain to industrial fish, which are then converted into food for salmon. It has been estimated that intensive salmon farming requires solar fixation by phytoplankton from a sea surface area approximately 50 000 times larger than the surface area covered by the cages. No matter how humans exploit salmon (fisheries, coastal sea ranching, or fish farming) the required ecosystem support areas is of similar size (about 1 km^2 per tonne harvest) (Folke and Jansson, 1992).

One of the most potentially serious effects of aquaculture on the environment is the introduction of exotic species into environments where they may compete with, or replace, native species. Such introductions are generally unintended, though some are the result of deliberate actions. Some countries, such as Denmark, have introduced legislation prohibiting the introduction of non-indigenous species.

For example, when the Virginia or American cupped oyster (*Crassostrea virginica*) was introduced into Europe, two other American molluscs were accidentally introduced with the oysters: the slipper limpet (*Crepidula fornicata*), which causes extensive damage by smothering oyster beds, and the oyster drill (*Urosalpinx cinera*), which preys on European flat oyster (*Ostrea edulis*) spat. The oyster drill has declined in numbers since the 1960s and is no longer a particular problem (Utting and Spencer, 1992).

A number of diseases of aquatic animals have been introduced to Europe along with imported fish and shellfish.

These include the crayfish plague that has decimated European crayfish populations and the sporozoan (*Bonamia ostreae*) which has been responsible for severe mortalities of European oysters along the Atlantic coast (Utting and Spencer, 1992).

Aquaculture has a range of other impacts on the environment. Increased production through aquaculture is brought about by medical products (antibacterial agents, anti-parasitic agents and sedatives). The treatment of nets with anti-fouling agents can increase water pollution. Aquaculture can occasionally compete with coastal landuse, for example, by blocking access to important recreational or ecologically sensitive areas.

It should also be added that water pollution can have significant impacts on aquaculture. For example, the toxic algae bloom off the Norwegian coast in spring 1988 caused losses to salmon and trout farms totalling about 480 tonnes (OECD, 1991).

FISHERIES POLICY

Fishing is of varying importance in its contribution to GNP in different European countries. For example, in Iceland in 1992 it comprised 15 per cent of GNP (Icelandic Ministry for the Environment, 1992). However, in most other Western European nations, fishing accounts for less than 1 per cent of GNP on average, ranging from approximately 0.02 per cent of GNP for Germany to 0.4 per cent for Portugal (1987 data) (ERM, personal communication). Although precise data are not available on the contribution of fishing to GNP for Central and Eastern European countries, it is important for employment and human consumption (European Parliament, 1992a).

Whether fishing is a major contributor to GNP or not, it remains a politically sensitive subject in most European countries because of the needs of areas that rely on fishing as the main source of income (eg, the Faeroes, the Shetlands and northern Norway), disputes over fishing areas and practices used, and overexploitation. These factors have led the EC and a number of European countries to develop policies aimed at ensuring the proper management of fisheries activities, which have met with varying degrees of success (outlined in Box 24B). It has been estimated that if the waters of the EU were properly regulated, they could yield a further ECU 300 million per year (CEC, 1991b).

Fisheries policies aim at maintaining the population size of target species at or above a level which is consistent with sustainable fishing and a maximum yield. In this respect fishing policies take account of the environment. However, few fishing policies or marine management regimes address the broad range of environmental objectives. In order to do this a shift of emphasis is needed from a focus on the users of the resource to an 'environmentally orientated' approach, in which policies are consciously designed to optimise sustainable resource use on the basis of sound scientific information about the *whole* ecosystem.

CONCLUSIONS

- Fisheries statistics for the Mediterranean are scarce and unreliable (however, there is evidence that overfishing takes place in this area). Difficulties with fisheries data also lie with the separating out of statistics for aquaculture from marine and freshwater catches. A major effort is required to improve the statistical basis upon which to base policy decisions.
- An excessive fishing effort, or inappropriate exploitation, can affect a whole marine ecosystem.
- The principal environmental impact associated with fisheries activities is the unsustainable harvesting of fish stocks and shellfish, which has led to the decline of many, and even the economic collapse of several fisheries in various European seas, and has potential consequences for the ecological balance of the aquatic environment.
- This overfishing is related to lack of adequate constraints on fishing effort. This in turn is exacerbated by the development of electronic fish-finding apparatus and other technological developments. It is likely to remain a serious problem for the short to medium term in Europe, particularly where regional fishing policies are not effective – for example, in the Mediterranean Sea.
- Beam trawling could be leading to serious impacts on benthic species.
- The level of impact of fishing activities on cetaceans is unclear, but there is concern that large numbers of dolphins are being killed as a result of fishing activities the Mediterranean, and that other marine mammals, such as monk seals, are seriously threatened. In the Northern European seas porpoises may also be under threat.
- There is some concern over the impact, through water pollution, from the growing aquaculture industry on river, lake, estuarine and coastal waters. Aquaculture, particularly that associated with the production of salmonid species, increased significantly in the 1980s in a number of Western European countries. There are signs that this growth is beginning to tail off, but similar developments are likely to occur in a number of Mediterranean and Central and Eastern European countries in the future.

REFERENCES

Backiel, T and Penczak, T (1989) The fish and fisheries in the Vistula River and its tributary, the Pilica. *River Symp Can Spec Publ Fish Aquat Sci* **106**, 488–503.

Bernes, C (Ed) (1993) The Nordic environment – present state, trends and threats. *Nord 1993: 12*. Nordic Council of Ministers, Copenhagen.

Brander, K (1981) Disappearance of the common skate *Raja batis* from the Irish Sea. *Nature* **290**, 48–9.

Caddy, J F (1993) Toward a comparative evaluation of human impacts on fishery ecosystems of enclosed and semi-enclosed seas. *Reviews in Fisheries Science* **1**(1), 57–95.

CEC (1991a) *The common fisheries policy.* European File, 3/91, Commission of the European Communities, Luxembourg.

CEC (1991b) *Report 1991 from the Commission to the Council and the European Parliament on the Common Fisheries Policy*, SEC (91) 2288 final, 18 December 1991, Commission of the European Communities, Luxembourg.

CEC (1992) Fisheries yearly statistics 1991. Eurostat, Commission of the European Communities, Luxembourg.

Central Bureau of Statistics of Norway (1993) *Natural resources and the environment 1992*. Central Bureau of Statistics of Norway, Oslo.

Couper, A D and Smith, H D (1989) *Times atlas and encyclopaedia of the sea*. Times Books, London.

Crivelli, A J and Labat, R (1992) Fisheries. In: *International Union for the Conservation of Nature and Natural Resources. Environmental Status Reports Vol IV: Conservation Status of the Danube Delta*, pp 71–4. IUCN Publications Unit, World Conservation Monitoring Centre, Cambridge.

Cross, D G (1989) A summary review of fisheries data and their shortcomings. In: Schrank, W E and Roy, E (Eds) *Economic Modelling of the World Trade in Groundfish*, NATO ASI Series E, Applied Sciences Vol 201, Kluwer Academic Publishers, Dordrecht.

Danish Ministry of the Environment (1991) *Environmental impacts of nutrient emissions in Denmark*. Danish Ministry of the Environment, Copenhagen.

European Parliament (1992a) *La situation du secteur de la pêche et de l'aquaculture dans les pays de l'Europe de l'Est*. Directorate-General for Research, European Parliament, Luxembourg.

European Parliament (1992b) *Small-scale fisheries in the EEC Member States: problems, prospects and measures at national and community level.* Directorate-General for Research, European Parliament, Luxembourg.

FAO (1981) *Atlas of the living resources of the seas.* FAO Fisheries Department, Food and Agriculture Organisation, Rome.

FAO (1990) *Inland fisheries of Europe,* European Inland Fisheries Advisory Commission (EIFAC) Technical Paper 52. Food and Agriculture Organisation Rome.

FAO (1991) *Environment and sustainability in fisheries.* Document COFI/91/3, Food and Agriculture Organisation, Rome.

FAO (1992a) *Review of the state of world fishery resources, Part 1: the marine resources.* Marine Resources Service, Fishery Resources and Environment Division, Fisheries Department, Food and Agriculture Organisation, Rome.

FAO (1992b) *Review of the state of world fishery resources, Part 2: inland fisheries and aquaculture.* Inland Water Resource and Aquaculture Service, Fishery Resources and Environment Division, Fisheries Department, Food and Agriculture Organisation, Rome.

FAO (1993a) *1991 FAO year book: fishery statistics – catches and landings,* Volume 72. Food and Agriculture Organisation, Rome.

FAO (1993b) *Recent trends in the fisheries and environment of the GFCM area.* General Fisheries for the Mediterranean, Twentieth Session, Valletta, Malta, 5–9 July 1993, Food and Agriculture Organisation, Rome.

FAO (1993c) *Inland fisheries of Europe.* European Inland Fisheries Advisory Commission (EIFAC) Technical Paper 52, Food and Agriculture Organisation, Rome.

FAO (1993d) *Aquaculture production 1985-1991,* Food and Agriculture Organisation, Rome.

Folke, C and Jansson, A M (1992) The emergence of an ecological economics paradigm: examples from fisheries and aquaculture. In: Svedin, U and Aniansson, B (Eds) *Society and the Environment,* 62–87. Kluwer Academic Publishers, Dordrecht.

Gowen, R J (1990) *An assessment of the impact of fish farming on the water column and sediment ecosystems of Irish coastal waters (including a review of current monitoring programmes).* Report prepared for the Irish Department of the Marine Environment, Dublin.

Gulland, J A (1983) *World resources of fisheries and their management.* In: Kinne, O (Ed) *Marine Ecology* 5(2), 839–1061.

Haslam, S M (1990) *River pollution: an ecological perspective.* Belhaven Press, London, 249 pp.

Hesthagen, T and Hansen, L P (1991) Estimates of the annual loss of Atlantic salmon, *Salmo salar L,* in Norway due to acidification. *Aquaculture and Fisheries Management* **22,** 85–91.

Hollis, G E and Jones, T A (1991) Europe and the Mediterranean Basin. In: Finlayson, M and Moser, M (Eds) *Wetlands.* Facts on File Inc, Oxford.

ICES (1992a) *ICES Fisheries statistics,* Vol 73. International Council for the Exploration of the Sea, Copenhagen.

ICES (1992b) *Report of the Study Group on Ecosystem Effects of Fishing Activities.* ICES, Copenhagen, 7–14 April 1992. International Council for the Exploration of the Sea, Copenhagen.

ICES (1993a) *Reports of the ICES Advisory Committee on Fishery Management 1992, Parts 1 and 2,* No 193, ICES Cooperative Research Report. International Council for the Exploration of the Sea, Copenhagen.

ICES (1993b) *Atlas of North Sea fishes: based on bottom-trawl survey data for the years 1985-1987,* ICES Cooperating Research Report No194, International Council for the Exploration of the Sea, Copenhagen.

Icelandic Ministry for the Environment (1992) *Iceland: national report to UNCED.* Iceland Ministry for the Environment, Reykjavík.

IUCN/UNEP/WWF (1991) *Caring for the earth.* International Union for the Conservation of Nature and Natural Resources, Gland, Switzerland.

IUCN (1990) *IUCN red list of threatened animals.* International Union for the Conservation of Nature and Natural Resources, Cambridge.

Jeppesen, E, Jensen, J P, Kristensen, P, Søndergaard, M, Mortensen, E, Sortkjer, O and Orlik, M (1990) Fish manipulation as a lake restoration tool in shallow, eutrophic, temperate lakes. 2; threshold levels, long-term stability and conclusions. *Hydrobiologia* **200/201,** 205–18.

Kangur, M (1993) Fishery in Estonian environment 1992, pp 43–8, Environment Data Centre, National Board of Waters and the Environment, Helsinki.

Lleonart, J (1993) Trends in Mediterranean fisheries yields. In: *Pollution of the Mediterranean Sea, Technical Annex to Working Document for the meeting 10–11 September 1993, Corfu, Greece.* Scientific and Technological Options Assessment Committee (STOA), European Parliament, Luxembourg.

Maitland, P S (1993) Conservation of freshwater fish habitats in Europe. *Convention on the Conservation of European Wildlife and Natural Habitats Standing Committee.* Council of Europe, Strasbourg.

Mann, R H K and Penczak, T (1984) The efficiency of a new electrofishing in determining fish numbers in a large river in central Poland. *J Fish Biol* **24,** 173–85.

Northridge, S (1991) *The environmental impacts of fisheries in the European Community waters, a Report to DG XI, CEC.* Marine Resources Assessment Group Ltd, London.

Northridge, S and Di Natale, A (1991) *The environmental effects of fisheries in the Mediterranean, a report to DG XI, CEC.* Marine Resources Assessment Group Ltd, London.

OECD (1991) *The State of the Environment.* Organisation for Economic Cooperation and Development, Paris.

Oldén, B, Peterz, M and Kollberg, B (1988) Sjöfågeldöd i fisknät i nordvästra Skåne. *Naturvardsverket rapport* 3414, Solna, Sweden.

Robinson, M A (1984) *Trends and Prospects in World Fisheries,* Fisheries Circular No 772. Fisheries Department, Rome.

Sokolov, L I and Berdichevskii, L S (1989) Acipenseriformes. In: Holcík, J (Ed) *The Freshwater Fisheries of Europe, I(II).* AULA-Verlag, Wiesbaden, Germany.

UNEP (1989) *Environmental data report* (2nd edn, 1989/1990). United Nations Environment Programme, Blackwell, Oxford.

UNEP (1991) *Environmental data report* (3rd edn, 1990/1991). United Nations Environment Programme, Blackwell, Oxford.

Utting, S D and Spencer, B E (1992) Introductions of marine bivalves into the United Kingdom for commercial culture – case histories. *ICES Marine Science Symposium* **194,** 84–91.

Wahlström, E, Hallanaro, E-V, Reinikainen, T and Environment Data Centre (1993) *The State of the Finnish Environment.* Environment Data Centre and Ministry of the Environment, Finland.

*Piazza San Marco,
Venice*
Source: G Arici,
Grazia Neri, Milan

25 Tourism and recreation

INTRODUCTION

In terms of employment, contribution to GDP, and as a growing item of consumer demand, tourism and recreation together have become one of the most important social and economic activities in Europe. These activities bring income and jobs, increased understanding of other cultures, preservation of cultural and natural heritage and investment in infrastructure, which in turn brings social and cultural benefits. On the other hand, some forms of tourism, and some recreational activities, can cause destruction of habitats, degradation of landscapes and competition for scarce resources and services, such as land, freshwater, energy and sewage treatment. In addition, host populations may suffer the loss of their traditions and become overdependent on tourism incomes.

These problems are exacerbated by the concentration of tourist activity in relatively short holiday seasons and in specific areas, often comparatively small, which are also subject to environmental pressures from other economic activities such as agriculture, industrial development, fishing and growing resident populations.

More than other human activities, tourism and recreation depend on the quality of the natural and cultural environment for their continued success. However, as countries or particular resort areas become attractive destinations for tourism and recreation, unmanaged environmental impacts may undermine future earnings. For example, it has been estimated that, in 1990, the algal plague in the Adriatic Sea cost an estimated ECU 1.5 billion in lost revenue from tourism and fishing (CEC, 1991). Thus, tourism and recreation can affect the natural environment to such an extent that they can threaten their own existence. This is particularly crucial in Central and Eastern European countries where the temptation might be to develop unsustainable tourism as a 'quick fix' to assist with economic recovery. Thus, like agriculture and forestry and fishing, tourism and recreation both affect, and are affected by, the environment.

The practice of the World Tourism Organisation (WTO) to use country groupings (Central and Eastern, Northern, Southern and Western Europe and the East Mediterranean) is adopted in this chapter (see Box 25A). The term 'tourism', as used by the WTO, includes all travel by people to destinations outside the place they normally live, for any purpose (including pleasure, professional, educational, health or other), but excluding excursionists (ie, those visiting for less than 24 hours). However, where data are available, this chapter also covers recreation and sporting tourism in the domestic market and includes, for instance, day-trip skiers, visitors to national parks, the countryside, historic sites or theme parks, and other activities that involve large but inadequately documented flows of people having potentially significant impacts on the environment. Box 25A describes the sources of available data on tourism and the environment.

In terms of arrivals, worldwide tourism almost tripled between 1970 and 1992, growing at nearly 5 per cent each year, whereas revenue from tourism increased almost sixteen-fold, and it is currently the third most important economic

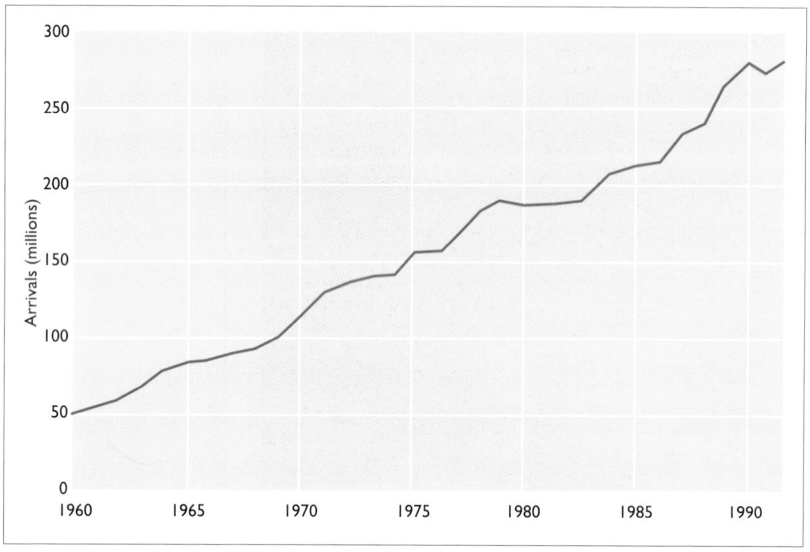

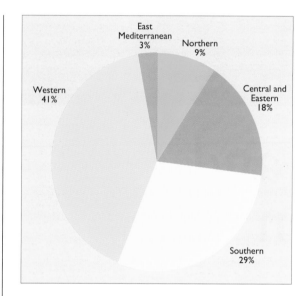

Figure 25.1
Trends in international tourist arrivals at frontiers in Europe, 1960–92
Source: WTO, 1993a; see also *Statistical Compendium*

Figure 25.2
International tourist arrivals at frontiers in Europe by region, 1992
Source: WTO, 1993a

KEY CONCEPTS: SUSTAINABLE TOURISM, INDICATORS AND CARRYING CAPACITIES

The need for sustainable tourism has been recognised at the international level in documents such as Agenda 21 and in the European Commission's Programme 'Towards Sustainability' (UNCED, 1992; CEC, 1993). Currently, there is no single accepted definition of sustainable tourism. However, a number of key elements might nevertheless be envisaged in its formulation (ERM and StfT, 1993):

- a longer-term perspective on policy making, as sustainable tourism is a goal which cannot be reached immediately;
- a recognition of the interdependence of economic and environmental systems; and
- a concern with biological limits within which human activities need to stay.

It has been suggested that it is necessary to balance five major elements to achieve sustainable tourism, without any single one becoming dominant: economic, tourist satisfaction,

Box 25A Sources of data and WTO country groupings

At the global level, the World Tourism Organisation (WTO) is the unique source of tourism data. The WTO reports its statistics in the following country groupings:

- Central and Eastern: Bulgaria, former Czechoslovakia, Hungary, Poland, Romania, former USSR;
- Northern: Denmark, Finland, Iceland, Ireland, Norway, Sweden, UK;
- Southern: Gibraltar, Greece, Italy, Malta, Portugal, San Marino, Spain, former Yugoslavia;
- Western: Austria, Belgium, France, Germany, Liechtenstein, Luxembourg, Monaco, The Netherlands, Switzerland;
- East Mediterranean: Cyprus, Israel, Turkey.

Most tourism statistics are available for Northern, Southern and Western Europe and the East Mediterranean. Comprehensive data are not available for tourism activity in Central and Eastern Europe. The WTO collects data on arrivals at frontiers, international and domestic tourist arrivals at accommodation establishments, international and domestic overnight stays, and the contribution to GDP from national tourist administrations (eg, WTO, 1993a). Recent trends on some of these indicators can be found in the *Statistical Compendium*. In addition, for most EU countries, Eurostat compiles data on tourism capacity, utilisation, visitor flows and employment in the tourism industry (eg, CEC, 1992a). The two principal indicators are tourist arrivals at accommodation establishments, and overnight stays. However, little information is available on the types of recreation and other activities that tourists undertake, their precise destinations (such as coasts, mountains or rural areas) within countries, or on seasonal flows. Further, as data are collected from national tourist bodies, they are not easily comparable across countries. As a result, much of this chapter is qualitative rather than quantitative.

international export earner in the world, surpassed only by oil and motor vehicles (WTO/UNCTAD data in WTO, 1993a). The majority of the Northern European countries are net spenders on tourism, but in a number of Mediterranean countries, including Spain, Portugal, Greece and Turkey, as well as some other countries such as Ireland, tourism makes a major contribution to the national economy. Thus in 1990 for the EU, tourism receipts as a contribution to GDP was 5.5 per cent, ranging from 1.3 per cent in The Netherlands to 9.4 per cent in Spain (CEC, DG XI, elaboration (1993) of WTO and Eurostat data). Few data are currently available on the contribution of tourism to the economy in Central and Eastern Europe. In Hungary, however, tourism accounted for about 3 per cent of GDP in 1992 (National Bank of Hungary, 1992).

According to the WTO, the number of international tourist arrivals to Europe increased from 190 million in 1980 to 288 million by 1992 (accounting for about 60 per cent of world tourism arrivals): an average annual growth rate of just over 3.5 per cent (Figure 25.1).

Figure 25.2 shows the variation in international tourist arrivals between different parts of Europe, with Central and Eastern Europe accounting for about 18 per cent (WTO, 1993a).

Box 25B
Measuring sustainable tourism: the need for better indicators

The total number of international arrivals at frontiers provides an impression of the overall tourist intensity in a given country, or group of countries, but this statistic is not specifically related to the environment. Arrivals statistics can also be useful because they are disaggregated by mode of transport. Number of overnight stays at accommodation establishments is probably the best existing indicator of local environmental pressure (ie, more useful than arrivals at accommodation establishments), although in many settings environmental impacts are associated with day visitors. However, all these indicators can at best be considered background indicators. Effective indicators of the real environmental impact of tourism and recreation should aim to provide measures which are exclusively related to physical and socio-cultural pressures from these activities. In practice the impacts of residents and other economic activities are difficult to separate from those of tourists. In addition, impacts are complex, cumulative and mostly site specific. Thus it has proved difficult to define even a few key, internationally comparable indicators. Environmentally relevant tourism data collected so far have tended to be on a localised case study basis (eg, those described in the national reports submitted to UNCED).

Indicators to monitor sustainable travel and tourism development are under consideration by organisations such as WTO, the World Tourism and Travel Environment Research Centre (WTTERC) and the European Commission (Eurostat). The WTO are developing measures at three levels: composite indices (eg, destination attractivity and site stress indices), national level indicators (eg, percentage of national area protected, endangered species), and site or destination specific indicators (eg, development density, area per tourist, or energy consumption per visitor-day). These types of indicators and indices should be determined by the needs of tourism sector managers, and hence reflect concerns such as the environmental, psychological or social carrying capacity of an area, the dependence on and control over tourism by local communities and the local or regional capacity for managing tourism impacts (eg, whether local tourism management plans are in place).

The WTO is now planning to test these indicators and indices on a voluntary basis in countries of different sizes, levels of development and tourism types. In this chapter, existing tourism statistics are used to make quantitative international comparisons, while case studies present more environmentally relevant data.

Source: WTO, 1993b

ecosystem, habitat or landscape can accommodate the various impacts of tourism and its associated infrastructure without damage being caused. It should also take account of seasonal variation.
2 Cultural and social capacity: the level beyond which tourism developments and visitor numbers adversely affect local communities and their ways of life, and any local identity is lost.
3 Psychological capacity: the level of tourism development or number of visitors compatible with the type of cultural/environmental experience that the visitors are seeking in relation to the particular destination. This depends on the type of tourism activity, and the expectations and actual experience that they have.

Attempts have been made in the literature to calculate these various carrying capacities for specific sites. However, as such analysis requires extensive data and local knowledge, it is not possible to include such material in this chapter. Indeed, the development of indicators and attempts to devise carrying capacities for different sites are long-term goals to achieve sustainable tourism. Currently, such tools are in their infancy in terms of conceptual development. Thus, policy makers concerned with the environmental impacts of tourism and leisure are guided mainly by practical experience and political choices.

ENVIRONMENT IMPACTS OF TOURISM AND RECREATION

Environmental impacts are experienced on several scales. At the local level, host communities suffer competition for scarce resources (particularly freshwater and land), air and water pollution, noise (eg, from excessive traffic), and environmental incidents such as avalanches. At the regional level, impacts may include loss of habitats and biodiversity, and air and water pollution. At the global level, emissions from road traffic and deforestation of large areas may contribute to climate change.

Since impacts are so largely dependent on the specific setting and type and scale of activity, six different settings for tourism and recreation, ranging from natural areas to completely built environments, are described in this chapter. This classification is used for indicative purposes only, and clearly there are overlaps in many cases. In each setting, case studies are used to provide examples of the range of impacts. Tourism and recreation vary from other human activities in that, being related to increased numbers of people present, existing pressures are generally exacerbated rather than new ones initiated. For these reasons, an overview table of significant environmental impacts is not presented for tourism.

National parks and protected areas

In Europe, a total of 2900 nationally protected areas cover 8.5 million km² (areas over 10 km² in IUCN management categories I to VI, here including sites in the former USSR and Turkey) (WCMC, personal communication,1994). Management categories, such as those devised by IUCN (see Chapter 9, for more details), can be designed and implemented to meet specific management objectives. Viewed in this way, conservation categories can become tools towards sustainable development (IUCN, 1990). Therefore, for example, in scientific reserves (IUCN category I) tourism and public access are generally forbidden. In categories II (national parks) and V (protected landscape or seascape), on the other hand, tourists are allowed access. However, the role that national parks play in tourism and recreation differs somewhat between countries.

social, cultural and environmental (Müller, 1993, in ERM and StfT, 1993). This implies a need for indicators to measure all these different aspects in order to highlight potential areas of concern and priorities for action (see Box 25B). Some principles of sustainable tourism are provided at the end of this chapter (Box 25G).

Indicators can be used to help identify problem or key impact areas, or can be combined to give some idea of carrying capacity (the number of tourists that an ecosystem can support indefinitely without running down the resource base). Ideally, carrying capacity should be assessed at three levels (ERM and StfT, 1993):

1 Environmental/physical capacity: the degree to which an

Tourism in protected areas has become increasingly popular in recent years. For instance, in the UK, 103 million visitor days are spent in national parks each year, equivalent to almost two days per person per year (FNNPE, 1993). Table 25.1 shows the popularity of selected destinations in IUCN categories II and V. Case studies indicate that the number of visitors, particularly to parks in Central and Eastern Europe, are increasing (Council of Europe, 1992; FNNPE, 1993).

National parks and protected areas are by their nature vulnerable to environmental degradation, and require careful environmental management. Some examples of environmental impact are provided in Box 25C. Problems are greatest in areas closest to urban centres and where arrivals are mainly by private car. However, increased tourism and recreation may also contribute to improved resource management, as a result of higher incomes for both parks and local people, and tourist interest in flora and fauna may help to safeguard biodiversity. For instance, in Majorca, 85 areas, representing 35 per cent of the land area, have been classified as protected areas since 1991 (Balearics Tourism Authority, 1992). This pattern is being observed elsewhere in Europe. Giving adequate levels of protection, the role of protected areas in tourism and recreation could be enhanced in the future. Guidelines are now being established for sustainable tourism in national parks and protected areas to supplement the already existing IUCN classification (WTO/UNEP, 1992).

Rural zones

Rural zones are popular tourist destinations throughout Europe. Many of the most attractive are described in Chapter 8. The most recent comprehensive survey of EU tourists in 1986 (European Omnibus Survey, 1987) suggested that, for their main holidays, 25 per cent of EU citizens prefer to visit the countryside. The types of impacts in rural areas are similar to those in protected areas, although they can affect wider areas, particularly so for IUCN category V areas (protected landscape or seascape).

Rural tourism in easily accessible areas is also popular in Central and Eastern Europe, with, for instance, the Plitvice Lakes area in Croatia attracting over 1 million visitors each year (Cambridge Economic Consultants, 1993). Throughout Central and Eastern Europe, countryside tourism has been sponsored by workplaces and trade unions. In Poland, for example, rural tourism has been popular for many years, as many urban dwellers still have relatives living in villages. Farmers used to host 'city people', giving them accommodation and food, receiving in return some help at harvest time. This form of tourism is at risk in Central and Eastern Europe, as many rural farmers are now elderly and cannot provide suitable accommodation. In parallel, uncontrolled building of weekend homes has proliferated, often in supposedly protected areas. For instance, in the

former Czechoslovakia, construction of weekend homes has led to overburdening of several easily accessible natural areas (eg, in the Jizera and Krkonoše mountains, and rivers Berounka and Dyje) (MECR and CAS, 1991).

Western European countries are witnessing the rapid development of rural tourism. This is caused partly by the necessity to diversify the rural economy, due to the declining role of agriculture. In France, countryside holidays are popular throughout the year (Figure 25.3). Rural tourism is well organised, with the 'Les Pays d'Accueil' programme catering for many different tastes, such as 'Tourisme à la ferme' and 'L'accueil au château' (Ministère de l'Environnement/Ministère du Tourisme, 1992; 1993).

Golf developments can cause environmental damage where they require removal of earth, take farm and forest land, destroy the natural landscape, disrupt existing hydrological patterns, and entail the draining of wetlands to create greens, fairways and artificial lakes. Erosion and flooding may result. In Ireland, new golf courses (which are normally exempt from planning permission) have been built on coastal dunes. The ploughing, planting with grasses and management activities (eg, mowing, fertilising and use of weed killers) have resulted in loss of floristic and faunistic diversity (Curtis, 1993). The endangered natterjack toad, *Bufo calamita*, was threatened in dune areas at Castlegregory, County Kerry, Ireland, in the early 1990s (Council of Europe, 1991). In humid climates, pesticide use is a particular concern due to leaching, especially from well-maintained greens. Thus it is important to avoid water sources, sandy areas, or areas with a shallow watertable when locating golf courses. In addition, water requirements for irrigation may put considerable strain on drinking water supplies (particularly a problem for golf courses in the Mediterranean such as along the Costa del Sol), and runoff may be polluted with fertilisers and pesticides. Floodlighting of courses for all-night play requires increased energy for lighting, may cause nuisance to local communities and disturbs wildlife. Golf courses also encourage increased development (eg, clubhouses and chalets) and road traffic. However, guidelines are being produced for sustainable construction and management of courses (eg, see Touche Ross, 1993), and some resorts such as those in Majorca now insist on the use of recycled water for irrigation.

Other recreational activities in rural areas include fishing (described in Chapter 24) and hunting (discussed in Box 25D).

In some parts of Europe, nature is still relatively pristine and untouched by tourism and recreation (eg, the northern part of Nordic countries and Russia). For example, there are at least 25 areas of 1000 km² or more without roads in the Scandinavian mountain chain (Bernes, 1993). However, the scale of tourism activities has increased rapidly, with tourist development centred around major resorts. The cumulative impacts of tourism and recreation, including transport, waste, noise and increased

Table 25.1
Visitors to a selection of European protected areas
Sources: FNNPE, 1993; IUCN, 1990

Country	IUCN category	Protected area ('000 ha)	No of inhabitants ('000s)	Annual visitor numbers ('000s)
Estonia	II (national park)	Lahemaa (65)	nd	1000
former Czechoslovakia	II (national park)	Tatra (74)	nd	4000
Slovenia	II (national park)	Triglav (85)	2	2000
Austria	V (protected landscape)	Hohe Tauern (179)	0	4000
England	V (protected landscape)	Lake District (229)	40	20 000
France	V (protected landscape)	Brotonne (40)	33	900
Poland	V (protected landscape)	Ojcow (2)	nd	250

Note: nd: no data available.

Box 25C Loving them to death: tourism and recreation in protected areas in Europe

New developments: nature and national parks are already under extreme pressure from the number of visitors, demand for outdoor activities and development of tourism facilities (eg, large hotels on the Mazurian Lakes, Poland).

Overcrowding occurs at peak periods (eg, in Hohe Tauern, Austria, visitors are concentrated in six months, while in Ojcow, Poland, 60 per cent are there during the three summer months). Most visitors are day trippers arriving by car (over 90 per cent to the Hohe Tauern), leading to traffic congestion, congestion of car parking space and litter problems.

Changes in habitats of native animals: wildlife attracts visitors, who disturb breeding patterns, leading to a fall in animal numbers; as species become rarer, more people come to 'see it while you still can'.

Path erosion and wear and tear arise from walking or mountain biking. Popular routes such as the Pennine Way in the UK require extensive maintenance and repair (see Chapter 7).

Introduction of exotic species: in Tenerife (Canary Islands) more than a dozen species have invaded the Teide National Park, having been carried as seeds on vehicles, clothes or tents of visitors.

Conflicts exist between tourism and nature conservation, traditional hunting and agriculture. For example, the Coto Donaña National Park in southwest Spain is an important breeding site for many of Europe's birds, and home to endangered species such as the imperial eagle, *Aquila heliaca*, and Spanish lynx, *Lynx pardina*, but is now threatened by water extraction for tourism and local agriculture (WWF-UK, personal communication, 1993). In Triglav National Park, Slovenia, 2 million tourists outnumber residents during the summer by almost 1000 to 1 (see Table 25.2).

Source: FNPPE, 1993, and ERM

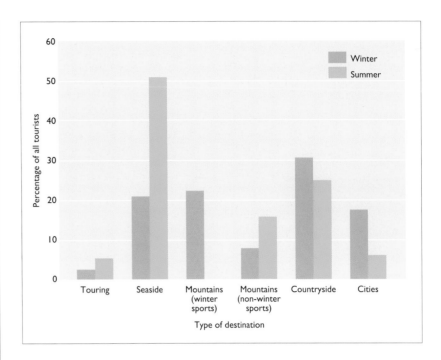

Figure 25.3
Types of domestic tourism in France, 1989
Source: Ministère de l'Environnement, 1991

population, could be substantial in certain areas. For example, wildlife species that frequent riparian habitats could be affected if large numbers of tourists are concentrated near rivers (Norwegian Directorate for Nature Management, 1993).

Mountains

Mountain areas are popular destinations for tourism in Europe. The majority of tourism and recreation in mountain areas in Europe is concentrated in the 190 000 km² of the Alps (as defined by the Convention on the Protection of the Alps) (Partsch, 1991). This area receives some 100 million tourists every year, of which about 40 per cent visit during the holiday season and the remaining 60 per cent at weekends (Partsch, 1991).

The four main destinations (France, Switzerland, Austria and Italy) account for some 92.5 per cent of the Alpine area, and 90 per cent of total visitor nights (Slovenia, Germany and Liechtenstein account for the remainder) (Jenner and Smith, 1992). The number of visitors to the Alps in the summer has stayed roughly constant over the last 20 years. However, winter visitors have increased considerably over the same period.

The extreme sensitivity of Alpine environments is

Box 25D
Hunting in Europe

A survey conducted by the French Comité National d'Information de Chasse-Nature (Pinet, 1987) found that in the mid-1980s there were over 6 million hunters in the EU creating an industry which employs approximately 100 000 people (including equipment manufacturers and retailers, gamekeepers, specialist magazine publishers, etc) and generated revenues of about ECU 5.6 billion (1 ECU = 6.94 French francs in May 1987). Four countries alone accounted for over 80 per cent of the EU's hunters: France (29 per cent), Italy (24 per cent), Spain (17 per cent) and the UK (13 per cent). In terms of hunting intensity (agricultural plus wooded areas divided by the number of hunters), Italy had the most intense index in Europe and Luxembourg the least (80). France alone had over 1.8 million hunters, who were estimated to have paid ECU 230 million for 'land access' rights alone, which covers the costs of gamekeepers, game rearing and use of the land. Much hunting is done on land associated with forests, as these provide the breeding, roosting and feeding grounds of much of the traditional game which is hunted.

Increased hunting pressure on some wildlife populations is a significant area of concern in many regions where industrial development or transport infrastructure provides greater access to wildlife areas. This can in turn contribute to habitat fragmentation, and terrain and forest damage.

Hunting also leads to effects on 'non-target' species. Lead poisoning of waterfowl through ingestion of lead shot is a phenomenon which has been recorded in at least 20 countries, and is now being investigated in depth in Europe. Shot ingestion levels show evidence of increasing over time as the density of available shot in wetlands increases. This is leading to increased bird mortality, a cause of some concern in many countries (Pain, 1992). There is also some evidence in the UK that lead weights used in recreational fishing are being ingested by waterfowl such as swans.

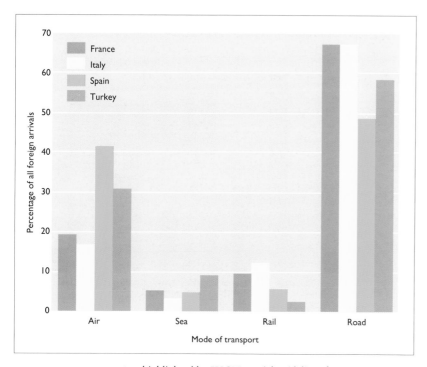

Figure 25.4
Foreign arrivals in
the Mediterranean by
transport mode, 1991
Source: WTO, 1993a

highlighted by IUCN special guidelines for managing
mountain protected areas and the Convention on the
Protection of the Alps agreed in 1991 by the International
Commission for the Protection of the Alps, which
encompasses a protocol on tourism and environment.

Skiing

In 1990 it was estimated that some 50 million people took an
Alpine ski holiday (Tourism Concern, 1992). This led to
over 40 000 ski runs and over 14 000 lifts by the late
1980s (CEC, 1992b). The trend in development of such
infrastructure is illustrated by the case of Switzerland,
where installations have grown from around 250 in 1954
to over 1900 by 1990 (BUWAL, 1992). The range of
environmental impacts arising from skiing is shown
in Box 25E.

In recent decades a number of completely new
developments have taken place, including major ones such as
that at Les Meunières in France for the Winter Olympics in
1992. The Three Valleys area, also in France, is now the
largest ski area in the world, with French skiers making up 80
per cent of the total (Tourism Concern, 1992).

The cumulative environmental impacts of skiing are
considerable. Construction of ski pistes and related
infrastructure, including access roads and parking, have required
the felling of forests, thereby removing habitats and natural
protection against avalanches and degrading landscapes (see Box
25E and Chapter 8). Most visitors travel by car, and exhaust
fumes lead to further forest damage and air pollution (BUWAL,
1992). In addition, during the last decade, resorts have

Table 25.2
Alpine tourism's
contribution to
polluting emissions in
Switzerland, 1987
Source: Partsch, 1991

Tonnes per year	Tourist traffic (tonnes/year)*	Central heating: hotels and second homes (tonnes/year)	Contribution of tourist traffic and central heating to Swiss total (%)
Sulphur dioxide	600	800	1
Nitrogen oxides	38 000	600	18
Hydrocarbons	24 000	500	7
Carbon monoxide	136 000	7000	23
Soot	50	100	1
Lead	160	–	24

Note: * Excluding public transport and air arrivals.

Box 25E Impacts of skiing in Europe

Mountains are the last remaining areas in Central and
Eastern Europe that are relatively untouched by human
activity. They support a rich biodiversity, although
relatively few species can survive in the harsh
environment. Recently introduced human activities related
to skiing threaten the balance of delicate ecosystems by
introducing pollution, deforestation, soil erosion and
human disturbance.

**Forest clearance and increased incidence of
avalanches:** some 100 km² of forest have been removed
throughout the Alps, which has led to higher incidence of
avalanches (CEC, 1992b). In Austria, the creation of
0.7 km² of ski runs in one year (1980) for the Winter
Olympics contributed to a major mudslide in the cleared
area in 1983 (ERM and StfT, 1993). New resort
construction may involve bulldozing, blasting and
reshaping of slopes, leading to a higher incidence of
avalanches.

Visual degradation of landscapes: natural forest
barriers may be replaced with unsightly concrete, plastic
and wooden barriers associated with developments.

Inappropriate development: the uncontrolled
development of ski-centres in some Central and Eastern
European countries (eg, in the High Tatra mountains,
Poland) are threatening the environment.

**Loss of habitats and disturbance of endangered
species:** operation of lifts, off-piste skiing, use of
all-terrain vehicles and compaction of snow on pistes
disturb rare species such as black grouse (*Tetrao tetrix*) in
the French Alps, which can be decapitated by overhead
wires, displaced from their breeding grounds and can face
competition for space with cable car installations (Jenner
and Smith, 1992).

Sewage disposal and water pollution: in the French
Pyrenees the sewage from summer tourist resorts
discharges directly to streams and leads to water
pollution. In the Alps chemicals used in preparing 36
glaciers for skiing have been reported to lead to increases
in nitrogen and phosphorus levels in drinking water
(Greenpeace data, quoted in Tourism Concern, 1992).

Exhaust from private cars and coaches have led to death
of trees, and to wildlife damage. In Switzerland 70 per
cent of domestic and foreign tourists arrive at their
destination by car (BUWAL, 1992). Emissions of
pollutants from traffic and heating in Switzerland are
shown in Table 25.2.

Unsustainable use of water: by 1992, 4000 snow
cannons were producing artificial snow to lengthen the ski
season in the Alps, using (and competing with other uses
for) 28 million litres of water per kilometre of piste. In
Les Meunières, France, 185 cannons installed for the 1992
Olympics were supplied by drinking water sources
(Tourism Concern, 1992). Artificial snow melts slowly,
reducing the short recuperation time for Alpine grasses
and flowers. Furthermore, skiing in sparse snow
conditions contributes to erosion and damages sensitive
vegetation. The result is a severe reduction in water
absorption and holding capacity of mountain slopes, and
increased risk of runoff and avalanches.

Overdevelopment of tourism: by 1991, employment
in the Swiss tourist industry in Alpine areas accounted for
probably one in every three jobs. Over 75 per cent of
accommodation is in the so-called para-hotel industry,
outside the regulated hotel sector (holiday apartments,
holiday houses and campsites) (BUWAL, 1992).

Source: ERM

experienced several poor snow seasons and have attempted to lengthen seasons using artificial snow machines, higher lifts and summer skiing on glaciers. These trends may compound environmental problems related to skiing and increase disturbances to natural areas (though lengthening the skiing season may also help reduce seasonal concentrations of tourists).

There are different problems associated with different forms of skiing. The more severe environmental impacts are normally associated with downhill skiing (eg, in terms of land taken, energy use for infrastructure and vegetation effects). Cross-country skiing tends to have less severe environmental consequences, and is increasingly popular as a form of winter tourism. Some Swiss and Austrian resorts (Zermatt, Saas Fee, Riederalp and Kühboden) have turned themselves into car-free zones or adopted voluntary capacity limitations on car parks, lifts (Grindelwald) or hotel space (see, for example, Krippendorf et al, 1992). However, these measures succeed at the cost of shifting the problem elsewhere, for instance to the Pyrenees or Central and Eastern Europe.

Summer tourism activities

Tourism and recreation in mountain areas in summer are less concentrated than, associated with skiing, in winter. Summer visitors tend to engage in walking and sporting pursuits. Along much-used long distance footpaths, constant wear in summer months can destroy the vegetation cover and leave deep grooves, leading to increased soil erosion. The most controversial aspect of outdoor leisure activities is that they have become increasingly motorised, resulting in noise and increased soil erosion – from snowmobiles in winter, and heavy duty vehicles and motorcycles in summer.

It is important to preserve mountain economies in order to develop sustainable tourism; one approach is to preserve the way of life of the local people. In Switzerland and Austria, a policy has been developed to subsidise local people who welcome tourists into their homes by financially supporting appropriate housing conversions. In this way, the local tourist economy benefits, and rural depopulation is reduced (Messerli, 1990).

Coastal areas

Coastal areas are popular tourist destinations all over Europe. Often, however, it is difficult to ascertain the specific role played by tourism in coastal zones, as other human activities (urban, agriculture, fishing and aquaculture, etc) are also prevalent. This is not helped by generally poor data availability (economic, social and environmental) for coastal areas.

The Mediterranean basin (here meaning all countries bordering on the Mediterranean Sea, but not the Black Sea, thus extending beyond the WTO definitions in Box 25A) is still the world's most important tourist destination,

Source: Michael St Muir Sheil

Box 25F Impacts of tourism in the Mediterranean

Overdevelopment: unplanned growth of hotels and tourism facilities with little regard to visual impacts or local architecture has led to visual degradation over vast areas. Land has been used for recreational facilities (eg, golf courses and theme parks), and major roads and airports encroach on protected areas (eg, Ria Formosa National Park in the Algarve).

Induced development pressures: for example, agricultural development aimed at meeting tourist (catering) needs around national parks, such as Coto Doñana in Southern Spain where wetlands are threatened by water abstraction and pesticide runoff.

Loss of habitats and of biodiversity: 75 per cent of dune systems from Gibraltar to Sicily have been lost since 1960 (CEC, 1992b), leading to loss of breeding grounds, such as that of the loggerhead turtle *Caretta caretta* (see the case study in Chapter 9, p 212). Over 500 Mediterranean plant species are threatened with extinction, and in France alone 145 species are on the verge of extinction or have already disappeared (CEC, 1992b).

Species impacts: tourism pressure on nesting sites of the loggerhead turtle (*Caretta caretta*) and green turtles (*Chelonia mydas*) led to curtailing the building of a hotel at Dalyan in Turkey in 1986. However, the very act of protecting the turtles has led to an increased influx of tourists – 5000 in summer – creating other environmental pressures such as waste dumps (WWF-UK, personal communication, 1993).

Lack of sewage and effluent treatment and disposal: only 30 per cent of municipal wastewater from Mediterranean coastal towns receives any treatment before being discharged (World Bank/European Investment Bank, 1990) and, as a result, some Mediterranean beaches failed EC bathing water quality tests (Spain, 7 per cent; France, 13 per cent; Italy, 8 per cent; and Greece, 3 per cent, in summer 1992). The total cost of developing the necessary level of sewage treatment was estimated at more than $10 billion. Spillages from pleasure boats are also a major source of pollution.

Unsustainable exploitation of natural resources, including excessive abstraction of drinking water and exploitation of fisheries resources. Overabstraction of water (for drinking, bathing, golf courses and water theme parks) has led to increased forest fires, with 2000 km^2 of Mediterranean forests destroyed each year by fire (CEC, 1992b).

Traffic congestion on coastal roads, and pressure to build new trunk roads through protected or unspoiled areas (eg, Southern Spain and the Algarve).

Changes in traditional lifestyles where local populations are outnumbered by tourists, particularly in poorer regions (the Balearics, Turkey, Croatia, Cyprus) with overdependence on tourist incomes (CEC, 1992b).

Sources: CEC, 1992b; World Bank/European Investment Bank, 1990; ERM

currently attracting some 35 per cent of international tourists worldwide. The Mediterranean basin's attractive landscapes, cultural heritage, traditional lifestyles and good climate and beaches have made it a very popular tourist destination. The area occupied by tourist accommodation and associated infrastructure in 1984 was approximately 4400 km^2 and 90 per cent of this was in three countries – Spain, France and Italy. This has been forecast to double to 8000 km^2 by 2025. The solid waste generated by tourists in the Mediterranean region, currently 2.9 million tonnes per year, would by this scenario increase to between 8.7 and 12.1 million tonnes by 2025, while wastewater would increase from 0.3 billion m^3 to between 0.9 and 1.5 billion m^3 (Grenon and Batisse, 1988).

The number of tourists to the Mediterranean basin tripled from 54 to 157 million between 1970 and 1990 (WTO, 1993a). In some countries, such as Greece, this number grew six-fold. In other countries, such as Turkey, there has been a four-fold increase in the number of tourists, and more than a three-fold increase in bed capacity between 1983 and 1991 (WTO, 1993a). In the worst affected areas the carrying capacity simply cannot accommodate the peak concentrations of tourists, which, for example, reached 4250 people/km^2 in the Tarragona area of Spain in August 1987 (CEC, 1992b).

Figure 25.4 on page 494 shows that, for the major Mediterranean destinations, most visitors arrive by road, and the next most by air.

Box 25F summarises the major environmental impacts of tourism and recreation in coastal areas, particularly the Mediterranean, which can lead to a downward spiral of environmental degradation. This has been experienced to some degree in the Algarve, along the Mediterranean coastline on the Spanish Costas, Cyprus and Turkey.

In Northern Europe, parts of the French and Belgian seaboard and traditional resorts in The Netherlands and UK and the Baltic Sea coasts still attract large numbers, but they do not suffer from the numbers or concentrations found in the Mediterranean. In some countries such as Denmark, the number of tourists has increased markedly in the last five years. The Black Sea is the major tourist destination for tourists in Central and Eastern Europe, currently attracting about 40 million tourists annually (GEF, 1993). However, declining environmental quality in the Black Sea is having serious negative consequences on tourism in the area. Poor planning, unsanitary conditions, insufficient drinking water and lack of user fees has led to increased environmental stress and severe health problems (including occasional outbreaks of cholera). The regular closure of some beaches has put pressure on previously clean ones. Furthermore, in the past, meagre user fees and tourist taxes gathered in countries with centrally planned economies were not usually reinvested in improving local services (Hey and Mee, 1993).

Cities and heritage sites

Many of the major tourist attractions in Europe are of historic, cultural or religious interest. Often these may be in or close to urban areas. Some 25 of the top 45 tourist attractions in Europe, each one attracting more than 750 000 visitors per year, are sites, museums and historic buildings in cities. In 1986, 19 per cent of EU citizens preferred to visit cities for their holidays (European Omnibus Survey, 1987). In Central and Eastern Europe, cities of cultural and historic interest, such as Prague, Budapest and St Petersburg, are increasingly attracting Western European tourists.

Tourism may act as an incentive for urban redevelopment. Large host populations and adequate infrastructures in major tourist cities easily absorb high-volume tourism. In smaller, confined areas, such as walled towns or historic sites (eg, Bath, UK, or Florence, Italy) the concentration of tourists gives rise to serious environmental management problems, either related to lack of infrastructure or to traffic congestion and emissions, or to wear and tear on buildings (Brady Shipman Martin, 1993).

The World Heritage List of UNESCO, currently with over 350 sites worldwide, includes cultural, natural and mixed sites, with the ultimate objective to ensure adequate protection of selected sites. Currently there are about ten times as many cultural sites (such as the 'leaning' Tower of Pisa, Italy) as natural sites (eg, Białowieża National Park/Puszcza Białowieska, Poland) in Europe. The designation of mixed sites (eg, Mont St Michel in France) has brought a new dimension to environmental management directed towards tourism by bringing the natural and cultural environments together (though these sites currently constitute less than 5 per cent of the list in Europe). The World Heritage List does not include each and every site worth protecting for future generations. However, because it provides a solid core of recognised sites, for which the World Heritage Convention facilitates international cooperation, the list provides a basis for coordinated action (Batisse, 1992).

Theme and leisure parks

There have been growing investments in theme parks throughout Europe (particularly in The Netherlands, Denmark, UK, France, Germany and Spain). Of the 45 most visited tourist attractions in Western Europe, some 20 were theme parks, attracting nearly 48 million visitors in 1991. There are now more than 50 major parks, each one attracting more than 750 000 visitors per year (Euromonitor, 1992). The so-called second generation parks are modelled on US-style parks providing accommodation in a 'good environment'. For instance, the Center Parcs chain (which started in The Netherlands) is based on holiday villages offering all-weather recreation in rural environments. These are designed to spread the tourist season throughout the year, and increase domestic tourism. However, the resource use of these parks is considerable. For instance, of the 13 parks run by Center Parcs in Northern and Western Europe, the average site covers 1.5 km^2 and, per person a day consumes 20 kWh of electricity, 6 m^3 of gas and 150 to 180 litres of water (depending on whether or not the park has a swimming pool) (Brettle, 1989)). Around the Mediterranean, water-based theme parks are being developed to attract visitors away from the congested coasts.

OUTLOOK

Main determining factors

Economic conditions

The future demand for tourism and recreation will be determined by a large number of factors, most importantly by national and regional economic trends. There is an almost constant elasticity of demand for leisure tourism (ie, greater disposable income means greater tourism and vice versa), while business tourism is less affected by economic conditions. The other major underlying factors determining the future demand for tourism are summarised below.

Increased leisure time

Shorter working weeks, flexible working patterns, and spreading of annual leave more evenly over the year are all expected to lead to an increased demand for non-sporting winter holidays and short breaks to leisure parks, national parks and cities. While throughout the EU an average of about one third of holidays are taken during the peak season (July to September), the figures are much higher for certain types of holiday, such as sun and beach, touring, countryside and sports holidays, particularly in certain countries, such as France (Institut für Planungskybernetik, 1989).

Demographic factors

At least as far as the EU is concerned, the forecast increase in Europe's population will have a negligible impact on tourism. However, changes in the population structure, with a greater proportion of pensioners, will increase demand for certain types of tourism and recreation (WTO, 1993c).

Socio-cultural factors

Trends towards later marriage, and more working women, means that many young couples have high disposable incomes; the increasing interest in education and different cultures is leading to a shift from package tourism to more environmentally sensitive types of tourism and recreation.

Increased environmental awareness

In a survey carried out in Germany in 1988, some 47 per cent of those interviewed reported serious environmental problems noticed at their holiday destination, compared with only 22 per cent in a comparable survey in 1985. A similar survey in 1992 shows that environmental quality was perceived as having deteriorated further and, in the case of Spain, over 70 per cent of those interviewed reported the Spanish environment as damaged by tourism (StfT, 1986; 1989; 1992). As a consequence of this new awareness, eco-tourism (see below is becoming increasingly important, although this is dependent on the country in question and on economic conditions.

Reduction in the cost of travel

Reduced travel costs and increased access to private transport, particularly in Central and Eastern and Southern Europe, will lead to more European travel, with large numbers of Central and Eastern Europeans already visiting the Mediterranean, and Western Europeans visiting Central and Eastern European countries. In addition, changes to the regulatory environment within the EU, such as deregulation of air transport and coaches, are expected to make short-break trips more accessible.

Types of tourism

These trends are leading to changes in the nature of tourism in Europe and its environmental consequences.

Mass tourism

Most of Europe will experience a gradual decline in mass tourism. Mass tourism exerts a heavy burden on the environment as it is concentrated not only in terms of absolute numbers of tourists, but also in both time and space. Although mass tourism is still popular for tourists from Central and Eastern Europe, in the longer term they are likely to take advantage of the opportunities for independent travel. Western and Northern Europeans are increasingly choosing self-catering accommodation and multi-centre or touring holidays based on fly-drive. However, there is expected to be continued growth in purpose-built villages, such as timeshare developments, retirement homes or weekend/second homes in apartment blocks. Typically these developments provide activity and health facilities such as golf courses, swimming pools and shops, meeting/conference centres and marinas.

Activity-based tourism

Activity-based tourism exerts specific stresses on the environment and is expected to grow faster than passive 'rest and relaxation' tourism. Golf and sailing are increasingly popular; for instance, in France, of a total 425 golf courses in 1991, some 300 were built in the previous ten years (Ceron, 1992). The most significant current development is the creation of golf resorts that incorporate large-scale developments into the course landscape. In Central and Eastern Europe international tourism operators are trying to develop golf courses and marinas in previously unspoiled areas such as the Lahemaa National Park in Estonia (FNNPE, 1993). Other adventure activities such as scuba diving, horse-riding, hot-air ballooning, canoeing, and off-road driving are all being developed as niche markets for both group and specialist tourism.

Eco-tourism

The World Tourism and Travel Environment Research Centre (WTTRC) has defined eco-tourism as:

implying a building of environmental considerations into all tourism and travel products and their consumption. This would suggest that sustainable tourism development is a level of activity at or below the level which will not result in environmental or socio-cultural deterioration or be perceived by tourists as depreciating their enjoyment and appreciation of the area.

(cited in ERM and StfT, 1993)

This, however, does not imply that eco-tourism does not have environmental impacts – only that such impacts are less than those of mass and activity-based tourism.

There is a wide variety of forms of eco-tourism, from village-type development to nature-based tourism, which might include walking and hiking, cycling in rural areas, staying in rural farmhouses, bird-watching, and other low impact sports. There is expected to be a growth in environmentally sensitive tourism and recreation, as a result of increasing environmental awareness. Areas such as the Covadonga Park (in the mountain region of Asturias and Coto Doñana, near Seville, Spain) and the Central Istria National Park (Croatia) are increasingly attracting tourists away from congested coastal strips to take advantage of unexploited inland areas. The Council of Europe has proposed a number of pilot projects on 'soft' tourism which comprise both natural and artificial sites, together with extensive agriculture and light industry (Lalumière, 1993).

Regional trends

How will trends in these types of tourism and recreation be manifest in different parts of Europe? Most forecasters of tourism trends expect that European tourism arrivals (both domestic and international) will continue to grow at an average rate of between 3 and 5 per cent each year in the 1990s (reaching 6 per cent each year between 1995 and 2000), which translates to 380 to 440 million arrivals each year by 2000 (eg, EIU, 1992). In favourable economic conditions, consumer purchasing power is likely to increase, and tourism expenditure will account for an increasing proportion of discretionary spending. The likely trends in international arrivals by each European region are summarized in Figure 25.5. For demographic reasons, the total number of domestic tourists is not likely to increase much, but growth is expected to result from an increased number of trips taken (eg, for second holidays and short breaks) with a broader seasonal spread (Euromonitor, 1992; Jenner and Smith, 1992). Such a development would allow operators to achieve better capacity utilisation and avoid the large-scale building of recent years.

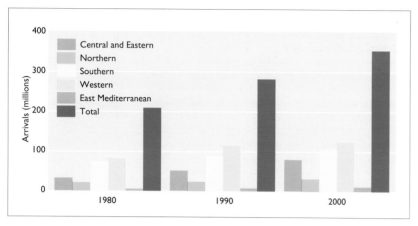

Figure 25.5
Trends in European
international
arrivals, 1985–2000
Source: WTO, 1993a; see
also *Statistical*
Compendium

Central and Eastern Europe

Growth in tourism to and from Central and Eastern Europe is expected to be greater than that for all Europe. By 1990, 50 million visitors were coming annually to Central and Eastern Europe, mainly from Western Europe. In a study carried out on tourism resources and strategies in Poland, Hungary, the former Czechoslovakia, the former Yugoslavia and Bulgaria, the development of cultural and nature-based tourism was seen as a major objective by all governments (Cambridge Economic Consultants, 1993). Measures to open up borders, deregulate exchange rates and provide quality tourist infrastructure, including roads, bed capacity and sporting facilities, are beginning to take effect. Developments are being focused on natural areas, cultural/health facilities such as spas, the coastlines of the Black Sea, and budget skiing resorts in Bulgaria and the Slovak Republic.

The Mediterranean basin

It has been forecast that by 2025 there will be about 380 million tourists in the Mediterranean basin (including all countries bordering the Mediterranean, but not the Black Sea) if economic growth is weak, and about 760 million tourists if it is strong. Almost half the tourists will be along the coastline. These figures are in addition to the growth of the local population, which has been forecast to rise from 350 million people in 1985 to between 530 and 570 million by 2025 (Grenon and Batisse, 1988). There is a trend towards environmental improvement in receiving areas, largely as a result of the demands made by tourists, particularly for good quality drinking and bathing waters. This is beginning to lead to better pollution monitoring and control through schemes such as Blue Flag beaches and Coastwatch, and the introduction of reduce, 're-use and recycle' policies by tourist operators (see below) (ERM and StfT, 1993).

Alpine areas

Tourist demand in the Alps is expected to stabilise by 2000 (Jenner and Smith, 1992), but to continue to grow in other mountain areas such as the Pyrenees, Greece (Mounts Olympus, Parnassus and Pindos) and Central and Eastern Europe, which will experience similar problems to the rest of Europe unless tourism development is planned and carrying capacity defined. In all major resorts, trends are away from reliance on skiing and snow and towards developing multi-sport facilities for the summer and spring seasons. These developments could be less environmentally damaging than prolonged skiing seasons and would make local communities less dependent on the short winter season.

Box 25G
Principles of sustainable tourism

Sustainable tourism, which includes tourism and associated infrastructures both now and in the future, is that which: operates within natural capacities for the regeneration and future productivity of natural resources; recognises the contribution that people and communities, customs and lifestyles make to the tourism experience; accepts that people must have an equitable share in the economic benefits of tourism; and is guided by the wishes of local people and communities in the host areas. This entails:

1 **Using resources sustainably:** the conservation and sustainable use of resources – natural, social and cultural – is crucial and makes long-term business sense.

2 **Reducing overconsumption and wastes:** avoids the costs of restoring long-term environmental damage and contributes to the quality of tourism.

3 **Maintaining diversity:** maintaining and promoting natural, social and cultural diversity is essential for long-term sustainable tourism, and creates a resilient base for the industry.

4 **Integrating tourism into planning:** tourism development which is integrated into a national and local strategic planning framework and management plans, and which undertakes environmental impact assessments of projects, plans and policies, increases the long-term viability of tourism.

5 **Supporting local economies:** tourism that supports a wide range of local economic activities, and which takes environmental costs and values into account, both protects those economies and avoids environmental damage.

6 **Involving local communities:** the full involvement of local communities in the tourism sector not only benefits them and the environment in general but also improves the quality of the tourism experience.

7 **Consulting stakeholders and the public:** consultation between the tourism industry and local communities, organisations and institutions is essential if they are to work alongside each other and resolve potential conflicts of interest.

8 **Training staff:** staff training which integrates sustainable tourism into work practices, along with recruitment of local personnel at all levels, improves the quality of the tourism product.

9 **Marketing tourism responsibly:** encouraging tourists to visit sites during off-peak periods to reduce visitor numbers, and when ecosystems are most robust. Marketing that provides tourists with full and responsible information increases respect for the natural, social and cultural environments of destination areas and enhances customer satisfaction.

10 **Undertaking research:** ongoing monitoring by the industry using effective data collection and analysis is essential to help solve problems and to bring benefits to destinations, the industry and consumers.

11 **Better information provision:** providing tourists with information in advance, and in situ (eg, through visitor centres), about tourist destinations.

Source: WWF, 1992

Box 25H Examples of legislation and policy tools being used to manage the impacts of tourism and recreation in Europe

Controls on landuse: in Cyprus, landuse controls have been introduced with zoning for protected areas, controlled development areas and transferable development rights which may be used in other uncontrolled zones. In Denmark a new 3 km coastal planning zone makes strong restrictions on tourist development on open coastlines. EC legislation on environmental impact assessment has influenced large-scale tourist projects in Europe.

Planning controls on new building: in Ria Formosa National Park (Portugal), planning permission for tourist facilities in the park is now dependent on operators adopting one of several national-park-approved designs.

Controls on bed capacity or infrastructure: in Malta, down-market bed capacity has been gradually eliminated and accommodation has been developed in collaboration with tour operators in 4- or 5-star hotels. In 1993, Majorca removed 25 000 bed spaces to ease tourism pressure. In Austria, voluntary restrictions on further building have been introduced by local communities in Kleinwalsertal (ERM and StfT, 1992).

Controls on the illegal building of second houses: in Sweden and Denmark, strict controls have been in place for several decades on the building of weekend homes. In Norway, a more restrictive approach to building second homes is now being adopted (Bernes, 1993).

Environmental standards for drinking water, bathing water, wastewater and air emissions are incorporated in EC Directives. Such standards were first enforced in Majorca in 1984, before Spain entered the European Community; guidelines have also been developed for open space and densities of new developments. Similar regulations have been introduced in Cyprus.

Establishment of protected areas and mixed areas around sensitive areas such as wetlands and dunes: for instance, the 'Green Lungs' example in Central and Eastern Europe described in Chapter 9.

Traffic management schemes and restrictions on the use of private cars. At the Glossglockner Hochalpenstrasse in Austria, stringent road pricing measures and conversion of car parks have been introduced to discourage car users. The target is to shift from 90 per cent of the 1.3 million visitors arriving by car to 70 per cent by coach and bus by 2000. Many towns and cities throughout Europe have introduced pedestrianised zones in the centres.

Limits to tourist numbers: function ratios, relating resident to guest populations, have been introduced in popular areas (eg, in Denmark, cooperation between planning authorities and the tourist industry takes place with the aim of controlling the development of tourism).

Economic instruments: where such measures have been applied in Europe, this has normally been for reasons other than environmental ones (ie, to raise revenue). However, some measures can indirectly help relieve environmental pressure (eg, the 10 per cent land levy on new developments in Cyprus can help reduce unfavourable developments).

Environmental awareness building: for tourists, host communities and tour operators. Numerous good-practice guides have been developed by hotel chains theme parks and automobile clubs. 'Blue Flags' awarded to beaches meeting certain standards (eg, EC regulations, provision of certain facilities and good beach area management) promoted by the Foundation for Environmental Education in Europe has affected tourist choice of beaches.

Training of those involved in the management of reception areas. For instance, partnerships have been developed between Central and Eastern European and Western European national parks such as Sumava in Bohemia and Bayerischer Wald in Germany and between Ojcow in Poland and the Peak District in the UK. A manual has been written of how to encourage ecological practices in hotels (ZAO, 1993). In addition, initiatives such as the 'Walled Towns Friendship Circle', an alliance of 109 towns in 15 countries, have produced *A Handbook of Good Practice for Sustainable Tourism in Walled Towns* (Bristol Business School, 1994).

Monitoring actions: it is vital to monitor the effects of tourism and of policy actions to refine and improve strategies (eg, the use of indicators to monitor effectiveness of tourism policies).

Northern European wilderness areas

Large areas with relatively untouched nature, such as found in the Nordic countries, in Russia, and some parts of Central and Eastern Europe, are also threatened. This is not because the number of visitors is too high, but because, relative to their 'untouched' status, almost any impact is too much. Thus, if these areas are to be protected and safeguarded, action must be taken to moderate their tourist development.

MANAGING FOR SUSTAINABLE TOURISM

Currently, in many countries there is an increased commitment to the concept of tourism based on sustainable development. This has led to international agreements on tourism and environment (eg, the Convention on the Protection of the Alps agreed in 1991, and the Euro-Mediterranean Declaration on Tourism within Sustainable Development in September 1993). However, unfortunately, no form of tourism is completely environmentally friendly. Tourism by its very nature is a burden on the environment, and therefore requires careful control and planning.

Some general principles that can be used to progress towards sustainable tourism are shown in Box 25G. In order to adhere to such principles, however, decision makers rely on a package of good practice tools such as:

- development of various codes of conduct by international and national bodies and the tourist industry; this type of self-regulation was an important concept recognised in Agenda 21, agreed at UNCED (UNEP, 1992);
- development of guidelines, such as those developed by WTO and UNEP for national parks and protected areas for tourism (WTO/UNEP, 1992);
- specific legislative acts and policy actions, which are being used by governments, conservation bodies, the tourist industry and tourists acting together or individually to reduce impacts on the environment; broad policies relating tourism to the environment are developed at the international or national level; planning is normally carried out at the regional level; while the actual management itself is rather site specific – some examples of specific actions are summarised in Box 25H.

CONCLUSIONS

- Tourism and recreation can threaten the environment if not well managed, and at the same time be affected by the environment. Thus, environmental degradation can affect holiday choice, and reduce future earning arising from tourism. Tourism and recreation can impact on the natural environment to such an extent that it can lead to a change in markets. This is of particular concern in Central and Eastern European countries, but probably also elsewhere in Europe, where the temptation might be to develop unsustainable tourism as a 'quick fix' to assist with economic recovery.

- Existing tourism statistics are inadequate for explaining environmental impacts. Indicators are being developed to measure sustainable tourism and to measure carrying capacities. Currently, the analysis and conclusions concerning the impacts of tourism on the environment are best carried out on a case-by-case basis.

- The undesirable impacts of tourism and recreation are largely site specific and vary according to the type of setting, size of area, the number of visitors, the seasonal concentration, the mode of transport used, development pressures, resource use, waste disposal and how well the area is managed. Broad policies, regional plans and local management actions are important in determining whether tourism and recreation lead to undesirable environmental impacts.

- Nature parks and protected areas – Numbers of visitors to environmentally sensitive sites are increasing, due to a growing interest in nature all over Europe. The main problems are caused by excessive use of passenger cars and the concentration of visitors in peak periods at popular sites that are easily accessible from cities. Different levels of protection can help in reducing an overload of visitors.

- Rural zones – Rural tourism has traditionally been popular all over Europe. It is currently expanding rapidly in Northern, Western and Southern Europe as an attempt to diversify the rural economy. Some rural tourism activities such as eco-tourism, which may have lower environmental impacts, are becoming increasingly popular. However, other recreational activities, such as golf and hunting, pose specific environmental problems, which require careful planning. Those remaining wilderness areas of Europe (such as the northern part of Nordic countries and Russia), which are still relatively untouched, require particular protection.

- Mountains – The cumulative environmental impacts from skiing are considerable. The main causes of environmental damage in mountain areas are the concentration of tourists in both time and space, deforestation and high levels of road vehicle usage. The development of ski infrastructure poses specific problems. Attempts to reduce environmental impacts in popular areas such as the Alps (eg, restrictions on car parking spaces) may succeed at the cost of shifting the problem elsewhere. Other impacts in mountains from tourism are significant (eg, erosion caused by trampling, and off-road driving) and need to be managed by zoning and other landuse planning measures.

- Coastal areas – Many coastal areas all over Europe are suffering from mass tourism overdevelopment. The greatest pressures are still felt in the Mediterranean – a situation likely to continue in future. Northern and Western European coastlines still attract large numbers but do not suffer from the environmental problems of the Mediterranean basin and Black Sea. Impacts arise from the sheer concentration of tourists, accommodation and infrastructure development, generation of solid and liquid wastes, and traffic. Such impacts could be mitigated by zoning and other landuse planning measures. The special environmental problems associated with coastal zones are discussed further in Chapter 35.

- Cities and heritage sites – The large number of visitors to many cities and heritage sites in Europe cause environmental problems such as traffic congestion and wear and tear on buildings and heritage sites. Visitor numbers in cities and heritage sites will increase in the future, as urban-based tourism becomes more popular.

- Theme and leisure parks – Theme and leisure parks are increasingly popular in Northern, Western and Southern Europe as a form of tourism and recreation. Their environmental impacts (eg, greater use of energy and water resources) tend to be site specific.

References

Balearics Tourism Authority (1992) Challenges and choices for the 90's, unpublished paper presented at English Tourist Board (ETB) *Tourism and the Environment* conference 16–17 November 1992, London.

Batisse, M (1992) The struggle to save our world heritage, *Environment* **34**(10), 13–20, 28–30 Heldref Publications, Washington DC.

Bernes, C (Ed) (1993) The Nordic environment – present state, trends and threats, *Nord 1993: 12*, Nordic Council of Ministers, Copenhagen.

Brady Shipman Martin (1993) Urban environment: the problems of tourism, final report. Study for CEC DG XI, Dublin (unpublished report).

Brettle, W (1989) 'Erfüllung moderner Urlaubswünsche', in the thesis 'Wirkungen des Wertewandels auf den Tourismus', Institute for Tourism, Technical University Berlin.

Bristol Business School (1994) *A handbook of good practice for sustainable tourism in walled towns*. Bristol Business School, University of the West of England. Study for CEC DG XXIII, Bristol (unpublished report).

BUWAL (1992) *The Swiss Report. UNCED 1992*. Swiss Federal Office of Environment, Forests and the Landscape (BUWAL), Berne.

Cambridge Economic Consultants (1993) *Tourism resources in Eastern Europe: problems and prospects for cooperation*. Study for CEC DG XXIII. PA Cambridge Economic Consultants, CEC, Luxembourg.

CEC (1991) *Europe 2000, outlook for the development of the Community territory*. Commision of the European Communities, Luxembourg.

CEC (1992a) *Tourism 1990 annual statistics*. Eurostat, Commision of the European Communities, Luxembourg.

CEC (1992b) *The state of the environment in the European Community*. COM(92) 23 final, Vol III, 27 March 1993, Commision of the European Communities, Luxembourg.

CEC (1993) *Towards sustainability, a European Community programme of policy and action in relation to the environment and sustainable development*. OJ No C, 138, 5–98, 17 May 1993, Commision of the European Communities, Luxembourg.

Ceron, J-P (1992) Parcours de golf. *L'environnement Magazine* no 1509, pp. 27–30, Paris.

Council of Europe (1991) *Recommendation No 33 (adopted on 6 December 1991) on the conservation of the natterjack toad (Bufo calamita) in Ireland*. Standing Committee of the Convention on the Conservation of European Wildlife and Natural Habitats (Berne Convention), Council of Europe, Strasbourg.

Council of Europe (1992) Specific environmental problems arising from the increase of tourism in Eastern and Central Europe. *Proceedings of the Budapest Colloquy, Tourism and Environment, 25–7 September 1991*, Strasbourg.

Curtis, T (1993) personal communication, Irish Office of Public Works, Dublin.

EIU (1992) *European tourist forecasts to 2005*. Special report 2454, Economist Intelligence Unit, London.

ERM and StfT (1993) *Environment and tourism in the context of sustainable development*, Report by ERM (Environmental Resources Management) and StfT (Studienkreis für Tourismus) for CEC, DG XI, December 1993, Brussels (unpublished).

Euromonitor (1992) *European tourism*. Euromonitor, London.

European Omnibus Survey (1987) *Europeans and their holidays*. Survey carried out in March–April 1986 for Eurobarometer, Surveys, Research and Analyses Unit, DG X, Commission of the European Communities, Brussels.

FNNPE (1993) *Loving them to death? The need for sustainable tourism in Europe's nature and national parks*. Final Report to the European Commission, FNNPE (Federation of Nature and National Parks of Europe), Grafenau, Germany.

GEF (1993) *Project document: environmental management and protection of the Black Sea*. Global Environment Facility, United Nations Development Programme, United Nations Environment Programme and The World Bank, New York.

Grenon, M and Batisse, M (Eds) (1988) *Le plan bleu: avenirs du Bassin Méditerranéen*. Report prepared for UNEP Mediterranean Action Programme, Economica, Paris.

Hey, E and Mee, L D (1993) The Black Sea Ministerial Declaration: an important step. *Environmental Policy and Law* **23**(5), October 1993, International Council of Environmental Law, Amsterdam.

Institut für Planungskybernetik (1989) *European travel monitor*. Institut für Planungskybernetik (IPK), Munich.

IUCN (1990) *1990 United Nations list of national parks and protected areas*. International Union for the Conservation of Nature and Natural Resources, Gland (Switzerland), and Cambridge (UK).

Jenner, P and Smith, C (1992) *The tourism industry and environment*. Economist Intelligence Unit, Special Report 2453, *The Economist*, London.

Krippendorf, J, Hamele, H and Zimmer, P (1992) *Reconciling tourist activities with nature conservation*. Report for the Council of Europe, 30 October 1992, Council of Europe.

Lalumière, C (1993) *Speech on the occasion of the Lucerne pan-European Ministerial conference* (Switzerland, 28 April 1993). Council of Europe, Strasbourg.

Messerli, P (1990) Tourismusentwicklung in einer unsicheren Umwelt: Orientierungspunkte zur Entwicklung angemessener Strategien, *Die Volkswirtschaft*. December 1990, 27–33, Berne.

MECR and CAS (1991) *Environment in the Czech Republic, part II*. Report prepared for presentation at the UNCED conference, Ministry of Environment of the Czech Republic and the Czechoslovak Academy of Science, Prague.

Ministère de l'Environnement (1991) *État de l'environnement*. Édition 1990, La Documentation française, Paris.

Ministère de l'Environnement/Ministère du Tourisme (1992) *Tourisme et environnement: 1ères rencontres internationales de La Rochelle*, 13–14 May 1991, La Documentation française, Paris.

Ministère de l'Environnement/Ministère du Tourisme (1993) *Tourisme et environnement: du tourisme de nature à l'ecotourisme*, hors série, 'Les cahiers Espaces', February 1993, La Documentation française, Paris.

Müller, H R (1993) Challenges in our time and their consequences for tourism management. Paper presented at seminar on *Tourism Management between Tradition and Sustainability*, 4 February 1993. Research Institute for Leisure Time and Tourism, University of Berne.

National Bank of Hungary (1992) Annual Report 1992. National Bank of Hungary, Budapest.

Norwegian Directorate for Nature Management (1993) *The state of habitat protection in the Arctic* (Canada, Denmark, Finland, Iceland, Norway, Russia, Sweden, USA). Conservation of the Artic fauna and flora, Report No 1, Second draft, Oslo.

Pain, D J (1992) Lead poisoning of waterfowl. In: *Lead Poisoning of Waterfowl, Proceedings of an International Waterfowl and Wetlands Research Bureau (IWRB) Workshop*, Brussels, 13–15 June 1991, IWRB, Slimbridge, UK.

Partsch, K (1991) *Rapport sur la situation des Alpes, Alpen-und Europabüro*. Karl Partsch, Member of the European Parliament, Sonthofen, Germany.

Pinet, J-M (1987) *L'économie de la chasse*. Comité National d'Informaton Chasse-Nature, Paris.

StfT (1986), (1989) and (1992) *Reiseanalyse*. StfT (Studienkreis für Tourismus), in cooperation with GFM-GETAS and BasisResearch, Starnberg, Germany.

Touche Ross (1993) *How green is the fairway? A common agenda for sustainable golf development and management*. Project undertaken for DG XXIII, CEC, Touche Ross, London (unpublished report).

Tourism Concern (1992) White gold. In: *Tourism in Focus*, Autumn 1992. RAP Ltd, Rochdale, UK.

UNCED (1992) *Agenda 21: Chapter 13 (Managing fragile ecosystems: sustainable mountain development) and Chapter 17 (Protection of the oceans, all kinds of seas, including enclosed and semi-enclosed seas, and coastal areas and the protection, rational use and development of their living resources)*, UNCED Secretariat, Conches, Switzerland.

UNEP (1992) Tourism and the environment: facts and figures. In: *UNEP Industry and Environment* July–December 1992 15(3–4), 3–5. UNEP Industry and Environment Programme Activity Centre, Paris.

World Bank/European Investment Bank (1990) *The environmental programme for the Mediterranean: preserving a shared heritage and managing a common resource*. World Bank, Washington DC/Luxembourg.

WTO (1993a) *Yearbook of tourism statistics, 45th edition*. World Tourism Organisation, Madrid.

WTO (1993b) *Basic data collection and techniques*. ENV/6/8(a), Environment Committee, Sixth Meeting, World Tourism Organisation, Madrid, 1–2 June 1993.

WTO (1993c) *First International Forum on Tourism for the Senior Citizen*. May 1993, World Tourism Organisation, Canary Islands, Spain.

WTO/UNEP (1992) *Guidelines: development of national parks and protected areas for tourism*. WTO/UNEP joint publication, Madrid/Paris.

WWF-UK (1992) *Beyond the green horizon: principles for sustainable tourism*. World Wide Fund for Nature, Godalming, UK.

ZAO (1993) *Ecologia in albergo, manuale per il recupero ambientale nei luoghi dell 'ospitalità'*. ZAO handbook partly funded by CEC DG XXIII, Selene Edizioni, Milan (also available in English and Greek).

Housing in the Isle of Dogs, London, UK
Source: Spectrum Colour Library

26 Households

INTRODUCTION

The domestic sector is an important area of the economy but is often unjustly overlooked as a source of environmental damage. Household spending accounts for a significant proportion of total industrial production, averaging just over 70 per cent in Europe as a whole and ranging from 56 per cent in the former USSR to 89 per cent in Greece in the late 1980s (ERM based on UN, 1992, and Euromonitor, 1992). The choices made by individuals concerning purchase of consumer items and how to run their homes can significantly influence environmental impacts from the domestic and other sectors. It is this range of behaviour and scope for choice which is addressed here. The main sources of data used in this chapter are described in Box 26A.

DEFINITION AND IMPORTANCE OF HOUSEHOLDS

A useful definition of 'households' for the purposes of this report is: a '... household denotes a person or group of persons, related or not to one another, who occupy the same accommodation and live there together' (CEC, 1993).

Households consume raw materials, electricity, other forms of energy, food and manufactured items while generating wastes which are released to land, water and air. They also require transport facilities and an infrastructure which may affect landuse, the landscape and natural resources. They therefore have direct impacts on the quality of the environment as described in Chapters 4 to 10.

Population growth, together with a trend towards fewer people per household, is contributing to a large rise in the number of households in the EU and EFTA countries (an increase of

around 10 per cent between 1980 and 1990) (Euromonitor, 1992). There is a similar trend in Central and Eastern Europe. In the former USSR, however, the rate of growth is constrained by the shortage of suitable housing. Smaller households use water and energy less efficiently and require more land per household member, so that these trends lead to greater per capita resource use.

Patterns of household consumption and consumer behaviour influence the environmental effects of households. At the same time households themselves are influenced by the physical and climatic conditions of their location and a variety of demographic, social and economic factors, including: household income; availability of goods; availability and quality of substitutes; environmental awareness of consumers; and culture and peer group pressure.

The direct environmental impacts of households arise from the consumption of resources, emissions and the other induced pressures on the environment described in Chapters 12 to 18. Use of resources and emissions which occur away from the household (for example, those associated with electricity generation, industrial or agricultural production of consumer items) is addressed under the specific sectorial chapters of interest (eg, Chapters 19, 20 and 22).

The importance of the household sector, therefore, lies in its demand for resources, the waste generated by consumption of those resources and its capacity to influence industrial and commercial activities through its spending power. This influence can be shown through the exercise of consumer choice, either in increased demand for perceived 'environment-friendly' products or in the avoidance of less 'friendly' products or manufacturers.

The significant environmental impacts stemming directly from the presence, consumption of resources and emissions of households are summarised in Table 26.1.

Box 26A Data sources and country groupings

In many cases data relating specifically to household consumption and emissions are not available. Households *per se* have not always been identified as a key sector by compilers of socio-economic and, particularly, environmental data. Information on households is, therefore, often included as part of a larger category (because it often cannot be identified separately).

Energy data are most often presented as combined figures for the domestic, commercial and service sectors, and these can be difficult to disaggregate because of the differing ways in which the various sectors are defined. Energy data for East Germany are included in Central Europe in this chapter, being pre-1991.

Socio-economic data may also pose problems of interpretation and comparison; for example, data on Central and Eastern Europe are often gathered under

different categories from EU and EFTA countries, making comparisons problematic. Also the severe currency fluctuations in 1992–93 and the different structure of the former centrally planned economies can make statistics relating to GNP (or GMP in the case of Central and Eastern countries) and spending difficult to interpret. Trends in population growth and trends in household size and structure and car ownership present fewer problems (see, eg, Euromonitor, 1992). Data for these parameters include East Germany in the EU group, as this is how the data were reported in the original Euromonitor source.

The country groupings used in this chapter to present data are as in Chapters 19, 20 and 21: EU, EFTA, Western, Central, former USSR and Central and Eastern Europe (see Box 19B).

Table 26.1
Overview of significant environmental impacts of main household requirements
Source: EEA-TF and ERM

Environmental media Household requirement	Air	Water	Soil/Land	Nature and wildlife landscapes
Land (living space)	• building activity → emissions of formaldehyde, radon	• gardening → leaching of pesticides and fertilisers	• renovations or extensions → building waste → disposal problems • housing, gardens, recreational areas → demands on space → land lost for other purposes	• housing, gardens → visual impacts on landscape
Energy (space heating/cooling, cooking, light and power)	• coal, gas and oil burning → emissions of CO_2, CO, NO_x, SO_2, black smoke, particulate matter and VOCs			
Domestic water supplies		• household sewage → emissions of organic matter, phosphates and nitrogen compounds • washing and cleaning (detergents) → organic matter and suspended solids • use of treated water for washing, cleaning, cooking, gardening, toilet flushing → pressures on drinking water supplies		
Consumption of goods including • food, beverages and tobacco • clothing, footwear • furniture, furnishings and household equipment • medicines and pharmaceuticals, • recreational and entertainment items	• aerosols, solvents, paints, refrigeration equipment → emissions of VOCs, CFCs • waste incineration → air pollution	• use of bleaches and disinfectants → chlorinated organic compounds • preparation of food → emissions of nutrients, organic matter	• domestic waste (including food waste, paper and cardboard, glass, ferrous waste (mainly cans) and textiles) → landfill requirements • → chemical wastes such as packaging, pesticides, oil, leftover paint, batteries, cosmetics, medicines, photographic developing fluids	
Transport and services	• combustion of petrol and diesel fuel → emissions of VOCs, NO_x, particulate matter, CO, CO_2 • car paint → emissions of VOCs • refuelling and maintenance → VOC emissions	• oils and lubricants washed into the sewer system from roads and driveways → water pollution	• → used tyres, used vehicles and vehicle components • land for roads, paths, and services including: water pipelines, telecommunications cables and electricity lines → land lost for other purposes	

Figure 26.1
Water consumption by households, Czech Republic, 1970–91
Source: MECR, 1992

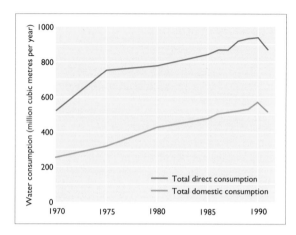

RESOURCE NEEDS OF HOUSEHOLDS

Land

There are few data available on the land area occupied by households. Eurostat estimates that the average floor area of households in the EU is the smallest in Greece ($80m^2$) and largest in The Netherlands ($105m^2$) and Denmark ($107m^2$) (CEC, 1993). Growth in population in the EU contributes to an increasing demand for land. On average the urban population in Europe is expanding while the rural population is contracting. Therefore urban households are on the increase, and pressure to urbanise rural land is further intensified.

Water

In most European countries, domestic users account for 5 to 25 per cent, and usually less than 20 per cent, of total water supplied for all purposes, although this proportion can be much higher in some countries (see Chapters 5 and 33 and the *Statistical Compendium*). The amount of water supplied to households is increasing as a proportion of total water supplied for all purposes – for example in The Netherlands (VROM, 1991) and the Czech Republic (see Figure 26.1). These increases are associated, at least in part, with changes in household structure and living patterns – described later in this chapter.

Almost all households in Europe receive high-quality water supplied directly through a distribution network. However, in rural locations in many parts of Europe there are households which still rely on public fountains or are self-supplied from their own wells.

In most European countries, the total water supplied per person ranges from roughly 150 to around 300 litres per day (only Spain exceeds this with 378 litres per capita per day – see *Statistical Compendium*). The variation may partly reflect the

different ways in which countries measure water use. Estimates of the average water consumption of various domestic appliances (UK DoE, 1992a) range from 100 litres for automatic washing machines (amount of water required each time the appliance is used), to 80 (baths), 50 (dishwashers), 30 showers and 10 (WCs). Details of the various uses to which this is put in all countries are not available, although statistics for The Netherlands show that the main uses are for flushing of water closets (WCs), bathing and laundry (see Figure 26.2). Average use per person in The Netherlands in 1986 stood at around 120 litres per day and this figure represents a 76 per cent increase since 1960. The increase is attributed to wider home ownership and use of showers, and to the less efficient use of water associated with the trend towards smaller but more numerous households (VROM, 1991).

Energy

European countries present data on energy consumption in various ways (Box 26A). Differences are particularly marked between the formerly centrally planned and the market economies of Western Europe. In this section the energy use of the domestic and commercial sectors are considered together (data exclude energy use for transport). For general energy consumption data see Chapter 19. Commercial energy use is normally a very small proportion of these combined sectors in Central Europe but typically accounts for up to about half the total of domestic and commercial energy use in Western Europe and the former USSR (WRI, 1992).

Final energy consumption by the domestic and commercial sectors has increased by almost 50 per cent for Europe as a whole (including the former USSR) from 1970 to 1990. On the other hand, total energy consumption has increased by 37 per cent over the same period (excluding the former USSR, the increases were 22 per cent and 20 per cent respectively). The domestic and commercial sectors accounted for 37 per cent (741 out of 2023 million tonnes of oil equivalent (Mtoe)) of total energy use in Europe (including the former USSR) in 1990, compared with 34 per cent in 1970 (496 out of 1465 Mtoe) (see *Statistical Compendium*). This increase cannot be attributed solely to population growth. There has been a general increase in domestic and commercial energy use per capita of 35 per cent across the whole of Europe in the past 20 years (from 0.74 in 1970 to 1.0 toe in 1990, including the former USSR), and this has been most marked in the former USSR, where the average per capita consumption rose by over 80 per cent (from 0.6 in 1970 to 1.1 toe in 1990). Both the absolute amount and intensity of energy use were extremely variable – ranging from 0.1 to 1.8 toe per capita; and consumption is lowest in Southern Europe (0.1 toe in Albania, and 0.2 toe per capita in Portugal and the former Yugoslavia) and highest in East Germany (1.8 Mtoe), and in Sweden, Finland, Switzerland and Luxembourg (1.5 Mtoe) (IEA, Eurostat and UN data; see *Statistical Compendium*).

Research in Norway has shown that energy use per capita is highest in single person households, and that for each additional person the additional heating required is less (Figure 26.3).

Trends in the proportions of domestic and commercial energy generated by fuel type are shown in Table 26.2. Throughout Europe over the past 20 years there has been a change in the mix of energy used in households. Coal and liquid petroleum products once widely used for space heating are now much less so, whereas gas, electricity and 'heat' (eg, the supply of hot water for central heating through a combined heat and power system) have all increased significantly. There were marked differences between Central and Eastern and Western Europe in 1990, with solid fuels still being the largest contributor in Central European homes (41 per cent of the sector as a whole), while in the EU and EFTA solid fuels are less than 5 per cent of total fuel consumption in this sector. The table also shows that natural gas began to

Figure 26.2
Main uses of household water consumption, The Netherlands, 1986
Source: VROM, 1991

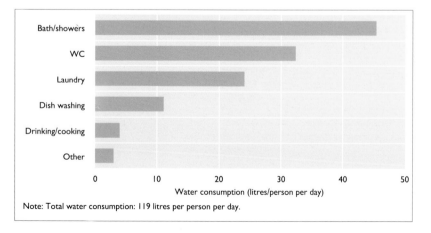

Note: Total water consumption: 119 litres per person per day.

make a significant contribution to energy supplies in Western Europe in the early 1970s and somewhat earlier in Central Europe and the former USSR.

Surplus heat from electricity generation, and from some industrial plants (eg, paper making), can be used for space heating in combined heat and power generation schemes with considerable gains in energy efficiency. Such schemes are, however, usually feasible only where there is a concentration of flats or small houses close to a plant. Central and Eastern Europe has a comparatively long history of district heating which almost doubled between 1970 and 1990. In 1990 this accounted for almost 20 per cent of final energy consumption of total domestic and commercial energy. Western Europe, however, has only recently begun to make significant use of district heating and only 3 per cent of Western Europe's final consumption for domestic and commercial energy is contributed from this source (in 1990). This contribution is supplied almost entirely by the Scandinavian countries. Iceland has been a pioneering country in the use of geothermal energy for space heating, and this source provides about 85 per cent of the energy requirements for space heating in Iceland. Electricity production using geothermal energy is limited, with only 6.4 per cent of total electricity in 1990 coming from this source (Icelandic Ministry for the Environment, 1992). In certain parts of Europe, solar and wind energy create energy self-sufficient houses, but this is confined mainly to demonstration projects.

EMISSIONS FROM HOUSEHOLDS

To air

Direct emissions of pollutants to the air by households arise principally through four sources:

- the use of solid fuels, heating oils and gas as domestic energy sources;
- the use of aerosols for cosmetics, cleaning agents etc;
- the use of paints and solvents; and
- fuelling and maintaining motor vehicles.

A list of key pollutants emitted by these activities is presented in Table 26.1. Indirect emissions come about from the disposal of consumer goods, CFCs from refrigerators being of particular note (see Chapters 14 and 15). Indoor air pollution is covered in Chapter 11.

The volume and nature of emissions associated with energy use are determined by: the amount of energy used; the fuel type; and the emissions control or treatment technology.

For 20 European countries in 1990, the combined combustion from the commercial, institutional and residential sectors accounted for 14 per cent of carbon monoxide and 17 per cent of carbon dioxide emissions (see Chapter 14).

More detailed data are available for some individual countries, which serve to illustrate the importance of the sector. In the UK, for example, coal burning in households is still a significant contributor to UK black smoke emissions, contributing to over one third of total black smoke emissions in 1991. Figure 26.4 shows the impact of the decline in domestic sector emissions on total emissions of black smoke in the UK. Although emissions from the sector have fallen both absolutely and as a proportion of the total in the past 20 years, it is clear that domestic coal users are responsible for a disproportionate fraction of black smoke emissions (for one kg of coal, the emission factor is 160 times that of a power station; Gillham et al, 1992). Carbon monoxide, sulphur dioxide and nitrogen oxides emissions by households on the other hand accounted for 5 per cent, 4 per cent and 3 per cent respectively of total UK emissions for the same year, and these amounts are roughly in accordance with the proportion of combustion of energy taking place in households (UK DTI, 1992; UK DoE, 1993).

Fuel type	EU 1970	EU 1980	EU 1990	EFTA 1970	EFTA 1980	EFTA 1990	Central Europe 1970	Central Europe 1980	Central Europe 1990	Former USSR 1970	Former USSR 1980	Former USSR 1990
Solid	22	8	4	8	3	2	53	44	41	37	27	16
Petroleum	53	45	33	66	54	36	21	18	12	33	29	24
Gas	12	28	36	2	4	5	4	9	14	15	21	29
Electricity	12	18	25	15	26	39	7	9	15	8	8	10
Heat[1]	1	1	2	1	6	10	15	20	18	7	14	21
Others[2]	0	0	0	8	7	9	0	0	0	0	0	0

Notes: 1 Refers to heat produced in combined heat and power (CHP) stations and, for non-EU countries, additionally includes heat generated in district heating plants (DHP). DHPs are generally less common in the EU. 2 Data under this heading are generally incomplete, but refer mainly to wood use in the domestic sector. A zero figure may thus reflect a lack of adequate data rather than zero consumption.

Table 26.2
Percentage of domestic energy use by fuel type, Europe, 1970–90
Source: See *Statistical Compendium*

Householders have little opportunity to treat emissions or to burn fuel at the same efficiency as an industrial user or a power station. Emissions from the domestic sector can be limited in a number of ways: by restricting the amount of energy used, by improving the efficiency with which it is used, by requiring the use of alternative fuels (eg, smoke control zones), or by encouraging consumption of a less polluting fuel type such as 'smokeless' or low sulphur fuels, and altering behaviour away from using in-house open fireplaces.

Significant emissions from sources other than combustion arise mainly from use of paints, solvents, aerosol sprays and refrigeration equipment. Few countries record these domestic emissions separately, but in The Netherlands it is estimated that around 20 per cent of all ozone depleting chemicals and 10 per cent of volatile organic compounds are emitted directly from households (VROM, 1991).

Sewage

By 1988 the great majority of European households were served with sewer connections, the only notable exceptions being in the former Yugoslavia and the rural areas of Greece, Portugal and Romania (WRI, 1992).

The most significant source of organic matter of anthropogenic origin in surface waters is human excreta. The range of per capita contributions to domestic wastewater is shown in Table 26.3.

Washing and cleaning also contribute organic matter and suspended solids to domestic sewage, but the most significant contribution – aside from excreta – is that of phosphates from detergents, originating mainly from households (UNECE, 1992). There is convincing evidence that the introduction of 'phosphate-free' detergents is changing this (see Box 26B).

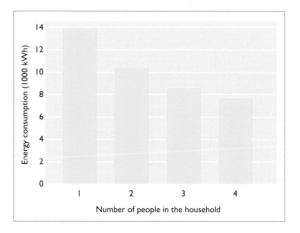

Figure 26.3
Average energy consumption per capita in households of different sizes, Norway
Source: CBS, 1991

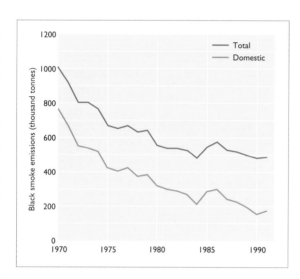

Solid waste

The main issues of concern with regard to domestic waste are:

The volume

The volume of waste produced by households, particularly in Western Europe, is increasing owing to increasing consumption and especially to increasing volumes of packaging material. Many products which could be bought loose in the past are now sold pre-packaged, and traditional packaging materials such as paper and cardboard have been replaced by various types of plastic packaging.

The recycleability

Many of the substantial contributors to the total volume of household waste are materials which could be recycled, including glass, plastics, paper, aluminium, textiles and ferrous waste (for example tin cans).

The presence of chemical pollutants

Domestic waste includes items which contain small quantities of hazardous materials which, if of industrial origin, would warrant special measures for safe disposal. Studies in The Netherlands have identified pesticides, oil, leftover paint, batteries, cosmetics, medicines and photographic developing

fluids as such items present in household waste (VROM, 1991).

Studies in Western Europe indicate that, although at least half of domestic waste, by weight, is recyclable, less than 10 per cent is actually recovered for recycling (ERM, 1992). Many cities, countries and pressure groups are now promoting recycling of household waste. However, recycling does not always yield clear environmental benefits, since the benefit derived needs to take into account the amounts of energy needed to collect and process the waste material as well as the nature of the waste streams generated by the recycling process itself (see also Chapter 15).

For recycling schemes to succeed where environmental benefits are clearly achievable, there needs to be: collaboration between governments to set up, promote and, where necessary, enforce schemes; an industry to collect and recycle wastes; adequate recycling facility capacity; and appropriate separation and disposal of waste. An example of such a collaboration is to be found in Germany where the introduction of the 'Green Dot' scheme for collecting and recycling packaging has not only resulted in large volumes of material being collected for recycling, but has also caused a decrease in packaging *usage* by manufacturers (ERM, 1992) (Chapter 15). This reduction may be due to manufacturers seeking to avoid recycling costs by eliminating unnecessary packaging.

In Norway a recent census found that the proportion of households sorting waste varies considerably with the type of household. The percentage is higher among multi-member households, particularly those with children (CBS, 1991).

DRIVING FORCES

The size and significance of the household sector's contribution to resource use and emissions to the environment is determined by the number and size of households and by their patterns of consumption and waste generation. The factors which have recently been most influential in altering the contribution to environmental impacts are:

- population growth, number of households, and household size;
- growth in disposable income and consumer spending;
- greater availability, affordability and sophistication of consumer items; and
- public attitudes to the environmental impacts of products.

Population growth, number of households and household size

Current trends indicate a growth in European population (see Chapter 12) and a reduction in average household size, which in 1990 ranged from 2.4 or less in Sweden, Denmark, Germany, Finland, Norway, The Netherlands and Switzerland, to more than 3.0 in Albania, Ireland, Italy, the former Yugoslavia and the former USSR (Euromonitor, 1992). (A figure of 3.9 quoted for the former USSR is distorted because the number of single member households was not recorded. This caveat applies to all the following section.)

During the period 1980 to 1990 the total number of households in Europe rose from 167 million to 183 million (excluding the former USSR). The average size of each household fell during the period from 2.9 to 2.7 people per household (excluding the former USSR) (see Figure 26.5). From the data available in Figure 26.6, the rate of household formation is shown to be fastest in Western countries and slowest in Central Europe. In Western Europe the number of households increased by almost 10 per cent between 1980 and 1990, but only about two thirds of this was due to the natural

Measurement	Per capita contribution (grammes per day)
BOD$_5$[1]	45–54
COD[2]	1.6–1.9 × BOD$_5$
Total solids	170–220
Suspended solids	70–145
Nitrogen	
Total (as N)	6–12
Organic	around 0.4 × total N
Free ammonia	around 0.6 × total N
Nitrate	around 0.0–0.05 × total N
Phosphorous	
Total (as P)	around 0.6–4.5
Organic	around 0.3 × total P
Inorganic (ortho- and polyphosphates)	around 0.7 × total P

Notes:
1 BOD$_5$: biological oxygen demand (over a 5-day period).
2 COD: chemical oxygen demand.

increase in population; the remainder was caused by the splitting of family units (Euromonitor, 1992). In Central and Eastern Europe on the other hand, the scope for the creation of new household units has been restricted largely by the scarcity of suitable housing, which probably explains the increases in household size.

With the exception of Ireland (where about one third of households have five or more members), the largest households are found in Central and Eastern Europe. In Hungary, Poland, the former Czechoslovakia and the former USSR, households with five or more members represent 17 to 24 per cent of the total number of households. Single member households appear to be far less common; in Poland and Romania for example, barely 14 per cent of households are single member units (Euromonitor, 1992).

Single member households are most common in Denmark and West Germany, where they account for about 34 per cent of the total. On the other hand, in Spain and Italy only about 13 per cent of households are single member units (one in eight) (Euromonitor, 1992). Households in the EU had an average size of 2.6 people in 1990; countries close to this average were Belgium, Greece, Luxembourg and the UK (Euromonitor, 1992).

The growth in the number of households is attributable largely to a changed social pattern towards more single member households and fewer children per family. It seems likely that this movement is largely complete in Western Europe, and that the rate of new household creations will decrease to that needed to accommodate population growth. In Central and Eastern Europe, however, where household creation has been constrained, there is a potential for household numbers to increase (Euromonitor, 1992).

Box 26B Experience of phosphate replacements in detergents

Germany

In West Germany in 1979, around 70 000 tonnes of phosphorus (P) were used in detergents; in 1989 the amount was a little over 5000 tonnes. The market share of phosphate-free detergents increased from 80 per cent in 1988 to more than 95 per cent in 1989–90, reducing further the amounts of phosphorus used.

In the river Rhine at Düsseldorf the annual phosphate load decreased by 64 per cent between 1979 – when measurements began – and 1989. This corresponds to a reduction of 28 400 tonnes of phosphorus per year. Phosphate concentrations were also found to have dropped by half between 1978 and 1987 in other important rivers (Ruhr, Main and Neckar).

The observed reduction in phosphate load is almost certainly due to the reduction in the use of phosphate in detergents, coupled with improved phosphorus removal during sewage treatment.

Switzerland

Phosphate content of detergents was reduced in a series of stages (1977, 1981 and 1983), until an outright ban on phosphate detergent came into force in 1987. Phosphorus concentrations declined very rapidly from 1986 onwards. In the river Rhine close to Basle, phosphorus concentrations fell from 170 µg P/l in 1979 to 130 µg P/l in 1985 and to 70 µg P/l in 1989. Similar reductions were found in lakes. The reductions are attributable to sewage treatment plants, as well as to the ban on phosphate in detergents.

Source: UNECE, 1992

Growth in disposable income and consumer spending

Assessments of comparative spending power are complicated by currency instabilities, and notably by the currency strength of the EFTA group. Subject to this qualification, however, according to Euromonitor data for 1990, average expenditure in Western Europe is about ECU 14 000 per capita, around four times that of Central and Eastern Europe and the former USSR where the average is about ECU 3400. Europe's most affluent consumers are found in Switzerland, where total consumer spending power is almost ECU 25 000 per capita and ECU 59 000 per household. West Germany, France, Belgium, Italy, Austria, The Netherlands, Norway and the UK all have expenditure levels in the region of ECU 13 700 to 16 200 per capita and ECU 33 000 to 47 000 per household. Denmark, Sweden, Iceland and Finland are all quoted as spending over ECU 17 000 per capita, but in the range ECU 39 000 to 46 000 per household, since households tend to be smaller (*Statistical Compendium*).

The countries with the lowest average consumer expenditure in Western Europe are Greece, Ireland and Portugal, with per capita expenditure of ECU 4500 to 8700 per year. Central European countries are at an even lower level, with average expenditure of just ECU 1500 per capita.

In the EU (excluding East Germany) GDP grew in the three decades between 1961 and 1991 by 4.8 per cent, 3.0 per cent and 2.3 per cent, respectively (average of annual percentage changes). (CEC, 1992a). During these periods, private consumption grew at approximately the same rate: 5.0 per cent, 3.4 per cent and 2.4 per cent, respectively (average of annual percentage changes) (CEC, 1992a). GDP data for Central and Eastern Europe and the former USSR are less reliable and hence conclusions cannot be drawn.

The way in which consumers allocate their expenditure on goods and services has implications for environmental impacts, and although the specific nature of impacts will depend very much on many interim factors, the proportion and volume of expenditure is of importance. Sectorial allocations of consumer spending vary widely in Europe, with the poorer countries spending the highest proportion of their income on essentials such as food and basic clothing (see Figure 26.7). Particularly in Central and Eastern Europe, the pattern has been distorted by heavy state subsidies on housing, transport and household fuels, but as these are reduced or removed, the proportion spent on essentials, especially on housing, will increase significantly (Euromonitor, 1992).

In Europe, as in other parts of the world, additional

Figure 26.5
European household sizes, 1980–90
Source: Euromonitor, 1992, and ERM

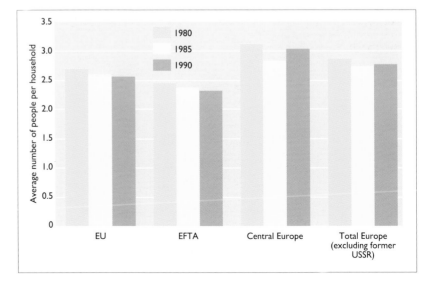

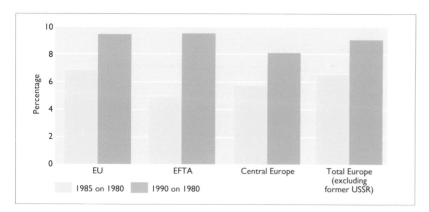

Figure 26.6
Percentage rise in number of European households since 1980
Source: Euromonitor, 1992, and ERM

income from increased economic prosperity is spent differently in poorer countries than in more affluent ones. Recent history has shown that, for Greece, Portugal, all of Central and Eastern Europe, and to a lesser extent Ireland, the first response to greater prosperity has been to use it primarily for the purchase of food and clothing. Once spending power rises above about ECU 5000 to 6000 per capita, the response generally is to reduce the proportion of the budget devoted to food and to spend more on other consumer goods, especially durables (Euromonitor, 1992).

Greater availability, affordability and sophistication of consumer items

Since World War 2, growth in personal incomes and technological development has made an ever wider range of consumer and leisure items available to households. These, together with associated social and cultural shifts, have transformed households resource needs and their patterns of consumption and waste generation.

The increased availability of plumbing, the fitting of baths and showers and the wider use of washing machines and dishwashers, have led to increased water use and volumes of sewage produced by households. Central heating (and cooling) systems increase the demand for energy from this sector. Similarly, leisure items such as television and stereo equipment or electrical appliances such as refrigerators, freezers, microwaves, ovens and other electronic equipment have all led to increased demand for electricity (see Chapter 19). Increased ownership of passenger cars is associated with changes in the number and nature of households (see below). The demand for resources and environmental impacts of road transport are considered in Chapter 21.

Although the initial effect of household possession of these items is greater consumption of resources, it is clear that technological development can, and generally does, improve the efficiency with which resources are consumed.

For example, a recent study investigating ecolabelling of washing machines found a difference of 45 per cent water use between various washing machines. The most water efficient used 13 l/kg of washing, the average being about 18 l/kg. The most energy efficient washing machine used only one third as much energy as the least energy efficient tested. The study also showed that detergent wastage could be reduced from 31 per cent to 1 per cent, with no significant effect on performance (UK DoE, 1992b). Similarly, increasingly more fuel efficient motor vehicles are being developed as efficiency of resource use becomes a major factor in consumers' decision making when they choose between products, particularly cars. Consumption of energy for heating can be reduced by better house insulation.

Vehicle ownership and use

Data on car ownership by households tend to be confused by multi-ownership, by the provision of company cars without an identifiable owner (particularly in the UK), and by numerous other factors. It is now certain, however, that Western Europe has more cars than households for the first time, following a dramatic expansion of sales in the late 1980s and early 1990s, and that about 75 per cent of all households have at least one vehicle at their disposal (Euromonitor, 1992). In Central and Eastern Europe, however, absolute levels of car ownership among households are lower, although rates of growth are higher than in Western Europe (see Figure 26.8 and also Chapter 21).

Public attitudes to the environmental impacts of products

There has been increased publicity in recent years given to the environmentally harmful effects of products and attempts by concerned groups to encourage consumers to find ways to reduce the impacts of their own activities and to influence the behaviour of other groups. Most industrial production supplies the domestic market and much of the remainder supplies manufacturers for domestic markets. There is, therefore, a great potential for consumers to exert pressure on industry as well as on governments through their exercise of choice. This pressure has been enhanced by the role played by non-governmental organisations. Furthermore, various surveys show that the vast majority of people are aware that they can protect the environment by changes to their own behaviour. Examples of consumer pressure successfully changing manufacturers' products include the appearance on the market and rapid success of phosphate-free detergents (see Box 26B) and the virtual elimination of CFC propellant aerosols from European markets (in advance of legislation banning them).

A 1992 survey in the EU asked people what actions they had taken or would consider taking to protect the environment. Figure 26.9 shows some personal actions claimed to have been taken by respondents to the survey. Seventy per cent of respondents said that they had bought or were prepared to consider buying 'environmentally friendly' products even if they were more expensive (CEC, 1992b).

The type of consumer pressure demonstrated by these results has initiated the proliferation of environmental information about products. Clear and accurate information is an essential tool to enable consumers to take account of the potential environmental impacts associated with the products they buy. Claims such as 'green', 'environmentally friendly', and 'ozone friendly' have been attached to many products and services. Occasionally these claims have been vague, meaningless and even misleading. While consumers undoubtedly need guidance in exercising choice, the unregulated profusion of claims and the lack of objective, independently verified information to support them can lead to confusion and ultimately to the frustration of consumer

Figure 26.7
Average household expenditure by product group (as % of income), Central and Eastern and Western Europe, 1990
Source: ERM based on Euromonitor, 1992

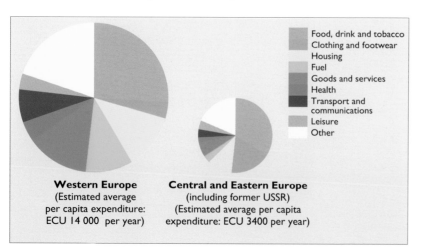

Western Europe
(Estimated average per capita expenditure: ECU 14 000 per year)

Central and Eastern Europe
(including former USSR)
(Estimated average per capita expenditure: ECU 3400 per year)

Food, drink and tobacco
Clothing and footwear
Housing
Fuel
Goods and services
Health
Transport and communications
Leisure
Other

choice. Schemes have, therefore, been set up which aim to impose standards or a code of practice regulating claims, such as the use of environmental logos. In the EU, these are being supplemented and may eventually be replaced by an EU 'ecolabel' (see Box 26C).

EXAMPLES OF CONTROL MEASURES

There are three main points at which governments can influence consumer behaviour in order to minimise environmental effects of household activities:

- by influencing purchase decisions – the kinds and numbers of products or resources that people buy;
- by influencing the use of products – the amounts of products or resources that are used;
- by influencing the disposal of products – the timing of replacement of goods and methods of disposal.

To influence behaviour in any of these categories, governments have at their disposal the following four complementary strategies (Cramer, 1993).

Coercive measures

These include legal measures to restrict activities or consumption of products. Examples include restrictions on emissions of smoke (for example, reducing air pollution from the domestic sector through smoke free zones in urban areas in the UK) and restrictions on the use of private motor cars (see Chapter 10) or banning of certain products (eg, drinks cans in Denmark since the early 1970s).

Financial regulation

Governments can ensure that the environmental costs of products throughout their entire life-cycle are included in the purchase price by using levies or taxation. The use of less harmful products can also be encouraged through subsidies or tax modifications (for example, the introduction of unleaded petrol at prices lower than those of leaded petrol has encouraged its use in many European countries).

In Denmark, disposable tableware is subject to a levy of 33 per cent of retail price or 50 per cent of import value, to discourage this form of waste generation, while pesticides for home use are taxed in a similar way (17 per cent of the retail price or 20 per cent of import value) to discourage their use in private gardens (ERM, 1992).

Information and training

To date, governments wishing to influence consumer behaviour both at home and in the workplace have provided information and education aimed at assisting or persuading consumers to modify their consumption decisions voluntarily in favour of less environmentally harmful alternatives. Examples include publicity and media campaigns, such as 'save energy' or 'energy efficiency' campaigns, and the regulation of product labelling, for example the EU ecolabelling scheme (Box 26C).

Institutional instruments

Governments can provide the facilities to enable consumers to adopt more environmentally responsible behaviour through a variety of institutional instruments, for example, by coordinating schemes to promote the separation and collection of recyclable components of domestic waste and measures to reduce household car use by promoting public transport systems.

CONCLUSIONS

- European households account directly for a substantial fraction of resource use: between 5 to 30 per cent of water supply, 15 to over 50 per cent of final energy consumption and a significant proportion of waste generation. In addition, most industrial production (70 per cent) supplies the domestic market, so that domestic consumption is responsible for a considerable proportion of industry's resource use and waste production.
- The significance of households as a major determinant of environmental quality, therefore, lies not only in their direct impacts but also in their ability to influence other sectors of the economy by exercising consumer choice.
- An increase in the total number of households throughout Europe, caused by growth in the population and decrease in average household size, is leading to increased per capita demand for resources. The rate of growth in Western countries is slowing but in Central and Eastern Europe – where growth in household numbers has been constrained by economic factors – there remains a substantial underlying demand for more land, water and energy for consumption by households.
- Similarly, the growth in consumer spending power, particularly in the West, has increased demands for goods and services, some causing significant environmental disruption. An example is the level of car ownership. Western European households now own on average at least one car each, an increase of over 20 per cent over the past ten years. This has led to severe urban congestion and air pollution in many major cities (Euromonitor, 1992). By contrast, fewer than one in three households in Central and Eastern Europe owns a car.
- Environmental management and control measures have achieved significant results in offsetting the adverse effects of increasing consumption in some countries. It has been demonstrated that there is substantial potential for more efficient use of resources in households, for example by metering of domestic water use and insulation of homes to reduce heating loss. In addition there are many examples of successful implementation of actions to reduce emissions, eg: the reduction of phosphates originating from detergents, the phasing out of domestic CFC usage prior to legislation, the introduction of smoke free zones and domestic recycling schemes.
- The successful elements of policy measures in Western Europe to deal with problems related to increased consumption will probably be useful in controlling the consequences of satisfying pent-up demand in Central and Eastern Europe.

Figure 26.8
Cars per household in Europe, 1980–90
Source: Euromonitor, 1992, and ERM

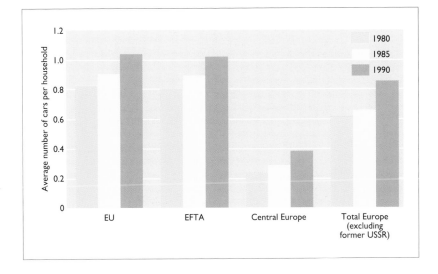

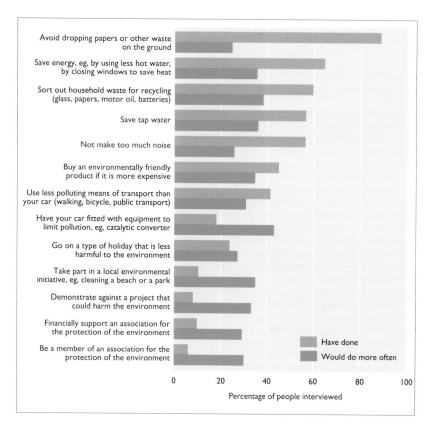

Figure 26.9
Personal actions
taken to improve the
environment, EU,
Spring 1992
Source: CEC, 1992b (see
Statistical Compendium)

- Scope for reducing the impact of the household sector on the European environment is, however, vested principally in life-style decisions of the individual consumer. Such decisions need to be educated, and would benefit from guidance from government and other relevant bodies, supported by a suitable policy framework. Past experience tends to show that these approaches are more successful than coercive measures, although these may also be effective.

Box 26C The EU ecolabel

The EU ecolabel award scheme is intended to promote products with a reduced environmental impact during their entire life-cycle. Products are to be assessed against environmental criteria specific to their product group, as agreed upon by Member States and the CEC. Those which can demonstrate compliance with the criteria will be permitted to use a distinctive logo – the ecolabel – on their packaging and promotional material informing consumers that they are among the least environmentally harmful products of that type. As well as providing authoritative guidance to consumers who wish to choose products for environmental reasons, the scheme aims to encourage the production of more environmentally benign products and to facilitate trade in these products. The scheme provides for an ecolabel to be awarded to those products which have the best environmental performance within particular product categories. The criteria will be set for each product category on the basis of a 'cradle-to-grave' life-cycle analysis of environmental performance (see Chapter 15), and only those products which meet these criteria will be awarded a label. There are five phases to the life-cycle analysis: pre-production; production; distribution (including packaging); utilisation; and disposal. Initially it is likely that around 10 to 30 per cent of products within a product category will qualify for a label.

Other key features of the scheme are that it will be voluntary, self-financing, and not include food, drink and pharmaceuticals (these are not covered as they are less clearly defined consumer products, and also impinge on health concerns).

REFERENCES

Arceivala, S J (1981) Wastewater treatment and disposal - engineering and ecology in pollution control. *Pollution Engineering and Technology* No 15. Marcel Dekker Inc, New York.

CBS (1991) *Natural resources and the environment*. Rapporter 92/1A, CBS (Central Bureau of Statistics of Norway), Oslo, Norway.

CEC (1992a) *European economy: the climate challenge, economic aspects of the Community's strategy for limiting CO2 emissions*. Statistical Annex 2, No 51, May 1992, DGII Economic and Financial Affairs, Commission of the European Communities, Luxembourg.

CEC (1992b) *Europeans and the environment in 1992*. Eurobarometer 37.0, prepared by INRA (Europe) European Coordination Office, Brussels, published by DG XI, Commission of the European Communities, Brussels.

CEC (1993) *Energy consumption in households*. Eurostat, Commission of the European Communities, Luxembourg.

Cramer, J (1993) *The contribution of consumers to sustainable development: seminar on low waste technology and environmentally sound products*, 24–28 May 1993, Warsaw, Poland, Senior Advisers to ECE governments on environmental and water problems, UNECE, Geneva.

ERM (1992) *Quantification, characteristics and disposal methods of municipal waste in the community: technical and economic aspects*. ERM (Environmental Resources Management), London.

Euromonitor (1992) The European Compendium of Marketing Information, Euromonitor. London.

Gillham, C A, Leech, P K and Eggleston, H S (1992) UK emissions of air pollutants 1970–1990, LR 887 (AP), Warren Spring Laboratory, UK Department of Trade and Industry, London.

Icelandic Ministry for the Environment (1992) *Iceland: 1992 national report to UNCED*. Skerpla, Reykjaviik.

MECR (1992) *Environmental yearbook of the Czech Republic, 1991*. Ministry for the Environment of the Czech Republic, in collaboration with the Czech Ecological Institute, Prague.

Meybeck, M, Chapman, D V and Helmer, R (Eds) (1989) *Global environment monitoring system: global freshwater quality, a first assessment*. Blackwell, Oxford.

UK DoE (1992a) The UK Environment, UK Department of the Environment, HMSO, London.

UK DoE (1992b) Ecolabelling criteria for washing machines. A study by PA Consulting for the UK DoE May 1992. UK Department of the Environment, London.

UK DoE (1993) *Digest of environmental protection and water statistics*, No 15, 1992. UK Department of the Environment, HMSO, London.

UK DTI (1992) *UK DTI digest of UK energy statistics*. UK Department of Trade and Industry, HMSO, London.

UN (1992) *National accounts statistics: main aggregates and economic tables*, United Nations, New York.

UNECE (1992) *Substitutes for tripolyphosphates in detergents*. United Nations Economic Commission for Europe, Geneva.

VROM (1991) *Essential environmental information 1991*. VROM (Ministry of Housing, Physical Planning and Environment), The Hague, The Netherlands.

WRI (1992) *World Resources 1992–93: a guide to the global environment*. World Rescources Institute, Oxford University Press, Oxford.

Part V
Problems

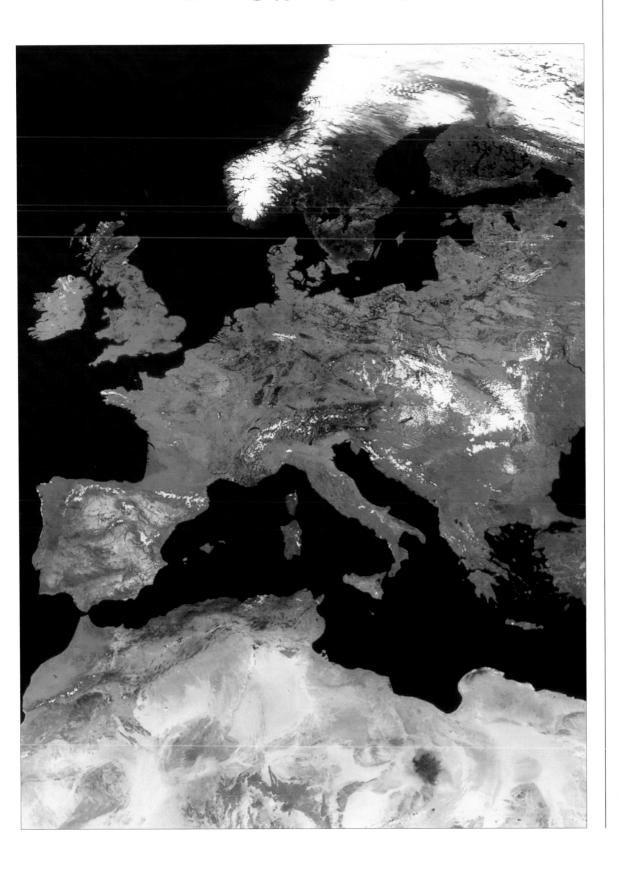

Introduction

One of the main goals for this report given by the ministers at the Dobříš Castle conference was to inform the public and raise awareness about environmental problems. Most environmental problems are interrelated. The flows of energy and the cycling of materials in the environment implies that most major problems appear not only in one natural medium, but may also have consequences for water, air, soil and living organisms.

Since there are no absolute criteria on which to base a selection of most relevant or prominent environmental problems, a method was sought which would combine the concerns of environmental ministers, expert international opinion as well as public perception of environmental risks. Thus work began by making a wide survey of the literature and analysing international initiatives by, among others, the EU, UNECE, UNEP, OECD and the US Environmental Protection Agency. Combining these results with the findings of the unfolding assessment itself, resulted in a list of 56 environmental problems or issues (see Appendix 1).

To identify the most prominent European environmental problems from out of this list of 56, reference was made to the original mandate and objectives of the report from which nine selection criteria were derived (see Chapter 1, page 5). A questionnaire was sent to more than 50 correspondents, mainly in public or independent environmental lines of work, which asked respondents to make scores against the nine criteria for each of the 56 problems in the consolidated list.

Overall, the results showed fairly consistent priorities among the various groups of respondents, but with NGOs being more preoccupied with local and ecological questions, and international groups giving more weight to global problems.

While recognising the relevance of all criteria used in the selection process, particular emphasis was given to the threat to sustainability, prominence of a European aspect, and long term character.

These results were analysed and compared with lists of priority problems made by the organisations mentioned above. The following 12 problem headings – which are not presented in any order of priority – were identified, over which there was widest agreement as to significance:

- climate change;
- stratospheric ozone depletion;
- the loss of biodiversity;
- major accidents;
- acidification;
- tropospheric ozone and other photochemical oxidants;
- the management of freshwater;
- forest degradation;
- coastal zone threats and management;
- waste reduction and management;
- urban stress; and
- chemical risks.

These problem headings were chosen to best represent composites of closely related or interconnected issues which combine to form prominent European environmental problems. This composite approach, combined with the selection process adopted, are the reasons for considering this list to be a robust representation of the 12 most significant environmental problems of concern to Europe. It is recognised, however, that other issues, particularly at regional level, may be considered by some of equal or even greater significant than these. The next most significant issues identified which were not chosen to be presented separately in this part of the report were soil and resource contamination, riverine inputs into seas and desertification, (see appropriate chapters for further details).

The systematic three-step presentation of environmental conditions, pressures and human activities presented in Parts II, III and IV of this report, has covered the basic facts of all of these problems. The aim of the following 12 chapters is to integrate these different components into clearly understandable stories of cause, effect and response for the 12 prominent problem headings selected. The presentation of each environmental problem therefore inevitably refers back to several previous chapters of the report, but the treatment of each problem can be read independently. The focus is on describing the problem, its origins and consequences. To the extent that observed trends are held to be a reliable guide for the near future, they are also presented. In a few cases, scenarios of the 'if-then' type are used. These are not forecasts of what will happen but instead, based on knowledge of natural or economic laws, describe possible consequences if such a course of action were to be followed. Even non-action, 'business-as-usual', is often the reference scenario making it clear that this choice of strategy also has its consequences like any other decisions.

For each problem presented, goals and strategies are indicated. Goals describe a future state of affairs, and strategies are the highways proposed or taken to reach these goals. Along the strategic roads lie the many concrete policy measures necessary in practical terms.

Photo overleaf shows a false-colour mosaic of satellite images. Red and brown colours represent different mixtures of visible and infrared reflections from the ground.
Source: Earth Satellite Corporation/Science Photo Library

*Storm brewing,
southern France*
Source: Frank Spooner
Pictures

27 Climate change

THE PROBLEM

A greenhouse effect has always existed, keeping the Earth warmer than it would be without an atmosphere. What is popularly referred to today as the 'greenhouse effect' is really the anthropogenically enhanced greenhouse effect by which an extra warming of the surface and the lower atmosphere is produced, leading to disturbances in the geosphere/biosphere system and, notably, an increase in the mean global surface temperature and in the mean sea level.

The natural presence of 'radiatively-active' gases in the atmosphere is essential for life: they trap heat in the lower atmosphere, thus creating – like a greenhouse – an environment which is far warmer (by about 33 °C) inside than out. By increasing the concentration of greenhouse gases, however, additional infra-red radiation, which otherwise would have been lost to space, is absorbed in the lower atmosphere and the Earth's radiation balance is upset. This energy is re-emitted in all directions, a large portion being sent back to the Earth's surface or elsewhere in the troposphere. This yields a radiative imbalance which can be restored only through a warming of the troposphere. On the other hand, the enhanced greenhouse effect will also cool the upper layers of the atmosphere (ie, the stratosphere, and above between 25 and 70 km).

Greenhouse gases in the atmosphere have increased since pre-industrial times by an amount that is radiatively equivalent to about a 50 per cent increase in carbon dioxide (CO_2), although CO_2 itself has risen by about 25 per cent; other gases have made up the rest. The global emissions of most greenhouse gases are expected to rise in the next decade: CO_2 at 0.5 per cent/year, methane (CH_4) at 0.9 per cent/year, and nitrous oxide (N_2O) at about 0.3 per cent/year. In contrast, CFC emissions are expected to decrease to near zero around the year 2000 following internationally agreed phase-out measures. As a result of these trace gas trends, an effective doubling of greenhouse gas concentrations (as commonly measured as 'CO_2-equivalents') is expected around 2030.

There has been much scientific activity aimed at identifying clues of the enhanced greenhouse effect by searching for trends in climatic and hydrological records (mainly temperature, precipitation, glaciers, runoff, freezing/melting date and sea level). Conclusions have so far been mainly tentative because definite signals from an enhanced greenhouse effect are weak and/or smaller than natural fluctuations, most of the homogeneous and comparable records, if any, are too short, and regional differences have added to the confusion (see Chapter 4). Our partial understanding of multimedia mechanisms involved has also been a limitation to making a safe forecast of future trends, although this has fostered large international research efforts such as the World Climate Research Programme (WCRP, of the World Meteorological Organisation (WMO), the International Council of Scientific Unions (ICSU) and the Intergovernmental Oceanographic Commission (IOC) of UNESCO) and the International Geosphere-Biosphere Programme (IGBP of ICSU), or international expert assessments such as under the Intergovernmental Panel on Climate Change (IPCC of WMO and UNEP).

At first glance, uncertainties about the distribution and timing of consequences of climate change may make some people feel that the cost of rapid implementation of actions intended to curb the man-made emissions responsible might be prohibitive. However, regardless of what happens to climate, application of relevant measures would bring its own benefits, such as longer availability of fossil fuels and other natural resources, as well as more rational and efficient energy use. Additionally, consequences such as stronger and more frequent storms, higher seas and less water in rivers are so far-

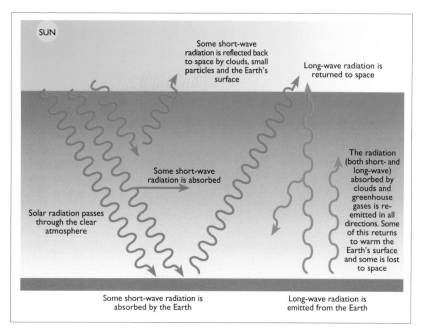

Figure 27.1
The greenhouse effect
Source: IPCC, 1992

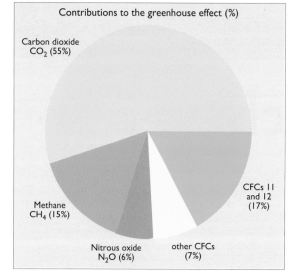

Figure 27.2
Current contributions of the different greenhouse gases to the enhanced greenhouse effect
Source: IPCC, 1990

Table 27.1
Mean atmospheric lifetimes of greenhouse gases with their global warming potentials (GWPs) relative to carbon dioxide
Source: IPCC, 1992

large-scale changes in the atmosphere (see Chapters 4 and 14). Conversely, climate change may greatly affect Europe's 680 million people and its natural environment.

THE CAUSES

Human exploitation of the world's natural resources, primarily fossil fuels – coal, oil and natural gas – but also biomass resources through agricultural and forestry practices, result in the release each year of over 20 000 million tonnes of carbon dioxide (CO_2) into the atmosphere. In addition, other atmospheric gases, notably methane (CH_4), nitrous oxide (N_2O) and chlorofluorocarbons (CFCs), are being released as a result of human activities. All these gases together with ozone (O_3) and water vapour are referred to as 'greenhouse gases', that is, they are relatively transparent to incoming short-wave radiation from the sun, but absorb long-wave infra-red radiation emitted from the Earth (Figure 27.1), and consequently entrap heat in the atmosphere. Ozone in the lower atmosphere (troposphere) is not emitted directly but is formed from anthropogenic emissions of nitrogen oxides (NO_x), volatile organic compounds (VOCs), methane and carbon monoxide. Particles (eg, sulphate) emitted into the atmosphere by anthropogenic and natural sources (mainly volcanoes) can also affect climate because they can reflect and absorb radiation.

Various physical and dynamic processes combine to redistribute heat horizontally and vertically in the atmosphere and the oceans. At the same time, they trigger various feedback effects that can either enhance or dampen (reduce) the initial trends. Most of these feedback processes are connected to the formation of clouds as well as to biospheric and oceanic responses.

Soils are an important reservoir for carbon; an estimate of the global soil carbon pool is 1500 gigatonnes (billion metric tonnes), or twice the atmospheric pool, excluding carbon in organic soils. Globally, the amount of carbon in decaying plant litter and soil organic matter may exceed the amount of carbon in living biomass by a factor of two or three. Soils also contribute to the natural fluxes of greenhouse gases; the contribution of the whole terrestrial biota (soil, vegetation and fauna) is 30 per cent for CO_2, 70 per cent for CH_4 and 90 per cent for N_2O.

Recent estimates suggest that 2 gigatonnes of carbon are absorbed by the oceans each year to replace the carbon taken by phytoplankton from sea water and subsequently deposited on the sea bed when the phytoplankton die.

The lifetime of greenhouse gases in the atmosphere is determined by their sources and sinks in the oceans, atmosphere and biosphere. Carbon dioxide, CFCs and nitrous oxide are removed only slowly from the atmosphere (see Table 27.1) and hence, following a change in emissions, their atmospheric concentrations take decades or centuries to adjust fully. Even if all human-made emissions of CO_2 had been halted in the year 1993, about half of the increase in carbon dioxide concentration caused by human activities would still be observable in the year 2100.

The concept of relative global warming potentials (GWPs) has been developed as an index of the relative radiative effect (and, hence, potential climate effect) of equal emissions of each of the greenhouse gases, to take into account the differing time that they remain in the atmosphere and their different absorption properties. The GWP defines the time-integrated warming effect due to an instantaneous release of unit mass (1 kg) of a given gas in today's atmosphere, relative to that of carbon dioxide. The relative contributions will change over time, and a period of 20 years will indicate the response to emission change in the short term (see Table 27.1). Anthropogenic CO_2 emissions are currently responsible for more than half of the enhanced greenhouse effect (Figure 27.2). However, the GWP concept

reaching that taking this 'precautionary' approach should be rewarding. This approach is also known as the 'no regrets' policy, because of the argument that preventing climate change would be far better than attempting to cure it, despite the uncertainties of how fast and how soon it might be happening.

Although Europe's land mass covers less than 10 per cent of the world's total area and its population is some 15 per cent of the world population, European countries contribute a significant amount to the global emissions of greenhouse gases and other substances which bring about

Gas	Lifetime (years)	GWP	Gas	Lifetime (years)	GWP
CO_2	50–200	1	HCFC-124	6.9	1500
CH_4	10.5	63	HCFC-141	10.8	1500
N_2O	132.0	270	HCFC-142	22.4	3700
CFC-11	55.0	4500	HFC-125	40.5	4700
CFC-12	116.0	7100	HFC-134	15.6	3200
CFC-113	110.0	4500	HFC-143	64.2	4500
CFC-114	220.0	6000	HFC-152	1.8	510
CFC-115	550.0	5500	CCl_4	47.0	1900
HCFC-22	15.8	4100	CH_3CCl_3	6.1	350
HCFC-123	1.7	310	CF_3Br	77.0	5800

Note: An integration time horizon of 20 years is assumed here.

Sources	CO_2	CH_4	N_2O	CFCs
Energy	80	26	9	–
Deforestation/land clearing	18	–	17	–
Other industry	3	–	15	100
Fertilised soils	–	–	48	–
Enteric fermentation	–	24	–	–
Rice cultivation	–	17	–	–
Landfills	–	11	–	–
Biomass burning	–	8	11	–
Animal waste	–	7	–	–
Domestic sewage	–	7	–	–

Table 27.2 Relative direct sources of global greenhouse gases (percentages of total anthropogenic emissions in 1990)
Source: IPCC, 1992

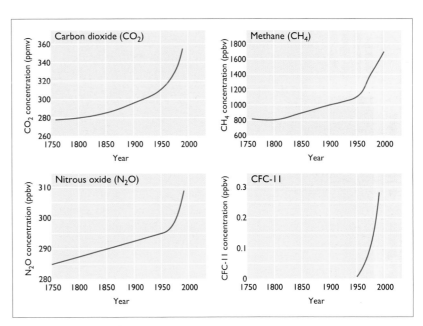

Figure 27.3
Concentrations of the greenhouse gases CO_2, N_2O, CH_4 and CFC-11 in the atmosphere since 1750
Source: IPCC, 1990

does not allow for the comparison of the relative effect of different greenhouse gases in terms of their contribution to climate change *per se*, such as may be measured in terms of the increase in the global mean surface temperature or sea-level rise. This comparison can only be made with the use of complete atmospheric models, and simple comparative indices are now being developed.

During the past century, the concentration of the most important greenhouse gases, and notably that of carbon dioxide, has been increasing steadily, showing that anthropogenic emissions have exceeded the natural capacity for their removal either through absorption at the Earth's surface or through chemical reactions in the atmosphere. Figure 27.3 illustrates the global CO_2, N_2O, CH_4 and CFC-11 concentrations since 1750.

The use of fossil fuels for energy production and transport is the most important source of global CO_2 emissions (Table 27.2). Emissions due to coal are about twice those of natural gas for the same amount of energy (oil-related emissions are about one-and-a-half times as much). Present conversion of tropical forests to agriculture and pasture releases up to 15 per cent of global emissions (approximately 7000 million tonnes of carbon in 1991).

Currently, European forests are a net sink for CO_2 as a result of afforestation and probably because of a CO_2-fertilisation and nitrogen-fertilisation effect. In most cases, the conversion of native forests to plantations will have resulted in a net loss of carbon to the atmosphere, including the carbon stored in timber products (Cannell et al, 1992).

Ordinary agricultural practices also play a role in the problem of climate change. The yearly burning of biomass releases an amount of carbon, in the forms of carbon

dioxide, carbon monoxide and methane, which is smaller than, but of the same order of magnitude as, the emissions from the burning of fossil fuels. The carbon released from biomass burning, however, like that from the decay of fallen leaves in temperate forests, does not correspond to a net release in the long term, since the same amount of carbon is removed from the atmosphere in the following growing season. Indeed, the oxidation of carbon at the higher temperature of burning tends to generate a higher percentage of carbon dioxide, which produces a smaller enhancement of the greenhouse effect, molecule by molecule. The burning of biomass, on the other hand, does affect the chemistry of the troposphere, notably by increasing the ozone concentration, which may also have local pollution effects.

Scientific assessments of global climate change have generally considered only long-lived greenhouse gases merged into one equivalent surrogate (CO_2), which is assumed to double homogeneously by the middle of the next century. The IPCC reports (IPCC, 1990 and 1992) mentioned the greenhouse property of ozone, but made no quantitative assessment of its warming potential. In effect, because ozone does not last long in the atmosphere, this assessment is more difficult since both temporal and spatial distribution should be taken into account. A recent estimate (Marenco et al, 1994) computes a relative radiative forcing for ozone at least 1200 times higher than for CO_2, that is, much higher than the other greenhouse gases except CFCs. With its current concentrations, ozone would thus contribute 22 per cent of the global warming in the northern hemisphere and 13 per cent in the southern hemisphere.

Table 27.3
Best estimates of climate change projections over the next 50 years
Source: Schneider, 1990

Indicators	Annual average change	Distribution of changes					Confidence of projection	
		Regional average	Change in seasonality	Interannual variability	Significant transients		Global average	Regional average
Temperature	+2 to +5°C	–3 to +10°C	Yes	Down?	Yes		High	Medium
Sea level	+10 to +100 cm	–	No	?	Unlikely		High	Medium
Precipitation	+7 to 15 per cent	–20 to +20 per cent	Yes	Up	Yes		High	Low
Direct solar radiation	–10 to +10 per cent	–30 to +30 per cent	Yes	?	Possible		Low	Low
Evapotranspiration	+5 to +10 per cent	–10 to +10 per cent	Yes	?	Possible		High	Low
Soil moisture	?	–50 to +50 per cent	Yes	?	Yes		?	Medium
Runoff	Increase	–50 to +50 per cent	Yes	?	Yes		Medium	Low
Severe storms	?	?	?	?	Yes		?	?

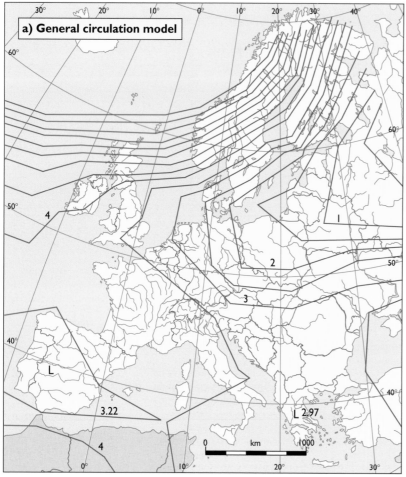

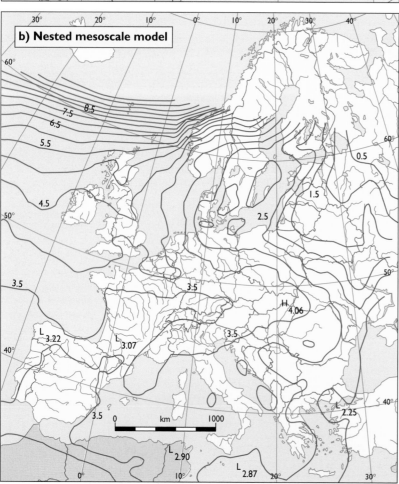

The increase of methane concentrations is correlated largely with increasing population, about 60 per cent of global emissions being associated with human activities. Methane from fossil origins is released by the exploitation of coal, oil (if it is not used or flared) and natural gases, and from distribution systems. As a product of the anaerobic digestion of organic material, it is also released from landfills, wetlands and rice paddies and by fermentation in the rumen of ruminants. Because of the potency of methane in the atmosphere and its relatively short lifetime, stabilising CH_4 concentrations will have a substantial impact on reducing potential warming.

Nitrous oxide emissions are produced mostly by denitrification processes in oxygen-free environments with a high nitrate load, such as soils and sediments in polluted waterbodies. N_2O is also released in limited quantities by the use of fossil fuels.

Carbon monoxide plays significant roles in controlling the chemistry of ozone production and hydroxyl radical OH destruction in the lower atmosphere. It directly affects the oxidising capacity of the lower atmosphere and thus influences the concentrations of other important trace gases such as methane, methylchloroform and the HCFCs.

Global anthropogenic emissions of sulphur gases have increased by about a factor of three during the past century, leading to increased sulphate aerosol concentrations, mainly in the northern hemisphere. Sulphate aerosols can affect the climate directly, by increasing the reflection of incoming solar radiation back to space in cloud-free air, and, indirectly, by providing additional cloud condensation nuclei. Model calculations show that the increase in sulphate aerosol concentration reaches a factor of 100 over Northern Europe in winter.

CONSEQUENCES OF CLIMATIC CHANGES

The primary consequence of increased radiation absorption is climatic change or, in terms of the most common climate indicators, rapid changes of surface temperature – about 0.3°C/decade (IPCC, 1992) – and precipitation changes. The consequent expected sea-level rise (3 to 10 cm/decade) and changes in hydrological and vegetation patterns may have serious effects on society, leading to high risks and to substantial costs. Current models predict average global changes, but the impacts of climate change will be most keenly felt on a regional and local scale where uncertainties are larger (see Table 27.3).

To evaluate the consequences of climate change, it is common to take as a reference case the doubling of CO_2 (or an equivalent amount of total greenhouse gases) in the atmosphere relative to its pre-industrial level (before the years 1750–1800). The IPCC business-as-usual scenario (which is considered a high emission scenario) suggests that this will occur by 2025. Under the IPCC low emission scenario, doubling will take place around 2060.

Map 27.1 Changes in January surface air temperature (°C) over Europe as a result of doubled CO_2:
a) *results from general circulation model (GCM) of the National Center for Atmospheric Research, Boulder, Colorado, USA, relative to the GCM control;*
b) *results from nested mesoscale model relative to the nested model control (Giorgi et al, 1991)*
Source: IPCC, 1992

Changes in climate patterns

General circulation model estimates of the increase in the global annual average temperature at the surface of the Earth vary from 1.5 to 4.5°C, given an increase in greenhouse gases equivalent to a doubling of the pre-industrial CO_2 concentration. The current 'best' estimate is 2.5°C (IPCC, 1990). The average global increase in precipitation and evaporation is estimated at 3 to 15 per cent. A recent reassessment of the IPCC90 climatic trends, which also takes into account the effects of CO_2 fertilisation, feedback from stratospheric ozone depletion and the counteracting radiative effects of sulphate aerosols, yields new projections for radiative forcing of climate and for changes in global-mean temperature and precipitation (Wigley and Raper, 1992). Changes in temperature and sea level are predicted to be less severe than those estimated previously, but are still far beyond (four to five times) the limits of natural variability.

The range of possible rise in temperature is determined largely by the uncertainty about the feedback processes. While the timing and extent of warming on a global basis is uncertain, still more indefinite is the change of climate on the regional level in Europe and elsewhere. The spatial resolution of models being still too coarse, regional estimates are less accurate than global ones. Models, however, consistently predict that temperature increases will be greater in the high latitudes than low latitudes. Hence the extent of temperature increases in Northern Europe are expected to be larger on average than those in the Mediterranean regions. Using the new technique of nesting computer models, improved estimates of possible temperature changes in Europe are becoming available (see Map 27.1).

The IPCC collected best model guesses as to the change in climate for Southern Europe from pre-industrial times to 2030, under the assumption that average global warming would be 1.8°C by that year. Temperature was estimated to be 2°C warmer in winter and 2 to 3°C warmer in summer, precipitation 0 to 10 per cent greater in winter and 5 to 15 per cent lower in summer, with a –5 to +5 per cent change in soil moisture in winter and –15 to –25 per cent change in summer. The ranges in these values indicate the disagreement between the three different general circulation models used for these calculations. As can be seen (Map 27.1), the models are relatively consistent, although in some cases they do not even agree on the direction of change.

Higher temperatures will yield more evaporation, resulting in more precipitation. This will intensify the hydrological cycles, with unknown consequences. Sudden regional or global climate changes following changes in the location and/or intensity of air and water flows cannot therefore be excluded.

Assessing climate impacts

Many studies on specific climate change impacts have not used a common framework, and some of their underlying assumptions may differ. As yet, there has been no comprehensive, systematic study of climate impacts in Europe. However, preliminary results from a still unpublished study of the European Commission (performed by the Climate Research Unit, East Anglia, UK; Environmental Resources Management Limited, London, UK; and the National Institute of Public Health and Environmental Protection (RIVM), Bilthoven, The Netherlands) indicate that the possible changes in the EU due to global warming would be increased agricultural yields and tourism, and reduced heating requirements. However, the study also points out that the monetary costs of protecting coastlines against sea-level rise are likely to far outweigh benefits.

Sea-level rise

A rise in the global mean sea level will be one important impact of global warming. It will result from:

1 the thermal expansion of sea water;
2 melting of mountain glaciers;
3 melting of inland ice caps, as in Greenland; and
4 changes in the Antarctic ice sheet.

According to IPCC best estimate scenarios, the global-mean sea level would be about 22 cm higher than today by 2050 and around 50 cm higher by 2100 (the low and high scenario estimates give 15 and 90 cm respectively for 2100). A rise of 10 to 20 cm over the last 100 years has been observed, due mainly to oceanic thermal expansion and retreating mountain glaciers.

The consequent environmental effects are many and varied: permanent inundation of low-lying land; increased frequency of temporary flooding from high tides or storm surges; changes in rates of beach, dune or cliff erosion; salinisation of estuaries, groundwater and surface water supplies, wetland ecosystems or agricultural soils; and effects on river hydrology, including inland flooding, through changes in river gradients. The most threatened areas in Europe, because of their location and elevation, are shown in Map 27.2.

Hydrological processes

Among the greatest potential impacts of climatic change will be the effects on the hydrological cycle. Increase in incidence of extremes, such as floods and droughts, would cause increased frequency and severity of disasters, while changes in precipitation, evapotranspiration and soil moisture would strongly alter agricultural practice and water management systems, and lead to severe land degradation. Forecasts of future changes of water resources attributable to human-

Map 27.2
European coastal regions threatened by a rise in sea level and by salt water intrusions into estuaries and groundwater
Source: ECGB, 1992

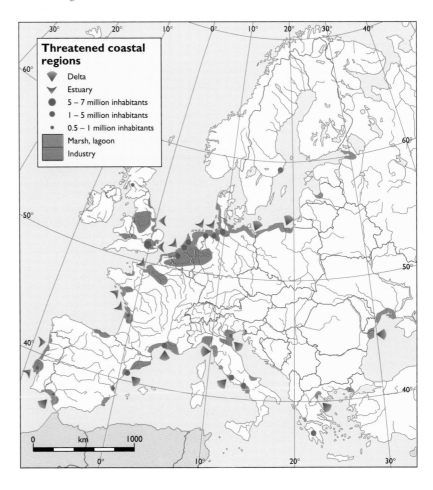

induced climate changes are still uncertain because of the unreliability of regional precipitation predictions. However, they do indicate that such changes would be fairly significant, even in the case of slight climate modification, and that not only river runoff but also water demand might be affected. For instance, a temperature rise of 1 to 2°C and a 10 per cent drop in rainfall would reduce river runoff in arid regions by 4 to 70 per cent (Shiklomanov, 1991).

Various case studies point out possible modifications in runoff and other processes that might occur if there were a doubling of CO_2. For example, a study of three catchments in Belgium (Bultot, 1988) indicated that winter floods may become more frequent in catchments that have low infiltration rates and unsteady base flows; however, in basins having high infiltration rates and more steady base flow, there might be a beneficial increase in base flow throughout the year. In a study of seven Norwegian stream basins, climate changes brought about by a doubling of CO_2 are estimated to increase runoff in mountainous catchments, while decreasing it in lowland basins because of greater evapotranspiration. The seasonal pattern of runoff will change markedly, particularly in catchments of intermediate elevation. Winter runoff will increase significantly, while summer runoff declines. Flooding will occur more frequently in autumn and winter, and the overall amount of flood damage is expected to increase. The net effect of changes in runoff patterns may also lead to a small increase in Norwegian hydropower production.

Preliminary model results have shown an increase in runoff (up to a doubling) of rivers discharging into the Arctic Ocean. Since freshwater runoff into the Arctic Ocean is a significant fraction of its total water mass, any substantial modification of the input of water as envisioned in model scenarios may change important thermal and salinity characteristics. This in turn could significantly modify currents, ice cover and atmospheric circulation. Changes in river flow also modify the delivery of dissolved constituents and particulates to the Arctic Ocean. Increases in nutrient inputs could increase net primary production and thereby sequester excess atmospheric CO_2.

Risks to ecosystems

Alteration of the hydrological cycle, with changes in runoff and moisture availability, will influence patterns of sedimentation and erosion and the recycling of organic matter and nutrients. These will in turn influence plant productivity, competition between species and biodiversity. The linkages are sometimes more subtle, with, for example, changed temperature and precipitation conditions combining to favour the outbreak of plant diseases or insect plagues. These complex linkages between different factors, together with uncertainties in predicted water availability, make it difficult to predict changes in natural ecosystems towards new equilibria. Moreover, changes may take place so quickly that plants have no time to spread to new habitats by natural dispersion mechanisms and the actual vegetation response could therefore differ markedly from equilibrium predictions.

Climate changes will influence ecosystems and agriculture since temperature and evapotranspiration/precipitation determine the major biogeographical zones (such as northern boreal, temperate, Mediterranean). Large regional differences may arise due to small changes in one of the critical factors, such as temperature, precipitation, evaporation and soil structure, yielding changes in vegetation structure and the range of species of a given zone. The potential impacts of climate change on agriculture may have dramatic consequences since Europe, as a main exporter of agricultural products, contributes significantly to the nutritional balance of non-European countries. This possibility for a major disturbance in agricultural trades

should be seen also in the context of a rapidly increasing world population.

A study of the potential range of maize-growing areas in Europe found that a 1°C increase in mean annual temperature in Europe would translate into a northward shift of approximately 200 to 350 km in Western Europe, and 250 to 400 km in Eastern Europe. New areas of potential maize production would be opened up in southern England, The Netherlands, Belgium, northern Germany and northern Poland. A mean annual increase of 4°C would move the boundary up into northern Russia and central Fennoscandia. However, this study did not take into account possible shifts in water availability, nor possible changes in the range of agricultural pests and diseases, nor other factors that might limit agropotential. Indeed, the IPCC estimated under their business-as-usual scenarios that Mediterranean countries which already depend heavily on irrigation would have 15 to 25 per cent less soil moisture in summer.

Although extensions of agricultural areas northward might prove to be beneficial, the same pressure on natural ecosystems to extend or shift their range may endanger these ecosystems. This is because of human or natural barriers that may prevent the migration of ecosystems as their climatic zones shift. The rate of change is also important because, in less than a hundred years, Europe's bioclimatic zones may shift hundreds of kilometres in latitude (or hundreds of metres in altitude in high Alpine regions), and it is generally accepted that this will place considerable stress on terrestrial ecosystems. In some cases it can be expected that flora and fauna will be unable to migrate fast enough to survive.

Terrestrial ecosystems are thought to play an important role in determining regional and global climate; two examples of this are in Amazonia, where destruction of the tropical rainforest leads to warmer and drier conditions, and the Siberian forests, which represent 40 000 million tonnes of stored carbon (ie, an amount equivalent to half of the forests of Amazonia). Boreal forest ecosystems may also affect climate: as temperatures rise, the amount of continental and oceanic snow and ice is reduced, so the land and ocean surfaces absorb greater amounts of solar radiation, reinforcing the warming in a 'snow/ice/albedo' feedback which results in large climate sensitivity to radiative forcings. This sensitivity is moderated, however, by the presence of trees in northern latitudes, which mask the high reflectance of snow, leading to warmer winter temperatures than if trees were not present. Results from the National Centre for Atmospheric Research (NCAR) global climate model show that the boreal forest warms both winter and summer air temperatures, relative to simulations in which the forest is replaced with bare ground or tundra vegetation (Bonan et al, 1992). This suggests that future redistributions of boreal forest and tundra vegetation (for example due to extensive logging in Russia/Siberia, or the influence of global warming) could initiate important climate feedbacks, which could also extend to lower latitudes. The position of the tree-line distinguishing boreal forest from tundra vegetation has altered in response to past climate changes and is likely to change with the warmer climate caused by increased atmospheric CO_2 concentrations. Results from the NCAR coupled-model raise the possibility of considerable climate changes caused merely by redistribution of boreal forest and tundra ecosystems. The decrease in snow-covered land surface albedo caused by northward migration of boreal forest into tundra zones in response to climate warming may produce further warming.

Plants are the essential basis of all terrestrial life. Any significant variation in the productivity and composition of plant life would initiate a cascade of changes affecting animal life. At first glance, elevated CO_2 levels might seem an agricultural blessing by enhancing plant growth. This 'CO_2 fertilisation effect', as it is called, is expected to be particularly pronounced if plants have plentiful supplies of nutrients, light and water. The CO_2 fertilisation effect also promises to provide a buffer for concerns about global warming by drawing more CO_2 from the atmosphere. However, recent studies (Bazzaz and Fajer, 1992) suggest that such assumptions about the benefits of a world replete with CO_2 may be overstated. An isolated case of a plant's positive response to increased CO_2 levels does not necessarily translate into increased growth for entire plant communities. Farming will also be adversely affected by the differential response of plants. Important crops, such as maize and sugarcane, may experience yield reductions because of the increased growth of weeds.

Even the notion that plants will serve as sinks to absorb ever-mounting levels of CO_2 is questionable. Increased competition between plants will also diminish the CO_2 enhancement of natural ecosystems such as meadows and forests, but also artificial ecosystems such as farms. Agricultural yields will improve in a CO_2-rich future only at the cost of large quantities of fertilisers, pesticides and water from irrigation.

Other organisms which depend on threatened plant species for food, shelter or mating sites may also become endangered. A strong reduction of species diversity would, in turn, undermine the integrity of natural ecosystems. Because individual plant and animal species supply a wealth of essential industrial, agricultural and medicinal products, the loss of diversity will have pervasive environmental and economic consequences. Changes in the nutritional quality of plant leaves could lower herbivore and predator populations within their habitats. If insect herbivores suffer population reductions in a world abundant with CO_2, many predators will have less prey. Some predatory insects, for example, feast on other insect pests that damage certain crops. Plant development and flowering times may be also altered unpredictably by elevated CO_2, thus disrupting pollination.

Shifts in climate and vegetation could also significantly affect future animal distribution, abundance and survival. Rapid rates of change, especially in combination with the existence of artificial urban and agricultural barriers, may affect the ability of many species to relocate to areas which are climatically and ecologically more favourable. Endangered or rare species would be particularly vulnerable to rapid change, especially if their distributions are spatially restricted and their niche-width is narrow. Rapid climatic change therefore becomes a threat for current biodiversity (see also Chapter 29).

Land degradation

The resulting change in atmospheric composition will affect soil biological processes and provoke, for instance, an increase in biomass and changes in composition of vegetation and organic matter. These biological transformations will have an unforeseeable impact on soils. For instance, there is some concern about anticipated effects on soil carbon storage in tundra boreal peatland. On the other hand, human interference in the land-cover brings about changes in the rainfall regime, in the hydrology of large river systems and in the Earth's albedo, all of which play a major role in the surface energy balance.

In the long term (over 50 to 100 years) a substantial increase in temperature and changes in rainfall patterns (perhaps slightly wetter winters) in Europe could modify landuse. Whole ecosystems may shift northwards, but this consequence of climate change will undoubtedly be modified by landuse policy (Cannell et al, 1992). Landuse changes can have such important impacts that soil carbon decreases in the top first metre by 40 to 60 per cent when forest or grassland is converted to cropland, and by 25 to 35 per cent when forest is converted to grassland. Overall, warmer temperatures are thought to result in more arid conditions. Increased aridity in Southern Europe could lead to a chain of events in the soil, beginning with the breakdown in supply of organic matter, continuing with salt accumulation and the formation of surface crusts, and ultimately ending in severe land degradation.

Loss of organic matter may result from removal of crops without the return of any material to compensate. It may also result from alteration of soil drainage, or tillage that accelerates oxidation of organic matter. Peaty soils tend to suffer oxidation of organic matter if drained, and can shrink alarmingly as a consequence. In drier climates, the loss of soil organic matter generally leads to a reduction in retention of soil moisture and a decline in vegetation cover, crops or natural plants, which in turn leads to increased erosion. A positive feedback can thus arise, as organic matter/moisture in soil falls, plant cover declines and thus the renewal of organic matter in soil is further reduced. But for extrapolation of data and making predictions, the underlying processes, which are largely microbial, need to be understood.

GOALS

Protecting the planet in the long term and maintenance of its ecological equilibrium will require joint efforts to reduce greenhouse gas emissions to a level corresponding to the natural absorbing power of the planet. Emissions should be reduced at such a rate that ecosystems can adapt naturally to climatological changes, and so that food production is not threatened. This requires the most rapid possible stabilisation of the concentration of greenhouse gases in the atmosphere.

One way to quantify the above premises is to determine the acceptable emissions on the basis of a maximum acceptable rate of change of temperature or sea level. Proposals have been made to derive such a limit from the rate of change in the past. The acceptable climatological changes can also be derived from the maximum rate at which ecosystems can adapt to changes in temperature and precipitation patterns. It is generally expected that, assuming a natural limit of the migration rate of terrestrial ecosystems of 50 km per century, a doubling of the CO_2 concentration before 2090 will lead to unstable ecosystems, particularly boreal forests (Map 27.3). This map shows the areas which do

Map 27.3
Adaptation of different ecosystems in Europe following a doubling of atmospheric CO_2 by 2090
Source: RIVM, 1991

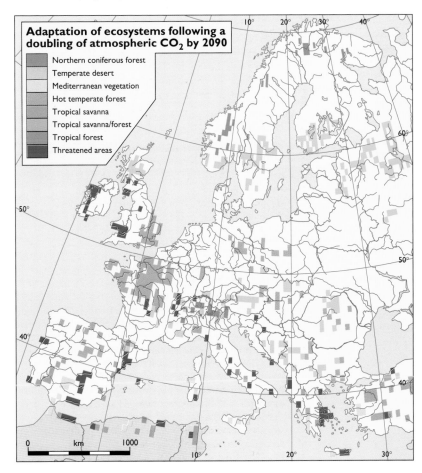

Adaptation of ecosystems following a doubling of atmospheric CO_2 by 2090

- Northern coniferous forest
- Temperate desert
- Mediterranean vegetation
- Hot temperate forest
- Tropical savanna
- Tropical savanna/forest
- Tropical forest
- Threatened areas

not have the same vegetation type across 50 to 80 km and which will be threatened if the CO_2 concentration doubles before 2090 and as climate zones shift by more than 50 km per century. Given these considerations, a temperature rise standard for sustainability of less than 0.1°C per decade has been proposed (Krause et al, 1990). Similarly, a provisional limit of a 2 cm rise in sea level per decade has been recommended to prevent damage to coastal zones, wetlands and coral reefs caused by too rapid climatological changes.

As discussed in previous chapters, the biosphere cannot operate outside certain limits. To remain within these, as yet unknown, limits, absolute limits on the acceptable rise in temperature or sea level could be set. Analyses of past temperature and rises in the sea level show that an absolute temperature change of 2.0°C relative to the pre-industrial era should be considered as a major risk. Considerable changes in ecosystems and sensitive coastal areas, as well as unexpected sudden changes in the climatic system, cannot be ruled out given a temperature change of more than 2°C.

There is a need to improve knowledge of the mechanisms and quantitative components of the budget of relevant compounds if the uncertainty of future projections for the climate and its related components is to be decreased. For example, there are certain greenhouse gases, such as CFC-14 and CFC-116, which have not been included in the IPCC assessment (Abrahamson, 1992). These two gases last for more than 10 000 years in the environment, and have large global warming potentials (being exceeded only by CFC-11, CFC-12, CFC-113 and HCFC-22). Both CFC-14 and CFC-116 are issued from aluminium smelting, but cement is the only primary commodity, other than fossil fuels, specifically included in conventional tabulations on greenhouse gases sources. However, it appears that, on a global basis, aluminium production is an even greater greenhouse contributor than cement production.

STRATEGIES

Despite the different assumptions behind the rapidly diverging scenarios for the rates of CO_2 emission, there are only very small differences in projected global warming rates over the next few decades. This indicates the commitment to global warming over this time created by past emissions and by current economic development trends. The projected warming to about 2050 is relatively insensitive to changes in global economic activity, or policy initiatives, which may be undertaken between now and 2050. The global warming under the different scenarios begins to diverge substantially by the second half of the next century, so that by 2100 there is a 100 per cent difference between the slowest (1.5°C) and most rapid (3°C) warming projection. This difference represents a significant opportunity for reducing the rate of global warming by the end of the next century, if certain economic and policy choices can be made over the next few years.

To adhere to the goal for sustainability of a maximum temperature rise of not more than 0.1°C per decade, it will be necessary to stabilise the concentration of greenhouse gases in the atmosphere at the lowest possible level, in the shortest possible time. According to the IPCC report the worldwide emissions of greenhouse gases such as carbon dioxide, nitrous oxide (N_2O) and CFCs should be reduced immediately by at least 60 per cent to stabilise the concentrations of these gases at the present level. Methane emissions should be reduced by 15 to 20 per cent, depending on the development of the emissions of other substances which are related to the decomposition of methane in the atmosphere, such as carbon monoxide and other hydrocarbons. At a conference in Toronto in 1988 a provisional guideline for the future worldwide development of CO_2 emissions was drawn up specifying reductions of 20 per cent by 2005 and 50 per cent by 2025 relative to 1988 (WMO, 1988). The Toronto

guideline has meanwhile been accepted as a policy basis by a number of countries.

Since 1988, there have been further advances in the international community's response to the threat of climate change. After 15 months work, the United Nations Framework Convention on Climate Change (FCCC) was adopted by the Intergovernmental Negotiating Committee (INC) on 9 May 1992, and was opened for signature during the United Nations Conference on Environment and Development (UNCED) in Rio de Janeiro from 4 June 1992. Over 160 countries have now signed the Convention, including all major industrialised countries and leading developing countries such as India, China, Brazil and Mexico. The Convention entered into force in March 1994.

The ultimate objective of the Convention is:

to achieve stabilisation of greenhouse gas concentrations in the atmosphere at a level that would prevent dangerous anthropogenic interference with the climate system ... within a timescale sufficient to allow ecosystems to adapt naturally to climate change, to ensure that food production is not threatened and to enable economic development to proceed in a sustainable manner.

Its main features are:

● a commitment by all parties to formulate, implement, publish and regularly update national programmes containing measures to limit emissions of greenhouse gases and protect sinks; and to report on these programmes and have them reviewed by the conference of the parties to the Convention;

● a commitment by developed countries (including the EU and its Member States) and other countries in Europe, including those undergoing the process of transition to a market economy, to take measures aimed at returning their emissions of CO_2 and of other greenhouse gases to 1990 levels by the year 2000;

● a commitment by OECD countries to provide new and additional funding to help developing countries draw up their national programmes, and to help meet the incremental costs of projects and measures which they identify in their programmes;

● a commitment by all parties to cooperate in further research into climate change and its effects, and to review regularly the working of the Convention, and the obligations to which countries have agreed.

A debate has recently started on when actions aimed at mitigating climatic change effects should be started. Some model results have been interpreted to suggest that waiting for another ten years in order to attain firmer scientific ground would imply at most a 5 per cent warmer global average temperature in 2100 than under an immediate action programme (Global Environmental Change Report, 1992). This argument may be short-sighted for several reasons:

● models do not allow for abrupt changes in the climate's response to global warming (such as shifts in ocean circulation, polar ice-cap changes or the release of methane from clathrates);

● taking only the global average as an indicator may overlook the possibilities for severe regional changes (eg, through interactions with the hydrological cycle) while the global average changes only slightly;

● non-greenhouse impacts (eg, formation of photo-oxidants, ozone depletion) may justify immediate action;

● it is unclear whether scientific knowledge or models will be more reliable after ten years and then still another ten years delay may be demanded; and

● if in any event reduction policy has to be started, it is most cost-effective to start as soon as possible in order to achieve the planned emission cuts.

If, for example, the world had immediately shifted in 1990 to a 20-year transition towards IPCC scenario B, CO_2 emissions would have had to be reduced by 7 per cent per year; waiting instead until 2000 to start the transition will require an annual CO_2 reduction of 20 per cent.

If CO_2-equivalent doubling in the next century is to be avoided, drastic reductions of global CO_2 emissions are primarily needed. In the IPCC business-as-usual scenario, an increase of 60 per cent is expected for these emissions by 2010. If no further measures are taken, only draconian reductions of 80 to 90 per cent within a few years after 2010 can postpone severe climate effects. If, by contrast, the emissions could be reduced by 2010, the most severe climate risks could be postponed beyond 2100. In that case, there would be time to halve CO_2 emissions by 2050 to reach a stabilisation level. This illustrates the need for rapid decisions.

The potential role of temperate forests as sinks for CO_2 has also been under scrutiny (see Box 23F). Germany has committed itself to a 25 to 30 per cent CO_2 emission reduction by 2005 (ie, 250 to 300 million tonnes per year), in which measures to enhance carbon sequestration by forests play an important role in the final goal. It has been estimated that expansion of forest area, a further increase of carbon storage by appropriate management, and the restoration and protection of forest health impaired by air pollution would result in an additional storage of 17 to 20 million tonnes of CO_2 per year, or 6 to 8 per cent of the reduction target. In order to store most carbon more rapidly, short rotation, fast-growing trees which provide products that last more than one rotation should be planted in soils with initially small organic carbon contents (Cannell et al, 1992). However, the role of forests as a sink for CO_2 may be minor and should be put into context since most of the planet's carbon is stored in sediments on the ocean floor.

Reductions of CO_2 emissions are technically feasible, even if consumer demands rise considerably, by increasing efficiency and by energy savings, by increased use of renewable energy sources (such as solar and wind energy) and – temporarily – by shifting fossil fuel energy generation from coal and oil to natural gas. More efficient use of nitrogen fertilisers should be stimulated to decrease N_2O fluxes. Environmental engineering solutions such as injection of carbon dioxide into the deep ocean has been shown to be theoretically feasible, but will obviously have limited application.

Because methane is a potent greenhouse gas, reduction in CH_4 emissions would be 20 to 60 times more effective in reducing the potential warming of the Earth's atmosphere over the next century than reductions in CO_2 emissions (IPCC, 1990). Reducing methane emissions will provide the additional benefits of reducing the potential for increasing tropospheric ozone (see Chapter 32). Furthermore, methane released by human activities can generally be viewed as a wasted resource, therefore reduction measures could be low cost, if not profitable (Hogan et al, 1991). It appears to be technically feasible to reduce anthropogenic CH_4 emissions by about 30 per cent (120 Tg per year). Over the next ten years, about a third to a half of these reductions would be needed to achieve the necessary 40 to 60 Tg reduction to stabilise atmospheric CH_4 concentrations. Some of the options are (Hogan et al, 1991): recovery systems in landfills; degasification before and during mining operations; improved handling and distribution in oil and natural gas systems (especially in Eastern Europe); use of specific feeds for ruminants; recovery systems for animal wastes and wastewater; improved management of rice cultivation; and fire-management and alternative agricultural practices.

Since the publication of the 1990 IPCC assessment, new evidence has emerged on the indirect effect of CFCs on global warming. By destroying ozone, the release of CFCs reduces the greenhouse effect. However, the problem is complicated by the fact that this process is strongly dependent on geographical location, especially latitude. Radiative calculations along a vertical column show that the negative effect of CFCs can cancel out the positive effect in some cases. The important conclusion is that, firstly, CFCs must be eliminated because of their role in the depletion of the stratospheric ozone layer, and, secondly, that their phase-out will not contribute significantly, if at all, to the alleviation of climate change, as previously thought.

According to the London agreement on substances that deplete the ozone layer, a phase-out of fully halogenated CFCs and halons will be achieved by the end of the century. However, envisaged substitutes will still provide some chlorine to the stratosphere and, like CFCs, have a high global warming potential. Consequently, depending on the type of halocarbons H(C)FC used to replace CFCs, implementation of the London agreement could result in even higher temperature forcing by the end of the next century. The unrestricted use of HCFCs has thus become an issue and, in a non-binding resolution, some parties of the Montreal Protocol call for a phase-out of HCFC by 2030. If such a phase-out leads to more abundant use of Class I HFCs (HFC-134a, -143a and -125), equilibrium temperature forcing by these compounds could even cause a greater warming (up to 55 per cent more) than if HCFC use were unrestricted (Kroeze and Reijnders, 1992). According to the same simulations, only a scenario involving virtual phase-out of CFCs by 2000, a phase-out of HCFC-22 by 1993, no use of Class I H(C)FCs to replace CFCs, no use of H(C)FCs for open foam, plus good housekeeping and maximum recycling or destruction of halocarbons to minimise emissions into the atmosphere, would not exceed unsustainable warming forcing of 0.08°C. It can therefore be argued that there is a case for the preference of non-climate forcing substitutes wherever possible, in order to minimise further global warming. For application such as aerosols, cleaning/drying, open and closed foam and fire extinguishers, non-H(C)FC substitutes are technically available. Most of the H(C)FCs to be used for refrigeration can probably also be replaced by non-H(C)FCs. Only for mobile air conditioning is a non-H(C)FC substitute currently unavailable.

Thus, there is a case for guiding the choice of H(C)FCs to replace CFCs by the use of both ozone depleting potentials and global warming potentials. The quest for alternatives by producers and users of halocarbons seems to be focusing on halocarbon effects on the ozone layer, so that the priority is given to HFCs over HCFCs or blends containing HCFCs. According to calculations (Kroeze and Reijnders, 1992), this results in an increase of the average global warming potential of the halocarbons used. If H(C)FCs are used without restriction to replace CFCs and halons, halocarbon applications that contribute most to temperature forcing are calculated to be refrigeration and mobile air conditioning. The gases that are expected to contribute most to calculated temperature forcing are HCFC-22 and HFC-134a. If HCFCs are phased-out to protect the ozone layer and replaced by HFCs, then HFC-134a, HFC-143a and HFC-125 will be the most important contributors to global warming, being emitted mainly from refrigeration and mobile air conditioning appliances.

International initiatives will certainly foster better harmonisation and emulations between countries in order to reach sustainable development levels. Steps towards this were taken at the UNCED in 1992 when nearly all European countries signed the Framework Convention on Climate Change. Furthermore, the European Commission is trying to reach an agreement on the use of tax instruments to stabilise CO_2 emissions at 1990 level by the year 2000.

REFERENCES

Abrahamson, D (1992) Aluminium and global warming. *Nature* **356**, 484.

Bazzaz, F A and Fajer, E D (1992) Plant life in a CO_2-rich world. *Scientific American,* January 1992, 18–24.

Bonan, G B, Pollard, D and Thompson, S L (1992) Effects of boreal forest vegetation on global climate. *Nature* **359**, 716–18.

Bultot, F (1988) Land-surface management and its impact on climate change and water balance. In: Cottenie and Teller (Eds) SCOPE Report on Belgian Research on Global Change. IGBP.

Cannell, M G R, Dewar, R C and Thornley, J H M (1992) Carbon flux and storage in European forests. In: Teller, A, Mathy, P and Jeffers, J N R (Eds) *Responses of Forest Ecosystems to Environmental Changes,* pp 256–71. Commission of the European Communities, EUR13902, Elsevier Applied Science, London.

ECGB (1992) Enquête Commission ('Protecting the Earth's Atmosphere') of the German Bundestag (Ed), *Climate change: a threat to global development.* Economia Verlag, Bonn.

Giorgi, F, Marinucci, M R and Viscanti, G (1991) A 2x CO_2 climate change scenario over Europe, generated using a limited-area model rested in a general circulation model. Part 11 Climate & Change Scenarios (unpublished manuscript).

Global Environmental Change Report (1992) Vol IV, no 75, pp 1-3. Cutter Info Corp, Arlington, Mass, USA.

Hogan, K B, Hoffman, J S and Thompson, A M (1991) Methane on the greenhouse agenda. *Nature* **354**, 181–2.

IPCC (1990) *Climate change: the IPCC scientific assessment.* Houghton, J T, Jenkins, G J and Ephraums, J J (Eds). Cambridge University Press, Cambridge.

IPCC (1992) *The supplementary report to the IPCC scientific assessment of climate change 1992:* Houghton, J T, Callander, B A and Varney, S K (Eds). Cambridge University Press, Cambridge.

Krause, F, Bach, W and Koomey, J (1990) *Energy policy in the greenhouse. Vol 1. From warming fate to warming limit. Benchmarks for a global climate convention,* International Project for Sustainable Energy Paths, El Cerrito, California.

Kroeze, C and Reijnders, L (1992) Halocarbons and global warming, II. *Science of the Total Environment* **112**, 269–90.

Marenco, A, Gouget, H, Nédélec, P, Pagès, J-P and Karcher, F (1994) Evidence of long-term increase in tropospheric ozone at mid-latitudes from Pic du Midi data series: positive radiative forcing. *Journal of Geophysical Research* 99(D8) 16617–16632.

RIVM (1991) *National environmental outlook 2, 1990–2010.* Rijksinstituut voor Volksgezondheid en Milieuhygiëne, Bilthoven, The Netherlands.

Shiklomanov, I A (1991) The world's water resources. *IHD/IHP 25 Years International Symposium,* 15–17 March 1990, 93–126, UNESCO, Paris.

Schneider, S (1990) Prudent planning for a warmer planet. *New Scientist* **17** November, 49–51.

Wigley, T M L and Raper, S C B (1992) Implications for climate and sea level of revised IPCC emissions scenarios. *Nature* **357**, 293–300.

WMO (1988) *Proceedings of the World Conference on the Changing Atmosphere: implications for global security.* Toronto 1988, WMO-Report 710. World Meteorological Organisation, Geneva.

Atmospheric monitoring
Source: Frank Morgan, Science Photo Library

28 Stratospheric ozone depletion

THE PROBLEM

The so-called ozone layer is located between 10 and 50 km above the Earth's surface and contains approximately 90 per cent of all atmospheric ozone. Under undisturbed conditions stratospheric ozone is formed as the result of a photochemical equilibrium involving oxygen molecules, oxygen atoms and solar radiation. The ozone layer protects life on the Earth's surface since ozone is the only efficient absorbent of the ultraviolet-B radiation (wavelengths 280 to 310 nm) from the sun. UV-B radiation is harmful to organisms in many ways (see Chapter 16).

The abundance of ozone is often measured through its total amount in an atmospheric column going from the ground to the top of the atmosphere. Ground-based and satellite observations have shown a decrease of total column ozone in winter in the northern hemisphere (an average depletion of the global ozone layer of 3 per cent in the period 1979–91, but with large latitudinal and seasonal differences). At mid-latitudes over Europe, the decline of the ozone column is about 6 to 7 per cent during the last decade. The decline is strongest in winter and early spring, when there is relatively little UV-B (Table 28.1). The 1991 Scientific Assessment (WMO, 1991) showed, for the first time, evidence of significant decreases in spring and summer in both the northern and southern hemispheres at middle and high latitudes, as well as in the southern winter. These downward trends were larger during the 1980s than the 1970s (see Figure 28.1). Recent evaluations show that seasonal averages of total ozone over Europe were 10 per cent lower

than long-term averages in winter 1991/92, and 13 per cent lower in winter 1992/93 (Bojkov et al, 1993).

The 'ozone hole' observed above Antarctica since the end of the 1970s is an extreme case of ozone layer depletion, with column reduction of about 55 per cent in October (ie, southern hemisphere spring) 1987 and 1989–93. At the 13 to 21 km level (lower stratosphere), the depletion was nearly complete, with loss of 95 per cent. There is great concern that substantial ozone decline could be produced in the northern high latitude regions, where the densely populated regions of Eurasia and North America could be under direct risk.

Continuous stratospheric ozone depletion trends, and the possible occurrence of an Arctic ozone hole, should be of great concern for Europe, with potential effects on human health, plants, animals and the food supply. Europe's responsibility is emphasised by the fact that Europe contributes approximately one third of the global annual emissions of ozone-depleting substances (see Chapters 4 and 14). The issue of stratospheric ozone depletion has stimulated

Table 28.1
Total ozone trends in per cent per decade (with 95% confidence limits) from satellite (TOMS) and ground-based observations
Source: WMO, 1991

Season	TOMS (1979–91)			Ground-based 26° N–64° N	
	45°S	Equator	45°N	1979–91	1970–91
December–March	−5.2 ± 4.5	+0.3 ± 4.5	−5.6 ± 3.5	−4.7 ± 0.9	−2.7 ± 0.7
May–August	−6.2 ± 3.0	+0.1 ± 5.2	−2.9 ± 2.1	−3.3 ± 1.2	−1.3 ± 0.4
September–November	−4.4 ± 3.2	−0.3 ± 5.0	−1.7 ± 1.9	−1.2 ± 1.6	−1.2 ± 0.6

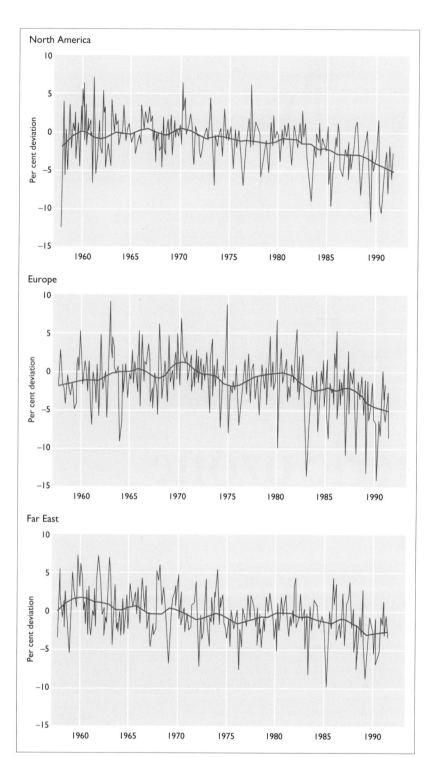

Figure 28.1
Changes in regional average ozone concentrations
Source: WMO, 1991

THE CAUSES

The increase of levels of chlorine (Cl) (Figure 28.2)and bromine (Br) has been shown to be the cause of the accelerating depletion of stratospheric polar ozone. The concentration of chlorine and bromine in the atmosphere was 1.5 to 2 parts per billion by volume (ppbv) chlorine equivalent in 1975, a few years before the 'ozone hole' started to appear (WMO/UNEP, 1989), while the present concentration is 3.9 ppbv. The artificial sources of chlorine and bromine are chlorofluorocarbons (CFCs or freons) and bromofluorocarbons (halons). These are almost exclusively artificial industrial products, first manufactured in the 1930s. CFCs are used as propellants in aerosols, coolants in refrigerators and air-conditioning units, foaming agents in the production of insulating and packaging materials, and cleaning agents. Halons are used particularly in fire extinguishers.

Direct emissions into the stratosphere of nitrogen oxides (NO_x) from aircraft also contribute to ozone layer depletion through NO_x-catalysed ozone removal and possible additional enhanced formation of polar stratospheric clouds that activate chlorine-catalysed destruction of ozone. Besides NO_x, engine emissions from aircraft include water vapour, unburned hydrocarbons, carbon monoxide, carbon dioxide and sulphur dioxide.

It has been established that important reactions involved in the destruction of stratospheric ozone are enhanced by the presence of solid particles such as ice crystals in polar stratospheric clouds, or small liquid droplets such as the sulphate aerosol particles observed in the lower stratosphere. As the results of heterogeneous reactions occurring on the surface of these particles, the unreactive and more abundant forms of chlorine molecules are partly converted into reactive species. The extreme ozone loss rates observed in the Antarctic lower stratosphere are due to suitable external conditions yielding high levels of active chlorine molecules during a sufficiently long time at the required low temperature for catalytic destruction of stratospheric ozone.

Extensive observation campaigns in the northern hemisphere during winter 1991–92 (the EC-sponsored European Arctic Stratospheric Ozone Experiment – EASOE) revealed low levels of NO_x (this helps to maintain high levels of active chlorine for ozone destruction), and enhanced levels of artificial chlorine compounds in the stratosphere in forms capable of destroying ozone. Many monitoring stations reported their lowest ever mean values of column ozone for winter months, this being explained partly by the

interest and dialogue among the scientific community, policy makers and the public. Scientific concern was already expressed in 1974, but effects stronger than predicted were observed in 1985, presenting the public with the alarming image of a hole in the sky. The first concrete international measures aimed at limiting ozone depletion were agreed in 1985 and 1987. Two important questions now are whether the agreed measures are sufficient for the atmosphere to reach its 'pre-hole' conditions, and what is the extent of any irreversible damage that has occurred to humans and ecosystems on Earth.

Figure 28.2 *Calculated atmospheric chlorine concentrations between 1950 and 1990*
Source: RIVM, 1991

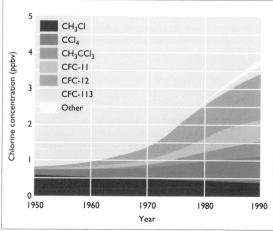

meteorological conditions. However, no feature was shown in the northern hemisphere which could appropriately be called an ozone hole. Calculated rates of ozone loss were at times of the order of 0.5 per cent per day (considered large), but the subsequent temperature rise, which led to a decrease in the levels of reactive chlorine compounds, precluded a major ozone loss. Nevertheless, measurements during the winter 1991–92 EASOE campaign indicated the potential of the chlorine already in the stratosphere to cause significant ozone loss.

The phenomenon of stratospheric ozone depletion by CFCs can be exacerbated by random increases in the presence of particles in the stratosphere following major volcanic eruptions such as those of El Chichon (Mexico, April 1982), Mount Pinatubo (Philippines, June 1991) and Mount Hudson (Chile, August 1991). In effect, the aerosol load in the atmosphere can be raised for a short time by one or two orders of magnitude after major volcanic eruptions, enhancing heterogeneous reactions which facilitate ozone destruction by reactive chlorine. For instance, ozone concentrations in the 3 to 6 months following the Pinatubo eruption were reduced by as much as 15 to 20 per cent in the altitude range 24 to 25 km, where the largest volcanic aerosol loading was observed (Grant et al, 1992).

Addition of NO_x to the atmosphere is expected to decrease ozone in the stratosphere and increase ozone in the troposphere. The observed increase in tropospheric ozone at lower altitudes in the northern hemisphere (see Chapters 4 and 32) cannot compensate for the ozone depletion in the stratosphere, since tropospheric ozone represents only about 10 per cent of the overall ozone column.

THE CONSEQUENCES

The main potential consequences of ozone depletion are:

- the disturbance of the thermal structure of the atmosphere, probably resulting in changes in atmospheric circulation;
- reduction of the ozone greenhouse effect (see Chapter 27);
- changes in the tropospheric ozone budget and, consequently, in the oxidising capacity of the troposphere (see Chapter 32); and
- increase of UV-B radiation at ground level.

Other effects might also occur. An unwanted surprise could be enhanced ozone depletion above the Arctic in connection with global stratospheric cooling, a feature known to accompany tropospheric greenhouse warming. Furthermore, even with the progressive phase-out of ozone-damaging chemicals, the high chlorine/bromine levels in the stratosphere, combined with lower temperatures, could enhance ozone depletion during the next 50 years. Moreover, this sequence can be exacerbated by the ozone loss itself, since the absorbing capacity of ozone has, so far, maintained the relatively high stratospheric temperatures.

All other factors being constant, there is no scientific doubt that stratospheric ozone depletion will increase UV-B radiation at ground level. Increased UV-B radiation could have effects on human health, ecosystems, materials, and physico-chemical processes in the lower atmosphere. Although more knowledge is needed to draw firm conclusions about several potentially important effects, the recognised deleterious effects are serious enough to justify action by the world's nations (UNEP, 1991).

Calculations based on the measured ozone trends indicate that they should be accompanied by increased UV-B over large areas of the Earth (see Figure 28.3), particularly at latitudes poleward of 30°. Evidence from direct observations is now emerging to confirm increased levels of UV radiation during episodes of ozone depletion in Antarctica (see Chapter 4). Smaller episodic increases have been measured in

Australia. Observations from the high-altitude observatory of Jungfraujoch in Switzerland indicate an increasing UV trend larger than expected (see Figure 4.8). Other factors may blur or even compensate increasing UV trends, such as cloud variability and increases in aerosol extinction and tropospheric ozone. This may be an explanation for decreasing UV trends observed in the USA.

The key areas of uncertainty in assessing the consequences of ozone depletion are:

- clarification and quantification of influences on human health, especially the immune system, and occurrences of melanomas and cataracts;
- effects on terrestrial and aquatic biota of increased UV radiation during enhanced ozone depletion episodes;
- quantification of the primary long-term effects on food production and quality, on forestry, and on natural ecosystems;
- the absence of a worldwide network of UV-B monitoring stations with suitable spatial and measurement resolution to support scientific and health assessments.

Effects on human health

In the northern hemisphere, the biologically active UV radiation (measured by the annual DNA-damage weighted dose) is estimated to have increased by 5 per cent per decade at 30°N and about 10 per cent per decade in the polar region. A few hours' exposure to UV radiation may result in sunburn and 'snow blindness' and hours to days of pain and irritation. Continuous exposure could result in skin ageing, skin cancer, a depression of the immune system and corneal cataracts. Irradiation of the skin with UV-B can suppress the immunological defence of the skin, and not only the exposed skin. Besides lowering resistance to infectious diseases, this could also reduce the effectiveness of vaccination programmes.

Skin cancer is probably one of the most common forms of cancer among white people. Most cases (90 per cent) are skin carcinoma. The most life-threatening form of skin cancer is melanoma (cancer of the pigment cells). It is estimated that a long-term reduction in the ozone concentration of 10 per cent will result in an increase of 26 per cent in the cases of non-melanoma skin cancer. Skin cancer deaths due to excess UV-B radiation at mid-latitudes are expected to rise to 2 per million inhabitants by 2030. The increase in visits to sunny

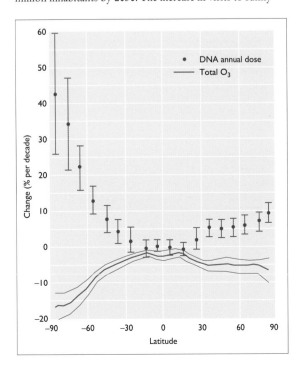

Figure 28.3
Trends in the annually integrated effective UV-B dose for DNA damage for cloudless skies, 1979–89
Source: UNEP, 1991

areas may have contributed to the worldwide rise in melanoma. Exposure to sunlight has now been associated with another cancer: that of the salivary gland. These findings suggest the possibility of a systemic effect of UV-B in humans, since the salivary gland is rarely, if ever, exposed.

As with skin carcinoma, it is the cumulative dose which determines the harmful effects of UV radiation on the eyes, for example, glaucoma and cataract. It is estimated that each 1 per cent depletion of the ozone column will increase occurrence of these afflictions by between 0.6 and 0.8 per cent. Worldwide, this would cause blindness in another 100 000 to 150 000 cases. Such concerns are not limited to fair-skinned populations but are also observed in deeply pigmented individuals, where pigment may be constitutive or acquired. On the basis that children spend considerably more time out of doors and have longer to live under a depleted ozone layer than adults, calculations for Britain show that a child's lifetime risk of developing non-melanoma skin cancer, under current ozone depletion rates, is 10 to 15 per cent higher than the risk under an intact ozone layer (Diffey, 1992). Under these conditions, an adult's risk is increased by less than 5 per cent. These predicted increases in risk can be minimised by changing behaviour during the summer months, for example spending less time outdoors, wearing wide-brimmed hats, applying tropical sunscreens, or a combination of these.

Effects on aquatic and terrestrial ecosystems

Effects on ecosystems due to an increase in UV radiation are difficult to quantify but are probably more important than those on human health (see Chapter 16). Prevention of these effects is the only possible way to avoid them, as protection of natural ecosystems from UV-B radiation is not possible. As with health effects, effects on ecosystems usually appear only after a delay. Through fisheries and crops these effects are also connected with the food supply and climate changes.

Marine phytoplankton produce at least as much biomass as all terrestrial ecosystems combined. Increases in UV-B radiation adversely affect the nitrogen metabolism, photosynthesis and vitally important adaptive functions such as orientation and mobility. As a result the survival and productivity of phytoplankton are threatened. Recent results (Smith et al, 1992) indicate a reduction by at least 6 to 12 per cent in phytoplankton productivity inside and outside the Antarctic ozone-hole region during the spring months in 1990. One consequence of losses in phytoplankton is reduced biomass production, which is propagated throughout the whole food web. According to a rough estimate, a long-term

16 per cent reduction in the ozone concentration may lead to a 5 per cent reduction in primary production and a reduction of fish stocks of 6 to 9 per cent. Phytoplankton fix a large proportion of the carbon dioxide in the oceans and are therefore important in relation to the greenhouse effect (see Chapter 27). A hypothetical loss of 10 per cent of the marine phytoplankton would reduce the oceanic annual uptake of carbon dioxide by about 5 gigatonnes (an amount equivalent to the annual global emissions of carbon from fossil fuel combustion). Also, phytoplankton production of dimethylsulphide (DMS), which acts as a precursor of cloud nucleation, would be reduced, with potential effects on the global climate.

There are wide differences in sensitivity to UV radiation between terrestrial species and varieties within species. In extensive field studies of crops, half showed a lower yield after exposure to increased UV-B radiation. A large number of the tested plants showed reduced growth, photosynthetic activity and flowering. Ecological equilibria will continue to be altered if depletion of the ozone layer continues.

Effects on materials

Damage to materials from increased UV radiation has not been much studied, with the exception of plastics, particularly PVC, which show more rapid yellowing and degradation. UV-B radiation particularly degrades wood and plastic products, leading to discoloration and loss of strength. These effects may result in increased costs of using higher levels of conventional light stabilisers, possible design of new stabilisers, and faster replacement of the affected products.

Effects on atmospheric chemical processes

Chemical reactivity in the troposphere is expected to increase in response to increased UV radiation. Tropospheric ozone concentrations could rise in moderately or heavily polluted areas, but should decrease in unpolluted regions (with low levels of nitrogen oxides), as recently confirmed by measurements in Antarctica. Other potentially harmful substances (hydrogen peroxide, acids and aerosols) are expected to increase in all regions of the troposphere due to enhanced chemical activity. These changes could exacerbate problems of human health and welfare, increase damage to the biosphere, and might make current air quality goals more difficult and expensive to attain.

Effects of future ozone depletion on climate change and global circulation cannot currently be predicted with reasonable certainty. However, results published in 1993 (Wang et al, 1993) indicate that the combined effect of stratospheric ozone decline and tropospheric ozone increase will significantly enhance radiative forcing in the northern hemisphere and contribute to surface warming.

GOALS

All leaks or emissions of substances causing stratospheric ozone depletion such as CFCs and halons which do not occur naturally in the environment should be avoided. First, a mix of compounds with allowed release rates and termination dates could be formulated, with the goal of minimising both short-term and long-term ozone losses. Additional important factors to consider include the feasibility, the economic and technological aspects of any measure, and the potential impact of the emitted compounds on greenhouse warming.

An effects-based approach could be to reduce the concentration of chlorine and bromine in the atmosphere to 1.5 to 2 parts per billion by volume (ppbv) chlorine equivalent (WMO/UNEP, 1989), the level measured in 1975

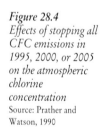

Figure 28.4
Effects of stopping all CFC emissions in 1995, 2000, or 2005 on the atmospheric chlorine concentration
Source: Prather and Watson, 1990

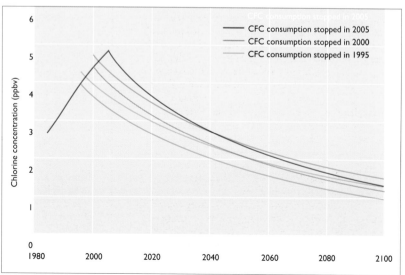

equivalent (WMO/UNEP, 1989), the level measured in 1975 just before the hole in the ozone layer was discovered. In fact, it will be almost impossible to reach 1.5 to 2 ppbv before 2080 due to the long atmospheric lifetime of the CFCs. Even if global consumption and emissions of CFCs and halons had been completely eliminated before 1995, the depletion of the ozone layer would still continue up to 2050 (see Figure 28.4). This is the earliest date at which chlorine concentrations previous to the discovery of the Antarctic ozone-layer 'hole' can be reached. Every five years' delay will lead to 0.5 ppbv more chlorine in the stratosphere and an 18-year extension of the depletion. The pre-industrial level of 0.6 ppbv could be a target for the centuries thereafter.

Another option is to establish a link with negligible or acceptable health hazards. At present it is not possible to draw up a standard for sustainability based on environmental and health impacts, due to scientific uncertainties. In the longer term it should be possible to formulate such a standard based on potential health hazards. It is clear, however, that stabilisation at the present concentration of 3.5 ppbv will result in unacceptable risks to human health and the health of ecosystems (RIVM, 1991).

Various indices have been developed for comparing the impact of one halocarbon source gas against others as scientific guidance to public and industrial policy makers when assessing the relative contribution of ozone-depleting emissions (Table 28.2). The simplest of these is the chlorine loading potential (CLP) representing the amount of total chlorine delivered from emission to the stratosphere, relative to that of the same mass of CFC-11. Ozone depletion is not only dependent on the amount of chlorine or bromine in the stratosphere but also on other physico-chemical processes and the atmospheric lifetime of the various molecules. Thus, the ozone depletion potential (ODP) represents the amount of ozone destroyed by emission of a gas over its entire atmospheric lifetime relative to that due to emissions of the same mass of CFC-11. As a relative measure, the ODP does reveal the trade-offs associated with the use of one compound rather than another. These indices have become increasingly important as the search for substitutes for the most damaging species has intensified, and concerns about increased short-term ozone losses have grown. Most emissions of CFCs reach the atmosphere some time after their production, given the way in which these products are used.

The 1989 Montreal Protocol to the Vienna Convention for the protection of the ozone layer (signed in 1987) requires the phase-out of fully halogenated CFCs. These will be replaced mostly with new chemicals not yet released into the environment, hydrochlorofluorocarbons (HCFCs) and hydrofluorocarbons (HFCs). However, any replacement compound should be scrutinised with respect to its possible detrimental effect on other environmental processes, such as global warming, photochemical pollution or other atmospheric changes (see also Chapter 27). The toxicity of these new chemicals and their degradation products need to be characterised not only by their effects on the ozone layer (expected to be less detrimental), but also by their possible effects on humans, fauna and flora (see Chapter 38).

Preliminary results based on a limited set of data (US EPA, 1990) on HFCs and HCFCs concluded that these chemicals could be used in a manner consistent with safety for humans and other biota. However, there are suggestions that HCFC-123 has induced an increase in non-malignant tumours in rats.

A drawback with substitute products for CFCs is that they often have a significant global warming potential, comparable with that of CFCs (see Table 27.1). Used in comparable quantities, their contribution to global warming will be substantial. Therefore use of HCFCs and HFCs is not a permanent solution, and their emissions and unnecessary releases should at least be restrained. In order to integrate these new constraints, the halocarbon global warming

potential (HGWP) index was introduced to compare the effects of the emissions of a given gas with the same emissions of CFC-11 (Fisher et al, 1990a, b).

A complementary goal is to minimise the risks of exceptional ozone depletion events from unforecastable physico-chemical processes (eg, in connection with unprecedentedly large halocarbon abundances or their coupling to unpredictable influences such as volcanic eruptions).

STRATEGIES

Prompt international actions for release reduction have been implemented as interim measures before the complete stop in use or production of compounds which lead to stratospheric ozone depletion. The Montreal Protocol states that the production of CFCs should be reduced to 50 per cent of the 1986 production level before 1999. It has been ratified by 32 European countries. The protocol was tightened further in London in June 1990, while provisions were also made to facilitate the participation of developing countries. This is

Species	Atmospheric lifetimes (years)	CLP	ODP
CFC-11	55	1.0	1.0
CFC-12	116	1.597	~1.0
CFC-113	110	1.466	1.07
CFC-114	220	2.134	~0.8
CFC-115	550	2.964	~0.5
HCFC-22	240	0.152	0.055
HCFC-123	47	0.0185	0.02
HCFC-124	129	0.0421	0.022
HCFC-141b	76	0.154	0.11
HCFC-142b	215	0.185	0.065
HCFC-225ca	60	0.0230	0.025
HCFC-225cb	120	0.0656	0.033
CCl$_4$	47	1.018	1.08
CH$_3$CCl$_3$	47	0.114	0.12
H-1301	67	0	~16
H-1211	40	0	~4
H-1202	33	0	~1.25
H-2402	38	0	~7
H-1201	58	0	~1.4
H-2401	46	0	~0.25
H-2311	29	0	~0.14
CH$_3$Br	35	0	~0.6

Table 28.2
Approximate atmospheric lifetimes of ozone depleting compounds, adopted chlorine loading potentials (CLP), and best estimate of ozone depletion potential (ODP) with CFC-11 as reference
Source: WMO, 1991

particularly important as forecast increases in CFC consumption, especially for refrigerators, in China and India would otherwise counteract all measures taken in the industrialised countries. According to the revised protocol, the production and use of all fully halogenated CFCs, and of the three most important halons and carbon tetrachloride (CCl$_4$), will be reduced to zero before 2000. For methyl chloroform (CH$_3$CCl$_3$), a 70 per cent reduction is aimed for by 2000 and 100 per cent by 2005. It was also decided to consider the contribution of CFCs and their alternatives to the greenhouse effect. Following the publication of new scientific evidence on enhanced ozone depletion, the EC decided to anticipate the London amendment with an 85 per cent cut by the end of 1993 and complete phase-out by the end of 1995 for CFCs and halons, as well as for carbon tetrachloride and methyl chloroform.

In 1992 the countries involved (65 signatory countries) considered the progress made, and agreed more stringent measures in the Copenhagen amendments to the Montreal Protocol. These will put a stop to global emissions of CFCs, carbon tetrachloride and methyl chloroform before 1996. The Copenhagen amendments regulate a limited consumption of

Figure 28.5
a) Annual
production and
emission of CFCs in
the world according
to four scenarios;
b) Chlorine
concentrations based
on these four
scenarios
Source: RIVM, 1991,
updated by Van Aalst,
personal communication,
1993

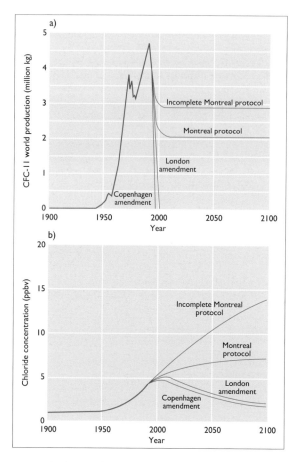

themselves be replaced. The next 20 years are likely to encompass the highest atmospheric halocarbon abundances that the Earth has presumably ever experienced, and therefore increased ozone depletion is to be expected.

Since the Montreal Protocol, there has been an intense search for substitutes for the CFCs harmful to the ozone layer. There are many alternatives to CFCs used in applications such as aerosol propellants, foam-blowing agents and cleaning of electronic equipment. However, in refrigerating, freezing and air conditioning equipment, the replacement will have to be made in two stages: replacement of the refrigerating CFC gases in existing equipment with minor modifications, and new equipment which can use non-halogenated gases or gas mixtures. Substitution products for the first stage are halocarbons with shorter tropospheric lifetimes and fewer chlorine atoms, for example, CFC-22.

Before the phase-out of compounds with large ODPs in the long term (ie, for chronic effects like the ozone hole), and reaching the time limit set on the widespread use of substitutes for the same reason, policy makers would probably consider how best to minimise the growing ozone losses that will occur during this transition (Solomon and Albritton, 1992). Therefore the critical issues involve identifying which substances are needed, for how long, and in what quantities.

Short term ODPs for the currently uncontrolled compounds could be a principal factor in those decisions. In those applications for which multiple substitutes exist (such as HCFC-141b and HCFC-123), industry may choose to invest in the compounds with the smaller short-term ODPs. Similarly, policy makers may set low ceilings on the production of compounds with larger projected ODPs in the next few decades, and higher ceilings on those with smaller short-term ODPs. The Copenhagen agreement set a ceiling for CFCs of 3.1 per cent of the 1989 levels, and 100 per cent of 1989 levels for HCFCs, expressed in ODP tonnes. A phase-out by 2030 was also decided.

The high global warming potential of HCFCs means that a shift to environmentally safer technologies should be implemented as soon as possible. The introduction of mandatory recycling for minor uses of CFCs and halons which may be deemed essential, and of legally binding schemes to recover CFCs and other ozone-depleting compounds held in current equipment (equivalent to several years' production), could also be envisaged as effective counteracting measures.

Finally, proposals for global environmental engineering have been made to help resolve the problem of stratospheric ozone depletion and attenuate its effects – for example, the injection of thousands of tonnes of alkanes (such as ethane or propane) into the polar stratosphere to prevent ozone depletion, and for climate change, iron fertilisation of the sea to enhance the oceanic fixation of atmospheric carbon. In order to avoid inherent risks involved in any large-scale intervention, including risks of unanticipated side-effects, proper scientific assessments of such proposals are required. First assessments using simple models show that the above proposals are not technically possible and may anyway not achieve their goals (Cicerone et al, 1991). Performing such assessments is clearly essential for a responsible exploration of the feasibility of such proposals. Indeed, pursuing such ideas can contribute to an improved understanding of environmental systems by exposing the weaknesses of the understanding of global systems and stimulating new research (Cicerone et al, 1992).

HCFC, leading to phase-out before 2030. Production of halons (notably for fire extinguishers) was phased out in 1994. However, because of the long lifetime of these compounds, consequent ozone depletion will continue well into the next century and beyond.

Figure 28.5 shows future CFC production and atmospheric chlorine concentrations according to four scenarios: an incomplete fulfilment of the Montreal Protocol, a full implementation of the Montreal Protocol, and the London and Copenhagen amendments. If the Montreal Protocol is not fully implemented, the atmospheric concentrations of chlorine will rise to more than 10 ppbv. In this event, a reduction of the ozone layer in the tropics of up to 4 per cent (and at higher latitudes of at least 4 to 12 per cent) is expected, particularly in winter. It is likely that the ozone depletion will be dramatically greater than this; the current models explain only half of the ozone depletion at mid- and high latitudes as measured by satellite observations. It will take at least 70 years before chlorine concentrations in the stratosphere decrease below 2 ppbv, at which no further depletion of the ozone layer is expected, even if the London amendment is strictly implemented. The unlimited production and use of CFC substitutes could of course prolong this 70-year period.

These calculations have shown that the rapid phase-out of CFCs, methyl chloroform, carbon tetrachloride and halons is the main factor in reducing ozone destruction over both short and long time horizons. If it is accepted that they are needed, substitute or better 'transitional substances' can help towards that goal. However, substitute products do still contain chlorine or bromine atoms and moreover have an elevated global warming potential, and thus should

REFERENCES

Bojkov, R D, Zierefus, C S, Balis, D M, Ziomas, I C and Bais, A F (1993) Record low ozone during northern winters of 1992 and 1993. *Geophysical Research Letters* **20**, 1351–4.

Cicerone, R J, Elliot, S and Turco, R P (1991) Reduced Antarctic ozone depletion in a model with hydrocarbon injections. *Science* **254**, 1191–4.

Cicerone, R J, Elliot, S and Turco, R P (1992) Global environmental engineering. *Nature* 356 (6369), 472.

Diffey, B L (1992) *Stratospheric ozone depletion and the risk of non-melanoma skin cancer in a British population*. Report prepared for Greenpeace UK, London.

Fisher, D A et al (1990a) Model calculations of the relative effects of CFCs and their replacements on stratospheric ozone. *Nature* **344**, 508–12.

Fisher, D A et al (1990b) Model calculations of the relative effects of CFCs and their replacements on global warming. *Nature* **344**, 513–16.

Grant, W B et al (1992) Observations of reduced ozone concentrations in the tropical stratosphere after the eruption of Mt Pinatubo. *Geophysical Research Letters* **19**(11), 1109–12.

Prather, M J and Watson, R T (1990) Stratospheric ozone depletion and future levels of atmospheric chlorine and bromine. *Nature* **344**, 729–34.

RIVM (1991) *National environmental outlook 1990–2010*. National Institute of Public Health and Environmental Protection, Bilthoven.

Smith, R C, Prézelin, B B, Baker, K S, Biligare, R R, Boucher, N P, Coley, T, Karentz, D, MacIntyre, S, Matlick, H A, Menzies, D, Ondrusek, M, Wan, Z, Waters, K J (1992) Ozone depletion: ultraviolet radiation and phytoplankton biology in Antarctic waters. *Science* **255**, 952–9.

Solomon, S and Albritton, D L (1992) Time-dependent ozone depletion potentials for short- and long-term forecasts. *Nature* **357**, 33–7.

UNEP (1991) *Environmental effects of ozone depletion: 1991 update*. United Nations Environment Programme, Nairobi.

US EPA (1990) *Hydrofluorocarbons and hydrochlorofluorocarbons: interim report*. Office of Toxic Substances, US Environmental Protection Agency, Washington DC.

Wang, W C, Zhvang, Y C and Bojkov, R D (1993) Climate implications of observed changes in ozone vertical distribution at middle and high latitudes of the Northern Hemisphere, *Geophysical Research Letters* **20**, 1567–70.

WMO (1991) *Scientific assessment of ozone depletion: 1991*. World Meteorological Organization, Global Ozone Research and Monitoring Project, Report 25.

WMO/UNEP (1989) *Scientific assessment of stratospheric ozone, Vol 1*, Report No 20.

Cactus and aloes in Almeria, Spain
Source: Spectrum Colour Library

29 Loss of biodiversity

INTRODUCTION

The Earth's genes, species, and ecosystems are the product of over 3000 million years of evolution, and are the basis for the survival of our own species. Biological diversity (often shortened to 'biodiversity') is a measure of the variation in genes, species and ecosystems. It is valuable because:

- diversity is the base of the stability and sustainable functions of natural systems;
- of its enormously wide range for potential and unexplored uses;
- here there is evidence that a removal of ecosystem components can have negative impacts; and
- variety is inherently interesting and more attractive.

Chapter 9 attempts to give a first broad assessment of nature and wildlife on a pan-European scale by examining the state and distribution of ecosystems and species.

A large variety of European landscapes, described in Chapter 8, forms the ecological base for a number of semi-natural habitats for many species of flora and fauna. By encompassing also many important natural habitats (relatively untouched areas), European ecosystems are composed of

Table 29.1
Distribution of higher plants by continent
Source: IUCN

Continent/region	Plant numbers
Latin America (Mexico through S America)	85 000
Tropical & subtropical Africa	45 000
Tropical & subtropical Asia	50 000
Australia	15 000
North America	17 000
Europe	12 500

more than 2500 habitat types (according to the CORINE Habitat Classification) (CEC, 1991) and include about 215 000 species, of which more than 90 per cent are invertebrates (animals without backbones).

Worldwide, about 1.4 million living species of all kinds of organisms have already been scientifically catalogued, but it is estimated that the actual number might be 20 to 30 times higher. Of the 1.4 million classified thus far, approximately 750 000 are insects, 40 000 are vertebrates and 250 000 are vascular plants and bryophytes (eg, mosses) (Parker, 1982). The remainder consists of a complex array of invertebrates, fungi, algae and microorganisms. Each species is the repository of an immense amount of genetic information. The number of genes range from about 1000 in bacteria and 10 000 in some fungi to 400 000 or more in many flowering plants and also in some animals (Hinegardner, 1976).

Although Europe appears to feature a rather modest range of flora and fauna, biodiversity is not defined by number alone. Each biogeographic region of the world has a different and typical set of environmental conditions and its own equilibrium or optimum. Hence, for example, it is impossible to base qualitative assessment of an oligotrophic bog in Ireland on the number of species: this highly specialised habitat is naturally poor in species, and species richness would indicate the presence of a 'sub-optimal' habitat condition probably due to eutrophication.

This example illustrates that biodiversity cannot be understood soley by comparing species; the 'optimal' state of individual habitats and their naturally occurring species must also be taken into account. Such a definition makes biodiversity as much a crucial principle of oligotrophic bogs as of tropical rainforests.

Within Europe, the distribution of species and ecosystems is widely variable, with the least richness found in the far north. Centres of biodiversity are found in the

the far north. Centres of biodiversity are found in the Mediterranean basin and on the margins of Europe in the Caucasus Mountains (Ukraine, Georgia, Armenia), with about 5000 plants being endemic to individual countries. Almost every European country has endemic species (that is, species found nowhere else), but many other species are found in more than one country and require international cooperation for their conservation. Since many of the domesticated plants of Europe originated in countries other than those where they are cultivated today, the genetic material that would be most useful in improving their characteristics is also found in other countries or even in other continents. This indicates the critical importance of developing productive links between and among countries.

GENETIC DIVERSITY

The number of species and the amount of genetic information in a representative organism constitute only part of the biological diversity on Earth. Each species is made up of many individuals. For example, the 10 000 or so ant species have been estimated to comprise 10 000 million million (10^{15}) living organisms, of which virtually no two members of the same species are genetically identical. At still another level, wide-ranging species consist of multiple breeding populations that display complex patterns of geographic variation in genetic polymorphism. Thus, even if an endangered species is saved from extinction, it may have lost much of its genetic diversity, requiring special efforts to restore the diversity.

Genetic diversity provides the variability within which a species can adapt to changing conditions. While this is important to all species, genetic variability in cultivated and domesticated species has become a significant socio-economic resource. Without the genetic variability which enables plant breeders to develop new varieties, food production in Europe would be far lower than it is at present, and far less able to adapt to the inevitable future environmental changes.

Further, biological resources – including genetic resources, organisms or parts of them, populations, or any other biotic component of any ecosystem with actual or potential use to humanity – are renewable and with proper management can support human needs indefinitely. These resources, and the diversity of the systems which support them, are therefore the essential foundation for sustainable development.

The emergence of biotechnology presents a certain potential for a productive link between conservation and sustainable utilisation of genetic diversity. Biotechnology can lead to new and improved methods of preservation of plant and animal genetic resources and speeding the evaluation of germ plasm collections for specific traits. Maintenance of a wide genetic base (one of the elements of biodiversity) is essential for providing biotechnology with the necessary resources, including both wild and domestic species. But biotechnology also poses ecological and economic risks that could ultimately undermine their potential contribution to the conservation of biodiversity. The release of genetically engineered organisms into the environment thus deserves the most careful supervision and monitoring (as discussed in Chapter 17).

TRENDS

The available evidence indicates that human activities are eroding biological resources and greatly reducing the planet's biological diversity. Estimating precise rates of loss, or even the current status of species, is challenging because no systematic monitoring system is in place and much of the baseline information is lacking. Few data are available on which genes or species are particularly important in the functioning of ecosystems, so it is difficult to specify the extent to which Europe is suffering from the loss of biological diversity. Since the ecological roles played by many species or populations are still only partly known, the wisest course – as recognised at UNCED (Rio de Janeiro, 1992) – is to follow the 'precautionary principle' and avoid actions that needlessly reduce biological diversity.

In Europe, the changes that have affected biodiversity over the past decades are closely linked to prevailing economic principles, such as ever increasing rates of production and consumption (the 'consumer lifestyle'). Traditionally, this economic system gives low values to biodiversity and ecological functions such as watershed protection, nutrient cycling, pollution control and soil formation. Being very much a cross-sectorial issue, biodiversity is geographically and functionally linked with many sectors of social, political and economic activities. No single sector can by itself ensure that biological resources are managed to provide sustainable supplies of products; rather, cooperation is required between the various sectors, ranging from research to tourism.

The decline of Europe's biodiversity in many regions (see, eg, Table 9.10 and Figure 9.7) derives mainly from highly intensive, partially industrial forms of agricultural and silvicultural landuse, from an increased fragmentation of remaining natural habitats by infrastructure and urbanisation and the exposure to mass tourism as well as pollution of water and air. Given the projected growth in economic activity (Chapter 2), the rate of loss of biodiversity is far more likely to increase than stabilise.

More generally, it has been estimated that almost 40 per cent of the Earth's net primary terrestrial photosynthetic productivity is now directly consumed, converted or wasted as a result of human activities (Vitoušek et al, 1986), and the percentage is likely to be even higher than this for Europe due to the pervasiveness of human influences on this continent.

The evaluation of representative sites in Chapter 9 indicates that all types of European ecosystems are facing severe stresses and are not sufficiently well managed to absorb these stresses (see Figure 29.1). The known effects of these stresses on wildlife as well as reports on various species groups (see also Chapter 9) point to a decline of diversity

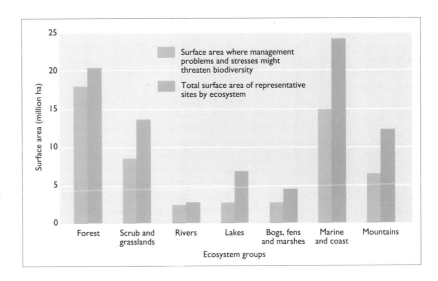

Figure 29.1

Representative sites of natural European ecosystem groups: aggregated total area, and area where management problems and stresses pose potential problems to biodiversity (see also Chapter 9)

Source: EEA-TF

Figure 29.2
Changes in plant diversity in Central Europe
Source: Kornas, 1983

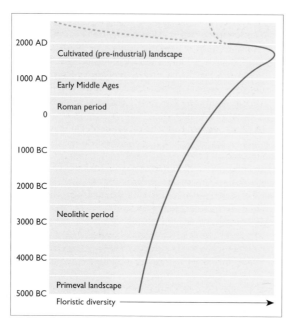

within ecosystems (loss of habitats), within habitats (loss of species) and among species (loss of species and decline of species abundance).

It might be concluded that major habitat changes and associated losses of biodiversity are the inevitable price paid for progress, as humans become an ever more dominant species. A very considerable body of work in the field of conservation biology over the past several decades has shown that reducing the area of habitat reduces not only the population of each species (and hence its genetic diversity), but also the number of species the habitat can hold. Many species that are not in immediate danger of extinction are suffering from declining populations and declining genetic variability. While some opportunist wild animal species – sparrows, pigeons, crows, starlings, hedgehogs – may be expanding their ranges and populations in Europe, far more are suffering population decline (see, eg, Figure 29.2). Low populations make species far more vulnerable to in-breeding, disease, habitat alteration and environmental stress.

The genetic variety of many cultivated species is being lost, thereby reducing ability to adapt to environmental change. The remaining gene pools in crops such as apples or wheat amount to only a fraction of the genetic diversity they harboured a few decades ago even though the species themselves are not threatened. If these trends continue, European agriculture will have a far narrower genetic base and potentially this will result in many fewer varieties of fruits and vegetables being on the market. Furthermore, these varieties will be less well adapted to local conditions, requiring larger investments in pesticides and fertilisers to maintain productivity. However, with changing patterns of agriculture in Europe, a remarkable potential opportunity exists to manage land in ways that are designed specifically for conserving biological diversity.

In the past few decades, scientists have significantly increased our understanding of the evolutionary processes that have created the biological diversity on Earth today and the contemporary factors that are leading to its reduction. The economic valuation of living natural resources has advanced greatly, and governments are seeking mechanisms for ensuring that the values of biodiversity are expressed in the cost of living and products. Present trends suggest that such principles are likely to find their way into the economic system, though greater investment in them would bring about faster progress. Greatly increased cooperation among European countries can be expected to help address the problems in a Europe-wide context.

SUSTAINABLE GOALS

To prevent an eventual decline in productivity and quality of life, the European landscape needs to be managed to conserve biodiversity, and biological resources have to be used sustainably. The following elements are needed to ensure this:

- Reliable, comprehensive and validated information on biological diversity relying on parameters such as soil, climate, topography, geology and landscape for monitoring status and trends of genes, species and ecosystems, and for predicting the impact of future changes.
- A well managed system of internationally and nationally protected areas established in each country, including representative ecosystems and the widest possible range of the country's biological diversity, taking into account Europe's role in global biodiversity.
- Forms of landuse management which guarantee the maintenance or reintroduction of biological diversity to all landscapes, by balancing all their functions – socio-economic as well as ecological – and by ensuring the wise use of their natural resources.
- A well-informed public aware of the status and trends of the ecosystems of their own country, in the rest of Europe and the world, being conscious of the impact made by their levels of resource consumption on biological resources.
- An appropriate body of legislation, economic incentives, and supporting regulations which encourage people to use biological resources sustainably and promote the conservation of biological diversity and the ethical and ecologically sound use of biotechnology.

STRATEGIES

Given that sustainable use of biological resources is the foundation of sustainable development, several strategic lines can be followed to bring about the goals outlined above:

1 Improve information on biodiversity. Decisions for planning conservation measures, identifying priorities and formulating management policies need to be based on a careful analysis of the most complete and up-to-date factual information. Baseline information on the status and distribution of landscapes, ecosystems and species that can serve as a benchmark for monitoring is available only for relatively few parts of Europe. Collaboration on a European scale is needed to develop data management standards for information on biodiversity (such as standard lists of species, inventory methods, etc), and jointly to determine the most efficient way to collect the European-wide information required.

2 Establish mechanisms to determine how the convention on biological diveristy can be most useful to each country and to Europe as a whole. All European countries and the European Community have signed the convention and, although the convention entered into force at the end of 1993, much remains to be done actually to implement it. A first step is to prepare national biodiversity strategies and action plans. Strategic issues to address include: means to regulate access to genetic resources, mechanisms for controlling import and possession of materials obtained contrary to the legislation of the originating country; mechanisms to identify the source of material from which any particular benefit is derived; funding mechanisms, intellectual property rights, and technology to transfer.

3 Elaborate further and implement the draft Action Plan for European Protected Areas prepared under the auspices of IUCN at Nyköping, Sweden, in June 1993.

4 Pay more attention to the economics of conserving biodiversity and using biological resources sustainably. In addition to the well-known values of agriculture, fisheries and forestry, the whole range of native species, including the genetic variation within species, makes contributions to agriculture, medicine and industry worth many billions of ECUs per year. Nevertheless, biodiversity tends to be perceived largely in scientific and conservationist terms rather than as a resource and as an economic good. This results partly from the difficulties in providing sufficient socio-economic insight to quantify the benefits of conserving biodiversity. Economics can help demonstrate the importance of biodiversity to human society by assigning values to the full range of goods and services it provides.

5 Concern for biodiversity needs to go beyond the creation of protected areas and conservation agencies. Given its cross-sectional linkages with almost all aspects of human life, biodiversity cannot be conserved through the artificial exclusion of economic forces. Rather, in the coming years, improvements in conserving biodiversity will increasingly need to come from improved management of land and water outside protected areas, and those sites which are given high levels of legal protection will need to be managed as part of wider landuse programmes. These wider programmes are likely to include diversification of agriculture and cropping strategies, encouragement of the spread of new varieties and non-traditional crops, rationalisation of cropping techniques so as to minimise ecological damage, and strategies for the rehabilitation and diversification of damaged habitats. In all cases, greater effort should be made to ensure that local communities are fully involved in the design and implementation of these measures.

6 The concept of biodiversity needs to be applied to ecosystems, species and genes (and perhaps even landscapes) over the whole of the European continent. The identification and preservation of the remaining valuable natural and semi-natural areas is of the same importance as the implementation of changes in landuse activities for areas where biodiversity has declined due to human impacts. Measures that point at regeneration, extensification and naturalisation of heavily overused landscapes have important functions for the realisation of biological networks (including corridors, buffer zones, etc) that encompass all biogeographic regions and landscapes of Europe.

7 Biodiversity needs to become a focus of research at national and European research centres, as a contributor (not an alternative) to practical action. Corporations with an interest in biodiversity, including those involved in biotechnology, pharmaceuticals, agrochemicals, and many other products, should be encouraged to combine forces to support the basic research required as a foundation upon which practical, applied research can be built.

REFERENCES

CEC (1991) CORINE Biotopes Programme. Manual: habitats of the European Community. Data specifications – part 2. EUR 12587. Commission of the European Communities, Luxembourg.

Hinegardner, R (1976) Evolution of genome size. In: Ayala, F J (Ed) Molecular Evolution, pp 179–99. Sinauer Associates, Sunderland, Mass.

Kornas, J (1983) Man's impact on flora and vegetation in Central Europe. Geobotany No 5, pp 277–86.

Parker, S P (Ed) (1982) Synopsis and classification of living organisms. McGraw-Hill, New York, 2 vols.

Vitoušek, P M, Ehrlich, P R, Ehrlich, A H and Matson P A (1986) Human appropriation of the products of photosynthesis. BioScience 36, 368–73.

The tanker Haven
sinking off Arenzo,
Gulf of Genoa, Italy
Source: Frank Spooner
Pictures

30 Major accidents

THE PROBLEM

Technology-related accidents are of major concern as sources of impacts on human health and the environment (Chapter 18). This concern arises from three interrelated characteristics: unpredictability of when and exactly how they will occur (and hence perceived lack of control), uncertainty over environmental pathways and impacts, and unforeseen interactions (human and technical) in the source facility. There is already substantial evidence of the short- and long-term level of human catastrophe which can result, for example, following the nuclear accident at Chernobyl and explosions such as at Los Alfaques and the release of dioxins at Seveso. The complex and possibly long-term damage to environmental resources (particularly soils and water) and dependent ecosystems is cause of increasing concern.

The causes of impacts from major accidents are distinguished by the fact that, although the source activities (eg, power generation, chemical processes and transportation) are planned and generally continuous, the hazards and environmental pressures associated with accidents are neither routine nor planned. Although statistics of past accidents provide an indication of what can be expected in the future, it is not possible to predict with absolute confidence either where or when an accident will happen. Combined with the significant uncertainty attached to the nature and magnitude of the resulting impacts, this justifies treating accidents as a significant source of societal 'risk' for the purpose of assessment and management. Risk, in this sense, can be characterised as the nature and magnitude of an undesired effect in relation to the probability of its occurring.

The industrial activities which give rise to these risks (primarily production and transport of chemicals) are increasing in intensity (see Chapter 20). In addition, interactions between human society and the natural environment are showing increasing signs of vulnerability to hazardous events (eg, the effects of Chernobyl and Seveso on agricultural activity, and the effects of hurricanes on the international insurance business). The prevention and control of such accidents to minimise the risks to within 'acceptable' limits can therefore be considered a priority. This chapter focuses on how appropriate goals can be set for a problem with such diverse elements, and how the risk can be managed by the parties concerned – industrial operators, regulatory and planning authorities and the public. Nuclear accidents are dealt with separately at the end of the chapter.

GOALS: SETTING ACCEPTABLE RISK LEVELS

Most attention, and therefore data collection and research, relating to accidents has traditionally focused on human health impacts. For this reason much of the following discussion describes this experience and looks at how it can serve as the basis for addressing environmental concerns. There are good reasons for doing this; the protection of human health and the environment from the consequences of accidents involve both complexity and uncertainty. This warrants the use of risk assessment and management for addressing both problems. Furthermore, it will increasingly be necessary to integrate human health and environmental issues into risk management of hazardous operations.

Human health risk

Most polluting activities can be characterised in terms of pollutant discharge levels (eg, sulphur dioxide emissions) over different time periods for different operating conditions. By

Box 30A Societal risk criteria: the Dutch national limits

Societal risk limits provide a convenient tool for setting criteria for the acceptability of hazardous activities to society as a whole rather than just individuals who may be affected. They take account of the fact that, for a particular industrial installation, a whole range of accidents are possible, each with a probability of happening and a range of possible impacts on the surrounding population. As with any hazardous event, a higher probability of occurrence can be tolerated for those events with lower resulting impacts. As the consequences become more serious the 'acceptable' frequency of occurrence becomes lower.

Figure 30.1 illustrates the societal risk limits, quantified in such terms, which have been adopted by The Netherlands. They specify, for example, that the probability of an accident resulting in ten deaths must not exceed 1 in 100 000 years (or 10^{-5} per year). A higher frequency is *unacceptable*. A frequency of 1 in 10 million years (10^{-7} per year) or less is *negligible*. An n-times larger impact should correspond with an n-squared times smaller probability of occurrence. Thus, the maximum permissible frequency for 1000 deaths is 10^{-9} per year, while a frequency of 10^{-11} or less is considered negligible.

A frequency in between these values is considered to warrant investigation of options to reduce the frequency. This large 'grey' area represents a margin between acceptability and unacceptability, to allow for the uncertainties associated with risk assessments, to allow for the consequences of cumulative exposure and to allow the two levels to be distinguished properly. This is in accordance with the overall approach taken towards setting risk limits in The Netherlands – the negligible risk level is considered to be 1 per cent of the maximum permissible level.

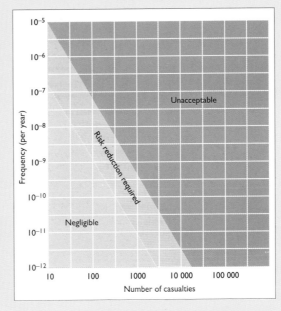

Figure 30.1 Societal risk limits adopted by The Netherlands
Source: VROM, 1990

linking these to associated impact levels (eg, acid deposition, ambient air quality) and setting impact criteria (eg, critical loads, air quality objectives), appropriate source-oriented targets can be set in terms of maximum permissible pollutant discharge levels. Plant operators can plan to achieve these targets through a range of technical and management measures. Success in achieving the targets can also be monitored relatively easily by the operator and the relevant authorities.

Goals with regard to accidents have been developed on a broadly similar basis, but with some important differences. Firstly, the emphasis to date has been on limiting direct impacts on human health and society, rather than on the environment, to acceptable levels. Mortality is usually the most frequently used measure of impact, taking account of the fact that for every death resulting from an accident there will also be a number of less seriously injured casualties, both in the short term (typically from an explosion) and in the long term (for example from radiation exposure or inhalation of toxic fumes). Secondly, because of the uncertainties associated with the occurrence of accidents and their consequences, impact targets have been expressed in terms of the probabilities of a range of impacts occurring, each with a different magnitude. This allows 'acceptable' risk to be defined as it affects both individuals and society as a whole.

Individual risk is usually defined as the probability per year that a certain individual will die as a result of an accident associated with a specific industrial plant or activity. This is because goals for the protection of individuals can be quantified as the maximum acceptable individual risk associated with a single plant. It is also possible to identify group risk, ie, the probability that a certain group of people (usually defined by size) outside an industrial facility will die due to an accident on site. Thus, goals for the protection of society can be quantified in terms of group risks by specifying the maximum acceptable probabilities per year of different numbers of people dying as a result of accidents at a particular plant.

Much of the initiative relating to identification of acceptable risks and risk assessment for planning purposes was taken in Europe by the UK Health and Safety Executive and the Dutch government. The cumulative risk assessment work for chemical complexes on Canvey Island in England (HSE, 1978; 1981) was a landmark in this respect, and subsequent investigations (HSE, 1989a; 1989b) have focused on development of appropriate risk criteria for landuse planning and their use with quantified risk assessments as an input to decision making (see below). The societal risk criteria adopted by the Dutch province of Groningen (Province of Groningen, 1979) laid the basis for Dutch national limits (VROM, 1990) and these are used as an example in Box 30A.

Risk criteria have traditionally been developed for application to potentially hazardous installations. However, they can in principle be applied to transport activities in order to limit the risks to which different groups located along transport routes should be exposed. This in turn would allow the application of risk assessment to route planning for the transport of hazardous substances. To allow for the exposure of population groups to hazards from many stationary and mobile sources, it is also possible to establish limits for cumulative risks. For example, the Dutch government has set the maximum permissible level of individual risk at 1 in 10^5 per year, ie, ten times the maximum level due to any one source.

The importance of risk perception

Setting risk targets for planning purposes involves three steps: deciding on a tolerable, or broadly acceptable, level of individual risk, deciding on the acceptability of accidents that affect numbers of people (societal risk), taking into account the possible aversion to large-scale and/or disruptive accidents, and a consideration of the achievability of the criteria.

Actual risks of different activities can be estimated in quantitative terms using a range of scientific approaches. However, setting appropriate criteria for acceptable or unacceptable risk is essentially a political issue. The human perception of risk is one of the most important factors which

Table 30.1
Annual mortality rate associated with certain occurrences and activities in The Netherlands
Source: VROM, 1990

Activity/occurrence	Annual mortality rate
Drowning as a result of dike collapse	1 in 10 million
Bee sting	1 in 5.5 million
Being struck by lightning	1 in 2 million
Flying	1 in 814 000
Walking	1 in 54 000
Cycling	1 in 26 000
Driving a car	1 in 5700
Riding a moped	1 in 5000
Riding a motorcycle	1 in 1000
Smoking cigarettes (1 packet a day)	1 in 200

the appropriate government authority needs to consider. At an individual level this depends on whether the risk is voluntary or involuntary, whether the individual perceives he or she can control the risk, the expected benefits of the activity in question and the nature and timing of possible deleterious effects.

There is growing evidence (Reed et al, 1992) that individual risk levels in the order of 1 in 1 000 000 and below do not generally give cause for concern and that levels in the order of 1 in 10 000 and above will often result in significant concern and a desire to see the risk reduced. Within this band the 'acceptability' of a given risk often varies, depending on the nature of the cause and the individual or group affected. The more an individual feels in control of a situation the more likely the risk will be judged acceptable (eg, driving a car along a motorway). If the exposure to the risk is considered involuntary, and/or the individual feels it is beyond his or her control, a similar level of risk is more likely to be judged unacceptable. For comparative purposes, Table 30.1 gives the mortality rate of a range of activities and occurrences in The Netherlands (VROM, 1990).

Notwithstanding this, the community's perception of risk will not be sufficiently wide-ranging to form the basis of acceptable risk levels. This must be the responsibility of the relevant government authority, which is in a position to take account of other important factors such as direct and indirect economic impacts (eg, through loss of local workforce or loss of confidence of investors). The feasibility of achieving different risk criteria is also an important factor to consider, if they are to be used as the basis of planning permission for hazardous, yet economically or socially important facilities (see below). A balance needs to be achieved. Thus, whereas the Dutch government has set an unacceptable group risk level of 1 in 2 billion (10^9) per year for 1000 deaths (see Box 30A), the Hong Kong limit is 1 in 2 million (10^6), reflecting the higher population density of Hong Kong.

It is important to appreciate that risk perception is subjective at the individual level and also, on a bigger scale, at societal level, where politicians must make judgmental decisions on behalf of the society they represent. For this reason, it is difficult to determine international goals for societal risk. The appropriate levels of acceptability or unacceptability are likely to vary on a country and even a regional basis.

Risk to the environment

Environmental impacts of accidents are not included in standard statistics on environmental quality. As discussed in Chapter 18, accidents should therefore be seen as an additional threat to the environment. However, the effects of accidents on the natural environment are usually complex and possibly far-reaching, and cannot be assessed easily by reference to a single indicator or parameter.

The impact of an accident usually has a widespread effect, for example, by damaging different parts of an affected

ecosystem successively, often via the food-chains. Several attempts have been made to model the environmental damage of accidents and to quantify it. An example is the Natural Resources Damage Assessment Model for Coastal and Marine Environments used in the USA and generally known by the acronym NRDAM/CME. It was developed in 1987 to assess damages to natural resources covered by the Comprehensive Environmental Response, Compensation and Liability Act (CERCLA) of 1980.

The NRDAM/CME contains submodels for physical fate, biological effects and economic damage, which are supplemented by a series of databases containing physical, chemical and biological information. The biological submodel uses lethal toxicity data for various compounds to assess the mortality of both adult and juvenile animals in the receiving habitat. However, there is a particular lack of information on the long-term effects of accidents on the environment. This is due particularly to the paucity of baseline information available. It is virtually impossible to assess the long-term ecological damage from a spill of toxic chemicals into a river if the original state of the affected ecosystem had not been previously examined.

Considering the complexity of the problem and also the substantial uncertainties involved, it is likely that the concept of environmental risk could be a potentially useful tool for setting goals in relation to the environmental impacts of major accidents. This is a relatively new concept and undeveloped compared with application of risk to human health impacts. However, it is now receiving serious attention, for example, in the Dutch Environmental Policy Plan. This provides for the setting of ecosystem risk limits for large accidents aimed at safeguarding populations of plants and animals rather than preventing damage to individual species. This might involve the setting of ecosystem protection levels (expressed as the percentage of species in an ecosystem) to act as maximum permissible impact levels. Setting ecosystem risk limits will require the development of reliable methods for assessing such risks in order that relevant standards can be set. Initially attention is being focused on the use of 'EC50' and 'LC50' values to allow the characterisation of disasters in relation to the likely exceedance of the critical limit for the most sensitive test organism or the loss of half a population, respectively.

A similar risk approach might be feasible for the protection of environmental functions of land and waters affected by major accidents (eg, the recreational value of natural areas, agricultural value of soils, use of surface waters for supply of drinking water). To some extent this might be assisted by expressing damage in monetary terms, though this has yet to be fully investigated and would be subject to the traditional judgmental problems associated with monetarisation of environmental resources.

MANAGING THE RISKS

Basic elements

The human health and environmental risks associated with accidents vary according to the location and inherent safety of industrial installations or other hazardous activities (eg, transport of dangerous substances over land or by sea). This reflects the variation of the two basic elements of risk: the magnitude of the consequences of an accident (strongly dependent on location) and the probability that these will occur (mainly dependent on the safety of the operation). Accident risk management therefore focuses on these two variables.

In practice, risk management requires a combination of source-oriented and impact-oriented measures, involving the operator of the hazardous activity, the relevant public authorities and services and members of the public. The basic elements are outlined below.

Source safety

The owners and managers of potentially hazardous operations are in a position to determine how likely it is that a particular accident, of a given type and size, will occur. This results from their influence over design technologies, management systems and human performance during normal and abnormal operations. The system design of a chemical process, the selection of appropriate plant, the incorporation of adequate safety measures into both automatic and manual (human) procedures and appropriate staff training at all levels will all contribute to reducing the probability of abnormal operating conditions occurring (eg, excessive operating temperatures or pressures). They will also help ensure that the magnitude of the immediate consequences of any such occurrence (eg, release of a toxic gas) will be minimised.

Since the 1950s the safety of industrial operations has been progressively improved by the introduction of a series of techniques relating to design standards, plant inspections, technical safety and safety audits (eg, HAZOPs). HAZOP (Hazard and Operability) studies are often undertaken on industrial sites to assess their safety and identify potential hazards. Under the group of EC Directives on Major Hazards (82/501/EEC, 87/216/EEC, 88/610/EEC), commonly referred to as the Seveso Directive, certain sites with an inherently high hazard potential need to prepare a safety study and have to do a HAZOP study.

Lately, industry has started to pay attention to 'human factors'. This relates not just to individual performance in doing a job or responding in an emergency situation, but also, and more importantly, to management, organisational and system aspects. It is now recognised that organisational and management failures are often an important contributory factor in major accidents, such as the Flixborough explosion in 1974 (Parker, 1975), the Piper Alpha offshore disaster in 1988 (Cullen, 1990) and Bhopal in 1984 (Bellamy, 1986). The report of the Zeebrugge ferry disaster (HMSO, 1987) identified a serious failure to appreciate responsibility for ship safety from director level downwards. These types of observations are leading to the development of integrated safety management systems and audit tools (eg, see Bellamy, 1986), which focus on all aspects of human factors in risk assessment to complement the existing technical focus.

However good such integrated approaches might be for assessing and reducing the impact from industrial accidents, any consideration of this topic cannot be separated from the socio-economic and political dimensions. For any safety management system to be wholly effective, two political/economic questions need to be addressed (Bourdeau and Green, 1989):

1. whether older plants, operating with outdated technologies, should be permitted to continue in production; and
2. whether plants, irrespective of age, that are so located in relation to vulnerable population groups as to present an irreconcilable conflict between risk and the cost of improvements should be relocated.

In the former case, particularly relevant to the countries of Central and Eastern Europe, there is an acute scientific/technological problem of assessing the hazards and quantifying the risks, especially where the plant has been modified over time with inadequate documentation of the changes from the original design. In both cases, the ultimate resolution of the question is likely to be encouraged by financial inducements by government.

Accident	Date	Size of spill	Clean-up costs[1]
Amoco Cadiz spill, France	16 March 1978	220 000 tonnes of crude oil	ECU199 million
Exxon Valdez spill, Alaska	24 March 1989	40 000 tonnes of crude oil	ECU2.4 billion
Haven incident, Italy	11 April 1991	10 000 tonnes of crude oil	ECU61 million (claimed)
Tanio incident, France	7 March 1980	13 500 tonnes of cargo oil	ECU42 million
Sandoz, Basle, Switzerland	1 November 1986	30 tonnes of various chemicals	ECU6 million

Note:
1 Converted to ECUs at October 1993 exchange rates.

Table 30.2
Clean-up costs following accidents
Source: Compiled by ERM

Emergency response planning

In some cases, appropriate measures on site or at the scene of an accident will ensure that pollutant releases are prevented or minimised and that significant impacts are avoided. In general, however, once an accident has happened, some level of impact is most likely, requiring clean-up measures in the case of environmental impacts, and possibly evacuation of the local population in the case of potential human health impacts. The scope for limiting impacts through response measures varies significantly, depending on whether the effect and impacts are instantaneous and simultaneous (eg, as with an LPG explosion following a road tanker crash), or whether there are delays between the time of the accident, the release of the pollutant and the impacts themselves (eg, as with an oil spill at sea).

Limiting impacts after an accident is usually very expensive (see Table 30.2) and, in the case of environmental impacts, rarely successful. For these reasons emphasis is put on prevention as the most cost-effective and efficient means of risk management. This involves both source safety measures (see above) and landuse and transport planning measures (see below).

Despite the necessary emphasis on prevention, it is recognised that accidents will continue to happen, often with considerable potential for disaster. In this context important initiatives are under way to coordinate international approaches to short-term accident response and follow-up activities with regard to clean-up of the environment and longer-term human health effects (eg, the UNECE convention on Transboundary Effects of Industrial Accidents, the UN Centre for Urgent Environmental Assistance).

The aim of an emergency or contingency plan is to localise any accident, to contain it and to reduce the harmful effects on health and the environment by rapid and responsible action. It should provide the necessary guidance to allow for a flexible response to a range of possible circumstances. Close coordination between the site operators and the local authorities and the strict definition of responsibilities in the case of an accident are essential (eg, Bourdeau and Green, 1989). In the EU, the Seveso Directive requires preparation of emergency plans and liaison between operators, local authorities and emergency services.

Emergency response plans must necessarily involve operators, local authorities and emergency services and members of the public who may be affected by an accident. They must start with an early warning of potential problems and provision of all relevant information to the relevant public authorities by the owner and/or manager of a hazardous operation. If the emergency plan is significantly to minimise the adverse impacts of any accidents, it needs to be continuously reviewed, taking into account any previous experiences and both the present as well as the future state of the industrial systems to which it is to be applied. Industrial audits serve to support the review of emergency plans.

As with other areas of environmental management, the transboundary aspect of this problem is becoming more important, as demonstrated by accidents such as Chernobyl and the Sandoz plant in Basle. This is emphasised by the UNECE (UNECE, 1993) along with the important role of industrial associations, such as CEFIC, in implementing programmes concerning awareness and preparedness for emergencies at local level (APELL) (UNEP, 1988).

Landuse and transport planning

Regardless of any steps taken by the owner and/or manager of an operation to ensure its safety and reduce the probability of a hazardous event occurring, there is an equal responsibility on the part of public authorities to ensure that potentially hazardous installations or traffic are separated adequately from population centres and environmentally sensitive areas. This will determine, ultimately, whether a human-made disaster is possible. As Reed et al (1992) have indicated, this underlies the very different casualty figures associated with the Flixborough explosion in the UK in 1974 and the isocyanate release at Bhopal, India, in 1984.

There are two complementary aspects to this element of risk management:

1 site and route selection for hazardous installations and for transport of hazardous substances (over land or by sea), respectively;
2 control over landuse in areas identified to be within the potential influence of existing or proposed hazardous activities.

If an operation is inherently highly hazardous, the decision on where to site it is of great importance. Environmentally sensitive as well as heavily populated areas should be avoided. There are similarities between this aspect of rural and urban planning and that concerned with protection of human health and the environment from routine industrial activities, involving the use of industrial zoning and environmental impact assessments (eg, the EC Directive on environmental assessment of projects, 85/337/EEC). This is an important feature of the Dutch environmental policy plan and is gaining favour elsewhere in Europe and beyond (eg, see Reed et al, 1992).

Within the context of landuse planning, industrial zones are often designated to avoid conflicts of use by predetermining the siting of industrial installations. Nevertheless, it is useful to quantify the off-site risks, particularly with regard to the cumulative risks associated with all facilities located in one area. Requiring new facilities to be subject to an environmental assessment will not ensure that cumulative environmental impacts will be avoided in an area of intensive development. For this reason the concept of strategic environmental assessment is gaining popularity as a tool for landuse planning. Similarly, managing the cumulative risks associated with accidents requires a more strategic approach than considering hazardous operations only on a case-by-case basis.

As for emergency response planning and environmental assessment of new projects, this area of risk management is likely to require more of an international focus in coming years, to take account of facilities developed close to national borders and particularly with regard to transport. In the light of recent oil spills in European waters (for example, the *Braer* off the Shetlands) the European Commission adopted a Communication entitled *A Common Policy on Safe Seas* (CEC, 1993). This proposes a package of measures intended to reduce the potential for major accidents at sea and the impacts which may arise as a result. In particular it identifies traffic restrictions in environmentally sensitive areas as a priority issue for international action. From the source safety point of view it makes a number of recommendations relating to ship inspection and training.

Integrated risk assessment and management

Integrated risk assessment and management attempts to provide a method for considering all the elements of risk management outlined above in developing the best overall means of ensuring that relevant risk criteria are satisfied for a particular installation or hazardous operation. The concept is based upon hazard identification, consequence/impact analysis, risk quantification and assessment against criteria and identification of mitigation options (either on-site or off-site).

Risk quantification is a technique which can be used to assess the implications of various siting and technical options for new developments. Frequencies of the initiating events identified during the hazard analysis (eg, a HAZOP) are normally quantified from the synthesis of operating data or equipment failure statistics (including those for the failure of safety barriers in the system). Logic diagrams, or 'fault trees', are then used to represent and quantify the sequence of events leading to a potentially hazardous outcome (eg, explosions, fires, oil spills, gas releases). Human performance data are incorporated into these analyses with an increasing degree of sophistication. Consequence models are used to assess the physical effects of the range of accidents. The consequences and frequencies for all events are then summed to give appropriate measures of individual or societal risk, which can be compared with the criteria.

There are inevitably inherent analytical uncertainties associated with this, for example, as a result of the imperfections in methodologies to model the behaviour of released chemicals under different conditions. These can be treated by undertaking probabilistic risk assessments. These are useful for understanding the confidence which should be attached to modelling outputs, but raise considerable problems as to how uncertain results should be compared with uncertain criteria.

Major industrial accident statistics in the EU are continually collected by MARS (Major Accident Reporting System) at the European Commission's Joint Research Centre (see Chapter 18). The availability of accident statistics which can be used as inputs to quantitative risk assessments is a key factor in the capability to reduce risks through safety management. The Seveso Directive puts an obligation on Member States to exchange information on major accidents in order to improve the basis of accident prevention and to ameliorate the consequences of such accidents.

An emerging aspect of this is the need to consider to what extent environmental impacts are associated with different types of accident, and therefore require different measures, from those relevant to direct human health impacts, for prevention or mitigation. One approach that is intuitively attractive is the preference for reduction of hazard rather than of exposure or impact: for example, reduce hazardous material inventories or improve safety systems rather than depend on locating plants away from housing or sensitive environments. It is likely that in future more explicit consideration of the environmental impacts of major accidents will feature in the requirements of international safety regulations, such as the Seveso Directive (which is being revised – CEC, 1994), and in combined health, safety and environment audits and management systems for industrial companies.

NUCLEAR ACCIDENTS

The use of radionuclides for industrial, medical and research purposes, as well as the operation of nuclear facilities, has to satisfy specific regulatory measures and nuclear legislation to ensure the protection of humans and the environment against the dangers of ionising radiation (see Chapter 16). Basic radiological protection requirements have been established worldwide and are kept under permanent review by the International Commission for Radiological Protection (ICRP). They are embodied in most national legislations and in EC legislation which lays down legally binding basic standards for the protection of the health of workers and the general public against the dangers arising from ionising radiation. The implementation of these standards is within the competence of each individual EU Member State.

Nuclear safety results from the existence and correct implementation of such a regulatory framework, as well as from adequate technologies and techniques, and from appropriate training and dedication of personnel. The opening of Central and Eastern Europe, and the break-up of the USSR, has disclosed marked differences compared with Western Europe in nuclear safety matters, and significant weaknesses have been identified. The main problem is to remedy past mismanagement and accidental environmental pollution, as well as preventing and mitigating possible new accidents.

Although the Chernobyl accident (see Box 18E) has not provided in itself any strong specific lessons of relevance to the designs and to the regulatory framework adopted for nuclear plants outside the former USSR, the very fact that it happened in a technologically advanced country, with all its heavy consequences on people and the environment, has posed a new challenge in Europe.

In essence the problem of nuclear accidents is how to deal with events which have potentially high consequences but which are expected to occur with only extremely low probabilities. Can further measures be taken to make their occurrence, for all intents and purposes, 'impossible', or can their potential consequences be reduced by, for example, improving containment of the plant?

Causes

A summary of the major nuclear accidents in Europe and their causes and consequences is presented in Chapter 18. Incidents which can occur in nuclear facilities may result in the release of radioactive materials (all plants) or primarily chemotoxic materials (especially uranium hexafluoride from enrichment plants). Possible underlying causes of such events include:

- loss of external power supplies;
- mechanical, electrical or electronic failures;
- earthquakes;
- fire;
- the impact of aircraft or other missiles;
- human error.

Such underlying causes may appear in combination and can give rise to intermediate situations such as nuclear criticality events (especially in chemical process plants) or loss of reactor cooling, which in turn lead to the release taking place.

Incidents are frequently related to handling radioactive liquids, sludge deposition in piping, storage of combustible or/and pyrophoric material, and the imperfect monitoring and control of these operations. Deficiencies in plant design and operation, mishandling of radioactive materials, human error, mismanagement and insufficient training, and neglect of appropriate regulatory frameworks, responsibilities and procedures are the main possible causes of abnormal events. In fact, according to the level of 'nuclear safety culture' of the country under consideration, only one or a few of them may be dominant.

Most abnormal events which have occurred at nuclear power plants in Western Europe have resulted from human error during operation of the plants. The operational and safety status of reactors of Soviet design in Central and Eastern Europe have been able to be assessed more precisely since the political changes there in the early 1990s. Nearly all the nuclear power plants of Soviet design show significant technical and operational weaknesses in various degrees according to the type of plant and the country under consideration. In the plants of the older types (RBMK and VVER 230 types) the design of the circuits and components is not satisfactory in terms of safety. Deficiencies in quality assurance and control and maintenance have been observed, as well as a scarcity of qualified and dedicated personnel. For some systems and some situations, the designers relied essentially on the operator's intervention, with a high probability of a crucial human error as a consequence.

Objectives

Since the end of the 1980s it has become fully accepted that, whereas severe accidents (ie, those having large-scale effects on humans and the environment) might occur in any nuclear power plant (even if the probability of occurrence is very low), the implementation of, and the priority given to, nuclear safety and environmental protection have been dealt with in a much less stringent manner in the former USSR and associated countries than in Western Europe. Two main objectives are therefore being pursued:

1 To reduce even further the probability of severe accidents in new nuclear power plants and, if one occurs, to limit its effect to the confines of the plant site. This objective is expected to be achieved mainly by means of new or improved design concepts or features, such as simplifying the ways and means of operating nuclear facilities, and the use of new containment and emergency cooling systems to counteract possible damage to the reactor core.
2 To establish general safety principles that are accepted by all countries and enforced by all of them, worldwide. This should include the promotion of a general and permanent awareness at all levels, including non-specialised workers, concerning matters of nuclear safety and environmental protection.

Strategies

Strategies for enhancing reactor safety are being carried out in several nuclear countries, mainly by means of research and development programmes. These are concerned with advanced reactor design, from the 'evolutionary', based on present experience, to the 'revolutionary', calling for new concepts. Virtually no substantial effort in advanced reactor design is occurring in isolation in a single country. Within the European Community, the cooperation between Framatome (F) and Siemens (D) on the future European Pressurised Water Reactor (EPR) is an example of evolutionary design. The EPR project will also provide a solid basis for the harmonisation of French and German safety requirements and possibly promote an enlarged European cooperation. Joint research in Europe should provide a good basis for future nuclear power plants and lead to the required common understanding of what should comprise basic safety features.

The new relationship established in the early 1990s between the countries of Central and Eastern Europe and the rest of the continent has created favourable conditions for the further development of the international dimension of nuclear safety. The International Conference on the Safety of Nuclear Power, convened by the IAEA in 1991 following a proposal by the European Commission, led to the conclusion that there was a need for an integrated international approach to all aspects of nuclear safety, and for the preparation of a formal framework. The international Convention on Nuclear Safety, which has arisen from this, aims mainly at achieving worldwide uniform, and higher, levels of safety at nuclear power plants. The convention was adopted on 17 June 1994 and will enter into force after ratification.

The specific problems of nuclear safety encountered in the countries of Central and Eastern Europe are taken into account by a strategy of assistance carried out by a group of 24 countries, including Western European countries, Canada, the US and Japan, with a financial commitment from the EC TACIS and PHARE programmes, between 1990 and 1994, of ECU416 million.

If, despite these measures, a nuclear emergency does occur, fast, reliable and appropriate information is required. In order to provide this at international level, both the International Atomic Energy Agency (IAEA) and the European Commission have set up special, mutually compatible communications systems ('Emercom' and 'ECURIE' respectively). The aim of these systems is to transmit urgent radiological information between the IAEA, the European Commission and their respective Member States. Exercises are regularly held between them to keep the systems prepared and operational.

REFERENCES

Bellamy, L J (1986) The Safety Management Factor: an Analysis of the Human Error Aspects of the Bhopal Disaster. Safety and Reliability Society Symposium, 25 September 1986, Southport, UK.

Bourdeau, P and Green, G (Eds) (1989) Methods for assessing and reducing injury from chemical accidents. SCOPE 40, IPCS Joint Symposia 11, SGOMSEC 6. John Wiley & Sons, Chichester, UK.

CEC (1993) A common policy on safe seas. COM(93)66 final. Commission of the European Communities, Luxembourg.

CEC (1994) Proposal for amendment of the Seveso Directive. COM(94)4 (OJC106). Commission of the European Communities, Luxembourg.

Cullen, Hon Lord (1990) The public inquiry into the Piper Alpha Disaster. Department of Energy Cm 1310, 2 vols. HMSO, London.

HMSO (1987) Merchant Shipping Act 1984. Mv. Herald of Free Enterprise formal investigation. Report of Court no. 8074. HMSO, London.

HSE (1978) Canvey: an investigation. UK Health and Safety Executive, HMSO, London.

HSE (1981) Canvey: a second report. UK Health and Safety Executive, HMSO, London.

HSE (1989a) Quantified risk assessment: its input to decision making. UK Health and Safety Executive, HMSO, London.

HSE (1989b) Risk criteria for land-use planning in the vicinity of major industrial hazards. UK Health and Safety Executive, HMSO, London.

Parker, R J (Chairman) (1975) The Flixborough disaster. Report of the Court of Inquiry, Department of Employment. HMSO, London.

Province of Groningen (1979) Pollution control and use of norms in Groningen. The Province of Groningen, The Netherlands.

Reed, S, Tromp, F and Lam, A (1992) Major hazards – thinking the unthinkable. Environmental Management 16, pp 715–22.

UNECE (1993) Policies and Strategies for the Prevention of, Preparedness for and Response to Industrial Accidents. Meeting of the Signatories to the Convention on the Transboundary Effects of Industrial Accidents (Second meeting, Geneva, 14–17 June 1993), ENVWA/WG.4/R.2., UNECE, Geneva.

UNEP (1988) APELL – Awareness and preparedness for emergencies at local level. A process for responding to technological accidents. Industry and Environment Office, United Nations Environment Programme, Tours, France.

VROM (1990) Premises for risk management, risk limits in the context of environmental policy. In: Dutch Environmental Policy Plan 'Kiezen of Verliezen' (Second Chamber of the States General, 1988–89 session, 21137). The Netherlands, Ministry of Housing, Physical Planning and Environment, The Hague.

Acid rain damage,
Slaska Poremba,
Poland
Source: C Martin,
The Environmental
Picture Library

31 Acidification

THE PROBLEM

Atmospheric emissions of acidifying substances such as sulphur dioxide (SO_2) and nitrogen oxides (NO_x), mainly from the burning of fossil fuels, can persist in the air for up to a few days and thus can be transported over thousands of kilometres, when they undergo chemical conversion into acids (sulphuric and nitric). The primary pollutants sulphur dioxide, nitrogen dioxide and ammonia (NH_3), together with their reaction products, lead after their deposition to changes in the chemical composition of the soil and surface water. This process interferes with ecosystems, leading to what is termed 'acidification'. The decline of forests in Central and Eastern Europe and the many 'dead' lakes in Scandinavia and Canada are examples of damage which are, in part, due to acidification. Modern forestry and agriculture contribute to but can also be affected by acidification. Acidifying substances also play a role in the greenhouse effect (see Chapter 27). Furthermore, nitrogen oxides contribute to the ozone problems (build-up of tropospheric ozone, depletion of stratospheric ozone; see Chapters 32 and 28), and, together with ammonia, contribute to the nitrogen fertilisation of natural terrestrial ecosystems; with phosphate they contribute to eutrophication in water (see Chapter 33).

By the end of the 1970s, acidification was widely recognised as a major threat to the environment. As a result large research programmes were set up to investigate the chain from emission to effects of acidifying substances and to indicate possible policy measures in this field. This has led to a much better understanding and modelling of the processes involved, which in turn has helped to formulate international agreements with explicit objectives for reducing emissions of pollutants leading to acidification.

The effects of these measures are now being evaluated in order to define sustainable releases of acidifying substances. This is based on studies which indicate that there are deposition loadings below which no harmful damage is observed, that is, the concept of 'critical loads'.

THE CAUSES

Anthropogenic sulphur dioxide emissions are due largely to the combustion of sulphur-containing fuels (oil and coal) used in power stations, other stationary combustion activities and process industries (refineries). Nitrogen oxides are emitted by combustion processes; transport, power generation and heating are the most important sources. Most of the ammonia in the atmosphere is due to the production and spreading of animal manure (see Chapter 22). Although ammonia is a base gas, it may lead to acidification after it reaches the soil and is nitrified. Nitrification is the bacteriological conversion in the soil of ammonium with oxygen to nitric acid. The acidifying substances (nitrogen oxides and sulphur dioxide), their conversion products and ammonium can generally be transported and spread over relatively long distances (thousands of kilometres), whereas ammonia is deposited relatively close to its source of emission (100 to 500 km).

Emissions of compounds leading to acidification increased considerably in Europe after the industrial revolution and especially after World War 2 (see Figure 4.3). For instance, sulphur dioxide emissions doubled between 1950 and 1970, but have generally levelled off since the early 1970s after the first oil crisis. Since 1980, European sulphur dioxide emissions in particular have been considerably reduced. Increasing oil prices following the oil crisis led to a full restructuring of the petroleum refining industry during the early 1980s, drastically reducing the output of heavy,

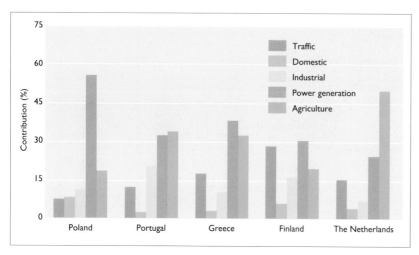

Figure 31.1
Relative contribution of source categories to the total potential acid deposition in selected countries: estimates for 1990
Source: Compiled by RIVM

residual fuel oil. Motor vehicles account for a considerable proportion of the total emissions of nitrogen oxides to the atmosphere in Europe (see Figure 31.1). Because of the general growth in private car use and road transport of goods, there are indications that emissions of nitrogen oxides will continue to increase. Increased emissions have been accompanied by increased deposition levels in Europe, with the relative contribution of nitrate increasing during the past 30 years.

The effects of emitted pollutants and the resulting transboundary air pollution on the environment within Europe have been assessed within the European Monitoring and Evaluation Programme (EMEP) under the UNECE Convention (see Table 31.1).

THE CONSEQUENCES

European ecosystems are sensitive to the deposition of sulphur and nitrogen. Sulphur dioxide and nitrogen oxides exert both a direct and an indirect influence on organisms and materials. In the Nordic countries, acidification of soils and surface waters is attributed mainly to sulphur deposition. However, nitrogen is estimated to contribute 10 to 30 per cent of the acidification in Denmark, southern Sweden and southern Norway. In some parts of Central Europe, nitrogen is responsible for a major part of acidification, and it seems likely that it will increase in importance as a source of acidification. This is explained by more stringent reduction of sulphur emissions as a consequence of international protocols, while nitrogen emissions seem to be continuously increasing due to insufficient measures to achieve reductions, which are largely counteracted by a relentless expansion in the numbers and use of the motor car.

Many of the processes resulting from the emission, transport and deposition of pollutants, the acidifying effects on soil and surface-water chemistry, and the link to biological

damage are now understood, particularly with regard to surface-water acidification. The effects of acidification result from indirect influences:

- organisms in water are affected when the hydrogen-ion concentration increases (pH decreases) and when toxic metals, leached from the ground, come into circulation;
- plants are affected when the chemistry and biology of the soil become altered by an increased concentration of hydrogen ions;
- people are affected through the consumption, in particular, of drinking water (surface and groundwater) which, due to the decreased pH, may have elevated metal concentrations.

The effects of acidification depend on the combination of two factors: the magnitude of deposition (wet and dry), and the natural, inherent sensitivity of the receiving media (soil and water) to acidification. Acidification effects can therefore occur both in the immediate vicinity of an emission source and at great distance from it. The earliest effects become apparent where both deposition and sensitivity are high. Most precipitation reaching land runs into watercourses or percolates through the soil and becomes groundwater. In most cases precipitation comes into contact with both the ground and vegetation, and can therefore have effects on both.

Lakes

In Scandinavia and North America lakes provided the first warning signals of acidification (see Chapter 5): the water became peculiarly clear; beds were gradually covered with ever thicker growths of species of *Sphagnum* mosses; and there was an increase in fish deaths and a decline in numbers of feeding fish and other aquatic animals. European rainwater unaffected by human activity should have a pH between 5 and 6. Today, increasing acidity has been observed over wide areas with notably lower pH – between 4 and 4.5, and even individual values as low as 3.

The process of surface water acidification essentially takes place in three stages:

1 Neutralisation of runoff water occurs in areas where the surrounding land and bedrock is easily weathered due to the generally high bicarbonate content. In due course, the content of bicarbonate ions in runoff declines, thus lowering the buffering capacity. There are no noticeable changes in lake biology during this stage.
2 At sufficiently low bicarbonate concentrations, large influxes of hydrogen ions can no longer be neutralised and the pH value begins to decrease. Periods of heavy rain are able to upset drastically the current balance, giving way to 'acid surges'. Large pH variations will cause direct and large-scale biological damage, such as mass death among fish and/or disruption to the reproductive capacity of fish species. The greater the influx of hydrogen ions at this stage, the more pronounced and prolonged will be the acidic periods of the lake. The second stage sets in if the pH of the water at any time during the year is lower than about 5.5, leaving the lake or watercourse moderately acidified.
3 The pH value stabilises around 4.5, even if precipitation is more acidic and the influx of hydrogen ions continues. At this point, humus substances and aluminium begin to buffer against further acidification. Aluminium in ion form increases drastically in acid surface water. This element has a strongly toxic effect on many organisms, and aluminium poisoning is becoming increasingly apparent as the real cause of the mass death of fish. This third stage has been reached in many Nordic lakes (see Chapter 5) and is marked by an entirely new ecosystem where fish life has for the most part completely disappeared. A severely acidified lake has characteristically

Table 31.1
Percentage of the deposited sulphur and nitrogen of indigenous origin in selected European countries in 1990 according to EMEP calculations
Source: Iversen et al, 1991

Country	Sulphur	Oxidised nitrogen	Reduced nitrogen
former Czechoslovakia	53	17	58
Denmark	25	6	77
France	42	38	85
West Germany	29	39	58
East Germany	80	15	68
Italy	61	37	75
Hungary	51	13	67
The Netherlands	21	15	82
Norway	5	6	28
Poland	53	21	67
Sweden	11	11	30
UK	87	46	86

unusually clear water, with the depth of visibility (Secchi depth) ranging from 4–5 m to 15–20 m, and *Sphagnum* moss spreading across its bed.

At present, severe acidification of freshwater is occurring over large areas of southern Norway, Sweden, Finland and some sites in Denmark (Brodin and Kuylenstierna, 1992). The most serious acidification of surface waters has been experienced by the Nordic countries, parts of Scotland and Germany and in Austria and the former Czechoslovakia (see Chapter 5). Less serious effects have also been observed in the Italian Alps and the French Vosges. In the Nordic countries, lake ecosystems seem to be in general more sensitive to acidification than forest ecosystems, forests generally being considered the most sensitive to nitrogen enrichment (see Chapter 34).

Soil

Compared with surface waters, soil has great buffering capacity. It requires a relatively heavy input of acid to produce appreciable acidification of soil and groundwater. However, the chemical stability of the soil system varies, and sensitivity to acidification is related to the type of bedrock, the kind of soil, landuse and the proximity to major emission sources. Regions with old lime-poor rocks resistant to weathering and overlaid by thin soil layers (such as Scandinavia and eastern North America) are the most sensitive to acidification. Over the greater part of Europe rocks have a higher content of lime (calcium carbonate), thereby providing a greater capacity to neutralise acid deposition or acids formed in the soil.

The natural production of acids from coniferous forest litter is normally neutralised by the buffering capacity of the soil. However, in places with poor buffering capacity and where the atmospheric deposition of acids exceed the natural acid production, complete neutralisation of acids can no longer take place. This leads to lower pH-values, increased leaching of base cations (for example, potassium, calcium, magnesium) and decreasing base saturation. At a later stage acidification can also lead to leaching of aluminium (Al^{3+}), which is harmful to organisms (eg, fine plant roots and aquatic organisms). Sulphur deposition is the key factor for soil acidification in the Nordic countries. Acid soils do not retain sulphur which is being leached to the groundwater as sulphate with equivalent amounts of base cations, thus further contributing to acidification. The critical load for sulphur deposition in areas with sensitive granite and gneiss bedrock has been estimated at 300 to 800 mg sulphur/m^2 per year (Christensen et al, 1993).

For nitrogen, the sequence is more complicated. Most forest ecosystems are nitrogen deficient and nitrogen addition (as atmospheric deposition or fertiliser application) will to a certain level act as a nutrient and increase tree biomass production. Beyond this level production will not increase and nitrate will start leaching, leading to acidification. Before this stage is reached other nutrients (eg, potassium, phosphorus, magnesium and micro-nutrients) can become growth limiting, leading to the well-known forest degradation symptoms such as discolouration of needles, decreased resistance to drought and frost, and increased sensitivity to insect plagues (see Chapter 34). Soil acidification caused by nitrogen leaching and associated problems are, in particular, a problem in Central Europe and The Netherlands (Grennfelt et al, 1993).

Evidence of anthropogenic acidification of soils is sparse due to lack of historical data; however, increase of soil acidity has been observed in southern Sweden (Tamm and Hallbäcken, 1988).

In agriculture, the increasing dependence on fertilisers over the last few decades (see Chapter 22) has entailed an increasingly serious acidification, especially with the use of ammonium-containing fertilisers and direct application of liquid ammonia. As the soil pH drops, there is an increase in the rate at which most trace nutrients and heavy metals are released from the soil and absorbed by plants. This can limit crop growth.

Forests

The susceptibility of forests (which generally are not fertilised – see Chapter 23) to acidification is dependent upon a number of factors such as: tree species, soil type, soil texture, soil depth and the thickness of the humus layer, the occurrence of bare rocks and of peatland, and soil moisture. The combined stress from acid deposition and photochemical oxidants may increase the potential for forest damage.

Nitrogen deposited in forest systems may have three main fates: uptake in stems and bark, accumulation in the living biomass or as organic compounds in soil, and leaching from the soil. The accumulation of nitrogen in the forest soil will change the availability of nitrogen in relation to other nutrients and to organic carbon. This may lead to increased nitrate leaching. Although the accumulation of nitrogen in soils is a natural process in northern forests, the deposition of nitrogen is far above the natural rate in Central Europe and Scandinavia. Nitrogen plays an important role in the acidification of soils in Central Europe but is of minor importance in Scandinavia, where sulphur is the main acidifying component. In areas with insignificant nitrate leaching, sulphur is also the main cause of lake acidification. The situation in Scandinavia may, however, change since the critical load for nitrogen is exceeded in many areas, and this will in the long run lead to nitrate leaching. The shift from a situation of minor to substantial nitrate leaching can occur over a very short time.

In coniferous forests of mountainous areas of Central Europe, particularly in the Czech Republic, Germany, Poland and the Slovak Republic, forest decline has occurred. This decline is believed to be caused by soil acidification and high concentrations of ozone and sulphur dioxide in ambient air (see Chapter 34). In the Nordic countries, no large-scale forest decline that can be attributed to anthropogenic acidification has been reported. However, substantial needle loss has been recorded in the coniferous forests of some areas, particularly in southern Sweden. In fact, Nordic forests are more productive than ever, but future effects that may be caused by long-term acidification cannot be discounted. Indeed, the areas now suffering in Central Europe also underwent an increase in yield prior to decline.

Other effects

Besides damage to ecosystems, acidification may in the long term reduce the quality of groundwater by leaching sulphur and nitrogen compounds and the mobilisation of aluminium and heavy metals in the soil. It may also damage materials (eg, buildings and fabrics) and crops. Acidification of the soil or the soil water may have a corrosive effect on materials, such as steel and cast-iron pipes, steel posts, concrete foundations and lead-jacketed cables buried in the ground. However, there are few data available to quantify the significance of these effects.

Current nitrogen deposition also causes environmental effects other than acidification. Nitrogen enrichment of soils by atmospheric deposition in large parts of Europe has caused substantial changes in natural vegetation (eg, heaths changed into grasslands). Direct deposition and leaching from soils are important sources of nitrogen to waterbodies, giving rise to, for example, eutrophication of coastal waters (see Chapter 6). The effects of acidification can also be enhanced by climate

Table 31.2
Importance of emission reductions to reduce acidification effects
Source: Grennfelt et al, 1993

Pollutants	Northern Scandinavia	Southern Scandinavia & UK	Central & Western Europe	Southern Europe
SO$_2$	XX	XXX	XXX	X
NO$_x$	0	X	XX	?
NH$_3$	0	X	XX	?

Notes:
XXX: very important; XX: important; X: some importance;
0: no importance.

change: increased soil temperatures can increase mineralisation and therefore leaching of nitrogen and sulphur from soils.

Regional pollution effects differ across Europe with the relative importance of the different precursors (see Table 31.2).

GOALS

Assessments of deposition loads to environmental media (surface waters, soil, vegetation) have led to the development of the 'critical-load' approach. The critical load to a medium is defined as a quantitative estimate of an exposure to one or more pollutants below which significant harmful effects on specified sensitive elements of the environment do not occur according to our present knowledge. Critical loads vary considerably from ecosystem to ecosystem and from site to site (see Chapter 4) such that specific methods of evaluation are required. Models developed to determine critical loads can differ as to their underlying assumptions.

The use of the critical-load concept was established by the UNECE nitrogen oxides protocol (see below) as the basis for future pollution control. Critical loads may be compared with current deposition rates to indicate the degree to which acidic deposition has to be reduced to maintain or reinstate sustainable environmental conditions. Previous attempts to reduce depositions have been based on only the vague understanding that damage was occurring and that emissions needed to be reduced, usually by suggesting national reductions.

The critical-load approach offers a useful method to allow a more rational abatement strategy to be carried out, based on the carrying capacity of ecosystems. Ecosystems are complex and therefore detailed dose–response relationships are difficult to obtain, but thresholds offer a simpler concept which may be defined and estimated from a broad appreciation of ecology. Ecosystem-based abatement strategies are an effective way to ensure long-term sustainable use of the environment.

In many areas of Europe, sulphur and nitrogen depositions currently exceed the critical-load values and in some areas are many times greater (see eg, Maps 4.10 and 4.11, and Chapters 5 and 7). In certain regions of Europe, estimated emission reductions of up to 90 per cent for sulphur dioxide and nitrogen oxides are needed to reach critical loads for nitrogen and sulphur, and more than 50 per cent for ammonia (Lövblad et al, 1992). For technical, social and economic reasons the emission reductions required to achieve critical-load values cannot realistically be implemented in a single step within a few years. The concept of 'target loads' is therefore used to indicate goals which countries may aim for within a given time frame. Target loads are nationally set values, which may vary within a country. They include a stepwise implementation of abatement strategies which may be set as a series of aims of which the ultimate objective is to reduce emissions to match critical loads.

Control measures need to take into account the regional variation of the contribution to acidification from sulphur and nitrogen throughout Europe. In acidified soils with substantial nitrate leaching, control measures need to be directed towards nitrogen as well as sulphur deposition. In soils without nitrate

leaching, but with ongoing sulphur acidification, a decrease in sulphur deposition may increase the resistance to nitrate leaching. In areas where sulphur and nitrogen both contribute substantially to acidification, it might be possible to use a general acidification concept (such as total potential acid deposition) where the contributions from the compounds are added. However, the relevance of the concept for all regions remains to be demonstrated scientifically.

The critical-load approach has clarified understanding of soil acidification and to a large extent also of freshwater acidification, eutrophication and direct impact on vegetation. Through the identification of receptors and damaging pollutants or pollutant loads, it is now possible to attribute deposition excess to emission sources in each European country with a significant level of confidence. The critical-loads approach will also improve quantification of environmental benefits in the future, especially through the widening of the scope of future protocols from single pollutants to a suite of pollutants and from single issues of concern to include several issues on a variety of scales. In terms of the cost–benefit comparison, each country can then estimate the cost of controls and evaluate the benefits of any improvement within its own territory. In the long-range transport context, the question is whether the benefits in other countries, when added up over a range of effects and a suite of pollutants, become significant (Grennfelt et al, 1993). Emission reduction of nitrogen oxides, ammonia and sulphur dioxide will be necessary to reduce exceedance above the critical loads for acidification and nitrogen deposition. Control of nitrogen oxides would reduce acidification but also eutrophication and photo-oxidant formation on both European and global scales; ammonia control would reduce acidification and eutrophication, while control of sulphur dioxide would reduce acidification on the regional scales.

If critical loads for acidification and nitrogen deposition can be reached within the next few decades, there will be substantial potential for:

- a considerable improvement in environmental conditions in coastal sea and brackish water areas in Europe;
- a considerable decrease in effects of ozone on human health, materials, crops and wild plants;
- a considerable reduction in erosion of buildings, metal constructions and other man-made artefacts; and
- an improvement of air quality on a local scale.

The last of these effects may form a basis for more stringent control than regional effects.

A decrease in sulphur emissions may result in an intensification of the greenhouse effect (via less solar reflection) and speed up the rate of temperature increase, whereas decreases in nitrogen emissions counteract this effect (see Chapter 27). This effect of reducing sulphur emissions should not constitute an excuse to prevent sulphur abatement, but instead increase the necessity to reduce greenhouse-gas emissions.

STRATEGIES
Legislation

The growing awareness of current and possible future effects of anthropogenic emissions of acidifying compounds and their long-range transport from country to country has already resulted in EC legislation and international negotiations on emissions reductions. A series of EC Directives have been developed to tackle emissions of acidifying substances from transport (70/220/EEC and 72/306/EEC) and industry, especially from large combustion plants (eg, 84/360/EEC, 88/609/EEC, 89/369/EEC and 89/429/EEC). At the international level, a sulphur protocol was developed within the framework of the convention on

Long-Range Transboundary Air Pollution (LRTAP) of the UNECE and adopted in 1985 in Helsinki. The protocol states that the signatories should reduce their sulphur dioxide emissions to at least 30 per cent below 1980 levels by 1993. By 1994, the sulphur protocol had been ratified by 20 European countries and Canada. In many Western countries, larger reductions have already been achieved, while the economic recession in Central and Eastern Europe has resulted in temporarily lower emissions. Total European emissions are expected to decrease by some 30 per cent during the period 1990 to 2000, if current policies are implemented.

The new sulphur protocol on further reductions of sulphur emissions was signed in 1994 by 25 European countries, the EC and Canada. Critical loads form an integral part of the new protocol, a significant new departure for such international agreements. The basic obligations require that sulphur emissions be controlled and reduced to protect human health and the environment and to ensure, in the long term, that deposition of oxidised sulphur compounds does not exceed critical loads.

In 1988 in Sofia, a further UNECE protocol was drawn up on the freezing of nitrogen oxide emissions. The Sofia protocol states that emissions of nitrogen oxides should not increase from the 1980–85 level until 1995. It has now been ratified by 22 European countries, the EC, the USA and Canada. Most Western European countries have declared that instead of freezing they will decrease by 30 per cent their nitrogen oxide emissions. However, the general growth in private car use and in road transport suggests that only a few countries will be able to reduce their national emissions by the announced 30 per cent.

All these agreements have the force of international law, and require their parties to monitor pollution levels, compute emissions, devise models for the dispersion of emissions, perform evaluations, compile statistics and report on results. A separate agreement, the Protocol on Long-Term Financing of the Co-operative Programme for Monitoring and Evaluation of the Long-Range Transmission of Air Pollutants in Europe (EMEP), was concluded in 1984 and was ratified by 30 countries. Today, some 100 monitoring stations in 25 countries are involved in the EMEP monitoring system.

Possibilities and limitations for emission reductions

Most SO_2 emission reductions which took place in Western Europe since 1980 were due to increased use of nuclear power, a switch from coal to oil and natural gas and technological improvements such as desulphurisation of petroleum products in the refining process, as well as a lower sulphur content in the crude petroleum intake. This development has also occurred partly because of increasing demand for low-sulphur products, following national and EC regulations. In Western countries especially, further reduction of sulphur dioxide emissions can be achieved not only by flue-gas desulphurisation, but also by use of advanced combustion technologies, energy conservation, increased energy efficiency of installations and increasing the share of non-fossil fuel power production. In Central and Eastern Europe, significant reductions of sulphur dioxide emissions are technically possible by switching to gas and low-sulphur coal, flue-gas desulphurisation, energy saving and implementation of energy-efficient installations.

According to petroleum companies, more than half of the sulphur content of crude oil is now removed in the refining process (and does not enter into the fuels). Accordingly, the sulphur content of residual fuel oil and middle distillates (diesel oil and light fuel oil) has also been reduced. Reduced output of heavy residual fuel oil from refineries also reflects the substitution of this fuel with lighter fuel products. A substitution of residual fuel oil containing 2.4 per cent sulphur with hard coal containing 1.6 per cent sulphur would lead to only a very slight reduction in sulphur dioxide

Source type	Typical SO₂ emissions (g/GJ)	Installed capacity (MWh)	Investment cost (ECU/kWh)	Operating cost (ECU/kWh)
Uncontrolled:				
Hard coal	500–1000			
Brown coal	500–4000			
Heavy oil	800–1500			
Controlled:				
Additive injection	140–1400	–	10–30	0.002–0.005
Wet scrubbing	<140	194 000	30–100	0.005–0.013
Spray dry absorption	<140	16 000	20–90	0.005–0.012
Ammonium scrubbing	40–400	2000	100	0.005–0.012
Wellman–Lord	<140	2000	150–250	0.008–0.018
Activated carbon	<140	700	100–120	0.006–0.007
Combined catalytic	<140	1300	80–120	0.002–0.003
Fluidised bed combustion	<400	30 000		

Table 31.3
Options for sulphur dioxide control in public power generation plants and other large combustion boilers
Source: Compiled by NILU

emissions because of the lower energy content of the coal.

Currently, more than 70 per cent of total sulphur dioxide emissions stems from coal combustion in thermoelectric power plants. Measures to reduce sulphur from this source include the removal of sulphur dioxide from stack gas (based on absorption in lime slurry or some other added absorbent) or recovery of sulphur as either sulphuric acid (the Wellman–Lord process) or gypsum. If not recovered, the absorbent may have to be disposed of in landfills.

For nitrogen oxides, a 60 per cent emission reduction could be achieved through the introduction of low-NO_x burners and modified combustion chambers, by new modes of operation (such as catalytic reduction techniques) in combustion installations for stationary combustion sources, as well as by Europe-wide implementation of improved catalysts in cars and improved diesel engines. Further reduction of emissions requires reaction of nitrogen oxides with ammonia to form elementary nitrogen, usually in the presence of a catalyst (selective catalytic removal, or SCR). Both flue-gas desulphurisation and denitrification are costly, and have only been implemented for a few large power plants and boilers.

The advanced combustion technologies of fluidised bed combustion and pressurised fluidised bed combustion (FBC and PFBC) are applicable for medium-sized boilers only. The fuel is mixed with lime to increase the retention of sulphur dioxide and the low combustion temperature reduces nitrogen oxide emissions. This technology is attractive for district heating and co-generation plants since construction costs are less than for conventional boiler designs. Drawbacks include waste disposal and possible N_2O formation because of the low combustion temperature.

Table 31.4
Typical emissions of nitrogen oxides from stationary combustion sources with and without nitrogen oxide emission control (g/GJ)
Source: Compiled by NILU

	Uncontrolled	Controlled		
		Process and combustion modifications	Non-catalytic selective removal	Catalytic selective removal
Investment (ECU/kW)		20–40	5–11	50–100
Coal, wet bottom	500–800	400–650	no data	<70
Coal, dry bottom	300–500	160–300	70–150	<70
Brown coal	160–270	70–100	<70	<70
Heavy fuel oil	200–530	100–150	50–70	<50
Distillate fuel oil	100–350	50–100	no data	<50
Gas	40–280	30–100	no data	<30
Fluidised bed combustion	70–250	60–140		no data
Pressurised fluidised bed combustion	90			no data
Turbine/boiler	1000–2500	25–200		7

Tables 31.3 and 31.4, which are based on information supplied to the ECE in connection with the Long-Range Transboundary Air Pollution (LRTAP) Convention, sum up typical performance data (emissions) and costs of these control options. The tables indicate that removal of sulphur and nitrogen oxides from the flue gases is technically feasible. However, only 50 000 MW electricity generating capacity in Europe applies the flue-gas desulphurisation and secondary denitrification control options. One of the problems associated with flue-gas desulphurisation is the disposal of waste products: gypsum or sulphuric acid. With the limited capacity now installed, gypsum is utilised in building materials, whereas the market for sulphuric acid is limited.

Motor vehicles' emissions of nitrogen oxides are the main cause of photochemical oxidant and ozone formation on a regional scale in Europe. Planned restrictions and introduction of technical control devices are expected to reduce total European nitrogen oxide emissions in 2000 by about 20 to 30 per cent with respect to 1990 emission levels. However, because of the general growth in private car use and in road transport of goods, there are indications that the actions taken will not be sufficient to reduce nitrogen oxide emissions (see Chapter 32).

Ammonia emissions from agriculture can in general be reduced by careful storage and application of manure and balanced animal feeding. These measures limit volatilisation of ammonia from manure and reduce the nitrogen content in animal fodder. This is difficult to achieve in areas of intensive animal husbandry, and may incur considerable investment costs for manure storage and handling equipment. European ammonia emissions can be expected to stabilise or even increase slightly during the next ten years, and the relative contribution of ammonia emissions to acidic deposition is expected to increase. So far, no internationally coordinated policy exists for this compound.

In summary, acid deposition in Europe is expected to decrease following reductions in sulphur and nitrogen emissions. However, these reductions are still insufficient, and in more than half of the European area the deposition loads are expected to remain in excess of critical loads, resulting in prolonged risk for ecosystems.

The future

Given a regional air pollution problem, the problem of emission reduction raises two important questions: which emissions should be reduced, and where should they be reduced, in order to provide the most cost-effective solution to the problem? The protocols to the LRTAP Convention specify national reductions as a percentage of national emissions in a given base year. While this approach has been practical as a first attempt to share the burden of reducing emissions equally between parties, it may not be the most cost-effective solution. Costs of reducing the emissions of a given pollutant may vary widely between countries, and the environmental effects of the releases may also be quite different. Both of these arguments are being addressed in the renegotiation of the protocols to the LRTAP Convention. A first assessment of the achievements of the LRTAP Convention under the sulphur protocol will be made in 1995.

There has been rapid acceptance of critical loads for use in negotiations because they form a bridge between scientists, policy makers and the public. Integrated assessment models have been developed that use information on emissions, atmospheric transport, costs of abatement, and critical or target loads to study the implication of different abatement strategies. These models can indicate the best abatement strategies necessary to achieve objectives such as reduction in the areas that exceed critical loads or reduction of the total deposition in excess of critical loads.

Studies indicate that very strong emission reductions of all acidifying compounds are needed to avoid exceedance of critical loads in Europe (about 90 per cent for sulphur dioxide and nitrogen oxides and more than 50 per cent for ammonia). Implementing critical loads for environmentally optimised and cost-effective abatement strategies means that some countries, for example in Central and Eastern Europe, may have to reduce emissions by more than 85 per cent. This has raised the question of creating a European environmental fund with the major task of distributing economic support and technology to Central and Eastern European countries where the economic situation is less favourable.

REFERENCES

Brodin, Y-W and Kuylenstierna, J C I (1992) Acidification and critical loads in Nordic countries: a background. *Ambio* **21**, 332–8.

Christensen, N, Paaby, H and Holten-Anderson, J (Eds) (1993) *Miljø og samfund – en status over udviklingen i miljøtilstanden i Danmark*. Faglig rapport fra DMU nr 93. Danmarks Miljøundersøgelser, Roskilde.

Grennfelt, P, Hov, Ø and Derwent, R G (1993) *Second generation abatement strategies for NOx, NH3, SO2 and VOC*. IVL Report B-1098, Gothenburg.

Iversen, T, Halvorsen, N E, Mylona, S and Sandnes, H (1991) *Calculated budgets for airborne acidifying components in Europe*. EMEP/MSC-W Report 1/91, The Norwegian Meteorological Institute, Oslo.

Lövblad, G, Amann, M, Andersen, B, Hovmand, M, Joffre, S and Pedersen, U (1992) Deposition of sulphur and nitrogen in the Nordic countries: present and future. *Ambio* **21**, 339–47.

Tamm, C O and Hallbäcken, L (1988) Changes in soil acidity in two forest areas with different acid deposition: 1920s to 1980s. *Ambio* **17**, 56–61.

Air pollution in Athens
Source: Frank Spooner Pictures

32 Tropospheric ozone and other photo-chemical oxidants

THE PROBLEM

Photochemical oxidants, especially ozone (O_3), are among the most important trace gases in the atmosphere with respect to their physical and chemical functions. Their distributions show signs of change due to increasing emissions of ozone precursors (nitrogen oxides, volatile organic compounds or VOCs, methane and carbon monoxide). World Health Organization (WHO) air quality guidelines for ozone, the one-hour short-term limit value of 75 to 100 ppb for health protection and the long-term value of 30 ppb over the growing season for protection of vegetation, are frequently exceeded in most parts of Europe. The EC has established a directive on air pollution by ozone (92/72/EEC) which sets thresholds for ozone concentrations in air to protect human health and vegetation and to inform or warn the public. There is no compound in the troposphere where the difference between actual atmospheric levels and toxic levels is so marginal as is the case for ozone. Tropospheric ozone is increasing particularly in the northern hemisphere due to the increases in anthropogenic emissions there. Comparisons of pre-industrial ozone data with modern data have shown that surface ozone has more

than doubled over the past century. Ozone concentrations in the middle troposphere over Europe and North America appear to have increased slightly over the last two decades. This increase has been witnessed both at ground level stations in the planetary boundary layer of the lower part of the troposphere (Bojkov, 1986; Volz and Kley, 1988; Anfossi et al, 1991) and at mountain tops in the elevated background troposphere (Marenco et al, 1994; see Figure 4.6). Episodic ozone peak hourly concentrations of 100 ppb and more occur frequently over Central Europe every summer. In the northern UK and Nordic countries such episodes very seldom occur and are mainly a consequence of long-range transport. In Southern Europe, local and subregional oxidant formation may be more important than regional long-range transport.

In the lower troposphere, close to the ground, ozone is a strong oxidant that at elevated concentrations is harmful to human health, materials and plants. In the upper troposphere, ozone is an important greenhouse gas and contributes greatly to the oxidation efficiency of the atmosphere. These effects are to be added to the detrimental effects of other pollutants such as sulphur dioxide, nitrogen oxides, carbon monoxide and particulates on human health in urban areas, and of acidifying compounds damaging ecosystems.

Sources		CO	CH$_4$	VOCs	NO$_x$
Global	Anthropogenic	62	63	10	76
	Natural	38	37	90*	24
Northern hemisphere	Anthropogenic	72	73	16	84
	Natural	28	27	85*	16

Note: * Including isoprene and terpenes.

Table 32.1
Relative natural and anthropogenic contribution (per cent) to tropospheric ozone precursor emissions in 1990
Source: RAS, 1993

Photochemical-oxidant precursor emissions are widely distributed spatially, and originate from various sectors of activity. Thus, control strategies have been more difficult to establish than for acidification, and ozone has proved to be one of the most difficult pollutants to control both in Europe and North America. The trend of increasing activity in the main sectors causing ozone formation (ie, transport and the petrochemical industry) means that this prominent environmental problem may worsen and could be around for a long time.

Long-term records of other photochemical oxidants, such as hydrogen peroxide or peroxyacetylnitrate (PAN), are extremely sparse. PAN measurements in The Netherlands show an increase of almost a factor of three over the past decade. It has been argued that the often observed spring ozone maximum at background stations, instead of a summer maximum concomitant with maximum photochemical activity implied by maximum sunlight in summer, is the result of accumulation of NO$_x$ and VOCs during winter, and the subsequent formation of ozone by photochemical activity when spring arrives (Penkett et al, 1986). A Greenland ice core record has shown evidence that hydrogen peroxide concentrations in the atmosphere have increased by 50 per cent over the past 200 years, with most of the increase occurring in the past 20 years, thus indicating that human activities may be responsible for such a clear change (Sigg and Neftel, 1991).

THE CAUSES

The formation of photochemical oxidants is much more complex chemically than the formation of acidifying compounds. Ozone is not directly emitted but is a secondary pollution component formed in the atmosphere by the action of sunlight on ozone precursors (nitrogen oxides, VOCs, methane and carbon monoxide) depending on many factors, including hydrocarbon reactivity, time-scale, amount of sunshine, ambient temperature, competition for hydroxyl radicals (OH) between hydrocarbons, concentration of nitrogen oxides and altitude. With the exception of VOCs, ozone precursor emissions are dominantly anthropogenic, and more so in the northern hemisphere where there is more industrial activity (see Table 32.1).

Regional ozone concentrations have been thoroughly investigated in northwestern Europe (eg, Derwent et al, 1991), whereas the situation in Southern and Eastern Europe still needs more investigation. It is noteworthy that most of the world's natural atmospheric ozone resides not in the troposphere but in the stratosphere, where it provides a crucial shield against biologically harmful ultraviolet radiation. The depletion, due to man-made CFCs, of this beneficial stratospheric ozone is another major environmental problem (treated in Chapter 28). However, there is no way in which the depletion of 'good' stratospheric ozone and the build-up of 'bad' tropospheric ozone can compensate each other.

The concentration of ozone in the troposphere is determined by the balance of downward transport of ozone-rich air from the stratosphere, photochemical production and loss within the troposphere, and removal by deposition to the ground and vegetation. Increasing anthropogenic emissions of ozone precursors have led to increased photochemical production of ozone in the troposphere. There is now strong evidence from observations that ozone concentrations have increased since the last century. This trend has to be related to a shift in the balance of ozone sources and sinks.

First assessment of the EUROTRAC/TOR ozone network (Beck and Grennfelt, 1993) has shown that the internal production of ozone within Europe contributes substantially to long-term mean concentrations.

Slow moving high-pressure systems with predominantly clear skies and elevated temperatures set the stage for the photochemical formation and accumulation of ozone and other oxidants over wide regions during episodes which last for several days. Such highly elevated ozone levels were first observed in the Los Angeles smog and were scientifically explained in the early 1950s. As a consequence of the rise of the background long-term ozone concentration, episodic values will inevitably reach higher values.

In the remote troposphere and clean continental sites where NO$_x$ levels are low, NO$_x$ is a limiting factor to ozone formation. Here the background levels of carbon monoxide, methane and natural hydrocarbons provide the fuel for ozone formation. Since carbon monoxide and methane have relatively long atmospheric lifetimes, they will determine the long-term average ozone concentration. At higher levels of nitrogen oxides (more than a few ppb), ozone formation is more dependent on the VOC availability.

Emissions of ozone precursors are increasing (see Chapter 14). Fuel combustion is the most important source of anthropogenic nitrogen oxides, while fuel combustion and evaporative emissions from motor vehicles and the use of solvents are the main sources of anthropogenic VOCs in Europe. Motor vehicles account for a considerable fraction of the total emissions of nitrogen oxides and VOCs to the atmosphere in Europe. VOC emissions are more uncertain than emissions of other major air pollutants, and the contribution of individual VOC sources to photochemical oxidant formation varies with their chemical composition (VOC speciation), and the time and place of release. Methane is released mainly from rumen fermentation in cattle, landfill sites, leaks in the distribution of natural gas and rice paddies, while carbon monoxide is emitted mostly from road transport.

VOCs may be classified according to their ability to form ozone. The concept of photochemical ozone creation potentials (POCPs) has been developed (Derwent and Jenkin, 1991) and recently refined (Andersson-Sköld et al, 1992) for taking into account several effects: the contribution to peak ozone values the same day and one day later; the maximum contribution to ozone formation; and the average contribution to ozone formation over 48 and 96 hours, including the amount of ozone deposited during transport. Under typical conditions for a high-pressure situation in southern Sweden, POCPs show that ethylene and acrolein are the most efficient ozone producers, followed by higher alkenes, aromatics, alkanes and ethers, while chlorinated species, alcohols and ketones are very weak producers of ozone. Nevertheless, the relative POCPs differ depending on the method of calculation and the chemical environment (Andersson-Sköld et al, 1992).

Following the Sofia Protocol to the Long-Range Transboundary Air Pollution (LRTAP) Convention (see Chapter 31), emissions of nitrogen oxides in Europe are expected to decrease in the next ten years by 20 to 30 per cent, but not homogeneously in every country. Implementation of the UNECE VOC protocol, which is now being ratified (signed in 1991 by 20 European countries, the EC, the USA and Canada, and ratified by ten by 1994), should lead to a 15 per cent European emission reduction of VOCs by 1999 with respect to values in the late 1980s. Reductions in Western Europe are expected to be larger than those in Central and Eastern Europe.

THE CONSEQUENCES

The increase of ozone levels is of concern because of ozone's adverse effects on human health (Lippmann, 1989), materials and ecosystems. The transient health effects are more closely related to length of exposure than to one-hour peak concentrations. The effects of long-term chronic exposure to ozone remain poorly defined but epidemiologic and animal inhalation studies suggest that current ambient levels are sufficient to cause premature ageing of lungs. The effects of ozone on transient functional changes are sometimes greatly enhanced by the presence of other environmental pollutants. Other photochemical oxidants cause a range of acute effects including eye, nose and throat irritation, chest discomfort, cough and headache. These have been associated with hourly oxidant levels of about 100 ppb.

Effects on the respiratory tract (impaired lung function, increased bronchial reactivity) are deemed the most important health effects of ozone. Pulmonary function decrements in children and young adults while undertaking exercise have been reported at hourly average ozone concentration in the range 80 to 150 ppb. Increased incidence of asthmatic attacks, and respiratory symptoms, have been observed in asthmatics exposed to similar levels of ozone. Maximum ozone concentrations are generally found in the afternoon.

Ozone contributes to damage to materials such as paint, textile, rubber and plastics. However, in general only little damage occurs to materials either because they are resistant enough after preventative measures (addition of anti-oxidant) or because their useful life is, in any event, rather short. However, works of art may be damaged after lengthy exposure to ozone.

In the case of crops and some sensitive natural types of vegetation or plant species, exposure to ozone will lead to leaf damage and loss of production. Generally ozone does not react with the surface of the plant but enters mainly through its open stomata (the pores on the underside of leaves). Ozone can affect all levels of biological organisation (Skärby and Selldén, 1984). At the cell level, by:

- increasing the permeability of plant cell membranes;
- inactivating enzymes;
- stimulating ethylene production as a stress reaction.

These changes, in turn, affect higher levels of plant organisation, leading to:

- symptoms on leaves (bleached, pigmented or necrotic spots);
- decreased growth of leaves and other plant organs;
- decreased yield and quality of individual plants;
- decreased yield and quality of crops and trees;
- predisposition of plants to fatal attacks from insects and/or diseases.

Tropospheric photochemical oxidants also need to be considered in relation to other environmental problems. Photochemical oxidants do not normally interact simply with acidification in ecosystems, since oxidants affect primarily the canopy, while the other damages are caused through soil changes (Grennfelt et al, 1993). There are, however, a number of investigations showing synergistic effects between ozone and acidifying compounds (Guderian and Tingey, 1987). It has been suggested that the combined stress from acid deposition and photochemical oxidants may increase the potential for forest damage. Acid fog deposition has been shown to interact with ozone, causing increased leaching from coniferous canopies.

As a second consequence of further increases in global trace gas emissions, a further decrease is expected to occur of the self-cleansing capacity of the troposphere (ie, the ability of the hydroxyl radical to react with most pollutants). This would result in longer atmospheric residence times of trace gases and, consequently, an enhanced greenhouse effect and an increased influx of ozone-depleting trace gases into the stratosphere.

GOALS

Observed and forecast trends in both ozone precursor emissions and tropospheric ozone concentrations may have serious consequences. For the assessment of environmental effects from ozone, control of both peak values and long-term values are important. To limit the effect of tropospheric photochemical oxidants on human health, vegetation, agricultural crops and forests, regional scale episodes of photochemical oxidant formation need to be controlled. This has been reflected in the UNECE development of critical levels for ozone, where short-term (one half to eight hours) as well as long-term (vegetation season) averages are considered. On the other hand, the stronger impact on human health of cumulative daily exposure, rather than one-hour peak concentrations, should be used as a guide when setting priorities.

The concept of critical level for ozone assumes a threshold below which no or small effects occur. The observed effects are related to the ozone dose expressed in ppb-hours above the threshold level. Recommendations have been made (UNECE, 1993) for critical levels for crops of 5300 ppb-hours above a reference level of 40 ppb (during daylight over three months) and for forests of 10 000 ppb-hours above 40 ppb (24 hours during six months). The reference level of 40 ppb is exceeded all over Europe, although the exceedance is much less in the north of Scandinavia than in Central Europe. The significance of ozone precursor emission reduction will thus vary within Europe (see Table 32.2).

The long-term average ozone concentrations at ground level are closely related to the ozone concentrations in the troposphere. Model calculations show that uniform percentage reductions in European emissions of nitrogen oxides and/or VOCs leads to stronger reduction of the high percentile concentration values (90, 95 and 98 percentiles) than of average ozone levels. In general the influence of European emissions on the 98 percentile oxidant values is two to three times as large as on the averaged concentrations. This again demonstrates that long-term averaged ozone concentrations depend more on the upper tropospheric oxidant (ozone, nitrogen dioxide, PAN, etc) concentrations, whereas high episodic peak concentrations are determined mainly by photochemical production over Europe (De Leeuw and van Rheineck Leyssius, 1989).

The global emissions of carbon monoxide, methane, and nitrogen oxides are expected to increase in the next decade by 0.9, 0.9 and 0.8 per cent per year respectively. As a consequence, and assuming that the chemical system stays linear, ozone concentrations in the northern-hemispheric troposphere are expected to keep rising at a rate of more than 1 per cent per year, leading to increased exposure of people and ecosystems to ozone. No explicit goals have yet been set with regard to tropospheric ozone. In order to halt these increases, it could be recommended that global nitrogen oxide emissions should at least be stabilised, since tropospheric

Table 32.2 Importance of nitrogen oxide and VOC emission reductions for oxidants reduction in different regions of Europe
Source: Grennfelt et al, 1993

	Northern Scandinavia	Southern Scandinavia	Central Europe	Southern Europe
NO$_x$	X	XX	XXX	XX
VOCs	0	X	XXX	XXX

Note: XXX very important; XX important; X some importance; 0 no importance.

Table 32.3
Importance of precursor changes for various oxidising and acidic compounds and visibility
Source: Grennfelt et al, 1993

Precursors	Episodic ozone	Average ozone	Average OH	Average H$_2$O$_2$	SO$_2$/SO$_4^{2=}$	NO$_x$/NO$_3^-$	NH$_3$/NH$_4^+$	Visibility
CH$_4$	+	+++	– – –	+++	weak	medium	0	+
CO	+	+	– – –	+++	weak	medium	0	+
NMVOC*	+++	+	weak	weak	++	medium	0	++
NO$_x$	strong(+/–)	+++	+++	– –	++	strong	medium	++
SO$_2$	0	0	0	–	strong	0	strong	strong
NH$_3$	0	0	0	0	weak	weak	strong	weak

Notes:

* NMVOC = non-methane VOC.

The sign of the relationship is indicated by +/– meaning that, when the precursor emission increases, the concentration of the different compounds or visibility increases/decreases.

concentrations of nitrogen oxides are a limiting factor for tropospheric ozone formation. Nitrogen oxide emissions from aircraft contribute strongly to tropospheric nitrogen oxide concentrations, and therefore to ozone formation; these aircraft emissions are expected to rise in the next decade by more than 3 per cent per year.

Control of background tropospheric ozone levels should be made with reference to their direct contribution to the mean level of ground level ozone and to the enhanced greenhouse effect. This latter effect cannot be appraised by just assuming an equivalent increase of carbon dioxide, as is normally done in various scenario assessments, because of the short residence time of ozone and its non-uniform distribution vertically, horizontally and temporally (Marenco et al, in press). This requires new research.

STRATEGIES

Because of the general growth in road transport (both private cars and other vehicles), there are indications that the actions already undertaken will not be sufficient to reduce photochemical oxidant formation in Europe. This is particularly the case for nitrogen oxides. For VOCs, the situation is more optimistic, but reductions are taking place mainly for those VOCs released by evaporation of solvents and other evaporative emissions, which are not necessarily the most photochemically active.

Agreed emission reductions will probably result in only a 5 to 10 per cent decrease in ozone peak concentrations, which is not sufficient to avoid exceeding current air quality standards. Part of this decrease could even be counteracted by the expected continuing increase in background tropospheric ozone. In order to avoid the present exceedances of the 75 ppb ozone level (see Chapter 4), simultaneous reductions in the European emissions of nitrogen oxides and VOCs of at least 70 per cent would be necessary. For nitrogen oxides, such reductions are technically feasible. For VOCs, such reductions could be achieved by Europe-wide introduction of improved three-way catalytic converters for cars, diesel engines with reduced VOC-emissions, reduced use and substitution of organic solvents and various add-on techniques in storage, transport and distribution of organic products (RIVM, 1992).

The further increase of global emissions of ozone precursors will also lead to global problems (such as decrease of the self-cleansing capacity of the troposphere, enhanced greenhouse effect, and increased influx of ozone-depleting trace gases into the stratosphere). In order to prevent or reverse this trend, global emissions of carbon monoxide, methane and low reactive hydrocarbons would need to be reduced. In Europe, transport emissions of carbon monoxide

and VOCs can be substantially reduced by further introduction of catalytic converters in cars. Industrial carbon monoxide emissions in Central and Eastern Europe can be reduced by implementing more advanced technology.

Since emissions of most compounds have common and linked effects, it is important to consider all effects together for setting emission reduction strategies. Through combined reductions in pollutant emissions, a wider range of adverse environmental impacts can be dealt with, a positive synergism in terms of reduction of adverse impacts can be obtained, and more flexible opportunities are made available (Grennfelt et al, 1993). For instance, a combined emission reduction programme may favour energy efficiency or fuel substitution, and could lead to a reduction in the amount of greenhouse gases emitted into the atmosphere. On the other hand, when it involves emissions of nitrogen oxides, a combined emission reduction can lead to locally increased ozone formation in areas of high emissions, or simply to transferring the location of peak ozone concentrations. Also, a reduction in global emissions of nitrogen oxides can lead to a global reduction in the hydroxyl radical and thus an increased lifetime of methane and CFC substitutes, with a consequent increase in the global warming potential of these species (see Chapter 27).

Table 32.3 shows the strength of the relationship between a change in the precursor emissions of ozone and acidifying compounds, and the concentration of different compounds or visibility effect. Concentrations of ozone, hydroxyl and hydrogen peroxide, intermediaries in the photchemical process, are particularly important for enhancing the rate of transformation of sulphur dioxide into sulphate, nitrogen oxide into nitrate and ammonia into ammonium. It appears from Table 32.3 that nitrogen oxide emissions stand out as a controlling factor for almost all parameters listed. Methane emissions are important for long-term changes in ozone, hydroxyl and hydrogen peroxide. Carbon monoxide has a controlling effect on the loss of hydroxyl. VOCs have a controlling effect on the peak episodic ozone concentrations, but otherwise VOCs (with the exception of methane) are thought to play only a minor role in atmospheric oxidation. All these relationships can be quantified through model calculations.

The reductions of nitrogen oxides and VOC emissions required to reduce ground-level and tropospheric ozone levels would also have additional environmental benefits:

- reduction in ambient concentrations of nitrogen dioxide and toxic hydrocarbons (such as benzene, PAH and organic particulates);
- reduction in acid deposition;
- reduction of the eutrophication of surface waters and soils;
- improvement of air visibility.

REFERENCES

Andersson-Sköld, Y, Grennfelt, P and Pleijel, K (1992) Photochemical ozone creation potentials: a study of different concepts. *Journal of the Air and Waste Management Association* **42**(9), 1152–8.

Anfossi, D, Sandroni, S and Viarenzo, S (1991) Tropospheric ozone in the nineteenth century: the Montcalieri series. *Journal of Geophysical Research* **96**, 17349–52.

Bojkov, R (1986) Surface ozone during the second half of the nineteenth century. *Journal of Climatology and Applied Meteorology* **25**, 343–52.

Beck, J P and Grennfelt, P (1993) Distribution of ozone over Europe. Proceedings of EUROTRAC Symposium 1992 (Borrel, P M et al, Eds) 43–58. SPB Academic Publishing, The Hague.

De Leeuw, F A A M and van Rheineck Leyssius, H J (1989) Calculation of long-term averaged and episodic oxidant concentrations for the Netherlands. Conference on the *Generation of Oxidants on Regional and Global Scale*. Norwich, 3–7 July 1989.

Derwent, R G, Grennfelt, P and Hov, Ø (1991) Photochemical oxidants in the atmosphere. *Nord 1991: 7*, Nordic Council of Ministers, Copenhagen.

Derwent, R G and Jenkin, M E (1991) Hydrocarbons and the long range transport of ozone and PAN across Europe. *Atmospheric Environment* **25A**, 1661.

Grennfelt, P, Hov, Ø and Derwent, R G (1993) Second generation abatement strategies for NO_x, NH_3, SO_2 and VOC. *IVL Report* **B-1098**, Gothenburg.

Guderian, R and Tingey, D T (1987) Notwendigkeit und Ableitung von Grenzwerten für Stickstoffoxide. UBA Berichte 1/87. Umweltbundesamt, Berlin.

Lippmann, M (1989) Health effects of ozone: a critical review. *Journal of the Air Pollution Control Association* **39**(5), 672–95.

Marenco, A, Gouget, H, Nedelec, Ph, Pagès, J-P and Karcher, F (1994) Evidence of a long-term increase in tropospheric ozone from Pic du Midi data series: Consequences for radiative forcing. *Journal of Geophysical Research* **99**(D8), 16617–32.

Penkett, S A and Brice, K A (1986) The spring maximun in photo-oxidants in the northern hemisphere troposphere. *Nature* **319** 655-7

RAS (Rapport de l'Académie des Sciences) (1993) *Ozone et pollution oxydante dans la troposphère (evaluation scientifique)*, Académie des Sciences, Paris (Ed).

RIVM (1992) The environment in Europe: a global perspective report 481505001. RIVM, Bilthoven.

Sigg, A and Neftel, A (1991) Evidence for a 50 per cent increase in H_2O_2 over the past years from a Greenland ice core. *Nature* **351**, 557–9.

Skärby, L and Selldén, G (1984) The effects of ozone on crops and forests. *Ambio* **13** (2), 68–72.

UNECE (1993) Workshop on Critical Levels for Ozone. United Nations Economic Commission for Europe, Berne.

Volz, A and Kley, D (1988) Evaluation of the Montsouris series of ozone measurements made at the nineteenth century. *Nature* **332**, 240–2.

Rapp-Bode dam,
Saxony-Anhalt,
Germany
Source: H Lange, Bruce
Coleman Ltd

33 The management of freshwater

THE PROBLEMS: CAUSES AND EFFECTS

A wide range of human activities can adversely affect the condition of the aquatic environment. Disturbances can result from a human activity in a seemingly unrelated area often far away from the impact site, and delayed in time. The great majority of issues involving European water availability and water quality are therefore most prominent in areas with high population densities, concentrated industrial activity and/or intensive agriculture. In addition, physical changes imposed on watercourses for construction of reservoirs, channelisation of rivers and improvement of land drainage have destroyed or are threatening many wetland habitats.

The close relationship between human activities and the condition and management of freshwater resources implies that conditions, except for widespread acidification, are better in the sparsely populated and humid Nordic countries than in the rest of Europe. Table 33.1 indicates that no marked differences separate the freshwater conditions in Eastern countries from those prevailing in Southern and Western Europe. However, it should be emphasised that there are areas of concern within all of the four European regions: Nordic, Eastern, Southern and Western countries. Habitat destruction through physical changes made to watercourses, groundwater overexploitation and contamination, eutrophication, toxic pollution and acidification are but a few water-related issues that can affect water use locally or regionally.

Water availability

The available freshwater in an area (or country) is the total amount moving in rivers and aquifers. It originates either from precipitation over the area itself, or from water that enters the area in rivers and regional aquifers.

Europe as a whole faces no water shortage problem but there exists a regional imbalance between supply and demand. The annual runoff in Europe is shown in Map 5.2, the total

Table 33.1
Summary of
freshwater-related
environmental
problems in different
parts of Europe
Source: Modified from
WHO/UNEP, 1991

Issue	Nordic	Western	Eastern	Southern
Pathogenic agents	0–1	0–1	1–2	1–2
Organic matter	0–1	1–2	1–3	2
Salinisation	0	0–1	2	1
Nitrate	0–1	1–3	1–2	1–2
Phosphorus	0–1	1–2	1–3	1–3
Heavy metals	1	0–2	1–3	0–2
Pesticides	0–1	2	2	1–3
Acidification	0–3	0–1	1	0–1
Radioactivity	0–1	1	1–3	1
Water availability	0	1–2	2	2–3
Reservoirs	1	2	2–3	2–3
River channelisation	0	2	1–2	1
Groundwater overexploitation	0–1	2	2	1–3

0 no problems or irrelevant; 1 some problems; 2 major problems; 3 severe problems.
Notes: See Figure 5.3 for definition of country groupings.

quantity for the continent being 3100 km³/year. The abstraction of water on a continental basis was 480 km³/year in the late 1980s, that is, about 15 per cent of available resources. In addition to the regional imbalance, the natural runoff variation from season to season, or from one year to the next, can set constraints on water usage.

The reliable runoff which can serve as a continuous supply source, even under drought conditions, is considerably less than average runoff, and depends on the local hydrological regime. As an illustration, the average annual 10-day minimum runoff is 1 to 3 l/s per km² for a portion of Western Europe. This is the amount of water which on average would be available during the 10 driest days of the year without artificial storage or transfers. For Europe as a whole, this corresponds to about 300 to 900 km³/year, compared with the present annual abstraction of 480 km³/year.

On top of the natural variation between seasons and years, long-term trends are possible. Multi-annual dry or wet periods can occur simultaneously over large regions in Europe. Climate change also may worsen the resource situation in many regions (see Chapter 27).

Around many large urban and industrial centres, of all Western European countries except Ireland, water abstraction has surpassed the natural recharge of groundwater and surface water resources. Overexploitation problems are, in particular, severe in coastal areas of the Southern European countries (eg, Spain, Italy, Greece), but areas of the Czech Republic, Hungary, Poland, the Russian Federation, Slovak Republic and Ukraine also face similar problems. In the case of groundwater abstraction, overexploitation leads gradually to lowered groundwater levels, which can in turn damage vegetation which is dependent upon shallow groundwater. It may further induce salt water intrusion in coastal aquifers, preventing their use for drinking water purposes. In urban areas, lowered groundwater levels can cause ground subsidence and damage to buildings.

The overexploitation of water resources, resulting from the regional imbalance between supply and demand, is clearly not sustainable. As long as it continues, overexploitation will lead to tension and hostility between water users, as well as expensive, and often futile, attempts to increase the resource base.

Water quality

Quality degradation further limits the availability of water for consumptive uses. Even water use for purposes other than human consumption is often prevented or limited because of poor quality. Thus, water quality needs to be considered as a broad term describing the suitability of water to sustain various human and ecological uses of the water resource. For example, high concentrations of nitrate or toxic substances restrict the use of water for drinking, and high concentrations of organic matter – and consequently low oxygen concentrations – in rivers set constraints on the survival of valuable fish, such as salmon and trout species. Nitrate occurring in high concentrations in groundwater and surface water can be a danger to human health, and it is considered to be the most important limiting nutrient for coastal and marine eutrophication.

Groundwater is polluted in numerous ways, including:

- fertiliser and manure surplus, leading to nitrate leaching;
- inappropriate use of pesticides;
- mining activities and deposition of hazardous wastes;
- atmospheric deposition of pollutants;
- salt water intrusion in coastal areas, caused by overexploitation of aquifers.

In many areas with intensive agriculture, in all parts of Europe, the quality of groundwater and surface water is threatened by nitrate leaching from the topsoil. Nitrate emissions from the domestic and industrial sectors are far less important than the contribution from agriculture.

Freshwater eutrophication is also a pan-European problem of major concern. The effects of eutrophication (eg, excessive growth of phytoplankton and filamentous algae, turbid water, oxygen deficiency and undesirable shifts in the biological structure of lakes) are caused by nutrient (especially phosphorus) enrichment of the waterbodies. Phosphorus emissions arise predominantly from point sources (domestic and industrial effluents), but the share of agriculture is not insignificant (see Chapter 14). If measures are not taken to reduce diffuse emissions from agriculture, this share is predicted to increase as point source contributions decrease due to improved sewage treatment and substitution of phosphorus-containing detergents.

Organic pollution of freshwaters with increased concentrations of ammonium and pathogenic microorganisms, oxygen deficiency and impoverished living conditions for biological communities, is a third major pan-European problem linked to the management of freshwater. These problems are, in particular, found in the more densely populated parts of the continent with no or insufficient treatment of sewage effluents. This is the case in the majority of Eastern and Southern European countries and in a few Western European countries as well. In other parts of Europe (Nordic, and the great majority of Western countries) use of more advanced sewage treatment technologies has led to significantly improved water quality in terms of lower organic matter and ammonium concentrations and better oxygen conditions.

Acidification of water affects its use for drinking and has profound impacts on the biological communities of rivers and lakes. The Nordic countries especially face severe acidification problems (see Chapters 5 and 31).

Discharge of saline waters from coal mines to surface waters is a severe environmental problem which particularly affects Poland, the Czech Republic and Ukraine. Such waters are highly corrosive if used for domestic or industrial purposes.

There is evidence that the scale of freshwater pollution caused by heavy metals, organic micropollutants and radionuclides is local and/or regional rather than pan-European, and the problems are predominantly related to untreated point-source emissions from a great variety of industries (see Chapter 20). In general, there have been too little monitoring data available for an overall assessment of the magnitude and extent of freshwater pollution by these substances. However, several severe cases of water pollution by heavy metals and radionuclides have been reported for the eastern countries of the Danube basin, the Russian Federation and Ukraine. For total pesticides, model calculations indicate that the quality of groundwater is threatened under most of the arable land in Europe (see Map 5.9).

Physical changes

Rivers and their floodplains interact in a number of ways. Undisturbed floodplains perform the following functions:

- as areas used by fauna for reproduction and as migration and travel corridors;
- provide shade, thus reducing water temperature and solar radiation reaching the surface of the river;
- reduce sediment and nutrient inputs to the river, thus improving water quality;
- increase water retention capacity during flood periods.

This natural balance between rivers and their floodplains has been disturbed by human intervention in almost all European countries.

In-river changes also take place as a result of regulation works, and these generally change the heterogeneous meandering river into a homogeneous straight channel with uniform flow and less habitat diversity than in the undisturbed situation. Upstream channelisation may also increase flood risks downstream. The most important in-river changes are:

- increased slope of the river bed;
- unstable sediment and increased transport of sediments;
- destruction of the pool-riffle sequence;
- reduced habitat and species diversity;
- reduced self-purification.

Major river regulation works have been carried out in almost all European countries for multiple purposes. The most common include those aimed at improving navigation and flood control, production of hydroelectricity, and improving drainage of adjacent farmland. Whatever the purpose of the regulation work, the natural interactions between the river and its floodplain are reduced or even destroyed. In addition the migration of fish, such as salmon, trout and sturgeon, to their spawning grounds in upstream river reaches is prevented because of barriers at locks, dams and weirs.

TRENDS AND SCENARIOS

Supply

On the water supply side, the most serious development seems to be the possible consequences of climate change. The societal impacts of climate change will be particularly striking through its consequences on the water cycle, whether as droughts and soil moisture deficit, or as sea-level rise, floods and increased erosion.

Southern Europe (35°–50°N, 10°W–45°E) was selected by the Intergovernmental Panel on Climate Change (IPCC) as one of five areas for estimating changes up to the year 2030 resulting from pre-industrial conditions. The estimated warming from a 'business-as-usual' scenario is about 2°C in winter, and varies from 2 to 3°C in summer. There is some indication of increased precipitation in winter. However, estimated summer precipitation decreases by 5 to 15 per cent, and summer soil moisture by 15 to 25 per cent, although the IPCC states that confidence in the regional estimates is low.

These scenarios show that the water availability could be particularly threatened in Southern Europe, where the

intensity of water use is already very high, and water scarcity has become a recurring problem.

Groundwater depletion occurs when annual abstractions exceed annual recharge. This is now taking place in many regions of Europe. As 65 per cent of European citizens depend on groundwater for drinking, any loss of these resources is very serious.

Demand

Whereas water supply is limited, this does not seem to apply to water demand. In Europe the demand for water has grown quickly and, although available statistics are difficult to interpret, some forecasts show continued growth (Figure 33.1).

The total per capita water abstraction varies between European countries from about 100 m³/year (Malta) to about 2000 m³/year (Estonia). The average water demand within EU countries increased by 35 per cent from 1970 to the end of the 1980s, and it is still increasing.

In order to understand the consumption trends, abstraction must be considered by sectors.

Domestic

Domestic water use is increasing in Central and Eastern European countries. In Western Europe it is rather stable or slightly growing (ie, few persons per household, many water appliances). Whereas household needs for drinking, washing and cooking could be adequately met by less than 100 l/capita per day or 35 m³/capita per year, present household consumption in Europe is roughly 150 to 300 l/capita per day. Furthermore, leakage from public distribution networks is a large problem in most countries. For example, 25 to 30 per cent losses are estimated for France, Spain and the UK, but may be 50 per cent in some cases.

Agricultural

The agricultural sector is a highly significant water user in most European countries. About a quarter of the water abstracted in Europe is used for agriculture as a whole, and irrigated agricultural land covers 18 million ha (1989) in Europe west of the former USSR. The irrigated area has increased by 70 per cent over 20 years in this part of Europe. It is particularly important (in terms of per cent of arable land) in the Balkan countries (excluding former Yugoslavia), Italy, the Iberian peninsula and The Netherlands. In the former USSR, more than 20 million ha were irrigated in the late 1980s. Agriculture counts for more than 50 per cent of all water abstraction in Albania, Bulgaria, Greece, Italy, Romania and Spain.

Some of the problems which can occur when irrigation is allowed to deplete river flows beyond sustainability are illustrated by the disastrous depletion of water in the Aral Sea, which has lost two thirds of its volume since 1965 (see Chapter 6).

Agricultural water use is generally increasing all over Europe, but is slower than the increase in irrigated areas. This probably reflects some improvements in irrigation technology. However, the efficiency of water use in irrigation can still be further improved (eg, by drip irrigation and tailwater re-use). On a global basis, irrigation comprises some three quarters of all water utilisation, of which 60 per cent is never used by the plants it is intended for. In the EU, it is estimated that roughly 50 per cent of irrigation water (35 km³/year out of 74.5) is lost through seepage or evaporation before it reaches the point of use.

It is difficult to foresee how the reform of the Common Agricultural Policy (CAP) will affect water consumption which, among other things, may lead to more bio-energy forests and increased agro-tourism in the EU (see Chapter

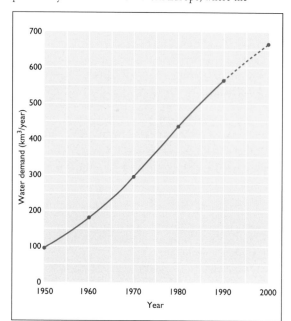

Figure 33.1
European water demand, 1950–2000
Source: WHO/UNEP, 1989

forests and increased agro-tourism in the EU (see Chapter 22). However, there is encouragement to add value to agricultural products, which can lead to growth in the food-processing industry, and an increasing demand for high quality water.

Industrial

Industry's water needs vary widely between countries. In Finland, Germany and Belgium, industry accounts for about 80 to 85 per cent of all water abstraction, whereas more agrarian countries (Greece, Portugal, Spain) abstract less than 30 per cent for industry. Industrial water use has for some time been stable or reduced in Western Europe, probably due to improved technologies. However, in some Central and Eastern European countries (eg, Bulgaria and the former USSR) industrial use of water is still increasing (UNECE, 1992).

The electricity-generating industry will probably increase its demand for cooling water, in contrast to other industrial sectors. For both process and cooling purposes, there seem to be significant opportunities for reducing water consumption through increased recycling.

Freshwater pollution

Water pollution levels and trends in Europe are, in general, not encouraging. However, in a number of countries (eg, Nordic and some Western European) legislation and action programmes, aimed at reducing point source emissions of contaminants, have in recent years halted pollution and improved surface water quality. However, all across Europe, the emissions from diffuse sources continue to threaten the quality of inland waters, examples being intensive farming and atmospheric deposition. This is due to the fact that many countries have invested large sums in improved treatment of industrial and domestic wastewater without, at the same time, paying sufficient attention to a simultaneous reduction of emissions of contaminants from other sectors such as agriculture. It should also be noted that, although the water quality of many larger European rivers may gradually improve, there are a multitude of small brooks and streams which continue to suffer because of neglect.

Organic waste from sewage and industry, causing loss of oxygen and severely degraded water quality, is now largely under control in Nordic and Western European countries. However, lack of sufficient sewage treatment in many Central, Eastern and Southern European countries as well as a few Western countries (see Chapter 14) still causes serious problems. Moreover, illegal or accidental spills of slurry and silage juice are occurring across Europe, spoiling, in particular, small streams in agricultural areas.

Nutrients from wastewater and agriculture cause widespread eutrophication of European rivers and lakes. Point source control measures are slowly reducing phosphate concentrations, although seldom to levels that reverse eutrophication, which is still often caused by release of sediment-bound phosphorus and phosphorus inputs from diffuse sources. Nitrate levels are still high, particularly in small rivers and groundwater in areas of Western and Central Europe with intensive agriculture and high application rates of fertilisers and manure. In many instances, nitrate concentration limits considered safe for human consumption are exceeded in European groundwater. Larger rivers and lakes normally have nitrate concentrations well below drinking water standards.

Heavy metal pollution of waters and sediments poses human health risks, in particular through fish consumption. Decreasing trends for mercury are now evident in some countries (eg, Norway, Sweden, Finland). Until the 1970s the Rhine was heavily polluted by mercury and cadmium, but levels have been significantly reduced since then because of improved wastewater treatment and replacement of these substances in industrial processes. However, water pollution from synthetic chemicals has become a new cause for concern. Organic micropollutants, including some much-used pesticides, are now widespread in the environment, with concentrations exceeding standards in many areas of intensive agriculture. In addition, their fate and ecological impact are often unknown.

Since the 1970s, acidification of European freshwaters, damaging aquatic life and changing the natural water quality, is known to have taken place in large areas where susceptible geology combines with deposition of acidifying compounds, mainly from fossil-fuel combustion (see Chapter 31). Another serious potential threat from airborne pollution is radioactivity from nuclear accidents, as witnessed after the Chernobyl accident (see Chapter 18).

Groundwater pollution is likely to become increasingly acute and widespread in coming years, particularly because of uncontrolled waste deposits, leaking petrochemical tanks, and ongoing percolation of pollutants into aquifers.

Groundwater pollution problems are exacerbated because natural regeneration of clean groundwater is extremely slow. The velocity of groundwater flow is in the order of a few metres per month or year in many aquifers. Purification techniques are developing, but are still very costly. Some soil contamination (eg, with chlorinated hydrocarbons or heavy metals) is for practical purposes irreversible today.

TRANSBOUNDARY RIVER MANAGEMENT

Several large European rivers have catchment areas shared between many countries (eg, Danube, Rhine, Elbe, Neman, Zapadnaya Dvina), and in some cases collaboration between riparian countries has been initiated to protect water resources. As examples of the importance for national water resources, The Netherlands receives 85 per cent of its water from transboundary rivers (eg, Rhine, Maas). Hungary receives an annual equivalent of nearly 1200 mm over the country from the Danube, whereas only 65 to 70 mm of runoff is actually generated within Hungary (see Figure 5.4). There are some 110 bilateral and multilateral agreements to regulate water quality, use and economy in Europe, thus helping to reduce international dispute over water supply sharing (UNECE, 1993). Other international agreements, such as the EC directives on water quality, also contribute to safeguarding European water resources for the future.

In recognition of the basin-wide, upstream–downstream and riparian interdependencies in the development, use and protection of the waters (particularly transboundary rivers), their basins and ecosystems, increasing effort in recent years has been devoted to integrated river basin management to ensure their sustainable development.

Under the terms of the Bucharest Declaration signed in 1985, the countries of the Danube basin began a programme of international joint sampling and analysis of water quality at border crossing points on the river Danube. This initiative was followed by the formulation of the Environmental Programme for the Danube River Basin, with the overall aim of strengthening collaboration between countries on the examination and development of solutions to environmental problems in the region. It was established for a three-year period at a 1991 meeting in Sofia of riparian countries, international and non-governmental organisations, and certain G-24 countries as financing partners, with the aim of developing a strategic action plan for the basin. This was signed by ministerial declaration in 1994.

The necessity for transboundary management of the river Rhine was recognised for shipping as early as the end of the eighteenth century. The transboundary pollution aspect was given attention in 1932, when the Dutch government protested against emissions of residual salt into the French part of the river. After World War 2, the pollution of the river increased and, in 1950, the International Rhine Commission started to study wastewater and water quality problems. The international cooperation against pollution was strengthened by the agreement, signed in Berne in 1963, regarding the International Commission for the Protection of the Rhine against Pollution. This was followed by Council Decision (77/586/EEC), to conclude the Convention for the Protection of the Rhine against chemical pollution to the Berne agreement of 1963. Especially to accelerate ecological improvements in the Rhine, ministers from riparian countries decided in 1986 to establish the Rhine Action Programme, with the following aim:

The ecosystem of the Rhine must become a suitable habitat to allow the return to this great European river of the higher species which were once present here and have since disappeared (such as salmon). (ICPR,1992)

In order to improve water quality ecological conditions in the Elbe (one of the most impacted larger rivers in Europe) and its tributaries, an agreement (Vereinbarung über die Internationale Kommission zum Schutz der Elbe – IKSE) was signed in 1990 by riparian countries and the EC.

Extensive international collaboration is now also taking place on the North and Baltic seas and the Mediterranean to reduce riverine emissions of nutrients and dangerous substances (see Chapter 6). Nevertheless, limited water resources, industrial pollution, mining activities, and/or dam construction plans regularly create some international tension. Two examples are disputes between Portugal and Spain on the sharing of the water of the Tajo river, and between Hungary and the Slovak Republic on the construction of the Gabcikovo dam on the Danube.

SUSTAINABLE GOALS

It is vitally necessary to balance Europe's water use with the long-term available resources. Resources are renewable, but limited, and scarcity is expected to increase in Southern and Eastern Europe. At the same time, pollution trends are constraining potential water use, whereas total water demand is increasing. Such trends are clearly not sustainable.

The following goals might be considered in management plans designed to improve the condition of freshwater resources:

- To provide Europe's population with sufficient water which is safe for drinking and general hygiene.
- Make available sufficient supplies of adequate quality water for other consumptive uses to meet realistic economic goals.
- Not to allow water abstraction to exceed natural recharge for long durations.
- To maintain adequate water and soil resources for biological protection, without resource depletion.
- To reduce nutrient emissions to limit eutrophication.
- To maintain or re-establish the self purification capacity of rivers.
- To improve the physical condition of European rivers to provide habitats for diverse flora and fauna.

STRATEGIES AND OPTIONS

Optional strategies and actions for reaching the above goals, and for adjusting Europe's water policies to maintain stability, include:

- Collecting of accurate information on the quantity and quality of water resources, as well as consumption patterns. Today, status assessments and prognoses are seriously hampered by lack of recent, accurate and comparable data. The need for an international and harmonised monitoring network for water resources has been recognised by several international organisations, eg, the GEMS/WATER global freshwater monitoring network established jointly in 1979 by UNEP, WHO, WMO and UNESCO (WHO/UNEP, 1989).
- On the supply side, giving priority to improving water quality, by:
  - preventing pollution at source;
  - restoring natural water resources to ecologically sound conditions; and
  - restoring the balance between the amounts of fertilisers added to soils, and the withdrawal of such substances by crops.

  Actions should be guided by cost-effectiveness.

- Continuing and intensifying efforts to introduce water-saving techniques to all sectors. There is a particular need for improving irrigation technology.
- On the demand side, regarding water as an economic good. Equitable allocation of water resources as a production factor should be based on market principles, however safeguarding drinking water may require special costing.
- Identifying the value of water resources (like other natural resources) in national economic accounting. Continued depreciation of resource stock value should be considered early in national policy making.
- Basing water resources management at all decision levels on an ecosystems approach. Sustainable water management requires a holistic view of the interrelations between:
  - groundwater, surface water, and the other elements of the water cycle;
  - landuse, economic and social activity, and water resource status; and
  - water quantity and water quality.

Water management should be linked to catchments, and be carried out with the participation of a well-informed public.

- Strengthening the provision of education and information.
- Further development of structures and procedures for solving conflicts between competing water users, at national as well as European level.

Consideration and implementation of such stategies and options in local, regional, national and international water policies would certainly help to fulfil the Dublin Statement on Water and Sustainable Development (ICWE, 1992), which calls for:

fundamental new approaches to the assessment, development and management of freshwater resources, which can only be brought about through political commitment and involvement from the highest levels of government to the smallest communities. Commitment will need to be backed by substantial and immediate investments, public awareness campaigns, legislative and institutional changes, technology development, and capacity building programmes. Underlying all these must be a greater recognition of the interdependence of all people, and of their place in the natural world.

REFERENCES

ICPR (1992) *Ecological master plan for the Rhine*. International Commission for the Protection of the Rhine against Pollution, Coblenz.

ICWE (1992) *The Dublin statement and report of the conference, 26–31 January 1992*. International Conference on Water and the Environment, Dublin.

UNECE (1992) *The Environment in Europe and North America: annotated statistics 1992*. United Nations, New York.

UNECE (1993) *Bilateral and multilateral agreements and other arrangements in Europe and North America on the protection and use of transboundary waters*. United Nations Economic Commission for Europe, ECE/ENVWA/32, Geneva.

WHO/UNEP (1989) *Global freshwater quality: a first assessment*, Meybeck, M, Chapman, D and Helmer, R (Eds). Blackwell, Oxford.

WHO/UNEP (1991) Earthwatch. GEMS. *Water quality: progress in the implementation of the Mar del Plata action plan and a strategy for the 1990s*. World Health Organisation/United Nations Environment Programme, London.

Dying forest of the Ore Mountains, northern Bohemia, Czech Republic
Source: Michael St Maur Sheil

34 Forest degradation

INTRODUCTION

Forest damage has long been recorded and is caused by natural phenomena (disease and pests, storms, fire, drought and other climatic stresses) or human-induced ones (eg, air pollution, fire, economic overexploitation, overgrazing), or by the interaction of human impacts and natural causes. This chapter focuses on the two most important causes of forest degradation across Europe: air pollution, which is a serious threat to the sustainability of forest resources in Central, Eastern and, to a lesser extent, Northern Europe, and fire, which is a major concern in Southern Europe. The degradation of forest quality in the sense of its authenticity is discussed in Chapter 9, where a non-exhaustive list of remaining semi-natural forests of ecological value in Europe is given. The overall changes in species composition, age structure and management are developed in Chapter 23.

DAMAGE ATTRIBUTED TO ATMOSPHERIC POLLUTION

Historical background

In 1981, public attention was drawn strongly to the problem of forest degradation when the press began to report that European forests were being damaged on an unprecedented scale. These reports in the general press as well as in the scientific literature triggered a lively and often controversial debate over the causes and extent of forest damage (Blank et al, 1988). There was particular concern over the fact that the first symptoms of damage were being found on fir (*Abies alba*) and, later on, on Norway spruce (*Picea abies*) – a widely distributed and commercially important species. The debate, however, was not always based on reliable information, and opinions were often characterised by preconceived ideas rather than facts. Confusion arose because, for the first time, it was speculated that air pollution was damaging trees in areas, such as Scandinavia, which were far from major pollution sources.

Forest damage is not a new phenomenon. Local problems have been reported since records began, but these were specific, resulting from definite causes, such as extreme weather conditions, insect attacks or local pollution. Unfortunately, in the early 1980s there was little knowledge about the effects of pollution on trees and the physical chemistry of forest soils in general. Furthermore, the symptoms observed – the lightening of tree crowns and the loss and yellowing of needles – were not characteristic of specific pollutants. It was widely accepted that acid rain, used as a synonym of air pollution in general, was the main culprit. This view dominated research on the problem and influenced political decision making for a long time; air pollution abatement legislation was introduced rapidly.

The problem

Widespread forest damage has been reported in most Central and Eastern European countries (as well as in the northeastern USA and Canada). This has stimulated a great deal of research into its possible causes, the greatest European research contribution coming from West Germany and the Nordic countries. Many of these research projects are now almost complete. Since 1985, results have been published in reports of the German Government's Advisory Council on Forest Decline/Air pollution, in *EC Air Pollution Research reports* (see, eg, CEC, 1990) and elsewhere. These results have greatly improved the understanding of the problem. The definition of forest damage has been refined, and more data have become available on the geographical distribution and severity of decline.

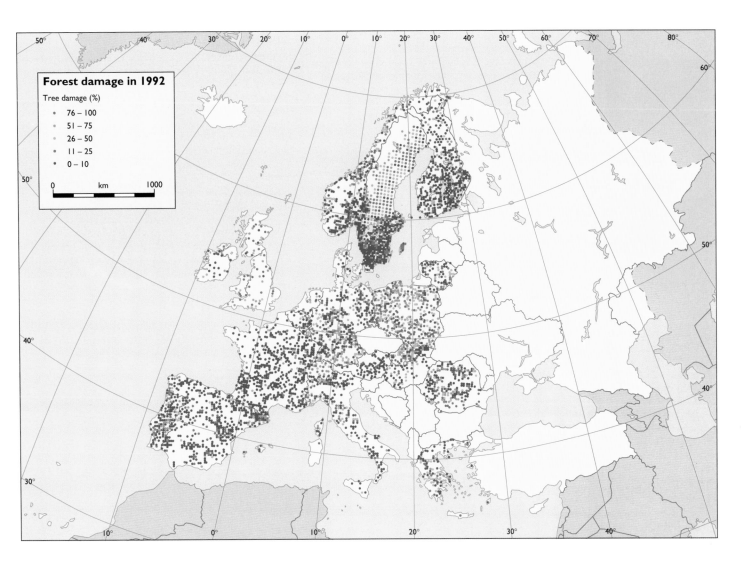

Forest damage in 1992

Tree damage (%)

- 76 – 100
- 51 – 75
- 26 – 50
- 11 – 25
- 0 – 10

0 km 1000

Launching of national and transnational surveys

In 1984, the West German authorities initiated annual standardised, nationwide forest damage surveys to provide a representative general account of the state of the forests. This initiative was soon adopted by other European countries. However, there were problems when comparing the results of one survey with another. In response, the International Cooperative Programme on Assessment and Monitoring of Air Pollution Effects on Forests (ICP-Forests) was established in 1985 by the Executive Body for the Convention on Long-Range Transboundary Air Pollution of UNECE. Its work was based on national monitoring inputs using conventional forest inventory methods, and was partly funded by UNEP between 1985 and 1990. Results from the national assessments are published in a series of *Forest Damage Reports* (see, eg, UNECE, 1991a).

A transnational EU survey was initiated under Council Regulation 86/3528/EEC. In 1987, the first survey was carried out, and in the successive years the practice was adopted by all Member States. The results of these standardised campaigns were published each year in the European Commission's forest health report (see, eg, CEC, 1989). To facilitate the comparison of forest damage surveys and national reports, a common methodology was used in accordance with the guidelines adopted by the parties to the UNECE Convention. Since 1991, the data from the UNECE and EU surveys have been published in a common annual report, *Forest Condition in Europe* (see, eg, CEC/UNECE,

1993). The report publishes transnational survey results from 23 European countries, based on the visual assessment of more than 90 000 trees in some 4400 standardised sampling plots defined by a uniform grid system (16 x 16 km), together with a summary of the 34 national survey results. Defoliation is based on an estimate of how much foliage has been lost compared with a reference tree, which seems to be healthy, in the same area. Discoloration is used as a secondary indicator, which is recorded in five classes of damaged trees. The classification system introduced by the EC and UNECE defines trees as damaged once the estimated defoliation exceeds 25 per cent. The results of this type of survey

Map 34.1
Forest damage in Europe, 1992
Source: CEC/UNECE, 1993

Healthy tree (defoliation <25%)

Damaged tree (defoliation >25%)

Source: Allgemeine Forst-Zeitschrift BLV Verlagsgesellschaft

(UNECE, 1991b), however, did not provide the explanation for the causal relationship between the observed damage and atmospheric pollution. Consequently, long-term observations and ecosystemic analyses on permanent plots will be carried out from 1994 onwards. (Soil and foliar analyses, deposition measurements and increment studies have been included in a sub-survey as amendment of Council Regulation 86/3528/EEC.)

Data analysis

About 184 million hectares of forests and wooded land were investigated in the 1992 survey, covering 34 European countries (EU, EFTA, the Czech Republic, Hungary, Lithuania, Poland, Romania and the Slovak Republic). In the transnational survey of 1992, which constitutes the largest sample since 1987, 24 per cent of European trees were considered to be damaged (that is, with defoliation greater than 25 per cent). This represents an increase of 2 per cent compared with 1991. Ten per cent of trees showed discoloration. Trees most affected are characteristically older age classes. As in the previous years' surveys, the areas of highest defoliation are located in Central Europe, but defoliation is also high in certain areas of Northern and southeastern Europe (Map 34.1). The majority of highly and critically affected forests occur in Bulgaria, the Czech Republic, Germany, Poland and the Slovak Republic, where severe conifer defoliation has been recorded. Of broadleaved trees, beech (*Fagus sylvatica*) in Denmark and birch (*Betula pubescens*) in Sweden are also suffering from defoliation.

Map 34.2
Climatic regions of Europe
Source: CEC/UNECE, 1993

Species

In the 1992 survey, 113 species were assessed, although the six most frequent species accounted for about two thirds of all trees. The six species are *Pinus sylvestris* (24 per cent), *Picea abies* (23 per cent), *Fagus sylvatica* (9 per cent), *Quercus robur* (4 per cent), *Pinus pinaster* (4 per cent) and *Quercus ilex* (3 per cent). On average, conifers are slightly more defoliated (24 per cent) than broadleaved trees (22 per cent), but less discoloured. Of the most common conifers found in Europe, fir (*Abies* spp) and spruce (*Picea* spp) are in the poorest condition (with 30 per cent and 26 per cent damaged trees respectively); *Picea sitchensis* was found to be the most damaged conifer species. Of the broadleaved trees, oak (*Quercus* spp) is the most damaged (24 per cent). The cork oak (*Quercus suber*) has the highest percentage of damaged trees of all species (33 per cent), particularly in Portugal, which may be the result of large-scale insect attacks and adverse climatic conditions, frequent forest fires, traditional landuse practices (cork exploitation) and atmospheric pollution. The holm oak (*Quercus ilex*) was the least damaged tree of all species.

According to the survey, defoliation seems to decrease with increasing diversity index, that is, higher defoliation is found more in monoculture than in mixed stands.

Climatic regions

To identify the relationship between damage and climate, Europe has been subdivided into nine climatic zones (Map 34.2). Comparison of survey results has shown that

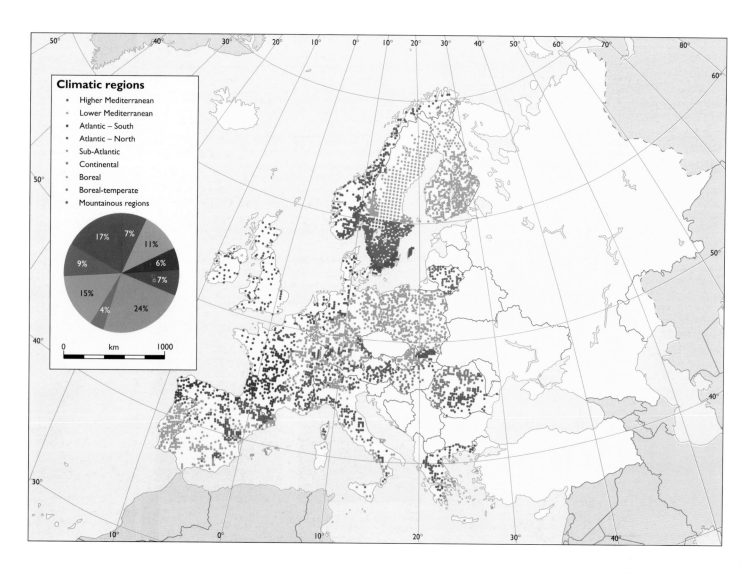

Box 34A Forest damage in the Krušné Hory, Czech Republic

The forest decline in northern and western former Czechoslovakia is unmistakable. This region emits a significant fraction of Europe's sulphur dioxide, nitrogen oxides and toxic metals as the result of an economy based heavily on mining and burning of lignite and coal, heavy industry and building material production (cement manufacturing). From the early 1960s, foresters first noted the decline and then the collapse of forests in the mountainous Erzgebirge region. Some 100 000 hectares of Norway spruce (*Picea abies*) forests have died, and it is estimated that a total of 54 per cent of the forests in the Czech Republic have suffered irreversible damage (Moldan and Schnoor, 1992). Dry deposition of sulphur dioxide and sulphate is aided by efficient collection in forest canopies. The problem is severe. Annual average sulphur dioxide gas concentrations in the Erzgebirge (100 μg/m³) are about ten times greater than those measured in the Bavarian and Black forests of Germany. Stands that have died are rapidly harvested, and much of the area has been replaced by grassy meadow. Norway spruce in the Erzgebirge is progressively replaced by more tolerant species of spruce (*Picea pungens*).

The longest legacy of the past 40 years may be the build-up of toxic constituents in the region's soils. Foresters have scraped away the top 15 cm of soil with bulldozers into wind rows in order to get seedlings to survive. They feed the soil with limestone and fertilisers, but regeneration remains poor. Metals (aluminium, arsenic, beryllium, cadmium, lead and zinc) make the soil toxic. Small-catchment research has demonstrated that the chemical weathering of calcium and magnesium cannot keep pace with the acidic deposition (Pačes, 1985). Aluminium that is liberated from the soil as the pH decreases is toxic to spruce trees. Changes in soil pH were recorded in more than 200 sites in the Czech Republic. A significant decrease in pH was found, from 3.8 in 1960 to 3.3 in 1985. This survey represents one of the few examples in the literature of relatively large-scale soil acidification caused by atmospheric deposition.

Dying forest
Source: Michael St Maur Sheil

defoliation of tree crowns has increased in all climatic zones except for the boreal region. The lower Mediterranean zone with the southern Atlantic and sub-Atlantic regions show forests in the worst condition. This is due particularly to the synergy between climate, management and local air pollution sources. Clear evidence of other environmental factors such as soil type and water availability on the intensity of defoliation was also shown.

Trends over time

The early projections (1980) suggested that forest damage would worsen and eventually destroy entire forests; so far this has not happened. Drastically defoliated and dead trees were, and are, scattered among less affected and unaffected trees, and there has been no large-scale deforestation, except in areas affected by exposure to high levels of sulphur dioxide gas in the mountainous regions of Central and Eastern Europe (Box 34A).

In order to compare results from year to year, a sub-sample of trees common to the different surveys has been studied since 1988. Since 1990 a substantial sub-sample of common sample trees (61 395 trees), which are preserved and repeatedly examined, has been established. It appears that there is a slight deterioration of the vitality of nearly all species of trees. The worsening of forest condition is happening particularly in the most damaged areas of Central Europe, but also in the Pyrenees, where no defoliation had been recorded previously.

Causes

Improved data and advances in methods of analysis and interpretation have shown that, because of the diversity of pollutants, forests and climates, the underlying assumption that air pollutants were exclusively to blame for forest decline was too simplistic. Recent research, as reported, for example, concerning Swiss forests (Buhrer, 1993), has cast doubt on the importance of the role of atmospheric pollution on forest degradation. Also, in the south of France, for instance, the cochineal *Matsucoccus feytaudi* was responsible for the defoliation of about 100 000 hectares of maritime pine forest (*Pinus pinaster*) during the late 1980s (Marchand, 1990). Overall, the most important probable causes of defoliation and discoloration are: adverse climatic conditions, insects, fungi, forest fires, improper management, game animals living in the forest and air pollution, most probably acting in combination.

Although very little direct impact of air pollution has been reported, it is a crucial weakening factor for the forest because of the interactions with nutrient uptake, soil acidification and toxification, tree physiology, etc. This makes it particularly difficult to understand the reasons for forest decline. Presently, five hypotheses dominate research: multiple stress; soil acidification and aluminium toxicity; interaction between ozone and acid mist; magnesium deficiency; and excess nitrogen deposition, as explained below.

1 The multiple stress hypothesis proposes that a number of stresses (air pollution, with its associated atmospheric deposition of nutrient, growth-altering, or toxic substances) impairs plant metabolism, making the tree more susceptible to other stress factors (such as climate, nutrient deficiencies, and secondary or contributing pathogens). This hypothesis is appealing, as it is vague enough to apply to any damage that cannot be explained in any other way. However, it does not relate to a single tree species nor to a specific damage type and, therefore, it does not allow an understanding of the causes and mechanisms of forest damage. Among the possible stresses, weather conditions are given more attention. But the lack of historical data implies that firm correlations cannot yet be made.

2 The soil acidification and aluminium toxicity hypothesis is based on a long-term nutrient-cycling study in the Solling area of Germany. It states that the natural acidification of forest soils is accelerated by the wet and dry deposition of sulphuric and nitric acids from anthropogenic sources, mainly the combustion of fossil fuels. Afforestation with conifers on unsuitable sites, and drainage of some wetland soils such as peat containing sulphides, can also lead to

acidification. Increased soil acidity, which leads to the breakdown of the soil buffering system and causes a leaching of aluminium ions and other heavy metals at levels toxic to plants, impairs the tree's root system and eventually harms the entire tree (see also Chapter 7). This hypothesis is not valid for trees growing on well-buffered soils such as the calcareous Alps, where damage still occurs. There is also evidence that the level of aluminium found in acid soils is likely to damage the tree roots and the fungi that live on them. Soil aluminium toxicity is, therefore, considered less as a self-standing explanation than as a contributory factor to damage, especially when nutrient deficiencies are involved.

3   The hypothesis that ozone interacts with acid mist and climatic factors to induce or exacerbate various nutrient deficiencies has been popular in recent years. The idea depends upon increasing ozone concentrations at medium to high altitudes (see Chapter 32). The major problem with this hypothesis is that symptoms produced by ozone in the laboratory are not the same as those in the field. Results are, therefore, difficult to extrapolate and it can be said that, where nutrient deficiencies are involved, the role of ozone has been overestimated. Nevertheless, the effects of ozone can be important in mountainous areas, especially in summer and on sensitive species such as pine.

4   Various reports have stressed the key role of magnesium deficiency in spruce yellowing damage. The sudden occurrence of these symptoms over a wide area is attributed to a series of dry years which reduced magnesium mineralisation and uptake. Magnesium depletion caused by tree harvesting (shorter rotation) and leaching of soils by acidic deposition, together with this reduced uptake of magnesium caused by drought, may explain why insufficient magnesium nutrition occurs now over a larger area than ever before.

5   The fifth and final hypothesis, concerning excess nitrogen deposition, has gained ground in the past few years. In Northern European countries, nitrogen (N) deposition can exceed 50 kg N per hectare per year, mainly in the form of ammonia released from animal waste. Most forest ecosystems are nitrogen deficient, and nitrogen will initially act as a nutrient. Later, nitrogen deposition will cause ecosystem changes and finally, after extensive periods at high nitrogen loads, nitrate leaching will occur and pollute the groundwater (see Maps 5.7 and 5.8). In most of the Nordic countries, nitrogen leaching is small (<1kg/ha/year), while in Western and Central Europe nitrogen leaching is common (>5kg/ha/year in Central Europe). Nitrogen uptake will contribute to acidification since it is associated with an uptake of alkaline ions, which will lead to losses of base saturation in the soil. Finally, the leaching of nitrogen in the form of nitrate is associated with leaching of cations and contributes to acidification in the same way as sulphate outflow from the forest soil. Local ammonium deposition may therefore be an important factor in forest degradation and, in excess, may increase susceptibility to climatic and biotic stresses and accelerate soil leaching (Grennfelt et al, 1993).

Economic effects

It is difficult to predict the economic effects of forest degradation; evaluations must include not only the loss to forestry, but also the decrease in recreational value and additional costs such as the reduced prevention of soil erosion and reduced protection from avalanches. In Poland, the economic losses resulting from declining forests have been estimated to be some 60 billion zlotys per year (Nilsson et al, 1992). The economic effects of forest damage are closely linked to projections of the course of that damage. So far, fears of severe economic consequences have not been realised.

There is no difference in the physical characteristics of timber from damaged and healthy trees; and although there has been an increase in the rate of unplanned salvage felling, the excess has been absorbed by the timber market.

Putting the regional differences to one side, it is important to resolve whether defoliated and healthy trees have significantly different growth patterns and, if so, to what extent growth and defoliation are correlated. The few data available show that coniferous trees grow more slowly if they lose more than 30 per cent of their foliage. This suggests a threshold rather than a linear relationship, but no reliable correlation has yet been established. These uncertainties rule out any accurate long-term projections of forest growth. Growth reductions since the 1950s or 1960s have been noted in some areas, but in other cases tree ring analysis has shown an impressive increase in growth rate during the last ten years (Becker et al, 1989). This contradictory phenomenon could be due to a good rain pattern of the past few years, combined with an increase in nitrogen deposition from the atmosphere, as for example in the Black Forest. An increase in tree growth rate does not necessarily exclude an impact of atmospheric pollution, the effects of which can be hidden. For example, nutrient excess promotes an increase in the height/roots ratio; this imbalance between above-ground (height) and below-ground (roots) development renders the trees more sensitive to wind. Nitrogen deposition on nutrient-poor sites can result in nutritional imbalance, leading to decreased yield and forest decline. In conclusion, it is most likely that atmospheric pollution in general renders trees more sensitive to any stress and diminishes their vitality. This will be all the more important as trees respond to any additional stress caused by global warming.

A model simulating forest damage due to air pollutants has been developed by the International Institute for Applied Systems Analysis (IIASA) (Nilsson et al, 1992) to predict growth losses for Western, Central and Eastern Europe. The loss of potential harvest caused by air pollutants is estimated to be about 85 million m³ per year averaged over 100 years. The regions most affected by forest decline attributed to air pollutants are the Eastern and Central regions of Europe. If, as is foreseen, Europe will face an annual roundwood deficit of some 40 million m³ per year by 2010, the deficit could amount to about 130 million m³ per year if decline caused by air pollutants is taken into account. Of course, this deficit would be partly compensated by the conversion of unused agricultural land to tree plantations, which was estimated at 8 per cent or 29 million m³ per year. This approach does not take into account the non-wood benefits of the forest and potential losses due to air pollution.

Options

A continuation of the present pollution load for extended periods of time is expected to threaten the vitality of forests over large areas of Europe. The atmospheric concentrations and the deposition of certain pollutants in many areas exceed the levels at which disturbance to forest ecosystems would be expected. In such situations, a reduction of the air pollution load should improve the condition of endangered forests and postpone further disruption to ecosystems. Sulphur dioxide, ammonia, nitrogen oxides (as a precursor of ozone and acid deposition) and other pollutants may all be important in particular areas. Reduction of emissions in accordance with the setting up of targeted critical loads is still the first priority for action (see Chapters 4 and 31).

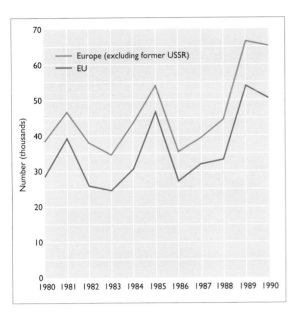

Figure 34.1 Number of forest fires in Europe, 1980–90
Source: FAO, 1992

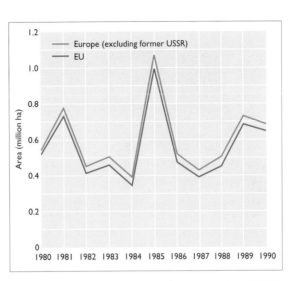

Figure 34.2
Forest area burnt in Europe, 1980–90
Source: FAO, 1992

Conclusions

Forest damage, as judged by European-wide surveys of tree crown condition, has increased during the last few years; however, this overall increase has not prevented some regional recovery of certain species, such as *Picea abies* and *Pinus sylvestris*. The most probable causes for the observed defoliation and discoloration have been reported to be adverse weather conditions, insects, fungi, forest fires and air pollution. Some countries, such as former Czechoslovakia, Germany and Poland, where several thousands of hectares of forests have been severely degraded, consider air pollution to be a factor leading to the weakening of forest ecosystems. However, recent research has cast doubt on the importance of the exclusive role of atmospheric pollution on forest degradation. In France, the programme DEFORPA (Dépérissement des Forêts et Pollution Atmosphérique), which was launched in 1984 to identify the causes of forest dieback and to further understanding of the effects of atmospheric pollution in general, has now been abandoned. The setting up of transnational surveys has provided a time series of large-scale spatial observations. However, they do not readily permit cause–effect relationships to be identified. In an attempt to improve the situation, a more detailed monitoring system, including the collection of information on soil, foliage, atmospheric deposition, local climate, tree growth, etc, has recently been established by the Eurpean Commission with UNECE.

DAMAGE DUE TO FIRE

Introduction

Research into the history of forest fires shows that fires have occurred as long as forest has existed. However, the causes of fires have changed over time. Originally, natural fires resulted from volcanic eruptions or lightning. Humans quickly made use of fire to clear and maintain land established as farmsteads and grasslands.

Forest after fire, Portugal
Source: Michael St Maur Sheil

Today, forest fires occur for many reasons, ranging from the heritage of past agricultural practices – firing land for grazing, for instance – to the consequence of modern development, such as clearing land for urbanisation and tourism, and criminal action (arson).

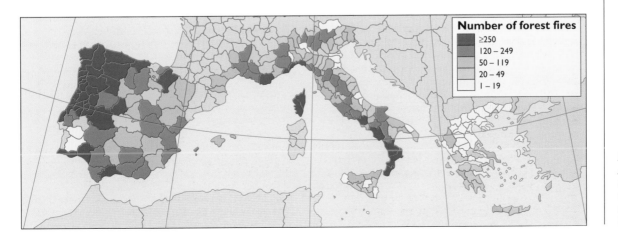

Map 34.3
Average annual number of forest fires, 1989–91
Source: Ministerial Conference on the protection of forests in Europe, 1993

Map 34.4
Burnt area for one year, as a percentage of overall forest area, 1989–91
Source: Ministerial Conference on the protection of forests in Europe, 1993

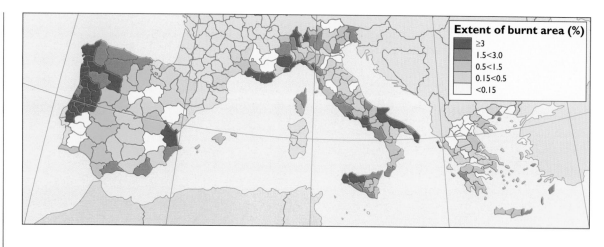

Map 34.5
Number of winter fires (January, February, March), as a percentage of the total number of fires, 1989–91
Source: Ministerial Conference on the protection of forests in Europe, 1993

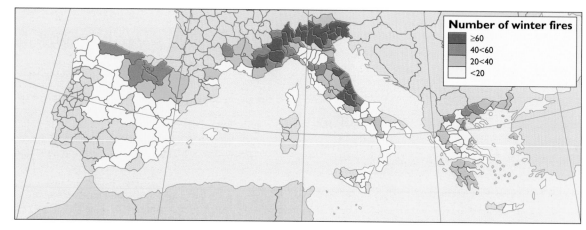

Map 34.6
Average number of fires of more than 50 ha, 1989–91
Source: Ministerial Conference on the protection of forests in Europe, 1993

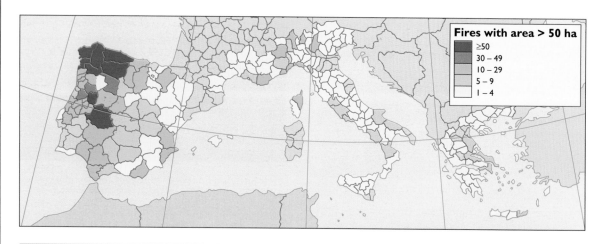

The problem

In Southern Europe and parts of the former USSR, fire is certainly the main problem for forest management, followed by drought and, to a lesser extent, air pollution. In Northern Europe, forest fires are sometimes used to protect heathland against tree invasion. A typical example is the management of the Lüneburger Heide in the north of Germany. In boreal forests, fire is used to maintain the diversity of the forest; enormous stands of rather pure *Picea abies* would develop in many areas of Sweden if burning did not occur, *Pinus sylvestris*, *Betula pubescens* and *Populus tremula* relying on disturbance for survival (Bradshaw and Hannon, 1992). Fire is then used as a management tool to maintain certain landscapes and ecological diversity.

The facts

In Europe, an average of 700 000 hectares (1980–90) of wooded land (the term 'wooded land' is used in a broad sense – it includes *maquis*, *garrigue*, heathland and even grassland from which trees have been removed) are burnt each year by a total of 60 000 fires (FAO, 1992). France, Greece, Italy, Portugal and Spain account for more than half of this total, with an average of 35 000 fires covering 500 000 ha of wooded land on average. During 1980–90, the number of fires in the five Mediterranean countries of the European Community increased, but the area burnt did not increase by the same proportion (Figures 34.1 and 34.2, p 563). In the EU, the recent trend has been towards many small fires occurring, which are then quickly extinguished.

Reliable data are difficult to find for the former USSR. One estimate was that a hundred times as much forest was destroyed by fire as was planted (WWF, 1992).

Box 34B EU work on prevention of forest fires (Regulation 92/2158/EEC)

A detailed analysis of the causes of fires and possibilities of controlling them, in connection with a study of socio-economic factors prevailing in regions at risk, is being undertaken by the EU. This investigation comprises a preliminary study of forest legislation and policies which have an impact on forest development (including town planning, environment policy, agricultural/animal husbandry policy) and which could indirectly cause fires. Attempts are also made to evaluate the effectiveness of awareness and information campaigns targeted at the public. Special attention is also given to economic incentives or disincentives taken to reduce the number of fires, such as town and country planning measures, support for agricultural/grazing activities, etc (CEC, 1991).

Since 1990, a pilot project for the development of an information system on forest fires has been under development in the EU (Council Regulations 86/3529/EEC and 92/2158/EEC). Data have been collected daily at regional level in France, Greece, Italy, Spain and Portugal to establish a geographical database on forest fires. Some of these data are reproduced in Maps 34.3 to 34.6 (pp 563-4). In the EU, it is estimated that 35 million ha of wooded lands are at risk from fire. In 1990, there were only two days with no fires in the whole of the EU. There is a seasonal pattern which shows that, in some regions, such as in the north of Italy, most fires occur in winter and, in others, such as the north of Spain, most occur in summer (winter fires are usually less damaging than summer fires). On the other hand, some regions, such as the northwestern part of Portugal and Spain and southern Italy, which have very different climates (Atlantic/Mediterranean) and vegetation types (high forest/shrubland), show similar distribution of fires, with a peak in summer. For the Mediterranean part of the EU, two thirds of the fires occur in summer and are responsible for three quarters of the total area burnt (CEC, 1994).

Socio-economic factors seem to lead to fires in these regions, where tensions arise from conflicts between, on the one hand, agriculturalists and pastoralists who want to clear the forest to use the land for other purposes and, on the other, foresters whose aim is to maintain the land for forestry. The size of the fires also varies between these regions. The largest fires (larger than 30 ha) occur in the north of Portugal and Spain, while in the south of Italy nearly all fires are smaller than 30 ha.

More than the number, it is the size of the fires which is determinant. The important factor, however, is the total area of forest affected, and small numbers of fires are damaging large areas of forest. For instance, in 1989 in France, 0.5 per cent of all fires damaged 40 per cent of the total area damaged, and 94 per cent of fires burnt less than 5 ha (Prometel, 1992).

The causes

Traditionally, forest fires have been ascribed to natural causes such as lightning, along with accidents, vandalism and arson. Causes are more diverse than is often assumed, and fire initiation is neither as random, nor, in some cases, as meaningless as some analyses have suggested. Understanding the reasons why fires start is very helpful when determining what to do to prevent or reduce their incidence (Box 34B).

Natural causes

Natural fires do not represent a large proportion of the total number of fires in Europe. Analysis of UNECE-FAO statistics suggests that, in 1987, natural fires made up 8 per cent in Spain, 3 per cent in Finland, 2 per cent in former Czechoslovakia, and less than 0.1 per cent in West Germany and Italy (WWF, 1992).

Accidents, negligence and arson

Many accidents leading to forest fires stem from negligence coupled with lack of understanding about the risks of, for example, dropping cigarette ends, or people starting fires for camping and barbecues. Much of the traditional understanding of fire management has also been lost because of the depopulation of many rural areas. The risk of fire has been further increased by the greater mobility of urban dwellers, who visit forest areas more regularly, but who may not understand the risks of starting a fire. In some areas, such as the Mediterranean, increases in coastal populations are creating additional stresses on forests through the sheer numbers of people visiting them.

Analysis of UNECE-FAO fire statistics suggests that negligence of various kinds is an important cause of fire in most European countries, but generally less so than arson (although the causes of many fires remain undetected). For example, in 1987 accidental fire accounted for some 59 per cent of fires in Turkey, 50 per cent in Austria, 32 per cent in Portugal and 25 per cent in Spain (UNECE/FAO, 1990).

Arson remains the most important known cause of fire in much of Europe. For example, combined figures for the whole of the Mediterranean for 1981–85 indicate that negligence was the cause of 23 per cent of fires, arson accounted for 32 per cent, while the cause of 40 per cent of fires was unknown (WWF, 1992).

Finally, other activities such as military exercises and waste disposal may also result in fire accidents if not well under control. Unprotected power lines running over forestry can create sparks, thus directly causing fires.

Management practices

Fire is traditionally used by farmers to clear the fields after harvesting and to provide nutrients to the soil through ashes. Fire is also used by shepherds to renew grassland. If these fires are not very well controlled (for instance, if they are initiated under weather conditions favourable to fire spreading quickly), they can spread to nearby forests and evolve into uncontrolled forest fires. Some other causes are mentioned by different authors (WWF, 1992): for example, fire for hunting, land speculation, wood speculation and private vengeance.

Consequences of fire on the environment

In the Mediterranean part of Europe, much of the landscape is the result of the frequent occurrence of fires. Fires have an important impact on the environment: burnt wood loses its scenic and recreational values; particularly intense and long-lasting forest fires can cause serious soil damage by slow combustion of organic matter and humus in the soil; serious erosion risks may appear if violent storms occur soon after a fire, especially on slopes and fragile soils. If fires occur too frequently, there may not be sufficient time for the forest to recover, which leads to its conversion into shrubland. Fires have an impact on forest composition, favouring species resistant to fires such as the holm oak (*Quercus ilex*), cork oak (*Quercus suber*) and some pines. All these species resprout and regenerate quickly after fire (sometimes within a year).

Fires are likely to affect the whole fauna and flora of an

ecosystem. Fires damage plants and threaten wildlife unable to escape. Some important animal species in the Mediterranean area have apparently suffered badly from forest fires, including wild boar (*Sus scrofa*), red deer (*Cervus elaphus*) and roe deer (*Capreolus capreolus*), badgers (*Meles meles*), porcupines (*Hystrix cristata*), and breeding birds (WWF, 1992). However, the effects on the dynamics of forest ecosystems are not always negative. For instance, a recent study has shown that the number of bird species is higher in burnt forests and remains higher for several years after a fire (Prodon, 1992). This is because the post-fire succession of growth offers a patchy vegetation (scattered bushes and young trees), which favours bird populations in general. This is also true for the diversity of forest undergrowth plant species.

Fires in radioactively contaminated forests pose a particularly serious problems (see Box 34C).

At the global level, forest burning contributes at least 20 per cent of the carbon dioxide released to the atmosphere each year from human activities. Many countries have emphasised the importance of the dry conditions in 1990 and 1991, both in terms of drought stress to trees and high frequency of forest fires. Any atmospheric changes that increase the frequency of dry conditions in Europe can have serious consequences for many forests, particularly in the south, by increasing the frequency of fires.

Economic costs

Costs are difficult to calculate. Burnt wood has little commercial value but burnt timber can often still be sold, and indeed forest fires are sometimes started deliberately as a means of wood speculation.

UNECE-FAO analyses (1990 and 1992) attempt to look at losses of both timber and other uses. In the particularly bad forest-fire summer of 1985 in Europe, Bulgaria lost 22 400 m³, Italy 2 million m³, Portugal 7 million m³, Turkey 556 300 m³, and former Yugoslavia 950 000 m³ of timber. However, costs of timber are only part of the problem, and additional costs for fire fighting and lost recreational use, all add to the total financial losses. For example, in 1992, France spent about ECU83 million for fire fighting. In 1990, Italy lost an estimated ECU68 million in terms of wood and other tangible losses and ECU3 million in other losses (FAO, 1992). Social values (loss of aesthetic and recreational resources) are not possible to quantify in financial terms. Local people suffer if fires eliminate traditional grazing lands within forests. Forest fires in residential areas, which have become increasingly frequent in the Mediterranean area, have immediate effects on people's quality of life through destruction of property and possessions. Fires also kill a number of people every year, including fire fighters.

Strategies for prevention and fighting fires

Any successful policy for the protection of forests against fires must have two objectives: to reduce the number of fire outbreaks; and to reduce the area burnt. The reduction of the number of fires has to be achieved by reducing the causes (in the Mediterranean region, 95 per cent of the fires are from human origin), while the reduction of the burnt-out area relies on prevention measures such as the installation of look-out structures, and on an active control and fire fighting once a fire has started. Since fire reduction is very much related to internal landuse conflicts, the discussion of fire protection measures below concentrates on those which help reduce the area burnt.

Box 34C Fires and radioactivity

After the Chernobyl nuclear accident in April 1986, large areas of Ukraine, Belarus and the Russian Federation were exposed to radioactive contamination, the main contaminants being the long-lived radionuclides caesium-137, strontium-90 and plutonium-239. In the most contaminated regions (Kiev, Zhitomir, Rovmo, Gomel, Mogilev and Bryansk regions), there are forests consisting of young and middle-aged pine and pine hardwood stands very sensitive to fire (because these forests are not thinned). In the year of the Chernobyl accident, a total of 1775 fires burned 2336 ha of forests in these regions. Fires in radioactively contaminated forests constitute a grave problem due mainly to the way in which they mobilise radioactivity into the air. Research shows that in 1990 the bulk of caesium-137 was concentrated in the forest litter and upper mineral layers of the soil. By 1992, vertical migration of radionuclides caused an increasing contamination of the upper soil layer so that, at present, surface fires (fires in contaminated litter and humus) and ground fires in contaminated peat pose a serious danger. Furthermore, incomplete combustion causes open sources of ionising radiation: in the zones where the density of radiocaesium contamination of the soil is 0.6 to 1.5 TBq per km² (TBq=10^{12} TBq) and higher, the specific radioactivity of ash and partly burned litter has been measured in the range of 180 to 1086 kBq per kg. The level of radioactive caesium in aerosols after a fire increases to ten times the level before the fire.

Another great danger is posed by high-intensity (crowning) fires which develop convective activity and lift radionuclides into the atmosphere. To obtain accurate quantitative assessment of the behaviour of radionuclides during forest fires, laboratory and field experiments are necessary. These experiments are needed to develop fire behaviour models, especially for heat and mass transfer in the near-ground layer. In addition, the contaminated forest environment has affected the working conditions of forest fire fighters and heavily contaminated fire-fighting equipment.

Source: Dusha-Gudym, 1992

Fire prevention

Prevention relates to the passive protection of the forest. Land management policies are an essential tool for minimising landuse conflicts and reducing fire risk. Forest management which takes fire into account should include regular thinning, brush clearance along roads, railways and power lines, and the plantation of stands/species which are less sensitive to fires. The most crucial infrastructure items to be developed for fire prevention are the creation of forest roads (which allow rapid intervention and act as fire-breaks), the provision of water supplies in the event of a fire and the establishment of fire-breaks.

Fire watch

In parallel with passive prevention, active protection measures include risk forecasting, involving analysis of meteorological data on a routine basis to foresee potential fire conditions, and regular watching, using fixed installations such as look-out towers or mobile ones (patrols and aircrafts), possibly with fire-fighting equipment. Since 1992, a daily exchange of information on the level of risk of forest fires during summer takes place between EU Member States and the European Commission; the objectives of this initiative were to provide the Southern countries with information on the hazard situation, and to realise a survey of the evolution of the risks.

Fire control

When a fire has started, how quickly it can be stopped will strongly depend on the promptness of its detection and how it is fought by the different land-based and airborne systems, their number and the effectiveness of their coordination. Therefore, sound deployment of capabilities and training of qualified personnel are factors in the control of forest fires. Since 1991, the European Commission (DGXI) has developed several actions to improve mutual assistance between Member States of the EU: a self-tuition programme, a group of cooperation officers, a register of fighting means and pilot projects.

Fire as a management tool

Prescribed burning can be used as a management tool to prepare land for replanting, to clear post-harvest debris, to destroy pests, etc, but also to prevent wild fires. Indeed, fire started under well-controlled conditions (in winter, with no wind) can be used to reduce the fuel load of the forest, that is, the litter undergrowth, over vast areas and at relatively low cost. This technique, which had fallen into disuse in much of Europe, has been reintroduced in some countries, where it proves to be very efficient (eg, in Portugal, France, Sweden). However, the conditions prevailing in Europe, that is, the high population density and the structure of land ownership – half of the forest is private – and the perception of the public are limiting factors for its extensive use.

Conclusions

Fires represent a major disturbance of forests in Europe and a threat to nearby populations. The belief that the number and extent of fires are increasing is much debated. Indeed, it is only recently that fires have been recorded systematically. Forest fires represent only a part of the total number of fires (less than 50 per cent in Italy). Changes in landuses which are occurring in Southern Europe (abandonment of agricultural land, depopulation) are increasing fire risk. The seriousness of the problem is due to the fact that natural causes have been compounded by human action (negligence, accidents, burning as an agricultural or grazing practice, and arson). The control of the causes is complex since they are often indicators of conflicts and tensions in the overall system of land management. Action measures often need to go far beyond forest activities, because the causes of fire can be related to socio-economic, legal, structural, and even cultural factors. Therefore only an integrated and progressive approach based on a sound knowledge of local problems, sustained coordination between the different involved parties, and political determination to succeed can be successful.

REFERENCES

Becker, M, Landmann, G and Nys, C (1989) Silver fir decline in the Vosges mountains (France): role of climate and silviculture. *Water, Air, and Soil Pollution* **48**, 77–86.

Blank, L W, Roberts, T M and Skeffington, R A (1988) New perspectives on forest decline. *Nature* **336**, 27–30.

Bradshaw, R and Hannon, G (1992). The disturbance dynamics of Swedish boreal forest. In: Teller, A, Mathy, P and Jeffers, J (Eds), *Responses of Forest Ecosystems to Environmental Changes*, pp 526–35. Elsevier Applied Science, London and New York.

Buhrer, J-C (1993) Des scientifiques suisses avouent s'être trompés sur 'la mort des forêts'. *Le Monde*, 3 September 1993, 10.

CEC (1989) *European Community forest health reports 1987–1988*. Commission of the European Communities, DG VI, Luxembourg.

CEC (1990) Air pollution and forest ecosystems in the European Community. *Air Pollution Report 29*. Ashmore, M R, Bell, J N B and Brown, I J (Eds) Directorate-General for Science, Research and Development, Brussels.

CEC (1991) *Rapport d'activité de la Commission dans le secteur régi par le Règlement (CEE) no 3529/86 du Conseil du 17 novembre 1986, et par le Règlement (CEE) no 1614/89 du Conseil du 29 mai 1989, relatifs à la protection des forêts contre les incendies*. Commission of the European Communities, DG VI, Brussels.

CEC (1994) *Action préparatoire au système d'informations sur les ince-dies de forêt*. Commission of the European Communities, DG VI, Brussels.

CEC/UNECE (1993) *Forest condition in Europe*. Report of the International Co-operative Programme on Assessment and Monitoring of Air Pollution Effects on Forests. Results of the 1993 survey. Commission of the European Communities, DGVI, Brussels.

Dusha-Gudym, S I (1992) Forest fires on the areas contaminated by radionuclides from the Chernobyl nuclear power plant accident. *International Forest Fires News* **7** (UN/ECE/FAO), August, 4–6.

FAO (1992) *Forest fire statistics 1990–1992*. ECE/TIM/58. Food and Agriculture Organisation, United Nations, New York.

Grennfelt, P, Hov, Ø and Derwent, R G (1993) Second generation abatement strategies for NO_x, NH_3, SO_2 and VOC. *IVL Report* B-1098, Swedish Environmental Research Institute, Gothenburg.

Marchand, H (1990) Les forêts méditerranéennes: enjeux et perspectives. *Les fascicules du Plan Bleu 2*, Programme des Nations Unies pour l'environnement. Plan d'action pour la Méditerranée. Economica, Paris.

Ministerial Conference on the protection of forests in Europe (1993) *Report on the follow-up of the Strasbourg Resolutions*. Helsinki, 16–17 June 1993, pp 93–122. Ministry of Agriculture and Forestry, Helsinki.

Moldan, B and Schnoor, J L (1992) Czechoslovakia: examining a critically ill environment. *Environmental Science and Technology* **26**, 14–21.

Nilsson, S, Sallnäs, O and Duinker, P (1992) *Future forest resources of Western and Eastern Europe*. IIASA, Parthenon Publishing Group – UK and USA.

Pačes, T (1985) Sources of acidification in Central Europe estimated from elemental budgets in small basins. *Nature* **315**, 6014, 31-6.

Prodon, R (1992) Animal communities and vegetation dynamics: measuring and modelling animal community dynamics along forest successions. In: Teller, A, Mathy, P and Jeffers, J (Eds), *Responses of Forest Ecosystems to Environmental Changes*, pp 126–41, Elsevier Applied Science, London and New York.

Prometel (1992) 36–15 Prometel (data base). Opération Prométhée CTID, Marseilles.

UNECE (1991a) *Forest damage and air pollution*. Report of the 1990 forest damage survey in Europe. UN/ECE/UNEP, Geneva.

UNECE (1991b) *Interim report on cause–effect relationships in forest decline*. UN/ECE/UNEP, Geneva.

UNECE/FAO (1990) *Forest fire statistics 1985–1988*. UNECE and FAO, United Nations, New York.

WWF (1992) *Forests in trouble: a review of the status of temperate forests worldwide*. Gland, Switzerland.

*Belgian coast at
Blankenberge*
Source: Eurosense UV

35 Coastal zone threats and management

INTRODUCTION

Human activities are often concentrated in coastal regions which are often least able to assimilate those activities, and where adverse effects are most apparent. Coastal zones are relatively fragile ecosystems, and disordered urbanisation and development of infrastructure, alone, or in combination with uncoordinated industrial, tourism-related, fishing and agricultural activities, can lead to rapid degradation of coastal habitats and resources. Mounting pressure on the coastal zone environment has, in several European countries, resulted in a rapid decline in open spaces and natural sites and a lack of space to accommodate coastal activities without significant harmful effects.

Not all effects on coasts are due to human activities; climate can have serious direct and indirect effects on the coastal environment. Effects of occasional storms can be disastrous, whereas the indirect effects of climate change in terms of sea-level rise (see Chapter 27) are predicted to cause potentially serious damage to unprotected low-lying areas along the European coast.

DEFINITION OF THE COASTAL ZONE

There is no common or unique definition of what constitutes a 'coastal zone', but rather a number of complementary definitions, each serving a different purpose. Although it is

generally intuitively understood what is meant by 'the coastal zone', it is difficult to place precise boundaries around it, either landward or seaward. For example, the coastal zone itself is an area considered in some European countries to extend seawards to territorial limits, while by others the edge of the continental shelf at around the 200 m depth contour is regarded as the limit. A general workable definition is:

*the part of the land affected by its proximity to the sea,
and that part of the sea affected by its proximity to the land
as the extent to which man's land-based activities have a
measurable influence on water chemistry and marine
ecology.*

(US Commission on Marine Science, Engineering
and Resources, 1969)

The landward boundary of the coastal zone is particularly vague, since oceans can affect climate far inland from the sea.

The coastal zone is the zone in which most of the infrastructure and human activities directly connected with the sea are located. An estimated 200 million of the European population (total 680 million) live within 50 km of coastal waters. This reflects partly the historical importance of coasts for human settlement for reasons of defence and for providing sources of food. Ports have generated industrial activity. Coastal zones are favoured areas for energy generation because of easy delivery of fuel for power stations and convenient disposal of cooling water. The landward part of the coastal zone plays an important role as a place for human settlement and tourism.

Human activity	Agents/consequences	Coastal zone degradation problems
Urbanisation and transport	Landuse changes (eg, for ports, airports); road, rail and air congestion; dredging and disposal of harbour sediments; spills at sea (oil, sewage garbage); water abstraction; wastewater and waste disposal	Loss of habitats and species diversity; visual intrusion; lowering of groundwater table; salt water intrusion; water pollution; human health risks; eutrophication; introduction of alien species
Agriculture	Land reclamation; fertiliser and pesticide use; livestock densities; water abstraction; river channelisation	Loss of habitats and species diversity; water pollution; eutrophication; reduction of freshwater inputs to coastal waters
Tourism, recreation and hunting	Development and landuse changes (eg, golf courses); road, rail and air congestion; ports and marinas; water abstraction; wastewater and waste disposal	Loss of habitats and species diversity; disturbance; visual intrusion; lowering of groundwater table; salt water intrusion in aquifers; water pollution; eutrophication; human health risks
Fisheries and aquaculture	Port construction; fish processing facilities; fishing gear; fish farm effluents	Overfishing; impacts on non-target species; litter and oil on beaches; water pollution; eutrophication; introduction of alien species; habitat damage and change in marine communities
Industry (including energy production)	Landuse changes; power stations; extraction of natural resources; process effluents; cooling water; windmills; river impoundment; tidal barrages	Loss of habitats and species diversity; water pollution; eutrophication; thermal pollution; visual intrusion; decreased input of freshwater and sediment to coastal zones; coastal erosion

Table 35.1
Relationships between human activities and coastal zone problems

In many cases, however, there has been overdevelopment of coastal zones, and this has led to degradation of the environment. This has in turn led to policies to rectify or reduce damage caused.

The variable coast

The European coastline (without lake shores) is approximately 143 000 km long including islands (estimated at a map scale of one to three million). This enormous distance (probably a minimum estimate), which includes arctic, temperate and subtropical climate zones, covers a large variety of geomorphological features, including ice barriers, rocks, cliffs, and shingly, sandy and muddy shores. The salinity of the water and concentrations of contaminants also show large variability. For example, the salinity ranges from almost zero in many estuaries with large freshwater input, up to ocean values of approximately 35 per thousand or even higher in saline Mediterranean lagoons. Because of the extreme conditions prevailing along the coast through tidal influence, strong winds, chemical composition of the water and salt spray, there is a clear zonation of biotopes from the sea landwards. This, in combination with climatic, geological and geomorphological differences across Europe, helps to create a large variety of natural coastal zone biotopes, including sea-bed communities of macroalgae and seagrasses, tidal mudflats, saltmarshes, salt steppes and gypsum scrubs, dune heaths, dune scrubs and natural or semi-natural dune woodlands.

COASTAL ZONE PROCESSES

Being the interface between sea and land (or between salt water and freshwater), the coastal zone acts as a very important buffer influencing the fate of riverine contaminants and direct discharges as they are transported from land-based sources to the sea (UNESCO, 1987). The gross flux of contaminants through the land/sea boundary is predominantly influenced by the geological, biogeochemical and landuse characteristics of the given catchment and its hydrology. Biogeochemical cycles in the seaward part of the coastal zone are controlled mainly by interactions between dissolved and particulate phases. Therefore, the processes that control net fluxes of contaminants traversing the boundary between coastal zone and ocean are basically those that

determine whether a contaminant is dissolved or adsorbed on particles or taken up by living organisms during transport in the coastal zone. Long-range transportation of nitrogen and persistent pollutants seems to be an increasing threat to the environmental conditions of coastal waters.

Sediments are an important component of the coastal environment. Only a small proportion of sediments coming from the continents reaches the open ocean, most being deposited in the shallow and calmer waters of the coastal zone, including estuaries, bays, fjords, and also in tidal

Map 35.1
Coastal evolution in the EU
Source: CORINE

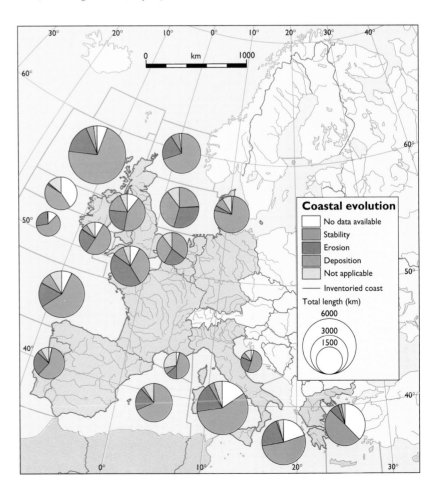

mudflats. Occasionally resuspension and water circulation patterns transport sediments to other areas along the coast. This balance between sedimentation and resuspension is treated further below under 'coastal erosion'.

Many environmental contaminants (eg, heavy metals, synthetic organic compounds, phosphorus, radionuclides) are adsorbed on suspended and settled particles. Sedimentation thus acts as a sink for these substances, if only temporarily, and plays an important role in determining the environmental impact of the contaminants. Sedimentation can lead to elevated concentrations of contaminants in the sediments of sheltered bays, harbours and tidal mudflats by accumulation. The water column may then become degraded by the release of sediment-bound contaminants over time, even when inputs from contaminant sources have stopped. The partitioning of chemicals between water, particles and settled sediment is therefore of critical importance in assessing or understanding the environmental behaviour of contaminants in the coastal zone, particularly at, and under the influence of, the salt water/freshwater boundary.

Although the seaward part of the coastal zone occupies only a small part of the world's oceans' water volume and surface area (0.5 per cent and 10 per cent, respectively), this zone (in particular the estuarine section) is functionally of great importance, with rates of production and consumption comparable with those found only in intensive agriculture and tropical and subtropical rainforests (Odum, 1971). It has been estimated that the coastal zone accounts for 30 per cent of ocean biological production (Mantoura et al, 1991) because of direct riverine nutrient input and upwelling along the shelf-edge. Most of the world's fish catch (70 to 80 per cent) comes from this zone.

THE PROBLEMS: CAUSES AND CONSEQUENCES

The most important underlying sources of pressure and the resulting problems (including physical, chemical and biological modifications) that these give rise to for the coastal environment are summarised in Table 35.1 (p 569). For more detailed information about the environmental pressures exerted by human activities, see the other chapters in Parts III and IV of this report.

The assessment of the state of the marine and coastal environment in Chapter 6 shows that, in all of Europe's seas, marine pollution of the coastal zone is a significant problem, whether locally (for example, in individual estuaries or bays) or regionally, such as in large enclosed or semi-enclosed seas. A similar conclusion emphasising marine pollution as an important coastal issue was made by the European Coastal Conservation Conference (ECCC, 1991). In particular, the enrichment of natural waters by nutrients, primarily nitrogen in marine waters but also phosphorus in low salinity waters, has been associated with increased primary productivity and nuisance algal growth (eutrophication) in the coastal zones and semi-enclosed and enclosed areas of seas. As well as being the major problem in coastal zones, eutrophication can also be a more widespread and general problem in the more enclosed seas of Europe, for example the Black Sea and Baltic Sea.

Physical modifications and habitat loss

In many European seas the coastal zone is an important area for human habitation, industry, location of centres of energy production, military activities, fisheries, bird life and recreation. This inevitably leads to a conflict of use over resources such as water (eg, for bathing and shellfisheries) but also for landuse (eg, harbours and marinas), which can have adverse effects on the quality of the coastal environment. The lack of effective coastal zone management

can lead to the loss of important components of the ecosystem and habitats (eg, dunes and wetlands). Physical modifications, such as the construction of ports and tourist facilities, result in the loss of habitats. The damming of rivers alters the hydrological regime, resulting in a reduction of freshwater flow. This can have serious consequences on the coast. Reducing freshwater flow generally means a reduced sediment load, which may induce coastal erosion.

Coastal erosion

Erosion of the shoreline and dunes, caused by ocean currents, tidal movements and wave and wind action, is a common natural phenomenon along many parts of the European coast. The result is the movement of vast amounts of sediment (mud, silt and sand). At present, approximately 70 per cent of all sandy coasts in the world are subject to coastal erosion (Hoogland, 1992). An inventory of coastal evolution in the EU (Map 35.1) undertaken within the EC CORINE programme, showed 55 per cent of the coastline (total length of 56 000 km) to be stable, 19 per cent to be suffering from erosion problems, and 8 per cent to be depositional.

The natural balance between erosion and sedimentation can be affected by those human activities which decrease the amount of sand available for erosion. Extraction of sand close to the coast weakens the coast locally, and dredging and construction of reservoirs on many rivers (see Chapter 5) have drastically decreased the riverine sand load to coastal waters, increasing the *in situ* demand for sand. Increased coastal erosion due to lack of riverine sand is reported to present major problems along the Romanian Black Sea coast and in the Caspian Sea (see Chapter 6). To protect the coast from erosion, various types of sea defences have been constructed in many countries with erosion problems. For example, in The Netherlands, where the largest proportion of the coast consists of dunes, 40 per cent of the coast is reinforced by defence works such as groynes, pile groynes and defences at the foot of dunes to protect them against the North Sea (Ministry of Transport and Public Works, 1990) and to avoid disastrous floods like that of 1953. Caution needs to be exercised in constructing defences designed to protect coasts since, by interfering in the natural erosion/deposition balance, these can create their own unwanted knock-on effects in neighbouring localities, leading to increased erosion or silting.

Habitat loss and degradation

Other physical modifications affect mainly the landward part of the coastal zone. The most important are those resulting from space-consuming developments of settlements, industries, agriculture (eg, reclamation of arable land), ports, military installations, tourism and recreation (eg, golf courses; see Chapter 25). Among other effects, these can lead to the creation of vast built-up areas at the expense of natural habitats (eg, dunes, saltmarshes) and, as a result, damage or destroy a substantial part of the natural coastline's habitats (CEC, 1991). In France, for example, 15 per cent of natural areas on the coast have disappeared since 1976 and are continuing to do so at the rate of 1 per cent a year. Italy, which had around 700 000 hectares of coastal marshes at the end of the last century, had no more than 192 000 hectares in 1972 and has fewer than 100 000 hectares today. Some estimates suggest that about one third of the coastal dunes in northwestern Europe and three quarters in the western Mediterranean have disappeared (Gehu, 1985). Such large-scale habitat destruction will inevitably lead to a decline in species distribution and abundance (see Chapters 9 and 29).

Contamination and coastal pollution

Excessive loads of contaminants are a direct result of humans being unable, or unwilling, to control or limit their polluting activities effectively. Pollution of the coastal zone is primarily a result of the contaminant load being discharged into receiving waters, resulting in such deleterious effects as harm to plants and animals, hazards to human health, hindrance to marine activities (including fishing), impairment of quality for use of sea water and reduction of amenities.

Type and origins of contaminants

Despite the coastal zone's ability to reduce the harmful effects of some contaminants (see the previous section), coasts are also vulnerable to pollution since wastewater is often discharged directly or indirectly into sheltered and shallow coastal waters with poor mixing. In such areas, large quantities are not needed to overwhelm the local receiving waters. The sea is often the ultimate recipient of many contaminants from many sources:

- directly from rivers bearing contaminants flowing into the sea;
- through direct discharges from outfalls or runoff from land into estuaries and coastal waters;
- through the dumping of waste such as sewage sludge, chemical waste and dredged spoil;
- from atmospheric deposition, which, in quantitative terms, can be as significant as that from other sources;.
- discharges from ships.

The most important contaminants in the coastal zone are organic matter, synthetic organic compounds (eg, PCBs and pesticides such as DDT and residues), microbial organisms, nutrients (mainly nitrogen and phosphorus), oil, litter, and, generally to a lesser extent, heavy metals (eg, cadmium, mercury and lead) and radionuclides.

Routine operations of ships, as well as shipping accidents (see Chapter 18), are regular, and often major, point sources of oil and risks for the introduction of unwanted organisms from ballast water. The contribution from oil and gas exploration and the routine operation of rigs is generally a minor source of contaminants in the coastal zone. While loads of contaminants from atmospheric deposition can be very significant over the whole surface area of individual seas (for example, inorganic nitrogen and some heavy metals, such as cadmium, in the Mediterranean and northern seas), their proportion of the total load to the coastal zone from all sources is likely to be relatively small.

Catchment management

In much of Europe there is a lack of effective catchment management, control and regulation through which a reduction of coastal zone pollution could be achieved (see Chapter 6). Many of Europe's seas have large multinational catchments (see Map 6.1), which may include states with no coast. Contaminants within the seas may therefore originate hundreds or even thousands of kilometres away from the sea. Hence, effective catchment management and pollution control measures require concerted international efforts and cooperation. This relates not only to water quality but also to water quantity, since the manipulation of river flows can have secondary effects on the receiving seas. There is also exchange of water, and hence contaminants, between many of the seas.

It is clear, therefore, that a significant contributory cause of coastal zone pollution in many of Europe's seas is inadequate control of diffuse sources of contaminants. Diffuse sources also include discharges of contaminants in industrial fume stacks and atmospheric emissions from some agricultural practices, such as ammonia arising from intensive livestock farming, which occur well inland of the coast.

Wastewater disposal

In many sea catchments there is inadequate treatment of domestic sewage, with crude, untreated sewage often being discharged directly to coastal waters, estuaries and rivers. In many of Europe's seas there are also uncontrolled and/or inadequately treated discharges of industrial wastes which can significantly add to the total loads of contaminants. Due to the breakdown of organic matter, such discharges can cause immediate problems to marine life from high oxygen demands in the water column and sediments and through toxic effects of ammonia. The introduction of sanitary litter to coastal waters with wastewater discharges often decreases considerably the amenity value of impacted areas.

Of immediate concern to humans is the presence of large numbers of pathogens that can cause illness and disease (eg, hepatitis and typhoid). Also associated with the presence of pathogens is the potential contamination of seafood, particularly shellfish, which in some seas can be commercially important as well as a source of food for fish and birds. There are many examples where the beaches of Europe's seas have been closed to bathers because of contamination by human pathogens arising from inadequately treated sewage. At present EU standards for bathing waters relate to contamination levels of indicator faecal bacteria (*E coli*) and other pathogens (*Salmonella* and entero-viruses) (CEC, 1993).

Sewage is also an important source of nutrients, and is therefore implicated in the problem of eutrophication. However, the major input of nutrients to coastal zones is often from tributary rivers. Nutrients in rivers arise from the application of nitrogen and phosphorus in fertilisers used in agriculture; phosphates are also used as 'builders' in detergents. Intensive farming practices and spreading of animal slurry contribute significant inputs of nutrients into rivers. Atmospheric inputs, soil erosion, release of nutrients from sediments and organic plant wastes also represent diffuse sources of nutrients. In urban catchments, point sources of nutrient input will be more important, and derive from sewage treatment works and industrial wastes. Another important point source of nutrients into the coastal zone of some seas is from fish-farm sites (eg, in the Baltic and Norwegian seas).

The fate and impact of nutrients

The environmental impacts of anthropogenic nutrients entering the sea have recently been subject to much public interest and scientific debate. Nutrients, however, are not pollutants *per se*, but stimulate and control plant growth when other environmental factors are optimal. Much evidence has accumulated to suggest that input of combined nitrogen (predominantly as nitrate) to the sea is the single most important factor contributing to 'cultural eutrophication', being the enrichment of a waterbody by nutrients of anthropogenic origin causing an accelerated production of plant material which during its presence and decay produces a series of undesirable phenomena that restrict the use of the water resource. Nitrogen is generally considered to be the limiting nutrient for primary productivity and biomass of algae in waters outside estuaries and inner coastal waters (Codispoti, 1989; Jickells et al, 1991). In estuaries, phosphorus, nitrogen or both can temporarily limit production.

There is scientific evidence to support the view that eutrophication-related phenomena in nearshore waters appear more frequently and are more serious than in the past, for example: the Adriatic Sea (Marchetti, 1990); Swedish west coast (Lundälv, 1990); Danish waters (Miljøstyrelsen, 1991); Black Sea (Mee, 1991); and in many areas of the Baltic Sea. The Paris Commission has also reported eutrophication problem areas within the North Sea, mainly along the eastern seaboard, such as within Belgian and Dutch coastal waters, and the German Bight (OSPARCOM, 1992). This is associated with increased inputs of anthropogenic nutrients from land into coastal waters.

Eutrophication has been associated with a number of effects, such as an increased prevalence of nuisance and toxic algal blooms within coastal zones. Physico-chemical factors other than nutrients (eg, salinity, temperature, light conditions and turbulence) can be involved in bloom formation, as can biological factors such as decreased zooplankton grazing on the algae. A recent study (Larsen et al, 1993) showed that zooplankton growth rates, and thereby the grazing potential, were markedly reduced by toxins released during an algal bloom, leading to self-propagation of the bloom. Dense blooms of colonial or chain-forming species (eg, species of *Nodularia*, *Phaeocystis* and *Chaetocerus*) can result in drifts of cells on the sea surface or on the beach, or slimy deposits on fish nets. Once extensive phytoplanktonic blooms die, they fall to the sea bed, where the resultant microbial degradation results in massive deoxygenation of the bottom waters causing wide-scale benthic mortality.

Some species of several phytoplankton groups (eg, diatoms, dinoflagellates, cyanobacteria) produce toxins, which are known to cause the death of marine invertebrates, for example mussels, that filter them from the water. In other cases the accumulated toxin may represent a hazard to organisms further up the food-chain (eg, fish, sea-birds, marine mammals and humans). The most widespread algal toxins are those causing paralysis (paralytic shellfish poisoning, PSP), diarrhoea (diarrhoeic shellfish poisoning, DSP), amnesia (amnesic shellfish poisoning, ASP), haemolysis and dermatitis. Although not yet documented for Europe, there are good reasons for awareness about ASP since, like the common PSP, it can cause very dangerous intoxications of humans. Amnesic shellfish poisoning was first discovered in Canada in 1987, where more than 150 people became seriously ill (with three deaths) as a result of eating infected mussels, *Mytilus edulis* (Quilliam et al, 1989).

There are also often changes in the composition of macroalgal community structure, from the perennial slower growing brown algae to the more ephemeral green algae such as *Enteromorpha* species.

The clouds of algal slime that occasionally arise in the northern Adriatic Sea (in particular, during the summers of 1988 and 1989, when they were washed up along tourist beaches) have received considerable public attention and are of great concern to the tourist industry. At present the underlying mechanisms forming and maintaining such large masses of slime are unknown and no rigorous hypothesis has yet been presented explaining this peculiar phenomenon (Battaglia, 1990). Eutrophication does not seem to be the prime factor involved and it is hypothesised that thermal anomalies following high insolation and calm wind conditions in the preceding spring may be of importance.

Factors affecting the fate of the riverine flux of dissolved nitrogen (mostly as nitrate) into the coastal zone are therefore very important for determining the eutrophic state of these waters. These factors include a series of complex biological and biochemical reactions of which denitrification is a very important process, transforming nitrate under anoxic conditions into the non-eutrophying gases N_2O and N_2. Estimates by Seitzinger (1988) suggest that about 45 per cent of the gross flux of total nitrogen to European estuaries is being lost in denitrification, thus reducing the risk of marine eutrophication. Denitrification in ocean shelf and estuarine systems of the world has been estimated by Christensen et al (1987) at 62 million tonnes of nitrogen per year. Although no estimate is available for the European coastal zone, it is clear that a significant part of an estimated European riverine gross flux of nitrogen (between 2.5 and 6.5 million tonnes per year) is lost to the atmosphere from the coastal zone before reaching the open oceans. However, this large loss of nitrogen, of primarily agricultural origin, does not stop many European coastal waters suffering from eutrophication.

Environmental impact of other contaminants

Synthetic organic compounds reach the coastal zone from a variety of sources, but in general riverine loads are higher than inputs from other sources. Marine organisms have the ability to accumulate some contaminants discharged into marine waters. In particular, some synthetic organic compounds such as organochlorine pesticides and PCBs have a high bioaccumulation potential and, though present at low concentrations in the water column, they can build up through the food-chain to form elevated concentrations in animals. It may be possible, therefore, for some contaminants to reach such high levels in food types as to pose a threat to human health. In some seas (eg, the Black Sea; see Chapters 6 and 24) the presence of pollutants has been linked with a decline in populations of commercial fish species, seals, porpoises, dolphins and whales. For example, studies on the three Baltic Sea seal species have shown pathological changes in the female reproductive tract, in adrenals, bone tissue and the alimentary tract of the animals (Olsson et al, 1992). On the basis of contaminant concentrations in the Baltic Sea, PCBs were suspected of being responsible for the occurrence of sterility among the seals. A similar suspicion has been reported for the observed decrease of the seal population and its reproductive failure in the Dutch Wadden Sea (Reijnders 1980; 1986). However, although elevated levels of chlorinated organics have been found in marine mammals in the North Sea, there is no direct evidence to relate levels found to any potential harm (Zabel and Miller, 1992).

Sediment may also be an important physical contaminant (eg, in the disposal of dredged spoil) through the restriction of light to, and the blanketing of, benthic communities.

Oil is a very noticeable contaminant which can adversely affect the quality of the coastal zone, in terms of both amenity and effects on organisms. Oil discharges from anthropogenic sources to the seas represent a significant coastal pollution problem despite the fact that global oil discharges decreased from an estimated 3.2 million tonnes in 1981 to 2.35 million tonnes in 1990 (GESAMP, 1993). This estimate is highly influenced by numbers and size of tanker accidents each year. The observed general decrease in the number of oil spills and the quantity of oil spilled (Figure 18.1, p 387) is due mainly to measures taken against oil spills from shipping by international conventions (see Chapter 6), whereas inputs of oil spills from land-based sources, in particular spills of petroleum hydrocarbons, are of increasing concern, causing sub-lethal effects in, for example, crustaceans and echinoderms. Sea-birds, mammals (eg, sea otters and polar bears) and caged fish in aquaculture farms are often the most conspicuous victims of oil pollutions. (See also Chapter 18.)

Metals also enter the marine environment from a variety of sources, with rivers generally predominating. Heavy metals occur naturally in sea water and sediments at levels reflecting local geology and geochemistry, and are generally present in such low concentrations in the sea that they do not constitute a major threat to marine organisms except at some contaminated sites.

Nuclear weapons testing, the dumping of radioactive waste, and discharges from nuclear power stations and reprocessing plants add to the radioactivity of the sea caused by naturally occurring radionuclides. The latter two artificial sources are the most important to the coastal zone. Anthropogenic inputs of radioactivity are generally considered to be less than 1 per cent of natural radioactivity, although the radionuclides involved are mostly of a different nature.

Litter

The Coastwatch Europe surveys (Coastwatch, 1994) have demonstrated that litter is a major problem within the coastal zone of all the seas surveyed (see Box 6A). Litter, in particular that made of synthetic materials such as ropes, nets,

Box 35A Main European initiatives concerning coastal zones

1 1973 Council of Europe – Resolution (73) 29 of the Committee of Ministers on the protection of coastal zones.

2 1973 Council Declaration on the European Community Environmental Action programme, recognising that conflicting economic development activities in coastal areas leads to the deterioration of resources having economic and ecological value.

3 1974 Helsinki Convention on the Protection of the Marine Environment of the Baltic Sea Area.

4 1974 Oslo Convention for the Prevention of Marine Pollution by Dumping from Ships and Aircraft.

5 1975 Council Directive 76/160/EEC concerning the quality of bathing water.

6 1975 Barcelona Convention for the Protection of the Mediterranean Sea against Pollution, and related protocols.

7 1976 OECD Recommendation of the Council on the principles relating to the administration of coastal zones.

8 1978 Paris Convention for Prevention of Marine Pollution from Land-Based Sources.

9 1979 Council Directive 79/923/EEC on the quality required of shellfish waters.

10 1979 Council Directive 79/409/EEC on the conservation of wild birds.

11 1981 European Coastal Charter, adopted by the Conference of Peripheral Maritime Regions of the EEC.

12 1982 Signing of the joint declaration on the Protection of the Wadden Sea by Denmark, Germany and The Netherlands.

13 1983 Council of Europe – Resolution of Planning Ministers on coastal areas.

14 1985 Council Directive 85/337/EEC on the assessment of the effects of certain public and private projects on the environment (Environmental Impact Assessment, EIA).

15 1988 Establishment of the North Sea Task Force (NSTF).

16 1991 Arctic Monitoring and Assessment Programme (AMAP) covering the Norwegian, White and Barents seas.

17 1991 Council Directive 91/271/EEC concerning urban wastewater treatment.

18 1991 Council Directive 91/676/EEC concerning the protection of water against pollution caused by nitrates from agricultural sources.

19 1992 Convention for the Protection of the Black Sea.

20 1992 Helsinki Convention on the Protection of the Marine Environment of the Baltic Sea Area (will upon entry into force supersede the 1974 Helsinki Convention) and the Baltic Sea Joint Comprehensive Programme.

21 1992 Paris Convention for the Protection of the Marine Environment of the North East Atlantic (will upon entry into force supersede the 1974 Oslo Convention and the 1978 Paris Convention).

22 1992 Council Resolution on intergrated management of coastal zones (OJ C 59, 6.3.1992).

23 1992 OECD Recommendation of the Council on the integrated management of coastal zones.

24 1992 Council Directive 92/43/EEC on the conservation of natural habitats and of wild fauna and flora.

25 1993 Odessa Declaration for improved assessment and monitoring of contaminants and the development of comprehensive plans for the restoration, conservation and management of the Black Sea.

26 1993 Resolution of the Council and the representatives of the Governments of the Member States, meeting within the Council, on a Community programme of policy and action in relation to the environment and sustainable development (EC Fifth Action Programme, 93/C 138/01).

plastic bags and packaging rings and straps, is increasingly being introduced into the marine environment, and is particularly noticeable in the coastal zone, where it tends to accumulate from the action of wind and waves. Such litter often floats in the sea, entangling and killing fish, birds and marine mammals. The presence of some forms of litter, such as sanitary, medical and glass, particularly on beaches, poses a direct threat to human users. The resultant decrease in amenity value of beaches also decreases the recreational attractiveness of the areas to both local people and visitors.

GOALS AND STRATEGIES FOR COASTAL ZONE MANAGEMENT

There is often a conflict of uses within the coastal zone, in particular where one use might have an adverse impact on another actual or potential use. Coastal defences, navigation, harbours and marinas, for example, would potentially affect the coastal zone aquatic ecosystem – directly through land and habitat loss and impacts on water quality, but also indirectly through the modification of water currents and sediment deposition patterns. There is therefore a need for strategic plans for coastal zones so that developments can proceed in an environmentally acceptable and predictable way. In many countries legislative and regulatory responsibilities for the coastal zone lie with a number of

bodies. Bringing these bodies together within a unified management plan would go some way to ensuring sustainable development of the coastal zone.

Therefore it is perhaps surprising that at present no comprehensive coastal zone management (CZM) scheme exists for Europe. Following the passing in the USA of the 1972 Act on CZM, several European maritime nations (eg, France, UK, The Netherlands, Denmark) have implemented national legislation for CZM. However, it is important to distinguish between measures which take an integrated view of planning and management at the coast, and those which cover only certain aspects of CZM (eg, coastal protection). Progress in the development of international or European CZM initiatives has, however, been slow. Important European initiatives are summarised in Box 35A.

Integrated coastal zone management has been extensively discussed in recent literature. Gubbay (1993) summarised CZM as 'the basic principles to promote the environmentally sensitive use of the coastal environment and to introduce an element of strategic planning for coasts'. For an integrated CZM process to be successful, three basic elements are required:

- development of an understanding of the coastal zone as a system made up of interlinked components and processes;
- using this knowledge to create a plan for its best use; and
- implementation and enforcement of the plan.

Management plans need to be designed to solve coastal problems through the achievement of a set of stated sustainability goals, which should include the maintenance and improvement of the usefulness of the coastal zone to humankind taking into account its use as habitat for plants and animals. The overall goals of integrated CZM need to promote sustainable use and respect the precautionary principle by including the following strategic principles (modified after Ketchum (1972) and Henningsen (1991)):

- determine human desires for using the coastal zone;
- determine the (carrying) capacity of the coastal zone to meet these desires (see Chapter 25);
- determine to what extent various uses balance capacities and how uses affect the coastal zone;
- solve eventual conflicts of uses and capacities pro-actively;
- rely on reliable, comparable and quantitative data for decisions and control measures;
- rely on active participation of a well-informed public;
- integrate environmental priorities into coastal development objectives at the local, national and international levels;
- integrate seaward and landward aspects; and
- encourage the establishment of CZM ecological or administrative units (eg, as already exists with the Wadden Sea collaboration between The Netherlands, Germany and Denmark).

Other, more specific goals might relate to particular problems of the individual coastal zones. Goals would have to be set in light of what would be achievable in the short and long terms with the likely resources (monetary and technical) available to the responsible authorities. Priority issues and problems should therefore be clearly defined within each coastal area, identifying those which would give relatively immediate environmental and economic returns and provide the best value for money. However, the severity of a problem should also help determine where actions should be taken even when the resolution of the problem is envisaged over a long time-scale.

The application of the strategies described in this section require a large amount of resources for implementing and monitoring. Without appropriate administrative arrangements in place, and without appropriate financial resources, there is little possibility of achieving any desired goal of reduced coastal zone pollution.

CONCLUSIONS

- For many coasts, the scale of environmental problems has not been fully quantified or understood. A first possible step towards changing that situation is to undertake detailed quality and status assessments. These are already periodically undertaken in many seas (eg, the North and Baltic seas). In other coasts and seas, for example in Eastern Europe, very few baseline data exist or, if they do, they have not been collated in a coordinated way. For some of Europe's seas therefore, the starting point for improvement would be to establish a baseline of quality in terms of input loads, contamination levels and biological status and effects. For other areas it is already clear what the problems are and the action required.
- Problems for priority action then need to be identified. They would have to be prioritised against agreed criteria and against the need for action on other environmental issues (eg, atmospheric pollution). Such criteria might include:
  - impact on human health;
  - the economic value of resources under threat;
  - the risk or closeness of the marine ecosystem to irreversible change or damage; and
  - the international importance of the habitat or species under threat.
- Real improvements in the environmental quality of coastal areas require the cause of the problems to be addressed: the direct pressures and human activities. Of particular concern are increased urbanisation, tourism and recreation. Public policies to help improve the compatibility of such human activities with the environment need to be carefully designed, taking local economic needs into account. Particular attention is needed where such public policies are not well established (eg, in Central and Eastern Europe).
- To achieve improvements in the environmental quality of coastal zones, there is a clear need for international collaboration and agreement. This is needed not only between the riparian states around a particular sea, but also of states within the catchment, and even those upwind, in the case of atmospheric transport of contaminants. Examples of such international cooperation, which already exists, can be found formalised in a number of conventions covering several of Europe's seas (eg, the Barcelona, Paris and Helsinki conventions).
- Associated with any action plan for improvement, there would, of course, be a monetary cost. Actions and priorities need to be assessed on the basis of both short-term and long-term goals, since, in many seas, improvements may not be detectable for many years. This should be reflected in timetables and goals, and in the assessment of cost-effectiveness.

REFERENCES

Battaglia, B (1990) Opening remarks. In: Barth, H and Fegan, L (Eds) *Eutrophication-related phenomena in the Adriatic Sea and in other Mediterranean coastal zones*, pp 9–11. Water pollution research report 16. Commission of the European Communities, Brussels.

CEC (1991) *Europe 2000: outlook for the development of the Community's territory*. Commission of the European Communities, Luxembourg.

CEC (1993) *Quality of bathing water, 1992*, Commission of the European Communities, EUR 15031 EN, Luxembourg.

Christensen, J P, Murray, J W, Devol, A H and Codispoti, L A (1987) Denitrification in continental shelf sediments has major impact on the oceanic nitrogen budget. *Global Biogeochemical Cycles* 1, 97–116.

Coastwatch (1994) *Coastwatch Europe: international results summary of the 1993 survey.* Coastwatch Europe Network, International Coordination, Trinity College, Dublin.

Codispoti, L A (1989) Phosphorus vs. nitrogen limitation of new and export production. In: Berger, W H, Smetacek, V and Wefer, D (Eds) *Productivity of the Ocean: Past and Present*, pp 377–94. Dahlem workshop report LS44. John Wiley & Sons, Chichester.

ECCC (1991) The golden fringe of Europe. Ideas for a European coastal conservation strategy and action plan. In: *Proceedings of the European Coastal Conservation Conference (ECCC) 19–21 November 1991*, pp 70–85. Scheveningen/The Hague.

GESAMP (1993) *Impact of oil and related chemicals and wastes on the marine environment*. IMO/FAO/UNESCO/WMO/WHO/IAEA/UN/UNEP Joint Group of Experts on the Scientific Aspects of Marine Pollution. Reports and Studies No GESAMP 50. London.

Gehu, J M (1985) *European dune and shoreline vegetation*. Nature and Environment Series no 32. Council of Europe, Strasbourg.

Gubbay, S (1993) *Coastal zone management and the North Sea ministerial conferences*. Report to the World Wide Fund For Nature from the Marine Conservation Society. UK.

Henningsen, J (1991) Integrated coastal management: challenges and solutions. In: *Proceedings of the European coastal conservation conference 19–21 November 1991*, pp 24–25. Scheveningen/The Hague.

Hoogland, J R (1992) Coastal defense and nature conservation: partners in development. In: *Proceedings of the European coastal conservation conference 19–21 November 1991*, pp 43–45. Scheveningen/The Hague.

Jickells, T D, Blackburn, T H, Blanton, J O, Eisma, D, Fowler, S W, Mantoura, R F C, Martens, C S, Moll, A, Scharek, R, Suzuki, Y and Vaulot, D (1991) Group report: What determines the fate of materials within ocean margins? In: Mantoura, R F C, Martin, J-M and Wollast, R (Eds) *Ocean margin processes in global change: report of the Dahlem Workshop on Ocean Margin Processes in Global Change, Berlin, 1990*, pp 211–34. John Wiley & Sons, Chichester.

Ketchum, B H (Ed) (1972) *The water's edge: critical problems of the coastal zone.* MIT Press, Cambridge, Massachusetts, and London.

Larsen, J, Hansen, P J and Ravn, H (1993) Giftige alger i danske farvande. Havforskning fra Miljøstyrelsen Nr 28. Miljøstyrelsen, Copenhagen.

Lundälv, T (1990) Effects of eutrophication and plankton blooms in waters bordering the Swedish west coast – an overview. In: Lancelot, C, Billen, G and Barth, H (Eds) *Eutrophication and algal blooms in North Sea coastal zones, the Baltic and adjacent areas: prediction and assessment of preventive actions*, pp 195–213. Water Pollution Research Report 12. Commission of the European Communities, Brussels.

Mantoura, R F C, Martin, J-M and Wollast, R (1991) Introduction. In: Mantoura, R F C, Martin, J-M and Wollast, R (Eds) *Ocean margin processes in global change: report of the Dahlem Workshop on Ocean Margin Processes in Global Change, Berlin, 1990*, pp 1–3. John Wiley & Sons, Chichester.

Marchetti, R (1990) Algal blooms and gel production in the Adriatic Sea. In: Barth, H and Fegan, L (Eds) *Eutrophication-related phenomena in the Adriatic Sea and in other Mediterranean coastal zones*, pp 21–42. Water Pollution Research Report 16, pp 21–42. Commission of the European Communities, Brussels.

Mee, L D (1991) *The Black Sea: a call for concerted international action*. Report of a UNEP/IOC/IAEA mission. IAEA-Marine Environmental Studies Laboratory, Monaco.

Miljøstyrelsen (1991) *Environmental impacts of nutrient emissions in Denmark*. Redegørelse fra Miljøstyrelsen, Nr 1. Copenhagen.

Ministry of Transport and Public Works (1990) *A new coastal defence policy for the Netherlands*. Ministry of Transport and Public Works, The Hague.

Odum, E P (1971) *Fundamentals of ecology*. 3rd edition. W B Saunders Company, Philadelphia, London and Toronto.

Olsson, M, Karlsson, B and Ahnland, E (1992) Seals and seal protection: a presentation of a Swedish research project. *Ambio* 21, 494–6.

OSPARCOM (1992) *Nutrients in the Convention Area*. Oslo and Paris Commissions, Paris.

Quilliam, M A, Sim, P G, McCulloch, A W and McInnes, G (1989) High-performance liquid chromatography of domoic acid, a marine neurotoxin, with application to shellfish and plankton. *International Journal of Environmental and Analytical Chemistry* 36, 139–54.

Reijnders, P J H (1980) Organochlorine and heavy metal residues in harbour seals from the Wadden Sea and their possible effects on reproduction. *Netherlands Journal of Sea Research* 14, 30–65.

Reijnders, P J H (1986) Reproductive failure in common seals feeding on fish from polluted coastal waters. *Nature* 324, 456–7.

Seitzinger, S P (1988) Denitrification in freshwater and coastal marine ecosystems: ecological and geochemical significance. *Limnology and Oceanography* 33, 702–24.

UNESCO (1987) Reports and Studies No 32. *Land/sea boundary flux of contaminants: contribution from rivers*. IMO/FAO/UNESCO/WMO/WHO/UN/UNEP Joint Group of Experts on the Scientific Aspects of Marine Pollution (GESAMP).

US Commission on Marine Science, Engineering and Resources (1969) *Our Nation and the Sea*. United States Government printing office, Washington DC.

Zabel, T F and Miller, D G (1992) Water quality of the North Sea: concerns and control measures. *Journal of the Institute of Water and Environmental Management* 6, 31–49.

Fly-tipped refridgerator, South Glamorgan, Wales
Source: Michael St Maur Sheil

36 Waste production and management

THE PROBLEM

The rapidly increasing quantities of waste generated in European countries are a major concern for Europe's environment (see Chapter 15). It is estimated that Europe produces annually over 250 million tonnes of municipal waste and more than 850 million of industrial waste. The annual average rate of increase of these wastes since 1985 in the OECD European area is estimated at around 3 per cent. Present disposal and processing capacity is probably not sufficient to deal with the expected growth. Often, existing facilities are not adequate to ensure acceptable environmental standards. The siting of new facilities usually encounters considerable opposition from local people concerned with the potential risks for their local communities.

Major constraints to safe management are imposed by significant changes in the quality of waste. Increasing amounts of discarded products contain substances now recognised to be toxic or highly toxic. Improper management and illegal dumping of waste, particularly hazardous and toxic waste, pose increasing threats to the environment and human health. Transfrontier movements of such waste from countries with strict regulations towards less-regulated countries increase the potential environmental risk of waste disposal in countries with insufficient control. There are increasing attempts to bring these problems under control by introducing national and international legislation.

Waste issues in Europe become apparent when the environmental impacts of waste management practices are examined:

- Disposal of waste by landfill, which is the main waste disposal route, if not properly managed, can cause leaching of contaminants into soil and groundwater.
- Landfill sites occupy considerable space with significant impacts on landuse and landscape, in some cases, however, landfilling can be used to restore derelict land, such as old mineral workings.
- Incineration of waste, unless properly regulated, leads to emissions of toxic substances into the atmosphere and to the production of large amounts of contaminated ashes.
- Recycling implies the least load of emissions and saves materials, but it involves considerable sorting and treatment during which pollutants present in waste may be transferred to the environment or incorporated into new products.

However, it is the production of waste in the first place which causes major environmental impacts. Waste production implies the use of material and energy and the depletion of the Earth's renewable and non-renewable resources (see Chapter 13). Waste issues and their solutions are inevitably linked to production and consumption throughout all stages of the life-cycle of materials and the use of energy (see Chapter 12).

Waste is also regarded as a 'second generation' pollution problem to which other problems are eventually reduced. Since the 1970s, European countries have achieved important progress in reducing emissions into air and water from production processes by imposing strict emissions standards

Box 36A Waste management hierarchy

1 Reduce generation of wastes, eg, by more efficient processes in manufacturing, reduction of disposable material in consumer goods or increase of durability in products.
2 Separate usable components of the waste at their source, eg, by more efficient control of effluents from manufacturing processes, separation of paper, glass, plastic and metals by householders, or concentration of used tyres or oil at collection centres.
3 Re-use of waste products directly, if possible, eg, return of materials to the production process as in steel-making or cement kiln operations, burning of household wastes to recover energy or exchange of material which is a waste from one process but may be a material useful for another process.

4 Transformation or other physical or chemical treatment in order to recycle usable materials from waste, eg, magnetic separation of ferrous scrap from household waste and subsequent use of the material to prepare ferrous products, reclamation of non-ferrous metals from mixed industrial wastes by thermal processes, re-refining of waste lubricating oils, or distillation and regeneration of spent solvents.
5 Destruction of the waste by physico-chemical treatment or incineration, eg, neutralisation by mixing alkaline and acid wastes or burning of pumpable liquid waste or solid wastes.
6 Permanent storage of the waste in or on land.
7 Dumping at sea (to be avoided as far as possible).

Source: OECD

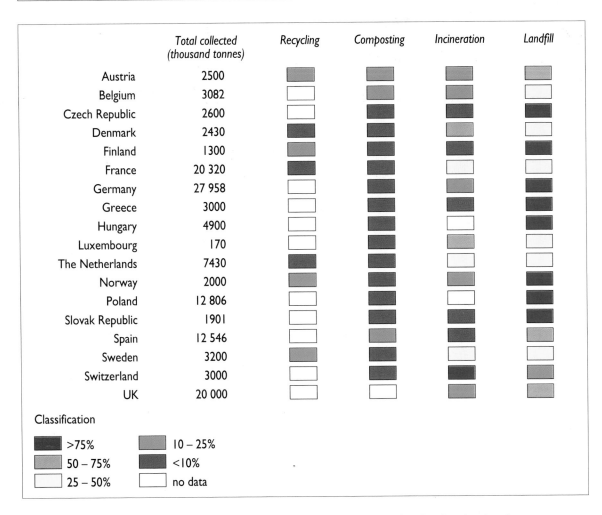

Table 36.1
Disposal of municipal waste in selected European countries, late 1980s
Source: OECD/Eurostat; EEA-TF; see *Statistical Compendium*

on conventional pollutants. However, as a result of their single-media mandate, environmental regulations have addressed air and water pollution problems separately. The result has been to move pollution problems to the least regulated environmental medium and the least controlled form of pollution. The implementation of emission control technologies has often resulted in increased amounts of solid waste from production processes. In addition, the concentration of hazardous substances in solid residues has increased.

European countries today recognise that reducing pollution loads requires an integrated pollution control strategy. A hierarchy of preferred options for waste management was adopted in 1976 by OECD countries (Box 36A). Reducing waste at source not only minimises the impact of waste treatment and disposal, it also enhances the efficient use of raw materials. However, despite the increasing emphasis on waste prevention, wastes have increased. Landfill and incineration, instead of recycling, are still the predominant practices in waste management, although differences exist between countries (Table 36.1). Waste often still escapes control or avoids strict regulations through transfrontier movement across European countries and from Europe to developing countries.

THE CAUSES

Waste production is one of the most revealing indicators of the interactions between human activities and environmental systems. Increased waste production in Europe follows production and consumption trends. Per capita production of municipal waste increases with living standards. More products and activities inevitably imply more waste. One of the most dramatic changes leading to increasing amounts of waste is the production and consumption of goods that last less time. However, the quantities of goods and services produced and consumed is only one side of the waste problem. Waste is often the result of inefficient energy and materials management by producers and consumers in all sectors of activity. Product design and manufacturing processes ultimately affect the quantity and quality of waste streams in all stages of the product life-cycle. In addition, consumer choice and product use play a major role. Finally, the environmental impacts of increasing quantities of waste

Table 36.2
Typical composition of uncleaned flue gases from waste incineration
Source: Barniske, 1989

Substance	Composition
H_2O	10 – 18 Vol. %
CO_2	6 – 12 Vol. %
O_2	7 – 14 Vol. %
CO	50 – 600 mg/m³
HCl	400 – 1500 mg/m³
HF	2 – 20 mg/m³
SO_2	200 – 800 mg/m³
NO_x	200 – 400 mg/m³
Dust	2000 – 15 000 mg/m³

Table 36.3
Solid residues from waste incineration
Source: Barniske, 1989

Residue	Specific amounts (dry) in kg/t waste	
Bottom ash	250 – 350	
Filter dust	20 – 40	
Residues from gas cleaning processes	Excluding dust	Including dust
– Wet process	8 – 15	30 – 50
– Semi-dry process	15 – 35	40 – 65
– Dry process	25 – 45	50 – 80

Figure 36.1
Waste disposal costs in selected European countries, 1992 ($ per tonne estimates)
Source: FEAD, personal communications, 1994

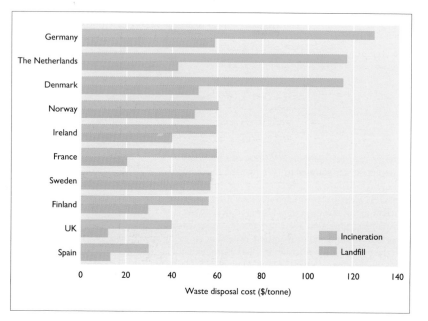

are strongly influenced by disposal methods and practices.

European countries rely predominantly on landfill for disposing of their waste. In the past, most landfills were rudimentary waste dumps without the necessary requirements to protect the soil and groundwater. This has frequently resulted in contamination where the adverse effects could be reversed only at extremely high costs. Landfills produce leachate that, if not collected, contaminate the groundwater with heavy metals such as arsenic, cadmium, copper, lead, manganese and zinc, as well as ammonia and organic compounds such as PCBs. In addition, landfills are one of the largest contributors to the total emissions of methane, a greenhouse gas, into the atmosphere. Measures to control landfill emissions are now specified in most regulations adopted by European countries. These normally require providing impermeable bottom-layers and building leachate collection systems and treatment plants as well as gas recovery and destruction systems. Continuous monitoring and care after closure are also recommended, although many European landfills currently in operation are not properly equipped for these activities. The discrepancy across European countries regarding the state of landfill sites is reflected in the scanty available information and low level of monitoring activities.

The second most common method of waste disposal after landfill is incineration. In some European countries, waste incineration increased considerably during the 1970s but has remained constant since the mid-1980s due to public opposition to new incineration plants combined with increasing capital and operation costs. Incineration is considered most suitable to reduce waste volume (by as much as 80 per cent) and to recover energy from solid residuals. However, concerns exist about the health and environmental effects of toxic substances that are released or formed during incineration (Table 36.2). Emissions from incinerators include metals such as cadmium, lead, mercury, chromium, tin and zinc. Organic compounds are released as uncombusted residues of chemical substances present in waste or generated during combustion. Incinerators also emit large quantities of compounds generally referred to as acid gases, including HCl, HF, SO_2 and NO_x. State-of-the-art incineration technologies can substantially reduce the emissions of such contaminants into air, but most of these substances will be found in solid residues from incinerator plants. Incineration produces vast quantities of fly and bottom ashes (Table 36.3) which contain hazardous metals (eg, cadmium and lead) and organic compounds (eg PAHs, PCBs, PCDDs and PCDFs).

Recycling capacity varies considerably across countries and between waste streams. In OECD Europe, one third of paper, glass and metal in municipal waste is submitted to recycling operations and one quarter of the organic fraction is incinerated to reclaim energy (Yakowitz, 1992). Recycling increased steadily between 1975 and 1985 but has remained constant or declined since 1985. The efforts to increase recycling rates have been strongly offset by increasing waste production trends and by the difficulties of finding a market for recovered materials. These difficulties have become evident in the most far-reaching German attempt to push the limits of recycling (see Chapter 15).

Economic factors play a critical role in choices to avoid generating waste or to manage waste production through one of the available options. Virgin materials are generally more competitive than secondary or recycled ones, although the price of virgin materials usually does not fully incorporate the environmental costs of extraction and processing nor the cost of their disposal. Different patterns of waste management across countries are also clearly linked to trends in the costs associated with different options for managing waste. In most European countries disposal of municipal waste in landfill is still the cheapest option (Figure 36.1). In countries where the cost of disposal on land is relatively lower, a greater proportion of waste goes to landfill. In other countries, tighter environmental standards have raised the price of waste disposal by landfill, particularly for toxic and hazardous waste, pushing industries towards treatment and recycling. However, the export of hazardous waste still remains a more competitive option than waste minimisation. According to OECD, the aggregate annual marginal savings to waste generators achieved in Europe through transfrontier waste movements represents roughly ECU200 to 250 million and the potential avoided costs average ECU250 per tonne, considering the overall cost of packaging, labelling and transportation. This is true even considering distances over 5000 km (OECD, 1991).

Different levels of regulatory strictness and disposal costs between European countries encourage increased waste movements towards less-regulated countries and particularly from Western to Eastern European countries. In the OECD European area there are some 100 000 annual transfrontier movements corresponding to more than 2 million tonnes of hazardous waste. In countries with stringent regulations, the wastes most likely to be exported, legally or illegally, are the highly hazardous ones, since these are the most expensive to treat domestically.

THE CONSEQUENCES

The lack of reliable data on the quantities and quality of various waste streams, their sources and disposal routes throughout Europe makes it impossible to quantify the overall environmental impact of waste production and management. However, indications of the dimension of the problems can be derived by the available data on waste production and management reported by national agencies and examined in Chapter 15. The assessment of the state of the environment (Part II of this report) shows that all environmental media are affected by increasing quantities of waste and improper disposal.

Past improper waste management practices have significantly affected soil and groundwater. All European countries are faced with the current and potential risk of old waste dumps. Although data on current soil contamination are not available throughout the whole European territory, some countries have started to register contaminated sites and to identify those which require immediate action. More than 55 000 are currently registered in only six countries of the European Union, more than 22 000 of which are in critical conditions (see Table 7.5). These sites are contaminated by heavy metals, organic chemicals, oils, pesticides, radioactive materials, asbestos and other hazardous minerals. Extrapolating to the whole of Europe, the total contaminated land area has been estimated to be between 47 000 and 95 000 km², between 1000 and 3000 km² of which originate from landfill (Franken, in press). The area of groundwater potentially polluted by point sources is estimated to be less than 1 per cent of the European territory. However, this area is often close to points of groundwater abstraction, causing significant threats to public water supply (Chapter 5) .

Potential costs of clean-up are extremely difficult to determine because they depend on the type and extension of pollution as well as on the characteristics of the site. They also depend on clean-up standards and techniques for soil decontamination. Estimated remediation costs for the most critical sites in the European Union are around ECU27 billion for a 15-year long-term plan (Carrera and Robertiello, 1993). However, the cost of remedial actions already carried out in countries such as Germany (ECU228 million) or the Netherlands (ECU1300 million) show that this figure underestimates the costs considerably (Table 7.5). Waste-related land and groundwater contamination is expected to be even more serious in Central and Eastern Europe due to past improper waste

management and the flows of toxic waste from Western to Central and Eastern European countries (see Chapter 15). When an accurate assessment of waste-related land contamination becomes possible, the true high costs of clean-up will be able to be assessed, which in many cases might prove to be prohibitive.

RESPONSES

European countries have adopted various regulatory measures to minimise waste and ensure its safe management. Different regulatory frameworks have emerged which establish:

1 programmes to encourage waste reduction, re-use and recycling;
2 standards and procedures to ensure safe storage, treatment and disposal; and
3 programmes to activate clean-up of contaminated sites.

In addition, international cooperation is being pursued to achieve control of transfrontier movement of hazardous waste.

Four guiding principles for the prevention and safe management of waste have emerged:

● the *prevention principle* establishes that generation of waste should be avoided at source;
● the *polluter pays principle* sets the cost of waste disposal to be borne by the waste generators;
● the *precautionary principle* suggests that measures should anticipate environmental harm to ensure that adverse environmental impacts do not occur;
● the *proximity principle* recommends that waste should be managed as near as possible to the source.

Table 36.4
Waste control options
Source: Alberti, 1992

Approaches	Measures	Examples of regulations
Effect-oriented measures	● Set limits on ambient concentration	Ambient quality standards Maximum Risk Limits
	● Clean-up of contaminated sites	Clean-up regulation Strict liability
Emission-oriented measures	● Restrictions on emissions from incineration	Emission standards Performance standards
	● Restrictions on landfill for hazardous substances	Specifications for landfill
	● Require adoption of Best Available Technologies for waste treatment	Integrated pollution control permit Best Available Technology
	● Restrictions on movement of hazardous waste	Manifest system for waste transport Transfrontier movement regulations
Source-oriented measures	● Set reduction targets for priority waste streams	Priority waste streams lists Waste streams reduction implementation plans
	● Require adoption of Best Available Technologies for production processes	Best Available Technology
Product-oriented measures	● Restrictions on products and packaging materials	Regulations on products – labelling – disposal
	● Ecolabelling of products	Regulation on packaging – products – waste
	● Ban or phase-out of certain hazardous substances	International conventions banning certain substances (eg, CFCs)
	● Restrictions on the use of certain substances in manufacturing processes	Phase-out of toxic substances from packaging materials

Control options

At each stage in the product life-cycle, from production to final disposal, various control options can be considered to reduce the amount and toxicity of waste. These measures are oriented to regulate products, sources, emissions and effects (Table 36.4). Priority is given to five strategies summarised by the so-called 5R approach recommended by the Senior Advisers to ECE Governments on Environment and Water Problems (UNECE, 1992). These are based on: reduction, replacement, recovery, recycling and reutilisation of industrial products, residues or waste.

Waste reduction

Waste can be avoided or reduced at source through the reformulation of products or processes (Box 36B). Clean products and technologies are meant to reduce waste from manufacturing as well as from final use of consumer goods. Product and process reformulation can also reduce the concentration of hazardous and toxic substances in waste streams by replacing them with other substances in the products and manufacturing process or by simply improving operational and maintenance procedures. Major results can be achieved in the chemical industry. Product reformulation can, for example, lead to the replacement of chlorinated solvents with water-based solvents. Process reformulation can simply require introducing closed-loop processes to replace cleaning systems. There are several examples of product, process and materials substitution among the various industrial sectors. Evaluating 500 case studies on the implementation of single or combined strategies adopted by industry, Huisingh (1989) found that product reformulation can result in 70 to 100 per cent reduction of certain hazardous waste streams, and can eliminate other emissions of toxic substances in air and water. Some of the most impressive results are among the producers of pharmaceuticals, organic chemicals, microelectronics, and automobiles, as well as photographic processing.

Preliminary results of a recent project conducted in The Netherlands – PRISMA: Project on Industrial Successes with Waste Prevention to assess waste prevention efforts by industries – show that successful reduction of hazardous substances in industrial waste streams can be achieved with substantial economic pay-back and other indirect benefits. An assessment of 45 prevention options showed that good house-keeping measures resulted in 25 to 30 per cent reduction in chemical substances, and technological change between 30 and 80 per cent reduction of waste and emissions. Substitution of raw materials resulted in the elimination of hazardous substances such as cyanide (zinc plating) and solvents (degreasing). Almost all cases (39) turned out to be cost-saving or neutral, with some extreme cases, for example, where a company saved more than Dfl1 million per annum (de Hoo et al, 1991).

Box 36B Clean production

Clean production is the application of an integrated strategy to reduce emissions from production processes while saving raw materials and energy. This concept views the industrial systems as analogues of ecosystems. Materials in an ideal industrial ecosystem are not depleted but after use they enter the production cycle again as raw materials. Manufacturing processes transform circulating material stocks from one form to another finding useful applications. There will always be unavoidable generation of waste. Industrial systems will never reach the recycling efficiency of biological systems, but it will be possible to develop a more closed industrial ecosystem through changes in product design, production processes and management practices.

Re-use, recovery and recycling

Once wastes have been generated, the preference accorded by the waste management hierarchy is for re-use, recovery and recycling. Re-use is effected by simple on-site or at-home operations to collect materials and put them back into the production and consumption process, instead of disposing of these substances within manufacturing and household waste streams. Re-use of industrial residues requires changes in in-plant practices plus source segregation and treatment in order to transform these wastes into secondary raw materials. The same concept applies to recycling, although this implies separation and treatment, which generally take place at off-site facilities, as well as setting up networks to exchange secondary materials among industries. According to the International Reclamation Bureau (IRB) – a non-profit organisation which represents 7000 national recycling organisations from 44 countries – recycling is the whole system in which obsolete or redundant products and materials are reclaimed, refined or processed, and converted into new, perhaps quite different, products. Thus, while re-use is performed on-site by the same waste generators, recycling requires a more complex organisational, economic and technological structure, the creation of which ultimately leads to significant economic impacts.

The difficulties in creating a stable market for secondary materials have limited the progress towards recycling hazardous waste in important industrial sectors. Among the constraints to increased recycling are: the limited quality of by-products, transportation costs, high recovery costs, and the low cost of alternative raw materials; there is also increasing concern for the environmental impact of recycling. On the other hand, several policy analysts dispute the desirability of creating such a stable market for toxic substances through recycling, pointing out that the new interests created by establishing such a market can indeed discourage hazardous waste reduction (Wynne, 1987; Caldart and Ryan, 1985).

Treatment

There is increasing concern in European countries to specify which waste disposal option is most appropriate for which waste stream. There is no single treatment option that can best ensure safe management of all different waste streams. The most common method of treating municipal waste is incineration. The future development of incineration as a method for disposing of waste greatly depends on the capability to reduce emissions by processing the waste submitted to incineration and adopting most up-to-date technologies for gas filtering and scrubbing. Methods for treating potentially hazardous waste range from neutralisation to incineration aimed at reducing the toxicity of waste through biological, chemical, physical and thermal processes. Biological methods include activated sludge and aerated lagoons. Chemical methods include oxidation, reduction, precipitation and other neutralisation technologies. Physical treatments include air stripping and carbon absorption. Thermal technologies include incineration and fluidised bed combustion. All these treatment technologies produce emissions and residuals which may still contain toxic substances. The extent to which these residuals can be avoided or contained depends upon the characteristics of specific substances and on the proper design and implementation of treatment plants.

Several European countries have now established emission limits for the incineration of municipal and hazardous waste. EC Directives 89/369/EEC and 89/339 on the prevention and reduction of air pollution from municipal waste incineration set standards for new and existing municipal waste incinerators. In a number of countries such as Austria, Germany and Sweden, new incineration plants should meet the best available technology (BAT) criteria in order to comply with emission limits. Existing plants need also to meet the same criteria to obtain renewal of operating permits. Rapid progress in the development of waste incineration technologies has outstripped the emission limit values set in the current regulations requiring systematic updating of such limits. The European Commission is now updating emission limit values for municipal waste and setting new limits for the incineration of hazardous waste (Table 36.5).

Landfill

Excluding sea dumping of wastes, which has been phased out by the Oslo Convention, the least preferred management alternative is waste disposal on land or under ground. Reducing the amount of waste disposed of in landfill is considered a priority in most European countries and listed in the waste management strategy of the EC Fifth Environmental Action Programme (CEC, 1993). Thus far, however, only The Netherlands has set reduction targets for landfill to 5 per cent of the overall solid waste by the year 2000. The Netherlands has also banned landfill for certain potentially hazardous wastes. The European Commision has proposed a Directive on landfills to harmonise reporting requirements (see also Box 36E).

Cleaning up contaminated sites

European countries have adopted various approaches to identify contaminated sites, assess potential risks and take the necessary remedial actions. In principle all costs for these actions are expected to be borne by the parties who have contributed to the improper disposal of contaminated waste at these sites. Even if no regulations concerning hazardous waste were in place at the time of such improper practices, acts that cause harm or damage are a sufficient ground for complaint and recovery. On this basis, a number of European countries such as France, Germany and the UK have instituted a 'strict liability' regime which implies that causation needs to be shown but negligence need not be proven. The regime of strict liability in these countries aims at ensuring that the high cost of remediation of the damage caused by an economic activity is borne by the operator. Denmark and The Netherlands, on the other hand, finance remedial actions through public funds. The European Commission is now working towards the harmonisation of liability regimes in EU Member States for all environmental damages caused by improper management of waste. A green paper on Remedying Environmental Damage has been issued by the European Commission as a basis for an EU-wide liability system in accordance with the prevention and polluter pays principles.

Preventing waste movements

The importance of a coordinated action at the international level to control waste movements and reduce the potential threats of improper waste management has become clear with the discovery of the damage caused by transfrontier movements of hazardous waste. The Basle Convention on the Control of Transboundary Movement of Hazardous Waste and their Disposal (UNEP, 1989), signed by 116 countries in 1989, allocates to the exporter states the responsibility for ensuring that exported waste is managed in a safe manner. The convention does not forbid waste shipments, but establishes that shipments must receive the written informed consent of importing states before they can take place. Twenty-nine European countries and the EU are party to the convention. Of those, 24 have signed and 15 have ratified the convention (Table 36.6).

In EU countries, the export of waste is controlled under Council Regulation 93/259/EEC, which updates and extends the controls originally introduced in 1984 and implements the provisions of the Basle Convention. Under the Regulation, all exports of waste for final disposal outside the Union and EFTA are banned. There are also specific controls on hazardous wastes being exported for recovery, and such exports from the EU, other than to OECD countries, will be banned after 1997 as a consequence of Decision II/12 taken at the Second Conference of Parties at Geneva in March 1994.

Transfrontier movements of wastes within Europe are influenced by a number of factors, including waste management capacity, regulatory standards and controls over transfrontier movements. Improvements in these respects across Europe, and particularly implementation of the Basle Convention, will help reduce these movements.

Since the movements of hazardous waste within European countries are strongly influenced by the different

Pollutants	EU proposal (mg/m3)	Germany (mg/m3)	The Netherlands (mg/m3)
Total dust	5	10	5
Total organic carbon	5	10	10
Hydrogen chloride (HCl)	5	10	10
Hydrogen fluoride (HF)	1	1	1
Sulphur dioxide (SO_2)	25	50	40
Carbon monoxide (CO)	50	50	50
Cadmium	0.05	0.05	0.05
Thallium	0.05	0.05	–
Mercury	0.05	0.05	0.05
Other heavy metals	0.5	0.5	1
Dioxins and furans	0.1 ng/m^3	0.1 ng/m^3	0.1 ng/m^3

Table 36.5
Emission limit values for waste incineration
Source: CEC, 1992

Table 36.6
European countries that are party to the Basle Convention as of 11 August 1994
Source: UNEP, 1994

Countries	Signature Final Act	Signature Convention	Ratification (r) Accession (a) Acceptance (A) Approval (AA)	Entry into force
Austria	√	√	12.1.1993 (r)	12.4.1993
Belgium	√	√	1.11.1993 (r)	30.1.1994
Croatia			09.5.1994 (a)	
Cyprus	√	√	17.9.1992 (r)	16.12.1992
Czech Republic	√		24.7.1991 (a)	1.1.1993
Denmark	√	√	6.2.1994 (AA)	
Estonia			21.7.1992 (a)	19.10.1992
Finland	√	√	19.11.1991 (A)	5.5.1992
France	√	√	7.1.1991 (AA)	5.5.1992
Germany	√	√		
Greece	√	√	4.8.1994 (r)	2.11.1994
Hungary	√	√	21.5.1990 (AA)	5.5.1992
Ireland	√	√	7.2.1994 (r)	8.5.1994
Italy	√	√	7.2.1994 (r)	8.5.1994
Latvia			14.4.1992 (a)	13.7.1992
Liechtenstein	√	√	27.1.1992 (r)	5.5.1992
Luxembourg	√	√	07.2.1994 (r)	8.5.1994
The Netherlands	√	√	16.4.1993 (A)	15.7.1993
Norway	√	√	02.7.1990 (r)	5.5.1992
Poland	√	√	20.3.1992 (r)	18.6.1992
Portugal	√	√	26.1.1994 (r)	26.4.1994
Romania	√		27.2.1991 (a)	5.5.1992
Russian Federation	√	√		
Slovak Republic	√	√	24.7.1991 (a)	5.5.1992
Slovenia			7.10.1993 (a)	5.1 1994
Spain	√	√	7.2.1994 (r)	8.5.1994
Sweden	√	√	2.8.1991 (r)	5.5.1992
Switzerland	√	√	31.1.1990 (r)	5.5.1992
UK	√	√	7.2.1994 (r)	8.5.1994
European Union	√	√	22.3.1989 (AA)	8.5.1994

regulatory requirements between countries for disposing of waste, it has become evident that the harmonisation of waste management standards across European countries is a critical step towards reducing the risk of transfrontier hazardous waste movements.

SUSTAINABLE GOALS

The challenge facing European countries in the next decade is to bring the use of materials and their management in balance with sustainability. Wastes should be minimised at such a level that natural resources and the environment are not threatened by their production and management. This requires changes in production and consumption patterns and modification in lifestyles (UNCED, 1992).

In order to quantify sustainability goals, biophysical limits to materials consumption and their management need to be established. This requires a better understanding of the interactions between ecological and economic systems as they are mediated by technological change and innovation. It is generally accepted, however, that current waste production trends and expected increases under current economic trends in European countries are unsustainable. To achieve sustainability entails both minimising the use of materials and reducing the impact of waste disposal.

Waste minimisation is the first priority. It requires the development of new resource-efficient methods and technologies. The capacity for re-use, recovery and recycling of waste needs to be ensured and sustained through positive incentives. Waste disposal needs to be reduced to a minimum while ensuring compliance of disposal and treatment facilities with health and environmental standards. This can be achieved only by providing the appropriate infrastructure, know-how and financial resources. In addition, the harmonisation of quality standards for waste management facilities is necessary if comparable levels of protection for the environment and human health are to be achieved throughout Europe.

STRATEGIES

The following strategies for waste production and management are crucial to move towards sustainable patterns of production and consumption:

1 **Reducing waste production per unit of GDP** is a priority in all European countries. Priority waste streams need to be identified according to their potential impact due to their quantities and hazard. Setting **reduction targets** for selected waste streams is critical to explore the options as well as for assessing the economic costs and their distribution between the various economic sectors. Measures to be considered range from adjusting the price of raw materials to regulations which specify waste minimisation requirements for production processes and products. Implementation plans could be developed through 'strategic discussion tables' with the target groups that directly influence the production and consumption patterns which generate waste (Box 36C).

2 Setting waste minimisation targets is also critical to **designing waste management plans** aimed at reducing consumption of resources and avoiding the impact of waste disposal. Most European countries require that plans should be developed in order to maximise re-use, recovery and recycling of useful materials and energy while ensuring safe management of waste that cannot be avoided. The design and implementation of integrated waste management plans entails identification of the most appropriate recycling, treatment and disposal

---

Box 36C
The EU strategic discussion plan

A new approach to waste prevention, re-use, and recycling as the priority options of the EU strategy on waste management is being considered by the European Commission (CEC, 1990a; 1990b). The aim of the new approach is to develop measures that help change the behaviour of the actors involved in the waste generation–handling–management chain. The involvement of these actors takes place through an interactive process called 'strategic discussion', bringing in independent facilitators to supervise the process. Participants in the strategic discussion include: the European Commission, representatives of Member State governments and regional local authorities, industrial and agricultural organisations, trade, consumers' and environmental groups, and research institutes. The aim of the discussion is to examine priority waste streams, identify the measures to be taken, and develop an implementation plan.

The new approach is currently being tested by the European Commission on a number of selected waste streams: used tyres, scrap cars, chlorinated solvents, electronic, construction, hospital waste and sewage sludge.

Some European countries have adopted a direct approach to waste minimisation in their national environmental policy by establishing targets and product regulations aimed at cutting priority waste streams. The Netherlands National Environmental Policy Plan (NEPP, 1990), for example, establishes that high priority waste streams are to be screened and measures will be drafted and implemented in partnership with target groups to reduce selected waste streams by specific deadlines.

---

options for specific waste streams (Box 36D). Strategic environmental assessment, eco-auditing and product life-cycle analysis could help explore waste prevention and management alternatives. In addition, environmental impact assessment procedures can help identify the proper design, technology and location for waste treatment facilities.

3 **Building the information base** on the sources, composition and management of various waste streams is necessary for developing and implementing waste minimisation and management strategies. Accurate and reliable data need to be made available in order to:
   – identify priority waste streams and minimisation targets;
   – establish integrated waste management systems;
   – optimise waste recovery and recycling;
   – develop ecological balances for assessing waste management alternatives;
   – implement the self-sufficiency and proximity principles.
   Categories of information on waste include:
   – waste production and management trends;
   – waste composition and potential hazard;
   – actual and potential waste minimisation;
   – waste movements and flows;
   – waste treatment/disposal capacity and costs;
   – environmental impacts of current waste management practices.

4 Establishing **monitoring networks** is needed to ensure early warning of potential contamination at waste sites and to assess the impact of waste management and major risk pathways, including discharge to air and groundwater (Box 36E).

Box 36D
Integrated waste management

Integrated waste management seeks waste minimisation through the various stages of the waste life-cycle. Waste streams and the options for their minimisation are seen as part of the entire cycle of production and consumption. The selection of waste treatment process and technology is part of a waste management strategy. Integrated waste management systems including plans, networks and facilities are one of the main objectives of the EC's Fifth Environmental Action Programme to implement a management strategy on waste. The development of ecological balances is suggested as the basis for evaluating waste management alternatives.

5 **Inventories of potential contaminated sites** need to be developed as a basis for setting priorities and planning clean-up actions. The present information is not adequate to make an assessment of potential risks and an evaluation of foreseeable costs of clean-up. This could require cooperation between Western and Central and Eastern European countries to share information, knowledge and financial resources.

6 The implementation of the principles established in the Basle Convention is necessary to ensure monitoring and control of waste shipments. The **harmonisation of hazardous waste management standards** across European countries will be necessary if measurable results are to be achieved to minimise transfrontier movements and their potential risks.

Box 36E Monitoring emissions from waste sites

The importance of monitoring activities at waste sites has gained attention from national authorities due to increased evidence of contamination from past improper waste management practices. Although regulations in several European countries require that monitoring at waste sites should be performed systematically, the knowledge of current and potential contamination of operating landfills in most European countries is still limited. Furthermore, little is known about the fate of pollutants at old contaminated sites.

In EU countries, the proposal for a Directive on landfill requires that countries should report on the capacity, operation and environmental conditions of landfills. Important considerations for the design of monitoring programmes are:

- consideration of major risk pathways, including discharge to air and groundwater;
- maximisation of early warning of adverse environmental effects;
- consideration of changes in natural background values;
- siting upstream and downstream monitoring wells for groundwater;
- intercepting a pollutant plume in the groundwater unsaturated zone before it reaches the aquifer.

In addition, encouragement is being given to associating monitoring programmes with contingency action plans. Major components of leachate to be monitored include: nitrogen species, chloride and the sodium/potassium ratio. Landfill gases to be monitored include carbon dioxide and methane.

REFERENCES

Alberti, M (1992) Minimization of hazardous waste in Western Europe: policy implementation and harmonization. PhD dissertation, MIT, Cambridge, Massachusetts.

Barniske, L (1989) Municipal solid waste incineration in the Federal Republic of Germany. *Environmental Impact Assessment Review* **9** (3), 279–90.

Caldart, C C and Ryan, C, W (1985) Waste generation reduction: a first step toward developing a regulatory policy to encourage hazardous substances management through production process change. *Hazardous Waste & Hazardous Materials* **2** (3), 309–31.

Carrera, P and Robertiello, A (1993) Soil clean up in Europe – feasibility and costs. In: Eijsackers, H J P and Hamers, T (Eds). *Integrated soil and sediment research: a basis for proper protection,* pp 733–53. Kluwer Academic Publishers, Dordrecht, The Netherlands.

CEC (1990a) *A Community strategy for waste management.* EC Council Resolution of 7 May 1990 OJ No C122.

CEC (1990b) *Analysis of priority waste streams: proposals to a new approach to prevention and recovery of specific waste streams.* Commission of the European Communities. DGXI, Brussels.

CEC (1992) *Proposal for a Council Directive on the incineration of hazardous waste.* COM(92) 9 Final - SYN 406. Brussels, 19 March 1992.

CEC (1993) *Towards sustainability: a European Community programme of policy and action in relation to the environment and sustainable development.* Commission of the European Communities, Luxembourg.

de Hoo, Brezet, H, Crul, M, and Dieleman, H (1991) *Prisma: Industrial Success with Pollution Prevention.* Nederlandse Organisatie voor Technologisch Aspectenonderzoek (NOTA), The Hague.

Franken, R O G (in press) *Point-sources of soil and groundwater pollution.* RIVM, Bilthoven.

Huisingh, D (1989) Cleaner technologies through process modifications, material substitutions and ecological based ethical values. *UNEP Industry and Environment* **12** (1), 4–8.

OECD (1991) State of the environment report 1991. Organisation for Economic Cooperation and Development, Paris.

UNCED (1992) *Agenda 21,* United Nations Conference on Environment and Development, Conches, Switzerland.

UNECE (1992) Recommendations to ECE Governments on the Five R Policies (Reduction, Replacement, Recovery, Recycling and Reutilisation of Industrial Products, Residues or Wastes). Fifth session of the Senior Advisers to ECE Governments on Environment and Water Problems. ECE/ENVWA/24, para.32. UNECE, Geneva.

UNEP (1989) *Basel Convention on the Control of Transboundary Movements of Hazardous Waste.* UNEP/IG, 80/3 (22 March 1989). United Nations Environment Programme.

UNEP (1994) *Parties to the Basel Convention on the Control of Transboundary Movements of Hazardous Waste and their Disposal.* United Nations Environment Programme, Geneva, 11 August 1994.

Wynne, B (1987) *Risk management and hazardous waste: implementation and dialectics of credibility.* Springer Verlag, Berlin.

Yakowitz, H (1992) Waste management in Europe. Paper presented at Globe Conference, Strasbourg, 17-20 May 1992. OECD, Paris.

*Housing in
Birmingham, UK*
Source: Spectrum Colour
Library

37 Urban stress

THE PROBLEM

Important changes in the quality of the urban environment
have occurred in Europe in the last few decades. Despite the
progress achieved in controlling local air and water pollution,
urban areas show increasing signs of environmental stress
(Chapter 10). Major concerns for European cities are the
quality of air, acoustic quality and traffic congestion. Open
spaces and green areas are under continuous threat due to
more competitive uses of limited land resources. The quality
of life in cities is also affected by the deterioration of
buildings and infrastructure and the degradation of the urban
landscape. On the other hand, cities absorb increasing
amounts of resources and produce increasing amounts of
emissions and waste, causing significant burdens on the
regional and global environment. A summary of major urban
environmental problems in Europe is presented in Box 37A.
These problems are warning signals of a more deep-seated
crisis, and call for a rethink of current models of organisation
and urban development.

Symptoms of environmental stress in cities become
evident when the quality of environmental components and
their effects on the health and quality of life of the urban
population are examined. However, the causes of urban
stress can be understood only when examining how cities
work and how their spatial organisation affects their
environmental performance. The quality of the urban
environment is a result of the interactions of many
interdependent variables and between urban activities and
city structure. The impact of urban activities on the local
environment is not the sum of each effect taken individually.
In cities, people are exposed simultaneously to the
concentrations of pollutants into the air, water and soil.
Synergies among these factors generate environmental stress.

Addressing the causes of environmental problems, instead of
treating their symptoms, is necessary if current efforts to
improve the quality of the urban environment are to succeed.

Urban environmental problems are often referred to as
local problems. The high concentration of people and
activities in cities is the cause of heavy pressures on the local
environment. Local environmental conditions affect the
health of the exposed population. However, environmental
problems affecting urban areas are closely linked with
regional and global problems by their common causes and
interdependent effects. Urban air pollution is linked to
acidification, photochemical smog and climate change
through the emissions of atmospheric pollutants from the
burning of fossil fuels. As cities deplete their local resources
and increase their dependence on imported global resources,
they become more vulnerable to the effects of global
environmental change. On the other hand, the
implementation of measures to improve the urban
environment have corresponding beneficial effects on the
regional and global environment.

CAUSES AND CONSEQUENCES

Urban air pollution, noise and traffic congestion are the
most recognised symptoms of urban environmental stress.
Causes of such stresses are clearly related to rapid changes in
urban lifestyles and increased urban activities which have
occurred in the last decades. Yet, their links with the patterns
of urban development are often not fully appreciated. The
interrelated nature of urban environmental problems can be
best explained by examining, for example, the relationship
between air pollution, energy consumption and
transportation trends.

Box 37A
Major urban environmental problems

Ambient quality
Exposure to air pollution and noise are increasing concerns in most European cities. These problems are indicated as causes of environmental stress and threats to public health. Despite the achievements in the reduction of traditional emissions (SO_2 and particulate matter) the majority of cities still exceed short-term WHO air quality guidelines (AQGs) at least once in a typical year. Compared with Eastern and Southern European cities, Northern and Western cities are better off as to long-term exposure to SO_2, but are now faced with other types of threats (eg, NO_x, VOCs). Energy production and industry are still major sources of air pollution in Eastern European cities. In Western cities the main source of air pollution is transport.

Urban space
The quality of urban space is dependent upon the relative share of built and open areas within cities. It also depends on the use of land and accessibility of open space and green areas achieved within the cities. Open and green space is an important component of the quality of urban life, but also for maintaining a balance of the city with the natural environment. In most European cities these elements are under threat as a result of depopulation of the urban centres and urbanisation of the urban fringe. Urban areas that have lost the functions for which they were designed and have not been renewed or re-used have become derelict land.

Urban traffic
Urban transportation is a major contributor to urban energy consumption, emission of air pollutants, traffic congestion and noise. European cities are all faced with increased urban mobility and change in travel mode from public transportation to private car, which have offset the improvements in car efficiency and the improvements in air quality achieved as a result of emission reduction measures. Car ownership has increased markedly during the last 20 years, particularly in Western Europe, and this trend is expected to continue. The share of traffic as a cause of urban air pollution is also expected to increase in Eastern European cities, where the number of cars per person have increased in the last few years.

Energy consumption
In addition to the share of energy consumption due to urban transportation, the amount of energy consumed by cities for heating and cooling offices and residential buildings in Western and Southern Europe has also increased significantly in the last two decades. In spite of the anticipated improvement in energy efficiency, an increase in energy consumption due to urban activities is expected to occur in Eastern European cities as a result of higher living standards.

Urban emissions and waste
Urban activities generate increasing amounts of emissions and waste. Urban waste has increased in volume and changed in composition in the last two decades. In spite of the progress made in recycling in several cities, waste is expected to increase further in the next decade. The disposal of municipal waste is still a major concern.

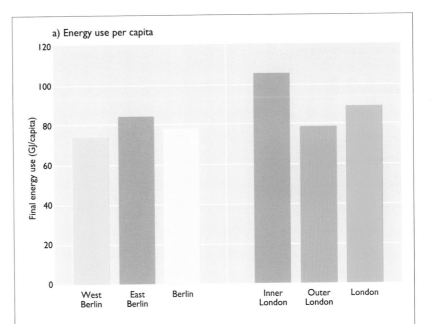

a) Energy use per capita

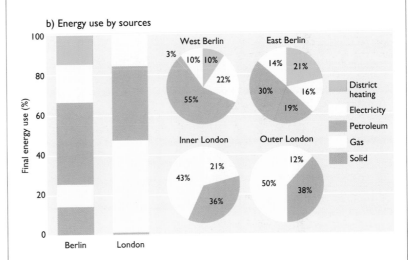

b) Energy use by sources

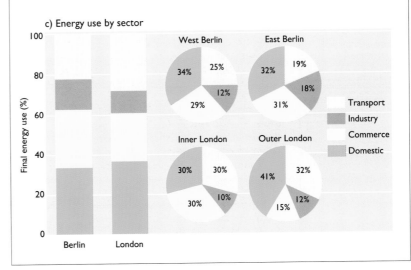

c) Energy use by sector

Major causes of air pollution in cities are the processes involving the combustion of fossil fuels. These include the production and consumption of energy for domestic and commercial building heating systems, industrial activities and transport. There are important differences in energy patterns among European cities and within cities' inner and outer areas, as illustrated by the examples of London and Berlin (Figure 37.1a, b and c). These are reflected in the levels of

Figure 37.1
Comparison of final energy use in Berlin and London
Source: LRC, 1993

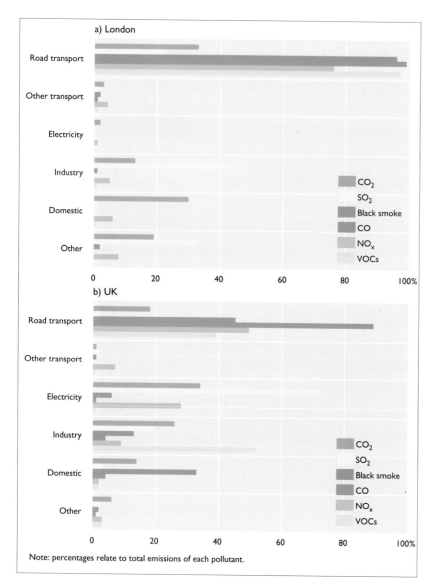

Note: percentages relate to total emissions of each pollutant.

Figure 37.2
Sources of principal air pollutants in London and the UK
Source: LRC, 1993

vehicles has increased substantially in the last decades despite the improvement achieved in their energy efficiency.

Current trends in urban mobility, and their increasing environmental impacts, are closely related to changes in the urban structure. Urban population density varies considerably from city to city, but almost all cities are experiencing decentralisation of population, whether they are growing or not (see Chapter 10). The decentralisation of urban activities traditionally located in urban centres as a result of economic specialisation, and the relocation of urban activities far from public transport links, have changed urban mobility patterns and increased car dependency of urban residents. Traffic flows in European cities have increased in terms of number and length of trips over the past decade. During the same period, the share of public transport of total motorised urban travel has fallen in most cities to about 20 per cent. The car's share of urban mechanised transport has reached over 80 per cent. These trends are major concerns in most European cities for their effects on human health and the environment.

The physical health of the European population is closely linked to environmental stress in urban areas (see Chapter 11). European urban areas are where environmental problems most affect the quality of life of citizens. Most air pollutants, such as sulphur dioxide, particulate matter, nitrogen oxides and carbon monoxide, directly affect human health. Other urban traffic-related factors of environmental stress in urban areas are noise, road accidents and congestion. Unacceptable noise levels of more than 65 dB(A) affect between 10 and 20 per cent of urban inhabitants in most European cities. Urban road accidents remain a significant problem despite the important progress achieved in the implementation of traffic safety measures. Traffic congestion has caused the reduction of average speed in urban centres by 10 per cent in the last two decades. These problems are among the most severe urban problems as perceived by people living in urban areas in a recent OECD survey of 132 cities (OECD/ECMT, 1993).

European cities contribute to local, regional and global environmental change. Major local environmental changes are the result of urbanisation and the transformation of the surface structure and thermal balance. Urban activities give rise to multiple environmental stress factors. Air pollution in cities combines with critical climatic conditions to generate high-level exposure episodes. Inner-city green areas are subject to high levels of pollution. Many historic monuments and buildings, especially those made from marble, calcareous sandstone, or other materials susceptible to damage, are affected by air pollutants. In urban air, soot and other primary particles are important contributors to visibility effects.

The state of the urban environment is inevitably linked to the state of the regional and global environment by the same causes of stress. Sulphur dioxide and nitrogen oxides emissions contribute to transboundary air pollution and increased acid deposition levels (Chapter 31). Nitrogen oxides are also precursors of ozone, which is the main constituent of photochemical smog (Chapter 32). Motor vehicles account for a considerable proportion of the total emissions of nitrogen oxides to the atmosphere in Europe, and their contribution is expected to increase in the long term following the growth in the use of the private car. Energy consumption and urban traffic are also major contributors to carbon dioxide emissions, which form the principal cause of the greenhouse effect (Chapter 27). Cities are also important sources of stratospheric ozone-depleting substances (Chapter 28).

emissions of air pollutants and air quality levels. Fuel substitution in domestic heating has reduced dramatically the emissions of sulphur dioxide and particulates per unit of energy consumed in Western European cities. The substitution of natural gas for coal and oil in European cities has also reduced emissions of carbon dioxide per unit of energy. However, the majority of cities still exceed short-term WHO-AQGs (air quality guidelines), and in most Eastern and Southern European cities long-term exposure to particulates and sulphur dioxide is still an important concern. Furthermore, as emissions from stationary sources of urban air pollution from domestic and industrial activities have receded, traffic-generated pollutants such as nitrogen oxides, carbon dioxide, particulates and volatile organic compounds have steadily increased.

Urban traffic is an increasingly important source of urban air pollution (Figure 37.2). Road transport in cities accounts for most of the summer smog in Europe and the exceedances of WHO-AQGs for ozone, nitrogen oxides and carbon monoxide. Urban transport as a whole represents around 30 per cent of total energy consumption in most European cities, second only to the domestic sector, which accounts for one third of the urban energy consumption. Energy consumption by urban transport has almost doubled in the last two decades, and its share of total urban energy consumption is expected to increase further following urban mobility trends. Most energy-intensive transport is road transport, which accounts for more than 85 per cent of total transport energy consumption. Total fuel consumption by

TRENDS AND SCENARIOS

The share of Europe's population living in urban areas is expected to increase over the next decade, with the largest cities housing a very significant proportion of the population. Urban lifestyles are already predominant in all European countries despite regional differences in the level of urbanisation and stage of urban development. Distinct phases of urban environmental problems in Europe can be associated with various stages in urbanisation. Urban air pollution in European cities can be related to the level of urban development as illustrated schematically in Figure 37.3. During the period of rapid urbanisation, air pollution levels rose rapidly, mainly from domestic and industrial sources. Switch in fuel use and the introduction of emission control measures have emerged as air pollution in European cities becomes a serious public concern. This has led to stabilisation of air quality condition and successive improvement. Suburbanisation and de-urbanisation, accompanied by a sharp decline in the manufacturing industry in the main Western European conurbations, have led to further improvement in air quality and a drastic reduction in emissions from industrial and domestic sources. However, the decentralisation of economic activities traditionally located in urban centres has contributed to the growth in travel demand and the use of the private car, which constitutes an increasing source of air pollution in cities. The current process of re-urbanisation is now accompanied by new air pollution concerns.

Future environmental conditions in European cities are highly dependent on the sustainability of economic development and the introduction of new technologies. While there is a paucity of information available to demonstrate the impact of regulations aimed at controlling individual emission sources, observed trends in sulphur dioxide, nitrogen oxides and lead concentrations show evidence of the effectiveness of these measures. Emission controls have been particularly successful in the reduction of dust and fly-ash emissions to the atmosphere, and in the reduction of the emissions of gases from process industries. To a large extent the exceedances of air quality guidelines for sulphur dioxide and suspended particles in European cities can be attributed to the burning of coal in small domestic stoves and boilers, and to uncontrolled emissions of dust from industrial sources. This is still a major problem in Central and Eastern European cities because of their heavy reliance upon coal – typically poor-quality brown coals and lignite. The supply of cleaner fuels, such as natural gas and low-sulphur distillate oil, to domestic heating applications and dust removal from industrial sources and larger boilers, together with increased energy efficiency, is expected to produce major improvements.

Although detailed emissions inventories for most European cities are not available, observed trends in national emissions inventories show that emissions of nitrogen oxides from stationary sources have decreased in many urban areas, while emissions from motor vehicles have increased, since the growth in traffic and road transport has been much larger than the reduction in emission factors. Motor vehicles are already the major source of urban air pollution in most Western European cities. Although the regulation on exhaust emissions of cars and lorries is expected to reduce substantially emissions of air pollutants from individual vehicles, the projected increase in car ownership and urban travel will largely offset the potential reductions. The relative contribution of mobile sources to total air pollution load is expected to increase substantially in Eastern European cities in the next decade.

Under current policies, European sulphur dioxide emissions are expected to decrease by about 30 per cent between 1990 and 2000. For aerosol particles, including both emitted dust and soot and secondary sulphate and nitrate aerosol, future developments are uncertain. No international

policy framework exists for these pollutants and even present levels are not well known. Following emission reductions of sulphur dioxide and nitrogen oxides, sulphate and nitrate aerosol levels are expected to decrease by 20 to 30 per cent. Concurrent with technological changes and shifts in fuel use, primary dust emissions from power generation and industry in Central and Eastern Europe are expected to decrease in the next 20 years by at least 30 per cent. While industrial particulate emissions are expected to decrease, the trend in future traffic emissions is uncertain, particularly with respect to soot.

Emission reductions in European cities will result in decreasing occurrence of winter smog and less pollution-induced visibility reduction. However, these reductions are not sufficient to prevent exceedance of the WHO-AQGs for sulphur dioxide and aerosol particles. This probably requires emission reductions of the order of 80 per cent, especially in hot-spot regions (eg, the Black Triangle). Significant emission reductions for dust and aerosol particles can be achieved by implementing various filtering techniques, such as electrostatic precipitators (ESP) on power plants and baghouse filters on industrial installations; fuel change from coal to gas in power production and residential heating; and energy saving.

SUSTAINABILITY GOALS

Sustainable urban patterns are essential to achieve sustainability in Europe. These imply bringing urban activities into balance with the capacity of ecosystems to provide life support services important for human health. These also require minimising the pressure of urban activities on the local, regional and global environment. Sustainable cities are cities that provide a livable and healthy environment for their inhabitants and meet their needs without impairing the capacity of the local, regional and global environmental systems to satisfy the needs of future generations. These challenges are interdependent and require coordinated efforts.

Making cities sustainable entails:

- minimising the consumption of space and natural resources;
- rationalising and efficiently managing urban flows;
- protecting the health of the urban population;
- ensuring equal access to resources and services;
- maintaining cultural and social diversity.

In order to achieve these goals, the design and management of urban systems need to be guided by urban sustainability principles. These include environmental capacity, reversibility, resilience, efficiency and equity.

The capacity of the environment to provide resources and absorb emissions and waste imposes absolute limits on human activities. The principle of environmental capacity requires that cities are designed and managed in order to deliver basic environmental, social and economic services

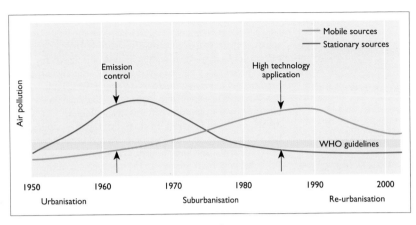

Figure 37.3
Schematic representation of urban air quality problems and urbanisation, 1950–2000
Source: Modified from UNEP/WHO, 1992

within the limits imposed by the natural environment. The concept of environmental capacity helps set the limits to the activities that a space can sustainably absorb. In ecology, carrying capacity is the maximum population size of any given species that a habitat can support indefinitely without permanently damaging the ecosystem upon which it is dependent. When applied to human systems, such limits are mediated by consumption patterns, technological development and lifestyles. In cities, this capacity applies to both the natural and built environments and the interactions between natural and human-made systems.

One important premise to the management of the urban environment is a view of cities as dynamic systems. Cities are continuously challenged by the need to adapt to new demands from changes in population and economic activities. The management of the urban environment needs to cope with such changes. The ability of cities to respond to these challenges without impairing environmental capacity is a measure of their vitality. Planning interventions in the urban environment should as much as possible be *reversible*. A crucial feature of the sustainable city is *resilience*. A resilient city is one able to respond and recover from external stresses.

Sustainable development is a much broader concept than environmental protection (CEC, 1994). Economic and social dimensions need to be integrated with the ecological dimension of urban systems. Two other notions relevant to urban sustainability are efficiency and equity. The *efficiency* principle requires getting the maximum economic benefit for each unit of resources used (environmental efficiency) and the greatest human benefit from each unit of economic activity (welfare efficiency). Equal access for urban inhabitants to resources and services (*equity*) is important to modify unsustainable behaviour exacerbated by inequitable distribution of wealth.

The stated objectives imply that a broad definition of urban health is adopted by urban planning and management. Cities are integrated systems that facilitate the delivery of a wide range of services and activities (Brugmann, 1992). Every service that is delivered in urban systems is dependent upon the productive capacity and the efficient management of many subsystems. A systems approach can help understand and manage the complex relationships that take place in cities.

To date, there is no such universally accepted definition of urban health that brings together all the significant aspects of urban life, nor such an approach that describes the relationships between urban life and the state of the urban environment. Different approaches focus on specific components of the urban environment, such as the air, land and water, or on the various functions of urban systems, such as housing, water and energy supply, or transport. In the same way, responses to environmental problems in urban areas have focused for a long time on the various symptoms, and problems have been dealt with in a compartmentalised fashion, often reflecting the structure of local authorities. Today, many European cities recognise that moving towards a sustainable city requires an integrated approach.

STRATEGIES

European cities have adopted various environmental strategies and pollution control measures to improve the urban environment. There is a wide range of planning and management options that can be classified under five main categories (Table 37.1): urban planning; urban management; economic instruments; standards; and public information. These are examined in a recent report on 'Sustainable Cities' prepared for the European Commission by the expert group on the urban environment (CEC, 1994).

This report stresses that there is no single set of actions which can apply equally to all European cities. Cities experience environmental problems in relation to their geographical profiles, demographic trends and economic assets. Cultural differences are also important determinants of the way people perceive urban problems. In addition, the organisational basis for implementing measures aiming at improving the urban environment varies between European countries according to the degree of decentralisation of competence and responsibilities. Nevertheless, a common European framework is important to identify and assess a broad range of options to operationalise sustainable development in urban areas.

The extent of the interrelationships within the urban system demonstrates the potential danger of *ad hoc* decision-making: the solution to one problem is often the cause of another. Effective management of urban air quality requires a strategy based on an overview of the urban system, with integrated decision-making in key areas. Few cities possess an administrative structure that can ensure such integration, most critically between landuse and transportation planning. While the city is the main focus of economic activity and the associated pressures on the environment, it cannot be analysed in isolation from the region within which it is located.

The range of actions of a European approach to urban areas is laid down in the *Green Paper on the Urban Environment* developed by the European Commission in

Table 37.1
Urban planning and management options
Source: EEA-TF

Problem areas	Urban planning	Urban management	Economic instruments	Standards	Public information
Ambient quality	• Landuse • Transport	• Integrated emission control	• Emission charges	• Air pollution • Drinking water • Noise exposure • Indoor pollution	• Monitoring • Reporting
Urban space	• Landuse • Urban renewal • Open space • Green areas • Housing	• Zoning • Designation of special areas	• Land pricing	• Green areas requirements • Building codes	• Protection and maintenance
Traffic	• Landuse • Transport • Parking control	• Traffic • Public transport • Traffic calming	• Petrol pricing • Road pricing	• Emissions • Performance • Petrol contents	• Public transport promotion
Energy	• Energy conservation programmes • Energy saving	• Energy efficiency	• Energy pricing	• Emissions • Emissions standards	• Energy saving
Waste	• Reduction targets • Integrated waste management	• Recycling schemes • Separate collection	• Taxation	• Products • Packaging • Disposal	• Ecolabelling • Recycling

1990 (CEC, 1990). Two major areas of actions are identified: policies which relate to the physical structure of cities (landuse, mobility, green areas and historical heritage); and policies which relate to reducing the impact of urban activities on the environment (energy, water and waste management). An important step forward in this direction is taken by the policy report 'Sustainable Cities' (CEC, 1994) which followed the publication of the Green Paper. The report explores the application of sustainable development in urban areas and strongly recommends a systems approach to achieve integration across the policy areas emphasised by the EC's Fifth Environmental Action Programme.

In the light of these recommendations, the following five strategic levels can be considered.

Sustainable urban planning

At the planning level, principles of urban sustainable design can be applied to urban landuse. A wide range of options can be considered to incorporate environmental considerations into urban planning, focusing on:

- selective densities (residential and jobs) around points of high access;
- location and proximity to public transportation nodes;
- energy-efficient planning;
- mixed-use development;
- landuse design considerations.

The scope for integrating environmental considerations into the design and management of urban systems varies in cities depending on their different planning systems. European cities have common objectives and targets as well as sets of procedures to influence the urban form and promote a balanced use of built-up and open areas.

Integrated transport management

Improved environmental quality in cities can be achieved by integrated transport strategies aimed at reducing the impact of urban traffic and the overall amount of fuel consumed by transportation. A wide range of measures are available in order to:

- ensure high accessibility of urban areas;
- reduce the level of car use and favour more efficient modes of travel;
- encourage a more efficient use of vehicles;
- facilitate walking and bicycling;
- improve the safety of pedestrians.

These measures include:

- traffic calming;
- traffic bans in designated areas;
- urban road pricing;
- integrated public transportation;
- vehicle priority schemes;
- parking controls.

Efficient management of urban flows

Urban design and renewal projects need to be oriented towards reducing the use of water, energy and materials. This can be achieved through the implementation of low-impact technologies and the rational allocation of resources. Reducing energy consumption in urban areas requires energy efficiency and energy conservation measures. Energy efficiency refers to measures aimed at improving the efficiency of energy transformation processes through technological improvements, switches in technology and

maintenance programmes. Energy conservation includes measures to improve the efficiency of end use and the rational allocation of energy sources. The visibility of ecotechnological solutions is particularly important for its high information value.

Targets and standards-setting

Achieving better urban environmental quality requires setting targets and implementing systematic monitoring to ensure that environmental quality standards and the objectives established by the international conventions are met in urban areas. This can be achieved by adopting add-on technologies for specific pollution sources or by measures aimed at preventing pollution at source (1, 2, 3 above). Environmental standards aim to protect human health as well as ecological health. Air quality standards established by national authorities or international bodies set both limit values (critical for human health) and guide values – values set as the objective for environmental improvement to prevent any long-term impact on health and the quality of life. Although lists of air quality guidelines published by WHO (WHO, 1987) or the EC (eg, 80/779/EEC) are already relatively extensive, there are many other chemicals in the air we breathe which pose a potential health hazard. Synergistic effects between compounds may also complicate the task of setting sound and safe guideline values (see also Chapter 38).

Urban environmental information

The collection of consistent information on urban environmental quality and the pressure of urban activities on the regional and global environment is required to establish clear relationships between changes in the state of the urban environment and patterns of urban development. The environmental performance of urban areas needs to be monitored through a consistent set of urban environmental

---

Box 37B Local Agenda 21

The important role of cities in achieving sustainable development has been recognised at the United Nations Conference on Environment and Development (Rio de Janeiro, 1992). Agenda 21, which commits 179 signatory states to a common programme of actions, assigns to cities the special task of developing Local Agenda 21. Chapter 28 sets the target that, by the year 1996, the majority of local authorities will have initiated a consultative process for developing an Agenda 21 for the community.

The development of Local Agenda 21 is an exploratory process of the opportunities for sustainable development at the local level (see, eg, ICLEI, 1993). Four main elements characterise this process:

1 the establishment of a local forum with the participation of all sectors of the urban community, such as local officials, urban planners, public utilities, local enterprises, experts, community and non-governmental organisations;
2 assessment of urban environmental problems through expert groups and workshops and identification of priority areas of actions to redirect urban development towards its sustainability;
3 development of local action plans towards sustainability through dialogue and partnership with the various actors involved; and
4 establishment of mechanisms for monitoring and reporting on the implementation of action plans.

monitored through a consistent set of urban environmental indicators to provide guidance in formulating and implementing urban environmental policies. Public information strategies are also important to provide the general public with immediate feedback on the state of the urban environment and the environmental effects of their actions. Information strategies include: public information systems, pollution alert systems and local environmental reporting.

Urban environmental stress is an increasing concern for the majority of European people. The importance of urban environmental problems in Europe is emphasised by the numerous links that can be made with almost all problems treated in the present report. Cities share an important role in achieving sustainable development. European cities are therefore challenged with an important responsibility. The Charter of European Cities and Towns Towards Sustainability adopted at Aalborg (Denmark, 27 May 1994) recognises this and provides the basis to implement the Local Agenda 21 process (Box 37B) in order to meet the mandate of the UN Conference on Development and the Environment (UNCED, 1992).

REFERENCES

Brugmann, J (1992) *Managing human ecosystems: principles for ecological municipal management.* International Council for Local Environmental Initiative, ICLEI, Toronto.

CEC (1990) *Green Paper for the urban environment.* COM(90) 218, Commission of the European Communities, Brussels.

CEC (1994) *European sustainable cities.* Draft report of an EC Expert Group on the Urban Environment, Sustainable Cities Project. Commission of the European Communities, Brussels.

ICLEI (1993) *The Local Agenda 21 Initiative – ICLEI Guidelines for Local Agenda 21 Campaigns.* International Council for Local Environmental Initiative, Toronto.

LRC (1993) *London energy study.* London Research Centre, London.

OECD/ECMT (1993) *Urban travel and sustainable development: an analysis of 132 OECD cities.* OECD, Urban Affairs Division, Paris.

UNCED (1992) *Agenda 21.* United Nations Conference on Environment and Development, Conches, Switzerland.

UNEP/WHO (1992) *Urban Air Pollution in Megacities of the World.* Blackwell, Oxford.

WHO (1987) *Air quality guidelines for Europe.* WHO Regional Publications, European Series No 23, World Health Organisation, Copenhagen.

*Toxic chemical
container*
Source: Benelux Press

38 Chemical risk

THE PROBLEM

In this present report it is shown that there are few environmental problems in Europe that cannot be traced back to some sort of excessive loading of chemicals. Economic development has been driven to a considerable extent by progress and innovation achieved by the chemical industry. This process has led to the marketing and use in different applications of ever-increasing numbers and quantities of chemical substances. More than 10 million chemical compounds (natural or man-made) have been identified. Of these, about 100 000 are produced commercially (200 to 300 new chemicals enter the market each year) and are potential subjects of concern. Estimates suggest that the current world production of chemicals is about 400 million tonnes (Lönngren, 1992). Many chemicals are applied directly to the environment or are discharged after use. Adequate toxicological and ecotoxicological data have been produced for only a very small fraction of the chemicals, and data on environmental pathways and ecotoxicological effects are even more sparse (see Chapter 17).

From being regarded mainly as a potential risk to humans, chemical impact on the environment became of increasing concern from the 1960s. Before then there was little appreciation of the atmosphere and hydrosphere having limited capacity to assimilate the ever-increasing releases of chemicals. The bioaccumulation of certain chemicals in the food-chain and its consequences for the normal function of ecosystems and for human welfare were not well understood either. Apart from a few early signs of chemical threats to the environment (eg, mercury pollution as evidenced by the Minamata disease in Japan, high levels of mercury in game and fish in Sweden, lake eutrophication, PCB and DDT pollution) it was only relatively recently recognised that chemicals which enter the environment can, in general, cause serious detrimental effects in all environmental compartments and to human health whether they be released through normal operation or as a result of accidents. Accidental discharges of chemicals and their environmental effects are treated in detail in Chapter 30.

The human-generated sources of dangerous chemicals which enter the air are widespread. The smokestacks of factories, power-generating stations, waste incinerators, motor vehicles and smelter discharge emit sulphur dioxide, nitrogen oxides, carbon monoxide, hydrocarbons and other combustion products often contaminated with substances such as heavy metals, dioxins, furans, etc.

In highly industrialised or populated areas, chemicals are often released into waterbodies in exceptionally large volumes through industrial discharge pipes and municipal sewage. Apart from large volumes of chemicals entering the environment in such areas, the environmental danger comes from the fact that these releases contain significant amounts of many types of chemicals which are commercially used.

The release of highly toxic chemicals (often as contaminants in high-volume production chemicals) also causes serious problems. Well known examples for these substances are chlorinated hydrocarbons, heavy metals and hydrocarbons (see Chapter 14). Once a chemical is emitted to the air it is usually deposited with rainwater and snow and eventually ends in runoff into rivers and seas. Another main source of chemicals in the environment is from the agricultural use of pesticides, which contain potentially dangerous chemicals that can leach into groundwater. Some non-degradable (heavy metals) and very slowly degradable substances (PCBs, dioxins) can be transformed into more toxic intermediate compounds (DDE from DDT or methylmercury from mercury). On the other hand, bioaccumulation may take place in the food-chain, which can result in concentrations toxic to biota, especially top predators.

Figure 38.1
Pathways for dispersion of chemicals in the environment
Source: VROM, 1991

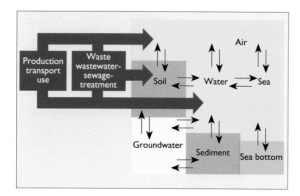

Natural transport from one compartment to the next further accelerates the dispersion of a given substance through the ecosystem (Figure 38.1). This eventually leads to accumulation of a substance and harmful effects in another compartment than that to which it was originally released. Most of the chemical compounds released to the environment are subject to biotic and/or abiotic degradation but some are persistent and therefore accumulate in the environment, leading to long-term exposure of organisms. Depending on the toxicity and persistence of the substance, exposure can lead to disorders, genetic mutation, adverse effects on reproduction, cancer, mortality and adverse effects on the nervous and immune systems. It may also cause effects on ecosystems.

A great variety of environmental perturbations caused by chemicals are considered in this report. In particular, the concentrations, trends and effects of chemicals in three environmental compartments: air, water (fresh and marine) and soil are thoroughly treated in Chapters 4 to 7. The human activities which promote the release of chemicals are described in Chapters 19 to 26, and the resulting loads to air and water with problem chemicals are the subject of Chapter 14. The health aspects and environmental impacts of a selection of chemicals considered to be particularly harmful are treated in Chapters 11 and 17. The environmental problems considered in this chapter of the report which relate to human activities and the emission of chemicals include: climate change, stratospheric ozone depletion, loss of biodiversity, major accidents, tropospheric photochemical oxidants, acidification and eutrophication.

To overcome the specific environmental problems caused by an identified chemical, remedial, preventative or legislative measures are taken in many industrialised countries (for example, controls on heavy metals, CFCs, PCBs, DDT, pentachlorophenol, etc). In the last few years, however, it has been generally recognised that such an approach is not sufficient. There must be an integrated concept which addresses the impact of all chemicals that may cause detrimental effects in the environment. However, such an integrated and systematic risk assessment approach is faced with a number of inherent difficulties, the most important being:

1 the large number of new and existing chemicals which have to be considered;
2 frequent lack of data which are needed for comprehensive risk assessment;
3 the possibility that initially innocuous chemicals may react in the environment and be converted to more toxic compounds and cause unforeseen secondary impacts;
4 the potential danger of almost any chemical to any kind of organism or ecosystem;
5 dangerous contaminants as by-products in synthesised chemicals produced in large quantities;
6 frequent lack of regional and global coordinated control measures; and
7 time delays between regulatory measures and beneficial effects.

GOALS

The overall goal in chemicals control is to reduce the amount of toxic substances in the environment and the risk of exposure of humans and ecosystems to a level where only negligible risks are to be expected (Figure 38.2).

The aim should be to reduce the concentration of each chemical to this level. To ensure this, the continuous monitoring and testing of each chemical released should be encouraged. When evidence arises of potential negative impacts, the exposure of the target environment has to be reduced by individual initiative and/or legislative measures.

Achievement of this overall goal of chemical risk control typically involves specifying the following subgoals, as proposed in the EC's Fifth Environmental Action – Programme (CEC, 1993):

1 Data collection on new and existing chemicals. This should include:
 – an effective notification procedure for all new chemicals containing a dossier of information about the chemicals' characteristics, hazards, uses and disposal, as well as a risk assessment;
 – an inventory of existing chemicals, concentrating on high-volume production chemicals and those substances considered to be most dangerous and in need of strict control; and
 – improved international coordinated data collection for chemicals.
2 Classification and labelling: the objective of classification is to identify all the physico-chemical, toxicological and ecotoxicological properties of chemicals. Having identified any hazardous property, the dangerous chemical is then labelled to indicate the hazard in order to protect the user, the general public and the environment, including water, air, soil and living organisms. The label provides the general public and persons at work with easy access to the identity of a dangerous chemical, its hazard and what to do in case of emergency.
3 Risk assessment: to combine information gathered from the classification procedure and knowledge obtained from research to provide assessment of dose/response effects and exposure to humans or ecosystems. This requires data about the level of exposure, the target populations, the biological, toxicological and ecotoxicological effects of a certain amount of pollutant, the threshold level. With this information the risk is defined and an evaluation made concerning the sustainable use of the substance.
4 Risk management: this has the overall aim of controlling emissions by banning highly hazardous chemicals, limiting the use of dangerous products, substituting dangerous with less dangerous products, limiting selling of dangerous chemicals to authorised purchasers, as well as informing consumers about hazardous chemicals and their alternatives.

Figure 38.2
Progressive achievement of reduced chemical risk to the environment
Source: VROM, 1991

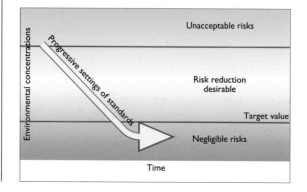

To make these goals operational it is important to realise their interconnectedness (Figure 38.3); tight international cooperation between scientists, environmental managers, policy makers and international organisations is absolutely crucial to improve environmental conditions by reducing the ambient load of chemicals.

STRATEGIES

The potential of chemicals to cause harm to life, health, the environment or property has few limits. Hazard control, therefore, aims to eliminate or reduce the levels of hazardous chemicals that can cause harm to humans or the environment, or that can trigger a fire or an explosion. Industry has gradually developed a risk management approach that takes into account the full life-cycle of chemicals. The manner in which these are handled determines whether their effects are beneficial or harmful. Strategies implemented by regulations, programmes and policies already exist internationally and within the European region (see Box 38A).

Policy coverage

Traditionally, risk containment regulations were restricted to chemical substances classified as dangerous. This changed, however, in the 1970s. In principle, two general strategies are available for addressing the control of chemical risk in the environment: a voluntary collaboration between industries and official institutions, or an approach based on legal obligations of industry. The first strategy has been practised by the UNEP and OECD, and, during its development, the EU legislative chemicals control system, which is detailed below, has drawn upon the programmes and experience gathered by these organisations.

Measures and initiatives from official bodies and industries exist in the following areas:

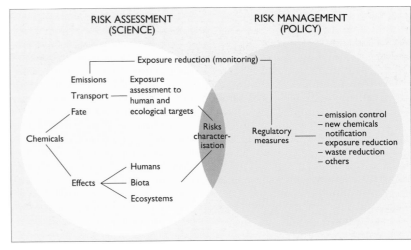

- establishing priorities on the basis of assessments of levels of exposure, effects, risk population, etc;
- reduction of emissions at all levels of the activity and treatment of effluents;
- compulsory testing for possible toxic and ecotoxicological properties of new and existing chemicals;
- establishing restrictions and control procedures for reducing hazards associated with the transport of dangerous chemicals;
- harmonisation of the regulations within Europe and between Europe and the rest of the world;
- containing the risks of major accidents;
- international research and development programmes;
- disseminating information to consumers about possible impacts of products;
- improving the interface between industry and environmental protection organisations;
- encouragement of improved management practices;
- encouragement of the development of substitutes for the most dangerous chemicals;
- alternative strategies for sectoral policies (eg, agriculture).

Figure 38.3
The relation between environmental protection research (risk assessment) and policy (risk management)
Source: Bourdeau, 1989

Figure 38.4
Comparison of international programmes on the control of chemical risk
Source: McCutcheon, 1993, personal communication

UN International Programme of Chemical Safety	Community Existing Regulation 93/793/EEC	OECD Programme on Existing Chemicals
	Industry data on HED-SET	
IRPTC Databank on Chemicals	EUCLID Database	
IPCS Working List	Priority Substances	OECD Working List current phase
	Review of EUCLID & Annex VII other data	Review of HPV and SIDS data
	Initial Risk Assessment	SIAM
EHCD	Risk Assessment Published in OJ	Post SIAM

Box 38A
International programmes on chemicals

UN, OECD and EU initiatives in the field of chemical risk control are compared in Figure 38.4. The horizontal arrows show the cooperations and exchange of data between the different organisations.

The left-hand column shows the UNEP dependent programmes. IRPTC was founded in 1975 to establish and operate a data bank on chemicals. IPCS deals with risk evaluation and development of methodologies for such evaluation and related matters based on these data (see Chapter 17).

The centre column shows the EU system, which is explained in detail in Figure 38.5.

The right-hand column contains the OECD strategy to overcome the lack of information and to find target chemicals.

Notes:
EHCD – Environmental Health Criteria Data.
EUCLID – European Chemical Information Database.
HED-SET – Harmonised Electronic Data Set.
HPV – High Production Volume Chemicals means chemicals produced in amounts exceeding 1000 tonnes per year.

IPCS – International Programme on Chemical Safety.
IRPTC – International Register of Potentially Toxic Chemicals.
OJ – Official Journal of the EU.
SIAM – SIDS Initial Assessment Meeting.
SIDS – Screening Information Data Set.

Figure 38.5
Main components of EU chemicals control strategy
Source: CEC, 1994

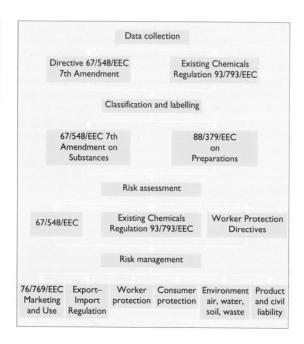

Chemical control in the EU

The basic legislation on the control of chemicals in the EU is EC Council Directive 67/548/EEC on the labelling, packaging and classification of dangerous substances – the first of a growing number of directives aimed at protecting human health and the environment. During the 1970s and early 1980s the Community's chemicals control legislation was basically a 'command and control' approach. This has developed considerably and the EC's Fifth Environmental Action Programme now makes it clear that future legislation should be based on a partnership between regulators, industry and the general public towards the attainment of sustainable development. Thus, the EC Regulation 93/1836/EEC, allowing voluntary participation by companies in a Community eco-management and auditing scheme, will be implemented as from June 1995. It aims to promote environmental performance improvements at industrial sites by committing them to establish environmental policies, programmes and management systems, to undertake environmental audits and supply information to the public. It provides the counterpart for product ecolabelling procedures which were introduced by EC Regulation 92/880/EEC.

Ecolabelling means the labelling with an EU-logo of products which have a reduced environmental impact compared with usual products, and is based on the development of a set of ecological criteria for groups of consumer products considered to raise significant environmental problems (eg, detergents, paper, washing machines, paints and packaging material). For products to qualify for the EU-logo they must comply with a set of specific ecological criteria. These are drafted to take account of the product's environmental impact at all stages of its life-cycle, from raw material through manufacture, distribution, use and final disposal. Currently, five Member States (Denmark, Germany, France, Italy and the UK) have started product life-cycle analysis of paper, detergents, paints, hair sprays, refrigerators and washing machines.

The main elements of chemicals control in the EU are shown in Figure 38.5, which indicates that legislation draws a line between 'existing' and 'new' chemicals. Existing chemicals are those which were on the EC market before 18 September 1981. The Commission made up an inventory listing these substances known as the European Inventory of Existing (commercial) Chemical Substances (EINECS). A chemical substance is 'new' and must be notified if it cannot be found among the 100 116 substances listed in EINECS. Substances in EINECS are exempt from notification, but if they are dangerous they must be classified and labelled by the manufacturer or importer.

Council Directive 92/32/EEC (Seventh Amendment of 67/548/EEC) regulates the testing and notification procedures for any new chemical put on the market in the EU. More than 800 chemicals have been notified by the EU since 1983 based on this system.

With the legal basis for the notification having been harmonised to a large extent within the EU, the chief goal for the years to come is to harmonise the assessment. This is done by the Risk Assessment Directive, 93/67/EEC, which requires an assessment of the chemical risks to humans/environment to be carried out, and establishes a 'Technical Guidance Document' which assists member countries to implement the environmental risk assessment of notified chemicals.

In contrast to these 'new' chemicals, many of the more than 100 000 chemicals in the EINECS inventory have never been tested for their effects on human health and the environment. To set up a long-term strategy to tackle this problem the EU adopted Council Regulation 93/793/EEC designed to identify and lead to the control of risks from

The UNCED (Rio de Janeiro, 1992) also decided in Chapter 19 of Agenda 21 (UNCED, 1992):

1   to expand and accelerate international assessment of chemicals;
2   to harmonise the classification and labelling of chemicals;
3   to exchange information on toxic chemicals and chemical risks;
4   to establish risk reduction programmes;
5   to strengthen the national capabilities and capacities for management of chemicals; and
6   to prevent the illegal international traffic in toxic and dangerous products.

It was underlined that a successful implementation of these six programme areas is dependent on intensive international work and improved coordination of ongoing international activities, as well as on the identification and application of technical scientific, educational and financial means. Collaboration on chemical safety between UNEP, ILO (International Labour Organisation) and WHO in IPCS (International Programme of Chemical Safety – for details see Chapter 17) should form the core for international cooperation on environmentally sound management of chemicals. In this field, Agenda 21 also stressed the need for cooperation between international programmes of IPCS, OECD and the EU (see Box 38A) and other regional and environmental programmes, as well as the involvement of NGOs.

An initiative in the USA has had an important impact on the use of chemicals. Here, companies with nine or more employees have to report on the use and disposal of 313 listed toxic substances. The report has to contain the release of these substances to air, water, or underground wells. This toxic release inventory (TRI) has now become an important source of public information, and introduction of this system has led to a rapid reduction of emission of these substances by one third in three years. Now a process of re-evaluating the TRI has started, since as currently formulated it incorporates only the amount of waste to the environment and not the hazardous material incorporated into the product. Furthermore, the list of toxic substances itself is incomplete. Several European countries (see Chapter 17) have recently adopted a similar approach for reducing chemical risk. Moreover, OECD/IPCS/UNEP/USEPA have initiated the pilot phase of a Pollutant Release Transfer Register based on the TRI.

some of the existing chemicals. According to this programme the main goal is the collection of basic information about existing chemicals, including their uses and characteristics, environmental fate and pathways, toxicity and ecotoxicity. This is to be reached through a three-step approach:

1 The initial phase (June 1993–June 1994) included the assessment of approximately 1800 chemicals produced or imported in amounts above 1000 tonnes per year. Fewer data are required for chemicals manufactured or imported in quantities between 100 and 1000 tonnes per year.

2 Phase two, beginning in 1994, aims to publish at regular intervals a priority list of substances 'requiring immediate attention because of their potential effects on human health and the environment' involving a cooperative procedure among the Member States. Special attention will be given to substances having chronic effects, those toxic to reproduction, and mutagens.

3 The third step includes risk assessment and the development of a control strategy. The priority chemicals will be divided up among Member States in order to perform individual risk assessment. The country acting as rapporteur for a particular chemical is able to ask for additional information from industry or from the importer before preparing the risk assessment. Based on this work, the rapporteur will propose and report a control strategy. The European Commission will make sure that the individual reports are comparable before final publication. On the basis of these reports the European Commission may propose restrictions on the use of the chemicals.

A further step in a chemical risk control strategy is to strengthen environmental research with the aim of improving, among other things: the understanding of the multitude of processes whereby chemicals are distributed in the different environmental compartments, their fate once there, how they affect the structure and functioning of ecosystems, the search for ways to prevent pollution effects, to restore damaged ecosystems, and to develop safer industrial and agricultural production processes. Equally important is research into the development of computer software for assessing the toxic properties of chemicals and for making maximum use of current information in mathematical models (eg, QSAR techniques; see Chapter 17) simulating environmental dispersion, fate and consequences of chemicals.

REFERENCES

Bourdeau, Ph (1989) Chemicals in the Environment. In: Marini-Bettòlo, G B (Ed) A Modern Approach to the Protection of the Environment. Pontificiae Acedemiae Scientiarum Scripta Varia 75, 457–76.

CEC (1993) Towards Sustainability. A Community Programme of Policy and Action in Relation to the Environment and Sustainable Development. Commission of the European Communities, Luxembourg.

CEC (1994) Chemical Risk Control. ECSC-EEC-EAEC, Commission of the European Communities, Luxembourg.

Lönngren, R (1992) International approaches to chemicals control. A historical overview. KEMI, Stockholm.

UNCED (1992) Agenda 21. United Nations Conference on Environment and Development, Conches, Switzerland.

VROM (1991) Essential environmental information. The Netherlands 1991. Ministry of Housing, Physical Planning and Environment, The Hague.

Part VI
Conclusions

There is no simple solution to the problem
Source: Spectrum Colour Library

39 General findings

There are no simple answers to the question: 'how healthy is Europe's environment?' The results of the assessment presented in the various parts of this report are necessarily detailed and specific. On several aspects, current knowledge of environmental processes and observations of facts lead with confidence to certain conclusions. On others, where data are insufficient or uncertainties are significant, best estimates can be derived on the basis of best available information. These need to be read in the context of current understanding, the sources of uncertainty, and the information and research needs for improving the assessment.

Highlights and responses

The main highlights and findings of the report are summarised below (Chapter 40). This also includes a summary of the most significant *policy responses* and/or *options* being taken or proposed. The *'findings'* focus on the main *facts* and *key trends* which require attention, and the *'responses'* are concerned either with management and policy actions or with monitoring and implementation. These results are presented according to the main themes of the assessment: by environmental media, pressures and the main human activities.

A full analysis of responses and the state of actions to protect the environment, their effectiveness and implementation was beyond the scope of the present exercise. This could be usefully pursued in parallel with future exercises reporting on the state of the environment.

Prominent environmental problems

Part V of the report (Chapters 27 to 38) highlighted the following 12 prominent European environmental problems derived out of a larger unconsolidated list of 56 found in Appendix 1:

- climate change
- stratospheric ozone depletion
- the loss of biodiversity
- major accidents
- acidification
- tropospheric ozone and other photochemical oxidants
- the management of freshwater
- forest degradation
- coastal zone threats and management
- waste reduction and management
- urban stress
- chemical risks

The 12 chapters covering these problems act as conclusions in their own right, where media pressures and reponses are integrated in assessments of particular problems of European concern. Thus, these are not treated again here.

The decision of which problems to highlight is a choice and depends upon the criteria used and how they are applied. Nevertheless, the current list appears robust and insensitive to major alterations when comparisons are made with important environmental problems identified in national reports and at international level by, for example, UNEP, OECD, the EC Fifth Environment Action Programme, and by the US Environmental Protection Agency. Additional issues may need to be highlighted – in particular, for example, soil erosion and desertification, which are of specific regional importance – but none would be deleted. This level of consensus on prominent European environmental problems is important to underline, since uncertainties in this field often cloud the real issues. Problems not directly highlighted in this list of 12 issues will normally be discussed within other relevant parts of the report.

The results in particular of the analyses of specific environmental problems demonstrate the importance of a multi-media, cross-sectorial integrated approach to environmental assessment. Integrated assessments can be made of different media, pressures, human activities and issues. They can benefit the understanding of the causes and effects of environmental problems and better clarify how to act and the consequences of such actions. Thus, to ensure coherent, continent-wide responses to environmental problems and threats, it is particularly important to perform integrated assessments at the European level. However, problems with data availability and incompatibility (see below) also make them particularly difficult to carry out. Future reports should aim to expand such integrated assessments to operationalise better the baseline information on the state of the environment.

Data and information

Throughout this report much mention has been made of problems with data and information. Although in recent years environmental data have increased in availability as environmental monitoring and research have developed, there are still significant data gaps as well as large discrepancies in quality. Furthermore, much information that is available is often inadequate to provide the basis for a rigorous assessment and accurate reporting on the state of Europe's environment. One major constraint at the European level is the lack of comparable, compatible and verifiable data across European countries due to existing differences in the methods of collection, the terminologies used and the degree of transparency. Improving the state of environmental information is a high priority in order to ensure that actions and decisions being taken by policy makers, the public and all sectors of society respond to the actual needs. The European Environment Agency has been established for this very purpose, and with its European environment information and observation network is now working to ensure the supply of objective, reliable and comparable information on the environment for the years to come. From the findings of this report a summary of the current state of environmental information, its strengths and weaknesses, is made below in Appendix 2. This includes an appraisal both of current data availability and existing gaps in the major areas of environmental concern.

Timeliness

The latest data presented in this report range from the late 1980s to 1993. As the report was being finalised more recent data became available which it was impossible to fully integrate into the assessment. This is a continuous process which can only be addressed by regular reporting and updating. Timely accurate reporting depends upon a better overall connection between data producer (monitoring), assessor (reporting) and consumer. A better understanding of the needs and uses of environmental information is essential to improve efficiency. Preparing this report was the first time that Europe's Environment had been assessed as a whole and in such a comprehensive manner. The difficulties encountered, described above, can largely be ascribed to this fact, which in turn contributed to the amount of time it has taken to prepare and complete this report.

Now that this exercise has been performed once over in a comprehensive way for the whole of Europe, future reporting at this level should be able to build on this, make comparisons and identify trends, overall facilitating a faster turn-around and a more regular reporting cycle.

Areas for future consideration

The report has attempted to provide a comprehensive analysis of the state of Europe's environment so as to constitute a strong basis for improving decision-making and raising environmental awareness across the board. Despite the comprehensive approach taken, the coverage shows some weaknesses, for example with regard to the assessment of rural environments, the service sector and the impacts of land disturbance. The state of non-renewable resources is not addressed and neither is the impact of military activities. Future reports might usefully consider these aspects in more detail.

Cooperation and information change

During the preparation process for this report, the opportunities for improved cooperation and coordination were explored. This process has involved, in particular, establishing contacts with a wide range of European data sources. Indeed, a network of contacts in more than 40 countries and international organisations was established to channel data and information and make assessments to form this report. This network included official national focal points and specialist institutes at national and international levels, both governmental and non-governmental, involving more than 250 individuals (see Contributions and Acknowledgements list at the front of the report). Preparing this report has shown the importance of a strong network for efficient environmental reporting. Maintaining and improving cooperation at all these levels is therefore of the utmost importance.

In addition to the report itself, this European cooperation facilitated the preparation of the *Statistical Compendium*, the companion volume to this report, the result of a first joint exercise of its kind between the statistical services of the different international organisations involved: Eurostat (the European Commission's statistical office), UNECE, OECD and WHO, and the European Environment Agency.

Additional products

A considerable effort was given to the design of the report's structure in order to provide a flexible, comprehensive document with multiple entry points to serve a wide variety of potential users. Other products are being developed to help disseminate and give access to the information more easily. In particular, a project transferring the report to CD-ROM is underway which will integrate into a single product some of the more detailed background information used to compile this report and especially the contents of the *Statistical Compendium*. A short 'popular' version of the report is also being commissioned, as well as an environmental atlas.

Future continuity

The European Environment Agency is the inheritor of this work. It will, among other things, produce three-yearly state-of-the-environment reports, which will provide the link with the activities in preparing this report, and assure continuity with future environmental reporting.

A timely response is essential in solving environmental problems
Source: Science Photo Library

40 Highlights and responses

AIR

Findings

1   Urban air quality is a continuing problem. Improvements have occurred since the 1970s for sulphur dioxide, particulate matter and lead. However, emissions of nitrogen oxides are not decreasing (due especially to road transport). Furthermore, tropospheric ozone is an increasing concern as a result of emissions of nitrogen oxides and volatile organic chemicals.

2   Short-term air pollution levels of one or more pollutants exceed the WHO Air Quality Guidelines (AQG) at least once in a typical year in about 70 to 80 per cent of the 105 European cities with more than 500 000 inhabitants.

3   Long-term exposure to sulphur dioxide and particulate matter exceeds WHO AQGs in 24 of the 61 cities studied monitoring these pollutants and included in the study for this report.

4   Short-term peak levels of ozone during summer photochemical smog episodes are estimated to exceed WHO AQG values in 60 large European cities, concerning over 100 million inhabitants. These episodes also affect areas outside the cities from where the pollution originates.

5   Europe accounts for about 25 per cent of global sulphur dioxide and nitrogen oxides emissions. Emissions of these gases from power plants contribute most to total acid deposition (30 to 55 per cent). Nitrogen oxides are providing a growing contribution to environmental acidification. The highest acid deposition levels are found in the highly populated and industrialised belt extending between Poland and the United Kingdom. Agriculture is the main source of ammonia emissions to the atmosphere, which also contribute to acid deposition. In 60 per cent of Europe's area, critical loads for acidification are exceeded. Very strong emission reductions of all acidifying compounds are needed to reduce deposition below these critical loads (about 90 per cent for sulphur dioxide and nitrogen oxides and more than 50 per cent for ammonia).

6   The European contribution to global emissions of chlorofluorocarbons (CFCs) and halons, which threaten stratospheric ozone, is 35 to 40 per cent.

7   Europe accounts for about 25 per cent of carbon dioxide and 16 per cent of methane emissions from human activities world wide. An increase in such greenhouse gases equivalent to a doubling of the carbon dioxide concentration is expected sometime during the 21st century and is estimated to produce warming exceeding 4 to 5 times the limits of natural variability. The rate of temperature increase would be about 0.3°C per decade. Overall precipitation and evaporation would increase by 3 to 15 per cent. Sea level would rise about 22 cm in 2050 and 50 cm by 2100. This would cause permanent flooding, and increased frequency of temporary flooding,

in many low-lying areas of Europe, as well as coastal erosion changes.

Responses/policy options

1   Promoting energy conservation, increased use of renewable energy resources, good housekeeping practices, reductions in road transport emissions, increased use of clean technologies.

2   Improving urban air quality through substitution of coal and heavy fuel oils and by controlling nitrogen oxides emissions from stationary sources and in motor vehicles and reducing emissions from diesel engines.

3   Implementation by all countries of existing international conventions such as the Geneva Convention on long-range transboundary air pollution.

4   Implementation of the recent volatile organic chemicals (VOCs) protocol under the Geneva Convention to tackle the problems of tropospheric photochemical oxidants.

5   Implementation by all countries of the Copenhagen Amendment to the Vienna Convention which will put a stop to global emissions of CFCs, carbon tetrachloride and methyl chloroform before 1996. A faster mandatory reduction scheme has already been adopted in the EU aiming at phase-out before 1995.

6   Implementing the Rio Framework Convention which sets the ultimate objective of stabilisation of greenhouse gas concentrations in the atmosphere at a level that would prevent dangerous anthropogenic interference with the climate system. Reductions of CO_2 emissions are technically feasible, by increasing efficiency and energy savings, by increased use of renewable energy sources and – temporarily – by shifting fossil fuel energy generation from coal and oil to natural gas. A tax on CO_2 energy, as proposed by the European Commission, would stimulate actions in this direction.

7   Development and implementation of integrated emission inventories and integrated pollution, prevention and control (as for water and soil).

INLAND WATERS

Findings

1   The temporal and regional distribution of water resources creates scarcity problems in several Central, Eastern and Southern European countries.

2   On average, 15 per cent (the world average is 8 per cent) of the total renewable resource is abstracted per year, ranging from 0.1 per cent in Iceland to more than 70 per cent in Belgium.

3   European industry abstracts more water (53 per cent) than agriculture (26 per cent) and the domestic sector (19 per cent). Water abstraction for domestic and agricultural purposes is increasing, whereas industrial abstraction is decreasing or has stabilised.

4   Sixty-five per cent of Europe's population is supplied from groundwater, the quality of which is threatened in many places. Overexploitation of groundwater is a serious problem in nearly 60 per cent of European industrial and urban centres, threatening also 25 per cent of major wetlands. In coastal regions of the Mediterranean, Black and Baltic seas, intrusion of saltwater into groundwater reservoirs is restricting the use of groundwater.

5   Nitrate concentrations in the soil water at one metre depth are estimated to exceed the EU guide level (25 mg NO_3/l) and the maximum admissible concentration (50 mg NO_3/l) in 85 per cent and 20 per cent, respectively, of agricultural land. Nitrate problems in groundwater are most conspicuous in the northwestern parts of the continent.

6   The EU standard for total pesticides (0.5 μg/l) is estimated to be exceeded in soil water in 75 per cent of agricultural land in EU-EFTA and in 60 per cent in Central and Eastern Europe.

7   River and lake eutrophication caused by excess phosphorus and nitrogen from agriculture, domestic and industrial effluents is a pan-European problem of major concern. The domestic sector and industry are the main contributors of phosphorus, agriculture the main contributor of nitrogen.

8   Nutrient concentrations in rivers and lakes are highest in a belt from the southern UK to the Balkans and Ukraine, with high population density and/or intensive agriculture. However, due to improved sewage treatment and substitution of phosphorus in detergents, contamination of large rivers with organic matter, ammonia and phosphorus has generally decreased over the last 10 to 15 years apart from in Central and Eastern Europe. Nitrate concentrations on the other hand continue to increase in most rivers. Water quality has improved markedly in several of the largest lakes in Nordic, Western and Southern European countries. Small shallow lakes, however, continue to be eutrophic due to high nutrient loadings.

9   Acidification of rivers and lakes on sensitive soils is causing widespread fish-kills in large parts of the Nordic countries. To a lesser extent, surface waters in Northwestern and Central Europe and in the northwestern part of the Russian Federation are also affected.

10   The concentration of heavy metals in the great majority of rivers, such as the Rhine, no longer exceeds the norms. However, concern still remains for the levels of metals, organic micropollutants, salts, radionuclides and pathogens, in particular in Central and Eastern Europe.

11   Channelisation, impoundment and other regulation works are major threats to the maintenance of natural physical conditions and functional integrity of river ecosystems. Highly intensive regulation works on a river system may increase the risk of flooding and thus add threats to human life and livestock.

Responses/policy options

1   Preventing leakage from water distribution systems. Reducing sectoral water demand by water-saving and improvement of reuse/recycling techniques.

2   Introducing or strengthening integrated water resource management.

3   Reducing the application of fertilisers (especially nitrogen as required by the EC Nitrate Directive) and pesticides to the precise level needed for optimal plant production and pest management to restrict the loss of these chemicals in order to safeguard drinking water and the ecological balance of surface waters.

4   Ceasing the dumping of mobile chemicals at sites overlaying important groundwater reservoirs. Initiation of clean-up operations at sites already severely contaminated and threatening the quality of groundwater.

5   Increasing and improving advanced sewage treatment (as asked for in the EC Directive on Urban Waste Water Treatment) and application of cleaner technologies in industry.

6 Development and implementation of integrated emission inventories and integrated pollution prevention and control (as for air and soil).

7 Establishment of improved set of base-line data for emissions of contaminants, contamination levels and ecological effects using appropriate methods producing reliable and comparable results.

THE SEAS

Findings

1 Europe's seas display a wide range of environmental conditions and impacts. The Black Sea, in general, is most affected by human activities and the northern seas (White, Barents, Norwegian seas and the North Atlantic Ocean) the least affected.
2 Lack of effective catchment management creates problems particularly in the Black and the Caspian seas, but the Mediterranean, the Baltic, the Barents and the White seas also face localised problems.
3 Coastal zone pollution and conflicting uses of the coastal zone are issues of growing concern in all areas. The biggest problems are found in the Black Sea, the Mediterranean, the North Sea and the Baltic.
4 All seas, except for the northern seas, are facing eutrophication problems in some areas with related adverse effects such as algal blooms. Nitrate has increased by a factor of two to three, and phosphorus by as much as seven in some coastal areas of the Black Sea and the Sea of Azov. In the Baltic, nitrogen and phosphorus have increased by a factor of four and eight, respectively.
5 Insufficient control of offshore activities is creating localised pollution problems in the Black Sea, the North Sea and the Caspian Sea.
6 Introduction of non-indigenous species (eg, comb jelly) has recently had severe ecological impacts on the Black Sea.
7 Contamination by organic micropollutants impacts the fauna in almost all Europe's seas. Concentrations of DDT residues and PCBs in fish are three to ten times higher in the Baltic than in the Northern Altantic. In the North Sea, high concentrations of PCBs and tributyl tin have impaired the reproductive success of seals and dogwhelks.
8 Overexploitation of fish and shellfish stocks is a common problem in all European seas.
9 The Mediterranean shows low biological productivity but high biodiversity (about 10 000 species) with many endemics, some of which (eg, monk seal and loggerhead turtle) are very endangered.
10 Of particular concern in the Barents Sea are the large amounts of radioactive waste dumped by the former USSR. So far, however, there have been no signs of radioactive contamination of water and biota.
11 Large changes in sea level have occurred over the last 60 years in the Caspian Sea, for which natural cycles are thought to be mainly responsible. Since 1977 sea level has risen by 1.5 metres.

Responses/policy options

1 Strengthening international cooperation by full implementation of existing conventions (eg, Barcelona, Helsinki, Oslo, Paris), and complementing these with others for those seas not yet covered.
2 Pursuing the responses and policy options for inland waters, which will have important impacts on the conditions of Europe's seas.

SOIL

Findings

1 Soil erosion is increasing in Europe. An estimated 115 million ha (12 per cent of the total land area is threatened by water erosion, 4 per cent by wind erosion) are affected. causing important loss of fertility and water pollution. More than 90 per cent of the eroded land is located in the Mediterranean zone.
2 Critical values for acidification are estimated to be exceeded in about 75 million ha of the forest soils of Europe.
3 Widespread overapplication of nitrogen and phosphorus fertilisers results in leaching and runoff, leads to contamination of drinking water and causes eutrophication.
4 It is estimated that 3.2 million ha suffer from organic matter loss in Europe and 42 million ha from compaction. Salinisation is estimated to affect 3.8 million ha. Waterlogging, which occurs mainly in the North, is estimated to involve 0.8 million ha.
5 There are 200 000 ha of derelict industrial land inventoried in the EU; remediation costs are estimated at ECU100 billion.

Responses/policy options

1 Implementing an integrated approach to fight soil erosion from improper farming or forestry practices, tourism, sport and other recreational activities, by linking socio-economic factors with soil vulnerability – technical solutions are well known but have often been unsuccessful because socio-economic factors are ignored. Giving preference to prevention methods.
2 Reducing acidic deposition to soils by lowering emissions of sulphur dioxide and nitrogen dioxides (implementing the Geneva LRTAP Convention).
3 Adopting measures and practices which mitigate effects of nitrogen leaching, eg, timing of fertiliser and manure application at correct doses; sowing winter crops; identification of vulnerable zones deserving protection; banning very mobile and broad-spectrum pesticides together with integrated pest management systems. Revising EC limit values of heavy metals in sludge and soils.
4 Compilation of a systematic register of contaminated sites, defining standardised limit values, and adopting appropriate legal and technical measures and clean-up plans for most contaminated sites.
5 Development and implementation of integrated emissions inventories and integrated pollution prevention and control (as for air and water).

LANDSCAPES

Findings

1 Europe's landscapes are diverse and rich in natural and cultural dimensions.
2 In many places European landscapes are undergoing changes or disappearing because of agricultural intensification or abandonment, urban expansion, infrastructure, transport, etc, adversely affecting important values and functions.
3 Six per cent of Europe's land area is under landscape protection, but generally with a weak legal status.

Responses/policy options

1 Making landscape considerations an important factor in promoting sustainable development and the conservation of biodiversity by complementing species and site-specific approaches. Involvement of farmers and landuse management authorities in landscape restoration
2 Reinforcement of international landscape protection.
3 Protecting outstanding landscapes through site designation and effective landuse planning mechanisms.
4 Improving information on landscapes and landscape changes.

NATURE AND WILDLIFE

Findings

Ecosystems

1 Europe's ecosystems comprise a large diversity of habitats and include a rich fauna and flora. Their distribution, size, ecological value and legal status are extremely varied.
2 A comprehensive and integrated database on the distribution of Europe's main natural or semi-natural ecosystems does not exist. The current assessment was thus based on 'representative' sites (1750 sites in total) identified and analysed for seven ecosystem groups: forests, scrub and grasslands, inland waters, wetlands, coastal and marine ecosystems, mountains, deserts and tundras.
3 Forests once covered 80 to 90 per cent of the territory and now account for 33 per cent of land cover. The 156 identified most natural sites cover about 2 per cent, and only a tenth of the larger examples (>10000 ha) can be found in Central and Western Europe. Most large forest and wetland ecosystems are located in the North and East.
4 Large and important (especially dry) grassland sites are concentrated in the South (mostly in Spain).
5 Natural river sites are small and threatened, and, since 1900, 75 per cent of sand dunes have been lost in France, Italy and Spain. Most river ecosystems lack naturalness due to regulation works, pollution and loss of riparian ecosystem functions.
6 Bogs, fens and marshes have disappeared in large numbers, mainly in Western and Southern Europe (Spain has lost 60 per cent). Oligotrophic and calcium-rich bogs and fens are highly endangered.

Fauna and flora

7 Many plant and animal species groups are currently declining and threatened with extinction: fish with 53 per cent under threat, reptiles 45 per cent, mammals 42 per cent, and amphibians 30 per cent. Marine mammals are affected by water pollution, fishing and hunting. The Mediterranean monk seal is among the 10 most threatened species in the world. More than 50 per cent of all threatened birds are coastal or wetland species.
8 21 per cent of Europe's 12 500 higher plant species are threatened; 27 species have actually become extinct. Aquatic, calcareous and acidic plants are at special risk. More than 30 per cent of previously rare, but not threatened, aquatic plants have become endangered.
9 Europe is a centre of international (often illegal) species trade.

Nature conservation

10 There has been a rapid increase in the total area of protected sites in Europe; two thirds have been added since 1972. However, most areas are small and fragmented: only 3000 (7 per cent) of the 45 000 European nature reserves are larger than 1000 ha. Landscape protection dominates (57 per cent), followed by national parks and nature reserves (32 per cent) and international designations (23 per cent) (with a 12.5 per cent overlap). The EC-Bird Directive covers 7 million ha, the UNESCO Biosphere Reserves about 10 million ha and Ramsar wetlands more than 5 million ha (some sites hold multiple designations).
11 Implementation and control is a problem in most international and national protection areas due most frequently to a lack of funds and staff.

Responses/policy options

1 Improvement of the implementation of existing protection schemes on national and international level, by monitoring threats and stresses in a systematic manner, and by providing funds and adequate management mechanism in protected areas. Developing strategies to integrate nationally protected areas into international nature conservation schemes.
2 Improvement and harmonisation of red species lists to achieve comparability on national and international levels.
3 Identification, maintenance and protection against impacts and outside stresses of all remaining large natural and semi-natural ecosystems, especially alluvial and lowland (ancient) forests, lowland dry grass- and heathlands, natural riparian corridors, wetlands and natural coastlines.
4 Establishment of ecological networks (buffer zones, corridors linkages) that integrate landuse activities outside protected areas.
5 Improvement of the ecological conditions of freshwater ecosystems (especially rivers) through adequate floodplain management, reduced emissions of pollutants within the catchment area, restoration of damaged components (eg, banks), reintroduction of native fish species, etc.
6 Application of an integrated approach to management of coastal zones, with special attention being given to overgrazing of saltmarshes, destruction of dunes and contamination of estuaries.

URBAN ENVIRONMENT

Findings

1 Two thirds of the European population live in urban areas, which cover about 1 per cent of the total land area. Almost all European cities are experiencing decentralisation of population, independent of population density and city growth.

2 The changes in urban structure have affected mobility patterns and trends in the last decades. Traffic flows have increased in all European cities in terms of the number and length of trips. The share of public transport has fallen in most cities by 20 per cent. However, some important initiatives have been taken in some cities (eg, Zurich) to reverse this trend.

3 Despite some improvements, urban air quality is frequently unsatisfactory in large and medium-sized cities due to sulphur dioxide, particulate matter and ozone. Most cities exceed short-term WHO Air Quality Guideline (AQG) levels (24-h) for sulphur dioxide and/or particulate matter/black smoke (winter smog). Most Central and Eastern European cities exceed WHO guidelines for combined exposure to sulphur dioxide and particulate matter of $50+50$ $\mu g/m^3$ (annual average) and $125+125$ $\mu g/m^3$ (24-h average).

4 Exceedances of 65 dB(A) noise level occur in most cities, affecting between 10 and 20 per cent of inhabitants in Western Europe and up to 50 per cent in some cases in Central and Eastern Europe.

5 A European city of one million inhabitants typically consumes on average each day about 11 500 tonnes of fossil fuels, 320 000 tonnes of water and 2000 tonnes of food.

It produces 300 000 tonnes of wastewater and 1600 tonnes of solid waste, plus 25 000 tonnes of carbon dioxide.

6 Urban water supply is not allocated and managed efficiently. Only 5 per cent of all piped water used in houses is for drinking. Leakages from the distribution system of 30 to 50 per cent are frequently reported.

Responses/policy options

1 Promoting sustainable urban patterns through a balanced landuse and efficient use of space.

2 Reducing demand for travel through landuse planning, focusing on encouraging development near points of high transport accessibility, proximity to public transportation nodes, mixed use development.

3 Promoting efficient transport management through urban road pricing, integrated public transportation, vehicle priority schemes, traffic calming, traffic bans in designated areas, parking controls.

4 Reducing energy consumption in urban areas through efficiency and conservation measures, including building design and materials, district heating and cogeneration, rational use of energy sources, efficient public transportation.

5 Reducing atmospheric emissions and exposure to noise pollution.

6 Orienting urban renewal projects towards improving quality of urban life, reducing the use of water, energy and materials.

7 Implementing programmes for separate waste collection, recovery and recycling.

HUMAN HEALTH

Findings

1 Excessive levels of air pollutants affect a significant proportion of the European urban population. Suspended particulate matter is estimated to pose the greatest burden to health, provoking asthma and obstructive airway disease.

2 Indoor exposure to radon is estimated to cause about 10 000 cancer deaths annually in Europe.

3 Contamination of drinking water and food by chemicals (eg, nitrates) and by microbiological agents is a problem in several areas. Bathing water contamination is estimated to result in over 2 million cases of gastro-intestinal diseases annually.

4 Lead remains a health problem. It may be responsible for mental impairment in at least 400 000 children in eastern parts of Europe.

5 The significant figures from road traffic accidents (over 350 people killed and 6000 injured *daily* in Europe) dominate accident casualty statistics, for which the road traffic environment is certainly partly responsible.

6 Major chemical accidents caused 1100 fatalities between 1980 and 1991.

7 Life expectancy at birth is several years lower and infant mortality rate higher in Central and Eastern Europe, but there is no clear evidence that this is related to environmental quality as such.

Responses/policy options

1 Performance of European-wide epidemiological studies to relate health to environmental conditions.

2 General reduction of population exposure to chemicals through all routes (air, water, food). Control of urban air pollution to respect the WHO Air Quality Guidelines.

3 Reducing nitrate in drinking water and treating sewage water.

4 Phasing out lead from petrol.

5 Cleaning up the most contaminated sites.

POPULATION – PRODUCTION – CONSUMPTION

Findings

1   Europeans represent 12.8 per cent of the world population (1990) and this proportion is decreasing. A strong decline in fertility has occurred in most countries.
2   Europe, and OECD countries, account for 82 per cent of world commercial energy consumption (40 per cent in Europe), and generate over 70 per cent of the world's industrial waste.
3   Between Eastern and Western Europe, adjusted living standards vary by more than a factor of 5.

4   Western Europe produces cereal surpluses but imports large quantities of animal feed from America and Asia. Most East European countries rely on food imports.

Responses/policy options

1   Aiming for greater convergence in the economies of Western and Eastern Europe, and for sustainable production and consumption.

EXPLOITATION OF NATURAL RESOURCES

Findings

1   Europe has 8 per cent of world wide renewable freshwater resources but accounts for 15 per cent of withdrawals. Some European countries are already using a significant proportion of their water resources.
2   Although Europe (excluding Iceland and the former USSR) had 1.9 million hectares more forest and wooded land in 1990 than in 1981, atmospheric pollution and uncontrolled logging (Russian Federation) are causing forest loss and damage.
3   Tropical deforestation is to some extent linked with Europe's demand and supply of natural resources. European countries contribute indirectly to deforestation in tropical regions through the use of wood and wood products from these regions.

4   Energy consumption in Europe (excluding the former USSR) is greater than production, implying that Europe contributes to the depletion of energy resources in other regions of the world.
5   Europe has depleted most of its high-grade mineral reserves and now relies mostly on imports, mainly from Africa. Unlike the rest of Europe, the former USSR is still a major producer and consumer of many minerals.

Responses/policy options

1   Reducing consumption of non-renewable resources and improved care of renewable ones to reach sustainability.

EMISSIONS

Findings

1   Emerging air and water pollution problems in the last two decades have been addressed separately. However, this approach has proven inadequate to face the cross-media nature of environmental problems.

Emissions to air (see also Air)

2   About 25 per cent of global sulphur dioxide and nitrogen oxides emissions arise from Europe.
3   35 to 40 per cent of global chlorofluorocarbon (CFC) and halon emissions come from Europe.
4   Europe accounts for 25 and 16 per cent respectively of global carbon dioxide and methane emissions arising from human activities world wide.

Emissions to water

5   Intensive agriculture and livestock farming, combined with increasing use of chemicals, add an enormous pressure on the aquatic environment from non-point source emissions.

6   In densely populated areas of Europe, 43 to 64 per cent of the phosphorus input to inland surface waters is related to sewage discharges. Industrial effluent makes up only 3 to 12 per cent of the total. In these areas 22 to 41 per cent of the phosphorus discharge stems from agricultural activities.

Responses/policy options

1   Development and implementation of integrated pollution and prevention control and integrated emissions inventories.
2   Implementation of national and international legislation and conventions for the reduction of emissions of certain contaminants.
3   Adoption of special measures and practices in the energy, industry, agriculture and transport sectors in particular for the reduction of sulphur dioxide and nitrogen oxides emissions to air, and nitrogen, phosphorus and pesticide emissions to water.

WASTE

Findings

1 Waste generation has increased in all European countries. On average about 350 kg of municipal waste are produced per capita per year, but there are large differences between countries. The rate of increase in OECD Europe was 3 per cent per year between 1989 and 1990. Waste composition varies considerably among countries.

2 Industrial waste production per year in the late 1980s reached 330 million tonnes per year in OECD Europe and 520 million tonnes in Central and Eastern Europe. An increasing amount of industrial waste is considered hazardous.

3 Most waste is disposed in landfills. In some countries, incineration increased considerably during the 1970s but has remained constant since mid-1985 due to public opposition combined with increasing capital and operation costs. Increased effort towards waste minimisation and recycling are still offset by high production rates.

4 A significant amount of hazardous waste is transferred across European countries and from Europe to developing countries.

5 All European countries are faced with the current and potential risk of old waste dumps, which are estimated to cover between 1200 and 2700 km^2, while there are 2000 km^2 of derelict industrial land.

Responses/policy options

1 Waste reduction through product re-design, process reformulation, substitution of raw materials, restrictions on products and packaging.

2 Materials reuse, recovery and recycling; closing material cycles; requirement of best available technologies for production processes; restrictions on products and packaging; separate collection.

3 Safe waste management through integrated waste management plans, restrictions on landfills, requirements of best available technologies for waste treatment facilities, emissions standards for incineration plants, systematic monitoring of disposal facilities.

4 Reduction of transfrontier movements of hazardous waste through full implementation of Basel convention, harmonisation of hazardous waste standards, monitoring transfrontier movements.

5 Clean-up of contaminated sites through inventories of potential contaminated sites, establishing legal frameworks, lists of priority clean-up sites and actions, design and implementation of clean-up plans.

NOISE AND RADIATION

Findings

1 It is estimated that 65 per cent of the European population is exposed to environmental noise levels exceeding L_{eq} 24-h 55 dB(A) (causing annoyance and disturbance of sleep), 17 per cent to levels exceeding L_{eq} 24-h 65 dB(A) (serious impact), and 1.4 per cent to levels above 75dB(A) (unacceptable).

2 There are indications of a 5 per cent increase in UV-B radiation in winter in the northern hemisphere; overexposure to ultraviolet radiation is so far due mostly to lifestyle (exposure to sun and sun lamps).

3 The average annual exposure from natural sources of ionising radiations to the average person varies in Europe between 2 and 7 mSv (17 Western European countries) compared with the world wide average of 2.4 mSv. Radon concentrations may vary dramatically from one locality to another and from house to house; in certain areas exposure is excessive (see Health).

4 Comparing doses from ionising radiation resulting from artificial human induced sources (such as medical uses, radioactive emissions and waste disposal) to natural radiation indicates that the increased exposure from such activities is, on the whole, not a great cause for concern. However, artificial radiation may contribute a significant fraction of the dose over which there is some control.

5 Improved working practices now means that the level of radiologically significant discharges to the Irish Sea is about 1 per cent of peak levels in the 1970s.

6 The Chernobyl accident had the most widespread effect in Europe for increasing exposure to ionising radiations. Areas of severe environmental contamination are found in the vicinity of the plant and also, from other sources, around the Techa river in the southern Ural Mountains. Potential unquantified threats of environmental contamination are found in the Barents and Kara seas from the dumping of radioactive objects by the former USSR.

Responses/policy options

1 Application of engineering and legal measures to reduce exposure to noise, supported by research and improved public information.

2 For UV exposure: promoting increased public awareness of the hazards of UV; implementing existing and stricter controls on the release of ozone-depleting substances leading to their eventual elimination; and 'active adaption' to an environment of greater exposure to UV radiation.

3 For ionising radiations: reducing all unnecessary exposures; controlling occupational safety and exposure to the general public according to international recommendations and against approved exposure limits; improving diagnostic medical applications; in high radon areas, application of the necessary measures to prevent accumulation in dwellings.

CHEMICALS AND GENETICALLY MODIFIED ORGANISMS

Findings

1 About 100 000 chemicals are marketed in the EU, and between 200 and 300 new ones appear each year.
2 The state of knowledge on the toxicity and ecotoxicity of chemicals in use and circulation is unsatisfactory.
3 Between 1991 and 1994, nearly 300 notifications of field releases of genetically modified organisms (GMOs) were made in the EU.

Responses/policy options

1 Implementation and coordination of international programmes for the data collection, risk assessment and risk management of new and existing chemicals (eg, EC Directives, IRPTC/UNEP, OECD, ECETOC, ECDIN, IARC, FAO).
2 Notification schemes for new chemicals (eg, EU legislative control system).
3 Classification and labelling of chemicals.
4 For GMOs: preventative approach (EC Directives) to establish a common set of environmental risk assessment requirements and safety measures, circulation of information and notification for deliberate releases.

NATURAL AND TECHNOLOGICAL HAZARDS

Findings

1 From a major industrial accident reporting system, it appears that most accidents occurred in the petroleum industry, and that highly inflammable gases and chlorine were the substances most often involved.
2 Tanker accidents contribute only about 10 to 15 per cent of all the oil that reaches the sea as a result of human activity.
3 Precise knowledge of the full consequences of the Chernobyl nuclear accident is unlikely, but some unexpected effects have already emerged.
4 Natural hazards are having an increased impact on human settlements, probably because of the greater number of settlements and their increased vulnerability due to their uncontrolled extension into high risk areas.

Responses/policy options

1 Good planning, management and control of routine activities, and implementation of codes of good practice (HAZOP studies, EC Seveso Directives).
2 Emergency response plans (fixed installations) and transport planning (routing, vehicle choice and timing) – implementation of EC Seveso Directives, UNECE Convention on Transboundary Effects of Industrial Accidents, and awareness and preparedness programmes such as APELL.
3 Improved monitoring and reporting of accidents and spills to assess long-term impacts.
4 Good landuse and emergency planning to reduce potential impacts of natural hazards and their interaction with human activities.
5 For enhancing the safety of nuclear reactors: the preparation of an integrated, international approach to all aspects of nuclear safety (possible international convention), research and development on advanced reactor design, and the implementation of a strategic international assistance programme to tackle the specific problems in Central and Eastern Europe.

ENERGY

Findings

1 In 1990, primary production of energy in Europe (excluding the former USSR) (1040 Mtoe) was 62 per cent of gross inland consumption (which was 37 per cent of total world gross consumption). Contribution to GDP is 5 per cent in the EU.
2 In 1990, for all Europe, energy use was split 41 per cent for industry, 22 per cent for transport, and 37 per cent in the domestic and commercial sectors. Transport and domestic/commercial energy use is increasing steadily.
3 In Europe as a whole, the mix of fuels has shifted away from solid fuels and nuclear towards increased use of gas. In Central and Eastern Europe, solid fuels are still the main indigenous energy resource. Renewable energy is a particularly significant contributor in EFTA countries.
4 Energy consumption and GDP have been decoupled since the energy crises of the 1970s. Large gains have been made in energy efficiency in Western Europe but little or no progress has been made recently; in Central and Eastern Europe high energy intensity is being reduced through industrial restructuring.
5 Electricity use is highest in Western Europe (especially EFTA). In the EU, 35 per cent is generated from nuclear power; EFTA relies for 63 per cent of electricity generation on renewables (particularly hydropower); in Central Europe, solid fuels provide nearly two thirds of electricity produced; in the former USSR, over one third of electricity was generated from gas in 1990.
6 The main impacts arising from energy use are local air pollution, acidification, tropospheric ozone and climate change. There are also more local impacts on water, soil and land. Energy consumption accounts for around 95 per cent of human induced sulphur dioxide emissions and 97 per cent for nitrogen oxides in 20 European countries in 1990.

Responses/policy options

1 Incorporating 'external' environmental costs into energy prices to make them better reflect real market prices (especially true in Central and Eastern Europe). Use of economic instruments, such as pollution charges, taxes (eg, carbon dioxide/energy tax), and tradeable permits. Removal of restrictive rules (inappropriate subsidies).

2 Energy efficiency programmes (for example, least cost planning, energy efficiency standards for appliances, products and vehicles); focus on operating efficiencies of power stations and energy distribution networks, and energy savings in the transport and household sectors.

3 Fuel switching to natural gas, which produces less carbon dioxide per unit energy generated.

4 Awareness building, education, and incentives aimed at sustainable energy use and behavioural change. Codes of conduct and agreements developed with energy industries.

5 Investments in research and development programmes, for example, 'clean coal technologies' and renewable energies.

INDUSTRY

Findings

1 Manufacturing industry contributes about 25 per cent of GDP in the EU and closer to 50 per cent in some Central and Eastern countries. Some of the main sectors having significant environmental impacts are chemicals, pulp and paper, cement, steel and non-ferrous metals.

2 There has been a reduction in the amount of energy consumed per unit output (particularly in Western Europe). For example, energy consumption per unit output in the chemical industry fell by 30 per cent between 1980 and 1989.

3 Industry leads to environmental impacts through emissions (to air, water, soil), waste production, and resource use (water, energy, raw materials).

4 For 20 European countries in 1990, 25 per cent of sulphur dioxide and 14 per cent of nitrogen oxides were emitted by industry. Dust emissions are a problem in Central and Eastern Europe. Industrial use of water (which accounts for 53 per cent of total water withdrawals in Europe) is declining through water recycling and process modification. There is a trend towards reduction of all emissions from industrial plants.

Responses/policy options

1 Developing integrated inventories of emissions to air and water, and wastes.

2 Introducing integrated pollution prevention and control systems in industries (eg, encouraging clean, low waste technologies). Applying Best Available Technology (BAT). Moving from 'end of pipe' solutions to the development of clean technologies.

3 Changing operating framework with the Polluter Pays Principle, licensing arrangements, emissions limits, tradeable permits and fiscal incentives (such as tax differentials). Creating more flexible regulatory frameworks for small and medium-sized enterprises.

4 Performance of environmental impact assessments (for projects, plans and programmes), eco-management and audit schemes, ecolabels; providing better consumer information.

5 Developing voluntary care agreements, and codes of good practice.

6 Implementation of industrial clean-up schemes (soil excavation, washing and disposal), adding immobilising additives, soil and groundwater protection and reducing air emissions.

TRANSPORT

Findings

1 The transport sector accounts for 7 to 8 per cent of GDP in the EU. The growth rate of passenger transport was 3.1 per cent per year between 1970 and 1990 (3.4 per cent for road passenger transport). There has been a doubling of private car ownership in Western Europe between 1970 and 1990 and a sixfold increase in Hungary over the same period . The annual average growth rate of freight transport was 2.5 per cent between 1970 and 1990 (4.1 per cent for road freight). A doubling of road transport for both passengers and freight is likely between 1990 and 2010 unless measures are taken.

2 The change in modal split towards road transport and the strong growth of this sector are exacerbating environmental impacts.

3 In the EU, transport accounts for about 25 per cent of energy-related carbon dioxide emissions, with road transport making up 80 per cent of the total. If current trends continue, carbon dioxide emissions from transport will increase by a further 25 per cent between 1990 and 2000. Relative increases will be even greater in Central and Eastern Europe.

4 In 1990, about 45 per cent of total nitrogen oxide emissions in 20 European countries were attributed to road transport. The greatest volumes of toxic emissions from road transport are of carbon monoxide, accounting for up to 90 per cent of total carbon monoxide emissions in some countries. VOCs from transport account for about 35 to 40 per cent of total VOC emissions in Europe. Particulate matter from diesel engines is also a cause for concern. The total emissions of pollutants will increase in the next few years, although emissions per vehicle will be reduced. Emissions from aircraft of nitrogen oxides and carbon monoxide are a cause for concern. The main source of noise is from road traffic.

5 Over the past 20 years, in the EU alone, more than 1 million people have died in traffic accidents, and more than 30 million have been injured and/or permanently handicapped.

6 The road network in the EU consumes 1.3 per cent of the total land area, compared with 0.03 per cent for the railway network.

7 The social and environmental costs of transport have been put at 2.5 per cent of GDP in Germany alone and near 5 per cent in the OECD area.

Responses/policy options

1 Investigating options for decoupling transport and economic growth, aimed at inhibiting increases in passenger and road freight transport. If this is not achieved, benefits arising from technological improvements are unlikely to occur at the same rate as disbenefits from increased transport growth.

2 Improving technical solutions for better vehicles (catalytic convertors and noise reduction measures), fuels (alternative and cleaner fuels, better efficiency) and final disposal.

3 Introduction of economic instruments to reduce transport demand and improve modal split (for example, road pricing).

4   Developing appropriate transport infrastructure (for example, combined transport) and landuse planning (for example, banning traffic in city centres).

5   Improvement of user behaviour (for example, better driver information on a more rational use of the car).

AGRICULTURE

Findings

1   Agriculture accounts for about 42 per cent of the total land area in Europe, varying between 10 per cent in the Nordic countries to 70 per cent in Hungary, Ireland, Ukraine and the UK. There is a wide diversity in farm holdings size, but the overall trend over time is an increase in average size. In most countries, the importance of agriculture in the economy is declining: in the EU, the average GDP share of agriculture was 2.8 per cent in 1991; it is lowest in former West Germany (1.1 per cent), and highest in Greece (16.1 per cent); in Eastern Europe the contribution from agriculture and forestry was between 10 and 16 per cent to Gross Material Product (GMP).

2   There have been consistent gains in agricultural production and labour productivity. Total European production of cereals increased by 42 per cent, from 199 million tonnes in 1970 to 283 million tonnes in 1990. As a result, surpluses have been created. However, much of Central and Eastern Europe remains a net importer of food.

3   The most important impacts of agricultural practices on the environment are water pollution, decline of soil quality, loss of biodiversity, and landscape changes. The loss of natural habitats such as hedgerows and wetlands has had significant effects on European landscapes and wildlife.

4   In some areas of Europe, agriculture is estimated to be responsible for up to 80 per cent of the nitrogen loading (mainly from fertilisers) and 20 to 40 per cent of the phosphorus loading (mainly from animal manure) of surface waters, threatening the quality of drinking water and triggering eutrophication. Imports of animal feed disrupts the nutrient cycle on European farms. Over one

third of the world's grain is now fed to livestock. This livestock excretes much more manure than the land can absorb. The average load of nitrogen has more than doubled since 1960, from 110 to 241 kg/ha per year in 1988. Ammonia emissions have become a serious problem.

5   There are considerable variations in application rates of pesticides (1000 types) with resulting impacts such as resistance build-up, pollution of surface and groundwater, reduction in biodiversity and ecotoxicological impact on non-targeted wildlife. Residues can be detected in food long after the use of the compound has been banned.

6   Agriculture is also a victim of environmental degradation: atmospheric pollution may reduce yield and contaminate food crops; erosion decreases productivity; agricultural soils can become contaminated by chemicals.

Responses/policy options

1   Reformulation of agricultural policies taking full consideration of environmental matters. Favouring extensification and reduction of surpluses by removing land from production, and by reducing applications of fertilisers to agricultural land in selected areas. Providing support to farmers for long-term land set-aside in environmentally sensitive areas and ensuring that agri-environmental measures are properly implemented.

2   Bringing into effect legal instruments and economic incentives aiming at lowering rates of fertiliser applications (good agricultural practices) and implementation of integrated pest management to optimise the efficiency of pesticide use.

FORESTRY

Findings

1   Forests cover 33 per cent of the total European land area; there has been a 10 per cent increase of forest cover over the last 30 years. The composition of the forest has changed, with the introduction of non-indigenous fast-growing species in pure plantations.

2   Wood production has increased in Europe since 1965 by a total of 18 per cent due to intensification of forest management and extension of the total forested area. Consumption has increased by 28 per cent over the same period (sawn-wood and panel products, pulpwood for paper and board, and fuelwood); only fuelwood has experienced a sharp decline over recent decades. In general, non-wood functions of the forest, such as recreation, water and soil protection, nature and landscape diversity have taken increasing importance.

3   In 1993, it was estimated that 24 per cent of European trees were measurably affected by adverse weather conditions, insects and fungi, air pollution and forest fires. Fires are increasing in Southern Europe (about 60 000 fires are damaging an average of 700 000 ha of forest each year).

Responses/policy options

1   Improvement of forest management favouring multiple use of forests and limiting environmental pressures. Achieving better balance between hardwood and conifers

2   Implementation of international agreements on critical thresholds of pollution (critical loads) and reduction of acidic deposition through coordinated abatement strategies (LRTAP convention and EC Directives)

3   Taking full consideration of socio-economic, legal, structural, and cultural factors in the formulation of action measures to combat fires and integrate them in a global land management policy.

4   Bringing into effect the four Helsinki Resolutions giving general guidelines for the sustainable management of forests in Europe based on the recommendations from the UNCED.

FISHING AND AQUACULTURE

Findings

1 Overfishing of certain stocks of fish and shellfish is occurring in the Northeast Atlantic (eg, Atlantic herring in the 1960s and 1970s) and Mediterranean (eg, swordfish), due to overcapacity of the fishing industry and use of certain fishing methods.
2 Fixed and drift nets have an impact on marine animals (eg, dolphins, monk seals, and possibly porpoises and seabirds).
3 Effluent from aquaculture causes water pollution. The introduction of exotic species through aquaculture may compete with or replace native species.

Responses/policy options

1 Adoption of more 'environmentally-orientated' approaches to fishing policy; use of Exclusive Economic Zones, Total Allowable Catches and other policy measures (eg, by EU); strengthening policy regime for the control of fishing methods (particularly the implementation of existing policies in the North Atlantic and the Mediterranean Sea and improving the regime for the Black Sea).
2 Reduction of the size of the fishing fleet. Restricting the length of surface drift nets, or banning them outright.

TOURISM AND RECREATION

Findings

1 Tourism, which represented about 5.5 per cent of GDP in 1990 in the EU, has become one of the most important social and economic activities in Europe.
2 The impacts of tourism are exacerbated by the concentration of tourist activity into short holiday seasons and in relatively small areas (natural parks, mountains, coastal areas and cities).
3 In mountain areas, the cumulative environmental impacts from skiing are considerable. The worst impacts are in the Alps, which receive 100 million tourists each year in an area of about 190 thousand km^2.
4 In coastal areas, the greatest pressures are still felt in the Mediterranean (157 million tourists in 1990).
5 In cities and heritage sites, visitor numbers will increase in the future, as urban based tourism travel becomes more popular.

Responses/policy options

1 Broad policies, regional plans and local management actions are important in determining whether tourism and recreation lead to unsustainable environmental impacts (for example, management of traffic flows, and investment in local infrastructures such as wastewater treatment plants).
2 Measures to reduce overload in fragile places (eg, different levels of protection in national parks).
3 Improving tourist behaviour through, for example, raising awareness to encourage a better seasonal spread of tourists; encouraging eco-tourism.

HOUSEHOLDS

Findings

1 Consumption by households accounts for 70 per cent of industrial production in Europe (ranging from 56 per cent in the former USSR to 89 per cent in Greece).
2 An increase in the total number of households (167 million in 1980 to 183 million in 1990 excluding the former USSR) combined with a reduction in household size is leading to increased demand for resources.
3 The growth in consumer spending power, particularly in Western Europe, has increased demands for goods and services, some of which have caused significant environmental disruption. Western European households now own on average at least one car each, an increase of 20 per cent between 1980 and 1990. In Central and Eastern Europe, fewer than one in three households owns a car.
4 The volume of waste from households continues to increase. Studies in Western Europe indicate that although 50 per cent of household waste could be recycled, less than 10 per cent is recovered for recycling.
5 European households account for about 20 per cent of total water supplied for all purposes. This fraction is increasing.

Responses/policy options

1 Legal measures: for example, to restrict activities or consumption of products (eg, reducing air pollution from the domestic sector).
2 Financial regulations: for example, imposing levies on certain products, to discourage their usage
3 Training and information: for example, environmental education, and better consumer information through eco-labels, publicity and media campaigns about insulating houses.
4 Institutional instruments: for example, establishing domestic recycling schemes and reducing the production of post-consumer waste, metering of domestic water and energy use.
5 Technical measures: for example, greater energy efficiency of household appliances.

Appendix 1
An inventory of 56 environmental problems

The following non-prioritised list of environmental problems or issues was compiled from the results of a general survey of the literature, reviews of international initiatives and the results of the present study. Problems were recorded as they were reported or identified in the analysis, and thus some are more general expressions of particular problems which are also listed. Obvious references to the same problem were combined, but references to linked problems or issues with a common cause were usually kept separate to maintain the important differences of perspective and foci of interest. There are obvious groups of related problems expressed here and few, if any, are exclusive to the rest.

The problems listed vary widely in character: some have broad significance, others have more confined relevance (in time or space); some are already having effects, others refer to potential effects or risks; some are uncontested, others are disputed. Thus, the inclusion of a problem in this list does not itself denote significance. The list is the result of a trawl performed widely enough, and on a sufficiently comprehensive foundation – as this current assessment represents – to expect confidently that the most significant environmental problems of concern to Europe are included.

The difficulties with an analysis identifying environmental problems can be best appreciated if environmental problems are seen to move through a development 'cycle', in regard to the way in which they are viewed and treated. First, there is the phase where awareness about the problem (or at this stage *issues*) begins and grows, and understanding increases. This often leads to a policy formulation stage followed by an implementation phase. Further evaluation of the problem and the actions taken to deal with it can result in modifications to the policy or its means of implementation and may give rise to insights into new problems. The level of recognition given to a problem, its mention in the literature and information about it by which it can be described, significantly depend upon its stage in this cycle of development. The list of 56 environmental problems identified below will therefore be strongly influenced by this, affecting both what is included and how it is expressed. For this reason also, despite the comprehensive approach taken for the analysis, and for this assessment in general, the list is unlikely to be complete. Problems most likely not to be included here are those which fall into the first 'awareness' stage of development. Conversely, the list is probably biased towards problems that fall into the policy formulation stage about which more is generally known and to which much attention is given.

This list of 56 broad environmental problems was used to help identify the 12 prominent European environmental problems, presented in Chapters 27 to 38 (see introduction to Part V). In defining the 12, a number of problems in the list below were consolidated so that more than 20 of the 56 are considered to be covered by the 12 prominent problems.

CONSOLIDATED LIST OF ENVIRONMENTAL PROBLEMS

- Nuclear accidents
- Stratospheric ozone depletion, increasing UV
- Loss of biodiversity and genetic resources
- Resources and quality of groundwater
- Acidification
- Hazardous waste (transport and storage)
- Climate change
- Forest degradation
- Waste disposal
- Nuclear waste
- Urban air quality
- Nature conservation and sensitive ecosystems
- Persistent air toxics
- Industrial accidents
- Direct pollution (discharges, dumping) to sea
- Soil contamination from waste disposal
- Conservation of threatened species
- Tropospheric ozone increase and episodes
- Waste production
- Soil and resource contamination
- Fragmentation and destruction of habitats
- Eutrophication of surface waters
- Riverine inputs into seas
- Changes in hydrological regime
- Management of large rivers and lakes
- Desertification
- Lack of water supply
- Stress and degradation due to tourism
- Food security
- Urban waste
- Growing vulnerability of complex systems
- Bioaccumulation (metals, POCs)
- Energy security
- Soil erosion
- Risks of biotechnology
- Microbiological pollution of surface water
- Oil spills
- Sea-level rise
- Intensification of land-use
- Introduction of new organisms
- Shortage of industrial raw materials
- Coastal erosion
- Natural radioactivity (radon)
- Loss of agricultural land
- Shift of biogeographical zones
- Draining wetlands
- Floods, droughts and storms
- Non-ionising radiation
- Societal health
- Landscape modification
- Noise
- Occupational health
- Loss of cultural heritage
- Seismic activity, volanoes
- Pests and locusts
- Thermal pollution of waters

Appendix 2
Information strengths and weaknesses

	Information strengths	*Information weaknesses*
Air	Meteorological data Monitoring networks of common pollutants (SO_2, NO_x, CO, O_3, lead) National emissions inventories European emissions inventories (eg, CORINAIR90) European contribution to global emissions	Detailed monitoring of toxic substances (eg, VOCs) Detailed emissions inventories of substances throughout Europe
Inland waters	National totals for water availability and abstraction available for most countries River runoff in large rivers well known River water quality in large EU rivers fairly well known (database exists covering the period since 1976) Efficient surface water monitoring networks in place in some countries Comparatively good data on basic water chemistry, eutrophication and acidification	Regional water resources statistics missing Present rates and trends of water abstraction by source and economic sectors poorly known Comparable and reliable data on groundwater quantity and quality almost completely lacking In general, comparison of surface water quality across Europe is very difficult due to lack of comparable and reliable data. In particular, there is a lack of data on small rivers and lakes Data on organic micropollutants, metals and radioactivity are patchy and incomplete Biological assessments of river quality carried out using a variety of methods, hence not comparable No pan-European water quality database exists Reporting schemes differ markedly between countries
The seas	Comprehensive surveillance of microbiological bathing water quality in EU waters Harmonised and efficient monitoring programmes for water quality, land-based emissions, and sea food contamination exist for a few seas, including the Baltic Sea and the North Sea, as a result of implementation of international conventions.	Very little comparable data on water quality and biology available for the Black Sea, the Caspian Sea, the White Sea and the Barents Sea Estimates of pollutant loads from different human activities and natural sources in general not available Unified procedures for estimating land-based emissions to seas missing Comparison of contaminant load estimates between different seas No pan-European marine water quality database exists Reporting schemes differ markedly between seas
Soil	Harmonised soil map of the EU at scale 1:1 million Global assessment of soil degradation map (1:10 million) for water and wind erosion and physical degradation Models for quantification of soil pollution Models for calculating critical loads and their exceedance for acidification of European forest soils Case studies for estimation of soil compaction effects	Updated soil map of Europe (1:1 million) harmonised with Eastern Europe. No European soil map at a scale of 1:250 000 is available for environmental assessment In general, lack of quantitative data on soil properties Soil survey, sampling, analytical methods and nomenclature vary between countries Soil and terrain attributes that influence environmental processes are often missing in soil surveys Data on soil fauna and flora, organic matter and pollutants are poor and inadequate Contaminated sites inventory lacking
Landscapes	Ground-based and remote sensed land use data National and regional reports about landscape types Database with protected landscapes (> 1000 hectares)	No international harmonisation of existing data Existing classification schemes are rudimentary and lack detail No information on distribution and quality of landscape types

	Information strengths	*Information weaknesses*
Nature and wildlife – ecosystems	Internationally harmonised classification systems and international inventories (of sites) exist for large parts of Europe (EU) Many countries and NGOs hold detailed information on habitats Existing data (eg, on natural potential vegetation, soil and land cover) are generally available for ecosystem assessments	No incorporation of data from Scandinavia and Central Europe into international systems (but in progress) Limited accessibility to East European data No data on ecosystem distribution for the whole of Europe Information on distribution and ecological quality of ecosystems groups is partly out of date, patchy and often incomplete
Nature and wildlife – species	International data on birds and mammals have relatively good coverage and are being regularly updated Red species lists on endangered species follow international standards Large amount of scientific work, literature, museum materials exist for many species and their populations	Internationally harmonised data on reptiles, amphibians and fish are incomplete, and lacking on invertebrates Insufficient large-scale monitoring schemes for internationally endangered, migratory and indicator species Insufficient data coverage for the supra-regional decline of unendangered species (eg, plants)
Urban environment	Census data (population, land area, housing, households, etc) Data on urban landuse, infrastructures, and public transport are available at the municipal level Data on services supplied at the municipal level (eg, drinking water, sewage, municipal waste, etc) Monitoring of concentrations of common air pollutants (SO_2, NO_x, CO, O_3, lead) Energy flows by sources and end-uses in a few cities Initiatives to create a common framework for developing urban environmental indicators	City boundaries definition is a major difficulty which undermines the meaning and comparability of data such as population, land use and infrastructure density Availability and comparability of environmental quality data (eg, noise) is very limited due to different methods of collection and classification Data on the use of resources (eg, energy and materials) are not available at the municipal level Information on environmental performance of urban planning and management are scarce
Emissions	National emission inventories of pollutants into the air containing detailed spatial information on human activities and their resulting (estimated) emissions of a wide range of pollutants Internationally coordinated inventories for emisions to the atmosphere	Direct emission measurements infrequent Emissions into water bodies not quantified in detail; almost complete lack of data on catchment scale Releases into or onto land incomplete and fragmentary (see Waste) Limited comparability of national emissions inventories Limited integration of inventories between emission to air, water, land, human activities and major environmental problems Timely emission statistics
Waste	National inventories of waste generation and management in several European countries International initiatives for the harmonisation of waste statistics (eg, European Waste Catalogue) Transfrontier movements of waste are being recorded by UNEP under the Basel Convention Data on most types of radioactive waste produced by civil activities available in most Western European countries National registers for contaminated sites in a few countries	Harmonised waste classification systems and harmonised inventories of waste generation and management are lacking Monitoring of landfills and emissions from waste treatment facilities and compliance with standards are insufficient Reliable data on transfrontier movement of hazardous waste (origin/destination) not available Data on radioactive waste handled at military sites are usually not available European and national inventories of contaminated sites still lacking
Noise and radiation	Some noise data available for transport (road, rail and air) Sources and effects of natural and artificial ionising radiation generally well known	Availability and comparability of noise exposure is poor due to different limit values, measurement techniques and diversity of descriptors used. Poor time series does not allow analysis Lack of data on non-transport noise (eg, recreational noise) Poor noise data in Central and Eastern Europe in general Lack of data on annoyance from noise Lack of data for Europe on UV-B reaching the Earth's surface Lack of data on artificial sources of UV-B Exposure data for electromagnetic fields lacking Contamination and effects from radiation 'hotspots' from spills, accidents and waste disposal by the former USSR lacking

	Information strengths	*Information weaknesses*
Chemicals and genetically modified organisms	Toxicity and ecotoxicity data for the small fraction of chemicals profiled and catalogued in international registers and newly coming on to the market	Toxicity and ecotoxicity data not satisfactory for most of the more than 100 000 chemicals in use and circulation, especially concerning potential impacts on human health and the environment Environmental pathways of chemicals
Natural and technological hazards	Impacts of individual accidents on human health usually well reported Location, nature and causes of major accidents reported on national basis and by a few international registers (eg, MARS) Industrial accidents in Northern Europe well covered Natural hazards widely reported on regional national and international levels	Data on environmental damage, eological effects, long-term recovery and clean-up actions not available Information on type and site of releases often very approximate; in general, accidents reported incompletely Paucity of data on industrial accidents in Eastern Europe; acute problem of hazard and risk quantification for older plants in Central and Eastern Europe Environmental impacts of natural hazards not routinely reported. Particular lack of information on potential interaction between natural hazards and human activities
Energy	Primary energy production Total energy consumption Forecasts of future energy consumption Spills from pipelines in Western Europe	Data on available resources of non-renewable energy sources (eg, coal and oil reserves) Sectoral fuel consumption by end-use Emissions to air originating from energy sources for the whole of Europe Detailed inventories of major energy using plant (their age, size, technology, etc) Data on renewable energy production and consumption Land area used by energy and electricity generation plants Energy conservation and efficiency measures Spills from pipelines in Central and Eastern Europe
Industry	Industrial production, employment and trade flows Energy use by industry as a whole Voluntary reporting by companies	Limited data on emissions to air from different industrial sectors for the whole of Europe In Central and Eastern Europe, industrial contribution to emissions not separated out from total figures Limited data on emissions to water Poor data on soil contamination from industrial activities Industrial waste data is very poor, especially in Central and Eastern Europe (volumes, composition, fate of various wastes) Poor industrial waste classification systems Limited data on resource use by industry (raw materials, energy, water)
Transport	Vehicle stocks and transport volumes in Western Europe Length of infrastructure network Energy use by transport Traffic accidents	Contribution of transport to GNP Vehicle stocks and transport volumes in Central and Eastern Europe Average distances travelled by different transport modes and data on 'alternative' modes of transport, such as walking and bicycling Emissions to air from road transport are difficult to estimate (CO_2, NO_X, VOCs etc) and off-road emissions lacking Emissions from aircraft, railway and water transport Transport waste (eg, old cars and tyres) Travel speeds, fleet composition, occupancy rates, load factors Noise from transport sources not collected on a systematic regular basis, using same methodology Routine and accidental releases from transport Land area used by different transport modes

	Information strengths	*Information weaknesses*
Agriculture	In general, adequate information on the structure of production, inputs to the land to maximise production, and production itself For EU and EFTA countries, comprehensive information on production, farm size, employment structure, fertiliser and pesticide use, and livestock	Data missing on the relative contribution to impacts on the environment of agriculture production and change in agriculture systems
Forestry	For Western Europe, a reasonable amount of data are available on area by broad forest type, rates of reforestation and afforestation, volumes of the standing stock of wood and annual growth, annual fellings and removals, production of roundwood, pulp and charcoal Surveys also exist on forest condition in a large part of Europe	Economic data on forest production, trade, and other factors, such as contribution to GDP/GNP, are either lacking or obsolete in Eastern Europe Few available inventory data for the Baltic states, Moldova and the European part of the Russian Federation Few data are available in general on specific environmental effects associated with forestry In general, accurate data are lacking on the location, extent and composition of afforestation and reforestation, planting with introduced tree species, ancient/natural forests
Fishing and aquaculture	Trend data on catches, fishing effort and fleet capacity (especially for the Northeast Atlantic) Stock data for the Northeast Atlantic	Contribution of fishing to GNP Catch data is not always reliable due to unreported catches or misreporting Estimates of maximum sustainable yields Data on fishing techniques, employment levels and productivity Catch and stock data for the Mediterranean and Black seas Catch and stock data for inland fisheries Data on aquaculture production Effluents arising from aquaculture

Tourism and recreation	Trend data on arrivals at frontiers, and tourism arrivals and nights stayed in accommodation establishments
	Arrivals at frontiers available by transport mode
		Contribution of tourism to GNP
		Visitor numbers for specific destinations or different tourism settings
		Relevant data on environmental impacts
		Emissions to air and water and waste production from tourist activities
		Use of different modes of transport used for tourism and recreation
		Pan-European attitude surveys on preferred tourism destinations
		Measures of the impact of tourism on the environment (indicators of carrying capacity), such as:
		  (i) composite indices of site attractivity and site stress index
		  (ii) national level (eg, endangered species)
		  (iii) site specific indicators (development density, area per tourist)

Households	Trend data on number of households and number of people per household (not good for former USSR) Trends in car ownership Trends in consumer spending Public attitudes	Energy data is not separated from the commercial and service sector End-use of energy by appliances Land area occupied by households Water supply to households, especially in Central and Eastern Europe Details on end-use of water by households Emissions to air and water (sewage production) and waste production (the volumes, recycling rates, amounts of hazardous pollutants) Resource use by households (eg, energy and water use, waste recycling rates)

Acronyms and abbreviation

ACP	African, Caribbean and Pacific (countries)
AMAP	Arctic Monitoring and Assessment Programme
ANC	acid neutralising capacity
AQG	air quality guidelines (of WHO)
BaP	benzo(a)pyrene
BAT	best available technology
BOD	biochemical oxygen demand
CAP	Common Agricultural Policy (EU)
CEC	Commission of the European Communities (now known as European Commission)
CEFIC	European confederation of chemical industries
CEI	Central European Initiative
CEN	European committee for standardisation
CFC	chlorofluorocarbon
CFP	Common Fisheries Policy (EU)
CH₄	methane
CHP	combined heat and power
CIPRA	Commission for the protection of the alps
CITES	Convention on International Trade in Endangered Species
CLP	chlorine loading potential
CMEA	Council for Mutual Economic Aid
CO₂	carbon dioxide
COD	chemical oxygen demand
CoE	Council of Europe
CORINE	CO-oRdination of INformation on the Environment (CEC)
CSD	Commission on Sustainable Development
CZM	Coastal Zone Management
DDT	dichloryldiphenyltrichorethane
DG	Directorate General (of CEC)
DMS	dimethylsulphide
DNA	deoxyribonucleic acid
EASOE	European Arctic Stratospheric Ozone Experiment
EBCC	European Bird Census Council
EBRD	European Bank for Reconstruction and Development
EC	European Community
ECCC	European Coastal Conservation Conference
ECDIN	Environmental Chemicals Data and Information Network (CEC)
ECE	Economic Commission for Europe (UN)
ECEH	European Centre for Environmental Health (WHO)
ECETOC	European Centre for Ecotoxicology and Toxicology
ECMT	European Conference of Ministers of Transport
EEA(-TF)	European Environment Agency (Task Force, CEC DG XI)
EEZ	Exclusive Economic Zone

EFTA	European Free Trade Association
EIA	Environmental Impact Assessment
EINECS	European Inventory of Existing Chemical Substances (CEC)
EMEP	European Monitoring and Evaluation Programme
EMF	electromagnetic field
ENCY	European Nature Conservation Year (1995)
EOAC	European Ornithological Atlas Committee
ESA	Environmentally Sensitive Area
EU	European Union
EWC	European Waste Catalogue
FAO	Food and Agriculture Organization (UN)
FCCC	Framework Convention on Climate Change (UN)
FGD	flue-gas desulphurisation
FOREGS	Forum of European Geological Surveys
FYROM	Former Yugoslavian Republic of Macedonia
gamma-HCH	lindane (organochlorine pesticide)
GATT	General Agreement on Tariffs and Trade
GDP	gross domestic product
GEF	Global Environmental Facility
GEMS	Global Environment Monitoring System (UNEP)
GESAMP	Group of Experts on the Scientific Aspects of Marine Pollution
GFCM	General Fisheries Council for the Mediterranean
GIS	geographical information system
GMO	genetically modified organism
GMP	gross material product
GNP	gross national product
GWP	global warming potential
H₂O₂	hydrogen peroxide
HCFC	hydrochloroflurocarbon
HELCOM	Helsinki Commission
HGWP	halocarbon global warming potential
IAEA	International Atomic Energy Agency (UN)
IARC	International Agency for Research in Cancer
IBA	Important Bird Area
IBSFC	International Baltic Sea Fishery Commission
ICBP	International Council for Bird Preservation (now BirdLife International)
ICC	International Chamber of Commerce
ICCAT	International Commission for the Conservation of Atlantic Tunas
ICES	International Council for the Exploration of the Sea
ICNIRP	International Commission on Non-Ionising Radiation Protection
ICPRP	International Commission for the Protection of the Rhine against Pollution
ICRP	International Commission on Radiological Protection

This list includes only acronyms and abbreviations commonly used throughout the report. See general index for a full listing.

ICSU	International Council of Scientific Unions
ICWE	International Conference on Water and the Environment
IEA	International Energy Agency (UN)
IEB	Initiative for Ecological Bricks
IGBP	International Geosphere-Biosphere Programme (ICSU)
IIASA	International Institute for Applied Systems Analysis
ILO	International Labour Organization
IMO	International Maritime Organisation
INES	International Nuclear Events Scale
IOC	Intergovernmental Oceanographic Commission (UNESCO)
IPCC	Intergovernmental Panel on Climate Change
IPCS	International Programme of Chemical Safety (WHO, UNEP, ILO)
IRB	International Reclamation Bureau
IRPA	International Radiation Protection Association
IRPTC	International Register of Potentially Toxic Chemicals (UNEP)
ISRIC	International Soil Reference and Information Centre
ISSG	Irish Sea Study Group
IUCN	World Conservation Union
IWC	International Whaling Commission
IWIC	International Waste Identification Code
IWRB	International Waterfowl and Wetlands Research Bureau
LCA	life-cycle analysis
LRTAP	Long-Range Transboundary of Air Pollution (Convention)
MAP	Mediterranean Action Plan
MARS	Major Accident Reporting System
MCBSF	Mixed Commission for Black Sea Fisheries
MEDPOL	Mediterranean Pollution Monitoring and Research Programme
MPC	maximum permissible concentration
MSY	maximum sustainable yield
NEAFC	Northeast Atlantic Fisheries Commission
NERI	National Environmental Research Institute, Denmark
NGO	non-governmental organisation
NILU	Norwegian Institute for Air Research
NMVOC	non-methane volatile organic compound
N_2O	dinitrogen oxide (traditionally called nitrous oxide)
NO	nitrogen oxide (traditionally called nitric oxide)
NO_2	nitrogen dioxide
NO_x	nitrogen oxides (non-specific, including NO and NO_2)
NP	National Park
NSTF	North Sea Task Force
OECD	Organization for Economic Cooperation and Development
OH	hydroxyl
ODP	ozone depletion potential
OSPARCOM	Oslo and Paris Commissions (established by the Oslo and Paris conventions on prevention of marine pollution)

PAH	polycyclic (or polynuclear) aromatic hydrocarbon
PAN	peroxyacetylnitrate
PCB	polychlorinated biphenyl
PCDD	polychlorinated dibenzodioxin
PCDF	polychlorinated dibenzofuran
PM	particulate matter
POC	persistent organic compound
POCP	photochemical ozone creation potential
POP	persistent organic pollutants
PPP	Polluter Pays Principle
RIVM	National Institute of Public Health and Environmental Protection, The Netherlands
RIZA	Institute for Inland Water Management and Waste Water Treatment, The Netherlands
SAC	Special Area of Conservation
SEA	strategic environmental assessment
SCOPE	Scientific Committee on Problems of the Environment
SNA	system of national accounts
SO_2	sulphur dioxide
SSB	spawning stock biomass
SPA	Special Protected Area
SPM	suspended particulate matter
TAC	total allowable catch
TBT	tributyl tin
toe	tonnes of oil equivalent
TSP	total suspended particulates
UNCED	United Nations Conference on Environment and Development (Rio de Janeiro, June 1992, 'the Earth summit')
UNCHE	United Nations Conference on the Human Environment
UNCHS	United Nations Centre for Human Settlements ('Habitat')
UNCTAD	United Nations Commission on Trade and Development
UNDP	United Nations Development Programme
UNECE	United Nations Economic Commission for Europe
UNEP	United Nations Environment Programme
UNESC	United Nations Economic and Social Council
UNESCO	United Nations Educational Scientific and Cultural Organization
UNIDO	United Nations Industrial Development Organisation
UNSCEAR	United Nations Scientific Committee on the Effects of Atomic Radiation
UNSTATO	United Nations Statistical Office (formerly UNSO)
VOC	volatile organic compound
WASH	Water and Sanitation for Health
WCED	World Commission on Environment and Development (UN)
WCMC	World Conservation Monitoring Centre
WCRP	World Climate Research Programme
WEC	World Energy Council
WHO	World Health Organization (UN)
WMO	World Meteorological Organization (UN)
WTO	World Tourism Organisation

List of illustrations

FIGURES

MAPS

TABLES

BOXES

General index

Geographical index

Sales

BELGIQUE / BELGIË
Moniteur belge/
Belgisch Staatsblad
Rue de Louvain 42/Leuvenseweg 42
B-1000 Bruxelles/B-1000 Brussel
Tél (02) 512 00 26
Fax (02) 511 01 84

Jean De Lannoy
Avenue du Roi 202/Koningslaan 202
B-1060 Bruxelles/B-1060 Brussel
Tél (02) 538 51 69
Fax (02) 538 08 41

Autres distributeurs/
Overige verkooppunten:

Librairie européenne/
Europese boekhandel
Rue de la Loi 244/Wetstraat 244
B-1040 Bruxelles/B-1040 Brussel
Tél. (02) 231 04 35
Fax (02) 735 08 60

Document delivery:

Credoc
Rue de la Montagne 34/Bergstraat 34
Boîte 11/Bus 11
B-1000 Bruxelles/B-1000 Brussel
Tél (02) 511 69 41
Fax (02) 513 31 95

DANMARK
J H Schultz Information A/S
Herstedvang 10-12
DK-2620 Albertslund
Tlf 43 63 23 00
Fax (Sales) 43 63 19 69
Fax (Management) 43 63 19 49

DEUTSCHLAND
Bundesanzeiger Verlag
Breite Straße 78-80
Postfach 10 05 34
D-50445 Köln
Tel (02 21) 20 29-0
Fax (02 21) 2 02 92 78

GREECE/ΕΛΛΑΔΑ
G C Eleftheroudakis SA
International Bookstore
Nikis Street 4
GR-10563 Athens
Tel. (01) 322 63 23
Fax 323 98 21

ESPAÑA
Boletín Oficial del Estado
Trafalgar, 27-29
E-28071 Madrid
Tel (91) 538 22 95
Fax (91) 538 23 49

Mundi-Prensa Libros, SA
Castelló, 37
E-28001 Madrid
Tel (91) 431 33 99 (Libros)
Tel (91) 431 32 22 (Suscripciones)
Tel (91) 435 36 37 (Dirección)
Fax (91) 575 39 98

Sucursal:

Librería Internacional AEDOS
Consejo de Ciento, 391
E-08009 Barcelona
Tel (93) 488 34 92

Fax (93) 487 76 59

Librería de la Generalitat
de Catalunya
Rambla dels Estudis, 118 (Palau Moja)
E-08002 Barcelona
Tel (93) 302 68 35
Tel (93) 302 64 62
Fax (93) 302 12 99

FRANCE
Journal officiel
Service des publications
des Communautés européennes
26, rue Desaix
F-75727 Paris Cedex 15
Tél (1) 40 58 77 01/31
Fax (1) 40 58 77 00

IRELAND
Government Supplies Agency
4-5 Harcourt Road
Dublin 2
Tel (1) 66 13 111
Fax (1) 47 80 645

ITALIA
Licosa SpA
Via Duca di Calabria 1/1
Casella postale 552
I-50125 Firenze
Tel (055) 64 54 15
Fax 64 12 57

GRAND-DUCHÉ DE LUXEMBOURG
Messageries du livre
5, rue Raiffeisen
L-2411 Luxembourg
Tél 40 10 20
Fax 49 06 61

NEDERLAND
SDU Servicecentrum Uitgeverijen
Postbus 20014
2500 EA 's-Gravenhage
Tel (070) 37 89 880
Fax (070) 37 89 783

ÖSTERREICH
Manz'sche Verlags-und
Universitätsbuchhandlung
Kohlmarkt 16
A-1014 Wien
Tel (1) 531 610
Fax (1) 531 61-181

Document delivery:

Wirtschaftskammer
Wiedner Hauptstraße
A-1045 Wien
Tel (0222) 50105-4356
Fax (0222) 50206-297

PORTUGAL
Imprensa Nacional
Casa da Moeda, EP
Rua Marquês Sá da Bandeira, 16-A
P-1099 Lisboa Codex
Tel (01) 353 03 99
Fax (01) 353 02 94

Distribuidora de Livros
Bertrand, Ld.a

Grupo Bertrand, SA
Rua das Terras dos Vales, 4-A
Apartado 37

P-2700 Amadora Codex
Tel (01) 49 59 050
Fax 49 60 255

SUOMI/FINLAND
Akateeminen Kirjakauppa
Akademiska Bokhandeln
Pohjois-Esplanadi 39/Norra esplanaden 39
PL / PB 128
FIN-00101 Helsinki / Helsingfors
Tel (90) 121 4322
Fax (90) 121 44 35

SVERIGE
BTJ AB
Traktorvägen 13
S-22100 Lund
Tel. (046) 18 00 00
Fax (046) 18 01 25
Fax (046) 30 79 47

UNITED KINGDOM
HMSO Books
(Agency section)
HMSO Publications Centre
51 Nine Elms Lane
London SW8 5DR
Tel (0171) 873 9090
Fax (0171) 873 8463

For this publication only:
Earthscan Publications
120 Pentonville Road
London N1 9SN
Tel (0171) 278 0433
Fax (0171) 278 1142

ICELAND
BOKABUD
LARUSAR BLÖNDAL
Skólavördustíg, 2
IS-101 Reykjavík
Tel 11 56 50
Fax 12 55 60

NORGE
Narvesen Info Center
Bertrand Narvesens vei 2
Postboks 6125 Etterstad
N-0602 Oslo 6
Tel (22) 57 33 00
Fax (22) 68 19 01

SCHWEIZ/SUISSE/SVIZZERA
OSEC
Stampfenbachstraße 85
CH-8035 Zürich
Tel (01) 365 54 49
Fax (01) 365 54 11

BĂLGARIJA
Europress Klassica BK Ltd
66, bd Vitosha
BG-1463 Sofia
Tel/Fax (2) 52 74 75

ČESKÁ REPUBLIKA
NIS ČR
Havelkova 22
CZ-130 00 Praha 3
Tel/Fax (2) 24 22 94 33

HRVATSKA
Mediatrade
P Hatza 1
HR-4100 Zagreb
Tel (041) 43 03 92
Fax (041) 45 45 22

MAGYARORSZÁG
Euro-Info-Service
Honvéd Európá Ház
Margitsziget
H-1138 Budapest
Tel/Fax (1) 111 60 61,
(1) 111 62 16

POLSKA
Business Foundation
ul Krucza 38/42
PL-00-512 Warszawa
Tel (2) 621 99 93, 628 28 82
International Fax&Phone (0-39) 12 00 77

ROMÂNIA
Euromedia
65, Strada Dionisie Lupu
RO-70184 Bucuresti
Tel/Fax 1-31 29 646

RUSSIA
CCEC
9,60-letiya Oktyabrya Avenue
117312 Moscow
Tel/Fax (095) 135 52 27

SLOVAKIA
Slovak Technical
Library
Nàm slobody 19
SLO-812 23 Bratislava 1
Tel (7) 52 204 52
Fax (7) 52 957 85

CYPRUS
Cyprus Chamber of Commerce
and Industry
Chamber Building
38 Grivas Dhigenis Ave
3 Deligiorgis Street
PO Box 1455
Nicosia
Tel (2) 44 95 00, 46 23 12
Fax (2) 36 10 44

MALTA
Miller Distributors Ltd
PO Box 25
Malta International Airport LQA 05 Malta
Tel 66 44 88
Fax 67 67 99

TÜRKIYE
Pres AS
Istiklal Caddesi 469
TR-80050 Tünel-Istanbul
Tel (1) 520 92 96, 528 55 66
Fax (1) 520 64 57

ISRAEL
ROY International
31 Habarzel Street
69710 Tel Aviv
Tel (3) 49 78 02
Fax (3) 49 78 12

Sub-agent
(Palestinian authorities):

INDEX Information Services
PO Box 19502
Jerusalem
Tel (2) 27 16 34
Fax (2) 27 12 19

MAGYARORSZÁG

EGYPT/MIDDLE EAST
Middle East Observer
41 Sherif St
Cairo
Tel/Fax (2) 393 97 32

UNITED STATES OF AMERICA/CANADA
UNIPUB
4611-F Assembly Drive
Lanham, MD 20706-4391
Tel Toll Free (800) 274 48 88
Fax (301) 459 00 56

CANADA
Subscriptions only
Uniquement abonnements

Renouf Publishing Co Ltd
1294 Algoma Road
Ottawa, Ontario K1B 3W8
Tel (613) 741 43 33
Fax (613) 741 54 39

AUSTRALIA
Hunter Publications
58A Gipps Street
Collingwood
Victoria 3066
Tel (3) 417 53 61
Fax (3) 419 71 54

JAPAN
Procurement Services Int (PSI-Japan)
Kyoku Dome Postal Code 102
Tokyo Kojimachi Post Office
Tel (03) 32 34 69 21
Fax (03) 32 34 69 15

Sub-agent:
Kinokuniya Company Ltd
Journal Department
PO Box 55 Chitose
Tokyo 156
Tel (03) 34 39-0124

SOUTH and EAST ASIA
Legal Library Services Ltd
Orchard
PO Box 0523
Singapore 9123
Tel 243 24 98
Fax 243 24 79

SOUTH AFRICA
Safto
5th Floor, Export House
Cnr Maude & West Streets
Sandton 2146
Tel (011) 883-3737
Fax (011) 883-6569

ANDERE LÄNDER
OTHER COUNTRIES
AUTRES PAYS
Office des publications officielles
des Communautés européennes
2, rue Mercier
L-2985 Luxembourg
Tél 29 29-1
Télex PUBOF LU 1324 b
Fax 48 85 73, 48 68 17

European Environment Agency

Europe's Environment: The Dobříš Assessement
Edited by David Stanners and Philippe Bourdeau

Luxembourg: Office for Official Publications of the European Communities

1995 – XXVI, 712 pp., num. tab., fig., map., pl. – 21.0 x 29.7 cm

ISBN 92-826-5409-5

Europe's Environment: The Dobříš Assessment covers the state of the environment in a Europe of nearly 50 countries, and presents comprehensive data which vary widely depending on topic and location from the late 1980s to 1993. It analyses different environmental themes relating to, inter alia, air, water, soil, nature, wildlife and urban areas, and looks at the pressures which impact on them, the human activities giving rise to those pressures and reviews the details of 12 prominent environmental problems of concern to Europe. In conclusion, the report summarises the main highlights and findings, the most significant policy responses and/or options being taken or proposed, as well as the strengths and weaknesses of the available data.

The report was requested by European environment ministers at the ministerial conference held at Dobříš Castle, near Prague in June 1991 and is intended to assist sound decision making and help raise public awareness about environmental problems. *Europe's Environment*, presented to the third conference of European environment ministers, should act as the basis for development of the regional 'Environment Programme for Europe'. It will be accompanied by a series of supporting documents including a separate 455 page *Statistical Compendium*, CD-ROM version, popular digest version, educational material etc.

Political

0   km   1000

50°   40°   30°   20°   10°   0°

GREENLAND

Greenland Sea

Arcti

Jan Mayen Island

Reykjavik • **ICELAND**

Arctic Circle

N o r w e g i a n Sea

N O R

Faeroe Islands

50°

Shetland Islands

Bergen
Lillehammer
Stavanger

Inverness

SCOTLAND Aberdeen

Glasgow
Edinburgh

North Sea

Alborg
Arhus
DENMARK
Esbjerg
Odense
Kiel
Ro

Newcastle upon Tyne

N IRELAND
Belfast

IRELAND

Castlegregory
Dublin • *Irish Sea*
Cork

Leeds
Liverpool Manchester
Sheffield

UNITED KINGDOM

WALES **ENGLAND**

Birmingham
Cardiff Oxford
Norwich
Delfzijl
Groningen
Hamburg
Bremen
GERMA

Exeter Bath
London
THE NETHERLANDS
The Hague
Rotterdam Amsterdam Osnabrück
Dortmund
Magd

English Channel
Blankenberge
Calais Antwerp
Essen
Düsseldorf Hanover
Lille
Charleroi **Brussels** Cologne
Leip

Roscoff
Le Havre
BELGIUM
Bonn
Frankfurt am Main

Rennes
LUXEM-BOURG
Pilsen

Nantes • Paris
Nuremberg

N o r t h

A t l a n t i c

O c e a n

40°

Tours
Strasbourg
Stuttgart

Dijon
Freiburg
Munich

Bay of Biscay

La Rochelle
FRANCE
Basle
Bern **SWITZERLAND**
Zurich **LIECHTENSTE**
Innsbruck

Vichy
Geneva

La Coruna
Bordeaux
Lyons
Chambéry
Aosta
Milan
Trento

Vigo
Santander
Bilbao
Gijon
Toulouse
Turin
Reggio Emilia
Genoa
Ferrara
SLO
Venice
Bologna
SAN MARI

Porto

Valladolid
Marseille Nice
MONACO
Ancona

PORTUGAL
Zaragoza
ANDORRA
Gulf of Lions
Toulon
Ligurian Sea
Florence

Gerona
CORSICA
ITALY

Lisbon
Estremoz
Madrid
Tarragona
Ajaccio
VATICAN CITY
Rome

Setubal Evora
SPAIN
Barcelona

Valencia
Balearic Islands
Palma
Sassari
Naples

Seville
MALLORCA
SARDINIA

Madeira
Granada
Cagliari
Tyrrhenian Sea

Malaga
Cartagena

Gibraltar
Almeria
M e d i t
Palerm

Strait of Gibraltar

30°
Algiers
SICIL

Rabat

CANARY ISLANDS
Tunis

Tenerife
Lanzarote
TUNISIA

Fuerteventura
MOROCCO
e

10°
0°
10°

ALGERIA